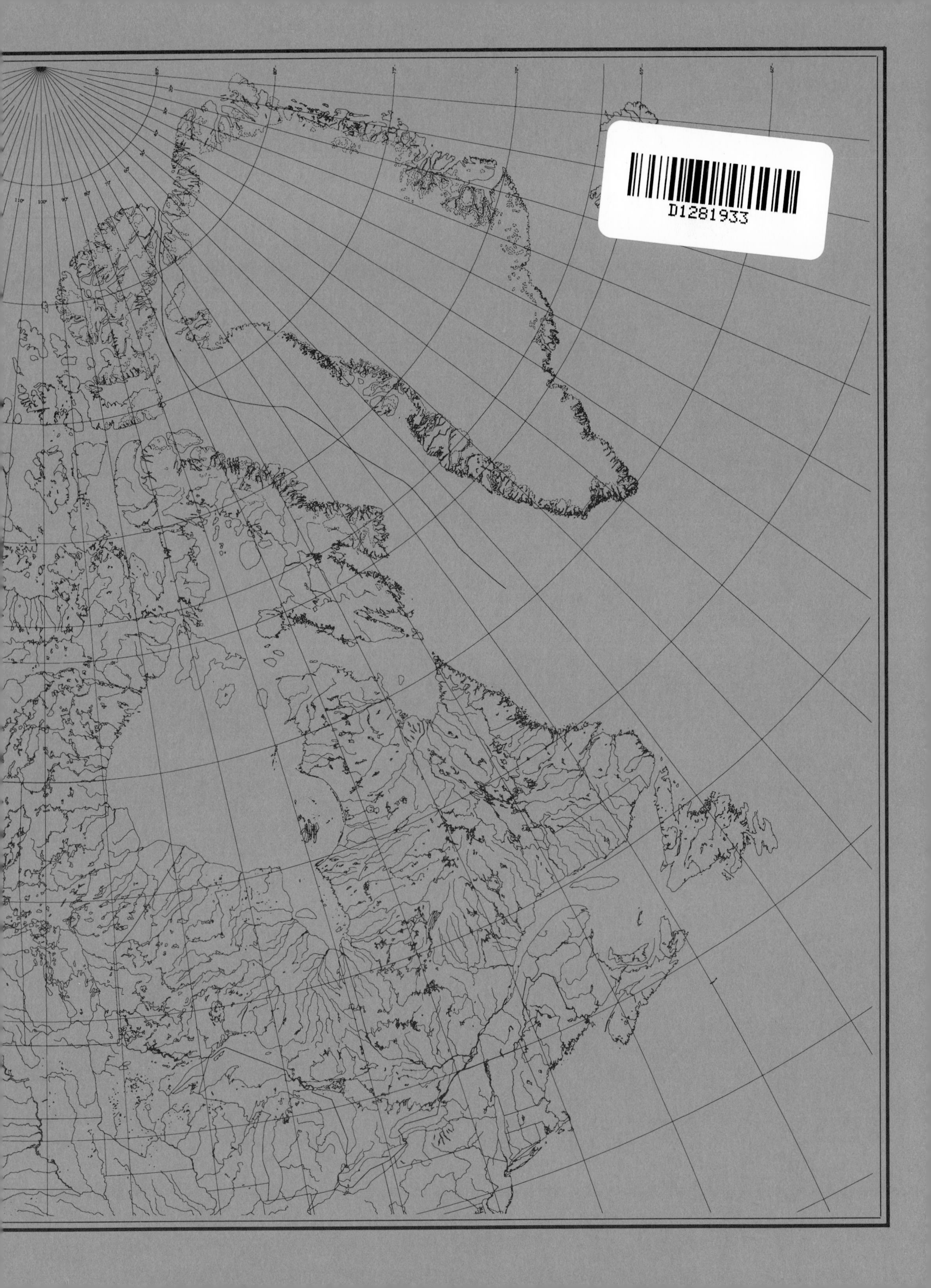
D1281933

The Decade of North American Geology
Summary Volume to Accompany the DNAG
Continent-Ocean Transect Series

Phanerozoic Evolution of North American Continent-Ocean Transitions

Edited by

Robert C. Speed
Department of Geological Sciences
Northwestern University
Evanston, Illinois 60208

1994

Acknowledgment

Publication of this volume, one of the synthesis volumes of *The Decade of North American Geology Project* series, has been made possible by members and friends of the Geological Society of America, corporations, and government agencies through contributions to the Decade of North American Geology fund of the Geological Society of America Foundation.

Following is a list of individuals, corporations, and government agencies giving and/or pledging more than $50,000 in support of the DNAG Project:

Amoco Production Company
ARCO Exploration Company
Chevron Corporation
Cities Service Oil and Gas Company
Conoco, Inc.
Diamond Shamrock Exploration Corporation
Exxon Production Research Company
Getty Oil Company
Gulf Oil Exploration and Production Company
Paul V. Hoovler
Kennecott Minerals Company
Kerr McGee Corporation
Marathon Oil Company
Maxus Energy Corporation
McMoRan Oil and Gas Company
Mobil Oil Corporation
Occidental Petroleum Corporation
Pennzoil Exploration and Production Company
Phillips Petroleum Company
Shell Oil Company
Caswell Silver
Standard Oil Production Company
Oryx Energy Company (formerly Sun Exploration and Production Company)
Superior Oil Company
Tenneco Oil Company
Texaco, Inc.
Union Oil Company of California
Union Pacific Corporation and its operating companies:
Union Pacific Resources Company
Union Pacific Railroad Company
Union Pacific Realty Company
U.S. Department of Energy

Published by The Geological Society of America, Inc.
3300 Penrose Place, P.O. Box 9140, Boulder, Colorado 80301

Printed in U.S.A.

Library of Congress Cataloging-in-Publication Data

Phanerozoic evolution of North American continent-ocean transitions / edited by Robert C. Speed.
p. cm.
At head of title: The Decade of North American Geology summary volume to accompany the DNAG continent-ocean transect series.
ISBN 0-8137-5305-8
1. Geology, Structural—North America. 2. Continental margins--North America. 3. Plate tectonics—North America. 4. Earth--Crust. I. Speed, Robert C. II. Decade of North American Geology Project.
QE626.P48 1994
551.7'0097—dc20 94-15452
CIP

Cover Photo: Map of North America showing the locations of the continent-ocean transects discussed in this volume.

10 9 8 7 6 5 4 3 2 1

Contents

NORTHERN NORTH AMERICA

TOPICAL SYNTHESES

Preface

The Geology of North America series has been prepared to mark the Centennial of The Geological Society of America. It represents the cooperative efforts of more than 2,000 individuals from academia, state and federal agencies of many countries, and industry to prepare syntheses that are as current and authoritative as possible about the geology of the North American continent and adjacent oceanic regions.

This series is part of the Decade of North American Geology (DNAG) Project, which also includes six wall maps at a scale of 1:5,000,000 that summarize the geology, magnetic and gravity anomaly patterns, regional stress fields, thermal aspects, and seismicity of North America and its surroundings. Together, the synthesis volumes and maps are the first co-ordinated effort to integrate all available knowledge about the geology and geophysics of a crustal plate on a regional scale.

The products of the DNAG Project present the state of knowledge of the geology and geophysics of North America through the 1980s, and they point the way toward work to be done in the decades ahead.

A. R. Palmer
General Editor for the volumes
published by The Geological
Society of America

J. O. Wheeler
General Editor for the volumes
published by the Geological
Survey of Canada

Foreword

The focus of this book is the structure and Phanerozoic evolution of the transitional region of the North American continent between its craton and bordering oceanic lithospheres (Fig. 1). The geological continent of North America here considered includes Canada, the United States except for Hawaii, Mexico, and northern Central America. The transitional region is the part of North America that is peripheral to the craton (zones 2–4, Fig. 1) and has undergone tectonism within a plate boundary zone at one or more times in the Phanerozoic. The transitional regions that are now active margins provide an understanding of tectonic phenomena that is highly resolved but of short observational period. In contrast, the ancient margins of North America—passive, active, and collisional—provide a long-term tectonic record although one of low resolution in time and space.

The objectives of this book are an up-to-date synthesis of the main structures in and sequences of passive, active, and collisional margins around Phanerozoic North America; its 600 m.y. history of growth and attrition; and interpretations of the processes of plate boundary tectonics.

The restriction of scope of this book to the Phanerozoic eon stems from the thought that at the end of Proterozoic time, the North American continent was mostly or wholly embedded within a supercontinent (Rodinia, McMenamin and McMenamin, 1990). A phase of widespread rifting and drifting at about the Proterozoic-Phanerozoic boundary blocked out the North American continent approximately in its current shape and surface area (Stewart, 1976; Bond and others, 1984). End-Proterozoic North America thus provides a moderately well-determined frame of reference with which to gauge the deformations, motions, additions, and attritions in Phanerozoic time, including the amalgamation of the transient supercontinent, Pangea.

This book is organized by chapters that have a mainly geographic basis. They are divided among regions that traditionally have been thought to have some degree of uniformity of Phanerozoic tectonic evolution. These are the eastern or Appalachian margin (Haworth and others; Rankin), southern or Ouachitan-Mexican margin (Buffler and Thomas; Ortega and others), western or cordilleran margin (Howell and Vedder; Saleeby and others; Monger and others; Moore and others); and northern or Arctic margin (Moore and others; Sweeney and others).

It has become evident in recent years, as discussed in these chapters, that heterogeneity is much greater than previously recognized; that long-distance correlations among rocks, structures, and events are commonly wrong or controversial; and that obliquity of velocity relative to the strike of the plate boundary zone is the rule. Nonetheless, the long-recognized diversity of Phanerozoic evolution among North American margins survives new data and analysis, and such differences are here employed in a hemispheric synthesis.

Besides the chapters with tectonic-geographic bases, this book contains topical chapters on modern active and passive margins of North America (Von Huene; Sawyer). These synthesize the recent findings of the processes and structures created at modern continent-ocean transitions.

Finally, a chapter on the Phanerozoic continent-ocean transitions of North America as a whole (Speed) offers an attempt to encapsulate the interactions, similarities, and differences around the margins of the continent.

BACKGROUND

This book is an outgrowth of a study program on continent-ocean transitions (The Transect Program) begun in about 1980 under the auspices of the U.S. Geodynamics Committee of the U.S. National Academy of Sciences. The ideas for such a study came from John C. Maxwell, then chairman of USGC, and the organization of the program was facilitated by Pembroke C. Hart, permanent coordinator of the USGC. The program expanded via the Canadian Committee on the lithosphere and the Institute of Geology, University of Mexico, to include the whole of North American margins and a group of about 200 geologists and geophysicists. The program received direct support from the National Science Foundation, the Geological Society of America, the University of Mexico, the Geological Survey of Canada, and the U.S. Geological Survey as well as indirect support from the many institutions of the participants.

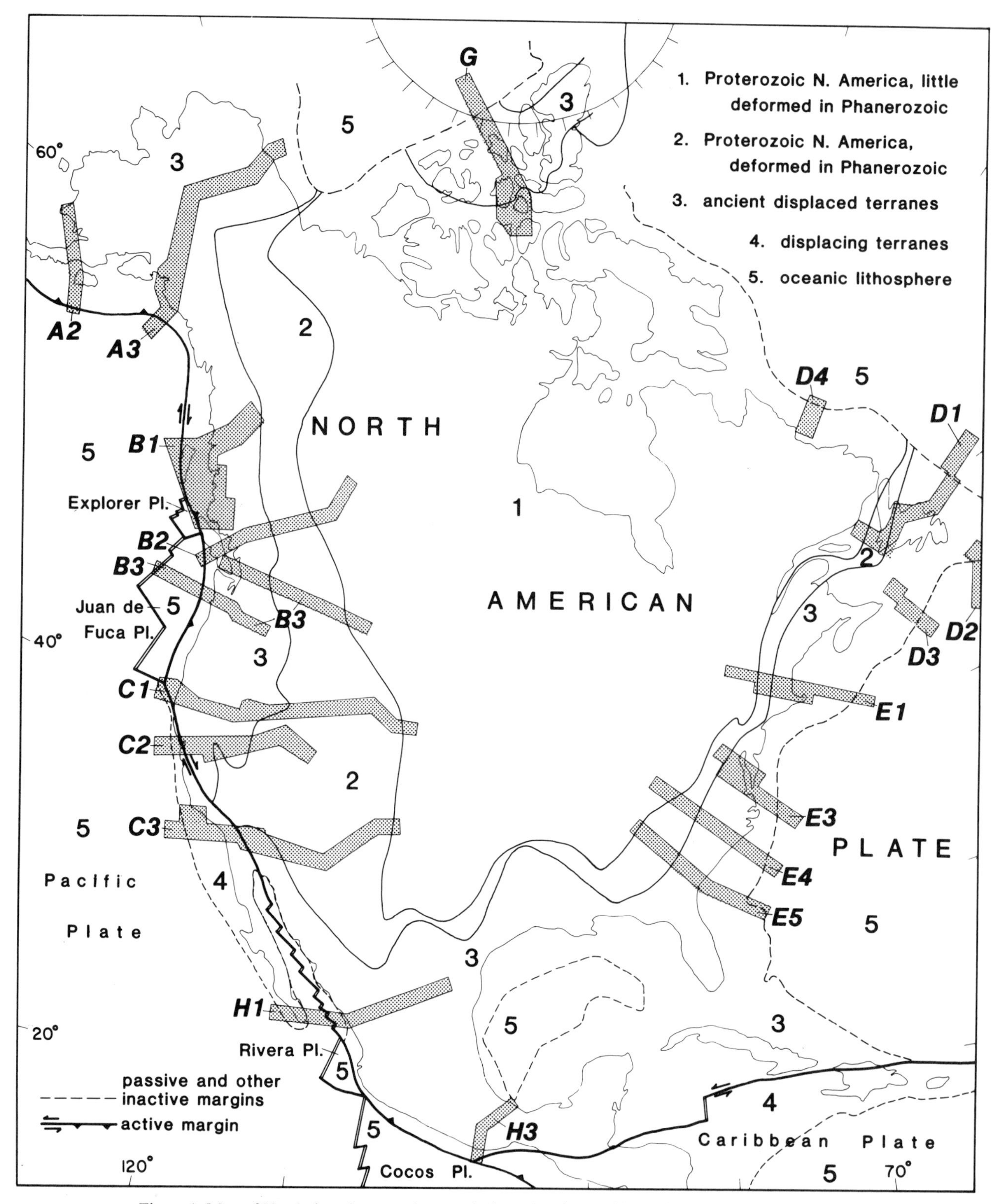

Figure 1. Map of North American continent and plate, showing positions of 19 published transects of the Centennial Continent-Ocean Transect Program.

The program was organized by concentration of study into 23 100-km-wide corridors or transects that cross the margin of North America, from craton to bordering oceanic lithosphere where possible (Fig. 1 and Appendix A). The transect positions were selected in recognition of important problems and by the concentration of data, especially of deep seismic profiling. The program employed existing data. Each transect was manned by a team of scientists that varied from 3 to 25.

The principal vehicle for synthesis and interpretation in the Transects Program was the preparation of graphic displays. These displays, called Continent-Ocean Transects, focus on tectonic cross sections and tectonic event-history diagrams, and include related maps, age-based sections, geophysical data profiles, and explanatory text. The tectonic sections are at 1:500,000 and extend down at least to the Moho. These are published by the Geological Society of America; Appendix A indicates the transect names and principal authors. The corridor abbreviations tie the transects to Figure 1. The tectonic sections of all 23 transects are compiled at 1:1,000,000 in a separate publication (Speed, 1991).

Most of the chapters in this book discuss the principal phenomena that occur in geographic groups of transects and frame them in a tectonic evolution. The reader will find data and discussion of local structures and tectonics in the individual transect's display sheets. Although each chapter should stand alone, much insight will be gained by reading the book with the transects in hand.

R. C. Speed
December 1993

REFERENCES CITED

Bond, G. C., Nickeson, P. A., and Kominz, M. A., 1984, Breakup of a supercontinent between 625 Ma and 555 Ma: New evidence and implications for continental histories; Earth and Planetary Science Letters, v. 70, p. 325–425.

McMenamin, M.A.S., and McMenamin, D.L.S., 1990, The emergence of animals; the Cambrian breakthrough: New York, Columbia University Press, 217 p.

Speed, R. C., 1991, Tectonic section displays, Centennial Continent-Ocean Transects: Boulder, Colorado, Geological Society of America, 2 sheets with text, scale 1:1,000,000.

Stewart, J. H., 1976, Late Precambrian evolution of North America; Plate tectonic implication: Geology, v. 4, p. 11–15.

APPENDIX A. BIBLIOGRAPHIC REFERENCES FOR PUBLISHED CONTINENT-OCEAN TRANSECTS

Speed, R. C., 1991, compiler, Tectonic section displays, Centennial Continent-Ocean Transects: Boulder, Colorado, Geological Society of America, 2 sheets with text, scale 1:1,000,000.

Transect corridor	Reference
A-2	Von Huene, R., and 5 others, compilers, 1985, A-2 Kodiak to Kuskokwim, Alaska: Boulder, Colorado, Geological Society of America, Centennial Continent-Ocean Transect no. 6, 1 sheet with text, scale 1:500,000.
A-3	Grantz, A., Moore, T. E., and Roeske, S., compilers, 1991, A-3 Gulf of Alaska to Arctic Ocean: Boulder, Colorado, Geological Society of America, Centennial Continent-Ocean Transect no. 15, 3 sheets with text, scale 1:500,000.
B-1	Yorath, C. J., and 7 others, compilers, 1985, B-1 Intermontane Belt (Skeena Mountains) to Insular Belt (Queen Charlotte Islands): Boulder, Colorado, Geological Society of America, Centennial Continent-Ocean Transect no. 8, 2 sheets with text, scale 1:500,000.
B-2	Monger, J.W.H., and 5 others, compilers, 1985, B-2 Juan de Fuca Plate to Alberta Plains: Boulder, Colorado, Geological Society of America, Centennial Continent-Ocean Transect no. 7, 2 sheets with text, scale 1:500,000.
B-3	Cowan, D. S., and Potter, C. J., compilers, 1986, B-3 Juan de Fuca Spreading Ridge to Montana Thrust Belt: Boulder, Colorado, Geological Society of America, Centennial Continent-Ocean Transect no. 9, 3 sheets with text, scale 1:500,000.
C-1	Blake, M. C., and 5 others, compilers, 1985, C-1 Mendocino Triple Junction to North American craton: Boulder, Colorado, Geological Society of America, Centennial Continent-Ocean Transect no. 12, 3 sheets with text, scale 1:500,000.
C-2	Saleeby, J. B., compiler, 1986, C-2 Central California Offshore to Colorado Plateau: Boulder, Colorado, Geological Society of America, Centennial Continent-Ocean Transect no. 10, 2 sheets with text, scale 1:500,000.
C-3	Howell, D. G., and 7 others, compilers, 1985, C-3 Pacific abyssal plain to the Rio Grande rift: Boulder, Colorado, Geological Society of America, Centennial Continent-Ocean Transect no. 5, 3 sheets with text, scale 1:500,000.
D-1	Haworth, R. T., Williams, H., and Keen, C. E., compilers, 1985, D-1 (West sheet) Grenville Province, Quebec, to Newfoundland: Boulder, Colorado, Geological Society of America, Centennial Continent-Ocean Transect no. 1, 1 sheet with text, scale 1:500,000. and Keen, C. E., and Haworth, R. T., compilers, 1985, D-1 (East Sheet) Rifted margin offshore northeast Newfoundland: Boulder, Colorado, Geological Society of America, Centennial Continent-Ocean Transect no. 1, 1 sheet with text, scale 1:500,000.
D-2	Keen, C. E., and Haworth, R. T., compilers, 1985, D-2 Transform margin south of Grand Banks: Offshore eastern Canada: Boulder, Colorado, Geological Society of America, Centennial Continent-Ocean Transect no. 2, 1 sheet with text, scale 1:500,000.
D-3	Keen, C. E., and Haworth, R. T., compilers, 1985, D-3 Rifted continental margin off Nova Scotia: Offshore eastern Canada: Boulder, Colorado, Geological Society of America, Centennial Continent-Ocean Transect no. 3, 1 sheet with text, scale 1:500,000.
D-4	Keen, C. E., and Haworth, R. T., compilers, 1985, D-4 Rifted continental margin off Labrador: Boulder, Colorado, Geological Society of America, Centennial Continent-Ocean Transect no. 4, 1 sheet with text, scale 1:500,000.
E-1	Thompson, J. B., Jr., Bothner, W. A., Robinson, P., and Klitgord, K. D., compilers, 1993, E-1 Adirondacks to Georges Bank: Boulder, Colorado, Geological Society of America, Centennial Continent-Ocean Transect no. 17, 2 sheets with text, scale 1:500,000.
E-3	Glover, L., III, Klitgord, K. D., and 15 others, compilers, 1994, E-3 Southwestern Pennsylvania to Baltimore

Canyon Trough: Boulder, Colorado, Geological Society of America, Centennial Continent-Ocean Transect no. 19, 2 sheets with text, scale 1:500,000 (in press).

E-4 Rankin, D. W. and 13 others, compilers, 1991, E-4 Central Kentucky to the Carolina Trough: Boulder,Colorado, Geological Society of America, Centennial Continent-Ocean Transect no. 16, 2 sheets with text, scale 1:500,000.

E-5 Hatcher, R. D., Jr., and 11 others, compilers, 1994, E-5 Cumberland Plateau to Blake Plateau: Boulder, Colorado, Geological Society of America, Centennial Continent-Ocean Transect no. 18, 2 sheets with text, scale 1:500,000 (in press).

G Sweeney, J. F., compiler, 1986, G Somerset Island to Canada Basin: Boulder, Colorado, Geological Society of America, Centennial Continent-OceanTransect no. 11, 2 sheets with text, scale 1:500,000.

H-1 Mitre-Salazar, L-M, and Roldan-Quintana, J., compilers, 1990, H-1 La Paz to Saltillo, Northwestern and northern Mexico: Boulder, Colorado, Geological Society of America, Centennial Continent-Ocean Transect no. 14, 2 sheets with text, scale1:500,000.

H-3 Ortega-Gutierrez, F., compiler, 1990, H-3 Acapulco Trench to the Gulf of Mexico across Southern Mexico: Boulder, Colorado, Geological Society of America, Centennial Continent-Ocean Transect no. 13, 1 sheet with text, scale 1:500,000.

Transects A-1, E-2, F-1, F-2, and H-2 were never completed as separate publications. Transects E-2, F-1, and F-2 are shown only on the Tectonic Section Displays (Speed, 1991).

Printed in U.S.A.

DNAG Continent-Ocean Transect Volume
Phanerozoic Evolution of North American
Continent-Ocean Transitions
The Geological Society of America, 1994

Chapter 1

North American continent-ocean transitions over Phanerozoic time

Robert C. Speed
Department of Geological Sciences, Northwestern University, Evanston, Illinois 60208

INTRODUCTION

This chapter presents a synthesis of the structure and Phanerozoic tectonic evolution of continent-ocean transitions around nearly the entire North American continent. Its objectives are to compare and contrast the modern transitions and Phanerozoic histories of specific margins of North America discussed in other chapters of this book and in the Transect displays (see Foreword) and to present ideas on processes in the evolution of continent-ocean transitions.

A word about the focus on the continent's margins and their place in the global scheme of tectonics may be warranted. The modern transitions between the North American continent and adjoining oceanic basins are where plate tectonics has most recently affected the continent aside from rigid translation. At active margins, where oceanic lithosphere underrides or slides along the continent, the development of the margin is ongoing and may have been continuous over a long time. In contrast, at passive margins where the continent has split and sea-floor spreading has occurred, the continent-ocean transition was developed in a discrete event and was discontinuous in time. At passive margins, the most recent North American transitions range from Holocene to mid-Mesozoic. At collisional margins, where the North American continent is or was on the underriding plate, the development of the margin was also discrete and discontinuous. As with passive margins, an existing collisional margin is generally the product of an ancient event.

During the development of all margin types, the motions between the North American plate and adjacent plates have commonly been broadly distributed and nonuniform across a zone of finite width inboard of the continental edge. Thus, a broad swath of continent must be studied to understand fully the tectonics of past and present plate boundary zones. Plate motions have caused the shapes and volumes of continents, as well as their positions, to evolve through time. The distributed motions in boundary zones of the North America plate have transferred large and small fragments of the North American continent and terranes of other origins from place to place along the margin of the continent and between North America and other continents. Such motions, moreover, have caused some loss of continental crust to the mantle by subduction and the addition of mantle material to the continent, mainly at its margins, by magmatism.

Perhaps most important, it is well known that most oceanic crust is returned to the mantle after a brief residence at the surface, certainly less than 200 m.y. In contrast, continental and other types of nonoceanic crusts survive, at least in some large proportion, over much longer increments of time, perhaps all of geologic history. The long-term record of continent-ocean interactions and of plate boundary tectonics is therefore in the continents, and the more recent, more interpretable chapters of the record are at the Phanerozoic continental margins.

This chapter begins with the North America plate, with goals of highlighting some modern tectonic effects on the North American continent and providing a reference for the likely complexity of past plate boundary zones. This is followed by discussions of the present-day and end-Proterozoic North American continents, the end states of North America's Phanerozoic tectonic evolution. Next considered are the Phanerozoic tectonic evolutions of North American continental margins, in the order eastern, southern, western, and northern. The orientations of ancient features are referred to in modern coordinates. Major points of the structure and processes of North American continent-ocean transitions and continental margin evolution are brought together at the end in a synopsis that is thoroughly mixed with interpretation and speculation.

NORTH AMERICA PLATE

Introduction

The North American continent is mainly but not completely in the North America plate, hereafter called NAm, whose relations with surrounding plates (Fig. 1) provide modern examples of complexities and processes of continent-ocean transitions. NAm contacts at least 10 other plates. Figure 1 shows most of the NAm boundaries, which are drawn as discrete faults

Speed, R. C., 1994, North American continent-ocean transitions over Phanerozoic time, *in* Speed, R. C., ed., Phanerozoic Evolution of North American Continent-Ocean Transitions: Boulder, Colorado, Geological Society of America, DNAG Continent-Ocean Transect Volume.

Figure 1. Map of Western Hemisphere showing North America plate and adjacent plates. Circled dots are rotation poles for plate pairs: NAm-Pa is North America–Pacific and NAm-SAm is North America–South America. Divergent plate boundaries: double lines; convergent and dominantly strike-slip boundaries: single heavy line. Regions of plate boundary zones (pbz) with broadly distributed relative plate motion shown for North America–Pacific and North America–South America (lined pattern). Abbreviations: QCF, Queen Charlotte fault; SAF, San Andreas fault; triple junctions: Mtj, Mendicino; Rtj, Rivera; Ttj, Tehuantepec; Atj, Azores. MAR, Mid-Atlantic Ridge; EPR, East Pacific Rise. Trenches: MA, Middle America; PR, Puerto Rico. C, Cayman Ridge; ABC, Alta-Baja California superterrane; TMVB, Trans-Mexican Volcanic Belt; Hisp, Hispaniola; S, Seattle; LA, Los Angeles; v pattern indicates active volcanic chains.

among rigid plates. Such faults are the loci of maximum rate of change of velocity across plate boundary zones of finite width. Boundaries of NAm extend west of Figure 1 to a triple junction with the Eurasia and Pacific plates near Japan (Fig. 2).

The major plates contacting NAm are the Pacific, South America, Africa, and Eurasia plates (Fig. 1), of which only the Pacific does not contain a continent. Minor plates, Explorer, Juan de Fuca, Rivera, and Cocos (Fig. 1), are all oceanic and are remnants of the trailing edge of the formerly extensive Farallon plate, which is almost fully subducted below western NAm (Atwater, 1970; Engebretson and others, 1985). The Caribbean, another small plate, is also probably wholly or partly of Farallon plate origin, but from an intraplate tract of the Farallon, not its trailing edge (Malfait and Dinkleman, 1972). The Caribbean-NAm boundary zone includes terranes of uncertain origin whose kinematic alliance may be principally with the Caribbean plate (see below).

The velocity vectors of bordering plates relative to NAm are shown along the edge of a rigid NAm in Figure 2. These are calculated from the angular velocities of the NUVEL 1 model (DeMets and others, 1990), which is a 12 plate global model with rates averaged over the past 3 m.y. It is evident that the NAm boundaries have highly varied motion, convergent and divergent with all degrees of obliquity, and that a few are nearly perfect strike slip. Obliquity is the ratio of magnitudes of the margin-parallel component of velocity to total velocity.

The motion of NAm relative to the mantle below it (Fig.

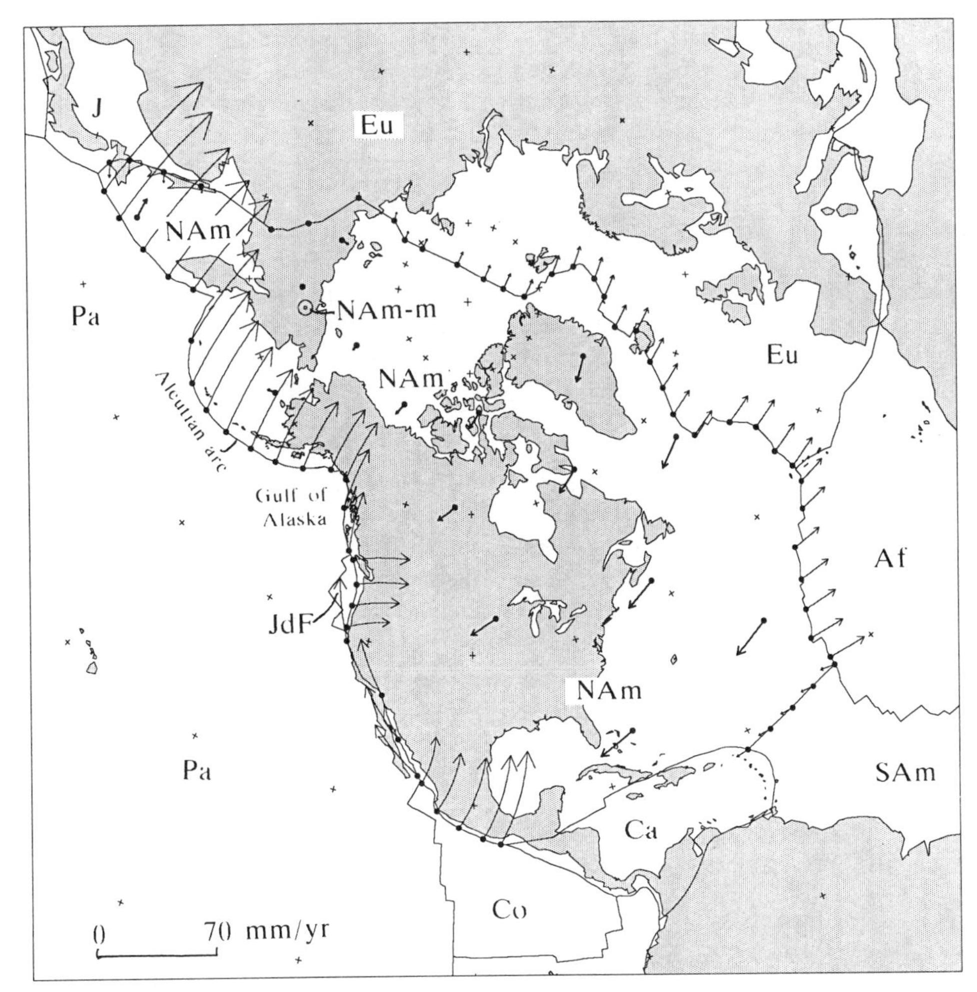

Figure 2. Map showing North America plate and adjacent plates; arrows on boundaries of North America plate indicate direction and magnitude of relative velocities of adjacent plates calculated at positions of dots; heavy arrows within North America plate indicate velocity of North America relative to the mantle about rotation pole, NAm-m. NAm, North America plate; SAm, South America plate; Eu, Eurasia plate; Af, Africa plate; JdF, Juan de Fuca plate; Pa, Pacific plate; Co, Cocos plate; Ca, Caribbean plate; J is Japan; scale bar indicates magnitude of velocity arrows.

2) can be calculated from an angular velocity vector of the HS2-NUVEL 1 model of Gripp and Gordon (1990), which employs nine Pacific hotspots and the Yellowstone hotspot of North America as mantle references and the NUVEL 1 model of relative plate motions. The NAm-mantle pole is in easternmost Siberia (Fig. 2), and the rate is 0.28°/m.y. clockwise around the pole. The linear velocity, which increases latitudinally away from this pole, is maximum in the southernmost reaches of NAm at about 30 mm/yr. This absolute plate velocity is slow relative to that of most plates that have either downgoing slabs or smaller proportions of continental lithosphere.

The principal variables among the plate boundary zones of NAm are the existence of convergent or divergent components, the obliquity of the velocity, the width, and the existence and migration rate of triple junctions.

Alaska and western Canada

The plate boundary zone of western North America extends in the area of Figure 1 from the Aleutian Trench east and southeast to the Tehuantepec triple junction with the Cocos and Caribbean plates. It is mainly convergent, but includes segments with either nearly pure strike-slip or divergent motions (Fig. 2). The western boundary zone is notable for the great width of NAm over which modern relative plate motions are taken up (Fig. 1).

The NAm-Pacific pole is in eastern Canada (Fig. 1) (DeMets and others, 1990), such that relative velocities are highly oblique along the northwest trend of the boundary zone from Mexico to the Gulf of Alaska. West of the Gulf of Alaska, the directions of velocity vary from normal to parallel around the south-convex arc of the Aleutian trench (Fig. 2). The plate boundary zone around the great bend of the Gulf of Alaska, the transition from nearly strike slip to strongly convergent motion, is structurally complex (Fig. 3) (Von Huene, this volume; Moore and others, this volume). The zone is narrow (a few tens of kilometers wide) in Canada where the velocity vector is nearly parallel to the Queen Charlotte fault, which is a few kilometers inboard from the continent-ocean boundary (Fig. 1), and where seismicity suggests an absence of active faults landward (Fig. 3). There is, however, a small convergent component across the Queen Charlotte fault (Fig. 2), indicated by plate motions (DeMets and others, 1990) and by active structures (Yorath and others, 1985). Northward toward Alaska, the boundary zone broadens into NAm with the bifurcation of the Queen Charlotte fault into several strands (Fairweather, Dalton, Cross Sound faults), all of which have major modern and ancient right-lateral components of slip. North of the corner in the plate boundary in the Gulf of Alaska, the active zone is at least 800 km wide (Fig. 3) and, according to Grantz and others (1991), may extend another 800 km north to Alaska's North Slope.

In the convergent realm of the boundary zone from the Gulf of Alaska west to the Aleutian trench (Fig. 3), NAm is fronted by a huge accretionary, mainly sedimentary, forearc (Von Huene, this volume) that overlies the subhorizontal Pacific plate. In the eastern forearc, the oceanic Yakutat terrane (Fig. 3) occupies the seaward portion and has detectable motions relative to both NAm and the Pacific plate (Grantz and others, 1991; Moore and others, this volume). The great breadth in NAm of the plate boundary zone north of the Gulf of Alaska seems related to the northeastern edge of the underlying Pacific plate, where its contact with NAm changes from a high to a low obliquity boundary below Alaska.

Landward of the forearc is an active magmatic arc that lies above the dipping slab of the Pacific plate. This is the Aleutian island arc west of about 170°W, where the volcanic chain lies between the forearc and an inactive oceanic marginal basin, the Bering Sea. The basin is probably either of Paleogene backarc spreading origin or a fragment of Pacific lithosphere trapped by a southward jump of the trench. East of 170°W, the magmatic arc is constructed on terranes of diverse infrastructure, but its tectonic affinity seems chiefly continental.

The Aleutian arc system changes trend by more than 70° in the area of Figure 2 west from the Gulf of Alaska, so that the obliquity of the NAm-Pacific velocity increases westward from 0 to 0.9. The bearings of slip vectors of thrust earthquakes below the arc system are between the velocity vector and trench normal and change westward with partial correspondence to the westward change in obliquity (DeMets and others, 1990; McCaffrey, 1992). If such bearings are taken as the minimum principal elongation direction, there is an evident partition of strain in the arc system between boundary-normal and boundary-parallel components (Walcott, 1978; Beck, 1983) that increases with obliquity. The partitioning in the Aleutian arc may reach a limit at obliquity of about 0.5 and be constant at higher values (McCaffrey, 1992). The Aleutian arc system, however, contains no evident major strike-slip faults, and the boundary-parallel compo-

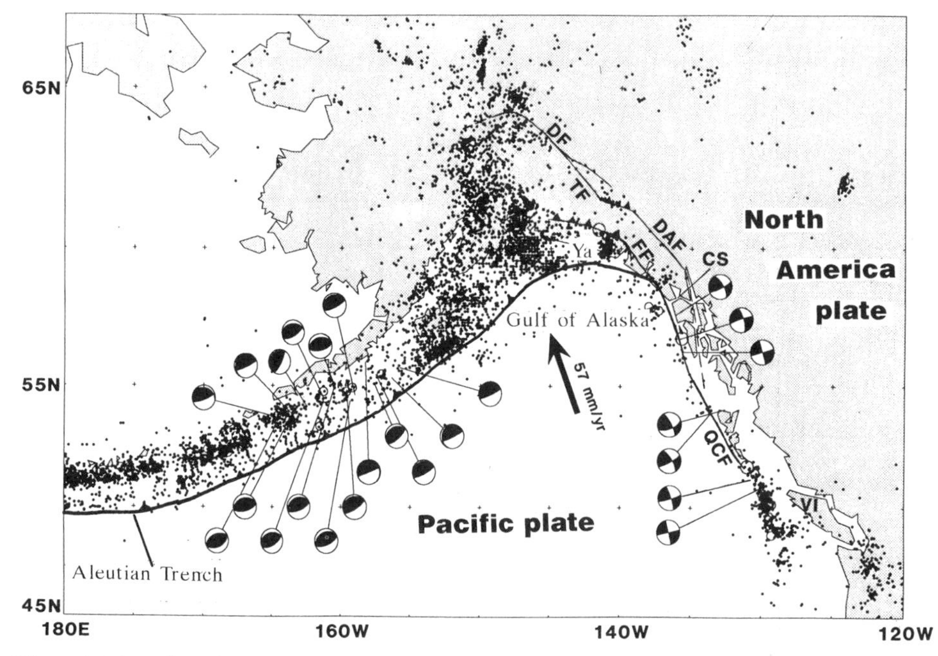

Figure 3. Map of southern Alaska and western Canada showing major active faults, shallow seismicity (epicenters 1963–1985, depths ≤50 km), focal mechanisms (lower hemisphere, compressional quadrants darkened), and Pacific–North America velocity in the Gulf of Alaska. Ya is Yakutat terrane; DF, Denali fault; DAF, Dalton fault; TF, Totschunda fault; QCF, Queen Charlotte fault; FF, Fairweather fault. VI is Vancouver Island; CS is Cross Sound; modified from DeMets and others (1990).

nent must be taken up in a broad, perhaps arcwide, zone of simple shear strain. Moreover, because the obliquity increases westward, there is also a westward increase in the boundary-parallel component of plate velocity. This implies that the arc is in extension parallel to its length, a prediction corroborated by normal faulting in the Aleutian arc (Geist and others, 1988).

Juan de Fuca

The Juan de Fuca and Explorer plates underride NAm at the Cascade forearc and continental magmatic arc after a short transit from the Juan de Fuca spreading ridge (Fig. 1) (Cowan and Potter, 1986; Monger and others, this volume; Saleeby and others, this volume). The Juan de Fuca plate velocity relative to NAm is N69°E, 42 mm/yr at Seattle (Fig. 1) according to the NUVEL 1 model. This direction is close to the N66°E bearing of minimum principal elongation of active deformation in the forearc obtained geodetically by Lisowski and others (1987). The direction suggests an absence here of partitioning of strain between boundary-normal and boundary-parallel components in spite of modest obliquity. The Juan de Fuca and Pacific plates and NAm are in contact at the Mendicino triple junction (Fig. 1), which is migrating northwesterly along the edge of NAm. The migration implies that the Juan de Fuca–Pacific spreading ridge collides progressively with NAm. The triple junction region shows large departure from rigid plate tectonics: thrusting of the Pacific plate above the Juan de Fuca, breakup of the Juan de Fuca plate along ridge-parallel faults, and northeast-vergent thrusting in NAm.

Southwestern plate boundary zone

Between the Mendicino and Rivera triple junctions, NAm and the Pacific plate are in contact, and their relative velocity, averaging 48 ± 1 mm/yr in this region, is distributed in a zone of remarkable width, about 2500 km. The velocity gradient is not constant across this plate boundary zone. The oceanic lithosphere of the Pacific basin has an apparently uniform velocity and undergoes no deformation. The leading edge of the continent (toe of slope) is attached to oceanic lithosphere, so that the velocity gradient exists fully within the continent. The maximum of the gradient is 100–300 km inland at the San Andreas fault, the late Neogene displacement rate of which is between 20 and 35 mm/yr as a function of position. The locus of the maximum is commonly taken to be the plate boundary in the context of classical rigid plate tectonics (Fig. 1). The region between the continent-ocean boundary and the San Andreas fault is the Alta-Baja California superterrane (Fig. 1). In the sense of rigid plate tectonics, this superterrane belongs to the Pacific plate while it is also part of the North American continent.

The Pacific-NAm velocity is highly oblique in general in southwestern North America, but the obliquity varies. It is slightly convergent at central California latitudes, more strongly convergent in the Los Angeles region, and slightly divergent in the Gulf of California.

The broad plate boundary zone is also notably high, about 1 km on average above the continent as a whole. Its elevation, together with high heat flow, Neogene volcanism over much of it, and velocity structure indicate that it overlies a dome of asthenosphere.

A principal structure of this segment of the boundary is the San Andreas fault system, a complex of anastomosing, mainly northwest-striking faults, with right-slip components in a zone about 100 km wide normal to the plate boundary zone. Of these, the San Andreas fault (Fig. 1) is the principal displacement-rate surface. East of the San Andreas system, the Sierra Nevada block (Fig. 4) is apparently translating northwest relative to NAm according to modern geodesy (Argus and Gordon, 1991). Farther east, the Basin and Range province of the United States and Mexico (Fig. 4) is undergoing heterogeneous extension (Saleeby and others, this volume; Howell and Vedder, this volume; Ortega and others, this volume). The bearings of maximum principal elongations, strain rates, and rotation rates in the Basin and Range are little known from direct observation. The Rio Grande rift on the eastern side of or east of the Basin and Range is a zone of localized intracontinental extension with active basalt venting and about 35 km total displacement (Fig. 4). The Colorado Plateau is a very broad, little-stretched block between the Rio Grande rift and the Basin and Range province.

The transtensional southern realm of this plate boundary zone, the Gulf of California, includes a series of short spreading segments that cause the development of oceanic lithosphere between conjugate edges of mainland Mexico and the Alta-Baja California superterrane. The superterrane split off at about 6 Ma

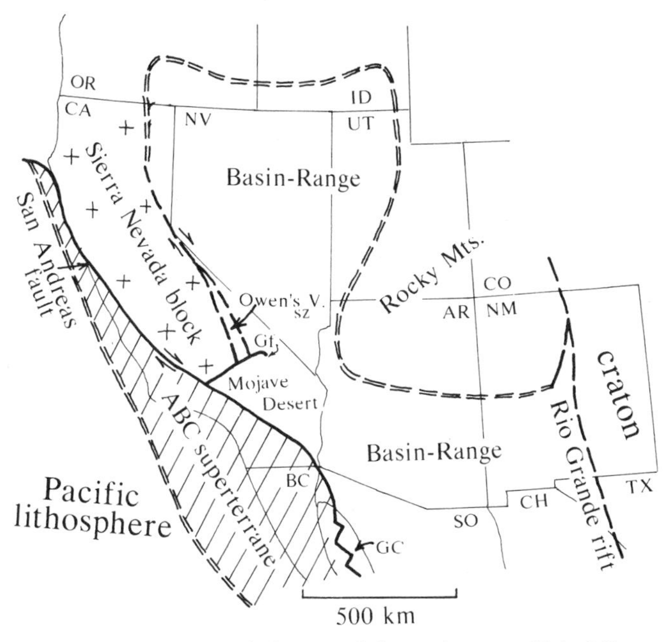

Figure 4. Some neotectonic features of the southwestern United States and northern Mexico. Heavy lines are active faults. Sierra Nevada block from Argus and Gordon (1991). Double dashed lines are tectonic province boundaries. SZ is shear zone; Gf is Garlock Fault; ABC is Alta-Baja California; GC is Gulf of California.

as the maximum in the velocity gradient jumped inboard of the ocean-continent boundary to the locus of a former magmatic arc, apparently selecting the weakest surface, according to a model by Lonsdale (1989) (modified by Sedlock and others [1993] and by Ortega and others [this volume]).

The modern kinematics and dynamics of this plate boundary zone are still controversial. In central California, the observed displacement rate on the San Andreas fault is about 35 mm/yr (Sieh and Jahns, 1984), and the strike of the fault is about 5° clockwise of the plate velocity vector, implying boundary-normal contraction (DeMets and others, 1990). The difference in boundary-parallel motions between the San Andreas fault and the plates at this latitude must be distributed among the faults of the San Andreas fault system, the Sierra Nevada block, the Basin and Range province, and the Rocky Mountains. The boundary-normal contraction in central California is apparently taken up by folding and thrusting at shallow seismic levels ($\leqslant$12 km) (Namson and Davis, 1988). This implies that the San Andreas system of oblique slip faults exists in a detached deforming brittle layer above a plastic zone where the shear direction and plate velocities may be parallel.

The Sierra Nevada block, now a large, seemingly unified rigid mass of former terranes and fault blocks, is an interesting contrast to the finely broken up Basin and Range (Fig. 4). A principal difference of the two regions is in heat flow, the Sierra Nevada low and Basin and Range high. This implies that the depth to the shallowest brittle-plastic failure transition and the relative strength of the upper brittle layer, both strongly controlled by temperature (Brace and Kohlstedt, 1980; Sawyer, this volume), are greater in the Sierra Nevada than in the Basin and Range. In turn, these factors may dictate the horizontal dimension of fault-bounded blocks in the brittle zone.

A second major question is the relation of the Basin and Range kinematics to the NAm-Pacific motions. Is the extension in the Basin and Range totally related to Neogene oblique plate motion, or is it a relict backarc zone of thinned lithosphere related to an earlier Farallon-NAm convergent margin, now mainly undergoing reactivation of inherited structures? Geodetic measurements point toward the first solution. A local network across Owen's Valley (Fig. 4), which is within or includes the boundary between the Sierra Nevada and Basin and Range province, implies boundary-parallel horizontal right-lateral simple shear (Savage and others, 1990). Furthermore, regional geodesy also suggests motion of the Sierra Nevada block as a whole relative to cratonal North America by right slip along its eastern boundary (Argus and Gordon, 1991). In Owen's Valley, the velocity direction from regional geodetics is N28W $\pm$ 3°, ~11 mm/yr. Given that plate boundary kinematics are steady over million-year increments and that geodesy gives valid samples of such motion, as argued by Argus and Gordon (1990), the following implications arise for the Basin and Range: (1) the Basin and Range undergoes right-lateral simple shear that is close to but does not exactly parallel the NAm-Pacific velocity vector; (2) Basin and Range shear takes up about a quarter of the plate motion; and (3) the maximum principal elongation is about N75W-S75E, a bearing about normal to the most frequent strikes of normal faults in the Basin and Range.

The Rivera triple junction is the intersection of NAm and the East Pacific Rise, the spreading ridge of the Cocos and Rivera plates on the east, and the Pacific plate on the west (Fig. 1). The triple junction is migrating southeast along the edge of NAm, causing a transition from low to high obliquity in motion along the boundary of NAm. A complex of grabens and volcanoes in Mexico just inland from the Rivera plate has been inferred to bound a nascent microplate that together with the Rivera plate may soon join the Pacific plate as the Pacific-NAm boundary jumps inboard, as it apparently did in the Gulf of California at 6 Ma (Allan, 1986; Ortega and others, this volume).

Southeast of the Rivera triple junction, the Cocos-NAm boundary is well defined by the Middle America Trench and by a Benioff zone with large thrust earthquakes (Fig. 5). The NUVEL 1 model indicates nearly normal convergence. The distribution of volcanoes in southern Mexico, however, is not easily related to the plate boundary; the only linear chain is the Trans-Mexican Volcanic Belt (Fig. 1), which crosses the Mexican mainland with discordance to the trench (Ortega and others, this volume). The Trans-Mexican Volcanic Belt may be a reactivated boundary of amalgamated Mexican terranes. The absence of a chain of volcanoes in NAm paralleling the trench is nonetheless a problem.

Southern plate boundary zone

The southern boundary of NAm is with the Caribbean (CA) and South America (SAm) plates and lies between poorly defined triple junctions, the Tehuantepec (NAm-CA-Cocos) and a nameless one at the Mid-Atlantic Ridge (NAm-SAm-Africa) (Fig. 1). The NAm-CA boundary (Molnar and Sykes, 1969) occupies a zone of finite but uncertain width that is commonly assigned to the Caribbean plate (e.g., Case and Holcombe, 1980; Burke and others, 1984). The zone includes terranes of northern Central America and the Greater Antilles and may extend south into oceanic lithospheres of the Caribbean basin. If the terranes in fact move mainly or wholly with the CA, the southern edge of NAm is the principal displacement surface, as shown in Figure 1. If so, the southern margin of the North American continent is rigid, unlike its western margin, which takes up plate boundary deformation over a broad zone. An explanation for the apparent difference in strength of the two continental margins is their difference in thermal history; the western margin was a broad continental arc system before intersection with the Pacific plate, and the southern margin is an old passive margin, fully cooled and probably webbed by dikes.

The NAm-CA plate velocity vector is nearly boundary parallel and left lateral, but the details are controversial (Jordan, 1975; Sykes and others, 1982; Stein and others, 1988; DeMets and others, 1990). Rates, about 15 mm/yr (Rosencrantz and others, 1988), can be determined only from a short spreading ridge, the Cayman Ridge (Figs. 1 and 5). The Cayman Trough is

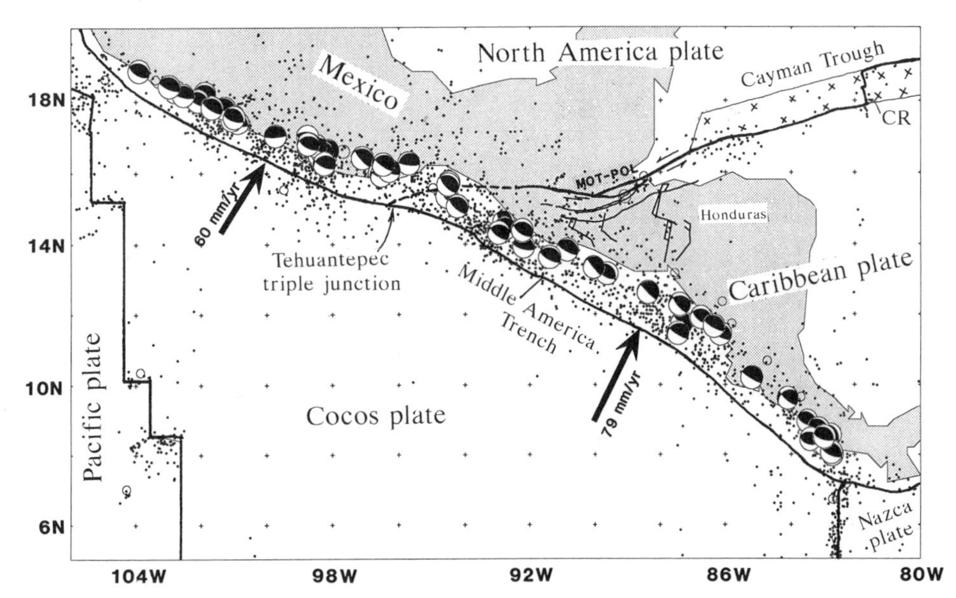

Figure 5. Plates around southern Mexico and Central America; shallow seismicity (1963–1985, depths ≤60 km) shown by black dots (magnitude 4.5 to 5.5) and circles (magnitude ≥5.5); focal mechanisms are lower hemisphere, compressional quadrants darkened; Cocos–North America and Cocos-Caribbean velocities shown at one site each; CR is spreading ridge in Cayman Trough. MOT-POL—Motagua and Polochic fault zones; modified from DeMets and others (1990).

a narrow oceanic pull-apart basin at a jog in the northern edge of the plate boundary zone. The floor of the trough is anomalously deep for its age, and its ridge produces anomalously thin oceanic crust; both peculiarities may be due to significant horizontal heat transfer into the walls of the trough relative to the short length of the ridge (Boerner and Sclater, 1987). The NAm-CA rate, however, may exceed that in the Cayman Trough by virtue of slip in the plate boundary zone to the south (Sykes and others, 1982).

The western reach of the NAm-CA boundary splays west into a system of faults (Motagua, Polochic, and others; Fig. 5), and south of this system, northern Central America is block faulted and undergoes active east-west extension (Burkart, 1983; Manton, 1987). The structure and kinematics of the intersection of this fault system with the Middle America Trench (Fig. 5), the Tehuantepec triple junction, are unclear because there are no evident morphologic or seismic manifestations of it.

The northern margin of the NAm-CA boundary zone east of Hispaniola is the Puerto Rico Trench, the locus of deepest ocean floor (10 km) in the Atlantic-Caribbean basin. It was regarded by McCann and Sykes (1984) as a subductional feature at a transpressional boundary and as a breakaway zone at a transtensional boundary by Speed and Larue (1991). Plate motion models can support either interpretation, depending on the kinematic data selected; use of slip vectors of thrust earthquakes around the eastern Caribbean implies transpression, whereas use only of strike-slip fault azimuth and Cayman Ridge data implies transtension (DeMets and others, 1990). Because of these uncertainties, NAm-CA velocity vectors are excluded from Figure 2.

The NAm-SAm plate boundary (Fig. 1) is especially mysterious because it is cryptic and apparently aseismic. Its existence is indicated by a nonzero NAm-SAm angular velocity calculated from (NAm-Africa)–(SAm-Africa); both terms are well known (Ladd, 1976; Pindell and others, 1988). The pole is in the central Atlantic between the Caribbean plate and the Mid-Atlantic Ridge (Fig. 1), according to NUVEL 1 and all other global plate models (DeMets and others, 1990). The most recent hypothesis for the NAm-SAm plate boundary (Argus, 1990) is that it occupies a wide deformation zone in the central Atlantic between about 18° and 25°N (Fig. 1). It is based on the existence of systematic departures in strikes of transform faults of the Mid-Atlantic Ridge from strikes predicted by a model using a single NAm-SAm boundary fault. The NAm-SAm velocity in this zone is transtensional in Argus's model (Fig. 2), but is probably highly nonuniform owing to the proximity of the pole to this boundary zone.

Eastern and northern boundaries

The eastern and northern boundaries of NAm are mainly divergent in the Atlantic basin and have velocity vectors that are boundary-normal (Figs. 1 and 2). The rates diminish northward along the spreading locus of the Mid-Atlantic Ridge from a maximum of 25 mm/yr to 0 at the Eurasia-NAm angular velocity pole in easternmost Siberia (Fig. 2). Within the central Atlantic, the boundary includes the Azores triple junction (Fig. 1) among the Eurasia and Africa plates and NAm. There, Africa converges northwesterly relative to Eurasia at a low rate, about 5 mm/yr. West of the pole in Siberia, the NAm-Eurasia boundary is convergent with little or no obliquity; the closure rate increases westward to a triple junction among the Eurasia, and Pacific plates and NAm (Fig. 2) (Chapman and Solomon, 1976).

NORTH AMERICAN CONTINENT—DEFINITIONS

The distinction between the North American continent and the North America plate (NAm) is first emphasized. The plate that carried the continent at any time in history is called the NAm. The word, North America, by itself, refers to the continent, not the plate.

By "North American continent," I mean the region that includes geographic North America and extends outward to an edge that is circumscribed by little-deformed oceanic lithosphere. The definition is elaborated below. Varied usages of the term "North American continent" are employed. The most frequent are present-day North America, referring to the continent today, and Proterozoic North America, referring to North America at the end of Proterozoic time. For ages between these end points of the Phanerozoic, the continent is referred to with an age modifier, for example, Late Ordovician North America.

Another usage inviting explanation is the "edge of the North American continent." The edge of present-day North America is today's continent-ocean boundary, or more exactly, the trace of the most inboard, little-deformed oceanic crust and its cover at the sea floor. The edge of Proterozoic North America, however, requires an age modifier because Proterozoic North America has had its margins excised to varied degrees at different times and places. Thus, "end-Proterozoic edge of Proterozoic North America" is used for the edge at the beginning of the span of inquiry. The "Middle Permian edge of Proterozoic North America" is, for example, a different line. Note that the Permian edge of Proterozoic North America is not the same as the edge of Permian North America.

Craton is used for the region of Proterozoic North America that is little modified by Phanerozoic tectonics. Continental margins are generally broad zones circumscribing the craton; all have been modified significantly by Phanerozoic tectonics. North American margins are divided among three principal classes, passive, active, and collisional, discussed below. Note that a margin refers to part of the continent, not the North America plate.

PRESENT-DAY NORTH AMERICA

Definition

An idea conveyed in the Introduction is that in a global historical-tectonic context, the continents are the ultimate repository at the earth surface of crustal materials other than oceanic crust of spreading origin. It is within this sense that I define the present-day North American continent. The usual criteria, bathymetry, size, and crustal thickness and composition, are inadequate. Thus, it is assumed here that the North American continent includes all contiguous ground from the continental interior out to boundaries with little disrupted oceanic crust, as illustrated in Figure 6. This means that the continent extends beyond the shoreline to the base of slope regardless of plate boundaries or crustal thickness or composition, except for a couple of assumed and arbitrary separations explained below.

At passive margins, the definition permits all degrees of crustal stretching and addition of basalt. There, the age of the edge of the present-day North American continent varies from Holocene to Jurassic, depending on the time at which drifting began at a nascent divergent boundary.

At active margins, which are at a convergent boundary with oceanic lithosphere descending below the continent, the edge of North America is the frontal emergent thrust of the forearc. There, the continent may include crustal fragments of any composition and thickness, including oceanic. The age of its edge is Holocene. For example, the oceanic Yakutat terrane, which is still being displaced relative to cratonal North America within the Alaskan forearc (Fig. 3), belongs to the continent because it is no longer little disrupted oceanic crust.

At active continental margins that contain a plate boundary zone with highly oblique motions, the continent extends seaward to the base of slope. Along much of western Canada (Fig. 1), for example, the continental edge is the trace of frontal thrusts of inactive accretionary forearcs and is slightly seaward of the Queen Charlotte fault, which is the locus of the velocity gradient maximum. Much of the San Andreas fault system (Fig. 1), in contrast, lies well within the continent. To the west of the San Andreas fault, present-day North America includes a wide sliver, the Alta-Baja California superterrane (Fig. 6), composed of rocks of varied kindred and ages as old as Proterozoic; the mean velocity of the terrane relative to the craton is close to that of the Pacific plate. Similarly, present-day North America, as defined, includes the nonoceanic Chortis terrane of southern Mexico and northern Central America (Fig. 6), although Chortis is apparently moving mainly with the Caribbean plate.

My definition of continental North America runs into snags in the Caribbean region because the circum-Caribbean island arcs (Central America, Greater Antilles) lie between evident continent and the little-disrupted oceanic crust of the Caribbean Sea (Fig. 6). If circum-Caribbean arcs are in discrete terranes within a broad plate boundary zone, they belong to continental North America; if, however, the arcs are built on oceanic crust contiguous with that of the Caribbean Sea, it can be argued that they represent the modern oceanic rather than continental realm. The second case probably applies to southern Central America and establishes the southern margin of the Chortis terrane as the edge of the North American continent. Evidence comes from seismic and onland studies, which strongly indicate that Caribbean oceanic crust underlies the region from southern Nicaragua to Panama (Bowland and Rosencrantz, 1988; Case and Holcombe, 1980; Frisch and others, 1992). It is not as clear that the second case applies to the Greater Antilles, but I assume that it does and that the plate boundary zone is intra-Caribbean and that the continental edge is the northern edge of that zone (Fig. 6).

Southern continental North America fully or partly surrounds three enclaves of little disrupted oceanic crust (Fig. 6). The Gulf of Mexico basin (Buffler and Thomas, this vol.) and the Yucatan Basin (Rosencrantz, 1990) are inactive, formed by short-lived(?) spreading episodes. The Cayman Trough (Holcombe and

Figure 6. Map showing North America continent defined and addressed in text (no pattern); solid lines are geologic boundaries; dash-dot lines are arbitrary boundaries used for restriction of discussion; v-dot pattern is oceanic lithosphere; lined pattern is nonoceanic crust not included in North America continent; GMB is Gulf of Mexico basin; YB is Yucatan basin; GA is Greater Antilles; L is Lomonosov Ridge; C is Cayman Trough.

others, 1973; Rosencrantz and others, 1988) is an actively spreading basin within the NAm-CA plate boundary zone. The northern and western margins of the Cayman Trough and part of its southern margin are taken as edges of present-day North America.

Further complications from my definition of present-day North America arise at its northern reaches (Fig. 6). There is no continent-ocean transition between Alaska and Siberia. The North American continent thus extends west through Eurasia to include western Europe. Here, we impose a nongeologic limit of consideration for present-day North America in the Bering Strait and south (Fig. 6). The Lomonosov Ridge (Fig. 6) is considered to be a sliver of the Eurasian continent (Sweeney and others, this vol.; Churkin and Trexler, 1981), that split off from the Barents shelf at the NAm-Eurasia plate boundary in the Cenozoic Eurasia basin. Although the Lomonosov is by my definition now a part of North America, it is arbitrarily excluded with Siberia from further consideration. Greenland is included in present-day North America by virtue of its contiguity with Arctic Canada north of the oceanic Baffin Bay (Fig. 6). Greenland, however, is little considered in this paper except for its role in North Atlantic spreading.

Zonation

Present-day North America can be divided into four tectonic zones (zones 1–4; Fig. 7). Zone 1 is the North American craton, and zones 2–4 constitute the transitional region of the continent, the part that has been shaped by continent-ocean tectonic interactions in Phanerozoic time. The transitional zones are surrounded by little disrupted oceanic crusts (zone 5; Fig. 7) that are products of ongoing spreading regimes (except for those of the Caribbean Sea).

Zone 1: Cratonal Proterozoic North America. This is the sialic nucleus of present-day North America that has behaved as a nearly continuous, little-deformed body since the Proterozoic. It

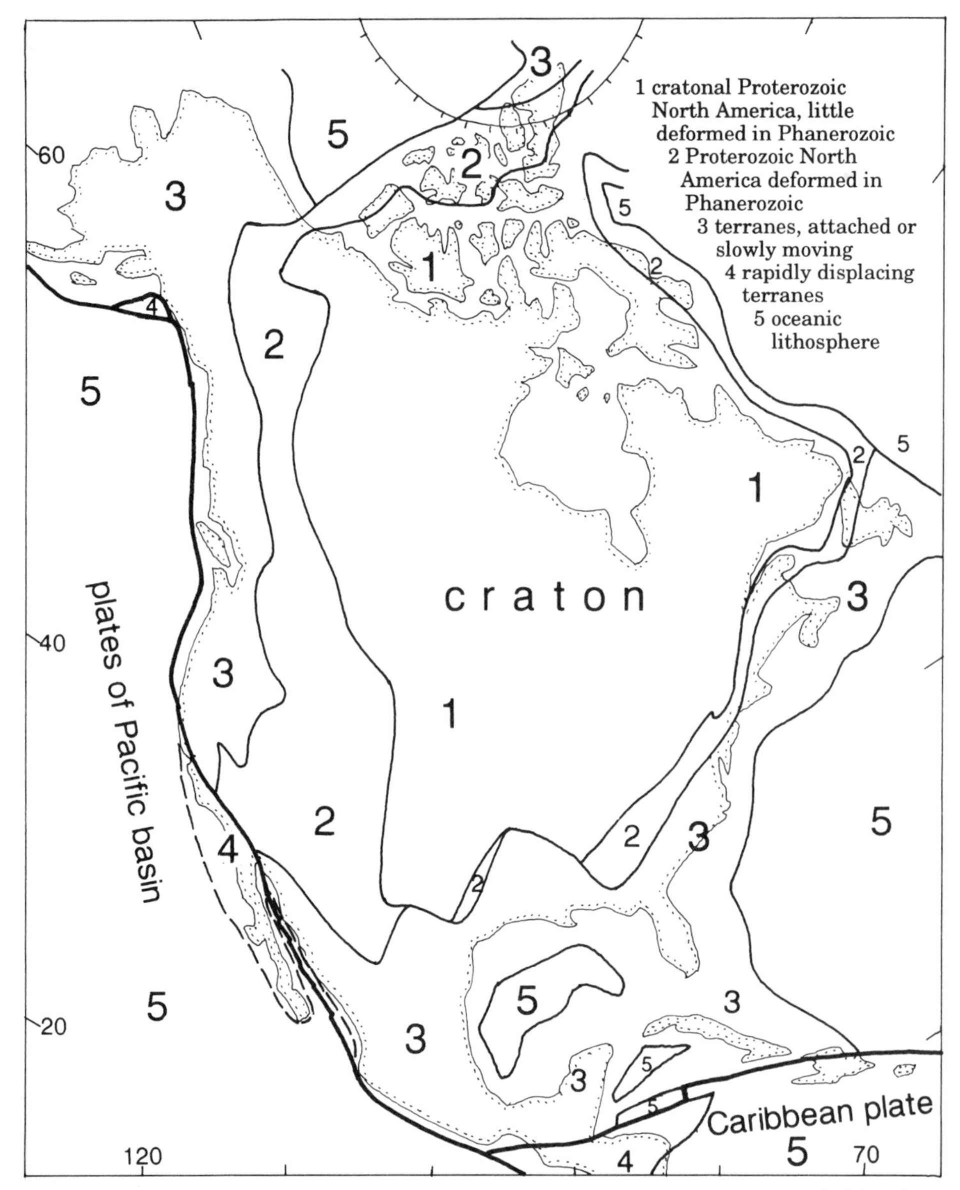

Figure 7. Zonation of present-day North American continent (zones 1–4). Zone 5 is oceanic lithosphere. Thickest lines are loci of maximum velocity in plate boundary zones.

existed within but did not compose all of Proterozoic North America. The craton is a single tectonic unit, and it provides the reference frame for the relative motions of units in the tectonically active outboard zones in the past 600 m.y. The periphery of zone 1 is a set of Phanerozoic deformation fronts.

Zone 2: Deformed Proterozoic North America. This zone contains the region of Proterozoic North America and its Phanerozoic cover that has undergone significant deformation related to Phanerozoic plate boundary tectonics. Unlike zone 1, zone 2 consists of multiple tectonic units, which presumably reflect different loci, times, and mechanisms of plate boundary events that affected the margins of Proterozoic and succeeding stages of North America. Although zone 2 is well fragmented, the pieces are generally thought not to have moved great distances with respect to one another or to zone 1. However, zone 2 may have been the source of continental terranes that have undergone large displacements and now exist in zones 3 or 4 or among accreted terranes of other continents. Examples of tectonic units in zone 2 are the foreland thrust and fold belts of the Valley and Ridge and Blue Ridge provinces of the Appalachian Mountains, and the Laramide belt of Mexico and the southwestern United States. Another region in zone 2 is the horizontally extended transitional crust that formed during rifting at the Atlantic margin in the early Mesozoic.

Zone 3: Displaced terranes. Zone 3 comprises terranes that are tectonic units that have certainly or probably undergone large displacement with respect to cratonal North America but are now fixed or moving slowly relative to the craton. The terranes of zone 3 are cratonward of faults that have the maximum velocity in present-day plate boundary zones. The word terrane implies a discrete and mappable fault-bounded block having a tectonic-stratigraphic history different from that of its neighbors, a designation first formally applied by Jones and others (1977). The concept that terranes had probably undergone substantial relative displacement came from recognition of the juxtaposition of long-ranging stratigraphic sequences and constituent fossils so different from one another that they could not be explained by local facies gradation (Monger and Ross, 1971; Jones and others, 1972). The identification of terranes also emerged from the discovery of differences in paleomagnetic inclination and declination relative to those of contemporaneous cratonal North America (Beck and Noson, 1972; Irving and Yole, 1972; Packer and Stone, 1972; Teissere and Beck, 1973; Beck, 1976). Terranes include fragments of continents, arcs, oceanic lithospheres, deformed oceanic strata, and rocks of uncertain origin. Zone 3 also includes cover strata and igneous rocks that formed in place after attachment of displaced terranes to North America or to earlier accreted units. Examples include the Avalon and Meguma terranes of the Canadian Appalachians, the Pericu and Zapoteco terranes of Mexico, and the Stikinia and Wrangellia terranes of the Pacific Northwest. Important considerations are the sources of such terranes, their transport histories, times of attachment, coalescence with other displaced terranes before accretion to North America, internal tectonic history, deep structure, and the deformation imposed on North America by their collision.

Zone 4: Displacing terranes. Zone 4 of present-day North America contains terranes whose velocities relative to the craton are close to those of adjacent plates. These are terranes whose positions in plate boundary zones are outside the faults that are the maxima in the velocity gradient. Examples are the Chortis terrane and the Alta-Baja California superterrane. The Yakutat terrane may belong to zone 4, although its velocity relative to the Pacific plate may be greater than that relative to NAm (Grantz and others, 1991).

PROTEROZOIC NORTH AMERICA

The North American continent at the beginning of the Phanerozoic era is here called Proterozoic North America (PNA). It contained Archean and Proterozoic basement and Proterozoic cover that exist in zones 1 and 2 of the present-day North American continent (Fig. 6). In today's craton (zone 1), such rocks exist with architecture almost identical to that at the beginning of the Phanerozoic. In zone 2 they have undergone deformation and moderate rearrangement by plate boundary zone tectonics during Phanerozoic time. Furthermore, fragments have been excised from Proterozoic North America and transferred as or within terranes to zones 3 or 4 of present-day North America, possibly to the margins of other continents, and to sites within oceanic lithosphere.

Figure 7 shows an estimated subsurface locus of today's edge of nearly continuous PNA. Put another way, Figure 7 shows the possible limits of continuous Proterozoic continental crust beyond zone 1 below Phanerozoic tectonic zones 2–4. Ages attached to segments of the modern edge of PNA indicate when the segment is thought to have formed by loss of tracts of preexisting continent. The Late Proterozoic–Cambrian segments thus have survived as edges of PNA since end-Proterozoic time, whereas the other segments mark sites where the edge of PNA has been whittled landward of its end-Proterozoic position. The bases for my definition of PNA and for the construction of Figure 7 are discussed below.

Origin and ages of end-Proterozoic breakup

The beginning of the Phanerozoic was a time of great consequence in North American history. It was then that PNA was blocked out, much in its present shape and size, from a preexisting Late Proterozoic supercontinent, Rodinia (Piper, 1983; Stewart, 1976; McMenamin and McMenamin, 1990). The process was one of rifting and the development of divergent plate boundaries that were fully or nearly fully circumferential to PNA at the end of the Proterozoic and were close (all or almost all) to today's edge of PNA (Fig. 8). The principal exception, where PNA at the end of the Proterozoic time might not have been bounded by a passive margin and may have extended greatly seaward of the present edge, is in Mexico and southern California. This possible exception, discussed later, is unlikely according to Ortega and others (this volume). Tracts of the former Rodinia external to such divergent boundaries drifted away, and passive continental margins and peripheral oceanic lithospheres developed around PNA over various durations into the Paleozoic. It is these passive margin phenomena, i.e., rift phase basins and sediment fill, normal faults, basaltic intrusions, thinned and subsided (transitional) crust and associated gravity effects, and postrift sediment wedges ("miogeoclines"), that record the place and time of the continent-shaping end-Proterozoic tectonics (Stewart, 1976; Rankin, 1976; Williams and Hiscott, 1987; Haworth and others, this volume; Bond and others, 1984).

The exact ages at which sea-floor spreading began to block out PNA along its various margins are difficult to determine. One difficulty is that the transition from rift to postrift phenomena at the continental margin, classically taken to be synchronous with onset of drift at the continent's edge, may be gradational or stratigraphically nonspecific through a major interval. This is because synrift and early postrift sediments (basal Sauk sequence of Sloss, 1963, 1988) are both commonly coarse and siliciclastic, and both are usually lacking in age-diagnostic fossils. The crystallization ages of basalt, assumed to be exclusively synrift, are difficult to date accurately and are poorly known. Another difficulty is that the hypothesis that the synrift-postrift transition is contemporaneous with drifting is questionable, according to recent ideas on the time-space migration of rifting and magmatism in continental breakup (Harry and Sawyer, 1992; Sawyer, this volume).

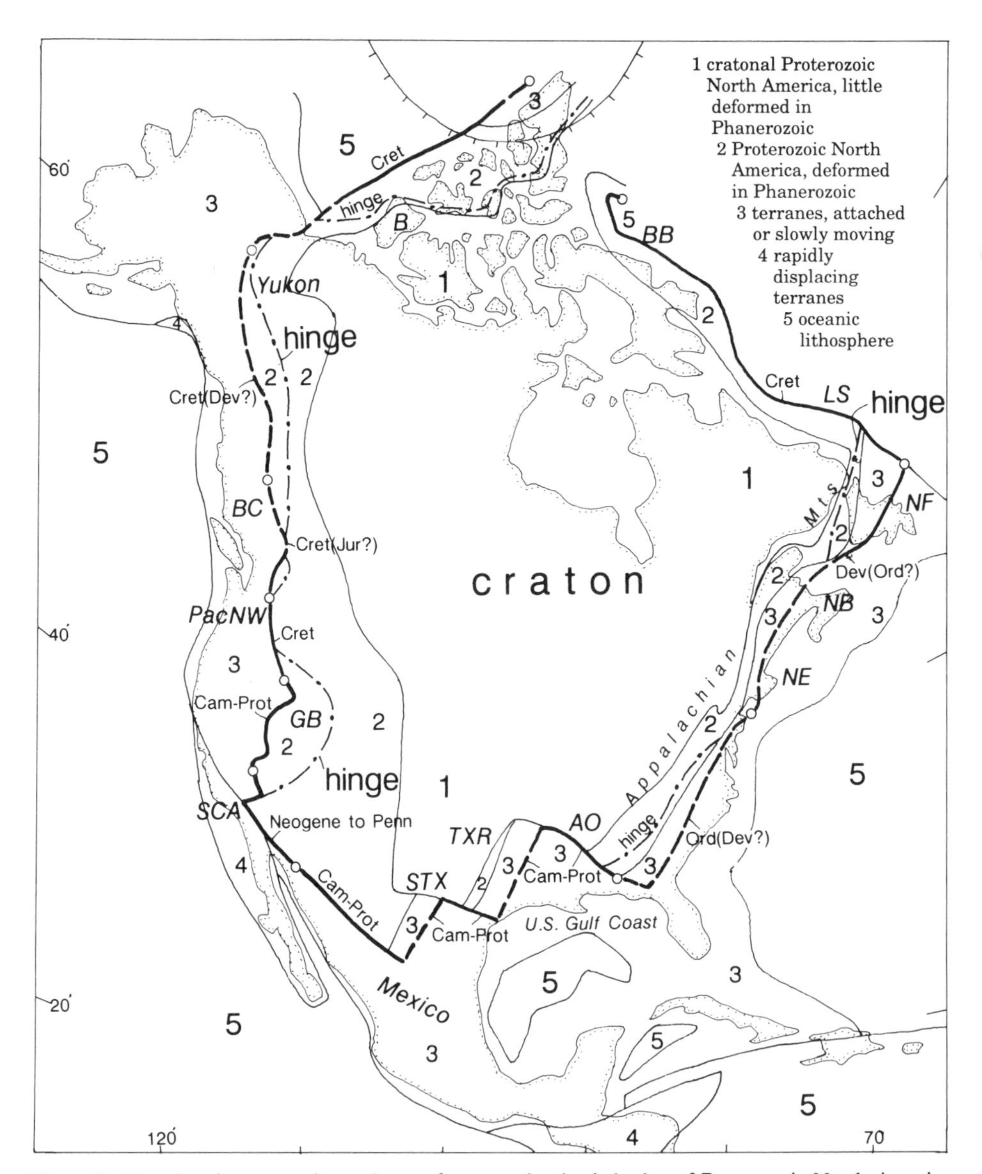

Figure 8. Map showing approximate locus of present-day buried edge of Proterozoic North America (heavy line; dashed where especially uncertain). Circles on buried edge separate segments of different age of formation; Cret, Cretaceous; Jur, Jurassic; Penn, Pennsylvanian; Dev, Devonian; Ord, Ordovician; Cam-Prot, end-Proterozoic. Dash-dot line is approximate locus of end-Proterozoic hinge. Numbers are tectonic divisions of the continent. Abbreviations: B, Banks Island; BB, Baffin Bay; LS, Labrador Sea; NF, Newfoundland; NB, New Brunswick; NE, New England; AO, Alabama-Oklahoma transform; TXR, Texas Rift; STX, southern Texas; SCA, Southern California; GB, Great Basin; PacNW, U.S. Pacific northwest; BC, British Columbia.

The onset of drift may occur during postrift events, during the subsidence of the passive margin wedge.

Some general data bearing on the ages of end-Proterozoic rifting are now reviewed. In the passive margins of eastern and western PNA, fossils in synrift strata (acritarch microbiotas) are Eocambrian (Vendian), and those in postrift strata are as old as Early Cambrian (Knoll and Keller, 1979, 1981; Simpson and Sundberg, 1987). Radiometric dates of synrift igneous rocks, almost all K-Ar, from sites all around PNA vary from 525 to 850 Ma, as reviewed by Bond and others (1984) and Rankin and others (1991); Bond and others (1984) noted a distinct concentration of dates between about 570 and 700 Ma, approximately Eocambrian. In a different approach, Bond and Komintz (1984) and Bond and others (1988) approximated the age of breakup in Late Proterozoic–Cambrian passive margin wedges of North America by fitting subsidence curves of postrift strata to lithospheric cooling curves, and extrapolated to the age of zero subsidence. They obtained 577 ± 23 Ma for sections in the Canadian

Rockies and the Great Basin and 600 ± 25 Ma for sections in the central Appalachians and in Newfoundland. All the data and models permit the synrift-postrift transitions at the western and eastern edges of PNA to have occurred within a short duration centered at the beginning of Phanerozoic time.

The edge of PNA in northern Canada probably formed at about this time because there is a lower Paleozoic (and Proterozoic?) passive margin wedge that thickens north to a Cretaceous edge of PNA (Fig. 8) (Sweeney and others, this volume). On Banks Island, clastic strata and basalt with 655–695 Ma K-Ar dates (synrift?) are overlain unconformably by (postrift?) Lower Cambrian quartz arenite and carbonate rocks (Dixon and Dietrich, 1990); such inferences suggest that the northern margin could have formed in the same interval as the eastern and western ones.

The southern edge of PNA in the U.S. Gulf Coast (Figs. 8 and 9) is interpreted to have been framed in Late Proterozoic–Cambrian time by long transform-dominated segments (Alabama-Oklahoma and South Texas transforms) between rift-dominated segments in the southern Appalachians and in Texas (Cebull and others, 1976; Thomas, 1976, 1991; Buffler and Thomas, this volume). The age of rifting in the southern Appalachians was probably latest Proterozoic, and the beginning of postrift subsidence was near the onset of the Cambrian Period, as it was to the north in the Appalachians. Ages of events in the Texas rift segment are uncertain, and are known only to be Middle Cambrian or older by the oldest known ages in the local Sauk sequence of Sloss (1963, 1988).

Along the Alabama-Oklahoma transform, however, there are no evident rift basins that precede the Sauk sequence, which is here platformal rather than a passive margin wedge and no older than late Middle Cambrian (Thomas, 1985; Buffler and Thomas, this volume). Another unusual feature of this transform segment is the occurrence of two major rift systems that are contemporaneous with the Sauk sequence and that strike about normal to the transform margin, the Mississippi Valley graben and the Birmingham fault system (Fig. 9). Both had Late Cambrian subsidence and sedimentation. Near the intersection of the Alabama-Oklahoma transform with the Texas rift segment in the Southern Oklahoma fault system (Wichita Mountains), there was abundant igneous activity whose rocks are dated between 577 and 525 Ma, suggesting a prolonged duration in latest Proterozoic and Cambrian time.

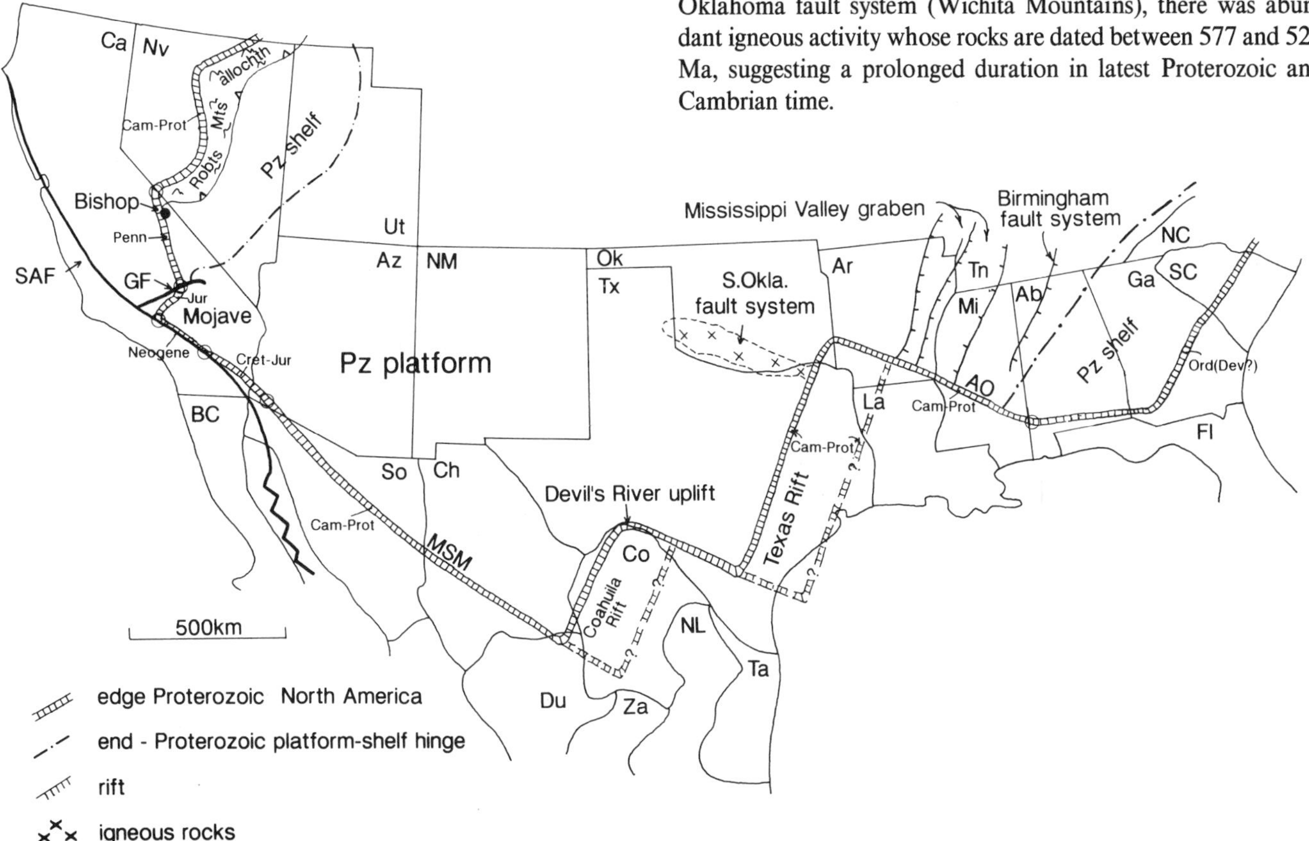

Figure 9. Map of southern edge of Proterozoic North America (PNA), shown as tracked double line except at Texas Rift and Coahuila Rift zones where pairs of double lines indicate hypothetical inner and outer (dashed) edges of stretched continental crust. Circles delimit segments of different age of edge of PNA: Cam-Prot, end-Proterozoic; Penn, Pennsylvanian; Jur, Jurassic; Cret, Cretaceous; Ord, Ordovician; Dev, Devonian. Abbreviations: SAF, San Andreas fault; PZ, Paleozoic; Robts Mts Alloch, Roberts Mountains Allochthon terrane; GF, Garlock fault; MSM, Mojave-Sonora megashear; AO, Arkansas-Oklahoma transform. Mexican states: BC, Baja California; So, Sonora; Ch, Chihuahua; Co, Coahuila; Du, Durango; Za, Zacatecas; NL, Nuevo Leon; Ta, Tamaulipais.

The seemingly aberrant features of the early southern edge of PNA may be explained by the 700–800 km migration of a spreading center along the Alabama-Oklahoma transform from the Texas rift segment to the rift segment in the southern Appalachians. Such passage may have taken 50 m.y. or more before North America was fully separated from relict Rodinia. This may account for the localized and prolonged igneous event near the continental corner in southern Oklahoma, at which the spreading center originated. The migration of and heating by the spreading center may have caused the Cambrian rifting within PNA north of the transform margin and the maintenance of a relatively high freeboard of adjacent tracts of southern PNA until late in Cambrian time. The South Texas transform may have acted similarly between the Texas rift and another postulated rift zone in northern Mexico (Figs. 8 and 9).

To conclude, existing data imply that continental North America was blocked out approximately in its current form in plan, with the possible exception of its southwestern reach in Mexico, by the breakup of Rodinia at about the beginning of Phanerozoic time. The duration of drifting to reach full isolation of North America from other continental fragments may have been locally significant (50 to 100 m.y.?) along long transform segments of the nascent system of divergent boundaries. There is no evidence for regional differences in breakup times or for long propagation of divergent boundaries along the edge of PNA. Furthermore, with the possible exception of the southwestern reach, PNA must have come from the interior of Rodinia, because it is otherwise fully enclosed by end-Proterozoic passive margins.

The name Laurentia has commonly been used for North America framed by the opening of the Iapetus Ocean along the Appalachian margin (Rankin, this volume). Laurentia is in my view synonymous with Proterozoic North America because there are no data that date any segment of the Appalachian margin as of post-Proterozoic breakup age. I choose to use PNA for the North American continent after the breakup of Rodinia because of its homology with names in a time series, such as Permian or present-day North America.

Edge of PNA and Phanerozoic changes

The features that could in principle be used to identify the position and character of the end-Proterozoic passive margin are the hinge, passive margin wedge, edge of cratonal crust, and the continent-ocean boundary (equivalently, the outer edge of transitional crust). The significance of such features comes mainly from observations and theory of the Mesozoic Atlantic passive margin, which provides an example with no superposed tectonism (Haworth and others, this volume; Rankin and others, this volume; Sawyer, this volume; see below).

The hinge is the locus of facies change in Paleozoic postrift cover, between thin platformal strata and the passive margin wedge that thickens rapidly seaward. The hinge is underlain by the contact between cratonal crust, with little thinning and scattered rift basins near the hinge, and transitional crust that is greatly thinned seaward. The outer edge of transitional crust is the locus of the oldest oceanic lithosphere produced by sea-floor spreading, possibly including a narrow gradational band called marginal oceanic crust.

Among these features, the end-Proterozoic hinge is the most accessible because it occurs widely at or near the surface in zone 2 (Fig. 8). Its occurrence in present-day North America is mainly in Phanerozoic foreland thrust and fold belts where the Paleozoic cover is detached from basement, imbricated, and translated significant distances, as much as several hundred kilometers, relatively continentward. The end-Proterozoic hinge shown in Figure 8 is its present-day locus and thus in a deformed position at most sites; its initial position was seaward. It is possible to restore the hinge in zone 2 to an initial locus because margin-normal foreland displacements are estimable. Zone 2 may, however, locally include significant margin-parallel displacements, and these may cast doubt on the accuracy of restorations.

The hinge is absent from the surface over long stretches of the margin of PNA (Fig. 8). The implications of such absence are, alternatively, that (1) a hinge never existed, or at least not in the form prescribed by the Mesozoic Atlantic rifted margin; (2) the hinge is tectonically buried below terranes (below zone 3; Fig. 8); and (3) the passive margin wedge with or without some craton was excised.

The seaward extent of thick continental crust, the subsurface edge of the craton, has been approximated at many transects by refraction and reflection depths to Moho, crustal reflectivity, and gravity (Monger and others, 1985; Cook and Oliver, 1981; Cook and others, 1981; Saleeby, 1986; Rankin and others, 1991; Haworth and others, 1985; Stewart and others, 1991; Hatcher and others, 1993). This edge is generally seaward of the hinge in zone 2 (Fig. 8). In southern British Columbia (Monger and others, 1985) and in Georgia (Hatcher and others, 1993), the positions of the edge of cratonal crust and the restored locus of the hinge are somewhat alike, implying that cratonal crust behaved rigidly during post-passive margin deformation. In general, however, the cratonal edge should be regarded as a minimum seaward locus for the hinge's initial position because the basement may have taken up some continuum contraction in later tectonism and because an outer fringe of cratonal crust may have been excised.

Transitional crust is rarely identified at the surface in tectonic zones 2–4, perhaps because no modern example exists in outcrop. The rare nappes and terranes of multiply tectonized, diked Precambrian crystalline rock are candidates, for example, the Monashee terrane of western Canada (Monger and others, 1985) and the external massifs of Grenvillean basement in the Appalachians (Rankin and others, 1991). The kinematic history of such bodies, however, is commonly unclear. Similarly, the ophiolites of zones 3 and 4 are not evidently related to the end-Proterozoic passive margin, with a possible exception in the Klamath Mountains of California (Wallin and others, 1988). An implication of the paucity of end-Proterozoic transitional crust and ophiolite at the surface in zones 2 and 3 is that such crusts were thoroughly detached from their cover and were overridden by terranes during the collisional phases of tectonics that appear to have widely followed the end-Proterozoic passive margin phase.

The existence of end-Proterozoic transitional crust in the subsurface is difficult to prove. High internal reflectivity is a predicted property, but where this occurs in zones 2 and 3, it is probably due to post-Paleozoic deformation. The isotopic compositions of Sr, Nd, Pb, and O in Phanerozoic magmatic rocks that were generated in or contaminated by cratonal and transitional crust, however, appear to permit the differentiation in the subsurface of such crusts from those of noncontinental terranes (Kistler and Peterman, 1973; Kistler, 1990; Armstrong, 1988). The isotopic compositions do not, however, allow discrimination between buried crust of PNA and terranes of continental origin.

The end-Proterozoic continent-ocean boundary is thus an elusive feature, perhaps because it may no longer exist along most of the PNA margins. Widths of modern passive margins between the rift zone and oceanic lithosphere vary from 50 to 500 km; this gives a range of reasonable values of distances seaward of the restored position of the end-Proterozoic hinge of the continent-ocean boundary at end-Proterozoic time along reaches where it has been excised.

The position of the present-day edge of PNA (Fig. 8) and the interpreted structure of the end-Proterozoic passive margin are now discussed as a function of position around North America.

In Arctic Canada (Fig. 8), a passive margin of approximately end-Proterozoic age is indicated by a hinge and a north-thickening lower Paleozoic passive margin wedge in zone 2. North American transitional crust probably extends north from the hinge to a Cretaceous edge of PNA (Sweeney, 1985). This edge is the product of spreading in the Early Cretaceous Canada basin, where the outer reaches of the end-Proterozoic passive margin and, probably, suprajacent Paleozoic collided terranes were excised and rafted away.

In eastern Canada in the Labrador Trough and Baffin Bay (Fig. 8), the edge of PNA is Late Cretaceous. It formed at a passive margin that developed on formerly cratonal North America. The rifting and drifting occurred at an early, failed arm of the North Atlantic (Srivastava and Tapscott, 1986; Keen and Haworth, 1985c).

In eastern Canada in Quebec and Newfoundland, the end-Proterozoic hinge is in zone 2 (Fig. 8), within the deformed foreland of the collisional Ordovician Taconian orogeny (Haworth and others, 1985, this volume). Here, zone 2 also includes to seaward a hinterland of thrust sheets of lower Paleozoic passive margin wedge and Proterozoic basement. Zone 3 contains terranes of Ordovician and Devonian emplacement ages (Williams and Hiscott, 1987; Dallmeyer and others, 1981). The edge of PNA may be Ordovician and a buried suture at which Ordovician terranes are attached (Fig. 8) (Stockmal and others, 1990). Alternatively, the edge of PNA may be Devonian and a contact with the Avalon terrane that wedged below the Ordovician terranes, as discussed later. The hinterland may include samples of the top of end-Proterozoic transitional crust clipped from the top of the overridden North American crust.

In the northern Appalachian region from south of Quebec to New England (Fig. 8), a hinge does not exist in zone 2, in contrast to Newfoundland. Here, the hinge and passive margin wedge underlie terranes of Ordovician emplacement age and occur in landward-vergent thrust duplexes that are seen in windows through the terranes (Thompson and others, 1993; Rankin and others, 1989; Rankin, this volume). The passive margin strata are metamorphosed and occur in duplexes together with nappes containing Proterozoic basement (Grenvillian) and relatively thin synrift strata of end-Proterozoic(?) age. The buried edge of PNA in New England (Fig. 8) is possibly a Devonian contact with Avalon-related terranes that wedged landward below Ordovician terranes, as may occur to the north in Newfoundland; this is discussed later.

In contrast with the northern Appalachians, the central and southern Appalachians contain extensive thick synrift strata of end-Proterozoic age in zone 2 in both the Ordovician foreland and hinterland (Blue Ridge) thrust stacks (Rankin and others, 1991; Hatcher and others, 1993; Transects E2 and E3 in Speed, 1991). The synrift successions of the hinterland include much volcanic rock, whereas those of the foreland do not. The most outboard thrust sheets of the hinterland include lower Paleozoic postrift-passive margin wedge strata, whereas landward the sheets contain only platformal postrift strata. The present position of the end-Proterozoic hinge is thus in outer zone 2 (Fig. 8). The edge of PNA is positioned by restorations of foreland and hinterland sheets of zone 2 and by geophysical suggestions of the extent of thick crust (Rankin and others, 1991; Hatcher and others, 1993). Both approaches give comparable results. This edge in the central and southern Appalachians is probably a suture with the Carolina terrane, which appears to have been thrust above North America. The age of the suture is uncertain, either Ordovician or Devonian (see below; Rankin, this volume).

To summarize, the Appalachian Mountains contain evidence that an end-Proterozoic passive margin existed along the entire seaboard from Newfoundland to Alabama. The passive margin was probably dominated by rifts rather than transforms because thick postrift wedge strata and local synrift deposits occur all along the margin. These and the hinge occur in Paleozoic nappes of the outer (hinterland) part of zone 2, except in Newfoundland, implying that the hinge lay well outboard of its current position at the beginning of the Phanerozoic. The present-day buried edge of PNA is possibly Devonian or Ordovician throughout or Devonian in the north and Ordovician in the south. The passive margin was deformed by margin-normal contraction and landward thrusting in the Ordovician, probably continuing throughout much of the rest of the Paleozoic. However, the possibility exists that at depth the current edge of PNA, and hence of the passive margin, is a surface of excision in later

Paleozoic time due to subduction or oblique plate boundary motions in the collisions of peri-Gondwanan (Avalon-related) terranes or the Gondwana continent with eastern North America.

The history of the southern edge of PNA, from Alabama to southern California, is especially difficult to determine because it is generally deeply buried and because its marginal orogenic belts consist mainly of terranes and synorogenic sediments (Transects F1 and F2 in Speed, 1991; Buffler and Thomas, this volume). The belts do not include parautochthonous imbricate stacks of North American basement or Late Proterozoic–early Paleozoic rift and passive margin wedge strata. Moreover, within exposed and drilled PNA rocks across this entire margin, beginning with Cambrian postrift equivalents (Sauk sequence; Sloss, 1963), strata are entirely of cratonal platformal facies. Geophysical and isotopic studies suggest, at least in the Gulf Coast segment of this edge, that the crust of PNA thins remarkably rapidly, over about 25 km, from normal cratonal values at the orogenic front southward to either greatly thinned continental or oceanic crust (Lillie and others, 1983; Keller and others, 1989).

All these features of the southern edge of PNA differ from those expected at an end-Proterozoic passive margin, using as reference the Mesozoic Atlantic margin of North America (Manspeizer, 1988; Sawyer, this volume). Aspects of the disparity may be explained in one of three ways: (1) original transform domination as discussed earlier; (2) tectonic excision and removal of a passive margin before Carboniferous and later collisions; or (3) complete tectonic burial of a passive margin by the later collisions.

In the U.S. Gulf Coast, the hypothesis of long transform segments (Thomas, 1976, 1985; Cebull and others, 1976), here accepted, accounts for the narrowness, for the lack of rift and postrift facies, and for the longevity of the base-Sauk unconformity. The hypothesis is supported by similar strike of short transform segments interpreted in the rift-dominated Appalachian end-Proterozoic margin and the long west-northwest–trending segments of the southern edge. It is also supported by the similarity of features in observations and model at the long Southern Grand Banks Fracture Zone continental transform and in the modern Atlantic basin (Keen and Haworth, 1985b; Todd and Keen, 1989). The hypothesis does not explain, however, the absence of North American synrift and passive margin wedge deposits in the foreland and hinterland imbricate stacks of the Ouachitan belt in the northeast-trending legs (eastern Texas and Coahuila, Fig. 9), which are presumably rift-dominated.

The Mexican leg of the southern edge of PNA, from Chihuahua to the Mojave Desert of southern California (Fig. 9), is where the end-Proterozoic configuration of PNA and Phanerozoic modifications to it are the most uncertain. Here, the current edge of PNA is fully bordered by terranes (Sedlock and others, 1993; Ortega and others, this volume). As in much of the U.S. Gulf Coast, there is no evidence for a broad zone of crustal thinning or rifting in PNA of end-Proterozoic age. Moreover, the Paleozoic cover of Proterozoic basement is entirely platformal, and there is not a foreland or hinterland belt of North American passive margin wedge strata thrust from outboard positions across the present edge. The Mexican leg differs from that of the U.S. Gulf Coast by the absence of a Paleozoic collisional belt equivalent to the Ouachitan belt of the Gulf Coast. Two ideas can explain these relations. First, the present edge of PNA in Mexico is, or is close to, an end-Proterozoic transform-dominated margin that is continuous eastward with that of the U.S. Gulf Coast (Stewart, 1988; Sedlock and others, 1993). Second, PNA may have extended far south and/or west of the present edge (Keller and Cebull, 1973; Dickinson, 1981), and Phanerozoic tectonics excised PNA back to the present edge. The first idea is somewhat more likely because the present edge of PNA does not contain structures indicative of excision, and terranes indicative of a large mass of PNA excised from Mexico have not been located. This question is taken up in a later section and by Ortega and others (this volume).

The present southern edge of PNA extends with similar properties from Mexico northwest into southern California. There, at approximately the Mojave Desert, it intersects the northeast-trending end-Proterozoic passive margin of the U.S. Cordillera (Fig. 9), which includes a well-defined hinge and very broad postrift sediment wedge. The orthogonality of these two edges of PNA with greatly different properties has long been interpreted to indicate that the southern California edge is a Phanerozoic truncation surface (Hamilton and Myers, 1966; Stewart and Poole, 1974). This is almost certainly true, because in southern California the San Andreas fault system cuts PNA, and there are structures in the PNA margin suggesting Paleozoic excision (Stevens and others, 1992). Furthermore, terranes outboard of PNA include continental fragments with both rifted margin and cratonal affiliations (Howell and Vedder, this volume). Therefore, unlike the Gulf Coast and Mexican segments of the southern edge of PNA, that in southern California has been substantially modified by Phanerozoic tectonics.

The position of the change in the age of the edge of PNA from probably end-Proterozoic in Mexico to late Phanerozoic in southern California is uncertain, but is possibly near the international boundary (Fig. 9). North of there to about Bishop, California (Fig. 9), ages of the edge vary from Pennsylvanian, which is probably the time of first excision of end-Proterozoic PNA in this region (Stevens and others, 1992), to Neogene where the Alta-Baja California superterrane contacts PNA. The Phanerozoic evolution of southern California is discussed more fully later in this chapter and by Howell and Vedder (this volume).

The Cordillera of Canada and the United States between Bishop and the Arctic (Figs. 8 and 9) contains an edge of PNA that is either deeply buried below terranes or is a vertical fault that extends down from surface contact between terranes and parautochthonous North American units of zone 2 (Fig. 8). The widespread existence of thick upper Proterozoic clastic strata and basalt and lower Paleozoic carbonate-rich passive margin wedge ("miogeoclinal") deposits over a generally broad width of zone 2 indicates that western end-Proterozoic North America was mainly rift dominated.

Northeast from the truncated edge of PNA near Bishop, California (Fig. 9), the region of the Great Basin contains a well-defined hinge and remarkably wide passive margin wedge outboard of the hinge (Stewart, 1972). The distance from the hinge to the edge of Precambrian crust, defined by isotopic (Sr, O) compositions of Phanerozoic plutons (Kistler and Peterman, 1973; Kistler, 1990; Elison and others, 1990) is as great as 700 km. One questions whether this breadth is the original or a deformed width of the end-Proterozoic passive margin. The principal Phanerozoic deformations are the approximately east-west late Mesozoic and early Cenozoic contraction of zone 2 foreland and hinterland (Fig. 7) and late Cenozoic Basin and Range extension. Estimates of total displacement in each by restorations are more than 200 km contraction (Elison, 1991) and 200–300 km extension (Wernicke and others, 1988). The small net displacement of these deformations implies a large initial width of the passive margin. The end-Proterozoic passive margin of the Great Basin may be comparable in original width to that of the Blake Plateau of the modern Atlantic basin.

The existing edge of PNA in the Great Basin is likely to be close to the end-Proterozoic edge (Fig. 9). This is suggested by facies changes across the lower Paleozoic passive margin wedge, including the preservation of shelf-edge or slope deposits at the outboard edge that appear to be pinned to continental (transitional?) crust (Speed and others, 1988). Both the original width and full preservation of the passive margin in the Great Basin are possibly unique along the western edge of PNA (Speed, 1983).

The edge of PNA in the Great Basin is a zigzag (Fig. 9). It is controversial whether this reflects end-Proterozoic rift-transform segmentation or Phanerozoic bending and faulting. There is evidence to support each origin (Oldow, 1984; Saleeby and others, this volume), and the effects of each are probably superposed.

The edge of PNA in the Pacific Northwest of the United States has a very different history from that of the Great Basin. Here, a sizeable fragment of PNA together with accreted terranes was probably excised and rafted away from eastern Oregon, southeastern Washington, and western Idaho in the Mesozoic (Speed, 1983), as discussed below. The fragment probably included the full width of the end-Proterozoic passive margin, because the end-Proterozoic hinge intersects the edge of PNA on both sides of the excised bight (Fig. 8). The far-traveled Wallowa terranes attached to the truncated edge of North America in the bight in mid-Cretaceous time, giving a minimum age of excision.

In northern Washington to central British Columbia, PNA extends seaward below the Mesozoic foreland thrust and fold belt of zone 2 (Fig. 8) (Monger and others, 1985, this volume; Cowan and Potter, 1985). The thrust belt includes the end-Proterozoic hinge and passive margin wedge strata whose restored width normal to the margin is at least 200 km (Price, 1986). The postrift strata are underlain by synrift deposits that include much basalt (Bond and Komintz, 1984). This reach of the western margin was probably rift dominated. Among terranes west of the deformed foreland (zone 3, Fig. 8) is the Monashee (Monger and others, this volume), which consists of tectonized Proterozoic crystalline rock plus sedimentary cover and basalt. It is a question whether such fragments, which are now structurally above the edge of PNA, are local thrust sheets derived from the subjacent passive margin or whether they are Mesozoic terranes that are far traveled and possibly exotic to North America.

The present-day edge of PNA in northern Washington to central British Columbia may be well landward of its end-Proterozoic locus. This is suggested by the greater restored width of passive margin wedge strata than the distance from edge of craton (zone 1) to terranes. However, the existence of continental crust below the terranes is indicated by isotopic compositions of Mesozoic plutons within the terranes (Armstrong, 1988) and by interpretation of deep seismic soundings (Cook and others, 1991). Such crust, however, may belong to the terranes, and continuity with PNA cannot be assumed without corroboration by suprajacent passive margin cover. Moreover, the reflections assigned to deep continental crust have no evident different character from those at shallower levels assigned to terranes. The potential for excision of outer passive margin features is indicated in this region by margin-parallel strike-slip faults of probable large individual and cumulative displacement from mid-Cretaceous through Paleogene time (Gabrielse, 1985) in zones 2 and 3. The age of possible excision is unclear; Jurassic and/or Cretaceous if the outer deformed foreland has moved substantially along the margin, or Cretaceous if the Jurassic thrusting of the outer foreland is local.

In northern British Columbia and the southern Yukon, the hinge of the passive margin is in the foreland thrust belt, as to the south in Canada (Fig. 8). The original margin of PNA in this reach was probably also excised in late Mesozoic time, but excision may have begun as early as the Devonian. The evidence is that in mid-Paleozoic time, the deposition of postrift carbonate-rich strata was interrupted by landward transport of clastic sediments (the Earn sequence; Gordey and others, 1987) across the shelf from sources in the outer passive margin wedge. The sediment source region, presumably deformed, does not now exist, at least at the surface, and its place is apparently taken by terranes of Mesozoic emplacement age (Monger and Berg, 1984).

Discussion

The North American continent was probably fully blocked out from a former supercontinent, much in its present size and shape, at approximately the end of the Proterozoic (Fig. 8). The breakup occurred at rift-dominated margins within a duration whose length is uncertain by about 70 m.y. The breakup may or may not have been synchronous at all positions at rift margins, but there is no evidence for diachroneity of breakup. The duration of continental separation was predictably longer at long intracontinental transforms, at least at the Alabama-Oklahoma transform.

The eastern and western end-Proterozoic margins of North America were apparently rift dominated and can be interpreted to have had west-northwest–east-southeast divergence (Fig. 10). A southern margin of PNA is inferred to have been wholly

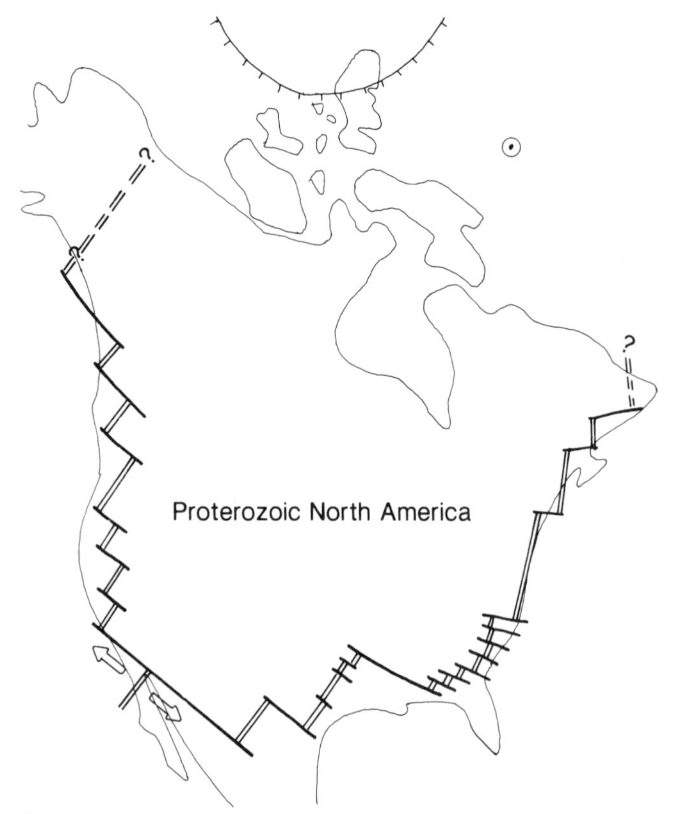

Figure 10. Conceivable edge of eastern, southern, and western North American continent at end of Proterozoic time due to breakup of Rodinia in a series of rift (double line) and transform (single heavy line) segments. Proterozoic North America (PNA) shown in a three-plate system with a single Euler pole (circled dot). Arrows show velocity directions of other two plates relative to PNA at boundary presumed to be most distant from Euler pole. Plate tectonics of northern margin of PNA uncertain.

created by end-Proterozoic breakup and to have been dominated by transform faults with west-northwest–east-southeast strike. If it is correct that today's southern edge of PNA is close to that of end-Proterozoic time, a contemporaneous rift-transform system of the eastern, southern, and western margins of PNA implies breakup and drift about a distant Euler pole (Fig. 10). This is difficult to confirm because of poor control of strikes of transform segments in the northern realm of PNA. In particular, the extent and orientation of the end-Proterozoic Arctic passive margin are little known. The width of the lower Paleozoic Arctic passive margin stratal wedge suggests, however, that it had at least local broad rift segments.

PHANEROZOIC EVOLUTION OF NORTH AMERICAN CONTINENTAL MARGINS

Introduction

The evolution of North American continental margins is now traced from the time of development of the probably encircling end-Proterozoic passive margin to the present day.

The generalized tectonic behaviors of regional lengths (1000 km) of the continental margin over substantial durations (≥50 m.y.) are classed here in three ways: passive, active, and collisional, the objective being to relate phenomena (structural, thermal, sedimentary) and plate boundary zone kinematics. A passive margin, as employed earlier and classically, implies horizontal extension, sea-floor spreading, and a regionally divergent component in the relative velocity across the plate boundary zone. At an active margin, oceanic lithosphere converged with the continent across a plate boundary and was subducted below the continent over a regional length of the margin and over a protracted period. Expected phenomena of an active margin are continental arc magmatism, forearc development, attachment of terranes, and widespread horizontal contraction, all with a significant lifetime. Collisional margins also evolved at convergent plate boundary zones, but where the North American continent belonged to the underriding plate. Here, tectonic activity affecting the North American continental margin presumably occurred in a discrete episode when a migrating trench intersected the margin. Following the collision, the plate boundary either jumped to a distant site or reconfigured locally, perhaps to develop an active margin on North America. Expected phenomena are horizontal contraction and the attachment of terranes (as large as continents).

The emphasis on regionality in the assignment of the three types of margins permits deviations in which relative motions and structures locally depart from those averaged over long lengths and times. Such deviations are predicted, for example, at geometric and rheologic irregularities in the continental edge. All three margin types may include segments with principally strike slip motion as a consequence of local margin orientation, and they may include segments where strike-slip motion is partitioned in different structures from those that take up contemporaneous contraction or extension.

The following discussions of Phanerozoic evolution are divided geographically around North America: eastern, southern, western, and northern. Such division is familiar, traditional, and tectonically meaningful because each of the four continental margins has an element of similarity of Phanerozoic evolution on strike and has differences in evolution from the others. The ever-improving resolution by new work on the structure and event history of the North American margin increasingly reveals the heterogeneity and localness of behavior. These advances, however, do not diminish the truth of the idea of regionality in margin evolution; rather, they provide evidence for the spectrum of processes which operate simultaneously or in sequence at a long segment of the continental margin.

Discussions below focus on the following topics: event histories of regional margin segments; configuration of the continental margin; breadth of the plate boundary zone; obliquity of the plate motion vector relative to the margin and its irregularities; fate of bordering oceanic lithosphere; magnitudes of horizontal contraction and extension; vertical displacements and timing and causes of unroofing; volume addition to the crust and circulation within the crust by magmatism; accretion against and below the continental margin by sediment and terranes; and margin-parallel transfer of sediment and terranes and tectonic attrition of the margin or its lower reaches.

Eastern North America

The eastern realm of North America extends from Florida to Labrador and embodies the Appalachian orogen and its foreland and the continental shelves and slopes of the Atlantic Ocean and Labrador Sea (Fig. 11). The Phanerozoic began with continued development of the end-Proterozoic passive margin of Proterozoic North America (Fig. 12) until mid-Ordovician time. The eastern margin then changed to a collisional one, which existed until about the middle of the Permian Period and culminated in the amalgamation of Pangea. There was no Middle Permian to Late Triassic North America, because the formerly discrete continent became a patch in the supercontinent of Pangea. In the Middle Triassic, a new passive margin began to develop with the onset of horizontal extension and rifting in a belt through central Pangea. The onset of rifting and spreading in the belt was diachronous from Middle Triassic to Paleogene time, blocking out the passive margins of eastern present-day North America and the oldest parts of the bordering oceanic lithospheres.

Early Paleozoic passive margin. The end-Proterozoic passive margin of eastern North America, now nowhere intact, continued development until mid-Ordovician time by thermal subsidence of the continental crust seaward of the hinge and deposition of a suprajacent carbonate-rich passive margin wedge. The bordering oceanic lithosphere, Iapetus, probably spread approximately normal to the continental margin. The first change in the mid-Ordovician transformation of the margin was the local uplift of the outer shelf of Ordovician North America and the onset of erosion and karst formation; these were followed by subsidence and deposition of black shale (Haworth and others, this volume; Rankin and others, 1989; Rankin, this volume). Such phenomena were brief and were succeeded by the onset of thrusting related to colliding terranes and the Taconian phase of the collisional Appalachian orogen at times between latest Early Ordovician and early Late Ordovician at varied positions. The switch from a passive to a collisional margin seems to have occurred in a short interval (20–30 m.y.) over a great length of margin, implying that Taconian collisions were the result of approximately margin-normal closure of Iapetus.

The small departures from the mean in the age of transformation are not systematically distributed along the margin. They are interpreted as due to a jagged edge of Ordovician North America, created in the breakup of Rodinia by rift-transform segments (Thomas, 1977, 1985) or by a series of triple armed rifts (Rankin, 1975, 1976). The irregular Ordovician edge is preserved in part in the structural salients and recesses of the Appalachian orogen.

Within this passive margin duration, intracontinental rifting occurred well inland of the hinge (Rankin, this volume; Thomas, 1985; Buffler and Thomas, this volume). This was probably related to the protracted development of the long Alabama-Oklahoma transform segment of the southern edge of early Paleozoic North America (Fig. 8). During this duration, active or collisional margin behavior is recorded within or among some terranes now within the Appalachian orogen (Penobscotian orogeny; Neuman and Max, 1989; Drake, 1987; Transect E2 in Speed, 1991; Virgillinian orogeny; Glover and Sinha, 1973; Transect E3 in Speed, 1991). Such events probably occurred during the transit of terranes across Iapetus, before their collision with North America.

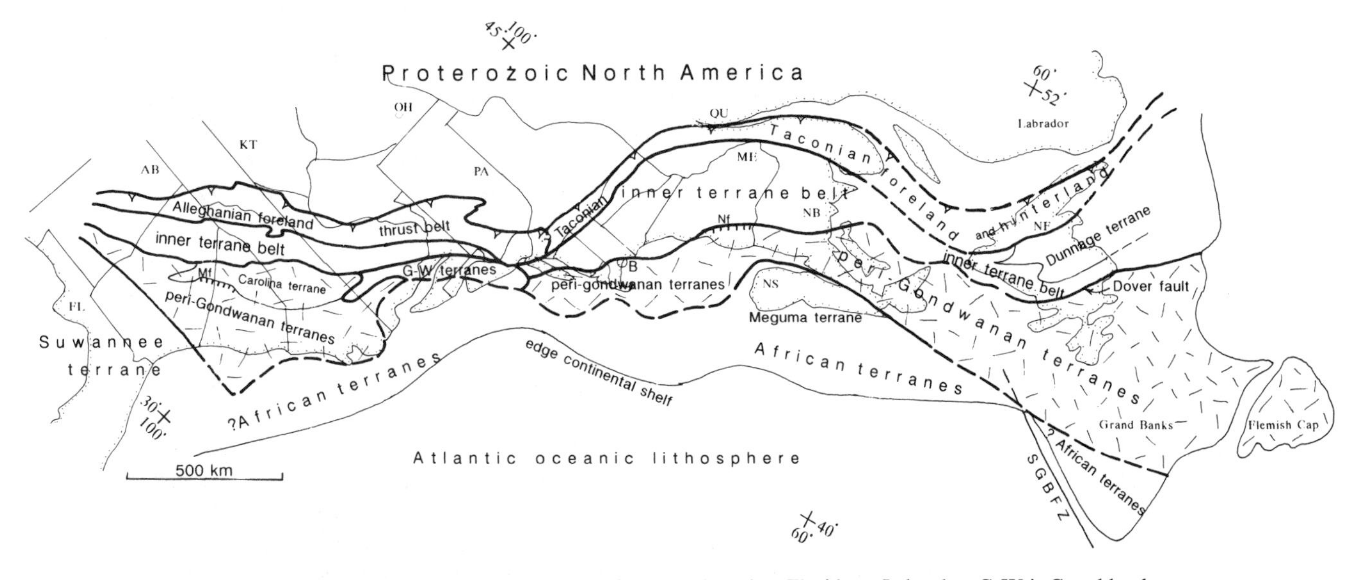

Figure 11. Paleozoic tectonic belts of eastern North America, Florida to Labrador. G-W is Goochland and Wilmington terranes; Mf, Modoc fault; Nf, Norumbega fault; B, Boston; NS, Nova Scotia; NB, New Brunswick; Qu, Quebec; SGBFZ, southern Grand Banks Fracture Zone; ME, Maine; FL, Florida; AB, Alabama; KT, Kentucky; OH, Ohio; PA, Pennsylvania.

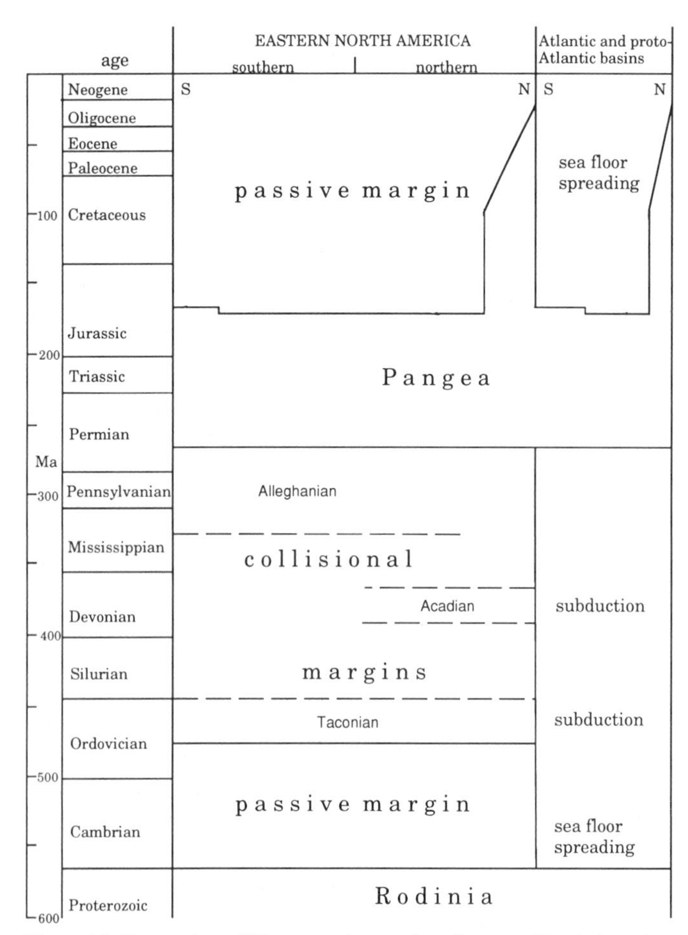

Figure 12. Succession of Phanerozoic margins of eastern North America and events in the Atlantic and proto-Atlantic basins. Late Mesozoic and Cenozoic passive margin has diachronous development, indicated by age of onset of drifting. The duration of Pangea indicates a period when no eastern margin existed..

Collisional margin. *Introduction.* The eastern margin of Early Ordovician North America extended from Alabama through northwestern Newfoundland to a broad shelf east of Greenland that included continent now in and west of northwestern Great Britain (Neuman and Max, 1989). The Iapetus oceanic lithosphere that was attached to this margin had been generated since the breakup of Rodinia, ~570 Ma. The subduction of part or all of Iapetus below terranes within or at the eastern edge of Iapetus took place before the first arrival of terranes at the North American continental margin in mid-Ordovician time. Collisions of terranes, including Gondwana, continued episodically through the Paleozoic Era, ending in the inclusion of North America within the Pangean supercontinent.

Today's general architecture of the Appalachian orogen that was produced by Paleozoic collisions is constituted by a series of margin-parallel tectonic belts: foreland and hinterland (both in zone 2 of Fig. 7) and terranes of several generations (Fig. 11). Such belts originated episodically in time and place, certainly not steadily, within the orogen. There was a general cratonward transport of older orogenic belts in the Paleozoic, such that early hinterlands became partly included in later forelands. Such transport is conspicuous in the southern Appalachians, but more subtle in the northern Appalachians. Moreover, the orogen has been modified to an uncertain extent by extension, unroofing, and magmatism of the rift phase of the Mesozoic passive margin.

Discrete tectonic pulses were long ago identified at different places in the Appalachian orogen (Rodgers, 1971; Zen, 1972) and were then assumed to have occurred along the full length of the orogen. The pulses are the Taconian (Ordovician), Acadian (Devonian), and Alleghanian (Carboniferous to Permian). It has been found, however, that, except for the Ordovician onset of collisional tectonics, tectonic phenomena vary in timing and character along the length of the orogen and are not easily grouped into three distinct phases (Rankin, this volume). Nonetheless, the names of the pulses are still used, mainly with a view toward temporal grouping of early, middle, and late Appalachian events.

The foreland comprises the thin-skinned fold and thrust belts and foreland basins. The fold and thrust belt overlies master detachments within Paleozoic strata, and these climb and had slip of their hangingwalls generally toward the continent. The foreland basins developed syntectonically on the landward side of the orogen and fold and thrust belts, subsiding probably in response to increasing vertical loads on the outer margin of North American lithosphere. Such basins were transient, their depoaxes migrating landward and their rearward portions progressively incorporated in the thrust belt or eroded from uplifts above blind thrust tips.

The hinterland is composed of fault-bound or shear-zone-bound packets of North American rocks that are mostly metamorphosed and were derived from more seaward sites and in part from deeper levels than those of the foreland (Rankin, this volume; Hatcher, 1978, 1987). The earliest hinterland (Ordovician) includes packets of Grenvillean Proterozoic basement and Late Proterozoic and lower Paleozoic passive margin strata from sites outboard of the hinge. The late Paleozoic hinterlands include terranes attached to North America in Ordovician collisions as well as rocks of the Ordovician hinterland. During first and subsequent collisional events, the hinterland packets underwent complex deformation, including thick-skinned thrust imbrication above intracrustal or deep supracrustal detachments, orogen-parallel strike slip, and horizontal extension with omission of vertical intervals. They have undergone plutonism and, at least locally, subsidence and deposition of cover at intervals during a collisional phase.

The belt of terranes consists of varied bodies thought to be non–North American at times preceding their attachment between mid-Ordovician and Permian time. Evidence that they are exotic includes the provinciality of early Paleozoic fossils (Wilson, 1966; Neuman and others, 1989), tectonostratigraphies incompatible with that of North America, and anomalous paleomagnetic latitudes relative to contemporaneous North American ones. As the collisional orogen developed through the

Paleozoic, early attached terranes were incorporated into younger hinterlands. Figure 11 divides terranes of all ages from early hinterland.

The crystalline basement of cratonal North America is thought by most investigators (Williams, 1975; Hatcher and others, 1989; Cook and Oliver, 1981; Cook and others, 1981; Rankin, 1975, this volume; Rankin and others, 1989; Stewart and others, 1991; Haworth and others, this volume; Transects D1–D4 and E1–E5 in Speed, 1991) to extend eastward below the Appalachian orogen, east of the inboard edge of terranes as drawn in Figure 11, to a locus shown as the present-day edge of PNA in Figure 8. The cross-sectional configuration of the orogen is thus a mammoth wedge above and offshore from the contiguous continent. This shape has almost certainly arisen from successive adjustments to accretion at the rear and front, underplating, horizontal contraction, and magma addition.

Plates and terranes. The plate tectonics of the collisional phases of the Appalachian orogen is not fully understood, mainly because of the uncertainties of the positions of major continents in the Paleozoic, the origins and affiliations of small terranes, and the difficulties of dating and correlation due to deformation and metamorphism. The large terranes that collided with Paleozoic eastern North American were Gondwana, Avalon, Carolina, and Baltica; the small terranes were a few of reasonably clear origin (e.g., Dunnage, Williams, 1975; Williams and Hatcher, 1983; Haworth and others, this volume) and many of less certain or problematic origin (e.g., Gander, Williams and Hatcher, 1983; Haworth and others, this volume; Goochland, central Maine, Secor and others, 1986; Horton and others, 1989; Rankin, this volume). All the terranes were in place by Permian time.

Baltica began closure relative to North America in mid-Ordovician time across the northern reach of Iapetus east of Newfoundland and completed its collision in the Devonian. Baltica and parts of Paleozoic North America now in the northwestern Eurasian plate, however, departed from Mesozoic North America with the opening of the Atlantic.

The pre-Permian relative motion of Paleozoic Gondwana (cored by Africa, South America, and Australia) relative to North America is controversial. Gondwana moved apparently far south of North America within the opening of Iapetus and perhaps other intervening oceans. In the Silurian, Gondwana may have moved north relative to North America, possibly colliding in the Early Devonian, and then drifting well south in the Devonian with the opening of a new intervening ocean before returning in the Carboniferous (Van der Voo, 1988). As one alternative, employing different data, the relatively northward Silurian motion of Gondwana may have continued monotonically until collision with eastern North America beginning in Early Mississippian time (Scotese and McKerrow, 1990).

The Avalon terrane of eastern Canada and small terranes of coastal New England thought to be related to, if not contiguous with, Avalon (Rankin, this volume), and the Carolina terrane (or Carolina composite terrane; Secor and others, 1986) (Fig. 11) all have certain pre-Middle Ordovician faunal and tectonostratigraphic similarities among themselves and with Gondwana. They differ from North America in the same properties. Their post–Middle Ordovician apparent polar wander paths, however, differ among themselves and from that of Gondwana, suggesting that they separated from Gondwana, perhaps were left behind in its wake, either in early Paleozoic oceans during the early southerly drift of Gondwana or after the mid-Paleozoic close approach of Gondwana with eastern North America, or both. To indicate the affinity with Gondwana of these terranes (Avalon, St. Croix, Ellsworth-Mascarene, and Carolina composite; Haworth and others, this volume; Rankin, this volume) but their disparate, movement histories, the terranes are called peri-Gondwanan.

If the idea of early separation of the peri-Gondwanan terranes from Gondwana applies, the collisions of such terranes with North America require convergent plate boundaries within Iapetus and/or later Paleozoic oceanic lithospheres. The existence of such oceans is indicated by the bodies of ultramafic rocks, some ophiolitic, that are within terranes in melanges or imbricated wedges or as an extensive sheet (e.g., Dunnage terrane, Fig. 11; Haworth and others, this volume).

The Carboniferous collision of nuclear Gondwana fronted by Africa with eastern North America was the climax of the Paleozoic opening and closing of oceans in the proto-Atlantic region (Wilson, 1966). This event probably had the maximum effect on shaping the Appalachian orogen as a whole (the Alleghanian-Hercynian belt), although the intensity of its effects certainly varied on strike. The Carboniferous suture between Africa and North America is difficult to delimit, either because it is mainly absent from present-day North America or because it is deeply buried below the Mesozoic-Cenozoic passive margin wedge. The Suwannee terrane of Florida and southern Georgia (Fig. 11) has strong West African correlatives and is almost certainly Gondwanan (Chowns and Williams, 1983; Dallmeyer, 1988); its suture with Carboniferous North America is a south-dipping zone and an inferred northerly overriding thrust (Tauvers and Muehlberger, 1987; McBride and others, 1989). The Meguma terrane (Fig. 11), which occupies outboard present-day North America from Newfoundland to offshore New England, is also thought to have belonged to pre-Carboniferous Africa (Schenk, 1981; Haworth and others, this volume). Between the Suwannee and Meguma terranes along the outer margin of today's North American continent and below the coastal plain sediments, there are several terranes (Charleston, Hatteras, Chesapeake) that are suspected by some to be Carboniferous Africa; others believe them to be part of or related to the Carolina composite terrane (Horton and others, 1989; Rankin, this volume; Secor and others, 1986). Thus, it is not yet clear whether the Jurassic breakup of Pangea in the Atlantic caused a net positive transfer of Paleozoic Africa to North America or an equal swap of territory for both present-day continents. It is possible, moreover, that the Jurassic breakup surface was exactly along the collisional suture at latitudes between the Meguma and Suwannee terranes.

Between PNA and the peri-Gondwanan terranes, the Appalachian orogen contains an inner belt of many small terranes

(Fig. 11). These are of varied kindred: magmatic island arc, forearc accretionary wedge, backarc, oceanic crust, continent; some are of mixed, disputed, and uncertain origin (Rankin, this volume; Williams and Hatcher, 1983; Zen, 1983; Horton and others, 1989; Secor and others, 1986). They have undergone varied degrees of deformation, metamorphism, and unroofing of preemplacement, synemplacement, and postemplacement age. Owing to such complexity, their boundaries and correlations are not easily established. The disagreements are mainly what is North American and what is alien continental basement or cover. It is also uncertain to what degree the small terranes are pieces of large terranes that were dispersed during postattachment deformation, chiefly by margin-parallel strike-slip and low normal faulting. It is also unclear whether the small terranes have attached underpinnings of crust and mantle, or whether the terranes at the surface are nappes or flakes and are unrelated to bodies at depth. It does appear, however, that most if not all the terranes inboard of the peri-Gondwanan terranes collided with North America and underwent major deformation, magmatism, and metamorphism in the Ordovician (Rankin, this volume). An important issue is whether the Goochland and Wilmington terranes (Fig. 11) of the central Appalachians are (1) autochthonous North America exposed through terranes in an extensive tectonic window (Glover and others, 1983); (2) North American, and transferred along the orogen by strike-slip faulting (Hatcher, 1987); or (3) bonafide exotic terranes (Rankin, this volume). Another issue is whether the Inner Piedmont terrane of the southern Appalachians contains Ordovician magmatic arcs equivalent to those of New England and Newfoundland. A problem in tectonic reconstructions of the Appalachians is whether microcontinental basements lie below certain terranes composed of exotic supracrustal rocks or whether such rocks are accretionary wedges derived from cover of now-subducted oceanic crust or from arc crust. Small Ordovician terranes of New England and eastern Canada (Central Maine, Dunnage, and Gander; Rankin, this volume; Haworth and others, this volume) are probably rootless, but they may be underlain by sialic crusts of uncertain heritage, as implied by isotopic and deep seismic studies (Stewart and others, 1991; Stockmal and others, 1990).

The attachment history of the peri-Gondwanan terranes to the Ordovician inner terrane belt is perhaps the most uncertain chapter of the evolution of the Appalachian orogen. The inboard edge of the Avalon terrane of Newfoundland and New Brunswick is a vertical, crust-penetrating surface (Dover and Turtle Head faults, Fig. 11; Haworth and others, this volume) that joined the terrane to Ordovician North America and had strike-slip motion until a time of attachment in the Devonian (Dallmeyer and others, 1981). Southwestward, the Avalon terrane is attached to other peri-Gondwanan terranes at its western edge (Rankin, this volume), and the onshore boundary of this terrane group dips below the edge of the inner terrane belt. Final attachment there occurred by Late Silurian or Devonian time. Farther southwest in southern New England, a set of peri-Gondwanan terranes (Fig. 11) underlies at least the eastern half of the inner terrane belt with subhorizontal contact (Getty and Gromet, 1992; Wintsch and others, 1992). Here, the latest movement on the boundary was Permian and caused the omission of a thick vertical section. Moreover, the crystalline basements of North America (Grenvillean crust) and the peri-Gondwanan terranes are in contact or close to it below the inner terrane belt of southern New England.

The collision of peri-Gondwanan terranes in the northern Appalachians thus had varied times within the middle and late Paleozoic and varied kinematics. It seems likely that the process involved many Gondwanan terranes rather than one (Avalonia) and that there may in fact have been a complex evolution of attachment, partial breakup, dispersal, and reattachment, much as with the Wrangellia terrane of western North America. It has long been recognized that the late Paleozoic phase (Alleghanian) of the Appalachian orogen was much more severe south of southern New England than to the north, presumably due to increasing proximity to the boundary of the colliding nuclear Gondwana.

The Carolina terrane and fully buried terranes on its outboard are the probable peri-Gondwanan affiliates of the southern Appalachian orogen (Fig. 11) (Secor and others, 1986; Rankin, this volume). The attachment of the Carolina terrane to the North American continent was at a time that is uncertain within the interval, Ordovician-Devonian. Its landward boundary is with the inner terrane belt and dips east, implying that the Carolina terrane overrode the other terranes. The boundary, however, has taken up a succession of motions: an earliest pre–Late Devonian ductile phase of uncertain direction, a hypothesized intermediate phase of flat faulting representing regional tectonic thinning in Silurian and/or Devonian time (Dennis, 1991), and a late Paleozoic (Alleghanian) phase with reverse, normal, and/or strike-slip movement (Horton, 1981; Gates and others, 1986; Hatcher and others, 1993). The arrival of the Carolina terrane is thought by Hatcher and others (1993) to record the final closing of the Iapetus ocean in the southern Appalachians. If Ordovician, the Carolina terrane was probably the principal collider in the development of the early southern Appalachian orogen and the inner terrane belt (Rankin, this volume; Secor and others, 1986). If later, the collision of the Carolina probably caused the redeformation, high-grade metamorphism, and plutonism of the hinterland and inner terrane belt that may be partly of mid-Paleozoic age and attributed to the Acadian phase of orogeny (Hatcher, 1987).

To conclude, the history of the peri-Gondwanan terranes of the Appalachian orogen is complex and still uncomfortably pliable. Such terranes may have arrived as a single composite giant in middle Paleozoic time and have undergone significant disruption and redistribution with varied kinematics by the late Paleozoic Alleghanian collision of nuclear Gondwana. Alternatively, these terranes may have been discrete fragments in the Iapetus and other early Paleozoic oceans, left behind from south-moving nuclear Gondwana, and have collided with North America with varied kinematic modes (underwedging, overriding, strike slip) at times throughout much of the Paleozoic.

Evidence for collision. Evidence that the Appalachian orogen evolved at a collisional rather than an active margin is strong for the initial (Ordovician) phase by virtue of the occurrence of early Paleozoic arc volcanic rocks in Ordovician terranes but not in the North American foreland or hinterland. Moreover, the terranes bearing Ordovician arc volcanic rocks are underlain and bordered landward by terranes that represent accretionary forearcs that include ophiolitic (or at least ultramafic) rocks, oceanic strata, and forearc basin deposits. Such forearcs imply that an ocean (Iapetus) existed between the edge of Ordovician North America and the magmatic arcs before juxtaposition. The evidence for a collisional rather than active margin in eastern North America in middle and late Paleozoic time is less strong than that for the early Paleozoic. The problem is the origin of indigenous magmatic rocks of middle or late Paleozoic age in the Appalachians. Plutonism of late Paleozoic and possibly mid-Paleozoic age occurred in the southern Appalachian in the inner belt of terranes (Dallmeyer and other, 1986). In New England and eastern Canada, Silurian and Devonian plutonism was widespread in the inner terrane belt, and there are a couple of late Paleozoic plutons in southern New England. All of these plutons are considered to be anatectic, generated within the crust, and a product of large tectonic thickening by Osberg and others (1989) and Sinha and others (1989). If the plutons are anatectic, evidence is lacking for the descent of post-Ordovician oceanic lithosphere below the North American continent. The eastern margin thus appears to have been collisional into the Permian (Fig. 12), unless former active margins have been excised and removed. The post-Ordovician closure of oceans east of North America probably occurred with the slabs of NAm dipping eastward. Paleozoic magmatic arcs that might be counterparts to such closure, however, have not been recognized in West Africa. The plate boundaries therefore probably existed within the oceanic realm between the continents.

Early collision. A history of the Paleozoic collisional margin that evolved as the Appalachian orogen begins in Middle Ordovician time, with the arrival at the edge of the North American continent of a forearc and magmatic arc above a seaward-dipping slab of Iapetus. The collision was of low obliquity (nearly head-on) from Georgia to Greenland. The forearc, composed of rock slices derived from oceanic realms and the continental rise, overrode the passive margin fully in the northern Appalachians but possibly only partly in the southern Appalachians before welding. Parautochthonous lower Paleozoic passive margin cover was accreted to the forearc during its transit across the shelf. One or a series of magmatic arcs were thrust westward above the forearc. It is disputed whether the arc in the southern Appalachians was in the Carolina terrane (Hatcher, 1987), as yet unrecognized, or was in terranes inboard of the Carolina (Rankin, this volume). On the seaward side of the colliding arcs there may have existed backarc regions with continuing magmatic activity, sedimentary basins, and microcontinental blocks (such as the Goochland-Wilmington terranes and a hypothesized basement in the northern inner terrane belt; Fig. 11; Rankin, this volume). These were accreted to North America in the Ordovician to form the inner belt of terranes before the arrival of the peri-Gondwanan terranes, with the possible exception of the Carolina terrane. The early collision was apparently completed by Late Ordovician or Early Silurian time. The collision interval included plutonism, deep-seated metamorphism, and progressive deformation in the hinterland and inner terrane belt. These events were apparently concomitant with large crustal thickening of North America adjacent to and/or below the overriding terranes. In southern New England, the thickening is attributed to breakback imbrication of the North American hinterland above a mid-crustal detachment (Stanley and Ratcliffe, 1985). Such structures attest to the continuing contraction of the North American margin after the initial collision with and attachment of terranes.

It is noteworthy that Ordovician plutonism, high-grade metamorphism, and progressive deformation are more intense in the Ordovician collision zone of southern New England than in areas just north and south of it, where collisional kinematics are thought to be more oblique than in southern New England. Such variability in phenomena was explained by Stanley and Ratliffe (1985) and Tremblay (1992) by collision with a zigzag passive margin, the promontories taking up the early collision by internal shortening as the colliding terranes slid obliquely along their flanks into the embayments. It is difficult to know whether this theory applies to the southern Appalachians because late Paleozoic contraction there has so severely reorganized the early structures.

Intermediate events. There seems to have been a true hiatus in orogeny at the continental margin between about earliest Silurian and mid–Early Devonian time in the Appalachian orogen as a whole (between Taconian and Acadian phases) (Glover and Sinha, 1973; Horton and others, 1989; Hatcher and others, 1993; Rankin, this volume). During this interval, thick sedimentary basins developed above or near the terranes that collided in the Ordovician in the northern Appalachians. Such basins were either successor to the earlier orogen, perhaps by cooling and thermal subsidence of collided arcs and thickened basement, or they may have preexisted back (seaward) of the collided Iapetan arcs. Equivalent basin fills probably exist in the southern Appalachians, on the basis of metasedimentary rocks with fossils thought to be of post-Ordovician age within the Taconian hinterland and the inner terrane belt. Such strata are strongly reworked by late and possibly mid-Paleozoic deformation (Tull and Groszos, 1990; Hatcher and others, 1993).

The character and cause of the mid-Paleozoic phase (Acadian) of the development of the Appalachian orogen are elusive. In the northern Appalachians it includes large margin-normal contraction, margin-parallel shear of uncertain sense, crustal thickening, and anatectic plutonism within the existing hinterland and the inner terrane belt, all of Devonian age. In addition, unroofing and cooling occurred in Devonian time (Osberg and others, 1989). In the northern Appalachians, the Ordovician hinterland and inner terrane belt and their cover underwent Devonian contraction as a deeply rooted triangle zone with westerly vergence on the landward side and easterly vergence on the seaward side. The eastern boundary of the triangle zone is a series of ductile and brittle faults that brought peri-Gondwanan terranes below or against the edge of mid-Paleozoic North America (Stewart and others, 1991; Rankin, this volume). The peri-Gondwanan terranes were not affected by the mid-Paleozoic penetrative deformation and metamorphism of the adjacent Appalachian orogen, but they do contain mid-Paleozoic plutons. The age and attitude of the final contact between the peri-Gondwanan terranes and the orogen to their west vary markedly with position in the northern Appalachians, from Silurian to Permian and horizontal to subvertical, but always with peri-Gondwanan terranes in the footwall.

Basic questions address the plate tectonic cause of the mid-Paleozoic (Acadian) orogenic phase and whether it was the collisions of the peri-Gondwanan terranes? It is appealing to consider that Acadian deformation was caused by such collisions, because no other cause is evident. The timing is permissible, but the idea requires postulation of postcollisional redistribution of the colliding elements to explain the range of times of final emplacement throughout the latter half of the Paleozoic and the varied structure of boundaries. Such variability can be reconciled with a Late Silurian–Devonian (Acadian) wedging of peri-Gondwanan terranes into the earlier terranes and/or between orogenic cover and Grenvillean basement of North America throughout the northern Appalachians. In Newfoundland, the Dover fault then took up late Acadian margin-parallel transport in the newly accreted margin, stranding a wedge of collided crust inboard of the Dover fault. Such a wedge, which could belong to the Avalon terrane, was imaged seismically by Stockmal and others (1990). The contact between the peri-Gondwanan terranes and inner terrane belt in Maine may be the initial one, whereas that in southern New England was truncated in Permian time by a flat shear zone with large omission of tectonic section and relatively eastward transport of the hangingwall (Wintsch and others, 1992). If this sequence is correct, Goldstein's (1989) evidence for late top-west transport in the shear zone separating the inner terrane belt and a peri-Gondwanan terrane in southern New England requires another, perhaps Mesozoic, reactivation.

In the southern Appalachians, a mid-Paleozoic collision comparable to that imagined for the northern Appalachians requires the arrival of the Carolina terrane in the Devonian. Even if this is so, the Carolina almost certainly was emplaced above inboard terranes, implying a change from wedging to overriding of colliding elements north to south along the margin of Devonian North America.

The ocean-continent configuration during the mid-Paleozoic is most uncertain. The absence of mid-Paleozoic ophiolites among terranes of the Appalachian orogen implies that oceanic lithosphere stayed deep below the colliding, relatively buoyant fragments of the peri-Gondwanan terranes. The Devonian plutonism in the Taconian hinterland and inner terrane belt of New England could suggest a local slab with landward dip, but the pluton compositions are commonly interpreted as indicating crustal anatectic sources (Osberg and others, 1989). Therefore, crustal thickening is indicated, and this accords with terrane wedging but provides no constraint on slab dip.

Late (Alleghanian) collision. The last phase (Alleghanian) of the collisional eastern margin of Paleozoic North America was the further deformation of the Appalachian orogen and the complete attachment of nuclear Gondwana fronted by Africa and/or terranes associated with it. These late Paleozoic motions caused the elimination of Paleozoic oceans between North America and Africa and the encasement of the eastern North American continent in the interior of the Pangean supercontinent. The Alleghanian suture lies outboard of the earlier attached peri-Gondwanan terranes, but its continuity is uncertain below the coastal plain and shelf of present-day eastern North America. The inner contacts of Meguma and Suwannee terranes in the northern and southern Appalachians, respectively (Fig. 11), are segments of the African suture, but it is controversial whether the central segment of the late Paleozoic suture lies in the North American or in the African continental shelf.

The Alleghanian phase was principally contractile, causing horizontal shortening and vertical thickening of much of the margin; it also included margin-parallel right-slip faulting, plutonism, unroofing in the Taconian hinterland and in the attached terranes, and local crustal extension, probably in pullaparts related to strike-slip faulting. The duration of the Alleghanian phase is said to have been 266 to 327 Ma (Secor and others, 1986), 270 to 330 Ma (Horton and others, 1989), and 250 to 330 Ma (Glover and others, 1983). Such estimates permit a 50 m.y. orogenic hiatus in Early Mississippian time between Acadian and Alleghanian phases.

The major unsolved problems of the Alleghanian phase are the origin of the extraordinary difference between the width and kinematics of the deformation zone in the southern versus northern Appalachians, and the configuration of plate boundaries during late Paleozoic continental closure.

In the southern Appalachian orogen, the width of the deformed margin of late Paleozoic North America and the net boundary normal displacement are remarkably large, at least 500 km and 300–350 km, respectively (Rankin and others, 1991; Hatcher and others, 1993). There, the orogen includes a major foreland basin, a broad thin-skinned thrust and fold belt that advanced relative to the Taconian foreland, and a very broad hinterland composed of pre-Alleghanian terranes. The hinterland underwent deep-seated contraction, plutonism, and highly differentiated unroofing. The Alleghanian foreland of the southern Appalachians in its most landward reach (Valley and Ridge province, Fig. 11) is perhaps the world's classic thin-skinned contractile belt. It includes stratigraphically controlled detachments that ramp up westward through Paleozoic platformal strata with break-forward sequence and margin-normal slip. The southern half of this belt is an imbricate of thrusts that were emergent or nearly so, whereas the northern half is a little-broken fold train that lies above blind thrusts and duplexes (Geiser, 1989). Late Paleozoic foreland basin deposits are deformed in their eastern reach, but undeformed west of the thrusts and the fold train. The horizontal contractile strain in the fold and thrust belt is heterogeneous and varies from 25% to 50%. The rear of the stratigraphically controlled foreland thrusting at the present surface within the Alleghanian foreland is the set of brittle thrusts that underlie the Blue Ridge sheet, which apparently acted as a huge (200 km wide), nearly coherent semirigid thrust packet with only minor imbrication in Alleghanian contraction (Hatcher, 1987). The Blue Ridge sheet, which contains the predeformed Ordovician foreland and hinterland (shown as western Blue Ridge in Fig. 11) and the landward portion of the inner terrane belt, translated northwest about 300–350 km above the pre–late Paleozoic continental platform (Rankin, this volume; Hatcher and others, 1993). One evident Alleghanian feature in the Blue Ridge sheet, however, is the bumps in the sheet's base caused by subjacent duplexes of overridden platformal strata (Hatcher and others, 1993).

East of the Blue Ridge sheet in the preexisting peri-Gondwanan terrane belt is the Alleghanian hinterland. Its late Paleozoic kinematics were complex and included partitioned margin-parallel right slip and margin-normal contraction. The hinterland contains belts of deeply unroofed rocks that underwent Alleghanian prograde high-grade metamorphism (upper amphibolite facies), belts of low-grade or retrograded rocks, and abundant late Paleozoic granitoid plutons. The relative ascent of the high-grade belts is attributed to climb on deep-seated east-facing ramps (Secor and others, 1989). In this view, the hinterland contains early thrust packets in the Alleghanian break-forward series that imbricated above a deep detachment, perhaps at the base of the brittle zone at mid-crustal depths (Hatcher, 1989). The detachment then ramped up westward to a preexisting brittle fault zone at the base of the Blue Ridge sheet. In addition to contractile ramping, late low-angle normal faulting probably caused local unroofing of deep-seated Alleghanian metamorphic rocks (Dallmeyer and others, 1986; Dallmeyer, 1988), although it may be difficult to differentiate Alleghanian unroofing from that due to regional extension during early Mesozoic time (Heck, 1989).

Alleghanian right-slip faulting is recognized in zones that had both brittle and ductile rheologies, implying deep penetration of such faults. The integral displacement of strike-slip faults across the hinterland is difficult to determine, but most workers think that strike slip was less than thrust heave across the Alleghanian orogen. Together, the two components of displacement imply either that the Africa–North America relative motion was right oblique, or that there was episodic nonoblique convergence and vertical axis rotation of one or both plates (Sacks and Secor, 1990).

The Alleghanian displacement zones of the hinterland are commonly reactivated older faults, especially Ordovician terrane boundaries. Moreover, the same zones underwent successions of slip with markedly different direction. For example, the Brevard zone may have been an Ordovician terrane boundary (Rankin, this volume; but see Hatcher, 1989) and may have had middle or late Paleozoic ductile right slip, late Paleozoic normal dip slip (Dennis, 1991), and finally, late Paleozoic brittle reverse dip slip.

Alleghanian plutonism in the hinterland of the southern Appalachians is of uncertain origin. Compositional data from Alleghanian plutons imply an eastward increase across the hinterland of noncontinental source rock (Sinha and others, 1989). This does not distinguish late Paleozoic subduction from anatexis in the earlier plate boundary zone that includes preexisting terranes with progressively more noncontinental affinity seaward.

The Alleghanian phase of the northern Appalachian orogen differs from the southern by the absence of foreland features, either a basin or thrust belt, and by the paucity of late Paleozoic deformation north of southern New England. Northeast of Boston, Alleghanian kinematics are limited to right-slip brittle faulting in the early and middle Paleozoic terrane assembly, commonly reactivating earlier terrane sutures, including the contacts of the peri-Gondwanan terranes (Rankin, this volume; Haworth and others, this volume). Carboniferous basins in this region are thought to be of pullapart origin (Bradley, 1982; Keppie, 1989). The implication is that final approach of Gondwana to the northern Appalachian margin was highly oblique.

In southeastern New England and coastal New Brunswick, however, the Alleghanian effects were more severe than to the north. Peri-Gondwanan terranes locally underwent Carboniferous extension and subsidence, possibly related to strike slip, as to the north. They also took up later Alleghanian contraction, metamorphism, and magmatism, unlike areas to the north. These responses to collision occurred during and/or before unroofing by low-angle normal faulting that omitted a substantial thickness of metamorphic section in southern New England (Wintsch and others, 1992; Getty and Gromet, 1992). It is possible that the peri-Gondwanan terranes in southern New England first arrived at and were wedged into the North American margin in late Paleozoic time, possibly as Meguma basement or as terranes associated with Meguma. Alternatively, they could have already existed from an Acadian collision and have been pushed below or farther below late Paleozoic North America's edge by the arrival of Meguma (Africa), which had a greater component of convergence in southern New England than farther north in the Appalachians. In either case, there seems to have been a change in obliquity of relative motion across the plate boundary zone that included southern New England to account for the gross differences of the northern and southern Appalachians in late Paleozoic time. Such differences are possibly explained by a change in trend of the late Paleozoic margin of North America from northeasterly north of southern New England to a more northerly trend to the south, a trend that may have existed before the large margin-normal displacements of the southern Appalachians.

Present passive margin. *Introduction.* The eastern margin of present-day North America from the Bahamas to Greenland is a passive margin that developed from Middle Triassic to the present (Fig. 12), including the onset of spreading of Atlantic oceanic lithosphere at times in the Jurassic, Cretaceous, and Cenozoic. The development of the passive margin (Fig. 13) was the

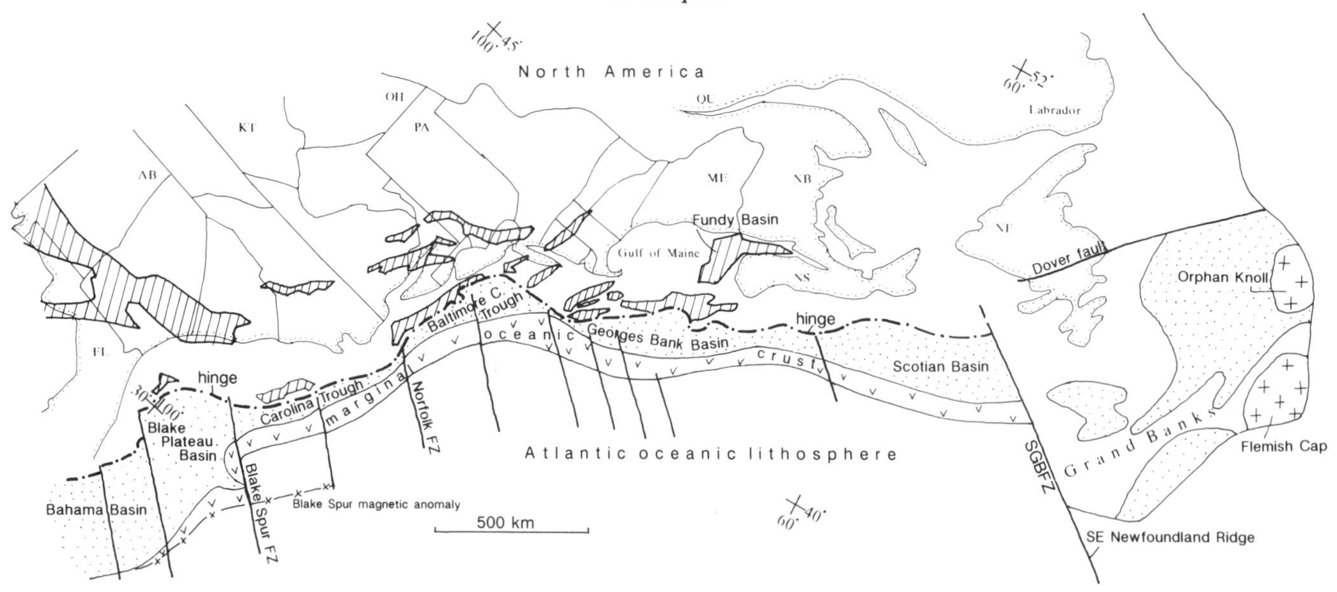

Figure 13. Tectonic features of the Mesozoic-Cenozoic passive margin of eastern North America. SGBFZ, Southern Grand Banks Fracture Zone; fz, fracture zone. Ruled pattern, rift basins; fine dot pattern, marginal basins; v pattern, marginal oceanic crust; cross pattern, continental plateaus. Abbreviations as in Figure 11.

result of the rifting and breakup of the Pangean supercontinent, the lifetime of which may have been as brief as 25 m.y. between final closure and first rifting and 90 m.y. between final closure and first spreading. The breakup of Pangea once again isolated the North American continent from others, but with modest rearrangements of its eastern edge relative to that of middle Paleozoic (pre-Alleghanian) North America. The change included acquisition from Africa of territory (the Suwannee and Meguma terranes and, possibly, terranes between them, as indicated in Fig. 11), and from the mantle, substantial volumes of basalt. The net growth of North America by the collision and withdrawal of Africa, however, is not clear, because the outboard terranes of the central Atlantic reach of North America are of uncertain Paleozoic provenance, African or North American, and it is possible that Jurassic Africa departed from the central Atlantic suture with territory acquired from Paleozoic North America.

The history of the Pangean breakup as related to North America begins with rifting in early Middle Triassic time (about 235–240 Ma) in the region of the Gulf of Maine–Nova Scotia shelf inland to the Fundy Basin (Fig. 13) (Keen and Haworth, 1985b). This was followed by rifting in Late Triassic time southward through the eastern Appalachian orogen to the Bahamas and, probably, across the present shelf over a width of about 300 km (deformed length) parallel to the northwest-southeast direction of maximum principal elongation (Thompson and others, 1993; Rankin and others, 1991; Hatcher and others, 1993; Rankin, this volume; Sawyer, this volume). Just north of Nova Scotia, the Triassic rifting ended at an intracontinental transform that, after drifting, became the Southern Grand Banks Fracture Zone (SGBFZ) transform, and to the east, the Southeast Newfoundland Ridge (oceanic fracture zone) and the Tethyan transform between Africa and Iberia (Fig. 13) (Haworth and others, this volume). The beginning of sea-floor spreading of normal oceanic crust in the rifted zone between Georgia (Blake Spur Fracture Zone, BSFZ) and northeastern Nova Scotia is estimated at about 175 Ma by marine magnetic anomalies (Klitgord and Schouten, 1986). South of the BSFZ (Fig. 13), which began as an intracontinental rift-rift transform, spreading of normal oceanic crust began at about 170 Ma, and the breakup locus was well east of the 175 Ma locus to the north of the BSFZ.

North of the SGBFZ transform as far as Labrador, rifting in North America of Maritime Canada occurred over a prolonged duration in the Jurassic and Early Cretaceous while North America was still connected to the Eurasian continent through the Grand Banks and while the central Atlantic was spreading to the south (Keen and Haworth, 1985a; Haworth and others, 1985). The breakup between the Grand Banks and Iberia in Eurasia occurred in mid-Cretaceous time, apparently when the Cretaceous Mid-Atlantic Ridge had reached today's eastern margin of the Grand Banks (Haworth and others, this volume). The breakup migrated north and west around the northern Grand Banks and into the Labrador Sea in the Late Cretaceous, causing the drift of Greenland and Eurasia from North America (Srivistava and Tapscott, 1986). In Eocene time, the Labrador Sea segment of Atlantic spreading was abandoned, and a new spreading arm began east of Greenland, separating it from Eurasia at what has become today's divergent boundary in the North Atlantic. In summary, rifting occurred over durations before breakup as short as 25 m.y. at some sites but as long as 100–150 m.y. at other places where spreading was delayed. The geography of Atlantic rifting and breakup suggests that the extension was imposed nearly synchronously over a very long reach, perhaps indicating motions in the mantle, whereas breakup occurred at disparate times, and was perhaps related to distance from the Euler pole of the diverging plates.

The present-day passive margin of eastern North America (Fig. 13) developed before and during the opening of the Atlantic Ocean with some similarity on strike, discussed presently, and with some differences on strike, discussed later. The development was generally two phase, rift and postrift. The rift phase is recorded by half grabens and their fill of sediments and basalt. The grabens occupy a belt (Fig. 13) that is onland in the eastern part of the Appalachian orogen and in at least the nearshore reach of the continental shelf, as inferred from magnetic anomalies and seismic profiles (Hutchinson and others, 1988). It is uncertain whether rift phase features like these occur farther outboard on the continental shelf because of the thick cover of postrift strata.

The postrift phase includes the following features: hinge, marginal sedimentary basins, marginal oceanic crust, transform faults, normal oceanic crust created by sea-floor spreading, and evaporite diapirism. Such phenomena occupy the outboard half of the passive margin, east from the hinge to Atlantic oceanic crust (Fig. 13) (Rankin and others, 1991; Rankin, this volume; Haworth and others, this volume). The hinge is the locus of major change in postrift subsidence, from a platform on the landward side to a marginal basin with deposition of a thick passive margin wedge on the seaward.

The development of transitional crust between the hinge and marginal oceanic crust and the occurrence of magmatism spanned both phases of passive margin evolution (Sawyer, this volume).

The continent-ocean boundary is zonally gradational from normal oceanic crust through a narrow belt of marginal oceanic crust (Fig. 13) to outer transitional crust that is probably rich in basalt dikes. This gradational boundary is generally marked by the East Coast Magnetic Anomaly between the SGBFZ transform and BSFZ, and by the Blake Spur Magnetic Anomaly south of there (Fig. 13). The source of the anomalies is probably marginal oceanic crust (Mutter, 1985). The seaward edge of carbonate wedges and of salt diapirs also marks approximately the continent-ocean boundary both north and south of the SGBFZ.

The rift-postrift transition is difficult to date with precision. In the Baltimore Canyon Trough segment (Fig. 13), it is thought to have followed rift magmatism dated 201 ± 1.3 Ma (Sutter, 1988) and to have preceded deposition of marginal basin strata at least as old as Early Jurassic (about 190 Ma) (Rankin, this volume). Without an unconformity between strata of the two phases, however, they are difficult to distinguish. In fact, rift and postrift phases appear to be widely continuous and marked mainly by increasing rates of subsidence outboard of the hinge. In the past, the rift-postrift transition has been assumed to date the onset of sea-floor spreading at the continent-ocean boundary. This is doubtful, as indicated by the 175 Ma age estimated for separation in the central Atlantic, which is at least 15 m.y. younger than strata apparently of postrift origin in a marginal basin.

The principal kinematic manifestations of passive margin development were the horizontal stretching of North American crust and, probably, lithospheric mantle, and the volume change of these layers by addition of basalt from the mantle. Both stretching and magmatism probably took place mainly before sea-floor spreading began. Another major kinematic manifestation was the subsidence of the surface of the crust relative to sea level, outboard of the hinge. The subsidence probably occurred throughout passive margin development, but most rapidly during the early postrift phase.

The width of the deformed passive margin varies enormously from about 150 to 1050 km, measured between the most landward rift basin and normal oceanic crust. The extension in the belt of exposed rift basins is probably small, ≤10%. Within the postrift belt, however, the extension is large and variable along strike, between 100% and 300%, as estimated by crustal thinning. The uncertainty of volume addition makes the values of calculated extension only minimums. The rifts, dikes, and margin as a whole are widely thought to have strikes normal to the maximum principal elongation, which was about N60W–S60E in the central Atlantic but more easterly in the younger North Atlantic (Klitgord and Schouten, 1986). Alternative views of the early kinematics exist, however, and propose regionally oblique extension and propagating rifting (Tankard and Balkwill, 1989).

The passive margin includes a regional substructure defined by promontories and embayments of the hinge and correlative differences in the postrift marginal basins (Fig. 13) (Rankin, this volume; Haworth and others, this volume; Sawyer, this volume). The promontories contain relatively wide marginal basins containing thick strata, whereas basins of the embayments are more narrow and have thinner strata. These alternating zones are connected across strike by transform faults or fault zones. It is disputed whether such zones are continuous with oceanic fracture zones such as the Blake Spur, Cape Fear, and Norfolk fracture zones (Fig. 13) (Klitgord and Schouten, 1986; Wernicke and Tilke, 1989).

Rift phase. The rift basins are half grabens, the master faults of which are listric and flatten to a mid-crustal detachment level. Such faults have both seaward and landward dips. The majority of dips are thought to be seaward along the entire margin, and where landward dips occur, they are mainly on promontories (Wernicke and Tilke, 1989). Exhumed normal ductile shear zones suggested to be of rift-phase age (Triassic) have been identified as deep principal displacement loci of rifting at mid-crustal levels. Examples are the Modoc faults of the southern Appalachians and Norumbega fault of the northern Appalachians (Fig. 13) (Secor and others, 1986; Goldstein, 1989; Heck, 1989). Such shear zones are thought to be reactivated listric thrusts and/or normal faults of the Appalachian orogen (Rankin, this volume).

Several features suggest, however, that the crust did not stretch significantly at levels below such shear zones in the passive margin landward of the hinge. These features are the unmodified cratonal thickness of crust (except for stepwise thinning seaward in Maine), the apparently minor volume of magmatic rock, and the paucity of lower crustal reflectors, all of which differ markedly from features of the crust seaward of the hinge (Ando and others, 1984; Green and others, 1988; Hutchinson and others, 1988; Stewart and others, 1991). This implies that stretching landward of the hinge occurred above a detachment. It is uncertain whether the large stretching of the lower and upper crusts in the passive margin seaward of the hinge began in the rift or postrift phase.

Synrift magmatism produced basalt dikes and lavas that are widely distributed but generally in small volume. The dikes strike mainly parallel to the rift axes and the continent-ocean boundary zone, implying minor, if any, margin-parallel shear during rifting. The timing of magmatism relative to rifting is varied. In the Newark basin (Fig. 13), magmatism occurred in a short pulse at about 201 Ma (Sutter, 1988) at the close of rift basin development. In onshore and offshore Georgia and South Carolina, widespread basaltic extrusion is interpreted to have occurred during and following the development of the postrift unconformity (Rankin, this volume; Hatcher and others, 1993). In contrast, basalt in the Labrador margin was extruded before faults bounding rift basins attained substantial throw (Haworth and others, this volume). For the passive margin as a whole, however, surficial magmatism seems to have been late in the rift phase.

Postrift phase. The postrift phase of passive margin evolution began with the development of the hinge (Fig. 13), which divided the margin into belts of large (5–15 km) and small (0–2 km) subsidence on the seaward and landward sides, respectively. The cessation of rifting was thus concomitant with a shift of all tectonic activity to the seaward belt and the development there of deep marginal basins with a seaward-thickening stratal wedge, greatly thinned transitional crust, and magmatism. The rift-postrift transition is marked at places by an unconformity indicated by the truncation of synrift strata. At other places, however, no unconformity or hiatus is evident, and the transition is marked only by sediment accumulation at increasing rates.

The marginal basins occur entirely in the subsurface of what is now the continental shelf (Fig. 13) from the Bahamas to Nova Scotia. These have different properties in segments along strike (Fig. 13): the Blake Plateau basin is wide and thick, the Bahamas and the Georges Bank–Scotia Basins are wide and thin, and the Baltimore Canyon Trough and the Carolina Trough are narrow and thick. Such differences indicate different magnitudes and mechanisms of horizontal stretching among the segments and the existence of extension-parallel displacement transfer zones between the segments (Bosworth, 1985). The continuity of such transfer zones with Atlantic transform faults is controversial (Wernicke and Tilke, 1989; Klitgord and Schouten, 1986).

In contrast to the passive margin from Nova Scotia south, the evolution of the margin of the Grand Banks (Fig. 13) was highly heterogeneous. There, it is hard to define a hinge that divides basins into rift and postrift phases. The SGBFZ transform south of the Grand Banks (Fig. 13) was the divider between these extensive regions of different Atlantic margin history. The Grand Banks seems to have undergone large stretching in local zones that isolated among them less-stretched microcontinental blocks such as Orphan Knoll and Flemish Cap (Fig. 13) (Haworth and others, this volume). Both the rift and postrift movement zones are thought to be reactivated old faults and are notably concentrated in the peri-Gondwanan (Avalon) terrane, which had a marked development of rift basins in Late Proterozoic and Cambrian time. The Labrador margin (Fig. 13), however, developed more typically by postrift subsidence with self similarity on strike in a narrow zone seaward of a hinge.

The transitional crust that developed below the marginal basins lies above an M-discontinuity that is well defined by a seismic refractor and by a base to a highly reflective lower crust (Stewart and others, 1991; Holbrook and others, 1992). The thickness of the transitional crust diminishes from normal continental values (40 km) at the hinge to a minimum of 10–15 km, generally at the boundary with oceanic crust. Values of horizontal extensional strain, based on integrated two-dimensional thinning at constant volume, range from 100% to 350%, suggesting large heterogeneity on strike (Haworth and others, this volume; Rankin, this volume; Royden and Keen, 1980; Sawyer, this volume). Possible factors in the cause of the apparent heterogeneity are the volume addition by diking and underplating of basalt from the mantle (Keen and deVoord, 1988) and of volume loss by dissolution of continental crust. The compressional wave velocity of the transitional crust (7.1 to 7.3 km/s) is higher than the mean velocity of cratonal crust, indicating either substantial intrusion by basalt or densification by loss or transformation of upper crust. The high reflectivity of transitional crust compared to that of normal continental crust in regions without Phanerozoic orogenesis is evidently related to one or both of these processes.

There are two margin-normal intracontinental displacement transfer zones that clearly continue seaward as major transform faults. The landward extension of the Blake Spur Fracture Zone (BSFZ, Fig. 13) separates two segments of markedly different transitional crust and marginal basins. North of the BSFZ extension, the marginal basin (Carolina Trough) is particularly narrow (50 km). In contrast, the marginal basin (Blake Plateau) south of the transfer zone is wide (500 km). Both basins, however, have approximately similar thicknesses of postrift strata (12 km) and thicknesses of subjacent crust (20 km) (Dillon and Popenoe, 1988; Rankin and others, 1991). Furthermore, the oceanward edges that mark the beginning of drifting of the two segments are probably of different age, the southern younger by about 5 m.y. (Klitgord and Schouten, 1986). The differences in deformed widths of the two marginal basins imply that there is much greater extensional displacement of the southern segment than the northern segment, and that the stretching of the southern segment went on longer before drifting than that of the northern segment. This may indicate that differential volume change also occurred across this transfer zone, and basalt underplating of the southern segment after intense stretching seems the most probable way to explain the observations. A margin-parallel gradient exists in the thickness of the Blake Plateau basin postrift strata; i.e., a rapid southwestward increase from the BSFZ to a maximum, and thereafter, thinning at a small rate farther southwest (Hatcher and others, 1993). This indicates that the BSFZ transfer zone was not a sharp discontinuity in strain.

The Southern Grand Banks Fracture Zone (SGBFZ, Fig. 13) is the continent-ocean boundary southwest of the Grand Banks: northwest, into the passive margin, it is the transfer zone between the Georges Bank–Scotia Basins segment on the south and the Grand Banks segment on the north. This displacement transfer zone took up the differential between substantial concentrated stretching, then drifting between North America and Africa on the south, and heterogeneous stretching without drift for about 60–70 m.y. on the north. During its evolution as a continent-ocean boundary, this transform was associated with the development of a major unconformity on its continental wall, a very narrow (50 km wide) and relatively shallow postrift basin across it, intraplate volcanism in both walls, and the anomalously thick layer 2 in the Atlantic crust adjacent to the transform (Todd and others, 1988; Haworth and others, this volume). The intracontinental transfer zone seems not to have followed any earlier structures.

The structures of the Blake Spur or Southern Grand Banks transfer zones within the passive margin are not known in detail. Moreover, the natures of the other transfer zones between postrift segments, or between promontories and embayments of the passive margin, are unclear.

Sea-floor spreading. The postrift phase of Atlantic passive margin development included continental separation and the probably contemporaneous cessation of horizontal stretching of the transitional crust and its cover. The occurrence of sea-floor spreading is registered by normal oceanic crust that is identified by thickness of 5–7 km, a rough upper crustal boundary to seismic reflection, and mainly horizontal sediment cover.

Dating of oldest normal oceanic crust by extrapolation of spreading rates from magnetic anomalies suggests 175 Ma from the BSFZ to the SGBFZ, 170 Ma south of the BSFZ (Klitgord and Schouten, 1986), 125 to 100 from the SGBFZ north to Greenland, and 80 to 50 in the southern Labrador Sea (Srivistava and Tapscott, 1986).

Between normal oceanic crust and clearly transitional crust below marginal basin strata, there is commonly a zone of crust about 25–50 km wide with unique properties that includes the continent-ocean boundary. This zone, called marginal oceanic crust (Fig. 13) (Mutter, 1985; Klitgord and Schouten, 1986), has an oceanic crustal velocity profile, but is thicker by a factor of two or more than typical oceanic crust. It is the source of the anomaly bands, the East Coast Magnetic Anomaly and Blake Spur Magnetic Anomaly, that mark the gradational continent-ocean boundary. The drilled upper reaches of marginal oceanic crust contain dipping stacks of basalt lava interleaved with marine sediments. Marginal oceanic crust probably represents loci of concentrated extrusion and intrusion that underwent substantial horizontal stretching and horizontal axis rotation before and/or during the onset of drift. As such, it may have been the first rift zone of the nascent spreading ridge. The timing of events in the development of marginal oceanic crust is unknown relative to stretching of transitional crust, however, and the two may have overlapped. Velocity distributions imply that magmatic rock of the marginal oceanic crust broadly extends below and/or within transitional crust as a landward-thinning wedge. This wedge may underplate the stretched continental edge and may have arisen by lateral flow from the arch of asthenosphere below the embryonic ridge crest (Keen and deVoord, 1988). Alternatively, it may be a zone of concentrated steep dikes that diverges upward from the finite width of the asthenospheric rise. The width of the initial zone of magmatism at the breakup locus was thus at least 50 km and may have been substantially greater.

Theory. A considerable body of theory of passive margin development has emerged from observations of the present-day margin of eastern North America. Passive margin tectonics began with horizontal extensional deformation confined within a broad band of Pangean lithosphere, perhaps 1000 km wide, and concentrated within the band's central third. The central third lay above an incipient rise of the asthenosphere, which either caused the deformation or was caused by it. In vertical section, the locus of displacements was either in a triangular zone widening upward from the incipient rise, or it was a single dipping zone of concentrated displacement on one side of the rise. The early extension (rift phase) caused lithospheric thinning and probable subsidence relative to sea level of a surface that was initially high from the Alleghanian collision. Magmas emerged from the asthenospheric rise through the deep region of the central zone and migrated laterally and upward through principal displacement zones to central and distal reaches of rift-phase deformation. Owing to dike concentration and thermal softening in the central zone, the stretching concentrated with time and at increasing rates toward the central zone interior. This caused kinematic abandonment of the distal rift zone(s), large subsidence of the flanks of the central zone and the establishment of the hinge, increasing concentration of magma toward the center, and ultimately, throughgoing fracture of the lithosphere above the asthenospheric rise (Harry and Sawyer, 1992; Sawyer, this volume). The transitional crust and marginal basins probably formed mainly during this postrift phase of thermal softening of Pangean lithosphere. The migration of stretching and magmatism toward the center perhaps led to lithospheric necking and concentrated initial extrusion at an embryonic ridge that became the marginal oceanic crust. Increasingly steady advection of magma in the central zone then led to spreading of normal oceanic crust.

Before breakup, the subsidence of continental lithosphere seaward of the hinge presumably reflected the net rate of thinning and isostatic adjustment relative to addition of sills and to thermal expansion. After breakup, the continuing subsidence was caused principally by thermal contraction (McKenzie, 1978; Royden and Keen, 1980). This exponentially decreasing rate with a time constant of about 62 m.y. can be recognized in marginal basin strata as young as Eocene. The switch in causes of subsidence during the postrift phase cannot be distinguished in the marginal basin strata that are a product of the centerward migration of extensional strain. The distal zone of extension landward of the hinge had only minor subsidence because of small stretching and magmatic heating.

The postrift unconformities that occur locally near the hinge are difficult to explain (Sawyer, this volume); an origin by flexure due to vertical loading by early marginal basin strata at more seaward sites is questionable, given the low flexural rigidity imagined for the early transitional crust.

A separate theory has been formulated for the evolution of continent-ocean transforms based on observations of the SGBFZ (Todd and Keen, 1989; Haworth and others, this volume). The theory explains the development of a postrift unconformity on the continental side that is narrowly confined, has substantial omission of section, and is diachronous on strike due to lateral heat conduction during the passage of the Mid-Atlantic Ridge. The theory does not include a reason for long maintenance of rift-phase tectonics north of the SGBFZ while breakup and spreading took place to the south of it. Harry and Sawyer (1992) proposed an origin for the BSFZ, by reactivation of a former plate boundary where the two walls had an inherent strength difference, a softer African lithosphere to the south and stronger North American lithosphere to the north.

Some other problems of the structure of the central Atlantic passive margin are controversial, such as the three that follow: (1) whether the principal displacement zones (faults and shear zones) of the passive margin are mainly reactivations of preexisting zones of the Appalachian orogen or mainly newly generated; (2) whether the principal displacement zone was initially a single dipping planar extensional fault-shear zone that crosscut the Pangean lithosphere or a conjugate pair of zones symmetric about a centerline, probably the crest of the rise of asthenosphere (Sawyer, this volume); and (3) whether the fundamental structure of an asymmetric margin oscillates on strike, as in the model of the East African rift zone (Rosendahl, 1987), or is constant on strike. As one example, it is proposed that at the beginning of extension, the Pangean lithosphere was crosscut by single planar dipping fault-shear zones, one to a strike segment about 500 km long and reversing dip segment to segment, as seems to occur in the East African Rift (Wernicke, 1985; Lister and others, 1986; Wernicke and Tilke, 1989). This theory accounts for the onstrike changes in marginal basin structure and the sinuosity of the hinge (Fig. 13) in the central Atlantic reach. Here, the segments of narrow marginal basins and transitional crust are thought to form in the footwall of the shear zone, and the broad basins and more extended crust are thought to form in the hangingwall. In this idea, the early planar fault-shear zone becomes greatly deformed by subsequent heterogeneous uplift and subsidence. Moreover, magmatism is not volumetrically correlated with the magnitude of vertically integrated strain, and the structural inheritance is minor.

There are substantially different models; for example, Harry and Sawyer (1992) employed the idea that structural inheritance controlled both the siting and development of the central Atlantic passive margin. Here, the principal locus of asthenospheric rise and partial melting was established by extensional thinning of the thickest crust, a result of the late Paleozoic superposition of Africa above North America, and was probably located close to the breakup locus. The early extension (rift phase) was taken up in weak regions: in the central zone that became progressively softened by magma advection, and in the distal zone at outward shallowing levels where fault and shear zones ramped up into the basal detachment and higher thrusts of the Appalachian foreland. The magma advection in the central zone caused progressive softening in and shift of all extension to that zone, the consequent breakup, and development of the hinge and thermally subsiding central zone crust. This proposal predicts that the locus of magmatism and the narrowing of the extensional realm from rift to postrift phase were consequent to preexisting structure, that the development may have been symmetric about the centerline, and that the onstrike changes were due to differences in inherited lithospheric strengths. The Harry and Sawyer (1992) model explains the geographic relation between Atlantic breakup and the North American–Gondwanan collision zone and provides a reasonable mechanical basis for the landward ramping up of the base of the belt of extensional strain. The rupture of a planar dipping fracture over a great extent of lithosphere is unlikely given the heterogeneity in strength expected in any lithosphere and, especially, in a former collision zone.

Southern North America

Introduction. The southern margin of the North American continent is between the Atlantic continent-ocean boundary east of Florida and plates of the Pacific basin (Fig. 6). The southern margin shifted southward during the Phanerozoic, from an early locus within the southern United States and northern Mexico (Fig. 8) to the present one that contacts terranes of oceanic affiliation in the Caribbean region (Fig. 6). The shift occurred by the accretion and stretching of terranes and in small part by internal sea-floor spreading. A time-space evolution of the southern margin is shown in Figure 14; its divisions are the bases for discussions below.

The eastern realm of southernmost PNA was a passive margin (Fig. 8) from end-Proterozoic to Mississippian time, when the collision of Gondwana began (Figs. 14 and 15). The collision migrated obliquely westward, embedding at least the eastern half of North America within Pangea (Fig. 16). Less clear, however, are the early Paleozoic configuration and tectonics of southern PNA in western Mexico and southern California, and the western limit of the late Paleozoic collision zone between Gondwana and North America. The principal alternatives for these variables in the western realm are displayed in two models (Fig. 16, a and b), the first of which is more likely.

In the late Paleozoic, at a time near that of the cessation of

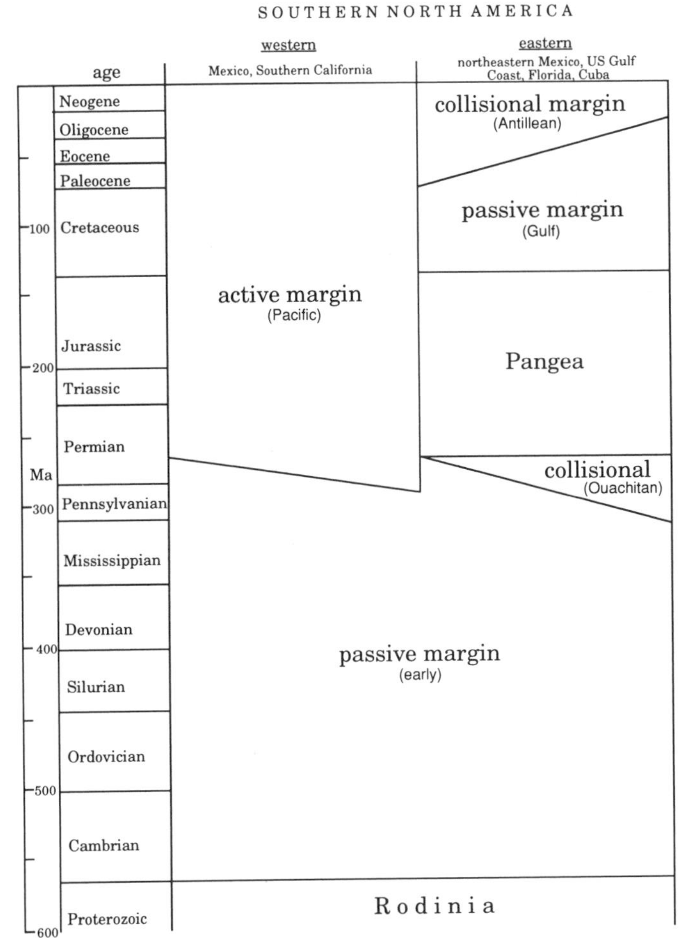

Figure 14. Succession of Phanerozoic margins of southern North America. Beginnings of passive margins taken as onset of drifting. No margin existed in eastern southern North America during duration of Pangea.

collision along the eastern southern margin, an active margin (here called Pacific) was initiated at the western southern margin along the west- or southwest-facing edge of Pangea shown in Figure 16a. The Pacific active margin evolved with subduction of oceanic plates of the proto-Pacific basin and the seaward migration of the continental edge by accretion of terranes (Fig. 17). Such evolution has continued to the present.

After a short duration of internment within Pangea during Late Permian and Early Triassic time, eastern southern North America regained its independence with the development of a passive margin as far west as eastern Mexico in early Mesozoic time (Fig. 18). This passive margin (here called Gulf) evolved in back of the Pacific active margin of western southern North America (Fig. 17). The stretching of the Gulf margin culminated in the Early Cretaceous with the breakup and drift of the North American and South American plates and sea-floor spreading in the Caribbean basin along a locus south of present-day Mexico, Cuba, and the Bahamas (Fig. 18) (Ladd, 1976; Pindell and others, 1988; Donnelly, 1989).

The Cretaceous southern continent-ocean boundary was then transformed diachronously into a collisional margin (Antillean) that took up left-oblique convergence in a plate boundary zone between NAm and the Caribbean plate. The collision began in southern Mexico in Late Cretaceous time and migrated northeastward through Paleogene time (Figs. 18 and 19). The present-day Caribbean–North American plate boundary zone (Figs. 6 and 18) occupies a zone of terranes on the south flank of this collisional margin and permits the continuing eastward transport of Caribbean lithospheres relative to North America.

Early passive margin. The southern edge of the North American continent evolved from the end-Proterozoic breakup of Rodinia as a passive margin in early Paleozoic time. The locus of this margin, moderately well known in its eastern reach (Thomas,

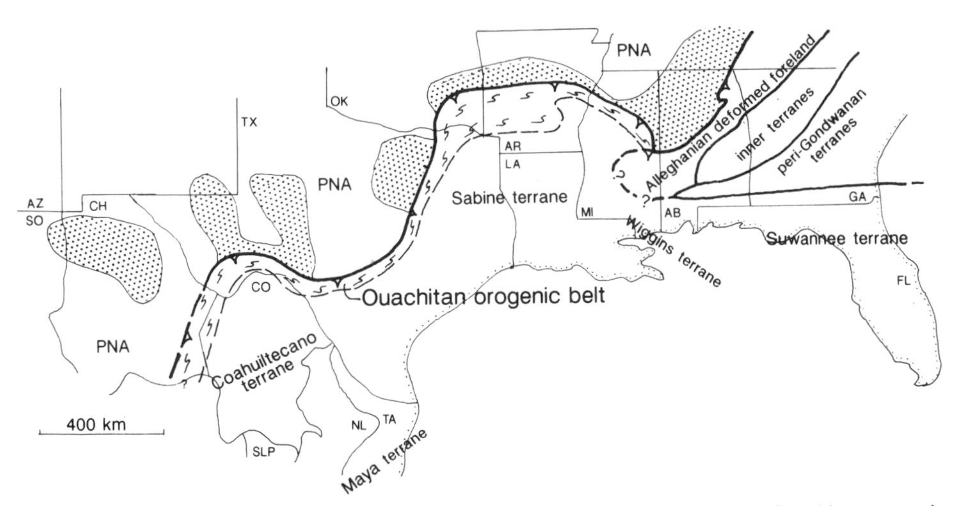

Figure 15. Late Paleozoic collisional zone of southern North America. S pattern is Ouachitan orogenic belt, comprising frontal imbricate and interior zones; dot pattern indicates autochthonous foreland basin strata deposited on Proterozoic North America (PNA). Coahuiltecano, Maya, Sabine, Wiggins, and Suwannee terranes are Gondwanan and were emplaced during late Paleozoic collision. Mexican state abbreviations: SO, Sonora; CH, Chihuahua; CO, Coahuila; SLP, San Luis Potosi; NL, Nuevo Leon; and TA, Tamaulipas.

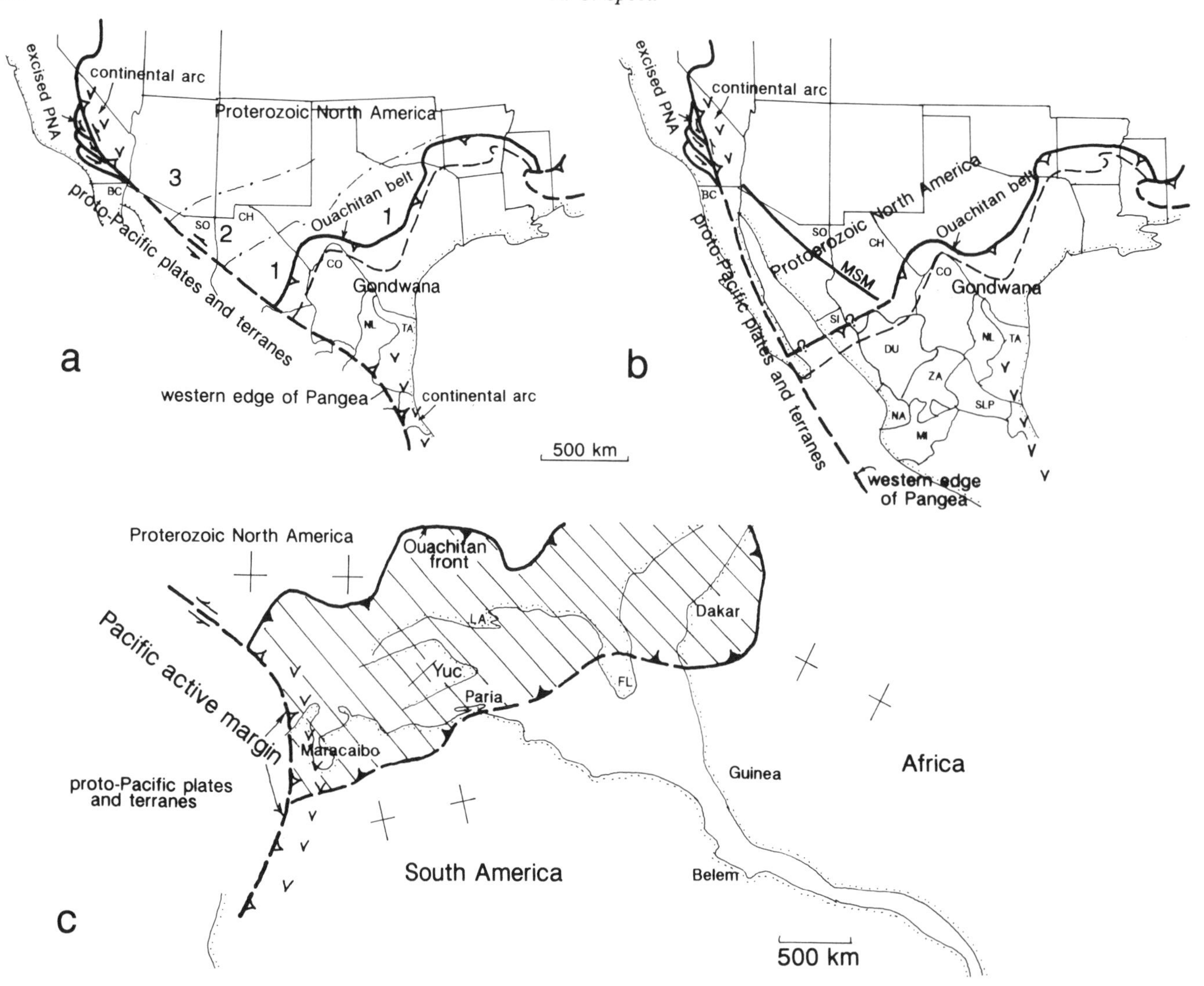

Figure 16. Models for configuration of Permian Mexico after collision of Proterozoic North America (PNA) and Gondwana and beginning of Pacific active margin (a and b) and model of continental collision zone (c). Modern shorelines shown for geography. Model a shows Permian edge of PNA same as present-day edge of PNA and as Permian edge of Pangea in northern and eastern Mexico. Numbers in PNA are basement age provinces after Bickford and others (1986): 1, Grenvillean (900–1200 Ma); 2, 1600–1700 Ma; 3, 1700–1800 Ma. v is volcano. Model b shows Permian edges of PNA and of Pangea an arbitrary large distance west of that in model a; position and extent of Ouachitan belt west of Chihuahua (CH) are arbitrary; MSM, Mojave-Sonora megashear. Abbreviations for Mexican states: BC, Baja California; SO, Sonora; CH, Chihuahua; CO, Coahuila; NL, Nuevo Leon; TA, Tamaulipas; SI, Sinaloa; DU, Durango; ZA, Zapateca; NA, Nayarit; SLP, San Luis Potosi; MI, Michoacan. Model c shows possible extent of collision zone of PNA and Gondwana affected by major deformation and heating (ruled pattern). Fit and orientation of the continents from Rowley and Pindell (1988); long direction of crosses in paleonorth. Yuc, Yucatan Peninsula; LA, Louisiana; FL, Florida.

1976, 1977, 1991; Buffler and Thomas, this volume) but less known in its western reach (Stewart, 1988; Ortega and others, this volume), is shown in Figure 8. This margin existed until a time between Mississippian and Permian, when it underwent convergent boundary tectonics (Fig. 14). In the Mesozoic, a passive margin once again developed in the same boundary zone.

The cryptic early Paleozoic passive margin of the U.S. Gulf Coast is characterized by steep gradients in gravity and magnetic anomalies that are taken to mark an abrupt transition from thick continent to Cambrian oceanic lithosphere (Lillie and others, 1983; Keller and others, 1989). The abrupt transition is also indicated by the tectonic juxtaposition of lower Paleozoic oceanic strata on North American lower Paleozoic cratonal strata in the Ouachitan orogen and by the absence of a passive margin wedge. Together with the inferred kinematics of end-Proterozoic plate motions along the southern breakup locus presented earlier, these

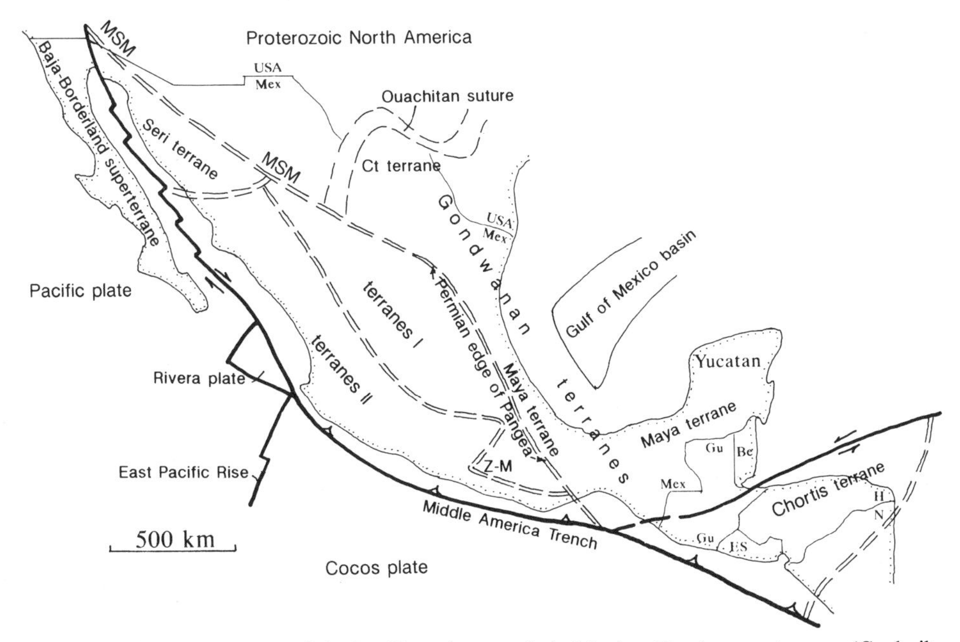

Figure 17. Tectonic features of the Pacific active margin in Mexico. Gondwanan terranes (Coahuiltecano [Ct] and Maya) are parautochthonous in Mesozoic Pangea and in the development of the Gulf passive margin. Sutures of post-Pangean terranes in groups (terranes I and II) and singly (Seri terrane) shown by double dashed lines. Terranes I are cored by Proterozoic basement and attached mainly in the Jurassic before terranes II; terranes II are of mainly Mesozoic arc origin in the proto-Pacific basin and have mainly Cretaceous attachment ages; Seri terrane came from southern California in the Jurassic; Chortis terrane and Baja-Borderland superterrane are currently on the move. Z-M, Zapoteco-Mixteco composite terrane; MSM, Mojave-Sonora megashear; Ct, Cuihuatecano terrane. Gu, Guatemala; Be, Belize; ES, El Salvador; H, Honduras; N, Nicaragua.

features suggest that the early passive margin of the Gulf Coast was transform dominated over much of its length, perhaps like the SGBFZ (Fig. 13) (Todd and others, 1988; Haworth and others, this volume). The model implies a generally abrupt transition from uplifted cratonal edge with prolonged postrift unconformity to oceanic lithosphere, and little or no development of stretched transitional crust and suprajacent marginal basins. This general style of early passive margin is inferred to have extended west as far as southern California (see below).

The Paleozoic oceanic lithosphere developed in a series of segments south of the passive margin, each of which younged eastward, as inferred from the left-stepping rift-transform segments (Fig. 8). Each segment presumably had isochrons at high angles to the continent-ocean boundary. The duration of eastward passage of spreading ridges against the southern continental edge may have been substantial, even 100 m.y., given slow spreading (2 cm/yr) and long transform segments. As a result of this process, early Paleozoic separation between southern North America and formerly conjoint major continents was probably large, as Gondwana migrated far from North America after the breakup of Rodinia (Van der Voo, 1988; Scotese and McKerrow, 1990).

The development of the early passive margin may have produced terranes of continental origin that migrated eastward and seaward relative to southern North America as the southern ocean spread in a largely margin-parallel direction. If so, such terranes may have been returned to the southern margin during the ensuing late Paleozoic left-oblique convergence. The Guachichil and Tepehuano terranes of eastern Mexico (in terrane belt I, Fig. 17) (Ortega and others, this volume), which have Grenvillean basement, and other basement-cored terranes in the internal zone of the Ouachitan orogen (Fig. 15) may have evolved in this way.

The ocean south of the early passive margin was probably the site of deposition of much or all of the allochthonous lower Paleozoic strata of the subsequent collision zone, the Ouachitan orogenic belt (Viele and Thomas, 1989). Such strata indicate early (Late Cambrian–Middle Ordovician) deposition of turbidites of continental provenance together with hemipelagic sediment and later (Late Ordovician–Devonian) pelagic and hemipelagic sediment (Lowe, 1989; McBride, 1989). This sequence of sedimentation is consistent with the idea that the diachronous migration of uplift along the long transform segments provided continental sediment for the relatively long duration of early turbidite deposition. The younger strata indicate that during the period Late Ordovician to Devonian, continental sediment did not reach the southern ocean basin, either because of distance, or perhaps because of the absence of significant seaward

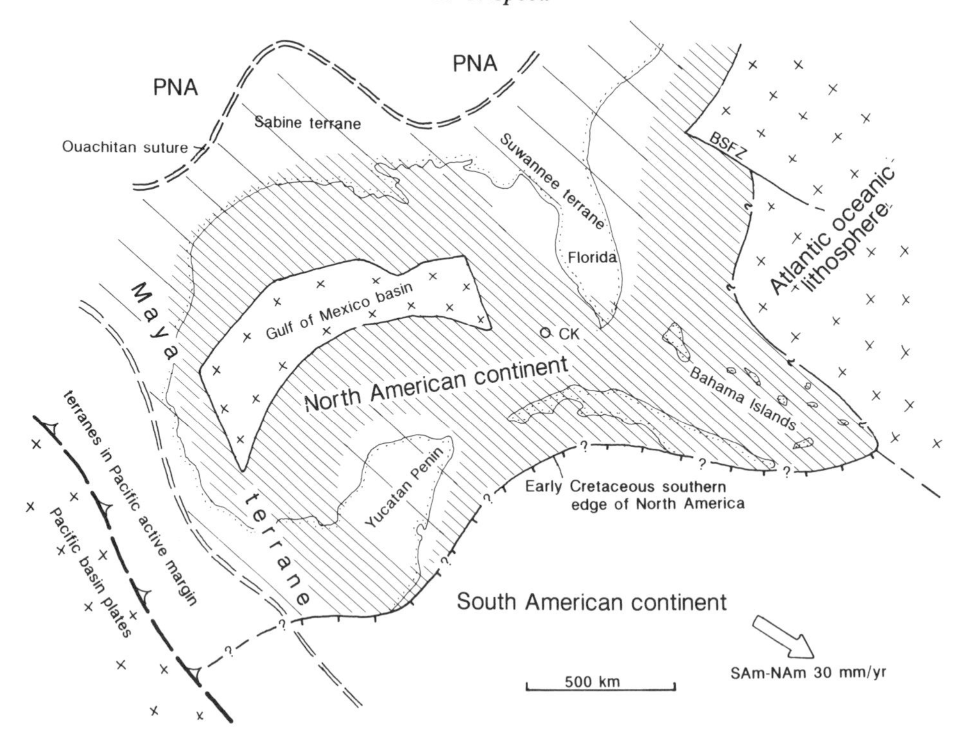

Figure 18. Eastern southern North America in Early Cretaceous time at initiation of drifting apart of North American and South American continents at highly approximate locus of queried line with tics. Gulf passive margin structures in region of North American continent with ruled pattern: close spacing for large postrift stretching and wide spacing for rifting only. x pattern is oceanic lithosphere; dashed double lines are sutures; heavy line with teeth is convergent boundary. Large arrow indicates direction of velocity of South America plate (SAm) relative to North America plate (NAm). Abbreviations: BSFZ, Blake Spur Fracture Zone; PNA, Proterozoic North America; CK, Catoche Knolls.

sediment transfer across Silurian and Devonian southern North America. The latter explanation may correlate with the idea that there was little or no subsidence due to stretching and thermal contraction at the long transform segments of the early Paleozoic southern edge.

Ouachitan collisional margin. The Ouachitan collisional margin occupies a belt from Alabama west to Chihuahua (Fig. 15) that is closely allied with the locus of the early passive margin (Viele and Thomas, 1989; Thomas, 1991; Buffler and Thomas, this volume). The belt's eastern end is an intersection in the subsurface with the contemporaneous Alleghanian collisional margin of eastern North America (Thomas, 1985). The western-most site at which the Ouachitan margin is recognized is in the subsurface of Chihuahua (Handschy and others, 1987), and it is controversial whether this margin extends or extended farther into Mexico. The issue is explored in the next section, with the conclusion that the present western end is probably the original one (Fig. 16a).

The Ouachitan margin was one segment of the zone of collision between North America and Gondwana that produced the supercontinent of Pangea (Fig. 16c). The part of Gondwana that collided with southern North America in southern Georgia and Florida is now in West Africa, except for the Suwannee terrane (Fig. 15), which was left attached to North America during continental separation of the Gulf passive margin phase. Between Florida and Chihuahua, late Paleozoic Gondwana was probably constituted by present-day northern South America and an active marginal belt that fronted it (Fig. 16c). Probable relics of Gondwana in this region of North America after the Gulf phase are the Coahuiltecano, Maya, and Sabine terranes, and possibly the Wiggins terrane (Fig. 15). The oldest record of the Ouachitan collision, Middle Mississippian, is in Alabama, and the youngest, Early Permian, is in western Texas, and intermediate dates are in between. This implies diachroneity and a right-lateral component of motion across the plate boundary.

The Ouachitan orogenic belt at the surface and near surface is a set of discontinuous boundary-parallel zones (Fig. 15), from the craton southward: zone of foreland basins, frontal imbricate zone, and interior zones of little metamorphosed and low-grade metamorphic rocks (Flawn and others, 1961; King, 1975; Viele and Thomas, 1989; Buffler and Thomas, this volume). South of the exposed orogenic belt below the Gulf Coastal Plain are other zones of the collisional margin that consist of Paleozoic igneous and metamorphic rocks and successor sedimentary basins (Nicholas and Rozendal, 1975; Pindell, 1985; Transects F1 and F2 in Speed, 1991). From another view, the allochthonous rocks of the Ouachitan collisional margin consist mainly of two components: terranes that were or arrived in front of Gondwana and upper

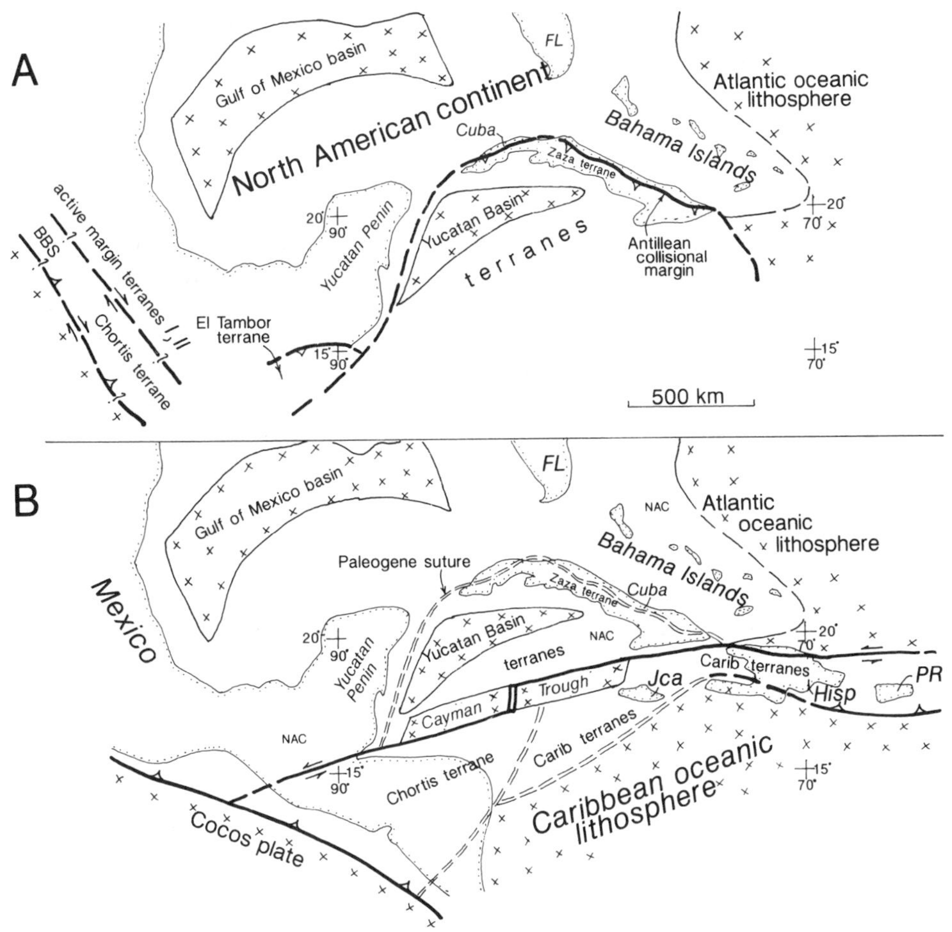

Figure 19. Margins of southern North America. A: Antillean collisional margin in middle Eocene time; x pattern is oceanic lithosphere. B: Present-day margin; North American continent extends south to Caribbean–North America plate boundary. Jca, Jamaica; Hisp, Hispaniola; PR, Puerto Rico; BBS, Baja-Borderland superterrane; FL, Florida; Carib, Caribbean; NAC, North American continent; broken double lines are sutures.

Paleozoic synorogenic North American sedimentary strata. The complex interface between the two components is within the frontal imbricate zone. The terranes overrode the North American passive margin at stratigraphic horizons mostly above the precollisional cratonal platform strata (Devonian and older). The early passive margin is thus probably fully intact, deeply tectonically buried below the Ouachitan margin (Keller and Cebull, 1973; Lillie and others, 1983; Buffler and Thomas, this volume).

The many foreland basins of the Ouachitan orogen (Fig. 15) are floored by PNA and lower Paleozoic platformal cover (Thomas, 1985). These basins, unlike those of the Appalachian orogen, are highly discontinuous parallel to the margin (Fig. 15), perhaps due to selective reactivation of faults of early passive margin origin. The Ouachitan foreland basins, like the Appalachian ones, contain surficial deformation fronts, dividing the basins into a cratonward autochthonous subzone and a subzone of imbrication within the frontal imbricate zone. The stratigraphic record of the Ouachitan foreland basins is of utmost importance to an understanding of the orogen because the Ouachitan deformed zones are so poorly exposed. Their fill records various events in the late Paleozoic: (1) the onset of Ouachitan subsidence, presumably due to vertical loading of North American lithosphere by terranes at seaward sites; (2) the first deposition of sediment derived from terranes to the south, signaling the minimum age of the near arrival and ramping up of Gondwana and/or terranes moving with it; (3) the durations of basement-rooted down to the south normal faulting within the basin, indicating the nature and timing of the flexure-accommodating mechanism; and (4) the migration within the foreland basin of the landward edge of the frontal fold and thrust belt. These data indicate foreland basin activity between Meremician and Wolfcampian in Alabama, Wolfcampian and Leonardian in west Texas, and durations in between commensurate with westward margin-parallel migration.

The frontal imbricate zone contains synorogenic strata (Middle Mississippian to Permian) that are deformed by breakforward emergent thrust imbricates and fault-related folds. Such strata are of varied facies that indicate tectonic juxtaposition of packets from oceanic to shallow-marine realms and telescoping of the sedimentary cover from varied sites ranging from well off the continent to the foreland basins. The zone includes only meager quantities of tectonic melange composed of fragments of North American lower Paleozoic platformal cover in upper Paleozoic matrix (Flawn and others, 1961; Viele and Thomas, 1989). In Texas, in a rift-dominated arm of the margin, the frontal imbricate zone includes two nappes of North American basement and Paleozoic platformal cover. Such nappes are minor components, but they show that the sole thrust of the collisional margin locally reached depths well below the early passive margin surface in the autochthon.

The interior zones of the exposed and nearly exposed Ouachitan orogen (included with frontal imbricate zone in Fig. 15) are constituted by strata at varied low grades of metamorphism up to greenschist facies, each in different subzones. In its eastern half (Arkansas and Oklahoma) (Fig. 15), the interior zone is entirely little-metamorphosed rocks, and is composed of folded nappes of two sediment types. One is preorogenic sedimentary rocks of Cambrian to Middle Mississippian age of offshelf, probably ocean-floor, origin. The other is suprajacent synorogenic Carboniferous sedimentary units, together with slices of serpentinite and gabbro. In its western half (Texas), the interior zone is like that of Arkansas and Oklahoma, but important differences exist in eastern Texas where interior zone nappes of different metamorphic grades are stacked. There, the lower nappes of little-metamorphosed oceanic preorogenic and synorogenic strata are directly thrust above shallow-marine foreland basin strata, and much of the frontal imbricate zone is either absent or tectonically covered. The higher nappes (Luling thrust sheet) contain more metamorphosed rocks in two tectonostratigraphic units, a lower one of probable North American basement and platformal cover (Nicholas and Waddel, 1989), and a higher one of pelitic and quartzose metasedimentary rocks of uncertain correlation. The nappe of North American platformal rocks in eastern Texas implies out-of-sequence thrusting within that part of the orogen.

South of the exposed Ouachitan zones of the eastern half of the orogen, below post-Paleozoic cover, is a thick basin (Fig. 15) of Carboniferous strata like those of the frontal and interior zones but much less deformed. South of this basin, the Paleozoic subcrop is made of volcanic rocks of late Paleozoic age and schist of late Paleozoic metamorphism, both of which are ascribed to terranes (Fig. 15) (Wiggins and Sabine; Buffler and Thomas, this volume). Similar igneous and metamorphic rocks occur in Coahuila (Fig. 15) outboard of the interior Ouachitan zone of Texas and Mexico (Nicolas and Waddell, 1989). In Coahuila, however, no late Paleozoic basin is known in between. The metamorphic rocks of the high nappes of the interior zones in eastern Texas may be contiguous with those farther outboard to the west and east.

The interior zones of the Ouachita collision zone and the Paleozoic rocks outboard of them are interpreted as a set of terranes that were transported to or near the continent-ocean boundary. Because the early transform-dominated passive margin was narrow relative to rift-dominated margins, the total width of stacking of crystalline crusts was perhaps small, perhaps between 0 and 50 km. The buried eastern Texas and Coahuila segments, presumably rift dominated, may have broader zones of stacking.

The Ouachitan orogen is viewed as having originated by the emplacement of a giant accretionary forearc on the early passive margin of North America (Graham and others, 1975; Thomas, 1976; Viele, 1979; Viele and Thomas, 1989; Houseknecht, 1986). The forearc amassed in front of and moved with Gondwana. The igneous and metamorphic rocks of the terranes were probably the northern fringe of the continental arc of Gondwana (Fig. 16c). The little-metamorphosed preorogenic strata of the interior zones are the inner forearc, accreted from ocean floor during early phases of North American–Gondwana convergence, during or before the onset of synorogenic sedimentation in Middle Mississippian time. The synorogenic sediments are thought to have been derived from the arc and the accretionary prism and to have been deposited in forearc basins as prism cover and as trench fill. With proximity to the continent, the encroaching arc system depressed the North American lithosphere, creating foreland basins and oceanward sediment transport from the continental interior to such basins and to the ocean floor beyond. The converging forearc accreted at least upper levels of Carboniferous sediment from the ocean floor and outer foreland basin fill before cessation of emergent imbricate thrusting inboard of the overridden continent-ocean boundary in Permian time.

The Ouachitan orogen was contemporaneous with the Alleghanian phase of the Appalachian collisional orogen around the corner of late Paleozoic North America, then in southeastern Georgia. Although the two orogenies were similar as continent-continent collisions, their manifestations differ in major respects. The Alleghanian of the southern Appalachians is characterized by large magnitudes of cratonward thrusting of the continental foreland and preexisting hinterland and by magmatism in and unroofing of the hinterland without apparent overriding of these zones by colliding terranes. The Alleghanian foreland basin fill was deformed against and overridden by nappes of pre-Alleghanian North American rock far from the edge of colliding terranes. Such effects imply that the eastern margin of North America underwent widespread horizontal contraction and thickening during Alleghanian collision. In contrast, the deformational belt of the North American continent in the Ouachita collision was almost completely overridden by far-traveled terranes. There is no suggestion of contraction and thickening of the preorogenic continent, and the terranes were driven directly into the foreland basin fill. There was no evident lithospheric unroofing or magmatism within the orogen; the local uplifts in the interior zone (Waco, Benton) may be due to out-of-sequence thrusting within the terranes.

At the junction of these two late Paleozoic collisional margins in Mississippi (Fig. 15), thrust sheets containing slate and believed to be Ouachitan lie below thrusts sheets of carbonate rock thought to be Alleghanian, with 90° differences in dip direction (Thomas, 1985). The timing of the structures is unknown, but the different attitudes of the Ouachitan and Alleghanian sheets imply that the displacement fields of the two orogens were not continuous. Perhaps the large contraction of the Alleghanian rift-dominated margin caused a prolonged migration of foreland thrust sheets landward above the corner with the Ouachitan transform-dominated margin, the collision at the corner having begun at the same time.

To speculate on the kinematics of the Ouachitan collision zone, the initial conditions were a long east-west–trending, narrow, relatively steep continental edge (except in eastern Texas and Coahuila) joined to the Paleozoic southern ocean at the early passive margin. Gondwana, fronted by a broad sedimentary forearc that may have been studded with terranes of crystalline rock, moved obliquely northwestward along this southern edge. The diachronous collision drove the forearc above the continental margin and caused large margin-normal contraction in the forearc. The heave of thick Gondwanan lithosphere or crystalline crust above North American continent, however, was apparently small or zero. The total thickening at the Ouachitan margin may thus have been relatively small, explaining the lack of anatectic plutonism and lack of large unroofing. The apparent absence of major east-west–striking strike-slip faults in the Ouachitan orogen and the paucity of North American continental rock in the orogen suggest that the oblique relative plate motion between Gondwana and North American was taken up mainly on the sole thrust of the overriding forearc. The long duration of thrusting (60 m.y.) within the Ouachitan orogen, however, implies continuing margin-normal contraction of the two lithospheres after initial contact, and this was taken up mainly within the forearc and continental arc of Gondwana, now in terranes in southern North America and in northern South America (Figs. 15 and 16c).

PNA and Pangea in Mexico. It is uncertain whether PNA extended well south and/or west of its current edge in Mexico (Fig. 8) in early Paleozoic time, and if it did, whether the Ouachitan belt, Gondwana, and Pangea occupied much of present-day southern and western Mexico at the end of the Paleozoic Era. Two models representing alternatives are shown in Figure 16. The first (Fig. 16a) assumes that the present edge of PNA in Mexico is or is close to the end-Proterozoic one, and that the site of westernmost recognition of the Ouachitan collisional belt, in Chihuahua, was at the western edge of Pangea (Van der Voo, 1983; Pindell, 1985; Stewart, 1988; Stewart and others, 1990; Sedlock and others, 1993; Ortega and others, this volume). The second model (Fig. 16b) assumes that PNA either exists well southwest of the edge shown in Figure 8 or that it existed there in the early Paleozoic. The second model also assumes that the Ouachitan belt extends or extended far west of its current end. In variants of the second model, PNA and the Ouachitan belt were thought by Keller and Cebull (1973) to underlie present-day central Mexico; alternatively, Silver and Anderson (1974), Dickinson (1981), and Anderson and Schmidt (1983) proposed that PNA and the Ouachitan belt existed there, but were removed by Mesozoic left slip along the Mojave-Sonora megashear (Fig. 16b).

Between Chihuahua and southern California, PNA was cratonal through Paleozoic time (Fig. 8). The craton includes Proterozoic belts of different ages (Fig. 16a) (Bickford and others, 1986). The current southern edge of the craton in Mexico, called the Mojave-Sonora megashear (MSM) (Silver and Anderson, 1974), is the intersection of PNA with terranes that compose Mexico to the south (Sedlock and others, 1993; Ortega and others, this volume). Silver and Anderson (1974) and Anderson and others (1979, 1991) argued that the MSM is an intracontinental fault of large left slip and Jurassic age, whereas Stewart and others (1990) and Sedlock and others (1993) thought that it is the end-Proterozoic edge of PNA and a surface of late Paleozoic and/or Mesozoic accretion.

The terranes are generally dissimilar in basement and cover to those of adjacent PNA, and paleomagnetic evidence suggests a latitude history for at least one terrane (Zapoteco; Ortega and others, this volume) that denies a source in northern Mexico (Sedlock and others, 1993). Furthermore, Proterozoic basement is apparently absent from many Mexican terranes, and Grenvillean basement exists only in small volume (Ortega and others, this volume). Such characteristics indicate that Mexican terranes as a whole are not products of local excision of the North American margin and, moreover, that PNA does not exist as an autochthonous or parautochthonous tract southwest of the Mojave-Sonora megashear.

The hypothesis that a large tract of PNA existed south and/or west of the MSM and has been excised and displaced from that region needs to explain where it went in order to be accepted. This requisite exists because of constraints on the Mesozoic breakup of Pangea and drift kinematics between North America and Gondwana (discussed presently). If such a tract exists, it or its fragments are presumably extensively floored by Grenvillean basement, according to the distribution of Proterozoic age belts north of the MSM (Fig. 16a). Two ideas for displacement are as follows. One is that the tract of PNA and the Ouachitan collision zone was displaced more than 700 km southeast on the MSM relative to PNA during mid-Mesozoic rifting and drifting within Pangea in the Gulf of Mexico region (Dickinson, 1981). The problem here is that the terranes of eastern Mexico and the Gulf region appear to be related to Gondwana and/or unrelated to PNA with Grenvillean crust (Sedlock and others, 1993). Another idea is that the tract of PNA lies in northwestern South America, transferred during the breakup of Pangea (Renne and others, 1989). Sialic crust with Grenvillean ages occurs in a belt from northern Colombia at the Caribbean plate boundary south at least 1500 km in the Andes (Goldsmith and others, 1971; Kroonenberg, 1982). This could be the missing tract. The idea implies that an ocean basin, perhaps a western extension of the Gulf of Mexico basin, opened up in Mexico during the breakup of Pangea and that this basin was filled by terranes. The hypothesis, however, has several difficulties: (1) no

Ouachitan collisional belt is recognized in South America between the Grenville-type rocks of the northern Andes and the interior shield (Priem and others, 1989); (2) cooling dates in much of the shield (Onstott and others, 1989) suggest unroofing related to a Grenville-aged orogeny, supporting a Late Proterozoic collision of North America and South America but not a mid-Mesozoic capture of Grenvillean rocks of the northern Andes; and (3) there are no recognized mid-Mesozoic ophiolites among terranes of Mexico (except for the Cuicateco terrane; Ortega and others, this volume) that would suggest a former oceanic basin south of the MSM. To conclude, there is no good evidence for terranes composed of a hypothesized former tract of PNA from south of the MSM in the Western Hemisphere.

In contrast, the hypothesis that the present-day southern edge of PNA and western end of the Ouachitan belt are the original ones (Fig. 16a) accommodates the diversity of terranes in Mexico and their tectonostratigraphic differences from PNA. The hypothesis also provides an explanation for the present-day existence in easternmost Mexico, taken as the western edge of Pangea, of features of the Pacific active margin that emerged in Permian time, apparently both north and south of the MSM. The characteristics of the edge of PNA in Mexico west of Chihuahua are much like those of the long Arkansas-Oklahoma transform of the Gulf Coast realm of the southern edge of PNA and can be rationalized as a northwestward continuation of the transform-dominated southern end-Proterozoic passive margin. In this idea, the western end of the Ouachitan belt marks the westward limit of Gondwana's migration during late Paleozoic collision with North America (Fig. 16a).

Pacific active margin. *Introduction.* An active margin developed on western Pangea, probably beginning in Permian time, in southern California, possibly as far south as the Mexican border, and in eastern Mexico and northwestern South America. This deduction is based chiefly on the loci of two belts of continental batholiths of long activity (Permian and Mesozoic). The probable Permian paleogeography of western Pangea, shown in Figure 16a, is interpreted partly from the existence of such a belt in terranes of Gondwanan origin in eastern Mexico (Ortega and others, this volume). The time and place of inception of this margin are indicated by the oldest plutons in the two belts (Lopez-Infanzon, 1986; Sedlock and others, 1993; Denison and others, 1971; Stewart and others, 1990; Gonsalez de Juana and others, 1980; Kellogg, 1984; Snow, 1992; Walker, 1988). The dating of such plutons, however, is not yet sufficient in precision or geographic coverage to interpret diachroneity in margin development or its temporal relations with the final amalgamation of Pangea in eastern Mexico.

The active margin that began in Permian time continued in the same two segments into the Triassic, during which the active margin in southern California extended northward in the western United States and in Jurassic time filled in the gap in northern Mexico (Fig. 16a). An active margin has existed along the western reach of southern North America from that time to the present.

The Pacific active margin records the subduction of oceanic lithospheres of the Pacific basin below western North America. Within this history, the Cenozoic and Late Cretaceous plate arrays and motions are resolved, whereas identification in the Pacific of earlier plates and motions is difficult (Engebretson and others, 1985; Stock and Molnar, 1988). The Cenozoic and Cretaceous reconstructions imply highly varied obliquity and speed of plate velocity with time and place across the Pacific active margin. Models using hotspots as a reference frame suggest a general left-lateral margin-parallel component in Jurassic and Early Cretaceous time and general right-lateral component since then (Engebretson and others, 1985; Sedlock and others, 1993). One and perhaps two phases of triple junction migration affected the Pacific active margin at Mexican latitudes. An ongoing phase began at about 25 Ma when the Pacific-Farallon divergent boundary intersected the trench along western Mexico (Atwater, 1970, 1989). Since then, a triple junction of North America–Pacific–Farallon plates has migrated southeast along Mexico to its current site in the Middle America Trench (Fig. 17). The once-extensive Farallon plate, now almost fully subducted and existing in relics as the Cocos and Rivera plates (Fig. 17), may have underridden North America since the Jurassic or before. An older triple junction may have formed off Mexico among NAm and the Farallon and Kula plates and migrated northward in Late Cretaceous time. The existence of this second triple junction is not in doubt, but it is unclear whether it was ever as far south as Mexico.

The existence of a gap in early (Permian-Triassic) continental arc magmatism between southern California and eastern Mexico can be explained by a plate velocity direction that was very highly oblique to that segment of the active margin (Stewart and others, 1990; Ortega and others, this volume). This implies approximately southeasterly migration of ocean lithosphere along the edge of PNA in northern Mexico in the late Paleozoic. In the Jurassic, a larger component of boundary-normal convergence evolved at this segment, filling in the former magmatic gap with continental arc.

The principal features of the long development of the Pacific active margin of southern North America are (1) the broadening of the active margin and the growth of most of Mexico by the accretion of terranes (Sedlock and others, 1993); (2) the probably large margin-parallel transport in and seaward of the active margin that brought in, transferred to, and removed terranes from Mexico (Beck, 1986; Hagstrum and others, 1987); and (3) the enormous volume of arc magma, of both continental and island-arc origins, that was added to the crust of this margin from Permian to Holocene time (Mitre-Salazar and others, 1990).

Mexico. The outgrowth of North America in Mexico from the late Paleozoic western edge of Pangea (Fig. 17) to the present has been by the attachment of terranes (Campa-Uranga and Coney, 1983; Sedlock and others, 1993; Ortega and others, this volume), together with volume addition of mantle-derived magma. At latest count (Sedlock and others, 1993), Mexico exclusive of PNA contains 16 terranes, several of which are composite. The attachments appear to have been largely one by one, not as large superterranes, and from inboard out, beginning at the edge of Pangea. The early-arriving terranes (those in terrane belt I; Fig. 17) probably attached in Late Triassic or Jurassic time, but possibly as far back as Permian in northwesternmost Mexico (Seri terrane). The attachment front migrated progressively outboard (southwestward) through the Mesozoic, and Mexico's current configuration was largely attained by mid-Cretaceous time with the attachment of terranes in belt II (Fig. 17). Boundary-parallel migration of terranes, involving excision, splitting, transport direction reversals, and reattachment, however, has continued at places along Mexico's active margin up to the present. Moreover, distributed strike-slip faulting of Mesozoic and Cenozoic(?) ages within interior Mexico has caused an uncertain but probably modest degree of rearrangement of terranes.

Terrane migration is active along two segments of the modern Mexican margin, both related to triple junctions. The Alta-Baja California superterrane was separated at about 6 Ma from earlier amalgamated terranes in western Mexico, and accelerated in the Pacific–North America plate boundary zone with northwestward motion relative to the North American plate (Fig. 1). The excision was caused by an inboard jump of the principal transform following the southeasterly passage along the margin of the NAm-Pacific-Farallon (Rivera) triple junction. The transform took up a new locus along a recently extinguished continental arc that lay above the Farallon-NAm convergent boundary (Stock and Hodges, 1989). Thus the process of terrane excision, in this case, was the establishment of a new locus of a maximum in the plate boundary zone velocity gradient along a belt of minimum strength within the active margin.

The Chortis terrane is currently part of the Caribbean plate (Fig. 17) and is moving eastward relative to southern Mexico of the North American plate. Chortis has migrated from an early Cenozoic site in the active margin of southwestern Mexico that was more than 1000 km west. The capture of the Chortis terrane and this latest episode in its complex kinematic history probably began near the end of Cretaceous time with the creation of today's Caribbean plate and its migration with the unstable NAm-Cocos (Farallon)-Caribbean triple junction (Fig. 1).

The inferred kinematic history of terranes in Mexico before the present phase (Mitre-Salazar and others, 1990; Ortega-Gutierrez, 1989; Ortega and others, this volume; Sedlock and others, 1993) includes several highlights, as follows. Terranes of belt I (Fig. 17) include Grenvillean basement. This implies they had a probable source region in the early passive margins of eastern or southern North America. In particular, one such terrane (Zapoteco) may have come from a site as distant as eastern Canada or its conjugate across the Paleozoic oceans. These terranes probably emerged from the end-Proterozoic breakup as microcontinents in a process like that of Orphan Knoll in the Atlantic off Newfoundland (Fig. 13) (Haworth and others, this volume), perhaps isolated within the Paleozoic oceanic crust(s) by ridge jumps. Succeeding margin-parallel transport implies net right-lateral displacement within the Paleozoic ocean basins of the proto-Atlantic, as well as within North America's Paleozoic collisional margins, before the final closure between Gondwana and North America in the Ouachitan orogen. The terranes with Grenvillean basement can be inferred either to have displaced relative to North America at the western front of Gondwana, or in its forearc, escaping westward into the proto-Pacific before cessation of collision. The Zapoteco terrane amalgamated with an ophiolite and subduction complex (Mixteco terrane) in Devonian time (Ortega-Gutierrez, 1989), implying that the Paleozoic proto-Atlantic contained at least one convergent boundary.

After dispersal in the late Paleozoic proto-Pacific, the terranes with Grenvillean basement and others began a southeastern migration relative to North America, culminating in attachments by the end of the Jurassic to early Mesozoic North America in Mexico, which consisted of the Gondwanan Coahuilatecano and Maya terranes (Fig. 17) (Sedlock and others, 1993). The other terranes in belt II (Fig. 17) are of varied kindred and probably came from sites within the proto-Pacific and western North America. They represent one or more major chunks of the Paleozoic passive margin wedge and continental basement, accretionary prisms of lower Paleozoic and upper Paleozoic ocean-floor strata, and of early Mesozoic island arcs. Their tectonic evolutions record the occurrence of Paleozoic and early Mesozoic plate boundary tectonics within the proto-Pacific fronting western and western southern North America. They also record the excision of tracts of the passive margin of western North America (Seri terrane) during the inception of a convergent plate boundary with collisional or active margin development, and the net left-lateral margin-parallel component of plate velocity fields across the southern California margin in latest Paleozoic and early Mesozoic time.

At about the beginning of the Late Cretaceous, some terranes on the outboard of the Mexican coalescence were excised and began northwestward migration (Fig. 19a), reflecting a switch in the sense of general obliquity of the plate velocity vector across the active margins.

The Chortis terrane is probably one of these that migrated northwesterly along the outer margin before being caught up in the Caribbean plate motion (Fig. 19, a and b). Such terranes also included those now in the late Neogene Baja-Borderland superterrane (Fig. 19a); many or all of these had undergone 1500 to 2500 km northwestward transport before amalgamating and lodging, apparently in Eocene time, against North America in northwestern Mexico and southern California (Hagstrum and others, 1987). Much of the Baja-Borderland superterrane was excised in late Neogene time as part of the Alta-Baja California superterrane (Fig. 20a). Moreover, many active strike-slip faults with at least tens of kilometers of slip dissect and deform the Baja-Borderland superterrane and other components of the Alta-Baja California superterrane. Such features make evident the progressive tectonic comminution and mixing of terranes at an active margin and the existence of a probable limit in the fineness with which terranes can be resolved in time and space.

The general right-lateral regime that began in the Late Cretaceous continues to the present. It now includes two triple junc-

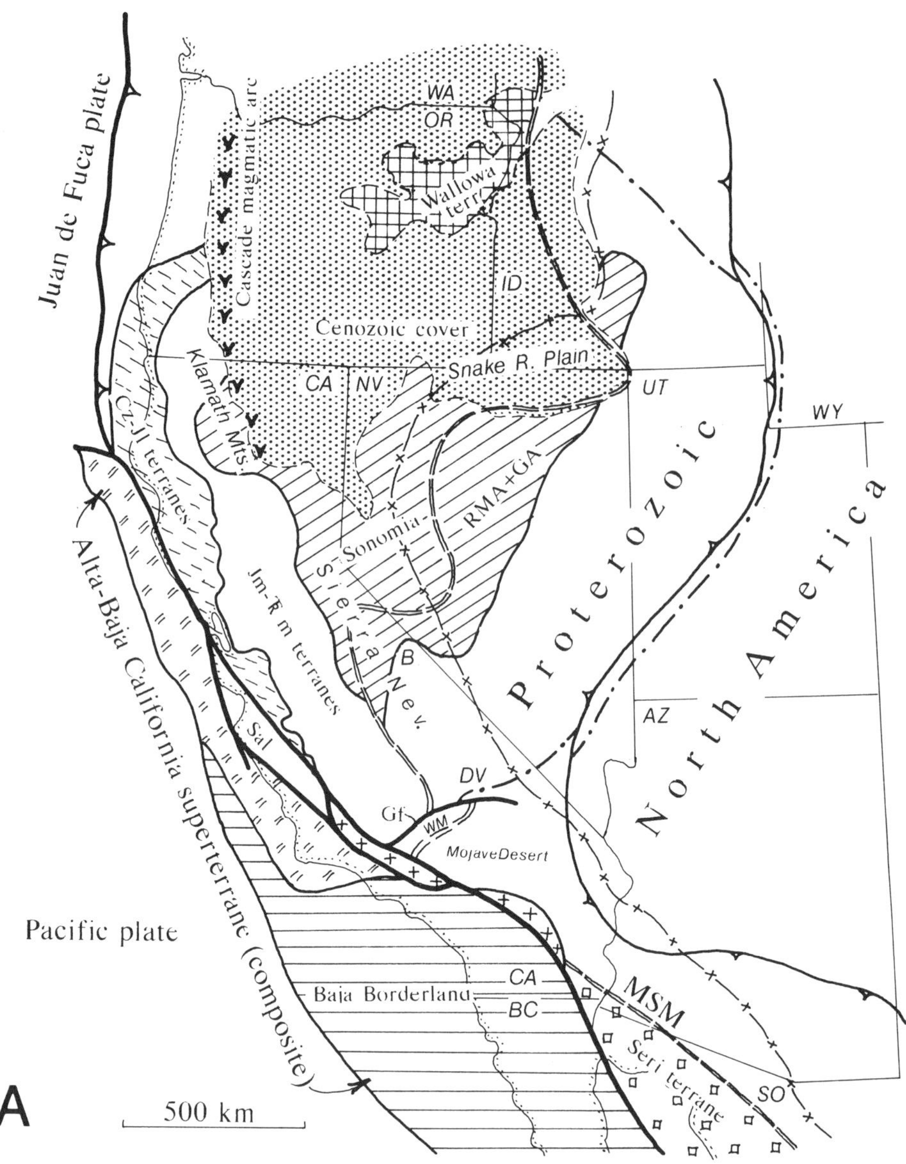

Figure 20. Margins of western North America in southwestern United States and southern North America in southern California and northwestern Mexico. A: Principal features of passive, collisional, and active margins; Gf, Garlock fault; DV, Death Valley; MSM, Mojave-Sonora megashear; B, Bishop; RMA + GA are Roberts Mountains and Golconda Allochthon terranes (legend on facing page); B: Map showing Rocky Mountain region of uplifted continent and basement-cored Cenozoic deformation and hinterland (lined region) of Mesozoic active margin; hinterland defined by patches of locally unroofed rocks that were deep-seated and metamorphosed in late Mesozoic time; fore is foreland west of hinterland.

tions and highly nonuniform relative motions along the active margin. The nonuniformity in the modern velocity field is certainly high compared to that interpreted from the record of ancient terrane migration. The difference is almost certainly due to the low resolution currently attainable for ancient kinematics.

Another regional feature of the active margin of western southern North America is the huge volume of magma added to the North American crust. Magmatism occurred in a continental arc that developed first in the Permian, above the western margin of Pangea, from southern California discontinuously to South America in Colombia and south (Fig. 16). Later, in Mesozoic time, it developed at more southwesterly loci above terranes, reflecting the growth of the active margin in that direction. For example, the early Cenozoic siliceous magmatism of the Sierra Madre Occidental of northwestern Mexico is said to be the largest discharge at the surface on Earth over a 10–20 m.y. duration (McDowell and Keizer, 1977). Moreover, many terranes evidently arrived in the Mesozoic with active or recently extin-

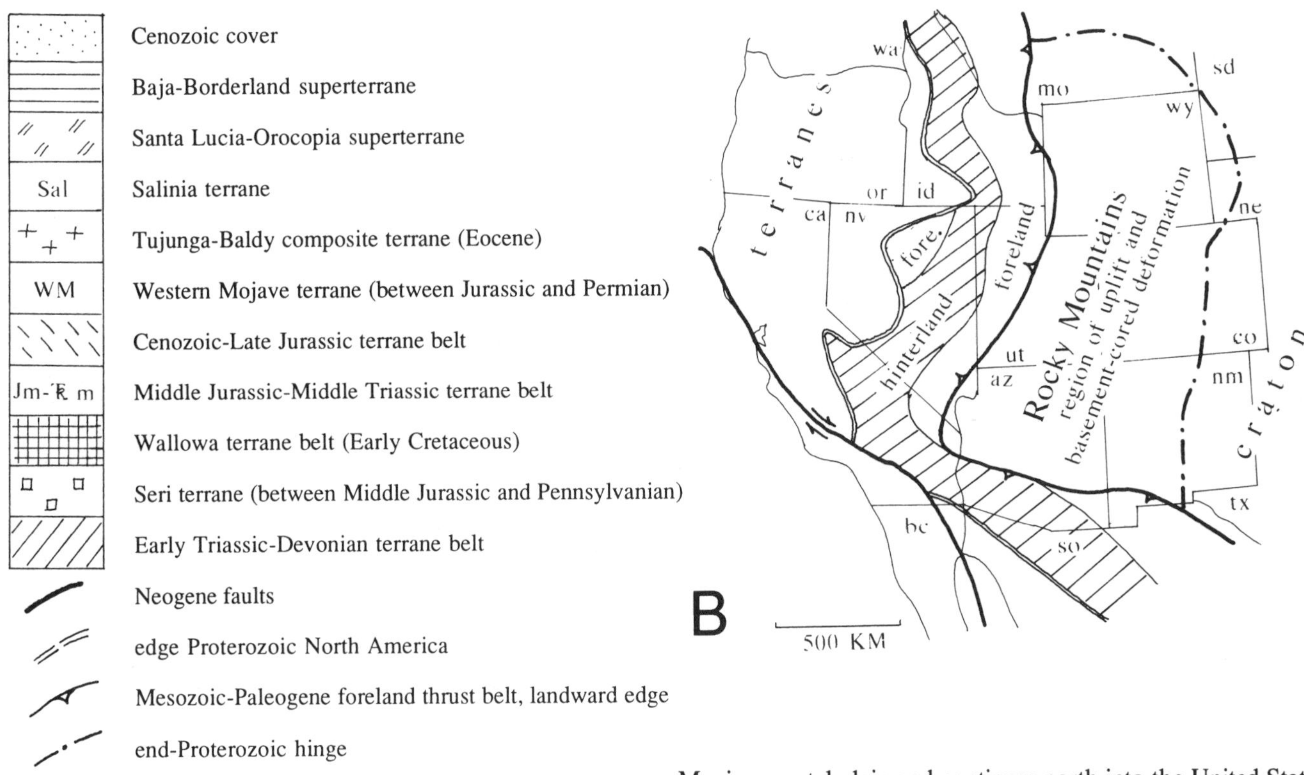

guished island arcs surmounting them, indicating that processes of magma generation occurred within the plate boundary zone seaward of the continental arc. Such processes may have related to local subduction, to local extension, or both.

The continental arc of Mexico, however, twice reversed the general southwesterly migration of its locus of principal magmatism, either with a duration of northeasterly migration of the locus or broadening of the existing locus in that direction. Such excursions occurred in the Early Cretaceous and in the Paleogene. They are explained by durations in which the slab of the incoming plate shallowed, followed perhaps by breaking across the slab and renewed subduction at steeper dips (Coney and Reynolds, 1977; Clark and others, 1982).

During Late Jurassic and Cretaceous time, basins developed in back of, perhaps partly within, the continental arc southward from the Mexico–U.S. border (Chihuahua basin; Dickinson, 1981) through central Mexico (Ortega-Gutierrez, 1989; Ortega-Gutierrez and others, this volume). The existence of earlier and later backarc basins is not evident, however, in the duration of active margin tectonism.

Another product of active margin tectonism is the Laramide orogenic belt, which occupies eastern Mexico west of the Gulf of Mexico coastal plain and continues north into the United States via New Mexico and Arizona (Suter, 1984; Padilla y Sanchez, 1986). It is a thin-skinned foreland thrust and fold belt composed of emergent thrust slices mainly of Mesozoic postaccretionary sedimentary cover. The thrusting in the well-studied part of the belt in eastern Mexico occurred from late Late Cretaceous to middle Eocene time, and proceeded with breakforward, top to the northeast propagation. The strikes and slip directions of thrusts vary greatly locally, however, and reflect irregularity in the attitudes of footwall ramps. The ramps were normal fault scarps of basement blocks that were created during early Mesozoic rifting when the Gulf of Mexico passive margin tectonics overran eastern Mexico in a brief campaign. Contractile deformations affecting late Mesozoic rocks in attached terranes west of the Laramide belt may represent earlier, more trenchward imbricates of that belt.

Southern California. The southern edge of PNA extends northwest from Mexico to southern California (Fig. 17), where it intersects the rift-dominated end-Proterozoic passive margin of western North America approximately at Death Valley (Fig. 20a). The edge in southern California is like that in northwestern Mexico by virtue of direct contact of terranes with the early Paleozoic craton (Howell and others, 1985; Howell and Vedder, this volume). It differs, however, because it has a more complex history of excision and attachment than that of Mexico, and because of its occurrence in the present-day Pacific–North America plate boundary zone (Fig. 20a). Questions concern the history of excision of the continent in southern California and the original position of a corner between the transform- and rift-dominated southern and western edges of PNA (Fig. 10).

PNA that was coherent in the Paleozoic extends northwest from Mexico along the northern wall of the Mojave-Sonora megashear (MSM) to a poorly defined intersection with the San Andreas fault system in California near the international border (Fig. 20a). Outboard of the MSM in Sonora is the Seri terrane (Ortega and others, this volume), which contains mainly Proterozoic continental crust and Paleozoic passive margin wedge strata like those of the rift-dominated passive margin from Death Valley north (Figs. 17 and 20a) (Stewart, 1988). The attachment of the Seri terrane at its present site occurred between late Paleozoic and Middle Jurassic time. An equivalent to the Seri terrane west of the San Andreas fault in southern California exists in the Alta-Baja California superterrane (Cortez terrane; Howell and others, 1985; Howell and Vedder, this volume).

North of the Seri terrane, PNA directly contacts the San Andreas fault as far as the Mojave Desert, except for a segment where sutured terranes (Tujunga and Baldy) intervene (Fig. 20a). The emplacement of these terranes on PNA was in Eocene time, perhaps related to the collision of a superterrane, the Santa Lucia–Orocopia allochthon (SLOA) (Fig. 20a) of Howell and Vedder (this volume), with PNA in southern California and with older terranes in central California. The SLOA had traveled far northwest from southern Mexican latitudes in the Late Cretaceous and Paleocene (Hagstrum and others, 1987) and attached to PNA in Eocene and Oligocene time. Most of the SLOA then reseparated, together with the Baja-Borderland superterrane, to become the Alta-Baja California (composite) superterrane in Neogene time. The Proterozoic rocks of the Tujunga terrane and the subjacent schists of ocean protoliths in the Baldy terrane may have traveled partly with the SLOA before Eocene collision, or they may be parautochthonous. If allochthonous, the Tujunga terrane may represent an excision of PNA from a more southeasterly site. If parautochthonous, the Baldy terrane may be the floor of a now-collapsed backarc basin within marginal Cretaceous PNA and the Tujunga terrane, the basin's outer margin.

In the Mojave Desert between the Garlock and San Andreas faults (Fig. 20a), PNA contacts one or more terranes of lower Paleozoic strata accreted from oceanic sites (Carr and others, 1984; Walker, 1988). The permissible age range of the attachment of this terrane(s) is Pennsylvanian to Jurassic.

From Death Valley north to Bishop, California (Fig. 20a), the present-day edge of PNA cuts orthogonally from the end-Proterozoic hinge across the passive margin wedge of the rift-dominated western edge of North America. This segment has long been recognized as a truncation (Hamilton and Myers, 1966), but good evidence for the Pennsylvanian age of truncation has only recently been deciphered (Stone and Stevens, 1988; Stevens and others, 1992). Bordering this segment to the west in the Sierra Nevada (Fig. 20a) are terranes, some of which contain lower Paleozoic passive margin wedge strata. They attached close to their present position following margin-parallel right slip in Jurassic or Cretaceous time (Schweikert and Lahren, 1990; Kistler, 1990; Saleeby, 1986; Stevens and others, 1992; Saleeby and others, this volume). The Sierra Nevada block, an amalgam of terranes, is now moving slowly northwest relative to North America in an interior locus of the broad Pacific–North America plate boundary zone (Fig. 5) (Argus and Gordon, 1991).

A possible kinematic evolution of the southern California margin of PNA is as follows. Early in Paleozoic time, a corner between the rift-dominated western edge and transform-dominated southern edge of PNA (Fig. 10) existed just west of southern California. A plate boundary zone with highly oblique velocity and left slip developed in the southern U.S. cordillera in late Paleozoic time (Speed, 1979; Stevens and others, 1992) and included this corner of PNA. The corner was excised and translated southeastward along the truncation and then along the end-Proterozoic edge of PNA (the MSM; Fig. 17) in Sonora as the Seri and perhaps other terranes. These attached to PNA in Sonora in late Paleozoic or early Mesozoic time. Terranes of oceanic and arc kindred at outboard positions probably migrated farther down the coast and accreted to the rapidly expanding North American continent in central Mexico. The southern California segment was transpressional and developed a continental arc by Permian time. By the middle of Cretaceous time, the sense of oblique motion switched to right slip, and previously accreted terranes were partly excised and translated northwest relative to North America. Thus, fragments of the passive margin wedge, other than the Seri terrane, may have moved well south and then returned to near their original positions by plate boundary zone tectonics.

The Neogene development of the San Andreas fault system excised a sizeable chunk of the North American continent in southern California and northwestern Mexico as the Alta-Baja California superterrane (Figs. 1 and 20a). The San Andreas fault has cut inboard through accreted terranes in Mexico to PNA in southernmost California and then outboard through accreted terranes in central California (Fig. 20a). Moreover, amalgamated terranes within the Alta-Baja California superterrane are being dismembered by faults of significant displacement within the San Andreas system (Sedlock and Hamilton, 1991).

Owing to the complexity of the tectonic history and the proximity of PNA to the continental edge in southern California, the edge of PNA has a dazzling span of ages, Pennsylvanian to Neogene, within a relatively small region (Fig. 8).

Gulf passive margin. A second Phanerozoic passive margin, here called the Gulf margin (Fig. 18), evolved in Mesozoic time in what is now the eastern southern North American continent (Salvador, 1987; Buffler and Sawyer, 1985; Dunbar and Sawyer, 1987; Buffler and Thomas, this volume; Ortega and others, this volume). The Gulf margin developed within Pangea, within the collision zone of Gondwana and Paleozoic North America. It is contiguous with the Mesozoic Atlantic passive margin at a junction in and southeast of southern Georgia, and the onset of rifting in both passive margins was at least approximately contemporaneous within later Triassic time.

The plate tectonics of the Gulf passive margin, as for the Atlantic one, was the divergence of the Gondwana plate and NAm. In the Gulf region, the locus of separation of the continents was well south of the surface trace of their Paleozoic suture. The initial configuration of Gondwana in Pangea may have been like that shown in Figure 16c, which is modified from Rowley and Pindell (1989). Analysis of Atlantic magnetic anomalies indicates that Gondwana moved southeast relative to North America steadily at 3 cm/yr, beginning about 175 Ma in the Gulf and central Atlantic margins (Ladd, 1976; Klitgord and Schouten, 1986; Pindell and others, 1988). The early divergent motions of the Gondwana and North America plates that occurred before the Late Cretaceous opening of the south Atlantic were thus taken up along a zone that extended west from the southern central Atlantic to a triple junction at the Pacific active margin, probably in or south of southern Mexico.

The divergence of Gondwana from the Atlantic margin caused sea-floor spreading at about 175 Ma north of the BSFZ (Fig. 18), whereas south of it, in the Atlantic and Gulf regions, the plate displacement continued to be taken up mainly by continental stretching and diking. The onset of sea-floor spreading in the southern central Atlantic was at ~170 Ma (Klitgord and Schouten, 1986), but separation of Gondwana from southern North America and the opening of a Caribbean basin between them did not begin until about Berriasian time (about 135 Ma). This age is estimated from the cessation of movement of blocks, notably the rotation and southward translation of Yucatan (Fig. 18), within the Gulf passive margin relative to PNA. The age is also indicated by the cessation of spreading of the Gulf of Mexico basin, which is internal to the passive margin (Buffler and Sawyer, 1985; Salvador, 1987; Pindell and Barrett, 1990; Buffler and Thomas, this volume).

The Caribbean basin presumably opened by sea-floor spreading, although this is difficult to prove because the oceanic lithosphere of the present-day Caribbean plate is partly or wholly far traveled (Malfait and Dinkleman, 1972; Sykes and others, 1982). The southern edge of the Gulf passive margin and of Cretaceous North America was thus probably a continent-ocean boundary (Fig. 18) within the present-day NAm-Caribbean plate boundary zone (Fig. 19b). The steady drifting of the South America plate from NAm ceased at a time between 84 and 110 Ma, when the south Atlantic spreading center was fully established between South America and Africa, and Gondwana was completely broken.

The Gulf passive margin extends from Florida west to eastern Mexico and south from the Ouachitan collision belt to the Paleogene Antillean collisional margin in southern Mexico and Cuba (Fig. 19a). Its maximum width in the direction of plate divergence (northwest-southeast) is about 1500 km. The bulk of the Gulf margin, at the surface at least, is stretched former Gondwanan crust ceded to the North American continent and oceanic crust of the Gulf of Mexico basin (Fig. 18). The spreading of the Gulf of Mexico basin in Oxfordian time produced about 450 km or one-third of the full deformed width of the passive margin in the drift direction. The development of the passive margin took about 100 m.y., from Late Triassic to the Early Cretaceous breakup of the North America and South America plates.

The prebreakup tectonics occurred in two phases (Buffler and Thomas, this volume). The early or rift phase of Late Triassic and Early Jurassic age produced grabens and basalt recognized onland in the U.S. Gulf Coast and eastern Mexico (Fig. 18). Rift phase structures and rocks of probable Jurassic age are also known farther outboard on the Gulf passive margin in Yucatan and Catoche Knolls (Schlager and others, 1984), in Cuba (Echevarria and others, 1991), and in the margin's presumed conjugate in northern South America in northern Colombia and Venezuela (Maze, 1984; Feo-Codocido and others, 1984). None of these rifts has a well established age range, and although it seems that rifting began first in the north, the migration history of stretching is not known. Where the Mesozoic rift structures occur onland today, crusts are thick, and total stretching was probably small. Examples onland are in the U.S. Gulf Coast, eastern Mexico, the Yucatan platform (Buffler and Sawyer, 1985; Transects F1 and F2 in Speed, 1991), and probably in northern South America.

The later or postrift phase, Middle and Late Jurassic, saw the migration of stretching to, or concentration in, the currently offshore Gulf of Mexico region and probably south to the future Caribbean breakup locus and to the present-day South American coastal borderland. The postrift phase followed the development of a widespread Middle Jurassic unconformity in the regions of preserved rift phase features on the fringes of the region of postrift tectonics (Transects F1 and F2 in Speed, 1991). The postrift tectonics caused horizontal extension that varied between 100% and 300% in the continental crust and the occurrence of local breakup and spreading in the Gulf of Mexico basin (Fig. 19). The postrift phase was followed by the birth and drift of the South American plate early in the Cretaceous.

The heterogeneity of extensional strain in the Gulf passive margin is also indicated by the existence of large blocks, such as the Yucatan Peninsula and southern Florida (Fig. 18), with apparently thick crust and small strain within a regional matrix of large, mainly postrift, stretching. The Gulf margin is thus like the Mesozoic passive margin of the Grand Banks of eastern Canada (Fig. 13) in its heterogeneity, long duration of extension, and inclusion of microcontinental blocks. Such blocks are likely candidates to be terranes in future convergent boundary tectonics, whereas the surrounding tracts of transitional crust may be dense and thin enough to be subductible.

Although the far-field kinematics and the plate tectonics of the Gulf passive margin are fairly well established, the movement histories of blocks within the margin are little known and probably complicated. For example, Gose and others (1982) and Molina-Garza and others (1990) showed that blocks in eastern Mexico and in Yucatan underwent substantial clockwise rotations during the rift phase. Moreover, the postrift spreading of the Gulf of Mexico basin probably reflects large vertical-axis rotation of its bordering blocks in Late Jurassic time (Buffler and Thomas, this volume). It is therefore unlikely that blocks within the passive margin displaced along hypothetical faults following small circle paths about the North American–South American Euler pole, as conceived by Klitgord and Schouten (1986). Rather, the velocity field of the blocks was probably highly nonuniform, and velocity gradients may have caused local spreading and convergence within the extending margin. The migration paths of blocks were probably turbulent and unpredictable.

The Gulf passive margin evidently began and evolved during the existence of the Pacific active margin at the western edge of Pangea. Two questions of importance about these margins are the characters of the boundary zone between them and their kinematic interactions. Regarding kinematics, two relations are evident: (1) the relative plate velocity vector across the Pacific active margin in Permian and Triassic time had at least a moderate degree of parallelism with the later northwest-southeast velocity trend between NAm and Gondwana at the Gulf passive margin; and (2) the broad area and large magnitude of stretching of the Gulf passive margin occurred in the immediate backarc of the active margin. These suggest a tectonic connection between the two margins. The reorganization of plate motions following the coalescence of Pangea may have begun by pulling a proto-Pacific slab(s) below the Mexican bight of Pangea (Fig. 16a) starting at about 280 Ma, followed by rifting within Pangea in the Gulf and central Atlantic starting some 50 m.y. later, and then separation of the North and South American plates at about 135 Ma. The unusual breadth of the Gulf passive margin in the direction of plate velocity perhaps records the juxtaposition of the Pangean rift zone with the active margin in Mexico and the combined effects of rifting and backarc circulation on the form of the asthenospheric rise.

The belt of plutons that marks the early (Permian-Triassic) continental arc of the Pacific active margin occupies the Coahuilatecano and Maya terranes that were attached during the NAm-Gondwana collision and formed the probable western edge of Pangea (Fig. 17). These terranes were also block faulted and rotated about vertical axes in the Late Triassic–Early Jurassic rift phase of Gulf passive margin development, and parts of the Maya terrane, including the Yucatan platform, migrated away from North America by lithospheric stretching and spreading of the Gulf of Mexico basin (Fig. 18). These same terranes in eastern Mexico later underwent further Mesozoic continental arc magmatism and foreland thrusting after Gulf stretching, implying an oscillation of tectonic regimes in the same ground.

The Gulf passive margin, like the Mesozoic Atlantic passive margin of eastern North America, was sited in regions of preexisting structures, evidently concentrating displacements in regions of low strength. Within the U.S. Gulf Coast (Fig. 18), early Gulf rifting was taken up in the Ouachitan collision zone. Although the average strike of such rifts parallels the east-west trend of the Ouachitan zone, there is large variance in the strikes (Pilger, 1978; Walper, 1980) that may be explained by reactivation of Ouachitan thrusts whose ramps had irregular dip directions. In contrast, the strike of rifts of the Gulf passive margin in eastern Mexico is mainly north-south and parallel to the Pacific active margin, on which and near which the rifts developed. There, passive margin extension was evidently taken up in a thermally softened zone. In both the U.S. Gulf Coast and eastern Mexico, rifts are oblique to the southeast direction of motion of Gondwana relative to North America, and the heterogeneity and rotation in Gulf passive margin deformation may be due to the zonal simple shear induced by preexisting structures.

The Gulf passive margin development also included major changes in place, time, and magnitude of stretching, from rift phase to postrift, with a general shift toward the Gulf of Mexico basin (Fig. 18). This might be explained by the localization of strain in the zone of greatest thermal softening reflecting maximum crustal thickness in the collision zone of Pangea (Fig. 16a), as conceived by Harry and Sawyer (1992) for the Atlantic passive margin. The mid-Jurassic spreading of the Gulf of Mexico basin (Salvador, 1987) supports the idea of relatively large mantle advection and crustal heating at this locus.

Antillean collisional margin. The eastern realm of the southern edge of the North American continent, between southernmost Mexico and the Bahamas (Fig. 19a), changed from a passive to a collisional margin at times between late Late Cretaceous (Campanian) and Eocene, younging west to east. The colliders were terranes that moved at least partly with the Caribbean plate northeast relative to the North American continent (Fig. 19a). Presumed oceanic lithosphere of the North American plate and the outer edge of the Gulf passive margin were overridden by such terranes. Since their attachment, the plate motions have continued, perhaps with modest reorientation, along a new, mainly strike slip boundary that cuts through the collision zone at loci toward the Caribbean plate (Fig. 19b) (Malfait and Dinkleman, 1972; Sykes and others, 1982; Pindell, 1985; Pindell and Barrett, 1990).

The idea of an Antillean collisional margin is pieced together from the following evidence. First, an ophiolitic subterrane (El Tambor) was thrust northward above the western southern edge of North America (Maya terrane) in late Late Cretaceous time (Rosenfeld, 1981) (Fig. 19a). The El Tambor is interpreted to have been oceanic crust attached to the Chortis terrane early in its current eastward motion relative to NAm or to an unidentified island arc. Second, in Cuba there exists a succession of thrust nappes covered by middle Eocene beds and fronted on the north offshore by an asymmetric Paleogene basin (Fig. 19a) (Pardo, 1975; Echevarria and others, 1991; Buffler and Thomas, this volume). The basin is interpreted as a foreland basin, related to the piling of nappes in Cuba. The nappes include, from top to bottom, Grenvillean basement with platformal Paleozoic cover (Renne and others, 1989), a complex of volcanic and ultramafic rocks from both island arc and oceanic realms, and Jurassic and Cretaceous carbonates. The carbonate nappes are thought to be North American passive margin strata and to be representative of the footwall of the nappe pile and the substrate of the foreland basin.

The nappe of Grenvillean basement may have come from southern or eastern PNA by way of excision from the end-Proterozoic passive margin, escape west from the Ouachitan collision zone, circulation south as a terrane through Mexico, as did the Zapoteco terrane (Sedlock and others, 1993), capture by the Caribbean plate, and transfer near its leading edge northeast to Cuba by middle Eocene time (Malfait and Dinkleman, 1972; Pindell and Barrett, 1990). Alternatively, this nappe may have come from Paleozoic Gondwana now in Colombia (Fig. 16b) and been left in the western Caribbean region during the breakup of Pangea.

The Chortis terrane, whose basement of probable Precambrian age is of uncertain affiliation but possibly Gondwanan, has also been swept east by the Caribbean plate (Figs. 17 and 19b) from a former position to the west at the Pacific active margin of southern Mexico, beginning early in the Cenozoic (Gose, 1983; Pindell and Barrett, 1990; Sedlock and others, 1993).

East of Cuba, the Antillean collision zone is defined by deformed island-arc rocks of Cretaceous and Cenozoic age in the Greater Antilles (Fig. 19b) (Donnelly, 1966; Mann and others, 1991). The arcs are presumably built on oceanic lithosphere that has fully or partly traveled with the Caribbean plate. They are in contact on the north with, and presumably are thrust above, the Gulf passive margin of North America. Since Eocene time, the North American–Caribbean plate motions have been taken up by distributed left-lateral strike slip through the Greater Antilles and Cayman Trough (Rosencrantz and others, 1988). This new plate boundary bypassed the collisional salient of Cuba (Fig. 19a), permitting its acquisition by the North American continent.

The Yucatan basin, floored by oceanic lithosphere of Paleogene or older age (Rosencrantz, 1990) occurs south of Cuba in the bypassed collisional salient (Fig. 19a). It is a question whether the Yucatan basin floor was part of the precollisional Caribbean plate or a product of local spreading related to the collision and/or the excision of the Cuban salient (Malfait and Dinkleman, 1972; Dillon and Vedder, 1973; Rosencrantz, 1990).

The locus of the Antillean collisional margin from southern Mexico to the Bahamas is presumed to lie above the Early Cretaceous edge of southern North America and oceanic lithosphere grown against the Gulf passive margin during drift of NAm and SAm. A question of the existence of this southern edge stems from the uncertainty of the origins of oceanic lithospheres that now occupy the Caribbean basin. The uncertainty arises from the lack of coherent magnetic anomalies in the basin (Donnelly, 1973) and from the existence of displacement zones fully around and within the basin (Burke and others, 1984; Donnelly, 1989; Pindell and Barrett, 1990). It is thus difficult to prove how much, if any, of today's Caribbean oceanic lithosphere is indigenous and due to spreading southward relative to an Early Cretaceous edge of southern North America. In support of Cretaceous sea-floor spreading in the Caribbean basin, however, it is doubtful that the 2000 km of separation of NAm and SAm was achieved solely by stretching of continental lithosphere.

Two groups of models of the evolution of the Caribbean basin and the Antillean collision zone have emerged. One assumes the Caribbean oceanic lithosphere is indigenous but developed with local convergent and other motion zones within the basin (Donnelly, 1989; Klitgord and Schouten, 1986). In such models, the Antillean collision zone contains the Gulf passive margin edge and intra-Caribbean arcs. The other assumes the Caribbean oceanic lithosphere to be wholly or partly a piece of the Farallon plate that moved northeast relative to the North American plate, beginning in the Cretaceous (Malfait and Dinkleman, 1972; Sykes and others, 1982; Pindell and Barrett, 1990). In this model, the Caribbean (Farallon) plate has overridden and consumed all or part of the indigenous Cretaceous Caribbean oceanic lithosphere. The Caribbean plate transported terranes from the Pacific and developing island arcs on its leading edges, and these collided with the passive margin of North America progressively northeastward toward the present.

Although the case is not closed, a Pacific origin of today's Caribbean plate seems better supported. The main factor is interpretation of large Caribbean–North America displacement (1100 km) at the Cayman Trough (Fig. 19) since 50 Ma (Rosencrantz and others, 1988). However, the paleomagnetism of a Cretaceous ophiolite of central America supports an indigenous origin; the ophiolite is assumed to be uplifted Caribbean plate, indicating the absence of a paleomagnetic inclination anomaly expected for the Farallon plate (Frisch and others, 1992).

Western North America

Introduction. The western margin of the North American continent here considered extends from southern California to central Alaska and east from the Pacific Coast as far as the craton below the Great Plains (Figs. 20, 21, 22, and 23). As in other North American margins, the western one has a fundamental structure of margin-parallel tectonic belts. The most landward of these belts is composed of deformed PNA (zone 2, Fig. 7). Seaward of deformed PNA, the belts consist of sets of terranes, each set differentiated mainly by age of attachment to North America and to some degree, by terrane kindred and postattachment deformation. In general, the age of attachment, terrane to terrane, is younger seaward.

In detail, however, the belt structure of the western margin varies markedly on strike, apparently more so than that of the eastern and southern margins. In the western margin, the total width of the terrane belts is highly nonuniform. Individual terrane belts are variably long and short on strike. Some seem to be continuous and composed of a unique terrane set along their full length, whereas others occupy several discrete segments along strike. The terranes have undergone synattachment and postattachment deformation that is heterogeneous parallel to the margin and has been variably contractile, extensile, and strike slip at different places and times. Furthermore, the width of, and structures developed in, deformed PNA of the western margin vary significantly on strike.

The western margin also differs from most of the others in its current activity as a plate boundary zone. In much of California and British Columbia, it includes the surfaces that accommodate the maximum rate of slip between the plates. These are the San Andreas and Queen Charlotte faults (Figs. 20 and 22). In contrast, in parts of the Pacific Northwest and in most of Alaska, the continent-ocean boundary is the surface of maximum slip rate.

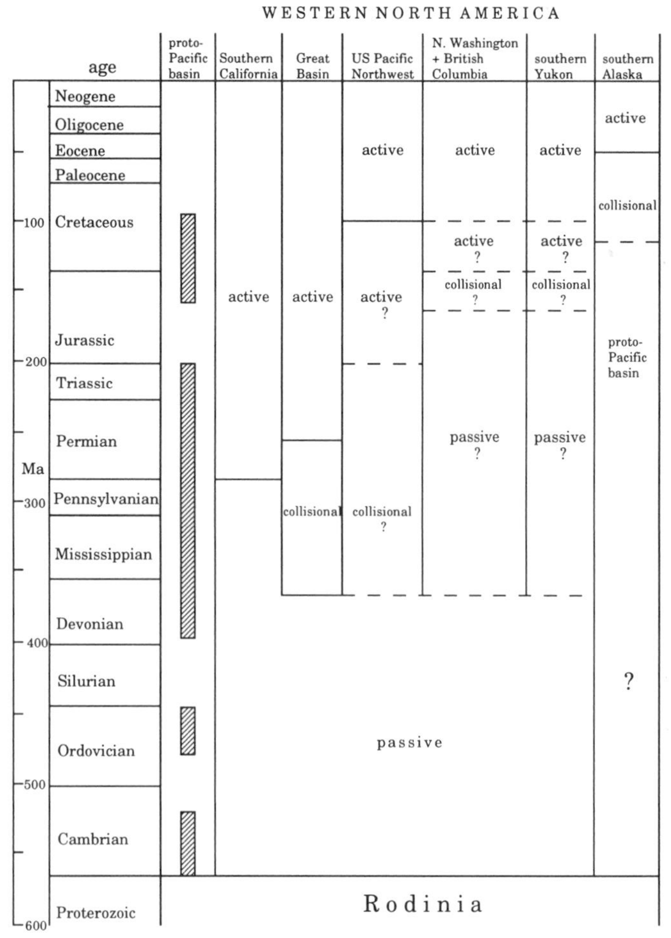

Figure 21. Successions of Phanerozoic margin types versus position in western North America and durations of tectonic activity in the proto-Pacific basins recognized in ophiolites and oceanic basalt in terranes.

A synthesis of tectonic regimes vs. time (Fig. 21) indicates a general evolution from passive to collisional to active for the western margin. It also indicates the existence of convergent boundary activity within the proto-Pacific ocean(s) for much of Phanerozoic time at unknown distances offshore western North America. Alaska is composed entirely of terranes and did not exist as a continent until mid-Mesozoic time (Moore and others, this volume).

Pacific basin lithospheres. An oceanic basin has probably existed along the western margin of North America throughout Phanerozoic time. The evolution of this basin can be viewed in three phases: (1) end-Proterozoic breakup of Rodinia and sea-floor spreading; (2) a Cambrian to Jurassic intrabasinal tectonic phase from which no plates survive; and (3) a Cretaceous-Cenozoic phase for which plates exist or for which their positions and kinematics can be reasonably well restored.

In the first phase, the end-Proterozoic breakup of Rodinia caused the development of the proto-Pacific oceanic basin and the growth of Paleozoic oceanic lithosphere against the newly generated passive margin of western North America. This oceanic basin has existed along the western margin of North America since end-Proterozoic time, although Phanerozoic lithospheres within the basin evolved markedly in this duration (phases 2 and 3). By Permian time, Paleozoic spreading had produced an oceanic basin, of which the proto-Pacific was a part, that was of global extent except for Pangea. The locations of the one or more pieces of Rodinia that drifted away from western North America are unknown. Dalziel (1991) suggested, however, that a part of Gondwana in western Antarctica and Australia was conjugate to western North America at the end of the Proterozoic.

In the second phase, plate boundaries were created within the proto-Pacific lithosphere, dividing it into plates. Moreover, boundaries that included convergent motions relative to proto-Pacific plates developed at various times and places along the western edge of the North American continent. However, convergence may not have occurred entirely along the western continental margin until mid-Cretaceous time. The continent also expanded basinward by accretion in phase 2. In post-Paleozoic time, there was large transit across the basin of oceanic lithosphere that was consumed below western North America. All of these events of the second phase are recorded within the western continental margin by rocks and structures in terranes and by continental arcs. The third phase, for which a Pacific record exists in or can be extrapolated from modern ocean-floor features, is a continuation of the second.

Glimpses of pre-Cretaceous plate tectonics within the proto-Pacific basin come from the tectonostratigraphic records of certain terranes transferred from proto-Pacific plates to North America at the active and collisional margins (Monger and Berg, 1984; Silberling and others, 1987). Times of oceanic spreading are indicated by the crystallization ages of ophiolites and oceanic basalts in terranes. In the Klamath Mountains of California (Fig. 20) (Blake and others, 1985), the Trinity terrane includes Lower Cambrian ophiolite that may record early ridge magmatism at the edge of newly born North America (Wallin and others, 1988), as well as Middle and Upper Ordovician ophiolitic rocks indicating divergent boundaries under oceanic or backarc conditions. Ophiolites of middle and late Paleozoic protolith ages in terranes in the Sierra Nevada (Fig. 20) (Feather River and Kaweah ultramafic bodies) (Saleeby, 1986; Saleeby and others, this volume) and in the Yukon (Devonian ultramafics imbricated with Yukon-Tanana terrane; Fig. 21; Mortensen and Jilsen, 1985) suggest the occurrence of Paleozoic spreading in the proto-Pacific basin. Moreover, terranes constituted by accretionary forearcs containing layered rocks that are probable oceanic deposits record the existence of Paleozoic oceans in the proto-Pacific basin. Examples of these are the Livengood and Angayucham terranes of central Alaska, which contain cover, respectively, from Ordovician or older and from Devonian or older ocean crust (Grantz and others, 1991), and the Northern Sierra and Roberts Mountains allochthon terranes of the U.S. Cordillera that were scraped off Devonian and older oceanic crust. The Cache Creek and Bridge River terranes of British Columbia were assembled in mid-Mesozoic time from strata on ocean crust at least as old as

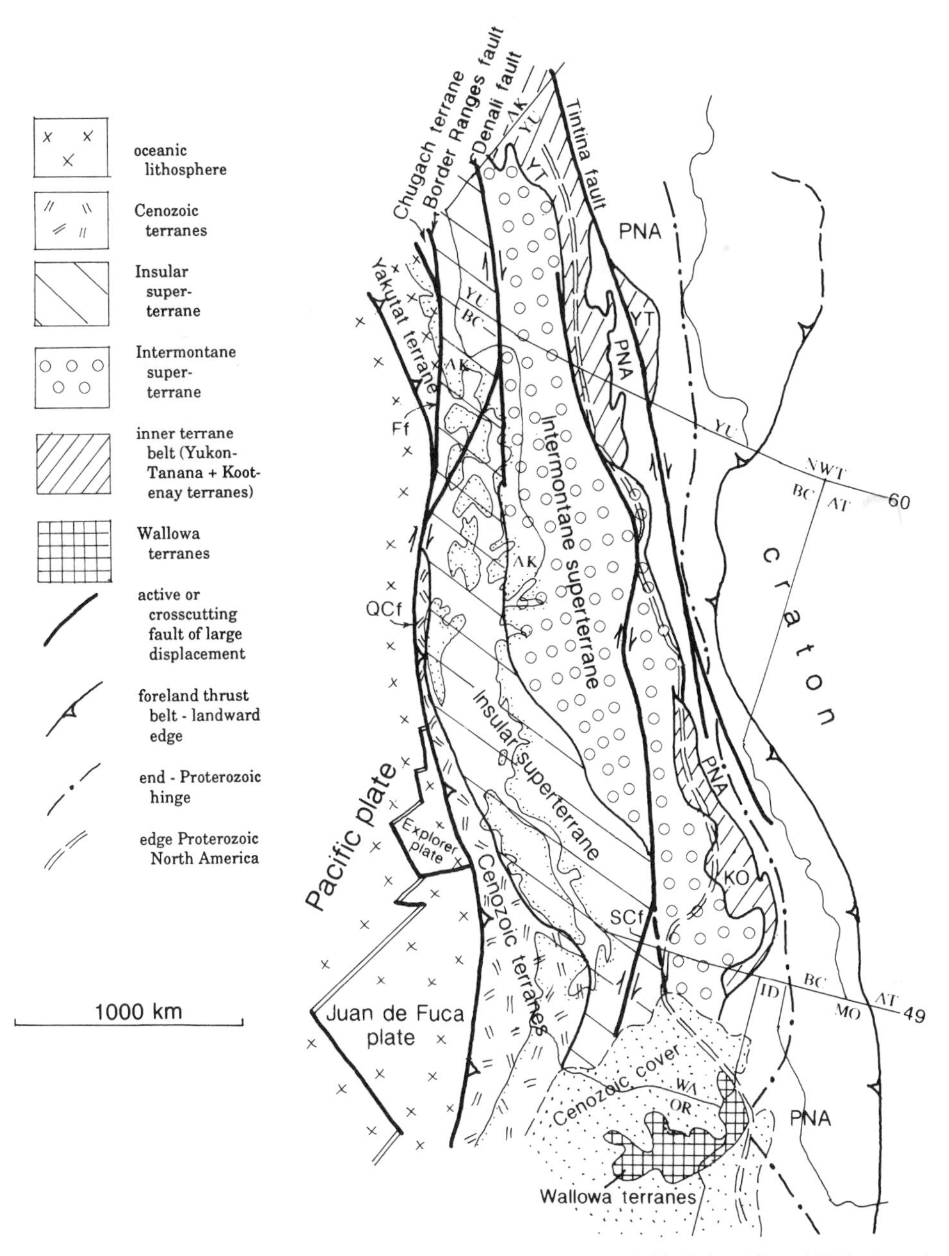

Figure 22. Margin of western North America in western Canada (British Columbia and Yukon) and Pacific Northwest of United States. Ff, Fairweather fault; QCf, Queen Charlotte fault; SCf, Straight Creek fault. YT, Yukon-Tanana terrane; KO, Kootenay terrane; PNA, Proterozoic North America; AK, Alaska; YU, Yukon; NWT, Northwest Territories; BC, British Columbia; AT, Alberta; WA, Washington; OR, Oregon; ID, Idaho; MO, Montana.

Mississippian (Monger and others, this volume). The 150 m.y. or so of depositional time represented by rocks of the Cache Creek and Bridge River indicate that the proto-Pacific crust on which they were delivered could have transited a very large oceanic basin, as suggested by J.W.M. Monger (1993, written commun.).

The occurrence of Paleozoic island arcs in terranes permits the inference of convergence within the proto-Pacific basin before the western margin of North America was fully active (Fig. 21). Examples of such island arcs are those representing each period from Ordovician to Permian in terranes of the eastern Klamath Mountains in California (Potter and others, 1990; Miller, 1987); Paleozoic arcs in the Sonomia, Turtleback, Quesnelia, and Stikinia terranes of the U.S. and Canadian margins (Saleeby and others, this volume; Monger and others, this volume), and the large Paleozoic arcs of the Wrangellia terrane of Alaska and Canada (Jones and others, 1977; Gehrels and Saleeby, 1987; Monger and others, this volume; Moore and others, this volume).

For all these examples of terranes of proto-Pacific origin, the

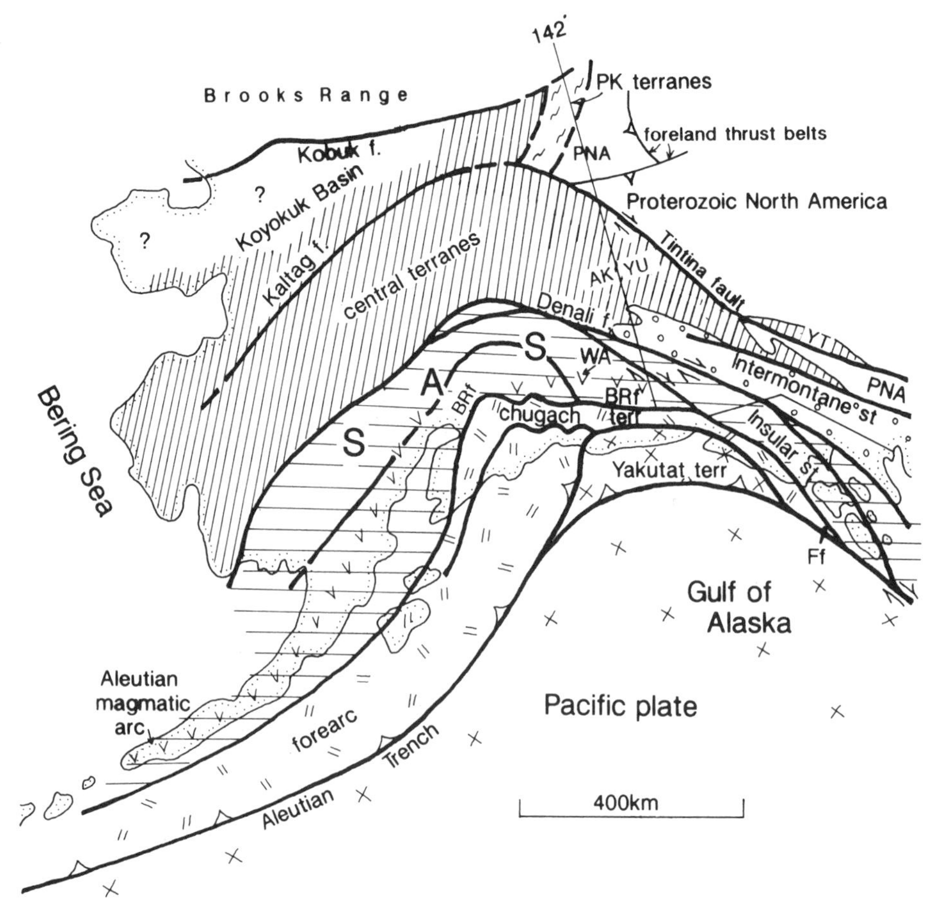

Figure 23. Major tectonic features of southern Alaska. SAS is Southern Alaska superterrane; st is superterrane; v is active volcanic chain. Heavy lines are faults or crosscutting Cenozoic faults of large displacement. x, oceanic lithosphere; PNA, Proterozoic North America; f, fault; BRf, Border Ranges fault; Ff, Fairweather fault; WA, Wrangell arc; PK, composite Porcupine and Kandik terranes.

nature of events can be inferred, but the position and orientation of plate boundaries at which they developed are generally unknown. The paleomagnetic latitude can be approximated, and the proximity to a continent, not necessarily North America, can be rudely inferred by sediment and fossils. The distance of transport along lines of latitude of terranes relative to North America, however, is always uncertain.

For the Cretaceous and Cenozoic, analyses of magnetic anomalies, fracture zones, hotspot tracks, paleomagnetic data, structures, and fossil provincialities provide a space-time picture of plate kinematics that is increasingly resolved toward the present (Atwater, 1970, 1989; Beck, 1976, 1983, 1991; Engebretson and others, 1985; Coney, 1972; Irving and others, 1985; Hyndman and others, 1990; Ross and Ross, 1983; Tozer, 1982; Von Huene and others, 1985; Cowan and Potter, 1986; Yorath and others, 1985; Grantz and others, 1991; Monger and others, 1985, this volume; Saleeby and others, this volume; Moore and others, this volume). They indicate the existence of three major oceanic plates, Pacific, Farallon, and Kula, with divergent boundaries among themselves that migrated toward and intersected the convergent boundary of the North American plate at varied times and places. The result of the intersections of such irregular plate boundaries has been the longshore migration of triple junctions and major obliquity changes along the western boundary, together with changing conditions along strike for the transport, excision, and attachment of terranes. In spite of the Cenozoic and probably older complexity of the field of relative velocities across the convergent plate boundary zones, there have been apparently long durations of margin-parallel components with a constant sense of transport along some segments of the boundary. These were generally left lateral in the southern Cordillera before about mid-Cretaceous time and right lateral since then (Avé Lallemant and Oldow, 1988; Beck, 1983; Engebretson and others, 1985). In particular, it appears that many of the terranes now in southern California, western Canada, and southern Alaska assembled with one another at far more southerly latitudes than their present ones and then translated with northward components within the plate boundary zone of North America and adjacent oceanic lithospheres (Packer and Stone, 1972; Thrupp and Coe, 1986; Hagstrum and others, 1987).

Passive margins. As noted earlier, the entire western margin of PNA, from Death Valley to the Yukon, was created by the breakup of Rodinia at the beginning of Phanerozoic time. From the Cambrian to times in the Paleozoic or Mesozoic that vary with position (Fig. 21), the western margin of PNA was a passive type. In the central western United States and in western Canada, this margin is marked at the surface mainly by a broad zone of thick Paleozoic passive margin wedge strata of postrift development and by subjacent clastics and basalt that are probably rift phase and at least partly of latest Proterozoic age. Such strata are entirely within the belt of Mesozoic and Cenozoic active margin deformation that is between the present-day craton and the terranes (zone 2; Fig. 7). A major variant to such passive margin structure exists in Idaho (Fig. 20), where the Paleozoic craton of North America extends to the zone of terranes (Cowan and Potter, 1986). I suggest below that the Oregon-Washington embayment is probably a product of excision of the continental edge during active margin tectonics rather than an end-Proterozoic feature, and that the western edge of end-Proterozoic North America was almost entirely rift dominated, as shown in Figure 10.

As noted above, the broad rifted margin of western PNA intersects the probably transform-dominated southern edge in southern California, where a former corner between these end-Proterozoic edges has probably been excised, fragmented, and dispersed mainly southeastward in terranes. In the Death Valley region (Fig. 20), the end-Proterozoic hinge between the craton on the southeast and the passive margin wedge on the northwest (Stewart, 1972) intersects the most inboard truncation surface. The occurrence of Late Pennsylvanian and Permian turbidite basins and thrusts in PNA at this boundary (Inyo Mountains) indicates that the truncation is probably late Paleozoic (Stevens and others, 1992; Snow, 1992).

The passive margin in the Great Basin (Fig. 20) is remarkable for its great width, up to 700 km, between the hinge and the existing edge of PNA (Blake and others, 1985; Saleeby, 1986; Saleeby and others, this volume). Its present width probably reflects a large original width because estimates of later deformation in the Great Basin are permissibly canceling: 180–290 km of Mesozoic contraction (Elison, 1991) and 200–300 km of Cenozoic extension (Wernicke and others, 1988). The great width and its thick passive margin stratal wedge suggest that the margin in the Great Basin is rift dominated. Rift domination must remain hypothetical, however, because the Great Basin is so little unroofed and the depths of exposure are small compared to most orogens. The broad structure of the margin outboard of the hinge is thought to be a greatly thinned wedge of transitional crust surmounted by rift-phase clastics and intruded by basalt (Stewart, 1972). This structural model comes mainly from the subsidence history of the Middle Cambrian to Late Devonian postrift wedge (Stewart and Suczek, 1977; Bond and others, 1988) and from the existence of diked Proterozoic and Archean crust in a few Mesozoic thrust nappes (Yonkee, 1992; Snoke and Miller, 1988). The Late Proterozoic(?) and Lower Cambrian clastic strata below the carbonate wedge, nowhere fully exposed, are poorly dated, and they contain no evidence of a rift-postrift transition or of contemporaneous magmatism.

The edge of PNA in the Great Basin is zigzag in plan (Fig. 20). It is a question whether the nonlinearity is primary, such as that produced by rift-transform-rift offsets, or is a product of Phanerozoic deformation. A primary origin is suggested by differences in the character and orientations of structures of the earliest-collided terrane, the Roberts Mountains allochthon, at its contact with edges of PNA of different orientation in the Great Basin (Oldow, 1984). The differences imply the existence of footwall ramps of substantially different attitude on which the terrane was emplaced. Alternatively, an origin of part of the zigzag pattern by younger deformation has been proposed on the basis of northwest-striking faults in the western Great Basin that produced heterogeneous right-lateral simple shear as far back as Late Cretaceous time or before (Speed, 1983; Saleeby, 1986; Saleeby and others, this volume). Both effects may be superposed.

In eastern Oregon, southern Idaho, and southeastern Washington, the passive margin outboard of the hinge is very narrow to absent (Fig. 20), in contrast to its width north and south. Where it is absent, the Paleozoic craton directly contacts terranes that did not originate in the North American continent. A likely explanation for this absence, for reasons given below, is that the passive margin was locally excised and moved away before collision in the Cretaceous (Davis and others, 1978; Speed, 1983; Cowan and Potter, 1986; Monger and others, this volume). Another is that the Idaho-Washington segment was an end-Proterozoic transform with a narrow, unrifted ocean-continent transition like that of the SGBFZ (Keen and Haworth, 1985b).

In northern Washington, British Columbia, and the southern Yukon Territory, the end-Proterozoic passive margin is moderately wide and is thought to have been rift dominated, on the basis of the great thicknesses of Late Proterozoic and Cambrian clastic sediments and basalt and an overlying wedge of Paleozoic and lower Mesozoic strata (Fig. 22) (Monger and others, this volume). The strata above the synrift rocks are carbonate rich and occupy a postrift wedge whose exponentially decreasing rate of subsidence can be tracked from Cambrian well into Ordovician time (Bond and others, 1988). Above the Ordovician, strata generally record shelf environments that subsided relative to the craton with only local and brief interruptions until the Middle Jurassic. This Cambrian to Middle Jurassic succession occurs entirely in thrust sheets within the late Mesozoic foreland fold and thrust belt, which extends well landward of the hinge. By restoration of strata of the fold and thrust belt, it is estimated that the transitional crust does or did extend at least 100 km seaward of the hinge (Monger and others, 1985, this volume).

It is a question whether the margin of PNA in western Canada was passive until the Late Jurassic onset of tectonism which led to the modern Canadian Cordillera, or whether plate boundary tectonics significantly affected the margin before then. The Cambrian to Middle Jurassic succession of PNA contains no structures that indicate compellingly the occurrence of major tectonism before late Mesozoic time. The succession, however, contains scattered features that suggest that a Late Devonian to Middle Mississippian tectonic event or events occurred at places within the passive margin. The features may be more prevalent and extensive northward from southern British Columbia to the southern Yukon Territory. They are local sedimentary basins, high-angle normal faults, clastic deposits with landward sediment transport, and minor alkalic magmatism (Struik, 1981; Gordey and others, 1987).

The clastic deposits (Earn Group, Exshaw Formation) are thought to have been derived from North American sediment sources of Paleozoic and Proterozoic ages that developed on the seaward side of the passive margin (Gordey and others, 1987). After the Middle Mississippian, the margin returned to a state of slow regional subsidence. The unroofed source rocks of the passive margin are nowhere exposed, implying that they were either covered by terranes in the Mesozoic or were excised and removed before the terranes collided.

The tectonics of the mid-Paleozoic events in western Canada are uncertain and have been interpreted disparately. One idea is that the features reflect margin-normal extension (Struik, 1981) with excision of outboard tracts of the continent (Templeman-Kluit, 1979). Eisbacher (1983) suggested that such extension was related to margin-parallel simple shear strain. Another hypothesis is that collision occurred along the outer margin, causing the development landward of foreland basins whose downflexing included crustal extension (Smith and others, 1993). A third idea is that mid-Paleozoic extension in the western Canadian margin may have been a far-field effect of Ellesmerian active margin tectonism at the northern margin of North America in the northern Yukon Territory, as discussed in a later section.

If the mid-Paleozoic tectonic events were caused by regional extension (hypotheses 1 and 3), they imply either an attempt at rifting and drifting in the Canadian margin or distal effects in a very broad plate boundary zone. In either case, the Canadian margin remained passive, and this interpretation is shown in Figure 23. The consequence of a possible mid-Paleozoic collision is considered in the next section. Whether extensional or collisional, the interpretations of the mid-Paleozoic features imply that an outboard reach of the Paleozoic margin is missing, either covered or excised.

Collisional margins. *Introduction.* The passive margin of western PNA collided with terranes at various places, beginning first in Late Devonian time and last in the mid-Cretaceous. By the Late Cretaceous, the western margin of the North American continent was entirely an active one (Fig. 21). The distinction between a collisional and an active margin in western North America is straightforward at some places but vague at others. The main criterion for an active margin is the existence of a continental magmatic arc. Collisional margins have no continental arc, although arc magmatic rocks may be thrust on them during collision. The continental arc criterion is not unique, however, because where plate boundary motion is highly oblique, a magmatic arc probably does not develop regardless of whether the slab dips toward North America (active margin) or away from it (collisional margin).

In general, some major kinematic questions of plate boundary zone evolution are alike for collisional and active margins in western North America. Are the terranes all or mostly far traveled or, at the other extreme, of only local transport, or a mix of both? How many times have the terranes attached to North America and been reseparated and/or divided as fragments of former individual terranes or terrane composites? What is the apportionment in terrane transport of the components of plate velocity between NAm and plates of the Pacific basin? What are the provenances of the terranes? Have they arisen entirely along the Phanerozoic edge of western North America, or do some come from cross-Pacific sites?

There are also questions specific to collisional margins. One is the configuration of plate boundary zones along western North America during collision. It can be presumed that the colliding terranes were positively buoyant relative to oceanic lithosphere, and that they overrode the ocean floor welded to North America at a seaward-dipping convergent plate boundary zone. Upon impact with or overthrusting of the rim of PNA or earlier attached terranes, the incoming terranes ceased closing relative to North America due to buoyancy arrest. A collision, so defined, had to be followed by a reorganization of the plate boundary zone for convergence to continue. Two such reorganizations can be imagined. One is that toward the Pacific, subduction zones all dipped seaward and upon each collision, the convergence jumped seaward of the newly attached terrane to a new or existing seaward-dipping zone. A second is that the plate boundary zone included a very broad region between a Pacific island arc whose subduction zone dipped toward North America and the edge of the North American continent. The broad backarc region gave rise to remnant island arcs, backarc oceanic lithosphere, and sedimentary basins, and it may have closed and reopened many times. In this idea, collisions with the adjacent continental margin took place by episodic thrusting of part or all of the backarc contents onto the continental edge while subduction was continuous at the more outboard locus. These alternative solutions were presented long ago (Burchfiel and Davis, 1972; Speed, 1977); there seems to be no resolution, in part because the polarity of collided arc terranes is so undetermined.

Early collisions in the Great Basin. The oldest recognized collision in the western North American continent emplaced the Roberts Mountains allochthon terrane (RMA) above the outer 100 km or so of the passive margin wedge in the Great Basin (Fig. 20) in Late Devonian and Early Mississippian time (Roberts and others, 1958; Stewart and Poole, 1974). This event, the Antler orogeny, has no counterpart elsewhere at a preserved strand of the Paleozoic western passive margin. This indicates either that the collision was only local, or if originally more far flung, the collided elements have been excised except for that of the Great Basin. The RMA terrane is interpreted as an accretionary forearc, constituted by sedimentary units, some of continental provenance, and basalt. These were scraped off an early Paleozoic ocean floor that had fronted western North America, presumably since the end of the Proterozoic (Johnson and Pendergast, 1981; Speed and Sleep, 1982). The RMA terrane probably moved a substantial distance along North America with high obliquity before collision. An island arc presumably fronted one side of the RMA, but this has not been identified. It was perhaps excised later from the site of collision or perhaps stayed attached to North America, but was overrun in a later collision. The effect on North America of the RMA collision was the development of a foreland basin (Poole, 1974; Speed and Sleep, 1982; Dickinson and others, 1983; Saleeby and others, this volume) that was progressively underwedged by the RMA (Jansma, 1988). There was, however, no thrust belt developed in the passive margin wedge below or ahead of the RMA and no magmatism associated with the collision.

The second collision recognized in the western margin of the North American continent was also in the Great Basin, along much the same reach as the first. This event, called the Sonoma orogeny by Silberling and Roberts (1962), was the emplacement of the Golconda allochthon terrane in Permian and early Early Triassic time above the RMA terrane (Speed, 1979; Saleeby, 1986; Saleeby and others, this volume). The Golconda allochthon, an accretionary prism of upper Paleozoic ocean-floor strata and basalt, was the forearc to the Sonomia terrane (Fig. 20), an island arc of late Paleozoic age built on fragments of one or more earlier arc terranes (Speed, 1979). This collision is thought to have brought lithospheres of Sonomia and North America into juxtaposition while the forearc overrode the outer edge of end-Paleozoic North America. Some differences in organization of terranes and in process were expressed by Silberling and others (1987) and by Miller and others (1984), respectively. The effects on the North American continent of the collision of the Sonomia and Golconda allochthon composite terrane were like those of the earlier collision of the RMA terrane. Thrusting and magmatism did not occur in the overridden continent; foreland basins subsided ahead of the Golconda allochthon, but these were small and discontinuous compared to the foreland basin of the RMA. Moreover, the contractile deformation of the Golconda allochthon terrane (Brueckner and Snyder, 1985; Babaie, 1987) and probably of the RMA terrane occurred almost entirely during growth of the accretionary forearcs; both were apparently relatively strong and coherent during collision with North America. The implications of both collisions in the Great Basin are that the overriding forearcs were massive enough to cause downflexing of the continent for foreland basins (Speed and Sleep, 1982), but that the thrusts below the forearcs were exceedingly weak compared to the strengths of the forearcs and to the upper crust of overridden North America. The weakness of the thrusts may have been due to the underplating of the forearcs by recently deposited oceanic sediments just before continental collisions and to the pressure of fluids, from dewatering sediments, that were channelled along the forearc bases, as is known in at least one modern forearc (Moore and others, 1988).

It is unknown how much, if any, of PNA was lost in the collisions in the Great Basin. The existence of lower Paleozoic strata indicative of continental slope or outer shelf below the RMA terrane and the absence of an evident duplex of thrust sheets of passive wedge strata below the terranes suggest that little or no continental crust was lost by subduction below the colliding crystalline lithosphere west of the forearcs.

The motion of the Sonomia and Golconda allochthon composite terrane relative to North America is thought to have had a left-lateral component (Speed, 1979). The plate boundary tectonics that caused their collision in the Great Basin may also have been responsible for the late Paleozoic deformation and excision of the corner of PNA in southern California (Fig. 17) (Stevens and others, 1992), which included the Seri terrane, now in northwestern Mexico (Fig. 17). Shortly after the attachment of the Sonomia and Golconda allochthon composite terrane, the western margin of the North American continent in the Great Basin became an active one, at about 230–240 Ma (Dilles and Wright, 1988; Speed, 1978). The change probably records a reversal in the dip of the slab bordering North America at Great Basin latitudes.

These early collisions in the Great Basin require special consideration because they have no preserved counterpart elsewhere along the western edge of North America. The Great Basin, moreover, is atypical because of the great width of its end-Proterozoic passive margin, not apparently due to later deformation, and of the lack of excisions within it (Speed, 1983). Two ideas to explain the tectonic behavior of the Great Basin are as follows. First, the end-Proterozoic passive margin may have had an odd configuration that trapped terranes moving along the coast, whereas at other reaches of the western margin, a different, perhaps linear, configuration prevented permanent attachments in collision mode before the Jurassic. Second, early collisions may have been widespread along the western margin, but their product is preserved only in the Great Basin. Although the anomalous breadth of its passive margin suggests that the Paleozoic Great Basin was peculiar to begin with, the second explanation that the early collided terranes were later removed elsewhere along the western margin seems more plausible. This is because of evidence for excisions and because terranes tectonostratigraphically similar to the RMA and the Sonomia-Golconda allochthon terranes exist with younger attachment ages at some other reaches of the western margin (see below).

Allochthons with tectonostratigraphies somewhat like that of the RMA were attached to the Seri terrane in Permian time (Stewart and others, 1990) and to the Paleozoic craton in the Mojave Desert (Fig. 20) at a time between Permian and Jurassic (Carr and others, 1984; Walker, 1988). The Stikine and Quesnell terranes of the Intermontane superterrane of British Columbia (Fig. 22), which might correlate with the Sonomia-Golconda allochthon terranes, attached to North America at their current sites no earlier than late Mesozoic (Monger, 1977, 1982, this volume). Thus, I propose that much of the western margin had a long collision history but that the early attached terranes were widely excised, probably together with an outer fringe of PNA, from most of western North America. The early collided terranes, or pieces of them, may have been episodically returned to coastwise flow in a broad plate boundary zone (Beck, 1983). The oddness of the Great Basin is its preservation of the Paleozoic margins (Speed, 1983).

Pacific Northwest of the United States. The western edge of PNA in much of the Pacific Northwest—western Idaho, Oregon, and southern Washington—differs decidedly from that of the Great Basin in its Paleozoic–early Mesozoic history. The unusually broad passive margin, the collisional terranes (RMA, Sonomia, Golconda allochthon), and the early Mesozoic active margin features of the Great Basin are absent (Fig. 20). Instead, the Paleozoic craton, which is typically inboard of the early Paleozoic passive margin of western North America, and which is here imbricated and intruded in a mid-Cretaceous active margin phase (Dover, 1980), extends to the edge of PNA (Speed, 1983; Cowan and Potter, 1986). Seaward of this edge are the Wallowa groups of terranes of noncontinental origin (Potter and others, 1990; Dickinson, 1979; Avé Lallemant, 1994). These attached to PNA in mid-Cretaceous time (Lund and Snee, 1988), probably at the beginning of continental arc plutonism in adjacent PNA.

This conspicuous change along the western margin has at least two competing explanations. First, the end-Proterozoic edge of PNA in the Pacific Northwest may have been very narrow, perhaps a long transform, and may have persisted as a passive margin until mid-Cretaceous time, in spite of evident convergent boundary tectonics between 350 and 100 Ma to its south in the Great Basin. Alternatively, some or all the attributes of the Paleozoic and early Mesozoic margin of the Great Basin existed in, but were excised and removed from, the Pacific Northwest. Such removal preceded the subduction of the deeply embayed edge of PNA in the mid-Cretaceous. The second explanation is probably correct (Davis and others, 1978; Speed, 1983) because the configuration of the edge of PNA (Figs. 20, 22) does not, at least at present, suggest a long transform in the Pacific Northwest relative to an orthogonal ridge in the Great Basin. In addition, to the north in western Canada, there are long margin-parallel faults of large right-slip displacement and mid-Cretaceous onset (e.g., Tintina, Straight Creek; Fig. 22; Gabrielse, 1985; Monger and others, 1985, this volume). Furthermore, the first hypothesis is implausible in its maintenance of extraordinary differences of tectonic evolution across a narrow band of the margin for as long as 250 m.y.

Therefore, it is likely that late Paleozoic collisions and an active margin of early Mesozoic age occurred in the Pacific Northwest. The relics of such events together with much of the end-Proterozoic rifted margin of PNA are probably elsewhere on the western margin, perhaps fully disrupted and widely distributed. The process of excision, unfortunately, is hard to read in the Pacific Northwest because so much of it is covered by Cenozoic volcanic rocks, including the 7-km-thick, Miocene Columbia River Basalt (Cowan and Potter, 1986). The process may have developed an oceanic pullapart basin in the place of the excised block of terranes and PNA (Davis and others, 1978; Speed, 1983). The inland edge of the basin and the present edge of PNA was perhaps a right-slip fault. The faulting was presumably caused by the margin-parallel component of the oblique plate velocity and may have begun with the onset of right-oblique convergence of plates of the Pacific basin relative to NAm (Engebretson and others, 1985). The excised terranes traveled northwest along the margin, and PNA of the Pacific Northwest was fronted by oceanic lithosphere that diminished in age northwest from a southern boundary near today's Snake River Plain (Fig. 20). The postulated oceanic lithosphere was later consumed below the Wallowa terranes now attached to PNA (Fig. 20), implying a collisional margin in mid-Cretaceous time in the Pacific Northwest. Thereafter, an offshore slab apparently began to descend with eastward dip, causing Late Cretaceous continental arc magmatism and foreland thrusting in Idaho (Fig. 20).

Western Canada and northern Washington. The western margin of the present-day continent in northern Washington, British Columbia, Alberta, and the southern Yukon Territory (Fig. 22) (hereafter, the Canadian Cordillera) has undergone convergent plate boundary tectonics from late Mesozoic time to the present (Fig. 23) (Monger and others, this volume; Cowan and Potter, 1986). Such tectonics have produced a deformed foreland, belts of terranes that are distinguished by tectonostratigraphic affiliations and amalgamation histories (Fig. 22), and by belts of concentrated plutonism and unroofing that occur chiefly along boundaries of the terrane belts.

By way of introduction, there are several major questions about the evolution of the Canadian Cordillera. First, were the terranes now in contact with PNA those that collided with PNA in the Jurassic, or has there been significant younger translation of terranes along the edge of PNA? Second, how much, if any, of the continental and transitional crust of end-Proterozoic North America was removed from the margin in the region now occupied by terranes (Fig. 22)? Third, what was the history of collisional versus active margins during the accretion of terranes before Late Cretaceous time? (From Late Cretaceous onward, it is clear that a Pacific basin slab underrode PNA.)

The deformed foreland of the Canadian Cordillera is a breakforward thin-skinned fold and thrust belt composed of passive margin wedge and cratonal strata (Price, 1981, 1986; Monger and others, 1985, this volume). It lies entirely above cratonal crust of normal thicknesses. Landward of the fold and thrust belt, the foreland includes a foreland basin of which the Upper Jurassic basal strata provide the earliest recognized deformation of PNA in the development of the Canadian Cordillera. The seaward contact of the deformed foreland is with terranes that are thought to have overridden the passive margin (Monger and others, 1982, 1985, this volume). It is unclear how much of the passive margin wedge is missing from the foreland thrust belt, due either to burial by terranes or to excision before collision. A restored section of the thrust belt in British Columbia by Price (1986) implied that today's edge of cratonal PNA was the contact with end-Proterozoic transitional crust and that the transitional crust extends or extended at least 100 km seaward. On the basis of deep seismic soundings, Cook and others (1991) interpreted thinned continental crust to extend about 400 km west of the cratonal crust of normal thickness. The reflection packets they chose to represent continental crust, however, differ little from supradjacent ones assigned to terranes. To conclude, the existence today of transitional crust below the terranes in western Canada is uncertain.

The contact zone of the deformed foreland and the terranes in southern British Columbia and northern Washington has a complicated deformation sequence: early breakforward (eastward) thrusting (Cretaceous?), intermediate backfolding and unroofing, late breakforward thrusting (Cretaceous and Paleocene) with the progressive development of the more landward realm of the thrust belt, and low-angle normal faulting and unroofing (Price, 1981, 1986; Brown and others, 1986). The sequence is interpreted in plate boundary tectonics by Monger and others (1985, this volume) to record (1) early collision of PNA with overriding terranes, (2) out-of-sequence faulting that permitted more outboard terranes to wedge below and backfold the early collision zone and the outer passive margin wedge, and then (3) general contraction and thickening of the collision zone. The fourth phase is thought to be due to regional extension and transtension, related to an increase in obliquity of relative velocity.

This sequence of deformations, which was perhaps continuous, evidently spanned the transition from a collisional to an active margin. The structures of the first three phases indicate different mechanisms of progressive contraction of a thickening, strengthening supracrust and give no clear signal of change in slab configuration. Plutons were intruded in a belt in the western reach of the deformed foreland between about 85 to 100 Ma and 60 Ma (Armstrong, 1988). These crosscut the contractile structures and, taken as a continental arc, imply that subduction of the continent began late in the tectonic sequence.

In Canada, the outboard zone of the deformed foreland and the belts of terranes broadly includes margin-parallel right-slip faults of long trace length (Fig. 22). Some of these began displacement as far back as mid-Cretaceous time. Estimates of displacement are varied; for example, 450 km (Mortensen and Jilsen, 1985) vs. 850 km (Gabrielse, 1985) for the Tintina fault (Fig. 22). The number of such margin-parallel faults (Fig. 22) indicates the possibility that cumulative right-lateral simple shear of large magnitude (thousands of kilometers) and Late Cretaceous–Paleogene age may exist within the Canadian Cordillera. If so, the outer reaches of the deformed foreland as well as the terranes include rocks and structures that were developed well south, perhaps at a succession of latitudes, and later translated northward in fault-bounded panels during collisional and/or active margin tectonics.

The innermost terrane belt comprises the Yukon-Tanana (Y-T) and Kootenay terranes (Fig. 22). These contain metamorphosed continent-derived clastic sedimentary strata of known and assumed early Paleozoic age and felsic extrusions and intrusions of Devonian and Mississippian age. In the Kootenay terrane, such rocks underwent deformation before late Paleozoic sedimentary overlap (Monger and others, this volume); in the Y-T, the deformation is known only to be pre–Late Triassic (Mortenson and Jilsen, 1985). It is clear that terranes of the innermost belt underwent anatectic or arc-type magmatism in mid-Paleozoic time, but it is unknown whether they were coherent stratal sequences (Mortenson and Jilsen, 1985) or tectonic complexes such as an accretionary forearc (Templeman-Kluit, 1979) before magmatism. It is debatable whether the premagmatic rocks are parautochthonous North American slope and shelf facies (Klepacki and Wheeler, 1985), perhaps analogous to those of the Taconian allochthon in New England, or are not representative of any nearby section of North America (Mortenson and Jilsen, 1985; Templeman-Kluit, 1979). The Kootenay terrane, moreover, is associated with several nappes of Proterozoic basement (Monashee terrane of Monger and others, this volume).

I suggest that the premagmatic protoliths of the Y-T and Kootenay terranes are accretionary complexes of lower Paleozoic continental and hemipelagic sediment deposited on oceanic crust that fronted the western (and perhaps, northern) passive margins of North America. Their early constitution may have been like that of the Roberts Mountains allochthon of the Great Basin and comparable terranes in the Mojave Desert, and part of the Seri terrane of northern Mexico (Fig. 20a), and like them, grew as accretionary prisms in Devonian and possibly earlier times. The Y-T and Kootenay terranes, however, underwent magmatism upon accretion, whereas the terranes to the south did not. The transport history of none of these terranes is understood. It is unclear whether they represent Devonian and earlier convergent boundary tectonics at many sites in the Pacific basin near western North America, or rather, a single site of convergence and later dispersion to far-flung sites as forearc fragments. If the suggestion by Smith and others (1993) that the Late Devonian–Mississippian basins on the margin of PNA in western Canada are foreland basins of collisional origin is correct, the hypothesis of widespread mid-Paleozoic convergence near or at the western margin is supported. In this case, the colliding elements have been excised or covered, and there is probably little chance of correlating such elements, now perhaps in Mesozoic terranes, with individual sites along the margin of mid-Paleozoic North America. If, however, the basins and other mid-Paleozoic tectonic features of PNA in western Canada are products of regional extension, the hypothesis of a restricted region of mid-Paleozoic convergence and later dispersion of the forearc is supported. In this case, the convergent boundary was probably near today's Great Basin edge of North America, because the collision of the Roberts Mountains allochthon and PNA probably began at the end of the Devonian. The other pieces may have been on the move until as late as the Cretaceous.

The Intermontane and Insular superterranes are seaward of the inner (Y-T and Kootenay) terrane belt (Fig. 22) (Monger and others, 1982, 1985, this volume). Between them, in British Columbia and at their southern ends in Washington, are many small terranes that probably represent sedimentary basins, island arcs, roots of accretionary forearcs, and pieces of continent that stuck to and moved with the superterranes (Cowan and Potter, 1986; Monger and others, 1985). On the seaward side of the Insular superterrane are small terranes that almost certainly accreted one by one in latest Cretaceous and Cenozoic time to an active margin of North America then fronted by the Insular superterrane (Fig. 22).

The Intermontane and Insular superterranes consist of terranes that did not evolve in the Paleozoic or early Mesozoic North American continent. The terranes are thus exotic to North America. Each superterrane arrived preassembled and added considerable girth to western Canada and northern Washington. The Intermontane includes island arcs and accretionary forearcs of Paleozoic and early Mesozoic age, including one with sedimentary rocks of probable continental provenance (Quesnel terrane). This superterrane is thought to have been assembled within the Jurassic (Cordey and others, 1987) outboard of the North American continent. Parts of it may be equivalent to the Sonomia and Golconda allochthon terranes of the Great Basin, either as fragments that emerged discretely from a long convergent plate boundary zone of NAm or that evolved by progressive fragmentation during margin-parallel transport of an originally larger edifice.

The Insular superterrane includes the Wrangellia terrane, which was the first terrane identified by tectonostratigraphy as being alien to North America and adjacent terranes. Paleomagnetism and fossils indicate that it was greatly displaced northward in latitude with respect to Triassic PNA (Jones and others, 1977; Hillhouse, 1977). The various bodies of Wrangellia distributed along 2500 km of the Canadian and Alaskan margins all have approximately the same Triassic paleomagnetic latitude, suggesting that the terrane has been redistributed from a more compact Triassic form (Yole and Irving, 1980). The Insular superterrane also includes the Alexander terrane (Berg and others, 1972), which is a succession of magmatic and sedimentary rocks of probable arc environment from Late Proterozoic to Permian age (Gehrels and Saleeby, 1987). The Alexander terrane may be the lower Paleozoic foundation for the strata of the Wrangellia terrane. Both terranes underwent extensive Triassic effusion of basalt that may have been associated with rifting. The Paleozoic tectonostratigraphy of the Insular superterrane (mainly Alexander terrane) is said to be unlike that of other terranes of western North America and its lower Paleozoic may be correlative with that of an orogen now in eastern Australia (Gehrels and Saleeby, 1987). The Alexander terrane thus provides the most evident example in the western margin of North America of possible transfer of terranes across the Pacific basin. If so, the Insular superterrane probably crossed near the equator, reaching the vicinity of western Pangea late in Paleozoic time, rifted in the Triassic, and then migrated northward along the western margin of North America in the Cretaceous (Gehrels and Saleeby, 1987).

A hypothesis of the convergent plate boundary zone evolution in southern Canadian Cordillera is given by Monger and others (1985, this volume), framed as a terrane attachment history and based on tectonostratigraphy. The Intermontane superterrane and the Kootenay terrane collided with PNA near their present positions above a west-dipping slab in the Middle Jurassic. The Insular superterrane was thrust against and below the edge of North America, which was the western edge of the Intermontane superterrane, in mid-Cretaceous time. In the view of Monger and others, the vergence of contractile structures may correlate in part with the slab attitude. In the collisional margin mode, generally top to the east shear occurred in the Jurassic thrusting of the Intermontane superterrane above North America. In the active margin mode with the slab directly underriding the continent, top to the west thrusting occurred in the cover and buoyant edifices during scraping off from the ocean floor in the Cretaceous emplacement of the Insular superterrane. The timing shown in Figure 21 is derived from this hypothesis.

A second hypothesis, based chiefly on paleomagnetic latitude histories of terranes, is that the Intermontane and Insular superterranes arrived together at their present positions at times in the Cenozoic (Avé Lallemant and Oldow, 1988; Sedlock and others, 1993). Here, both superterranes existed well south of Canada, amalgamated in the Cretaceous, and migrated northwestward relative to North America in the plate boundary zone to begin attachment in the Late Cretaceous and Paleogene. The margin-parallel right-slip component, however, has persisted through the Cenozoic and has been taken up in a broad shear zone whose boundaries have migrated seaward with time. Thus, the Tintina and nearby faults (Fig. 22) took up some of the early motion in this shear zone, and these were abandoned in the Paleogene as shear shifted to or was concentrated among more seaward faults. In this second hypothesis, Mesozoic convergent boundary effects on PNA (foreland deformation and sediment input from exotic terranes) were caused by collision of terranes that are no longer present, have moved farther along the margin, and/or have been buried tectonically. Such rearrangement of margin-hugging terranes is permissible because none of the boundaries between terranes and PNA has direct evidence of pre-Paleogene stitching (stratal overlap or postkinematic intrusion).

Paleomagnetic inclinations of the Intermontane superterrane indicate no Triassic latitude anomaly relative to North America (Irving and others, 1985; May and Butler, 1986), but mid-Cretaceous plutons in the Intermontane superterrane indicate about 10° of southward latitude relative to contemporaneous North America (Irving, 1979; Irving and others, 1985; Beck, 1983; May and Butler, 1986). Such data imply southward transport of some or all terranes of the Intermontane superterrane between Triassic and mid-Cretaceous, then northward transport thereafter (Avé Lallemant and Oldow, 1988), notwithstanding arguments that the plutons have tectonically flattened inclinations (May and Butler, 1986). In the second hypothesis, the amalgamation of the Insular and Intermontane superterranes occurred during the northward transport and perhaps was related to the mid-Cretaceous plutonism in the Intermontane superterrane.

The second hypothesis for convergent-boundary tectonics in British Columbia provides a comfortable explanation for Cretaceous excision in the U.S. Pacific Northwest: the right-lateral shear stepped inboard of the early Mesozoic margin north of a rift in southern Idaho(?) (now the Snake River Plain; Fig. 20a), extracting early terranes and much of the passive margin. If true, pieces of the outer deformed foreland, the Kootenay terrane, and the Intermontane superterrane of British Columbia came from the U.S. Pacific Northwest.

Avé Lallemant and Oldow (1988), on the basis of several interpreted structures, suggested that left-lateral margin-parallel motion mainly or wholly affected the western margin from Alaska to the Great Basin before mid-Cretaceous time through the Jurassic and perhaps back into the Permian. This motion history seems probable, if difficult to document, and may include the late Paleozoic excision of the corner in southern California between the western and southern margins.

Alaska. Alaska consists totally of terranes except for a possible fringe of PNA at the Canadian border (Figs. 23 and 24) (Churkin and Trexler, 1981; Stone and others, 1982; Grantz and others, 1991; Silberling and others, 1987; Jones and others, 1977, 1982; Moore and others, this volume). The terranes may be divisible into four groups, each having an approximate east-west trend (Figs. 23 and 24). The northern group is north of the Kobuk fault and includes the present-day physiographic Brooks Range and North Slope. The northern edge of the northern group is part of the present-day northern edge of North America. The northern group terranes assembled at their current site in mid-Mesozoic time (Grantz and others, 1991), possibly beginning earlier (Avé Lallemant and Oldow, 1988), and concurrent with or before the arrival of terranes of the central group.

The central group extends south to the Denali-Fairweather fault (Figs. 23 and 24). It is a mix of large and small terranes of varied origin and mid-Cretaceous assembly. There is no direct evidence for or against latitudinal transport of the central group terranes before the Late Cretaceous, and paleomagnetic data suggest no latitudinal shift since the Late Cretaceous (Harris and others, 1987).

The Southern Alaskan superterrane (SAS) is between the Denali-Fairweather fault and the Border Range fault (Fig. 23). It includes the Wrangellia and Alexander terranes that continue west from the Insular superterrane of Canada, together with other terranes (Grantz and others, 1991). The SAS underwent large northward transport relative to North America after Late Triassic time (Stone and others, 1982), arriving at its current latitude, perhaps in discrete fragments, at times from Late Cretaceous to Eocene (Panuska, 1985; Thrupp and Coe, 1986).

The southernmost group of terranes, those south of the Border Ranges fault (Fig. 23), compose the forearc of today's active margin between North America in Alaska and the Pacific plate (Von Huene and others, 1985; Von Huene, this volume). The forearc has broadened progressively southward from its northern edge by accretion mainly of sediment from early in the Cretaceous to the present. Moreover, in Paleocene time, the forearc edifice evidently translated northward relative to North America together with parts of the SAS. The velocity of these terranes relative to North America, however, may have been less than that of the adjacent oceanic Farallon (or Kula) plate because convergence was probably widely distributed in the plate boundary zone. The forearc probably arrived at its current position by the end of Eocene time, after which it broadened by progressive accretion.

Two major post-accretionary structures affect much of Alaska. One is a set of faults of very great trace length that are west-opening splays from the margin-parallel faults of the western margin in Canada (Figs. 23 and 24). These are right lateral or right transpressional and of Late Cretaceous and Cenozoic age; some are still active (Grantz and others, 1991). Such faults in part bound terranes and terrane groups and accommodated principal displacements of terranes. Some of them, however, cut through terranes and have caused further division, dispersion, and westward transport relative to North America of terranes after they reached Alaskan latitudes. The second structure is an oroclinal bend of terrane groups in southern Alaska and the Aleutian Trench about a northerly axial trace through the Gulf of Alaska (Fig. 23). Paleomagnetic declinations indicate counterclockwise rotation of the western relative to the eastern limb of the orocline (Grantz and others, 1991) at a time between Late Cretaceous and middle Eocene. The orocline may be a plate boundary zone deformation of the Paleogene collision between Eurasia and North America in Siberia (Fig. 1) (Coe and others, 1985).

Alaska seems to have attained its current configuration by the end of the Eocene, and since then, to have had a southern active margin consuming plates of the Pacific basin below its mainland. The evolution before the Eocene is less certain.

An early event, perhaps the earliest, was the Mesozoic development of a northern margin to North America in Canada and Alaska (see below) and the oceanic spreading of the Canada basin (Fig. 24). This apparently caused the translation of the northern group of terranes southward from a position in the present Canada basin (Moore and others, this volume; Sweeney and others, this volume).

The migrating northern group of terranes probably collided with PNA at or near the northern end of its western margin, now in the northern Yukon Territory (Fig. 24). This event may have caused the excision of parts of outboard Paleozoic North America, which then may have extended farther west into Alaska.

The northwest-striking Cordilleran foreland fold and thrust belt of southern western Canada curves to a westerly strike in the Yukon Territory (Figs. 22 and 23). In the central Yukon, it intersects another Cretaceous and Cenozoic fold and thrust belt (the northern Yukon belt) that srikes north to northeast and extends north to the shelf of the Canada basin (Figs. 23 and 24) (Norris and Yorath, 1981; Norris, 1985). The two fold and thrust belts are at least partly contemporaneous, but their kinematic relations are not yet clear. The position of the northern Yukon

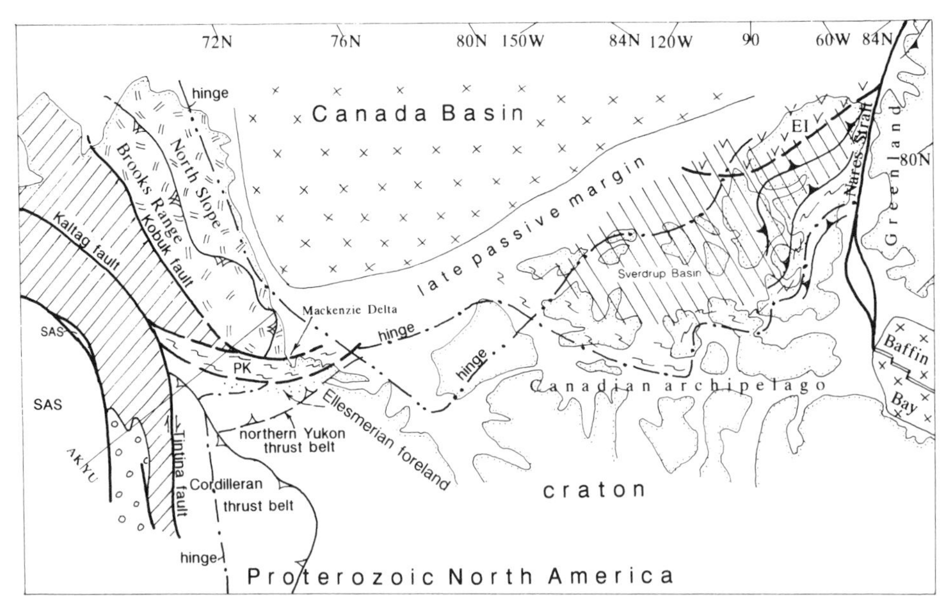

Figure 24 (Legend on facing page). Margin of northern North America in Canada and Alaska. EI, Ellesmere Island; SAS, Southern Alaska superterrane; PK, composite Porcupine and Kandik terranes; AK, Alaska; YU, Yukon.

belt implies that related to collision and progressive contraction of the northern group of terranes with PNA. The kinematics of the two thrust belts, however, are almost certainly related to the large contemporaneous right-lateral slip on the Tintina fault that is the southwestern boundary of the thrust belts (Fig. 23).

The northern group of terranes in the Brooks Range is in a north-vergent thrust complex of Late Jurassic through Tertiary age, to which the North Slope is the foreland (Fig. 24). The foreland includes a major foreland basin of Cretaceous and Tertiary age and blind thrusts that extend well north of the mountains. The thrust complex of the Brooks Range evolved in a mainly breakforward sequence, except for some out of sequence thrusting that brought internal nappes relatively forward, local duplexing, and generation of culminations. The thrusting achieved large contraction (500 km; Oldow and others, 1987; 800 km; Grantz and others, 1991).

The tectonostratigraphy of the terranes and nappes or subterranes of the northern group provide important clues to pre-Mesozoic continent-ocean transitions and to the collisional character of the Brooks Range thrust complex. The highest and southernmost terrane of the complex (Misheguk Mountain; Grantz and others, 1991) was emplaced as a hot ophiolite of probable island-arc origin in the Jurassic above the Angayucham terrane, an accretionary prism of oceanic sedimentary units and lava flows of Jurassic to Devonian age. Below the Angayucham are nappes of the extensive Arctic-Alaska terrane, which underlies most of the Brooks Range and extends north across the North Slope to the Cretaceous northern margin with the Canada basin (Fig. 24). The Arctic-Alaska terrane (actually a superterrane) contains PNA plus Paleozoic terranes and was contiguous with Arctic Canada in early Mesozoic time (see below). The nappes of the Brooks Range indicate a preexisting structure of the Arctic-Alaska terrane downward (northward) in the stack, from pre-Mississippian terranes to North American early passive margin. The Angayucham terrane can be interpreted as a Jurassic forearc scraped from a possibly extensive ocean basin of early Paleozoic spreading age (Fig. 25) (Grantz and others, 1991). The postulated Angayucham oceanic lithosphere was subducted southward at a Jurassic convergent boundary of which the Mishiguk Mountain terrane represents the magmatic arc. The northern end of the Angayucham slab was evidently attached to the Arctic-Alaska terrane, perhaps at a Mississippian suture, which also entered the subduction zone but became jammed shortly thereafter in Jurassic or Early Cretaceous time. Further convergence was taken up at least in part by the contraction of the Brooks Range.

The southern side of the northern group of terranes thus can be taken as a collisional zone. This zone may be substantially displaced from its original locus.

The central group contains terranes of diverse origin (Silberling and others, 1987; Grantz and others, 1991): continent, including known or suspected Proterozoic crust (Nixon Fork, Ruby); oceanic lithosphere of diverse age ranges (Livengood—Paleozoic; Toznita—Mississippian and Triassic; Kanuti—Jurassic); and accretionary complexes of sediments of arc and/or continental provenance and Paleozoic to Cretaceous depositional ages (Figs. 23 and 24). The central group also contains the large Yukon-Tanana terrane, which extends well southeastward into Canada (Fig. 22) where it is the innermost of terranes at the western margin.

The continental Nixon Fork terrane contains lower Paleo-

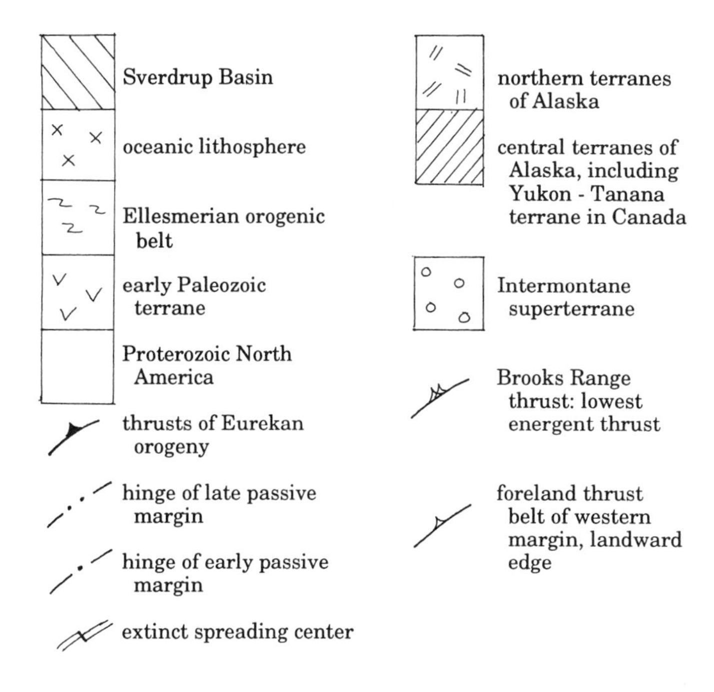

zoic platformal strata above basement. It is thought to have been overridden from the south by the Yukon-Tanana and Livengood terranes in Cretaceous time (Grantz and others, 1991). It is a question whether the Nixon Fork is a sliver of PNA projecting westward between the collision zones of western Canada and the northern group of terranes in Alaska, or is far traveled. The platformal origin of its lower Paleozoic strata suggests the latter, because, by virtue of its position, a parautochthonous Nixon Fork would be expected to contain a thick Paleozoic passive margin wedge (Fig. 24). The continental Ruby terrane differs from the Nixon Fork; its stratal content suggests that it could have been derived from an end-Proterozoic passive margin, and its Devonian rift features and plutons suggest a northern North America derivation.

The central group of terranes was interpreted by Grantz and others (1991) to be a Cretaceous north-vergent imbricated stack that overran oceanic lithosphere of the Koyukuk basin, the northern side of which was underthrust by the Angayucham slab. The multiple occurrence of oceanic above nonoceanic terranes and the lack of evident major magmatic arcs of Jurassic or Cretaceous age in the central group suggest a general collisional origin. The many Mesozoic convergent plate boundaries that must be represented in the central group thus can be interpreted to have existed in the ocean basins, which separated the nonoceanic terranes and whose sediment fills now occur as terranes within the imbricated array. If island arcs existed in the ocean basins, they must have been small and/or eroded upon obduction.

The central group probably records the accumulation of tectonic fragments of highly varied protolith age and origin, perhaps amalgamated over much of Phanerozoic time. The right-slip faults and active seismicity of central Alaska (Fig. 4) imply that the translation and rearrangement of terranes continues today. The structural relation of the central group and the SAS is uncertain, in part because the Denali fault system, thought to have about 500 km of right slip, separates the two.

The SAS is an amalgam of terranes of Paleozoic arc and Mesozoic arc and rift-related rocks with generally distinctive tectonostratigraphy (Jones and others, 1977; Nokelberg and others, 1985), as in its counterpart in western Canada, the Insular superterrane (Fig. 23). There are several important properties of the SAS. First, geologic evidence (intrusions, overlapping strata) indicates local linkage among blocks of the SAS main terrane constituents by Late Jurassic time and local contact of the SAS with both adjacent terrane groups (central and southern) by Late Cretaceous time. Second, the SAS is bounded by faults of pre-Eocene origin that are at least locally still active. Third, the SAS as a belt has apparently not exchanged terrane fragments across its boundaries since mid-Cretaceous time or before. Fourth, a continental magmatic arc, now the active Aleutian-Wrangell arc (Fig. 23), has occupied the SAS since the Eocene. Fifth, paleomagnetic inclinations indicate large northward transport of Mesozoic rocks of SAS relative to North America, as in the Insular superterrane, but differences of up to 30° of paleolatitude exist among volcanic rocks of Paleocene age (Moore and others, this volume). Those Paleocene strata with little or no latitude anomaly occur at the landward reach of SAS, whereas those with moderate and large anomalies are on the seaward side.

This set of properties is not easy to reconcile in a kinematic history. Perhaps the most apt solution is that right-lateral boundary-parallel motion strongly rearranged the SAS after attachment of its landward edge, as alluded to by Moore and others (1983) and Thrupp and Coe (1986). One might conceive that the SAS and Insular superterranes lay well offshore and were underridden by Pacific basin slabs in the Cretaceous, then closed with respect to North America and attached at latitudes that include the present and more southerly ones in the Late Cretaceous. During and after first attachment, the SAS and forearc as a whole took up heterogeneous simple shear on vertical planes in the right-oblique boundary zone between the Pacific basin and North America plates. In this story, the simple shear zone migrated seaward with time, permitting large northward transport of the seaward reach of the SAS relative to the landward in the Paleocene, and then transferring all the shear seaward of the SAS by Eocene time. The hypothesis requires the postulated simple shear planes to have always paralleled the SAS boundaries, even as the SAS changed orientation around the Gulf of Alaska (Fig. 23). The structures in the SAS that could have taken up large boundary-parallel simple shear have not been identified.

Active Margin. *Introduction.* Considered here are margins of western North America where the downgoing oceanic slab at a convergent boundary belongs to a plate adjacent to the North America plate and where the slab moves against the contiguous continent, causing the development of a continental magmatic arc and related phenomena. Such margins have also been classed as Andean, employing western South America as the global example. Whereas modern active margins are generally easy to spot by

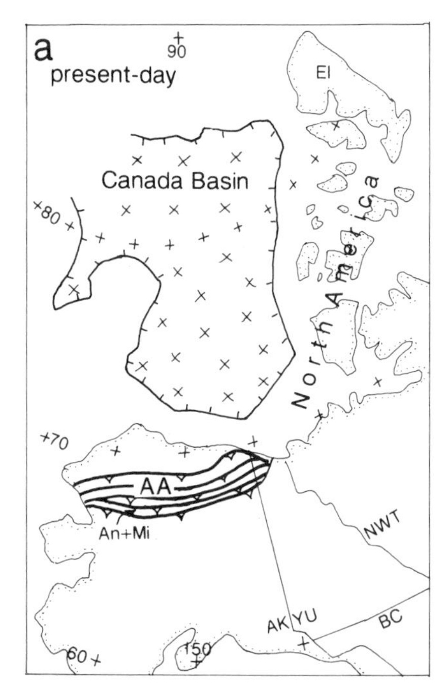

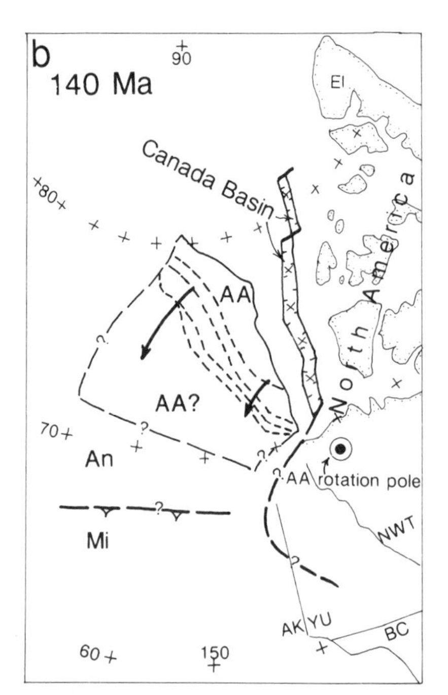

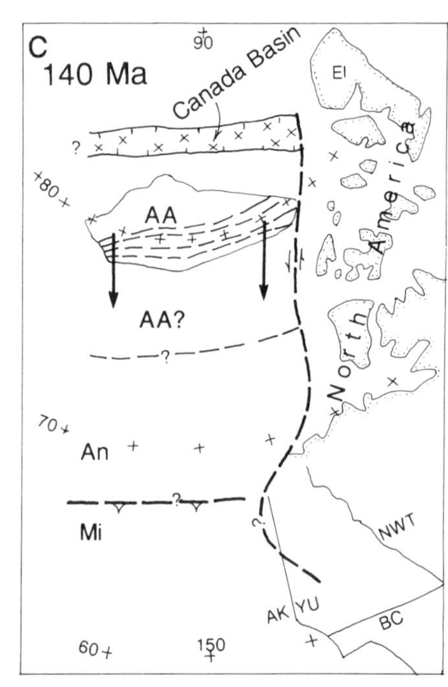

Figure 25. Alternative hypotheses of provenance and path of Arctic-Alaska terrane (AA) and related opening of the oceanic Canada Basin (x pattern) in Early Cretaceous time. a. present-day geography showing approximate subaerial perimeter of AA terrane and traces of major thrusts. AK, Alaska; YU, Yukon; NWT, Northwest Territory; BC, British Columbia. b. Rotation model, showing rotation pole and restored orientation of present-day perimeter of and thrust traces (short dashes) in AA terrane; longer dashed lines on southern AA terrane indicate possibly greater initial extent of AA if thrust imbrication began during drift; heavy arrows in AA indicate velocity paths of AA relative to North America; An is Angayucham ocean basin; Mi is Micheguk island arc; position shown of Mi is arbitary; it may have been south or north. C. Strike-slip model; notations as in b.

seismicity and volcanoes, ancient ones are not easy to distinguish from other convergent boundary configurations. The principal criterion is a plutonic-volcanic belt of sustained magmatism (tens of millions of years) that may be taken as a continental arc. This interpretation may be supported by backarc-foreland phenomena that differ from those of collisional margins. The forearc-terrane zones of an active margin are between a magmatic arc and oceanic slab, whereas at collisional margins they are directly emplaced against PNA or older attached terranes. There are uncertainties in the interpretation of ancient active margins in both western and eastern North America: whether a set of plutons is slab related or anatectic, the duration of magmatism, whether an existing continental arc is autochthonous or a terrane, whether continental arcs have been excised from some stretches of the margin, and whether an arc in a terrane predates or postdates attachment to North America.

In general, the age of beginning of active margin tectonism in western North America and the certainty of that age both seem to diminish from south to north. An active margin first developed in the Permian in southern California and in the Middle Triassic at central Great Basin latitudes. In the U.S. Pacific Northwest, a continental arc certainly existed as early as Late Cretaceous time and exists today above the Juan de Fuca slab (Fig. 20a). The Mesozoic evolution of active versus collisional margins in most of western Canada, however, is less certain. Southern Alaska includes an active margin that has existed approximately in place at least since the Eocene.

Continental arcs. The Permian and Early Triassic active margin of southern North America exists in southern California as far northwest as the Mojave Desert, as indicated by scattered plutons in PNA (Fig. 20a) (Miller and Bradfish, 1980; Walker, 1988). These were succeeded by younger sets of pluton, for which dating (zircon U-Pb and hornblende K-Ar) indicates Jurassic and Cretaceous intrusion ages, and by Jurassic and Cenozoic volcanism. This long-lived stretch of active margin probably began with the excision of the postulated corner of PNA in southern California between the Paleozoic rift-dominated western margin and the transform-dominated southern margin.

In the western Great Basin (Fig. 20a), which is thought to have been parautochthonous since Early Triassic time (Speed, 1978), a continental arc emerged in the Middle Triassic (Dilles and Wright, 1988). This stretch of magmatic arc continued sporadically to the Neogene with possible lapses in parts of the Early and Late Cretaceous and Paleogene (Saleeby and others, this volume). The late Neogene extinction of this arc is apparently related to the transition in the active margin to highly oblique motion and with the northward migration of the Mendicino triple junction, followed by the propagating San Andreas ridge-ridge transform. The Triassic and Jurassic continental arc of the western Great Basin includes abundant volcanic rocks that emerged on the eastern side of the arc and perhaps on the arc crest. The western edge of this arc may have been excised, or it may exist in the Sierra Nevada, engulfed in Cretaceous plutons.

There are huge granitic batholiths in the central and western Sierra Nevada and in the eastern Peninsular Ranges of California that were produced by continental arc magmatism in Jurassic and Cretaceous and Early Cretaceous time, respectively (Saleeby and others, this volume; Silver and others, 1979). Both ranges are probably out of place relative to their sites in the Mesozoic active margin (Tesierre and Beck, 1973; Frei, 1986; Schweikert and Lahren, 1990; Argus and Gordon, 1991; Howell and Vedder, this volume). Both probably incorporate slices of the Mesozoic active margin that were transferred northwest along the margin well inboard in the mainly right-oblique, mid-Cretaceous to Cenozoic plate boundary zone. The pre–mid-Cretaceous plutons of these batholiths may have undergone earlier southeasterly transport, according to Avé Lallemant and Oldow (1988), who suggested large left-oblique motion in the plate boundary zone in mid-Mesozoic time. More generally, it seems probable that active or recently active continental arcs are likely foci for concentration in an active margin of margin-parallel motion because of thermal weakness at all lithospheric depths and possible existence of quartzose rocks to deep levels. If true, the outboard halves of continental arcs are prime candidates to be terranes.

In the U.S. Pacific Northwest, the continental arc is Late Cretaceous (Lund and Snee, 1988) and exists chiefly in PNA (Fig. 20a). The older active margin of the Great Basin may have originally extended north into the Pacific Northwest, been excised and drifted away, and a new, more inboard active margin generated later, when margin-normal motion became sufficient (Speed, 1983).

The large chains and regions of Mesozoic batholiths in western Canada and central Alaska are difficult to interpret for active margin processes. The Canadian Cordillera contains two long pluton chains that occupy important tectonic boundary zones: the mid-Cretaceous Coast Plutonic Complex exists in the suture of the Insular and Intermontane superterranes, and the Omineca Plutonic Belt of Jurassic and Cretaceous age is in the boundary zone of the Intermontane superterrane–Kootenay terrane and PNA (Monger and others, 1982, 1985, this volume). There are also scattered mid-Cretaceous postkinematic plutons in the western zone of the foreland thrust belt; these may or may not be comagmatic with the Omineca belt. The central group of terranes in Alaska, which continues southeast into the Canadian Cordillera east of the Insular superterrane, contains an abundance of batholiths of Cretaceous and Tertiary age (Grantz and others, 1991). These were apparently emplaced after the juxtaposition and stacking of terranes.

At issue is whether such plutonic belts in Canada and Alaska are slab induced or are anatectic and a product of partial melting of deep levels of multiply imbricated crustal rocks in collisional zones. The latter process is conceivable, if not probable, because (1) many of the terranes, near or far traveled, include an abundance of felsic sedimentary or igneous rock of varied ages that may provide plutons of compositions like those of continental arcs; (2) such plutons apparently were emplaced over short durations, in contrast to the roughly 150 m.y. recorded in the bona fide continental arc of the western Great Basin; and (3) an analog of short spurts of anatectic plutonism exists in the eastern Great Basin due to tectonic thickening of the crust well inboard of the contemporaneous continental arc (Speed and others, 1988).

The Alaskan continental arc of the present-day active margin occupies the SAS (Fig. 23). The magmatism of the modern chain of volcanoes began in Eocene time and has evolved in three major pulses of between three and seven million years. Such timing implies that magmatism related to downgoing slabs of Pacific basin plates commenced as the component pieces of the SAS and of the inner terranes of the forearc had reached their present latitudes relative to North America.

To review, some principal attributes of the Alaskan active margin are as follows. (1) The accretion of the forearc began in Cretaceous time (Moore and others, 1983) and has been active since then by southward progradation above Pacific basin plates. (2) The forearc began to suture locally to the moving SAS in Late Cretaceous time. (3) The SAS probably acted as a backstop within the forearc for margin-normal motion of rock transferred from Pacific basin slabs to the overriding edifice. (4) The Cretaceous forearc and the SAS migrated northward relative to the western edge of North America, perhaps from Mexican latitudes and perhaps mainly in the Late Cretaceous, first arriving in southern Alaska late in Cretaceous time. (5) The outer realm of the forearc, including the outer SAS, continued relative northward translation with margin-parallel simple shear of large transport lengths. (6) The position of the trench, taken as the front of the forearc, is unknown for Cretaceous and Paleogene time; both the trench and forearc, as a growing edifice, may have migrated northward together within the Pacific basin, or pieces of forearc may have migrated northward within the active margin after accretion at a trench off Canada and Alaska fixed in time near the present one. Both of these hypotheses invoke a large fraction of Pacific-NAm velocity to the outer forearc. The first hypothesis, however, implies the existence of one or more convergent boundaries between the forearc (including the SAS) and northern Alaska, taken to be the Arctic-Alaska terrane (Fig. 24), which was in place early in the Cretaceous. The second hypothesis requires no such convergent boundaries in back of the migrating forearc. The choice is unclear—the first hypothesis is supported by the wide deformed tract of the central group of terranes with Cretaceous ages of imbrication and of varied terrane origin, including oceanic. The second hypothesis is favored by the immense cross-sectional area of the forearc south of the SAS, implying that there was consumption of a great length of Pacific basin lithosphere and that this was the locus of principal convergence. The absence of a conspicuous, discrete Cretaceous continental arc in southern Alaska may also point to the second hypothesis.

Deformed foreland. The region of western PNA between the terranes and the present-day craton is here called the deformed foreland (zone 2; Fig. 7). It contains belts of different styles of deformation that developed during Mesozoic and Cenozoic intervals of convergent plate boundary tectonism (Fig. 20a). The best defined and most extensive of these is the foreland thrust and fold belt, which exists continuously from southern California to the Yukon. A seaward zone of the fold and thrust belt that locally contains Mesozoic plutons has been called hinterland by some geologists. Landward of the fold and thrust belt in much of the United States is a belt of basement-cored, reverse-fault–bounded blocks that forms the Rocky Mountains (Fig. 20b). The deformation in the three belts is mainly sequential, younging landward, but overlapping in time across adjacent belts.

The Mesozoic evolution of the deformed foreland in southern California is less evident than that to the north because it has undergone so many superposed tectonic episodes (Howell and Vedder, this volume) and, probably, excisions. The onset of folding and thrusting of Paleozoic continental strata in southern California appears clearly to be older, Permian and Early Triassic, than that (Jurassic) in the Great Basin (Walker, 1988; Snow, 1992). The difference correlates with the earlier onset of active margin phenomena in southern California than in the Great Basin. It is difficult to be sure, however, that the southern California contraction was in a foreland rather than within a continental arc.

From the Great Basin to the Arctic (Figs. 20a, 22, and 24), the fold and thrust belt is a thin-skinned breakforward imbricate stack of stratal cover of PNA of Late Proterozoic, Paleozoic, and early Mesozoic age. The belt developed between Middle or Late Jurassic and Paleogene time (Royse and others, 1975; Cowan and Potter, 1986; Price, 1981, 1986; Monger and others, 1985; Saleeby, 1986; Blake and others, 1985; Heller and others, 1986). The fold and thrust belt includes faulted foreland basin sediments, and it is bounded landward by autochthonous foreland basin fill. The fold and thrust belt occupies various positions on strike relative to the hinge of the end-Proterozoic passive margin. Its landward front is approximately coincident with the hinge in the Great Basin, well inboard of a former hinge in the U.S. Pacific Northwest, and significantly inboard of the hinge in British Columbia and the Yukon (Figs. 20a and 22).

The fold and thrust belt took up between 150 and 200 km of boundary-normal contraction within Phanerozoic strata between the northern Great Basin and central British Columbia. North and south of there, the magnitude is less (Elison, 1991). The outer realm of the deformed foreland in Canada north of southern British Columbia evolved during—and has been rearranged by—margin-parallel right simple shear of substantial magnitude (simple shear strain of 4 or more), implying partitioning of oblique slip over at least some time intervals. There is a possible periodicity in rates of thrusting of 40–50 m.y. along the entire belt, reaching maxima of three times the mean value (Elison, 1991). It is uncertain whether such periodicity is in or out of phase with rates of margin-parallel shear, but there does seem to be harmony in contraction rates and changes of plate velocity across the active margin (see below).

The degree of plutonism in the deformed foreland varies substantially with position. Plutons are abundant in the foreland of the U.S. Pacific Northwest, whereas they occur only moderately in the seaward fringe of the fold and thrust belt in Canada and hardly at all in the Great Basin. These differences may be related to the proximity of a slab in active margin configuration. In Idaho, a slab may have shallowly underridden the edge of PNA after a rim of continent, including the hinge, had been excised in the mid-Cretaceous. Such tectonics presumably caused thin-skinned contraction in the former craton that was immediately followed by plutonism (Idaho batholith) in the Late Cretaceous (Speed, 1983). In the Great Basin, in contrast, the slab was apparently at great depth (~500 km) below the thin-skinned fold and thrust belt, at least in the Mesozoic. In the Great Basin, contraction was taken up in the upper brittle zone by imbrication of thick passive margin wedge strata and by the piling up of these strata against the hinge, which acted as an inherited backstop of relatively strong and cold basement. The fold and thrust belt in Canada is in yet another configuration, complicated there by strain partitioning and by the uncertainty of slab position and orientation in the Cretaceous and Paleogene. The foreland deformation in Canada, however, clearly migrated far landward of the hinge relative to that of the Great Basin at about the same total displacement. This suggests that in Canada there was either a narrow passive margin or the prior excision of a fringe of continent.

The hinterland of the western margin is not a sharply defined concept. In Canada, hinterland was earlier used for territory now called terranes by Monger and others (1985, this volume; Kootenay and Monashee). In the Great Basin, a hinterland may exist as shown in Figure 20b, based on the occurrence within the passive margin wedge of Mesozoic plutons and of deep-seated synkinematic metamorphism in local sites of large Cenozoic unroofing (core complexes). Such characteristics were ascribed to a hinterland in the Great Basin by Allmendinger and Jordan (1984), whereas the same region was entirely called foreland by Speed (1983) and Saleeby and others (this volume). The evolution of this belt is now examined.

The Great Basin contains thick Paleozoic passive margin wedge and successor strata and terranes emplaced at the early collisional margin (Fig. 20a). These were imbricated mainly by thin-skinned thrusting in Jurassic and Cretaceous time. Such thrusting concentrated in several discrete subbelts that are ramps to the surface. The subbelts are separated by broad tracts of little deformation at shallow levels. Such tracts, however, are thought to lie above a deep detachment with probable duplexing that crossed the Great Basin from the continental arc to the foreland basin (Speed and others, 1988). That is, the entire broad passive margin wedge of the Great Basin is detached above some stratigraphic level and brittlely contracted. In cross section, the detached strata are in a giant triangle zone, flanged westward above the arc on the west and eastward against the craton on the east. It is inferred that a subjacent plastic zone of strata and transitional crust underwent greater horizontal contraction and thickened substantially (Speed and others, 1988). Magmatism of minor volume at shallow levels occurred in the thickened zone in two brief spurts, at 165 and 80 Ma (Saleeby and others, this volume), while the continental arc on its west was active and while thrusting was ongoing at places within it. This implies that the magmatism of the deformed foreland was anatectic, probably related to thickening and melting of transitional crust sufficient to create enough magma for advection to the moderate depth levels now exposed. The unroofing of the deeper levels occurred mainly during Cenozoic extension (Snoke and Miller, 1988), and not during Mesozoic thrusting. In the Mesozoic, therefore, the full Great Basin looked like a foreland, with thin-skinned brittle thrusting at shallow depths; however, this developed above a plastically deforming basement and lower cover, which upon exhumation reveals the duality of deformation regimes with depth. The region of later exhumation is what is called hinterland.

The belt in the deformed foreland of basement-cored thrust blocks (Rocky Mountains) east of the fold and thrust belt (Fig. 20b) developed during and after the most landward deformation in the fold and thrust belt, in Paleogene and Oligocene time. This thick-skinned belt occurs entirely within formerly cratonal PNA. In Wyoming, the total contraction is perhaps 50 km, and the magnitudes to the south around the Colorado Plateau are unknown. An origin given for this belt is the shallowing of the slab during the Paleogene, causing contraction and magmatism far inboard of their Mesozoic loci (Coney and Reynolds, 1977).

It is a continuing question how foreland (including hinterland) contraction along the western margin of North America related to plate boundary kinematics. In the Great Basin there seems an evident relation of long-term subduction below the continent and contraction continentward of the arc. A lag between earliest recognized arc plutonism and first contraction at least at shallow depths in the deformed foreland was between 30 and 50 m.y. Contraction was probably continuous piecewise thereafter for at least 120 m.y., until about 60 Ma at the eastern front of the foreland (the fold and thrust belt), and for another 20 m.y. after that to the landward in the basement-cored thrust belt. The lag may represent a duration for the reduction of viscosity by thickening, heating, and melting in the basement of the continental margin before strain rates became large enough to cause perceivable brittle contraction of the upper zone.

The correlation between plate velocity changes in the model of Engebretson and others (1985) with possible changes in strain rates of the fold and thrust belt noted by Elison (1991) may record the imposition of new fields of motions in the plastic basement that decayed with time.

If the Great Basin story indicates that the foreland contraction of long duration is due to an active margin configuration rather than collision, the foreland of western Canada can be inferred to have been an active margin since the Jurassic. If correct, one could infer, as did Monger and others (1982, this volume), that the inner terranes (Intermontane superterrane and Kootenay and Yukon-Tanana terranes) collided in Jurassic time, after which an active margin developed, including the Omineca plutonic belt as a continental arc. Alternatively, the relation with foreland thrusting could imply that the active margin that existed along mid-Mesozoic western Canada was excised before the attachment of current terranes as a whole late in Cretaceous time.

Forearc. The discussion of the forearc of the active margin of western North America is begun in an abstract manner with a view toward illustrating complexities. At the simplest, the forearc is the tectonic edifice that lies seaward of the continental arc. Its constitution is complex. Components are bodies of a wide range of sizes that have been transferred from adjacent plates to the North America plate above a downgoing slab. The smallest bodies are probably kilometer-wide thrust-bound packets of sedimentary rocks and the largest are arc terranes and oceanic plateaus hundreds of kilometers wide. The transfer may occur at North America's leading edge or below the forearc, underplating it after some distance of transit inboard of the leading edge with the downgoing plate. Another component is the postaccretionary sedimentary cover and magmatic rock that are parautochthonous within the forearc. The sediment cover is varied. The forearc basins that occur along the backside of the forearc, between its structural high and the front of the continental arc, have moderately long lives. Other small tectonic basins develop seaward on the forearc, and these seem to be more transient. Magmatic rocks of the forearc, generally rare, appear to form by local rifting, perhaps mainly in the form of pullapart basins, and by anatexis where the forearc is very thick or passes above an anomalous heat source, such as a mid-ocean ridge.

The forearc of western North America is mostly, if not wholly, within the present-day plate boundary zone and therefore takes up part of the motion between oceanic plates and NAm. From Mexico to the Gulf of Alaska, such motion imposes on forearcs a heterogeneous right-lateral simple shear strain together with contraction, and these vary in proportion and magnitude with position on and across strike (Fig. 2). West of the Gulf of Alaska, heterogeneous right-lateral simple shear also occurs in the forearc. The oblique plate boundary velocity fields that affect western North America today have probably been partly taken up in forearcs of the active margin since their inception in the Mesozoic. It follows, therefore, that the forearc of a given latitude today is unlikely to have accreted there except for its youngest tracts and possibly, its most inboard tract, which may have stuck to the edge of pre–active-margin North America. Moreover, the concept of margin-parallel shear in the plate boundary zone predicts that the components in a section across a given forearc are continually changing. It further implies that bodies within a forearc were not necessarily transferred entirely from adjacent plates, but may include the edge of North America as a source. Added to the transience of forearc structure and position imposed by plate boundary zone motions, there is the continual heterogeneous deformation of forearcs as a whole due to volume changes, contraction, basal shear, and progressive strengthening.

From the foregoing theory of forearc evolution, it is evident that the analyses of the history of a given forearc are difficult at best and that such difficulty increases with age of forearc. One or a set of bodies of certain rocks incorporated in the forearc by a restricted set of processes presumably becomes a terrane; for example, an island arc, a collection of seamounts, or hundreds of slices of offscraped ocean-floor strata each may make up a terrane. Such terranes later fragment and disperse, and perhaps join with fragments of other like terranes within the forearc over time. There would seem to be some minimum in the size of bodies with discretely resolvable histories. Thus, the division of a forearc into component parts is a basic problem, as is their history of motion relative to one another and to North America. Especially interesting problems are how the forearcs take up large rigid-body rotation that can be detected in them by paleomagnetic declination anomalies relative to North America (Beck, 1976, 1991), the depth-dependent structure of surfaces that take up margin-parallel simple shear, and the controls in the partition of shear and contraction.

Noted earlier, several properties in terranes of western North America, whether of forearc (active margin) or collisional origin, suggest large latitudinal transport relative to the craton: fossils (Monger and Ross, 1971; Ross and Ross, 1983), paleomagnetic inclination (Beck, 1976, 1991; Yole and Irving, 1980), and margin-parallel faults of large displacement and varied age. The large magnitude and uniformity of sense of Late Cretaceous and Cenozoic paleomagnetic declination anomalies relative to the craton also imply a consistent direction of margin-parallel transport over that time. An alternative view, that except for the Neogene, the forearc and terranes in general of western North America have had only margin-normal displacement, was expressed by Butler and others (1989, 1991). The alternative seems unlikely in light of the evidence for large transport of terranes. Moreover, it seems unlikely from the evidence that western North America has been a broad plate boundary zone for much of Mesozoic and Cenozoic time and that velocity direction across the plate boundary has been at least moderately oblique for much of that period.

The forearc in California lies between the toe of the continental slope and sutures within the Sierra Nevada and Klamath Mountains (Fig. 22) (Saleeby, 1986; Blake and others, 1985; Saleeby and others, this volume). The sutures mark the loci of Triassic accretion of sediment and seamounts from the first slab to subduct beneath western North America north of southern California. It is questionable, however, whether the Sierra Nevada and Klamath Mountains have moved significantly since early Mesozoic time and whether the sutures and batholiths they contain developed at least partly up or downcoast from their present sites. The paleomagnetic inclinations of Jurassic and Cretaceous plutons in the Sierra Nevada and Klamath Mountains indicate between 0 and 1000 km displacement relative to North America (Frei, 1986); furthermore, the Sierra Nevada block is currently moving northwestward with about a quarter of the total Pacific–North America velocity magnitude (Argus and Gordon, 1991), and at least three Mesozoic margin-parallel strike-slip faults of possibly large (hundreds of kilometers) displacement exist within the Sierra Nevada (Speed, 1978; Saleeby, 1986; Schweickert and Lahren, 1990; Kistler, 1990; Saleeby and others, this volume). It thus seems improbable that the inner edge of the forearc is in its Triassic position.

Seaward from the sutures of the Sierra Nevada and Klamath Mountains are terranes of Jurassic age of accretion whose fault boundaries were at least partly reactivated in later time. These terranes include Jurassic ophiolites and island arcs, which are respectively interpreted as products, of spreading in and collapse of pullapart basins along margin-parallel faults of the active margin (Saleeby, 1982; Saleeby and others, 1982, this volume). An alternative hypothesis is that such terranes traveled far on incoming plates (Ingersoll and Schweikert, 1986).

The western fringe of the Jurassic terranes of the foothills of the Sierra Nevada is covered by Upper Cretaceous turbidites of the Great Valley. Such strata are considered to be forearc basin deposits. The provenance of such sediment was a continental arc, the roots of which could be the batholiths of the Sierra Nevada (Dickinson, 1976; Ingersoll, 1983). To the west, the strata of the Great Valley include Cretaceous and Jurassic beds that record basinal sedimentation and deformation in the forearc before the forearc basin came into existence.

Farther west, the forearc is composed chiefly of terranes of Jurassic to Paleogene clastic strata and basalt, collectively called Franciscan, that were offscraped and underplated from downgoing Pacific basin plates. The margin-parallel transport of those terranes is probably large and varied, as implied by paleomagnetic inclinations (Alvarez and others, 1980; Tarduno and others, 1986) and by the fragmentation and redistribution of terranes by postaccretionary margin-parallel faulting (Blake and others, 1985).

The movement histories of two large sets of terranes, the Santa Lucia–Orocopia allochthon (SLOA) and the Baja-Borderland superterrane (Fig. 20a) (Beck, 1991; Howell and Vedder, this volume) are of particular interest in forearc evolution. Both have large paleomagnetic inclination anomalies relative to Cretaceous North America (Champion and others, 1986; Hagstrum and others, 1990). The SLOA includes terranes of diverse origin: continental arc, ophiolite, older collisional margin fragments, and Cretaceous sedimentary rocks. These assembled in the Late Cretaceous at latitudes of southern Mexico and migrated north in the forearc to attach against PNA in southern California in Eocene time (Howell and others, 1985). The Baja-Borderland superterrane moved northward from southern Mexico in the Cenozoic and attached outboard and south of the SLOA in southern California. Both the SLOA and Baja-Borderland were then excised together as the Alta-Baja California superterrane from PNA with the late Miocene birth of the San Andreas system of faults (Fig. 5). An exception is a fringe of the SLOA left on the margin of PNA (Tujunga and Baldy terranes; Fig. 20a). Moreover, the excision included a large sliver of presumed PNA (Salinia terrane) (Ross, 1985) with the Alta-Baja California (composite) superterrane. The Alta-Baja California superterrane has migrated about 350 km northwest within the forearc relative to its pre–late Miocene position. It is important to note that this giant composite superterrane is undergoing major reorganization by margin-parallel right-slip faults of the San Andreas system within it, as well as by thrusting and wedging (Namson and Davis, 1988). If the San Andreas system continues for a long time to come, the simple shear imposed on the Alta-Baja California superterrane may stretch it out, probably in fragments, over a length of margin tens of times longer than its original (Miocene) length. This kinematic fate may, therefore, be like that of the SAS-Insular (Wrangellia-Alexander) superterrane at the beginning of the Cenozoic.

The forearc of the U.S. Pacific Northwest extends from the base of continental slope east to the edge of PNA, at which mid-Cretaceous collision of terranes was probably concurrent with the beginning of subduction beneath the continent. The forearc evidently prograded seaward by accretion in Late Cretaceous and Paleogene time and became the basement for the Cascade continental arc (Fig. 20a) and its backarc that evolved in mid-Cenozoic time (Cowan and Potter, 1986). The forearc has continued to accrete seaward to the present (Fig. 22). The principal constituents are sedimentary rocks of continental provenance from the Farallon, Kula, and Juan de Fuca plates and tracts of basalt that may be seamounts and/or oceanic plateaus. The existence of margin-parallel transport in this segment of the forearc is indicated mainly by clockwise paleomagnetic declination anomalies in terranes (Simpson and Cox, 1977). Faults of large trace length cutting the forearc have not been recognized, but margin-parallel faults of Late Cretaceous and Paleogene age (Straight Creek, Tintina, and others; Fig. 22) may extend south from Canada below the widespread Neogene Cascade cover of southern Washington and Oregon.

In western Canada, it is difficult to discern what part of the terrane belt is forearc to an active margin and what part is collisional and pre-active margin, as noted earlier. At one extreme, the Insular superterrane and terranes seaward of it may compose a Late Cretaceous and Cenozoic active margin, as suggested by Monger and others (1982, 1985, this volume). In this hypothesis, the extensive Insular plus SAS composite superterrane arrived at the edge of North America on a downgoing oceanic plate, detached and accreted to the forearc, became dismembered by margin-parallel faults, and migrated relatively north in proportion to distance from the continent. Such faults also existed in the continental arc and foreland of this model.

An alternative model for western Canada is that all terranes from the Insular superterrane eastward arrived together late in the Cretaceous at a collisional margin with North America, inboard of a subduction zone for Pacific basin plates. In this case, the forearc is restricted to terranes on the seaward flank of the Insular terrane (Cowan and Potter, 1986; Monger and others, 1985).

The forearc of southern Alaska is as wide as 300 km between its backstop, the Southern Alaska superterrane, and the downgoing Pacific plate (Fig. 23). The forearc is perhaps a northward continuation of the narrow belt of Cenozoic (and latest Cretaceous) terranes that extends well south along western North America (Fig. 22). The Alaskan forearc consists mainly of accreted sedimentary rocks whose age range of deposition youngs oceanward from early Mesozoic at the inner boundary to Quaternary at the modern trench. The forearc is divided among terranes (including Chugach and Yakutat; Fig. 23) by the age range of accretion, proportions and types of sediment and of basalt, and by deformation and metamorphism (Moore and others, 1983; Von Huene and others, 1985; Grantz and others, 1991; Moore and others, this volume).

Sediments enter the forearc at the Aleutian Trench by accretion of the top fraction of strata from the Pacific plate; by underplating in duplexes the base of the forearc from the residual strata on the downgoing plate; and by deposition on the top of the forearc. The main deposition is in a forearc basin at the rear of the forearc. The accreted strata consist of contemporaneous coarse trench wedge fill and of sections of turbiditic and hemipelagic strata that may have accumulated on the Pacific plate over its lifespan. The turbidites include continent-derived sediment, probably shed westward from sources in Mexico to British Columbia during the northward movement of Pacific basin plates relative to North America in Late Cretaceous and Cenozoic time. The turbidite blankets evidently extended greatly seaward, probably crossing filled trenches at the active margins south of Alaska.

As noted, the history of movements of the southern Alaska forearc is very complex. The early-accreted (Paleocene and Late Cretaceous) parts (Chugach terrane) evidently attached to the SAS backstop at varied distances well south of southern Alaska. These moved north by arc-forearc translation and/or by intra-arc right-lateral simple shear to their current configuration. During the Cenozoic, accretion continued at the front of and below this translating and deforming edifice. The kinematics of the forearc also include internal motions such as progressive horizontal contraction and probable excisions at the base by reattachment of forearc to the downgoing slab (Von Huene, this volume). They also include progressive unroofing at the structural high near the rear of the forearc due to thickening and underplating.

The Yakatat terrane is an extensive chunk of oceanic crust that began to detach from the Pacific plate and to imbed in the forearc at about 6 Ma (Fig. 23).

The forearc of southern Alaska provides a probable modern example of the history of ancient forearcs that populate the terrane belts throughout western North America. The main points are as follows. First, ocean-floor sediment, including much of continental provenance, and lava were swept up from thousands of kilometers of sea floor into a compact edifice. Second, the edifice itself was greatly translated and elongated by margin-parallel shear as it grew by processes of accretion. Third, the edifice stabilized ultimately and has grown most rapidly in a configuration normal to relative plate velocity. The implication of the third point is that forearcs that grow at plate boundaries with high obliquity probably migrate as edifices along the margin, perhaps to a site where the obliquity is a minimum.

Northern North America

Introduction. The northern margin of the North America continent is bordered by the oceanic Canada basin of the Arctic Ocean off the Canadian Archipelago and Alaskan North Slope (Fig. 24). The margin evolved in Phanerozoic time (Fig. 26) as the Innuitian province (Trettin and Balkwill, 1979). It includes an end-Proterozoic passive margin, early Paleozoic to Devonian active and collisional(?) margins, and a late passive margin, which exists to the present. Rifting of the late passive margin began either in mid-Mesozoic or early Carboniferous time. Drifting and sea-floor spreading in the Canada basin began in the Early Cretaceous (Grantz and May, 1983; Sweeney, 1985; Sweeney and others, 1986, this volume).

The northern margin also includes an extinct plate boundary zone of northern Baffin Bay and Nares Strait (Fig. 24), in which Greenland moved with the Eurasia plate away from NAm in Late Cretaceous and Paleogene time before the onset of spreading in the North Atlantic.

An important controversy exists whether the late passive margins of the Canadian archipelago and of the Alaskan North Slope are conjugate rifted margins (Moore and others, this volume; Sweeney and others, this volume), or whether they are a right-angle rift and transform system (Avé Lallemant and Oldow, 1988). Figure 25 depicts the alternative hypotheses.

Early passive margin. The end-Proterozoic passive margin is indicated by lower Paleozoic facies changes and mafic dikes in the northern realm of the Canadian archipelago (Fig. 24). Platformal cover of the craton grades northward across a hinge to a north-thickening passive margin wedge (Fig. 24). The outer wedge includes Silurian and Lower Devonian clastic rocks with northerly provenance (Trettin and Balkwill, 1979).

The deposition of the early passive margin wedge was apparently interrupted at one or more times in the early Paleozoic by the collisions of terranes now exposed in the northernmost Canadian Archipelago (Figs. 24 and 26). The early passive margin of the Archipelago was fully disrupted by the development of an active margin in the Ellesmerian orogeny, which occurred or peaked in Late Devonian time (Fig. 26). Thus, the end-Proterozoic passive margin of northern North America was short-lived, like that of eastern North America, but in contrast with the relatively long lived western margin.

In the westernmost Canadian Archipelago, the early passive margin hinge is crossed perpendicularly by the hinge and west-thickening sedimentary wedge of the late passive margin (Fig. 24). It is unknown whether the early passive margin continues on a westerly trend in the subsurface to a truncation at the continent-ocean boundary of the Canada basin, or whether it bends to a

NORTHERN NORTH AMERICA

age (Ma)		central Alaska	northern Alaska	Canada
	Neogene	collisional	passive	spreading (Baffin Bay) and transpression (Eureka)
	Oligocene			
	Eocene			
	Paleocene			
100	Cretaceous			passive: sea floor spreading (Canada Basin)
	Jurassic	proto-Pacific basin		passive: rifting (Sverdrup Basin)
200	Triassic			
	Permian			
300	Pennsylvanian			
	Mississippian			
	Devonian	?		active (possibly collisional) (Ellesmerian)
400	Silurian			
	Ordovician			
500	Cambrian			passive
600	Proterozoic			Rodinia

Figure 26. Succession of Phanerozoic margin types in northern North America.

southerly trend below the late passive margin deposits. The latter trend is shown in Figure 25, according to inferences by Dixon and Dieterich (1990).

West of the Canadian Archipelago on the mainland in the northern Yukon Territory, strata representing the end-Proterozoic passive margin of PNA are widespread (Fig. 24) (Norris and Yorath, 1981; Norris, 1985). They comprise thick upper Proterozoic clastic rocks and lava that are probably synrift and Cambrian to Devonian carbonate and shale that are probably a postrift, west-thickening passive margin wedge. Such strata may occupy a strand of little disturbed PNA, but they occur chiefly in Cretaceous and Cenozoic thrust imbricates of the deformed foreland (Fig. 24) (Norris, 1985).

It is unclear whether the early passive margin strata of the northern Yukon are parautochthonous and reflect end-Proterozoic rifting and drifting approximately in place. Alternatively, they may have arrived as terranes before Cretaceous thrusting, and the early Mesozoic edge of PNA in the northern Yukon may have been a cratonal truncation. The second idea comes from the existence of terranes containing similar rocks to the west near the

Alaska border (Howell and Wiley, 1987) and from the interpretation by Norris and Yorath (1981) that strike-slip faulting preceded thrusting near the boundary of the deformed foreland and the craton. The evidence that an inner fringe of the passive margin wedge adjoins on the craton in the northern Yukon (Norris, 1985) is more influential, and I assume that the wedge is nearly in place: its trend north of the Yukon shoreline is uncertain.

The end-Proterozoic tectonics of the early passive margin of northern North America are not easy to decipher. In the Canadian Archipelago, the early passive margin has been extensively modified by convergent-boundary tectonics and is poorly exposed. The much greater thickness of lower Paleozoic strata over an apparently broad region north of the hinge compared to that on the craton suggests rift domination. The lithology and thickness of upper Proterozoic and lower Paleozoic strata in the northern Yukon make it clear that they evolved in a rift-dominated margin. It seems probable, however, that a corner of end-Proterozoic PNA existed between the apparently west-facing passive margin in the northern Yukon and the north-facing one in the Canadian Archipelago. This possible corner may have been excised in the migration of the Arctic-Alaska terrane and the opening of the Canada basin (Fig. 25). The Yukon margin may have been a continuation of the rift-dominated western passive margin and may have differed substantially in tectonics from the margin in the Archipelago. To conclude, the role of the early northern margin in the breakup of Rodinia (Fig. 10) is uncertain.

The continental Arctic-Alaska terrane of the Brooks Range and North Slope of Alaska (Fig. 24) includes subterranes that were derived from the early passive margin and the Ellesmerian active margin at more northerly sites of northern North America (Grantz and others, 1979, 1991). Competing hypotheses, however, present substantially different restored positions of the Arctic-Alaska terrane (discussed below and in Fig. 25).

The Arctic-Alaska terrane is overlain in the southern Brooks Range by the Angayucham terrane, a Jurassic accretionary wedge of oceanic rocks of known Early Devonian through Jurassic ages (Grantz and others, 1991). The Angayucham was probably accreted from now-subducted oceanic lithosphere that was Devonian or pre-Devonian. A pre-Devonian age is permitted by the observation at the fronts of modern accretionary wedges that only an upper fraction of the sedimentary sequence is scraped off (Moore and others, 1988). The role of the Angayucham lithosphere in the evolution of the northern margin is an interesting problem. There are three conceivable origins. First, it might have been a product of the end-Proterozoic breakup of Rodinia and spreading at the edge of the early passive margin. Second, it might have been a product of early Paleozoic rifting and spreading adjacent to or within the early passive margin: this hypothesis opens the possibility of an intermediate passive margin phase in northern North America. Third, it may have developed far from North America and been translated to an adjacent position in mid-Paleozoic or later (Ellesmerian time or later). The first idea seems denied by the expected consumption of oceanic lithosphere of the early passive margin at the Late Devonian Ellesmerian active margin or pre-Ellesmerian collisional margin. The second hypothesis is also denied by the expectation of collapse of the ocean basin by Ellesmerian orogeny. The third origin is permissible and discussed below.

Active and collisional(?) margins. The earliest evidence of tectonics associated with a convergent plate boundary is in northern Ellesmere Island (Fig. 24), where there is a tectonic assemblage of Ordovician felsic and basic volcanic rocks, granite, and ophiolite. The deposition of Silurian and Lower Devonian flysch from this source within the lower Paleozoic passive margin wedge testifies to early Paleozoic emplacement of the exotic components of this northerly terrane (Trettin and Balkwill, 1979). It is not clear whether the Ordovician rocks of northern Ellesmere Island are a terrane at a collisional margin or mix of terranes and arc rocks at an active continental margin.

Volcanic rocks of probable arc origin and lower Paleozoic age also exist locally in the Arctic-Alaska terrane (Moore, 1987). These may suggest a brief active margin or collision of an exotic arc in early Paleozoic time and may correlate with Ordovician rocks on Ellesmere Island.

The Ellesmerian orogeny deformed the early passive margin of the Canadian Archipelago in Late Devonian and Early Mississippian time, transforming the margin to an active one (Fig. 24). The passive margin strata were folded and thrust with southerly vergence, widely intruded by granitic rocks, and metamorphosed (Trettin and Balkwill, 1979; Sweeney and others, this volume).

In the northern Yukon Territory, the Porcupine and Kandik terranes, which occur between the deformed foreland and the Arctic-Alaska terrane (Fig. 24), contain lower Paleozoic passive margin strata and earlier synrift deposits (Howell and Wiley, 1987). Parts of this terrane pair underwent contractile deformation and intrusion by granite in the Devonian (Norris and Yorath, 1981), almost certainly an effect of the Ellesmerian orogeny. The Porcupine and Kandik terranes may be of local origin or may have accompanied the Arctic-Alaska terrane during part or all of its Mesozoic displacement. Both the deformed foreland and the Porcupine and Kandik terranes contain Late Devonian and Early Mississippian granite-bearing clastic rocks that underwent continentward sediment transport and probably accumulated in foreland basins (Nilsen and others, 1976; Gordey and others, 1987; Howell and Wiley, 1987). This orogenic sediment wedge implies that the Ellesmerian orogeny occurred in the northern Yukon, whether or not the Ellesmerian belt in the adjacent terranes is parautochthonous.

Within the Arctic-Alaska terrane of the Brooks Range, a Late Devonian angular unconformity has been recognized as evidence of deformation correlative with that of the Ellesmerian orogeny (Grantz and others, 1991). In the North Slope (Grantz and others, 1991) and in the northeastern Brooks Range (Oldow and others, 1987), Devonian deformation occurs below the regional unconformity. Oldow and others (1987) found evidence for north over south shear and suggested that tectonic transport of Ellesmerian age is southward all across the Canadian and Alaskan northern margin.

The plate tectonics of the Ellesmerian orogeny are thought to be the convergence, perhaps left oblique, of some or all of the Siberian shield with the early passive margin of North America (Grantz and others, 1991; Sweeney and others, this volume). If so, the climax was a collision. The plate closure preceding the collision of the continents, however, was accommodated by the descent of a slab below North America, as implied by the widespread granitic intrusions. The margin was thus an active one.

The timing and plutonism of the Ellesmerian orogeny suggest a possible origin for the extensive Yukon-Tanana and Kootenay terranes of the western margin. Those terranes are apparently accretionary complexes that include much lower Paleozoic clastic sediment of continental provenance and that underwent Devonian deformation and plutonism. The terranes may have originated in the Ellesmerian forearc and then later swept south along the western margin of North America during the sustained phase of left-lateral transport envisioned by Avé Lallemant and Oldow (1988).

Late passive margin. The Mississippian and younger development of the northern margin (Fig. 26) can be separated into the (1) early formation of the Sverdrup basin, (2) the continued rifting of the continental Sverdrup basin and the opening of the oceanic Canada basin, and (3) the related Cretaceous and Cenozoic tectonics of northern Baffin Bay, the Nares Strait, and the Eurekan orogeny of the eastern Canadian Archipelago (Fig. 24).

The Sverdrup basin is an intracontinental sedimentary basin that evolved from Mississippian to Late Cretaceous time in three evident pulses of rifting and diking (Balkwill, 1978; Sweeney and others, this volume). The basin developed above the Ellesmerian orogenic belt of the Canadian Archipelago (Fig. 24) and contains up to 13 km of clastic sedimentary rocks of continental provenance, together with evaporites and basalt.

The early Sverdrup basin developed in left-stepping rift basins that were intruded by basalt dikes from Mississippian to Early Permian time. The rifts are reactivations of Ellesmerian structures. The tectonics of the early basin are not known; hypotheses include the onset of regional extension related to the opening of the mid-Mesozoic Canada basin, thermal subsidence of the Ellesmerian orogenic belt, and an Ellesmerian foreland basin.

In the Jurassic, the Sverdrup basin underwent a second pulse of rifting with the development of half grabens, dikes, and salt intrusions. Late in the Jurassic, rifting jumped northwest to the northwestern edge of the Sverdrup basin, and the oceanic Canada basin lithosphere was emplaced against this margin. The age of spreading of the Canada basin is probably Early Cretaceous, according to indirect evidence; magnetic anomalies in the basin are poorly lineated (Grantz and others, 1979; Sweeney, 1985; Sweeney and others, this volume).

The tectonics of the opening of the Canada basin are disputed. Two main hypotheses are illustrated in Figure 25. One idea is that the Alaskan and Canadian margins are conjugate rift margins (Fig. 25a) and that the Arctic-Alaska terrane was excised from the Canadian margin and rotated counterclockwise (Fig. 25b) about a pole at the intersection of the two margins (Grantz and others, 1979; Howell and Wiley, 1987; Grantz and others, 1991; Moore and others, this volume). This idea is supported by tectonostratigraphic matches across the presumed conjugate margins and by paleomagnetic directions in the Alaskan North Slope that imply large counterclockwise rotation in mid-Cretaceous time relative to cratonal North America. An alternative idea is that the Canadian passive margin is transform dominated and left lateral (Fig. 25c), such that the excised Arctic-Alaska terrane translated southwest about 800 km to a position along the northernmost western margin of North America (Nilsen and others, 1976; Dutro, 1981; Avé Lallemant and Oldow, 1988). Evidence for top to the south tectonic transport in the Devonian deformation at a site in the Arctic-Alaska terrane (Oldow and others, 1987) supports the left-slip hypothesis. The approximate parallelism of Ellesmerian-age transport in the Arctic-Alaska terrane and in the Canadian Archipelago implies that large rotation did not occur in migration of the terrane. At issue between these hypotheses is the credibility of the paleomagnetic data indicating rotation (Avé Lallemant and Oldow, 1988), on one hand, and the pervasiveness and tectonics (possible local backthrusting?) of top to the south Devonian transport in the Arctic-Alaska terrane on the other.

The alternative origins of the Arctic-Alaska terrane illustrated in Figure 25 are drawn with an end-Jurassic intraoceanic convergent boundary approximately where northern Alaska is today. The island arc represented by the Misheguk Mountain terrane, the highest of the northern group of terranes stacked in the Brooks Range, is shown to be the leading edge of the overriding plate at this boundary. This plate boundary may in fact have been well south or north of the position drawn. The Angayucham oceanic lithosphere was downgoing at this boundary, and the Angayuchum terrane evolved as the accretionary forearc at this boundary. The Mesozoic thrusts of the Arctic-Alaska terrane developed during its collision with the Misheguk arc and Angayucham forearc. The thrusting accounts for a substantial contraction of the Arctic-Alaska terrane (Grantz and others, 1991) from its precollisional width (Fig. 25, b and c). The northern side of the Angayucham oceanic lithosphere at the Late Jurassic onset of the migration of the Arctic-Alaska terrane is shown attached to the leading edge of the Arctic-Alaska terrane.

The Angayucham lithosphere may have belonged to the incoming plate of the Ellesmerian active margin, perhaps together with pieces of or all of the Siberian continent. The arrest of subduction by continent-continent collision along part of the Ellesmerian active margin left the Angayucham lithosphere joined to North America along other parts of the margin, as well as outboard of Ellesmerian-age terranes that are now included in the Arctic-Alaska terrane. Figure 25, b and c, shows the incipient spreading of the Canada basin following the separation of the Arctic-Alaska terrane.

The eastern third of the Sverdrup basin was affected by the Eurekan orogeny from Late Cretaceous to Miocene time (Sweeney and others, this volume) (Fig. 24). This event caused about 60 km of contraction in a northwest-southwest direction, and included uplift of Campanian age at its southeastern reach on the northwestern flank of the Nares Strait (Fig. 24) and later folding and thrusting with southeast vergence in the northwestern realm of the orogen (De Paor and others, 1989). The thrusts may be reactivated Paleozoic faults (Balkwill, 1978). Farther northwest is the Eurekan hinterland, composed of Sverdrup basin rocks that underwent no deformation or thermal effects. The heave on the Eurekan thrusts is thought to diminish southwest to zero in southern Ellesmere Island, and grabens south of there imply Eurekan-age northwest-southeast extension.

The timing, position, and limited extent of the Eurekan orogeny has indicated to many workers a relation with displacements between Greenland and North America and sea-floor spreading in Baffin Bay–Labrador Sea and in the Nares Strait (Fig. 24) (Miall, 1983). The Eurekan orogeny may have begun with uplift during the northward Late Cretaceous–Paleogene propagation of the tip of the rift-drift system of the Labrador Sea and Baffin Bay and then taken up local contraction north of a Greenland–North America plate pole in middle Cenozoic time. The Eurekan orogenic region is thought to be part of a zone of right-oblique transpression that includes the Nares Strait and northern Greenland and that accommodated the drift of Greenland (De Paor and others, 1989). The Eurekan orogeny provides a fine example of kinematic complexity at the corner of a basically simple intracontinental propagating rift-transform system, wherein contractile deformation occurred locally in a system that led to a passive margin.

SYNOPSIS, INTERPRETATION, SPECULATION

Overview

North America first emerged as a discrete continent, much in its present-day form, at the beginning of the Phanerozoic Eon, about 545 Ma. The birth of the continent was caused by the development of an encircling network of divergent plate boundaries within a Proterozoic supercontinent. Such tectonics also created the North America plate, and the initial continent and plate were the same tract. Oceanic lithosphere accreted to the edges of the early Phanerozoic North American continent as it was progressively displaced from the formerly joined continental fragments and as the plate grew.

Since 545 Ma, the North American continent has undergone geographic translation and vertical-axis rotation as part of the North America plate. It has also undergone deformation by reshaping and volume change. Such deformation has taken place mostly or entirely in regions of kinematic and physical transition between North America's craton, which is its Proterozoic continental nucleus, and bordering oceanic lithospheres. Both the craton and the oceans are extensive realms of little Phanerozoic deformation, or equivalently, of nearly uniform relative velocity fields. The regions of transition are the margins of the present-day continent. They are the boundary zones between the North America plate and other plates bordering the continent where the relative plate motions have been distributed sporadically through Phanerozoic time. The view that continental margins are present and past plate boundary zones contrasts with early plate theory that embraces rigid plates and discrete boundaries.

The transitional regions contain the record of the growth and distortion of the North American continent over Phanerozoic time. They contain, moreover, the only record of Phanerozoic continent-ocean tectonic interactions before those of the present phase. To expand on these concepts, continents and ocean floors are the fundamentally different elements of the lithosphere, the Earth's strong outer shell. The oceanic lithosphere is subductible (negatively buoyant relative to the upper mantle) and circulates vertically with the mantle. The residence time at Earth's surface of a unit of ocean floor is short (<200 m.y.). Continents, in contrast, circulate only minimally or perhaps, not at all, with the mantle. They are for the most part permanent occupants of the Earth's surface. They are the ultimate repositories of nonsubductible (positively buoyant) materials transferred to them by converging oceanic lithospheres and by advections of magma from the mantle. The continents thus contain the only record at the surface of global tectonics that occurred before spreading of existing ocean floor, before about 200 Ma.

Present-day North American continent

Employing the foregoing concept of continental evolution, the present-day North American continent is defined to include all contiguous crust out to little-deformed oceanic lithosphere of the bordering major ocean basins: Pacific, Atlantic, Caribbean, and Arctic. Thus defined, the continent contains all of the contiguous regions in which plate boundary motions are occurring or have occurred, including nonoceanic terranes that move wholly or partly with adjacent plates relative to the craton. Examples of such terranes are the Alta-Baja California and Chortis. A difficulty presented by this definition, however, is the inclusion of the Eurasian continent within North America; to avoid such claims, an arbitrary nongeologic division of continents is erected between Alaska and Siberia.

The present-day North American continent is zoned tectonically in a somewhat concentric pattern from the craton out to the contact with oceanic lithospheres of the Pacific, Caribbean, Atlantic, and Arctic basins. The craton is Proterozoic North America and its cover that have been little deformed since the Proterozoic (zone 1, Fig. 7). The craton is surrounded, except in southern California, by a zone of Proterozoic North America and its cover (zone 2; Fig. 7) that has been deformed substantially by one or more tectonic events in Phanerozoic time. Zone 2 includes parautochthonous tracts that have undergone major extension (rifted margins) and/or contraction (deformed forelands) and other tracts that may have been displaced substantial distances within this zone parallel to the cratonic margin. Farther outboard is a zone of terranes (zone 3, Fig. 7) that are either fixed or currently in small magnitudes of motion relative to the craton.

The terranes consist of fault-bounded blocks that have been displaced large or unknown distances from their places of origin, which may have been but were generally not Proterozoic North America. The terranes with large current velocity relative to the craton are in the most outboard zone in western North America (zone 4; Fig. 7).

The zones of Phanerozoic deformation (zones 2, 3, and 4; Fig. 7) together constitute the continental margins and can be classed in three ways: passive, active, and collisional. The defining variables are the sign of the margin-normal component of plate velocity and the direction of dip of a slab relative to the North American continent. All three classes are generalized over a long length of margin and long duration and may include local departures from the defining characteristics. The velocity fields between the North American craton and plates bordering the North America plate have been almost entirely oblique to the continental margins. The obliquity, defined as the departure from normality of the total velocity vector, varies along the margin with changes in trend of the margin, changes in radius relative to Euler poles, and jumps across triple junctions. The sign of the margin-normal component of velocity vanishes at 90° obliquity (pure strike slip), an orientation that exists in only minor length around present-day North America.

Passive margins arise by divergent margin-normal components of plate velocity and include no slabs. The northern and eastern margins of present-day North America are passive. The width of the zone of plate boundary deformation in passive margins of present-day North America varies from 50 to 1050 km (after deformation). The strain has a positive horizontal principal stretch, which increases in magnitude oceanward. At the surface, the edge of the continent is the contact with the zone of marginal oceanic crust, which was probably the founding spreading ridge of the adjacent ocean floor. The development of a passive margin occurs as a series of events in a finite duration.

The episode of development of present-day North American passive margins at central Atlantic latitudes began with the onset of rifting in the Late Triassic and ended about 150 m.y. later with cessation of detectable subsidence. At the Labrador and Greenland passive margins, the beginning was in the Cretaceous and the duration relatively short, probably commensurate with concentration of stretching in a narrower zone than to the south. At the Arctic passive margin, the onset of rifting may have been as early as late Paleozoic; the duration is unknown, and subsidence may still be occurring.

An active margin has predominantly convergent normal velocity components relative to adjacent oceanic lithospheres, and the slabs of the converging plates dip below North America. The entire western margin of present-day North America (Mexico to Alaska) can be considered active. The velocity obliquity varies markedly along this margin, and segments of pure strike slip (Parkside zone of the San Andreas fault) and even slight divergence (Gulf of California) occur within it. The active margin is notable for its width, as great as 2500 km, although its width in historic time as indicated seismically is highly varied, from about 50 to 2500 km inboard from the continent-ocean boundary. Where the velocity direction is highly oblique, the relative plate motions appear to be taken up entirely within the continental margin. In contrast, where the velocity is close to margin-normal (southern Mexico, Gulf of Alaska), a significant proportion of the relative motion is accommodated at the continent-ocean boundary (toe of the forearc). The converging plates of oceanic lithosphere seem not to deform as part of the plate boundary zone with North America, except for downbending to subduct and at triple junctions.

The active margin of present-day western North America probably first developed in Permian time in eastern Mexico and southern California, when North America temporarily was a territory of Pangea. The first appearance of the active margin was younger northward, perhaps as late as mid-Cretaceous in northwestern Canada, implying northward propagation.

An active margin may operate continuously, even steadily, over a long duration, in contrast to passive and collisional margins. The continental girth is generally increased at an active margin by accretion of terranes and sediment in the forearc, but volume loss may occur locally by excision or by tectonic erosion at the base of the forearc.

A collisional margin has generally convergent margin-normal components of relative plate velocity and a slab that dips away from the continental margin. The slab is oceanic lithosphere that either formed in the North America plate by spreading during the development of an earlier passive margin or was attached during an earlier collision. The collision is the arrest of convergence when the leading edge of the North American continent attempts to subduct and the plate motions are taken up at new, probably outboard loci. The southern margin of present-day North America is collisional.

The process of collision causes accretion to the former edge of the North American continent. The zone of former relative motion either migrates outboard in the newly attached terranes or jumps far outboard to new or existing plate boundaries. The duration of activity of a collision is finite, measured between the first deformation of North America and quiescence after suturing.

The southern edge of the present-day North American continent is the Antillean collisional margin. There, the Caribbean plate with a leading edge surmounted by island arcs is thought to have obliquely overridden oceanic lithosphere of the North America plate. The collision has been diachronous from west to east, beginning in the Cretaceous. Upon suturing, the plate boundary zone has migrated toward the Caribbean, leaving collided elements such as southern Cuba and northern Hispaniola as fixed additions to the North American continent.

Proterozoic North America

North America first became a discrete continent in nearly its current form at the end of the Proterozoic Eon. The continent at this initial stage is called Proterozoic North America (PNA). The end points of the Phanerozoic tectonic evolution of North America are PNA at its beginning, and the present-day continent, discussed above.

PNA was blocked out in the breakup of the supercontinent, Rodinia, and was wholly or largely circumscribed by passive margins. Oceanic spreading at the edges of the new continent began between 545 and 600 Ma. The eastern and western passive margins of PNA (orientations in present coordinates) were rift dominated. The southern margin, in contrast, was transform dominated and probably extended from Georgia to southern California, and had rift segments in eastern Texas and Coahuila. In this view, the Mojave-Sonora megashear is a segment of the end-Proterozoic transform system. These long transform segments may have been partly intracontinental for long durations (50–100 m.y.) before southern PNA was fully bounded by ocean.

The orientations of transform segments of the reconstructed eastern, southern, and western margins suggest that the fragmentation of Rodinia that created PNA was highly coordinated. The newly formed plates around these margins moved about a single, distant Euler pole.

The northern (Arctic) margin of PNA is less-well understood than the others. It was almost certainly of passive type, and it developed at or near the onset of Phanerozoic time. Its characteristics suggest, however, that it was rift dominated and north facing. If so, the end-Proterozoic plate motions relative to northern North America had a different Euler pole (or poles) than that of the other margins.

PNA is embedded within terranes of present-day North America, except along Labrador and Greenland, where the modern continent-ocean boundary is also the edge of present-day PNA. Elsewhere, today's edge of PNA is the contact of the deformed foreland (zone 2; Fig. 7) with the zone of terranes (zone 3; Fig. 7) that accreted to North America through Phanerozoic time.

The extent of PNA is somewhat less today than at the end of the Proterozoic. The major loss was in the Mesozoic breakup of Pangea when northeasternmost PNA was taken off as part of Eurasia. Apart from this, the changes in PNA over Phanerozoic time can be gauged by the additions, losses, and transfers of volume in the region between the present-day craton and the outer edge of zone 2.

The craton is slightly smaller in area today than it was at the end of the Proterozoic, when it was the region bounded by the passive margin hinge. The inward migration has been by the propagation of Phanerozoic deformation inboard of the end-Proterozoic hinge and by excisions that extended into the end-Proterozoic craton.

PNA in zone 2 has undergone processes that have both decreased and increased its surface area. Zone 2 has been narrowed by margin-normal contraction in the deformed forelands that circumscribe almost all of the present-day craton. The inward total displacement of such contraction from the end-Proterozoic continent-ocean boundary in western, eastern, and perhaps northern North America has been as much as 500 km; 200 km is perhaps a hemispheric average. The principal constituent of the contractile realms of zone 2, however, is detached Phanerozoic cover to PNA, not the Proterozoic crust itself. The crust of PNA probably extends outboard of the surface perimeter of zone 2 below the terranes. It is difficult to know just how far out parautochthonous Proterozoic crust extends. I consider it doubtful that such crust exists in its original width out to an end-Proterozoic continent-ocean boundary at margins where Phanerozoic collisions have occurred, with a couple of exceptions. In much of the eastern, western, and northern margins, losses of crust from the periphery of PNA are identified or inferred. A principal process for such loss is excision of PNA and transfer into the zone of terranes by components of motion parallel to the plate boundary zone. Another process may be the subduction and breaking off of a rim of transitional crust that stays attached to the North American slab during collision.

The full extent of end-Proterozoic PNA may, however, be preserved in southern North America and in the Great Basin of western North America. In southern North America, such preservation in spite of major collision may be explained by the nonsubductibility of the southern North American crust because it exists at a transform-dominated passive margin and because of its greater strength than that of the leading edge of the colliding Gondwana. In the Great Basin, preservation might be explained by an unusual circumstance of early armoring by terranes of the end-Proterozoic edge.

Two processes have caused increases in area of PNA in parts of zone 2 over Phanerozoic time. One was horizontal extension in passive margin development in the Mesozoic in northern, eastern, and southern North America. Another was the intrusion of magma as dikes and plutons from the mantle, in both passive and active margins.

What has been the net change of PNA through Phanerozoic time as a result of all these processes? Other than the loss in North Atlantic breakup, the area of the craton has diminished only a little. In surficial extent, zone 2 as a whole has almost certainly undergone loss of area. This is because contraction and excision have strongly affected zone 2, whereas Mesozoic rifting and diking have only little extended PNA (in contrast to the zone of terranes). I suspect that there is also a net loss in subsurface extent. Zone 2 has also been reshaped by margin-parallel components of strain. To conclude, PNA today is a little smaller than at birth and its evolution has been entirely at its margins: the similarities are more impressive than the differences.

Terranes

Terranes constitute the extensive peripheral belts of present-day North America that are between PNA and oceanic lithosphere, except in the North Atlantic. Those in zone 3 (Fig. 7) are either fixed or slowly moving relative to the craton, whereas those of zone 4 are moving with velocity close to that of the adjacent plate. As noted, the terranes have displacement histories that differ relative to that of adjacent PNA and to those of their neighbors. Terranes were incorporated in North America when and where the continent's margins were plate boundary zones of active and collisional types.

Terranes consist of protoliths of diverse origin and age range. They are composed of oceanic magmatic arcs, seamounts, continental arcs, and accreted wedges of sedimentary strata derived from varied sites: ocean-floor turbidites and pelagic beds, continental rise, and intra-arc basins. The sedimentary strata are of continental, oceanic, and arc provenance. Some terranes include continental crust, only some of which can be related to PNA. Notably scarce among terranes, at least at the surface, are oceanic crust and mantle and transitional (stretched, diked) continental crust. Together, the rocks of terranes mainly indicate the selective transfer of nonsubductible materials to North America across the boundaries of converging plates.

The transfer of terranes to North American margins has occurred throughout the Phanerozoic at least as far back as Ordovician time, which was apparently the age of the first convergent plate boundary zone at a North American margin. Following their initial incorporation, almost all terranes have been reactivated in succeeding episodes of movement in the same or later margins. For example, terranes in the southern Appalachian orogen near the edge of PNA probably were first attached to North America in the Ordovician, underwent margin-parallel transport in middle Paleozoic time, and were thrust substantially farther continentward in late Paleozoic time.

In the present-day active margin of the southwestern United States, the full width of the zone of terranes is in motion relative to the craton, even though many of the constituent terranes attached to their neighbors in the Cretaceous. There, the full Pacific–North America plate velocity appears to be taken up across zones 2, 3, and 4 between a welded continent-ocean boundary and a fault zone at the edge of the craton. The velocity distribution across this 1500-km-wide zone is nonuniform; the maximum in the horizontal gradient is across the 100-km-wide San Andreas system of faults, which is conventionally taken as the plate boundary. The kinematic histories of terranes in this and perhaps all margins are extraordinarily complex, including transient amalgamations and attachments to PNA, excision and reentry into the margin-parallel flow, vertical-axis rotation, and slicing and dispersal of pieces of former terranes and terrane groups.

There has clearly been an increase in the area of the North American continent over Phanerozoic time due to accretion of terranes. There seems to have been rather little loss of terranes attached at any margin by subsequent divergent boundary formation, except in the north Atlantic and, conceivably, in the central Atlantic. In general, collisions in the eastern, southern, and western margins have contributed material to North America at the expense of the overriding plates, which later pulled away. The exchange budget of the northern margin is less clear.

It is uncertain how much of PNA has gotten into the zone of terranes. That some transfer has occurred is indicated by evidence that plate boundary zone motions were able to extract major tracts from the edge of PNA. It is also indicated by tectonostratigraphic alliances of certain terranes to sequences of PNA. For example, the Seri terrane of northern Mexico and the Salinia terrane of coastal central California can be related to source regions of PNA in southern California. As a whole, however, there seems to be rather little clearly identified PNA among terranes, perhaps because deformation and metamorphism have greatly obscured many displaced fragments of PNA.

Evolution of Phanerozoic margins: General

The Phanerozoic margins of North America began as a set of end-Proterozoic passive margins. Thereafter, they underwent moderately to greatly different tectonic evolutions in the transformation to subsequent collisional, active, and passive margins. The major contrasts are along segments of the modern cardinal directions: east, south, west, and north.

Despite the independence in evolutions of these segments, there is a similarity of basic structure around almost the entire present-day continent. The similarity is in the existence and features of belts peripheral to the craton. The belts are a deformed foreland on the inboard, followed outward by one or more belts of terranes distinguished by the ages of attachment one to another. Such ages diminish outward. This basic pattern presumably is a feature of the secular growth of continents and develops regardless of the timing and kinematic sequence in plate boundary zone evolution.

There are two exceptions to this basic structural pattern. One is the North Atlantic margin. There, a divergent plate boundary cut through the craton, failed after a short duration of spreading (Labrador Sea), and finally found success with a jump and sustained spreading east of Greenland. This event is distinctive because it is the only post-Proterozoic boundary to cross a great length of craton. The Labrador and Greenland boundaries cut the craton without evident control by preexisting structure, either by anisotropic, weak, or hot zones, unlike the central Atlantic, which had such structural controls.

A second exception exists in southern California, where there have been successive excisions of the cratonal edge, together with passage and transient attachment of terranes. This circumstance can be explained by the hypothesis that a corner or salient in the edge of PNA has existed in southern California from the end-Proterozoic breakup. The corner has been subject to progressive rounding in Phanerozoic plate boundary zones.

There are also many differences in structure among the principal margins of today's continent. In the following, the evolution of the eastern, southern, western, and northern margins of North America are reviewed and some major differences and ideas on their causes are presented.

Eastern North America

The end-Proterozoic plate divergence that created PNA caused the development of a passive margin in eastern North America from Alabama to Newfoundland. This early passive margin existed until mid-Ordovician time, when eastern North America underwent the first of a series of collisions. The collisions continued episodically through Paleozoic time, culminating in the arrival of nuclear Gondwana (Africa). The collisional margin existed as a deforming region until late in Permian time. For a while thereafter, eastern North America was embedded in the Pangea supercontinent. In the Middle Triassic, a new (Atlantic) passive margin began to develop in the earlier collisional margin. The onset of continental drift within the new passive margin varied geographically between Early Jurassic and early Cenozoic time.

The end-Proterozoic passive margin was rift dominated: its development included an unusually broad zone of rifting that extended far inland of the hinge (Rome Trough, Mississippi Valley graben) and continued activity well after drifting began. The breadth and duration of the zone of rifting are attributed to tectonics along the southern margin of North America.

The beginning phase of the collisional margin is noteable for the nearly simultaneous Middle Ordovician arrival of arc terranes along the full length of the early passive margin. This event (Taconian) records the subduction of ocean lithosphere that existed in the North America plate due to spreading against the early passive margin. The simultaneity of collision along strike suggests that closure of this ocean basin employed the same Euler pole as did the basin's opening. The Taconian terranes include accretionary forearcs, magmatic arcs, and arc crusts, and possibly major backarc basins. It is a question whether continental fragments exist in any of them, and if so, whether they are of Gondwanan derivation.

The Taconian collision caused the development of a deformed foreland that extended well landward of the early passive margin hinge. The terranes were thrust landward above North American crust, but it is difficult to separate the early overlap from the total overlap due to sequential collisions. At least locally, the overriding terranes were themselves imbricated, and out of sequence breakback duplexing occurred in North American crust and cover below the terranes. Major heterogeneity of deformation along the early collision margin may relate to nonuniform obliquity due to the zigzag geometry of the early passive margin.

The later collisions occurred in two phases, one broadly mid-Paleozoic (Acadian), and another in the late Paleozoic (Alleghanian).

The colliding terranes in the Acadian event are apparently continental and are related tectonostratigraphically to Gondwana (hence, peri-Gondwanan terranes). They may have been fragments in the oceans east of Paleozoic North America that were dispersed from the trailing edge of Gondwana as it drifted relative to the former Rodinia, and then were swept in one by one to eastern North America. Alternatively, such terranes may have arrived as one large peri-Gondwanan fragment that was divided and dispensed tectonically within the collision belt. The latter idea is supported by the existence of Acadian margin-parallel faults. A remarkable range of ages of attachment and attitudes of contacts exists between peri-Gondwanan terranes and preexisting North America, which was the outer edge of the Taconian belt of terranes. Wedging of peri-Gondwanan terranes below the earlier terranes is widespread.

The Alleghanian orogen is a type example of an ancient collision zone between two large continents, now greatly exhumed and its deeper structure exposed. A possible modern analogy is the India-Eurasia collision, where India is the North American craton and the Himalayas are the Alleghanian plate boundary zone. Accepting the analogy, the late Paleozoic Alleghanian Mountains of southeastern North America may have been the continent's highest ever.

The Alleghanian collision was right oblique and had apparently decreasing obliquity toward the south. The collisional plate boundary zone occupied both North America and presumably the leading edge of Gondwana, which is now mostly deeply buried on both sides of the Atlantic. The sole thrust zone was apparently a deep detachment that lay below the strong crustal layer of both continents and below the suture between them. Gondwana did not apparently override late Paleozoic North America significantly at near-surface levels. The sole thrust ramped up through pre-Alleghanian North American terranes to follow ever-more-brittle zones landward. The principal ramp may have been the deeply buried seaward face of the Paleozoic craton of PNA or a suture zone among pre-Alleghanian terranes. The ramp caused the marked elevation of high-grade tectonites of the hangingwall, and probably greatly promoted mountain building within North America during continentward transport. Forward of the main ramp, the sole thrust probably followed a thin zone of reactivated low-angle faults in the earlier terranes and foreland, and farther landward, to newly formed detachments and ramps to the surface within the cover of the preexisting craton.

The hangingwall of this thrust system was only little or moderately imbricated where it was thick, and consisted of deformed rock (as far west as the Blue Ridge), whereas it underwent strong imbrication where it was thin and consisted of well-layered cover. The front of the deformed foreland moved substantially landward into the craton relative to its pre-Alleghanian locus. The total margin-normal displacement within North America on and in the Alleghanian hangingwall in the southern Appalachians may be as great as 350 km.

Kinematic partitioning occurred within the Alleghanian hangingwall by the late development of discrete right-lateral margin-parallel faults between panels with margin-normal contraction. It is not clear whether the partitioning reflects an increase in velocity obliquity, a strengthening of the sole thrust zone that may have initially accommodated the full oblique velocity, or a progressive weakening of loci in the hangingwall.

Similarity in at least a couple of major features indicates the Alleghanian and the modern India-Eurasia collisions are comparable processes. The suture zones are both apparently steeply dipping. The overriding plates did not overrun a great width of the underriding continents. Rather, both sole thrusts extended below the suture into or below the underriding continental crust at depth, and at least much of the subsequent convergence was taken up by imbricate thrusting totaling hundreds of kilometers in the upper crust of the underriding plate. In both cases, the imbricate zone emerged on the craton, causing and then deforming a linear foreland basin. Almost certainly, convergence was also taken up within the leading edge of the overriding plates, but this is little studied in either case. In the India-Eurasia collision, the transitional crust that originally fronted northern India now underlies Eurasia and may in the future be lost to the mantle. In eastern North America, it is unclear whether the end-Proterozoic transitional crust partly exists today below the outer terranes, now underplates West Africa, or is in the mantle.

Some major questions persist on the collisional margins of eastern North America as a whole. Did the subduction zones in fact dip away from North America throughout the Paleozoic? What was the distribution, sense, and magnitude versus time of margin-parallel shear? How far seaward below the terranes does PNA actually extend?

The Atlantic margins of eastern North America are probably the world's prime examples of modern passive margins, including both rift- and transform-dominated types. Passive margin development at central Atlantic latitudes took place entirely within and perhaps coextensively with the Alleghanian collision zone. It began with the initiation of rifting as little as 25 m.y. after the cessation of thrusting associated with that collision. The principal structure of the central Atlantic passive margin is thought to be an extensive normal detachment that stepped down eastward through the Appalachian orogen and below the continental shelf. The downdip end of the detachment probably reached the asthenosphere below a locus that was the maximum crustal thickness of the Alleghanian collision zone and, coincidentally, the minimum thickness of mantle lithosphere. This same locus, during and after stretching, was the maximum in the rate of magmatism, and ultimately it became the drift boundary. The updip path of the detachment probably followed preexisting shear zones and faults within the unroofing Appalachian orogen. Rifting occurred in the hangingwall where it was thin, and probably all the way across the hangingwall at shallow levels. After some duration of margin-normal extension and magmatic heating that probably increased eastward, stretching and probably intrusion concentrated in the eastern realm of the zone of extension. There, 200%–300% margin-normal elongation of the lithosphere was achieved, and transitional continental crust was created. The shift caused high rates of subsidence in the highly stretched realm relative to the rates in the little extended landward zone of rifting only, and caused the development of a hinge between the two. An eastward-thickening passive margin sediment wedge was deposited above the transitional crust. After about 50 m.y. from the beginning of rifting, the highly stretched zone in the central Atlantic broke apart at the protoridge of marginal oceanic crust, sea-floor spreading began, and thermal contraction probably continued the high subsidence rate of lithosphere east of the hinge.

The history of passive margin development of eastern North America in the north Atlantic differed from that of central Atlantic latitudes just discussed. The two regions are divided by the transform Southern Grand Banks Fracture Zone (SGBFZ). Although rifting began in both regions at about the same time, the northern region (Grand Banks) stretched heterogeneously and probably with small total displacement for at least 100 m.y. while the central Atlantic on its south was spreading. The SGBFZ permitted Gondwana (Africa) to translate far to the east of North America in the Grand Banks, and the Mid-Atlantic Ridge to migrate 2000 km east of the central Atlantic protoridge before north Atlantic drifting began in the Cretaceous. Studies of the SGBFZ transform-dominated margin indicate that it was an abrupt transition over about 50 km from little-stretched and presumably little-thinned continent south to ocean floor. The transform faulting was accompanied by a propagating thermal uplift and erosion along the continental wall of the SGBFZ, linked to the passing spreading ridge. Moreover, the SGBFZ has relatively thin sediment cover. All of these characteristics contrast markedly with those of the rift-dominated central Atlantic passive margin. The SGBFZ seems not to follow a preexisting structure.

Southern North America

Southern North America took its initial shape as a transform-dominated passive margin early in Phanerozoic time. This passive margin persisted until the right-oblique collision of Gondwana in late Paleozoic time encased eastern southern North America within the Pangean supercontinent. The western front of Gondwana migrated as far west as present-day eastern Mexico, resulting in a large embayment in the western side of Pangea between Coahuila and southern California. The collisional Ouachitan orogen of southern North America was contemporary and contiguous with the Alleghanian of eastern North America around the corner of the late Paleozoic continent in Alabama. The two collision zones, however, differ remarkably in their structure. During the late stages of, or after, the Ouachitan collision, an active margin developed at places along the embayment of western Pangea in the late Paleozoic. This active margin has continued to the present. In the Middle Triassic, the region of Pangea within the earlier collisional zone began rifting and the development of the Gulf passive margin. Large extension continued at places in this passive margin until about Early Cretaceous time, when Gondwana broke away from southern North America at a locus along the southern edges of Yucatan, Cuba, and the Bahama Platform. The last and continuing tectonic phase of eastern southern North America is its left-oblique collision with oceanic lithosphere and island arcs of the Caribbean plate.

The end-Proterozoic passive margin comprised three long transform and two shorter rift segments between corners of PNA in southern California and Alabama. The predominant transform segments probably bounded little-stretched, thick continental crust with a relatively precipitous slope to adjacent ocean floor, by analogy with the modern SGBFZ. The southern ocean that formed against southern North America probably spread at ridges that intersected the transform segments at high angles. Small ridges presumably developed at the toes of the left-stepping rift segments in east Texas and Coahuila. At least one longer ridge system probably developed farther west in Rodinia or beyond and south of the westernmost transform segment (the Mojave-Sonora megashear). This longer ridge was presumably connected by the transform to the ridge system offshore of and parallel to western North America. The transform segments of the developing margin kept southern North America in contact with continents on its south well into the Paleozoic (50 to 100 m.y.?) before oceanic lithosphere fully lined the margin. Lateral heating from the migrating oceanic ridges probably kept the freeboard high, perhaps even erosional, and prevented the deposition of marine sediment cover on the continental transform segments until late in Cambrian time. The transform motion and ridge migration may also have caused protracted early Paleozoic rifting at high angles well to the north in the craton in the Mississippi Valley and Rome grabens.

The Paleozoic oceans east and south of North America contained plate boundaries, and there was oblique convergence on one or more of the boundaries. This is implied by terranes now in Mexico that include Grenvillean continental crust from a northern end-Proterozoic site and another indicating early Paleozoic ophiolite–Grenvillean crust collision.

The Ouachitan collisional margin developed along the southern edge of Paleozoic North America in a right-oblique convergent plate boundary zone with Gondwana. The deformation began late in the Mississippian in Alabama and in the Middle Permian in Coahuila, clocking the east to west migration of deformation front. The Permian age probably marks the time at which collision ceased in the Ouachitan orogen. West of Florida, Gondwana was fronted by a forearc composed mostly of older and contemporary Paleozoic sedimentary rocks, together with bits of oceanic crust. The forearc also included terranes with nonoceanic crustal underpinnings, some of which were swept up as microcontinents from the Paleozoic oceans and from the early North American passive margin. South of the forearc, Gondwana was a continental arc, and south of that, the Guyana shield. The front of southern North America at the onset of collision contained a series of discontinuous foreland basins, unlike the Alleghanian foreland basin, which was continuous on orogenic trend in the southern Appalachians. The Ouachitan foreland basins presumably developed because of loading of North American lithosphere by the encroaching Gondwana, but they did so with subsidence rates that varied markedly, somewhat periodically, with position along the continental edge. Such variations are not predicted by loading theory; they may reflect control of subsidence rates either by reactivation of preexisting faults (heterogeneous flexural rigidity) or by oblique flexure due to lateral migration of the collision.

The Ouachitan collision drove the Gondwanan forearc above the North American continent-ocean boundary at generally high, upper Paleozoic horizons in North American cover. The sole thrust emerged within the foreland basins, perhaps no more than 100 km inland from the continent-ocean boundary. The width of the zone of imbrication of, and the heave in, North American rocks are thus remarkably small in the Ouachitan zone. The Gondwanan continental arc and crystalline crust apparently did not override the southern edge of the North American continent. The pre-Ouachitan North American basement and cover were apparently not detached and displaced landward above the sole thrust, except in the east Texas rift segment of the southern margin. In most of the exposed Ouachitan zone, moreover, there has been no significant unroofing, no magmatism, and no evident strike-slip faulting on discrete surfaces.

Such characteristics differ markedly from those of the contemporaneous Alleghanian collision of Gondwana and North America. I suggest that the differences are due to three factors: (1) the relative subductibility of the crust of the two preexisting passive margins; (2) the relative strengths of the colliding continental edges; and (3) the strength of the colliding edifices relative to the sole thrust zone.

First, the crust of the transform-dominated passive margin of southern North America was probably less subductible than the transitional crust of the eastern margin, assuming that the end-Proterozoic crust of the rift-dominated eastern margin still existed in late Paleozoic time below Taconian and Acadian terranes. The presumably tabular, thick, little-diked southern crust must have been more positively buoyant than the wedge-shaped, thin, heavily diked crust presumed for the eastern margin. If so, the southern margin's crust descended less far, if at all, below Gondwana before buoyancy arrest of the local convergence. This explains the zero or minimal overlap of crystalline crusts in the Ouachitan zone. In contrast, the transitional crust of eastern North America may have descended well below Gondwana's leading edge, permitting large cratonward transport of terranes and cover above it.

Second, southern North America as an old, cold passive margin was probably strong relative to the front of colliding Gondwana, which was an active margin. Thus, the deformation that accommodated continuing oblique convergence after initial suturing probably migrated south within Gondwana by contraction in the magmatic arc, backarc, and foreland, and by right-lateral faulting within the arc. The former edge of North America behaved more rigidly. The Alleghanian collision, in contrast, was the union of two weak margins, the Gondwanan front presumably carrying a continental arc and eastern North America, the warm Acadian collisional margin that had ceased deformation only 50 m.y. earlier. The Alleghanian convergence thus was probably distributed in both colliding edifices.

Third, the apparent lack of contraction and strike-slip faults in the North American basement and pre-Mississippian cover of much of the Ouachitan zone may be due to the greater strength and adhesion of such rocks than those of the sole thrust zone. The sole thrust was weak because it probably contained poorly lithified sediments and high fluid pressure throughout the base of the Gondwana forearc and within North American foreland basins. In this view, the oblique velocity was taken up in the thrust zone until suturing, after which margin-parallel components may have been partitioned during further convergence in the weak Gondwana magmatic arc.

The Pacific active margin began in the Permian, causing the

formation of continental arcs in eastern Mexico and southern California at the north-trending arms in the bight of western Pangea. The absence of a North American magmatic arc in the intervening arm of the bight suggests that the arm's trend was subparallel to the velocity vector between Pangea and the adjacent plate of the Permian Pacific basin. By Jurassic time, the continental arc occupied the full bight, suggesting that obliquity decreased with time. Terranes from the western Gondwana forearc, and perhaps those extruded from the Ouachitan collision zone, were the earliest to accrete at the Pacific active margin in eastern and central Mexico. Terranes that later attached in Mexico probably came from sites off western North America.

The rift-dominated Gulf passive margin began development in the Middle Triassic within the earlier Ouachitan collision zone of eastern Mexico and the southern United States. The region of extension and diking then apparently expanded southward in Pangea, and by Jurassic time occupied what is now the Gulf of Mexico, Cuba, and northern Venezuela. This region included the backarc of the Pacific active margin. Although the divergence of the North America and Gondwana plates was apparently uniform (northwest-southeast, 3 cm/yr) over at least much of the duration of rifting and drifting, the deformation in the plate boundary zone is notably heterogeneous and includes local vertical-axis rotation and area change and motion of blocks on paths that were not small circles about the Euler poles of the major plates. The early extension (rift phase) over the full region was probably small, but later stretching of large magnitude, together with magmatism and local spreading, concentrated in the Gulf of Mexico basin in the Jurassic. The heterogeneity and migration of deformation in the Gulf passive margin may be related to several factors, for example, the loci of greatest crustal thickening in the prior collision, interfering structures of the collision and the backarc, and the geography of backarc advection. It is a puzzle why the final breakup of Pangea and probable spreading in the Caribbean basin occurred at the beginning of the Cretaceous so far south of the region of very large Jurassic stretching in the Gulf of Mexico.

The Antillean collisional margin developed with left-oblique convergence along the rift-dominated Gulf passive margin, beginning in southern Mexico late in Cretaceous time. The Caribbean plate and its island arcs overrode North America to some unknown extent and caused imbricate thrusting of North American cover, at least in Cuba. The deformation that permitted continued motion in this plate boundary zone seems to have been in the Caribbean plate above or south of the suture. The velocity gradients migrated or concentrated in the overriding plate because the arc-fronted Caribbean was weak compared to the North American passive margin, which had cooled and strengthened through the Cretaceous. The idea that the eastern realm of the Caribbean–North America plate boundary zone (the Puerto Rico Trench) is transtensional can be explained by intra-Caribbean motions (different Euler poles of the western and eastern Caribbean relative to North America) or by a recent small change of Caribbean–North America velocity.

Western North America

Western North America evolved with a passive margin from the end-Proterozoic breakup of Rodinia. The passive margin existed until times between the Late Devonian and Jurassic, when the first collisions occurred at different sites. The collisional margins later changed to active margins, first in southern California in the Permian and later to the north, possibly as late as Cretaceous time in Canada. The active margin continues to the present.

Western North America differs from the eastern and southern continental margins because it has always faced out to Phanerozoic oceanic lithosphere. Its collisions have been with terranes of small or moderate size, and it has not been encased in a continental amalgam like Pangea. Nonetheless, western North America contains long, distinctive tectonic belts like those of the other margins: a deformed foreland of PNA and, outboard of that, terranes whose ages of attachment to one another generally decrease toward the ocean.

The present-day plate boundary zone between North America and adjacent plates of the Pacific basin occupies most of the active margin. The principal exception may be the eastern Canadian Cordillera, which appears to be currently (temporarily?) excluded from the velocity gradient zone. The continental magmatic arc is currently active in segments where the plate velocity is of low to moderate obliquity. Such segments occur exclusively in the zone of terranes, but batholiths representing earlier activity of the continental arc occur in either the deformed foreland or the zone of terranes at different positions along the margin and, occasionally, straddling both belts. Where the inactive arc exists in terranes, it has almost certainly been displaced from its original positions. Today's forearc is exclusively within the terranes. The zone of terranes represents the preserved portion of accumulated forearcs that formed at convergent plate boundary zones (collisional and active margins) along the western North American continent since Late Devonian time.

The end-Proterozoic passive margin is a rift-dominated type. Its passive margin wedge is remarkably broad in the Great Basin, about 700 km and as wide as the Blake Plateau, the widest rift-dominated segment of the modern Atlantic margin. The original widths of other segments of this passive margin are unclear because their outer reaches were probably excised.

The earliest recognized collision began in Late Devonian time in the Great Basin with the emplacement of a mainly oceanic sedimentary forearc (Roberts Mountains allochthon terrane) above the outer third of the former passive margin. The collision brought an already deformed, thickened, and strengthened edifice above the passive margin wedge. There was little contemporary deformation of either wall and no magmatism or unroofing. A foreland basin of modest amplitude developed progressively at the toe of the allochthon and was finally underwedged by the allochthon. Such features imply that the sole thrust was weak relative to its walls and that the overriding mass was thin. The first collision in the Great Basin was followed by another at the end of the Paleozoic with almost identical characteristics, which stacked collided forearcs. In both collisions, no continental crust appears to have been subducted, implying that buoyancy arrest and suturing occurred as soon as the arc and continental crusts came in contact. This is puzzling because the continental crust of the Great Basin is thought to be of transitional type. Perhaps the colliding arc crusts were extraordinarily weak, and progressive convergence was taken up within the newly attached arc terranes.

North and south of the Great Basin in western North America, the edge of PNA is lined by terranes whose ages of attachment (or at least, encroachment) are Mesozoic and Cenozoic. Many of them include tectonostratigraphies that ally them with forearcs of offshore arc systems of early Paleozoic or late Paleozoic development, like those of Paleozoic collision age in the Great Basin. These Paleozoic forearcs, moreover, include sedimentary rocks of continental provenance, implying that the forearcs are partly sweepings from the ocean floor(s) that fronted the former passive margins.

In northern British Columbia and the Yukon Territory, the terranes of Paleozoic forearcs (Yukon-Tanana, Kootenay) include mid-Paleozoic magmatic rock that can be interpreted as evidence of a broader realm of convergent boundaries. These terranes perhaps originated in the mid-Paleozoic Ellesemerian active margin of northern North America and later, migrated south along the western edge of PNA. The Paleozoic tectonostratigraphic record in the early terranes as a whole, however, suggests that plate boundaries, convergent and divergent, existed within the Pacific basin off western North America at many times in the Paleozoic. This implies that many discrete arc systems may have existed and that each may have collided independently with the western continental margin.

It is a question whether the terranes of Paleozoic rocks and forearc construction, other than those of the Great Basin, remained within the Pacific basin until the Mesozoic or whether they had collided earlier at one or more times over a long stretch of the margin and been excised and reset in motion in a later plate boundary zone. Some evidence points to multiple collisions and excisions. The structure expected for an excised terrane composed of a forearc thrust above PNA, however, is not commonly found. This may deny the multiplicity of early collisions, or imply either that most terranes are too little unroofed to see underpinnings of PNA, or that excision detached the forearc, leaving PNA in place.

By Late Cretaceous time, the entire length of western North America was an active margin. Since then, oceanic lithospheres of the Pacific basin have descended below the continent, except where local high obliquity caused transient cessation of subduction. The hallmark for an active margin is a continental magmatic arc of sustained existence. The beginning of continental arc magmatism in southern California in the Permian and in the Great Basin at times in the Triassic marks the diachronous onset of the active margin in southern western North America. In the northwestern United States and Canada, however, it is difficult to prove whether the pre–Late Cretaceous margins were active or collisional, because the batholiths older than that are in terranes, and the terrane attachment history is uncertain. It is conceivable that an early Mesozoic continental arc existed in PNA fully up and down the coast, but that such rock has been largely excised and removed by later margin-parallel transport. Such loss seems almost predictable because the locus of minimum strength through an active margin should be the hot arc. The margin-parallel velocity component across an oblique plate boundary zone can be expected to be taken up along the continental arc. An example of an excision through an arc is the jump at 6 Ma of the maximum in the plate velocity gradient inboard across Baja California to the Gulf of California along the locus of a Neogene continental arc. As further support for this possibility, it is now well established that there is a partition of velocity gradient components into one or more narrow zones of strike-slip separated by broad panels of margin-normal strain in many plate boundary zones worldwide.

Another example of inboard jump of the maximum of the velocity gradient and excision and the removal of territory from the continent is western Idaho and adjacent Oregon and Washington in mid-Cretaceous time. There, an extensive tract of the continent was extracted, comprising terranes and PNA inboard beyond the passive margin hinge. The causative mechanism may have been a right-stepping pullapart; the excised tract probably moved northward along the coast, and the gap left behind it may have been filled by local sea-floor spreading. In the Late Cretaceous, the postulated oceanic lithosphere was overridden by incoming terranes, and an active margin commenced or resumed. The new coastal PNA, which has been cratonal before the excision, then became the infrastructure for a new continental arc (the Idaho batholith) and deformed foreland.

A basic question is: once entered into the belt of terranes along western PNA, whether in a collisional or active margin, what further transport does a terrane undergo within the belt? Paleomagnetics, paleontology, structures, and tectonostratigraphy argue for large margin-parallel flow, probably at nonuniform rates versus time and with increasing mean rates outward. The distances of transport along the margin are probably as much as thousands of kilometers. There is evidence that the flow was generally left-lateral in the late Paleozoic and early Mesozoic and right lateral after about mid-Cretaceous time. The potential for large margin-parallel transport is incontestable, because the relative plate velocity across the terrane belt has probably always had some degree of obliquity. Furthermore, the existence in western North America of long margin-parallel faults with large offset back at least to the Jurassic indicates that such transport has occurred. It is still an issue, however, how rotated blocks within the zone of terranes with multiple strike-slip faults accommodate their motion relative to neighboring blocks.

The margin-normal components of the plate velocities are taken up at shallow levels by thrusting and folding or normal faulting in panels between the strike-slip faults. Within such panels there must be thickening and thinning, including volume changes, that cause mountain belts and large degrees of unroofing, such as the tens of kilometers undergone by the Catalina terrane and the Coast Range batholith. Where the unroofed deep-seated rocks were relatively cold, volume addition below them by wedging or underplating of newly arrived terranes is one solution. Where the unroofed rocks were hot, advection of anatectic magma and softened walls within a contractile belt is a solution.

The extraordinary differences in collisional and active margin evolutions versus position can be glimpsed by comparisons of southern California, where oblique excisions seem to have been the rule, and Alaska, which can be taken as a progressively growing, deforming coagulation.

Southern California apparently emerged at the breakup of Rodinia as a corner, perhaps as a peninsula, of PNA between the southern and western passive margins. Its first recognized reentry into a plate boundary zone was in the Pennsylvanian, with excisions of PNA that cut into both craton and passive margin. Newly oriented or continuing transpression caused the development of a continental magmatic arc in Permian time. The arc continued sporadically or continuously throughout Mesozoic time with attachment of terranes, including one consisting of a lower Paleozoic forearc. Parts of the southern California margin were further excised, such that ever more interior regions of PNA fronted or were within the active margin. A large amalgamated group of terranes, the Santa Lucia–Orocopia allochthon (SLOA), attached to the southern California edge of PNA in Eocene time, and a second superterrane (Baja Borderland) attached outboard and south of SLOA in the Miocene. Both superterranes had evidently migrated north from latitudes of southern Mexico before attachment. They underwent deformation together by right-slip faulting in the Neogene, culminating in excision as a giant superterrane (Alta-Baja California). The maximum velocity shifted around in the plate boundary zone, apparently settling at the San Andreas fault in the past 6 m.y., but also causing the Alta-Baja California superterrane to be fragmented. The excision left small bits of the superterranes (SLOA) still attached to PNA and sent slivers of PNA (Salinia) into the coastwise flow. One main point of this history is the perpetuity of excision of PNA at an initial salient; another is that, although PNA is being whittled back, the flow of terranes continues on its outboard. The flow is inhomogeneous and nonsteady, including the transient attachment of component terranes to one another and to PNA, and fragmentation, rotation, deformation, and dispersal.

Alaska differs from the rest of western North America by the very large width of the region that consist of terranes—virtually the whole state and surrounding marine regions. Among Alaskan terranes, the earliest to attach to PNA (Early Cretaceous) was apparently the giant Arctic-Alaska terrane (actually superterrane). This terrane, which consists of PNA and terranes of early Paleozoic collision age, was excised from northern (Arctic) North America in the Jurassic and traveled south around the corner between the northern and western margins of early Mesozoic North America to attach at its present site. The oceanic Canada basin formed by sea-floor spreading at the trailing edge of the Arctic-Alaska terrane. The terrane composite grew by accretion at its leading edge and deformed internally to yield the Brookian thrust belt and North Slope foreland basin. Upon attachment to PNA in the northern Yukon Territory, the Arctic-Alaska terrane provided a backstop for terranes that swept northwest along the margin of western North America in Late Cretaceous and Cenozoic time. The result has been a vast coalescence of terranes in Alaska, now as much as 1600 km wide, and still rapidly building southward by accretion.

At the present, the Pacific–North American plate velocity passes through an obliquity minimum in the Gulf of Alaska that includes a stretch of nearly head-on convergence. The minimum has existed for at least much of the 100 m.y. over which the Alaskan coalescence has grown. The relation between the rate of terrane accumulation and obliquity is evident. The large margin-parallel components of velocity along western North America to the south of the Gulf of Alaska have kept that margin in a flow and delivered terranes progressively to the region where such components minimize or disappear.

Within the zone of terranes of western North America near its outer margin, there is a belt, remarkable for its length of 2000 km from Vancouver to central southern Alaska, that contains the distinctive composite Wrangellia and Alexander terranes. This terrane composite apparently now occurs in large discrete blocks, but earlier these probably existed in a single continuous edifice, which was at equatorial latitudes early in the Cretaceous. This original edifice translated rapidly north and collided in Late Cretaceous time in southern Canada, where it broke up, and blocks of it migrated north along the western margin as far northwest as central southern Alaska. The times of attachment of Wrangellia-Alexander blocks to their neighbors seem to young outward across the belt from late Late Cretaceous through Paleocene time, about 20 m.y. The restriction of the blocks to a belt and the long duration of attachment implies that transport was channelized and that accretion occurred progressively along the inner wall. Such orderly flow seems in contrast to the inhomogeneity of the flow in southern California.

The belt of Wrangellia-Alexander blocks and other terranes, upon assembly in southern Alaska (the Southern Alaska superterrane), became the foundation for the Alaskan continental arc early in the Cenozoic. Moreover, it became the backstop for the accretionary forearc of the Gulf of Alaska, possibly the world's broadest. The forearc contains ocean-floor sedimentary rocks, much of continental provenance, and basalt that have been transferred from thousands of kilometers of downgoing oceanic crust of plates of the Pacific basin. The transfer occurs at imbricate thrusts that emerge from a detachment extending from the forearc base into the adjacent plate at or above the base of its sedimentary cover. Transfer also occurs below the forearc, by underplating its bottom. The inner realm of the forearc may have first accreted at more southerly latitudes and migrated north on the exterior of the belt containing the Wrangellia-Alexander blocks.

The Canadian Cordillera presents important insights and uncertainties in the evolution of collisional and active margins. There, the zone of terranes comprises several belts, including the one with Wrangellia-Alexander blocks. Each belt contains somewhat distinctive terranes and had probably undergone assembly before arrival at its current positions. All of the belts are cut by margin-parallel strike-slip faults. A problem in kinematic interpretation is whether the terrane belts came in one by one or as a group. Another problem is how much transport and reshaping occurred after arrival. The Wrangellia-Alexander example indicates a great deal. Finally, although it is clear that terranes collided with PNA in British Columbia in Late Jurassic time, perhaps for the first time, it is not certain that the terranes currently contacting PNA are the same ones.

The history of terrane migration at Great Basin latitudes also offers interesting insights. There, the terranes that accreted to PNA in mid-Paleozoic and end-Paleozoic time are still apparently in their place of original attachment. This is in a recess in the edge of PNA that may have been an end-Proterozoic ridge-transform elbow and/or a product of early Mesozoic deformation. There seems to be no comparable recess preserved along the rest of the western margin. Outboard of the recess, terranes have been in the margin-parallel flow, probably with increasing velocity outward. The full width of the active margin at Great Basin latitudes, including the deformed foreland, however, currently is moving relative to the craton. The modern rate east of the San Andreas system is as great as 10 km/m.y. If such movements have occurred as a simple-shearing continuum for 100 m.y., substantial displacement and reshaping of the inner terrane zone and deformed foreland has taken place without destroying the systematics of terrane belts and age trends.

The deformed foreland of western North America is another belt that evokes interest by its enormous length and restricted duration of activity. Its deformation began in the Jurassic, apparently everywhere except in southern California, where the onset was Triassic or possibly Permian. Its timing and deformation clearly reflect convergence across the plate boundary zone of western North America, although it is not certain whether such deformation occurred in both the later phase of collisional margins or exclusively in the ensuing active margin. The process has been the imbrication of detached passive margin wedge and succeeding cover, which piled up at or near the slope break between transitional crust and the craton. At many places, however, the front of thrusting migrated beyond the hinge onto the craton. This belt includes local zones, called hinterlands, where younger extensional deformation has unroofed nappes and rocks from a deeper ductile realm of foreland contraction. In spite of the varied plate kinematics along the western margin over 150 m.y. and the evidence for margin-parallel terrane migration, the foreland seems to reflect generally uniform total margin-normal displacement of the cover over about 2500 km on strike. The displacement rates, however, have varied with position and time. The uniformity of total displacement along the belt thus implies that foreland processes have tended to even out the contraction along the edge of thick, relatively cold cratonic crust. This phenomenon might be explained by plastic horizontal shortening of the deep crust below the passive margin wedge. The continuum deformation of the substrate might tend to balance over time the cumulative displacements in the heterogeneous, layer-dominated brittle zone.

The Laramide zone of basement-cored thrusts is east of the classic belt of deformed foreland containing thin-skinned stratal imbricates in the central western United States. The Laramide movements were generally in succession, but overlapped in Paleogene time with the latest thrusting of the thin-skinned belt. The Laramide zone may owe its existence to the fracture and contraction of thick, brittle cratonal hanging wall above a deep detachment that propagated eastward in the brittle-ductile transition of the sialic crust from below the thin-skinned belt.

Northern North America

The Arctic region of North America began with an end-Proterozoic passive margin, which was succeeded by an active margin that climaxed in Late Devonian and Early Mississippian time by the Ellesmerian orogeny. This active margin may have begun as early as Ordovician time, or was preceded by a discrete Ordovician convergent plate boundary zone. After the Ellesmerian active margin, northern North America has existed as a passive margin from the Mississippian to the present. The northward propagation of divergence and sea-floor spreading of the Labrador Sea into Baffin Bay in the Late Cretaceous and early Cenozoic imposed heterogeneous deformation, the Eurekan orogeny, around the tip of this failed arm.

The end-Proterozoic Arctic passive margin was probably rift dominated: if true, the drift of continental fragments away from northern PNA followed paths about Euler poles different from that of fragments relative to other North American margins. A corner may have existed between the northern and western edges of PNA in the northern Yukon Territory.

The Ellesmerian active margin may have climaxed in the Early Mississippian collision of a Siberian continent with northern PNA. Northern North America may have been fronted by an extensive accretionary forearc that underwent anatectic magmatism before and/or during collision. The collision may have extended around the corner in PNA along northern western North America. Bodies of the forearc may have escaped toward the south and collided with PNA at varied sites. The widespread Yukon-Tanana and Kootenay terranes may have come from this postulated forearc. The Monashee terrane of western Canada may be Proterozoic crust clipped off the Yukon corner of PNA by the collision.

The later passive margin of northern North America may have begun with Mississippian rifting of the extensive Svedrup basin. If so, the rifting occurred for 200 m.y. before the Jurassic–Early Cretaceous onset of spreading in the Canada basin and the drifting away of the Arctic-Alaska terrane, the excision of which took away much of the passive margin of northern PNA.

The Arctic-Alaska terrane and allied small terranes may

have rotated out of Arctic North America to their present position in northern Alaska about a pole near its present eastern end. Alternatively, the Arctic-Alaska terrane may have slid rectilinearly on a left-slip boundary from an initially more polar site. The Arctic-Alaska terrane was fronted by and moved with the Angayucham oceanic lithosphere, the southern end of which subducted below an island arc to the south. This convergence climaxed in a Late Jurassic collision of the arc and its forearc (the Angayucham terrane) and the Arctic-Alaska terrane. The collision caused extensive thin-skinned contraction by imbrication of the Arctic-Alaska terrane, estimated between 500 and 800 km displacement in the cover. The Brooks Range thrust belt of northern Alaska is the product of this contraction.

REFERENCES CITED

Allan, J. F., 1986, Geology of the northern Colima and Zacoalco grabens, southwest Mexico: Late Cenozoic rifting in the Mexican volcanic belt: Geological Society of America Bulletin, v. 97, p. 473–485.

Allmendinger, R. W., and Jordan, T. E., 1984, Mesozoic structure of the Newfoundland Mountains, Utah: Horizontal shortening and subsequent extension in the hinterland of the belt: Geological Society of America Bulletin, v. 95, p. 1280–1292.

Alvarez, W., Kent, D. V., Silva, I. P., Schweickert, R. A., and Larson, R. A., 1980, Franciscan limestone deposited at 17° south paleolatitude: Geological Society of America Bulletin, v. 91, Part I, p. 476–484.

Anderson, T. H., and Schmidt, V. A., 1983, The evolution of Middle America and the Gulf of Mexico–Caribbean Sea region during Mesozoic time: Geological Society of America Bulletin, v. 94, p. 941–966.

Anderson, T. H., and Silver, L. T., 1981, An overview of Precambrian rocks in Sonora, México: Universidad Nacional Autónoma de México, Instituto Geología, Revista, v. 5, p. 131–139.

Anderson, T. H., Eells, J. L., and Silver, L. T., 1979, Precambrian and Paleozoic rocks of the Caborca region, Sonora, México, *in* Anderson, T. H., and Roldán-Quintana, eds., Geology of northern Sonora (Field Trip Guidebook 27): Hermosillo, Sonora, Instituto de Geología, Universidad Nacional Autonoma de Mexico, p. 1–22.

Anderson, T. H., McKee, J. W., and Jones, N. W., 1991, A northwest trending, Jurassic fold nappe, northernmost Zacatecas, México: Tectonics, v. 10, p. 383–401.

Ando, C. J., and 11 others, 1984, Profiling in New England Appalachians and implications for architecture of convergent mountain chains: American Association of Petroleum Geologists Bulletin, v. 68, p. 819–837.

Argus, D. F., 1990, Current plate motions and crustal deformation [Ph.D. thesis]: Evanston, Illinois, Northwestern University, 163 p.

Argus, D. F., and Gordon, R. G., 1990, Pacific–North American plate motion from very long baseline radio interferometry compared with motion determined from magnetic anomalies, transform faults, and earthquake slip vectors: Journal of Geophysical Research, v. 95, p. 17,315–17,324.

——, 1991, Current Sierra Nevada–North America motion from very long baseline interferometry: Implications for the kinematics of the western United States: Geology, v. 19, p. 1085–1088.

Armstrong, R. A., 1988, Mesozoic and early Cenozoic magmatic evolution of the Canadian Cordillera, *in* Clark, S. P., Jr., Burchfiel, B. C., and Suppe, J., eds., Processes in continental lithospheric deformation: Geological Society of America Special Paper 218, p. 55–91.

Atwater, T., 1970, Implications of plate tectonics for the Cenozoic tectonic evolution of western North America: Geological Society of America Bulletin, v. 81, p. 3513–3536.

——, 1989, Plate tectonic history of the northeast Pacific and western North America, *in* Winterer, E. L., Hussong, D. M., and Decker, R. W., eds., The Eastern Pacific region: Boulder, Colorado, Geological Society of America, The Geology of North America, v. N, p. 21–72.

Avé Lallemant, H. G., 1994, Pre-Cretaceous tectonic evolution of the Blue Mountains Province, northeastern Oregon: U.S. Geological Survey Professional Paper 1438 (in press).

Avé Lallemant, H. G., and Oldow, J. S., 1988, Early Mesozoic southward migration of Cordilleran transpressional terranes: Tectonics, v. 7, p. 1057–1075.

Babaie, H. A., 1987, Paleogeographic and tectonic implications of the Golconda allochthon, southern Toiyabe Range, Nevada: Geological Society of America Bulletin, v. 99, p. 231–243.

Balkwill, H. R., 1978, Evolution of the Sverdrup Basin, Arctic Canada: American Association of Petroleum Geologists Bulletin, v. 62, p. 1004–1028.

Beck, M. E., Jr., 1976, Discordant paleomagnetic pole positions as evidence of regional shear in the western Cordillera of North America: American Journal of Science, v. 276, p. 694–712.

——, 1983, On the mechanism of tectonic transport in zones of oblique subduction: Tectonophysics, v. 93, p. 1–11.

——, 1986, Model for late Mesozoic–early Tertiary tectonics of coastal California and western Mexico and speculations on the origin of the San Andreas fault: Tectonics, v. 5, p. 49–64.

——, 1991, Case for northward transport of Baja and coastal southern California: Paleomagnetic data, analysis, and alternatives: Geology, v. 19, p. 506–509.

Beck, M. E., Jr., and Noson, L., 1972, Anomalous paleolatitudes in Cretaceous granitic rocks: Nature Physical Science, v. 235, p. 11–13.

Berg, H. C., Jones, D. L., and Richter, D. H., 1972, Gravina-Nutzotin belt: Tectonic significance of upper Mesozoic sedimentary and volcanic sequence in southern and southeastern Alaska: U.S. Geological Survey Professional Paper 800D, p. D1–D24.

Bickford, M. E., Van Schmus, W. R., and Zietz, I., 1986, Proterozoic history of the midcontinent region of North America: Geology, v. 14, p. 492–496.

Blake, M. C., Jr., Bruhn, R. L., Miller, E. L., Moores, E. M., Smithson, S. B., and Speed, R. C., 1985, C1 Mendicino triple junction to North American craton: Boulder, Colorado, Geological Society of America, Centennial Continent/Ocean Transect no. 12, scale 1:500,000.

Boerner, S., and Sclater, J. G., 1987, The two-dimensional infinite dike problem: Approximate solutions for the heat loss in small marginal basins, *in* Wright, J. A., and Louden, K. E., eds., Boca Raton, Florida, CRC Press, 30 p.

Bond, G. C., and Kominz, M. A., 1984, Construction of tectonic subsidence curves for the early Paleozoic miogeocline, southern Canadian Rocky Mountains: Implications for subsidence mechanisms, age of breakup, and crustal thinning: Geological Society of America Bulletin, v. 95, p. 155–173.

Bond, G. C., Nickeson, P. A., and Kominz, M. A., 1984, Breakup of a supercontinent between 625 Ma and 555 Ma; new evidence and implications for continental histories: Earth and Planetary Science Letters, v. 70, p. 325–345.

Bond, G. C., Grotzinger, J. P., and Komintz, M. A., 1988, Cambro-Ordovician eustacy: Evidence from geophysical modeling of subsidence in Cordilleran and Appalachian passive margins, *in* Kleinspen, K. L., and Paola, C., New Perspectives in basin analysis: New York, Springer Verlag, p. 129–160.

Bosworth, W., 1985, Geometry of propagating continental rifts: Nature, v. 316, p. 625–627.

Bowland, C. L., and Rosencrantz, E., 1988, Upper crustal structure of the western Colombian Basin, Caribbean Sea: Geological Society of America Bulletin, v. 100, p. 534–546.

Brace, W. F., and Kohlstedt, D. L., 1980, Limits on lithospheric stress imposed by laboratory experiments: Journal of Geophysical Research, v. 85, p. 624–625.

Bradley, D. C., 1982, Subsidence in late Paleozoic basins in the northern Appalachians: Tectonics, v. 1, p. 107–123.

Brown, R. J., Journeay, M., Lane, L. S., Murphy, D. C., and Rees, C. J., 1986, Obduction, backfolding, and piggyback thrusting the metamorphic hinterland of the southeastern Canadian Cordillera: Journal of Structural Geology, v. 8, p. 255–268.

Brueckner, H. K., and Snyder, W. S., 1985, Structure of the Havallah sequence, Golconda allochthon, Nevada—Evidence for prolonged evolution in an accretionary prism: Geological Society of America Bulletin, v. 96, p. 1113–1130.

Buffler, R. T., and Sawyer, D. S., 1985, Distribution of crust and early history, Gulf of Mexico basin: Gulf Coast Association of Geological Societies Transactions, v. 35, p. 333–344.

Burchfiel, B. C., and Davis, G. A., 1972, Structural framework and evolution of the southern part of the Cordilleran orogen, western United States: American Journal of Science, v. 272, p. 97–118.

Burkart, B., 1983, Neogene North America–Caribbean plate boundary across northern Central America—Offset along the Polochic fault: Tectonophysics, v. 99, p. 251–270.

Burke, K., Cooper, C., Dewey, J. F., Mann, P., and Pindell, J. L., 1984, Caribbean tectonics and relative plate motion, *in* Bonini, W. E., Hargraves, R. B., and Shagam, R., eds., The Caribbean–South American plate boundary and regional tectonics: Geological Society of America Memoir 162, p. 31–63.

Butler, R. F., Gehrels, G. E., McClelland, W. C., May, S. R., and Klepacki, D., 1989, Discordant paleomagnetic poles from the Canadian Coast Plutonic Complex: Regional tilt rather than large-scale displacement?: Geology, v. 17, p. 691–694.

Butler, R. F., Dickinson, W. R., and Gehrels, G. E., 1991, Paleomagnetism of coastal California and Baja California: Alternatives to large-scale northward transport: Tectonics, v. 10, p. 561–576.

Campa-Uranga, M. F., and Coney, P. J., 1983, Tectonostratigraphic terranes and mineral resources distributions of México: Canadian Journal of Earth Sciences, v. 20, p. 1041–1051.

Carr, M. D., Christiansen, R. L., and Poole, F. G., 1984, Pre-Cenozoic geology of the El Paso Mountains, southwestern Great Basin, California: A summary, *in* Lintz, J., Jr., ed., Western geological excursions, Volume 4, (Geological Society of America Cordilleran Section meeting guidebook): Reno, Nevada, Mackey School of Mines, p. 84–93.

Case, J. E., and Holcombe, T. L., 1980, Geologic-tectonic map of the Caribbean: U.S. Geological Survey Map MI 1-1100, scale 1:,500,000.

Cebull, S. E., Shurbet, D. H., Keller, G. R., and Russell, L. R., 1976, Possible role of transform faults in the development of apparent offsets in the Ouachita–southern Appalachian tectonic belt: Journal of Geology, v. 84, p. 107–114.

Champion, D. E., Howell, D. G., and Marshall, M., 1986, Paleomagnetism of Cretaceous and Eocene strata. San Miguel Island, California Borderland, and the northward translation of Baja California: Journal of Geophysical Research, v. 91, p. 11,557–11,570.

Chapman, M. E., and Solomon, S. C., 1976, North American–Eurasian plate boundary in Northeast Asia: Journal of Geophysical Research, v. 81, p. 921–930.

Chowns, T. M., and Williams, C. T., 1983, Pre-Cretaceous rocks beneath the Georgia Coastal Plain—Regional implications, *in* Gohn, G. S., ed., Studies related to the Charleston, South Carolina earthquake of 1886—Tectonics and seismicity: U.S. Geological Survey Professional Paper 1313, p. L1–L42.

Churkin, M., Jr., and Trexler, J. H., 1981, Continental plates and accreted oceanic terranes in the Arctic, *in* Nairn, A.E.M., Churkin, M., Jr., and Stelhi, F. G., eds., The ocean basins and margins, Volume 5, The Arctic Ocean: New York, Plenum Press, p. 1–20.

Clark, K. F., Foster, C. T., and Damon, P. E., 1982, Cenozoic mineral deposits and subduction-related magmatic arcs in México: Geological Society of America Bulletin, v. 93, p. 533–544.

Coe, R. S., Globerman, B. R., Plumley, P. W., and Thrupp, G. A., 1985, Paleomagnetic results from Alaska and their tectonic implications: American Association of Petroleum Geologists Circum-Pacific Earth Science Series, v. 1, p. 85–108.

Coney, P. J., 1972, Cordilleran tectonics and North American plate motions: American Journal of Science, v. 272, p. 603–628.

Coney, P. J., and Reynolds, S. J., 1977, Cordilleran Benioff zones: Nature, v. 270, p. 403–406.

Cook, F. A., and Oliver, J. E., 1981, The late Precambrian–Early Paleozoic continental edge in the Appalachian orogen: American Journal of Science, v. 281, p. 993–1008.

Cook, F. A., Brown, L. D., Kaufman, S., Oliver, J. E., and Petersen, T. A., 1981, COCORP seismic profiling of the Appalachian orogen beneath the Coastal Plain of Georgia: Geological Society of America Bulletin, v. 92, Part 1, p. 738–748.

Cook, F. A., Varsek, J. L., and Clowes, R. M., 1991, LITHOPROBE reflection transect of southwestern Canada: Mesozoic thrust and fold belt to midocean ridge: American Geophysical Union Geodynamics Series, v. 22, p. 247–256.

Cordey, F., Mortimer, N., DeWever, P., and Monger, J.W.H., 1987, Significance of Jurassic radiolarians from the Cache Creek terrane, British Columbia: Geology, v. 15, p. 1151–1154.

Cowan, D. S., and Potter, C. J., 1986, B3 Juan de Fuca spreading ridge to Montana thrust belt: Boulder, Colorado, Geological Society of America, Centennial Continent/Ocean Transect no. 9, scale 1:500,000.

Dallmeyer, R. D., 1988, Late Paleozoic tectonothermal evolution of the western Piedmont and eastern Blue Ridge, Georgia: Controls on the chronology of terrane accretion and transport in the southern Appalachian orogen: Geological Society of America Bulletin, v. 100, p. 702–713.

Dallmeyer, R. D., Blackwood, R. F., and Odom, A. L., 1981, Age and origin of the Dover fault: Tectonic boundary between the Gander and Avalon zones of the northeastern Newfoundland Appalachians: Canadian Journal of Earth Sciences, v. 8, p. 1431–1442.

Dallmeyer, R. D., Wright, J. E., Jr., Secor, D. T., and Snoke, A. W., 1986, Character of the Alleghanian orogeny in the southern Appalachians: Part II, Geochronological constraints on the tectonothermal evolution of the eastern Piedmont in South Carolina: Geological Society of America Bulletin, v. 97, p. 1329–1344.

Dalziel, I.W.D., 1991, Pacific margins of Laurentia and East Antarctica–Australia as a conjugate rift pair: Evidence and implications for an Eocambrian supercontinent: Geology, v. 19, p. 598–601.

Davis, G. A., Monger, J.W.H., and Burchfiel, B. C., 1978, Mesozoic construction of the Cordilleran "collage," central British Columbia to central California, *in* Howell, D. G., and McDougall, K. A., eds., Mesozoic paleogeography of the western United States: Society Economic Paleontologists and Mineralogists, Pacific section, Pacific Coast Paleogeography Symposium 2, p. 1–32.

DeMets, C., Gordon, R. G., Argus, D. F., and Stein, S., 1990, Current plate motions: Geophysical Journal International, v. 101, p. 425–478.

Denison, R. E., Burke, W. H., Jr., Hetherington, E. A., and Otto, J. B., 1971, Basement rock framework in parts of Texas, southern New Mexico, and northern Mexico: West Texas Geological Society Publication 71-59, p. 3–14.

Dennis, A. J., 1991, Is the central Piedmont suture a low-angle normal fault?: Geology, v. 19, p. 1081–1084.

DePaor, D. G., Bradley, D. C., Eisenstadt, G., and Phillips, S. M., 1989, The Arctic Eurekan orogen: Geological Society of America Bulletin, v. 101, p. 952–967.

Dickinson, W. R., 1976, Sedimentary basins developed during evolution of Mesozoic-Cenozoic arc-trench system in western North America: Canadian Journal of Earth Sciences, v. 13, p. 1268–1287.

——, 1979, Mesozoic forearc basins in central Orogen: Geology, v. 7, p. 166–170.

——, 1981, Plate tectonic evolution of the southern Cordillera: Arizona Geological Society Digest, v. 14, p. 113–135.

Dickinson, W. R., Harbaugh, D. W., Saller, A. H., Heller, P. C., and Snyder, W. S., 1983, Detrital modes of upper Paleozoic sandstones derived from the Antler orogen in Nevada: Implications for the nature of the Antler orogeny: American Journal of Science, v. 283, p. 481–509.

Dilles, J. A., and Wright, J. E., 1988, The chronology of early Mesozoic arc magmatism in the Yerington district of western Nevada and its regional implications: Geological Society of America Bulletin, v. 100, p. 644–652.

Dillon, W. P., and Popenoe, P., 1988, The Blake Plateau Basin and Carolina Trough, *in* Sheridan, R. E. and Grow, J. A., The Atlantic continental margin, U.S. Boulder, Colorado, Geological Society of America, The Geology of North America, v. I-2, p. 291–328.

Dillon, W. P., and Vedder, J. G., 1973, Structure and development of the continental margin of British Honduras: Geological Society of America Bulletin, v. 84, p. 2713–2732.

Dixon, J., and Dietrich, J., 1990, Canadian Beaufort Sea, *in* Grantz, A., Johnson, L., and Sweeney, J. F., eds., The Arctic Ocean region: Boulder, Colorado, Geological Society of America, The Geology of North America, v. L, p. 239–256.

Donnelly, T. W., 1966, Geology of St. Thomas and St. John, U.S. Virgin Islands: Geological Society of America Memoir 98, p. 85–176.

—— , 1973, Magnetic anomaly observations in the eastern Caribbean sea, *in* Edgar, N. T., and Saunders, J. B., eds., Initial reports of the Deep Sea Drilling Project, Volume 15: Washington, D.C., U.S. Government Printing Office, p. 1023–1030.

—— , 1989, Geologic history of the Caribbean and Central America, in Bally, A. W., and Palmer, A. R., eds., The geology of North America—An overview: Boulder, Colorado, Geological Society of America, The Geology of North America, v. A, p. 299–321.

Dover, J. H., 1980, Status of the Antler Orogeny in central Idaho—Clarification and constraints from the Pioneer Mountains, *in* Fouch, T. D. and Magathan, E., eds., Paleozoic paleogeography of the west-central United States: Society of Economic Paleontologists and Mineralogist, Rocky Mountain Paleogeography Symposium 1, p. 371–386.

Drake, A. A., Jr., 1987, Pre-Taconian deformation in the Piedmont of the Potomac Valley—Penobscottian, Cadomian, or both?, *in* Contributions to Virginia geology, Volume V: Charlottesville, Virginia, Division of Mineral Resources Publication 74, p. 1–18.

Dunbar, J., and Sawyer, D. S., 1987, Implications of continental crust extension for plate reconstruction: An example from the Gulf of Mexico: Journal of Geophysical Research, v. 93, p. 9075–9092.

Dutro, J. T., Jr., 1981, Geology of Alaska bordering the Arctic Ocean, *in* Nairn, A.E.M., Churkin, M., Jr., and Stelhi, F. G., The ocean basins and margins, Volume 5, The Arctic Ocean: New York, Plenum Press, p. 21–36.

Echevarria, G., and eight others, 1991, Oil and gas exploration in Cuba: Journal of Petroleum Geology, v. 14, p. 259–274.

Eisbacher, G. H., 1983, Devonian-Mississippian sinistral transcurrent faulting along the cratonic margin of western North America: A hypothesis: Geology, v. 11, p. 7–10.

Elison, M. W., 1991, Intracontinental contraction in western North America: Continuity and episodicity: Geological Society of America Bulletin, v. 103, p. 1226–1238.

Elison, M. W., Speed, R. C., and Kistler, R. W., 1990, Geologic and isotopic constraints on crustal structure in the northern Great Basin: Geological Society of America Bulletin, v. 102, p. 1007–1022.

Engebretson, D. C., Cox, A., and Gordon, R. G., 1985, Relative motions between oceanic and continental plates in the Pacific basin: Geological Society of America Special Paper 206, 59 p.

Feo-Codocido, G., Smith, F. D., Jr., Aboud, N., and Di Giacomo, E., 1984, Basement and Paleozoic rocks of the Venezuelan Llanos basin: *in* Bonini, W. E., Hargraves, R. B., and Shagam, R., eds., The Caribbean-South American plate boundary and regional tectonics: Geological Society of America Memoir 162, p. 175–187.

Flawn, P. T., Goldstein, A., Jr., King, P. B., and Weaver, C. E., 1961, The Ouachitan system: Austin, University of Texas Publication, 401 p.

Frei, L. S., 1986, Additional paleomagnetic results from the Sierra Nevada: Further constraints on Basin and Range extension and northward displacement in the western United States: Geological Society of America Bulletin, v. 97, p. 840–849.

Frisch, W., Meschede, M., and Sick, M., 1922, Origin of Central American ophiolites, evidence from paleomagnetic results: Geological Society of America Bulletin, v. 104, p. 1301–1314.

Gabrielse, H., 1985, Major dextral displacements along the Northern Rocky Mountain Trench and related lineaments in north-central British Columbia: Geological Society of America Bulletin, v. 96, p. 1–14.

Gates, A. E., Simpson, C., and Glover, L., 1986, Appalachian Carboniferous dextral/strike-slip faults: An example from Brookneal, Virginia: Tectonics, v. 5, p. 119–133.

Gehrels, G. E., and Saleeby, J. B., 1987, Geologic framework, tectonic evolution, and displacement history of the Alexander terrane: Tectonics, v. 6, p. 151–174.

Geiser, P. A., 1989, Description of structures: Central Appalachians, *in* eds., The Appalachian-Ouachita orogen in the United States: Hatcher, R. D., Jr., Thomas, W. A., and Viele, G. W., Boulder, Colorado, Geological Society of America, The Geology of North America, v. F-2, p. 362–375.

Geist, E. L., Childs, J. R., and Scholl, D. W., 1988, The origin of summit basins of the Aleutian Ridge: Implications for block rotations of an arc massif: Tectonics, v. 7, p. 327–341.

Getty, S. R., and Gromet, L. P., 1992, Geochronological constraints on ductile deformation, crustal extension, and doming about a basement-cover boundary, New England Appalachians: American Journal of Science, v. 292, p. 359–397.

Glover, Lynn, III, and Sinha, A. K., 1973, The Virgilina deformation, a late Precambrian to Early Cambrian(?) orogenic event in the central Piedmont of Virginia and North Carolina: American Journal of Science, v. 273-A, p. 234–251.

Glover, Lynn, III, Speer, A., Russell, G. S., and Farrar, S. S., 1983, Ages of regional metamorphism and ductile deformation in the central and southern Appalachians: Lithos, v. 16, p. 223–245.

Goldsmith, R., Marvin, R. F., and Mehnert, H. H., 1971, Radiometric ages in the Santander Massif, Colombian Andes: U.S. Geological Survey Professional Paper 750D, p. 44–49.

Goldstein, A. G., 1989, Tectonic significance of multiple motions on terrane-bounding faults in the northern Appalachians: Geological Society of America Bulletin, v. 101, p. 927–938.

Gonzalez de Juana, C., Iturralde, J. M., and Picard, X., 1980, Geologia de Venezuela y de sus cuencas petroliferas: Caracas, Venezuela, Ediciones Foninves, 1031 p.

Gordey, S. P., Abbott, J. G., Templeman-Kluit, D. J., and Gabrielse, H., 1987, Antler clastics in the Canadian Cordillera: Geology, v. 15, p. 103–107.

Gose, W. A., 1983, Late Cretaceous–early Tertiary tectonic history of southern Central America: Journal of Geophysical Research, v. 12, p. 10,585–10,592.

Gose, W. A., Belcher, R. C., and Scott, G. R., 1982, Paleomagnetic results from northeastern México: Evidence for large Mesozoic rotations: Geology, v. 10, p. 50–54.

Graham, S. A., Dickinson, W. R., and Ingersoll, R. V., 1975, Himalayan-Bengal model for flysch dispersal in the Appalachian-Ouachita system: Geological Society of America Bulletin, v. 86, p. 273–286.

Grantz, A., and May, S. D., 1983, Rifting history and structural development of the continental margin north of Alaska, *in* Watkins, J. S. and Drake, C. L., eds., Studies in continental margin geology: American Association of Petroleum Geologists Memoir 34, p. 77–100.

Grantz, A., Eittreim, S., and Dinter, D. A., 1979, Geology and tectonic development of the continental margin north of Alaska: Tectonophysics, v. 59, p. 262–291.

Grantz, A., Moore, T. E., and Roeske, S. M., 1991, A3 Gulf of Alaska to the Arctic Ocean: Boulder, Colorado, Geological Society of America, Centennial Continent/Ocean Transect no. 15, scale 1:500,000.

Green, A. G., and 9 others, 1988, Crustal structure of the Grenville front and adjacent terranes: Geology, v. 16, p. 788–792.

Gripp, A. E., and Gordon, R. G., 1990, Current plate velocities relative to the hotspots incorporating the NUVEL 1 global plate motion model: Geophysical Research Letters, v. 17, p. 1109–1112.

Hagstrum, J. T., Sawlan, M. G., Hausback, B. P., Smith, J. G., and Gromme, C. S., 1987, Miocene paleomagnetism and tectonic setting of the Baja California peninsula, México: Journal of Geophysical Research, v. 92, p. 2627–2640.

Hagstrum, J. T., McWilliams, M., Howell, D. G., and Grommé, S., 1990, Mesozoic paleomagnetism and northward translation of the Baja California Peninsula: Geological Society of America Bulletin, v. 96, p. 1077–1090.

Hamilton, W., and Myers, W. B., 1966, Cenozoic tectonics of the western United States: Geophysical Review, v. 4, p. 509–549.

Handschy, J. W., Keller, G. R., and Smith, K. J., 1987, The Ouachita system in northern México: Tectonics, v. 6, p. 323–330.

Harris, R. A., Stone, D. B., and Turner, D. L., 1987, Tectonic implications of paleomagnetic and geochronologic data from the Yukon-Koyukuk province, Alaska: Geological Society of America Bulletin, v. 99, p. 362–375.

Harry, D. L., and Sawyer, D. S., 1992, A dynamic model of extension in the Baltimore Canyon Trough: Tectonics, v. 11, p. 420–436.

Hatcher, R. D., Jr., 1978, Tectonics of the western Piedmont and Blue Ridge, southern Appalachians: Review and speculation: American Journal of Science, v. 278, p. 276–304.

——, 1987, Tectonics of the southern and central Appalachian internides: Annual Review of Earth and Planetary Sciences, v. 15, p. 337–362.

——, 1989, Tectonic synthesis of the U.S. Appalachians, *in* Hatcher, R. D., Jr., Thomas, W. A., and Viele, G. W., eds., The Appalachian-Ouachita orogen in the United States: Boulder, Colorado, Geological Society of America, The Geology of North America, v. F-2, p. 511–535.

Hatcher, R. D., Jr., Thomas, P. A., Geiser, P. A., Snoke, A. W., Mosher, S., Wiltschko, D. V., 1989, Alleghanian orogen, *in* Hatcher, R. D., Jr., Thomas, W. A., and Viele, G. W., eds., The Appalachian-Ouachita orogen in the United States: Boulder, Colorado, edited by Geological Society of America, The Geology of North America, v. F-2, p. 233–318.

Hatcher, R. D., Jr., and 11 others, 1993, E-5 Cumberland Plateau (North American craton) to Blake Plateau basin: Boulder, Colorado, Geological Society of America, Centennial Continent/Ocean Transect no. 18, scale 1:500,000.

Haworth, R. T., Williams, H., and Keen, C. E., 1985, D-1 Northern Appalachian Mountains across Newfoundland to south Labrador Sea: Boulder, Colorado, Geological Society of America, Centennial Continent/Ocean Transect no. 1, scale 1:500,000.

Heck, F. R., 1989, Mesozoic extension in the Southern Appalachians: Geology, v. 8, p. 711–714.

Heller, P. L., Bowdler, S. S., Chambers, H. P., Coogan, J. C., Hagen, E. S., Shuster, M. W., Winslow, N. S., and Lawton, T. F., 1986, Time of initial thrusting in the Sevier orogenic belt, Idaho-Wyoming and Utah: Geology, v. 14, p. 388–391.

Hillhouse, J. W., 1977, Paleomagnetism of the Triassic Nikolai Greenstone, McCarthy quadrangle, Alaska: Canadian Journal of Earth Sciences, v. 14, p. 2578–2592.

Holbrook, W. S., Reiter, E. C., Purdy, G. M., and Toksoz, M. N., 1992, Image of the Moho across the continent-ocean transition, U.S. east coast: Geology, v. 20, p. 203–206.

Holcombe, T. L., Vogt, P. R., Matthews, J. E., and Murchison, R. R., 1973, Evidence for seafloor spreading in the Cayman Trough: Earth and Planetary Science Letters, v. 20, p. 357–371.

Horton, J. W., Jr., 1981, Shear zone between the Inner Piedmont and Kings Mountain belts in the Carolinas: Geology, v. 9, p. 28–33.

Horton, J. W., Jr., Drake, A. A., Jr., and Rankin, D. W., 1989, Tectonostratigraphic terranes and their boundaries in the central and southern Appalachians *in* Dallmeyer, R. D., ed., Terranes in the circum-Atlantic Paleozoic orogens: Geological Society of America Special Paper 230, p. 213–245.

Houseknecht, D. W., 1986, Evolution from passive margin to foreland basin: The Atoka Formation of the Arkoma basin, south-central U.S.A.: International Association of Sedimentologists Special Publication 8, p. 327–345.

Howell, D. G., and Wiley, T. J., 1987, Crustal evolution of northern Alaska inferred from sedimentology and structural relations of the Kandik area: Tectonics, v. 6, p. 619–631.

Howell, D. G., Gibson, J. D., Fuis, G. S., Knapp, J. H., Haxel, G. B., Keller, B. R., Silver, L. T., and Vedder, J. G., 1985, C3 Pacific abyssal plains to Rio Grande rift: Boulder, Colorado, Geological Society of America Centennial Continent/Ocean Transect no. 5, scale 1:500,000.

Hutchinson, D. R., Klitgord, K. D., Lee, M. W., and Trehu, A. M., 1988, U.S. Geological Survey deep seismic reflection profile across the Gulf of Maine: Geological Society of America Bulletin, v. 100, p. 172–184.

Hyndman, R. D., Yorath, C. J., Clowes, R. M., and Davis, E. E., 1990, The northern Cascadia subduction zone at Vancouver Island, seismic structure and tectonic history: Canadian Journal of Earth Sciences, v. 27, p. 313–329.

Ingersoll, R. V., 1983, Petrofacies and provenance of late Mesozoic forearc basin, northern and central California: American Association of Petroleum Geologists, Bulletin, v. 67, p. 1125–1142.

Ingersoll, R. V., and Schweikert, R. A., 1986, A plate-tectonic model for Late Jurassic orogenesis, Nevadan orogeny, and forearc initiation, northern California: Tectonics, v. 5, p. 901–912.

Irving, E., 1979, Paleopoles and paleolatitudes of North America and speculations about displaced terranes: Canadian Journal of Earth Sciences, v. 16, p. 669–694.

Irving, E., and Yole, R. W., 1972, Paleomagnetism and the kinematic history of mafic and ultramafic rocks in fold mountains belts: Ottawa, Ontario, Earth Physics Branch Publication 42, p. 87–95.

Irving, E., Woodsworth, G. J., Wynne, P. J., and Morrison, A., 1985, Paleomagnetic evidence for displacement from the south of the Coast Plutonic Complex, British Columbia: Canadian Journal of Earth Sciences, v. 22, p. 584–598.

Jansma, P. E., 1988, The tectonic interaction between an oceanic allochthon and its foreland basin during continental overthrusting; the Antler orogeny, Nevada [Ph.D. thesis]: Evanston, Illinois, Northwestern University, 315 p.

Johnson, J. G., and Pendergast, A., 1981, Timing and mode of emplacement of the Roberts Mountains allochthon, Antler orogeny: Geological Society of America Bulletin, v. 92, p. 648–658.

Jones, D. L., Irwin, W. P., and Ovenshine, A. T., 1972, Southeastern Alaska—A displaced continental fragment?: U.S. Geological Survey Professional Paper 800-B, p. 211–217.

Jones, D. L., Silberling, N. J., and Hillhouse, J., 1977, Wrangellia, a displaced terrane in northwestern North America: Canadian Journal of Earth Sciences, v. 14, p. 2565–2577.

Jones, D. L., Silberling, N. J., Gilbert, W., and Coney, P., 1982, Character, distribution, and tectonic significance of accretionary terranes in the central Alaskan Range: Journal of Geophysical Research, v. 87, p. 3709–3717.

Jordan, T. H., 1975, The present-day motions of the Caribbean plate: Journal of Geophysical Research, v. 80, p. 4433–4439.

Keen, C. E., and deVoorgd, B., 1988, The continent-ocean boundary at the rifted margin off eastern Canada: New results from deep seismic reflection studies: Tectonics, v. 7, p. 107–124.

Keen, C. E., and Haworth, R. T., 1985a, D4 Rifted continental margins off Labrador: Boulder, Colorado, Geological Society of America, Centennial Continent/Ocean Transect no. 4, scale 1:500,000.

——, 1985b, D-2 Transform margin south of Grand Banks: Offshore eastern Canada: Boulder, Colorado, Geological Society of America Centennial Continent/Ocean Transect no. 2, scale 1:500,000.

——, 1985c, D-3 Rifted continental margin off Nova Scotia: Offshore eastern Canada: Boulder, Colorado, Geological Society of America Centennial Continent/Ocean Transect no. 3, scale 1:500,000.

Keller, G. R., and Cebull, S. E., 1973, Plate tectonics and the Ouachita system in Texas, Oklahoma, and Arkansas: Geological Society of America Bulletin, v. 84, p. 1659–1666.

Keller, G. R., Kruger, J. M., Smith, K. J., and Voight, W. M., 1989, The Ouachita system: A geophysical overview, *in* Hatcher, R. D., Jr. Thomas, W. A., and Viele, G. W., eds., The Appalachian-Ouachita orogen in the United States: Geological Society of America, The Geology of North America, v. F-2, p. 689–693.

Kellogg, J. N., 1984, Cenozoic tectonic history of the Sierra de Perija, Venezuela-Colombia, and adjacent basins, *in* Bonini, W. E., Hargraves, R. B., and Shagam, R., eds., The Caribbean-South American plate boundary and regional tectonics: Geological Society of America Memoir 162, p. 239–261.

Keppie, J. D., 1989, Northern Appalachian terranes and their accretionary history, *in* Geological Society of America Special Paper 230, p. 159–192.

King, P. B., 1975, The Ouachita and Appalachian orogenic belts, *in* Nairn,

A.E.M., and Stehli, F. G., eds., The ocean basins and margins, Volume 3, The Gulf of Mexico and Caribbean: New York, Plenum Press, p. 201–241.

Kistler, R. W., 1990, Two different lithosphere types in the Sierra Nevada, California *in* Anderson, J. L., ed., The nature and origin of Cordilleran magmatism: Geological Society of America Memoir 174, p. 271–281.

Kistler, R. W., and Peterman, Z. E., 1973, Variations in Sr, Rb, K, Na, and initial $^{87}Sr/^{86}Sr$ in Mesozoic granitic rocks and intruded wall rocks in central California: Geological Society of America Bulletin, v. 84, p. 3489–3512.

Klepacki, D. W., and Wheeler, J. O., 1985, Stratigraphic and structural relations of the Milford, Kaslo, and Slocan Groups, Goat Range, British Columbia: Geological Survey of Canada Paper 85-1A, p. 277–286.

Klitgord, K. D., and Schouten, H., 1986, Plate kinematics of the central Atlantic, *in* Vogt, P. R., and Tucholke, B. E., eds., The western North Atlantic region: Boulder, Colorado, Geological Society of America, The Geology of North America, v. I-2, p. 351–375.

Knoll, A. H., and Keller, F. B., 1979, Late Precambrian microfossils from the Walden Creek Group, Ocoee Supergroup, Tennessee: Geological Society of America Abstracts with Programs, v. 11, p. 185.

Kroonenberg, S. B., 1982, A Grenvillian granulite belt in the Colombian Andes and its relation to the Guiana Shield: Geologie et Mijnbouw, v. 61, p. 325–333.

Ladd, J. W., 1976, Relative motion of South America with respect to North America and Caribbean tectonics: Geological Society of America Bulletin, v. 87, p. 969–976.

Lillie, R. J., Nelson, K. D., de Voogd, B., Brewer, J. A., Oliver, J. E., Brown, L. D., Kaufman, S., and Viele, G. W., 1983, Crustal structure of Ouachita Mountains, Arkansas: A model based on integration of COCORP reflection profiles and regional geophysical data: American Association of Petroleum Geologists Bulletin, v. 67, p. 907–931.

Lisowski, M., Savage, J. C., Prescott, W. H., and Dragert, H., 1987, Strain accumulation along the Cascadia subduction zone in western Washington [abs.]: Eos (Transactions, American Geophysical Union), v. 68, p. 1240.

Lister, G., Etheridge, M. A., and Symonds, P. A., 1986, Application of the detachment fault model to the formation of passive continental margins: Geology, v. 14, p. 246–250.

Lonsdale, P., 1989, Geology and tectonic history of the Gulf of California, *in* Winterer, E. L., Hussong, D. M., and Decker, R. W., eds., The eastern Pacific region: Boulder, Colorado, Geological Society of America, The Geology of North America, v. N, p. 499–521.

Lopez-Infanzon, M., 1986, Petrologia y radiometria de rocas igneas y metamorficas de Mexico: Asociacion Mexicanos Geologicos Petroleros Boletin, v. 38, p. 59–98.

Lowe, D. R., 1989, Stratigraphy, sedimentology, and depositional setting of pre-orogenic rocks of the Ouachita Mountains, Arkansas and Oklahoma, *in* Hatcher, R. D., Jr., Thomas, W. A., and Viele, G. W., eds., The Appalachian-Ouachita orogen in the United States: Boulder, Colorado, Geological Society of America, The Geology of North America, v. F-2, p. 575–590.

Lund, K., and Snee, L. W., 1988, Metamorphism, structural development, and age of the continent–island arc juncture in west-central Idaho, *in* Ernst, G., ed., Metamorphism and crustal evolution of the western United States (Rubey Volume 7): Englewood Cliffs, New Jersey, Prentice-Hall, p. 296–331.

Malfait, B. T., and Dinkelman, M. G., 1972, Circum-Caribbean tectonic and igneous activity and the evolution of the Caribbean plate: Geological Society of America Bulletin, v. 83, p. 251–272.

Mann, P., Draper, G., and Lewis, J., 1991, Overview of the geologic and tectonic development of Hispaniola *in* Mann, P., Draper, G., and Lewis, J., eds., Geologic and tectonic development of the North America-Caribbean plate boundary in Hispaniola: Geological Society of America Special Paper 262, p. 1–29.

Manspeizer, W., 1988, Triassic-Jurassic rifting and opening of the Atlantic, an overview, *in* Manspeizer, W., ed., Triassic-Jurassic rifting: Amsterdam, Elsevier Scientific Publishers, Developments in Geotectonics, v. 22, p. 41–80.

Manton, W. I., 1987, Tectonic interpretation of the morphology of Honduras: Tectonics, v. 6, p. 633–651.

May, S. R., and Butler, R. F., 1986, North American Jurassic apparent polar wander: Implications for plate motion, paleogeography, and Cordilleran tectonics: Journal of Geophysical Research, v. 91, p. 11,519–11,544.

Maze, W. B., 1984, Jurassic La Quinta Formation, Sierra de Perija, northwestern Venezuela, igneous petrology and tectonic environment *in* Bonini, W. E., Hargraves, R. B., and Shagam, R., eds., The Caribbean-South American plate boundary and regional tectonics: Geological Society of America Memoir 162, p. 263–282.

McBride, E. F., 1989, Stratigraphy and sedimentary history of pre-Permian Paleozoic rocks of the Marathon uplift, *in* Thomas, W. A., and Viele, G. W., eds., The Appalachian-Ouachita orogen in the United States: Boulder, Colorado, Geological Society of America, The Geology of North America, v. F-2, p. 603–620.

McBride, J. H., Nelson, K. D., and Brown, L. D., 1989, Evidence and implications of an extensive early Mesozoic rift basin and basalt-diabase sequence beneath the southeast Coastal Plain: Geological Society of America Bulletin, v. 101, p. 512–520.

McCaffrey, R., 1992, Oblique plate convergence, slip vectors, and forearc deformation: Journal of Geophysical Research, v. 97, p. 8905–8921.

McCann, W. R., and Sykes, L. R., 1984, Subduction of aseismic ridges beneath the Caribbean plate: Implications for the tectonics and seismic potential of the northeastern Caribbean: Journal of Geophysical Research, v. 89, p. 4493–4519.

McDowell, F. W., and Keizer, R. P., 1977, Timing of mid-Tertiary volcanism in the Sierra Madre Occidental between Durango City and Mazatlan, México: Geological Society of America Bulletin, v. 88, p. 1479–1487.

McKenzie, D. P., 1978, Some remarks on the development of sedimentary basins: Earth and Planetary Science Letters, v. 40, p. 25–32.

McMenamin, M.A.S., and McMenamin, D.L.S., 1990, The emergence of animals: The Cambrian breakthrough: New York, Columbia University Press, 30 p.

Miall, A., 1983, The Nares Strait problem: A reevaluation of the geological evidence in terms of a diffuse, oblique-slip plate boundary between Greenland and the Canadian Arctic Islands: Tectonophysics, v. 100, p. 227–239.

Miller, C. F., and Bradfish, L. J., 1980, An inner Cordilleran belt of muscovite-bearing plutons: Geology, v. 8, p. 412–416.

Miller, E. L., Holdsworth, B. K., Whiteford, W. B., and Rogers, D., 1984, Stratigraphy and structure of the Schoonover sequence, northeastern Nevada: Implications for plate-margin tectonics: Geological Society of America Bulletin, v. 95, p. 1063–1076.

Miller, M. M., 1987, Dispersed remnants of a northeast Pacific fringing arc: Upper Paleozoic terranes of McCloud faunal affinity, western U.S.: Tectonics, v. 6, p. 807–830.

Mitre-Salazar, L.-M., and eight others, 1990, H-1 La Paz to Saltillo, northern Mexico: Boulder, Colorado, Geological Society of America, Centennial Continent/Ocean Transect no. 13, scale 1:500,000.

Molina-Garza, R. S., Van der Voo, R., and Urrutia-Furugauchi, J., 1990, Paleomagnetic data from Chiapas, southern México: Implications for tectonic evolution of the Gulf of Mexico: Eos (Transactions, American Geophysical Union), v. 71, p. 491.

Molnar, P., and Sykes, L. R., 1969, Tectonics of the Caribbean and middle America regions from focal mechanisms and seismicity: Geological Society of America Bulletin, v. 80, p. 1639–1684.

Monger, J.W.H., 1977, Upper Paleozoic rocks of the western Canadian Cordillera and their bearing on Cordilleran evolution: Canadian Journal of Earth Sciences, v. 14, p. 1832–1859.

Monger, J.W.H., and Berg, H. C., 1984, Lithotectonic terrane map of western Canada and southeastern Alaska: U.S. Geological Survey Open-File Report 84-523, p. B1–B31.

Monger, J.W.H., and Ross, C. A., 1971, Distribution of Fusulinaceans in the western Canadian Cordillera: Canadian Journal of Earth Sciences, v. 8, p. 259–278.

Monger, J.W.H., Price, R. A., and Tempelman-Kluit, D. J., 1982, Tectonic accretion and the origin of the two major metamorphic and plutonic welts in the Canadian Cordillera: Geology, v. 10, p. 70–75.

Monger, J.W.H., Clowes, R. M., Price, R. A., Simony, P. S., Riddihough, R. P., and Woodsworth, G. J., 1985, B2 Juan de Fuca plate to Alberta plains: Boulder, Colorado, Geological Society of America, Centennial Continent/Ocean Transect no. 7, scale 1:500,000.

Moore, J. C., Byrne, T., Plumley, P. W., Reid, M., Gibbons, H., and Coes, R. S., 1983, Paleogene evolution of the Kodiak Islands, Alaska: Consequences of ridge-trench interactions in a more southernly latitude: Tectonics, v. 2, p. 265–293.

Moore, J. C., and 16 others, 1988, Tectonics and hydrogeology of the northern Barbados Ridge: Results from Ocean Drilling Program leg 110: Geological Society of America Bulletin, v. 100, p. 1578–1593.

Moore, T. E., 1987, Geochemistry and tectonic setting of some volcanic rocks of the Franklinian assemblage, central and eastern Brooks Range: Alaska Geological Society Bulletin, v. 2, p. 691–710.

Mortensen, J. K., and Jilsen, G. A., 1985, Evolution of the Yukon-Tanana terrane: Evidence from southeastern Yukon Territory: Geology, v. 13, p. 806–810.

Mutter, J. C., 1985, Seaward dipping reflectors and the continent-ocean boundary at passive continental margins: Tectonophysics, v. 114, p. 117–131.

Namson, J. S., and Davis, T. L., 1988, Seismically active fold and thrust belt in the San Joaquin Valley, central California: Geological Society of America Bulletin, v. 100, p. 257–273.

Nelson, K. D., and eight others, 1985, New COCORP profiling in the southeastern United States, Part I: Late Paleozoic suture and Mesozoic rift basin: Geology, v. 13, p. 714–718.

Neuman, R. B., and Max, M. D., 1989, Penobscottian-Grampian-Finnmarkian orogenies as indicators of terrane linkages *in* Dallmeyer, R. D., ed., Terranes in the circum-Atlantic Paleozoic orogens: Geological Society of America Special Paper 230, p. 31–45.

Neuman, R. B., Palmer, A. R., and Dutro, J. T., Jr., 1989, Paleontological contributions to Paleozoic paleogeographic reconstructions of the Appalachians, *in* Hatcher, R. D., Jr., Thomas, W. A., and Viele, G. W., eds., The Appalachian-Ouachita orogen in the United States: Boulder, Colorado, Geological Society of America, The Geology of North America, v. F-2, p. 375–384.

Nicholas, R. L., and Rozendal, R. A., 1975, Subsurface positive elements within Ouachita foldbelt in Texas and their relation to Paleozoic cratonic margin: American Association of Petroleum Geologists Bulletin, v. 59, p. 193–216.

Nicholas, R. L., and Waddell, D. E., 1989, The Ouachita system in the subsurface of Texas, Arkansas, and Louisiana, *in* Hatcher, R. D., Jr., Thomas, W. A., and Viele, G. W., eds., The Appalachian-Ouachita orogen in the United States: Boulder, Colorado, Geological Society of America, The Geology of North America, v. F-2, p. 661–672.

Nilsen, T. H., Brabb, E. E., and Simoni, T. R., Jr., 1976, Deep sea fan deposition of the Devonian Nation River Formation, Yukon-Kandik area, Alaska, *in* Miller, T. P., ed., Recent and ancient sedimentary environments in Alaska: Fairbanks, Alaska Geological Society Symposium proceedings for 1975, p. E1–E20.

Nokelberg, W., Jones, D. L., and Silberling, W. J., 1985, Origin and tectonic evolution of the Maclaren and Wrangellia terranes, eastern Alaska Range, Alaska: Geological Society of America Bulletin, v. 96, p. 1251–1270.

Norris, D. K., 1985, Eastern cordilleran foldbelt of northern Canada: Its structural geometry and hydrocarbon potential: American Association of Petroleum Geologists Bulletin, v. 69, p. 778–808.

Norris, D. K., and Yorath, C. J., 1981, The North American plate from the Arctic Archipelago to the Romanyof Mountains, *in* Nairn, A.E.M., Churkin, M., Jr., and Stelhi, F. G., eds., The ocean basins and margins, Volume 5, The Arctic Ocean: New York, Plenum Press, p. 37–103.

Oldow, J. S., 1984, Spatial variability in the structure of the Roberts Mountains allochthon, western Nevada: Geological Society of America Bulletin, v. 95, p. 174–185.

Oldow, J. S., Avé Lallemant, H. G., Julian, F. E., and Seidensticker, C. M., 1987, Ellesmerian (?) and Brookian deformation in the Franklin Mountains, northeastern Brooks Range, Alaska, and its bearing on the origin of the Canada basin: Geology, v. 15, p. 37–41.

Onstott, T. C., Hall, C. M., and York, D., 1989, $^{40}Ar/^{39}Ar$ thermochronometry of the Imataca complex, Venezuela: Precambrian Research, v. 42, p. 255–292.

Ortega-Gutierrez, F., 1989, H3 Acapulco Trench to Gulf of Mexico across southern Mexico: Boulder, Colorado, Geological Society of America, Centennial Continent/Ocean Transect no. 12, scale 1:500,000.

Osberg, P. H., Tull, J. F., Robinson, P., Hon, R., and Butler, J. R., 1989, The Acadian orogen, *in* Hatcher, R. D., Jr., Thomas, W. A., and Viele, G. W., eds., The Appalachian-Ouachita orogen in the United States: Boulder, Colorado, Geological Society of America, The Geology of North America, v. F-2, p. 179–232.

Packer, D. R., and Stone, D. B., 1972, An Alaskan Jurassic paleomagnetic pole and the Alaskan orocline: Nature, v. 237, p. 25–26.

Padilla y Sanchez, R. J., 1986, Post-Paleozoic tectonics of northeast Mexico and its role in the evolution of the Gulf of Mexico: Geofisica Internacional, v. 25, p. 257–206.

Palmer, A. R., compiler, 1983, The Decade of North American Geology 1983 geologic time scale: Geology, v. 11, p. 503–504.

Panuska, B. C., 1985, Paleomagnetic evidence for a post-Cretaceous accretion of Wrangellia: Geology, v. 13, p. 880–883.

Pardo, G., 1975, Geology of Cuba, *in* Nairn, A.E.M., and Stehli, F. G., eds., The ocean basins and margins, Volume 3, The Gulf of Mexico and Caribbean: New York, Plenum Press, p. 553–615.

Pilger, R. H., 1978, A closed Gulf of Mexico, pre-Atlantic Ocean plate reconstruction and the early rift history of the Gulf and North Atlantic: Gulf Coast Association of Geological Societies Transactions, v. 28, p. 385–393.

Pindell, J., 1985, Alleghenian reconstruction and subsequent evolution of the Gulf of Mexico, Bahamas, and proto-Caribbean: Tectonics, v. 4, p. 1–39.

Pindell, J. L., and Barrett, S. F., 1990, Geological evolution of the Caribbean region: A plate-tectonic perspective, *in* Dengo, G., and Case, J. E., eds., The Caribbean region: Boulder, Colorado, Geological Society of America, The Geology of North America, v. H, p. 405–432.

Pindell, J. L., Cande, S. C., Pitman, W. C., III, Rowley, D. B., Dewey, J. F., Labrecque, J., and Haxby, W., 1988, A plate-kinematic framework for models of Caribbean evolution: Tectonophysics, v. 155, p. 121–138.

Piper, J.D.A., 1983, Proterozoic paleomagnetism and single continent plate tectonics: Geophysical Journal, v. 74, p. 163–197.

Poole, F. G., 1974, Flysch deposits of the Antler foreland basin, western U.S. *in* Dickinson, W. D., ed., Tectonics and sedimentation: Society of Economic Mineralogists and Paleontologists Special Publication 22, p. 58–82.

Potter, A. W., and 10 others, 1990, Early Paleozoic stratigraphic, paleogeographic, and biostratigraphic relations of the eastern Klamath belt, northern California *in* Harwood, D. S. and Miller, M. M., eds., Paleozoic and early Mesozoic paleogeographic relations; Sierra Nevada, Klamath Mountains and related terranes: Geological Society of America Special Paper 255, p. 57–74.

Price, R. A., 1981, The Canadian foreland thrust and fold belt in the southern Canadian Rocky Mountains, *in* McClay, K. R., and Price, N. J., eds., Thrust and nappe tectonics: Geological Society of London Special Publication 9, p. 427–448.

——, 1986, The southeastern Canadian Cordillera: Thrust faulting, tectonic wedging, and delamination of the lithosphere: Journal of Structural Geology, v. 8, p. 239–254.

Priem, H.N.A., Kroonenberg, S. B., Boelrijk, N.A.I.M., and Hebeda, E. H., 1989, Rb-Sr and K-Ar evidence for the presence of a 1.6 Ga basement underlying the 1.2 Ga Garzon-Santa Marta granulite belt in the Colombian Andes: Precambrian Research, v. 42, p. 315–324.

Rankin, D. W., 1975, The continental margin of eastern North America in the southern Appalachians: The opening and closing of the proto-Atlantic Ocean: American Journal of Science, v. 275-A, p. 298–336.

——, 1976, Appalachian salients and recesses: Late Precambrian continental breakup and the opening of the Iapetus Ocean: Journal of Geophysical Research, v. 81, p. 5605–5619.

Rankin, D. W., Hall, L. M., Drake, A. A., Jr., Goldsmith, R., Ratcliffe, N. M., and Stanley, R. S., 1989, Proterozoic evolution of the rifted margin of Laurentia, *in* Hatcher, R. D., Jr., Thomas, W. A., and Viele, G. W., eds., The Appalachian-Ouachita orogen in the United States: Boulder, Colorado, Geological Society of America, The Geology of North America, v. F-2, p. 10–42.

Rankin, D. W., and 13 others, 1991, E-4 Central Kentucky to Carolina trough: Boulder, Colorado, Geological Society of America, Centennial Continent/Ocean Transect no. 16, scale 1:500,000.

Rast, N., and Skehan, J. W., 1993, Mid-Paleozoic orogenesis in the North Atlantic: The Acadian orogeny *in* Roy, D. C. and Skehan, J. W., eds., The Acadian Orogeny; recent studies in New England, Maritime Canada, and the autochthonous foreland: Geological Society of America Special Paper 275, p. 1–25.

Renne, P. R., Mattinson, J. M., Hattan, G. W., Somin, M. L., Onatott, T. C., Millan, G., and Linares, E., 1989, $^{40}Ar/^{39}Ar$ and U-Pb evidence of Late Proterozoic (Grenville-age) continental crust in north-central Cuba and regional tectonic implications: Precambrian Research, v. 42, p. 325–341.

Roberts, R. J., Hotz, P. E., Gilluly, J., and Ferguson, H. G., 1958, Paleozoic rocks of north-central Nevada: American Association of Petroleum Geologists Bulletin, v. 42, p. 2813–2857.

Rodgers, John, 1971, The Taconic orogeny: Geological Society of America Bulletin, v. 82, p. 1141–1178.

Rosencrantz, E., 1990, Structure and tectonics of the Yucatan Basin, Caribbean Sea, as determined from seismic reflection profiles: Tectonics, v. 9, p. 1037–1059.

Rosencrantz, E., Ross, M., and Sclater, J., 1988, Age and spreading history of the Cayman Trough as determined from depth, heat flow, and magnetic anomalies: Journal of Geophysical Research, v. 93, p. 2141–2157.

Rosendahl, B., 1987, Architecture of continental rifts with special reference to East Africa: Annual Reviews of Earth and Planetary Science, v. 15, p. 445–503.

Rosenfield, J. H., 1981, Geology of the western Sierra de Santa Cruz, Guatemala, Central America, an ophiolite sequence [Ph.D. thesis]: Binghamton, State University of New York, 313 p.

Ross, C. A., and Ross, J.R.P., 1983, Late Paleozoic accreted terranes of western North America, *in* Stevens, C. H., ed., Pre-Jurassic rocks in western North America suspect terranes: Los Angeles, Pacific section, Society of Economic Paleontologists and Mineralogists, p. 23–35.

Ross, D. C., 1985, Basement rocks of the Salinian block and southernmost Sierra Nevada and possible correlations across the San Andreas, San Gregorio–Hosgri, and Rinconada-Reliz-King City fault zones: U.S. Geological Survey Professional Paper 1317, 37 p.

Rowley, D. B., and Pindell, J. L., 1989, End-Paleozoic and early Mesozoic western Pangea reconstruction and its implications for the distributions of Precambrian and Paleozoic rocks around meso-America: Precambrian Research, v. 42, p. 411–444.

Royden, L., and Keen, C. E., 1980, Rifting process and thermal evolution of the continental margin of eastern Canada determined from subsidence curves: Earth and Planetary Science Letters, v. 51, p. 343–361.

Royse, F., Jr., Warner, M. A., and Reese, D. L., 1975, Thrust belt structural geometry and related stratigraphic problems, Wyoming–Idaho–northern Utah, *in* Bolyard, D. W., ed., Symposium on deep drilling frontiers in the central Rocky Mountains: Denver, Colorado, Rocky Mountain Association of Geologists, p. 41–54.

Sacks, P. E., and Secor, D. T., Jr., 1990, Kinematics of late Paleozoic continental collision between Laurentia and Gondwana: Science, v. 250, p. 1702–1705.

Saleeby, J. B., 1982, Polygenetic ophiolite belt of the California Sierra Nevada—Geochronological and tectonostratigraphic development: Journal of Geophysical Research, v. 87, p. 1803–1824.

——, 1986, C2 Central California offshore to Colorado Plateau: Boulder, Colorado, Geological Society of America, Centennial Continent/Ocean Transect no. 10, scale 1:500,000.

Saleeby, J. B., Harper, G. D., Snoke, A. W., and Sharp, W. D., 1982, Time relations and structural-stratigraphic patterns in ophiolite accretion, west-central Klamath Mountains, California: Journal of Geophysical Research, v. 87, p. 3831–3848.

Salvador, A., 1987, Late Triassic–Jurassic paleogeography and origin of Gulf of Mexico Basin: American Association of Petroleum Geologists Bulletin, v. 71, p. 419–451.

Savage, J. C., Lisowski, M., and Prescott, W. H., 1990, An apparent shear zone trending north-northwest across the Mojave Desert into Owens Valley, eastern California: Geophysical Research Letters, v. 17, p. 2112–2116.

Schenk, P. E., 1981, The Meguma zone of Nova Scotia—A remnant of western Europe, South America, or Africa *in* Kerr, J. W., and Ferguson, A. J., eds., Geology of the north Atlantic borderlands: Canadian Society of Petroleum Geologists Memoir 7, p. 119–148.

Schlager, W., and 16 others, 1984, Deep Sea Drilling Project Leg 77, southeastern Gulf of Mexico: Geological Society of America Bulletin, v. 95,, p. 226–236.

Schweikert, R. A., and Lahren, M. M., 1990, Speculative reconstruction of the Mojave–Snow Lake fault: Implications for Paleozoic and Mesozoic orogenesis in the western United States: Tectonics, v. 9, p. 1609–1629.

Scotese, C. R., and McKerrow, W. S., 1990, Paleozoic revised world maps and introduction, *in* McKerrow, W. S., and Scotese, C. R., eds., Paleozoic biogeography and paleogeography: Geological Society of London Memoir 12, p. 1–21.

Secor, D. T., Jr., Snoke, A. W., and Dallmeyer, R. D., 1986, Character of the Alleghanian orogeny in the southern Appalachians: Part III, Regional tectonic relations: Geological Society of America Bulletin, v. 97, p. 1345–1353.

Secor, D. T., Jr., Murray, D. P., and Glover, L., III, 1989, Geology of the Avalonian rocks, *in* Hatcher, R. D., Jr., Thomas, W. A., and Viele, G. W., eds., The Appalachian-Ouachita orogen in the United States: Boulder, Colorado, Geological Society of America, The Geology of North America, v. F-2, p. 57–85.

Sedlock, R., and Hamilton, D. H., 1991, Late Cenozoic tectonic evolution of southwestern California: Journal of Geophysical Research, v. 96, p. 2325–2352.

Sedlock, R. L., Speed, R. C., and Ortega, F., 1993, Tectonic evolution of Mexico: Geological Society of America Special Paper 278, 130 p.

Sieh, K. E., and Jahns, R., 1984, Holocene activity of the San Andreas fault at Wallace Creek: Geological Society of America Bulletin, v. 95, p. 883–896.

Silberling, N. J., and Roberts, R. J., 1962, Pre-Tertiary stratigraphy and structure of northwestern Nevada: Geological Society of America Special Paper 72, 58 p.

Silberling, N. J., Jones, D. L., Blake, M. C., Jr., and Howell, D. G., 1987, Lithotectonic terrane map of the western coterminous United States: U.S. Geological Survey Map MF 1874C, scale 1:2,500,000.

Silver, L. T., and Anderson, T. H., 1974, Possible left-lateral early to middle Mesozoic disruption of the southwestern North American craton margin: Geological Society of America Abstracts with Programs, v. 6, p. 956.

Silver, L. T., Taylor, H. P., and Chappell, B. W., 1979, Some petrological, geochemical, and geochronological observations of the Peninsular Ranges batholith near the international border of the USA and Mexico, *in* Abbot, P. L., and Todd, V. R., eds., Mesozoic crystalline rocks, Peninsular Ranges batholith, pegmatites, and Point Sal ophiolite (Guidebook Geological Society of America annual meeting): San Diego, California, San Diego State University, p. 83–110.

Simpson, E. L., and Sundberg, F. A., 1987, Early Cambrian age for synrift deposits of the Chilhowee Group of southwestern vergence: Geology, v. 15, p. 123–126.

Simpson, R. W., and Cox, A., 1977, Paleomagnetic evidence for tectonic rotation of the Oregon Coast Range: Geology, v. 5, p. 585–589.

Sinha, A. K., Hund, E. A., and Hogan, J. P., 1989, Paleozoic accretionary history of the North American plate margin (central and southern Appalachians): Constraints from the age, origin, and distribution of granitic rocks, *in* Hillhouse, J. W., ed., Deep structure and past kinematics of accreted terranes: American Geophysical Union Geophysical Monograph 50, p. 219–238.

Sloss, L. L., 1963, Sequences in the cratonic interior of North America: Geological Society of America Bulletin, v. 74, p. 93–114.

—— , 1988, Tectonic evolution of the craton in Phanerozoic time, *in* Sloss, L. L., ed., Sedimentary cover—North American craton; U.S.: Boulder, Colorado, Geological Society of America. The Geology of North America, v. D-2, p. 25–51.

Smith, M. T., Dickinson, W. R., and Gehrels, G. E., 1993, Contractional nature of Devonian-Mississippian Antler tectonism along the North American continental margin: Geology, v. 21, p. 21–24.

Snoke, A. W., and Miller, D. M., 1988, Metamorphic and tectonic history of the northeastern Great Basin, *in* Ernst, W. G., ed., Metamorphism and crustal evolution in the western United States (Rubey Volume 7): Englewood Cliffs, New Jersey, Prentice-Hall, p. 606–648.

Snow, J. K., 1992, Large-magnitude Permian shortening and continental margin tectonics in the southern Cordillera: Geological Society of America Bulletin, v. 104, p. 80–105.

Speed, R. C., 1977, Island arc and other paleogeographic terranes of late Paleozoic age in the western Great Basin, *in* Stewart, J. H., Stevens, C. H., and Fritsche, A. E., eds., Paleozoic paleogeography of the western United States: Society of Economic Paleontologists and Mineralogists, Pacific Section, Pacific Coast Paleogeography Symposium 1, p. 349–362.

—— , 1978, Paleogeography and plate tectonic evolution of the early Mesozoic marine province of the western Great Basin, *in* Howell, D. G., and McDougall, K. A., eds., Mesozoic paleogeography of the western United States: Society of Economic Paleontologists and Mineralogists, Pacific Section, Pacific Coast Paleogeography Symposium 2, p. 253–270.

—— , 1979, Collided Paleozoic microplate in the western United States: Journal of Geology, v. 87, p. 279–292.

—— , 1983, Evolution of the sialic margin in the central-western United States *in* Watkins, J. S., and Drake, C. L., eds., Studies in continental margin geology: American Association of Petroleum Geologists Memoir 34, p. 457–470.

—— , 1991, Tectonic section display: Boulder, Colorado, Geological Society of America, Centennial Continent/Ocean Transects, scale 1:1,000,000 (2 sheets).

Speed, R. C., and Larue, D. K., 1991, Extension and transtension in the Puerto Rico Trench: Geophysical Research Letters, v. 17, p. 461–465.

Speed, R. C., and Sleep, N., 1982, Antler orogeny and foreland basin, a model: Geological Society of America Bulletin, v. 93,, p. 815–828.

Speed, R. C., Elison, M. W., and Heck, F. R., 1988, Phanerozoic tectonic evolution of the Great Basin, *in* Ernst, G., ed., Metamorphism and crustal evolution of the western United States (Rubey Volume 7): Englewood Cliffs, New Jersey, Prentice-Hall, p. 572–605.

Srivastava, S. P., and Tapscott, C. R., 1986, Plate kinematics of the North Atlantic, *in* P. R. Vogt and B. E. Tucholke, eds., The Western North Atlantic region: Boulder, Colorado, Geological Society of America, The Geology of North America, v. M, p. 379–404.

Stanley, R. S., and Ratcliffe, N. M., 1985, Tectonic synthesis of the Taconian orogeny in western New England: Geological Society of America Bulletin, v. 96, p. 1227–1250.

Stein, S., and nine others, 1988, A test of alternative Caribbean plate relative motion models: Journal of Geophysical Research, v. 93, p. 3041–3050.

Stevens, C. H., Stone, P., and Kistler, R. W., 1992, A speculative reconstruction of the middle Paleozoic continental margin of southwestern North America: Tectonics, v. 11, p. 405–419.

Stewart, D. B., Wright, B. E., Unger, J. D., Phillips, J. D., Hutchinson, D. R., and others, 1991, Global Geoscience Transect 8, Quebec–Maine–Gulf of Maine transect, southeastern Canada, northeastern United States of America: U.S. Geological Survey Open-File Report 91-353, 1 plate, 65 p., scale 1:1,000,000.

Stewart, J. H., 1972, Initial deposits in the Cordilleran geosyncline: Evidence of a late Precambrian (<850 m.y.) continental separation: Geological Society of America Bulletin, v. 83, p. 1345–1360.

—— , 1976, Late Precambrian evolution of North America: Plate tectonics implications: Geology, v. 4, p. 11–15.

—— , 1988, Latest Proterozoic and Paleozoic southern margin of North America and the accretion of México: Geology, v. 16, p. 186–189.

Stewart, J. H., and Poole, F. G., 1974, Lower Paleozoic and uppermost Precambrian miogeocline, Great Basin, western United States, *in* Dickinson, W. R., ed., Tectonics and sedimentation: Society of Economic Paleontologists and Mineralogists Special Publication 22, p. 28–57.

Stewart, J. H., and Suzcek, C. A., 1977, Cambrian and latest Precambrian paleogeography and tectonics in the western United States, *in* Stewart, J. H., Stevens, C. H., and Fritsche, A. E., eds., Paleozoic paleogeography of the western United States: Society of Economic Paleontologists and Mineralogists, Pacific Section, Pacific Coast Paleogeography Symposium 1, p. 1–17.

Stewart, J. H., Poole, F. G., Ketner, K. B., Madrid, R. J., Roldán-Quintana, J., and Amaya-Martínez, R., 1990, Tectonics and stratigraphy of the Paleozoic and Triassic southern margin of North America, Sonora, México, *in* Gehrels, G. E., and Spencer, J. E., eds., Geologic excursions through the Sonoran Desert region, Arizona and Sonora: Tucson, Arizona Geological Survey Special Paper 7, p. 183–202.

Stock, J., and Molnar, P., 1988, Uncertainties and implications of the Late Cretaceous and Tertiary position of North America relative to the Farallon, Kula, and Pacific plates: Tectonics, v. 7, p. 1339–1384.

Stock, J. M., and Hodges, K. V., 1989, Pre-Pliocene extension around the Gulf of California, and the transfer of Baja California to the Pacific plate: Tectonics, v. 8, p. 99–116.

Stockmal, G. S., Colman-Sadd, S. P., Keen, C. E., Mariller, F., O'Brien, S. J., and Quinlan, G. M., 1990, Deep seismic structure and plate tectonic evolution of the Canadian Appalachians: Tectonics, v. 9, p. 45–62.

Stone, P., and Stevens, C. H., 1988, Pennsylvanian and Early Permian paleogeography of east-central California: Implications for the shape of the continental margin and the timing of continental truncation: Geology, v. 16, p. 330–333.

Stone, D. B., Panuska, B. C., and Packer, D. R., 1982, Paleolatitudes versus time for southern Alaska: Journal of Geophysical Research, v. 87, p. 3697–3707.

Struik, L. C., 1981, A reexamination of the type area of the Devono-Mississippian Cariboo orogeny, central British Columbia: Canadian Journal of Earth Sciences, v. 18, p. 1767–1775.

Suter, M., 1984, Cordilleran deformation along the eastern edge of the Valles–San Luis Potosí carbonate platform, Sierra Madre Oriental fold-thrust belt, east-central México: Geological Society of America Bulletin, v. 95, p. 1387–1397.

Sutter, J. F., 1988, Innovative approaches to the dating of igneous events in the early Mesozoic basins, *in* Froelich, A. J., and Robinson, G. R., Jr., eds., Studies of the early Mesozoic basins of the eastern United States: U.S. Geological Survey Bulletin 1776, p. 194–200.

Sweeney, J. F., 1985, Comments about the age of the Canada Basin: Tectonophysics, v. 114, p. 1–10.

—— , 1986, G Somerset Island to Canada Basin: Boulder, Colorado, Geological Society of America, Centennial Continent/Ocean Transect no. 11, scale 1:500,000.

Sykes, L. R., McCann, W. R., and Kafka, A. L., 1982, Motion of Caribbean plate during last 7 million years and implications for earlier Cenozoic movements: Journal of Geophysical Research, v. 87, p. 656–676.

Tankard, A. J., and Balkwill, H. R., 1989, Introduction, *in* Tankard, A. J., and Balkwill, H. R., eds., Extensional tectonics and stratigraphy of the North Atlantic margins: American Association of Petroleum Geologists Memoir 46, p. 1–6.

Tarduno, J. A., McWilliams, M., Sliter, W. V., Cook, H. E., Blake, M. C., Jr., and Premoli-Silva, H., 1986, Southern Hemisphere origin of the Cretaceous Laytonville limestone of California: Science, v. 231, p. 1425–1428.

Tauvers, P. R., and Muehlberger, W. R., 1987, Is the Brunswick magnetic anomaly really the Alleghanian suture?: Tectonics, v. 6, p. 331–342.

Teissere, R. F., and Beck, M. E., Jr., 1973, Divergent Cretaceous paleomagnetic pole position for the Southern California Batholith, U.S.A.: Earth and Planetary Science Letters, v. 18, p. 296–300.

Tempelman-Kluit, D. J., 1979, Transported cataclasite, ophiolite and granodiorite in Yukon: Evidence of arc-continent collision: Geological Survey of Canada Paper 79-14, 27 p.

Thomas, W. A., 1976, Evolution of Ouachita-Appalachian continental margin: Journal of Geology, v. 84, p. 323–342.

—— , 1977, Evolution of Appalachian-Ouachita salients and recesses from reentrants and promontories in the continental margin: American Journal of Science, v. 277, p. 1233–1278.

—— , 1985, The Appalachian-Ouachita connection: Paleozoic orogenic belt at the southern margin of North America: Annual Review of Earth and Planetary Sciences, v. 13, p. 175–199.

—— , 1991, The Appalachian-Ouachita rifted margin of southeastern North America: Geological Society of America Bulletin, v. 103, p. 415–431.

Thompson, J. B., Jr., Bothner, W. A., Robinson, P., Isachsen, Y. W., and Klitgord, K. D., 1993, E-1 Adirondacks to Georges Bank: Boulder, Colorado, Geological Society of America, Centennial Continent/Ocean Transect no. 17, scale 1:500,000.

Thrupp, G. A., and Coe, R. S., 1986, Early Tertiary paleomagnetic evidence and the displacement of southern Alaska: Geology, v. 14, p. 213–217.

Todd, B. J., and Keen, C. E., 1989, Temperature effects and their geological consequences at passive margins: Canadian Journal of Earth Sciences, v. 26, p. 2591–2603.

Todd, B. J., Reid, I., and Keen, C. E., 1988, Crustal structure across the Southwest Newfoundland Transform Margin: Canadian Journal of Earth Sciences, v. 25, p. 744–759.

Tozer, T. T., 1982, Marine Triassic faunas of North America: Geologische Rundschau, v. 71, p. 1077–1104.

Tremblay, A., 1992, Tectonic and accretionary history of Taconian oceanic rocks of the Quebec Appalachians: American Journal of Science, v. 292, p. 229–252.

Trettin, H. P., and Balkwill, H. R., 1979, Contributions to the tectonic history of the Innuitian province, Arctic Canada: Canadian Journal of Earth Sciences, v. 16, p. 748–769.

Tull, J. F., and Groszos, M. S., 1990, Nested Paleozoic "successor" basins in the southern Appalachian Blue Ridge: Geology, v. 18, p. 1046–1049.

Van der Voo, R., 1983, A plate-tectonics model for the Paleozoic assembly of Pangea based on paleomagnetic data *in* Hatcher, R. D., Jr., Williams, H., and Zietz, I., eds., Contributions to the tectonics and geophysics of mountain chains: Geological Society of America Memoir 158, p. 19–32.

—— , 1988, Paleozoic paleogeography of North America, Gondwana, and intervening displaced terranes: Comparisons of paleomagnetism with paleoclimatology and biogeographical patterns: Geological Society of America Bulletin, v. 100, p. 311–324.

Viele, G. W., 1979, Geologic map and cross section, eastern Ouachita Mountains, Arkansas: Geological Society of America Map and Chart Series MC-28F, 8 p.

Viele, G. W., and Thomas, W. A., 1989, Tectonic synthesis of the Ouachita orogenic belt, *in* Hatcher, R. D., Jr., Thomas, W. A., and Viele, G. W., eds., The Appalachian-Ouachita orogen in the United States: Boulder, Colorado, Geological Society of America, The Geology of North America, v. F-2, p. 695–728.

Von Huene, R., Box, S., Detterman, B., Fisher, M., Moore, C., and Pulpan, H., 1985, A2 Kodiak to Kuskowin, Alaska: Boulder, Colorado, Geological Society of America, Centennial Continent/Ocean Transect no. 6, scale 1:500,000.

Walcott, R. I., 1978, Geodetic strains and large earthquakes in the axial tectonic belt of North Island, New Zealand: Journal of Geophysical Research, v. 83, p. 4419–4429.

Walker, J. D., 1988, Permian and Triassic rocks of the Mojave Desert and their implications for timing and mechanisms of continental truncation: Tectonics, v. 7, p. 685–709.

Wallin, E. T., Mattinson, J. M., and Potter, A. W., 1988, Early Paleozoic magmatic events in the eastern Klamath Mountains, northern California: Geology, v. 16, p. 144–148.

Walper, J. L., 1980, Tectonic evolution of the Gulf of Mexico, *in* Pilger, R. H., Jr., ed., The origin of the Gulf of Mexico and the early opening of the central north Atlantic Ocean: Baton Rouge, Louisiana State University, p. 87–98.

Wernicke, B., 1985, Uniform-sense normal simple shear of the continental lithosphere: Canadian Journal of Earth Sciences, v. 22, p. 108–125.

Wernicke, B., and Tilke, P. G., 1989, Extensional tectonic framework of the central Atlantic passive margin *in* Tankard, A. J., and Balkwill, H. R., Extensional tectonics and stratigraphy of the North Atlantic margins: American Association of Petroleum Geologists Memoir 46, p. 7–21.

Wernicke, B., Axen, G. J., and Snow, J. K., 1988, Basin and Range extensional tectonics at the latitude of Las Vegas, Nevada: Geological Society of America Bulletin, v. 100, p. 1738–1757.

Williams, H., 1975, Structural succession, nomenclature, and interpretations of transported rocks in western Newfoundland: Canadian Journal of Earth Sciences, v. 12, p. 1874–1894.

Williams, H., and Hatcher, R. D., Jr., 1983, Appalachian suspect terranes *in* Hatcher, R. D., Jr., Williams, H., and Zietz, I., eds., Contributions to the tectonics and geophysics of mountain chains: Geological Society of America Memoir 158, p. 33–53.

Williams, H., and Hiscott, R. N., 1987, Definition of the Iapetus rift-drift transition in western Newfoundland: Geology, v. 15, p. 1044–1047.

Wilson, J. T., 1966, Did the Atlantic close and then re-open? Nature, v. 211, p. 676–681.

Wintsch, R. P., Sutter, J. F., Kunk, M. J., Aleinikoff, J. N., and Dorais, M. J., 1992, Contrasting P-T-t paths: Thermochronologic evidence for a late Paleozoic assembly of the Avalon composite terrane in the New England Appalachians: Tectonics, v. 11, p. 672–689.

Yole, R. W., and Irving, E., 1980, Displacement of Vancouver Island: Paleomagnetic evidence from the Karmutsen Formation: Canadian Journal of Earth Sciences, v. 17, p. 1210–1228.

Yonkee, W. A., 1992, Basement-cover relations, Sevier orogenic belt, northern Utah: Geological Society of America Bulletin, v. 104, p. 280–302.

Yorath, C. J., and seven others, 1985, B1 Intermontane Belt (Skeena Mountains) to Insular Belt (Queen Charlotte Islands): Boulder, Colorado, Geological Society of America Centennial Continent/Ocean Transect no. 8, scale 1:500,000.

Zen, E-An, 1972, The Taconide zone and the Taconic orogeny in the western part of the northern Appalachian orogen: Geological Society of America Special Paper 135, 72 p.

—— , 1983, Exotic terranes in the New England Appalachians—Limits, candidates, and ages: A speculative essay *in* Hatcher, R. D., Jr., Williams, H., and Zietz, I., eds., Contributions to the tectonics and geophysics of mountain chains: Geological Society of America Memoir 158, p. 55–81.

Manuscript Accepted by the Society October 25, 1993

ACKNOWLEDGMENTS

I am indebted to L. L. Sloss, D. W. Rankin, J.W.M. Monger, and D. W. Howell for painstaking reviews of this manuscript. Their many comments have been incorporated extensively and have improved the exposition. I also thank P. J. Hart, J. Oliver, and J. C. Maxwell for their active participation in the execution of the Transects Program and A. R. (Pete) Palmer for his guidance and encouragement in the production of this book. Alice Gripp was a great help with computer graphics.

Printed in U.S.A.

DNAG Continent-Ocean Transect Volume
Phanerozoic Evolution of North American
Continent-Ocean Transitions
The Geological Society of America, 1994

Chapter 2

Transects of the ancient and modern continental margins of eastern Canada

Richard T. Haworth* and Charlotte E. Keen
Atlantic Geoscience Centre, Geological Survey of Canada, Bedford Institute of Oceanography, P.O. Box 1006, Dartmouth, Nova Scotia B2Y 4A2, Canada
Harold Williams
Department of Earth Sciences, Memorial University of Newfoundland, St. John's, Newfoundland A1B 3X5, Canada

INTRODUCTION

The continental margin transect program evolved through a desire to define the specific characteristics of North America's modern continental margins and to compare the modern margins with Paleozoic examples, now found within the continent. Thus, the features of modern margins can be used to locate and elucidate the structures of ancient analogues. Conversely, the exposed rocks of ancient margins provide an insight into the evolution and deep structures of modern examples. Ocean-continental transitions, rifting mechanisms, development of sedimentary basins, processes of continental breakup, and ancestral controls are all features that warrant comparison between modern and ancient examples.

The Atlantic coast has the best-known passive margin of North America. Inboard of it is the Appalachian Mountain chain, which records a clear history of late Precambrian rifting and the passive development of a Paleozoic continental margin. Transects across both these modern and ancient examples provide information on scale and detail that is useful for comparisons.

The modern continental margin of eastern Canada (Fig. 1a, 1b) has a wider variety of ages and styles than that of its eastern United States counterpart. Segments of the margin southeast of Nova Scotia and east of Labrador are typical of rifted margins, though of different ages. Between them lies the tortuous segments of the northern and eastern Grand Banks where rifting and subsidence occurred over a broad area of the continental shelf. In contrast, the southern boundary of the Grand Banks is a transform segment. Rocks of the Appalachian Paleozoic margin are best preserved and exposed in western Newfoundland.

The Newfoundland–northern Grand Banks transect (A to I) crosses the entire Appalachian Orogen and the modern margin. It can be used for ancient-modern comparisons, or viewed as two separate sections: one, an ancient continental margin now incorporated in the Appalachian Orogen (A to F); the other, a severely extended and rifted modern margin (G to I). The Nova Scotia (M to O) and Labrador (P to Q) transects show the characteristics of the modern rifted continental margin. These transects are not extended westward because onshore rocks are structurally complex and of inappropriate age and affinity to the theme of this chapter. The southern Grand Banks transect (J to L) is chosen to demonstrate the characteristics of a transform margin segment.

Much of the information used in the preparation of these transects is published elsewhere, and some of the transects are updates of earlier syntheses (Williams and Stevens, 1974; Haworth and Keen, 1979; Keen and Hyndman, 1979). We therefore stress the comparative aspects and regional implications of the transects, and make no attempt to describe the details of each.

We hope our contribution will be of benefit to other "transectuals" (an apt label coined by Richard Buffler and his coworkers), and we wish to acknowledge the cooperation of our peers, graduate students, and colleagues to whom we owe a continuing debt.

PHANEROZOIC EVOLUTION OF EASTERN CANADA: A TRANSECT OVERVIEW

The Phanerozoic evolution of eastern Canada involved the rifting and passive development of an early Paleozoic continental margin that bordered the Iapetus Ocean (Fig. 2). The closing of Iapetus and the destruction of its North American margin led to the development of the Appalachian Orogen. Opening of the modern Atlantic began in Jurassic time. Its axis split the Appalachian orogen longitudinally, and it lay well east of the Iapetus suture. Thus, a variety of suspect terranes were left clinging to the North American Paleozoic margin (Williams and Hatcher, 1983).

*Present address: Geological Survey of Canada, 601 Booth St., Ottawa, Ontario K1A 0E8, Canada.

Haworth, R. T., Keen, C. E., and Williams, H., 1994, Transects of the ancient and modern continental margins of eastern Canada, *in* Speed, R. C., ed., Phanerozoic Evolution of North American Continent-Ocean Transitions: Boulder, Colorado, Geological Society of America, DNAG Continent-Ocean Transect Volume.

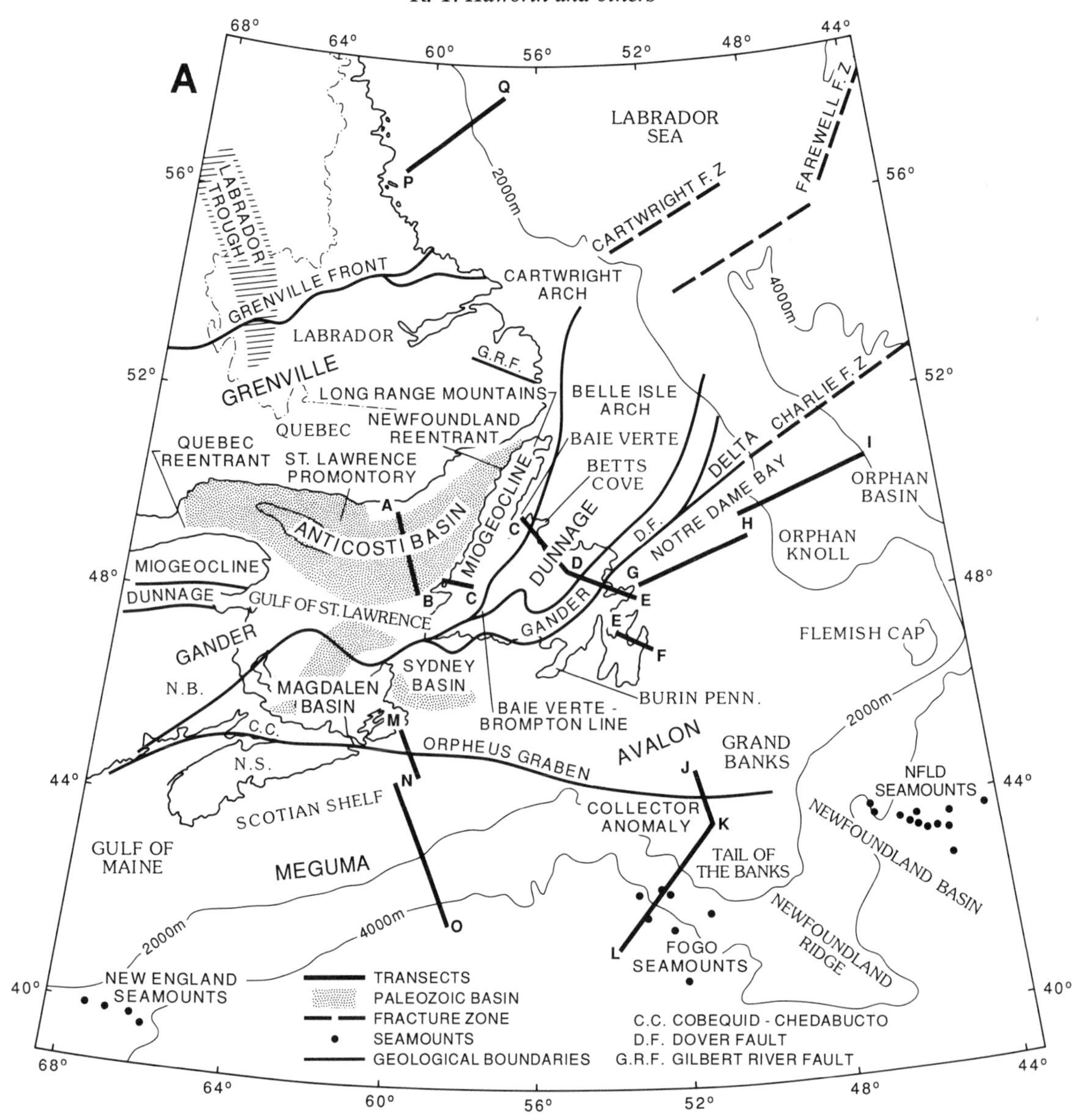

Figure 1a, b. Location of transects and the main tectonic elements of the eastern Canadian margin. Figure 1a (above) illustrates Paleozoic and older features, and Figure 1b (facing page) illustrates the Mesozoic-Cenozoic features. Transects are shown as solid lines, designated by letters for convenience in identifying changes in direction or offsets in the line of section. Transect D-1 corresponds to the composite section A to I, which crosses Newfoundland and the Orphan Basin. Transect D-2 corresponds to J to L, which crosses the southern margin of the Grand Banks. Transect D-3 corresponds to M to O across the Nova Scotian margin. Transect D-4 corresponds to P to Q across the Labrador Margin. The lines of crosses denote major pre-Mesozoic geological boundaries between structural provinces or terranes (Fig. 1a); the Grenville Front, and boundaries between the Appalachian Dunnage, Gander, Delta, Avalon, and Meguma terranes. The Labrador Trough is indicated by horizontal shading. Offshore, the distribution of Triassic rift basins and of major depocentres along the present margin are illustrated (Fig. 1b) as stippled areas. The location of major oceanic structures such as seamounts, fracture zones, and linear ridges, as well as the 2,000 and 4,000 m bathymetric contours are shown.

The record of the Paleozoic margin of Iapetus is contained in rocks of western Newfoundland and their southern correlatives. Development was initiated by the rifting of Grenvillian basement with coeval mafic dike intrusion, mafic volcanism, and the accumulation of thick clastic sequences (Williams and Stevens, 1974). These events began in the late Precambrian, as indicated by isotopic ages of mafic dikes (Pringle and others, 1971; Stukas and Reynolds, 1974).

A thinner sequence of mainly carbonate rocks formed at the continental shelf from Early Cambrian to latest Early Ordovician time (Rodgers, 1968). These sediments thicken eastward and record an upward transition from immature arkosic sandstones to

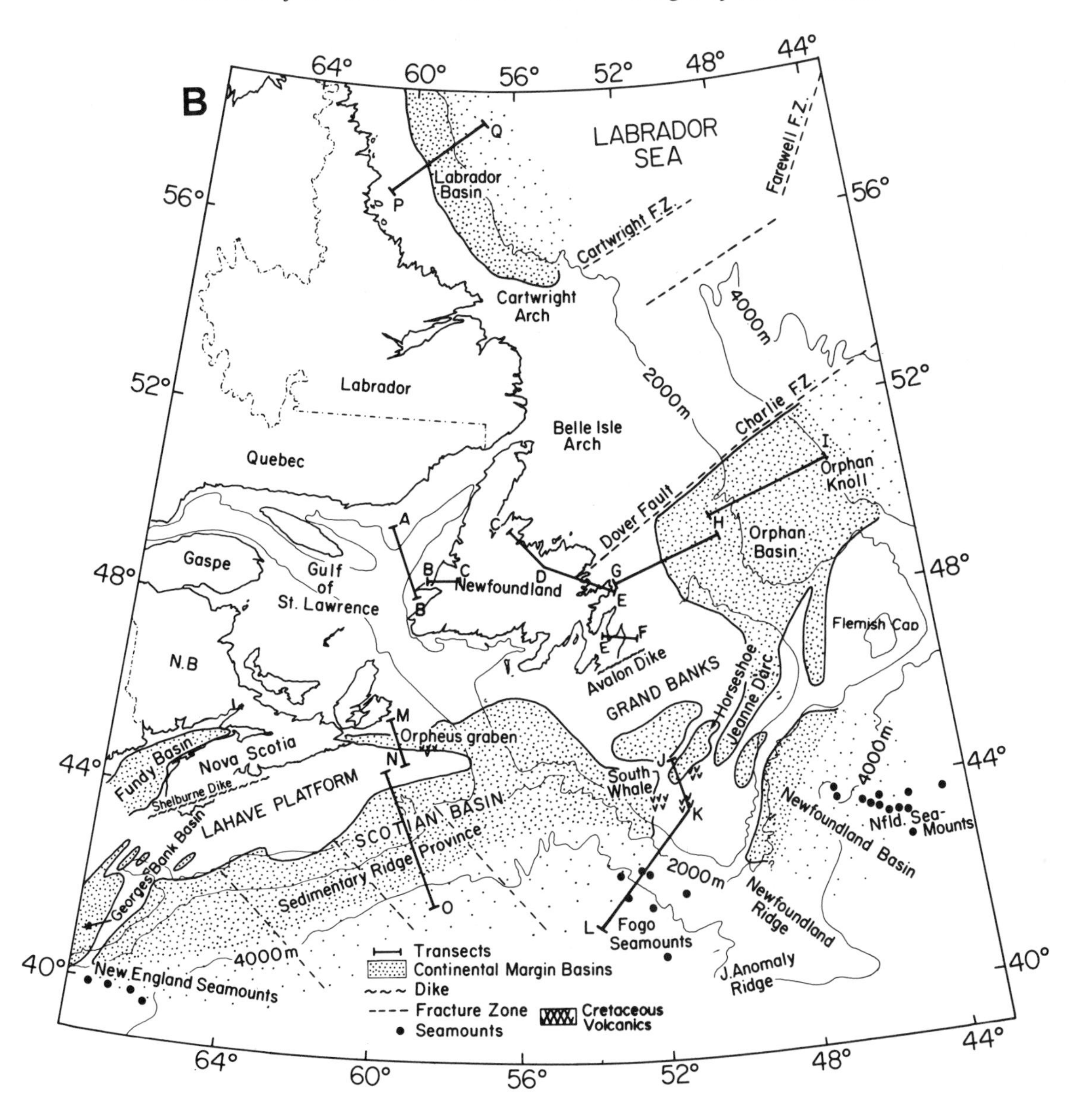

mature quartz sandstones, and then limestones and dolomites. Rocks deposited at the continental slope and rise are now preserved in allochthonous sequences structurally above the carbonate shelf.

Destruction of the ancient stable margin began toward the end of Early Ordovician time with a series of accretionary events that lasted throughout the Paleozoic (see Fig. 26). Development of an ancient karst topography across the carbonate shelf may reflect upwarp of the area as a precursor to collision. Later subsidence is recorded first by the deposition of deeper water carbonates across the disturbed bank, then by a flood of terrigenous clastic rocks from the east. These were structurally overridden by a sequence of contrasting rock assemblages in separate slices. The structurally lowest slices consist of sedimentary rocks from the nearby continental margin, and the highest ophiolitic slices represent farther-travelled oceanic crust and mantle. The structural pile was assembled from the east, possibly through peeling of successively landward sections from the leading edge of the sinking continental plate.

In Newfoundland, the North American margin of Iapetus is succeeded to the east by three large composite terranes (Williams and Hatcher, 1982, 1983). All are of unknown paleogeography with respect to North America. From west to east, these are the Dunnage, Gander, and Avalon terranes. An additional, more easterly terrane, the Meguma, occurs in mainland Nova Scotia.

Island-arc rocks of the Dunnage terrane overlie an ophiolitic substrate. Accordingly, the Dunnage is an example of a lower Paleozoic island arc built upon the crust of Iapetus. Terranes east of the Dunnage are of continental affinity.

The Gander terrane contains a thick clastic sequence of Ordovician and earlier age that might be viewed as a miogeoclinal apron along the west flank of the Avalon terrane, although adjacent Avalonian volcanics seem an unlikely source for Gander graywackes and quartzites. The Gander terrane may have once

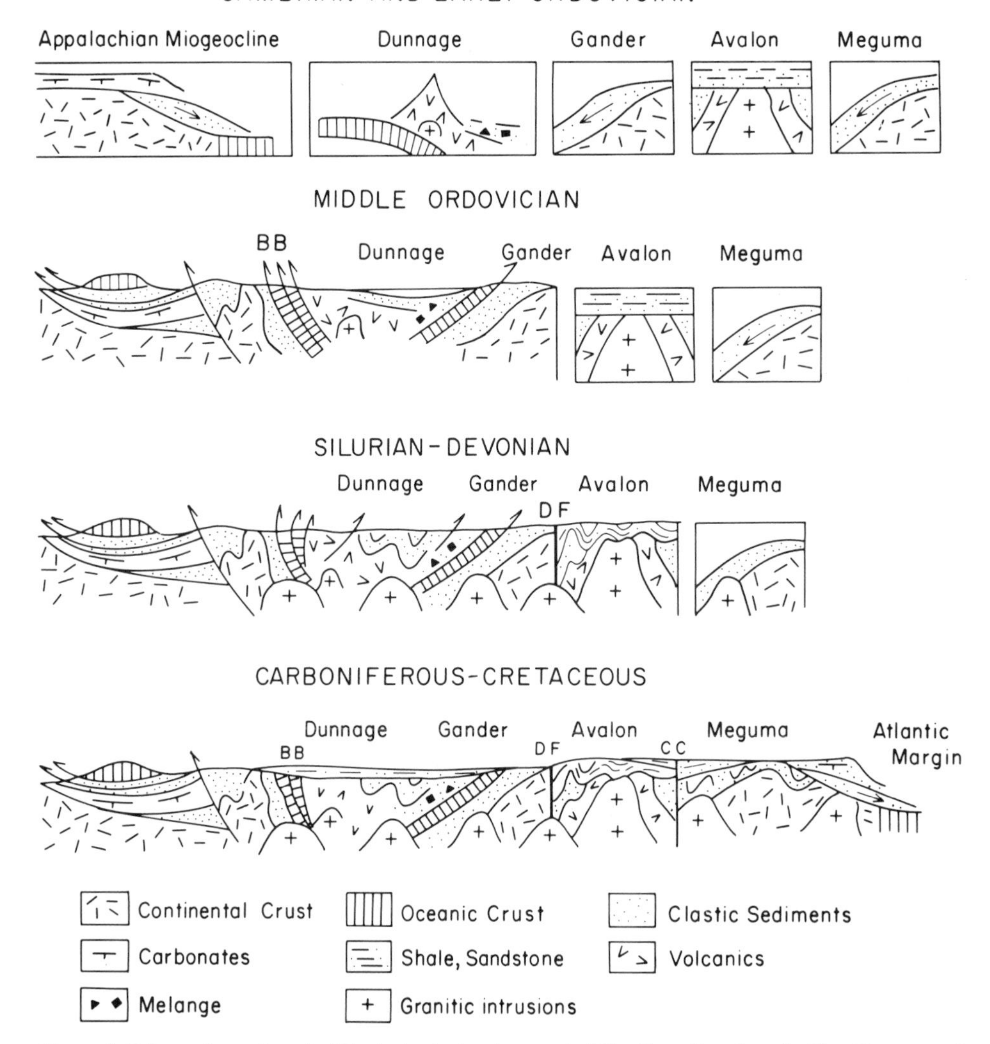

Figure 2. Schematic portrayal of the tectonic development of the Canadian Appalachian Orogen and Atlantic margin. BB, Baie Verte–Brompton Line; DF, Dover Fault; CC, Cobequid-Chedabucto fault (or Glooscap Fault).

bordered a continental craton but it is nowhere linked to a craton now. Possibly the offshore Delta terrane (Fig. 1) is a continental remnant linked to the Gander terrane.

The Avalon terrane is defined on its late Precambrian thick sedimentary and volcanic sequences, and overlying Cambrian shales with Atlantic or European faunal affinities. Its basement is probably Grenvillian (Olszewski and Gaudette, 1982; Farrar and Glover, 1983). The stratigraphy and structure of the Avalon terrane are very well known in places, but its significance in the Appalachian orogen remains a mystery. Its setting in the late Precambrian has been interpreted as either an archipelago of volcanic islands related to subduction and a Pan-African orogenic cycle that predated the opening of Iapetus (Rast, 1980), or volcanos and sedimentary basins related to rifting and the initiation of Iapetus (Strong, 1979).

The Meguma terrane has 13 km of marine graywackes and shales of Early Ordovician and Cambrian age. These sedimentary rocks were derived from a gneissic source of continental dimensions that lay to the southeast (Schenk, 1981). The terrane has a miogeoclinal aspect and is underlain by gneisses, which exhibit an isotopic age of 1180 Ma (Keppie and others, 1983).

The appearance of cosmopolitan faunas in Middle Ordovician rocks of the Appalachians, and in Middle Ordovician rocks of the Caledonides of Scandinavia on the opposite side of Iapetus, indicates a breakdown of preexisting faunal barriers (McKerrow and Cocks, 1976). This implies convergence between terranes previously separated by Iapetus and is compatible with the Ordovician age for destruction of parts of its continental margins. Widespread occurrences of Late Silurian and Devonian terrestrial rocks throughout the Appalachian-Caledonide Orogen, coupled with the wide extent of middle Paleozoic orogenesis, indicate the almost complete destruction of Iapetus at this time. Restricted late Paleozoic marine deposits and the limited extent of late Paleozoic deformation in the Appalachian-Caledonide Orogen signify the narrowness of the remaining seaways and perhaps the final structural tightening of these and other weak crustal areas.

Stratigraphic and sedimentologic analyses of the Appalachian Orogen indicate that it built outward from the miogeocline in three accretionary events. These are dated stratigraphically as Early to Middle Ordovician, Silurian-Devonian, and Carboniferous-Permian; and these times of accretion coincide with the main orogenic episodes affecting the system, respectively the Taconian, Acadian, and Alleghanian (Williams and Hatcher, 1982, 1983). The boundaries between the first accreted western terranes are subhorizontal to moderately dipping zones marked by ophiolites and mélanges; implying head-on subduction and obduction. Later boundaries between eastern terranes are steep ductile shears and brittle faults, implying transcurrent movements.

In Triassic time, rift basins developed within the Bay of Fundy area and the Orpheus Graben area of the Scotian Shelf. These basins represent the initial phase of opening of the present Atlantic Ocean. By Late Triassic–Early Jurassic time, the locus of rifting shifted to the present edge of the continental margin off Nova Scotia. The rifted margin off Nova Scotia continued to develop while the margin south of the Grand Banks was an active transform fault during the Jurassic and Early Cretaceous (Fig. 3). Rifting between Iberia and the Grand Banks was also initiated in the Early Cretaceous. As a consequence of this, extensive Early Cretaceous volcanism is detected on the continental margin in wells on the Grand Banks and Scotian Shelf (Jansa and Pe-Piper, 1987), and also in the Fogo Seamounts, Newfoundland Ridge, and Newfoundland Seamounts (Fig. 1a, 1b). Although the presence of Jurassic granites and a series of lamprophyre dikes in central Newfoundland have been proposed as evidence that rifting began northeast of Newfoundland during Jurassic time, no Mesozoic sedimentary rocks older than Early Cretaceous have yet been sampled in that region except on Orphan Knoll. Rifting progressed to the Labrador Shelf during the Early Cretaceous, as shown by the development of normal faults, volcanism, and the deposition of sedimentary rocks. Active drifting as demonstrated by magnetic anomalies continued from the Late Cretaceous through to the Early Oligocene in the Labrador Sea. Elsewhere along the southeastern Canadian margin it has continued to the present day.

Transect A to F (D–1 West), Figure 4

The transect crosses the entire Appalachian orogen from Precambrian Grenville crust in the west to Precambrian "Avalon" crust in the east. Grenville crust underlies at least the westernmost 200 km of the transect (A to C) within the Gulf of St. Lawrence and western Newfoundland. The crust thins to the east, and has a thickness of between 30 and 40 km. In the Gulf of St. Lawrence, an intermediate-velocity lower crustal layer underlies the Paleozoic sedimentary basin. The thickness of the crust increases significantly beneath the exposed edge of the Appalachian miogeocline and the Dunnage terrane. Beneath the Dunnage terrane, the lower crustal intermediate velocities reappear, increasing the crustal thickness to approximately 45 km. Similar velocities also occur at shallow depths of less than 10 km, where they are associated with the surface outcrop of ophiolite suites. The crustal thickness abruptly decreases, and the lower crustal intermediate velocities disappear east of the Dover Fault at the Gander-Avalon boundary. However, the shallow intermediate velocity layer still persists within the Avalon terrane of the Grand Banks. Regional continuity of the surficial ophiolite sheet within the Dunnage terrane can be demonstrated geophysically (Haworth and Miller, 1982). However, continuity of the Dunnage ophiolite into the Avalon terrane is neither demonstrable nor expected, and the intermediate-velocity layer there may have a different significance. The detached ophiolites above the miogeocline of western Newfoundland and the geometry of the Dunnage Zone ophiolite sheet are compatible with models that invoke eastward subduction of North America (Fig. 2; Church and Stevens, 1971; Strong and others, 1974; Strong, 1977).

The sharp surficial contact between the Gander and the Avalon zones at the Dover Fault is emphasized by the abrupt change in crustal thickness across the fault. These contrasts support the idea of major transcurrent displacement during accretion of the Avalon terrane. Alternating volcanic and sedimentary belts of the Avalon terrane are expressed by high and low magnetic anomalies, respectively.

The northeast Newfoundland margin: Transect G to I (D-1 East), Figure 5

Plate motions between northwest Europe and North America created the modern margin northeast of Newfoundland in Late Cretaceous time. Rifting may have started in the Early Cretaceous or the Late Jurassic, perhaps in response to plate motions between Iberia and North America to the south (Fig. 3). In general, the Mesozoic history of this area is complex because of the variety, proximity, and geometry of plates involved in continental breakup during the Cretaceous. Rifting appears to have taken place over a long time interval, about 50 m.y. Jurassic intrusions found in northeast Newfoundland and beneath the shelf may be the products of the early rift phase.

The region shown on transect D1 East is underlain by rocks of the Avalon terrane (Haworth, 1977). Faults between the volcanic and sedimentary belts of the Avalon terrane, initially active in late Precambrian time, have subsequently been reactivated during the Paleozoic and Mesozoic. As a result, Mesozoic basins have developed on the Grand Banks with the Precambrian sedimentary basins as their loci.

The Orphan Basin is a broad depression, partly filled with sediments, which is bounded to the east by Orphan Knoll, a continental fragment, and to the north by a transform fault, created along the seaward continuation of the Dover Fault into the Charlie Fracture Zone. To the south lies the Flemish Cap, a continental fragment similar to that of Orphan Knoll. Thus, the fragmented nature of this margin segment is quite different from

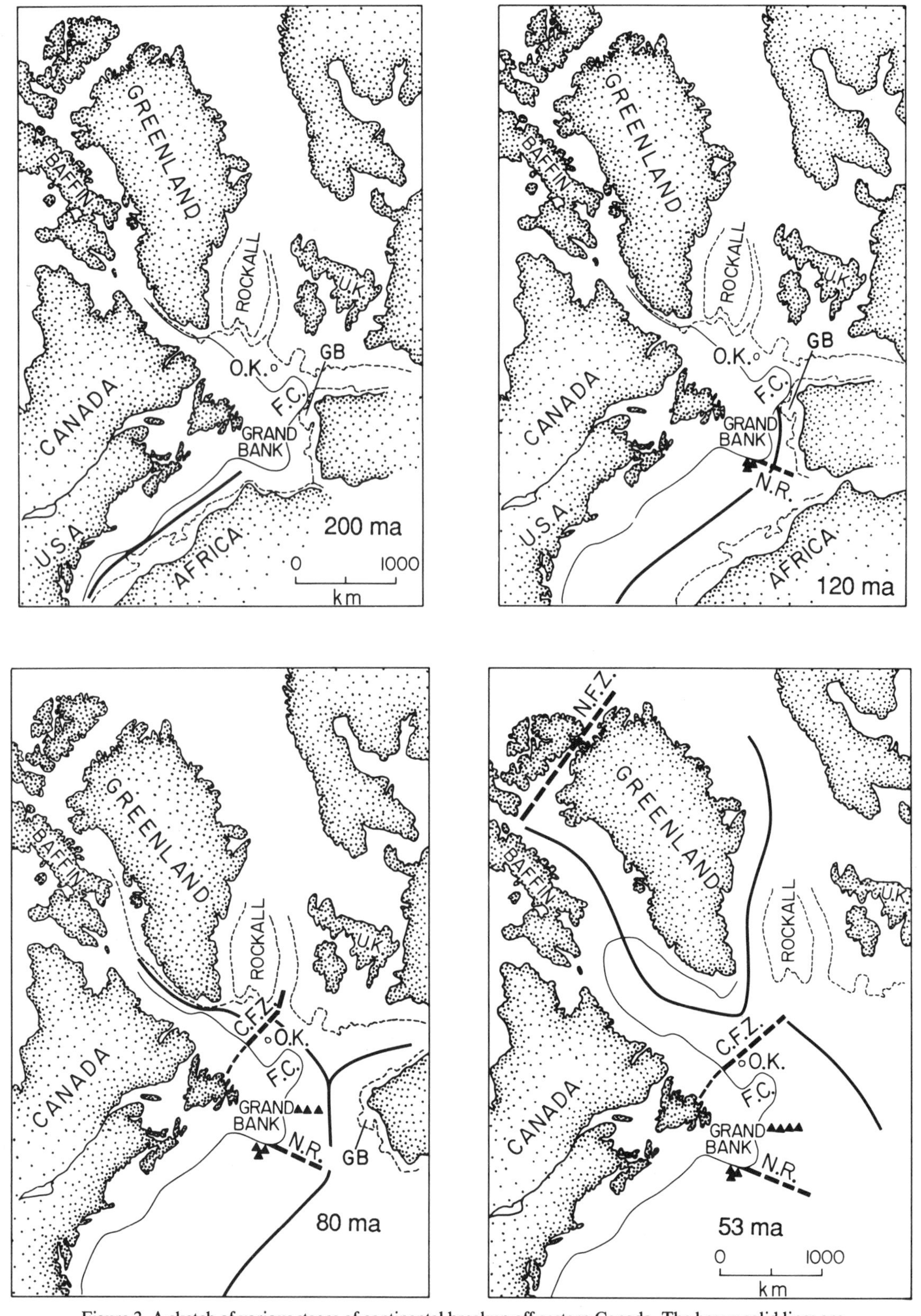

Figure 3. A sketch of various stages of continental breakup off eastern Canada. The heavy solid lines are incipient or active mid-ocean ridges. The dashed lines are fracture zones. Triangles are seamounts. The lighter, solid and dashed lines outline the approximate edge of shelf on the western and eastern sides of the Atlantic, respectively. N.R., Southeast Newfoundland Ridge; F.C., Flemish Cap; O.K., Orphan Knoll; C.F.Z., Charlie Fracture Zone; GB, Galicia Bank; N.F.Z., Nares Strait Fracture Zone.

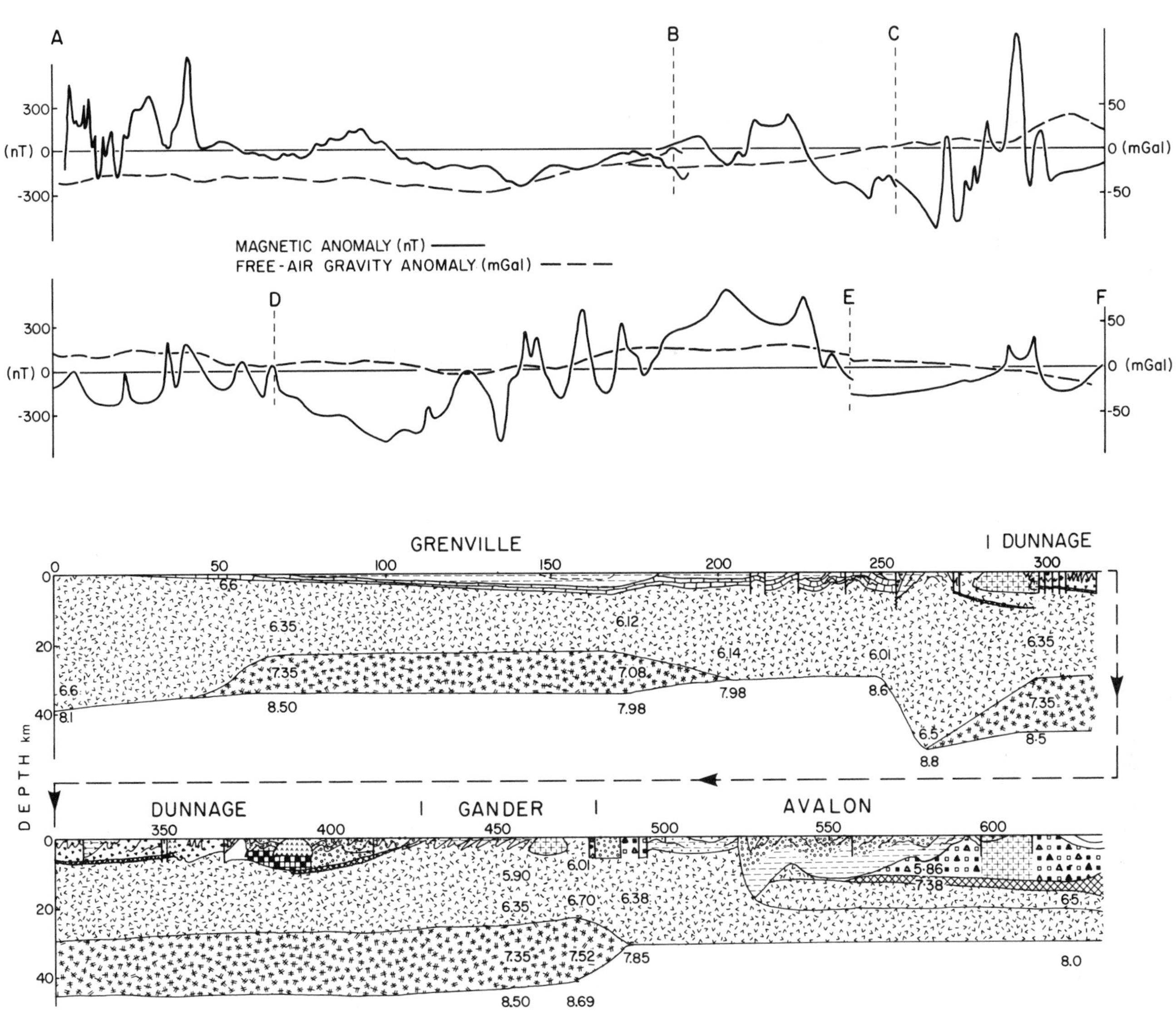

Figure 4. The western part of Transect D-1 (A to F, Fig. 1) crossing the Appalachian Orogen in Newfoundland, shown in two sections with no vertical exaggeration. Although much of the detailed surface geology can only be found on the full-size transect, the correlation between terrane boundaries defined by surface geology and significant changes in crustal structure are still evident. A deep layer of anomalously high velocity is found beneath the miogeocline in the Gulf of St. Lawrence, and also beneath the Dunnage and Gander terranes where it is abruptly truncated at their boundary with the Avalon terrane. A shallower layer of high-velocity rocks is apparently continuous with exposed ophiolites occupying a synformal structure beneath the Dunnage terrane. A similar layer beneath the Avalon terrane has no demonstratable continuity with the Dunnage layer. The deep crustal structure was derived from seismic refraction data reported by Press and Beckman (1954), Dainty and others (1966), Ewing and others (1966), Sheridan and Drake (1968), Fenwick and others (1968), Berry and Fuchs (1973), and Hobson and Overton (1973). The gravity and magnetic data were obtained from contour maps by Weaver (1968), Haworth and MacIntyre (1975), Hood and Reveler (1977), and Miller and Deutsch (1978). Patterns are explained on Figure 5.

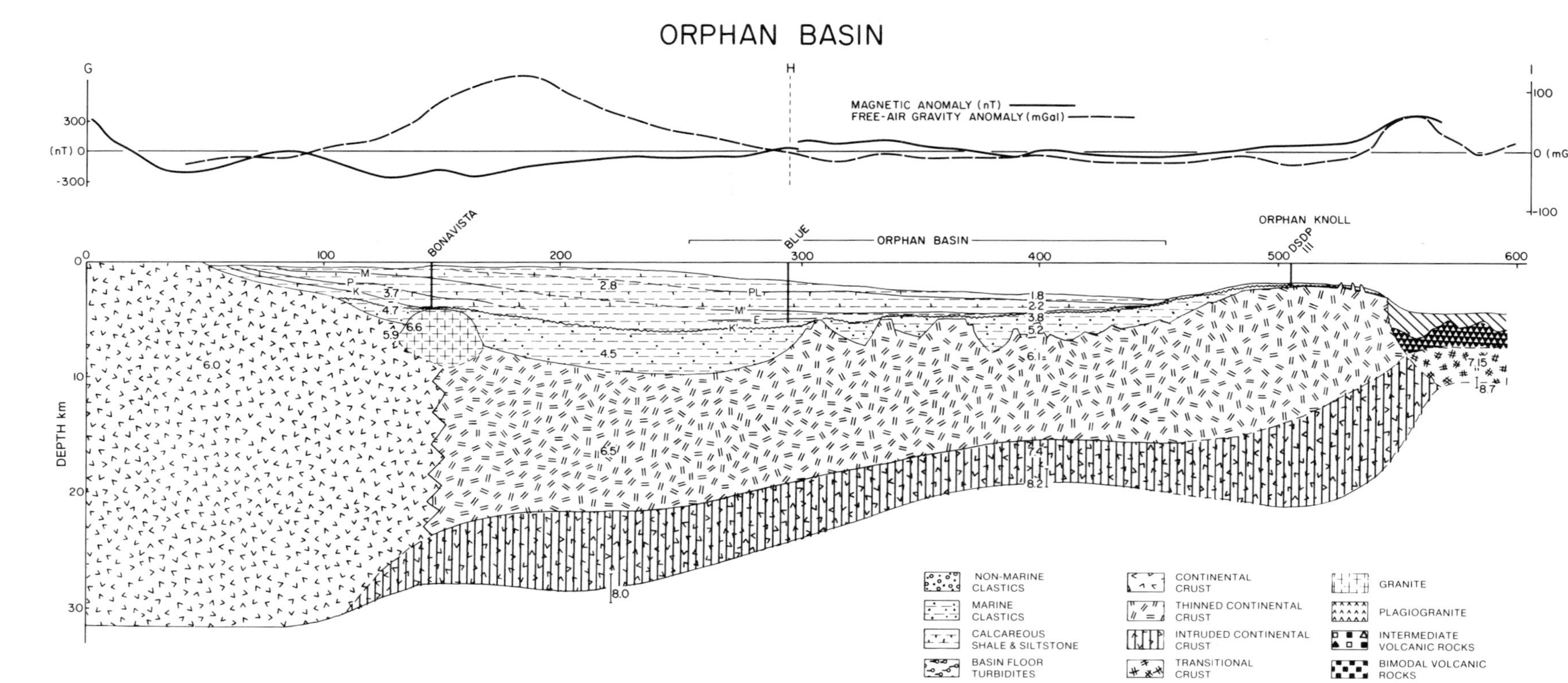

Figure 5. The eastern part of Transect D-1 (G to I, Fig. 1). The transect crosses the shelf northeast of Newfoundland and the Orphan Basin. It terminates in the southern Labrador Sea just east of Orphan Knoll, a continental fragment. Two deep exploratory wells, Blue and Bonavista, are shown as vertical lines. The sedimentary stratigraphy from these wells and from DSDP Site 111 were extrapolated to other parts of the transect using seismic reflection data (Gradstein and Williams, 1981; Ruffman and van Hinte, 1973; Grant, 1975). Within the sediments the following horizons have been marked: PL, Pliocene; M, Miocene; E, Eocene; P, Paleocene; and K, Cretaceous. The Cretaceous marker corresponds to a mid-Cretaceous unconformity in the Blue well. The deep crustal structure of this region was derived from Keen and Barrett (1981), Sheridan and Drake (1968), Fenwick and others (1968), and Dainty and others (1966). Gravity and magnetic data were compiled from Haworth (1977). An explanation of the patterns used in Figures 4 to 8 is given.

the relatively sharply bounded margin to the south (transects J to L and M to O).

Both seismic and magnetic data indicate that Orphan Belt is underlain by continental crust (Haworth, 1977; Keen and Barrett, 1981). Normal faults within the basement suggest extension during rifting. The syn-rift sediments are Lower Cretaceous shallow-marine clastics. They are separated by a regional unconformity of mid–Late Cretaceous age from overlying latest Cretaceous and younger sediments that were deposited after rifting. The subsidence history of this margin is difficult to interpret according to a simple cooling model. Very rapid subsidence occurred in the Eocene, with little or no tectonically controlled subsidence thereafter. This rapid subsidence may be related to decoupling between Rockall Plateau and Orphan Basin as their common transform margins separated (Fig. 3).

The subsedimentary structure of Orphan Basin is characterized by crustal thinning, and by the presence of a high-velocity lower crustal layer (Fig. 5). Its thin crust is consistent with extension of the lithosphere during rifting. The lower crustal layer may be a product of magma, generated during rifting, which has intruded or underplated the continental crust. Orphan Knoll separates the Orphan Basin from the deep ocean. The existence of this positive feature has not been explained satisfactorily. The simplest explanation is that the locus of rifting jumped, from a point within the basin to a point east of Orphan Knoll, but supporting evidence is lacking. Whatever mechanism caused the Knoll to persist as a positive feature must also account for the sharp ocean-continent contact that occurs to the east.

Margin south of the Grand Banks: Transect J to L (D-2), Figure 6

The transform margin south of the Grand Banks was formed by shearing between the African and North American plates. The age of its independent development progresses from Early-Middle Jurassic in the northwest where it joins the rifted Nova Scotian margin, to Early Cretaceous at its southeastern terminus. Continental basement beneath the southern Grand Banks comprises rocks of the Meguma and Avalon terranes (Haworth, 1975;, Haworth and Keen, 1979). Elongate half-grabens occur on the Grand Banks. They are filled with Triassic nonmarine and paralic clastic sediments, Triassic and Lower Jurassic salt, and carbonates. The geometry of plate movement resulted in these rift basins remaining active after sea-floor spreading had begun to the southeast. Rifting did not stop until the final decoupling of Africa from North America in Early Cretaceous time. In Early to Late Cretaceous time, the Grand Banks were uplifted, perhaps in response to the early development of rifting between Iberia and the eastern Grand Banks (Fig. 3). This produced a major unconformity ("U"), above which latest Cretaceous and younger sediments blanket the rift basins and basement. Except within the rift grabens, sediments are thin over much of the Grand Banks, and a deep sedimentary basin, such as that along the rifted margin to the south, did not develop along the transform margin.

The contact between the Avalon and Meguma terranes appears to have been reactivated either by rifting or by plate reordering on a more global scale. Early to Middle Cretaceous volcanism occurred near this boundary on the Grand Banks and in the Orpheus graben (Fig. 1a, 1b). Volcanism also occurred on oceanic crust (Tucholke and Ludwig, 1982; Sullivan and Keen, 1978) and formed the Fogo Seamounts and the J-Anomaly Ridge along the southeastern part of the transform margin. Farther north and east, the Newfoundland Seamounts of mid-Cretaceous age formed after the separation of Iberia from North America (Fig. 3). The relationship between these regions of volcanic activity is unclear. The Newfoundland Seamounts may be the seaward prolongation of the Avalon-Meguma boundary.

The ocean-continent boundary appears to be much sharper across the transform margin than across the rifted margins (Keen and Hyndman, 1979). There is no broad zone over which the crystalline continental crust has been thinned. The oceanic crust adjacent to the transform margin is anomalous: it consists of a very thick layer 2, while layer 3 is either thin or absent. This may be related to the excess volcanism of the region, but the magmatic processes involved are unclear. The narrow ocean-continent transition at the transform margin is compatible with the hypothesis that extensional forces caused the crustal thinning observed at the rifted margins. The absence of a large sedimentary basin further supports the inferred absence of extension and thinning along the transform margin during rifting.

Margin off Nova Scotia: Transect M to O (D-3), Figure 7

The contemporary rifted margin off Nova Scotia was formed in response to plate motions between Africa and North America. Rifting of the Meguma and Avalon terranes began in the Late Triassic and produced rift basins beneath the present continental shelf. These are similar to those on the Grand Banks, but active rifting ceased in mid-Jurassic time when sea-floor spreading began. The rift basins are not uniformly distributed; they are more frequent in the Avalon terrane where Precambrian grabens are their locus. Triassic volcanism is exemplified by the North Mountain basalts and the Shelburne dike in Nova Scotia (Fig. 1), and most of the known volcanic rocks associated with rifting appear to lie well inland of the present ocean-continent transition. Rifting in the Early Jurassic (Wade, 1981) was accompanied by the deposition of thick salt beds beneath much of the present shelf and slope. Early to Middle Jurassic deepening led to a less restricted marine environment and the development of carbonate platforms. It was during this time that sea-floor spreading began east of the rifted North American continent. This rift-drift transition does not appear to be marked by a period of uplift, erosion, or the development of a clear breakup unconformity.

After the onset of sea-floor spreading, the margin continued to subside, and over 10 km of Jurassic and younger sediments were deposited on the outer shelf and slope. Variations in morphology, sediment type, and deposition rate throughout the post-

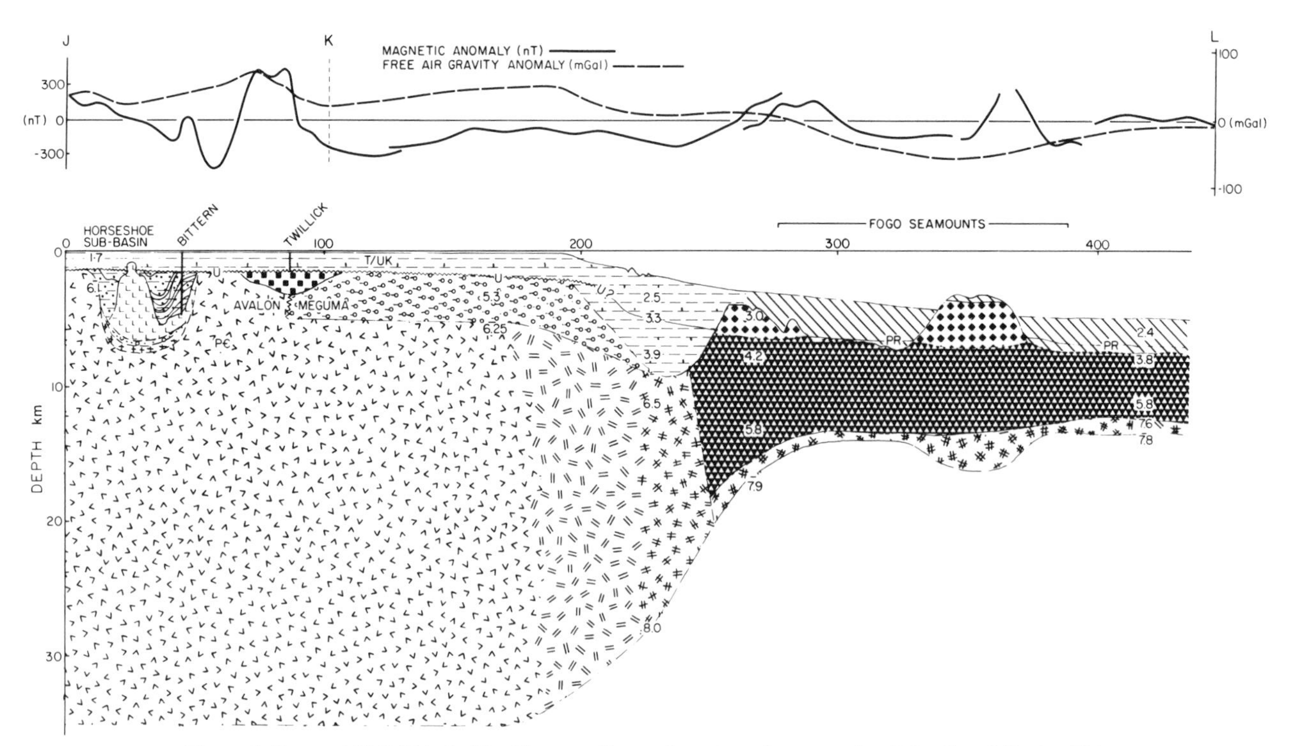

Figure 6. Transect D-2 (J to L, Fig. 1) across the southern transform margin of the Grand Banks. This transect crosses Horseshoe Sub-Basin, one of the deep, linear basins, that are common on the Grand Banks. The Avalon-Meguma boundary occurs near the Twillick well, where mid-Cretaceous volcanics have been sampled. Two of the Fogo Seamounts, one of which is now buried by sediments, occur on the oceanic portion of the profile. The sedimentary stratigraphy was derived from studies of the well data and from seismic reflection profiles (Barss and others, 1979; Barss and others, 1980; Jansa and Wade, 1975; Wade, 1981; Grant, 1977). The wells are shown as solid vertical lines. "U" is the Lower Cretaceous unconformity discussed in the text. Above it lie Upper Cretaceous and younger sediments. The deep structure along this transect was derived from seismic refraction data reported by Press and Beckman (1954), Bentley and Worzel (1956), Jackson and others (1975), and Keen and Keen (1974). Gravity and magnetic data were taken from Shih and others (1981) and Haworth and MacIntyre (1975). Patterns are explained in Figure 5.

rift history of the margin were largely controlled by the relative rates of subsidence, sediment influx, and eustatic sea-level changes (Jansa and Wade, 1975). These processes caused a major change in the nature of sedimentation, from carbonate deposition to clastic influx in Late Jurassic–Early Cretaceous time; they produced several regional unconformities, and they controlled the position of the edge of the shelf. Salt diapirism has further altered the stratigraphy of the marginal sedimentary basin.

The deep structure of the margin (Keen and Hyndman, 1979) is characterized by extensive thinning of the crystalline continental crust by factors of 2 to 3. The thinning intensifies toward the ocean-continent transition. High seismic velocities of 7.4 km s^{-1} are found within the transition zone, similar to the lower crustal velocities in Orphan Basin. The transition is also associated with the East Coast Magnetic Anomaly, a prominent marker that can be traced southward as far as the Blake Plateau. The thinned crust may be a measure of the amount of extension that occurred during rifting. The cooling of the lithosphere after the rifting episode can explain the shape of the subsidence history curves determined from deep borehole data.

Labrador Margin: Transect P to Q (D-4), Figure 8

The Labrador margin is a result of continental breakup between Greenland and North America in Late Cretaceous time. Unlike the other contemporary margins described here, this margin formed within exposed Precambrian crystalline rocks. A long rift interval of about 40 to 50 m.y. is recorded in the sediments and volcanic rocks, spanning Early to Late Cretaceous time (Umpleby, 1979). The rift phase can be divided into several episodes. The first involved the emplacement of subaerial volcanic rocks in Early Cretaceous time, followed by a second episode in which normal faults developed in the basement and nonmarine clastic sediments were deposited in a subsiding rift basin. A third episode involved the development of a regional unconformity during mid–Late Cretaceous time. Subsequently,

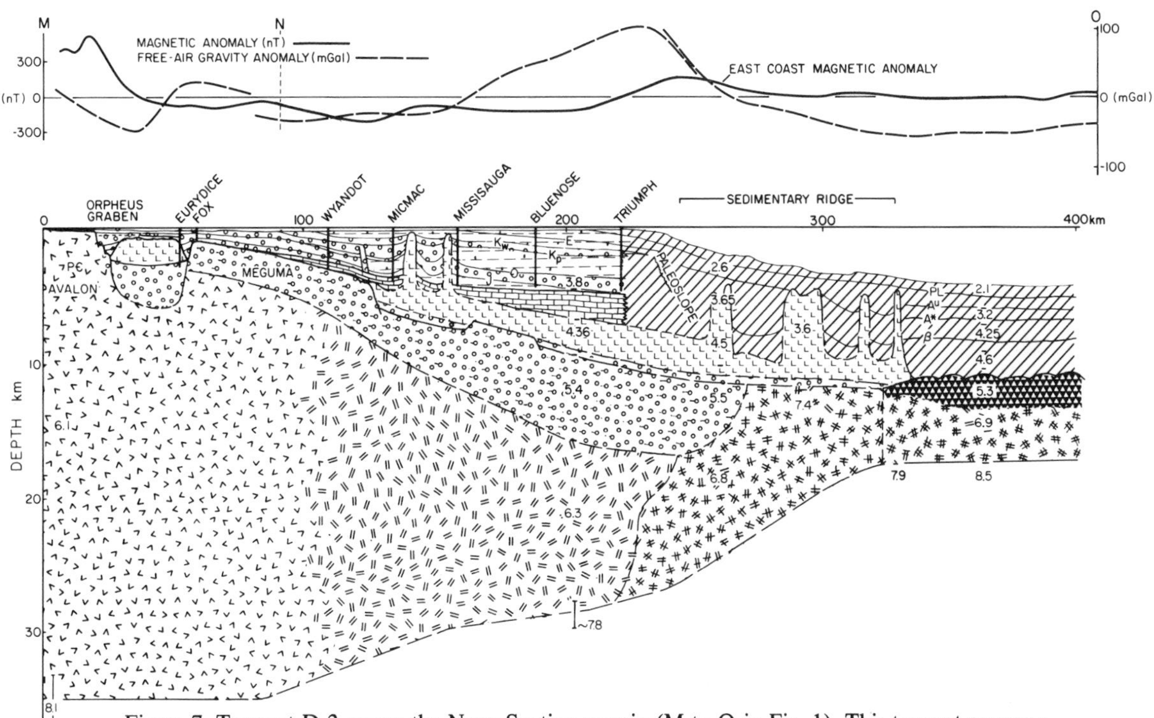

Figure 7. Transect D-3 across the Nova Scotian margin (M to O in Fig. 1). This transect crosses a mature rifted margin, now occupied by over 10 km of sediment. The Orpheus Graben, which developed at the Avalon-Meguma boundary, is one of the few well-developed Triassic grabens on the Nova Scotian Shelf. Farther seawards the Sedimentary Ridge Province, encompassing numerous salt diapirs, lies beneath the continental slope, near the ocean-continent transition. Deep exploratory wells (e.g., Bluenose) are shown by vertical lines. The sedimentary stratigraphy used to construct the transect is controlled by data from these wells and by seismic reflection data (Jansa and Wade, 1975; Wade, 1981; Barss and others, 1979; Barss and others, 1980). The sedimentary horizons marked are: PL, Pliocene (?); E, Eocene; K_w, Upper Cretaceous Wyandot Chalk marker; K_p, Upper Cretaceous Petrel limestone; O, Neocomian; J, top Jurassic; P€, Precambrian Oceanic sedimentary horizons A^u (Eocene), A* (Upper Cretaceous), and β (Lower Cretaceous) are also shown. The deep structure of the margin was deduced from seismic refraction data reported by Barrett and others (1964), Berger and others (1966), Keen and others (1975), Keen and Cordsen (1981), and Dainty and others (1966). Gravity and magnetic anomalies were obtained from the data of Shih and others (1981), Haworth and MacIntyre (1975), and Stephens and Cooper (1973). Patterns are explained in Figure 5.

continental separation was completed and postrift sediments, mainly fine-grained clastics, were deposited on the margin.

Interpretation of the deep structure of this margin and of the nature of the ocean-continent transition has been the topic of much debate. Thinned and intruded continental crust occurs beneath the outer shelf. Deep boreholes have bottomed in Precambrian rocks, Paleozoic sediments, and Cretaceous volcanics. The lower crust is again characterized by velocities of 7.3 km s^{-1}, and may have the same origin as the comparable layer within the Orphan Basin. Because of the similarity between the velocities of the deep crustal layer and that of oceanic layer 3, it is difficult to use the velocity structure of the margin to define the ocean-continent transition (van der Linden, 1975). The best estimate for the position of the oldest oceanic crust is where the top of oceanic basement can be seen as a clear reflector. However, there remains a broad region beneath the edge of the shelf and the slope where crustal affinities are unclear.

THE MODERN TRANSITION

The characteristic signatures of gravity and magnetic anomalies, the deep crustal structure, and the geologic record of margin subsidence and sedimentation are the primary data that allow us to reconstruct the evolution of the modern margin and to assess the processes likely to have caused continental breakup. The properties of the sedimentary basins are discussed in later sections. In this section, the geophysical characteristics of the margin are summarized. More detailed descriptions can be found in Haworth and Keen (1979), Keen and Hyndman (1979), Srivastava (1978), Srivastava and others (1981), Grant (1975), Haworth

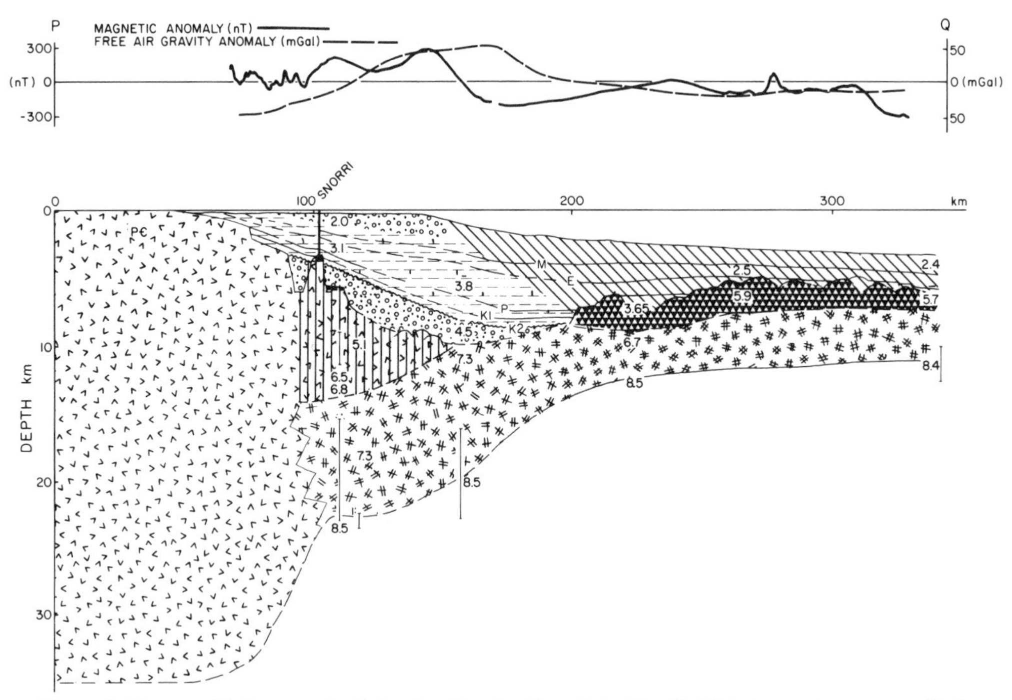

Figure 8. Transect D-4 across the Labrador Margin (P to Q in Fig. 1). This transect crosses a rifted margin, which is somewhat younger than that off Nova Scotia. The sedimentary stratigraphy is derived from studies of material from the deep exploratory wells such as Snorri, shown by a vertical line, and from seismic reflection data (Gradstein and Williams, 1976; Umpleby, 1979; Hinz and others, 1979). Specific sedimentary horizons are designated with the following symbols: M, Miocene; E, Eocene; P, Paleocene; K1, top Cretaceous; and K2, mid-Cretaceous unconformity; PЄ, Precambrian; The deep crustal structure was derived from seismic refraction data reported by van der Linden (1975). Gravity and magnetic data were obtained from the compilations of Srivastava (1978, 1979). Patterns are explained in Figure 5.

(1977, 1981), Keen and others (1977), and Keen and Barrett (1981).

Oceanic crust adjacent to the margin

The identification of oceanic crust near the continental margin is generally based on one or more of the following three characteristics: (1) seismic velocities typical of oceanic layers 2 and 3, and a relatively thin oceanic crust by comparison with the adjacent continental crust; (2) the seismic definition of the top of oceanic basement, which is a rough, semireverberant reflector in most areas and is usually overlain by subhorizontal reflectors that denote flat-lying sediments; and (3) the presence of lineated oceanic magnetic anomalies. All three of these markers are indistinct toward the ocean-continent boundary. Volcanism in continental crust near the ocean-continent transition and the presence of high-velocity sediments over basement confuses identification of crustal type on the basis of seismic velocity structure. The large thickness of sediments on the continental slope tends to mask the top of oceanic basement (Fig. 9). The oceanic magnetic anomalies become weak, in part because of the greater depth to magnetic basement, and perhaps also because the mechanism of earliest emplacement of oceanic lithosphere against the continent was somewhat different from the processes that led to magnetic stripes at mid-ocean ridges. These are the main reasons for disagreement on the nature or position of the ocean-continent boundary, even from the same set of observations. All of these difficulties exist in defining the eastern Canadian margin. There are no rigorous parameters by which oceanic crust can be identified; each situation must be considered independently. The best that we can do is to identify truly oceanic and truly continental crust, the distance between the two being up to 200 km in the case of the Nova Scotian margin or 400 km for the Orphan Basin margin.

The ocean crust south and east of the Grand Banks and Nova Scotia is characterized by the presence of a Jurassic Quiet Magnetic Zone, in which oceanic magnetic anomalies are very weak (Barrett and Keen, 1976), and by the presence of the East Coast Magnetic Anomaly, a prominent magnetic marker that trends subparallel to the edge of the shelf (Fig. 10; Keen, 1969; Keen and others, 1975; Emery and others, 1970). The origins of the Quiet Magnetic Zone and of the East Coast Magnetic Anom-

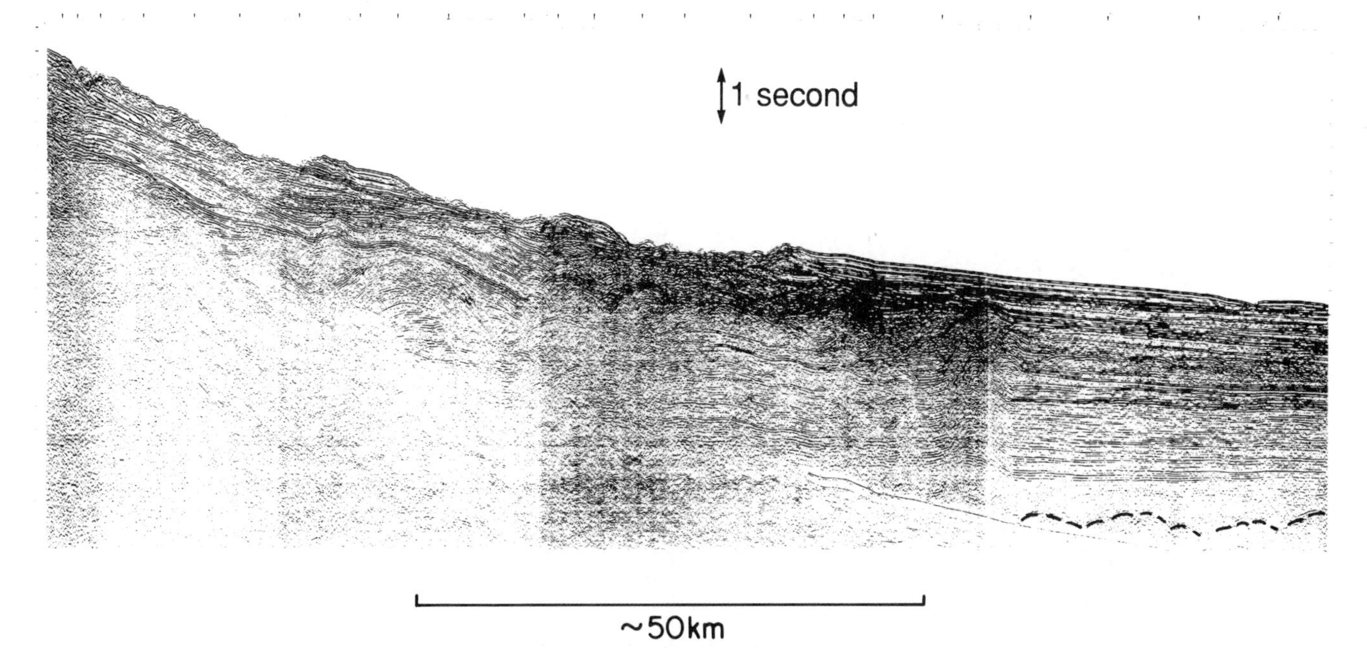

Figure 9. Part of Seiscan-Delta Line 108, a multichannel seismic reflection line along Transect D-3 (N–O in Fig. 1). The vertical scale is two-way travel time (sec). This portion of the line shows oceanic basement on the seaward part of the line dipping beneath the margin. Ocean basement (heavy dashed line) and the overlying flat sediments are interrupted by the confused sedimentary strata beneath the continental slope and upper rise. Much of this complexity is due to the presence of salt diapirs that form part of the Sedimentary Ridge Province.

aly are still uncertain (Lowrie, 1979; Klitgord and Behrendt, 1979; Vogt and others, 1970; Barrett and Keen, 1976). However, it is generally accepted that the Quiet Zone is underlain by oceanic crust (but see Gradstein and others, 1977), and that the East Coast Magnetic Anomaly marks, or lies within, the region of the ocean-continent transition.

Offsets in the East Coast Magnetic Anomaly (Fig. 10) have been interpreted as fracture zones (Fig. 1). These offsets in most places correspond to boundaries between basins and platforms; for example, the boundary between the Scotian Basin and the LaHave Platform, and between the LaHave Platform and Georges Bank Basin. The East Coast Magnetic Anomaly is very weak just north of Transect D3 (Fig. 7) and is apparently absent on the northern Nova Scotia margin. The reasons for this are unclear. Barrett and Keen (1976), on the basis of calculated spreading rates within the Quiet Magnetic Zone, estimate the age of the ocean crust in the vicinity of the East Coast Magnetic Anomaly to be 175 Ma. Despite the uncertainties, this is the best estimate that can be obtained for the age of the oldest oceanic crust adjacent to the Nova Scotia margin.

East of the Quiet Magnetic Zone lies the Keithley sequence of oceanic magnetic anomalies, which correspond to oceanic crust between 108 and 153 Ma (anomalies M0 to M25; Larson and Hilde, 1975). The youngest of these anomalies, M0–M2 (J-Anomaly), are found within the Newfoundland Basin, east of the Grand Banks (Keen and others, 1977). The J-Anomaly (~108 Ma) represents the oldest sea floor that can be identified in the Newfoundland Basin (Fig. 10). West of this anomaly, there appears to be a deep sedimentary trough, probably underlain by continental crust (Fig. 1; Sullivan, 1978, 1983). The age of the Newfoundland Seamounts is compatible with the age for the oldest sea floor. One of these seamounts yields an Ar^{39}/Ar^{40} age of 97 Ma (Sullivan and Keen, 1977).

The southern Grand Banks margin exhibits evidence of extensive volcanism. The Fogo Seamounts may be evidence of a leaky transform. The J-Anomaly Ridge, a basement high underlain by a thickened section of oceanic crust, may be analogous to the present Iceland–Reykjanes Ridge system. At DSDP site 384 on the J-Anomaly Ridge, shallow-water limestones of mid-Cretaceous age lie above basaltic basement, suggesting that the ridge was once at sea level (Gradstein and others, 1977). Similarly, the Newfoundland Ridge may be a transverse aseismic feature analogous to the Iceland–Faeroes Ridge (but see Grant, 1977; and Gradstein and others, 1977 for other proposals). This hypothesis is based on the assumption that a large magma chamber could have been tapped to create these features (Stewart and Keen, 1978; Tucholke and Ludwig, 1982). Farther north, the Newfoundland Seamounts may represent the trace of a leaky transform fault, extending from the intersection of the Meguma-Avalon boundary with the continental margin, and formed during complex plate motions between Iberia and North America (Haworth, 1975; Sullivan, 1983).

The early sea-floor spreading history of the Labrador Sea, east of Orphan Basin and the Labrador Shelf, has been described

Figure 10. Magnetic anomalies, shown as contours where data density is sufficient and in profile form where data are sparse. The contoured data are derived from the *Magnetic Anomaly Map of Canada* (McGrath and others, 1977), from Haworth and MacIntyre (1975), and Haworth and Jacobi (1983). The profiles were obtained from data reported by Shih and others (1981), Srivastava (1978, 1979), and Keen and others (1977). The positions of the East Coast Magnetic Anomaly (ECMA) and the J-Anomaly (J) are indicated. In the Labrador Sea the trends of lineated oceanic anomalies are shown along with magnetic anomaly numbers (after Srivastava, 1978). The position of the extinct mid–Labrador Sea Ridge is shown by a heavy dashed line. Fracture zones are shown as lighter dashed lines. Off Nova Scotia, the latter have been inferred from gravity as well as magnetic anomalies.

by Srivastava (1978) and revised recently by Srivastava and Tapscott (1986). The oldest oceanic magnetic anomaly that can, in general, be identified in these regions is anomaly 33 (80 Ma). However, the oldest sea floor east of Orphan Basin and the Labrador Shelf may be older than anomaly 34 (84 Ma). Off central Labrador in the vicinity of transect D4 (Fig. 8), anomaly 31 (68 Ma) is the oldest anomaly that can be identified. These complexities in identification of magnetic anomalies arise because the anomalies are weak and confusing near the margin, leading some to speculate that transitional or continental crust underlies a large part of the Labrador Sea (van der Linden, 1975; Grant, 1980). Near the continental margin, flat-lying basement reflectors exhibit internal reflections (Hinz and others, 1979). Again the oceanic nature of these reflectors has been questioned, but we believe that they represent oceanic basaltic flows interbedded with sediment. There are no equivalents of the East Coast Magnetic Anomaly north of Nova Scotia, with the exception of the region bounded by the Charlie and Cartwright Fracture Zones. In

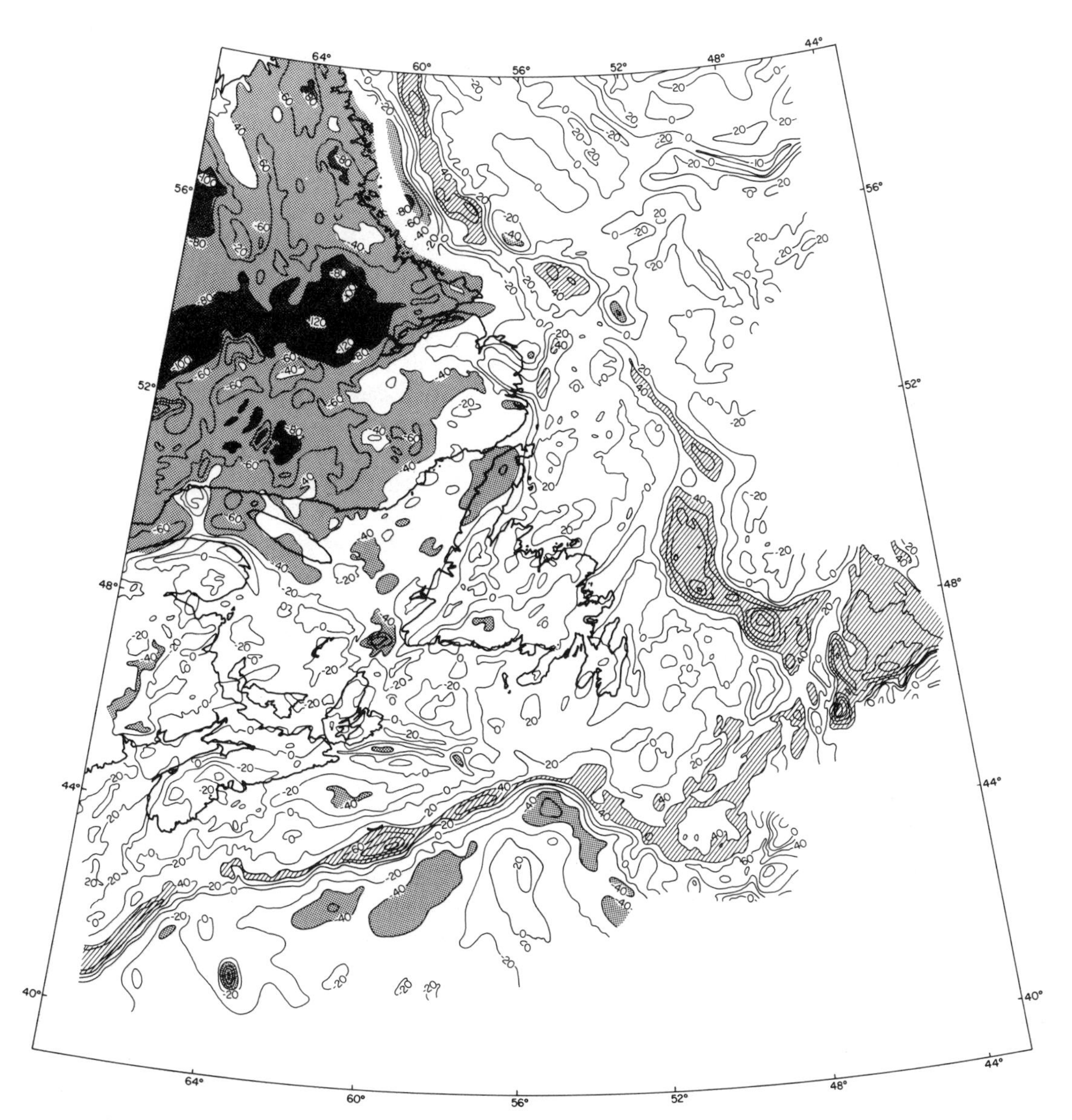

Figure 11. Gravity Anomaly Map: free air at sea, Bouguer on land. Contour interval is 20 mgal. Anomalies larger than 40 mgal are shown by the hatchured pattern. Anomalies less than –40 mgal are shaded, while anomalies less than –80 mgal are the darkest areas. Note the positive anomalies along the edge of the shelf, and the corresponding negative anomalies on the slope and rise. The transition to large negative anomalies in the northern Gulf of St. Lawrence and in western Newfoundland appears to be associated with the ancient passive margin at the edge of the Grenville Province. Data are from the *Gravity Map of Canada* (Earth Physics Branch, 1980), Shih and others (1981), Srivastava (1979), and Haworth and Jacobi (1983).

this region, a magnetic anomaly exceeding 500 nT in amplitude lies along the outer shelf and slope, and hypotheses similar to those for the East Coast Magnetic Anomaly have been suggested for its significance and origin (Fenwick and others, 1968).

Gravity anomalies

Prominent free-air gravity anomalies are observed on many rifted margins (Fig. 11). A large positive anomaly lies over the outer continental shelf, and a corresponding negative anomaly lies over the continental slope or rise. These anomalies do not coincide with the East Coast Magnetic Anomaly or its equivalents, but rather they follow the morphology of the present margin in most places. In the Orphan Basin region (D-1 East; Fig. 5), the positive gravity anomaly is located landward of the shelf edge, while the ocean-continent transition lies several hundred kilometers farther east. Such anomalies might not therefore be diagnostic of the location of ancient ocean-continent transitions.

The gravity anomalies have two causes: (1) changes in crustal thickness, sediment thickness, and water depth near the

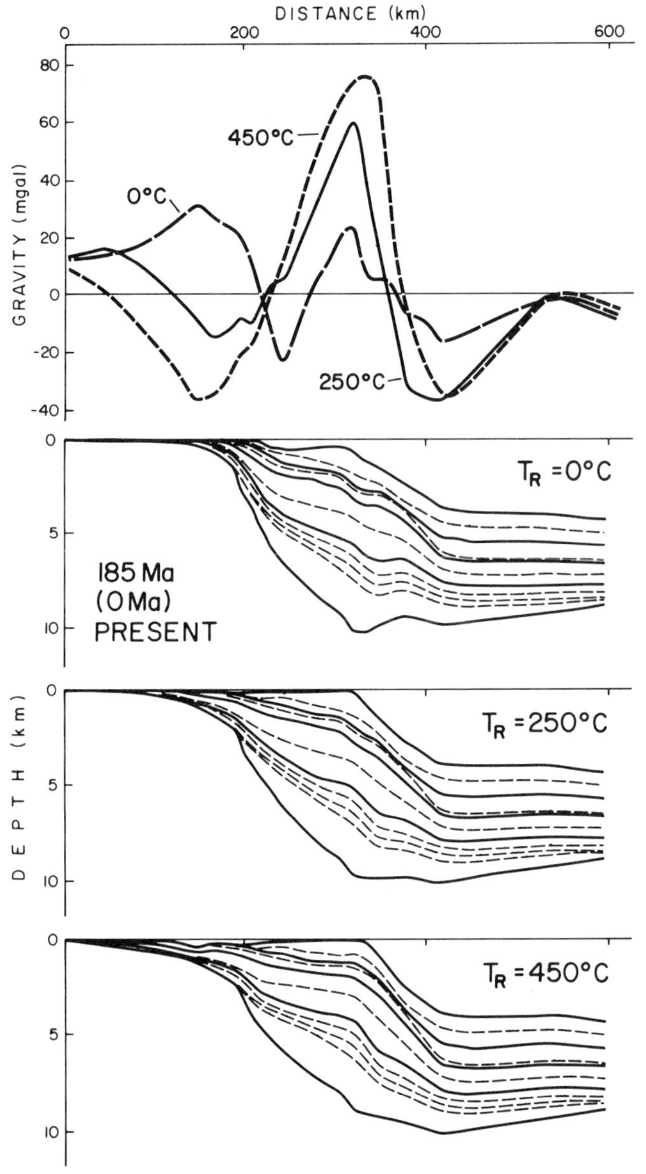

Figure 12. The variations in free-air gravity anomaly (top) produced along transect M to O (Fig. 1) for different modes of isostatic adjustment to the load imposed by the development of the sedimentary basin. The load of the basin is supported by flexure of an elastic plate whose thickness is defined by the depth to a particular isotherm T_R. The T_R = 0°C gravity curve corresponds to the simple Airy model of isostacy. The resulting sedimentary basin is shown below. The 200°C and 400°C curves correspond to progressively stronger lithospheres. The gravity anomalies are quite different for these three loading responses. Conversely, the basins produced by these three loading mechanisms are similar. An important exception occurs landward of the hinge line where a broader "coastal plain" is produced by the stronger lithosphere. After Beaumont and others (1982).

shelf edge; and (2) isostatic compensation for the load of sediment and water by flexure in the underlying lithosphere. In most models of the density structure that satisfy the gravity anomalies across continental margins, the relative emphasis given these two causes is arbitrary and variable. On the Nova Scotian margin, models have shown that the means by which isostatic compensation is achieved affects the amplitudes by 50 percent or more (Fig. 12; Keen and others, 1981; Beaumont and others, 1982; Karner and Watts, 1982).

The free-air gravity anomaly map exhibits a discontinuous pattern of highs along the shelf edge, separated by regions where the highs are less pronounced (Fig. 11). The corresponding lows on the slope and rise exhibit a similar pattern. In general, the largest anomalies are associated with the deepest sedimentary basins, while smaller anomalies correspond to platform regions. Such a pattern is consistent with a significant contribution to these anomalies from the effects of loading and isostatic adjustment. The gravity anomalies are considerably smaller across the transform margin segment, south of the Grand Banks.

Deep structure of the margin

The deep structure of the rifted margin off eastern Canada exhibits significant thinning of the continental crust out to the ocean-continent boundary. This thinning, by factors of 2 to 4, over distances of several hundred kilometers, is the dominant characteristic of the transects (Figs. 5 to 8). Furthermore, along each transect the regions exhibiting the thinnest crust correspond to the regions of maximum subsidence, suggesting that there may be a fundamental relationship between the deep crustal structure and sedimentary basin development.

The truly transitional crust has probably been thinned and intruded, and in such regions the properties of the lithosphere are probably different from typical ocean or continent. These zones are in many places characterized by lower crustal velocities of about 7 km s^{-1} and are about 100 km wide off Labrador and Nova Scotia, and about 400 km wide beneath Orphan Basin. The 7 km s^{-1} velocities are typical of basaltic intrusive rocks such as gabbros, and they are observed beneath continental crystalline basement. We tentatively interpret this to be evidence for the intrusion of basaltic magma, which underplates or intrudes the thinned continental crust during rifting. However, similar velocities are also present at depth under the stable parts of the continent (e.g., within the Appalachians, Fig. 4), and the significance of these high velocities in the lower crust remains uncertain.

Finally, the transform margin segment south of the Grand Banks exhibits a much narrower region of thinned continental crust. There is no major sedimentary basin in this region, only a narrow sedimentary trough beneath the slope, which is bounded to the southeast by the oceanic Fogo Seamounts. Within the oceanic region, layer 3 appears to be very thin, and most of the oceanic crust appears to consist of basaltic layer 2. The absence of extensive thinning of the continental crust suggests the absence of extensional forces, and the peculiar oceanic crustal structure may reflect excessive volcanism.

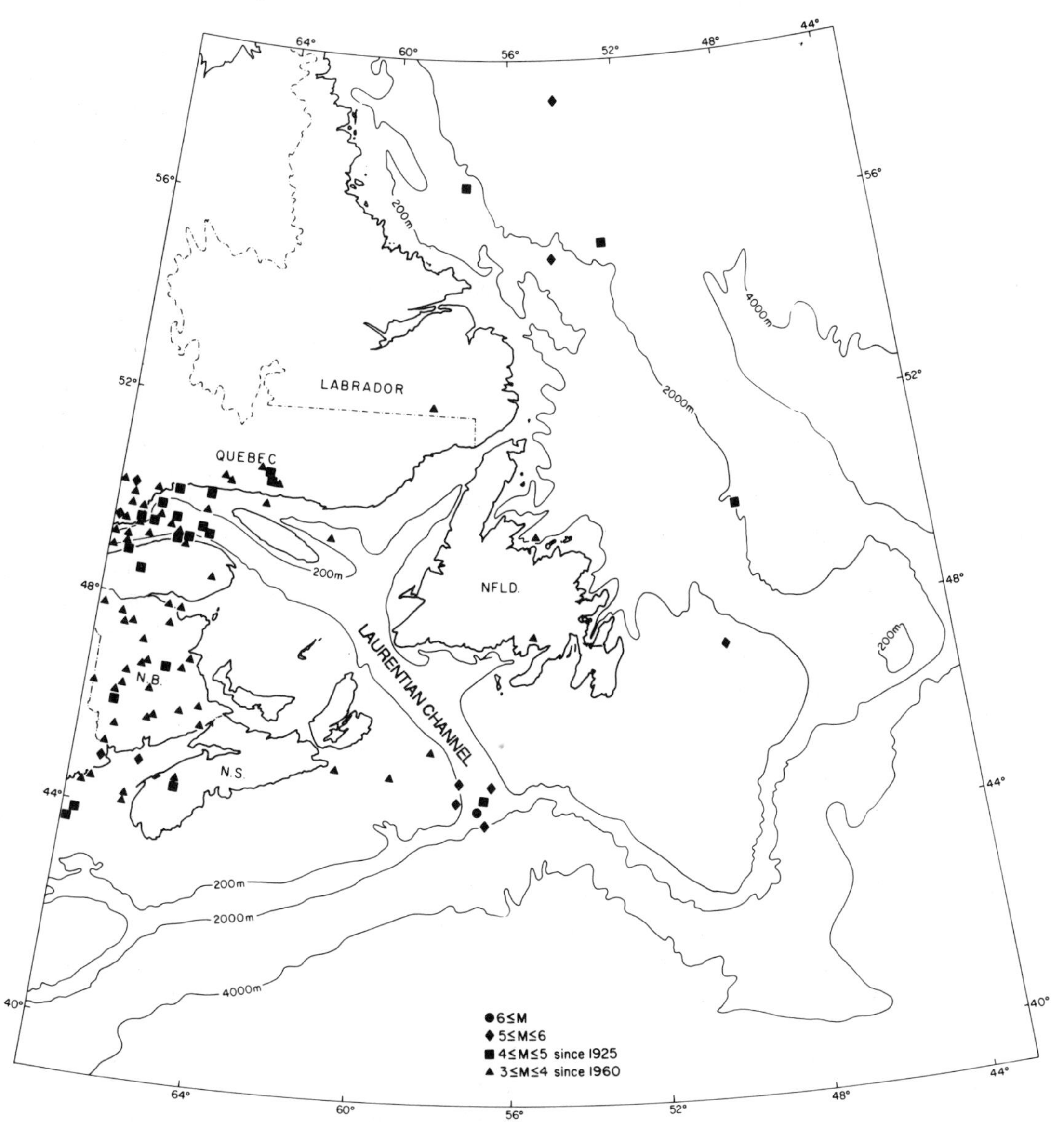

Figure 13. Location of earthquake epicenters off Atlantic Canada. After Basham and others (1979, 1983).

Neotectonics

There is evidence for the occurrence of major earthquakes on the eastern Canadian margin within historic time (Basham and Adams, 1982; Basham and others, 1983; Basham and others, 1979). The best known of these is the magnitude 7.2 "Grand Banks" earthquake of 1929. This earthquake occurred near the junction of the Nova Scotian and southern Grand Banks margin segments, at the mouth of the Laurentian Channel (Fig. 13; Doxsee, 1948). Our present knowledge of the distribution and frequency of earthquakes occurring offshore (Fig. 13) is inferred from land-based seismometer stations and may be biased by their locations. Major centers of seismicity occur at the mouth of the Laurentian Channel and at the mouth of the St. Lawrence River. In addition, a relatively high level of seismicity is observed in New Brunswick, by contrast with the low levels in Nova Scotia and Newfoundland.

The cause of these events and their localization is unknown. Stein and others (1979) suggested that much offshore seismicity relates to the reactivation of old faults under stress caused by isostatic rebound of the lithosphere following the last deglaciation. Furthermore, the eastern St. Lawrence River region and the Laurentian Channel region of high seismicity lie near major geologic boundaries (Fig. 1). However, the irregular distribution of seismically active regions and the extent of paleotectonic control remain unresolved, but environmentally important, problems.

DEVELOPMENT OF MESOZOIC RIFT BASINS

The beginning of Mesozoic continental breakup is marked by the development of rift basins with concomitant volcanism and mafic dike intrusion. These basins generally formed within half-grabens or within a series of half-grabens bounded by normal faults. They contain nonmarine and shallow-marine sediments whose thickness ranges from about 1 to 6 km. Volcanism is associated with rifting but it appears to have been sporadic and localized. The stratigraphic sections of some offshore rift basins contain a well-developed breakup unconformity, while deposition in others was continuous throughout both the rift and postrift phases.

The configuration of offshore rift basins (Fig. 1), their stratigraphy, and depositional environment of the sediment fill (Fig. 14) is derived mainly from seismic reflection and deep borehole data. Most of these data were obtained from regional syntheses published by McWhae (1981), Umpleby (1979), and Cutt and Laving (1977) for the Labrador margin; by Jansa and Wade (1975), Amoco and Imperial (1973), Given (1977), and Wade (1981) for the Grand Banks and Nova Scotia margins; and by Wade (1981), Scholle and Wenkam (1982), and Ballard and Uchupi (1975) for the Gulf of Maine. McWhae (1981) gives a regional overview for the entire Canadian Atlantic margin. The biostratigraphy and depositional environment are described by Gradstein and others (1975), Gradstein and Williams (1976, 1981), Gradstein and Srivastava (1980), and Barss and others (1979).

Triassic–Early Jurassic rifting: Grand Banks to Gulf of Maine

The Late Triassic was a time of rifting, which affected the eastern Canadian margin from the northern Grand Banks to the Gulf of Maine, a margin nearly 2,000 km long. Narrow, elongate basins, many of which are asymmetric half-grabens, formed on the Grand Banks, on the Nova Scotian margin, and in the Gulf of Maine (Fig. 1). Many appear to be underlain by Permian to Devonian sediments, suggesting that their development was controlled by Late Paleozoic structures. The oldest sediments of rift origin are Upper Triassic nonmarine clastics that have been sampled in the Bay of Fundy rift, the Gulf of Maine, the Orpheus graben, and in the subbasins on the Grand Banks. Nonmarine deposition was followed by the development of an extensive salt basin, which expanded from north to south, as the Tethys flooded the region. On the Grand Banks the oldest salt is Late Triassic, while on the Nova Scotian margin it is earliest Jurassic (Jansa and others, 1980). The salt is over 2 km thick on the Grand Banks and Nova Scotia margin, but is thin beneath Georges Bank Basin, which appears to mark the southwestern extent of these deposits. No salt occurs in the Bay of Fundy, which instead received over 4 km of nonmarine clastic sediments. The most northern occurrence of salt is in the region of the Jeanne d'Arc basin on the Grand Banks.

In most areas the salt is conformably overlain by carbonate, which developed as more open-marine conditions were established. Carbonate deposition persisted throughout most of the Jurassic. Continental clastics of the Middle Jurassic Mohican Formation in the Scotian Basin are an exception. Both the change from salt to carbonate deposition and the Mohican interruption in carbonate deposition have been interpreted to mark the onset of sea-floor spreading between Africa and North America in mid-Jurassic time. In either case, no breakup unconformity occurs between the syn-rift and postrift sediments beneath Georges Bank or the Nova Scotian margin.

Upper Triassic mafic volcanics and dikes are observed in southern Nova Scotia and in eastern Newfoundland, and can be traced offshore by geophysical methods. In southeastern Nova Scotia the tholeiitic North Mountain basalts rest conformably on nonmarine Upper Triassic sediments (Poole and others, 1970). These basalts are dated at about 195 Ma, and have been traced offshore beneath the Bay of Fundy and Gulf of Maine (King and MacLean, 1976). Volcanic rocks of unknown age were sampled in the Montagnais well on the LaHave Platform. In southwestern Nova Scotia, the Shelburne Dike is the most prominent of the northeast-trending Triassic dikes found throughout Atlantic Canada. It can be traced offshore and extends some 300 km subparallel to the present margin (Fig. 1). Papezik and Hodych (1980) proposed that it may be continuous with the Avalon Dike in eastern Newfoundland, which itself can be traced magnetically for more than 100 km. Both trend northeasterly, compatible with many of the other magnetic trends indicative of dikes on the northeastern Newfoundland Shelf (Haworth and Keen, 1979). Elsewhere, particularly within the deeper sedimentary basins, there is no evidence for Triassic volcanic rocks, but in most cases they would be too deeply buried to detect.

Jurassic-Cretaceous rifting: Grand Banks

While rifting ended in the Gulf of Maine and Nova Scotia regions with the onset of Early to Middle Jurassic sea-floor spreading, the development of rift basins persisted on the Grand Banks until Cretaceous time. During the Jurassic and Early Cretaceous, the southern margin of the Grand Banks was an active transform fault, and the eastern margin experienced continental rifting between Iberia and North America. This complex situation culminated in the mid-Cretaceous, as sea-floor spreading began east of Newfoundland, and Africa decoupled from the North American plate (Fig. 3).

The southwestern corner of the Grand Banks experienced a similar rift-drift history to the Nova Scotian margin; postrift subsidence was coeval with sea-floor spreading in the Early Jurassic and continued to the present. On the remainder of the Grand Banks, the rift basins continued to subside during the Jurassic, but in Early Cretaceous time a major erosional unconformity developed (Fig. 14) in response to uplift and rifting between Iberia and North America. Earlier Mesozoic sediments were eroded, except in some deep sub-basins. The locus of maximum disturbance appears to have been along the Avalon Uplift, which trends northwestward across the Grand Banks from the Newfoundland Ridge. The unconformity occurs as far north as the southern part of the Jeanne d'Arc basin. North of the Avalon Uplift the development of the margin is markedly different.

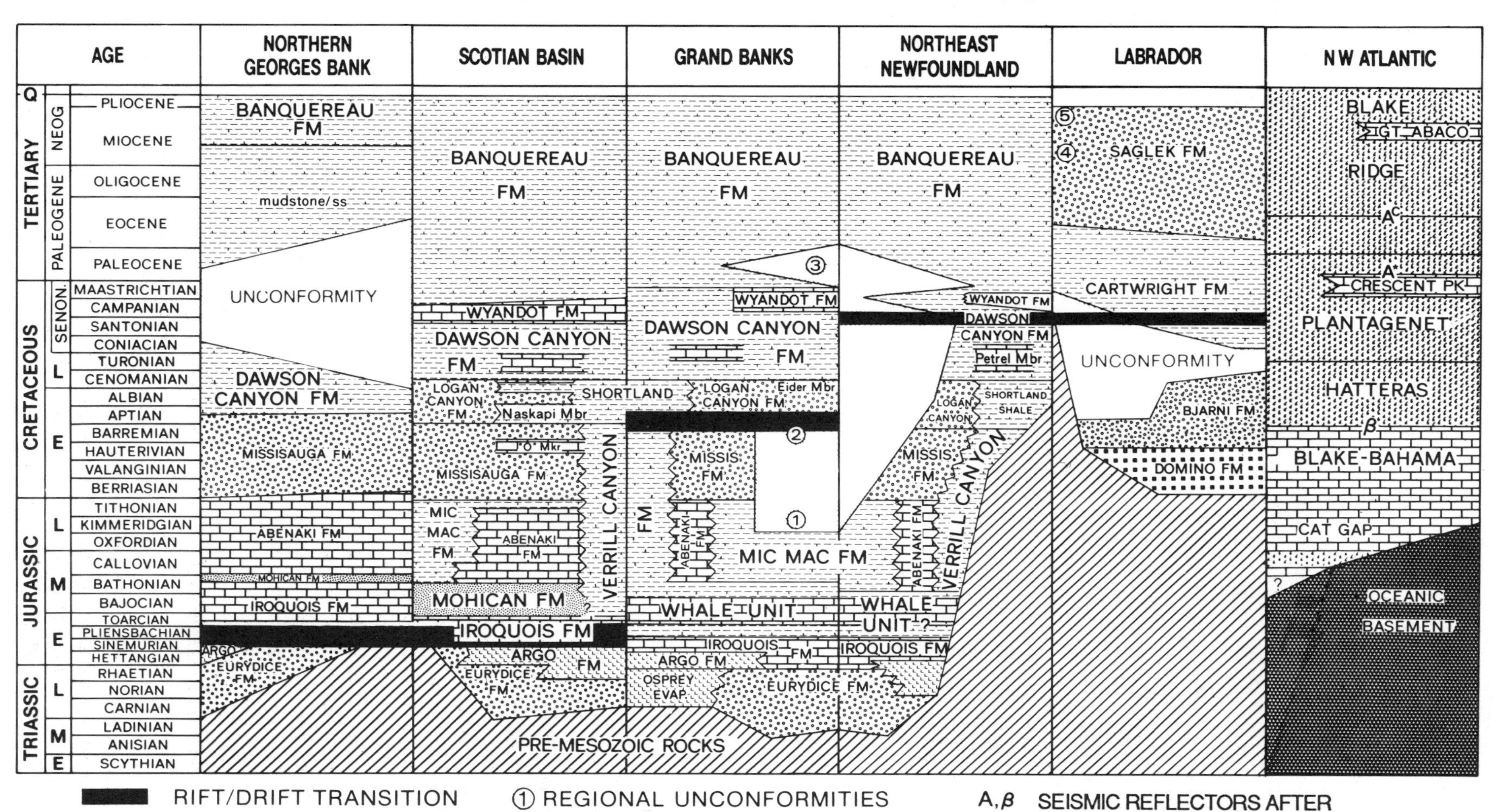

Figure 14. Generalized stratigraphic columns for the Mesozoic-Cenozoic rocks on the Canadian Atlantic margin. Major unconformities are shown as white areas. The numbers refer to the five regional unconformities of McWhae (1981). The heavy black lines are the inferred times of the rift-drift transition. Patterns are those explained in Figure 5. The column for the northwest Atlantic shows the stratigraphic positions of prominent deep ocean seismic reflectors A^c, A^*, and β. This diagram was derived from the data of Wade (1981), Umpleby (1979), Jansa and others (1979), and Scholle and Wenkam (1982).

Early Cretaceous alkaline volcanism has been detected in many of the wells on the Grand Banks (Pe-Piper and Jansa, 1987; Gradstein and others, 1977). These volcanics lie near the Avalon-Meguma boundary, suggesting reactivation along that zone of weakness. Oceanic volcanism of comparable age is observed in the Fogo Seamounts, the Newfoundland Ridge, and the Newfoundland Seamounts, which collectively encircle the southeastern tip of the Grand Banks (Sullivan and Keen, 1977, 1978; Tucholke and Ludwig, 1982; Fig. 1).

Jurassic–Early Cretaceous rifting: Northeast Newfoundland and Labrador

The Grand Banks marks the transition between Triassic rifting to the south and Early Cretaceous rifting to the north. In the Orphan Basin, basement faulting, subsidence, and sedimentation appear to have started in the Early Cretaceous, although rifting may have started as early as Jurassic time (Umpleby, 1979). While nonmarine Jurassic sediments were sampled at DSDP site 111 on Orphan Knoll, no Mesozoic sediments older than Early Cretaceous have been observed elsewhere in the region. In the Blue well (Fig. 5), shallow-marine, Lower Cretaceous sediments overlie Paleozoic sediments, and there is a marked Upper Cretaceous breakup unconformity.

Jurassic rifting is suggested by the occurrence of Jurassic lamprophyre dikes and mafic intrusions in the Notre Dame Bay area of northeast Newfoundland (Strong and Harris, 1974). Offshore, the Bonavista well bottomed in Jurassic aplite. However, further drilling is necessary to delineate the extent and development of Jurassic rift basins in the region.

On the Labrador margin, Precambrian and Paleozoic basement rocks were disrupted by normal faults that developed during the Early Cretaceous. Subaerial volcanic rocks and terrestrial clastic sediments filled the rift basins (Fig. 14), which lie beneath the present outer shelf and slope. The volcanism appears to have occurred first, in Berriasian to Hauterivian time, although minor volcanic activity may have persisted until Aptian time. The distribution of the volcanic rocks is unknown, but they were encountered in five of the wells drilled on the Labrador Shelf. The overlying Bjarni Formation consists of up to 2,000 m of Hauterivian to Early Cenomanian sediments. These strata are tilted by normal faults that were active during their deposition. Most of the Lower Cretaceous sediments occupy half-grabens, and they are thin or absent on basement highs. This may reflect a major unconformity in mid–Late Cretaceous time, separating syn-rift and postrift sediment.

Rift basin geometry; Relationships to Paleozoic and Precambrian lineaments

Major structural boundaries within the Precambrian and Paleozoic basement were reactivated during rifting and in places are mimicked by the geometry of the rift system that developed off southeastern Canada in Triassic time and off northeastern Canada in Early Cretaceous time (Fig. 1). The continental margin is segmented into a series of platforms and basins, which developed during rifting. Many of the boundaries between these elements coincide with major Paleozoic or Precambrian lineaments. The Labrador rift basins are bounded to the south by the Cartwright Arch, which appears to be an offshore extension of the Grenville Front. Early Cretaceous volcaniclastic rocks near the Grenville Front, on the Labrador coast at Fords Bight, suggest reactivation along that structural boundary. Offshore, the Cartwright Arch appears to have acted as a positive element throughout the Mesozoic development of the margin. Furthermore, an oceanic fracture zone, the Cartwright Fracture Zone, appears to have developed along the seaward extension of the Grenville Front–Cartwright Arch system (Fig. 1).

The Labrador basins are separated from the Orphan Basin by a platform region, the Belle Isle Arch. The southern terminus of this platform coincides with the Dover Fault, separating the Avalon and Gander terranes. The seaward extension of the Dover Fault is the Charlie Fracture Zone, a major North Atlantic transform (Haworth, 1977), which is also the northern boundary of the Orphan Basin.

These two examples of platforms and basins, the nature of the boundaries between them, and their relationship to oceanic fracture zones typify the geometric development of rift basins. Another example of this kind of segmentation occurs off Nova Scotia, where the LaHave Platform and Yarmouth Arch separate the Scotian and Georges Bank Basins (Fig. 1). Although the boundaries of the LaHave Platform are not clearly related to older continental lineaments, they are related to the location of oceanic fracture zones, as observed in offsets in the East Coast Magnetic Anomaly (Fig. 10).

The Bay of Fundy and Orpheus Triassic rift basins formed along the Avalon-Meguma boundary. Late to mid-Cretaceous volcanism on the Grand Banks and in the Orpheus graben was also centered near this boundary. The seaward prolongation of the Avalon-Meguma contact coincides with the Newfoundland Seamounts (Haworth, 1975). These have a mid-Cretaceous age and may represent volcanic activity along a leaky transform that developed during changes in plate motions between Iberia and North America (Sullivan and Keen, 1977).

Most of the basement faults that developed during Mesozoic rifting occur in the Avalon terrane, rather than Meguma terrane. This reflects the relative degree of Precambrian and Paleozoic faulting of the two terranes. Furthermore, all the rift basins on the Grand Banks overlie Avalon basement and occur between Avalonian structural highs, which like the rift basins, exhibit a northeasterly trend (Haworth, 1981).

During rifting, the asymmetry of listric normal faults, which produces a series of half-grabens with a single polarity, may be controlled by earlier faulting (Bally, 1982), perhaps thrust faults that developed during Appalachian orogenesis. This asymmetry is apparent in the basins off eastern Canada, but their ancestral controls are unclear. Sedimentary deformation by salt diapirism, and the depth of basement structures obscure many aspects of basement geometry. An exception occurs, in part, on the Labrador margin (McWhae, 1981). Basement structure in that region shows both northwest-trending normal faults, stepping down toward the margin, and northeast-trending "transform" faults, perpendicular to the margin. These trends are compatible with lineaments in the adjacent Precambrian Shield. However, the available data are inadequate for definition of listric faults below the basement surface.

Rather ironically, the important southern transform margin of the Grand Banks appears not to have developed along a preexisting lineament. Furthermore, normal faults within the Orphan Basin appear to be perpendicular to the northeastward trend of Avalon structures. These exceptions suggest that extensional forces during rifting controlled the locus of continental breakup, and that although preexisting boundaries were convenient, they were not essential for the development of Mesozoic rift structures.

EVOLUTION OF SEDIMENTARY BASINS ALONG THE PRESENT OCEAN-CONTINENT TRANSITION

Subsidence history

After sea-floor spreading began, the sedimentary basins along the ocean-continent transition continued to subside, in some cases after a period of uplift marked by a breakup unconformity. Their development is characterized by subsidence and almost continuous sedimentation. This is demonstrated by the subsidence curves for the Labrador Shelf, Orphan Basin, and Nova Scotian Shelf (Fig. 15 to 17). The subsidence history, de-

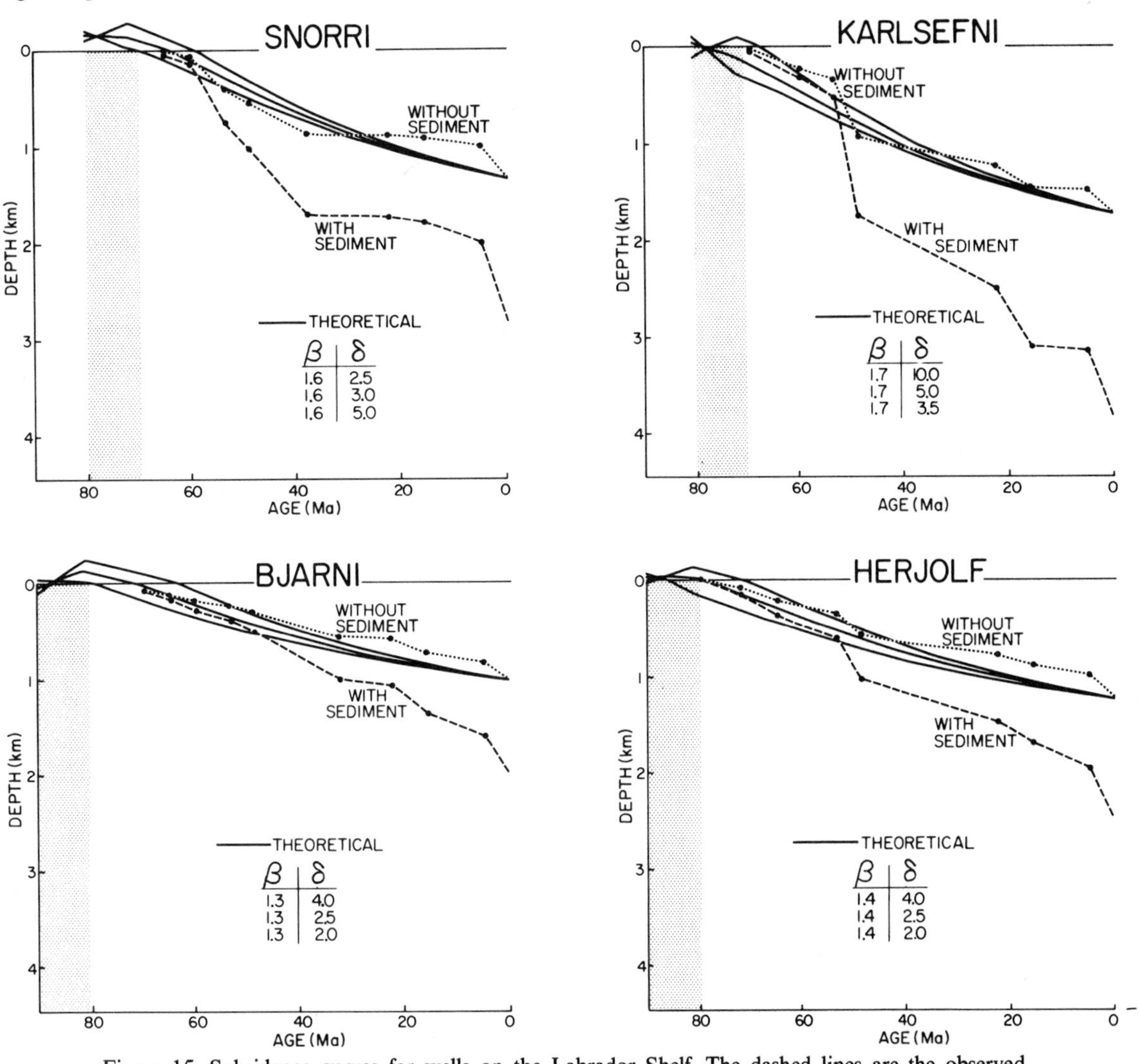

Figure 15. Subsidence curves for wells on the Labrador Shelf. The dashed lines are the observed subsidence curves. The curves labelled "without sediment" have been corrected to remove the effect of sediment loading so that they show the subsidence due only to tectonic processes. The solid curves are theoretical predictions of the tectonic subsidence for the indicated amounts of stretching. The theoretical model corresponds to the depth-dependent extension model discussed in the text in which the upper lithosphere is stretched by β and the lower lithosphere by δ (see Fig. 19). The theoretical curves with the maximum early uplift correspond to the largest values of δ, as given by the tabulated β, δ values. After Royden and Keen (1980).

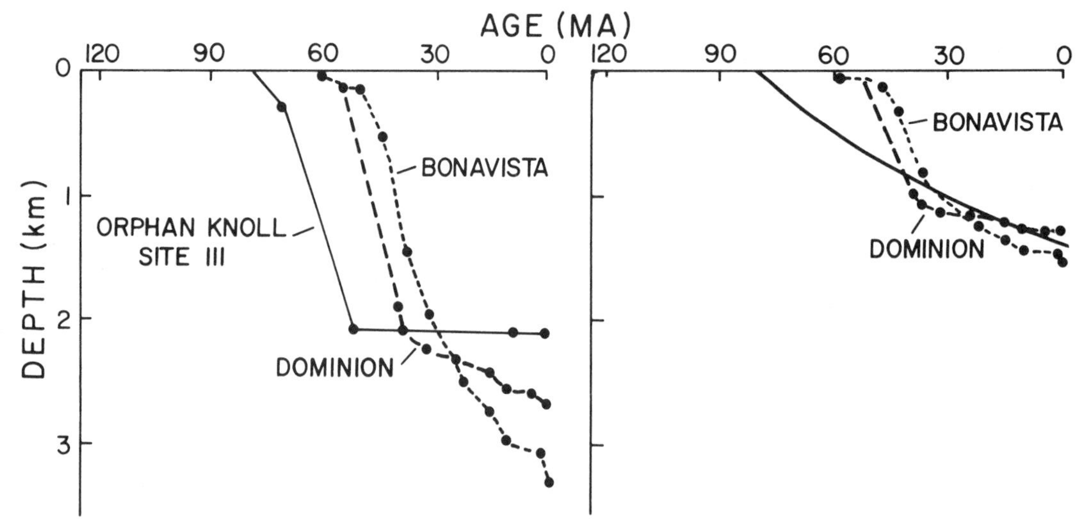

Figure 16. Subsidence curves for the Orphan Basin region. The left-hand side shows subsidence at DSDP Site 111 on Orphan Knoll, and for the Bonavista and Dominion wells on the northeast Newfoundland Shelf. The right-hand side shows the subsidence curves for Dominion and Bonavista after the correction has been applied to remove the effect of sediment loading. The solid line is theoretical subsidence predicted by the depth-dependent extension model. After Keen and Barrett (1981).

rived from the biostratigraphy for selected deep boreholes, is best defined for the Nova Scotian Shelf (Royden and Keen, 1980; Keen, 1979). The subsidence curves exhibit the familiar exponential shape, indicative of cooling and thermal contraction of the lithosphere. The subsidence of the Labrador margin can be described in a similar manner, although the curves (Fig. 15) show considerable scatter, probably because of variations in paleowater depth, which have not been considered. Postrift sedimentation of the Labrador margin was delayed for about 10 m.y. after the beginning of sea-floor spreading, probably because this margin was uplifted just before the commencement of sea-floor spreading.

The subsidence of Orphan Basin (Fig. 16) is anomalous, in that the subsidence history is not compatible with a simple model of a cooling lithosphere (Keen and Barrett, 1981). The region did not undergo significant subsidence until the Eocene, when subsidence was extremely rapid. This was followed by anomalously slow subsidence in the later Tertiary. This anomalous behaviour may be the result of the northern boundary of the basin being a transform margin, the Charlie Fracture Zone. During continent-continent transform motion, the two plates would remain coupled so that the basin would continue to be actively rifted until continental separation permitted rapid foundering of the basin. This hypothetical scenario is analogous in some respects to continued rifting of the Grand Banks until Cretaceous time. However, the mechanical and thermal processes involved are unclear.

Sedimentary processes

On the Labrador margin and in Orphan Basin the postrift sediments consist largely of fine-grained clastic material. On the Scotian margin, however, salt diapirism and carbonate sedimentation have significantly affected the patterns of terrigenous clastic sedimentation. During Jurassic and earliest Cretaceous time, carbonate platforms and banks developed across the margins from the Grand Banks to the Gulf of Maine (Fig. 14; Jansa, 1981). Their development was primarily determined by the rate of carbonate production compared with the rates of subsidence, sea-level change, and clastic sediment input (Jansa, 1981). The rate of carbonate production appears to depend primarily on climate (Jansa, 1981). The development of Late Jurassic carbonate platforms and Early Cretaceous carbonate banks now provide convenient markers for the location of the paleoshelf edge. This edge occurs under the middle shelf in parts of the Scotian Basin, and beneath the continental slope off the LaHave Platform. Carbonate deposition was terminated in Late Jurassic to Early Cretaceous time, corresponding to an influx of terrigenous clastic sediments that was sufficient to bury the carbonate banks. Terrigenous clastic sedimentation has predominated from Early Cretaceous time to the present. An example of the termination of carbonate buildup is shown on transect D3 (Fig. 7), where influx of the deltaic Verrill Canyon shales overstepped the carbonate bank, presumably at the time of climatic cooling and decreased carbonate production. This caused a seaward progradation of the continental shelf edge in Early Cretaceous time.

Numerous diapiric structures presently occur beneath the Grand Banks and Nova Scotian margin (Wade, 1981). These diapirs, which appear to consist of salt of Early Jurassic age, occur beneath the shelves, but are most numerous beneath the continental slope where they form the Sedimentary Ridge Province (Fig. 1). The main phase of diapirism occurred in Early Cretaceous time because the diapirs pierce sediments of that age.

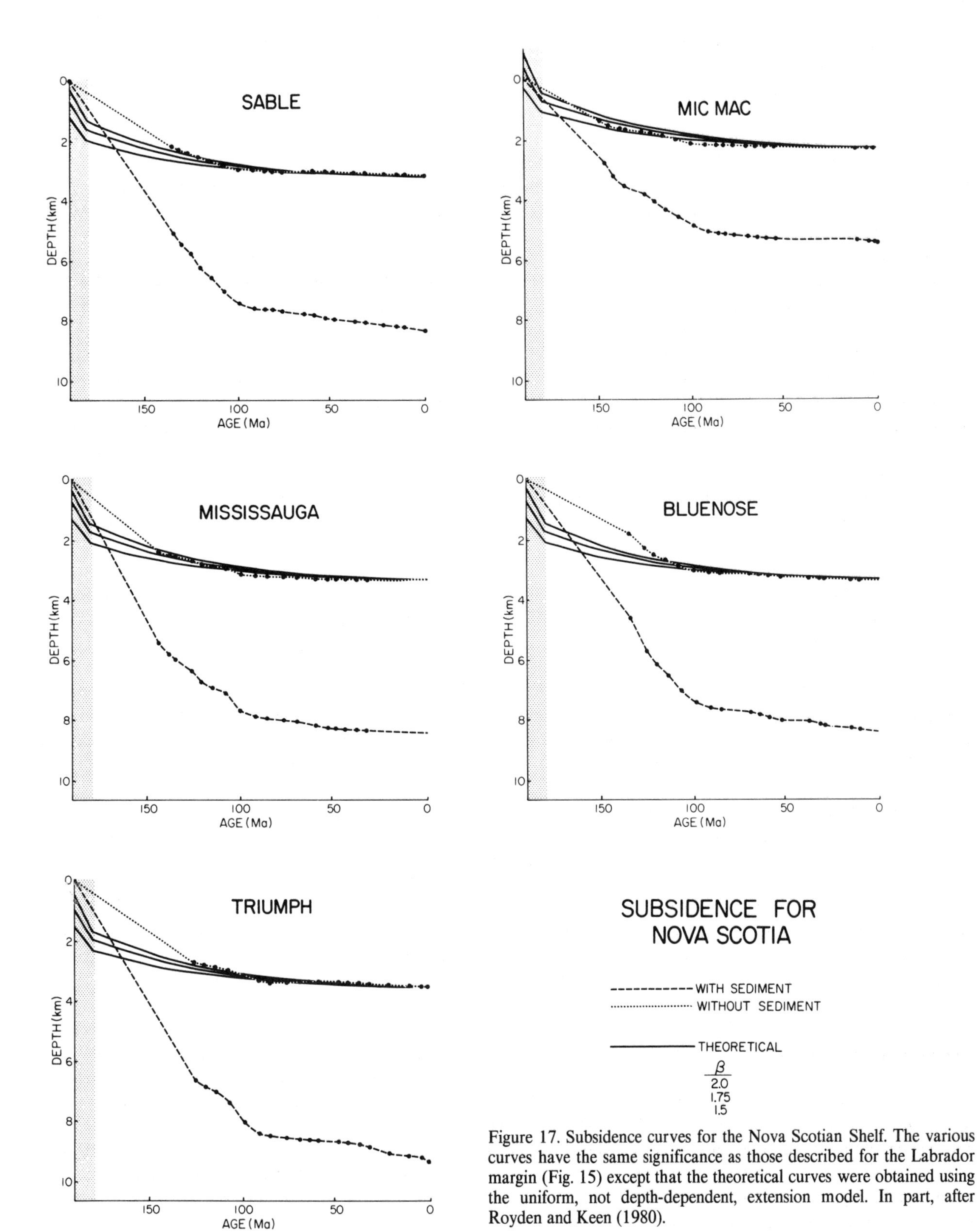

Figure 17. Subsidence curves for the Nova Scotian Shelf. The various curves have the same significance as those described for the Labrador margin (Fig. 15) except that the theoretical curves were obtained using the uniform, not depth-dependent, extension model. In part, after Royden and Keen (1980).

A late phase of motion occurred in Late Cretaceous–Early Tertiary time, when contemporary sediments were thinned and elevated above the diapirs.

The position of the edge of the continental shelf has varied significantly as the margins evolved, depending on the balance between subsidence and sediment influx. This balance controls the development of carbonate reefs, which act as barriers to slope sedimentation, and also determines the stability of sediments deposited on the outer shelf and slope. The nature of the sediments beneath the present continental slope and rise is unknown because of the absence of boreholes. It is generally thought that the sediments on the slope are fine-grained equivalents of those on the shelf (McWhae, 1981). However, the nature of the transition from continental shelf to ocean basin sediments is unknown.

Regional stratigraphic studies of the western Atlantic ocean basin, using DSDP results and seismic reflection profiles (Jansa and others, 1979; Tucholke and Mountain, 1979) provide a framework within which shelf and deep ocean sediments can be compared (Fig. 14). Carbonate followed by clastic sedimentation is seen in both shelf and ocean regions. Prominent Upper Cretaceous carbonate markers (Wyandot Chalk, and Crescent Peaks Member) occur in both regions. However, neither the shelf nor the deep-ocean stratigraphy is likely to be representative of the continental slope and rise facies.

King (1975) described the Mesozoic-Cenozoic sediments on the continental shelf and rise as potential miogeoclinal and eugeoclinal sedimentary units, respectively. These units meet on the continental slope, where a series of paleoslopes can be recognized. The development of these paleoslopes depends on the relationship between rates of subsidence, sediment input, and eustatic sea-level changes (King and Young, 1977). On older margins, such as the Nova Scotian margin, subsidence rates are slow, and King and Young (1977) observed "destructional" paleoslopes, because the shelf sediments had been cut back by slumping and erosion. On younger margins, such as the Labrador margin, subsidence is relatively rapid, and constructional paleoslopes predominate, because the shelf sediments prograde over the slope. Recognition of these paleoslopes is important because their development disrupts the apparent stratigraphic continuity across the continental slope. Therefore, correlation of stratigraphic units from the shelf to the rise using seismic reflectors can be misleading.

Unconformities and sea-level changes

Five regional unconformities within the marginal sedimentary basins have been recognized by McWhae (1981; Fig. 14). These occur along the entire eastern Canadian margin, although they are not uniformly well developed. Some unconformities can be related to major tectonic events, such as rifting on a regional scale, or orogenesis on a global scale (McWhae, 1981). However, no comprehensive studies have been conducted of eustatic or relative sea-level changes as recorded in the seismic stratigraphy of the margin (Vail and others, 1977). Estimates of paleowater depth have been reported by Gradstein and others (1975) and Gradstein and Williams (1976, 1981). Watts and Steckler (1979) derived a eustatic sea-level curve, using biostratigraphic data for wells on the Nova Scotian Shelf. However, major problems in separating the effects of tectonic subsidence, sediment loading, and sea-level changes (Watts and Steckler, 1981) must be resolved before the origin of unconformities can be properly understood.

PROCESSES OF CONTINENTAL BREAKUP AND EVOLUTION OF THE CONTEMPORARY MARGINS

The underlying tectonic processes that caused continental breakup are largely responsible for the later evolution of the margins. These processes affected the entire lithosphere, and therefore, observations of both deep crustal structure and the geologic record of subsidence and sedimentation should be examined for clues to their nature. Some of the more important aspects of these observations, reviewed or alluded to in the preceding sections, include the following (Fig. 18):

(1) Rift basins form and are filled with nonmarine and shallow-marine sediments. The basement faulting forms a series of asymmetric half-grabens.

(2) The interval of rifting varies from about 20 to 25 m.y. off Nova Scotia to about 50 m.y. in Orphan Basin.

(3) The rift phase can in some cases be divided into early and late stages. The early stage is accompanied by volcanic activity. The late stage is in some cases characterized by the development of a breakup unconformity. There may be a correlation between the development of breakup unconformities and duration of the rift phase (Fig. 18).

(4) The margin is dissected into basins and platforms, their boundaries coinciding with oceanic fracture zones.

(5) Crustal thinning is observed at rifted but not at transform margins.

(6) Post-rift subsidence follows an exponential pattern of decay.

There is no consensus on the primary tectonic processes that occur during rifting, and a large number of different models have been proposed (Falvey, 1974; Bott, 1971; Sleep, 1971; McKenzie, 1978). This proliferation of models is comparable to the situation for contemporary continental rift zones (e.g., Mohr, 1982). The models are of two basic types: those which require the asthenosphere to provide energy to the base of the lithosphere, and those which involve stretching or extension within the lithosphere, perhaps because of global plate tectonic controls. The former implies dynamic rifting as the result of convection in the asthenosphere beneath the rift zone, while the latter implies a passive role for the asthenosphere. It appears that extension can explain most of the characteristics of the rifted margin, and the application of this model to the eastern Canadian margins is described below.

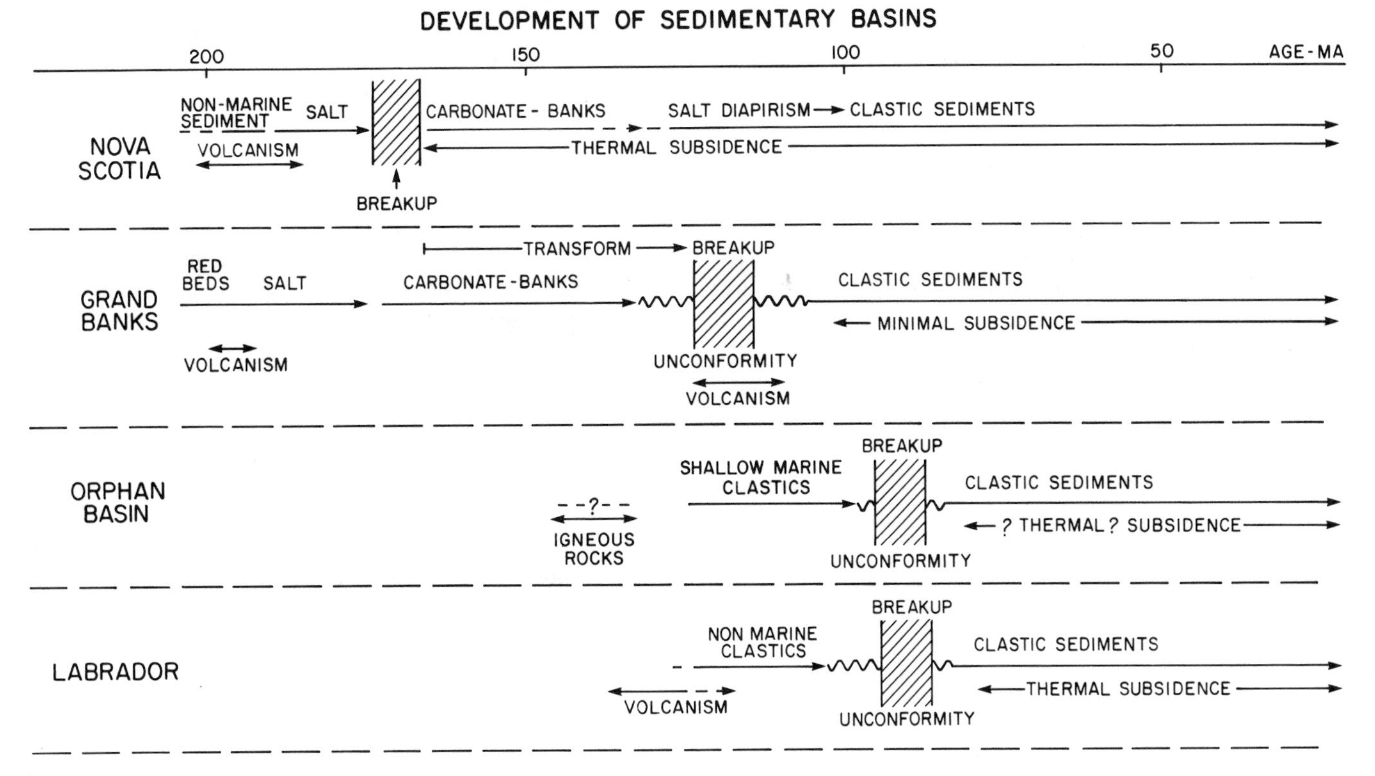

Figure 18. Schematic illustration of the evolution of sedimentary basins on the margins of eastern Canada. The evolution is depicted for four segments of the margin; Nova Scotia, Grand Banks, Orphan Basin–northeast Newfoundland Shelf, and Labrador. The times of final continental breakup are shown by diagonal shading, and the presence of a breakup unconformity is indicated by a wavy line. Vertical arrows denote the approximate onset of rifting. Sediment types, presence of volcanism, and subsidence are indicated.

The extension model

This model has been applied to a number of rifted margins. Its numerical and physical basis has been described by McKenzie (1978), Royden and others (1980), Jarvis and MacKenzie (1980), Royden and Keen (1980), and Foucher and others (1982). The essential elements of this model are the following (Fig. 19):

(1) During rifting the lithosphere is extended by an amount, β, which may vary across the rift zone.

(2) Extension causes thinning of the lithosphere, including the crust, to a thickness $1/\beta$. The asthenosphere rises passively to fill the space created by thinning. The geothermal gradient and heat flow are increased.

(3) During extension, the surface elevation changes in response to two opposing factors. The thermal expansion tending to cause uplift is counteracted by the subsidence resulting from the replacement of crustal material by denser mantle material. The net effect is generally subsidence, called initial subsidence.

(4) After extension, lithospheric cooling and contraction occurs. This also produces subsidence; the thermal subsidence. Its time dependence is approximately exponential, with a time constant of about 60 m.y.

Such a model predicts the existence of rift basins due to the initial subsidence during rifting. Also, the shape of postrift subsidence curves is compatible with the thermal subsidence predicted by this model (Figs. 15 and 17). Normal faults in basement and thinning of the continental crust also result from extension.

The predictions of the model can be tested quantitatively against observational data. In simplest terms, the amount of crustal thinning is proportional to the total subsidence observed in the overlying basins. Estimates of the amount of extension, β, can therefore be obtained by two independent methods if data are available. First, crustal thinning gives estimates of β, usually based on seismic refraction data. Second, β can be obtained from subsidence curves for the basin, estimated from biostratigraphic studies of the well samples. Royden and Keen (1980) obtained estimates of β for various parts of the Nova Scotian and Labrador margins using these two methods. They fitted theoretical curves to the observed subsidence, thereby choosing the best fitting values of β (Figs. 15 and 17). They then compared these estimates of β with those obtained from seismic observations of crustal thinning (Fig. 20). This study showed that quantitative predictions of the extension model provide a good fit to both the observed deep crustal structure and to the subsidence on the Nova Scotian margin. On the Labrador margin, however, the apparent uplift of that region just before the onset of sea-floor spreading required a modification of the extension model. This variant of the model, the depth-dependent extensional model, requires the lower lithosphere to be stretched and thinned more than the upper lithosphere, in order to produce the observed uplift (Fig. 19). The failure of the simple extension model to predict uplift and the development of breakup unconformities is one of its major deficiencies.

The subsidence of the margin is greatly amplified by the isostatic response of the lithosphere to the sediment load. This effect can be estimated if all parts of the margin are assumed to be in isostatic equilibrium according to the simple Airy model of compensation (Watts and Steckler, 1979; Royden and Keen, 1980). However, explanation of some properties of the margin, such as the large free-air gravity anomalies and the coastal onlap of sediments landward of the hinge zone, require a more realistic model for the mechanical response of the lithosphere to surface loads. The model that has been applied to the Nova Scotian margin consists of an elastic plate, which supports the loads by flexure (Fig. 21; Keen and others, 1981; Beaumont and others, 1982; Watts and Steckler, 1981). The thickness of the plate is a measure of its flexural rigidity and is allowed to vary according to the temperature distribution in the lithosphere. The base of the plate is therefore defined as an isotherm (T_R in Fig. 21). This treatment of the mechanical behavior of the lithosphere allows the lithosphere to become more rigid as it cools, and also allows the rigidity to vary across the margin, from cool, rigid continent to hot, less rigid, new oceanic lithosphere.

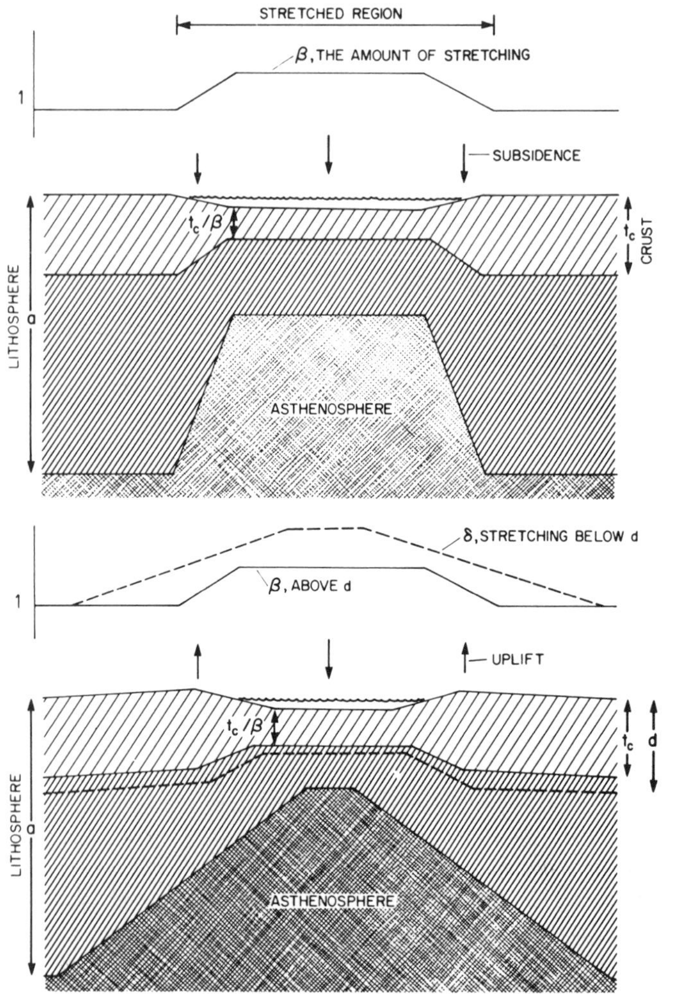

Figure 19. Illustration of the extension process. In the upper part of the figure the whole lithosphere is extended and thinned by an amount β. The crust whose original thickness is t_c is thinned to t_c/β. Hot asthenosphere rises beneath the thinned, hot region. For the model shown there will be subsidence during rifting indicated by the arrows. The lower part of the figure illustrates what happens if stretching is not constant through the whole lithosphere. Instead the upper lithosphere is stretched by β and the lower lithosphere by δ. Uplift would then occur as illustrated by the arrows. This is the depth-dependent extension model.

Keen and others (1981) and Beaumont and others (1982) used this mechanical model to trace the evolution of the Nova Scotian margin, assuming extension during rifting. The modelling included: (1) estimates of β obtained from the transect shown in Fig. 7; (2) a value of T_R appropriate to the rigidity of the lithosphere; and (3) the sediment input as a function of time and position, obtained from the observational data. The models were stepped through time, and the thermal properties of the lithosphere, basin shape, stratigraphy, and gravity anomalies were predicted at each time step (Fig. 22). This study showed that the extension model quantitatively predicts the basic properties of rifted margins. It also showed that sediment onlap landward of the hinge zone is due mainly to flexure of the lithosphere during sediment loading and that gravity anomalies across the margin (Fig. 12) are a sensitive indicator of plate rigidity.

The advantage of the extension model is that it allows a number of disparate observations to be integrated within a single

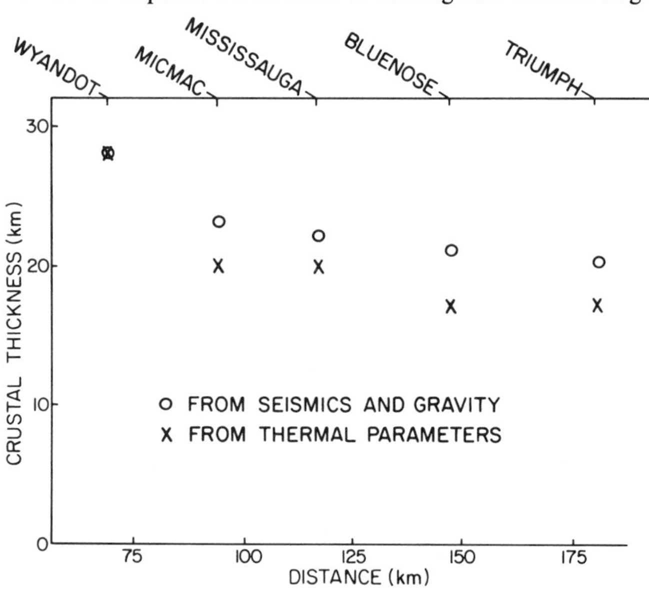

Figure 20. A comparison of two independent estimates of the amount of extension, β, across the Nova Scotian Shelf, as indicated by crustal thickness. The crosses represent crustal thickness inferred from the best fit to the tectonic subsidence calculated for each well named at the top of the diagram. The open circles correspond to measurements of crustal thickness obtained from seismic refraction experiments, supplemented by gravity data. After Royden and Keen (1980).

concept. For example, gravity anomalies, deep crustal structure, and basin evolution can all be quantitatively estimated. Furthermore, the thermal history of the sediments can be predicted, allowing their thermal maturity to be estimated (Fig. 23). This is an important element in petroleum source-rock assessment (e.g., Keen, 1979; Beaumont and others, 1982).

Volcanism, uplift, and breakup unconformities

The simple extension model does not predict volcanism during rifting, and as noted above for the Labrador margin, the model has to be modified to predict uplift. Two important aspects of the extension process require further assessment: the mechanical response of the lithosphere to finite rates of extension, and the production of basaltic melt within the upwelling asthenosphere.

The rheological properties of the lithosphere are strongly temperature dependent, and its viscosity changes with depth by many orders of magnitude (e.g., England, 1983; Beaumont, 1979). Furthermore, the uppermost crust will extend by brittle failure, deeper crustal levels may extend by ductile shearing, while the remainder of the lithosphere deforms through viscous flow. The deformation of the lithosphere under extension will therefore vary with depth. This mechanical behavior will be highly dependent on strain rates during extension, and it should be noted that uplift is exhibited at margins where the rift phase was prolonged. Numerical modelling of these mechanical properties is required to further refine quantitative models of rifting.

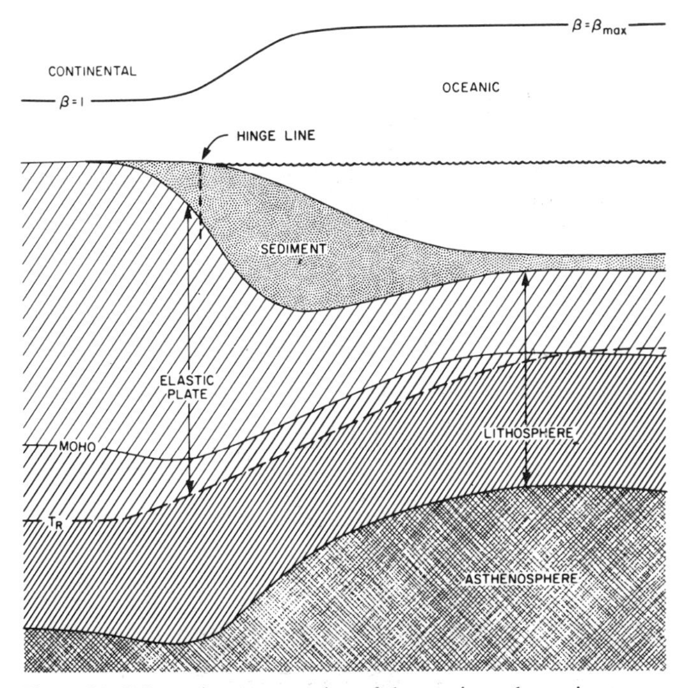

Figure 21. Schematic representation of the continental margin at some time during its evolution after rifting. The variation in the amount of extension across the margin is shown at the top. The lithosphere (and crust) have been thinned during rifting, but the lithosphere has cooled sufficiently since rifting to allow sediments to accumulate. The sediment load also produced subsidence, which depends on the rigidity of the lithosphere. These mechanical properties of the lithosphere are included in the model by proposing that the sediment load is supported by an elastic plate whose thickness and rigidity varies with temperature. After Beaumont and others (1982).

A second important consequence of extension is that partial melting will occur in the upwelling asthenosphere. Aspects of this process have been discussed by Foucher and others (1982) and by Beaumont and others (1982). Melt may rise through the lithosphere and intrude or underplate the crust, or remain in the asthenophere, and reduce the initial subsidence because of its low density. Some of the possible consequences of melt production are illustrated schematically in Figure 24. Of these, the more important are the relative uplift that will occur as stretching proceeds and partial melting is increased (Foucher and others, 1982); and the crustal underplating by basaltic magma. The latter could explain many of the observations of high seismic velocities in the lower crust near the ocean-continent transition (Figs. 5 to 8), and the relatively smooth transition from continental to oceanic crustal thicknesses (Beaumont and others, 1982). This model might also explain the extensive uplift and volcanism on the Grand Banks in mid-Cretaceous time. It is of interest to note that excess volcanism is also seen in oceanic crust around the Grand Banks; in the Fogo and Newfoundland seamounts, and in the J-Anomaly and Newfoundland ridges (Tucholke and Ludwig, 1982). However, the alternative possibility that a mantle plume existed beneath this region cannot be discounted (Morgan, 1981).

Platforms, basins, and conjugate margins

The extension process can be visualized as a kind of "continental spreading" in which transform offsets separate basins of large extension from platforms of limited extension (LePichon and Sibuet, 1981). The transition from basin to platform is quite sharp, with a change in depth to basement of 4 to 6 km occurring over horizontal distances of 20 to 40 km. Such features suggest that fundamental offsets in the locus of maximum extension exist from the earliest phase of basin development (Fig. 25). The geometric evolution of basins and platforms, and the observation that conjugate margins are not symmetrical (basins appear to develop conjugate to platforms; Jansa and Wiedmann, 1982) are consistent with this. It is suggested, following Bally (1982), that in each region along the margin there is a dominant polarity for listric normal faults, (i.e., they are either rotated toward or away from the continent). This polarity may be inherited from older fault systems. Polarity reversals and the offsets in the margins are consistent with the observation that Mesozoic rift geometry mimics promontories and re-entrants in the Appalachian system (Williams, 1979). It argues against the conventional concept of a rift system which is symmetrical with respect to a central rift valley.

SIALIC TERRANES AND THEIR PALEOZOIC ACCRETIONARY HISTORY

Four major orogenic episodes (Fig. 26) have affected the sialic crust crossed by the transects: The Grenvillian in Late

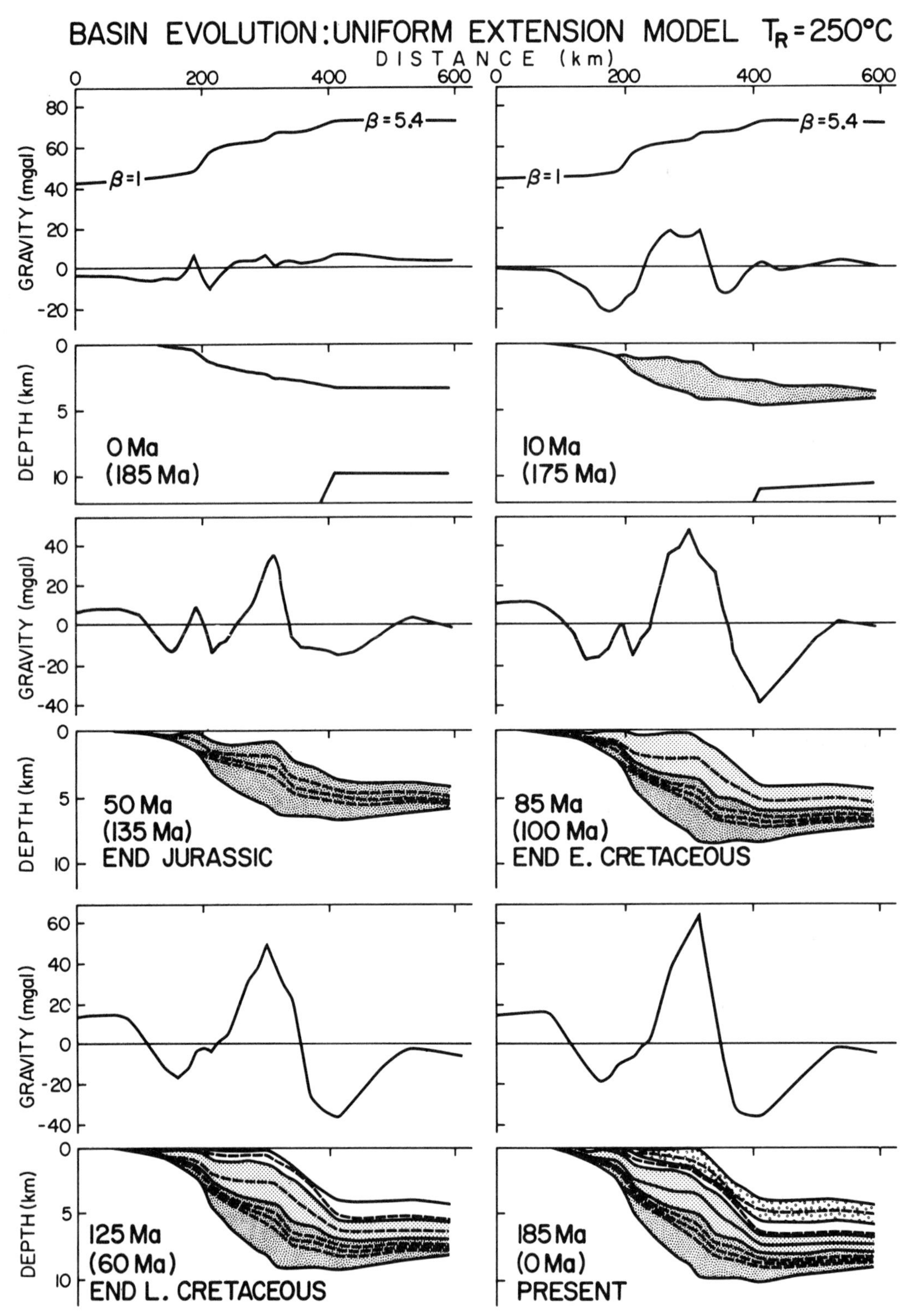

Figure 22. The evolution of the sedimentary basin along transect D-3 (Fig. 7), and its associated gravity anomaly as predicted by the uniform extension model. The uppermost curves are the amounts of extension, β, used in the model. These were obtained from observations of crustal thinning. The thickness of the elastic plate supporting the load of sediment and water on the subsiding margin is dictated by the choice of the 250°C isotherm to correspond to the bottom of the plate. Each panel shows the basin at a different time, and the corresponding free-air gravity anomaly. After Beaumont and others (1982).

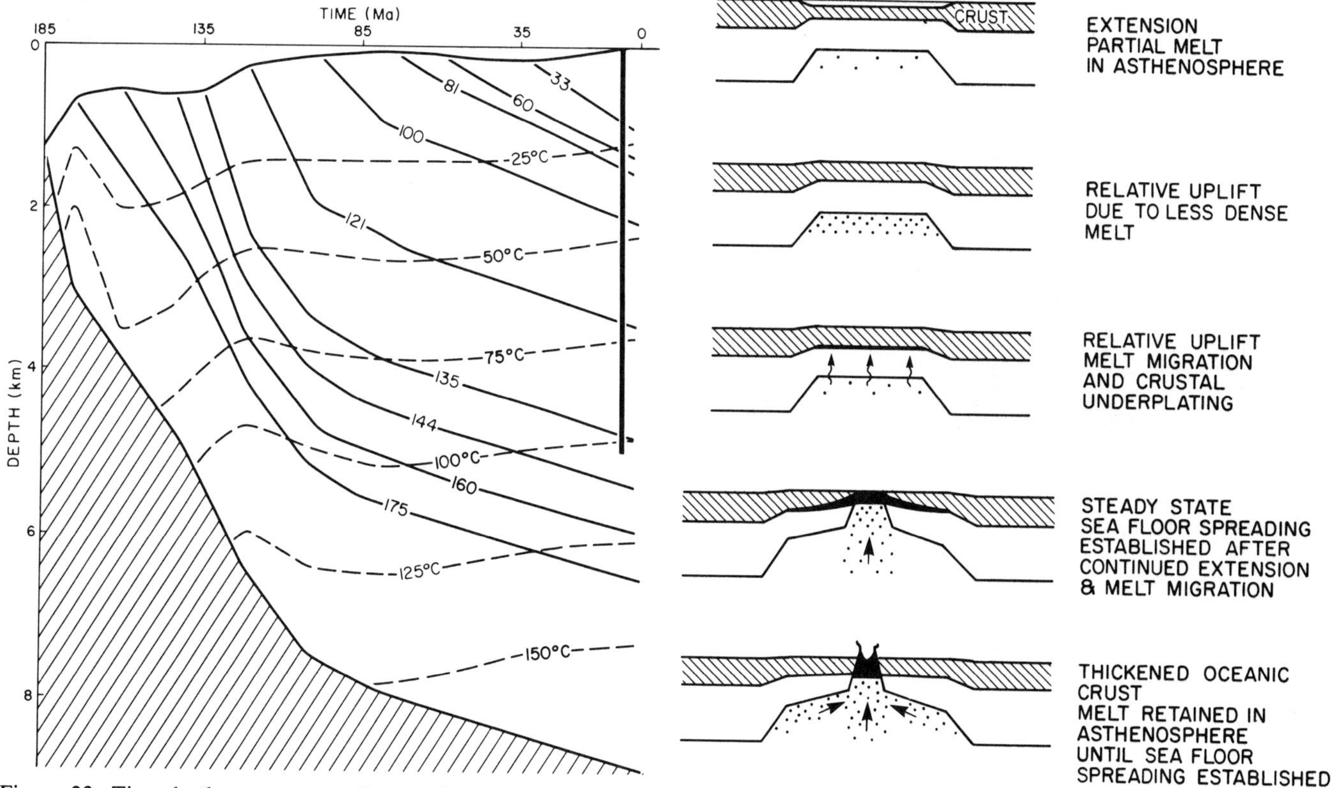

Figure 23. Time-depth temperature diagram for the Triumph well (Fig. 7). Solid lines are the subsidence curves for strata of various ages. The ages are shown in million years (Ma). Isotherms are shown as dashed lines every 25°C. Paleotemperatures can be determined; for example, sediments deposited at 121 Ma reached 25°C about 104 Ma, and 50°C about 59 Ma. Their present temperature is about 70°C. The hachured region is crystalline basement. After Beaumont and others (1982).

Figure 24. Sketch of the possible consequences of melt production in the asthenosphere during extension and rifting. The dots represent partial melt in the asthenosphere, and black regions consist of crystallized products of the melt. Note that two scenarios are suggested. First, the melt may intrude the thinned lithosphere during extension and underplate or intrude the lower continental crust. Alternatively, the melt may be largely retained in the asthenosphere until continental separation is complete. Only then does the melt rise to the surface, creating thickened oceanic crust near the ocean-continent transition.

Proterozoic time; the Taconian in Ordovician time; the Acadian in Devonian time; and the Alleghanian (Hercynian or Variscan) in Permo-Carboniferous time.

The Grenville orogenic front, crossing central Labrador almost perpendicular to the Atlantic Coast, marks the northern limit of Late Proterozoic deformation and intrusion of an area of rocks of Helikian or earlier age. The tectonic controls of the Grenville structural province are still the subject of debate, but one model for the Grenvillian Orogeny is that of a Himalayan-type continent-continent collision in the Late Proterozoic (Dewey and Burke, 1973). According to this model, the apparently over-thickened Grenville crust would be analogous with that of the Tibetan Plateau. A major gravity gradient associated with the Grenville Front (high to the south, low to the north; Thomas and Tanner, 1975) can be traced to the coastline. It is less pronounced offshore, but its trend projects toward an offset in the continental margin, at the seaward end of which is the Cartwright Fracture Zone (Fig. 1).

Trending northward, oblique to the Grenville Front, is the Labrador Trough containing Precambrian sedimentary and volcanic rocks (Fig. 1). The trough rocks lie unconformably on Archean crust of the Superior Province and they may relate to a Proterozoic continental margin (Dewey and Burke, 1973). Sediments within the Labrador Trough can be traced, with some westward offset, into the Grenville Proavince. The trend of the trough, projected across the Grenville Province, coincides with the major Quebec Reentrant of the Appalachian Orogen within the Gulf of St. Lawrence (Williams, 1978). Appalachian structural development may therefore have been controlled by the older feature.

The sinuous course of the west flank of the Appalachian Orogen (expressed in the Quebec Reentrant, St. Lawrence Promontory, and Newfoundland Reentrant of Williams, 1978) is interpreted to reflect the orthogonal geometry of rift and transform segments of the late Precambrian–early Paleozoic continental margin (Thomas, 1977; Williams and Doolan, 1979). Several important depositional and structural features appear to be controlled by this orthogonal form. Ancient reentrants are the main sites of rift-related dikes and volcanic rocks, they contain the thickest and best preserved rift-related clastic sequences, they are

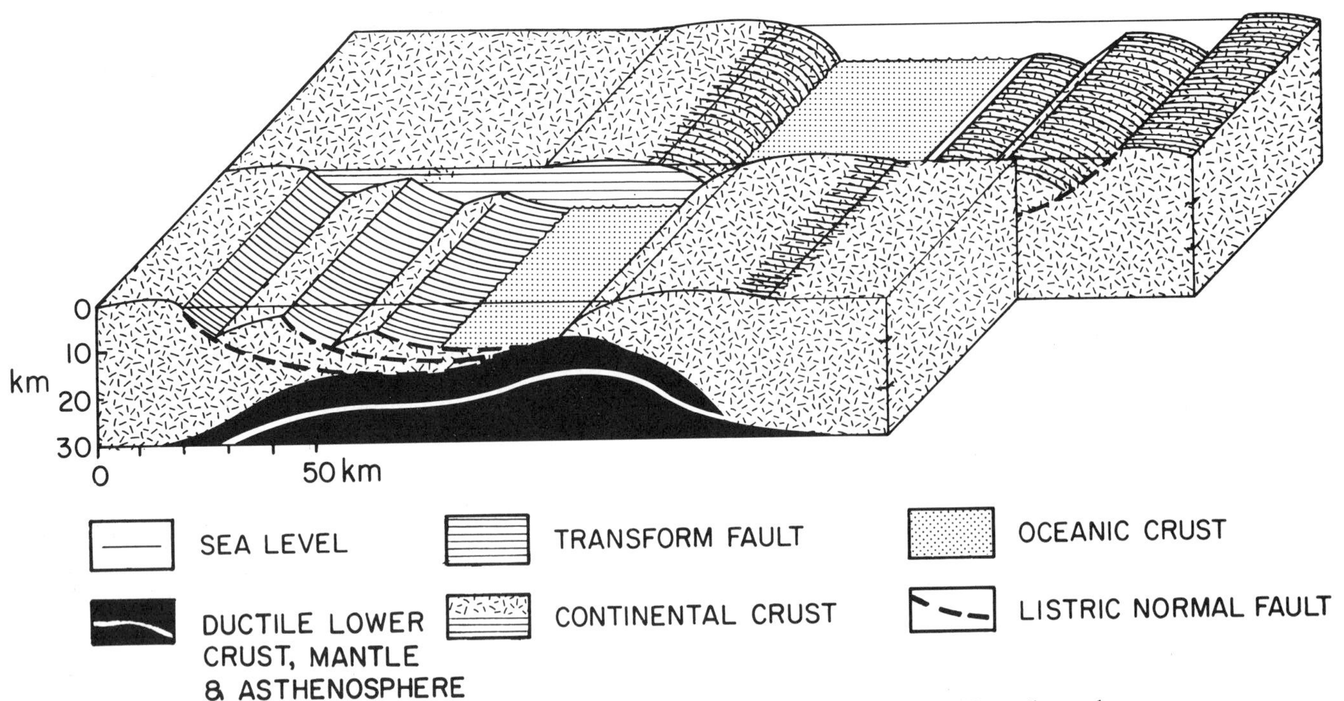

Figure 25. Illustration of offsets in rifted margins, separating basins and platforms. The sediments have been removed for clarity. Note that the listric faults in any one margin segment have only one polarity, creating asymmetry in the rift basin geometry. After Bally (1982).

the locus of clastic wedges shed toward the craton during destruction of the margin, they harbor most of the allochthons of sedimentary rocks and accompanying far travelled ophiolite suites, and they are marked by wide zones of thin-skinned deformation in cover rocks. In contrast, ancient promontories lack most of the above features and they are marked by narrow zones of thick-skinned deformation involving Grenvillian basement.

It is postulated that the edge of the Grenville craton is marked by a rapid increase in gravity from the low values that dominate the Grenville Province (Fig. 11). This line of gravity gradient can be traced the entire length of the Appalachians (Haworth and others, 1980). The COCORP line across the southern Appalachians has shown that the line of maximum gravity gradient occurs close to a major change in the nature of the subsurface reflectors. This has been interpreted as indicating that the gravity gradient marks the ancient continental margin (Cook and Oliver, 1981), or the boundary between continental and transitional crust (Cook, 1984). The situation resembles that of a modern margin, and suggests that the Appalachian gravity gradient resulted from late Precambrian rifting perhaps modified by Paleozoic collisional tectonics.

The line of maximum gravity gradient follows the Quebec Reentrant northward toward the St. Lawrence Promontory and Newfoundland Reentrant, although the effect of the sedimentary basin in the central Gulf of St. Lawrence locally obscures the pattern. The gradient is further obscured by the high gravity anomalies associated with the ophiolite suites at Bay of Islands and Hare Bay. However, in general, the gravity field reflects the major offset in the Appalachian margin within the Gulf of St. Lawrence (Fig. 1). North of Baie Verte (Fig. 1a), the gravity gradient follows the east coast of the Great Northern Peninsula and Grenvillian rocks of the Long Range Mountains (Williams, 1978). Rocks of the Paleozoic margin continue north-northeastward off southern Labrador (Haworth and others, 1976). Geophysical trends suggest that the Paleozoic margin trends eastward near the offshore extrapolation of the Gilbert River Fault (Fig. 1). The present continental margin truncates the Paleozoic margin close to an unnamed fracture zone, just south of the intersection between the Grenville Front and the present margin (Fig. 1).

The Paleozoic margin in Newfoundland

The history of the Paleozoic continental margin of eastern Canada is contained in the rocks along the west flank of the Appalachian Orogen in western Newfoundland (Transect A to F, Fig. 4). Inliers of Grenvillian gneisses are recognized as uplifted basement blocks. Thick clastic sequences overlie the gneisses and these are coeval with mafic dikes and flows, all related to initial rifting. These rocks are overlain, in turn, by a thick Cambrian-Ordovician carbonate sequence that records the passive development of the margin (Williams and Stevens, 1974).

In western Newfoundland, Grenvillian basement gneisses of the Long Range Inlier are cut by a swarm of mafic dikes of late Precambrian age that trend northeastward parallel to the ancient margin. These dikes and coeval mafic flows are interpreted as related to rifting prior to continental breakup. Toward the west, the flows are thin, columnar basalts of terrestrial deposition. They either lie directly on Grenvillian gneisses or are separated from the gneisses by thin units of arkosic sandstone. Eastern correlatives, now within Taconic allochthons, are in places pillowed, indicating submarine eruptions, and they are underlain by thick marine clastic sequences. Locally at Belle Isle, the proportions of black mafic dikes and intervening pink Grenville gneisses are roughly equal. The dikes are everywhere steep to vertical, but sedimentary rocks, cut by the dikes, vary in attitude from subhorizontal to steep, implying tilting of the sedimentary rocks either before or during mafic dike intrusion.

Late Precambrian to Early Cambrian carbonatites and lamprophyres along the St. Lawrence River are roughly coeval with mafic dikes and flows of the Long Range Inlier. The St. Lawrence intrusions are interpreted as the result of rifting (Doig, 1970) within the failed arm of an Iapetus triple junction (Burke and Dewey, 1973). Persistent alkali igneous activity in the St. Lawrence region (Currie, 1970) possibly relates to subsequent abortive attempts at separation during the opening of the present Atlantic Ocean.

Pillow lavas and breccias associated with slope/rise clastics in allochthonous sequences have chemical affinities to continental margin tholeiites farther west (Williams and Smyth, 1983; Strong and Williams, 1972). Other allochthonous volcanic rocks are alkali basalts, possibly once forming part of a seamount chain near the ancient continental margin (Jamieson, 1977; Baker, 1978).

Monotonous clastic sequences, several kilometers thick and forming the lowest stratigraphic units of Taconic allochthons, were derived from the ancient continental slope and rise. The rocks are poorly sorted, with conglomerate, graywacke, siltstone, shale, feldspathic sandstone, and arkose predominating. Blue quartz like that in the Grenvillian basement is abundant, and granite and metamorphic rock fragments are common in coarser beds. These rocks are interpreted as deep-water turbidites of late Precambrian to Early Cambrian age. Autochthonous correlatives to the west are thinner quartz sandstones with well-sorted, rounded grains and even beds. Cross-beds are common, and the sediments represent shallow-water deposits in areas of low relief. An increasing maturity in the clastic sequences from east to west and bottom to top implies rugged terrane with sediment accumulation in deep water at the onset of deposition followed by gradual erosion and infilling of depositional troughs, concluding with the deposition of Lower Cambrian mature beach sands.

The surface onto which the Cambrian seas transgressed cannot everywhere be described as a peneplain, for there is considerable local relief. Since there is worldwide evidence for an early Paleozoic marine transgression interpreted to be the result of a prolonged eustatic rise in sea level (Vail and others, 1977), it is difficult to separate the local tectonic and regional eustatic influences on sedimentation during this early development of the ancient continental margin.

At the north end of the Long Range Inlier, the Grenvillian basement and its cover rocks are separated by a sharp unconformity. Dips are subhorizontal to gentle in the west, increasing to moderate and steep toward the east. Locally, there is imbrication of basement gneisses and cover rocks along the east side of the inlier (Williams and Smyth, 1983). East of the Long Range Inlier at the Baie Verte Peninsula, Grenvillian basement rocks are deformed with the cover sequences, and mafic dikes are represented by amphibolites. The exposed edge of the ancient continental

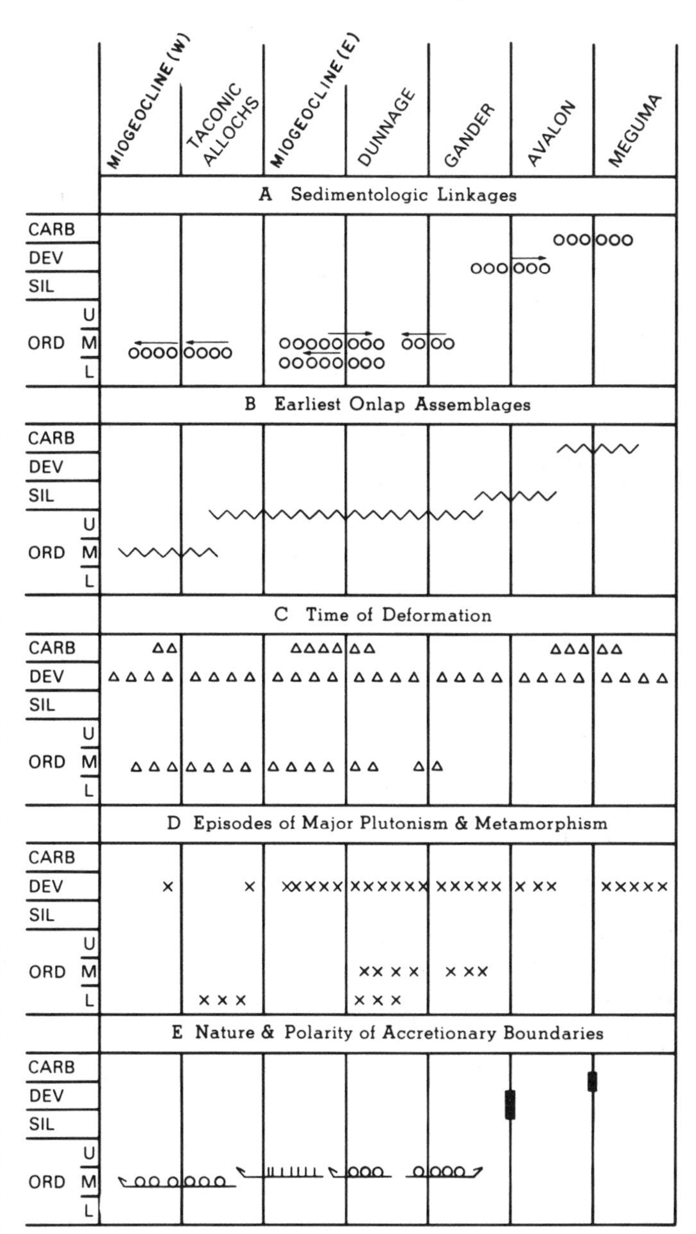

Figure 26. Sedimentologic and stratigraphic analyses of accretionary history of the Canadian Appalachians compared to time of deformation, plutonism, and metamorphism. Arrow in A indicates direction of sedimentary transport and sedimentologic linkage gives the time at which sediments were first deposited across a terrane boundary. In E, arrow with circles indicates thrust zone marked by mélange and ophiolitic rocks, arrow with vertical lines indicates sharp thrust contact, and thick vertical bar indicates steep fault zone, commonly with mylonites. After Williams and Hatcher (1983).

margin is defined by a steep zone of ophiolite occurrences, the Baie Verte–Brompton Line (Fig. 27; Williams and St. Julien, 1982).

The west Newfoundland carbonate sequence ranges in age from Early Cambrian to Middle Ordovician. It thickens eastward across the Appalachian deformed zone from less than a kilometer to 2 km or more. The carbonate sequence rests conformably above Cambrian clastic rocks, and its Middle Ordovician top grades upward and eastward into black graptolitic shales overlain in turn by a westward transgressive clastic wedge (Fig. 27). An erosional disconformity is recorded at the Lower Ordovician–Middle Ordovician contact, but the Ordovician-Cambrian boundary is gradational. Algal mounds, desiccation cracks, and local erosional disconformities, combined with cross-bedding in interlayered mature sands, all indicate that the carbonate rocks accumulated in shallow water. The thickness of the carbonates across western Newfoundland indicates that the continental margin subsided much more rapidly in the east, although shallow-water conditions were also maintained there. Across much of the northern Gulf of St. Lawrence, the carbonate units are thin and dips are consistently shallow, as seen on the seismic reflection profiles accompanying the transect A to F (D-1 West) and on the seismic reflection line across the equivalent section in Quebec (Ministère des Richesses Naturelles du Quebec, 1979; Seguin, 1982).

Allochthonous sequences above the carbonates in western Newfoundland originated east of the bank sequence. They include Middle Cambrian to Lower Ordovician shales with limestone breccia units most likely built up offshore just beyond the bank edge. The spectacularly coarse limestone breccias at Cow Head form part of a thin (300 m) condensed sequence that spans the same interval as the carbonate sequence. Thinner, finer, and shalier units of the same age in nearby transported sequences are interpreted as sediments formed in deeper water and away from the bank edge.

Rodgers (1968) delineated the eastern edge of the carbonate bank and interpreted it as an abrupt declivity, like the margin of the present Bahama Bank. This explains the marked facies change and the lack of interlayering between eastern exposures of the carbonate sequence and clastics and volcanics farther east. He also pointed out that the eastern carbonate exposures closely follow the locus of Precambrian inliers along the western margin of the Appalachians, so that the edge of the carbonate bank may itself have been localized by the original eastern extent of Grenvillian basement. However, the carbonates in western Newfoundland are slightly west of the easternmost exposures of Grenvillian basement in gneiss domes of the Baie Verte Peninsula. The situation suggests that in western Newfoundland the rifted continental edge lay to the east of the carbonate bank edge.

Destruction of the Paleozoic margin took place during the Ordovician Teconian orogeny. Its onset is signalled by the pre-Middle Ordovician unconformity and the appearance of carbonate breccias, black shales, and clastics above the carbonates (Stevens, 1970). The unconformity reflects uplift and warping, possibly in response to eastward subduction and migration of the continental margin across the peripheral bulge of an offshore contemporary island arc (Jacobi, 1981). The carbonate bank then subsided rapidly with deeper water facies migrating landward. Where post-orogenic Middle Ordovician shallow-water carbonates overlie the Humber Arm Allochthon in western Newfoundland, subsidence of the bank may have been as much as the total structural thickness of the assembled allochthon. This implies that the former bank sank to depths of a kilometer or more.

Easterly derived clastic rocks, contemporaneous with the collapse of the bank, are first recorded in Lower Ordovician parts of transported sequences and they transgressed westward across the Cow Head breccias. Farther west, they are of early Middle Ordovician age where they overlie the carbonate bank sequence. The clastics contain chromite and volcanic grains, suggesting that oceanic crust was already exposed to the east. The clastics become coarser upward, probably reflecting the gradual encroachment of the Humber Arm Allochthon.

The lowest sedimentary slices of the Humber Arm Allochthon contain the youngest parts of the offshore slope/rise strati-

CAMBRIAN TO EARLY ORDOVICIAN

Carbonate bank
Bank-foot breccias
Rift facies sedimentary rocks
oceanic crust
Grenvillian basement

EARLY ORDOVICIAN

EARLY TO MIDDLE ORDOVICIAN

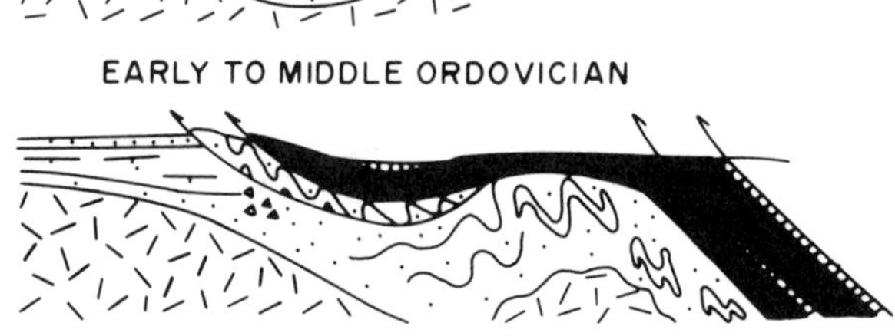

EARLY TO MIDDLE ORDOVICIAN

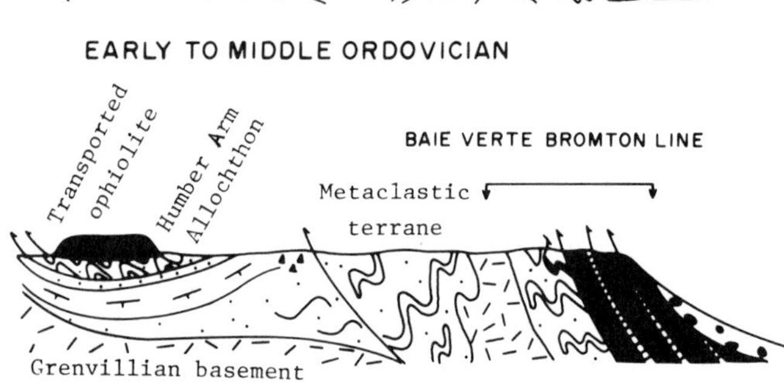

Figure 27. Model for the destruction of the Paleozoic margin in Newfoundland (after Williams and St. Julien (1982).

graphic sequence. The structural slices are underlain by mélange with a black shale matrix. Similar shaly mélanges occur between successively higher slices and they represent the combined effects of mass wasting and tectonic mixing at the soles of the advancing slices. The local occurrence of large serpentinite, gabbro, diorite, and volcanic blocks within the structurally lowest mélanges clearly attests to the proximity of ocean-floor rocks during formation of the mélanges. This implies that the lowest mélange zones are the surfaces of latest movement and that the transported rocks were emplaced along them as an already assembled allochthon. The Bay of Islands ophiolite slice, now the structurally highest, is interpreted as the first to have moved. Assembly and transport of the slices was a relatively slow process that spanned about 5 graptolite zones or approximately 20 m.y. A model for the destruction of the Paleozoic continental margin of western Newfoundland is depicted in Figure 27.

Suspect terranes of eastern Canada

Rocks of the Canadian Appalachians east of the Baie Verte–Brompton Line constitute a number of suspect terranes that were accreted to the Paleozoic margin of eastern Canada.

The *Dunnage terrane* is widest (150 km), and its rocks are best preserved in northeast Newfoundland. Its width decreases rapidly southwestward, but it increases to more than 200 km offshore and maintains that width to the present Atlantic margin northeast of Newfoundland. High seismic velocities observed offshore indicate that its ophiolitic basement has a synclinal form (Haworth and others, 1978).

Deformation across the Dunnage terrane is uniformly mild compared with that in rocks of the bordering North American margin (to the west) and Gander terrane (to the east). Ophiolitic complexes at the Baie Verte–Bromptom Line are imbricated and steeply dipping to overturned, but the sequences of units everywhere face eastward. Farther west, the ophiolitic complexes are structurally and metamorphically gradational with ophiolitic mélanges and metaclastic rocks that are an integral part of the continental margin sequence. Deformation and metamorphism are interpreted as Ordovician and related to westward transport of ophiolitic complexes, such as the Bay of Islands, from a Dunnage root zone to their present positions above the continental margin or miogeocline.

The age and chemistry of volcanic rocks in the Dunnage terrane indicate that an island arc was active during the passive development of the North American margin to the west (Kean and Strong, 1975). A cover of Middle Ordovician (Caradocian) shales above the island arc volcanic rocks indicates that the evolution of the volcanic arc ceased at the time of destruction of the miogeocline and final emplacement of the Humber Arm Allochthon (Dean, 1978). The relationships also imply that the continental margin was destroyed by collision with an island arc and that the Dunnage terrane was accreted to the North American margin in Early to Middle Ordovician time (Fig. 26).

Dunnage ophiolitic rocks are also thrust eastward upon clastic rocks of the Gander terrane. The ophiolitic rocks are overlain by polymictic conglomerates and Caradocian shales, implying a Middle Ordovician disturbance. Nearby olistostromes contain huge rafts of volcanic rocks surrounded by Caradocian shales; these are interpreted as marine slides that slumped westward, back toward the basin from which the ophiolitic rocks originated (Pajari and others, 1979). These relationships are somewhat analogous to those of the Baie Verte–Brompton Line, where olistostromes and coarse conglomerates, above the ophiolite complexes, are thought to have slid eastward.

Palinspastic restoration of ophiolitic suites of the Dunnage terrane in northeast Newfoundland indicates that the ocean in which they originated had a width of more than 1,000 km (Williams, 1980).

Basement to the *Gander terrane* is interpreted as continental, although this is equivocal. In light of the ophiolitic Dunnage terrane and its restored width, it is implied that the thick graywacke sequence of the *Avalon terrane,* indicated by characteristic gravity and magnetic anomalies, and seismic velocities, makes it one of the most extensive of all Appalachian terranes. Southward, the Avalonian magnetic and gravity highs can be traced to the highlands along the Fundy coast of New Brunswick, and thence offshore and parallel to that coast into Massachusetts, Connecticut, and Rhode Island. These geophysical connections provide continuity to previously known exposures. Paleogeographic reconstructions of the North Atlantic indicate that the geophysical characteristics of the Avalon terrane continue across the European margin and provide connections between Avalonian rocks of North America and correlatives in Europe and northwest Africa (Haworth and Lefort, 1979; Haworth, 1981).

Faults throughout the Avalon terrane, possibly as old as late Precambrian, are the locus for later reactivation leading to the superposition of Mesozoic basins above basins of late Precambrian age (Fig. 1).

Although absent on land, an additional suspect terrane is suggested on the continental margin northeast of Newfoundland where geophysical markers associated with the Gander terrane veer northward away from the northeastward extension of the Dover Fault. Haworth (1980) named the intervening area the *Delta zone* because of its triangular shape. It appears to have been linked with the Gander terrane before the major transcurrent movement between the Gander and Avalon terranes. It could therefore represent a remnant of the craton at the margin of which the rocks of the Gander terrane developed.

The *Meguma* terrane comprises sedimentary rocks of Cambrian through Silurian-Devonian age that are unique to Nova Scotia (Schenk, 1981). The oldest rocks are interpreted to have deep-water, turbidite affinities, and paleocurrent analysis indicates that their source lay to the east. The youngest (Devonian) rocks are terrestrial. The Avalon-Meguma terrane boundary is overlapped by Carboniferous sediments (Fig. 26), but no older linkages have been established. Even the Devonian intrusions of the Meguma terrane are distinct from nearby plutons in the Avalon terrane (Clark and others, 1980).

The Cobequid-Chedabucto fault separates the Avalon and Meguma terranes in Nova Scotia. It is marked by major anomalies in the magnetic and gravitational fields. The magnetic high on the Avalon side of the boundary (the Collector Anomaly of Haworth, 1975) can be followed eastward from the Cobequid Highlands of Nova Scotia across the northern edge of the Mesozoic Orpheus graben on the Scotian Shelf to the Grand Banks, suggesting that the Tail of the Banks is part of the Meguma terrane. The steep, straight nature of the fault boundary implies transcurrent movement between the two terranes.

Terrane accretion and associated tectonic history

The accretionary history of the Canadian Appalachians is summarized in Figure 26. Sedimentologic linkages indicate that the Dunnage and Gander terranes were in juxtaposition with each other and with the Appalachian miogeocline during the Middle Ordovician. Subsequent Ordovician-Silurian sedimentation onlapped the entire accretionary terrane as best seen in Quebec and New Brunswick. Accretion of the Avalon terrane cannot be established prior to Silurian-Devonian time. The Meguma may have arrived still later but before Carboniferous time. Accretion of the orogen therefore appears to have progressed eastward outward from the North American margin (Williams and Hatcher, 1982).

The earliest accretionary event corresponds to the time of Taconic orogeny and associated deformation, plutonism, and metamorphism and is interpreted as related to the main closure of Iapetus. Subsequent Acadian orogeny (Devonian) affected almost the whole orogen. Its effects are most intense east of the Taconic deformed zone, and its timing corresponds to the best estimate of the time of Avalonian accretion. Large granitic plutons accompany Acadian deformation and occur in all the terranes east of the North American margin. Their chemistry is incompatible with island-arc plutonism, so that subduction is not considered as a means for their generation. In New Brunswick, Fyffe and others (1981) considered them to have been generated by melting of a compressed or overthickened crust. In Newfoundland, Strong (1980) suggests that they were generated within a "megashear," which allows the simultaneous production of different granitoid types in locally different tectonic settings because of differing degrees of partial melting. Hanmer (1981) attributes the fabric of the plutons to the shearing process. Duration of shearing may have spanned Middle Ordovician through Triassic time (Strong, 1980), or it may have been confined to the late Paleozoic (Arthaud and Matte, 1977).

Paleomagnetic control from which to deduce major displacements between terranes remains equivocal. Morel and Irving's (1978) interpretation suggests that Iapetus did not close until Early Devonian time, followed by the appearance of another ocean basin before closure and transcurrent movement in mid- to Late Carboniferous time. Kent and Opdyke (1978) have renewed interest in the earlier suggestion of Roy and Robertson (1968) that major transcurrent movement occurred along the axis of the Appalachian system in Devonian-Carboniferous time. Van der Voo (1982) believes that Appalachian displacements are related to Hercynian movements determined paleomagnetically in Europe, thus providing support for the idea of a late Paleozoic megashear (Arthaud and Matte, 1977). The level of paleomagnetic control in Newfoundland during the Ordovician through Carboniferous does not permit sharp discrimination among these hypotheses.

A lack of obvious offsets or repetitions of Appalachian terranes indicates that any major movements must have been confined to terrane boundaries.

Transcurrent movement between the Avalon and Gander terranes is most likely because of their contrasting characteristics. However, Blackwood (personal communication, 1982) believes that the *final movement* between the terranes was primarily vertical, possibly as part of a thrust fault, with the Gander metamorphic rocks thrust eastward over Avalon sandstones and volcanics. The refraction data emphasize the contrast between the terranes by revealing a change in crustal thickness approximately coincident with the Dover Fault (Fig. 4, transect A to F). However, whereas the Gander terrane thickness was determined close to northeast Newfoundland (Dainty and others, 1966), the Avalon terrane thickness is extrapolated from a determination near southeastern Newfoundland (Jackson and others, 1975).

The Dover Fault is extrapolated geophysically across the northeast Newfoundland Shelf to the southwest end of the Charlie Fracture Zone (Haworth, 1977). Major differences in plate motion of the present North Atlantic have occurred on opposite sides of this feature (Srivastava, 1978). The eastern end of the fracture zone also appears to correlate with a major zone of breakup at the Irish continental margin (Haworth, 1980). It therefore seems that the Gander/Avalon terrane boundary (or the Delta/Avalon terrane boundary at the edge of the continental shelf) had a major influence on subsequent plate geometry.

The Paleozoic North American margin, identified by the major eastward increase in gravity, is the location of a significant change in crustal thickness. A 15- to 20-km-thick, lower crustal layer, with seismic velocities in the range of 7.4 to 7.5 km s^{-1}, increases the crustal thickness to about 47 km beneath the Dunnage and Gander terranes (Fig. 4). The coincidence of this thicker crust with the ophiolitic rocks at the surface suggests that the deep layer might be a consequence of ocean closing. However, the Dunnage terrane includes a layer with similarly high velocities at depths of less than 8 km, as determined by short refraction profiles. This shallow layer is in apparent continuity with exposed ophiolites, and occupies a synformal structure beneath Notre Dame Bay (Haworth and Miller, 1982). Recent suggestions that the Dunnage terrane has no roots (Karlstrom and others, 1982; Karlstrom, 1983) imply that the shallow high-velocity layer is an allochthonous slice of oceanic crust. The deeper high-velocity layer may therefore result from metamorphism of the lower continental crust as it sank beneath the imposed allochthonous load. Unfortunately, Karlstrom's (1983) model was prepared without the benefit of deeper crustal information, and it has been challenged by Miller (1984). It is puzzling that anomalous lower crust, like that beneath the Dunnage terrane, also occurs beneath the Gander terrane and within the Gulf of St. Lawrence.

The southwestward extension of the Avalon/Meguma boundary from Nova Scotia is the locus for the Triassic development of the Fundy Basin. Triassic activity is also recorded at its eastern extension in the Orpheus Graben (Fig. 7, transect M to O). This and the existence of the Newfoundland seamounts along the extrapolation of the trend of the Avalon/Meguma boundary across the Trail of the Banks (Haworth, 1975) indicate that it remained a zone of weakness. Geometrical constraints must have determined that it did not, however, act as the locus for development of the southern transform margin of the Grand Banks.

Most geological models for the development of the Canadian Appalachians focus on the late Precambrian through Ordovician history of the evolution and destruction of Iapetus. The Silurian through Carboniferous history is less well known. The conflict between geological and paleomagnetic models is rather severe, and the relatively few mid-Paleozoic poles from the Canadian Appalachians only add to the confusion. Whereas paleomagnetic data favor a wide Ordovician ocean, the appearance of cosmopolitan Middle Ordovician faunas and synchronous ophiolite emplacement suggests narrowing and destruction of the ocean at that time. Possibly a Middle Paleozoic ocean existed to the east of the Dunnage terrane, separating the Gander/Avalon and/or Avalon/Meguma terranes.

It is not until the Devonian that the paleomagnetic data favor the total closure of oceans between North America and Africa. The pervasive plutonism and the widespread deformation during Devonian time as part of the Acadian orogeny is geologically indicative of the culmination of important movement. Most geological models infer the closure of only one ocean spanning the Taconian and Acadian Orogenies. Many consider the Taconian as representing the onset of oceanic closure, and the Acadian as the final phase. Colman-Sadd (1982) has pointed out that a 40 to 60 m.y. quiescent period between successive orogenies is a characteristic feature of collisional zones such as the Zagros and Himalayas. However, his model still favors the almost complete closure of Iapetus during the Ordovician and remains in apparent conflict with the paleomagnetic evidence.

In Late Devonian through Carboniferous time, the deposition of primarily terrestrial sedimentary rocks was confined to rift basins. Volcanic rocks have been drilled in one such basin—the Carboniferous basin in the Gulf of St. Lawrence—at a depth of approximately 3,000 m (Howie and Barss, 1975). This basin developed in a "rift" that crosscuts most early Paleozoic features. Recent attempts to model the evolution of the basin invoke "pull-apart" tectonics and the creation of a depocenter by transcurrent movement along a sigmoidal fault (Bradley, 1982). Whatever the cause of the depocenter, the deposition is episodic (G. Quinlan, personal communication, 1982).

In Newfoundland, Carboniferous rocks are found mainly within a narrow rift basin close to and paralleling the Paleozoic miogeocline and the Baie Verte–Brompton Line. The rocks continue offshore with fold trends that become progressively more oblique to the Paleozoic margin and more nearly parallel to the present margin of Labrador. Fauna from the basin have been interpreted to suggest that a Carboniferous ocean existed in that area (Jansa and others, 1978), but anything more than a local seaway is difficult to reconcile with other data. A depositional gap during most of the Permian marks a geologically quiet period prior to the opening of the present Atlantic Ocean.

TECTONIC HEREDITY AND REPETITION

Complementary to the foregoing chronologic history of the development and destruction of Phanerozoic continental margins of eastern Canada, we conclude by highlighting two topics that have permeated that history: inheritance and passive-margin development.

Inheritance

Several geologic boundaries in eastern Canada have either been the locus for continued tectonic activity throughout the Phanerozoic or have propagated as a result of orogenesis. Their history can be described in the context of the closing of Iapetus and opening of the North Atlantic (Williams, 1984). The closing of Iapetus created the Caledonian-Appalachian orogen including accretion of a variety of suspect terranes. The opening of the North Atlantic split the orogen longitudinally.

South of Newfoundland, early Mesozoic opening of the North Atlantic parallels the axis of the latest compressional event related to Iapetus closure. North of the Grand Banks, the axis of opening developed two branches at an Early Tertiary triple point in the southern Labrador Sea. The northwestern arm of this triple junction cuts obliquely across older structures.

Where Mesozoic-Cenozoic rifting cuts across the Appalachian Orogen in northeast Newfoundland and the Grenville Structural Province of Labrador, some Paleozoic terrane boundaries and Precambrian structural features propagated eastward, and their prolongations coincide with offsets in the present margin and major oceanic fracture zones. Most obvious is the prolongation of the Dover Fault into the Charlie Fracture Zone. The importance of the former as a Paleozoic terrane boundary is apparent from the contrasting surface geology of the Avalon and Gander terranes as well as their deep crustal contrasts. Similarly, the prolongation of the Precambrian Grenville structural front is expressed in the Cartwright Fracture Zone. Surprisingly, the fundamental Avalon-Meguma terrane boundary is not expressed in a rift or transform margin, although it localized the Fundy rift basin and the Orpheus depression.

The Tail of the Bank at the Grand Banks transform margin is a modern promontory that parallels the Paleozoic St. Lawrence Promontory of the Appalachian Orogen. Furthermore, the "transform" linking the St. Lawrence Promontory to the Quebec Reentrant lies along the prolongation of the older Labrador Trough, which may represent a much older (Aphebian) continental margin. Although inheritance and ancestral controls are implied among these features, there are no actual structural breaks that can be traced from one to another.

Finally, there are features that have maintained tectonic activity over long periods but which never developed into continental margins. For example, the St. Lawrence Valley shows evidence for incipient rifting prior to the opening of Iapetus and the Atlantic, and continues to be a zone of high earthquake activity.

On a more global scale for the entire North Atlantic region, there is an obvious spatial relationship between the distribution of rocks affected by Grenvillian orogeny, those affected by Paleozoic orogeny, and the location of present continental margins (Williams, 1987). Thus, tectonic patterns in the North Atlantic may have been determined by events that began more than 1,000 m.y. ago.

Overview of passive-margin development

A continent-wide evaluation of the role that contemporary passive continental margins have played in the evolution of North America necessarily awaits publication of all contributions to the Transects Program. Nevertheless, the widespread similarities seen in eastern Canada between development of the late Precambrian/early Paleozoic margin of Iapetus and the Mesozoic margin of the North Atlantic warrant specific comment.

The breakup of the Grenville Craton is heralded in the late Precambrian by lamprophyres in the St. Lawrence estuary much as the Jurassic lamprophyres in Notre Dame Bay herald the opening of the North Atlantic. Both occurrences are well away from the respective rift zones. The late Precambrian dikes of the Long Range Inlier in Newfoundland are roughly parallel to the margin of Iapetus and have that similarity with the Triassic dikes of the Atlantic margin (e.g., the Shelburne and Avalon Dikes). However, the Long Range dikes form a higher proportion of the crust (as seen at Belle Isle) than do the Triassic dikes. They may have hidden analogues at the edge of the present margin and provide an exposed example of an intruded crust similar to that anticipated beneath the Orphan Basin.

At the modern Atlantic margin, rifts developed subparallel to the dikes, and transforms developed orthogonal to the rifts. Some of the transforms are located along earlier discontinuities and they divide the margin into zones that evolved separately as platforms or basins. Listric normal faults in the rift basins produced rugged topography that led to the deposition of coarse clastic sediments. The duration of this phase of development is between 25 m.y. (Nova Scotia) and 50 m.y. (Orphan Basin).

The sinuous course of the west flank of the Appalachian Orogen is interpreted to reflect ancient promontories and reentrants that were controlled by orthogonal rifts and transforms. Considerable relief is indicated by the thickness and coarseness of clastic sediments above the Grenvillian basement, although listric normal faults are neither recorded nor emphasized in the ancient example. The duration of ancient rifting is estimated at 50 m.y. but it could have been longer.

Extremely thick sedimentary sequences at the junction between rifted and transform segments of the margin off Nova Scotia and excess volcanism at the Grand Banks transform are characteristics compatible with those of ancient reentrants of the Appalachian Orogen.

The drift phase at the Paleozoic margin is marked by the deposition of thick, shallow-water carbonates from Early Cambrian through Early Ordovician (about 100 m.y.). This is analogous with carbonate buildup of similar duration at the Mesozoic margin of Nova Scotia. In the ancient example, coarse breccia was shed seaward from an oversteepened bank edge, much as it is shed today from the Bahama Bank. A breakup unconformity is not observed at either the ancient example or the modern example of Nova Scotia, although other segments of the present margin exhibit a well-developed unconformity between sediments of rift and drift phases.

Evolution of the carbonate bank is the last phase of margin development for which an ancient-modern comparison can be made in eastern Canada. Growth of a bank results from a precarious balance among thermal subsidence of the margin, eustatic sea-level changes, climate, and sediment input. Carbonate growth at the modern margin was terminated by thermal subsidence and an influx of sediment from the west. In contrast, the ancient carbonate bank was destroyed by tectonic loading and covered by an influx of orogenic sediment from the east.

The continent-ocean transition at the modern margin is in places wide, varying between 30 and 400 km, with variable crustal properties. It neither coincides with the morphologic shelf edge nor the line of maximum gravity gradient. We emphasize these important observations and caution the use of these markers in palinspastic restorations of ancient transitions.

OUTSTANDING PROBLEMS

One aspect of the development and destruction of passive margins that has received relatively little attention in this contribution is the nature of the geodynamic processes involved in the various stages of evolution. Quantitative models of the evolution of contemporary rifted margins have been developed, which depend on stretching of the lithosphere during rifting and on the resulting generation of magma and its migration to crustal levels. These concepts have not yet been fully tested, and their confirmation and refinement can only come from comparison of model predictions with observations. The requisite observational data are still lacking. Furthermore, our present understanding of transform margins is primitive, with few of the necessary observations being available.

The Paleozoic ocean closure, thrusting, and accretion of suspect terranes has only been defined on the basis of near-surface geology. The geodynamic processes that involve the whole lithosphere are not well documented. This is highlighted by our present inability to relate the variations in deep crustal structure to near-surface geology beneath the Appalachian Orogen. To illustrate this deficiency more clearly, three fundamental problems are briefly described here.

First is the lack of evidence of appreciable extension within the Canadian Appalachians. If there was major transcurrent faulting in the region in the late Paleozoic, and if many of the major terrane boundaries, such as the Dover Fault and Avalon-Meguma boundary, reflect transcurrent motion along the sinuous boundaries between terranes, large regions of extensional tectonics should result. These regions are prevalent throughout the western Cordillera, and join transcurrent fault zones. The Basin and Range is a prime example (Bally and others, 1981). However, in the Canadian Appalachians, Late Paleozoic extensional features appear to be confined to parts of the Gulf of St. Lawrence and to small Carboniferous basins in western Newfoundland. The role that the Gulf of St. Lawrence has played in orogenesis has, in fact, been largely ignored. If the southern Gulf is an extensional feature, possibly the lower crustal layer there is the equivalent of those beneath rifted margins, which were formed by basaltic intrusion or underplating of stretched continental crust.

Extension within the Cordillera has also led to the exposure of metamorphic core complexes (Davis and Coney, 1979). These are interpreted as the product of crustal extension and faulting of the cover rocks. Core complexes are unknown in the Appalachians. This lack of evidence for extensional tectonics, with some exceptions such as the Gulf, reinforces our belief that present models for the evolution of the orogen through Late Paleozoic time are incomplete or incorrect. It is conceivable that extensional tectonics played a major role in parts of the orogen but the evidence was destroyed in later compressional phases of development. Detailed geophysical studies in the Gulf of St. Lawrence might help clarify some of these problems, as the stratigraphy and underlying basement structures should record the tectonics of Late Paleozoic events. The importance of late-phase transcurrent and extensional tectonics cannot be overemphasized. Present cross sections through the orogen, including the relationship of deep structure to near-surface geology, may be confused by these late motions.

Second, the timing, location, and intensity of thrusting appear to vary along the length of the Appalachian Orogen. This affects the development of the foreland basin, which occurs west of the deformed zone as a result of downward flexure of the lithosphere by the load of the thrust sheets (Quinlan and Beaumont, 1984). Quinlan and Beaumont (1984) have used the post–mid-Ordovician stratigraphy of the Appalachian foreland basin in the United States to establish the history of thrusting and the size of the thrust loads. A similar analysis is required for the Anticosti Basin, west of the Canadian Appalachians. Such studies are important to present concerns regarding the size and assemblage of the Humber Arm Allochthon, which differs significantly from the large, 5- to 15-km-thick, allochthonous slices of basement, which were overthrust to the west in the southern Appalachians. In general, such studies are pertinent in analyzing mechanisms, geometry, and time of terrane accretion in the northern Appalachians.

Thirdly, what processes control the initiation of collisional tectonics and the siting of subduction zones? Is an imminent phase of North Atlantic closing dependent on global plate reorganizations? Or are processes operating at the present rifted margin, such as sediment loading, contributing to the geometry of a future collisional scenario? The traditional question of the siting of rifts, "geosynclines," and orogens is explained in a simple and admirable way by plate tectonics, but the specific controls on plate kinematics remain a mystery.

REFERENCES CITED

Amoco-Imperial Oil, 1973, Regional geology of the Grand Banks: Bulletin of Canadian Petroleum Geology, v. 21, p. 479–503.

Arthaud, F., and Matte, P., 1977, Late Paleozoic strike-slip faulting in southern Europe and northern Africa; Result of a right-lateral shear zone between the Appalachians and the Urals: Geological Society of America Bulletin, v. 88, p. 1305–1320.

Baker, D. F., 1978, Geology and geochemistry of alkali volcanic suite (Skinner Cove Formation) in the Humber Arm Allochthon, Newfoundland [M.S. thesis]: St. John's, Memorial University of Newfoundland, 314 p.

Ballard, R. D., and Uchupi, E., 1975, Triassic rift structure in the Gulf of Maine: American Association of Petroleum Geologists Bulletin, v. 59, p. 1041–1072.

Bally, A. W., 1982, Musings over sedimentary basin evolution: Philosophical Transactions of the Royal Society of London, series A, v. 305, p. 325–338.

Bally, A. W., Bernoulli, D., Davis, G. A., and Montadert, L., 1981, Listric normal faults: Oceanologica Acta, Special Publication, Geology of Continental Margins Symposium, p. 87–101.

Barrett, D. L., and Keen, C. E., 1976, Mesozoic magnetic lineations; The Magnetic Quiet Zone and seafloor spreading in the northwest Atlantic: Journal of Geophysical Research, v. 81, p. 4875–4884.

Barrett, D. L., Berry, M., Blanchard, J. E., Keen, M. J., and McAllister, R. E., 1964, Seismic studies on the eastern seaboard of Canada: The Atlantic coast of Nova Scotia: Canadian Journal of Earth Sciences, v. 1, p. 10–22.

Barss, M. S., Bujak, J. P., and Williams, G. L., 1979, Palynological zonation and correlation of sixty-seven wells, eastern Canada: Geological Survey of Canada Paper 78-24, p. 1–118.

Barss, M. S., Bujak, J. P., Wade, J. A., and Williams, G. L., 1980, Age, stratigraphy, organic matter type and colour, and hydrocarbon occurrences in 47 wells, offshore eastern Canada: Geological Survey of Canada Open-File Report 714, 6 p.

Basham, P. W., and Adams, J., 1982, Earthquake hazards to offshore development on the eastern Canadian continental shelves: Proceedings Second Canadian Conference on Marine Geotechnical Engineering, Dartmouth, Nova Scotia, 6 p.

Basham, P. W., Weichert, D. H., and Berry, M. J., 1979, Regional assessment of seismic risk in eastern Canada: Seismological Society of America Bulletin, v. 69, no. 5, p. 1567–1602.

Basham, P. W., Adams, J., and Anglin, F. M., 1983, Earthquake source models for estimating seismic risk on the eastern Canadian continental margin: Fourth Canadian Conference on Earthquake Engineering, Vancouver, British Columbia, p. 495–508.

Beaumont, C., 1979, On the rheological zonation of the lithosphere during flexure: Tectonophysics, v. 59, p. 347–366.

Beaumont, C., Keen, C. E., and Boutilier, R., 1982, On the evolution of rifted continental margins; Comparison of models and observations for the Nova

Scotian margin: Royal Astronomical Society Geophysical Journal, v. 70, p. 667–715.
Bentley, C. R., and Worzel, J. L., 1956, Geophysical investigations in the emerged and submerged Atlantic coastal plain; Part X, Continental slope and continental rise south of the Grand Banks: Geological Society of America Bulletin, v. 67, p. 1–18.
Berger, J., Cok, A. E., Blanchard, J. E., and Keen, M. J., 1966, Morphological and geophysical studies on the eastern seaboard of Canada; The Nova Scotian Shelf: Royal Society of Canada Special Publication 9, p. 102–113.
Berry, M. J., and Fuchs, K., 1973, Crustal structure of the Superior and Grenville Provinces of the northeastern Canadian Shield: Seismological Society of America Bulletin, v. 63, p. 1393–1432.
Bott, M.H.P., 1971, Evolution of young continental margins and formation of shelf basins: Tectonophysics, v. 11, p. 319–327.
Bradley, D. C., 1982, Subsidence in late Paleozoic basins in the northern Appalachians: Tectonics, v. 1, p. 107–123.
Burke, K., and Dewey, J. F., 1973, Plume generated triple junctions; Key indicators in applying plate tectonics to old rocks: Journal of Geology, v. 81, no. 4, p. 405–433.
Church, W. R., and Stevens, R. K., 1971, Early Paleozoic ophiolite complexes of the Newfoundland Appalachians as mantle-oceanic crust sequences: Journal of Geophysical Research, v. 76, no. 5, p. 1460–1466.
Clark, D. B., Barr, S. M., and Donohoe, H. V., 1980, Granitoid and other plutonic rocks of Nova Scotia, *in* Wones, D. R., ed., Proceedings, The Caledonides in the U.S.A.: Blacksburg, Virginia Polytechnic Institute and State University Memoir 2, p. 107–116.
Colman-Sadd, S. P., 1982, Two stage continental collision and plate driving forces: Tectonophysics, v. 90, p. 263–282.
Cook, F. A., 1984, Harmonic distortion on a seismic reflection profile across the Quebec Appalachians; Relation to Bouguer gravity and implications for crustal structure: Canadian Journal of Earth Sciences, v. 21, p. 346–353.
Cook, F. A., and Oliver, J. E., 1981, The late Precambrian–early Paleozoic continental edge in the Appalachian orogen: American Journal of Science, v. 281, p. 993–1008.
Currie, K. L., 1970, An hypothesis on the origin of alkaline rocks suggested by the tectonic setting of the Monteregian hills: Canadian Mineralogist, v. 10, pt. 3, p. 411–420.
Cutt, B. J., and Laving, J. G., 1977, Tectonic elements and geologic history of the south Labrador and Newfoundland continental shelf, eastern Canada: Canadian Petroleum Geology Bulletin, v. 25, p. 1037–1058.
Dainty, A. M., Keen, C. E., Keen, M. J., and Blanchard, J. E., 1966, Review of geophysical evidence on crust and upper mantle structure on the eastern seaboard of Canada, *in* Steinhart, J. S., and Smith, J. J., eds., The earth beneath the continents: American Geophysical Union, p. 349–369.
Davis, G. H., and Coney, P. J., 1979, Geologic development of the Cordilleran metamorphic core complexes: Geology, v. 7, p. 120–124.
Dean, P. L., 1978, The volcanic stratigraphy and metallogeny of Notre Dame Bay, Newfoundland: Memorial University of Newfoundland, Geology Report 7, 204 p.
Dewey, J. F., and Burke, K.C.A., 1973, Tibetan, Variscan, and Precambrian basement reactivation; Products of continental collision: Journal of Geology, v. 81, p. 683–692.
Doig, R., 1970, An alkaline rock province linking Europe and North America: Canadian Journal of Earth Sciences, v. 7, no. 1, p. 22–28.
Doxsee, W. W., 1948, The Grand Banks earthquake of November 18, 1929: Publications of the Dominion Observatory, v. 7, p. 323–335.
Emery, K. O., Uchupi, E., Phillips, J. D., Bowin, C. O., Bunce, E. T., and Knott, S. T., 1970, Continental rise off eastern North America: American Association of Petroleum Geologists Bulletin, v. 54, p. 44–108 and 1120–1139.
England, P. C., 1983, Constraints on extension of continental lithosphere: Journal of Geophysical Research, v. 88, p. 1145–1152.
Ewing, G. N., Dainty, A. M., Blanchard, J. E., and Keen, M. J., 1966, Seismic studies on the eastern seaboard of Canada; The Appalachian System, I: Canadian Journal of Earth Sciences, v. 3, p. 89–109.
Falvey, D. A., 1974, The development of continental margins in plate tectonic theory: Australian Journal of Petroleum Exploration, v. 14, p. 95–106.
Farrar, S. S., and Glover, L., III, 1983, Grenville basement in the Piedmont east of the pre-Appalachian (pre-Caledonian) edge(?) of the North American craton: Geological Society of America Abstracts with Programs, v. 15, no. 3, p. 123.
Fenwick, D.K.B., Keen, M. J., Keen, C. E., and Lambert, A., 1968, Geophysical studies of the continental margin northeast of Newfoundland: Canadian Journal of Earth Sciences, v. 5, p. 483–500.
Foucher, J.-P., LePichon, X., and Sibuet, J.-C., 1982, The ocean-continent transition in the uniform lithospheric stretching model; Role of partial melting in the mantle: Philosophical Transactions of the Royal Society of London, v. A305, p. 27–43.
Fyffe, L. R., Pajari, G. E., Jr., and Cherry, M. E., 1981, The Acadian plutonic rocks of New Brunswick: Maritime Sediments and Atlantic Geology, v. 17, p. 23–36.
Given, M. M., 1977, Mesozoic and early Cenozoic geology of offshore Nova Scotia: Bulletin of Canadian Petroleum Geology, v. 25, p. 63–91.
Gradstein, F. M., and Srivastava, S. P., 1980, Aspects of Cenozoic stratigraphy and paleoceanography of the Labrador Sea and Baffin Bay: Palaeogeography, Palaeoclimatology, and Palaeoecology, v. 30, p. 261–295.
Gradstein, F. M., and Williams, G. L., 1976, Biostratigraphy of the Labrador Shelf: Geological Survey of Canada Open-File Report 349, 39 p.
——, 1981, Stratigraphic charts of the Labrador and Newfoundland Shelves: Geological Survey of Canada Open-File Report 826, 5 p.
Gradstein, F. M., Williams, G. L., Jenkins, W.A.M., and Ascoli, P., 1975, Mesozoic and Cenozoic stratigraphy of the Atlantic Continental Margin, eastern Canada, *in* Yorath, C. T., Parker, E. R., and Glass, D. J., eds., Canada's continental margins and offshore petroleum exploration: Canadian Society of Petroleum Geologists Memoir 4, p. 103–131.
Gradstein, F. M., Grant, A. C., and Jansa, L. F., 1977, Grand Banks and J-Anomaly Ridge; A geological comparison: Science, v. 197, p. 1074–1076.
Grant, A. C., 1975, Structural modes of the western margin of the Labrador Sea, *in* van der Linden, W.J.M., and Wade, J. A., eds., Offshore geology of eastern Canada; Volume 2, Regional geology: Geological Survey of Canada Paper 74-30, p. 217–231.
——, 1977, Multichannel seismic reflection profiles of the continental crust beneath the Newfoundland Ridge: Nature, v. 270, p. 22–25.
——, 1980, Problems with plate tectonics; Labrador Sea: Bulletin of Canadian Petroleum Geology, v. 28, p. 252–275.
Hanmer, S., 1981, Tectonic significance of the northeastern Gander Zone, Newfoundland; An Acadian ductile shear-zone: Canadian Journal of Earth Sciences, v. 18, p. 120–135.
Haworth, R. T., 1975, The development of Atlantic Canada as a result of continental collision; Evidence from offshore gravity and magnetic data, *in* Yorath, C. J., Parker, E. R., and Glass, D. J., eds., Canada's continental margins and offshore petroleum exploration: Canadian Society of Petroleum Geologists Memoir 4, p. 59–77.
——, 1977, The continental crust northeast of Newfoundland and its ancestral relationship to the Charlie Fracture Zone: Nature, v. 266, p. 246–249.
——, 1980, Appalachian structural trends northeast of Newfoundland and their trans-Atlantic correlation: Tectonophysics, v. 64, p. 111–130.
——, 1981, Geophysical expression of Appalachian-Caledonide structures on the continental margins of the north Atlantic, *in* Kerr, J. W., and Fergusson, A. J., eds., Geology of the North Atlantic borderlands: Canadian Society of Petroleum Geologists Memoir 7, p. 429–442.
Haworth, R. T., and Jacobi, R. D., 1983, Geophysical correlation between the geological zonation of Newfoundland and the British Isles, *in* Hatcher, R. D., Jr., Williams, H., and Zietz, I., eds., Contributions to the tectonics and geophysics of mountain chains: Geological Society of America Memoir 158, p. 25–32.
Haworth, R. T., and Keen, C. E., 1979, The Canadian Atlantic margin; A passive continental margin encompassing an active past, *in* Keen, C. E., ed., Crustal properties across passive margins: Tectonophysics, v. 59, p. 83–126.

Haworth, R. T., and Lefort, J. P., 1979, Geophysical evidence for the extent of the Avalon zone in Atlantic Canada: Canadian Journal of Earth Sciences, v. 16, no. 3, pt. 1, p. 552–567.

Haworth, R. T., and MacIntyre, J. B., 1975, The gravity and magnetic fields of Atlantic offshore Canada: Marine Sciences Paper 16, Geological Survey of Canada Paper 75-9, 22 p.

Haworth, R. T., and Miller, H. G., 1982, The structure of Paleozoic oceanic rocks beneath Notre Dame Bay, Newfoundland, *in* St. Julien, P., and Beland, J., eds., Major structural zones and faults of the northern Appalachians: Geological Association of Canada Special Paper 24, p. 149–173.

Haworth, R. T., Grant, A. C., and Folinsbee, R. A., 1976, Geology of the continental shelf off southeastern Labrador: Geological Survey of Canada Paper 76-1C, p. 61–70.

Haworth, R. T., Lefort, J. -P., and Miller, H. G., 1978, Geophysical evidence for an east-dipping Appalachian subduction zone beneath Newfoundland: Geology, v. 6, p. 522–526.

Haworth, R. T., Daniels, D. L., Williams, H., and Zietz, I., 1980, Bouguer gravity anomaly map of the Appalachian orogen: Memorial University of Newfoundland Map no. 3, scale 1:1,000,000.

Hinz, K., Schluter, H. V., Grant, A. C., Srivastava, S. P., Umpleby, D., and Woodside, J., 1979, Geophysical transects of the Labrador Sea; Labrador to southwest Greenland, *in* Keen, C. E., ed., Crustal properties across passive margins: Tectonophysics, v. 59, p. 151–183.

Hobson, G. D., and Overton, A., 1973, Sedimentary refraction seismic surveys, Gulf of St. Lawrence, *in* Hood, P. J., eds., Earth science symposium on offshore eastern Canada: Geological Survey of Canada Paper 71-23, p. 325–336.

Hood, P. J., and Reveler, D. A., 1977, Magnetic anomaly maps of the Atlantic Provinces: Geological Survey of Canada Open-File Report 496, scale 1:1,000,000.

Howie, R. D., and Barss, M. S., 1975, Upper Paleozoic rocks of the Atlantic Provinces, Gulf of St. Lawrence and adjacent continental shelf, *in* van der Linden, W.J.M., and Wade, J. A., eds., Offshore geology of eastern Canada; Volume 2, Regional Geology: Geological Survey of Canada Paper 74-30, p. 35–50.

Jackson, H. R., Keen, C. E., and Keen, M. J., 1975, Seismic structure of the continental margins and ocean basins of southeastern Canada: Geological Survey of Canada Paper 74-51, 13 p.

Jacobi, R. D., 1981, Peripheral bulge; A causal mechanism for the Lower/Middle Ordovician unconformity along the western margin of the northern Appalachians: Earth and Planetary Science Letters, v. 56, p. 245–251.

Jamieson, R. A., 1977, A suite of alkali basalts and gabbros associated with the Hare Bay Allochthon of western Newfoundland: Canadian Journal of Earth Sciences, v. 14, p. 346–356.

Jansa, L. F., 1981, Mesozoic carbonate platforms and banks of the eastern North American margin: Marine Geology, v. 44, p. 97–117.

Jansa, L. F., and Wade, J. A., 1975, Geology of the continental margin off Nova Scotia and Newfoundland, *in* van der Linden, W.J.M., and Wade, J. A., eds., Offshore geology of eastern Canada; Volume 2, Regional geology: Geological Survey of Canada paper 74-30, p. 51–105.

Jansa, L. F., and Wiedmann, J., Jr., 1982, Mesozoic-Cenozoic development of the eastern North American and northwest African continental margins; A comparison, *in* von Rad, U., Hinz, K., Sarnthein, M., and Seibold, E., eds., Geology of the northwest African continental margin: New York, Springer-Verlag, p. 215–269.

Jansa, L. F., Mamet, B., and Roux, A., 1978, Visean limestones from the Newfoundland Shelf: Canadian Journal of Earth Sciences, v. 15, p. 1422–1436.

Jansa, L. F., Enos, P., Tucholke, B. E., Gradstein, F. M., and Sheridan, R. E., 1979, Mesozoic-Cenozoic sedimentary formations of the North Atlantic basin; Western North Atlantic, *in* Talwani, M., Hay, W., and Ryan, W.B.F., eds., Deep drilling results in the Atlantic Ocean; Continental margins and paleoenvironment: American Geophysical Union Maurice Ewing Series 3, p. 1–57.

Jansa, L. F., Bujak, J. P., and Williams, G. L., 1980, Upper Triassic salt deposits of the western North Atlantic: Canadian Journal of Earth Sciences, v. 17, p. 547–559.

Jarvis, G. T., and McKenzie, D. P., 1980, Sedimentary basin formation with finite extension rates: Earth and Planetary Science Letters, v. 48, p. 42–52.

Karlstrom, K. E., 1983, Reinterpretation of Newfoundland gravity data and arguments for an allochthonous Dunnage zone: Geology, v. 11, p. 263–266.

Karlstrom, K. E., van der Pluijm, B. A., and Williams, P. F., 1982, Structural interpretation of the eastern Notre Dame Bay area, Newfoundland; Regional post–Middle Silurian thrusting and asymmetrical folding: Canadian Journal of Earth Sciences, v. 19, p. 2325–2341.

Karner, G. D., and Watts, A. B., 1982, On isostasy at Atlantic-type continental margins: Journal of Geophysical Research, v. 87, p. 2923–2948.

Kean, B. F., and Strong, D. F., 1975, Geochemical evolution of an Ordovician island arc of the central Newfoundland Appalachians: American Journal of Science, v. 275, no. 2, p. 97–118.

Keen, C. E., 1979, Thermal history and subsidence of rifted continental margins; Evidence from wells on the Nova Scotian and Labrador Shelves: Canadian Journal of Earth Sciences, v. 16, p. 502–522.

Keen, C. E., and Barrett, D. L., 1981, Thinned and subsided continental crust on the rifted margin of eastern Canada; Crustal structure, thermal evolution, and subsidence history: Geophysical Journal of the Royal Astronomical Society, v. 65, p. 443–465.

Keen, C. E., and Cordsen, A., 1981, Crustal structure, seismic stratigraphy, and rift processes of the continental margin off eastern Canada; Ocean bottom seismic refraction results off Nova Scotia: Canadian Journal of Earth Sciences, v. 18, p. 1523–1538.

Keen, C. E., and Hyndman, R. D., 1979, Geophysical review of the continental margins of eastern and western Canada: Canadian Journal of Earth Sciences, v. 16, p. 712–747.

Keen, C. E., and Keen, M. J., 1974, The continental margins of eastern Canada and Baffin Bay, *in* Burk, C. A., and Drake, C. L., eds., The geology of continental margins: New York, Springer-Verlag, p. 381–389.

Keen, C. E., Keen, M. J., Barrett, D. L., and Heffler, D. E., 1975, Some aspects of the ocean-continent transition at the continental margin of eastern North America: Geological Survey of Canada Paper 74-30, p. 189–197.

Keen, C. E., Hall, B. R., and Sullivan, K. D., 1977, Mesozoic evolution of the Newfoundland Basin: Earth and Planetary Science Letters, v. 37, p. 307–320.

Keen, C. E., Beaumont, C., and Boutilier, R., 1981, Preliminary results from a thermo-mechanical model for the evolution of Atlantic-type continental margins, *in* Proceedings of the 26th International Geological Congress, Geology of Continental Margins Symposium, Paris, 1980: Oceanologica Acta, p. 123–128.

Keen, M. J., 1969, Magnetic anomalies off the eastern seaboard of the United States; A possible edge effect: Nature, v. 222, p. 72–74.

Kent, D. V., and Opdyke, N. D., 1978, Paleomagnetism of the Devonian Catskill redbeds; Evidence for motion of the coastal New England–Canadian Maritime region relative to cratonic North America: Journal of Geophysical Research, v. 83, p. 4441–4450.

Keppie, J. D., Odom, L., and Cormier, R. F., 1983, Tectonothermal evolution of the Meguma terrane; Radiometric controls: Geological Society of America Abstracts with Programs, v. 15, no. 3, p. 136.

King, L. H., 1975, Geosynclinal development of the continental margin south of Nova Scotia and Newfoundland, *in* van der Linden, W.J.M., and Wade, J. A., eds., Offshore geology of eastern Canada; Volume 2, Regional geology: Geological Survey of Canada Paper 74-30, p. 199–206.

King, L. H., and MacLean, B., 1976, Geology of the Scotian Shelf: Marine Science Paper 7, Geological Survey of Canada Paper 74-31, 31 p.

King, L. H., and Young, I. F., 1977, Paleocontinental slopes of the east coast geosyncline (Canadian Atlantic Margin): Canadian Journal of Earth Sciences, v. 14, p. 2553–2564.

Klitgord, K., and Behrendt, J. C., 1979, Basin structure of the U.S. Atlantic margin, *in* Watkins, J. S., Montadert, L., and Dickenson, P., eds., Geological and geophysical investigations of continental margins: American Association

of Petroleum Geology Memoir 29, p. 85–112.
Larson, R. L., and Hilde, T.W.C., 1975, A revised time scale of magnetic reversals for the Early Cretaceous, Late Jurassic: Journal of Geophysical Research, v. 80, p. 2586–2594.
LePichon, X., and Sibuet, J. -C., 1981, Passive margins; A model of formation: Journal of Geophysical Research, v. 86, p. 3708–3720.
Lowrie, W., 1979, Geomagnetic reversals and ocean crust magnetization, *in* Talwani, M., Harrison, C. G., and Hayes, D. E., eds., Deep drilling results in the Atlantic Ocean; Ocean crust: American Geophysical Union Maurice Ewing Series 2, p. 135–140.
McGrath, P. H., Hood, P. J., and Darnley, A. G., 1977, Magnetic anomaly map of Canada: Geological Survey of Canada Map 1255A, 3rd edition, scale 1:1,500,000.
McKenzie, D. P., 1978, Some remarks on the development of sedimentary basins: Earth and Planetary Science Letters, v. 40, p. 25–32.
McKerrow, W. S., and Cocks, L.R.M., 1976, Progressive faunal migration across the Iapetus Ocean: Nature, v. 263, p. 304–306.
McWhae, J.R.H., 1981, Structure and spreading history of the northwestern Atlantic region from the Scotian Shelf to Baffin Bay, *in* Kerr, J. W., and Fergusson, A. J., eds., Geology of the north Atlantic borderlands: Canadian Society of Petroleum Geologists Memoir 7, p. 299–332.
Miller, H. G., 1984, Reinterpretation of Newfoundland gravity data and arguments for an allochthonous Dunnage zone (comment): Geology, v. 12, p. 60–61.
Miller, H. G., and Deutsch, E. R., 1978, The Bouguer anomaly field of the Notre Dame Bay area, Newfoundland, with map 163, Notre Dame Bay: Earth Physics Branch Gravity Map Series 163, scale 1:250,000.
Ministère des Richesses Naturelles du Quebec, 1979, Interpretation geologique de la ligne sismique 2001: Dossier Publique 721.
Mohr, P., 1982, Musings on continental rifts, *in* Palmason, G., ed., Continental and oceanic rifts: American Geophysical Union Geodynamic Series, v. 8, p. 293–309.
Morel, P., and Irving, E., 1978, Tentative paleocontinental maps for the early Phanerozoic and Proterozoic: Journal of Geology, v. 86, p. 535–561.
Morgan, W. J., 1981, Hotspot tracks and the opening of the Atlantic and Indian oceans, *in* Emiliani, C., ed., The oceanic lithosphere: New York, John Wiley and Sons, The sea; ideas and observations on progress in study of the seas, v. 7, p. 443–487.
Olszewski, W. J., Jr., and Gaudette, H. E., 1982, Age of the Brookville Gneiss and associated rocks, southeastern New Brunswick: Canadian Journal of Earth Sciences, v. 19, p. 2158–2166.
Pajari, G. E., Pickerill, R. K., and Currie, K. L., 1979, The nature, origin, and significance of the Carmanville ophiolitic mélange, northeastern Newfoundland: Canadian Journal of Earth Sciences, v. 16, p. 1439–1451.
Papezik, V. S., and Hodych, J. P., 1980, Early Mesozoic diabase dikes of the Avalon Peninsula, Newfoundland; Petrochemistry, mineralogy, and origin: Canadian Journal of Earth Sciences, v. 17, p. 1417–1430.
Pe-Piper, G., and Jansa, L. F., 1987, Early Cretaceous volcanicity on a northeastern American margin and its implications for plate tectonics: Geological Society of America Bulletin, v. 99, p. 803–813.
Poole, W. H., Sanford, B. V., Williams, H., and Kelly, D. G., 1970, Geology of southeastern Canada, *in* Douglas, R.J.W., ed., Geology and economic minerals of Canada: Geological Survey of Canada Economic Geology Report 1, 5th edition, p. 229–304.
Press, F., and Beckmann, W., 1954, Geophysical investigations in the emerged and submerged Atlantic coastal plain; Part 8, Grand Banks and adjacent shelves: Geological Society of America Bulletin, v. 65, p. 299–314.
Pringle, I. R., Miller, J. A., and Warrell, D. M., 1971, Radiometric age determination from the Long Range Mountains, Newfoundland: Canadian Journal of Earth Sciences, v. 8, p. 1325–1330.
Quinlan, G., and Beaumont, C., 1984, Appalachian thrusting, lithospheric flexure, and the Paleozoic stratigraphy of the eastern interior of North America: Canadian Journal of Earth Sciences, v. 21, no. 9, p. 973–996.
Rast, N., 1980, The Avalonian plate in the northern Appalachians and Caledonides, *in* Wones, D. R., eds., Proceedings, The Caledonides in the U.S.A.: Blacksburg, Virginia Polytechnic Institute and State University Memoir 2, p. 63–66.
Rodgers, J., 1968, The eastern edge of the North America continent during the Cambrian and Early Ordovician, *in* Zen, E-an, White, W. S., Hadley, J. B., and Thompson, J. B., Jr., eds., Studies of Appalachian geology: Northern and Maritime: New York, Wiley Interscience, p. 141–149.
Roy, J. L., and Robertson, W. A., 1968, Evidence for diagenetic remanent magnetization in the Maringouin Formation: Canadian Journal of Earth Sciences, v. 5, no. 2, p. 275–285.
Royden, L., and Keen, C. E., 1980, Rifting process and thermal evolution of the continental margin of eastern Canada determined from subsidence curves: Earth and Planetary Science Letters, v. 51, p. 343–361.
Royden, L., Sclater, J. G., and Von Herzen, R. P., 1980, Continental margin subduction and heat flow; Important parameters in formation of petroleum hydrocarbons: American Association of Petroleum Geologists Bulletin, v. 64, p. 173–187.
Ruffman, A., and Van Hinte, J. E., 1973, Orphan Knoll, a 'chip' off the North American "plate," *in* Hood, P. J., ed., Earth science symposium on offshore eastern Canada: Geological Survey of Canada Paper 71-23, p. 407–449.
Schenk, P. E., 1981, The Meguma Zone of Nova Scotia; A remnant of western Europe, South America, or Africa, *in* Kerr, J. W., and Fergusson, A. J., eds., Geology of the North Atlantic borderlands: Canadian Society of Petroleum Geologists Memoir 7, p. 119–148.
Scholle, P. A., and Wenkam, C. R., eds., 1982, Geological Studies of the Cost Nos. G-1 and G-2 wells, United States north Atlantic outer Continental Shelf: U.S. Geological Survey Circular 861, 193 p.
Seguin, M. K., 1982, Geophysics of the Quebec Appalachians: Tectonophysics, v. 81, p. 1–50.
Sheridan, R. E., and Drake, C. L., 1968, Seward extension of the Canadian Appalachians: Canadian Journal of Earth Sciences, v. 5, p. 337–373.
Shih, K. G., Macnab, R., and Halliday, D., 1981, Multiparameter survey data from the Scotian margin: Geological Survey of Canada Open-File Report 750.
Sleep, N. H., 1971, Thermal effects of the formation of Atlantic continental margins by continental break-up: Geophysical Journal of the Royal Astronomical Society, v. 24, p. 325–350.
Srivastava, S. P., 1978, Evolution of the Labrador Sea and its bearing on the early evolution of the North Atlantic: Geophysical Journal of the Royal Astronomical Society, v. 52, p. 313–357.
—— , 1979, Marine gravity and magnetic anomalies maps of the Labrador Sea: Geological Survey of Canada Open-File Report 627.
Srivastava, S. P., and Tapscott, C. R., 1986, Plate kinematics in the North Atlantic, *in* Vogt, P. R., and Tucholke, B. E., eds., The western North Atlantic region: Boulder, Colorado, Geological Society of America, The Geology of North America, v. M, p. 379–404.
Srivastava, S. P., Falconer, R.K.H., and MacLean, B., 1981, Labrador Sea, Davis Strait, Baffin Bay; Geology and geophysics, a review, *in* Kerr, J. W., and Fergusson, A. J., eds., Geology of the North Atlantic borderlands: Canadian Society of Petroleum Geologists Memoir 7, p. 333–398.
Stein, S., Sleep, N. H., Geller, R. J., Wang, S. C., and Kroeger, G. C., 1979, Earthquakes along the passive margin of eastern Canada: Geophysical Research Letters, v. 6, p. 537–540.
Stephens, L. E., and Cooper, R. V., 1973, Results of underwater gravity surveys over the southern Nova Scotia continental shelf: Earth Physics Branch Gravity Map Series, no. 149, p. 1–10.
Stevens, R. K., 1970, Cambro-Ordovician flysch sedimentation and tectonics in west Newfoundland and their possible bearing on Proto-Atlantic Ocean, *in* Lajoie, J., ed., Flysch sedimentology in North America: Geological Association of Canada Special Paper 7, p. 165–177.
Stewart, I.C.F., and Keen, C. E., 1978, Anomalous upper mantle structure beneath the Cretaceous Fogo Seamounts indicated by P-wave reflection delays: Nature, v. 274, p. 788–791.
Strong, D. F., 1977, Volcanic regimes of the Newfoundland Appalachians, *in*

Baragar, W.R.A., Coleman, L. C., and Hall, J. M., eds., Volcanic regimes in Canada: Geological Association of Canada Special Paper 16, p. 61–90.

——, 1979, Proterozoic tectonics of northwestern Gondwanaland; New evidence from eastern Newfoundland: Tectonophysics, v. 54, p. 81–101.

——, 1980, Granitoid rocks and associated mineral deposits of eastern Canada and western Europe, *in* Strangway, D. W., ed., The continental crust and its mineral deposits: Geological Association of Canada Special Paper 20, p. 741–769.

Strong, D. F., and Harris, A., 1974, The petrology of Mesozoic alkaline intrusives of central Newfoundland: Canadian Journal of Earth Sciences, v. 11, p. 1208–1219.

Strong, D. F., and Williams, H., 1972, Early Paleozoic flood basalts of northwest Newfoundland; Their petrology and tectonic significance: Geological Association of Canada Proceedings, v. 24, p. 43–54.

Strong, D. F., Dickson, W. L., O'Driscoll, C. F., Kean, B. F., and Stevens, R. K., 1974, Geochemical evidence for eastward Appalachian subduction in Newfoundland: Nature, v. 248, p. 37–39.

Stukas, V., and Reynolds, P. H., 1974, $^{40}Ar/^{39}Ar$ dating of the Long Range dikes, Newfoundland: Earth and Planetary Science Letters, v. 22, p. 256–266.

Sullivan, K. D., 1978, Structure and evolution of the Newfoundland Basin [Ph.D. thesis]: Halifax, Nova Scotia: Dalhousie University, 294 p.

——, 1983, The Newfoundland basin; Ocean-continent boundary and Mesozoic seafloor spreading history: Earth and Planetary Science Letters, v. 62, p. 321–339.

Sullivan, K. D., and Keen, C. E., 1977, Newfoundland seamounts; Petrology and geochemistry, *in* Baragar, W.R.A., Coleman, L. C., and Hall, J. M., eds., Volcanic regimes in Canada: Geological Association of Canada Special Paper 16, p. 461–476.

——, 1978, On the nature of the crust in the vicinity of the Newfoundland Ridge: Canadian Journal of Earth Sciences, v. 15, p. 1462–1471.

Thomas, W. A., 1977, Evolution of Appalachian-Ouachita salients and recesses from re-entrants and promontories in the continental margin: American Journal of Science, v. 277, p. 1233–1278.

Thomas, M. D., and Tanner, J. G., 1975, Cryptic suture in the eastern Grenville Province: Nature, v. 256, no. 5516, p. 392–394.

Tucholke, B. E., and Ludwig, W. J., 1982, Structure and origin of the J-Anomaly Ridge, western North Atlantic Ocean: Journal of Geophysical Research, v. 87, p. 9389–9407.

Tucholke, B. E., and Mountain, G. S., 1979, Seismic stratigraphy, lithostratigraphy, and paleosedimentation patterns in the North American basin, *in* Talwani, M., Hay, W., and Ryan, W.B.F., eds., Deep drilling results in the Atlantic Ocean; Continental margins and paleoenvironments: American Geophysical Union Maurice Ewing Series 3, p. 58–86.

Umpleby, D. C., 1979, Geology of the Labrador Shelf: Geological Survey of Canada Paper 79-13, 34 p.

Vail, P. R., Mitchum, R. M., Todd, R. G., Widmier, J. M., Thompson, S., Sangree, J. B., Bubb, J. N., and Hatlelid, W. G., 1977, Seismic stratigraphy and global changes of sea level, *in* Payton, C. E., ed., Seismic stratigraphy; Applications to hydrocarbon exploration: American Association of Petroleum Geology Memoir 26, p. 49–212.

van der Linden, W.J.M., 1975, Crustal attenuation and sea-floor spreading in the Labrador Sea: Earth and Planetary Science Letters, v. 27, p. 409–423.

Van der Voo, R., 1982, Pre-Mesozoic paleomagnetism and plate tectonics: Annual Review of Earth and Planetary Science, v. 10, p. 191–220.

Vogt, P. R., Anderson, C. N., Bracey, D. R., and Schneider, E. O., eds., 1970, North Atlantic magnetic smooth zones: Journal of Geophysical Research, v. 75, p. 3955–3968.

Wade, J. A., 1981, Geology of the Canadian Atlantic margin from Georges Bank to the Grand Banks, *in* Kerr, J. W., and Fergusson, A. J., eds., Geology of the North Atlantic borderlands: Canadian Society of Petroleum Geologists Memoir 7, p. 447–460.

Watts, A. B., and Steckler, M. S., 1979, Subsidence and eustasy at the continental margin of eastern North America: American Geophysical Union Maurice Ewing Series 3, p. 218–234.

——, 1981, Subsidence and tectonics of Atlantic-type continental margins: Oceanologica Acta, v. 4 (supp.), Geology of continental margins, p. 143–153.

Weaver, D. F., 1968, Preliminary results of the gravity survey of the island of Newfoundland: Dominion Observatory Gravity Map Series no. 53-57, scale 1:500,000.

Williams, H., compiler, 1978, Tectonic-lithofacies of the Appalachian orogen: St. John's, Memorial University of Newfoundland Map no. 1, scale 1:1,000,000.

——, 1979, The Appalachian orogen in Canada: Canadian Journal of Earth Sciences, v. 16, p. 792–807.

——, 1980, Structural telescoping across the Appalachian orogen and the minimum width of the Iapetus Ocean, *in* Strangway, D. W., ed., The continental crust and its mineral deposits: Geological Association of Canada Special Paper 20, p. 421–440.

——, 1984, Miogeoclines and suspect terranes of the Caledonian-Appalachian orogen; Tectonic patterns in the North Atlantic region: Canadian Journal of Earth Sciences, v. 21, p. 887–901.

Williams, H., and Doolan, B. L., 1979, Evolution of Ouachita salients and recesses from re-entrants and promontories in the continental margin (comment): American Journal of Science, v. 279, no. 1, p. 92–95.

Williams, H., and Hatcher, R. D., Jr., 1982, Suspect terranes and accretionary history of the Appalachian orogeny: Geology, v. 10, p. 530–536.

——, 1983, Appalachian suspect terranes, *in* Hatcher, R. D., Jr., Williams, H., and Zietz, I., eds., Contributions to the tectonics and geophysics of mountain chains: Geological Society of America Memoir 158, p. 33–53.

Williams, H., and Smyth, W. R., 1983, Geology of the Hare Bay Allochthon, *in* Geology of the Strait of Belle Isle area, northwestern insular Newfoundland, southern Labrador, and adjacent Quebec: Geological Survey of Canada Memoir 400, part 3, p. 109–132.

Williams, H., and Stevens, R. K., 1974, The ancient continental margin of eastern North America, *in* Burk, C. A., and Drake, C. L., eds., The geology of continental margins: New York, Springer-Verlag, p. 781–796.

Williams, H., and St. Julien, P., 1982, The Baie Verte–Brompton line; Early Paleozoic continent-ocean interface in the Canadian Appalachians, *in* St. Julien, P., and Beland, J., eds., Major structural zones and faults of the northern Appalachians: Geological Association of Canada Special Paper 24, p. 177–207.

Manuscript Accepted by the Society December 28, 1987

ACKNOWLEDGMENTS

We are grateful to Chris Beaumont, Michael Keen, Louise Quinn, Bill Dillan, and Bob Speed for their constructive comments on various versions of this paper. Bill Kay prepared the diagrams and Sharon Hiltz patiently typed the many drafts of this manuscript. The study was financially supported by the Geological Survey of Canada, and by Canadian Natural Science and Engineering Research Council Grants to one of the authors (H.W.).

Printed in U.S.A.

DNAG Continent-Ocean Transect Volume
Phanerozoic Evolution of North American
Continent-Ocean Transitions
The Geological Society of America, 1994

Chapter 3

Continental margin of the eastern United States: Past and present

Douglas W. Rankin
U.S. Geological Survey, MS 926, National Center, Reston, Virginia 22092

INTRODUCTION

The continental margin of eastern North America has been studied as part of the North American Continent-Ocean Transect Program by synthesizing existing data along six corridors, each at least 100 km wide, between Newfoundland and Georgia that extend from the Phanerozoic craton to the oceanic crust. The term Phanerozoic craton is used here for a cratonal area that has not undergone significant deformation since the beginning of Phanerozoic time. One of these corridors is in Canada (D1; Haworth and others, 1985), and two additional Canadian transects (D2 and D3; Keen and Haworth, 1985a, 1985b) cross the present continental margin, but do not extend westward to the craton. Herein I analyze the evolution of the eastern continental margin as depicted by the five corridors or transects, E1 through E5, prepared or in preparation for the United States. The five transects in eastern United States are outlined in Figure 1 and are as follows.

E1, Adirondacks to Georges Bank (Thompson and others, 1993). The corridor length is 1000 km.

E2, New York Appalachian basin to the Baltimore Canyon trough (Drake and others, unpublished). The corridor length is 860 km.

E3, Pittsburgh to Baltimore Canyon trough (Glover, 1989; Glover and others, unpublished). The corridor length is approximately 850 km.

E4, Central Kentucky to the Carolina trough (Rankin and others, 1991). The corridor length is 1100 km.

E5, Cumberland Plateau to Blake Plateau (Hatcher and others, 1994). The corridor length is approximately 1280 km.

Because of the time constraints, E2 and E3 will not be published under the auspices of The Decade of North American Geology.

These transects were selected to utilize seismic reflection profiles, either existing or in progress, and to illustrate commonality and differences in rocks, structure, and tectonics along the eastern margin of the continent. The individual transect displays and descriptions generally focus on features of that specific transect. Features in common along the margin, as well as differences, are brought out here and in the tectonic section display (Speed, 1991).

After this program was well underway, study of another transect was undertaken by an international team in the gap between the E and D series transects. The Quebec–Maine–Gulf of Maine transect extends from the exposed Grenvillian rocks of the Canadian shield near the city of Quebec, through central Maine and across the Gulf of Maine to the same southeastern end as transect E1 (Fig. 1). This second-generation transect differed from those under the North American Continent-Ocean Transect Program in that the authors profited from the earlier experience, and it was designed to acquire and integrate several complementary new data sets. No discussion of the continental margin in the eastern United States is adequate without recognition and use of that extensive data set (Spencer and others, 1989; Stewart and others, 1986, 1991).

Several other major synthesis efforts overlap the North American Continent-Ocean Transect Program. One of the more recent includes four volumes published by the Geological Society of America under the aegis of the Decade of North American Geology (DNAG): *The Atlantic Continental Margin* (Sheridan and Grow, 1988a), *The Appalachian-Ouachita Orogen in the United States* (Hatcher and others, 1989), *Geology of the Continental Margin of Eastern Canada* (Keen and Williams, 1990), and *Precambrian: Conterminous United States* (Reed and others, 1993). The products of two International Geological Correlation Program (IGCP) projects should also be mentioned: the volume edited by Harris and Fettes (1988) from Project 27, the Caledonide orogen, and the volume edited by Dallmeyer (1989a) from Project 233, Terranes in the Circum-Atlantic Paleozoic orogens. Many of the same investigators have participated in more than one of these efforts and, indeed, in more than one transect. To some this may seem like overkill, but it has resulted in a great volume of overview literature and maps in addition to the publication explosion resulting from research on the eastern continental margin in the past 20 years.

This chapter, in a volume dedicated to the continental margins of North America, is intended for those readers who

Rankin, D. W., 1994, Continental margin of the eastern United States: Past and present, *in* Speed, R. C., ed., Phanerozoic Evolution of North American Continent-Ocean Transitions: Boulder, Colorado, Geological Society of America, DNAG Continent-Ocean Transect Volume.

Figure 1. (Explanation follows these two pages.) Map of eastern United States and adjacent Canada showing the location of transects E1 to E5 and the Quebec-Maine-Gulf of Maine transect, as well as the location of geographic, geologic, and geophysical features discussed in the

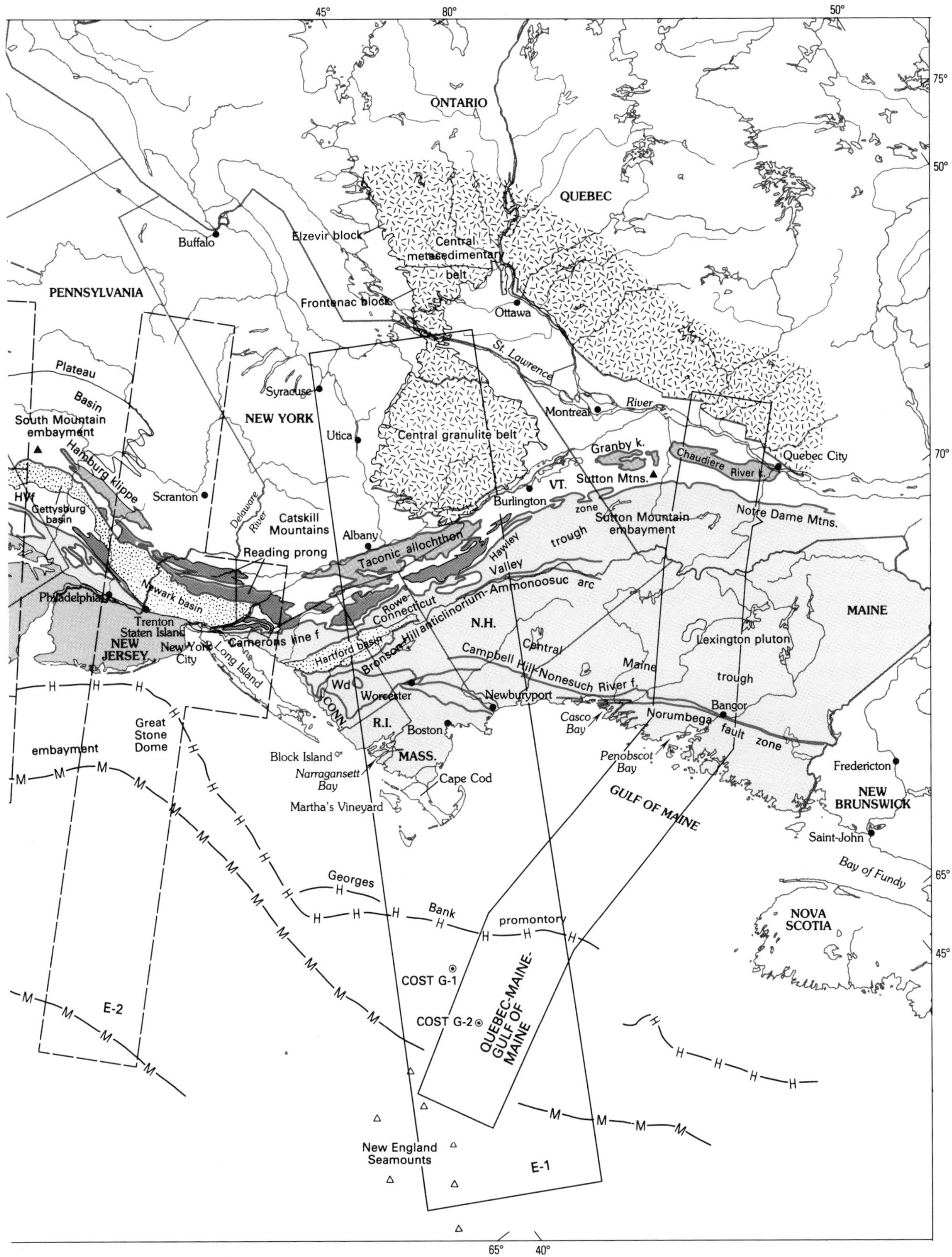

text. Data from Austin and others (1980), Klitgord and others (1988), Rankin and others (1993a), Slivitzky and St. Julien (1987), and Wernicke and Tilke (1989).

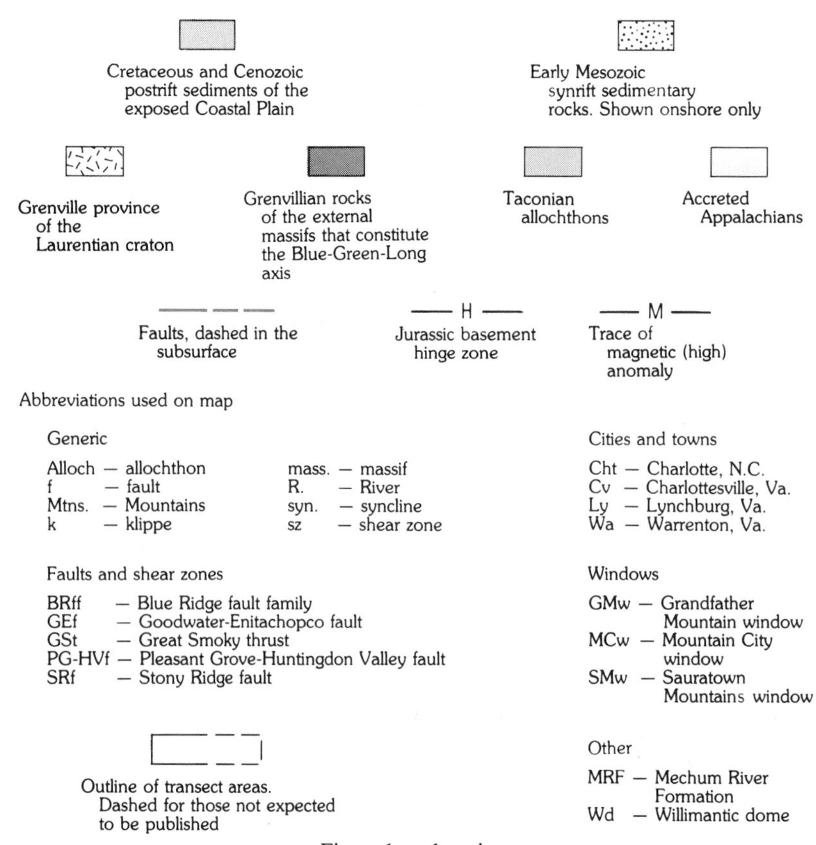

Figure 1 explanation.

wish to have much information about the eastern margin presented as a smaller package than two DNAG volumes. The chapter is not a consensus of the transect teams or their leaders. It reflects my attempt to ferret out an interpretation as internally consistent as possible. At the time of writing, I had available "in press" copies of E1, E4, and the Quebec–Maine–Gulf of Maine transects as well as a prereview version of parts of E2, E3, and E5, plus, of course, the vast published literature. Where appropriate, alternate interpretations are presented or referenced, but the interpretation preferred here is generally stated. Many problems are clearly unresolved.

Various terms are used in this paper for areally extensive structural and tectonic units, such as trough, belt, zone, block, terrane, and province—all spelled with lower case first letters. In general, the terms are descriptive and follow traditional or local usage. Province is used for large subdivisions of the Phanerozoic craton or Appalachian orogen whether based upon physiography or bedrock geology. King (1955) distinguished between physiographic provinces and more or less coextensive geologic belts. That distinction is not followed here, but King's (1955) subdivision of the southern Appalachian Piedmont into belts has become traditional. Trough is used as a nongenetic term for the region of an elongate depression filled with stratified rocks. Belt and zone as large-scale structural-tectonic units have been used interchangeably by some, but I generally use zone for relatively narrow units, as in the Brevard zone (Brevard fault zone). With exceptions as noted, terrane is used for distinct accretionary units, probably alien and far traveled (tectonostratigraphic terranes), in the plate tectonic sense. Block is used in the tectonic sense, which means a crustal unit bounded completely or partly by faults.

The geologic time scale of the Decade of North American Geology (Palmer, 1983) is used throughout this chapter, except for the subdivisions of the Ordovician. On the advice of Anita G. Harris (U.S. Geological Survey, 1991, oral commun.), I used the correlation chart for the Ordovician System in Canada (Barnes and others, 1981) and graptolite zonation of Finney (1986). In this scheme the Caradocian, critical to interpreting the Taconian orogeny, spans the upper half of the Middle Ordovician and extends into the Upper Ordovician (Blackriveran to Edenian). The Llandeilian was of short duration. The Ordovician time scale of Tucker and others (1990), based upon U-Pb zircon dating of stratotypes in Britain, is in reasonable agreement with the time scales of Barnes and others (1981) and Finney (1986).

The E transects are all within the North American plate (see Vogt, 1986, Fig. 1). The modern eastern plate boundary is the Mid-Atlantic Ridge, which is about 2600 km east of the southeast end of E1 and about 3000 km east of the end of E5 along spreading slip lines (fracture-zone traces) (Klitgord and Schouten, 1986). The present continental margin is thus a passive or trailing margin; it is, in fact, part of the type passive or "Atlantic type" margin (see Grow and Sheridan, 1988). On the craton side, transects E1, E2, E3, and E5 begin either over subsurface or exposed rocks of the Grenville province (Fig. 1). These rocks are dominantly of Middle Proterozoic age, experienced Middle Proterozoic orogenies, and constitute the youngest orogenic belt of the Phanerozoic craton (Hoffman, 1988). Transect E4 begins just west of the Grenville front (the western limit of significant Grenvillian deformation) in the eastern granite-rhyolite province, which is composed of 1.5–1.4 Ga anorogenic igneous rocks (Denison and others, 1984; Lidiak and Hinze, 1993).

The Grenville province is an orogenic belt more than 1000 km wide in which Middle Proterozoic thrusting was directed northwestward toward the craton, and the entire crust was imbricated. The orogen is the result of a continental collision (the Grenvillian orogeny) about 1.1 Ga between the North American craton and a continent which lay to the southeast (present direction) in the area now partly occupied by the Appalachian orogen (Moore, 1986). The Grenvillian orogeny may have been the final stage in the assembly of a Proterozoic supercontinent (Piper, 1982; Dalziel, 1991) for which Thompson and others (1993) proposed the name Panchaos.

The Grenville province has been divided into numerous belts, zones, and terranes (see Moore, 1986). Which are discrete accretionary units in the plate tectonic sense is still debated. Culotta and others (1990), Lidiak and Hinze (1993), and McLelland and others (1993) extrapolated the subdivisions of the cratonal Grenvillian province of Canada into exposed and buried parts of the craton in eastern United States.

Continental terranes appear to have been accreted to ancestral North America during the Grenvillian orogeny, although much of the colliding material was probably carried away during the early Paleozoic opening of the Iapetus Ocean. The North American paleocontinent by the late Early Cambrian, therefore, included only those parts of the Grenville orogen that were left behind during the opening of the Iapetus. The Cambrian eastern margin of the North American paleocontinent was thus a rifted margin facing the Iapetus Ocean. In accordance with the terminology of Scotese and others (1979), the term Laurentia is used here for this paleocontinent, as opposed to Proterozoic North America, in order to emphasize that the late Early Cambrian paleocontinent did include those parts of the Grenville orogen that were left behind after the opening of the Iapetus. For the area discussed in this chapter, except possibly for the very northwestern end of E4, all of the paleocontinent Laurentia is thought to consist of rocks that experienced the Grenvillian orogeny. Use of the term "Grenvillian crust" emphasizes that that crust belongs to the youngest orogenic belt of the Phanerozoic craton; "Laurentian crust" emphasizes that it is part of the Early Cambrian North American paleocontinent.

The Phanerozoic tectonic history of the eastern continental margin includes Late Proterozoic to Early Ordovician development of the Iapetan passive margin of Laurentia, the complex multiphase Middle Ordovician to Permian collision and accretion that formed the Appalachian orogen through the closing of Iapetus and other paleooceans, and the Mesozoic and Cenozoic development of the present margin of North America (the opening of the Atlantic Ocean). Within the Appalachian orogen those parts that include the former Laurentian margin are referred to as the Laurentian Appalachians (Rankin, 1993a) and those parts thought to consist of accreted terranes are here called the accreted Appalachians (Fig. 1).

The timing of orogenies in the Appalachians is still under study; some argue that there was a continuum of deformation. Three major orogenic episodes have traditionally been recognized. These are listed below along with my prejudices for the time of the orogenies developed from information discussed in the text (see Figures 5 and 7; also Horton and others, 1989).

Alleghanian orogeny: about 330 to 270 Ma. Accretion of easternmost terranes, collision with Gondwanaland, and westward thrusting of terranes as far east as the Piedmont onto autochthonous North America.

Acadian orogeny: about 400 to 380 Ma in New England. Accretion of previously amalgamated terranes?

Taconian orogeny: about 470 to 440 Ma. Closing of the Iapetus Ocean and accretion of previously amalgamated terranes and possibly other terranes to Laurentia.

The above synopsis presents the Appalachians as seen from the North American perspective. It does not take into account events that affected terranes prior to their accretion to North America. The Penobscottian orogeny is now recognized to have affected areas from Virginia to Newfoundland, and is interpreted to be the amalgamation of some exotic terranes about 520–480 Ma, prior to their accretion to Laurentia in the Taconian orogeny (See Neuman, 1967; Colman-Sadd and others, 1992). A Silurian orogeny, unnamed at present, probably represents the amalgamation of terranes farther outboard in New England and the Canadian Appalachians (O'Brien and others, 1991) prior to their accretion to North America in the Acadian orogeny. It may be that this Silurian orogeny in the Canadian Appalachians was time transgressive into the Acadian orogeny in New England.

It is convenient to divide the Appalachian orogen in the United States longitudinally into northern or New England, central, and southern segments separated by the New York and Roanoke structural recesses. As noted by Rodgers (1970), the New York recess marks the more profound changes along strike. One of the more important differences across the New York recess is that the great thickness of fossiliferous Silurian and Devonian sedimentary and associated volcanic rocks of the northern Appalachian internides (hinterland) are, as far as we know, absent in the central and southern Appalachians. It may be more than a coincidence that the Devonian Acadian orogeny was a major event in the northern Appalachians, but that evidence for that orogeny south of New York is equivocal.

A number of themes are developed in this chapter. One is that different parts of the continental margin along strike had different histories, and one should be cautious in extrapolating features or events from one part of the margin to another. All major Phanerozoic events that affected the eastern continental margin are diachronous, even the Mesozoic opening of the Atlantic Ocean. Rifting propagated along strike; collisions were typically oblique. Both of these are a consequence of composite interaction of plate motions on a spherical earth (Dewey, 1975). A subset of observations related to diachronism is that events affected the continental margin with different intensities. For example, the Acadian orogeny appears to have been a major event only in the northern Appalachians, whereas the Alleghanian orogeny, except for coastal New England, had more pronounced effects in the central and southern Appalachians.

Terranes have different neighbors at different times. This is perhaps best illustrated by terranes characterized by a peri-Gondwana fauna (Neuman and others, 1989). These terranes may have once been part of a coherent high-latitude paleocontinent (Neuman and Harper, 1992) that was fragmented in the early Paleozoic. The various pieces, here called proto-Gondwana terranes, amalgamated with parts of other plates, themselves different, and were assembled with the growing North American plate at different times.

An argument is developed herein that the west-sloping Bouguer gravity gradient (the Appalachian gravity gradient) (Fig. 2) separating paired anomalies in the western Appalachian orogen has different sources along strike. The gravity high east of the gradient in the northern Appalachians is interpreted to mark a failed continental Iapetan rift in autochthonous Grenvillian crust analogous to the early Mesozoic Hartford basin (Fig. 1). A consequence of this interpretation is that slope-rise deposits of the Taconian Granby and Chaudière River klippen of southern Quebec (Fig. 1) are rooted southeast of the rift, but are still considered to be part of the Laurentian margin. South of Maryland the gradient forms a shallow arc concave toward the foreland outboard of the internal massifs, and is interpreted to mark the crustal edge of Laurentia at depth.

The theme of tectonic heredity appears throughout the discussion. Tectonic heredity has been a major influence on the geometry of the continental margin. The western Mesozoic continental rift system within the Appalachian orogen appears to follow the Iapetan continental rift system (Rankin, 1976). The locations of several oceanic fracture zones of the present Atlantic Ocean basin appear to be controlled by earlier features on the continental margin (Klitgord and others, 1988). Once a zone of weakness (fault and/or shear zone) is formed in the upper crust, it is commonly reactivated in subsequent events. Another recurring theme in this chapter is the importance of foreland studies (both stratigraphic-sedimentologic and structural) in unraveling the history of the hinterland. For example, the dating of fossiliferous clastic wedges in the foreland provides some of the best time constraints on the age of orogenic pulses in the Appalachians, whereas the absence of a convincing Devonian clastic wedge in the foreland of the southern Appalachians argues against a major Acadian orogeny in the southern Appalachians.

A major theme is the effect of Mesozoic extension on levels of exposure in parts of the Appalachian orogen and in reconstituting the lower crust and the Moho. Southeast of the Chain Lakes block in Maine (see Fig. 3) (Stewart and others, 1991), the Connecticut Valley in E1 (Ando and others, 1984), Arvonia, Virginia, in E3 (Pratt and others, 1988), and the Lowndesville shear zone in E5 (Cook and others, 1979), the Moho is reflective to seismic energy and interpreted to be of Mesozoic age. To the northwest of these regions, the Moho, as determined by seismic reflection studies, is poorly defined and may have grown diffuse with time. Stewart and others (1991) showed that above the reflective Moho, lower crustal seismic reflections extend undisturbed beneath Paleozoic terrane boundaries, indicating that lower crust was also reconstituted during the Mesozoic. Much of the area of tectonostratigraphic terranes thought to have been accreted to Laurentia during the Paleozoic lies above the region of reflective Moho interpreted to have been formed through Mesozoic reconstitution of the lower crust. Thus, although these terranes as seen today may be thin relative to their areal extent, they do preserve the record of the Paleozoic development of the Appalachian orogen. One conclusion from the theme of Mesozoic extension is that an understanding of the Paleozoic history of much of the Appalachian internides must be deduced from surface geology and the upper crust in the region of accreted terranes.

I begin with a description of the eastern continental margin from the craton and Appalachian foreland on the west, across the crystalline Appalachians including the accreted terranes, and across the Mesozoic (modern) continental margin to oceanic crust east of the East Coast magnetic anomaly. This largely two-dimensional description is followed by a discussion of some aspects in three dimensions, such as the source of the Appalachian gravity gradient, crustal thickness, Alleghanian shortening, and early Mesozoic extension. Next, the evolution through time from the Late Proterozoic to the Quaternary of the eastern continental margin is reviewed. The chapter ends with a discussion of the major differences between the transects.

EASTERN CONTINENTAL MARGIN IN SPACE

Mid-continent to Laurentian Appalachians

In the eastern United States the autochthonous Proterozoic rocks of the Phanerozoic craton are exposed only in the Adirondack massif (transect E1). These rocks are part of the Grenville province of the Canadian shield, with which they are continuous across the St. Lawrence River through the Frontenac arch (Fig. 2). The Adirondack massif is a structural dome defined by the radial dips of the surrounding nonmetamorphosed Upper Cambrian and Lower Ordovician sedimentary rocks that overlie crystalline rocks metamorphosed to upper amphibolite and granulite facies about 1.1 Ga, during the Ottawan phase of the Grenville orogeny (McLelland and others, 1993). The doming is certainly post-Carboniferous, and the initiation of the doming may be as young as Tertiary (Isachsen, 1975, 1976).

South of the Adirondacks the craton is covered by nearly flat lying Paleozoic platform rocks that are within the Interior Lowlands of King (1977). Despite the relative stability of that cratonal area, the region is characterized by broad warps of the basement surface that are commonly 300 km across and have structural relief of a few kilometers. One of these, the long Cincinnati-Findlay arch (Fig. 2), separates the Illinois basin on the west from the Appalachian basin on the east. It has been interpreted as a peripheral bulge activated and reactivated in response to lithospheric downwarping of the Appalachian basin under loads of successive Taconian, Acadian, and Alleghanian overthrusts (Quinlan and Beaumont, 1984; Beaumont and others, 1988). Paleozoic sedimentary rock cover on the basement is as thin as less than 0.5 km on parts of the Cincinnati arch, as much as 4 km in the Rome trough (Fig. 2), and thicker than 10 km under the imbricate thrust sheets of the Valley and Ridge province (Fig. 1). Southeast of the Cincinnati-Findlay arch rocks of the Acadian or Alleghanian clastic wedges (molasse) form surface exposures, elevations are slightly higher, and the area is known as the Appalachian Plateaus province (including the Allegheny and Cumberland Plateaus) (Fig. 1). The Appalachian Plateaus province is generally taken as the westernmost part of the Appalachian foreland as far north as the Mohawk Valley (Fig. 4).

Anyone who has tried to describe an orogenic belt is familiar with the conflicting nomenclature applied to the various parts of it, whether those parts are physiographic, structural, sedimentologic, lithotectonic, or other. Boundaries between the parts may be fuzzy, either in concept or because the parts are gradational into one another. Nonetheless, it is almost impossible to describe an orogenic belt without segmenting it into parts and relating those parts to one another. The terms externides and internides as here applied to the Appalachians are essentially interchangeable with the sedimentary Appalachians and crystalline Appalachians, and are separated approximately by the basement external massifs. The Appalachian basin as used by Colton (1970)—from the crest of the Cincinnati-Findlay arch southeastward to the Talladega-Blue Ridge fault, Shenandoah massif, Reading prong, and Logan's line—but not the Appalachian basin as used by Hatcher (1989a), is virtually coextensive with the sedimentary Appalachians.

Blind thrusts extend a considerable distance westward under the Appalachian Plateaus (Geiser, 1989), and thrusts intersect the present erosion surface as far west as the Sequatchee Valley and Pine Mountain fault (E5 and E4). Geiser (1989) also pointed out that deformed fossils, the most widespread expression of layer-parallel shortening, extend as much as 200 to 300 km west of the Allegheny structural front (Fig. 2), which is the northwestern limit of folded rocks, in New York and Pennsylvania. In Ohio, West Virginia, and Virginia, blind thrusts under the Appalachian Plateaus carry deformed nonmarine rocks of the Dunkard Group, the youngest rocks affected by Appalachian (Alleghanian) deformation. The age of the Dunkard Group is uncertain, within the range Late Pennsylvanian to Early Permian (Secor and others, 1986). The Allegheny structural front is a sharp demarcation and is traditionally taken as the boundary between the Appalachian Plateaus on the northwest and the Valley and Ridge province on the southeast. The Valley and Ridge province is the classic foreland fold and thrust belt of the Appalachians. The Valley and Ridge province changes along strike, from a zone of imbricate thrust sheets at the present level of erosion in the southern Appalachians (E4 and E5) to a fold belt draped above a set of duplexes, which end in blind thrusts in the central Appalachians (E2 and E3). Geiser (1989) argued that the lateral difference is real, and not just that the southern Appalachians are more deeply eroded. The youngest deformed strata belong to the late Paleozoic Alleghanian clastic wedge (molasse), hence the age of deformation is Alleghanian.

The southeastern part of the Valley and Ridge province is typically a broad valley having a number of local names, such as the Great Valley of Pennsylvania (E2), the Shenandoah Valley of Virginia (E3), and the Tennessee Valley of Tennessee (E5). Strata include basal Cambrian clastic rocks, Cambrian and Ordovician shelf carbonates, and Middle and Upper Ordovician flysch deposits that are synorogenic with the Taconian orogeny. The latest deformation of these rocks is Alleghanian because in some areas Carboniferous strata are deformed. Clastic wedges are not present in the foreland north of Mohawk Valley, and the classic Valley and Ridge topography dies out in the lower Hudson Valley (Fig. 1). The geomorphic lowland underlain by Cambrian and Ordovician rocks continues north from the Great Valley of Pennsylvania into the Hudson-Champlain Valley, and the entire linear topographic low along the Appalachians has been called the Appalachian Valley (Hack, 1989).

The Hudson-Champlain Valley is a narrow and topographically subdued foreland fold and thrust belt. The northwestern limit of Appalachian deformation, called Logan's line (E1), is the Champlain thrust or later normal faults that offset it. Along Logan's line deformed rocks of the Appalachian orogen are thrust to within 5 km across strike of Middle Proterozoic rocks of the craton exposed in the Adirondack massif (Fig. 2). Farther northeast, beyond the city of Quebec, these deformed Appalachian rocks are thrust over Grenvillian rocks of the exposed craton (see Riva and Pickerill, 1987). In the Hudson Valley, a thrust interpreted to be of the Champlain thrust family (Taconic frontal thrust system) is unconformably overlain by Lower Devonian rocks (Bosworth and others, 1988; Thompson and others, 1993).

Emplaced within the miogeoclinal and synorogenic flysch deposits of the Appalachian Valley during the Taconian orogeny are the Taconic allochthon (E1) and the Hamburg klippe (E2) (Fig. 1). These are composed of Late Proterozoic rift facies rocks and Cambrian and Ordovician slope-rise strata, and are here considered to be part of Laurentia. Stanley and Ratcliffe (1985), on the basis of sedimentary facies and structural arguments, concluded that the lowest and westernmost of the Taconic allochthons are the first moved, farthest traveled, and were rooted east of the external massifs and probably east of the internal massifs (also see Drake, 1989b).

East and southeast of the Appalachian Valley and the external massifs lie the internides or crystalline Appalachians, which continue at the surface to the Atlantic shoreline in New England and are overlapped by the Atlantic Coastal Plain south of New

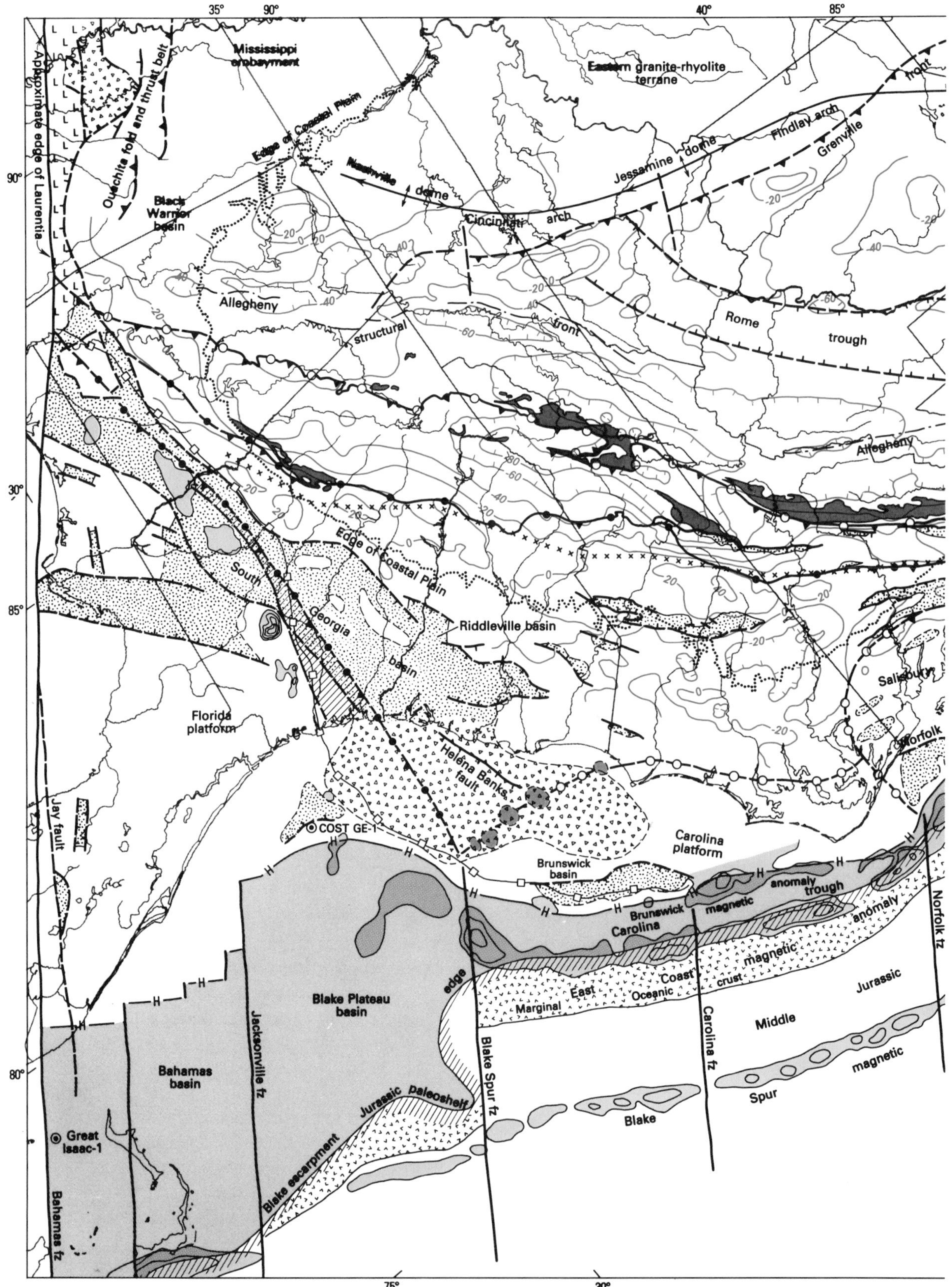

Figure 2. (Explanation follows these two pages.) Features of the continental margin of the eastern United States: Past and present. Data from Austin and others (1980), Gravity Anomaly Map Committee (1987), Klitgord and others (1988), Magnetic Anomaly Map Committee (1987), McBride and Nelson

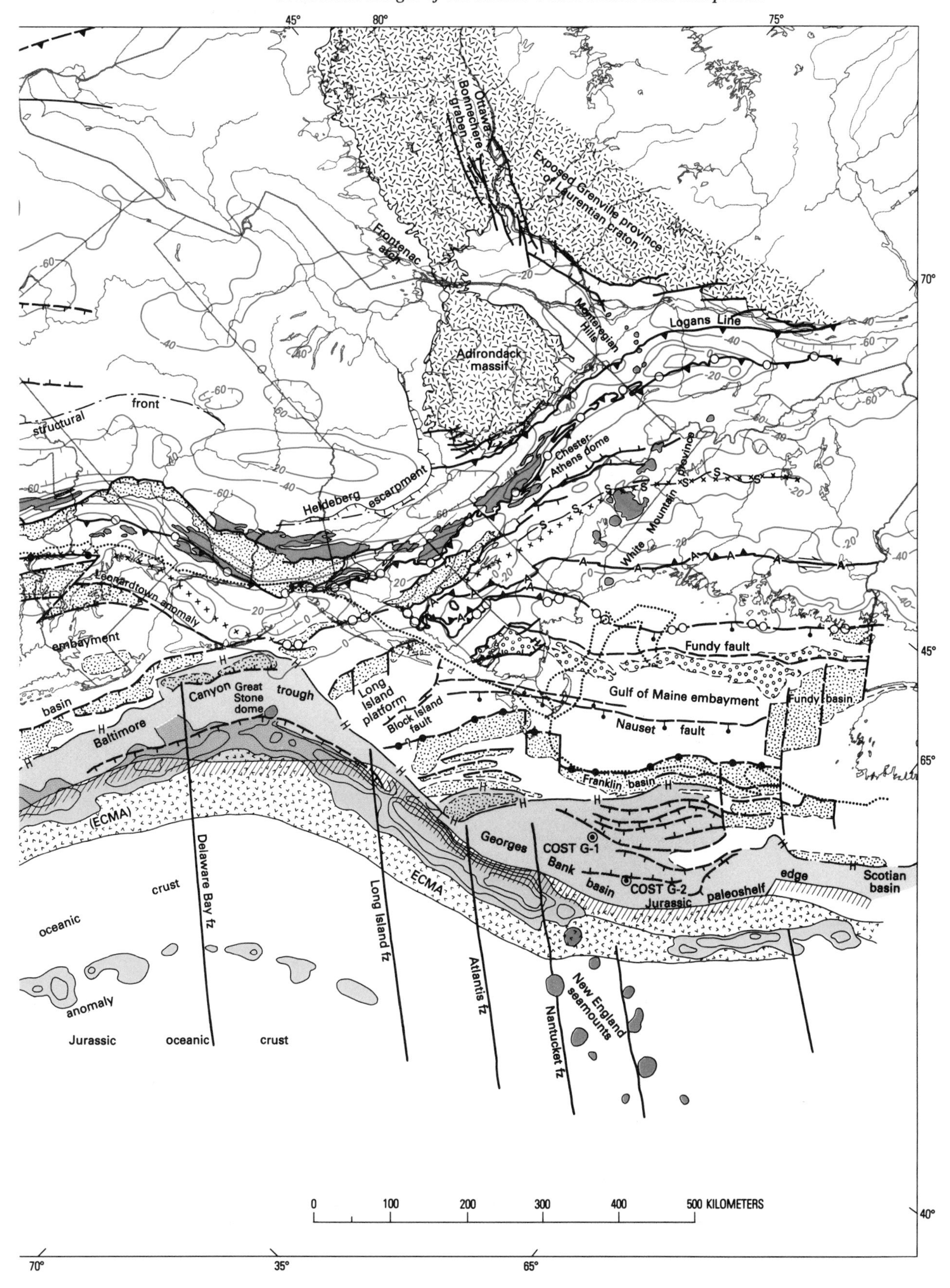

(1988), McBride and others (1989), Moench and St. Julien (1989), Rankin and others (1993a), Trehu and others (1989). Abbreviation fz equals oceanic fracture zone.

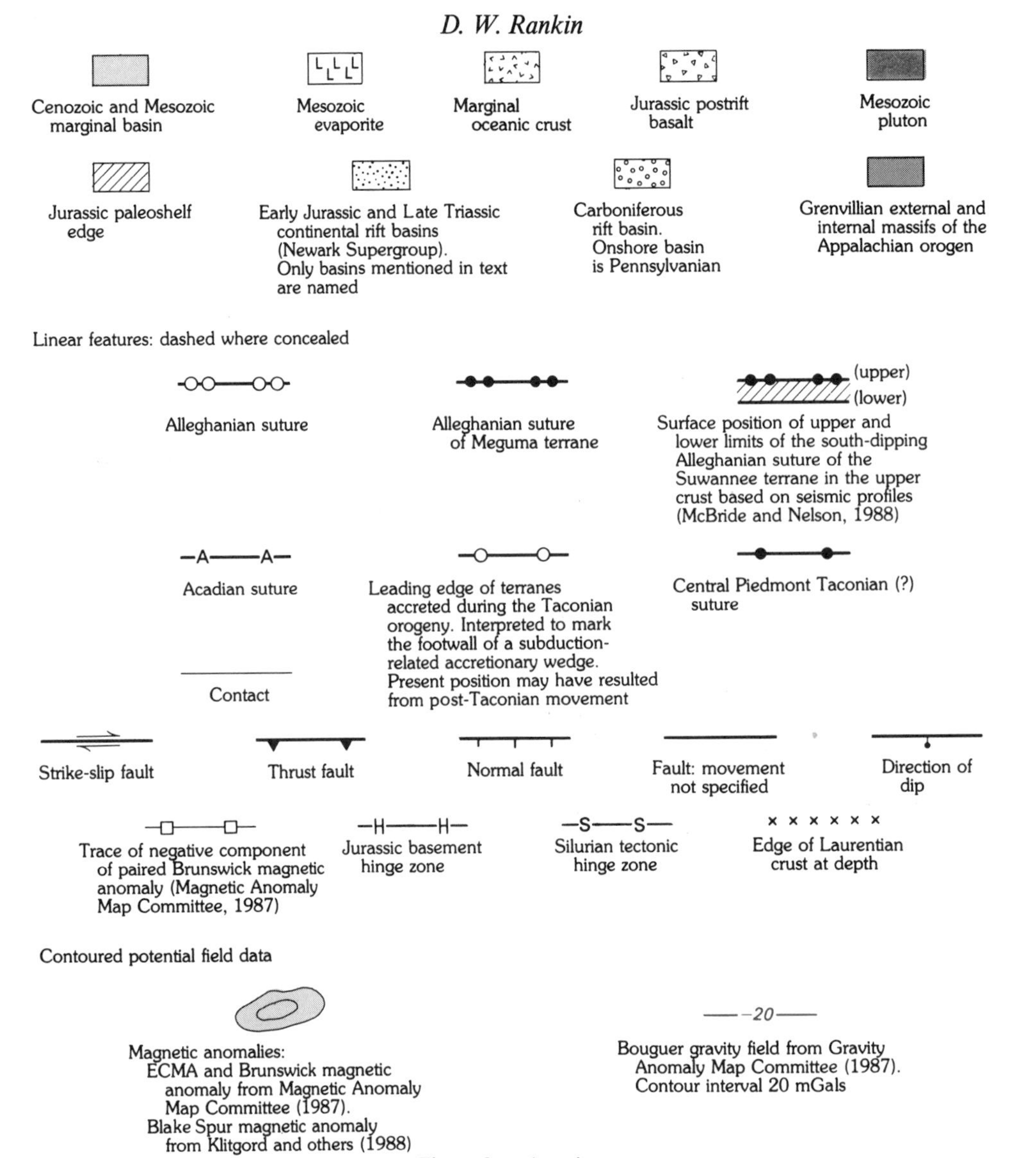

Figure 2 explanation.

York City. In the subsurface, except possibly for transect E5, the internides probably continue out to the continental margin. The internides are complex and have many differences along the length of the orogen. Only generalities can be treated here.

Laurentian basement crops out within the orogen in a chain of en echelon external massifs of Grenvillian crust along the southeast side of the Appalachian Valley (Fig. 1). The trace of the long dimension of these external massifs forms the Blue-Green-Long axis (named for the Blue Ridge, Green Mountains, and the Long Range; Rankin, 1976). All of the external massifs are now known to be allochthonous, and some are part of large, thin, far-traveled crystalline thrust sheets. Most of the westward transport of the Berkshire massif occurred late in the Taconian orogeny (E1) (Sutter and others, 1985). Timing of the emplacement of the Green Mountain massif is less well constrained. On the basis of regional considerations, Stanley and Ratcliffe (1985) argued for emplacement late in the Taconian orogeny, but left open the possibility of later movement on the faults. Ratcliffe and others (1988) documented Acadian east-west or northwest-southeast shortening on faults within the Green Mountain massif.

Grenvillian basement also crops out in a series of gneiss domes or internal massifs, well within the Paleozoic metamorphic core of the orogen (Figs. 1 and 2). The larger of these internal massifs are the Chester-Athens dome (E1), a series of domes in southeastern Pennsylvania and adjacent Delaware and Maryland, which expose Baltimore Gneiss (E3), the Sauratown Mountains anticlinorium (E4), and the Pine Mountain belt (southwest of E5). Most of these domes (foliation domes) are now interpreted to result from refolding of crystalline nappe systems, duplexing of underlying thrust sheets, or both. The internal massifs are the southeasternmost exposures of Grenvillian crust; hence, the rifted margin of Laurentia must have been at least as far southeast as they are. Glover (1989) argued that the Goochland terrane (Fig. 3) in E3 is also an internal massif of Grenvillian crust. That interpretation has been questioned (Horton and others, 1989; Rankin and others, 1989).

The external and internal massifs are collectively referred to as Grenvillian outliers. All underwent metamorphism and/or plutonism from 9.5 to 1.3 Ga, and all, except the smallest, are tied stratigraphically to a Laurentian cover sequence (Rankin and others, 1989).

Rock assemblages in the Grenvillian outliers, or external and internal massifs, are not all the same. Rankin and others (1993a, 1993b) outlined some of the differences between groups of outliers and suggested correlations with subdivisions of the cratonal Grenville province and new subdivisions where appropriate. Protoliths in the outliers in New England, New Jersey, and Pennsylvania north of the Pleasant Grove–Huntingdon Valley fault (Fig. 1) include much stratified rock that closely resembles the central granulite belt of the Adirondack Highlands. Drake (1990), however, suggested that areas rich in marble, such as the Franklin–Sterling Hill district along the northwestern flank, but within the Reading prong, might be part of the Frontenac block exposed in the Adirondack Lowlands. Protoliths of most of the Grenvillian outliers south of Maryland include a higher percentage of plutonic rocks than their northeastern counterparts and were collectively called the Blue Ridge plutonic belt by Rankin and others (1993b). They also noted that the southern part of the French Broad massif (Fig. 1) and the Sauratown Mountains massif appear to contain a higher percentage of stratified rocks. Horton and McConnell (1991) and McConnell and others (1992) interpreted most of the stratified rocks in the Sauratown Mountains massif to be younger cover rocks. Rankin and others (1973), however, interpreted the stratified rocks to be part of the basement and intruded by Grenvillian plutonic rocks. I still favor that interpretation. The internal massifs of Maryland, Pennsylvania, and Delaware south of the Pleasant Grove–Huntingdon Valley fault are thought to contain a higher percentage of metavolcanic rock, do not resemble either the central granulite belt or the Blue Ridge plutonic belt, and were placed in a new subdivision, the Baltimore belt. The rocks of the Goochland terrane do not resemble those of any of the other Grenville subdivisions and may never have been part of Laurentia.

Cover rocks of rift facies and of Late Proterozoic and earliest Cambrian age that were transported in thrust sheets along with the Grenvillian basement provide much of the evidence for continental rifting and subsequent opening of the Iapetus Ocean. Anorogenic plutons and dikes of Late Proterozoic and Early Cambrian age that are within the Grenvillian massifs between New York City and northwestern North Carolina (E2, E3, and E4) are correlated with coeval and more extensive volcanic rocks in the cover sequence and with continental rifting. In some areas mafic dikes are so abundant (as much as 50% of areas as large as 8 km^2; Rankin, 1975) that this allochthonous Laurentian continental crust is probably comparable to the transitional crust of the Mesozoic Atlantic margin.

The external massifs are near the western margin of the metamorphic core of the orogen. For most of the orogen, the westernmost episode of Paleozoic metamorphism was Taconian, although evidence for the age of the westernmost metamorphism in the central Appalachians is equivocal (Dallmeyer, 1975; Russell and others, 1984; Sutter and others, 1985; Laird, 1989; Drake, 1989b). Taconian metamorphism is thought to have attained the granulite facies in the Jefferson terrane (Fig. 3) just east of the southwestward projection of the Blue-Green-Long axis in southwestern North Carolina (Absher and McSween, 1985) (E5). An early, 500 Ma, medium-high- and high-pressure metamorphism affected rocks on both sides of the boundary between the Laurentian and accreted Appalachians in northern Vermont and southern Quebec. This metamorphism may have been subduction related and earlier than the collision of accreted terranes (Laird and others, 1984; Sutter and others, 1985). The age of the westernmost metamorphism at the southern end of the orogen in the Talladega block (Fig. 1) is controversial and is discussed more fully later. Briefly, however, Lower Devonian fossiliferous strata were metamorphosed to the greenschist facies and thrust over nonmetamorphosed rocks on Alleghanian thrusts (Tull and Stow, 1980; Kish, 1990). This metamorphism has been thought to be Acadian, but I wonder if it might not be Alleghanian.

Accreted Appalachians

Thus far, discussion of the Appalachian orogen has focused on those rocks that were intruded into Laurentian basement or deposited on it, even if later transported, i.e., on the Laurentian Appalachians. Now those parts of the orogen are discussed that appear to have been accreted to the eastern margin (present geography) of Laurentia during one of the Paleozoic collisional events that formed the Appalachian orogen.

Wilson (1966) suggested that juxtaposed dissimilar early Paleozoic faunal provinces in New England and the Maritime Provinces of Canada and in northwestern Europe could be accounted for if an ocean basin separated them from late Precambrian to the end of Middle Ordovician time. Thus, the concept evolved that a proto-Atlantic Ocean (Wilson, 1966) or Iapetus Ocean (Harland and Gayer, 1972) existed approximately, but not precisely, in the same location as the present North Atlantic. As we have learned more about the Appalachian-Caledonide orogen and plate tectonic processes, Wilson's initial suggestions are being modified continuously into more and more complex models. Similarities in the early Paleozoic eastern or Atlantic province fauna, their host rocks, and the underlying basement in southeastern New England and Maritime Canada led to the idea of a large coherent terrane that collided with Laurentia during the closing of "the" Paleozoic Ocean. The name Avalon is commonly associated with this large terrane, after the classic locality in southeastern Newfoundland. Avalon-like rocks, also bearing an Atlantic province fauna, crop out in the Carolina slate and Belair belts (discussed below), and some have extended the Avalon terrane to include these (Rodgers, 1972; Rast and Skehan, 1983; Williams and Hatcher, 1983). My favored interpretation is that of Secor and others (1989); i.e., although a large coherent terrane with an Atlantic province fauna may have existed, it was fragmented and the pieces collided with ancestral North America at times as different as the Taconian and Alleghanian orogenies. The large coherent terrane was probably proto-Gondwana, and the fragments now preserved along the eastern margin of North America have what Neuman and others (1989) characterized as a peri-Gondwanan fauna. Neuman and Harper (1992), in discussing primarily Early and Middle Ordovician brachiopod faunas, also recognized a third fauna called Celtic, which they interpreted as being more closely related to the peri-Gondwana fauna than to Laurentia; they interpreted the fauna to have lived near islands, some of which were volcanic, in the Iapetus Ocean.

The terms Avalon terrane or Avalonia should be and are here restricted to the assemblage of rocks that most resemble those of the Avalon Peninsula, Newfoundland. The essential features of this terrane include extensive Late Proterozoic magmatism and Cambrian platformal sedimentary rocks, bearing an Atlantic fauna, that did not experience an early Paleozoic orogeny. In the United States the Avalon terrane includes the Hope Valley and Esmond-Dedham blocks and their offshore extensions.

Between the clearly exotic terranes that contain an Atlantic province fauna and the Laurentian eastern margin, numerous additional terranes that may be exotic have been suggested. Their fossil characterization is poor to nonexistent. These are the "problematic terranes" of Goldsmith and Secor (1993). They are problematic because of major disagreements over the existence, delineation, time of docking, and names of these terranes. It is reasonably certain, however, that the Appalachian orogen east of the Laurentian margin is a collage of terranes accreted at different times through the closing of more than one ocean. One possible assemblage of terranes of the accreted Appalachians is presented in Figure 3 (taken with modification from Rankin and others, 1993a), which serves as a basis for the following discussion. The assemblage is the one I favor, but much of the evidence is conflicting and parts of the assemblage are highly speculative.

The westernmost of the accreted terranes as portrayed in Figure 3 are the Brompton-Cameron terrane of New England and the Jefferson terrane of the central and southern Appalachians. The two terranes have much in common and are thought to include subduction-related accretionary wedges. The two terranes are discussed in the remainder of this section and reasons for considering them to include exotic accretionary wedges are outlined.

A series of faults from Vermont to Alabama is interpreted to be the western boundary of accreted terranes. The faults separate metamorphosed stratified rocks that contain bodies of ultramafic rocks on the east from metamorphosed stratified rocks stratigraphically tied to Laurentian basement and without ultramafic rock. From north to south, these faults are the Hazens Notch, Whitcomb Summit, Cameron's line, and Pleasant Grove–Huntingdon Valley faults; hypothetical faults in Virginia; and the Hayesville (but not the Fries), Allatoona, Goodwater-Enitachopco, and Hollins line faults (see Figs. 1 and 4).

The ultramafic bodies are commonly associated with metamorphosed basalts and gabbros, and together these have been interpreted as fragmented ophiolites representing pieces of normal oceanic crust. Isotopic studies suggest, however, that some of the assemblages have continental contamination and may be other than of mid-ocean origin (Shaw and Wasserburg, 1984). Of the areas studied by Shaw and Wasserburg (1984), the best candidates for normal oceanic crust are in the Jefferson terrane near the North Carolina–Georgia border (Fig. 3) and in the Hazens Notch slice of northern Vermont (Fig. 4). The ultramafic bodies are thought to be slivers along imbricate faults and/or olistoliths in a sedimentary melange. They were interpreted by Glover (1989) as intrusive bodies (sills and dikes) related to late Iapetan rifting in the region of the E3 transect.

The series of faults named above is interpreted to mark the subduction zone above which developed a subduction-related accretionary wedge containing ophiolite fragments and along which an ocean closed. There is not agreement as to the size of that ocean: suggestions include a back-arc ocean by Hatcher (1978a), a relatively narrow ocean by Robinson (*in* Tucker and Robinson, 1990), and the large Iapetus Ocean by Stanley and Ratcliffe (1985), Horton and others (1989), and Rankin and others (1993a), among others. The subduction zone is thought to have dipped to the east or southeast because of the absence of appropriate arc-related rocks west of the line. In Rankin and others (1989), I argued that the west-directed accretionary wedge developed east of the subduction-related trench and possibly east of the Iapetus spreading center and was related to the eastern margin of Iapetus. That wedge would be exotic to Laurentia. In the discussion that follows and in the accompanying illustrations, I continue to support that interpretation, but acknowledge that its validity is far from established. The interpretation here is that the subduction complex marks the disappearance of a large amount of oceanic crust (Stanley and Ratcliffe, 1985, estimated the Taconian consumption of about 1000 km of oceanic crust).

In New England the accretionary wedge is composed of deep-water sedimentary and volcanic rocks that constitute the Hazens Notch slice and the Rowe-Hawley zone of Stanley and Hatch (1988) along the western margin of the Brompton-Cameron terrane (Figs. 3 and 4). Rocks of the Hazens Notch slice and Rowe-Hawley zone are thought to be more oceanward facies than those of the Taconic allochthon. The Whitcomb Summit–Hazens Notch thrust is thought to be the root zone of the Taconic allochthon, which carries slope-rise deposits of the Laurentian margin over early Paleozoic shelf rocks of the foreland (Stanley and Ratcliffe, 1985, modified in Rankin and others, 1989). Rocks of the Hazens Notch slice bear the closest resemblance to the slope-rise deposits of the Taconic allochthons and were included in the Laurentian margin by Stanley and Ratcliffe (1985). However, they are distinct from the Taconic allochthons in that they include abundant ultramafic rocks and the allochthons contain none. In the interpretation of Zen and others

(1986) the similarities between the accretionary wedge and Laurentian slope-rise deposits suggest that the accretionary wedge is a deep-sea facies of pre-Silurian Laurentian cover rocks obducted onto Laurentia. The westward vergence of the wedge presumably resulted from underplating an obducting oceanic slab and somehow mixing ophiolitic fragments within the underplating material. In the interpretations favored here, the Taconic allochthons are derived from slope-rise deposits of the Laurentian margin, whereas the Hazens Notch slice is included in west-verging accretionary wedge of the Brompton-Cameron terrane.

In the southern Appalachians, many of the sedimentary rocks of the accretionary wedge are coarser grained and may not be deep-water facies (such as the Ashe Formation and Lynchburg Group as redefined by Rankin and others, 1993b). Depending upon where the accretionary wedge formed relative to Laurentia and the Iapetus spreading ridge, the rocks of the wedge may or may not be exotic to Laurentia. At the risk of oversimplification, however, the ground-surface trace of the faults that carry the accretionary wedge is taken to a first approximation to mark the Iapetus suture, and the rocks of the wedge to be the westernmost of the accreted terranes. Some of these faults have clearly been reactivated, and the accretionary wedge was transported farther west by later faulting (Alleghanian in the southern Appalachians). The last major movement on some faults may be Alleghanian, but the faults are interpreted to mark the site of offscraping of Iapetan sedimentary cover and the subducted edge of Laurentian cover. Hence, the inferred continuous trace of these faults is called the Iapetus or Taconian suture. This surface is shown as the Iapetus suture in section BB[1] of E1, in E2, and in E4, and as an important terrane boundary in E5. Section AA[1] of E1 shows the Rowe-Hawley zone being a suture zone without identifying any specific fault as the suture. In this model the Grenvillian internal massifs would be exposed in windows through the suture and overlying accretionary wedge.

Alternative models are used in E3 and E5. Although the discussion here may be a little premature, it is given here to alert the reader to a more complete discussion later in the chapter. Glover (1989) included the Jefferson terrane (a term not needed in his model) and some rocks east of that (Evington Group rocks that are here included in the Potomac composite terrane; Fig. 3) in Laurentia, included the Chopawamsic terrane (a Cambrian volcanic arc; Fig. 3) in the Carolina terrane, and interpreted the Taconian suture to be a broad melange in a collision zone between the Evington Group of Laurentia and his Carolina terrane. In the Glover model the Taconian suture is repeated by thrusting, and its easternmost exposure is on the southeast side of the Goochland terrane (his Laurentia) beneath the overthrust Carolina terrane. Hatcher (1989b) interpreted the Hayesville fault to mark the Taconian closing of a back-arc basin along the Laurentian margin and interpreted the central Piedmont suture (Fig. 1) to mark the closing of Iapetus, which could have been during either the Taconian or Acadian orogeny.

East and southeast of the Brompton-Cameron and Jefferson terranes, much of the commonality of the Appalachians breaks down. It is perhaps better to outline the geology of the northern, central, and southern parts of the accreted Appalachians separately. The focus of the discussion is on those parts of the orogen crossed by transects E1 to E5 and the Quebec–Maine–Gulf of Maine transect.

Northern Appalachians. Figure 5 is a terrane diagram that summarizes many of the features of the terranes of New England and environs.

Brompton-Cameron and Central Maine terranes and their relation to the Canadian Appalachians. The first two terranes east of the Iapetus suture are the Brompton-Cameron terrane (Zen, 1989) (which includes the Hazens Notch slice for purposes of generalities) and the Central Maine terrane (Fig. 4) (Zen, 1989). The oldest rocks in both terranes are penetratively deformed and variably metamorphosed clastic sedimentary and volcanic rocks of probable Cambrian through Early or Middle Ordovician age. Most rocks are poorly dated. The interpretation followed here, not without misgivings, is that the Central Maine terrane was amalgamated with the Brompton-Cameron terrane following the closing of an eastern ocean during the Late Cambrian to Early Ordovician Penobscottian orogeny, and that both terranes were accreted to Laurentia during the Taconian closing of Iapetus in the Middle to Late Ordovician (Boone and Boudette, 1989; Moench and St. Julien, 1989).

In New England the older rocks of the Brompton-Cameron terrane are exposed along the western margin of that terrane and constitute the accretionary wedge and included ultramafic bodies discussed previously. Along strike to the northeast in Quebec the zone of the accretionary wedge becomes the Baie Verte–Brompton line and the St. Daniel Melange (Fig. 4) (Cousineau and St. Julien, 1992; Williams and St. Julien, 1982). In the region of the Quebec–Maine–Gulf of Maine transect, the Thetford Mines ophiolite, which is in the Brompton-Cameron terrane, crops out along the northwestern edge and is overlain conformably on the southeast by the St. Daniel Melange (St. Julien and Hubert, 1975; Moench and St. Julien, 1989). Neither the ophiolite nor the melange is fossiliferous. Sm-Nd isotopic studies by Shaw and Wasserburg (1984) demonstrated that the Thetford Mines ophiolite did not have the signature of normal oceanic crust formed at a mid-ocean spreading center. Further geochemical studies combined with the isotopic data suggest that the ophiolite is typical of oceanic crust created by fore-arc rifting in a subduction-zone environment (Laurent and Hébert, 1989). Plagiogranites of the ophiolite complex are 478 ± 3 Ma (U-Pb zircon ages; G. Dunning, personal commun., *in* Laurent and Hébert, 1989) and the final stage of thrust emplacement of the ophiolite is 456 ± 4 Ma (Rb-Sr isochron; Clague and others, 1981). The St. Daniel Melange is overthrust from the southeast by a graphitic slate, tuffaceous sandstones, and chert assemblage of the Beauceville Formation of early Caradocian age (St. Julien and Hubert, 1975). The

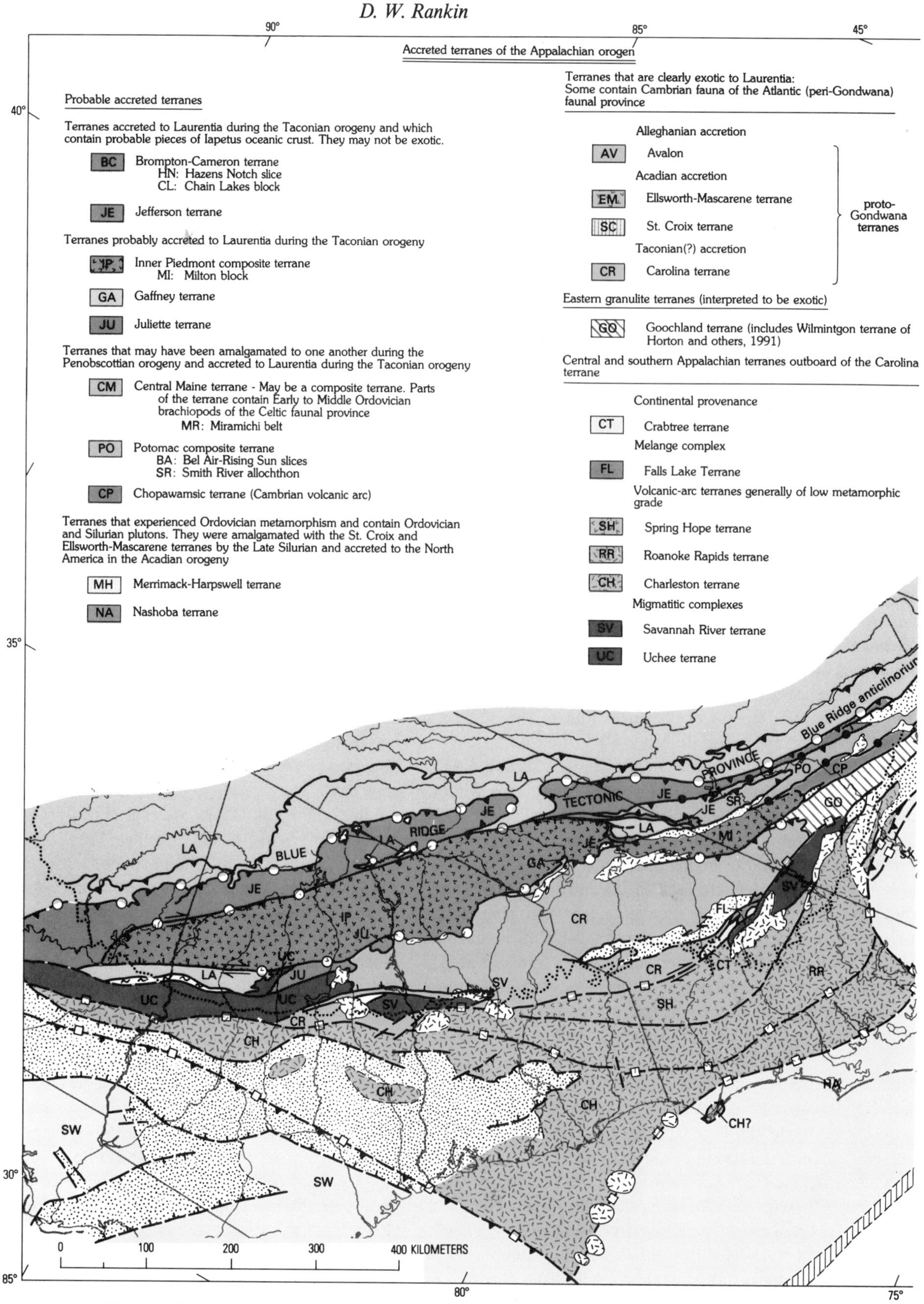

Figure 3. Tectonostratigraphic terranes of the Appalachian orogen. Modified from Horton and others (1991) and Rankin and others (1993a), with additional data from Berquist and Marr (1992), Daniels and Leo (1985), Stewart and others (1991), and Zietz and Gilbert (1980).

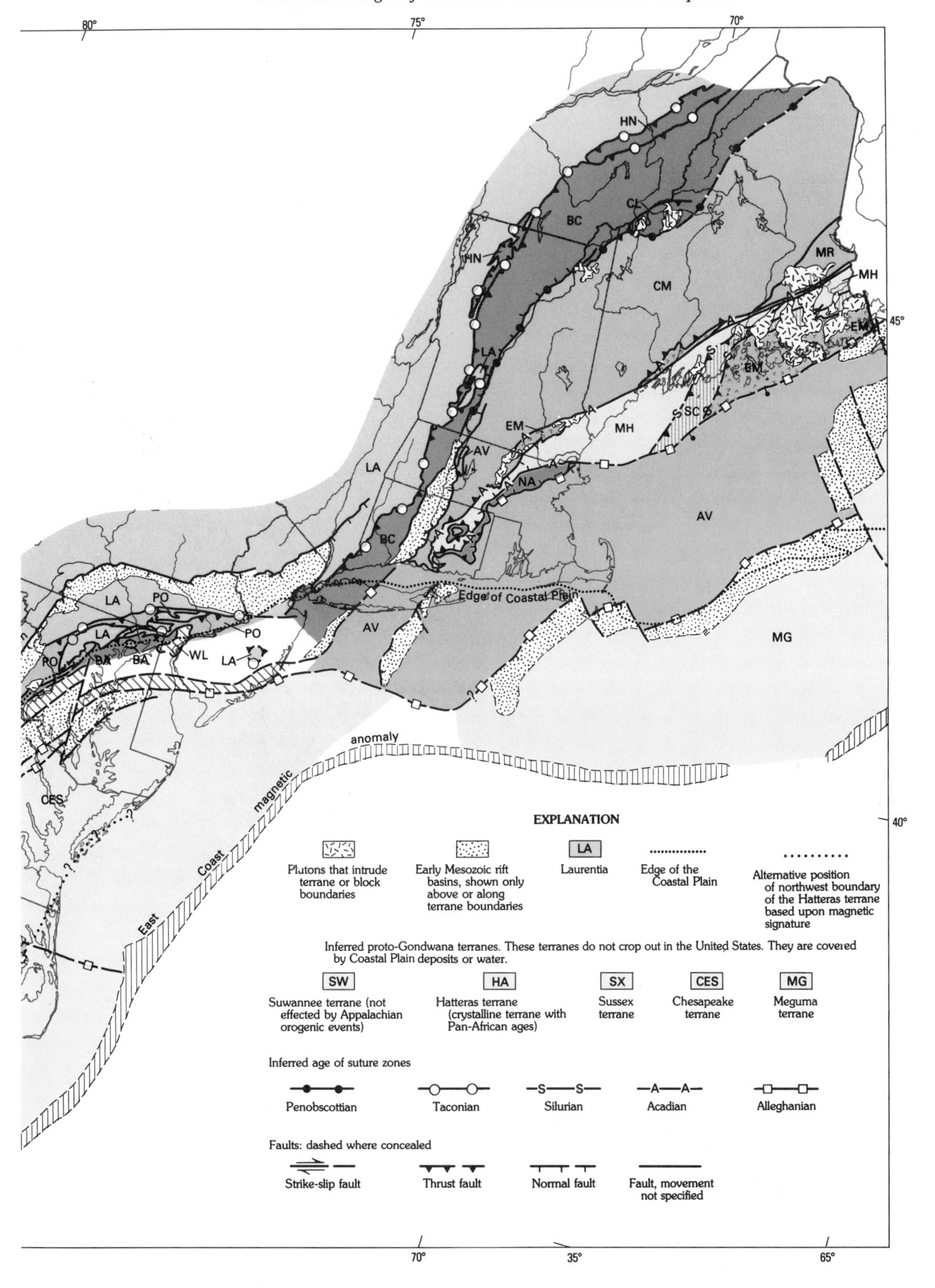
80°
75°
70°
HN
BC
CL
HN
CM
MR
MH
45°
EM
LA
EM
S
SC
EM
MH
AV
LA
NA
AV
BC
LA
PO
Edge of Coastal Plain
AV
MG
LA
PO
PO
BA
BA
WL
LA
anomaly
magnetic
Coast
East
CES
40°
EXPLANATION
Plutons that intrude terrane or block boundaries
Early Mesozoic rift basins, shown only above or along terrane boundaries
LA
Laurentia
Edge of the Coastal Plain
Alternative position of northwest boundary of the Hatteras terrane based upon magnetic signature
Inferred proto-Gondwana terranes. These terranes do not crop out in the United States. They are covered by Coastal Plain deposits or water.
SW
Suwannee terrane (not effected by Appalachian orogenic events)
HA
Hatteras terrane (crystalline terrane with Pan-African ages)
SX
Sussex terrane
CES
Chesapeake terrane
MG
Meguma terrane
Inferred age of suture zones
Penobscottian
Taconian
Silurian
Acadian
Alleghanian
Faults: dashed where concealed
Strike-slip fault
Thrust fault
Normal fault
Fault, movement not specified
70°
35°
65°

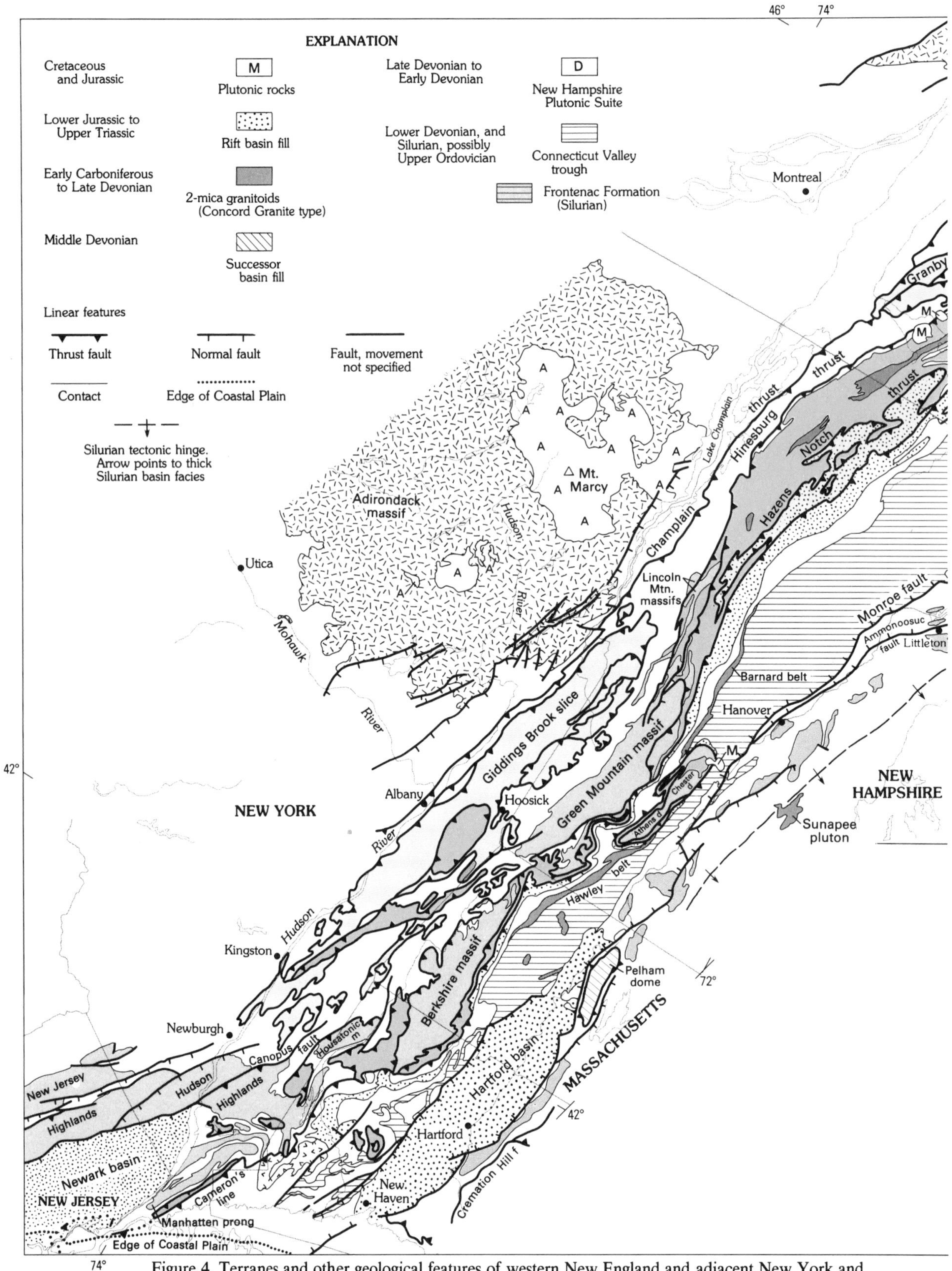

Figure 4. Terranes and other geological features of western New England and adjacent New York and Quebec. Data from Boone and others (1989), Doll and others (1961), Hatch and Stanley (1973), Lyons and others (1994), Moench, 1990, Osberg and others (1985), Panish and Hall (1985), Rankin and others (1993a), Rodgers (1985), Slivitzky and St. Julien (1987), Thompson and others (1993),

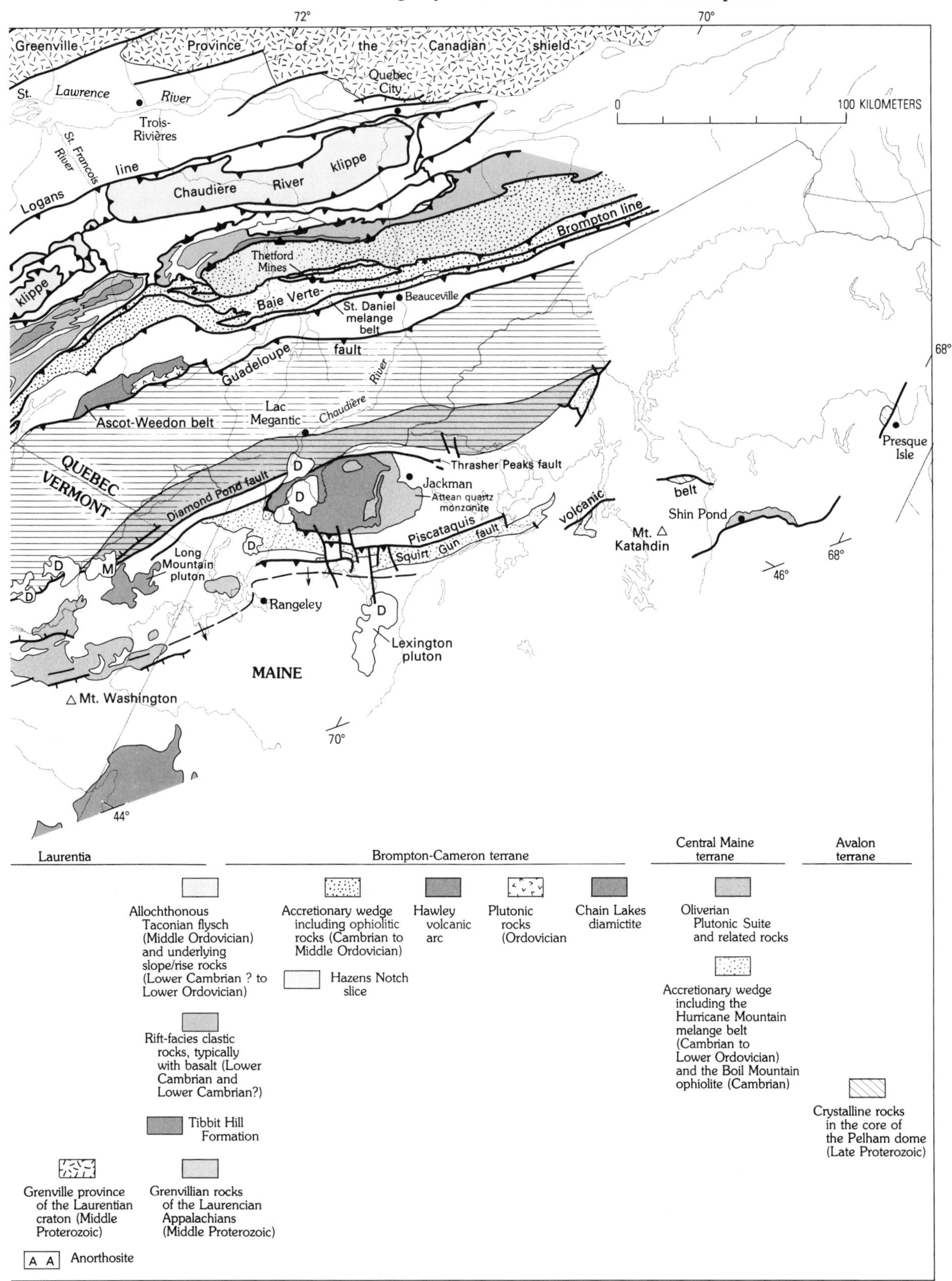

Trzeinski and others (1992), and Zen (1983). Abbreviations used: d, dome; m, massif. Plutons shown are Ordovician plutons, plutons that intrude terrane boundaries, and those that are referred to by name in the text.

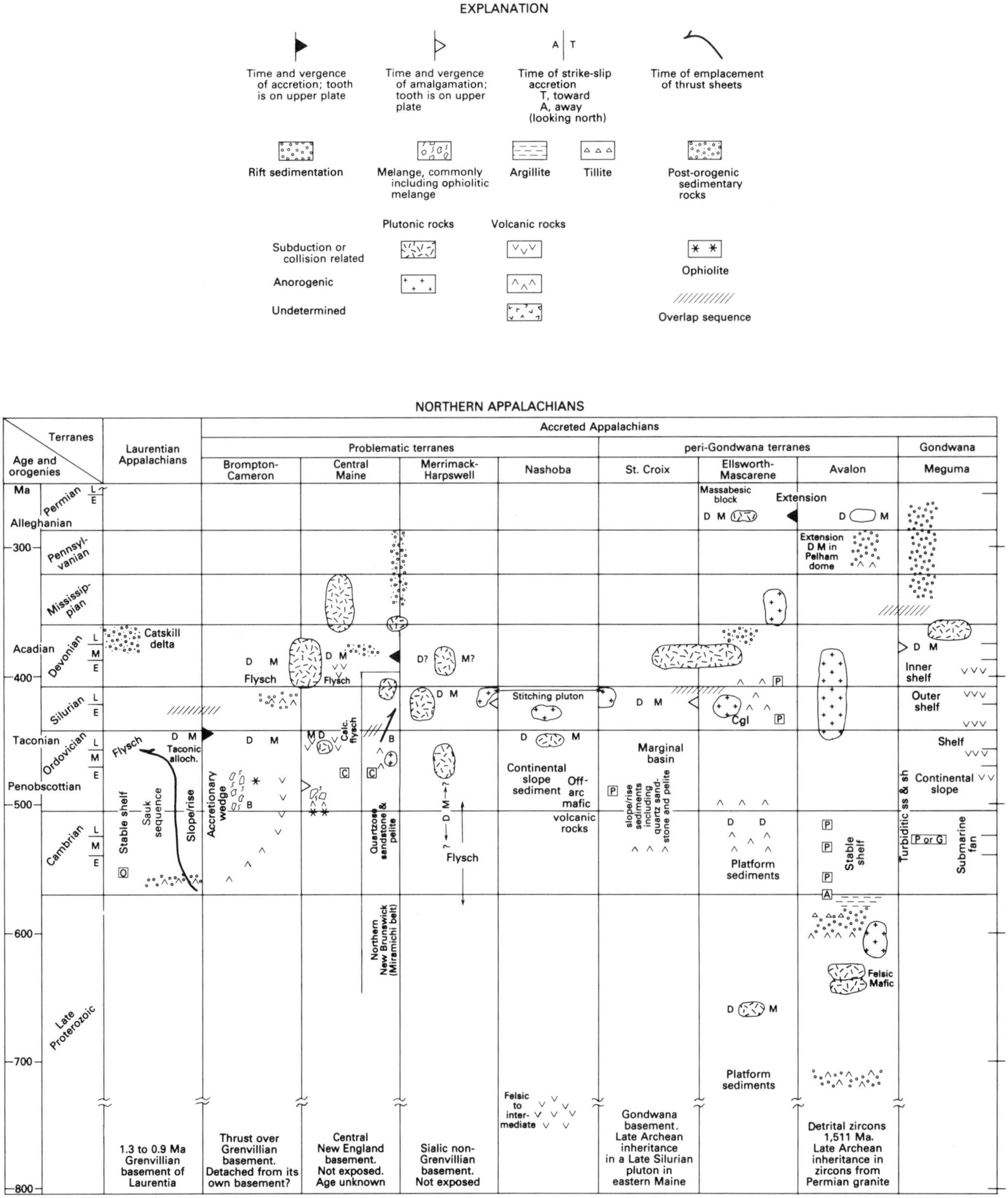

EXPLANATION

Time and vergence of accretion; tooth is on upper plate
Time and vergence of amalgamation; tooth is on upper plate
A | T Time of strike-slip accretion T, toward A, away (looking north)
Time of emplacement of thrust sheets

Rift sedimentation
Melange, commonly including ophiolitic melange
Argillite
Tillite
Post-orogenic sedimentary rocks

Plutonic rocks
Volcanic rocks
Subduction or collision related
Anorogenic
Undetermined
Ophiolite
Overlap sequence

NORTHERN APPALACHIANS

Terranes
Age and orogenies
Laurentian Appalachians
Accreted Appalachians
Problematic terranes
peri-Gondwana terranes
Gondwana
Brompton-Cameron
Central Maine
Merrimack-Harpswell
Nashoba
St. Croix
Ellsworth-Mascarene
Avalon
Meguma

Ma
Permian L E
Alleghanian
300
Pennsylvanian
Mississippian
Acadian
Devonian L M E
400
Silurian L E
Taconian
Ordovician L M E
Penobscottian
500
Cambrian L M E
600
Late Proterozoic
700
800

Massabesic block
D M
Extension
D M
Extension D M in Pelham dome
Catskill delta
D M
Flysch
D M
Flysch
D? M?
D M
Stitching pluton
D M
Cgl
P
Inner shelf
Outer shelf
Shelf
Continental slope
Turbiditic ss & sh
P or G
Submarine fan
D M
Flysch
D M
Taconic alloch.
Calc. flysch
B
D M
D M
Continental slope sediment
Off-arc mafic volcanic rocks
Marginal basin
Stable shelf
Sauk sequence
Slope/rise
Accretionary wedge
B
C
C
Quartzose sandstone & pelite
D M
Flysch
slope/rise sediments including quartz sandstone and pelite
P
D D
Platform sediments
P
P
P
Stable shelf
O
A
Northern New Brunswick (Miramichi belt)
Felsic
Mafic
D M
Platform sediments
Felsic to inter-mediate

1.3 to 0.9 Ma Grenvillian basement of Laurentia
Thrust over Grenvillian basement. Detached from its own basement?
Central New England basement. Not exposed. Age unknown
Sialic non-Grenvillian basement. Not exposed
Gondwana basement. Late Archean inheritance in a Late Silurian pluton in eastern Maine
Detrital zircons 1,511 Ma. Late Archean inheritance in zircons from Permian granite

Figure 5. Terrane diagram for New England and adjacent Canada. Fauna: O, Olenellus; C. Celtic; P. peri-Gondwana; A, acritarchs; E, Ediacarian; G, Gondwanan. Other abbreviations: M, metamorphism; B, blue-schist metamorphism; D, deformation. Sources of data are given in appropriate places in text. Additional source: van Staal and others (1990) for northern New Brunswick.

Beauceville is interpreted to be unconformably above the melange, however, because it also crops out in a small, apparently synclinal autochthonous body northwest of the thrust. These relations are interpreted to date the accretion of the Brompton-Cameron terrane to Laurentia as a Taconian event.

The suture between the Brompton-Cameron and Central Maine terranes in the area of the Quebec–Maine–Gulf of Maine transect is a Penobscottian thrust, the Squirtgun fault, which was later reactivated in the Acadian (Fig. 4). Elsewhere the suture is covered by younger rocks. The ophiolitic Boil Mountain Complex, metavolcanic Jim Pond Formation, and the overlying melange of the Hurricane Mountain Formation of the Central Maine terrane form the hanging wall of the Squirtgun fault (Boone and Boudette, 1989). The lower part of the Jim Pond volcanic sequence contains a large proportion of felsic rocks, and the assemblage has the geochemical signature of island-arc rocks, suggesting that the ophiolite is a suprasubduction ophiolite; the upper part of the volcanic sequence is mid-ocean ridge basalt–like (Coish and Rogers, 1987). The Jim Pond Formation is 500 ± 10 Ma (U-Pb zircon age by Aleinikoff and Moench, 1985). Boone and Boudette (1989) and Boone and others (1989) presented evidence that the melange of the Hurricane Mountain Formation underwent pre-Taconian deformation and metamorphism; i.e., a pre-Middle Ordovician unconformity and an isotopic apparent age of 485 Ma on metamorphic calcic amphibole in a metagabbro intrusive into the Hurricane Mountain Formation. Boone and others related that deformation and metamorphism to the Late Cambrian to Early Ordovician Penobscottian orogeny, which represents the amalgamation of the Central Maine terrane to the Brompton-Cameron terrane.

Observations that first led to the concept of a Penobscottian orogeny ("disturbance") came from the occurrence in the Central Maine terrane near Shin Pond, Maine (Fig. 4) (Neuman, 1967), of late Arenigian to early Llanvirnian (near the Early and Middle Ordovician boundary) strata unconformably above previously deformed Lower Cambrian(?) quartzite and slate of the Grand Pitch Formation. The benthic shelly fossils belong to the Celtic faunal province, and are interpreted to have lived adjacent to volcanic islands at middle to high latitudes within the Iapetus Ocean (Neuman, 1984; Neuman and Harper, 1992). In the interpretation here, that ocean was probably not Iapetus, which was west of the Brompton-Cameron terrane, but a more easterly ocean.

Neuman (1984) noted that the Celtic provincialism ended in the late Llanvirnian, perhaps coinciding with the subduction of hundreds of kilometers of oceanic crust. Late Arenigian to early Llanvirnian strata containing a Celtic fauna also crop out in northern New Brunswick, Canada (see Fig. 8), and are above quartzite and slate of the Miramichi Group (Neuman, 1984; van Staal and Fyffe, 1991). The New Brunswick locality is within the Miramichi belt, commonly considered a terrane correlative or continuous with terranes south of the Norumbega fault (Zen, 1989; Stewart and others, 1991; Rankin and others, 1993a). The Miramichi is a belt of intensely deformed Cambrian and Ordovician volcanic and sedimentary rocks and Silurian sedimentary rocks (Ruitenberg and others, 1977; van Staal and others, 1990). The Miramichi is included here in the Central Maine terrane largely because of the Celtic fauna and subsequent history, which contrasts with terranes south of the Norumbega fault. The Matapedia cover sequence, which is possibly as old as Late Ordovician and certainly as old as Late Silurian, is common to the Miramichi belt and the adjacent Central Maine terrane in New Brunswick (Fyffe and Fricker, 1987). Helderbergian (earliest Devonian) fossils of North America affinity are also present within the Miramichi in western New Brunswick (Fyffe and Fricker, 1987).

The ophiolite of the Boil Mountain Complex, at the leading edge of the Central Maine terrane, was thrust northwestward onto feldspathic paragneiss and granofels of the Chain Lakes massif or block of the Brompton-Cameron terrane (Fig. 4). Pb/Pb ages on zircon of about 1.5 Ga (Naylor and others, 1973) plus relict, possibly Precambrian, sillimanite-grade metamorphism led to the interpretation that these rocks constituted non-Laurentian basement somewhat older than Grenvillian. This was the "Craton X" of Zen (1983) and "Basement B" of Osberg (1978). The potential existence of a non-Grenvillian Middle Proterozoic basement in the Appalachian internides has strongly influenced thinking about Appalachian terranes since 1973. Much of the Chain Lakes block consists of diamictite. Clasts of the Chain Lakes block, including the diamictite, up to a few kilometers in size are in the St. Daniel Melange north of Beauceville (Fig. 4) (Cousineau, 1991). Recent work suggests that the Chain Lakes is not a Middle Proterozoic massif, but a Late Proterozoic to Lower Cambrian clastic deposit containing clasts of several ages and history (Dunning and Cousineau, 1990; Trzcienski and others, 1992a). Whether those deposits were derived from Laurentia (far traveled across the St. Daniel Melange, as favored by Stewart and others, 1991) or from an exotic continental mass to the southeast is not now known. A concordant U-Pb age on monazite from the Chain Lakes diamictite of 468 ± 2 Ma indicates that the age of high-grade metamorphism of the southern part of the block is Ordovician, probably Middle Ordovician (Dunning and Cousineau, 1990). The monazite is part of a high-temperature–low-pressure sillimanite-bearing assemblage that was interpreted by Trzcienski and others (1992a) to be a contact-metamorphic assemblage adjacent to plutons of the Highlandcroft Plutonic Suite (correlated with the Oliverian Plutonic Suite, which is shown in Fig. 4), even though reported zircon ages from the Highlandcroft are somewhat younger (453 to 443 Ma; Lyons and others, 1986). In fact, the Attean Quartz Monzonite, which is intrusive into the Chain Lakes diamictite (Fig. 4), has an U-Pb zircon age of 443 ± 4 Ma (Lyons and others, 1986). Because the Boil Mountain Complex did not undergo that metamorphism, Trzcienski and others (1992a) argued that the Boil Mountain was not juxtaposed to the Chain Lakes block until after the intrusion of the Highlandcroft Plutonic Suite (Late Ordovician), which casts doubt on the suggested Penobscottian age of obduction of the Boil Mountain Complex.

At least two belts of Middle to Late Ordovician subduction-related volcanic rocks are within the Brompton-Cameron and Central Maine terranes (Fig. 4). The western belt, the Hawley arc, is in the Brompton-Cameron terrane and includes the Hawley Formation in Massachusetts, the Barnard Volcanic Member of the Missisquoi Formation in Vermont, and the Ascott-Weedon sequence in Quebec. The eastern, more extensive belt is the Ammonoosuc arc (including the Oliverian and Highlandcroft Plutonic Suites) and is more or less coextensive with the Bronson Hill anticlinorium in the western margin of the Central Maine terrane. The arcs must have been built over an east-dipping subduction zone or zones as noted in the previous section. The collision of the arcs with Laurentia during the closing of the Iapetus Ocean was the proximate cause of the Taconian orogeny. Osberg (1978), Hall and Robinson (1982), and Stanley and Ratcliffe (1985) thought that the Ammonoosuc arc developed on non-Laurentian basement.

The arcs developed during subduction, perhaps related in time to the high-pressure metamorphism in the Hazens Notch slice of the Brompton-Cameron terrane in northern Vermont (Fig. 4) and prior to the collision with Laurentia that resulted in the emplacement of the Taconic allochthon. Tucker and Robinson (1990) demonstrated that the igneous activity of the Ammonoosuc arc dates from about 454 to 442 Ma (which agrees well with the 453 to 443 Ma ages of the Highlandcroft Plutonic Suite), that the arc was built partly or entirely on continental crust, and that the igneous activity was coeval with or slightly younger than the Caradocian emplacement of lowest slice (Giddings Brook slice) of the Taconic allochthon (Fig. 4). Thus, although the Ammonoosuc arc at the surface is in the Central Maine terrane, it formed over a subduction zone active under the Brompton-Cameron terrane after amalgamation with the Central Maine terrane. The magmas may have originated in the Brompton-Cameron terrane. It is not clear at present whether there were one or two Ordovician subduction zones (the Hawley arc could be a distal part of the Ammonoosuc arc) or two sequential arcs over one subduction zone. Thompson and others (1993) and Tucker (*in* Tucker and Robinson, 1990) argued that the Hawley arc was a separate arc and the initial colliding arc. If the Hawley arc proves to be older than the Ammonoosuc, that could resolve the dilemma of the Ammonoosuc possibly being younger than the emplacement of the earliest transported slice of the Taconic allochthon.

It is important to remember that the Brompton-Cameron and Central Maine terranes are identified and distinguished on the basis of their pre-Silurian rocks, which make up less than half of the surface area of the former, and much less than half of the surface area of the Central Maine terrane. The pre-Silurian rocks of the Central Maine terrane include, among others, the Boil Mountain Complex–Hurricane Mountain melange belt, the rocks of the Shin Pond area containing the Celtic fauna, and, as interpreted here, the Miramichi belt, which also contains Lower and Middle Ordovician Celtic fauna. It is likely that the pre-Silurian rocks of the Central Maine terrane are an assembly of older continental and oceanic island terrane fragments brought together or amalgamated in pre-Silurian time.

Some observations critical to understanding the pre-Silurian assembly come from the Canadian Appalachians. Williams (1978, 1979) divided central Newfoundland into the Dunnage zone, consisting of vestiges of the Iapetus Ocean, and the Gander zone, consisting of the eastern margin of the Iapetus Ocean. Williams and Hatcher (1983) considered these zones to be suspect terranes, but Williams and others (1988) reverted to the term zone because of the controversial nature of these boundaries. Williams and others (1988) also divided the Dunnage zone into an eastern Notre Dame and western Exploits subzone along the Red Indian line. The Notre Dame subzone is characterized by a widespread sub-Silurian unconformity and, at least along its northwestern margin, northwest-directed structures of Taconian age (Williams and ohers, 1988; Colman-Sadd and others, 1992). The Exploits subzone is characterized by a Lower and Middle Ordovician Celtic brachiopod fauna (Neuman, 1984; Williams and others, 1988), a sedimentary record that is continuous from the Ordovician into the Silurian (Williams and others, 1988), and southeast-directed thrusting of late Arenigian age in the Penobscottian orogeny (Colman-Sadd and others, 1992). These relations suggest that in the United States the Brompton-Cameron terrane is correlative with the Notre Dame subzone and the Central Maine terrane is correlative with the Exploits subzone.

The Gander zone in Newfoundland consists of a monotonous sedimentary succession characterized by quartz-rich sandstone of apparent continental provenance and the paucity of volcanic rocks (Williams and others, 1988; Colman-Sadd and others, 1992). Sedimentation began at least as long ago as the latest Proterozoic and continued into the late Arenigian. Southeast-directed thrust faults carry rocks of the Exploits subzone (including ophiolite) over the Gander zone (locally exposed in windows through the Exploits). Thrust movement is dated at about 474 Ma (late Arenigian) and is interpreted to be an event in the Penobscottian orogeny (Colman-Sadd and others, 1992). A sedimentary linkage indicates that the Exploits subzone and Gander zone were juxtaposed by the Llandeilian (van Staal and Williams, 1991).

Recent work in the Miramichi belt of northern New Brunswick (van Staal and Fyffe, 1991; van Staal and others, 1990; van Staal and Williams, 1991) showed that rocks of the lower part of the Tetagouche Group (bimodal but dominantly felsic volcanic and volcaniclastic rocks) stratigraphically overlie the Cambrian and Lower Ordovician Miramichi Group (quartzite and quartz wacke–shale rhythmite). Calcarenite, calcareous siltstone, and conglomerate, which interfinger with tuffaceous sandstone and dark shale at the base of the Tetagouche volcanic rocks, contain middle Arenigian to earliest Llanvirnian brachiopods (Celtic fauna) and conodonts (Neuman, 1984; van Staal and Fyffe, 1991). To the northwest, across a thrust-fault zone that includes a narrow blueschist belt traceable for 70 km along strike, volcanic rocks of the Fournier Group (correlative with the Tetagouche

Group) overlie a Middle Ordovician ophiolitic substrata (van Staal and others, 1990; van Staal and Fyffe, 1991; van Staal and Williams, 1991). Middle Ordovician volcanic and sedimentary rocks thus span an ocean-continent boundary that was later imbricated (van Staal and Williams, 1991). The bimodal volcanic rocks of the Tetagouche Group were interpreted by van Staal and others (1990) to be a rifting suite (472–466 Ma) formed during the early stages of the opening of a back-arc basin eastward of the Taconian magmatic arc or arcs. The 464 to 460 Ma ophiolitic substrata under the Fournier Group are slightly younger than the bimodal rocks of the Tetagouche and were interpreted by van Staal and others (1990) to be oceanic crust formed during the more advanced stages of back-arc spreading. The closing of this back-arc basin, which was called Iapetus II by van der Pluijm and van Staal (1988), on a northwest-dipping subduction zone produced the blueschists, and continued thrusting placed the Fournier Group over the Tetagouche Group (van Staal and others, 1990). The age of blueschist is about 447 Ma and the thrusting is thought to have continued into the Silurian (van Staal and others, 1990).

Evidence for a Penobscottian orogeny in the Miramichi belt is therefore circumstantial. Van Staal and Williams (1991) and van Staal and Fyffe (1991) pointed out that Middle Ordovician volcanic rocks of the Miramichi belt have been assigned to the Dunnage zone (Exploits subzone), but that they are better interpreted as an overstep sequence on the Dunnage and Gander zones. A true distinction between the two zones, in their interpretation, can only be made where Lower Ordovician oceanic crust is preserved and juxtaposed against rocks of the Gander zone, as is the case in Newfoundland.

Important aspects of the above review of the Canadian Appalachians in terms of our understanding of the U.S. Appalachians are given below. (1) The Gander zone rocks of the Miramichi belt were attached to the Central Maine terrane by the Middle Ordovician. (2) Evidence from Newfoundland suggests that the east-directed forces were operative along the eastern margin of a terrane correlated with the Central Maine terrane as long ago as late Arenigian. (3) A back-arc basin well to the east of the Taconian magmatic arc or arcs opened in the Middle Ordovician in the Miramichi belt. (4) The back-arc basin closed on a west-dipping subduction zone that generated blueschist about 447 Ma. (5) The closing of the back-arc basin led to collision of the main mass of the Central Maine terrane and rocks of the Gander zone (terrane), and eastward thrusting of the blueschists over the Gander terrane and some of the Middle Ordovician volcanic overstep sequence. (6) The east-directed deformation continued into the Silurian, resulting in the generation of Late Silurian to earliest Devonian (417 to 402 Ma U-Pb ages; Bevier and Whalen, 1990) granitoid plutons in the Miramichi belt in New Brunswick.

These relations raise problems, as yet unresolved, that will become better defined as the discussion continues. Neuman (1984) noted the similarity between the Grand Pitch Formation of the Shin Pond, Maine, area and the Miramichi Group in New Brunswick. If that correlation is correct and if the Miramichi Group and hence the Grand Pitch Formation are part of the Gander zone, we do not yet have an adequate understanding of the pre-Silurian terrane assembly in New England. Van Staal and Fyffe (1991) included the St. Croix terrane of coastal Maine (described below) in the Gander zone, and that correlation is accepted here. However, the St. Croix terrane as discussed below was amalgamated with the Ellsworth-Mascarene terrane in the Silurian and the faunal assemblage in the Ellsworth-Mascarene terrane was distinctly non-North American into the earliest Devonian. Perhaps the Gander zone, as part of proto-Gondwana, was fragmented in the early Paleozoic and different pieces of it docked with North America at different times.

Extensive Silurian through Lower Devonian sedimentary (some fossiliferous) and volcanic rocks, up to several kilometers thick and metamorphosed to various degrees, constitute the overlap sequences of the Brompton-Cameron and Central Maine terranes. It is these sequences in the internides of the northern Appalachians and their absence to the south that makes the contrast with the internides of the central and southern Appalachians so striking. These sequences include the calciferous schists (Thompson and others, 1993) and associated rocks of the Connecticut Valley trough, a much thinner sequence in the Bronson Hill anticlinorium, and a tremendous volume of strata in the Central Maine trough (Fig. 1). The base of the sequence northwest of the Bronson Hill anticlinorium in northern New Hampshire, Quebec, and Maine includes well-stratified pale green or greenish-gray phyllite, tuffaceous metasandstone, and a suite of bimodal volcanic rocks of the Frontenac Formation of probable Silurian age (Fig. 4). These rocks are thought to have been deposited in an incipient rift (Stewart, 1989; Moench, 1990; Moench and others, 1992).

The east boundary of the Connecticut Valley trough and the Bronson Hill anticlinorium toward the north is the Monroe fault (Hatch, 1988), but toward the south middle Paleozoic rocks of the Connecticut Valley trough are juxtaposed against middle Paleozoic rocks of the Bronson Hill anticlinorium. The nature of that contact is equivocal (Thompson and others, 1993), but a recent consensus is that the Bronson Hill rocks are above an erosional unconformity on the Connecticut Valley rocks (Trzcienski and others, 1992b). Mesozoic movement has been identified on a number of en echelon normal faults near the eastern border of the Connecticut Valley trough, including the Monroe fault. These are roughly on strike with the eastern border faults of the Hartford Mesozoic basin to the south in Massachusetts and Connecticut (Figs. 1 and 4).

The middle Paleozoic strata are thin and unconformably above the volcanic-arc rocks in the Bronson Hill anticlinorium, but thicken markedly across a synsedimentary tectonic hinge line to the east in the Central Maine trough (Figs. 2 and 4) (Hatch and others, 1983; Moench and Pankiwskyj, 1988). As noted by Stewart (1989), this Silurian tectonic hinge line is essentially at the same position as the hypothesized southeastern limit of Grenvillian crust at depth (Fig. 2). Stewart (1989) estimated that the Central Maine trough may once have contained as much as 2200 km^3/km of strike length of strata, mostly of Silurian and Devonian age. The Central Maine trough also contains some of the highest grade metamorphic rocks in the Appalachians (sillimanite–K-feldspar zone in pelitic rocks). Plutons, many in the form of large gently dipping sheets, are abundant. Plutons are mostly of Devonian age, but some may be latest Silurian, and others are Carboniferous, in the range 330 to 315 Ma. Osberg and others (1989) thought that the Carboniferous plutons were late Acadian. Fossiliferous rocks as young as Emsian (late Early Devonian) were metamorphosed during the Acadian orogeny. Small successor basins in Maine (Fig. 4) contain post-Acadian plant-bearing strata of Middle Devonian age.

As succinctly described by Thompson and others (1993), "The Bronson Hill zone (anticlinorium) is characterized by westward-verging fold and thrust nappes, many of which carried hotter rocks westward over colder ones during the Acadian orogeny. Continued deformation folded the axial surfaces of the nappes and culminated in the rise of a roughly en echelon series of gneiss domes constituting the Bronson Hill anticlinorium . . . Nappes near the eastern margin of the Central New England zone, on the other hand, have been regarded as eastward-verging. . . ." The grade of Acadian metamorphism drops off sharply to as low as chlorite grade westward into the Connecticut Valley trough from the Bronson Hill anticlinorium. Where Mesozoic normal faults such as the Ammonoosuc fault (Fig. 4) are present, the drop in grade is across this fault. The grade rises again to the west to grades as high as sillimanite-muscovite, which overprints Taconian metamorphism and then diminishes farther west.

The identity of the basement, at least to mid-crustal level, under the Central Maine trough and the Bronson Hill anticlinorium is uncertain, but, as indicated above, a number of workers have suggested that it is non-Laurentian and distinct from the proto-Gondwana terranes farther southeast. Among them are Osberg (1978), Zen (1983), Williams and Hatcher (1983), Zartman (1988), and Spencer and others (1989). The presence of 613 Ma rocks in the core of the Pelham dome, Massachusetts (refer to Figs. 4 and 6), raises the possibility that proto-Gondwanan basement may extend as far west as the southern part of the Bronson Hill anticlinorium. This will be discussed more fully in later sections. Ayuso and others (1988) have shown through Pb and O isotopic studies of granite plutons in Maine that Grenvillian crust underlies the Chain Lakes block at depth. Harrison and others (1987) also concluded that Grenvillian crust underlies the western part of the Central Maine trough at depth, based upon the age of inherited zircons in Acadian granitoids. More recent work (Ayuso and Bevier, 1991) suggests that the source region for plutons in the Central Maine trough in Maine may be a mixture of Grenvillian and Avalonian (proto-Gondwanan) basements.

As noted above, Acadian structures verge westward along the Bronson Hill anticlinorium and eastward toward the eastern side of the Central Maine terrane. In Massachusetts and southern New Hampshire in the area of E1, where strain is extreme, west-verging and east-verging nappes may be the result of upward flow in the intervening area of high-grade and migmatitic rocks caught in a tectonic vise during the final stages of the Acadian orogeny (Thompson and others, 1993). In the area of the Quebec–Maine–Gulf of Maine transect, where the metamorphic grade is lower and the strain is less, Acadian structures in Silurian rocks along the southeastern side of the Central Maine terrane also verge southeastward. If older southeast-directed structures, such as the thrusting in the Miramichi belt of northern New Brunswick or the Exploits subzone of Newfoundland, existed in New England, the evidence has been obscured by Acadian and later events.

The Norumbega–Hackmatack Pond–Black Pond–Cremation Hill fault system appears to be a major structural break marking the southeast side of the Central Maine terrane. To the southeast of that fault system, the bedrock is very faulted. Except for Pennsylvanian fossils in an areally restricted coal near Worcester, Massachusetts, Silurian fossils near Newburyport, Massachusetts, Late Proterozoic, Cambrian, and Pennsylvanian fossils in the Esmond-Dedham block (see below), and a number of localities in coastal eastern Maine (Billings and Lyttle, 1981), all of this broken up territory is nonfossiliferous. Correlations are by lithology and stratigraphic succession constrained by isotopic ages of plutonic and volcanic rocks. There are about as many ways of making tectonic subdivisions of this area southeast of the Cremation Hill–Norumbega fault system and relating the resulting pieces to each other as there are geologists who have tried to do so. My interpretation of the assembly of terranes in coastal and offshore New England is shown in Figures 5 and 6 along with selected plutons that are mentioned in the discussion that follows or that stitch the boundaries between the terranes.

The thesis that I develop in the following discussion is that the Ellsworth-Mascarene, St. Croix, Nashoba, and Merrimack-Harpswell terranes were amalgamated on northwest-directed thrust faults in the Silurian, and that the assembled package, here called the Merrimack-Mascarene composite terrane, was accreted to the Central Maine terrane on a southeast-directed thrust-fault system during the Acadian orogeny. The collision of that assembled package with the Central Maine terrane may have been the cause of the Acadian orogeny. The docking of the Avalon terrane with North America on southeast-directed thrust faults (Massachusetts and Connecticut) was presumably the cause of the Alleghanian orogeny in New England. The outboard Meguma terrane was probably amalgamated with the Avalon terrane prior to the Alleghanian orogeny, as determined from the Early Carboniferous overlap sequence in Nova Scotia.

Stewart and others (1991) portrayed the Avalon terrane boundary in the Gulf of Maine as the northwest-directed Gulf of Maine thrust fault. This fault is in a location where there is no seismic control and the dip of the fault is not constrained. To the northeast, in coastal New Brunswick, the exact position of the Avalon terrane boundary is controversial and complicated by the presence of the very large Devonian Saint George–Mount Douglas batholith (Currie, 1988). Many faults are steeply dipping and strike-slip motion is deduced for many. Currie concluded that the Avalon terrane is uplifted relative to rocks to the north, and that Avalonian crust may extend far to the northwest in the subsurface, possibly as far as the Iapetus suture. A northwest-directed Gulf of Maine thrust fault carrying the Avalon terrane over terranes northwest of it is difficult to reconcile with the late Paleozoic history deduced in southern New England. I have not indicated any sense of movement on the Gulf of Maine fault in Figure 6. The accretion of the Avalon terrane appears to have been diachronous: Acadian in Newfoundland (Dallmeyer and others, 1981), Acadian in New Brunswick, if the Saint George–Mount Douglas batholith intrudes the terrane boundary, and Alleghanian in southern New England.

Stewart and others (1991) interpreted the Central Maine terrane to be thrust southeastward over the Falmouth-Brunswick block (their Nashoba-Casco-Miramichi composite terrane) on the Hackmatack Pond fault (Fig. 6). This fault can be seen in the seismic record to dip about 25° northwest and can be followed in the records to depths of 10 km, albeit with some complications. A comparable fault, the west-dipping Black Pond thrust fault (Fig. 6) (Peper and Pease, 1976) and its continuation to the south as the Cremation Hill fault, was suggested by Zen (1989) to bound the east side of the Central Maine terrane in Massachusetts, and that interpretation is shown in Figures 3 and 6. Zen, however, portrayed these thrusts as east dipping. In the intervening region in New Hampshire, the Central Maine terrane is bounded on the southeast by the steeply northwest dipping or vertical Campbell Hill–Nonesuch River fault, along which the Massabesic block is upthrown on the southeast side (Fig. 6).

Rocks exposed in the core of the Pelham and Willimantic domes (Fig. 6) are interpreted to be inliers of the Avalon terrane exposed in windows through structurally higher allochthonous terranes. The Massabesic and Falmouth-Brunswick blocks are also thought to be fragments of another terrane, possibly the Ellsworth-Mascarene terrane, along vertical or steeply southeast dipping faults that show either right-lateral strike-slip or normal movement, down to the southeast. These proto-Gondwanan inliers are discussed in a separate section following the section on the Avalon terrane.

Merrimack-Harpswell terrane. A belt of flysch and calcareous flysch of a wide range of metamorphic grades crops out from south-central Connecticut to southeastern Maine. The belt includes volcanic rocks locally (generally minor) as well as minor carbonate and quartz arenite. Some pelitic rocks are sulfidic. This belt is here designated the Merrimack-Harpswell terrane (Fig. 6). Major stratigraphic units include the Hebron Formation and Scotland Schist Member of the Oakdale Formation in Connecticut, the Paxton Group (as used by Pease, 1989) and Oakdale Formation in Massachusetts, the Merrimack Group in northern Massachusetts, New Hampshire, and southwestern Maine, and the Saco-Harpswell sequence (Hussey, 1988) in Maine. The basement upon which these strata were deposited is nowhere exposed, but Pb isotopic studies of plutons in Maine show that the underlying crust is a sialic non-Grenvillian basement (Ayuso and Bevier, 1991). No fossils have been recovered from the Merrimack-Harpswell terrane.

The terrane is bounded on the northwest by the fault system that extends from the Cremation Hill fault to the Norumbega fault (Fig. 6). The Black Pond fault is an Acadian east-directed thrust, the Wekepeke and Flint Hill faults are probably Mesozoic faults, down to the east (Rodgers, 1970; Goldsmith, 1991), and the Nonesuch River and Norumbega faults have complicated histories that include major right-lateral, strike-slip movement (Hussey and Newberg, 1978; Wones and Thompson, 1979; Hubbard and others, 1991). The Merrimack-Harpswell terrane is bounded on the southeast by a fault system that includes the west-dipping Clinton-Newbury fault and the west-directed Sennebec Pond thrust fault. The assertion that the contact between the Merrimack-Harpswell and Nashoba terranes in Connecticut is a fault that is an extension of the Clinton-Newbury fault of Massachusetts (Fig. 6) and is located west of the outcrop belt of the Tatnic Hill Formation follows the interpretation of Barosh (1982), Pease (1982), Wintsch and Aleinikoff (1987), Gromet (1989), and Goldsmith (1991). Most published quadrangle maps along this boundary in Connecticut portray the boundary as a stratigraphic contact (see Rodgers, 1985). Zen (1989) placed the terrane boundary along the Tatnic Hill fault east of the Tatnic Hill Formation. In part because of this, the lithologic assemblages of our adjacent terranes are different and I have used different terrane names from those used by Zen (1989).

To the extent that the stratified units of the terrane are correlative, the stratified rocks and their deformation are pre-Middle Ordovician. Deformed and metamorphosed strata of the Merrimack Group are intruded by the nonfoliated Exeter Diorite (473 ± 37 Ma, Rb-Sr whole-rock age; Hon and others, 1986). Rb-Sr ages, interpreted as resetting ages of stratified rocks in the Merrimack-Harpswell terrane from Massachusetts to Maine, range from 540 to 480 Ma, suggesting that the protoliths are of Cambrian and Ordovician or older age (Olszewski and Gaudette, 1988).

Metamorphism of the Merrimack-Harpswell terrane produced assemblages ranging from the greenschist to upper amphibolite facies. Parts of the Merrimack Group adjacent to the Wekepeke fault are chlorite grade, but grades in the terrane are high as sillimanite-orthoclase along the central Maine coast and in southern Connecticut. A Silurian thermal event produced calc-alkaline magmas from Massachusetts to Maine in the Merrimack-Harpswell terrane as well as in the adjacent Nashoba terrane. The age of the higher grades of regional metamorphism is thought to be Acadian superimposed on the earlier episodes (Eusden and Barreiro, 1988; Guidotti, 1989; Osberg and others, 1989). The nonfoliated Exeter Diorite is not in an area of higher grade superimposed Acadian metamorphism.

The plutonic history of the Merrimack-Harpswell terrane in Massachusetts and adjacent states is long and complex (Wones and Goldsmith, 1991). An early cycle of calc-alkaline magmatism changed over time from mafic and intermediate mantle-derived magmas to felsic I-type magmas derived from continental

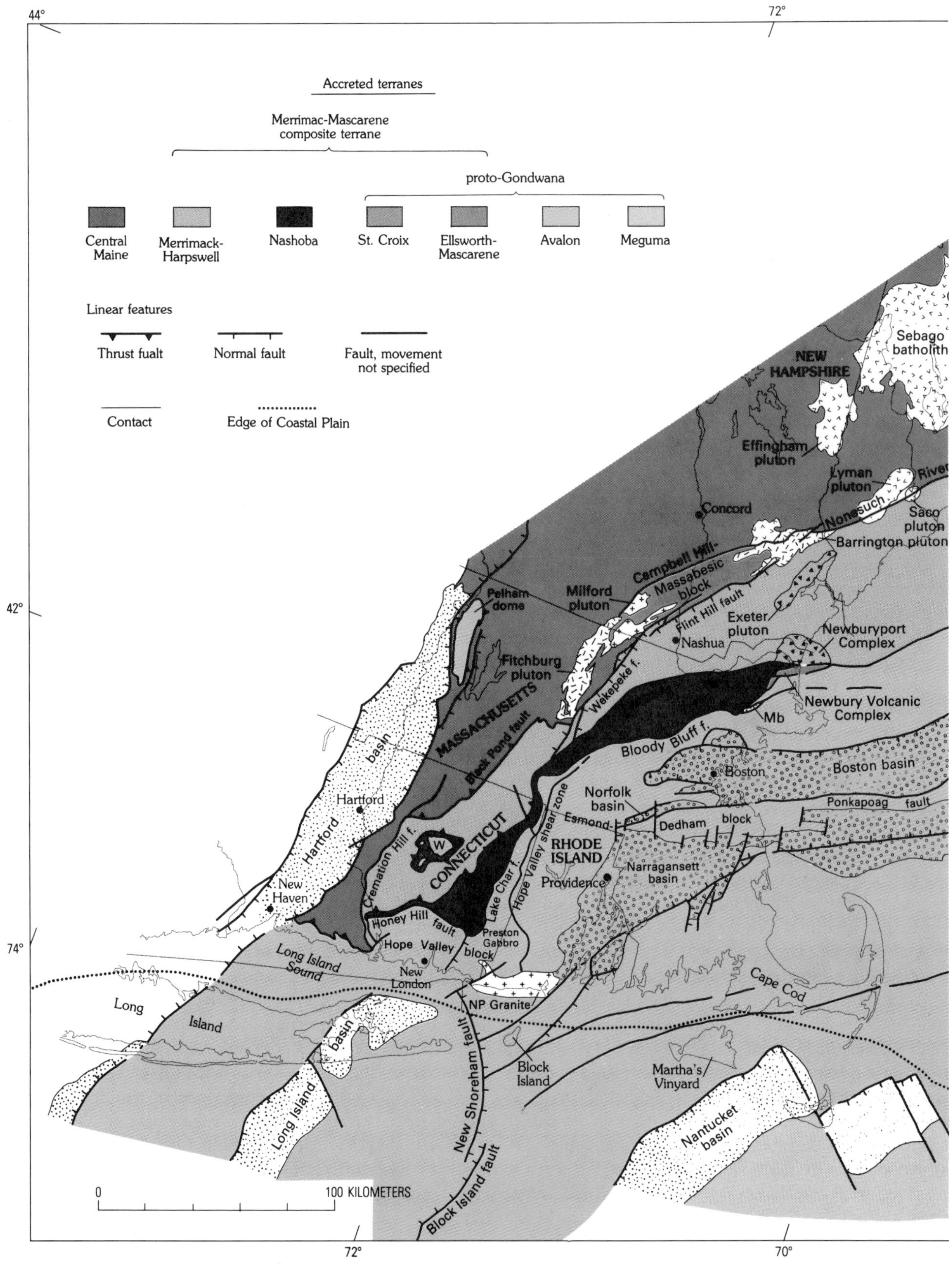

Figure 6. Terranes and other geological features of eastern New England, Long Island, and the Gulf of Maine. Date from Hopeck (1991), Hutchinson and others (1985), Lyons and others (1993), McMaster and others (1980), Osberg and others (1985), Rankin and others (1993b), Rodgers (1985), Thompson

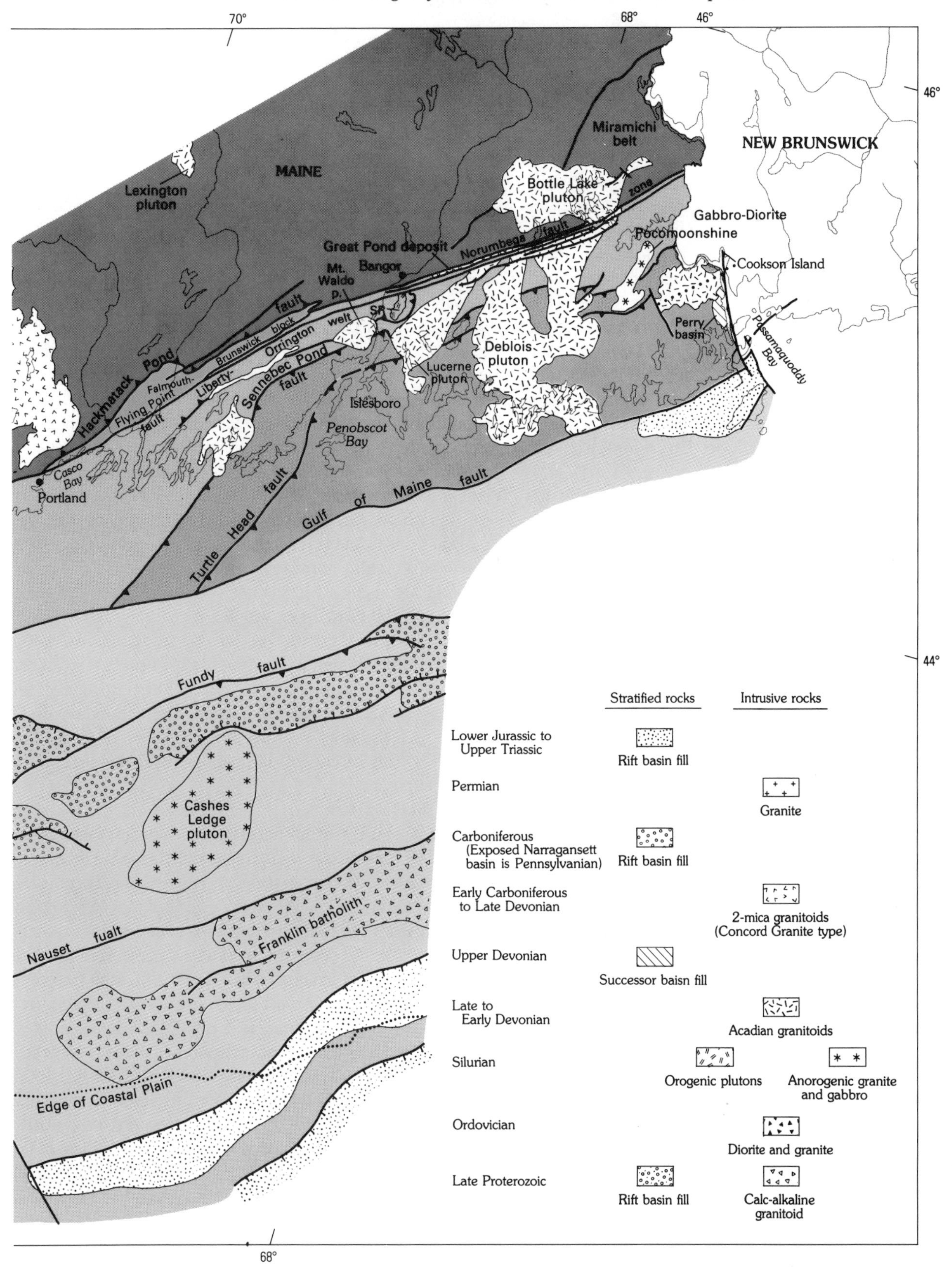

and others (1993), Stewart and others (1991), Zen and others (1983). Abbreviations used: Mb, Middleton basin; NP, Narragansett Pier; SR, Strickland Ridge pluton; W, Willimantic dome. Plutons shown are those that intrude terrane boundaries and those that are referred to by name in the text.

crust. The Exeter Diorite and Newburyport Complex (455 ± 15 Ma, U-Pb zircon age by Zartman and Naylor, 1984) were succeeded in adjacent parts of the terrane by sheet-like gneissic bodies of the Silurian Ayer Granite (433 ± 5 Ma, U-Pb zircon age by Zartman and Naylor, 1984). Hepburn and others (1987) interpreted the assemblage in Massachusetts to represent a volcanic-plutonic arc developed near or on a continental margin and to be related to a convergent plate boundary. Younger two-mica peraluminous granites such as the Chelsmford Granite in Massachusetts (389 ± 5 Ma, Pb/Pb zircon age) and the Cantebury Gneiss in Connecticut (395 ± 10 Ma, U-Pb zircon age) were emplaced during the Acadian orogeny (Zartman and Naylor, 1984). Thus the Merrimack-Harpswell terrane must have been adjacent to the Central Maine terrane at least by the late Early Devonian. The Barrington pluton (Fig. 6) intrudes the Campbell Hill–Nonesuch River fault, but has also been affected by minor motion in zones parallel to and along the boundary (Eusden and Barreiro, 1988). The Barrington pluton has a 364 Ma Pb/Pb monazite age (Eusden and Barreiro, 1988).

One more point needs to be considered concerning the Merrimack-Harpswell terrane. Several workers (Zen, 1989; Stewart and others, 1991; Goldsmith and Secor, 1993; Rankin and others, 1993a) included rocks of the Miramichi belt, north of the Norumbega fault in eastern Maine and adjacent New Brunswick, with rocks of the Merrimack-Harpswell terrane and other rocks of New England. As stated earlier, I have not done so here because of the presence of a Celtic fauna in the Miramichi of northern New Brunswick, the abundance of volcanic rocks in the Miramichi in contrast to the Merrimack-Harpswell terrane as defined here, and because the overstepping Matapedia cover sequence in New Brunswick indicates that the Miramichi was part of North America by the Late Ordovician to Early Silurian (Fyffe and Fricker, 1987). If the Miramichi were an extension of the Merrimack-Harpswell terrane, it would be difficult to reconcile the presence of Late Ordovician to Early Silurian North American fossils in Matapedia with the Late Silurian to Early Devonian Atlantic fossils in the St. Croix and Ellsworth-Mascarene terranes and the Silurian (prior to the intrusion of the Pocomoonshine Gabbro-Diorite) assembly of the Merrimack-Harpswell and St. Croix terranes (amplified below). As discussed earlier, however, the Miramichi belt of New Brunswick has now been interpreted to include rocks of both the Exploits subzone (Central Maine terrane?) and the Gander zone. The Merrimack-Harpswell terrane, as defined in this chapter, has many of the characteristics of the Gander zone. It could have been part of that zone, perhaps like the St. Croix terrane, that was separated from the rest of the Gander zone in the early Paleozoic and later amalgamated as part of the Merrimack-Mascarene composite terrane.

Nashoba terrane. In Massachusetts and Connecticut, rocks southeast of the Clinton-Newbury fault (Fig. 6) have a greater volcanic component than those of the Merrimack-Harpswell terrane and are of higher metamorphic grade. The rocks may belong to a separate terrane called the Nashoba block by Thompson and others (1993) and Hepburn and others (1987), the Putnam-Nashoba zone by Gromet (1989), and the Nashoba terrane by Rankin and others (1993a) (Fig. 6). The terrane in its type area in Massachusetts consists of a thick pile of mafic volcanic and volcanogenic sedimentary rocks along with some pelitic and calcareous rocks metamorphosed and migmatized under amphibolite facies conditions (Hepburn and others, 1987). The mafic volcanic rocks are primarily on the southeast side of the terrane (the Marlboro Formation) and are metamorphosed basalts, including some high-alumina types that are most compatible with an arc or marginal basin setting. The basalts have Nd model ages that suggest that they are older than 450 Ma (DiNitto and others, 1984). The oldest rocks are felsic to intermediate metavolcanic wackes of the Fish Brook Gneiss (730 ± 26 Ma U-Pb zircon age by Olszewski, 1980), but it is possible that the Fish Brook is unconformable beneath other units (Hepburn and others, 1987). Hepburn and others pointed out the similarity in composition, position within the orogen, and permissible range of ages between the Marlboro Formation and the Chopawamsic Formation of Virginia (Chopawamsic arc of Fig. 1).

There is not agreement as to what constitutes the Nashoba terrane in Connecticut (except that the mafic volcanic Quinebaug Formation is roughly correlative with the Marlboro Formation). Part of the argument is whether less volcanic-rich stratified rocks of the Tatnic Hill Formation, which is structurally above (stratigraphically above according to Dixon and Lundgren, 1968; Rodgers, 1985) and west of the Quinebaug Formation, correlate with rocks of the Nashoba terrane west of the outcrop belt of the Marlboro, or whether they are in a different terrane and separated from the Quinebaug by a terrane boundary fault as favored by Zen (1989). The Tatnic Hill Formation is included here in the Nashoba terrane.

The Nashoba terrane is bounded on the east and southeast by a fault system that from northeast to southwest includes the Bloody Bluff, Lake Char, and Honey Hill (Fig. 6). The Bloody Bluff fault dips moderately to steeply northwest. Dips become less steep to the southwest and the Honey Hill fault dips gently north. The faults have been interpreted as west-over-east thrust faults, but Goldstein (1989) showed that different segments of the fault system have different histories. The Bloody Bluff fault bears evidence for the earliest movement. Goldstein (1989) summarized the evidence for Late Silurian to Early Devonian left-oblique thrusting. Some later normal movement down to the northwest is indicated. Zones of mylonitization as thick as 3 km mark the Lake Char and Honey Hill faults. Kinematic indicators show that the Nashoba terrane moved down and to the northwest relative to the Hope Valley block of the Avalon terrane regardless of the strike of the faults (Goldstein, 1989). Evidence for earlier thrust(?) movement is not preserved along the Lake Char and Honey Hill faults. The Lake Char fault is interpreted to reappear as the outer frame of the Willimantic window so that the Nashoba terrane is an intermediate-level tectonic sheet through which the Avalon terrane is exposed in an inner window. The

Nashoba terrane is structurally truncated in south-central Connecticut and offshore northeastern Massachusetts.

Plutonism in the Nashoba terrane is most thoroughly studied in Massachusetts and ranges in age from Ordovician to Early Devonian. It began with prekinematic or synkinematic peraluminous granite such as the early phase of the Andover Granite, followed by postkinematic calc-alkaline diorite and tonalite, such as the Sharpners Pond Diorite, succeeded by less-mafic tonalite, granodiorite, and monzogranite, and ended with peraluminous granite and pegmatite represented by rocks also mapped as Andover Granite (Wones and Goldsmith, 1991). The early Andover Granite is about 450 Ma, the Sharpners Pond Diorite has a U-Pb zircon age of 430 ± 5 Ma, and the younger Andover Granite is about 408 Ma (Zartman and Naylor, 1984). The Preston Gabbro within the southeastern angle of the Nashoba terrane in Connecticut (Fig. 6) is 424 ± 5 Ma (U-Pb zircon age by Zartman and Naylor, 1984), and is thus part of the Silurian magmatic cycle that distinguishes the Nashoba and Merrimack-Harpswell terranes from others. Wones and Goldsmith (1991) characterized both the older and younger phases of the Andover Granite as S-type granite derived from deep-seated sedimentary material in a high-temperature regime. They also noted that the older phase of the Andover intrudes metamorphosed slope-facies sediments and off-arc volcanic rocks of the Nashoba and Marlboro Formations and is not represented in the Merrimack-Harpswell terrane to the west. The Silurian cycle, however, has characteristics of lower pressure I-type intrusives that formed at convergent plate boundaries and, because of the isotopic heterogeneity, is intepreted to have assimilated crustal material (Hill and others, 1984; Wones and Goldsmith, 1991).

Proto-Gondwana terranes. Terranes of New England southeast of the Honey Hill–Sennebec Pond fault system include a number of localities where Early Cambrian to earliest Devonian (Gedinnian) fossils belong to an Atlantic faunal province distinct from the North American faunal province of coeval rocks in the Laurentian Appalachians and from the Celtic faunal province of a more restricted age range in the Central Maine terrane (Neuman and others, 1989). The Late Proterozoic fossils are not province specific. Neuman and others (1989) noted that the Atlantic fauna of New England and the Maritime Province of Canada have affinities closer to the peri-Gondwana terranes than with Baltica. Those affinities constitute some of the evidence for a Tornquist's Sea between Baltica and Western Europe (Cocks and Fortey, 1982) that opened along a Late Proterozoic and Early Cambrian rift system that may have joined the Iapetan rift system at a triple junction in the present North Sea (Rankin and others, 1988).

Atlantic fauna are found in four terranes that in this chapter constitute the Gondwana terranes of New England. Two terranes, the St. Croix and Ellsworth-Mascarene, crop out in Maine and are interpreted to be cut off by the Gulf of Maine fault, outboard of which is the Avalon terrane (Fig. 6). That terrane is exposed in coastal southeastern New England as well as in inliers northwest of the Clinton-Newbury fault. The Meguma terrane, which is the fourth proto-Gondwana terrane, is not exposed in the United States but is interpreted to underlie the outer part of the Gulf of Maine.

St. Croix terrane. The older rocks of the St. Croix terrane are characterized by abundant black pelite and include conglomerate, quartzite, impure carbonate, and minor bimodal volcanic rock. Fossils are rare, but this older package of rocks is thought to be of Cambrian and Middle Ordovician or possibly Late Ordovician age (Fyffe and others, 1988; Neuman, 1973). Graptolites on Cookson Island, New Brunswick, north of the head of Passamaquoddy Bay (Fig. 6) are of Tremadocian (Early Ordovician) age and of the Atlantic faunal province (Fyffe and Riva, 1990). Brachiopods in staurolite-grade rocks in the St. Croix terrane west of Penobscot Bay are probably of Arenigian to Caradocian age (Neuman, 1973). Because of the age uncertainty, they could not be assigned to a province. The St. Croix terrane is interpreted to consist of a continental slope-rise sequence that passes upward into a marginal basin formed on rifted continental crust (Fyffe and others, 1988; Stewart and others, 1991). The continental basement is thought to be proto-Gondwanan and slope-rise sediments to have been deposited on its northwestern side (Stewart and others, 1991). The St. Croix terrane may be equivalent to the Gander zone of Newfoundland (van Staal and Fyffe, 1991), but with a different accretionary history.

The St. Croix terrane is bounded on the southeast by the Turtle Head fault, a northwest-directed thrust fault with a component of right-lateral strike-slip movement (Fig. 6) (Stewart and others, 1991). Late Silurian roundstone conglomerates deposited on both sides of the Turtle Head fault and unconformably above the Cambrian and Ordovician strata suggest that the St. Croix terrane was amalgamated to the Ellsworth-Mascarene terrane by the Late Silurian (Gates, 1989; Stewart and others, 1991). Rocks of the southwestern part of the St. Croix terrane west of Penobscot Bay were metamorphosed to grades as high as the upper amphibolite facies prior to the final movement on the Turtle Head fault; regional metamorphism in the Ellsworth-Mascarene terrane southeast of the Turtle Head fault did not exceed greenschist facies (Guidotti, 1989; Stewart and others, 1991). The age of metamorphism of the high-grade rocks west of Penobscot Bay has generally been assumed to be Acadian (Guidotti, 1989). D. P. West (1992, written commun.) has indicated that on the basis of hornblende cooling ages (about 414 Ma), the metamorphism is Silurian. That is consistent with the observation that the contact aureole of the Mt. Waldo pluton (about 382 Ma; Hogan and Sinha, 1989) is superimposed on the regional metamorphism (albeit lower grade) of the St. Croix terrane (D. B. Stewart, 1992, personal commun.). Note that with the establishment of a Silurian age for the metamorphism of the St. Croix terrane, it can be said that no part of the proto-Gondwana terranes in New England was significantly affected by Acadian regional as opposed to contact metamorphism.

The St. Croix terrane appears to have been amalgamated to

the Merrimack-Harpswell terrane in the Silurian. The Pocomoonshine Gabbro-Diorite is a postkinematic complex pluton that intruded in two separate episodes across the boundary (Sennebec Pond fault) between the Merrimack-Harpswell and St. Croix terrane (Fig. 6). The northern mafic part of the pluton, which is within the Merrimack-Harpswell terrane, has an $^{40}Ar/^{39}Ar$ hornblende crystallization age of 422.7 ± 3 Ma (West and others, 1992a). An $^{40}Ar/^{39}Ar$ muscovite plateau age of 414.2 ± 3 Ma from the adjacent contact aureole in regional greenschist facies rocks of the Merrimack-Harpswell terrane gives a minimum age for the intrusion of the central and southern dioritic parts of the pluton (West and others, 1992a). This age is close to the age of the synkinematic two-mica granite of Strickland Ridge (412 ± 16 Ma) (Fig. 6) and the quartz diorite of Winterport (412 ± 14 Ma) that intrude higher grade rocks of the Liberty-Orrington welt in the Merrimack-Harpswell terrane north of Penobscot Bay (U-Pb zircon ages by Zartman and Gallego, 1979).

The contacts between the Merrimack-Harpswell and St. Croix terranes (Sennebec Pond fault) and the St. Croix and Ellsworth-Mascarene terranes (Turtle Head fault) are cut by Early Devonian plutons. These include the metaluminous and postkinematic Mt. Waldo, Lucerne, and Deblois plutons (Fig. 6), which were intruded about 390 to 380 Ma (see Hogan and Sinha, 1989). The plutons may be orogenic plutons related to the Acadian orogeny (Osberg and others, 1989) rather than the more inclusive extension-related Maine magmatic province proposed by Hogan and Sinha (1989).

Ellsworth-Mascarene terrane. The Ellsworth-Mascarene terrane has a Proterozoic continental basement (protoliths were platform sediments cut by mafic dikes and sills) that was metamorphosed to amphibolite facies about 670 to 650 Ma ($^{40}Ar/^{39}Ar$ hornblende ages, Stewart and Lux, 1988) and intruded by Late Proterozoic granitic pegmatites, as evidenced from exposures near Islesboro in Penobscot Bay (Fig. 6) (Stewart and others, 1991). Basement is not exposed elsewhere in the Ellsworth-Mascarene terrane in the United States.

Three different volcanic-rich sequences occupy much of the terrane in the Penobscot Bay area. The sequences range in age from probable Cambrian to Early Devonian, and the three sequences are probably separated by deformational events. As summarized by Stewart and others (1991), the oldest sequence is nonfossiliferous and consists predominantly of bimodal continental rift-facies basalt and rhyolite interbedded with siltstone, quartz arenite, and carbonate. On Isleboro, volcanogenic sedimentary rocks interpreted to be a distal facies of the older volcanic sequence are separated from the Proterozoic basement by a thin platform sequence of conglomerate, quartz arenite, impure carbonate, and pelite. The rocks of the oldest sequence are Cambrian or Ordovician and were metamorphosed to greenschist facies prior to their erosion and incorporation as detritus in unconformably overlying Ordovician conglomerate that is basal to a bimodal volcanic suite (Castine Volcanics). The youngest sequence is relatively nonmetamorphosed, is above conglomerates as old as late Llandoverian (Early Silurian), and is interlayered with or above sedimentary rocks as young as Gedinnian (Early Devonian) (Brookins and others, 1973). The fossils are of the Atlantic faunal province. Volcanic rocks of the youngest sequence are mostly a bimodal basalt-rhyolite suite (Stewart and others, 1991). A recent isotopic age of about 500 Ma (McLeod and others, 1991) for the nonfossiliferous Castine Volcanics, which had been considered part of the latest volcanic sequence (Brookins and others, 1973), raises the probability that this volcanic sequence is intermediate in age between the other two. D. B. Stewart (1992, oral commun.) stated that the Castine Volcanics are less deformed than the Ellsworth Schist (part of the oldest volcanic sequence) in the same structural block. These relations suggest that the deformation of the oldest volcanic suite is Early Ordovician or older.

Rocks correlative with the Ellsworth Schist of the oldest volcanic sequence crop out extensively northeast of Penobscot Bay, and rocks correlative with the youngest sequence, which are bimodal (Gates and Monech, 1981), cover large areas in eastern coastal Maine. Interlayered strata contain an Atlantic fauna of Silurian to Early Devonian age. The abundance of volcanic protoliths in the Ellsworth-Mascarene terrane has led to the characterization of that area as the Maine coastal volcanic belt (see Brookins and others, 1973).

The Ellsworth-Mascarene terrane, like the St. Croix terrane, is intruded by voluminous mafic and felsic plutons of Silurian to Early Carboniferous age. The lumping of all plutons into a single province (the coastal Maine magmatic province of Hogan and Sinha, 1989) is probably an oversimplification, as noted earlier. Early Devonian granites such as that of the Lucerne pluton are metaluminous (the type IIb plutons of Hogan and Sinha, 1989). The granites that are clearly A type, such as those of the Deer Isle and Cadillac Mountain plutons, are Early Carboniferous. Stewart and others (1991) pointed out that at least some of the plutons are consanguineous with and intrude the volcanic cover of the youngest volcanic sequence. The coastal Maine magmatic province was characterized by Hogan and Sinha (1989) as consisting of bimodal plutonism with alkalic tendencies, properties that are compatible with an extensional tectonic regime. Those properties may well apply to the Silurian and Carboniferous plutons. The Silurian plutons and associated volcanic rocks do have some similarity in chemistry to, and overlap in age, the Ordovician through Devonian anorogenic plutons of the Avalon terrane (see below).

The Newbury Volcanic Complex in northeastern Massachusetts is a fragment of the Ellsworth-Mascarene terrane caught in faults between the Merrimack-Harpswell terrane and the Esmond-Dedham block of the Avalon terrane at the northeastern termination of the Nashoba terrane (Fig. 6). Basalts, andesites, and rhyolites of the Newbury are little metamorphosed and are interlayered with and beneath mudstone, siltstone, and sandstone containing an Atlantic fauna of Late Silurian to Early Devonian age (Shride, 1976). Hepburn and others (1987) pointed out the similarity in age and composition between the intermediate and granitic plutons of the Nashoba terrane and the Newbury Volcanic Complex. Hall and Robinson (1982) and Hepburn and others (1987) suggested that the Newbury is a down-faulted remnant of a cover sequence on the Nashoba block. That interpretation is difficult to reconcile with the terrane history favored here. The present position of the Newbury Volcanic Complex can be accounted for by about 160 km of right-lateral displacement on the Gulf of Maine fault; however, if the Massabesic block is an uplifted fragment of the Ellsworth-Mascarene terrane, that lateral displacement may not be necessary.

Avalon terrane. Scattered outcrops between Boston and coastal Rhode Island are of little metamorphosed Lower, Middle, and possibly Upper Cambrian rocks that contain an Atlantic fauna. Recognition of the distinctiveness of this fauna and the faunal correlation of those rocks with those in coastal New Brunswick and Avalon of Newfoundland goes back to Walcott (shown as the Atlantic Coastal provinve on a map in Walcott, 1891). The significance of this fauna was first related to sea-floor spreading by Wilson (1966). The Cambrian rocks of southeastern New England are now interpreted to be in a distinctive terrane, the boundaries of which are still debated. Rocks exposed in southeastern New England southeast of the Honey Hill–Lake Char–Bloody Bluff fault system are here included in the Avalon terrane. It includes both the Esmond-Dedham and Hope Valley terranes of O'Hara and Gromet (1985), both of which are here referred to as blocks of the Avalon terrane (Fig. 6).

The oldest rocks in the Avalon terrane occur as deformed roof pendants in areally more extensive plutons. They record bimodal volcanism and quartz-rich sedimentation; perhaps they formed in a rifting environment (Hepburn and others, 1987). They have not been dated but may be related to an older group (pre–700 Ma) of quartzites, carbonates, and oceanic basalt in the Avalon terrane of Canada (see O'Brien and others, 1983) and/or to the basement of the Ellsworth-Mascarene terrane near Isleboro. U-Pb ages of detrital zircons from one of these older quartzites north of Boston (the Westboro Formation) are discordant with an upper intercept of 1511 ± 22 Ma (Olszewski, 1980). This age is interpreted as the age of the provenance of the zircons, which may have been northwest Africa.

A younger Late Proterozoic igneous cycle began with a largely mafic plutonic-volcanic complex, the Salem Gabbro-Diorite, located mostly north of Boston in the Esmond-Dedham block (Wones and Goldsmith, 1991). The early mafic complex as well as the older metavolcanic and metasedimentary rocks of the terrane were intruded by ~630 Ma batholiths of calc-alkaline granite and granodiorite such as the Dedham Granite (Esmond-Dedham block) (see Wones and Goldsmith, 1991) and the Sterling Plutonic Group of Connecticut (Hope Valley block) (Zartman and others, 1988). The early Salem Gabbro-Diorite and later granitoid batholiths are thought to be part of a subduction-related volcanic-plutonic arc formed at the edge of a continental mass (Wones and Goldsmith, 1991). Slightly younger Late Proterozoic granites in the Esmond-Dedham block are associated with an episode of rifting that produced the Boston basin and a suite of bimodal volcanic rocks in and near that basin.

Within the Boston basin (Fig. 6), dated basal volcanic rocks (602 ± 3 Ma U-Pb age by Kaye and Zartman, 1980) as well as a Late Proterozoic to earliest Cambrian microfauna in thick argillite in the upper part of the Boston Bay Group (Lenk and others, 1982) establish the age of the group. The basal volcanic rocks include high-silica ash-flow tuffs in terrestrially exposed collapsed calderas (Thompson and Hermes, 1990). Sedimentary rocks of the Boston Bay Group include deposits of proximal debris flows and high- and low-density turbidity currents on a marine slope/fan (Smith and Socci, 1990). Diamictites and laminated pebbly mudstones in the middle of the group (Squantum Member of the Roxbury Conglomerate) are thought to be of glaciogenic origin (Smith and Socci, 1990). Brittle faults and basal and marginal conglomerate lenses indicate deposition in a region of high relief in the Boston basin (Billings and others, 1939; Kaye, 1982).

Nance (1990) noted that plate consumption as represented by arc magmatism such as the Dedham Granite did not lead to any significant collision. Strata within the Boston basin show little evidence of a compression. He suggested that subduction was oblique and was terminated by transform movement.

Following the volcanism of the Boston Bay Group, igneous activity ceased and the area became a stable shelf (the Boston platform) for the next 100–150 m.y. Lower and Middle Cambrian platformal sedimentary rocks with an Atlantic fauna overlie the Boston Bay Group with apparent conformity, and south of Boston the fossiliferous Lower Cambrian strata are nonconformably above the Dedham Granite. Clasts in Pennsylvanian conglomerate of the Narragansett basin (Fig. 6) contain Late Cambrian fossils.

Rocks of the Avalon terrane underwent sporadic Late Ordovician through Devonian peralkaline plutonism and, in the Esmond-Dedham block, local felsic volcanism. The magmatism was anorogenic and mostly felsic, but includes some gabbro. As noted above, this magmatism overlaps in age some of the anorogenic plutonism of the Ellsworth-Mascarene terrane.

The Norfolk and Narragansett fault-bound basins within the Esmond-Dedham block (Fig. 6) are filled with Carboniferous (Westphalian and Stephanian; Lyons, 1979) alluvial strata that include some coal and, near the bottom, bimodal volcanic rocks. Fragments of much smaller basins, including a coal-bearing one near Worcester, Massachusetts, are preserved along the Clinton-Newbury fault. These are too small to show in Figure 6 (see Zen and others, 1983). These basins are similar to more extensive ones in the Maritime Provinces of Canada that contain strata ranging in age from Late Devonian into the Permian (Bradley, 1982). Although the basins in Maritime Canada are thought to be pull-apart basins along dextral strike-slip faults (Bradley, 1982), the Norfolk and Narragansett basins are more likely the result of extension with little strike-slip motion (Snoke and Mosher,

1989). The Carboniferous basins have no counterparts south of New England.

Following Alleghanian compressive deformation and Barrovian prograde metamorphism (thermal peak about 275 Ma according to Zartman and others, 1988; Dallmeyer and others, 1990), the peraluminous Narragansett Pier Granite intruded southwestern coastal Rhode Island about 273 Ma (Zartman and Hermes, 1987) and stitches the Hope Valley and Esmond-Dedham blocks. Zartman and Hermes (1987) reported a Late Archean inheritance in zircons from the Permian Narragansett Pier Granite, and speculated that this part of the Avalon terrane was obducted onto the West African shield during Permian collision. The presence of the small Late Triassic Middleton basin along a strand of the Bloody Bluff fault in northeastern Massachusetts (Fig. 6) (Kaye, 1983) adds an extensional stage to the long and complicated history of fault movement in coastal New England.

O'Hara and Gromet (1985) recognized a high-grade ductile shear zone, the Hope Valley shear zone, that divides the Avalon terrane into two structural blocks (Fig. 6). They suggested that the two blocks, the Hope Valley and Esmond-Dedham, were distinct terranes. The Hope Valley terrane underwent amphibolite facies metamorphism in the late Paleozoic and has a strong tectonic fabric. Except in western Rhode Island, the Esmond-Dedham block has been metamorphosed in the Paleozoic to grades no higher than the lower greenschist facies, and brittle deformation dominates. Other reasons for suggesting two distinct lithotectonic terranes are that the Late Proterozoic to Earliest Cambrian Boston Bay Group, the fossiliferous Cambrian platform strata, and the Ordovician to Devonian anorogenic plutonism were thought to be restricted to the Esmond-Dedham block. Some workers would not include the Hope Valley block in Gondwana because of the absence of Atlantic fauna–bearing strata. Zen (1989), for example, included the Hope Valley terrane in his Nashoba-Casco-Miramichi terrane. The Hope Valley and Esmond-Dedham blocks are here thought to be part of the Avalon terrane because they have a common Proterozoic history, as discussed earlier (except for the absence of latest Proterozoic and Cambrian strata in the Hope Valley), they do share Ordovician to Devonian anorogenic plutonism, and they were both metamorphosed and deformed during the Alleghanian orogeny and intruded by a late Alleghanian pluton. Work by Zartman and others (1988) argued that the Joshua Rock Granite Gneiss in the Hope Valley block of southern Connecticut is part of the Paleozoic anorogenic suite. The Joshua Rock is mildly peralkaline, contains sodic pyroxene, and is intrusive into the Late Proterozoic New London Gneiss. Wintsch (1992) demonstrated that in western Rhode Island and eastern Connecticut Alleghanian (Early Permian) metamorphic isograds are not offset by the Hope Valley shear zone and that the cooling histories of rocks on both sides of the zone are indistinguishable. He concluded that the Hope Valley and Esmond-Dedham blocks behaved as a single thermal unit and that any independent history of them must predate apparently uniform Early Permian metamorphism.

Inliers of proto-Gondwana terranes. Isolated fragments of proto-Gondwana terranes occur in four areas northwest of the Clinton-Newbury-Sennebec Pond fault zone. They crop out in windows in the cores of the Willimantic and Pelham domes and as upraised blocks in the Massabesic and Falmouth-Brunswick blocks (Fig. 6). Observations that led to this interpretation for each of these areas include at least one of the following: lithologic correlation, documented Late Proterozoic plutonic and/or volcanic rocks, evidence for a Late Pennsylvanian to Early Permian thermal event, and documented Permian plutons. In addition, all except the Massabesic block lack Acadian plutons. All four areas experienced a Late Pennsylvanian to Early Permian thermal event that may be the same thermal event that produced the Alleghanian regional metamorphism of western Rhode Island (Zartman and others, 1988; Eusden and Barreiro, 1988; West and others, 1988; West and Lux, 1989; Dallmeyer and others, 1990; Gromet and Robinson, 1990; Tucker and Robinson, 1990; Getty and Gromet, 1992a; Robinson and others, 1992; Wintsch and others, 1992). The core rocks of the Willimantic and Pelham domes are thought to belong to the Avalon terrane. Rocks of the Massabesic and Falmouth-Brunswick blocks may belong to the Ellsworth-Mascarene terrane.

Rocks in the core of the Willimantic dome are exposed in an inner window framed by a thrust sheet of the Nashoba terrane, which, in turn, is in a window framed by the Merrimack-Harpswell terrane. The rocks have been correlated with those in the Hope Valley block, include metavolcanic rocks (Wintsch and Fout, 1982; Ambers and Wintsch, 1990) and are about 625 Ma (Zartman and others, 1988). The Late Proterozoic Avalon rocks in the Willimantic dome underwent gneissic deformation and metamorphism (temperatures of 500 to 550 °C) about 304 Ma (U-Pb sphene ages by Getty and Gromet, 1992a). The metamorphic fabric of rocks of the Avalon and Nashoba rocks is cut by pegmatites and small granitic bodies indistinguishable from the Permian granitic magmatism related to the 273 Ma Narragansett Pier Granite (Getty and Gromet, 1992a).

Rocks in the core of the Pelham dome are exposed in a window framed by rocks of the Bronson Hill anticlinorium of the Central Maine terrane (Tucker and Robinson, 1990; Wintsch and others, 1992). Protoliths of the stratified gneisses are interpreted to be immature feldspathic wackes with a quartz-rich continental source interlayered with volcanic rocks, mostly felsic, deposited in an intracratonic basin or along a rifted continental margin (Tucker and Robinson, 1990). One sample of alkali rhyolite gneiss is about 613 Ma (U-Pb zircon) and another shows strong evidence of zircon inheritance with a minimum age of 1402 Ma (Tucker and Robinson, 1990). Rb-Sr, Sm-Nd, and U-Pb isotopic analyses of metamorphic minerals indicate that rocks at deep structural levels in the Pelham window underwent medium-grade metamorphism and penetrative deformation about 290 to 295 Ma (Gromet and Robinson, 1990; Robinson and others, 1992).

The Massabesic block is an upraised block in southern New Hampshire between the Campbell Hill fault on the northwest and the Mesozoic Wekepeke and Flint Hill faults on the southeast (Fig. 6). The block consists of the Massabesic Gneiss Complex (middle- to high-grade metasedimentary and metavolcanic rocks and orthogneisses) intruded by younger plutonic rocks. Orthogneisses of the complex are 600 to 620 Ma (U-Pb zircon ages; Besancon and others, 1977), and metavolcanic rocks are no younger than 646 Ma (Aleinikoff and others, 1979; both interpreted to be protolith ages). Other orthogneisses of the complex that crosscut and are folded with the layered gneisses yield protolith ages of about 475 Ma (Aleinikoff and others, 1979). The Massabesic block is thus distinctive in that it experienced a pre-Middle Ordovician orogenic episode reminiscent of the deformation history of the St. Croix and Ellsworth-Mascarene terranes. It is more like the Ellsworth-Mascarene terrane in the prominence of metavolcanic rocks. The Massabesic Gneiss Complex was intruded by Early Permian granite of Milford, New Hampshire (275 Ma, U-Pb zircon age; Aleinikoff and others, 1979).

The Falmouth-Brunswick block is separated from the Central Maine terrane by the northwest-dipping Hackmatack Pond thrust fault and from the Merrimack-Harpswell terrane by the southeast-dipping Flying Point fault (part of the Norumbega system; Fig. 6), which has a large component of dip-slip movement up to the northwest. The block consists of uniformly strongly metamorphosed and extensively migmatized quartzofeldspathic gneiss and amphibolite that are lithologically distinct from other units of the Casco Bay Group southeast of the Flying Point fault (Hussey, 1988). The ages of the protoliths are poorly constrained. A Rb-Sr whole-rock resetting age of the Cushing Formation from the block of 495 ± 23 Ma (Olszewski and Gaudette, 1988) suggests that the protoliths are at least as old as Cambrian. The metamorphic character of the Falmouth-Brunswick block is distinguished from terranes on both sides by the presence of early kyanite overprinted by Acadian Buchan metamorphism (Guidotti, 1989). The assemblages indicate that rocks of the block underwent pressures as much as 4 kbar higher than rocks of the Merrimack-Harpswell terrane southeast of the Flying Point fault (Stewart and others, 1991). The history of the Falmouth-Brunswick block is clearly polymetamorphic. The Rb-Sr whole-rock ages may represent metamorphic resetting, which if true, would suggest a pre-Middle Ordovician metamorphic episode similar to that of the Massabesic block and the St. Croix and Ellsworth-Mascarene terranes. Because of this and the metavolcanic component, the Falmouth-Brunswick block is thought to be a fragment, albeit from greater depth, of the Ellsworth-Mascarene terrane.

Both the Massabesic and Falmouth-Brunswick blocks are now uplifted relative to the Merrimack-Harpswell terrane. If those two blocks are part of the Ellsworth-Mascarene terrane, those relations are difficult to reconcile with the interpretation of Stewart and others (1991) that the Avalon terrane is thrust over the northern terranes on a northwest-directed Gulf of Maine thrust fault. As noted earlier, however, the dip and sense of motion on the Gulf of Maine fault are not really known. It is not certain, of course, that the Massabesic and Falmouth-Brunswick blocks are pieces of the Ellsworth-Mascarene terrane. Other correlations, however, seem even more unsatisfactory. Others (Gromet, 1989; Getty and Gromet, 1992a, 1992b; Wintsch, 1992) have correlated the Massabesic with the Avalon terrane. The described lithologies and geologic histories of the blocks, including the presence of Acadian metamorphism and plutonism, are different enough from the Avalon to make this correlation unlikely. The two blocks contain abundant pre–Middle Ordovician volcanic rocks, as does the Nashoba terrane. However, the Massabesic block (Eusden and Barreiro, 1988) and the southwestern part of the Falmouth-Brunswick block (West and others, 1988; West and Lux, 1989) underwent Alleghanian metamorphism, whereas the Nashoba terrane did not. The sparsity of volcanic protoliths in the Merrimack-Harpswell terrane adjacent to the Massabesic block makes correlation across the Flint Hill fault unlikely. The portrayal in Figure 6 of the Massabesic and Falmouth-Brunswick blocks as inliers of the Ellsworth-Mascarene terrane is, thus, a best guess, but a tentative one.

Recent studies of the thermal history of southeastern New England have yielded some exciting and unexpected results. On the basis of isotopic studies of hornblende ($^{40}Ar/^{39}Ar$) and titanite (U-Pb), the Avalon terrane of southeastern Connecticut (Hope Valley block) and the two Avalon inliers reached at least 500 °C during the Alleghanian orogeny, but adjacent and structurally overlying terranes were affected by Alleghanian regional metamorphism only close to the inliers (Gromet and Robinson, 1990; Tucker and Robinson, 1990; Getty and Gromet, 1992a; Robinson and others, 1992; Wintsch and others, 1992; Wintsch, 1992). Samples from the Avalon terrane and its inliers show hornblende and titanite ages ranging from about 290 to 260 Ma, or Late Pennsylvanian to Early Permian. Hornblende and titanite ages from adjacent terrances are as follows: for the Bronson Hill anticlinorium (Central Maine terrane) adjacent to the Pelham dome, 357 Ma (Tucker and Robinson, 1990), and for the Nashoba terrane in Connecticut, 351 Ma (Wintsch and others, 1992). U-Pb monazite ages for the Nashoba terrane in the Willimantic window are about 400 Ma (Getty and Gromet, 1992a). The conclusion from these data is that the structurally lower Avalon terrane was not in its present position relative to the adjacent overlying terranes until after the Alleghanian metamorphic peak. Getty and Gromet (1992b) proposed a model in which tectonic convergence in the Pennsylvanian (about 305 to 290 Ma) culminated with the underthrusting of Avalonian crust deeply beneath a previously metamorphosed Acadian metamorphic belt with which the Avalonian crust is currently in contact. The deeply buried Avalonian crust was deformed and metamorphosed in the process. Wintsch and others (1992) made a similar proposal, but argued for an earlier convergence.

Goldstein (1989) documented that the last movement on the Clinton-Newberry and Lake Char fault systems was normal faulting, the west side down. Getty and Gromet (1992a, 1992b) documented evidence for late extensional tectonics in the Willimantic dome. The following is a summary of their conclusions (also see Wintsch and others, 1992; Wintsch, 1992). Mylonitic amphibolites of the Avalon terrane along a low-angle shear zone that frames the Avalon terrane of the Willimantic dome (the "basement-cover contact" of Getty and Gromet, 1992a, 1992b) or the Willimantic fault of Wintsch and Fout (1982) were totally recrystallized during lower amphibolite-facies northwest-southeast–oriented ductile stretching about 275 Ma. A disrupted zone of heterogeneous deformation extends upward from the Willimantic shear zone (interpreted as a detachment fault) into the overlying high-grade pelitic gneisses, which were metamorphosed during the Acadian orogeny, and consists of stretched lozenges bounded by a network of anastomosing normal-shear-sense zones that formed in the Permian. Kinematic indicators in the cover rocks on the west flank of the Willimantic dome indicate a relative movement of tops to the northwest. Following the deep underthrusting of Avalonian crust in the Pennsylvanian, extensional collapse about 280 to 265 Ma led to extreme attenuation and/or removal of considerable structural section between the underlying Avalonian rocks and the Acadian cover. The sequence of events ended with the present juxtaposition of terranes with different and apparently unrelated deformational and thermal histories. Thus the present position of the Avalonian inliers may be accounted for by an extensional mechanism analogous to that which produced the Cordilleran metamorphic core complexes. The Willimantic detachment fault or zone may or may not connect with the Lake Char–Honey Hill fault, although they reflect the same regional extensional event.

Several points or problems need to be mentioned before leaving the discussion of the inliers. The Falmouth-Brunswick and Massabesic blocks, here suggested to be fragments of the Ellsworth-Mascarene terrane, differ in important aspects from the Avalon inliers. They both experienced a Devonian (Acadian?) thermal event, and they are bounded on the south side by faults that had Mesozoic movement. The northeastern part of the Falmouth-Brunswick block yields Devonian cooling ages for hornblende and muscovite (West and others, 1988, 1992a), and the Massabesic block is cut by Devonian plutons. Map relations permit the interpretation that the Massabesic terrane was welded to the Central Maine terrane by the Fitchburg Complex (390 ± 15 Ma; Zartman and Naylor, 1984) and to the Merrimack-Harpswell terrane by the Barrington pluton (364 Ma Pb/Pb monazite age; Eusden and Barreiro, 1988). The presence of the Mesozoic faults, together with the distribution of metamorphic mineral assemblages and the cooling ages, leads to the suggestion that the present position of the blocks is due to uplift in response to Mesozoic extension. Muscovite and biotite $^{40}Ar/^{39}Ar$ cooling ages indicate that a thermal contrast existed across the Flying Point fault between the Falmouth-Brunswick block and the Merrimack-Harpswell terrane as late as 250 Ma (West and Lux, 1989).

If the hypothesis of Pennsylvanian west-directed underthrusting of proto-Gondwana is correct, the proto-Gondwanan and Grenvillian crust are juxtaposed west of the Pelham dome (a point also made by Getty and Gromet [1992b]). Whether there is a medial New England basement in Maine, as favored by Zen (1983) and Osberg (1989), or whether Grenvillian and Gondwanan crust are juxtaposed at depth, is unresolved. The suggestion by Bothner and others (1984) that the Massabesic Gneiss Complex is the higher grade equivalent of the Merrimack-Harpswell terrane is at odds with the interpretation presented here.

Offshore extensions including the Meguma terrane. Rocks of the Avalon terrane and the Carboniferous Narragansett basin extend offshore beneath the Gulf of Maine and Long Island Sound (Fig. 6) (Hutchinson and others, 1988; Stewart and others, 1991; Thompson and others, 1993; McMaster and others, 1980). Stewart and others (1991) speculated that in the region crossed by the Quebec–Maine–Gulf of Maine transect (Fig. 1) the belt of highly magnetic rocks between the Gulf of Maine fault and the Fundy fault may be high-grade Late Proterozoic gneisses like those of the Hope Valley block. Note, however, that the high metamorphic grade of the Hope Valley block is of Alleghanian age. Hutchinson and others (1988) hypothesized that the Gulf of Maine fault might be the Variscan front (northwest limit of Alleghanian deformation). The Gulf of Maine fault is extrapolated from the Bloody Bluff fault in Massachusetts to mapped faults in New Brunswick along a band of strong magnetic highs (Hutchinson and others, 1988). Strong seismic reflectors in the Quebec–Maine–Gulf of Maine transect that dip south 25° to 30° at mid-crustal to deep crustal levels project to the surface near the band of magnetic highs and are interpreted to mark this Alleghanian fault (Stewart and others, 1991). Another band of magnetic highs and south-dipping seismic reflections marks the Fundy fault, which has had a complex history of movement.

Hutchinson and others (1988), mostly on the basis of geophysics, divided the offshore extension of the Esmond-Dedham block south of the Fundy fault into a central plutonic zone and a southern plutonic zone separated by the Nauset anomaly (fault?). They delineated the large Cashes Ledge pluton, which has been sampled, in the central plutonic zone and the large granitic Franklin batholith in the southern plutonic zone (Fig. 6). The Cashes Ledge pluton, which forms a large magnetic high, is a peralkaline granite of the Ordovician to Devonian anorogenic suite of Avalon. Zircons from several samples have slightly discordant U-Pb ages that have an upper intercept of 422 Ma (Hermes and others, 1978). The Franklin batholith may be a Late Proterozoic calc-alkaline pluton like the Dedham Granite.

As portrayed by Thompson and others (1993) and as shown in Figure 6, the largest part of the Late Proterozoic Boston basin is offshore. This rift basin is, thus, the same order of magnitude in size as all but the largest of the onshore Mesozoic rift basins. It would be fascinating to know if the offshore part of the Boston basin is as little deformed and metamorphosed as the onshore part. Photomicrographs of the Cashes Ledge pluton in Hermes and others (1978) do not have an apparent tectonic fabric. Note also the large interpreted area of the Carboniferous basins offshore.

The southern boundary of the Avalon terrane was interpreted by Stewart and others (1991) to be the Nauset fault. Leo and others (1991) dated granitic rocks from drill core on both sides of the Nauset fault on Cape Cod; Avalonian age (about 584 Ma) granitic rocks occur south of the Nauset fault. I have followed, therefore, the interpretation of Hutchinson and others (1988) and placed the boundary between the Avalon and Meguma terranes along the Nantucket and Franklin early Mesozoic rift basins southeast of the Gulf of Maine (Fig. 6). The Avalon-Meguma terrane boundary has considerable right-lateral offset south of Cape Cod and comparable left-lateral offset in the eastern Gulf of Maine (Fig. 3).

The farthest outboard continental rocks under the Gulf of Maine probably belong to the Meguma terrane (Figs. 3 and 6), which is exposed only in southern Nova Scotia. The Meguma terrane also occupies a large area on the Scotian shelf and is thought to extend seaward to the continental margin. Keppie (1989) stated that in Nova Scotia the Meguma terrane is bounded on its north side by a major east-west dextral shear zone that swings into a more westerly directed listric thrust with a dextral component of movement.

In Nova Scotia the Meguma terrane is dominated by Upper Cambrian to Lower Ordovician clastic metasedimentary rocks of the Meguma Group. It is intruded by granitic plutons of Devonian to Carboniferous age (375 to 355 Ma) that are compositionally distinct from plutons in the adjacent Avalonian composite terrane (Clarke and others, 1980). Recently described Middle Cambrian trilobites from the Meguma Group in Nova Scotia confirm that the Meguma terrane is a proto-Gondwana terrane (Pratt and Waldron, 1991). The Meguma Group is overlain conformably by paralic Silurian and Lower Devonian strata. Schenk (1980, 1991) estimated the thickness of these Upper Cambrian to Lower Devonian siliciclastic and volcaniclastic sedimentary rocks of the Meguma terrane to exceed 14 km and the volume to be tremendous. The strata were deposited on the continental margin of proto-Gondwana adjacent to what is now northwestern Africa. They were interpreted by Schenk (1991) to represent four major transgressive-regressive sedimentary cycles, each cycle being terminated by igneous activity, typically in the form of subaerial volcaniclastic rocks. Rocks as young as Emsian (Early Devonian) were deformed and metamorphosed prior to the intrusion of the South Mountain batholith (about 370–360 Ma). Early Carboniferous strata overlap the Meguma-Avalon terrane boundary and clasts from plutons of the Meguma are found in Mississippian rocks of the overlap sequence. Thus, the Meguma terrane must have been juxtaposed to the Avalon terrane between the Early Devonian and the deposition of the Carboniferous overlap sequence.

White Mountain Plutonic-Volcanic Suite and related rocks. A unique feature of the New England Appalachians is that plutonic and locally associated volcanic rocks from Middle or Late Triassic to Middle Cretaceous age occur in a broad, northwest-trending belt extending from the autochthonous Grenville province near Montreal across the Appalachian orogen of central New England to oceanic crust well beyond the southeast end of transect E1 (McHone and Butler, 1984). In Quebec these include the intrusives of the Monteregian Hills; in New England, the White Mountain Plutonic-Volcanic Suite; and offshore, the New England Seamount Chain. The volcanic rocks of the White Mountain Plutonic-Volcanic Suite occur as down-dropped masses in calderas. These igneous rocks have been attributed to the passage of a hot spot (Crough, 1981; Morgan, 1981), but because the ages of the intrusions do not become younger systematically along the trend (Foland and Faul, 1977), it is more likely that their emplacement is related to Mesozoic extension leading to the opening of the Atlantic Ocean (McHone and Butler, 1984; Thompson and others, 1993) and, for some, emplacement along a leaky transform fault. I return to discussion of these igneous rocks in the section on continental rifting and the opening of the Atlantic.

Central Appalachians. The accreted terranes of the central Appalachians are crossed by transects E2 and E3. In the region crossed by E2 most of the area of accreted terranes is covered by early Mesozoic rocks of the Newark basin and/or the emerged and submerged Atlantic Coastal Plain. Most of the discussion of the accreted terranes in the central Appalachians, therefore, is with reference to the area of E3, except for the following paragraph. Figure 7 is a terrane diagram that summarizes many of the features of the terranes of the central and southern Appalachians.

Rocks of the Brompton-Cameron terrane, including the large ultramafic body on Staten Island, crop out near New York City and rocks of the Potomac composite terrane (more later) are exposed south of the Pleasant Grove–Huntingdon Valley fault between Trenton, New Jersey, and Philadelphia, Pennsylvania (Figs. 1 and 3). Volkert and Drake (1991) suggested that the Brompton-Cameron and Potomac composite terranes are continuous in the subsurface, on the basis of potential field data and rock types from drill holes. Because the Potomac terrane clearly underwent Penobscottian deformation and the Brompton-Cameron terrane did not, this seems unlikely. A terrane boundary separating them is probably present in the subsurface. The Jefferson terrane (discussed below) is a more likely southern counterpart to the Brompton-Cameron terrane.

Laurentian margin and the Jefferson terrane. E3 crosses the Piedmont of northern and central Virginia, an area for which there is debate over fundamental geologic relations. A place to start is with a discussion of the northern termination of the Jefferson terrane. Except where interrupted by the Mesozoic Gettysburg basin, the Lower Cambrian clastic Chilhowee Group and the underlying rift-related metabasalts of the Catoctin Formation wrap around the Grenvillian core of the northeast-plunging Blue Ridge anticlinorium in Maryland and Pennsylvania (see Rankin and others, 1989, Plate 2). In northern Virginia, clastic rocks of the Fauquier Group (name revised by Rankin and others, 1993b), devoid of ultramafic rocks, intervene between the Catoctin and Grenvillian basement of the southeast limb of the anticlinorium. South of Warrenton, Virginia (Fig. 1), the ultramafic-bearing Lynchburg Group (name revised by Rankin and others, 1993b) crops out between the Fauquier Group and a metabasalt traditionally mapped as the Catoctin Formation. For the E3 transect, Glover (1989) interpreted that the largely alluvial Fauquier Group grades laterally across a shoreline oblique to the present strike of the Blue Ridge into the marine Lynchburg Group; that the ultramafic rocks in the Lynchburg are intrusive and rift related; and that the Evington Group, physically overlying the Catoctin or where the Catoctin is absent, the Lynchburg, is a deeper water distal facies equivalent of the Catoctin Formation, Chilhowee Group, and Lower Cambrian Shady Dolomite. The Evington Group is considered by Glover to be the youngest Laurentian sequence known to the Piedmont of Virginia.

Rankin (*in* Rankin and others, 1989, 1993b) noted that the alluvial to marine transition documented by Wehr (1985) was vertical, from the Fauquier Group up into the Lynchburg Group. The contact between them locally truncates units in the Fauquier and could be interpreted as a thrust fault that neatly separates units devoid of ultramafic rocks from units bearing ultramafic rocks. This unnamed thrust fault is part of the fault family that, in the interpretation of Rankin and others (1989, 1993b) and Horton and others (1989), is the leading edge of terranes, including the Jefferson, accreted to Laurentia during the closure of the Iapetus Ocean. The Lynchburg Group is thus part of the Jefferson terrane and the Fauquier Group is part of Laurentia. On the basis of the mapping of Furcon (1939) and the aeromagnetic pattern (Zietz and others, 1977), the Catoctin Formation above the Fauquier is not continuous along strike to the southwest with the Catoctin Formation above the Lynchburg south of Warrenton. The unnamed thrust fault may pass through this gap. The Lynchburg Group and the Jefferson terrane are here interpreted to terminate at tectonic contacts under the Mesozoic Culpeper basin near Warrenton, Virginia.

South of the Culpeper basin, Pavlides (1989) mapped Laurentian continental margin deposits (including his informal formation of True Blue) west of the Mountain Run fault (Fig. 1) as the westernmost sediments of a back-arc (Chopawamsic arc) basin unconformably above the "Catoctin" greenstone belt south of Warrenton. Pavlides interpreted that greenstone to be the Catoctin Formation, as had earlier workers, and along with the underlying Lynchburg to be part of Laurentia. Pavlides correlated these continental margin deposits with those shown by Horton and others (1989) as slope-rise deposits of the Westminster terrane (allochthonous offshore sequence) to the north in Maryland. The Westminster terrane, as a native terrane in the sense of Gray (1986), is included here in Laurentia, whereas the formation of True Blue and other "continental margin deposits" mapped by Pavlides (1989) west of the Mountain Run fault are included here in the accreted Jefferson terrane as shown in Figures 1 and 3.

The continental margin deposits west of the Mountain Run fault may trace south into some of the proximal facies of the Evington Group as mapped by Patterson (1989). From a study of textures of perthite in arkoses of the Fauquier and Lynchburg Groups and the continental margin deposits west of the Mountain Run fault, and the comparison of these with perthites from granulites of the Grenvillian basement of the Blue Ridge, Kline (1991) concluded that these arkoses were all derived from the adjacent Grenvillian basement. He questioned the exotic nature and, therefore, the validity of the Jefferson terrane. The solution of Hatcher and others (1990), of thrusting the Lynchburg Group over the interpreted southern extension of the Laurentian margin Catoctin Formation and the formation of True Blue does not answer Kline's objection and is inconsistent with actual map relations. Continuing work may demonstrate that, in fact, the Jefferson terrane represents rocks close to the Laurentian margin rather than the eastern margin of Iapetus. Nonetheless, I find the continuity along strike of this part of the orogen impressive. It is unlikely that the origin of the ultramafic rocks in the Lynchburg Group of E3 is different from those elsewhere in the Jefferson terrane and its New England counterpart, the Brompton-Cameron terrane. The Jefferson terrane remains a concept to be tested by careful study of the ultramafic rocks, the provenance of detrital material, and the contact between the Fauquier and Lynchburg Groups.

Potomac composite terranes. The Potomac composite terrane is a stack of thrust sheets containing tectonic and olistostromal melange complexes as well as fragments of probable oceanic crust and submarine-fan turbidites. The Bel Air–Rising Sun and Smith River terranes of Horton and others (1989, 1991) are here included as thrust sheets in the Potomac composite terrane.

East of the Mountain Run fault and the Culpeper basin, Drake (1987) mapped a stack of three lithotectonic units in northern Virginia overlain unconformably by the turbiditic (metasiltstone and phyllite) and tuffaceous Popes Head Formation, all intruded by the 494 ± 14 Ma (Rb-Sr whole rock; Mose and Nagel, 1982) or 560 Ma (U-Pb zircon; Seiders and others, 1975) Occoquan Granite. Each lithotectonic unit consists of a precursory melange (wildflysch) and an overriding allochthon for which Drake (1985) coined the term tectonic motif. The tectonic motifs are thought to be terrane fragments that collided, were overlapped by the Popes Head Formation, and then were deformed together and intruded by the Occoquan Granite to form the Potomac composite terrane (Fig. 3). The lowest tectonic motif includes deposits of a submarine fan without ophiolitic debris; the middle tectonic motif includes a submarine fan that contains ophiolitic debris overlying the Sykesville Formation (precursory melange); and the upper tectonic motif includes a fragment of a large dismembered ophiolite, the Piney Branch Complex. The mafic-ultramafic Bel Air–Rising Sun slices near Baltimore, Maryland (Fig. 3), are considered here to be additional fragments of oceanic crust in the Potomac composite terrane. The three tectonic mofits are more complexly deformed than the overlying Popes Head Formation. Drake (1987) concluded that the Occoquan Granite was synkinematically intruded during the oldest fold phase in the Popes Head Formation and suggested a Late Cambrian age for the intrusion because of the possibility of inherited zircons in the granite (see Higgins and others, 1977).

The terrane-amalgamating deformation is clearly pre-Taconian if the Occoquan is Late Cambrian, and Drake (1987) suggested that it probably is the Penobscottian orogeny. The age of the pre–Popes Head deformation that amalgamated the tectonic motifs is not well constrained. Drake considered it to be Late Proterozoic to earliest Cambrian and suggested that the deformation might be Cadomian (a Late Proterozoic orogeny affecting proto-Gondwana terranes in southern United Kingdom and France). The Cadomian orogeny overlaps in age the suite of 650–600 Ma calc-alkaline plutons that intrude deformed rocks in the Avalon terrane of southeastern New England. Note, however, that these plutons were considered by Wones and Goldsmith (1991) to be subduction related, not collision related. The Potomac composite terrane is inboard of probable proto-Gondwana terranes and also inboard of the Chopawamsic terrane (see below). The younger age limit of the deformation in terms of isotopic ages is still the age of the Occoquan Granite.

The Potomac composite terrane and the Jefferson terrane are thought to have been thrust onto the Laurentian margin during the Taconian orogeny. In other words, the collision that caused that thrusting is the Taconian orogeny. In Maryland, internal Grenvillian massifs near Baltimore are exposed in windows through the Potomac composite terrane. The evolution of the Potomac composite terrane as determined by Drake (1987), with major reservations about correlating the pre–Popes Head deformation with the Cadomian orogeny, is accepted here as a working model.

Pavlides (1989), working in the same belt of rocks in central Virginia but extrapolating his interpretations to the Potomac Valley and the Baltimore area, mapped a stack of thrust slices that contain melange. He interpreted the melange to have formed in a Cambrian and Ordovician back-arc or marginal basin that lay on

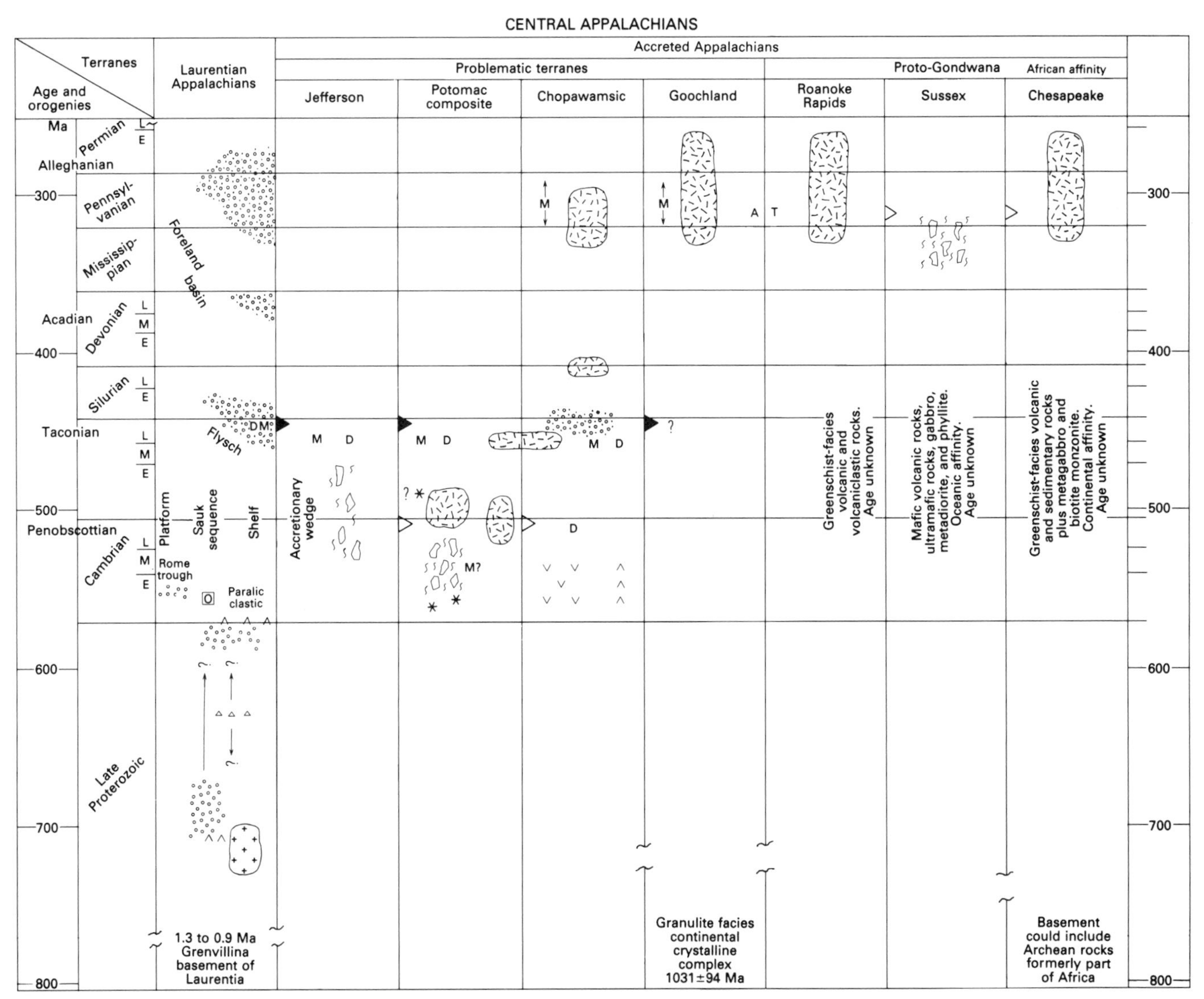

Figure 7. Terrane diagrams for the central and southern Appalachians (on following page spread). See Figure 5 for explanation. Sources of data are given in appropriate places in text. Additional source: Hutson and Tollo (1992) for central Virginia.

the continent side of the Central Virginia volcanic-plutonic belt (an island arc that formed in Cambrian time) that constitutes the Chopawamsic terrane of Figure 3. The back-arc basin is interpreted to have formed initially by the rifting away from Laurentia, near the beginning of the Cambrian, of a piece or pieces of continental crust that became basement for the island arc. After the rifting away of the continental fragments, a transform-segmented spreading ridge underlay the marginal basin in its early history. This ridge was oriented perpendicular to the long axis of the basin and was the source of ultramafic rocks in the melanges. The melanges are thought to have formed as sedimentary slump or slide aprons on the slope of the rifted landmass or by along-trough basin sedimentation. The island-arc terrane was subsequently thrust westward during the closing of the marginal basin in Cambrian and Ordovician time.

Another interpretation for rocks here included in the Potomac composite terrane was offered by Glover (1989). He recognized a zone of melange bordering the island-arc terrane (Chopawamsic terrane) that was thrust westward over a lower grade deep-water graded metagraywacke sequence, which contains clasts of the island-arc material. Glover stated that these rocks in the James River area (Fig. 1) are undoubtedly part of the complex melange sequence of Drake (1987) in northern Virginia, but considered the complex deformational history to represent

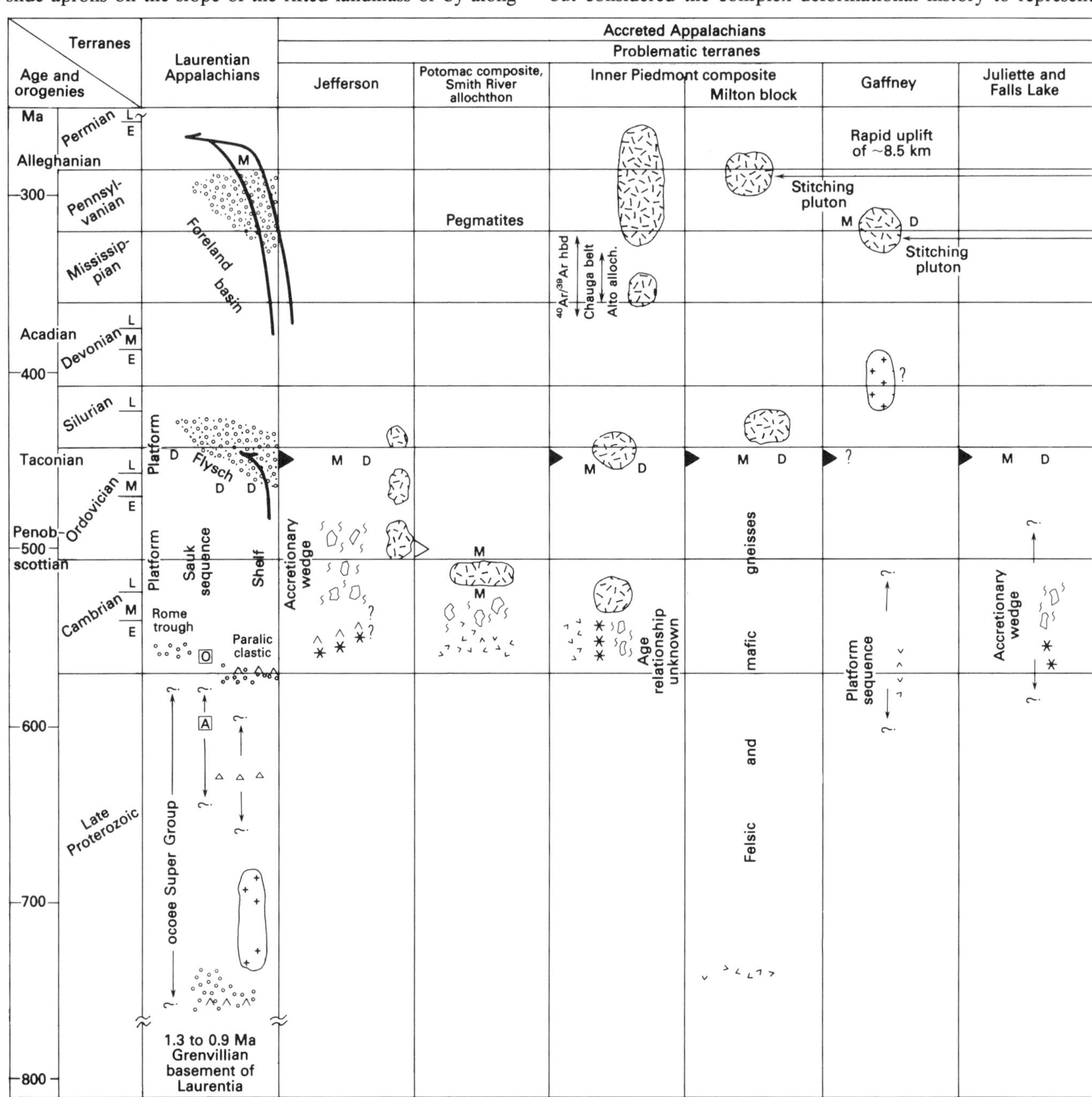

Figure 7 (continued). Southern Appalachians.

different stages in the deformation of a complex accretionary wedge that developed over a long interval of time, beginning offshore from Laurentia. As discussed below, Glover (1989) interpreted the island arc to be part of the Carolina terrane (his Carolina). The thrusting of the accretionary wedge onto the Laurentian margin was interpreted to be a Taconian event and the melange zone to mark the Taconian suture.

In the interpretation of Drake (1987) (and in this chapter), both the Jefferson and Potomac composite terranes are exotic. They were amalgamated in the Penobscottian orogeny and accreted to Laurentia during the Taconian closing of an ocean basin. The allocthonous internal Grenvillian massifs of Baltimore Gneiss and the stratigraphically attached cover sequence are exposed in windows through the Potomac composite terrane. Neither Pavlides nor Glover considered the rocks of the Jefferson and Potomac composite terranes to be exotic, but rather thought them to be part of the Laurentian margin. Pavlides (1989) invoked a spreading ridge in a back-arc basin between rifted continental fragments (the Baltimore Gneiss) and Laurentia oriented perpendicular to the axis of the basin. The Chopawamsic arc formed on Grenvillian basement (Baltimore Gneiss) of the continental fragments in response to a farther-outboard, west-dipping subduction zone. Glover (1989) considered the Chopawamsic arc to be part of the Carolina terrane. Both Pavlides and Glover interpreted the me-

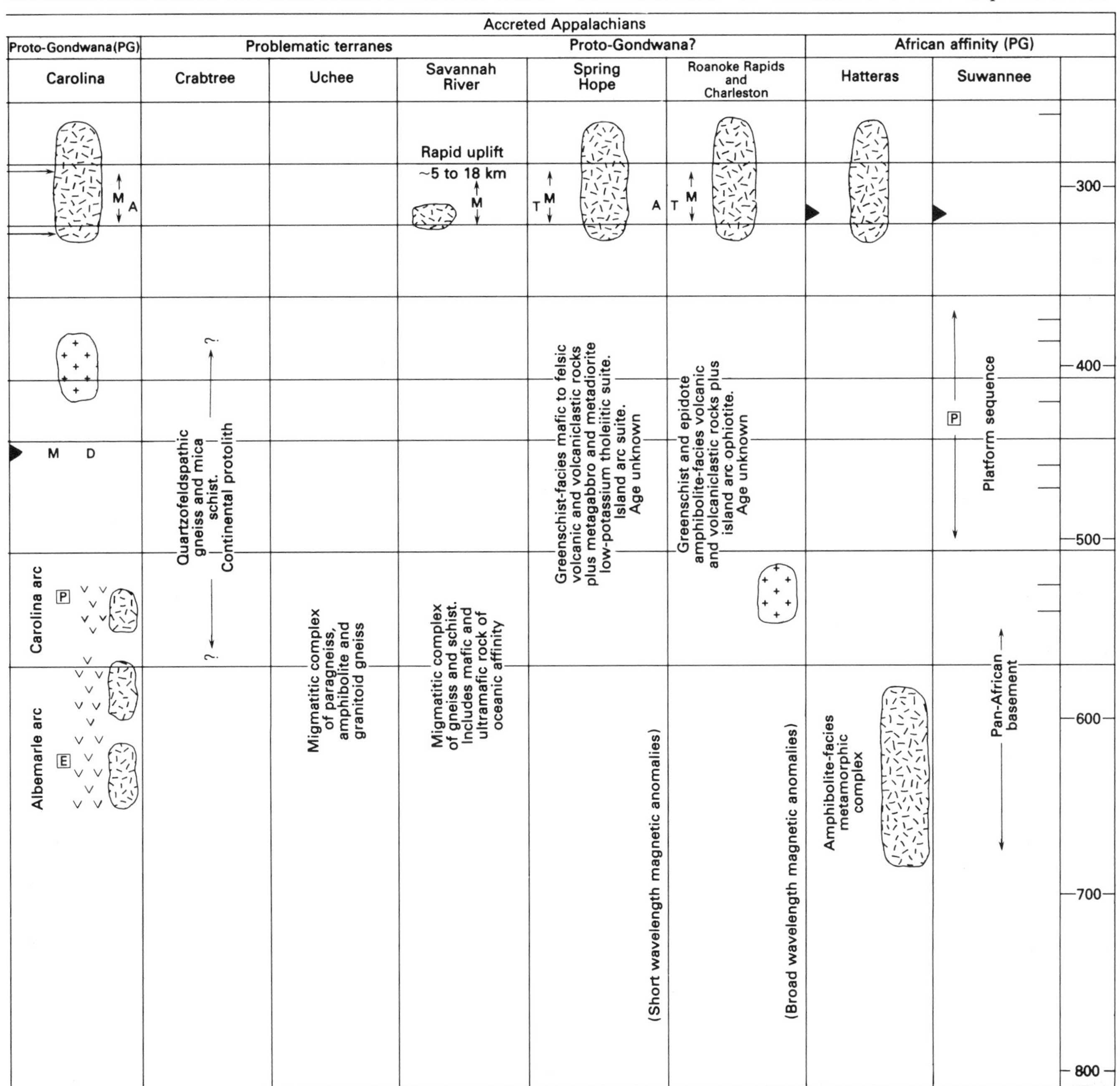

Figure 7 (continued). Southern Appalachians.

lange of the Potomac composite terrane to form in response to the westward movement of the Chopawamsic arc.

The Smith River allochthon (Fig. 1) of Conley and Henika (1973) is mostly in the region between E3 and E4. The Smith River allochthon was portrayed as a separate terrane by Horton and others (1989, 1991). It has many of the characteristics of the Inner Piedmont composite terrane to the southwest and the Potomac composite terrane to the northeast. It is here tentatively considered to be one of thrust slices of the Potomac composite terrane. Glover (1989) and Gates (1987), however, considered rocks of the Smith River terrane to be part ofthe Laurentian margin.

The Smith River allochthon occupies an area more than 200 km long and as wide as 35 km, and lies in a synformal klippe that is above the Jefferson terrane at its southwestern end in E4 and above the rest of the Potomac composite terrane (as interpreted by Rankin, 1993b) at its northeastern end, just barely within the area of, but not shown on, E3. The synform tightens and the bounding faults converge southwestward into the Brevard fault zone wherein the allochthon pinches out. The metamorphosed stratified rocks include debris of ultramafic and mafic rocks and may include an ophiolitic melange. Gneissic diamictite (melange?) occurs in the upper of two major units of the allochthon. Regional amphibolite facies metamorphism and deformation predate the Late Cambrian Leatherwood Granite (516 Ma U-Pb zircon age by Sinha and others, 1989) and the emplacement of the allochthon (Conley and Henika, 1973). The age of the metamorphism has been interpreted as Taconian (see Gates and Spear, 1991), but as pointed out by Drake (1989b), the prograde amphibolite facies was followed by retrogression (during transport?) and then a second prograde event (Rankin, 1975; Espenshade and others, 1975) much like parts of the Potomac composite terrane. The first prograde metamorphism and deformation could well be Penobscottian, particularly because it predates the Leatherwood Granite. As discussed by Horton and others (1989), the aeromagnetic character of the Smith River allochthon is similar to that of the Potomac composite terrane to the north and dissimilar to the Inner Piedmont and Milton terranes. Future work is needed to confirm that the Smith River allochthon is a thrust slice of the Potomac composite terrane rather than a continuation across the Sauratown Mountain window of the Inner Piedmont terrane. The latter is the interpretation of Conley and Henika (1973) and an earlier interpretation of Rankin (1975). The Smith River allochthon is here included in the Potomac composite terrane because of aeromagnetic and lithologic similarities.

Chopawamsic terrane. The Chopawamsic terrane (Fig. 3) consists of the remnants of a Cambrian volcanic arc (Fig. 1) and is equivalent to the Central Virginia volcanic-plutonic belt of Pavlides (1981). The use of the terrane name and its outlines follow that of Horton and others (1989, 1991). Pavlides (1981) described the Chopawamsic Formation in Virginia as a continentward facies of mixed mafic and felsic arc volcanic rocks interlayered with nonvolcanic metagraywacke and metapelite intruded by plagiogranite and trondhjemite. The Ta River Metamorphic Suite in the Chopawamsic terrane is thought to be an eastern, more oceanic (mafic rocks are more abundant and more MORB-like than those of the Chopawamsic Formation) facies of the arc consisting of amphibolite interlayered with biotite schist and gneiss. These relations led Pavlides (1981) to propose that the arc was built on sialic crust above a west-dipping subduction zone. The Chopawamsic Formation and its probable correlative to the northeast in Maryland, the James Run Formation, are of Cambrian, possibly Early Cambrian age, on the basis of discordant zircon data. Pavlides's regional interpretation, invoking parautochthoneity, is of a Cambrian continental margin sequence (formation of True Blue and related rocks) bordering Laurentia, grading eastward into back-arc basin fill (now the melange zone), east of which is an island arc, and east of that, the Iapetus Ocean. For transect E3, Glover (1989) argued that because the Cambrian shelf deposits to the west in the Valley and Ridge province do not contain pyroclastic material, the Chopawamsic arc could not have been close to Laurentia. In accordance with Glover's observation, the arc is considered here to be an accreted terrane. I concur with Drake's (1989b) observation that structurally an east-dipping subduction zone is likely.

Volcanic rocks of the Chopawamsic terrane have been correlated with volcanic rocks of the Carolina terrane by many, including Higgins (1972), Rodgers (1972), Rankin (1975), and Glover (1989). I now accept arguments of the Pavlides (1981), that the two volcanic terranes are probably not related. The Carolina terrane (Fig. 3) is complex and may be a composite of at least two arcs (Fig. 1); it is discussed more fully in the section on the accreted terranes of the southern Appalachians. The Chopawamsic terrane differs from the Carolina in gravity and magnetic expression, in lacking the bimodal volcanic suite and the voluminous felsic volcanic rocks that characterize parts of the Carolina terrane (Pavlides, 1981), and in the absence of an Atlantic (or any other) fauna in the Chopawamsic.

The Chopawamsic terrane is overlain unconformably by the Upper Ordovician metamorphosed clasitc rocks of the Arvonia Slate and the Quantico Formation, which are preserved in tight en echelon synclines (Fig. 1). The Arvonia Slate is also unconformably above the Columbia Granite of Jonas (1928), dated at 454 Ma (Rb-Sr whole-rock age by Mose and others, 1980). If the isotopic age of the Columbia Granite is correct, it is unlikely that the Arvonia Slate is older than Late Ordovician (Fig. 7A). The basal unconformity is interpreted by numerous workers to represent the time interval of the Taconian orogeny, i.e., it is the Taconian unconformity. The fossils preserved in the Arvonia and Quantico are the only known pre-Mesozoic fossils in the Appalachian internides south of the Potomac River, except for those in the Carolina terrane, the Talladega block (Fig. 1), and possibly in the Ocoee Supergroup. The Arvonia and Quantico, which are preserved only in those tight synclines, are also the only known post-Cambrian Paleozoic strata in the internides of the Appalachians south of the Potomac River, except for those in the Talladega block and possibly (but not proven) in the Murphy syncline (Fig. 1). This is in marked contrast with the internides of the New England Appalachians, where numerous units ranging from Ordovician to Pennsylvanian age contain fossils. Ordovician to Pennsylvanian strata probably cover about 90% of the internides of New England (plutons included in the area of adjoining country rock).

Goochland terrane. In the interpretation of Glover (1989), Grenvillian rocks crop out in the eastern Piedmont in a region called the Goochland granulite terrane by Farrar (1984) and the Goochland terrane (Fig. 3) by Horton and others (1989, 1991). The terrane covers a sizeable area bounded on the west by the Spotsylvania lineament, which separates it from the Chopawamsic terrane, and on the east by the Hylas fault zone (Fig. 1) near the Coastal Plain overlap. The lithic sequence from lowest to highest consists of, in the usage of Farrar (1984), the State Farm Gneiss (felsic to intermediate gneiss of both igneous and sedimentary protoliths), the Sabot Amphibolite (mafic volcanic protolith), and the Maidens Gneiss (a heterogeneous unit that is probably a stratified metavolcanic and metasedimentary unit). Farrar (1984) reported granulite facies assemblages in all three units. A later amphibolite facies metamorphism has been superimposed on the terrane. Farrar's (1984) interpretation that all units have been intruded by a small anorthosite body is not supported by subsequent mapping. Marr (1985) reported that the metaanorthosite intrudes only the Sabot Amphibolite, but the Sabot as mapped by Marr is more extensive than that portrayed by Farrar (1984). Weems (1985) interpreted the Maidens Gneiss to unconformably overlie the State Farm Gneiss and the meta-anorthosite. Gates and Glover (1989) reported that the meta-anorthosite intrudes the State Farm Gneiss and possibly the Sabot Amphibolite. The State Farm Gneiss has a Rb-Sr whole-rock age of 1031 Ma, but a large uncertainty of ±94 m.y. (Glover and others, 1982). The State Farm Gneiss and Sabot Amphibolite are exposed in three en echelon, doubly plunging antiforms surrounded by the Maidens Gneiss. Glover (1989) reported a low-angle discordance between the Sabot and State Farm that could be a fault or unconformity.

The Maidens Gneiss is continuous to the north with the Po River Metamorphic Suite and has been inferred to be correlative to the south with the Raleigh Gneiss of Farrar (1985) in North Carolina. Berquist and Marr (1992), however, interpreted the Nutbush Creek fault in North Carolina to connect with Hylas fault zone in Virginia (Fig. 1) and thus to separate the Raleigh Gneiss from the Goochland terrane. This interpretation is shown in Figure 3. The Raleigh Gneiss is here included in the Savannah River terrane, which is discussed in the section on the southern Appalachians.

Considerable controversy exists over the origin of the Goochland terrane. Glover (1989) interpreted it to be Grenvillian Laurentian basement brought up on faults. Hatcher (1989b) agreed that it is probably Grenvillian basement but emplaced by major dextral strike-slip movement. Horton and others (1989, 1991) and Rankin and others (1989) leaned toward the interpretation that the Goochland terrane is an exotic (non-Grenvillian) accreted terrane, rather than a slice of Laurentia. The rather simple structure and stratigraphy of the doubly plunging antiforms is not like any of the Grenvillian outliers. Furthermore, the assemblage of rock types is different from that in nearby Grenvillian outliers (but the anorthosite has a counterpart in the Shenandoah massif in the Roseland Anorthosite), and the great uncertainty in the only age determination is bothersome.

Coastal Plain subcrop; Chesapeake and Sussex terranes. The extrapolation of the Goochland terrane, or any rocks, under the emerged or submerged Coastal Plain is speculative. Several recent subcrop maps have been attempted, including that for E3 (Glover, 1991, written commun.), Horton and others (1989, 1991), Rankin and others (1993a), Thomas and others (1989), and Daniels and Leo (1985). All are interpretive because of insufficient data. Glover (1991, written commun.) portrayed sizeable areas underlain by Alleghanian granitic plutons. Justification for identifying them as Alleghanian comes from their eastern position in the orogenic belt and analogies with other parts of the orogen. Alleghanian isotopic ages have been obtained for some plutons partly exposed along the Fall Line. The Petersburg Granite near Richmond has a U-Pb zircon age of 330 ± 8 Ma (Wright and others, 1975), and Pavlides and others (1982) reported U-Pb zircon and Rb-Sr whole-rock ages of 300 to 325 Ma from small intrusive bodies farther north near Fredericksburg. The Petersburg Granite is strongly deformed along the Hylas ductile shear zone, which has a dextral shear sense (Gates and Glover, 1989). The granites near Fredericksburg are locally strongly foliated. In contrast to the sizeable areas of plutonic rocks shown in the coastal plain subcrop shown in E3, the compilations by Horton and others (1989, 1991) and Rankin and others (1993b) and Figure 3 portray smaller areas underlain by Alleghanian granite and relatively more area underlain by early Mesozoic basins based on the work of Daniels and Leo (1985), Hansen and Edwards (1986), and Hansen (1988).

The rocks into which the Alleghanian plutons are intruded are the subject of considerable debate, as outlined in the next paragraphs.

Crystalline rocks beneath the Coastal Plain in the area of Chesapeake Bay may be a fragment of the Paleozoic African continent (proto-Gondwana) accreted to North America during the Alleghanian orogeny and left attached to North America when the present Atlantic Ocean opened. This area is shown as the Chesapeake terrane in Figure 3 (Chesapeake block of Horton and others, 1991), and the concept follows the ideas of Lefort (1984) and Lefort and Max (1991). Sparse drill-hole data summarized by Daniels and Leo (1985) and Lefort and Max (1991) indicate that the area is underlain by greenschist facies metavolcanic and metasedimentary rocks, including laminated siliceous metavolcanic rocks and calcareous metasiltstone, biotite-epidote-garnet-amphibole schist and gneiss, metagabbro, and biotite quartz monzonite. Daniels and Leo (1985) and Glover (1989) interpreted the low-grade metamorphic rocks to be part of the Carolina terrane. Hansen and Edwards (1986) concluded that some of the rocks in the area of the Chesapeake terrane on the Delmarva Peninsula are not easily correlated with other rocks of the Appalachians, which opens the possibility that they constitute a new terrane. Lefort and Max (1991) pointed out similarities of these Delmarva rocks with specific rocks of the Reguibat uplift belt in West Africa. If the Chesapeake terrane, which probably continues out to the present continental margin, is a piece of Africa, it could include Archean to Middle Proterozoic basement like that in the Reguibat uplift and/or Pan-African supracrustal rocks formerly of the Mauritanide belt.

The Chesapeake terrane is bounded on the west by the Sussex terrane of Horton and others (1989) (Fig. 3). Rocks of the subcropping Sussex terrane produce a positive Bouguer gravity anomaly that has the plan of a gentle arc concave to the east, east of Richmond, Virginia (Fig. 2). This anomaly was called the Sussex-Leonardtown anomaly by Daniels and Leo (1985) and the Salisbury anomaly by Sheridan and others (1991). It coincides roughly with a series of magnetic highs. The amplitude of the gravity anomaly is greater than that associated with the greenstone of the Catoctin Formation, which suggests that mafic rocks constitute a major part of the source (Horton and others, 1989). Drill holes within the envelope of the anomaly have penetrated foliated mafic volcanic rocks, ultramafic rocks, gabbro, metadiorite, and phyllite. Lefort and Max (1991) postulated that the mafic rocks of the Sussex-Leonardtown anomaly (Sussex terrane as used here) are oceanic crust preserved along the Alleghanian suture (their Chesapeake Bay suture). Horton and others (1989, 1991), because of the mixture of rock types recovered from drill holes, suggested that terrane might be a melange. The Sussex terrane is considered here to be a melange along the suture of Alleghanian age between the Chesapeake terrane and the Roanoke Rapids (discussed later) and Goochland terranes. Glover (1989) also interpreted the western border of the Sussex terrane to be a suture, but one of Taconian age between the Carolina terrane and the Goochland nappe of Laurentia.

Lefort and Max (1991) further postulated that the arcuate shape of the Chesapeake Bay suture is caused by the indenting during Alleghanian collision of the Carboniferous margin of North America by a projection of the West African craton, the Reguibat uplift. That geometry, of course, could result from the accommodation of "the indenter" into a preexisting embayment in the North American continental margin. The proposed suture changes from a thrust near Richmond to high-angle shear zones to the north and south that allowed translation of the African salient westward (misprinted "eastward" in Lefort and Max, 1991). Part of the indenter may have been left behind during the Mesozoic opening of the present Atlantic Ocean and may be present in the subsurface beneath the Coastal Plain of the Chesapeake Bay area. This interpretation was portrayed by Horton and others (1991) and accepted by Rankin and others (1993b) (see Fig. 3).

Southern Appalachians. The accreted terranes of the southern Appalachians are crossed by transects E4 and E5. The geology and tectonics of this part of E4 and E5 are sufficiently alike that they can be dealt with together. The accreted terrane adjacent to Laurentia on both transects is the Jefferson terrane, which occupies the eastern part of the Blue Ridge physiographic province. The area of low relief underlain by crystalline rocks east of the Blue Ridge is called the Piedmont province. The name Piedmont province, although commonly used for a physiographic province, is used here in a geologic sense to include the pre-Mesozoic rocks between the Brevard fault zone (Fig. 1) on the northwest and the present continental margin on the southeast, including those beneath coastal plain sediments. The Piedmont province in this sense also includes the area of E3 roughly from the Mountain Run fault zone (Fig. 1) to the continental margin. In the area of E4 the Piedmont province includes, from northwest to southeast, the Inner Piedmont composite, the Smith River allochthon of the Potomac composite (already discussed), and the Gaffney, Carolina, Spring Hope, Charleston, and Hatteras terranes. E5 includes the Inner Piedmont composite, Juliette, Uchee, Carolina, Savannah River, Roanoke Rapids, and Charleston terranes. Where appropriate, the following summaries are taken from the E4 explanatory text (Rankin and others, 1991) and from Horton and others (1989, 1991).

Jefferson terrane. Interlayered clastic metasedimentary and metavolcanic rocks together with mafic-ultramafic complexes constitute the Jefferson terrane. The terrane extends for more than 1200 km along strike from beneath the Gulf Coastal Plain in Alabama to its northern termination under the early Mesozoic Culpeper basin in northern Virginia (Figs. 1 and 3). The Jefferson terrane is thrust over the Laurentian margin to the northwest on a series of faults, from the Hollins line fault in Alabama to an unnamed fault in Virginia, interpreted to mark the site of closing of the Iapetus Ocean in the Taconian orogeny. The terrane is bounded on the southeast by the Brevard fault zone and its probable continuation to the northeast as the Bowens Creek and Mountain Run faults (Fig. 1).

Protoliths of the Jefferson terrane are graywacke, pelite, calcareous sandstone, basalt, and mafic-ultramafic complexes, minor conglomerate and quartz arenite, and rare marble. The sedimentary rocks are thought to be marine, and the mafic-ultramafic complexes are thought to be dismembered ophiolites (Hatcher and others, 1984; Lipin, 1984), at least some of which represent oceanic crust (Shaw and Wasserburg, 1984). Metamorphosed basalts, now amphibolites, of the Jefferson terrane that are not obviously part of mafic-ultramafic complexes have chemical characteristics of oceanic crust generated at a spreading center (Stowe and others, 1984; Misra and Conte, 1991). Parts of the Jefferson terrane have been interpreted as melange (Abbott and Raymond, 1984). The Jefferson terrane is thought to constitute an accretionary wedge developed over an east-dipping subduction zone in the Iapetus Ocean analogous to the western part of the Brompton-Cameron terrane of New England (Rankin and others, 1989).

The age of the strata of the Jefferson terrane is poorly known and model dependent; no fossils have been recovered. The age in the model preferred here is bracketed between the time of marine sedimentation in the Iapetus Ocean and regional metamorphism. The maximum age of the strata is thus Early

Cambrian. The minimum age of the strata is not known because the age of regional metamorphism of the Jefferson terrane is not known. My prejudice (discussed below) is that the age of metamorphism of the Jefferson terrane (eastern Blue Ridge) is the same as that of adjacent Laurentia (western Blue Ridge), which is Taconian (Butler, 1973; Dallmeyer, 1975). Hence the age of the strata in the Jefferson terrane could range from Early Cambrian to Middle Ordovician.

Rocks of the Jefferson terrane were metamorphosed to garnet grade or higher (see compilation by Laird, 1989). Granulite facies conditions were attained in the area of E5, and interpreted to be of Taconian age (Absher and McSween, 1985). Butler (1991) noted that mineral dates from kyanite and higher grade rocks in the central and eastern Blue Ridge (Jefferson terrane) are mostly 340 to 420 Ma, whereas the mineral ages from the western Blue Ridge (Laurentia) reflect a metamorphic peak of 460 to 480 Ma. The timing of metamorphism(s) in the Jefferson terrane is not yet resolved.

Rocks of the Jefferson terrane are intruded by numerous granitoid plutons in a broad zone, generally toward the southeast side of the terrane, from the coastal plain overlap in Alabama to near Mt. Airy, North Carolina (Fig. 1). The plutons appear to be oldest and largest near the coastal plain overlap in Alabama. Russell and others (1987) reported a whole-rock Rb-Sr isochron age of 490 ± 28 Ma supported by zircon data that approximate a chord with an intercept age of 496 ± 74 Ma for quartz diorite of the very large Elkahatchee pluton. They also reported a U-Pb zircon age of 460 ± 19 Ma for the Kowaliga pluton and associated Zana pluton. These plutons (not illustrated) are of an approximate age and location to be related to subduction during the closing of the Iapetus. The older and more northwestern Elkahatchee pluton has a low initial $^{87}Sr/^{86}Sr$ of 0.7036, which indicates no significant involvement of mature upper crustal material or sediments in the formation of the magma. The initial ratios of 0.7044 and ~0.7061 for the younger Kowaliga and related pluton suggest that those magmas may have intruded crust that was already heated and thus partly melted and amenable to assimilation (Russell and others, 1987).

Other granitic plutons in this Alabama to North Carolina belt yield isotopic ages ranging from 460 to 380 Ma, although inherited zircons and low Rb/Sr ratios hamper precision (Butler, 1973; Miller and others, 1983; McSween and others, 1991). It is not certain that all of these bodies are related, but many are characterized by having low potassium contents, being peraluminous, synkinematic, or late kinematic, and in not being associated with contemporaneous mafic to intermediate igneous rocks. Perhaps the best dated of the granitoids, albeit different from many in being more potassic and less peraluminous, is the Austell Gneiss of Georgia. Arth (1987, written commun.) determined a whole-rock Rb-Sr isochron age of 432 ± 8 Ma and slightly discordant U-Th-Pb ages on zircon for the Austell. The $^{207}Pb/^{206}Pb$ age is 430 Ma. Butler (1973) reported that replotting U-Pb data on a concordia diagram yields an age for the Spruce Pine pegmatites of 430 ± 20 Ma, although he favors an age of about 380 Ma for the pegmatites. Rankin and others (1989) pointed out that except near Spruce Pine (Fig. 1), the regions of the thermal peak of regional metamorphism (thought to be Taconian) in the Blue Ridge tectonic province—including the granulites—lie northwest of the granitoid belt, and contain no sizeable bodies of Paleozoic granitoids. Rankin and others (1989) hypothesized that these granitoids are related and suggested a Silurian age for them.

Inner Piedmont composite terrane. The Inner Piedmont composite terrane (Fig. 3) is a stack of nappes or thrust sheets of stratified rocks intruded by a variety of plutonic rocks of several ages, some as old as Cambrian. The stratified rocks are metasedimentary and metavolcanic rocks mostly of medium metamorphic grade (up to sillimanite-muscovite), but there is a zone of low-grade rocks, the Chauga belt, along the northwestern margin of the terrane. The Inner Piedmont appears to contain metamorphic complexes of different tectonic affinities and may be an amalgamation of several disrupted terranes. The Milton block (Fig. 3) was portrayed by Horton and others (1989, 1991) as a separate terrane. It is here considered to be part of the Inner Piedmont composite terrane, although it is structurally isolated from the main body of the terrane (see below). Charnockitic gneiss in North Carolina suggests that some old continental fragments, perhaps Grenvillian basement, may be present (Goldsmith and others, 1988). Scattered ultramafic bodies and mafic-ultramafic complexes indicate that some oceanic material may be present. Inequigranular biotite gneiss, reminiscent of diamictite, in the South Mountains of North Carolina (Fig. 1) (Goldsmith and others, 1988) may represent metamorphosed melange. Higgins and others (1984, 1988) interpreted the Clairmont Formation near Atlanta (southwest of E5) to be a metamorphosed melange. Other parts of the Inner Piedmont appear to have a coherent stratigraphy of metasedimentary and metavolcanic gneiss and schist reminiscent of the Jefferson terrane.

Horton and McConnell (1991) summarized and compared the stratigraphic packages within the Inner Piedmont. Rocks along the northwestern margin of the Inner Piedmont consist of phyllite, graphitic phyllite, feldspathic arenite, marble, amphibolite, and Cambrian Henderson Gneiss, a granitic augen orthogneiss. These rocks constitute the Chauga belt (Fig. 1) of Hatcher (1972), and are generally of low metamorphic grade. Those within the Brevard fault zone have a mylonitic overprint. The Chauga belt is overthrust from the southwest by a series of nappes of higher metamorphic grade made up of various combinations of schist (much is sillimanite bearing), mica gneiss, hornblende gneiss, and amphibolite. The contact between the Chauga belt and the nappes is coincident with a metamorphic gradient and appears to be a thrust fault or tectonic slide in some areas and a gradational contact in others (Horton and McConnell, 1991). The Alto allochthon (Fig. 1) is a klippe of about 50 km × 8 km of schist (sillimanite common), gneiss, and amphibolite that structurally overlies the Chauga belt.

The Inner Piedmont composite terrane is separated from the Jefferson terrane by the Brevard fault zone, a major crustal boundary, which is probably now a splay off the sole thrust on which the Blue Ridge crystalline thrust sheet moved (Cook and others, 1979). Williams and Hatcher (1983) and Hatcher (1989a, 1989b) did not consider the Brevard to be a terrane boundary because of similarities between rocks of the Tallulah Falls Formation (herein of the Jefferson terrane) and rock of the Alto allochthon of the Inner Piedmont (Hatcher, 1978b; Hopson and Hatcher, 1988). They considered the Inner Piedmont, Jefferson, and Potomac terranes to be part of a single large Piedmont terrane. Higgins and others (1988) also considered similar packages of rocks on both sides of the Brevard fault to be correlative and in the same thrust sheets, although the sheets are not continuous across the Brevard. Nelson and others (1989), however, did not recognize a stratigraphic correlation between the Tallulah Falls Formation and rocks of the Alto allochthon. They considered the Alto allochthon to be a klippe of more southeasterly Inner Piedmont nappes. Furthermore, the Henderson Gneiss (Cambrian) is the dominant rock just southeast of the Brevard for more than 120 km along strike, but no rock of similar petrography or age has been found in the Jefferson terrane (Reed and Bryant, 1964). The Rb-Sr whole-rock age of the Henderson is 525 ± 27 Ma (Odom and Fullagar, 1973) and a Pb/Pb zircon age of 593 Ma was reported by Sinha and Glover (1978). On the basis of zircon morphology (Lemmon, 1981), the Henderson Gneiss is of intrusive origin. Such an areally extensive granitic pluton is not what one would expect to be intruded adjacent to the Cambrian passive margin of Laurentia. The presence of the Henderson Gneiss argues that the Inner Piedmont is not continental slope-rise material of Laurentia, but is exotic with respect to Laurentia (D. T. Secor, Jr., 1991, written commun.).

The age of regional metamorphism in the Inner Piedmont is Taconian. The most compelling evidence for this is in the relation between metamorphism, thrusting, and plutonism in South Carolina. There, a major Inner Piedmont thrust sheet composed of sillimanite-muscovite–grade rocks is thrust over kyanite-grade rocks (see A. E. Nelson and others, 1987, and references therein). The Caesars Head Granite, which has a 435 Ma $^{207}Pb/^{206}Pb$ zircon age, intrudes across this thrust fault (Horton and McConnell, 1991). The age of accretion of the Inner Piedmont to the Jefferson terrane is also thought to be Taconian (see Horton and others, 1989; Horton and McConnell, 1991). Zircons from mylonitized Henderson Gneiss in the Brevard fault zone at Rosman, North Carolina, yield a U-Pb upper intercept age of 456 Ma, which was interpreted by Sinha and Glover (1978) to be the age of mylonite formation during the Taconian orogeny.

$^{40}Ar/^{39}Ar$ hornblende plateau ages in northwestern Georgia provide some evidence for an episode of Devonian metamorphism. Dallmeyer (1989b) reported similar hornblende ages in thrust sheets of contrasting metamorphic grade; he reported hornblende plateau ages of 332 to 359 Ma from the higher grade Alto allochthon and 322 to 362 Ma from the lower grade Chauga belt. As noted by Glover and others (1983), the available evidence could be interpreted to indicate a period of Devonian heating (which they called Acadian) or a continuum of the thermal peak from the Taconian orogeny into the Devonian Period.

The Brevard fault zone has had a long and complex history consisting of several periods of ductile deformation followed by an episode of brittle deformation. Exotic slices of little metamorphosed platformal carbonates were emplaced by thrusting along the brittle Rosman fault of the Brevard fault zone during the Alleghanian orogeny (Horton and Butler, 1986; Hatcher, 1987).

The Inner Piedmont composite terrane is bounded on the southeast by a series of faults or ductile shear zones discussed more fully in the section on the Carolina terrane. That boundary was interpreted as a fundamental plate boundary between Laurentia and exotic terranes by Glover and Sinha (1973), and was named the Central Piedmont suture (Fig. 1) by Hatcher (1987). In those interpretations, the central Piedmont suture marks the disapperance of the Iapetus Ocean in contrast to the site favored here, which is the Hollins Line–Hayesville fault family, as discussed earlier.

The Milton block is a highly deformed assemblage of interlayered felsic and mafic gneiss. The dominant structure is a nearly recumbent antiformal nappe rooted just west of the boundary with the Carolina slate belt of the Carolina terrane (Tobisch and Glover, 1971; Tobisch, 1972). Tobisch and Glover (1971) and Glover and Sinha (1973) included the nappe in the Charlotte belt (Fig. 1), and considered it to be a higher grade and stratigraphically older part of the Carolina slate belt (U-Pb zircon ages may be as old as 740 Ma). Hatcher (1989b) also included the Milton block in the Carolna terrane. The Milton block is distinguished from the Charlotte belt, which is more or less along strike to the southwest, by having a smaller area occupied by plutonic rocks (Butler and Secor, 1991). It is separated from and welded to the Charlotte belt by the postmetamorphic Alleghanian Churchland pluton (282 ± 6 Ma Rb-Sr whole-rock age by Fullagar and Butler, 1979). Lithologically, the rocks of the Milton block are much like the Inner Piedmont and, as pointed out by D. L. Daniels (1987, written commun., *in* Horton and others, 1989), the aeromagnetic character of the Milton block and main Inner Piedmont terrane are similar. Hibbard (1991) determined that the contact between the Milton block and the Carolina slate belt is tectonic. He likewise correlated the Milton block and Inner Piedmont terrane and proposed that the Milton block–Carolina terrane contact is a continuation of the central Piedmont suture. The Milton block is separated from terranes to the west for the most of its length by the Dan River–Danville Triassic basin (Fig. 1). Sill-like granitic sheets (429 ± 4 Ma Rb-Sr whole-rock age; Kish, 1983), thought to be syntectonic, suggest that the Milton block of the Inner Piedmont composite terrane was emplaced during the Taconian orogeny, or at the latest, in the Silurian.

Gaffney and Juliette terranes. Two unconnected and probably unrelated packages of rocks occupy relatively small areas southeast of the central Piedmont suture partly within E4 or E5. These were interpreted as probable terranes by Horton and others (1989, 1991) and are the Gaffney and Juliette terranes (Fig. 3).

The Gaffney terrane in the Carolinas consists of the Blacksburg Formation on the western flank of what is traditionally known as the Kings Mountain belt. This formation consists of pelite, which is commonly graphitic, interbedded with quartzite, marble, calc-silicate, and metabasalt (amphibolite) (Horton, 1984). The age of the Blacksburg Formation is uncertain and the tectonic setting is controversial. Rankin (1975) pointed out that these rocks are similar to the platformal sequence that overlies the Grenvillian basement rocks of the Pine Mountain and Sauratown Mountain internal massifs. The Gaffney terrane could be a similar internal continental terrane of Laurentia, but one in which the basement is not exposed. The terrane, however, is southeast of the central Piedmont suture (here the Kings Mountain shear zone, Fig. 1), and the presence of amphibolite does not support the platformal association unless they are rift volcanics (the petrogenetic character is unknown). Supplee (1986) argued that the Blacksburg Formation is conformably above rocks of the Carolina terrane to the southeast, and is not separated from them by a fault of any magnitude. The Blacksburg Formation is portrayed as the Gaffney terrane in Figure 3, but the validity of that designation is uncertain.

The Juliette terrane, south of the Lowndesville shear zone in E5, is a region underlain by rocks containing abundant ultramafic and mafic fragments in a matrix of schistose metamudstone and metagraywacke intruded by variably sized granitic bodies (Figs. 1 and 3). Farther southeast, outside E5, a separate area of the Juliette terrane includes a sizeable mafic-ultramafic body called the Gladesville Complex by Hatcher and others (1984). Higgins and others (1984, 1989) called the Juliette terrane the Juliette slice of the much more extensive Macon melange and considered the Gladesville complex to be an ophiolitic fragment. Hatcher and others (1984) interpreted the Gladesville to be part of a mafic arc complex, but also to be tectonically emplaced. Hatcher (1989b) and Hooper and others (1991) did not consider the Juliette to be a separate terrane, but rather part of the Carolina (their Avalon) terrane. The interpretation here follows that of Higgins and others (1988), that the Juliette terrane is a melange of an accretionary wedge into which fragments of ophiolite have been tectonically sliced or imbricated.

Carolina terrane. The Carolina terrane (Fig. 3) consists of metamorphosed Late Proterozoic to Middle Cambrian or younger volcanic, volcaniclastic, and epiclastic rocks, and coeval intrusive rocks, plus a variety of younger intrusive rocks. The terrane is an island-arc terrane that may also be a composite terrane.

Considerable impetus was given to the tectonic modeling of the Southern Appalachians by the description of the first fossil recovered from Appalachian Piedmont south of the Arvonia syncline by St. Jean (1973) as ?*Paradoxides Carolinaensis* (from the Carolina slate belt east of Charlotte, North Carolina; Fig. 1). It is ironic that subsequent work showed that that fossil was not a Middle Cambrian Atlantic province fauna tribolite, but a Late Proterozoic *Pteridinium* (Gibson and others, 1984). Nonetheless, once it was demonstrated that fossils were there, other fossil localities were discovered. It is now well established that rocks of the Carolina slate and related belts have an Atlantic province fauna (Neuman and others, 1989). Rodgers (1972) was certainly one of the first, if not the first, to imply a correlation of rocks in eastern Newfoundland with the Carolina slate belt and to suggest that the latter was a southern continuation of the Avalon or Avalonian belt. Secor and others (1983; also see Samson and others, 1990), while confirming the presence of a Middle Cambrian Atlantic province tribolite fauna in the Carolina slate belt of South Carolina, pointed out significant differences between the Carolina slate belt and the Avalonian terranes of Canada and New England in terms of paleontology, stratigraphy, and the deformational and displacement history. They recommended that the Carolina slate belt and the adjacent Charlotte belt (Fig. 1) be identified as a separate exotic terrane called the Carolina terrane. This usage was followed with slight modification by Horton and others (1989, 1991), Rankin and others (1993b), and in this chapter. The modification is that the Belair belt and part of the Kings Mountain belt (Battleground Formation, which is intruded by Late Proterozoic epizonal tonalite; Horton, 1984) are also included in the Carolina terrane.

The Carolina terrane is thus one of the proto-Gondwana terranes of the Appalachians. Other terranes of the southern Appalachians outboard of the Carolina terrane, such as the Roanoke Rapids, Spring Hope, Charleston, and Hatteras terranes (Fig. 3), may also be proto-Gondwana terranes, but they lack diagnostic (or any other) fossils.

On the basis of stratigraphy, fossils, isotopic ages, and geochemistry, at least two sequences of rocks have been identified that are not correlatable and may even be separate terranes (Secor and others, 1989). The older sequence, called the Albermarle arc (Fig. 1) (Horton and others, 1991) in North Carolina and southern Virginia, is of Late Proterozoic age, ranging from 690 to 600 Ma, and contains Ediacarian or Vendian macrofossils. Only the uppermost unit could be as young as Cambrian. The younger sequence, in South Carolina and Georgia, is called the Carolina arc (Fig. 1) (Horton and others, 1991). It lies unconformably above a Cambrian epizonal intrusive complex (U-Pb zircon age of 550 ± 4 Ma; Dallmeyer and others, 1986) and contains fossils as young as middle Middle Cambrian (Samson and others, 1990). Volcanic products of both arcs are dominated by fragmental material as opposed to flows. Feiss (1982) summarized existing geochemical data for both areas and pointed out the difficulty of arriving at definitive petrogenetic conclusions because of extensive hydrothermal alteration. Most of the data are of major elements. Nonetheless, Rogers (1982) and Feiss supported the traditional interpretation that the arcs are subduction related, in part, because of low TiO_2 contents of the more mafic rocks. The Albermarle arc is decidedly bimodal, whereas the Carolina arc is not (Butler and Secor, 1991). Butler (1979) noted that the evolution of the Albemarle arc toward a bimodal suite with abundant rhyolite may indicate crustal thickening. Shelley and others (1988) and S. A. Shelley, J. W. Shervais, and D. T. Secor, Jr. (1991, written commun.) indicated that early volcanism in the Carolina arc was transitional between tholeiitic and calc-alkaline. These rocks have characteristics of a mature orogenic arc built either on tectonically thinned continental crust or older, intraoceanic-arc terrane. Mafic volcanic rocks with tholeiitic affinities that may be younger than the mature arc may represent an intra-arc rift basin within the Carolina arc (Dennis and Shervais, 1991). Both the mature arc and tholeiitic basalt, however, show depletion of compatible major and trace elements relative to MORB, enrichment of light rare earth elements (REEs) relative to MORB, and depletion of heavy REEs relative to MORB (S. A. Shelley and others, 1991, written commun.).

Glover and Sinha (1973) interpreted relations in the Albermarle arc near the Virginia–North Carolina border between E3 and E4 as evidence for the "Virgilina deformation, a late Precambrian and Early Cambrian(?) orogenic event," which would be roughly contemporaneous with the Avalonian event to the north. Horton and others (1989) reviewed the evidence for this, and concluded that at present the evidence for the deformation in the Carolina terrane is equivocal.

The Paleozoic structure of the Carolina slate belt in the area of E4 is dominated by a series of open folds with north-northeast–south-southwest axes; the metamorphic facies is greenschist. Rocks in the amphibolite facies Charlotte belt (Fig. 1) are west of the Carolina slate belt and are separated from the slate belt by the Gold Hill shear zone, a zone of strong shearing but undetermined displacement (Schroeder and Nance, 1987). The Charlotte belt differs from the Carolina slate belt in that it contains very little metasedimentary rock. The Charlotte belt contains an abundance of plutonic rocks, both those probably coeval with the presumably Late Proterozoic arc volcanic rocks, and younger middle and late Paleozoic plutons. The Charlotte belt apparently represents the axial zone of the arc. The polarity and original location of the arc are unknown.

The boundary between the Carolina terrane and the Inner Piedmont composite terrane to the west is the central Piedmont suture, which is a chain of faults or ductile shear zones that include, from the southwest (E5) to the northeast (E4), the Lowndesville and Kings Mountain shear zones and the Eufola fault (Fig. 1). At least the Eufola fault and the Towaliga fault (southwest of E5) experienced some Mesozoic dip-slip normal movement.

The age and even the nature of the emplacement of the Carolina terrane, however, are in dispute. Hatcher and Odom (1980), Williams and Hatcher (1983), and Hatcher (1987) equated the Carolina terrane with the "Avalonian terranes" of the northern Appalachians, and suggested accretion during the Devonian Acadian orogeny. Considerable evidence, however, points to Ordovician metamorphism of the Albemarle arc (Kish and others, 1979; Sutter and others, 1983; Noel and others, 1988). Recent paleomagnetic studies of the Albemarle Group of the Albemarle arc where the age of magnetization, deformation, and metamorphism are constrained by 455 Ma $^{40}Ar/^{39}Ar$ plateau ages (Noel and others, 1988) indicate a Late Ordovician paleolatitude of about 22°S (Vick and others, 1987). This paleolatitude is consistent with those expected for the eastern margin of Laurentia in the Ordovician. Vick and others (1987) concluded that the metamorphism, deformation, and magnetization (acquired during folding) were coeval with docking and the accretion of the Carolina terrane onto Laurentia, i.e., Taconian accretion.

Dennis (1991) outlined a challenging new hypothesis for the postaccretion history of the central Piedmont suture. He pointed out that lower grade rocks of Carolina terrane form the hanging wall of the southeast-dipping Lowndesville shear zone, whereas higher grade rocks of the Inner Piedmont terrane form the footwall. $^{40}Ar/^{39}Ar$ mineral ages young from west to east across the Inner Piedmont near the Savannah River (Fig. 1), consistent with an eastward-progressing unroofing of the Inner Piedmont. He suggested that normal dip-slip movement on the central Piedmont suture could account for these observations, although no mineral fabric has been reported to support a normal shear sense. Dennis argued for Devonian timing of the extensional faulting because of the presence of a linear belt of Silurian or Early Devonian subalkaline to alkaline granitoids and gabbro-norites in the Charlotte belt. These may represent mantle-derived magmatism along an axis of rifting contemporaneous with major crustal extension. These plutonic rocks are the Salisbury and Concord Plutonic Suites (Goldsmith and others, 1988), which Wones and Sinha (1988) also related to extensional tectonics. These ideas remain to be tested but have considerable appeal. Except for the presence of the Silurian or Devonian plutons, however, Mesozoic extension might explain equally well the distribution of metamorphic rocks.

The Carolina terrane, in addition to the Avalon terrane of southeastern New England and the Avalon terrane of Newfoundland, underwent a thermal event that generated calc-alkaline granitoids in the Late Proterozoic. Significant differences between the areas, however, have led a number of workers to argue that there was probably not a single continuous terrane. No preplutonic basement has been identified in the Carolina terrane. In New England and on the Avalon Peninsula, but not in the Carolina terrane, Lower Cambrian strata rest nonconformably on granites as young as about 610 to 540 Ma. In fact, no Lower Cambrian fossils have been recovered from the Carolina terrane. On the one hand, Lower Cambrian limestones and red beds characteristic of the Avalon terrane are apparently absent in the Carolina terrane. On the other hand, the thick sequences of Cambrian felsic volcanic and volcaniclastic rocks of the Carolina terrane are not found in the Avalon terrane (Secor and others, 1983). The times of docking of the various proto-Gondwana terranes characterized by an Atlantic fauna are different, ranging from Ordovician(?) for the Carolina terrane, to Acadian for a Silurian-amalgamated package of the Merrimack-Harpswell, St. Croix, and Ellsworth-Mascarene terranes, to Alleghanian for the Avalon terrane in southern New England. As currently configured, the indenter, or Chesapeake terrane, isolates completely the Carolina terrane from the New England proto-Gondwana terranes. Paleomagnetic studies indicate that the Albemarle arc was at a low paleolatitude (22°S) about 455 Ma, whereas the Avalon terrane in Nova Scotia was at a higher paleolatitude (42°S) in the Late Ordovician (Van Der Voo and Johnson, 1985, Fig. 4).

Outer migmatitic terranes. Higher grade migmatitic complexes appear at the coastal plain overlap along E5. Those include the Uchee terrane (Fig. 3) southeast of the Pine Mountain window (Fig. 1), and the Savannah River terrane (Fig. 3), formerly the Kiokee belt (Fig. 1), southeast of most of the Carolina terrane.

The Uchee terrane is a migmatitic complex of paragneiss, amphibolite, and granitoid gneiss (Hanley, 1986). The Phenix City Gneiss of Bentley and Neathery (1970), which is one of the principal rock units, has U-Pb zircon and Rb-Sr whole-rock data that strongly suggest a Late Proterozoic to Early Cambrian age (Russell, 1985). The Uchee terrane is separated from the Laurentian rocks of the Pine Mountain window by the Bartletts Ferry fault. The nature of the boundary between the Uchee and Carolina terranes is not determined. Higgins and others (1984, 1988), who included the Uchee terrane in the Macon melange complex, inferred a major thrust fault along this boundary. The southeast border of the Uchee terrane is covered by coastal plain sediments.

The Savannah River terrane is an antiformal belt of amphibolite facies rocks beneath greenschist facies rocks of the Carolina arc part of the Carolina terrane. The ages of the protoliths are not known, but they are intruded by penetratively deformed Alleghanian plutons. The terrane in the type area, which is the Kiokee belt (Fig. 1), is exposed beneath the coastal plain overlap from Columbia, South Carolina, to central Georgia (Fig. 3). It has been interpreted by Snoke and others (1980) and Secor and others (1986) as an Alleghanian infrastructure. Maher (1987) first made the analogy to metamorphic core complexes of the Cordillera, an analogy amplified by Snoke and Frost (1990) and adopted here.

The central and southeastern parts of the Savannah River terrane of the Kiokee belt consist of a migmatitic assemblage of gneiss and schist. The assemblage of rocks includes biotite-amphibole layered gneiss containing lenses and pods of sillimanite schist, amphibolite, serpentinite, leucocratic paragneiss, and quartzite; calc-silicate nodules are volumetrically minor but widespread (Maher and others, 1991). The rock units are disrupted and lack stratigraphic continuity. Zones of variably serpentinized mafic and ultramafic pods are thought to have originated in a mid-ocean ridge setting (Sacks and others, 1989), and their present setting as zones and pods is a predynamothermal or syndynamothermal phenomenon (Maher and others, 1991). Sacks and others (1989) and Higgins and others (1984, 1989) hypothesized that the Savannah River terrane (Kiokee belt) originated as a subduction-related accretionary complex. The oceanic geochemical affinity of these mafic and ultramafic rocks contrasts with the volcanic arc geochemical affinity of mafic and ultramafic rocks of the Carolina terrane (Dennis and Shervais, 1991; Shelley and others, 1988), and led to the suggestion by Maher and others (1991) that the Kiokee belt is a separate lithotectonic terrane that they named the Savannah River terrane.

At the northeastern end of the Kiokee belt, the Savannah River terrane includes pelitic schist and subordinate quartzo-feldspathic gneiss and quartzite plus quartzo-feldspathic orthogneiss and amphibolite thought to be intrusive into the stratified sequence (Snoke and Frost, 1990). The metamorphic assemblage in the pelitic schist includes kyanite, staurolite, garnet, and biotite. Snoke and Frost (1990) estimated that the schist was metamorphosed at temperatures between 645 and 695 °C and pressures of 7.2 to 8.2 kbar.

The time of the peak metamorphism, including the crystallization of kyanite, of the Savannah River terrane and the adjacent Carolina terrane is dated between 315 and 295 Ma by Dallmeyer and others (1986). $^{40}Ar/^{39}Ar$ cooling ages of hornblende and biotite indicate that the Savannah River terrane underwent relatively rapid uplift between about 300 and 280 Ma (Dallmeyer and others, 1986).

The Savannah River terrane in the Kiokee belt is exposed in the core of an east-northeast–trending foliation arch. The terrane is bounded on the northwest by the northwest-dipping Modoc shear zone and on the southeast by the southeast-dipping Augusta fault (Fig. 1). The Modoc shear zone is actually a steeply dipping transition zone, 4 to 5 km thick, characterized by intense plastic strain, lenticular granitic bodies, and a steep metamorphic gradient (Snoke and Frost, 1990, and references therein). $^{40}Ar/^{39}Ar$ plateau ages of 268 Ma from biotite from ductile shear zones in a granite (crystallization age of about 315 Ma) in and adjacent to the Modoc shear zone date cooling following dynamic recrystallization associated with shear-zone development (Dallmeyer and others, 1986). Sacks and Secor (1986) concluded that the movement history of the Modoc shear zone includes a major episode of hanging-wall-down movement on a gently northwest dipping zone that was later tilted to the observed steep dip. The Augusta fault is a plastic to brittle fault with oblique-normal movement (southeast side down) and is marked by an about ~250-m-thick zone of mylonitic granitoid rocks in the footwall (Maher, 1987). Granitic veins in the mylonite are synkinematic to late kinematic. On the basis of muscovite plateau ages, Maher and others (1990) concluded that the Savannah River terrane cooled as a relatively intact block through the muscovite blocking temperature (about 350 °C) just prior to juxtaposition against the southeastern part of the Carolina terrane along the Augusta fault about 274 Ma. They estimated a net dip-slip movement component on the Augusta fault of 8 km.

The accumulated data indicate that the Savannah River terrane is a mid-crustal block brought up by crustal extension late in the Alleghanian orogeny (Maher, 1987; Atekwana and others, 1989; Snoke and Frost, 1990; Maher and others, 1991). On the basis of aluminum-in-hornblende barometry, Vyhnal and McSween (1990) estimated the amount of uplift to be about 18.5 km across the Modoc shear zone and about 5.1 km across the Augusta fault. Analogies were made by Maher (1987), Snoke and Frost (1990), and Maher and others (1991) to Cordilleran core complexes, but these authors also noted differences between the core complex model and antiformal character of the Savannah River terrane. Snoke and Frost (1990) suggested that the uplift of the Savannah River terrane may have been a composite feature reflecting crustal arching during Alleghanian strike-slip faulting and plutonism (Gates and others, 1988) as well as a local isostatic effect during broadly synchronous crustal extension.

Could a significant amount of the uplift of the Savannah River terrane be due to early Mesozoic extension? (This question was also raised by Maher [1987]). COCORP data indicate that the Augusta fault becomes listric at mid-crustal levels (Cook and others, 1981) and that the western border fault of the Riddleville sub-coastal plain half graben flattens into the extrapolated Augusta fault at depth (Petersen and others, 1984). Heck (1989) also speculated that the extensional movement on the Augusta fault was Mesozsoic rather than Alleghanian. The published cooling ages, however, would require the Mesozoic movement to be in the brittle realm.

The Savannah River terrane is here interpreted to reappear in northern North Carolina in the Raleigh belt (Raleigh terrane of Stoddard and others, 1991). Rocks of the Raleigh belt (Fig. 1) are predominantly gneiss and schist of middle and upper amphibolite facies exposed in an antiform that plunges southwestward (Stoddard and others, 1991). The rocks are much like those in the Kiokee belt and those in the Goochland terrane in Virginia. The Raleigh belt was included in the Goochland terrane by Horton and others (1989, 1991). Rocks of Raleigh belt differ from those in the Goochland in not having definitive granulite facies mineral assemblages (sillimanite is present, however) (Stoddard and others, 1991), and in having no example of a simple stratigraphic sequence such as found in the antiforms in the Goochland terrane west of Richmond. Berquist and Marr (1992) determined that the Nutbush Creek fault traces northeastward into the Hylas fault (Fig. 1), which would separate the type Goochland terrane in Virginia from the Raleigh belt. The Raleigh belt is thought to be structurally beneath the Spring Hope terrane (Horton and others, 1989). Vyhnal and McSween (1990) determined that the Raleigh belt underwent uplift between 312 and 285 Ma of as much as 8 to 15 km relative to terranes on either side (aluminum-in-hornblende barometry). The analogy with the Kiokee belt is obvious, and I suggest that the Raleigh belt is part of the Savannah River terrane.

Terranes of the Appalachian orogen outboard of the Carolina terrane. The remainder of the Appalachian orogen crossed by E4 and E5 is covered by strata of the Atlantic Coastal Plain. Except for the coastal North Carolina area, most of the rocks recovered from the subcrop by drilling are greenschist facies mafic to felsic volcanic, volcaniclastic, and epiclastic rocks of unknown age as well as red beds of probable early Mesozoic age. The pre-Mesozoic rocks are interpreted to be in one of three volcanic arc terranes; the Spring Hope, Roanoke Rapids, and Charleston (see Horton and others, 1989, 1991). It is not entirely clear that there are three distinct arc terranes, or that some or all of these rocks should not be included in the Carolina terrane. Nonetheless, two terranes are extrapolated into the subsurface in the area of E4 from exposures in the eastern Piedmont near the Virginia–North Carolina state line: the Spring Hope terrane on the west and the Roanoke Rapids terrane on the east (Fig. 3).

The Spring Hope terrane was interpreted by Boltin and Stoddard (1987) to be a volcanic arc on oceanic crust or very thin continental crust. Some justification for separating these rocks from the Carolina terrane is the presence in the Raleigh, North Carolina, area of the intervening Savannah River terrane and a melange complex (Falls Lake terrane) adjacent to part of the Nutbush Creek fault zone west of the Savannah River terrane (Fig. 1). The Nutbush Creek fault zone is an Alleghanian strike-slip fault with right-lateral movement, and its trace beneath the coastal plain is hypothesized from an arcuate linear magnetic high (Hatcher and others, 1977; Horton and others, 1991; Stoddard and others, 1991).

The Roanoke Rapids terrane differs from the Spring Hope terrane by its content of a mafic-ultramafic complex (informally named the Halifax County complex by Kite and Stoddard, 1984) in northern North Carolina and a small body of peralkaline granite (whole-rock Rb-Sr "scattercron" age of 531 ± 60 Ma by Mauger and others, 1983). The Roanoke Rapids terrane is separated from the Spring Hope by the Alleghanian Hollister fault zone (Fig. 1) with dextral strike-slip movement (Stoddard and others, 1991, and references therein). Broad-wavelength magnetic highs attributable to crystalline rocks under the coastal plain imply the presence of mafic material similar to the Halifax County complex at middle to upper crustal levels. The Halifax County complex was thought to be an island-arc ophiolite by Kite and Stoddard (1984), indicating that the terrane originated at least in part in an oceanic setting.

The third terrane, the Charleston terrane, is nowhere exposed and much of it is not only covered by Cretaceous and younger coastal plain strata, but also by rocks of the large South Georgia basin of probable Triassic and Jurassic age (Fig. 2). As defined by Horton and others (1989), the Charleston terrane is equivalent to the "Charleston magnetic terrane" of Higgins and Zietz (1983) and the "Brunswick terrane" of Williams and Hatcher (1983), except for some boundary differences. The Charleston terrane should not be confused with the "Charleston block" of older usage (Rankin, 1977) for what is now generally known as the South Georgia basin. The Charleston terrane is distinguished from the Carolina and Spring Hope terranes solely on the basis of broad, generally low amplitude magnetic anomalies (Higgins and Zietz, 1983; Williams and Hatcher, 1983). Higgins and Zietz (1983) and Daniels and Leo (1985) explained the broad-wavelength character of the magnetic anomalies as the result of eastward dampening of magnetic anomalies caused by the increasing thickness of early Mesozoic and coastal plain cover. Regardless of the wavelength of the anomalies, however, the Charleston terrane stands out as a magnetically high domain, generally more than 200 gammas higher than adjacent areas of the Carolina terrane (Horton and others, 1991).

Widely separated boreholes to basement in coastal North Carolina have penetrated an amphibolite facies metamorphic complex containing plutonic rocks that range in composition from granite to diorite. These rocks, which are nowhere exposed, constitute the Hatteras terrane of Horton and others (1989) or the Hatteras belt of Thomas and others (1989). Its western border is merely bracketed between boreholes, and it is thought to extend offshore to the continental margin. Samples of the plutonic rocks from the Hatteras terrane north of the area of E4 have yielded Rb-Sr ages of 685 to 583 Ma (Denison and others, 1967; Russell and others, 1981). The known ages of these rocks are thus comparable with those from the Carolina terrane.

The configuration of terrane boundaries in the subsurface outboard of the Spring Hope terrane portrayed in Figure 3 is different from that portrayed by Horton and others (1991), E4, and Rankin and others (1993a); that in Figure 3 is based largely on a different interpretation of the aeromagnetic pattern (Zietz and Gilbert, 1980). The projection of the Hollister fault boundary between the Spring Hope and Roanoke Rapids terranes in Figure 3 more closely follows the arcuate trend, convex to the southeast, of the magnetic anomaly pattern than does the north-south projection portrayed by Horton and others (1991). The boundary chosen generally separates the broader wavelength highs of the Roanoke Rapids terrane from the shorter wavelength highs of the Spring Hope terrane, and implies that in South Carolina, the Roanoke Rapids terrane intervenes between the Spring Hope and Charleston terranes. There is no clear distinction between the Roanoke Rapids and Charleston terranes based upon the magnetic field. The delineation of two terranes is maintained because no mafic-ultramafic complex rocks or peralkaline granite have been recovered from the subsurface in the Charleston terrane (see Chowns and Williams, 1983). Figure 3 also shows an alternate interpretation of the northward projection of the Charleston-Hatteras terrane boundary, and is also based upon the magnetic fabric (Zietz and Gilbert, 1980)—borehole data there are nonexistent.

Suwannee terrane. It is now generally accepted that the Suwannee terrane of Thomas and others (1989) and Horton and others (1989), which includes most of the Tallahassee-Suwannee terrane of Williams and Hatcher (1983), beneath the coastal plain of Florida and southern Georgia and Alabama is a piece of proto-Gondwana that did not undergo significant Appalachian deformation. The terrane contains fossiliferous Ordovician to Devonian sandstones and shales that are nonmetamorphosed, little deformed, and contain European or African marine faunas unlike those of similar age in rocks deposited on Laurentia (Pojeta and others, 1976; Neuman and others, 1989). The Suwannee terrane also includes an assortment of igneous and metamorphic rocks, some of which yield Pan-African ages (about 675 to 550 Ma) (see discussions by Thomas and others, 1989; Dallmeyer, 1989b; Horton and others, 1989, 1991, and references therein). These rocks are everywhere covered by 1 to 3 km of coastal plain strata. All available data substantiate Wilson's (1966) conclusion that the Florida subcrop represents a detached fragment of the African plate, which was accreted to North America during the Alleghanian orogeny. The Alleghanian suture is buried beneath the South Georgia basin and coastal plain strata in southern Georgia, perhaps close to the onshore part of the Brunswick magnetic anomaly (Fig. 2). Lively debate continues about the exact location and nature of the suture. The configuration as a south-dipping suture across Alabama and Georgia shown in Figure 2 is constructed from the seismic profiles shown by McBride and Nelson (1988). The only potential encounter of the E transects with the Suwannee terrane is in a small area of the offshore part of E5.

Given our present state of knowledge, it is difficult to relate terranes outboard of the Carolina terrane to the rest of the Appalachians or to construct a coherent terrane history. My sense is that they all reached their present position in the late Paleozoic Alleghanian orogeny. They could, of course, have been amalgamated earlier. Deduced dextral movement on the Nutbush Creek and the Hollister faults (Stoddard and others, 1991) suggests that these outboard terranes arrived from the northeast by strike-slip movement. The terranes have no counterparts in the northern Appalachians and may be separated from the northern Appalachians by a fragment of the Paleozoic African continent (the Chesapeake terrane).

The present Atlantic continental margin

In previous sections, west to east near-surface changes across the Appalachian orogen were discussed, particularly with reference to each of the transects. The rocks of the orogen as far as the continental margin were discussed, whether they were exposed or covered by strata in Mesozoic rift basins, the Atlantic Coastal Plain, the Atlantic Ocean, or combinations of the three. The discussion now shifts to rocks and structures that formed during the evolution of the present continental margin. Because these younger stratified rocks are little deformed and many are essentially flat lying, techniques other than direct observation of outcrops, whether subaerial or subaqueous, are necessary in order to understand the geologic history of the margin. These include the use of borehole data, seismic profiles, and potential field data.

Because the continental margin is similar enough along the U.S. East Coast, we consider the margin and the changes across it from west to east as a whole, rather than discussing each transect individually. Differences between the transects are noted where appropriate.

A brief outline of the present continental margin and its evolution serves as a frame of reference for the discussion that follows. Evolution of the present margin began in the early Mesozoic when, as part of Pangea, geologic features in the eastern United States were continuous with those now in West Africa. The area of the E transects was roughly from Senegal to Morocco. Early Mesozoic crustal extension, ultimately roughly normal to the present shorelines, resulted in continental rifting across a broad area in which tilted fault blocks of continental crust were covered with varying thicknesses of fluvial and lacustrine sediments and rift-related volcanic rocks. By the Middle Jurassic, the rate and amount of stretching was sufficient for the development of a spreading center and the creation of oceanic crust between North America and Africa (Schlee and Klitgord, 1988). Sediment accumulation, following continental separation on a rapidly cooling and subsiding basement, produced a shelf, slope, and rise morphology on the edges of the Atlantic Ocean basin; i.e., two passive continental margins were formed. Our concern here is the North American passive margin west of the oceanic crust. A marginal zone along which the depth to basement increases abruptly from a depth of about 2 to 4 km to more than 8 km is known as the basement hinge zone (Figs. 1 and 2). Seaward of the hinge zone there is also an abrupt shallowing of the mantle as well as a change in properties of the crust giving rise to transitional crust. Transitional crust has seismic velocities intermediate between typical continental crust and oceanic layer 3 crust. It is typically 100 to 200 km wide and is thought to consist of extended and thinned continental crust intruded by many mafic dikes. The boundary between transitional crust and true oceanic crust is roughly along the East Coast magnetic anomaly (ECMA) (Figs. 1 and 2). Recent data collected by wide-angle seismic reflection and refraction studies have identified a layer with a seismic velocity of 7.2 to 7.5 km/s and a thickness that may in places exceed 10 km. This layer extends beneath both the marginal basins (thought to be floored by transitional crust) and oceanic basement (layer 3) (Trehu and others, 1989, and references therein; Holbrook and others, 1992). The 7.2 to 7.5 km/s layer is thought to be a zone where anomalously large volumes of magma accumulated beneath or underplated, and perhaps intruded, highly stretched continental crust and early oceanic crust during latest rifting or earliest sea-floor spreading. The 7.2 to 7.5 km/s layer has been identified in transects E4 and E2, and is assumed to be present under Georges Bank (Fig. 1).

The oldest sedimentary strata related to the evolution of the present continental margin in the United States are of early Mesozoic age and are most thoroughly studied where preserved in exposed grabens or half grabens in crystalline rocks of the accreted Appalachians (Figs. 1 and 2). The continental rift basins are filled with synrift, mostly terrestrial deposits, which are faulted and tilted. The synrift deposits of basins offshore and east of the Fall Line are overlain unconformably by postrift, commonly marine deposits of the drifting phase. This unconformity is known as the postrift or breakup unconformity, and extends beyond the borders of the rift basins as an erosional surface on crystalline rocks. Grow and others (1988a) pointed out that the erosional surface may represent a long time of nondeposition, commonly well into the Late Jurassic: they therefore preferred the term postrift to breakup unconformity; postrift is adopted here. The extent to which early Mesozoic synrift basins west of the Fall Line were once covered by postrift sediments is the subject of ongoing discussion.

Hutchinson and Klitgord (1988a) noted that buried rift basins landward of the basement hinge zone are characterized by tilted, subparallel strata (seismic reflectors), whereas those seaward of the hinge zone are characterized by a wedge-shaped reflector geometry. They attributed this change in reflector geometry across the hinge zone to differential erosion in response to uplift landward of the hinge and continued subsidence seaward of the hinge. Thus, the postrift unconformity is an angular unconformity across the rift basins landward of the hinge, whereas seaward of the hinge it is close to a disconformity. The exposed rift basins and the buried basins on both sides of the hinge probably had similar early histories involving their formation as half grabens along listric or planar normal faults. Hutchinson and Klitgord (1988a) suggested that uplift along and shoreward of the hinge zone and continued subsidence (rifting) seaward of the hinge zone produced the observed differences. The uplift may be a flexural bulge caused by sediment loading seaward of the hinge and/or a thermal bulge caused by lateral heat flow early in postrift history. The data do not allow distinction between the mechanisms.

Landward of the hinge zone, postrift strata are relatively thin and constitute the Atlantic Coastal Plain whether west of the present shoreline (emerged coastal plain) or east of the shoreline (submerged coastal plain). Large areas of the coastal plain are devoid of synrift sediments except in narrow rift basins, and are called marginal platforms. The platform areas are interrupted by areas underlain by somewhat thicker postrift sediments, which may be underlain by thin synrift deposits, and which extend farther inland relative to the hinge zone than the marginal platforms. These are called marginal embayments (Klitgord and others, 1988).

The hinge zone has a sinuous trace along the margin. Wernicke and Tilke (1989) referred to those segments of the hinge zone that are convex toward North America as salients and those that are concave as embayments. Promontories (adopted here) rather than salients would have been a better term for the convex portions as used by Thomas (1977) for the Iapetus margin. Except for the Gulf of Maine embayment (Fig. 2), the marginal platforms and marginal embayments are opposite the promontories and embayments, respectively, of the hinge zone. Wernicke and Tilke (1989) noted that continental rifting tends to be asymmetrical in cross section and that the sense of asymmetry shifts along strike. They concluded that the inflections in the sinuous trace of the hinge zone define segments of the rifted margin with opposing asymmetry.

Along the continental margin outboard of the hinge zone and hence above transitional crust lies a chain of deep marginal basins in which as much as 13 km of postrift sediments accumulated above earlier synrift sediments (Fig. 2). A carbonate-bank, paleoshelf-edge complex developed along the outer edge of the marginal basins and separated shelf deposition from slope-rise deposition (Fig. 2) (Klitgord and others, 1988). The discontinuous reefal bank at the paleoshelf edge developed during the Middle and Late Jurassic and continued into the Early Cretaceous (Poag and Valentine, 1988). The miogeocline of the continental margin thus extends from the coastal plain to the slope-rise sediments and straddles three different kinds of crust: continental, transitional, and oceanic (Sheridan and Grow, 1988b).

The Cenozoic history of the continental margin is one of continued crustal subsidence caused by thermal and sediment loading effects and the resulting accumulation of sediments. The sedimentary record also reflects repeated changes in sea level, shifting depocenters, and the advent of the Gulf Stream as a major ocean current in the early Tertiary. As noted by Thompson and others (1993), the entire sedimentary package seaward of the paleoshelf edge consists of postrift material of the continental slope-rise and abyssal plain; the package thins from thicknesses of as much as 5 km at the continental margin to zero at the Mid-Atlantic Ridge (see Tucholke, 1986).

Early Mesozoic continental rift basins. The following description of early Mesozoic continental rift basins is largely adopted from the synthesis by Olsen (1989). Mesozoic strata within each continental rift basin typically dip 10° to 15° toward the border fault near which they are commonly warped into broad folds or, in places, tighter en echelon folds (Manspeizer and deBoer, 1989). Synrift deposits within the basins belong to the Newark Supergroup (Olsen, 1989). The deposits are dominantly fluvial and lacustrine, but include eolian deposits in Canada and coal in some of the basins in Virginia and North Carolina. The oldest strata and hence the earliest known rifting is Middle Triassic (Anisian) in the Fundy basin on the Canadian border (Fig. 2). In the United States the initiation of rifting was Late Triassic (Carnian). Basins from Culpeper northward contain Upper Triassic to Lower Jurassic sedimentary rocks and thin tholeiitic basalt flows of earliest Jurassic age. The southern exposed basins contain only Upper Triassic sedimentary rocks, and no Jurassic sedimentary rocks. They contain no basalt flows, although the northern part of the Deep River basin is intruded by sizeable diabase bodies, and diabase dikes cut the basin fill of that and other southern basins.

Most Triassic sequences in basins from Nova Scotia to North Carolina exhibit a tripartite division of vertically isolated facies beginning with a lower, dominantly fluvial interval, a middle "deep-water" lacustrine interval that is commonly gray or black, and an upper, mostly lacustrine interval that is generally red. Conglomerates are prominent toward the border faults. Jurassic strata show no clear vertical facies patterns, but this may be due to the limited age range of preserved Jurassic rocks. Only the Hartford basin, which is crossed by E1, preserves any strata younger than earliest Jurassic. In the Hartford basin strata are as young as late Early Jurassic (Toarcian, which according to Palmer [1983], ended 187 Ma; see Manspeizer and others, 1989, Fig. 4). Lacustrine rocks of the Newark Supergroup exhibit simple and compound sedimentary cycles about which there is a large body of literature (see Olsen, 1989).

Atlantic Coastal Plain. The Atlantic Coastal Plain extends from the Fall Line to the basement hinge zone, a breadth of more than 300 km in E4 but less than 100 km in E2. It is underlain by an eastward-thickening wedge of little deformed, gently seaward dipping Mesozoic (post–Middle Jurassic) and Cenozoic strata. In general, clastic sediments prograde seaward into carbonate-rich sediments. The coastal plain wedge accumulated as the postrift unconformity tilted seaward in the Late Jurassic and the shoreline migrated landward primarily during the Early Cretaceous. Steckler and others (1988) concluded that landward of the hinge zone the tilting of the unconformity was caused by flexural downwarping of the continental lithosphere by sediment loading. Sea-level fluctuations superimposed on this continuous seaward-dipping flexure produced recurring migration of the strandline (transgression and regression) across the coastal plain (summarized by Sheridan and Grow, 1988a, 1988b). The present emerged and exposed coastal plain narrows from a width of about 200 km in North Carolina to zero on Long Island. Coastal plain sediments are present in the subsurface beneath the Pleistocene drift on Long Island, possibly on Cape Cod, and on a number of islands in between (Weed, 1991). Small exposures of Upper Cretaceous and Tertiary sedimentary units are present on Block Island, Martha's Vineyard, and coastal Massachusetts south of Boston (Woodworth and Wigglesworth, 1934; Frederiksen, 1984). The Martha's Vineyard and Block Island exposures are in deformed blocks and slabs emplaced by ice-driven Pleistocene thrust faults (Woodworth and Wigglesworth, 1934).

The Atlantic Coastal Plain in our area of consideration consists of the Gulf of Maine embayment (E1), Long Island platform (E1 and E2), Salisbury embayment (E2 and E3), Carolina platform (E4), and Southeast Georgia embayment (E5). Klitgord and others (1988) postulated that the borders of the platforms and embayments are inherited from preexisting Paleozoic structures. By analogy, the inflections in the trace of the basement hinge zone would also be inherited. The outline of the Salisbury embayment and the southeastern border of the Long Island platform coincide in general with the interpreted position of the Alleghanian suture bordering the Chesapeake terrane (Figs. 2 and 3) (Lefort and Max, 1991). The Southeast Georgia embayment lies over the South Georgia rift basin and the Alleghanian suture between the Suwannee terrane and the Appalachian orogen. The northern border of the South Georgia basin coincides roughly with the interpreted northern boundary of the Charleston terrane (Figs. 2 and 3) (Horton and others, 1991).

The Gulf of Maine embayment (E1), recessed into the continental margin between Nova Scotia and Cape Cod, is the only embayment on the U.S. Atlantic margin not covered by a thin veneer of Mesozoic (post–Early Jurassic) sedimentary strata (Austin and others, 1980). It is covered by as much as 200 m of Quaternary material, including glaciomarine deposits (Poag and Valentine, 1988). It is possible that a Mesozoic veneer was removed by glacial erosion (Klitgord and others, 1988); however, the Gulf of Maine embayment is part of a promontory in the analysis of Wernicke and Tilke (1989). Perhaps it never was covered by Jurassic and Cretaceous sediments.

Transect E4 lies along the Cape Fear arch, a major basement warp with an axis roughly perpendicular to the coastline in the Carolina platform (Gohn, 1988). The general configuration of the arch is depicted by structural contours on the pre-Cretaceous basement surface in E4 (Rankin and others, 1991). The arch was present as early as the middle part of the Cretaceous. Even middle Cretaceous strata (Albian and Cenomanian) are missing from the arch, although they are present to the north and south (Gohn, 1988). Santonian (Upper Cretaceous) rocks are the oldest beds that extend across the Cape Fear arch.

Basalt flows in the subcrop that lie on the eroded surface of Appalachian crystalline rocks underlying the Carolina platform (Fig. 2) in E4 have been traced in seismic profiles to near Charleston, South Carolina, where they have been sampled by drilling (Rankin and others, 1991, and references therein). The basalts at Charleston were erupted subaerially (Gohn and others, 1983). They have been dated at 184 ± 33 Ma (Jurassic) ($^{40}Ar/^{39}Ar$ ages; Lanphere, 1983), which is close to the age given for the first oceanic crust or the rift-drift transition (175 Ma, according to Klitgord and Schouten, 1986). However, the compositions of the basalts are similar to those of the continental eastern North American basalt province (Gottfried and others, 1983). Because the basalt appears on seismic profiles to be only slightly faulted, major block faulting of the platforms and rift-basin filling had nearly ceased by the time these flows were extruded (Dillon and Popenoe, 1988; Klitgord and others, 1988). The erosion surface on which the basalt lies is continuous across the Triassic(?) Brunswick basin and is considered to be the postrift unconformity (Hutchinson and others, 1983b; Dillon and others, 1983). The existence of the Brunswick basin has been questioned by Austin and others (1990), who interpreted weak landward-dipping seismic reflectors in the position of the Brunswick magnetic anomaly to be defraction tails on small normal-fault steps on the basalt. Hutchinson and others (1983b), however, concluded that a wedge of low-density rock is necessary at that position to fit the gravity field; the Brunswick basin is shown in Figure 2.

McBride and others (1989) suggested that the Jurassic basalt on the postrift unconformity on the Carolina platform is continuous with an areally extensive basalt flow and/or diabase sill sequence that they identified in the area of the buried Triassic and Jurassic(?) South Georgia basin. Seismic profiles indicate that basin fill beneath the mafic sequence ("J" reflector), locally as thick as 6 km, is complexly dipping and is presumably fault bounded. The "J" reflector is broken and tilted where crossed by only one of several seismic profiles. The "J" reflector is overlain by essentially flat lying coastal plain sediments. McBride and others (1989) interpreted the basaltic interval above the South Georgia basin as marking the postrift unconformity and a time of a more widespread subsidence accompanied by relatively minor normal faulting. McBride and others (1989) noted that if the identification and correlation of these two basalt layers or sequences (the Carolina platform and South Georgia basin) are correct, these basalts collectively constitute one of the largest outpourings of basalt on continental crust in the world.

Marginal basins. As reviewed by Grow and Sheridan (1988), the collection in the late 1960s of proprietary multichannel seismic profiles along the U.S. Atlantic continental margin by the petroleum industry revealed the presence of as much as 10 km of sedimentary strata beneath the continental shelf of New Jersey. Further work led to the definition of five deep Mesozoic and Cenozoic basins along the U.S. Atlantic margin now considered to be of first-order geologic importance. From northeast to southwest, these are the Georges Bank basin (E1), the Baltimore Canyon trough (E2 and E3), the Carolina trough (E4), the Blake Plateau basin (E5), and the Bahamas–South Florida basin. Although a number of petroleum exploration boreholes and stratigraphic test boreholes have penetrated to Jurassic strata in the basins, only two boreholes seaward of the hinge zone have penetrated the older synrift sequence (the COST G-2 well on Georges Bank and the Great Isaac-1 well in the Bahamas; Fig. 2) (Sheridan and Grow, 1988b). Stratigraphic control for the basins comes from extrapolation from these boreholes and others outside the basins along regional seismic reflection lines.

The generalized stratigraphic framework and depositional history of the marginal basins were summarized by Poag and Valentine (1988) and are paraphrased below. Synrift graben-fill deposits, interpreted to be coarse terrigenous siliciclastics of Triassic age, are the oldest rocks in the marginal basins. The synrift section is thickest in the Baltimore Canyon trough (5 km). Restricted marine carbonates and evaporites were deposited as the initial postrift deposits following a major erosional hiatus. The rifting to sea-floor–spreading transition was in the Early Jurassic. Shallow-water limestones and shelf-edge reefs characterized the outer parts of the basins as sea-floor spreading proceeded. By Late Jurassic to Early Cretaceous time a discontinuous reefal bank formed the eastward margin of the basins. Jurassic strata form most of the postrift deposits and are thickest in the Baltimore Canyon trough (more than 12 km). Terrigenous siliciclastic sediments buried the shelf-edge barrier in the Early and middle Cretaceous and terminated the carbonate regime. Upper Cretaceous deposits consisted of a fining-upward sequence from fine-grained sandstones to shales and chalk beneath a steadily deepening shelf sea. The Cretaceous section is thickest in the Baltimore Canyon trough (more than 2 km). An especially widespread interval of nondeposition and erosion was followed by renewed carbonate deposition in the Paleogene. Neogene deposition consisted of deltaic progradation of siliciclastics. Quaternary deposits are also dominantly siliclastic, including glaciomarine deposits in the Georges Bank basin. The maximum thickness of the Cenozoic deposits is also in the Baltimore Canyon trough (more than 2 km).

A number of aspects of Poag and Valentine's (1988) summary are worth amplifying. The ages of rocks beneath the postrift unconformity in the two boreholes that penetrated them are not well controlled. The synrift strata of the Great Isaac-1 well (Bahamas) are arkose and volcaniclastic rocks that are nonfossiliferous (Sheridan and others, 1988b). In the COST G-2 well, the biostratigraphy and ages are controversial, and the location (depth) of the postrift unconformity is in dispute. The COST G-2 well bottomed in salt at a depth of about 6.7 km, and the lower 2 km of the section in the hole are nonfossiliferous, except for some poorly preserved Carnian to Norian (Late Triassic) palynomorphs at a depth of about 6.3 km (Cousminer and Steinkrauss, 1988). Seismic evidence indicates that the basement surface is at least 500 m below the well bottom (Schlee and Klitgord, 1988). Cousminer and Steinkrauss (1988) interpreted the lower nonfossiliferous strata of the well to be of Late Triassic age on the basis of sparse palynomorphs and dinoflagellates recovered from the core more than 2 km above the salt. Manspeizer and Cousminer (1988) interpreted a "subtle sequence boundary" (Sheridan and Grow, 1988) above the Late Triassic fossils at a depth of about

4.15 km as the postrift unconformity. Hutchinson and Klitgord (1988a), however, argued that the COST G-2 stratigraphy of Cousminer and Steinkrauss (1988) and Manspeizer and Cousminer (1988) is inconsistent with regional seismic stratigraphy. In the nearby COST G-1 well, on the northwest flank of the Georges Bank basin (Figs. 1 and 2), the postrift unconformity has been identified in the well, and can be traced acoustically to a little above the top of the salt in the COST G-2 well. The stratigraphy in the two wells is correlated by reflectors tied to fossiliferous Middle Jurassic strata at the base of the COST G-1 well (see Schlee and Klitgord, 1988). Hutchinson and Klitgord (1988a) and Schlee and Klitgord (1988) concluded that the nonfossiliferous lower 2 km of the COST G-2 well are of Lower to Middle Jurassic age and that the Late Triassic fossils above that are reworked, perhaps eroded from basins uplifted along the hinge zone. Their conclusion is accepted here.

Marginal basins evolved by continuing crustal subsidence from rift basins that were within or seaward of the hinge zone. Rift basins landward of the hinge zone are deeply eroded, and preserve only the older synrift deposits (Klitgord and others, 1988; Hutchinson and Klitgord, 1988a). Klitgord and others (1988) were able to identify the postrift unconformity in the regions of the hinge zone where it is an erosional unconformity, and traced it across the marginal basin where it is a more conformable surface. The identification of the postrift unconformity, however, may be controversial (see above discussion of the COST G-2 well). Dillon (*in* Rankin and others, 1991) stated that the age and tectonic significance of the horizontal acoustical reflector taken to define the floor of the Carolina trough (contour shown on the geologic strip map of E4) are not precisely known. This reflector is continuous with reflections interpreted to be the postrift unconformity on the sides of the trough, but the horizontal reflections on the floor of the trough may be caused by synrift or very early postrift deposits, including salt.

Evidence for salt is found in three basins: the Georges Bank basin, the Baltimore Canyon trough, and the Carolina trough (Sheridan and Grow, 1988b, and references therein). At the end of the rifting phase, the basin probably flooded periodically, resulting in evaporite (including salt) deposits. Initially, salt was probably deposited with continental rift-facies clastic rocks, and deposition continued until continental breakup and the first formation of oceanic crust (Dillon, *in* Rankin and others, 1991). Salt was penetrated in the COST G-2 well of the Georges Bank basin, and may be of Carnian to Norian (Late Triassic) age (Cousminer and Steinkrauss, 1988). As reported in Schlee and Klitgord (1988), salt interfingers with Upper Triassic red beds landward of the hinge zone on the Canadian margin; seaward of the hinge zone, the salt is overlain by lowermost Middle Jurassic limestone. For the Baltimore and Carolina troughs, salt deposition is inferred from a series of probable salt diapirs along the seaward edge of the basins (also present along the seaward edge of the Georges Bank basin). These diapirs are most extensively developed in the Carolina trough, where a string of 28 diapirs is aligned along the axis of the East Coast magnetic anomaly (ECMA). This alignment probably indicates basement control of the location of the diapirs. It is possible that the continent-ocean boundary formed the southeast boundary of the early evaporite basin, and/or a ridge or scarp of basement at the boundary controlled the flow of salt (Dillon, *in* Rankin and others, 1991). At such a ridge, the flow of salt seaward, caused by the weight of the continental margin sediments in the marginal basin, might be interrupted and the salt might begin to rise. In the area of E4, disruption of the sea floor above the domes and slumping in adjacent sediments, probably caused by oversteepening of the uppermost strata during diapiric flow, show that salt flow continues.

In the Carolina trough, a normal-fault system is parallel to and on the landward side of the string of salt diapirs (Dillon and Popenoe, 1988, and references therein; Dillon, *in* Rankin and others, 1991). These faults have throws that increase progressively with depth, showing that they are growth faults contemporaneous with deposition. Seismic profiles show that these faults do not flatten with depth, as do ordinary down-to-basin faults. The faulting is interpreted to result from the movement of salt into the diapirs, causing a nearly vertical subsidence of the stratigraphic section in the trough. As with the diapirs, the growth faults break near-surface sediment layers and are thus active.

The marginal basins differ in details, some of which are reviewed below. Sediments of the Georges Bank basin accumulated in one main depocenter and several ancillary depressions above a complexly block faulted basement (Schlee and Klitgord, 1988). Series of half grabens that step down seaward (identified in seismic sections and by distinctive magnetic anomaly patterns) mark the hinge zone in the basement. Synrift sediments accumulated in these and other restricted elongate basins that trend northeast beneath the postrift strata. The synrift sediments may exceed 3 km in thickness at the axis of the main basin. The trend of the half grabens is parallel to the hinge zone, but the continental margin at the southwest end of the basin cuts east-west across the trend of the grabens. Postrift deposits are as thick as 7 km in the main basin and more than half of these are Middle and Upper Jurassic sedimentary strata. The Georges Bank basin is relatively broad. The across-strike distance from the hinge zone to oceanic crust as interpreted in E1 is nearly 200 km (Thompson and others, 1993). In E1, the paleoshelf edge is well landward (more than 30 km) of the ECMA (Fig. 1).

The Baltimore Canyon and Carolina troughs are probably the simplest of the marginal basins; both are oriented perpendicular to the oceanic fracture zones, and are probably normal pull-apart basins (Sawyer, 1988). Both are relatively long, narrow, and deep basins. The Baltimore Canyon trough is the best studied and drilled of the marginal basins because of petroleum exploration, although exploratory drilling has probed only the top third of the basin fill (Grow and others, 1988a). The width of the Baltimore Canyon trough between the hinge zone and the ECMA ranges from 60 to 100 km. The trough has a distinctive segmented structure. It is broken into three parts of different widths and basement depth at the intersection of oceanic fracture zones (Klitgord and others, 1988). These are a northern section (E2), central section (E3), and a southern section, which is the shallowest (E3).

The thickest accumulation of synrift and postrift sediments along the U.S. Atlantic margin is in the northern Baltimore Canyon trough, which contains as much as 18 km of Mesozoic and Cenozoic sedimentary deposits (Grow and others, 1988a). About 70% (9 km) of the postrift deposits are Middle and Upper Jurassic, 15% are Cretaceous, and 15% are Cenozoic. Grow and others (1988a) attributed the great thickness of postrift sediments in the Baltimore Canyon trough to a tremendous influx of clastic material during the Middle Jurassic derived from the nearby young mountains of the Appalachian orogen both west and north of the trough. They also pointed out that where the postrift sediments are thickest off New Jersey, the Late Jurassic shelf edge migrated eastward over oceanic crust, which is consistent with the voluminous sediment supply (Fig. 2). This is the only area along the U.S. Atlantic margin where the paleoshelf edge migrated over oceanic crust.

The Great Stone dome is interpreted to be a mafic intrusion of Early Cretaceous age about 20 km in diameter and about 40 km landward of the ECMA in the Baltimore Canyon trough (Fig. 2) (Grow and others, 1988a). This is the only major intrusion to penetrate high into postrift strata along the U.S. Atlantic margin landward of the ECMA. Much deeper intrusives may underlie the Georges Bank basin, and the Cretaceous New England Seamount Chain reaches the continental rise south of that basin (Schlee and Klitgord, 1988; Thompson and others, 1993). It is also possible that the circular magnetic and gravity anomalies under the Carolina platform off South Carolina are Mesozoic intrusives (Horton and others, 1991; Rankin and others, 1993a).

The Carolina trough is the narrowest of the marginal basins; the ECMA is typically 50 to 80 km east of the hinge zone. The salt diapirs and the related growth faults have been discussed. Sedimentation in the Carolina trough was most rapid in the Jurassic, during which about 7.5 km of 12 km of postrift sediments in the trough accumulated (Dillon and Popenoe, 1988; Klitgord and others, 1988). The paleoshelf edge grew vertically above and perhaps eroded back slightly from the landward edge of the ECMA (Klitgord and others, 1988). Dillon and Popenoe (1988) noted that no clearly defined reefs appear in seismic profiles of the Carolina trough. The reefal structures at the Jurassic and Cretaceous paleoshelf edge, common along the other marginal basins, if they existed along the Carolina trough, are largely eroded away.

The ECMA stops at the north edge of the Blake Plateau basin (Fig. 2) (Klitgord and others, 1988). The Blake Plateau basin extends seaward of the projection of that anomaly to the Blake Escarpment, a massive carbonate escarpment located landward of the Blake Spur magnetic anomaly (Figs. 1 and 2).

The Blake Plateau basin is the widest of the marginal basins in the U.S. Atlantic margin. The basin is more than 400 km wide, which is about 6 times as wide as the Carolina trough just to the north (Dillon and Popenoe, 1988). It is inferred that rifting occurred over a wider zone for a longer period of time than for the other basins, and was accompanied by widespread mafic intrusions (Dillon and Popenoe, 1988; Klitgord and others, 1988). Up to 13 km of sediments fill the Blake Plateau basin. These are dominated by carbonates and constitute a huge carbonate platform in contrast to the basins to the north, which contain considerable clastic material (Dillon and Popenoe, 1988). No salt diapirs are known. The shelf margin reef of the basin died at the beginning of the Late Cretaceous and was never reestablished. During most of the Late Cretaceous and Paleogene, flow of the Gulf of Mexico water to the Atlantic was across the southeast Georgia embayment (Suwannee strait) and northward across the Carolina trough shelf. This flow presumably prevented deposition of continental sediment on the more seaward Blake Plateau (Dillon and Popenoe, 1988). The plateau at that time was covered by marl that accumulated in water deposits of several hundred meters. The presence of that marl also means that the Gulf Stream was not operative across the Blake Plateau until the end of the Paleocene. The Gulf Stream became a major ocean current at the end of the Paleocene or beginning of the Eocene, when the Gulf of Mexico water could no longer pass through the Suwannee strait, but began the flow around the Florida peninsula (Dillon and Popenoe, 1988). The Gulf Stream has over time migrated back and forth across the Blake Plateau, scoured its surface, and limited progradation of the Florida-Hatteras Shelf. A new deepwater flow at the beginning of the Oligocene eroded and steepened the continental slope, and created the steep cliff of the Blake Escarpment off the Blake Plateau basin.

Atlantic Ocean Basin. The upper part of the crystalline basement of the Atlantic Ocean basin is oceanic crustal layer 2, which is composed almost entirely of extrusive basalt (Bryan and Frey, 1986). Layer 2 is 2 to 3 km thick, and has seismic velocities that are variable both laterally and vertically. The velocity generally increases downward from about 4 to 6 km/s (Trehu and others, 1989). Unusually thin upper crust that has anomalously low velocity has been associated with oceanic fracture zones (Trehu and others, 1989). The basement surface is commonly characterized by a series of hyperbolic echoes or irregular diffractions due to rough basement topography (Klitgord and others, 1988; Grow and others, 1988a, 1988b). This surface generally appears in the seismic record tens of kilometers seaward of the ECMA and gradually rises seaward. The most landward part of this diffracting surface is thought to be the top of early Middle Jurassic oceanic crust (Klitgord and Grow, 1980). Because continental rise sediments onlap oceanic basement surface seaward, it is interpreted that the oceanic crust is younger seaward (Klitgord and Grow, 1980).

The ECMA (Fig. 2), a prominent positive magnetic anomaly, is taken as roughly marking the landward edge of oceanic crust from the Blake Spur fracture zone northeast into the Canadian margin (Grow and Sheridan, 1988). Details of the source of the ECMA remain controversial (see below). The Blake Spur magnetic anomaly is a prominent positive magnetic anomaly of lower amplitude than the ECMA, and is located about 150 to 250 km seaward of it (Klitgord and others, 1988). It can be traced from the Bahamas fracture zone to the Long Island fracture zone (Fig. 2). North of the Blake Spur fracture zone it is flanked on both sides by oceanic crust. The Blake Spur magnetic anomaly is interpreted to mark a major shift in the location of the spreading center in the Middle Jurassic (Vogt, 1973). Series of landward-facing scarps between the ECMA and the Blake Spur magnetic anomaly are thought to mark down-to-the west faults (Klitgord and others, 1988). The Blake Spur anomaly coincides with the most seaward of these basement steps.

As noted above, on most of the multichannel seismic-reflection profiles, there is a landward limit of well-defined oceanic crust characterized by hyperbolic echoes. Along most of the U.S. margin, this characteristic signature of oceanic crust is lost landward beneath a highly reflective sedimentary reflector (J3) at the most landward of the landward-facing scarps (J3 scarp) (Klitgord and Grow, 1980). Seaward of Georges Bank (E1), however, well-defined oceanic crust extends landward to a zone of seaward-dipping reflectors, which are in turn seaward of the Jurassic paleoshelf edge (Klitgord and others, 1988; Thompson and others, 1993). The seaward-dipping reflectors are above a poorly defined basement surface. The zone of seaward-dipping reflectors here is about 25 km wide and in E1 is coextensive with the ECMA. Analogous zones of seaward-dipping reflectors occur on other continental margins at the very edge of oceanic crust. On some of those margins where they have been drilled, they are composed of basaltic flows and dikes interbedded with a small amount of sediment (Klitgord and others, 1988, and references therein). The zones of seaward-dipping reflectors have been attributed to the earliest formation of oceanic crust or final formation of rift-stage crust (Mutter, 1985). This crust, characterized by seaward-dipping reflectors along the landward edge of oceanic crust, was called marginal oceanic crust by Klitgord and others (1988) (Fig. 2).

The identification of a crustal layer with velocities of 7.2 to 7.5 km/s, interpreted to be a zone of magmatic underplating beneath the marginal basins, suggests that the ECMA may be a boundary in the upper crust only (Trehu and others, 1989). The ECMA "may represent a lateral change from older, very extended continental rocks to younger volcanic rocks in the upper crust overlying a continuous lower crust formed by magmatic processes during the late rifting and early seafloor-spreading phases of margin evolution. The implications for thermo-mechanical models of the rifting process of a thick underplated layer are only beginning to be explored" (Trehu and others, 1989, p. 377).

CRUSTAL PROPERTIES ACROSS THE EASTERN CONTINENTAL MARGIN

Three topics illustrate aspects of crustal properties across the eastern continental margin of the United States. These are (1) the Appalachian gravity gradient and gravity high, (2) the crustal thickness and Alleghanian shortening, and (3) the early Mesozoic extension. For a more thorough discussion of crustal properties, see Pakiser and Mooney (1989), Hatcher and others (1989b), Sheridan and Grow (1988a), and Keen and Williams (1990).

Appalachian gravity gradient and gravity high

The west-sloping Appalachian gravity gradient is a major feature of the Appalachians from Newfoundland to Alabama and is crossed by all of the E transects and the Quebec–Maine–Gulf of Maine transect (Fig. 2). The gradient is between a trough in the Bouguer gravity field on the northwest and a ridge in the field to the southeast. The trough is typically less than –50 mGal and the ridge is typically greater than 20 mGal, although the crest of the ridge is less for much of the length of Vermont and between the Chaudière River (Fig. 1) and the Shickshock Mountains of Quebec (refer to Fig. 8) (Society of Exploration Geophysicists, 1982; Haworth and others, 1980). The greatest magnitude of the gradient as well as the maximum range of the Bouguer anomalies are within or adjacent to the E4 transect; lows of less than –105 mGal occur over the Blue Ridge and highs of greater than +45 mGal occur over the Carolina slate belt (Kumarapeli and others, 1981, reported a somewhat greater maximum of +54 mGal over the Sutton Mountains of southern Quebec). The Appalachian gravity gradient has been discussed by numerous investigators who have constructed transverse and longitudinal profiles and modeled the feature in various ways. These investigators include Diment (1968), Best and others (1973), Long (1979), Cook and Oliver (1981), Kumarapeli and others (1981), Hutchinson and others (1983a), Karner and Watts (1983), Thomas (1983), Cook (1984), Pratt and others (1988), and Thompson and others (1993).

Interpretations of the Appalachian gravity gradient include changes in crustal thickness (caused by the edge of Laurentia at depth or thinning of Laurentian crust toward a continental margin; i.e., transitional crust or Mesozoic thinning of North American crust), continental rifting, dense bodies in the upper crust, a dipping slab of dense material (suture zone), or some combination of these interpretations.

Although regional gravity maps show a continuous paired Bouguer anomaly and accompanying gradient along the length of the Appalachians, it is obvious that several factors contribute to these features, and that these factors are not the same along the length of the orogen. Symptomatic of these differences is the observation that much of the discussion of these features focuses on the gravity high in the northern Appalachians and on the gravity gradient in the central and southern Appalachians.

First, consider the paired anomaly and gradient in New England and adjacent Quebec. Kumarapeli and others (1981) presented a convincing case for the Bouguer positive anomaly (amplitude of +54 mGal; magnitude of gradient 90 mGal) over the Sutton Mountains (Fig. 1) being caused by two components—a nonmagnetic, dense, deep-seated source (mid-crustal and lower crustal levels) and a dense, near-surface magnetic source. They modeled the near-surface magnetic source as a pile about 8 km thick of mostly mafic volcanic rocks that crop out at the present level of erosion as the relatively narrow belt of the

Tibbit Hill Formation (also called the Tibbit Hill Volcanic Member of the Pinnacle Formation in the United States) (Fig. 4) in the core of the Sutton Mountain anticlinorium. They concluded that the deepest reflector on three southeast-trending seismic reflection lines that partly transect the area of the gravity high is the surface of Grenvillian basement. This reflector slopes gently southeast from depths of 2 to 3 km near the St. Lawrence River toward the Sutton and Notre Dame mountains (Fig. 1). The available space above the reflector under the Sutton Mountains could accommodate an 8-km-thick pile of Tibbit Hill, which is the oldest unit of the cover sequence in the region (Lower Cambrian). The deep-seated source is not discussed in detail, but the assertion that the Tibbit Hill volcanic rocks are rift volcanics at a rift-rift-rift (RRR) triple junction (see also Burke and Dewey, 1973; Rankin, 1976; Kumarapeli and others, 1989) suggests that the deep-seated source is caused by crustal thinning and/or mafic intrusions also related to rifting. To the extent that that is true, the exposed Tibbit Hill Formation cannot have moved significantly from its original position above autochthonous Grenvillian basement.

Some of the transported rocks of the external domain such as the Granby and Chaudière River nappes (Figs. 1 and 4), which are now northwest of the Sutton Mountain and Notre Dame Mountain anticlinoria, are allochthons composed of Lower Cambrian slope-rise sedimentary and minor volcanic rocks and were emplaced, at least in part, above Middle Ordovician wildflysch (St. Julien and Hubert, 1975; Slivitzky and St. Julien, 1987). These rocks are less deformed than those of the internal domain including the Tibbit Hill Formation and other rocks in the core of the Sutton Mountain and Notre Dame Mountain anticlinoria. Perhaps the transported rocks, like the Taconic allochthons, were emplaced early in the Taconian deformational sequence across the Tibbit Hill Formation and associated rocks. If so, their root zone probably lies between the continental rift-facies volcanic rocks of the Tibbit Hill Formation and the accretionary wedge of the accreted Hazens Notch slice of the Brompton-Cameron terrane. Such an interpretation is permissive of relatively minor displacement of the Tibbit Hill volcanic sequence from its underlying Grenvillian crust.

Rooting the Granby and Chandière River nappes along the northwest boundary of the Hazens Notch slice and thus requiring them to have passed over the continental rift-facies rocks (Tibbit Hill Formation) may account for the presence of medium-high–pressure minerals in the metabasalt. Trzcienski (1976) identified crossite in the Tibbit Hill Formation in the area of the gravity high over the Sutton Mountain anticlinorium. He estimated a crystallization pressure of 6 to 7 kbar (20 to 25 km depth of burial) based on the sodium content of the crossite. Medium-high–pressure Ca-rich amphibole also occurs in biotite-grade rocks (about 375 °C at 5 kbar) along a relatively narrow zone 40 km long in mafic rocks of the Tibbit Hill Formation in northern Vermont (Laird and Albee, 1981; Laird and others, 1984). Perhaps Taconic-like allochthons, now completely eroded, once buried the Tibbit Hill Formation of northern Vermont. Note that the glaucophane-bearing, high-pressure rocks in northern Vermont occur only in a restricted area of the Hazens Notch slice, where Bothner and Laird (1991) estimated recrystallization temperatures as high as 550 °C and a pressure of 12.5 kbar (depth of burial of about 45 km). The high-pressure rocks of the Hazens Notch slice are separated from the medium-high–pressure rocks of the Tibbit Hill Formation by a belt of medium-pressure rocks. The extreme conditions of the high-pressure rocks in the Hazens Notch slice are interpreted to have occurred on the downgoing slab of a subduction zone about 500 Ma, prior to collision (Laird and others, 1984; Sutter and others, 1985). This was followed by rapid ascent on thrust faults (representing collision) and partial retrogression of the Hazens Notch slice at the same time as the Taconian lower temperature and medium-high–pressure metamorphism of the Tibbit Hill Formation (about 470 Ma) (Laird and others, 1984; Bothner and Laird, 1991).

It is important to keep in mind for later discussion that early Paleozoic rocks west and northwest of the Green Mountain–Sutton Mountain anticlinorium are in three distinct packages: (1) an autochthonous cover sequence on Grenvillian basement; (2) imbricately faulted platform-shelf rocks bounded on the west and northwest by the Champlain thrust and Logan's line; and (3) slope-rise rocks of the Taconic allochthon and the Granby and Chaudière River nappes. In Quebec the first of these is the autochthonous domain of St. Julien and Hubert (1975), and the second and third are their external domain.

The gravity high in central and southern Vermont shows a marked correlation with the exposed Grenvillian basement of the Green Mountain massif, and has an amplitude of 85 mGal above the gravity trough to the west under the Taconic allochthon (Thompson and others, 1993). The exposed basement rocks, which are of relatively low density, cannot produce the gravity high. Modeling by Diment (1968) led to his suggestion that the anomaly was caused by the uplift of a formerly laterally continuous layer of mid-crustal rocks along a narrow zone of deformation (faulting) at the western edge of the high. Modeling by Thompson and others (1993) along the northern line of section in E1 and along the nearly coincident COCORP line from the Adirondacks across the Green Mountains massif permits the presence of an east-dipping slab of transitional or diked crust (density about 2.99 gm/cm^3) that extends to within 8 to 10 km of the surface. That slab is overlain by a higher slab containing the less-dense Grenvillian basement that is exposed at the present erosion level. In their model the slab of buried transitional crust as well as the autochthonous, nondiked Grenvillian crust immediately to the west (which is exposed as the Adirondack massif) extend to depths slightly greater than 40 km beneath the Green Mountains. The crust under the Adirondacks is somewhat thinner (36 km) and has a gradational contact with the mantle (Taylor, 1989). Note that at this latitude in Vermont, no thick pile of Lower Cambrian rift volcanics is present and the gravity high does not coincide with a strong magnetic high. This is in strong contrast to the coincident gravity and magnetic high over the Sutton Mountains (Kumarapeli and others, 1981).

The New York–New England COCORP data (see Ando and others, 1984) as reprocessed by Phinney and Roy-Chowdhury (1989) reveal a reflector at about 0.3 s two-way traveltime (about 0.6 km depth) under the Taconic allochthon, which was interpreted as the sole thrust of the Taconic allochthon, and a reflection at about 3 s, which was interpreted as the top of the Grenvillian basement or bottom of the autochthonous Paleozoic platform rocks. The lower surface reappears east of a data gap as a subhorizontal reflector under the Green Mountain massif. Beneath the eastern part of the exposed massif this reflector dips eastward, reaching a depth of at least 4 s at the Connecticut River (the end of that segment of the profile). The Green Mountain massif thus appears to be thrust westward over platform sedimentary rocks and, along with the basement of the Chester and Athens domes, rooted at least as far east as the Connecticut Valley.

The similarity of the gravity field over the Green Mountain massif with that over the Sutton Mountains suggests that they have similar causes. An argument was made above that the gravity high over the Sutton Mountains is caused by a pile of rift-facies mafic volcanic rocks overlying diked Grenvillian crust, both related to crustal extension prior to the opening of the Iapetus Ocean. It is an enigma that the Green Mountain gravity high so closely mimics the shape of the exposed massif, which does not contain abundant dikes suggestive of rifted crust. I favor the interpretation that the Green Mountain–Sutton Mountain–Notre Dame Mountain gravity highs are caused by an Iapetus rift zone in Grenvillian continental crust that is essentially in place, in which mafic dikes or intrusives are an important constituent. Perhaps the different crustal properties of the rift zone compared with the nonrifted crust on either side (my interpretation) caused the root zone of the imbricate thrust stack (east-dipping reflectors of basement slices) to be located beneath and east of the eastern part of the Green Mountain massif (see Ando and others, 1984, Fig. 3). A culmination in the overriding duplex stack of basement slices to produce the Green Mountain anticlinorium (see Stanley and Ratcliffe, 1985) could place the low-density, dike-free basement of the Green Mountain massif over autochthonous high-density, dike-rich crust. Resolution of the seismic reflection data is not adequate in the vicinity of the Green Mountain massif to confirm or deny such a possibility.

A recent seismic refraction–wide-angle reflection study across the Adirondacks and central Vermont to western Maine by Hughes and Luetgert (1991) revealed a major lateral velocity change from higher velocity Grenvillian crust of the Adirondacks to lower velocity and layered upper and middle crust of the Appalachians. The velocity change occurs along a ramp-like boundary that projects to the surface under the Champlain Valley and dips eastward to a depth of about 25 km beneath the eastern edge of the Connecticut Valley trough in Vermont. The profile crosses the Green Mountain–Sutton Mountain anticlinorium at the northern point of the Lincoln Mountain massifs (Fig. 4). No Grenvillian basement is specifically identifiable by the seismic study above the ramp. The profile crosses the saddle in the Appalachian gravity high between the Green Mountain massif and the Sutton Mountains. The higher velocity crust (6.55 to 6.65 km/s) beneath the ramp was interpreted by Hughes and Luetgert (1991) to be a subsurface continuation of the Marcy Anorthosite. The rift zone of highly diked crust proposed above should presumably be part of the higher velocity Grenvillian crust below the ramp. J. H. Luetgert (1991, oral commun.) stated that such a zone of diked crust would not be distinguishable from the Marcy Anorthosite by the refraction–wide-angle reflection study.

From the Green Mountain massif northward to the Notre Dame Mountains (Fig. 1) the linear nature of the gravity high suggests that this zone of Iapetus continental rifting and diking was relatively narrow and, based upon evidence presented below, a long way inboard from where the continental margin ruptured to form Iapetus oceanic crust (Fig. 2). Grenvillian crust is thought to extend in the subsurface along the Quebec–Maine–Gulf of Maine transect as far southeast as the Lexington pluton (Fig. 4), which is about 170 km southeast of the trace of the gravity high (Stewart and others, 1991). As noted earlier, Harrison and others (1987) concluded that two Concord Granite–type plutons (two-mica, undeformed granitoid) in western New Hampshire were probably generated from Grenvillian crust. The northern of those two plutons (Long Mountain) is about 110 km southeast of the trace of the gravity high over the Sutton Mountains, and the southern one (Sunapee) is about 70 km east of the Green Mountain gravity high (Fig. 4). Along the U.S. Geological Survey marine line 36 on the Long Island platform, Phinney and Roy-Chowdhury (1989) interpreted a crustal block of Grenvillian basement and rocks here included in the Brompton-Cameron terrane to extend eastward beneath a wedge-shaped upper crustal mass. That mass is between the east-dipping downward continuation of the border fault of the Mesozoic Long Island basin and the west-dipping Block Island fault (Fig. 6) (Hutchinson and others, 1985). The crest of the gravity high is not well defined in southeastern Connecticut (Fig. 2), but by this interpretation Grenvillian crust may extend as much as 150 km east-southeast of the Manhattan prong (a Grenvillian massif in and north of New York City).

South of the Green Mountain massif, the gravity high is broader, more diffuse, and displaced progressively farther east or southeast of the external Grenvillian massifs (Blue-Green-Long axis). The axis of the Berkshire massif is about 20 km west of the crest of the gravity high (Thompson and others, 1993), and the French Broad massif is about 200 km northwest of the high in North Carolina (Fig. 2) (Haworth and others, 1980). The Reading prong and the French Broad massif are, in fact, essentially over the trough of the gravity low. Note also that from Newfoundland to Maryland the trace of the gravity gradient and the crest of the high follow bends in the Appalachian structural trends (salients and recesses) or, in other terms, mimic the Blue-Green-Long axis (Rankin, 1976; Gravity Anomaly Map Committee, 1987). South of Maryland, the generalized trace of the gradient maximum and high is a shallow arc concave toward the foreland. The gradient is deflected around the locus of the Sauratown

Mountains anticlinorium and passes along the length of the Grenvillian external massif exposed in the Pine Mountain window, Georgia and Alabama (Fig. 2) (Rankin, 1975).

A number of workers have made the assumption that the source of the high is the same along the length of the orogen and have noted the southward increase in the displacement of the external massifs from the trace of the gravity high. Some of these workers suggested that the external massifs have been detached from their high-density roots by Alleghanian thrusting of increasing displacement southward in the central and southern Appalachians (e.g., Rodgers, 1970; Bryant and Reed, 1970; Rankin, 1975; Thomas, 1983; Thompson and others, 1993). Rodgers (1970) and Bryant and Reed (1970) used this apparent displacement in support of the hypothesis of an allochthonous crystalline Blue Ridge. Support for the general concept comes from the dying out of the Alleghanian foreland fold and thrust belt in southern New York State, and the change in geometry of the gradient and high from following salients and recesses in the north to a broad arc in the south.

In the southern Appalachians, the proximity of the gravity gradient to the Sauratown Mountains and Pine Mountain internal massifs, which are the southeasternmost exposures of convincing Grenvillian basement, raises the possibility that the gradient is an expression of the southeastern limit of Laurentia (Fig. 2). The Goochland terrane, which is here not considered to have been part of Laurentia, lies southeast of the gradient. At the latitude of Richmond, Virginia, the Goochland terrane coincides with the gravity high (Fig. 2).

If we accept the evolutionary history of the Appalachians as deduced here from surface geology, the southeastern limit of Laurentia should be a suture zone between Laurentia and a volcanic arc or continental plate, but perhaps including rocks of an accretionary wedge. The suture zone should, in turn, reflect the rifted margin of Laurentia but, of course, the shape of the rifted margin could have been modified prior to or during continental accretion. A modeling of the gravity gradient by Hutchinson and others (1983a) permitted the interpretation that the gradient marks the position of the Iapetus suture. This interpretation was adopted for transect E4. The position of the suture on the E4 cross section under the Charlotte belt was constrained by an attempt to construct a balanced cross section for rocks exposed at the surface as far southeast as the Brevard zone (Rankin and others, 1991). Thomas (1983) presented a similar interpretation for the gravity gradient along the Savannah River COCORP line.

A few other points should be made briefly. Several workers, to a greater or lesser degree, attribute the gravity gradient and high to eastward thinning of the continental crust during Mesozoic extension (Nelson and others, 1986; Glover and others, 1987; Pratt and others, 1988; Costain and others, 1989b). Mesozoic extension is here considered to be an important aspect of the eastern continental margin, and we return to that topic later. Mesozoic extension may have modified the slope and magnitude of the gradient, but the sizes of the Mesozoic basins along the E4 transect do not seem adequate to account for all of the 25% crustal thinning (Hutchinson and others, 1983a) under the Piedmont, nor is extension the major cause of the gradient and high in the northern Appalachians, if the model presented above is valid. Wernicke and Tilke (1989), however, pointed out that if there is significant foowall upward motion on a listric fault, the actual separation between the hanging-wall cutoff and the footwall cutoff may be much larger than the slip deduced from the thickness of strata in the basin. This possibility remains to be tested in the southern Appalachians.

The gravity gradient appears to be truncated by the Alleghanian suture beneath the coastal plain of Alabama. Hence the gradient probably predates the latest stages of the Alleghanian orogeny. In Maryland and northern Virginia the gradient and high are west of the internal massifs of the Baltimore Gneiss. Perhaps the gravity high there is related to a subsurface Iapetan rift analogous to the Green Mountain–Sutton Mountain high, rather than the Laurentian continental margin. The residual Bouguer gravity field for wavelengths less than 250 km (Hildenbrand and others, 1982) indicates a bifurcation of the Bouguer gravity high in northern Virginia. If the gravity gradient marks the eastern border of Laurentia in the central and southern Appalachians, it is clearly a different feature from the gradient (and high) in the northern Appalachians.

Crustal thickness and Alleghanian shortening

In addition to the individual transects, summary papers (Sheridan and others, 1988a; Braile and others, 1989; Costain and others, 1989b; Taylor, 1989; Trehue and others, 1989) discussed crustal thickness, velocity structure, and the data on which they are based. Figure 3 of Braile and others (1989) portrayed crustal thickness as contoured data, and those data are reproduced in Figure 8. Here I summarize, without defending, crustal-thickness data from the transects along with some other information as appropriate.

Profile AA′ of E1 (Thompson and others, 1993) portrays the crust as 35 km thick under the Adirondacks, thickening to 40 km under the Green Mountain gravity high, then gradually thinning across central New England to about 35 km at Cape Ann, Massachusetts. Offshore, the depth to the Moho decreases gradually to a broad zone of constant depth at about 28 km between the Nauset fault and the basement hinge zone (upper surface of the basement) under the Franklin basin (Fig. 5). From the hinge zone the depth to the Moho decreases across the zone of transitional crust to about 15 km over oceanic crust. Here, as in the other transects, layer 2 of the oceanic crust is about 2 km thick and layer 3 is 5 to 6 km thick. Gravity modeling of onshore data along profile BB′ of E1 is consistent with a broad root as much as 45 km thick and including higher density (Iapetan transitional?) crust from the Berkshires to the vicinity of the Clinton-Newbury fault. This root is bordered on either side (Mohawk Valley–Taconic allochthon on the west and southeastern Massachusetts on the east) by continental crust about 35 km thick.

The depth to the Moho along the Quebec–Maine–Gulf of

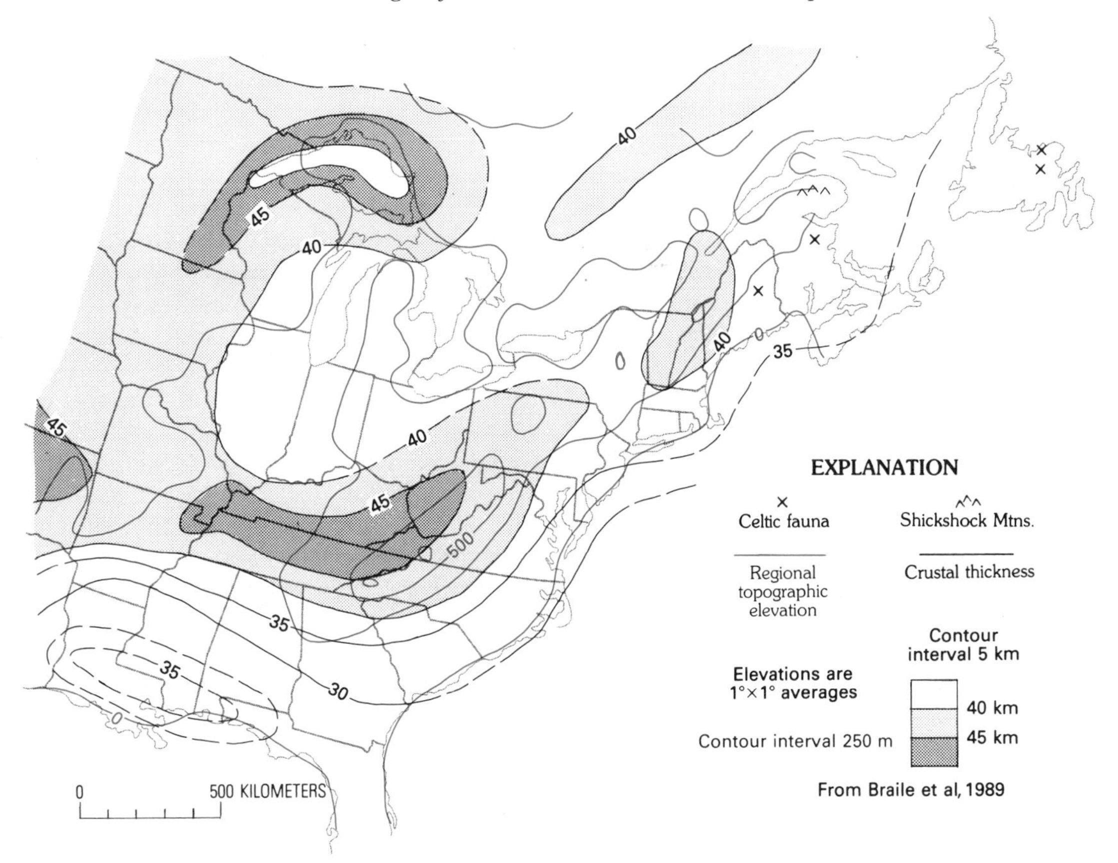

Figure 8. Map showing regional topographic elevation superimposed on contoured crustal thickness for part of eastern North America (from Braile and others, 1989).

Maine transect is somewhat better controlled and offers some new insight. The crust is about 40 to 42 km thick in northwestern Maine and thins to about 34 km under coastal Maine (Stewart, 1989). The crust appears to thin across two steps about 15 km wide in which the Moho rises 3 to 4 km across each step. As noted by Stewart (1989) the steps strike parallel to regional structural trends and are parallel to, but about 15 km southeast of, the maximum gradients in the gravity field. The Moho continues across the Gulf of Maine at a uniform depth of about 33 to 36 km (Hutchinson and others, 1988). Both across Maine and the Gulf of Maine, the Moho passes undisturbed under several major faults (Norumbega, Sennebec Pond, and Turtle Head; Fig. 6) that are thought to be Paleozoic terrane boundaries. Stewart (1989) and Hutchinson and others (1988) concluded from this that the Moho here is a Mesozoic feature, and suggested that the same may be true for much of the Appalachian orogen or at least the eastern Appalachian orogen. Hutchinson and others (1988) defined the Moho as the surface separating a zone of abundant, generally subhorizontal reflectors above from the region below that produced essentially no reflections. The zone of abundant reflections varies in thickness from less than 1 s under the northern Gulf of Maine to as much as 5 s beneath the central and southern part, suggesting that the lower crust–Moho transition is variable in thickness along strike (Hutchinson and others, 1988). Phinney and Roy-Chowdhury (1989) made a similar observation in the reprocessing of line 36 on the Long Island platform. They proposed that the Moho be considered a package of rocks in which the velocity increases rapidly, rather than an abrupt boundary; evidence from refraction studies restricts the thickness of the package to only 1 to 2 km.

Farther south along transect E3 and specifically along profile AB of Pratt and others (1988) along the Interstate 64 U.S. Geological Survey seismic reflection profile, a crustal root under the eastern part of the Valley and Ridge province may be more than 50 km thick (modified from data of James and others, 1968). The crust thins eastward across the Blue Ridge and Piedmont to a little less than 35 km under Richmond, Virginia. The profile of Pratt and others (1988) shows a direct correspondence between the rise of the Moho and the rise in the Bouguer gravity field (Appalachian gravity gradient). The crust then thickens abruptly to a root 45 km deep under Williamsburg (Fig. 1) before thinning toward the continental margin. In gravity modeling along profile AB, Pratt and others (1988) incorporated a slab of denser mafic material dipping eastward about 45° beneath the Sussex-Leonardtown anomaly (Fig. 2) that penetrates the entire crust. This slab intersects the Moho at the eastern hinge of the 45-km-deep root near Williamsburg. It is difficult to relate Figure 6 of Pratt and others (1988) to surface geology because some of the vibration points on their Figure 7 are mislabeled and because the Mesozoic basin identified beneath the coastal plain cover in their Figure 6 is not shown on E3 (Glover, 1991, written commun.). Pratt and others (1988) also incorporated in their model an eastward-thickening wedge of less-dense crustal material southeast of the subsurface Mesozoic basin. In the interpretation favored here and elsewhere (Lefort and Max, 1991; Horton and others, 1991), the bottom of the east-dipping mafic slab marks the Alleghanian suture. The less-dense mantle material east of the slab may be a distinctive feature of the Chesapeake terrane, because such a feature has not been identified in other Appalachian gravity models.

The crustal thickness in E4 (Rankin and others, 1991) as adopted from the gravity model of Hutchinson and others (1983a) increases southeastward from about 35 km near the Cincinnati arch to a maximum of about 45 km under the Inner Piedmont. The Moho rises abruptly eastward under the gravity gradient to about 35 km and maintains that depth farther east, except for some possible crustal thinning under the Piedmont in the vicinity of abundant Mesozoic dikes and the Wadesboro–Deep River basin (Fig. 1), out to the basement hinge zone. The zone of transitional crust is about 100 km wide, and the Moho under the ocean crust is at a depth of about 15 km below sea level. Similar crustal thicknesses are reported for E5, which includes the COCORP Savannah River seismic reflection transverse (Cook and others, 1981), except that no crustal root is shown under the western Appalachians. The Moho is portrayed at a depth of 45 km from the Cumberland Plateau to the Inner Piedmont (but shown in Speed, 1991, as only a little more than 30 km).

Figure 8 shows the regional topographic elevation ($1^\circ \times 1^\circ$ averages) superimposed on a contour map of crustal thickness for part of eastern North America (both from Braile and others, 1989, Figs. 3 and 10). The crustal thickness is measured from the surface to the inferred Moho and is based upon seismic refraction data. The Bouguer gravity field of the eastern United States is shown in Figure 2. From these data it can be seen that there is a crustal root under the western Appalachians, and that it is, in general, under or offset somewhat to the northwest from those areas of highest average elevation and lowest regional Bouguer gravity. The crustal root of Braile and others (1989) shown in Figure 8A is also offset somewhat to the northwest from the thickest crust shown in E3 and E4.

At least three mechanisms have been suggested to account for the thicker continental crust, or the negative Bouguer gravity field under the Blue Ridge (Fig. 2), or the area of high average elevation. Cook (1984) as well as Hutchinson and others (1983a) and James and others (1968) attributed the negative gravity field to a root of continental crust caused by the stacking of thrust sheets, including the 6- to 10-km-thick Blue Ridge crystalline thrust sheet. Karner and Watts (1983) demonstrated by modeling that flexure of continental crust under the load of thrust sheets and obducted material could, over time, produce the Appalachian gravity field. Nelson (1992) argued that lithospheric delamination following collision and crustal thickening could cause crustal thinning under the internides accompanied by magma production and regional metamorphism. Sacks and Secor (1990) also proposed a delamination model and applied it to the southern Appalachians. Their mechanism, however, is actually slab pull and detachment of an oceanic lithosphere slab, and hence is not delamination in the sense of Nelson (1992) or Bird (1978), who first proposed the idea. The observed features attributed to the stacking of thrust sheets (Karner and Watts, 1983) or delamination (Nelson, 1992) are results of the Alleghanian orogeny.

The third mechanism involves Mesozoic extension. Wernicke and Tilke (1989) noted that the areas of high Appalachian summits correlate with promontories of the offshore basement hinge zone, and that areas of low Appalachian summits correlate with embayments in the hinge zone. The areas of high summits correspond roughly to the areas of average high elevation shown in Figure 8, except that the summit elevations better portray the Adirondack–New England area as being a high area within the Georges Bank promontory. Wernicke and Tilke (1989) suggested that the areas of high Appalachian summits are remnants from Jurassic long-wavelength arching due to synrift crustal thickening by underplating by mafic magmas, and are perhaps analogous to the broad uplift of the Arabian shield adjacent to the Red Sea.

It is possible that each of these mechanisms contributed to the observed relations. As discussed in the next section, I am convinced that Mesozoic extension played a major role in the present configuration of the eastern part of the Appalachian orogen. The extent to which mafic rocks underplated the region of the topographic high and crustal roots should be tested by gravity modeling. It appears to me that the crustal root in the central and southern Appalachians is best explained by the stacking of thrust sheets. Support for this comes from the observation that the position of the crustal root, high average elevation, and gravity low all migrate from under the Appalachian Plateaus in West Virginia and Virginia to under the Blue Ridge in North Carolina, paralleling the migration of the gravity gradient. The direction of the migration is consistent with increasing northwestward displacement on Alleghanian thrusts.

Ductile deformation during compression may have contributed to crustal thickening and the formation of the crustal root. This was suggested by Iverson and Smithson (1982) as one option to resolve the dilemma predicted by Elliott's (1973) observation that palinspastic reconstructions across orogens commonly restore cover sequences to positions outboard of their presumed autochthonous sialic basement. Rankin and others (1991) incorporated the Iverson and Smithson suggestion in the E4 discussion. Retrodeformation of the E4 cross section, which is balanced only as far southeast as the Brevard fault zone, restores Grenvillian basement of the Blue Ridge crystalline thrust sheet southeastward (misstated as southwestward in Rankin and others, 1991) to the vicinity of Lumberton, North Carolina (Fig. 1). That amounts to a shortening of 300 km during the Alleghanian if the thrusting is indeed Alleghanian. The amount of shortening required is controlled in part by the assumed stratigraphy of duplexes under the structural culminations of the Mountain City and Grandfather Mountain windows. The 300 km shortening is a conservative interpretation in which the duplex at depth under the Grandfather Mountain window is assumed to be composed of Grenvillian basement and Late Proterozoic cover rather than Paleozoic cover alone. If the duplex is made up of younger rocks (Paleozoic), the amount of crustal shortening would be greater. In the stated conservative interpretation, the position of the restored Grenvillian basement near Lumberton is about 200 km southeast of the hypothesized position of the southeasternmost autochthonous Grenvillian crust at depth, and hence is an example of dilemma raised by Elliott (1973). The southeastern limit of Grenvillian crust is interpreted to be the Iapetus suture. If the crustal root predicted at depth by the gravity model of Hutchinson and others (1983a) is caused in part by subdecollement thickening of the lower crust, its retrodeformation would yield autochthonous Grenvillian crust and displace the Iapetus suture about 120 km to the southeast, nearly two-thirds of the distance needed to accommodate the allochthonous cover sequences and its attached basement. This is not intended to be a rigorous treatment, but only to show that the crustal root of the gravity model approaches the size needed to restore allochthonous Grenvillian crust over autochthonous Grenvillian crust.

The above discussion is clearly oversimplified. First, the reconstructions deal only with Alleghanian crustal shortening, because we do not have adequate constraints to make reconstructions of Taconian shortening. Second, the reconstruction of the balanced cross section does not take into account the effects of penetrative deformation, which includes considerable bulk shortening within the Blue Ridge thrust sheet and shear strain adjacent to the thrusts. The reconstructions do not account for pressure solution in carbonates of the Valley and Ridge (see Woodward and others, 1986). The estimates of shortening are therefore most certainly minimums. Hatcher (1989b) estimated the amount of Alleghanian shortening of the southern Appalachians to be between 350 and 500 km, which, in the vicinity of E5, would restore the leading edge of the Blue Ridge crystalline thrust sheet to a position immediately southeast of the present surface position of his central Piedmont suture (Lowndesville shear zone) (Fig. 1). Hatcher also cautioned that these estimates were based only on Alleghanian brittle nonpenetrative deformation in the plane of the sections he used, and that the strike-slip component was ignored as it was in the E4 discussion.

Mesozoic extension as deduced from crustal studies

In recent years numerous seismic reflection and refraction studies from nearly every continent have shown that areas of Mesozoic to recent crustal extension are characterized by a highly reflective Moho, whereas little-deformed Precambrian crational areas of continental interiors have a poorly defined Moho across which the number of reflections decreases gradually (Allmendinger and others, 1987, and references therein). This change in the character of the Moho can be seen in all of the seismic studies across the U.S. Appalachians from Maine to Georgia. On the Quebec–Maine–Gulf of Maine transect, the reflective Moho extends northwest nearly to the Chain Lakes block (Fig. 4) (Stewart and others, 1991). On the New England COCORP profile, the reflective Moho extends west to the Bronson Hill anticlinorium (Fig. 1) (Ando and others, 1984). On the I-64 line in Virginia, the reflective Moho extends northwest to the Chopawamsic terrane (Pratt and others, 1988). Along the Savannah River (east Georgia) COCORP line, the reflective Moho extends northwest to the vicinity of the Lowndesville shear zone (Fig. 1) (Cook and others, 1979), and on the west Georgia COCORP line, to the Pine Mountain window (Fig. 1) (Nelson and others, 1985). Because the highly reflective Moho passes undisturbed beneath dipping reflectors thought to be Paleozoic terrane boundaries, it is interpreted to be significantly younger than those boundaries and to have formed, or more properly reformed, in the Mesozoic as the continental crust was heated and extended. There is no evidence on the seismic profiles that major faults cut the reflective Moho as portrayed by Gates and others (1988). At least most of the continental extension ceased at the time of continental separation, so the extension probably dates from Late Triassic to Middle Jurassic.

Stewart and Luetgert (1990) concluded from seismic reflection and refraction studies in New England that the lower crust above the reflective Moho was also reconstituted during Mesozoic extension. The lower crust is characterized by numerous strong subhorizontal reflectors that are not interrupted by through-going faults. Major upper crustal structures imaged in seismic reflection profiles are lost in the top of the lower crust as defined by seismic refraction data. These include the southeast- or east-dipping decollement surface under the Chain Lakes block and Green Mountains massif, and Mesozoic faults that become listric at the top of the lower crust. Stewart and Luetgert (1990) cautioned that beneath the extended part of the orogen, the present lower crust cannot be unambiguously related to the overlying terranes in the upper crust.

Results from a recent short seismic reflection profile across part of the coastal plain of New Jersey also indicate that the lower crust continues uninterrupted under a prominent linear positive gravity anomaly called the Sussex-Leonardtown (Fig. 2) or the Salisbury anomaly (Sheridan and others, 1991). The anomaly was interpreted by Sheridan and others (1991) to be a faulted repetition of the Iapetus suture exposed in the Piedmont about 100 km to the northwest (the Pleasant Grove–Huntingdon Valley fault; Fig. 1, not the Martic line). The Sussex-Leonardtown anomaly, however, is interpreted here and by Horton and others (1991) to be the Alleghanian suture as proposed by Lefort and Max (1991). The important point is the continuity of the lower crust beneath the possible suture, again suggesting reconstitution of the lower crust during Mesozoic extension. Note, however, that not all lower crust is reconstituted with time. Presumably the temperature to which that crust was heated and the duration of that heating as well as mechanical changes during deformation determines the degree to which the lower crust is reconstituted. Very old features may also be preserved unless they have been modified by an important (significantly hot and long) younger process. For example, the 1.0 to 1.3 Ga Grenville front tectonic zone has been imaged through the entire crust under Lake Huron by seismic reflection profiling (Green and others, 1988).

Let us discuss briefly some of the surface expressions of the extension. The most obvious are the early Mesozoic basins and the Mesozoic dikes. The implications of dike orientation will be discussed later along with emplacement of the plutons of the White Mountain Plutonic-Volcanic Suite. There is considerable literature on the pure-shear vs. simple-shear mechanisms for the formation of rift basins during extension. This is not reviewed here; see Klitgord and others (1988), Manspeizer and others (1989), and Crespi (1988) for an introduction to the literature as applied to eastern United States. Although current interest focuses on the simple-shear mechanism for basin formation, Keen and others (1987) concluded from crustal studies off northeastern Newfoundland that the data better fitted the pure-shear model. Bell and others (1988), however, concluded that positive Bouguer gravity anomalies over the hanging walls of the early Mesozoic half grabens in eastern North America favor the simple-shear model over the pure-shear model for lower crustal extension. The interpretation of the asymmetry of rifting and footwall uplift by Wernicke and Tilke (1989) also implies simple shear.

Many of the border faults of the early Mesozoic basins are thought to be reactivated Paleozoic structures, commonly thrust faults, and many become listric at depth (Ratcliffe, 1971; Ratcliffe and others, 1986; Bobyarchick and Glover, 1979; Petersen and others, 1984; Hutchinson and Klitgord, 1988b). To a first approximation the amount of extension by the listric faults is at least as much as their throw and may be considerably more (Wernicke and Tilke, 1989).

The distribution of Paleozoic metamorphic grades in the Appalachians is such that there are alternating belts for low- and high-grade rocks (see Morgan, 1972). Some of this undoubtedly is because of the shifting of the loci of thermal maxima with time. The grade of metamorphism at the surface also reflects the depth of recrystallization of the exposed crust. Some of the differences in levels of crustal exposure are because of later folding of isogradic surfaces, as in the Blue Ridge in North Carolina (Rankin, 1975). Evidence is accumulating, however, that at least some of the belt pattern of metamorphic grades is the result of early Mesozoic extension and the isostatic rise of deeper crustal material on the footwalls of normal faults as the fault blocks are rotated. We may ultimately learn that early Mesozoic extension is a major cause of the belt pattern of metamorphic grades. Stewart (1989) suggested that the notched or lobed map pattern of isograds on the geologic map of Maine (Osberg and others, 1985) might originate from the erosional planing of an array of rotated extensional fault blocks along old faults reactivated in the early Mesozoic. Alternatively, the three lobes of metamorphic highs in western Maine are spatially related to concentrations of Devonian and Carboniferous plutons (sumarized by Guidotti, 1989). The Ammonoosuc fualt, now recognized as a west-dipping Mesozoic normal fault, juxtaposes chlorite-grade rocks of the hanging wall against kyanite-staurolite-grade rocks of the footwall in central western New Hampshire (Thompson and Norton, 1968; Rodgers, 1970; Spear and Rumble, 1986). The Ammonoosuc fault is on strike with the Hartford basin farther south in the Connecticut Valley (Fig. 4). The Hartford basin is a half graben with an eastern border fault, a geometry consistent with the west-side-down movement of the Ammonoosuc fault.

In an earlier section it was pointed out that both the Massabesic and the Falmouth-Brunswick blocks are bounded on their southeast sides by young, possibly Mesozoic, faults, i.e., the Wekepeke–Flint Hill and Flying Point faults, respectively (Fig. 6). Movement on the Wekepeke–Flint Hill system is up to the west, and in places juxtaposes kyanite-staurolite assemblages of the Massabesic block against chlorite-grade rocks of the Merrimack-Harpswell terrane (Zen and others, 1983). West and Lux (1989) showed that a thermal contrast existed across the Flying Point fault as late as 250 Ma.

In the southern Piedmont, the southeast-dipping ductile Augusta fault, which places greenschist facies rocks of the Carolina terrane over amphibolite facies rocks of the Savannah River terrane, has normal dip-slip movement (Maher, 1987). Earlier, in the discussion of the Savannah River terrane, I reviewed the evidence for 8 km of dip-slip movement in the late Paleozoic on the Augusta fault (5.1 km of differential uplift; Vyhnal and McSween, 1990). I also wondered, as had Heck (1989), if some of the extensional movement might be Mesozoic. Maher's evidence for extension on the Augusta fault is based on fabrics in syntectonic granitic veins and dikes, which are pre-Mesozoic. The cooling ages, discussed earlier, would require any Mesozoic movement to be in the brittle realm. The Augusta fault becomes listric at depth and may merge with the western border fault of the Riddlesville sub–coastal plain basin (Petersen and others, 1984). As noted earlier, Dennis (1991) suggested that the southeast-dipping Lowndesville shear zone, which places lower grade rocks of the Carolina terrane over higher grade rocks of the Inner Piedmont, could have normal dip-slip movement, although his own observations of the fabrics along this boundary offer no support for such a model. COCORP data for Georgia line 1 as reprocessed by Phinney and Roy-Chowdhury (1989) suggest that the Lowndesville becomes listric at a depth of about 12 km. Dennis (1991) suggested Devonian extension, but other than the previously discussed Silurian or Early Devonian plutons (see discussion of the Carolina terrane), we have no evidence for Devonian extension. Could the extension be Mesozoic? It is roughly at the position of the Lowndesville shear zone that the Moho becomes reflective, a property correlated above with extension (Allmendinger and others, 1987; Taylor, 1989).

South of E5, the northwest-dipping Towaliga fault (Fig. 1), which bounds the Pine Mountain window on its northwest side, was interpreted by Clarke (1952) to be a northwest-verging thrust fault that had transported the Inner Piedmont allochthon over the

Pine Mountain belt. Schamel and Bauer (1980) concluded that a large segment of the Towaliga was a postmetamorphic brittle normal fault with zones of silicified breccia. The COCORP west Georgia line (K. D. Nelson and others, 1987) shows that the sole thrust of the Inner Piedmont terrane or allochthon does not continue beneath the Pine Mountain belt. Instead, the reflections terminate at the downward projection of the normal-fault segment of the Towaliga. Thus Grenvillian basement appears to be offset as much as 9 km on a northwest-dipping brittle normal fault that may be Mesozoic.

Boyer and Elliott (1982) suggested that the Brevard zone may have been reactivated as an extensional fault during the Mesozoic largely because Triassic faults can be traced into it and because of the linearity of the zone (David Elliott, ca. 1982, oral commun.) The Stony Ridge fault zone (Fig. 1), which is within the Sauratown Mountains anticlinorium, was portrayed by Espenshade and others (1975) as connecting with the Brevard to the southwest. The Stony Ridge is characterized by silicified microbreccia and was interpreted by Butler and Dunn (1968) and Espenshade and others (1975) as a Mesozoic fault. Fullagar and Butler (1980) reported a Rb-Sr whole-rock age of 180 ± 3 Ma on the silicified microbreccia. Jurassic dikes are common in the Piedmont but rare northwest of the Brevard zone, suggesting that the Brevard marks a distinct change in crustal properties (see Costain and others, 1989b, Fig. 3), and is a likely candidate for reactivation. The southeast-dipping Brevard fault zone becomes listric with depth (Cook and others, 1979; Costain and others, 1989a), and places greenschist facies rocks of the Chauga belt in the hanging wall over amphibolite facies rocks of the Jefferson terrane in the footwall. The Brevard appears to be a splay off the Blue Ridge thrust fault, and Phinney and Roy-Chowdhury (1989) endorsed the interpretation that the Lowndesville shear zone is also a splay off the Blue Ridge thrust. Early thrust movement has been documented for the Brevard fault zone as well as later dextral strike-slip and brittle dip-slip movement (Vauchez, 1987; Edelman and others, 1987; Horton, 1982). Horton (1982) found evidence only for compressional movement (high-angle reverse) on the late brittle fault for the Brevard at Rosman, North Carolina (Fig. 1), but it has not been documented that all brittle structures in the Brevard zone are related to the brittle fault at Rosman (Bobyarchick and others, 1988).

Virtually all of the Piedmont faults have long, complicated histories, including early thrust movements. I do not dispute the early history of these faults. My point is that the present distributions of metamorphic zones are not explained easily by thrust movements, or the transpressional model of Gates and others (1988), and that perhaps significant reactivation during Mesozoic extension has not been given adequate consideration.

Before leaving the topic of Mesozoic extension, one must acknowledge that in the Canadian Maritime Appalachians and coastal New England, pull-apart basins were formed by the late Paleozoic by dextral movement on wrench faults (Bradley, 1982). These basins, including the Narragansett basin (although Snoke and Mosher, 1989, argued for a dominantly extensional origin for the Narragansett), contain mainly nonmarine, clastic strata, up to 9 km thick, that may range in age from Late Devonian to Permian (Bradley, 1982; Thomas and Schenk, 1988). Perhaps the forces that opened these basins also account for the late Paleozoic extensional movement on faults around the Willimantic dome (Getty and Gromet, 1992a, 1992b; Wintsch, 1992) and on the Clinton-Newbury and Lake Char fault systems (Goldstein, 1989). At least for the Narragansett basin, however, the forces that opened the basin in the Carboniferous predate the Alleghanian orogeny and Narragansett Pier Granite (273 Ma), which in turn predate the extensional fabrics of the Willimantic dome. How the late Paleozoic extension in New England relates to the late Paleozoic uplift of the Savannah River terrane (Snoke and Frost, 1990) is not known.

EASTERN CONTINENTAL MARGIN IN TIME: A HISTORY

Most of the remainder of this chapter is a review of the Phanerozoic evolution of the eastern continent-ocean transition as it developed through time. The preaccretion history of non-Laurentian terranes is not emphasized.

Continental rifting, opening of Iapetus, and formation of the Laurentian passive margin of North America

Long after the Grenvillian orogeny of about 1.1 to 1.0 Ga, a Proterozoic supercontinent (Panchaos of Thompson and others, 1993) ancestral to eastern North America underwent a prolonged period of extension resulting in the formation of continental rifts, intrusion of anorogenic mafic and felsic magmas and accompanying volcanism, and finally rupture to form an ocean basin. Uplift and erosion exposed granulite terrane upon which rift-related sedimentary and volcanic rocks were deposited or erupted. Data indicate that there were two pulses of continental rifting in the central and southern Appalachians separated by about 200 m.y. The first, about 760 to 700 Ma, did not proceed to continental separation; the second one, about 570 to 560 Ma, did (Badger and Sinha, 1988; Aleinikoff and others, 1991; J. Aleinikoff, 1992, written commun.; Rankin and others, 1993b, and references therein). The two pulses spacially overlapped between North Carolina and northern Virginia; older volcanism dominated in the south and younger volcanism dominated in the north. Rocks associated with the younger pulse of rifting occur locally along the western Appalachian orogen all the way to Newfoundland. In the Sutton Mountains of Quebec (Fig. 1) these include the youngest synrift volcanic rocks (554 Ma; Kumarapeli and others, 1989) along the orogen. The older, unsuccessful pulse of continental rifting does not appear to have extended north of Virginia.

Isotopic ages are not yet abundant enough for us to be sure which rocks are related to each of the pulses of rifting. Rift-facies clastic rocks were deposited in local basins along the length of the Appalachians, except in Pennsylvania and New Jersey. Perhaps the great thickness of the Ocoee Supergroup in Tenneessee, North Carolina, and Georgia (Fig. 1) is because those deposits span both pulses of rifting. Basalt is interlayered with most rift-facies sedimentary rocks, except for the nearly volcanic-free Ocoee Supergroup. Rhyolite is mostly restricted to salients (embayments) in Appalachian structural trends, although the rhyolite of the Mt. Rogers embayment, and hence the embayment, appears to be older than the South Mountain and Sutton Mountain embayments (Fig. 1) (Aleinikoff and others, 1991). Basement rocks of the Grenvillian massifs are intruded by gabbros, mafic and felsic dikes, and anorogenic granitoids, all thought to be genetically related to the Late Proterozoic and Early Cambrian crustal extension. The intrusive rocks in the south date mostly from the earlier pulse of extension, and at least some of the granitoids in the central Appalachian also date from the earlier episode of rifting (Tollo and others, 1991). The granitoids of the northern Appalachians (New York and Newfoundland) date from the younger pulse of extension (see Rankin and others, 1993b). Dikes, principally mafic, in parts of the French Broad and Shenandoah massifs and the Reading prong (Fig. 1) form a significant volume. The massifs are probably exposed examples of transitional (extended and intruded) crust.

The Iapetus Ocean opened somewhere east of the Grenvillian massifs along an irregular margin that in part is reflected in the present structural salients and recesses of the Appalachian orogen (Rodgers, 1975; Rankin, 1976; Thomas, 1977). The trace of the Blue-Green-Long axis (Fig. 1) mimics the embayments and promontories of the rifted margin of Iapetus. The jagged nature of the rifted margin has been interpreted as originating from the offset of an oceanic spreading ridge by transform faults (Rodgers, 1975; Thomas, 1977, 1991) or the intersection of rifts at triple junctions (Rankin, 1976). The external massifs may be the locus of continental rifts, but are not the locus of an oceanic ridge-transform system (Rankin, 1983). A consequence of the rift triple-junction model is that one of the rifts at a triple junction could fail; i.e., not produce continental separation and oceanic crust, but leave a failed arm or aulacogen pointing into the continent at an embayment (Burke and Dewey, 1973). In the Appalachians the best candidate for an Iapetan failed arm is the Ottawa-Bonnechère graben, which trends west-northwest from the Sutton Mountain embayment (Fig. 2) (Rankin, 1976; Kumarapeli and others, 1989). Documentation that promontories evolved from rift triple junctions is more difficult because the failed arm could be carried off with the departing continent. Promontories would probably undergo greater strain than embayments during subsequent collisions and may be destroyed or obscured. Palinspastic reconstruction by Stanley and Ratcliffe (1985), for example, indicates that the original position of the Laurentian margin may have been as much as 1000 km east of the present position of the Berkshire massif and may have been a promontory now destroyed by the collision of an uneven margin (Ratcliffe, 1988).

Once the crustal fracture pattern was established, it is reasonable that one of the two remaining sets of rift systems (the third being the failed arm) could evolve into transform faults, leaving the other set to become the oceanic ridge. Thus, the rift-rift-rift system could evolve into a ridge-transform system. The Iapetan crustal fracture system probably influenced the location and geometry of the Mesozsoic rift system of eastern North America (Rankin, 1976), and some of the fractures became the locus of North Atlantic transform faults, such as the Pico fracture zone southeast of the Gulf of St. Lawrence.

A possible explanation for the concentration of rhyolite at major Appalachian bends (now preserved only in embayments) is that the bends represent triple junctions generated over hot spots or mantle plumes. A possible source for silicic melts of peralkaline affinity is continental crust that has already undergone a major episode of dehydration and melting (D. R. Wones, 1984, oral commun.). Certainly the granulitic Grenvillian crust with its abundant metaaluminous (calc-alkaline) plutonic rocks would provide such a source for melts.

The time of continental breakup, i.e., rift to drift transition, is near the Late Proterozoic–Cambrian boundary. For much of the Laurentian margin, continental separation was in the Early, even earliest, Cambrian (Simpson and Sundberg, 1987; Williams and Hiscott, 1987). Ratcliffe (1987) placed the rift to drift transition in the latest Proterozoic. This is in part based upon deductions from a palinspastic reconstruction of rocks in thrust slices to give a lateral transition from continental alluvial facies on the west to marine slope-rise deposits and finally, farthest east, deep-marine deposits which now host the ultramafic rocks. Rocks interpreted to be marine occur in the Taconic allochthon well beneath the *Elliptocephola asphoidec* faunal zone, but the age of the deep-marine deposits palinspastically farther east is poorly constrained. *Elliptocephola asphoidec* is an olenellid trilobite that characterizes the oldest Cambrian fauna of the Taconic allochthon, but is nonetheless thought to be of late Early Cambrian age (Palmer, 1971). The span of time involved in the disagreement (latest Proterozoic vs. earliest Cambrian) over the age of continental separation may not be large, and the age of separation may have been diachronous.

Continental extension apparently continued in the craton into the Cambrian (the age constraints are poor), as indicated by graben development in the basement and the accumulation of as much as 2 km of Lower Cambrian(?) strata in the Rome trough (Fig. 1) and related structures (Ammerman and Keller, 1979; Read, 1989; Rankin and others, 1991). Possibly some of the normal faults in the southeastern Adirondacks (Fig. 4) were initiated during Iapetan rifting. Seismic reflection profiles across the southern Appalachians show dipping reflectors at depth beneath the Blue Ridge and Inner Piedmont that have been interpreted as Iapetan rift-stage grabens within autochthonous Laurentian basement beneath the early Paleozoic cover sequence and the Blue Ridge crystalline thrust sheet (Harris and others, 1981; Costain and others, 1989a). One wonders, however, whether such Late Proterozoic–Cambrian rift basins would be preserved in a recognizable state given the amount of crustal thickening due to the load of overriding thrust sheets (model of

Karner and Watts, 1983) or ductile deformation of autochthonous basement in the model adopted for transect E4 from Iverson and Smithson (1982).

The width of the Late Proterozoic–Cambrian extensional system from the Rome trough across the Sauratown Mountains internal massif to the Appalachian gravity gradient and the interpreted present position of the Iapetan suture in E4 is about 400 km. This compares conservatively with the 550 km width of the early Mesozoic zone of extension of E4 from the Dan River–Danville Triassic basin along the southeast margin of the Sauratown Mountains window to the Mesozoic (modern) continental margin east of the Carolina trough (Fig. 1).

With the opening of the Iapetus Ocean and the thermal subsidence of the continental margin, the sea transgressed onto Laurentia beginning in the Early Cambrian along the Appalachian Valley, but not until Late Cambrian across the site of the present Adirondack massif. Platform sediments, mostly carbonates, were deposited across the Interior Lowlands and a passive margin carbonate bank developed along the Laurentian margin. The carbonate sedimentation was interrupted at the end of the Early Cambrian by an influx of west-derived clastic sediments (Rome, Waynesboro, and Monkton Formations). An erosion surface on which karst topography developed marks the top of the Knox and Beekmantown Groups (Lower Ordovician to Lower Middle Ordovician) and essentially the end of stable platform and shelf deposition (end of the Sauk sequence of Sloss, 1963).

Any analysis of the accreted terranes of the Appalachians depends upon the interpretation of the eastern or southeastern margin of Iapetus. The initial proposition of Wilson (1966) concerning the closing and subsequent reopening of the Atlantic Ocean depended heavily upon juxtaposed contrasting early Paleozoic faunal realms. As evidence developed for the opening cycle of this early ocean (Bird and Dewey, 1970; Rankin, 1975), a question arose as to what constituted the eastern margin of Iapetus and what the source was for the terranes of exotic fauna accreted to Laurentia during the Paleozoic. The eastern rifted margin of Iapetus is preserved in the Caledonides of Scandinavia (see Gee, 1975), but south of Tornquist's line (Baltic Sea to Black Sea), we don't know where the eastern rifted margin of Iapetus was or is now. Dalziel (1991) revived a suggestion of Bond and others (1984), that eastern Laurentia may have been rifted from the proto-Andean margin of South America. It seems almost certain, however, that what collided with Laurentia as preserved in eastern United States was not the eastern rifted margin of Iapetus (see Zen and Palmer, 1981; Rankin and others, 1988; Secor and others, 1989). Most of the arguments focus on the proto-Gondwana terranes because those are the best characterized and are clearly exotic. In addition to the faunal arguments, the evidence is as follows. (1) Basement, where it exists, is not of Grenvillian age. (2) Some of the proto-Gondwana terranes underwent a compressional episode at the end of the Late Proterozoic (the Avalonian or Cadomian orogeny) that did not affect Laurentia but was older than the Early Cambrian opening of Iapetus. (3) Other terranes, such as the Massabesic and Nashoba, contain rocks of widely ranging but pre-Avalonian ages and intervene between Laurentia and proto-Gondwana terranes.

Penobscottian amalgamation and Taconian accretion

Neuman (1967) mapped an unconformity in north-central Maine near Shin Pond (Fig. 4) between the Lower Cambrian(?) Grand Pitch Formation (pelite and quartzose sandstone) and the late Early Ordovician Shin Brook Formation (tuffs and tuffaceous sandstone). Because of the strong lithologic contrasts across the unconformity, and because the unconformity is older than the Taconian orogeny (see below), Neuman proposed the name Penobscot disturbance (now Penobscottian orogeny) for the event that produced the unconformity. The unconformity has now been recognized in other parts of Maine (Boone and Boudette, 1989), New Brunswick (Neuman, 1984), and Newfoundland (Colman-Sadd and others, 1992). In the central Appalachian Piedmont, the thrust stack of the Potomac composite terrane (Fig. 3) was assembled during the Penobscottian from east to west and from the top down (thrusts propagated toward the foreland). The thrust stack was unconformably overlain by the Popes Head Formation prior to the intrusion of the Occoquan Granite (600 to 500 Ma; Drake, 1989b). If the Smith River allochthon (Fig. 3) is another thrust slice of the Potomac composite terrane, the Penobscottian orogeny affected rocks as far south as North Carolina (Horton and others, 1989). As noted by Drake (1989b) and Hatcher (1989b), no unconformity is recognized in the platform deposits of the Appalachian foreland between the Middle Cambrian and late Early Ordovician. Therefore, the Penobscottian deformation must have occurred at a place remote from the Laurentian margin.

We can only speculate, and not very satisfactorily, on the broader role of the Penobscottian orogeny in the evolution of the Appalachians. Neuman and Max (1989) discussed the evidence that the Penobscottian orogeny, the Grampian orogeny in the British Isles, and the Finnmarkian orogeny in Scandinavia are all Late Cambrian to Early Ordovician events that characterize terranes having similar predeformational and postdeformational features. These orogenies were interpreted by them to record a major orogen-wide closing of the Iapetus Ocean basin in the Cambrian. They argued that the relevant sedimentary sequences, when they were deformed, occupied the margins of a very large Early Ordovician Iapetus Ocean. This seems somewhat at odds with the observation by Hatcher (1989b) that the Penobscottian, Grampian, and Finnmarkian orogenies did not affect their respective forelands. An interpretation that has application to the Appalachian synthesis is that the Penobscottian orogeny represents closing of an eastern ocean or eastern arm of Iapetus east of the Chain Lakes block and that the ophiolite of the Boil Mountain Complex (Fig. 4) is the oceanic crust of that ocean basin (Boone and Boudette, 1989). The Penobscottian, in this interpretation, resulted from the amalgamation of the Central Maine terrane with the Brompton-Cameron terrane, and the Potomac composite terrane (including the Smith River allochthon) with the

Jefferson terrane (Fig. 3). Subsequently, the assembled terranes collided with Laurentia, resulting in the Taconian orogeny. The model assumes that the Boil Mountain Complex–Hurricane Mountain Formation package is older than the Thetford Mines ophiolite–St. Daniel melange package, and that the Central Maine terrane crust (Craton X of Zen, 1983) is distinct from Laurentia. An unresolved problem with this model is the west-directed collision of the Carolina terrane, which was outboard of the Potomac composite terrane during the Taconian.

The Exploits subzone of the Dunnage zone in Newfoundland (the probable correlative of the Central Maine terrane) has a recognizable southeastern margin. The Exploits subzone is of oceanic affinity and includes Lower to Middle Ordovician (late Arenigian to early Llanvirnian) rocks containing a Celtic brachiopod fauna (Williams and others, 1988; Neuman, 1984). The Gander zone is interpreted to be the eastern edge of an ocean basin (here, the eastern ocean of the Central Maine terrane) or the western slope-rise margin of proto-Gondwana. The Exploits subzone, including fragments of oceanic crust (ophiolites), was thrust eastward over the Gander zone at the time of closure of the ocean, about 474 Ma (late Arenigian) (Colman-Sadd and others, 1992). That event is interpreted to be the Penobscottian orogeny. A similar east-directed thrusting and ocean closure may have occurred in the Miramichi belt of New Brunswick, but evidence for that has been obscured by later events.

The classic area for the study of the Taconian orogeny is in the Hudson Valley of New York (Fig. 4), where both the Taconian unconformity (separating Middle Ordovician rocks from Lower Silurian strata) and the Taconic allochthon are present. The Taconian orogeny is thought to have been caused by attempted subduction of the Laurentian margin in an oceanic trench. As pointed out by Drake (1989b), this interpretation is more straightforward in the northern Appalachians, where there is an Ordovician island arc or perhaps two Ordovician island arcs (Ammonoosuc and Hawley; Fig. 4). The stratigraphy and history of these arcs are complex and are still being investigated. Trzcienski and others (1992b), for example, considered the Barnard Volcanic Member of the Missisquoi Formation to be stratigraphically below the Hawley Formation, whereas Aleinikoff and Karabinos (1990) reported a 418 ± Ma age (U-Pb zircon) for the Barnard in central Vermont.

Tucker and Robinson (1990) showed that the Ammonoosuc arc in southern New England was active from about the middle of the Caradocian until latest Ashgillian and possibly earliest Llandoverian time (454 to 442 Ma). They therefore argued that the Ammonoosuc arc was too young to be the locus of subduction-related volcanism before the closure of Iapetus and collision that resulted in the emplacement of the Giddings Brook slice (Fig. 4) of the Taconic allochthon (early Caradocian or possibly Llandeilian, which on the Ordovician time scale of Tucker and others [1990] is about 462 to 455 Ma; see below). Tucker (*in* Tucker and Robinson, 1990) suggested that the Hawley arc might be older than the Ammonoosuc arc and might be the suprasubduction-zone arc related to the closing of the Iapetus. However, the Ammonoosuc arc in northern New Hampshire and adjacent Vermont contains two intrusive tonalite bodies, thought to be comagmatic with the adjacent volcanic rocks, that have U-Pb zircon ages of about 469 Ma (Aleinikoff and Moench, 1992). The problem is unresolved.

The best candidate for an Ordovician volcanic arc of possible Taconian affinity in the central and southern Appalachians is the Hillabee greenstone (informal name) (along the Hollins line fault, which is the boundary between the Talladega block of the Laurentian Appalachians and the Jefferson terrane in Alabama; Fig. 1). The age and structural position of the greenstone are controversial (Horton and others, 1989). Tull and Stow (1980) interpreted the greenstone to be within the Talladega block, stratigraphically above the Devonian Jemison Chert and overthrust by the Jefferson terrane. Such an age and position would be unique for a proximal volcanic deposit in the Laurentian Appalachians. Higgins and others (1988), however, placed the Hillabee above the Hollins line fault and in rocks here included in the Jefferson terrane. Rocks of the Hillabee have geochemical characteristics suggestive of an immature volcanic-arc setting (Tull and Stow, 1980; Stow, 1982). Quartz dacite in the Hillabee greenstone has yielded zircon ages of 462 to 456 Ma (Russell, 1978; Russell and others, 1984).

With the possible exception of the Hillabee greenstone, subduction-related volcanic rocks of Ordovician age have not been identified in the central and southern Appalachians. The known arc rocks (Fig. 1: Chopawamsic and Carolina terranes, Cambrian or possibly Early Cambrian, and Late Proterozoic to Middle Cambrian, respectively), are older and are outboard of terranes thought to have been amalgamated in the Penobscottian. Furthermore, these arcs do not correspond in age to a clastic wedge in the foreland. Nonetheless, the commonality of some Taconian features along the Laurentian margin in the United States is striking. These include a belt of ultramafic rocks thought to be fragments of Iapetus oceanic crust, the Ordovician clastic wedges, and the Caradocian age bentonites in the foreland. Note that the Tucker and Robinson (1990) age for the Ammonoosuc arc of 454 to 442 Ma is in good agreement with the age of the eastern North America bentonites, about 454 Ma ($^{40}Ar/^{39}Ar$ age spectrum by Kunk and Sutter, 1984).

In the interpretation favored here, the Taconian orogeny marked the closing of Iapetus and the collision of the passive margin of Laurentia with an island arc bordering previously assembled terranes. Subsequent orogenies must have involved other collisions and closing of ocean basins outboard of Iapetus.

The unconformity at the top of the Knox and Beekmantown Groups is the first record of the Taconian orogeny on the craton. The age and magnitude of the unconformity vary along and across strike (see Drake, 1989a). Jacobi (1981) and Mussman and Read (1986) proposed that this unconformity was caused by the upward flexure of the shelf of the downgoing slab as it passed over the peripheral bulge a few hundred kilometers in front of the subduction zone within the closing Iapetus Ocean. A necessary condition of this proposal is that the subduction zone was east dipping, a conclusion also reached from the distribution of igneous rocks in the internides. By analogy with modern passive margin–arc collisions, a flexural foredeep in front of the arc migrates with time toward the craton as a result of plate convergence. This results in the drowning of the carbonate platform and may be recognized along ancient margins by a facies change from carbonate to shale. Closer to the trench axis, shale grades into flysch. The age of the shelf drowning (carbonate to shale transition), which with luck may be determined paleontologically, is diachronous across strike, and that diachroneity may be used to determine the rate of plate convergence. If either or both the passive margin and the arc are irregular or the collision oblique, the age of shelf drowning will also be diachronous along strike.

It has long been known that the shelf drowning during the Taconian orogeny was diachronous along and across strike (Rodgers, 1971). The main phase of the Taconian orogeny produced a clastic wedge in the foreland basin of Upper and Middle Ordovician flysch (the Martinsburg Formation and related rocks) centered in Pennsylvania and southeastern New York. The main clastic wedge in the foreland basin of the southern Appalachians consists of Middle Ordovician flysch (Sevier Shale and related rocks) and is centered in eastern Tennessee. Rodgers (1953) referred to the episode that produced the Tennessee wedge as the Blountian phase of the Taconian orogeny. Flysch deposits of both clastic wedges are succeeded by postorogenic molasse. Molasse of the main Taconian pulse extends from New York to Alabama and is of Late Ordovician to Early Silurian age. Molasse of the Sevier basin is restricted to eastern Tennessee and environs, and is of Middle Ordovician age (Bays Formation and related rocks) (see Drake, 1989a). Because the Bays molasse is overlain by the southern extension of the Martinsburg flysch, except where the latter is cut out by an unconformity, Drake (1989a) considered the Blountian to be a separate orogenic episode.

Bradley (1989), however, considered the diachroneity to be a continuum. In his application of the shelf-drowning model to Taconian plate kinematics, he noted that in the eastern United States and southern Quebec, shelf drowning began earliest in Alabama and Georgia (Llanvirnian) and latest in the Mohawk Valley and southern Quebec (possibly as young as Ashgillian). He estimated that the rate of plate convergence during the Taconian collision was 1 to 2 cm/yr, and that the minimum width of ocean that closed was 500 to 900 km.

The across-strike diachroneity of shelf drowning is well documented in the classic Hudson Valley area. There the picture is both complicated and augmented by the Taconic allochthon, which transports slope-rise deposits and synorogenic flysch onto the platform. The youngest transported flysch in the Taconic allochthon (Pawlet Formation of the Giddings Brook slice) is correlative with the *C. bicornis* Subzone, which in the zonation of Finney (1986) is early Caradocian or possibly Llandeilian. The oldest autochthonous deep-water shale representing shelf drowning is also the easternmost (Walloomsac Formation near Hoosick, New York, Fig. 4) and also correlates with the *C. bicornis* Subzone (Finney, 1986). Hames and others (1991) estimated the absolute age of the Walloomsac to be about 460 to 455 Ma. The age of shelf drowning decreases to the west and may be as young as Ashgillian near Utica, New York, in the Mohawk Valley (Fig. 4).

Surface-breaking thrusts of Caradocian age in the frontal zones of the Taconic allochthon transported masses of sedimentary rocks over detritus shed from the advancing thrust fronts. Successively higher, and ultimately the last-moved, slices were transported by a fold-thrust mechanism. The highest carried Grenvillian basement and its already metamorphosed stratified cover (Ratcliffe and Harwood, 1975). The age of peak Taconian metamorphism east of Albany, New York, is 450 to 440 Ma (Hames and others, 1991), which is younger than the initial movement of the Taconic allochthon, as it should be from the above field relations. Also late in the Taconian sequence of deformation were the faults of the Champlain thrust family (Taconic frontal thrust system) that locally displace the leading edge of the Taconic allochthon (Stanley and Ratcliffe, 1985; Bosworth and others, 1988).

A frontal zone of imbricate Taconian thrusting also developed in the southern Appalachians. It is less-well documented because a number of thrusts were reactivated during the Alleghanian orogeny, and because much of the Taconian thrust belt was probably overridden by the ~300 km of movement of the Blue Ridge crystalline thrust in the Alleghanian sheet. We can, however, document that movement on some of the major thrusts in the Blue Ridge, such as the Greenbrier fault (Fig. 1), was prior to the peak of Taconian metamorphism (Hadley and Goldsmith, 1963; Milton, 1983).

The Jefferson terrane and pre-Silurian rocks of the Brompton-Cameron terrane (Fig. 3) are interpreted to be an accretionary wedge containing ophiolite fragments formed in association with an east-dipping subduction zone, down which large volumes of oceanic crust were consumed. The site of deposition of the strata of the accretionary wedge is undetermined, although I argue for an eastern site. It may have been east of and above the subduction zone, possibly even outboard (east) of the Iapetus spreading ridge, and thus on the east side of the Iapetus Ocean (Rankin and others, 1989), or it may have been deposited close to the slope-rise sequence of the Laurentian margin, as was suggested for the Rowe-Hawley zones (Fig. 4) of New England (Stanley and Ratcliffe, 1985; Zen, 1989). If the eastern (outboard) site is correct, the Jefferson and Brompton-Cameron terranes are exotic and their western boundary (the family of faults extending from Hollins line fault, Alabama, to the Hazen's Notch fault, Vermont; Figs. 1 and 4) marks the leading edge of the accreted terranes and the closure of the Iapetus Ocean. If the western (inboard) site is correct, the terranes are not in the strict sense exotic, nor is their western boundary the Iapetus suture. To a first approximation, sole thrust of the accretionary wedge (the Hollins line to Hazen's Notch fault) is accepted here as the surface expression of the Iapetus suture and is so shown in Figure 2.

Taconian plutons of the northern Appalachians are largely restricted to the Ammonoosuc arc and the vicinity of Cameron's line (Figs. 1 and 4) in Connecticut and New York (Wones and Sinha, 1988; Sinha, 1988). In the southern Appalachians, pretectonic or syntectonic gneissic granites, typically garnet bearing and in the age range of 460 to 420 Ma (Harper and Fullagar, 1981), are common in the Inner Piedmont composite terrane (Fig. 3). These are good candidates for subduction- and/or collision-generated Taconian plutons. The somewhat younger (Silurian?) two-mica, epidote-bearing granodiorite and quartz diorite of the Blue Ridge (Spruce Pine Plutonic Suite) may be the result of decompression melting after stacking of the Taconian thrust sheets (Wones and Sinha, 1988). As noted by Rankin and others (1989), the regions of highest grade Taconian metamorphism in the Blue Ridge tectonic province lie northwest of the Silurian(?) granitoid belt and contain no sizeable bodies of Paleozoic granitoids, except near Spruce Pine, North Carolina (Fig. 1).

The southeastern extent of the effects of the Taconian orogeny is unknown. The Taconian orogeny is thought to have affected terranes in the southern Appalachians as far southeast as the Carolina terrane (Fig. 3), on the basis of studies of metamorphism, deformation, and magnetization (Vick and others, 1987). The Upper Ordovician Arvonia Slate and Quantico Formation are preserved in tight en echelon synclines within the Chopawamsic terrane of the central Appalachians (Figs. 1 and 3). These formations may have been deposited in a euxinic, intra-arc successor basin, subsequent to the suturing of the Chopawamsic terrane with terranes to the west (Drake, 1989a). Hence, the Taconian orogeny affected rocks of the central Appalachians as far southeast as the Chopawamsic terrane. In following sections, I argue that the Black Pond–Campbell Hill–Hackmatack Pond fault family is the Acadian suture between the Central Maine terrane and the Merrimack-Harpswell terrane, which was already amalgamated with the St. Croix and Ellsworth-Mascarene terrane, and probably the Nashoba terrane (Fig. 6). In this interpretation, no rocks of the northern Appalachians southeast of the Black Pond–Campbell Hill–Hackmatack fault family were affected by the Taconian orogeny.

To review, both the St. Croix and Ellsworth-Mascarene terranes are proto-Gondwana terranes. Fossils of the Atlantic faunal province as young as Gedinnian (Early Devonian) are present in the Ellsworth-Mascarene terrane and its outlier in Massachusetts, the Newbury Volcanic Complex (Fig. 6). The St. Croix and Ellsworth-Mascarene terranes were amalgamated by the Late Silurian, as evidenced by fossiliferous Upper Silurian roundstone conglomerates unconformably above Cambrian and Ordovician strata on both sides of the mutual terrane boundary, the Turtle Head fault (Fig. 6). The St. Croix terrane is stitched to the Merrimack-Harpswell terrane by the Silurian Pocomoonshine Gabbro-Diorite (Fig. 6). The Merrimack-Harpswell and Nashoba terranes were probably also amalgamated in the Late Silurian. Although their early histories were different (Early to Middle Ordovician postkinematic intermediate composition plutons of mantle derivation in the Merrimack-Harpswell vs. prekinematic or synkinematic peraluminous granites of Middle to Upper Ordovician age in the Nashoba), they both underwent Silurian plutonism, as did the St. Croix and Ellsworth-Mascarene terranes. In summary, an amalgamated block of the Merrimack-Harpswell, Nashoba, St. Croix and Ellsworth-Mascarene probably lay somewhere southeast of the Central Maine terrane in the Early Devonian, far enough away to maintain a distinctive non–North American fauna. This amalgamated block is here called the Merrimack-Mascarene composite terrane.

Acadian Orogeny

The effects of the Acadian orogeny are most obvious in New England, where fossiliferous Lower Devonian strata of central Massachusetts, Vermont, New Hampshire, and Maine are deformed and metamorphosed, whereas isolated deposits of fossiliferous Middle and Upper Devonian strata in Maine are little deformed. Most of this discussion focuses on New England. As addressed below, and contrary to Osberg and others (1989), I find little evidence to support a Middle or Late Devonian orogeny south of the area of Fredricksburg, Virginia.

As an introduction to discussing the Acadian orogeny, it is important to consider the setting in which it developed. In the interval between the Taconian and Acadian, a considerable volume of molasse from Taconian highlands was deposited in the foreland basin of the central and southern Appalachians. Taconian molasse (Shawangunk Formation) wedges out north of Kingston, New York (Fig. 4). Zen (1991) used that observation to argue that the bulk of sediment from the denudation of the Taconic highlands in New England was shed eastward.

Silurian and Lower Devonian strata are thin across the Bronson Hill anticlinorium, but thicken eastward into the Central Maine trough and westward into the Connecticut Valley trough (Figs. 1 and 4). Some of the post-Taconian strata in the troughs may be Upper Ordovician. The thickening of the sedimentary section into the Central Maine trough is abrupt across a tectonic hinge zone (Fig. 4) and begins with the westerly derived clastics, some conglomeratic, of the Lower Silurian Rangeley Formation (Hatch and others, 1983). As noted earlier, the Central Maine trough may have once contained as much as 2200 km^3 of strata per kilometer of strike length of the trough (Stewart, 1989). Zen (1991) pointed out that the thickness, extent, and nature of the pre-Acadian rocks in the Central Maine trough are similar to those of the post-Acadian Upper Devonian Catskill delta. Stewart (1989) noted that the volume of sediments per strike kilometer inferred to be in the Central Maine trough is comparable to that of the post–Lower Jurassic slope-rise sediments of the present Atlantic margin from central Long Island to central Nova Scotia.

The Connecticut Valley trough is bordered on its southeastern side in northern New Hampshire and adjacent Quebec by the Silurian Frontenac Formation (Fig. 4), which includes a bimodal volcanic suite. These rocks are thought to represent a rift (Stewart, 1989; Moench, 1990, and references therein; Moench and others, 1992). Pre-Frontenac rocks of the Bronson Hill anticlinorium in a number of areas are intruded by bimodal but largely mafic swarms of dikes and sheeted mafic intrusives (Moench, 1990). According to Moench (1990), at least one felsic dike in the bimodal suite is dated as Silurian (418 ± 4 Ma, U-Pb zircon), and they are interpreted to be comagmatic with the Frontenac volcanic rocks, also dated as Silurian (418 ± 4 Ma, U-Pb zircon). North-northeast–trending dikes are particularly abundant immediately east of the Monroe fault from Hanover, New Hampshire, to northern New Hampshire (Fig. 4) (e.g., see White and Billings, 1951; Lyons, 1955). Eric and Dennis (1958) noted that along the Connecticut River and about 15 km west of Littleton, New Hampshire (Fig. 4), mafic dikes constitute 80% of the exposures in an area 3.2 km wide. Recent work by Rankin (unpublished field data) confirms that this is a sheeted dike complex. White and Billings (1951) reported that a few mafic dikes cut rocks of the Connecticut Valley trough adjacent to the Monroe fault (Fig. 4), but that they are uncommon farther west. It is not established how much of the sedimentary section exposed in the trough is younger than the dike swarm. Hence, it is not known how much of the trough could be underlain by extended crust containing abundant dikes. It is intriguing that the U-Pb age (418 Ma) of the Barnard Volcanic Member on the west margin of the trough in central Vermont (Fig. 4) is the same as that of the Frontenac Formation and the associated felsic dike on the east side of the trough in northern New Hampshire. Seismic reflection and refraction studies show no evidence of oceanic crust beneath the Connecticut Valley trough (Ando and others, 1984; Spencer and others, 1989; Hughes and Luetgert, 1991).

In a thorough discussion of the Acadian orogeny in New England, Osberg and others (1989) noted that the record includes flysch (Connecticut Valley and Central Maine troughs) and molasse (Catskill delta) deposits, volcanism, several episodes of folding and faulting, syntectonic and late-tectonic plutonism, and several pulses of metamorphism. In New England, as summarized by Osberg and others (1989), thermal effects of the Acadian orogeny began in the Silurian and continued at least to the end of the Devonian. Volcanic activity peaked from the Pridolian to the Emsian time. The main phase of plutonic activity occurred in the interval 410 to 370 Ma. In southern New England early plutons either predate or coincide with the main phase of Acadian dynamometamorphism. Plutons toward the north were intruded at successively higher structural levels and in Maine clearly postdate the main phase of Acadian deformation. At deep structural levels metamorphism occurred over a time span of 30 to 40 m.y. and lasted until the end of the Devonian. At higher structural levels, metamorphism occurred in pulses (Osberg and others, 1989). The maximum age of metamorphism is documented by the metamorphism to chlorite grade of fossiliferous Emsian age strata near Littleton, New Hampshire (Fig. 4). Correlative strata in adjacent areas are isoclinally folded and metamorphosed to sillimanite grade.

In northern Maine, Middle Devonian fossiliferous strata of successor basins (Fig. 4) lie unconformably on folded sandstone and volcanic rocks in small basins, but are themselves broadly folded and faulted. In southeasternmost Maine (Fig. 6), coarse conglomerates of the Upper Devonian Perry Formation contain clasts of an underlying Acadian granite. Detritus eroded from Acadian highlands formed the marine and nonmarine Catskill delta complex of New York and adjacent Pennsylvania (Fig. 1) that prograded westward onto the platform during the Middle and Late Devonian (Woodrow and Sevon, 1985). On the basis of the relationships in the successor basins of Maine, the Eifelian and younger strata in the delta postdate Acadian orogenic events (Osberg and others, 1989).

Even in New England, where the evidence for the Acadian orogeny is indisputable, the plate motions responsible are in dispute. In contrast with the evidence developed for the Taconian orogeny, there is no convincing belt of ultramafic bodies (obducted oceanic crustal fragments) related to the Acadian orogen, no good evidence of subducted oceanic crust, no convincing island arc, no identified continental margin, and no foreland basin. Osberg and others (1989) reviewed the arguments that the thickening Silurian section east of the Bronson Hill anticlinorium represents either a continental margin or the filling of a broad extensional basin. At present we have no evidence for oceanic crust under this basin. Rankin (1968) proposed that the Piscataquis volcanic belt (Fig. 4) of north-central Maine was an island arc, and Rodgers (1981), Hon and Roy (1981), and Bradley (1983) have incorporated this island arc in their models. Stewart (1989) argued that garnet phenocrysts in many of the rhyolites from the Early Devonian Piscataquis volcanic belt indicate that they originated by partial melting of felsic rocks at lower crustal or upper mantle pressures equivalent to depths of greater than 50 km. Such a deep source supports the overthickening he postulated along the buried Taconian suture over the edge of Laurentia, but does not rule out the possibility that Acadian compression could have caused or contributed to the crustal thickening. Stewart (1989) also noted that the seismic refraction study of Luetgert and Bottcher (1987; see also Hughes and Luetgert, 1991) found no evidence of oceanic crust in Maine, nor did the Pb-isotope studies of plutonic feldspars by Ayuso and others (1988) suggest a major contribution from oceanic crust at the time those plutons sampled the crust.

Evidence of plate movements that produced the Acadian orogeny is circumstantial. The classic early Acadian fold nappes of the Bronson Hill anticlinorium involved tens of kilometers of east-over-west shortening, as well as slightly younger west-directed thrust faults in the Central Maine trough (see review by Robinson and others, 1991). The time of nappe emplacement was estimated to have been about 400 Ma by Chamberlain and England (1985). Acadian nappes and structures verge east along the eastern margin of the Central Maine terrane (Hall and Robinson, 1982; Stewart and others, 1991; Thompson and others, 1993). Segments of the eastern boundary of the Central Maine terrane that are thrust faults (Black Pond and Hackmatack Pond faults; Fig. 6) are east directed. Hubbard and others (1991) indicated that the major dextral strike-slip deformation on the Norumbega fault occurred prior to Middle Devonian time.

Earlier I argued that the Merrimack-Mascarene composite

terrane, consisting of the Merrimack-Harpswell, Nashoba, St. Croix, and Ellsworth-Mascarene terranes, as well as the Massabesic and Falmouth-Brunswick blocks of the Ellsworth-Mascarene terrane, was amalgamated by the Late Silurian. It was situated far enough from ancestral North America so that it maintained a distinctive non–North American fauna until the Gedinnian. I would like to be able to argue that it was the underthrusting of ancestral North America by this composite terrane that caused the Acadian orogeny. Among the numerous problems with this model is the timing of events. The Central Maine terrane was stitched to the Massabesic block and Merrimack-Harpswell terrane by about 364 Ma (Barrington pluton; Fig. 6) and possibly as early as 390 Ma (Fitchburg Complex; Fig. 6). It is difficult to reconcile the 400 Ma emplacement of nappes of the Bronson Hill anticlinorium driven by collision with a terrane amalgamated by the Late Silurian, but which maintained a non–North American fauna until the Gedinnian (408 to 401 Ma; time scale of Palmer, 1983). Furthermore, in northern New Brunswick, rocks of the Central Maine terrane (Exploits subzone) that contain the Celtic fauna were deposited on rocks thought to be of the Gander zone (marginal to proto-Gondwana) (van Staal and Fyffe, 1991; van Staal and Williams, 1991). This suggests that the Central Maine terrane and proto-Gondwana were already linked in the Middle Ordovician. A back-arc basin subsequently opened, and oceanic crust formed. This back-arc basin closed beginning in the Late Ordovician on a west-dipping subduction zone in which blueschist formed. Subsequent collision led to east-directed thrusting of the blueschists over the Gander zone rocks and generation of Silurian and earliest Devonian plutons (see van Staal and others, 1990; Bevier and Whalen, 1990). This deformation may have been continuous into the Acadian orogeny. The problem of fitting all of these observations and interpretations into a coherent history remains unresolved.

Evidence for a significant Devonian compressional event (the Acadian orogeny) in the central and southern Appalachians is marginal, although some have argued strongly for it (see Osberg and others, 1989). The best evidence for Devonian reginal metamorphism and penetrative deformation is in the Talladega block in Alabama southwest of transect E5. In the Talladega block (Fig. 1), which is part of Laurentia, an Early to Middle Devonian dynamothermal event is bracketed by Early Devonian fossils (Siegenian; Tull and others, 1988) in the greenschist facies Jemison Chert, and K-Ar whole-rock slate and phyllite ages that suggest a thermal peak no later than 399 ± 17 Ma (1 σ) (Early Devonian) (Kish, 1990). Osberg and others (1989) suggested that the Hillabee greenstone, which they interpreted to be stratigraphically above the Jemison Chert, could be the remnant of a volcanic arc, somewhat similar to the Piscataquis volcanic arc. (The controversy over the age [Ordovician isotopic zircon ages] and structural position of the Hillabee was reviewed in the previous section.) The Jemison Chert is underlain by the Erin Slate Member of the Lay Dam Formation as used and mapped by Cook (1982). Carboniferous lycopods were reported from the Erin by Smith (1903). The original fossils have been reexamined and their ages confirmed, but attempts to refind and collect the fossil have been unsuccessful (Gastaldo and Cook, 1982). The Erin is a thinly foliated, variably carbonaceous phyllite with a dominant slaty cleavage (Gastaldo and Cook, 1982). Cook (1982, Fig. 6) showed the Erin in his study area to be almost entirely separated from the rest of the Lay Dam Formation by a thrust fault. Could the stratigraphic order be reversed and the metamorphism (here chlorite grade) be Alleghanian rather than Acadian?

A number of studies have suggested a Devonian thermal event in the Piedmont. Dallmeyer (1978) reported postmetamorphic $^{40}Ar/^{39}Ar$ cooling ages of hornblende of about 355 Ma in the western Piedmont near Atlanta, Georgia. Dallmeyer (1988) also reported that high-grade rocks of the Alto allochthon at the northwestern edge of the Inner Piedmont in South Carolina (Fig. 1) (E5) underwent diachronous postmetamorphic cooling through hornblende closure temperatures between 335 and 360 Ma, which suggests a Late Devonian or earlier thermal peak. Hopson and Hatcher (1988) interpreted the age of uplift and emplacement of the Alto allochthon to be later than the thermal peak and it could coincide with the hornblende ages. A Late Devonian–Early Mississippian (ca. 360 to 340 Ma) regional thermal event may be indicated by $^{40}Ar/^{39}Ar$ spectra of hornblende in the southeastern Charlotte belt of South Carolina (Dallmeyer and others, 1986). Hatcher (1989b) cautioned that these ages are equivocal and may represent post-Taconian uplift. Note that the ages reviewed in this paragraph are all younger than the Acadian orogeny of the northern Appalachians.

Several features, admittedly in the form of negative evidence, argue against an Acadian orogeny in the central and southern Appalachians. Distal parts of the Catskill delta (Acadian molasse) extend into the central Appalachians (Pennsylvania and Virginia), but there is no comparable clastic wedge of Acadian molasse to the south. There is no Acadian flysch in the foreland basin of the central or southern Appalachians, and no identified Acadian flysch in internal basins of the internides comparable to that in the Connecticut Valley and Central Maine troughs. In fact, the only known post-Cambrian fossiliferous rocks in the Appalachians south of the Potomac River and east of west border of the Jefferson terrane are the Upper Ordovician Arvonia Slate and Quantico Formation of central and northern Virginia (Fig. 1). Both of these formations were probably metamorphosed during the Alleghanian orogeny, as opposed to the Acadian orogeny (Flohr and Pavlides, 1986). One could argue that Acadian flysch of the internides has been removed by erosion, but the distribution of metamorphic assemblages (Morgan, 1972) suggests that, if anything, a deeper crustal level is exposed in southern New England than in the Piedmont. Furthermore, the Silurian-Devonian Salisbury and Concord Plutonic Suites (discussed below) of the Charlotte belt (Fig. 1) include hypabyssal rocks (Goldsmith and others, 1988). There is nothing in either the central or southern Appalachians comparable in volume to the extensive suite of Acadian granite (largely early peraluminous

S-type plutons), granodiorite, and tonalite of New England (Wones and Sinha, 1988; Sinha, 1988). The southernmost good candidate for an Acadian granite is the Falls Run Granite Gneiss near Fredericksburg, Virginia (Fig. 1): that hornblende and biotite granite gneiss, which ranges in composition from monzonite to granite, is about 410 Ma, based on U-Pb and Rb-Sr studies (Pavlides and others, 1982).

The Salisbury Plutonic Suite (granitoids) and Concord Plutonic Suite (gabbro-diorite-syenite) are mostly in the Charlotte belt in and near E4 (Fig. 1). Ages range from about 415 to 362 Ma (Sinha and others, 1989). Goldsmith and others (1988) interpreted the Salisbury and Concord Plutonic Suites to be syntectonic, but Wones and Sinha (1988) and Sinha and others (1989) related these rocks to extensional tectonics. Wones and Sinha (1988) made a tentative but tantalizing correlation of the Concord Plutonic Suite with the 370 Ma alkalic intrusives of the Esmond-Dedham block of the Avalon terrane in southeastern New England. The implication of this is that the extensional episode, if it occurred, took place in the homeland of the Carolina and Avalon terranes (proto-Gondwana?) before the fragmentation of proto-Gondwana. Attributing the plutons in both the Carolina and Avalon terranes to a Silurian-Devonian episode of extension in the proto-Gondwana homeland is difficult to reconcile with the Taconian accretion of the Carolina terrane and the Alleghanian docking of Avalon. Sinha and others (1989) proposed that the Silurian-Devonian plutons of the Charlotte belt were generated in a transtensional environment during strike-slip accretion of the Carolina terrane

Alleghanian orogeny

The Alleghanian orogeny was the last of the Paleozoic compressional events that formed the Appalachian orogen, and for many years was called the Appalachian orogeny. It is also the last orogeny that eastern North America experienced, and marked the final stage of accretionary history in the Appalachian-Caledonide orogen. The Alleghanian was a continent-continent collision between North America (here, the eastern United States) and proto-Gondwana (here, West Africa) that closed the Theic-Rheic Ocean and formed the supercontinent of Pangea. The collision was oblique and not only affected southeastern New England, but was responsible for the classic foreland fold and thrust belt of the central and southern Appalachians. The amount of Alleghanian crustal shortening increases southward to Tennessee from a hinge near New York City. The closing of the Theic-Rheic Ocean involved the accretion of a succession of outboard terranes in New England (Avalon and Meguma) and the Chesapeake "indenter" terrane (Fig. 3). What was accreted in the south is problematic, but it certainly includes the Suwannee terrane and, I argue, the Hatteras terrane. It is puzzling that we can identify several terranes that were probably accreted in the Alleghanian in New England where Alleghanian effects do not extend far inland, whereas in the southern Appalachians the Alleghanian foreland fold and thrust belt lies as much as 500 km west of the Alleghanian suture (Fig. 2).

Secor and others (1986) summarized the evidence for the timing of the Alleghanian in the central and southern Appalachians. The onset of the orogeny is marked in the foreland by the transition from shallow-water shelf carbonate of the upper Mississippian to uppermost Mississippian marine or brackish shale (flysch). Petrographic studies and facies analyses of Pennsylvanian rocks in the ensuing clastic wedge indicate that there was progressive unroofing of a batholithic source terrane to the southeast during the Namurian and Westphalian. The youngest deformed strata of the clastic wedge are nonmarine rocks of the Dunkard Group, which are carried by blind thrusts in Ohio, West Virginia, and Pennsylvania. Secor and others (1986) estimated the duration of the Alleghanian from evidence in the clastic wedge to be from about 40 to 60 m.y. (327 to 286 or 266 Ma), depending upon whether the Dunkard was deposited in the Pennsylvanian or Permian.

These ages agree well with the ages of deformation, metamorphism, and plutonism determined from the internides of the central and southern Appalachians. Numerous plutons—mostly granite to monzogranite, but with locally conspicuous gabbroic rocks—in the Piedmont southeast of the Brevard zone, or its extrapolated extension, between Georgia and Maryland are Carboniferous to Permian in age (330 to 260 Ma) (Sinha and others, 1989). Barrovian metamorphism in the Gaffney and adjacent Carolina terranes in the central Piedmont of North and South Carolina ($^{40}Ar/^{39}Ar$ hornblende plateau ages of 323 to 318 Ma from amphibolites in the country rock) surrounds the High Shoals Granite (dated at 317 Ma, U-Pb on zircon; Horton and others, 1987).

Three distinct Alleghanian deformational events, summarized by Dallmeyer (1989b), affected the eastern Piedmont of Georgia and South Carolina. The first accompanied regional metamorphism to grades as high as kyanite or sillimanite and emplacement of granitoid plutons at mid-crustal levels between 315 and 295 Ma. The second (about 295 to 285 Ma) folded the early isothermal surfaces. The final phase produced dextral shear zones between about 290 and 265 Ma. A comparable complex series of Alleghanian events (deformation, amphibolite facies metamorphism, and plutonism) affected the eastern Piedmont of North Carolina near Raleigh (Stoddard and others, 1991).

The volume of Alleghanian granitoids in the Piedmont is huge (Snoke and Mosher, 1989). These granitic rocks are of two major varieties that are spatially intermixed (Hermes and Murray, 1988). The more common variety is metaluminous biotite granite that contains minor hornblende and generally has relatively low initial strontium ratios (0.702 to 0.708). These are similar to the Caledonian I-type granite of Pitcher (1983). Less common are peraluminous two-mica granites with higher initial strontium ratios that resemble the S-type granites of Chappell and White (1974). Many of the Alleghanian plutons are clearly postkinematic and impose contact aureoles on adjacent regionally metamorphosed rocks. Others are deformed and associated with Alleghanian plastic shear zones or are within broader zones of higher grade Alleghanian metamorphism and deformation (Gates and others, 1988; Snoke and Mosher, 1989). Sinha and Zietz (1982) summarized chemical and isotopic variations of the plutons that suggest an eastward-decreasing contribution from a continental source. They argued that an arc, convex to the southeast, of at least 50 Alleghanian plutons roughly along the eastern edge of the Piedmont is a continental magmatic arc developed over a west-dipping subduction zone. Pindell and Dewey (1982), in contrast, argued for an east-dipping Alleghanian subduction zone because of the westward vergence of the Alleghanian thrust sheets. Nelson (1992) also questioned the west-dipping subduction zone because seismic reflection data show a predominantly east- (or south-) dipping fabric throughout the crust of the eastern Piedmont, coastal plain, and adjacent continental shelf. Pindell and Dewey (1982) and Nelson (1992) suggested that the Alleghanian granites might have been generated during continental collision rather than during subduction. The coincidence of the range of ages of the plutons with the time spanned by the Alleghanian foreland clastic wedge supports this idea (Nelson, 1992).

Faults and folds of the thin-skinned foreland fold and thrust belt (Valley and Ridge province) were driven by the very large Blue Ridge crystalline thrust sheet. The Valley and Ridge province changes from a stack of imbricate thrust sheets in the southern Appalachians, to a folded roof thrust series above duplexes that end in blind thrusts in the central Appalachians (Geiser, 1989). The thrusting was northwest directed and probably propagated toward the foreland as the thrust front ramped upward into progressively younger strata (see Boyer and Elliott, 1982). Along the southeastern margin of the Valley and Ridge province in E5, the Great Smoky thrust (Fig. 1) carries the Blue Ridge crystalline thrust sheet over Mississippian rocks. Pennsylvanian rocks of the clastic wedge are carried by and are overridden by thrusts in E4. The present position of the Blue Ridge thrust sheet is due to movement on postmetamorphic (Taconian metamorphism) Alleghanian thrusts that at least locally formed duplexes and structural culminations. These were subsequently eroded to create windows such as the large Grandfather Mountain and Mountain City windows in E4 (Fig. 1). Some of the Alleghanian thrusts on which parts of the Blue Ridge sheet moved were probably reactivated Taconian thrusts (Rankin and others, 1989).

Numerous Alleghanian faults have been identified within the internides. For some, such as the Brevard zone, a long and complicated history has been documented. Rb-Sr whole-rock dating of ultramylonites from the Brevard indicate an episode of fluid-enhanced deformation at 273 Ma (Sinha and others, 1988). Earlier thrusting was followed by right-lateral strike-slip movement as deduced from S-C fabric studies (Vauchez, 1987; Edelman and others, 1987) and subsequently by late brittle reverse faulting along the northwestern boundary of the zone (the Rosman fault of Horton, 1982). It was probably during the last brittle episode that pieces of platform carbonate and shale were brought up from beneath the Blue Ridge crystalline thrust sheet (Hatcher, 1987). For many of the Alleghanian faults, from the Brevard southwest to the Nutbush Creek fault (Fig. 1), late right-lateral movement has been documented. Large strike-slip displacements have been proposed (Druhan and others, 1988), but not proven. This late dextral shear is consistent along the eastern margin of the Appalachians (Gates and others, 1986) and has led to the interpretation, preferred here, that the last stage of the Alleghanian orogeny after the closing of the Theic-Rheic Ocean (Van der Voo, 1983) was dextral plate motion between mid-Paleozoic North America and proto-Gondwana (Secor and others, 1986). Hatcher (1989b) speculated that the dextral motion might have been caused by the Reguibat indenter.

Carboniferous sedimentary rocks do not crop out in the foreland north of Pennsylvania, so there is no record of an Alleghanian clastic wedge in the northern Appalachians. Nor is there a record of Alleghanian deformation in the foreland. The effects of the Alleghanian orogeny in the U.S. northern Appalachians include plutonism, metamorphism, and deformation. These effects, once thought to be restricted to coastal New England (reviewed by Snoke and Mosher, 1989), are now known to extend inland almost as far as southwestern New Hampshire (Robinson and others, 1992). Although Late Devonian to Carboniferous basins of Maritime Canada appear to be pull-apart basins initiated as a result of dextral strike-slip movement (Bradley, 1982), the Norfolk and Narragansett basins (Fig. 6) probably formed by extension with little strike-slip movement (Snoke and Mosher, 1989).

High-temperature, ductile Alleghanian deformation in New England is most obvious in the Narragansett basin, where well-dated Pennsylvanian strata are involved in polyphase deformation and metamorphism (Snoke and Mosher, 1989). Deformation of the basin can be separated into four distinct phases. The first two phases produced folds and were accompanied by regional metamorphism to grades as high as sillimanite in the southwestern part of the basin. Kyanite of this regional metamorphism crystallized in some Pennsylvanian rocks of the basin (Murray, 1988), indicating that parts of the basin were at mid-crustal levels prior to the intrusion of the Narragansett Pier Granite. Deformation of the second two phases can be mostly attributed to discrete movement along a complex system of nonparallel, intersecting sinistral and dextral, near-vertical strike-slip faults. In general, sinistral movement (D_3) preceeded dextral movement (D_4). This late-stage faulting (D_3 and D_4) was coincident with the intrusion of the S-type Narragansett Pier Granite (272 ± 4 Ma U-Pb zircon age by Zartman and Hermes, 1987), which truncates the regional metamorphic isograds.

As reviewed earlier in the section on the Avalon terrane and inliers thereof, the Avalon terrane was not affected by Acadian metamorphism, nor was it intruded by Acadian plutons. Furthermore, rocks of the Avalon terrane were not in their present position relative to Late Devonian North America at the time of peak Alleghanian metamorphism (Getty and Gromet, 1992a, 1992b; Robinson and others, 1992; Wintsch and others, 1992). It is tentatively concluded, therefore, that the Avalon terrane did not dock with North America until the Alleghanian, and that the initial collision and peak metamorphism were with and adjacent to rocks other than their present neighbors.

A number of Carboniferous two-mica granites (Sebago, Effingham, and Lyman plutons; Fig. 6) intrude the Central Maine terrane. Their ages are in the range of 325 to 320 Ma (Aleinikoff and others, 1985; Hayward and Gaudette, 1984; Gaudette and others, 1982). Osberg and others (1989) included these with others dating back to about 350 Ma (365 Ma is probably a better number; Lyons and others, 1994) and suggested that they formed a group of postkinematic Acadian granites generated at depth by partial melting of tectonically thickened crust, and ultimately emplaced at higher crustal levels. The younger granites (325 to 320 Ma), however, are about 25 m.y. younger than the rest of the group and could be Alleghanian plutons. The Narragansett Pier Granite and the granite near Milford in southern New Hampshire (Fig. 6) (about 275 Ma; Aleinikoff and others, 1979) are clearly Alleghanian granites of Permian age.

The Avalon terrane is interpreted to be exposed in windows through the Central Maine and Merrimack-Harpswell terranes (the Pelham and Willimantic domes, respectively). That configuration indicates that the Alleghanian collision involved the underthrusting of the Avalon terrane beneath the Late Devonian North American margin. A similar configuration was argued for the earlier Acadian collision of the Merrimack-Mascarene composite terrane with the Central Maine terrane. As determined from the position of the Pelham dome, the amount of underthrusting was about 100 km. The main body of the Avalon terrane is bounded on the northwest by the Honey Hill, Lake Char, and Bloody Bluff faults (Fig. 6). Early movement on these faults has been interpreted to be northeast- and east-directed thrusting, primarily because they placed higher grade rocks over lower grade rocks (Snoke and Mosher, 1989, and references therein).

Work by Goldstein (1989), Getty and Gromet (1992a, 1992b), Wintsch and others (1992), and Wintsch (1992) indicates that, except for the Pelham dome region, southeastern New England underwent a period of late, major east-west extension, which accounts for observed fabrics and the discordance in cooling ages across the bounding faults of the Avalon terrane. Late extensional fabrics have not been observed near the Pelham dome. Instead, the Avalon core rocks and their Central Maine cover have a strong north-south–trending mineral lineation of Pennsylvanian age that indicates a top-to-the south sense of shear (Robinson and others, 1992). Data indicate that the extension identified southeast of the Pelham dome was late Paleozoic, between 280 and 265 Ma (Getty and Gromet, 1992a). The realities of these conclusions and their implications are still being investigated. The Narragansett basin in Rhode Island contains fossiliferous strata as young as Stephanian (Lyons, 1979) (minimum age of 286 Ma; time scale of Palmer, 1983). These sediments were metamorphosed at mid-crustal levels (kyanite is present in some rocks) prior to the intrusion of the Narragansett Pier Granite at about 272 Ma and prior to extensional movement on the Lake Char fault and around the Willimantic dome. We do not know the timing of the opening of the interpreted very extensive continuation of the Narragansett basin offshore (Fig. 6) or its relation to Alleghanian deformation.

Evidence is accumulating that the U.S. Appalachians underwent late orogenic or postorogenic extension and that a significant part of that extension took place in the late Paleozoic, prior to the well-documented Late Triassic to Early Jurassic rifting. Whether the cause of the late Paleozoic extension was the same in the northern and southern Appalachians is not known. It is probably premature to attempt to correlate the details of sequences of events in the different parts of the orogen. We reviewed the data that indicate that the Kiokee belt of the Savannah River terrane, bounded by the Modoc shear zone and Augusta fault (Fig. 3), underwent rapid uplift about 300 Ma relative to the Carolina terrane both northwest and southeast of it (Maher, 1987; Atekwana and others, 1989; Snoke and Frost, 1990; Maher and others, 1991). The amount of that uplift was estimated from aluminum-in-hornblende barometry to have been about 18.5 km across the Modoc shear zone and about 5.1 km across the Augusta fault (Vyhnal and McSween, 1990). The aluminum-in-hornblende barometry also indicated that the Raleigh belt of the Savannah River terrane (Fig. 3) was uplifted as much as 8 to 15 km relative to terranes on either side between 312 and 285 Ma (Vyhnal and McSween, 1990). The parts of the Gaffney and adjacent Carolina terranes (Fig. 5) for which Alleghanian metamorphism has been documented (Horton and others, 1987) have been uplifted as much as 8.5 km relative to the Carolina terrane 7.5 km to the east (Vyhnal and McSween, 1990).

Continental rifting, opening of the Atlantic, and formation of the modern passive margin of North America

The present tectonic configuration of eastern North America is a consequence of the breakup of Pangea and the opening of the present Atlantic Ocean. The cycle began in the northern and Maritime Appalachians with Carboniferous (even Late Devonian) and Permian transtensional basin formation. Transtensional forces in the north overlapped in time compressional forces of the Alleghanian orogeny in the south.[1] It continued with Late Triassic to Early Jurassic continental rifting across a zone as wide as 500 km. It concluded with Middle Jurassic continental breakup, formation of oceanic crust and the subsequent subsidence of the new continental margin, and deposition of onlapping sediments. New oceanic crust formed at the Mid-Atlantic Ridge. The Atlantic Ocean widened and North America and Africa drifted apart. The continental margin of North America was an active extensional margin during the rifting phase. It evolved into a passive or trailing-edge margin during the drifting phase.

[1]Note added in press. Speed, this volume, attributes the Carboniferous basins, which in Maritime, Canada were initiated as long ago as the Upper Devonian, to oblique collision in the consolidation of Pangea rather than to early stages of Mesozoic rifting. I concur with that interpretation.

Crustal extension produced continental rifting that resulted in a disconnected series of half grabens both along and across the strike. The border faults are listric normal faults, and many of these are reactivated Paleozoic thrust or strike-slip faults. Ratcliffe and others (1986) demonstrated that the northwestern border fault of the Newark basin along the Delaware River is a reactivation of low-angle Paleozoic thrust and mylonite zones that emplaced the Musconetcong thrust sheets of Grenvillian basement over the Paleozoic shelf sequence. They postulated that the Mesozoic fault pattern formed by passive response to extension rather than by active wrench-fault tectonics, which was proposed by Manspeizer (1980). Offshore seismic studies show that listric border faults may merge at depth with low-angle detachment surfaces that decoupled the upper and lower crust (Hutchinson and Klitgord, 1988b). For most basins the major border fault is on one side, but antithetic faults may dip into the border fault. The major border fault may be on either side of the basin. Some basins appear to be in pairs, such as the Dan River–Danville vs. Wadesboro–Deep River and the Newark vs. Hartford (Fig. 1).

Rifting of Pangea along the broad zone within which the North American–African split would occur (central Atlantic zone) propagated southward during the Middle and Late Triassic from the Arctic–North Atlantic zone, where rifting began as early as Permian (Klitgord and others, 1988). This is consistent with the Fundy basin (Fig. 2) preserving the earliest fossils (Middle Triassic). McHone and Butler (1984) identified a coastal New England igneous province of Early to Late Triassic (240 to 210 Ma) syenite to alkali granite complexes and dikes that suggest extension in a northwest-southeast direction consistent with the orientation of the Fundy basin. During this early phase, active rifting was inboard of the two major African fragments, the Meguma and Suwannee terranes. Major strike-slip fault zones connected the central Atlantic rift system with the Arctic–North Atlantic system (see Klitgord and others, 1988, Fig. 19).

In the Early Jurassic the central Atlantic system became linked to the Ligurian Tethys rift system in the Mediterranean rather than to the Arctic–North Atlantic system (Klitgord and others, 1988, Fig. 19). This change was accompanied by a tremendous pulse of magmatism and changes in the direction and distribution of rifting. Rifting in the onshore basins waned, the basement hinge zone formed, and the focus of rifting shifted to the marginal basins. Rifting inboard in the Fundy and South Georgia basins (Fig. 2) ceased, but continued or was initiated in the Scotian and Bahamas basins (Fig. 2), leaving the Meguma and Suwannee terranes attached to North America (Fig. 3). Gulf of Mexico water, however, continued to flow to the Atlantic across the Suwannee strait (south Georgia rift) until the Paleocene (Dillon and Popenoe, 1988). As noted earlier, McBride and others (1989) hypothesized that one of the largest outpourings in the world of basalt on continental crust was on the postrift unconformity in the south Georgia rift and on the Carolina platform (Fig. 2). They suggested that the south Georgia rift came close to resulting in continental separation and the creation of oceanic crust.

The magmatism relating to the Early Jurassic rifting phase is of two quite different types. The first constitutes the eastern North America (ENA) tholeiitic dolerite (diabase) province of basalt, diabase sheets, and dikes; the province is part of a vast magmatic superprovince that surrounds most of the central and northern Atlantic. Olsen (1989) reported that more than 100 K-Ar and $^{40}Ar/^{39}Ar$ ages from basalt of the Newark Supergroup (recalculated with new decay constants where possible) give a mean age of 203 Ma with the median at 198 Ma. Sutter (1988) concluded from $^{40}Ar/^{39}Ar$ age spectra that the igneous activity (lavas, dikes, and sills) of the ENA province was a short-lived event spanning no more than a few million years about 201.2 ± 1.3 Ma, toward the end of the continental-rifting episode. Several different parental magmas are necessary to produce the varieties and distributions of ENA magmas. The ENA dikes may change character, from quartz-tholeiitic through olivine-tholeiitic to alkalic basalts, with increasing distance from zone of actual continental breakup (see McHone and Butler, 1984).

The other pulse of Jurassic magmatism constitutes a north-northwest–trending zone of alkalic mafic to felsic, but dominantly felsic, plutonic-volcanic complexes in central New England (Fig. 2). These are the main components of the White Mountain Plutonic-Volcanic Suite, which has generally been thought of as consisting of all Mesozoic plutons in central New England (the White Mountain magma series of Billings, 1956). The Mesozoic plutons of central New England are one of the components of a generally northwest trending zone of Mesozoic plutonism extending from the Cretaceous Monteregian Hills, Quebec, to the Cretaceous(?) New England Seamounts in the Atlantic Ocean, and including the Triassic plutons of coastal New England. There has been a tendency to think of these as all genetically related. One explanation has been that they represent the track of one or more hot spots (Crough, 1981; Morgan, 1981), but because there is no systematic trend of ages of the plutons (Foland and Faul, 1977), the hot-spot interpretation is suspect (see Thompson and others, 1993). McHone and Butler (1984) argued with considerable reason that the "White Mountain magma series" province, perhaps better called the White Mountain province as in Figure 2, consists of a more coherent grouping if one considers only the Jurassic plutons of central New England (mostly in New Hampshire) as composing the province. They suggested that the emplacement of the complexes was controlled by a series of north-trending faults and rift valleys active in the Jurassic and roughly comparable to fault zones in the Connecticut and Champlain valleys. The general north-northwest trend of the White Mountain province, however, seems to be significantly different, especially considering its location between the north-trending Hartford basin and the northeast-trending basins in the Gulf of Maine. The White Mountain province seems to indicate incipient extension in a northeast-southwest direction, as discussed later. Most of the complexes of the White Mountain province are of Middle Jurassic age, but they range from Early to Late Jurassic (see McHone and Butler, 1984). The White Mountain province is within the area of the ENA dolerite province.

Continental separation took place during the evolution of the marginal basins. Poag and Valentine (1988) interpreted that more than 5 km of synrift terrigenous siliciclastic rocks of Triassic age in the Baltimore Canyon trough underlie the postrift unconformity and marine carbonates and evaporites. Klitgord and others (1988), extrapolating from available data in the COST wells of the Georges Bank basin, concluded that the lowest strata of the postrift sections in the marginal basins are probably late Early Jurassic age (about 190 Ma, according to the Palmer [1983] time scale). Basalts above the postrift unconformity at Charleston, South Carolina, were dated at 184 ± 33 by Lanphere (1983). The youngest sediments preserved in the onland early Mesozoic rift basins are the Toarcian age (minimum age about 187 Ma) rocks in the Hartford basin. Klitgord and Schouten (1986) estimated the beginning of sea-floor spreading as 175 Ma based upon the assumption of a constant spreading rate prior to the Tithonian. With the initiation of sea-floor spreading, crustal extension ceased on the continent.

The initiation of sea-floor spreading was not synchronous along the central Atlantic system. Rifting continued in the Blake Plateau and Bahamas basins (the ECMA does not extend south of the Blake Spur fracture zone) while sea-floor spreading was creating typical ocean crust east of the Carolina trough to the Scotia basin (Fig. 2). Klitgord and others (1988) suggested that the zone of seaward-dipping reflectors east of the ECMA was generated during this early stage of sea-floor spreading. About 170 Ma the central Atlantic spreading center jumped eastward to the Blake Spur magnetic anomaly (Vogt, 1973), and oceanic crust was being generated along the entire central Atlantic rift system, in the Ligurian Tethys, and in the Gulf of Mexico. "By 170 Ma, the central Atlantic Ocean basin was 200 km wide. . . .This margin was cooling and subsiding, and substantial sediment was accumulating on the shelf behind the paleoshelf edge. This was the start of the period of rapid subsidence on the margin as the spreading center and active plate boundary shifted farther and farther to the east" (Klitgord and others, 1988, p. 50–51).

In the Cretaceous, magmatism recurred along the general trend of the White Mountain province but with a more northwesterly or even west-northwesterly trend, rather than north-northwesterly. A chain of alkalic mafic plutons (now the Monteregian Hills) intrude rocks in Quebec between the Late Proterozoic to Cambrian Ottawa-Bonnechère graben (in part coextensive with the Monteregian Hills) and the White Mountain province. Rocks of the Monteregian Hills are 130 to 110 Ma, as summarized by McHone and Butler (1984). Early Cretaceous plutonic complexes, some with lavas and pyroclastics in down-dropped masses, occur in the southeastern part of the White Mountain province. The New England Seamounts extend more than 1300 km from the upper continental rise southeast of Cape Cod southeasterly across the Atlantic Ocean floor. The northwesternmost seamounts are just landward of the ECMA in E1 at a 45 km right-lateral offset in the anomaly (Klitgord and Schouten, 1986). The seamount chain ends to the southeast on about 85 Ma oceanic crust. The seamounts are thought to range in age from about 100 to 83 Ma (all Early Cretaceous) (see McHone and Butler, 1984). The seamounts may be along a leaky transform fault (and the trend is parallel to the fracture zones as portrayed by Klitgord and Schouten, 1986). However, Thompson and others (1993) suggested that the general northwest trend from the New England Seamounts to the Monteregian Hills represents an unsuccessful rift that ceased activity with the opening of the subparallel and more successful rifts that led to the opening of the Davis Strait and the Greenland and Norwegian seas.

Present tectonic stress field

Prowell (1989) reviewed the distribution and characteristics of known faults in eastern United States that have Cretaceous or younger movement. Because the recognition of such faults is generally dependent upon observed displacement of geologic contacts of appropriate age, most of the known faults are in coastal plain deposits or along the Fall Line. These faults are mostly northeast-trending reverse-fault zones and fault systems up to 100 km long. Vertical displacements of as much as 76 m have occurred since the Early Cretaceous, and progressively smaller offsets have occurred in strata spanning the Cenozoic. A component of strike-slip motion has been reported for some of the reverse faults, but dip-slip motion is dominant. Along the Atlantic margin, fault zones in the northeast tend to strike more northerly than those in the southwest.

These observations plus a compilation of focal mechanisms for earthquakes in the New York–New Jersey area led Zoback and Zoback (1980) and Wentworth and Mergner-Keefer (1983) to suggest an Atlantic seaboard stress province with a northwest-oriented maximum horizontal compression. New data compiled by Zoback and Zoback (1989)—including breakouts in boreholes both on land and on the continental shelf, hydraulic fracturing studies at several localities, and better constrained earthquake focal mechanisms, particularly for the New York–New Jersey and Charleston, South Carolina, areas—are incompatible with the generalization of northwest compression. Zoback and Zoback (1989) concluded that most of the central and eastern United States is part of a broad "midplate" stress province, which includes most of Canada and possibly the western Atlantic basin, characterized by northeast- to east-northeast–oriented maximum horizontal compression. They noted that many of the young reverse faults with northeast strikes are in eastern Virginia and that this area seems to be anomalous. For the province as a whole, however, the northeast to east-northeast orientation of compression coincides with that inferred from both the absolute plate motion and ridge-push directions for North America.

DIFFERENCES AMONG THE TRANSECTS

Important if not unique features of transects or groups of transects are listed below. This list is intended to point out differences between the transects. It is not a catalogue of the features of each transect. Details of terranes, many of which are controversial, will not be repeated, because they were covered in the preceeding parts of this chapter. Only those features actually traversed by one of the E transects are listed.

Transect E1

Exposed autochthonous cratonic basement in the Adirondack massif which is also within a post-Carboniferous, possibly Tertiary dome.

A buried Iapetan continental rift zone indicated by a linear Bouguer positive gravity anomaly.

Aspects of the Taconian orogeny: (1) foreland fold and thrust belt; (2) the Taconic allochthon; (3) one or more Ordovician volcanic arcs (Hawley arc and Ammonoosuc arc); (4) essentially no Taconian molasse in the foreland.

Extensive Silurian and Devonian strata in the internides forming overlap sequences on terranes joined in the Penobscottian and Taconian orogenies: (1) the accumulation east of a Silurian tectonic hinge zone of a tremendous volume of clastic sedimentary rocks in the Central Maine trough that probably represent the erosion of Taconian mountains; (2) Silurian rifting west of the tectonic hinge zone.

Aspects of the Acadian orogeny: (1) deformed and metamorphosed Silurian and Devonian strata of the internides including nappes and mantled gneiss domes of the Bronson Hill anticlinorium; (2) extensive plutonism and overthickened crust; (3) the Catskill delta (Acadian molasse) in the foreland.

Terranes composing the Merrimack-Mascarene composite terrane that underwent Silurian plutonism and magmatism, were amalgamated in the Silurian, and accreted to North America in the Acadian.

Late Proterozoic extension in the southeastern New England Avalon terrane to produce the Boston basin.

Strata with an Early Cambrian Atlantic province fauna nonconformably above Late Proterozoic calc-alkaline plutons (Avalon terrane).

Pennsylvanian extensional and transtensional basins.

Aspects of the Alleghanian orogeny: (1) a Permian pluton intruding regionally metamorphosed Pennsylvanian strata; (2) a terrane (Avalon) now part of the eastern Appalachian orogen that was not affected by either the Taconian or Acadian orogenies; (3) windows (Pelham and Willimantic) exposing Late Proterozoic rocks metamorphosed in the Pennsylvanian, but not during the Acadian orogeny, framed by rocks that were metamorphosed during the Acadian, but not the Pennsylvanian.

The early Mesozoic rift basin containing the youngest preserved strata (Toarcian strata in the Hartford basin).

Mesozoic plutonic-volcanic complexes (Monteregian Hills, White Mountain Plutonic-Volcanic Suite, and New England Seamounts) ranging in age from Early Triassic to Early Cretaceous. Many of those of Jurassic age are aligned along a northwest-trending failed rift.

The Gulf of Maine embayment in which Paleozoic strata are covered only by local early Mesozoic basins and Pleistocene deposits.

A broad, structurally complex Mesozoic-Cenozoic marginal basin (Georges Bank basin).

A paleoshelf edge well landward of the ECMA.

Features shared by transects E2 to E5 that distinguish that group from E1 (some exceptions noted)

The reader is reminded that E2 and E3, unfortunately, will not be published by the Decade of North American Geology.

Late Proterozoic to Early Cambrian anorogenic plutons in the Laurentian Appalachian (except for E5).

E3 and E4 cross the South Mountain and Mt. Rogers Iapetan embayments and their attendant thick piles of anorogenic rhyolites that appear to be as much as 200 m.y. different in age (about 760–700 and 600–550 Ma).

Extensive Taconian molasse in the foreland.

The rarity of post-Cambrian stratified rock in the internides south of the Potomac River. A possible interpretation of this is that most of the internides are composed of Cambrian and older strata plus younger plutons.

Alleghanian molasse in the foreland in which rocks as young as Pennsylvanian and possibly Permian are deformed.

Alleghanian foreland fold and thrust belt: (1) for E2 and E3 the belt is a fold belt draped above a set of duplexes that end in blind thrusts; (2) for E4 and E5 the belt is an imbricate thrust belt.

Extensive Alleghanian plutonism (except for E2).

A prominent west-facing Bouguer gravity gradient in the Appalachian internides that is progressively farther outboard from the external massifs from New York toward Georgia.

Extensive but thin (up to a few kilometers) Cretaceous and Cenozoic coastal plain deposits between the Fall Line and basement hinge zone offshore.

Transect E2

Taconian Hamburg allochthon.

Narrowest internides of the Appalachian orogen.

The deepest marginal basin. The northern Baltimore Canyon trough contains as much as 18 km of Mesozoic and Cenozoic synrift and postrift sediments.

The only major intrusion to penetrate high into postrift strata landward of the ECMA (Early Cretaceous Great Stone dome).

The paleoshelf edge is over oceanic crust.

Transect E3

A Cambrian volcanic arc (Chopawamsic terrane).

Fossiliferous Taconian successor basins in the internides (Arvonia Slate and Quantico Formation).

The enigmatic Middle Proterozoic(?) Goochland granulite terrane.

The subsurface Chesapeake terrane—a possible fragment of African crust that could contain Archean rocks.

Transect E4

The only transect that extends westward to the subsurface Grenville front and the Paleozoic peripheral bulge of the Cincinnati arch.

Cambrian rift basin (Rome trough) in the subsurface of the Appalachian basin.

Large windows in the Blue Ridge crystalline thrust sheet.

An area of Alleghanian regional metamorphism in the central Piedmont. Could this be a metamorphic core complex?

The largest internal Grenvillian massif crossed by a transect (Sauratown Mountains massif, exposed in a window).

Silurian and Devonian anorogenic plutonic complexes (Concord and Salisbury Plutonic Suites) in the Charlotte belt.

Exposed early Mesozoic rift basins that contain only Late Triassic strata and no volcanic rocks.

A medium-grade metamorphic terrane (Hatteras terrane) in the subsurface along the coastline that includes Late Proterozoic Pan-African–age rocks.

The narrowest of the Mesozoic-Cenozoic marginal basins (Carolina trough).

Extensive salt diapirs along the continental margin in the Carolina trough and related growth faults due to progressive withdrawal of salt.

No clearly defined reef at the paleoshelf edge.

Features shared by transects E4 and E5

The Pine Mountain thrust sheet at the leading edge of the Alleghanian foreland fold and thrust belt.

The Brevard fault zone and central Piedmont suture (Kings Mountain shear zone in E4 and Lowndesville shear zone in E5).

One or more Late Proterozoic to Cambrian volcanic arcs of proto-Gondwana (the large composite Carolina terrane). Largely Late Proterozoic with Ediacarian fossils in E4; largely Middle Cambrian with Atlantic province fauna in E5.

Transect E5

Extensive deposits of Late Proterozoic clastic rocks that are unlike their probable correlatives in other transects in being volcanic free (Ocoee Supergroup). The recent report in the Walden Creek Group of fossils that could be as young as Silurian (Unrug and Unrug, 1990) is still being evaluated in terms of reproducibility and age.

Paleozoic(?) slope-rise deposits and/or Taconian successor basin deposits (Tull and Groszos, 1990) of the Murphy syncline.

A number of small windows through the Blue Ridge crystalline thrust sheet including some rimmed by ultramafic rocks.

A belt of Alleghanian regional metamorphism near the Fall Line possibly exposed in a metamorphic core complex.

A small part of the Suwannee terrane of African affinity.

The paired high-low Brunswick magmatic anomaly that roughly marks the Alleghanian suture between the Suwannee terrane and the Appalachian orogen of North America.

The subsurface early Mesozoic South Georgia basin that now covers the Alleghanian suture between the Suwannee terrane and the Appalachian orogen.

Extensive early Middle Jurassic basalt on the postrift unconformity above the South George basin. The hypothesized volume of basalt suggests that rifting here came close to resulting in continental separation and the formation of oceanic crust.

Marine deposits in the Suwannee Strait above the South Georgia basin, which indicate that the ancestral Gulf Stream flowed through the Suwannee Strait until the Paleocene.

The widest of the Mesozoic-Cenozoic marginal basins (the Blake Plateau basin). This basin extends across the southern termination of the ECMA and rifting is thought to have continued in this basin until continental crust separated along the Blake Spur anomaly, forming the east side of the basin about 170 Ma.

ACKNOWLEDGMENTS

This synthesis, in addition to the voluminous published research, has drawn upon many years of participation in field trips throughout the Appalachians and attendance at meetings and conferences. It would be almost impossible to acknowledge the many colleagues who have contributed to my knowledge of eastern geology. I wish to acknowledge the following for specific helpful discussions during preparation of this manuscript: W. A. Bothner, D. L. Daniels, W. P. Dillon, A. A. Drake, Jr., A. J. Froelich, W. F. Hanna, A. G. Harris, J. W. Horton, Jr., J. H. Luetgert, R. B. Neuman, N. M. Ratcliffe, J. E. Repetski, B. D. Stone, J. B. Thompson, Jr., R. P. Tollo, W. E. Trzcienski, D. P. West, Jr., and R. P. Wintsch. I have also benefitted from many discussions, encouragement and futile advice to be brief from D. B. Stewart. The manuscript has been improved by reviews by W. P. Dillon, A. E. Gates, J. W. Horton, Jr., W. Manspeizer, D. T. Secor, Jr., R. C. Speed, and D. B. Stewart. I thank A. R. Palmer and R. C. Speed for their patience and exhortations to persevere.

REFERENCES CITED

Abbott, R. N., Jr., and Raymond, L. A., 1984, The Ashe metamorphic suite, northwest North Carolina: Metamorphism and observations on geologic history: American Journal of Science, v. 284, p. 350–375.

Absher, B. S., and McSween, H. Y., Jr., 1985, Granulites at Winding Stair Gap, North Carolina: The thermal axis of Paleozoic metamorphism in the southern Appalachians: Geological Society of America Bulletin, v. 96, p. 588–599.

Aleinikoff, J. N., and Karabinos, P., 1990, Zircon U-Pb data for the Moretown and Barnard Volcanic Members of the Missisquoi Formation and a dike cutting the Standing Pond Volcanics, southeastern Vermont, *in* Slack, J. F., ed., Summary results of the Glens Fall CUSMAP project, New York, Vermont, and New Hampshire: U.S. Geological Survey Bulletin 1887, p. D1–D10.

Aleinikoff, J. N., and Moench, R. H., 1985, Metavolcanic stratigraphy in northern New England—U-Pb zircon geochronology: Geological Society of America Abstracts with Programs, v. 17, p. 1.

——, 1992, U-Pb zircon ages of the Ordovician Ammonoosuc Volcanics and related plutons near Littleton and Milan, New Hampshire: Geological So-

ciety of America Abstracts with Programs, v. 24, no. 3, p. 2.
Aleinikoff, J. N., Zartman, R. E., and Lyons, J. B., 1979, U-Th-Pb geochronology of the Massabesic Gneiss and the granite near Milford, south-central New Hampshire: New evidence for Avalonian basement and Taconic and Alleghenian disturbances in eastern New England: Contributions to Mineralogy and Petrology, v. 71, p. 1–11.
Aleinikoff, J. N., Moench, R. H., and Lyons, J. B., 1985, Carboniferous U-Pb age of the Sebago batholith, southwestern Maine: Metamorphic and tectonic implications: Geological Society of America Bulletin, v. 96, p. 990–996.
Aleinikoff, J. N., Zartman, R. E., Rankin, D. W., Lyttle, P. T., and Burton, W. C., and McDowell, R. C., 1991, New U-Pb zircon ages for rhyolite of the Catoctin and Mount Rogers Formations—More evidence for two pulses of Iapetian rifting in the central and southern Appalachians: Geological Society of America Abstracts with Programs, v. 23, no. 1, p. 2.
Allmendinger, R. W., Nelson, K. D., Potter, C. J., Barazangi, M., Brown, L. D., and Oliver, J. E., 1987, Deep seismic reflection characteristics of the continental crust: Geology, v. 15, p. 304–310.
Ambers, C. P., and Wintsch, R. P., 1990, Subsurface identification of Merrimack, Putnam-Nashoba, and metavolcanic Avalon terrane rocks, and ductile to brittle fault rocks in the 1.45-kilometer-deep research hole, Moodus, Connecticut, *in* Socci, A. D., Skehan, J. W., and Smith, G. W., eds., Geology of the composite Avalon terrane of southern New England: Geological Society of America Special Paper 245, p. 171–186.
Ammerman, M. L., and Keller, G. R., 1979, Delineation of Rome trough in eastern Kentucky by gravity and deep drilling data: American Association of Petroleum Geologists Bulletin, v. 63, p. 341–353.
Ando, C. J., and 11 others, 1984, Crustal profile of a mountain belt: COCORP deep seismic reflection profiling in New England Appalachians and implications for architecture of convergent mountain chains: American Association of Petroleum Geologists Bulletin, v. 68, p. 819–837.
Atekwana, E., Sutter, J., and Schwartzman, D., 1989, Thermochemistry of the Pennsylvanian Liberty Hill, Pageland, and Lilesville plutons and the Wadesboro basin, North and South Carolina: Geological Society of America Abstracts with Programs, v. 21, no. 3, p. 2.
Austin, J. A., Uchupi, E., Shaughnessy, D. R., III, and Ballard, R. D., 1980, Geology of New England passive margin: American Association of Petroleum Geologists Bulletin, v. 64, p. 501–526.
Austin, J. A., Jr., and 7 others, 1990, Crustal structure of the Southeast Georgia embayment–Carolina trough: Preliminary results of a composite seismic image of a continental suture(?) and a volcanic passive margin: Geology, v. 18, p. 1023–1027.
Ayuso, R. A., and Bevier, M. L., 1991, Regional differences in Pb isotopic compositions of feldspars in plutonic rocks of the northern Appalachians Mountains, U.S.A. and Canada: A geochemical method of terrane correlation: Tectonics, v. 10, p. 191–212.
Ayuso, R. A., Horan, M. F., and Criss, R. E., 1988, Pb and O isotopic geochemistry of granitic plutons in northern Maine: American Journal of Science, v. 288-A, p. 421–460.
Badger, R. L., and Sinha, A. K., 1988, Age and Sr isotopic signature of the Catoctin volcanic province: Implications for subcrustal mantle evolution: Geology, v. 16, p. 692–695.
Barnes, C. R., Norford, B. S., and Skevington, D., 1981. The Ordovician System in Canada: Correlation chart and explanatory text: International Union of Geological Sciences Publication 8, 27 p., 2 charts.
Barosh, P. J., 1982, Structural relations of the junction of the Merrimack province, Nashoba thrust belt and the southeast New England platform in the Webster-Oxford area, Massachusetts, Connecticut, and Rhode Island, *in* Joeston, R., and Quarrier, S. S., eds., Guidebook for field trips in Connecticut and south-central Massachusetts (New England Intercollegiate Geological Conference, annual meeting, 74th): Connecticut Geological and Natural History Survey Guidebook 5, p. 395–418.
Beaumont, C., Quinlan, G., and Hamilton, J., 1988, Orogeny and stratigraphy: Numerical models of the Paleozoic in the eastern interior of North America: Tectonics, v. 7, p. 389–416.
Bell, R. E., Karner, G. D., and Steckler, M. S., 1988, Early Mesozoic rift basins of eastern North America and their gravity anomalies: The role of detachments during extension: Tectonics, v. 7, p. 447–462.
Bentley, R. D., and Neathery, T. L., 1970, Geology of the Brevard fault zone and related rocks of the Inner Piedmont of Alabama: Alabama Geological Society, Eighth Annual Field Trip, Guidebook, 119 p.
Berquist, C. R., Jr., and Marr, J. D., 1992, Faulted terrane boundaries in the southeastern Piedmont of Virginia: Geological Society of America Abstracts with Programs, v. 24, no. 2, p. 4.
Besancon, J. R., Gaudette, H. E., and Naylor, R. S., 1977, Age of the Massabesic Gneiss, southeastern New Hampshire: Geological Society of America Abstracts with Programs, v. 9, p. 242.
Best, D. M., Geddes, W. H., and Watkins, J. S., 1973, Gravity investigation of the depth of source of the Piedmont gravity gradient in Davidson County, North Carolina: Geological Society of America Bulletin, v. 84, p. 1213–1216.
Bevier, M. L., and Whalen, J. B., 1990, Tectonic significance of Silurian magmatism in the Canadian Appalachians: Geology, v. 18, p. 411–414.
Billings, M. P., 1956, The geology of New Hampshire—Part II, Bedrock geology: Concord, New Hampshire State Planning and Development Commission, 200 p.
Billings, M. P., and Lyttle, P. T., 1981, Paleontological control of Paleozoic stratigraphy, New England: U.S. Geological Survey Miscellaneous Field Studies Map MF-1302, scale 1:1,750,000.
Billings, M. P., Loomis, F. B., Jr., and Stewart, G. W., 1939, Carboniferous topography in the vicinity of Boston, Massachusetts: Geological Society of America Bulletin, v. 50, p. 1867–1884.
Bird, J. M., and Dewey, J. F., 1970, Lithosphere plate-continental margin tectonics and the evolution of the Appalachian orogen: Geological Society of America Bulletin, v. 81, p. 1031–1060.
Bird, P., 1978, Initiation of intracontinental subduction in the Himalaya: Journal of Geophysical Research, v. 83, p. 4975–4987.
Bobyarchick, A. R., and Glover, L., III, 1979, Deformation and metamorphism in the Hylas zone and adjacent parts of the eastern Piedmont in Virginia: Geological Society of America Bulletin, v. 90, p. 739–752.
Bobyarchick, A. R., Edelman, S. H., and Horton, J. W., Jr., 1988, The role of dextral strike-slip in the displacement history of the Brevard zone, *in* Secor, D. T., Jr., ed., Southeastern geological excursions (Geological Society of America Southeastern Section field trip guidebook): Columbia, South Carolina Geological Survey, p. 53–154.
Boltin, W. R., and Stoddard, E. F., 1987, Transition from Eastern slate belt to Raleigh belt in the Hollister quadrangle, North Carolina: Southeastern Geology, v. 27, p. 185–205.
Bond, G. C., Nickeson, P. A., and Kominz, M. A., 1984, Breakup of a supercontinent between 625 Ma and 555 Ma: New evidence and implications for continental histories: Earth and Planetary Science Letters, v. 70, p. 325–345.
Boone, G. M., and Boudette, E. L., 1989, Accretion of the Boundary Mountains terrane within the northern Appalachian orthotectonic zone, *in* Horton, J. W., Jr., and Rast, N., eds., Melanges and olistostromes of the U.S. Appalachians: Geological Society of America Special Paper 228, p. 17–42.
Boone, G. M., Doty, D. T., and Heizler, M. T., 1989, Hurricane Mountain Formation melange: Description and tectonic significance of a Penobscottian accretionary complex, *in* Tucker, R. D., and Marvinney, R. G., eds., Structure and stratigraphy: Augusta, Maine Geological Survey, Studies in Maine Geology, v. 2, p. 33–83.
Bosworth, W., Rowleg, D. B., Kidd, W.S.F., and Steinhardt, C., 1988, Geometry and style of post-obduction thrusting in a Paleozoic orogen: The Taconic frontal thrust system: Journal of Geology, v. 96, p. 163–180.
Bothner, W. A., and Laird, J., 1991, Iapetan crustal edges and the preservation of high pressure metamorphism in Vermont: Geological Society of America Abstracts with Programs, v. 23, no. 1, p. 10.
Bothner, W. A., Boudette, E. L., Fagan, T. J., Gaudette, H. E., Laird, J., and Olszewski, W. J., 1984, Geologic framework of the Massabesic anticlinorium and the Merrimack trough, southeastern New Hampshire, *in* Hanson, L. S., ed., Geology of the coastal lowlands, Boston [Mass.] to Kennebunk,

Maine (New England Intercollegiate Geological Conference annual meeting, 76th): Salem, Massachusetts, Salem State College, p. 186–206.

Boyer, S. E., and Elliott, D., 1982, Thrust systems: American Association of Petroleum Geologists Bulletin, v. 66, p. 1196–1230.

Bradley, D. C., 1982, Subsidence in late Paleozoic basins in the northern Appalachians: Tectonics, v. 1, p. 107–123.

——, 1983, Tectonics of the Acadian orogeny in New England and adjacent Canada: Journal of Geology, v. 91, p. 381–400.

——, 1989, Taconic plate kinematics as revealed by foredeep stratigraphy, Appalachian orogen: Tectonics, v. 8, p. 1037–1049.

Braile, L. W., Hinze, W. J., von Frese, R.R.B., and Keller, G. R., 1989, Seismic properties of the crust and uppermost mantle of the conterminous United Staes and adjacent Canada, *in* Pakiser, L. C., and Mooney, W. D., Geophysical framework of the continental United States: Geological Society of America Memoir 172, p. 655–680.

Brookins, D. G., Berdan, J. M., and Stewart, D. B., 1973, Isotopic and paleontologic evidence for correlating three volcanic sequences in the Maine coastal volcanic belt: Geological Society of America Bulletin, v. 84, p. 1619–1628.

Bryan, W. B., and Frey, F. A., 1986, Petrologic and geochemical evolution of pre–1 Ma western North Atlantic lithosphere, *in* Vogt, P. R., and Tucholke, B. E., eds., The western North Atlantic region: Boulder, Colorado, Geological Society of America, The Geology of North America, v. M, p. 271–296.

Bryant, B., and Reed, J. C., Jr., 1970, Geology of the Grandfather Mountain window and vicinity, North Carolina and Tennessee: U.S. Geological Survey Professional Paper 615, 190 p.

Burke, K., and Dewey, J. F., 1973, Plume-generated triple junctions: Key indicators in applying plate tectonics to old rocks: Journal of Geology, v. 81, p. 406–433.

Butler, J. R., 1973, Paleozoic deformation and metamorphism in part of the Blue Ridge thrust sheet, North Carolina: American Journal of Science, v. 273-A, p. 72–88.

——, 1979, The Carolina slate belt in North Carolina and northeastern South Carolina: A review: Geological Society of America Abstracts with Programs, v. 11, p. 172–173.

——, 1991, Metamorphism, *in* Horton, J. W., Jr., and Zullo, V. A., eds., The geology of the Carolinas: Knoxville, University of Tennessee Press, p. 127–141.

Butler, J. R., and Dunn, D. E., 1968, Geology of the Sauratown Mountains anticlinorium and vicinity, *in* Guidebook for field excursions (Geological Society of America Southeastern Section meeting guidebook): Southeastern Geology, Special Publication 1, p. 19–47.

Butler, J. R., and Secor, D. T., Jr., 1991, The central Piedmont, *in* Horton, J. W., Jr., and Zullo, V. A., eds., The geology of the Carolinas: Knoxville, University of Tennessee Press, p. 59–78.

Chamberlain, C. P., and England, P. C., 1985, The Acadian thermal history of the Merrimack synclinorium in New Hampshire: Journal of Geology, v. 93, p. 593–602.

Chappell, B. W., and White, A.J.R., 1974, Two contrasting granite types: Pacific Geology, v. 8, p. 173–174.

Chowns, T. M., and Williams, C. T., 1983, Pre-Cretaceous rocks beneath the Georgia Coastal Plain—Regional implications, *in* Gohn, G. S., ed., Studies related to the Charleston, South Carolina earthquake of 1886—Tectonics and seismicity: U.S. Geological Survey Professional Paper 1313, p. L1–L42.

Clague, D. A., Rubin, J., and Brackett, R., 1981, The age and origin of the dynamothermal aureole underlying the Thetford Mines ophiolite: Canadian Journal of Earth Sciences, v. 18, p. 469–486.

Clarke, D. B., Barr, S. M., and Donohoe, H. V., 1980, Granitoid and other rocks of Nova Scotia, *in* Wones, D. R., ed., Proceedings of the Caledonides in the U.S.A., I.G.C.P. project 27, Caledonide orogen: Blacksburg, Virginia Polytechnic Institute and State University, Department of Geological Sciences, Memoir 2, p. 107–116.

Clarke, J. W., 1952, Geology and mineral resources of the Thomaston quadrangle, Georgia: Georgia Department of Mines, Mining and Geology Bulletin 59, 103 p.

Cocks, L.R.M., and Fortey, R. A., 1982, Faunal evidence for oceanic separations in the Paleozoic of Britain: Geological Society of London Journal, v. 139, p. 465–478.

Coish, R. A., and Rogers, N. W., 1987, Geochemistry of the Boil Mountain ophiolite complex, northwest Maine, and tectonic implications: Contributions to Mineralogy and Petrology, v. 97, p. 51–65.

Colman-Sadd, S. P., Dunning, G. R., and Dec, T., 1992, Dunnage-Gander relationships and Ordovician orogeny in central Newfoundland: A sedimentary provenance and U/Pb age study: American Journal of Science, v. 292, p. 317–355.

Colton, G. W., 1970, The Appalachian basin—Its depositional sequences and their geologic relationships, *in* Fisher, G. W., Pettijohn, F. J., Reed, J. C., Jr., and Weaver, K. N., eds., Studies of Appalachian geology—Central and southern: New York, Interscience Publishers, p. 5–47.

Conley, J. F., and Henika, W. S., 1973, Geology of the Snow Creek, Martinsville East, Price, and Spray quadrangles, Virginia: Virginia Division of Mineral Resources Report of Investigations 33, 71 p.

Cook, F. A., 1984, Geophysical anomalies along strike of the southern Appalachian Piedmont: Tectonics, v. 3, p. 45–61.

Cook, F. A., and Oliver, J. E., 1981, The late Precambrian–early Paleozoic continental edge in the Appalachian Orogen: American Journal of Science, v. 281, p. 993–1008.

Cook, F. A., Albaugh, D. S., Brown, L. D., Kaufman, S., Oliver, J. E., and Hatcher, R. D., Jr., 1979, Thin-skinned tectonics in the crystalline southern Appalachians; COCORP seismic-reflection profiling of the Blue Ridge and Piedmont: Geology, v. 7, p. 563–567.

Cook, F. A., Brown, L. D., Kaufman, S., Oliver, J. E., Petersen, T. A., 1981, COCORP seismic profiling of the Appalachian orogen beneath the Coastal Plain of Georgia: Geological Society of America Bulletin, Part 1, v. 92, p. 738–748.

Cook, T. A., 1982, Stratigraphy and structure of the central Talladega slate belt, Alabama Appalachians, *in* Bearce, D. N., Black, W. W., Kish, S. A., and Tull, J. F., eds., Tectonic setting in the Talladega and Carolina slate belts, southern Appalachian orogen: Geological Society of America Special Paper 191, p. 47–59.

Costain, J. K., Hatcher, R. D., Jr., and Coruh, C., 1989a, Appalachian ultradeep core hole (ADCOH) project site: Regional seismic lines and geologic interpretation, *in* Hatcher, R. D., Jr., Thomas, W. A., and Viele, G. W., eds., The Appalachian-Ouachita orogen in the United States: Boulder, Colorado, Geological Society of America, The Geology of North America, v. F-2, plate 12.

Costain, J. K., and 6 others, 1989b, Geophysical characteristics of the Appalachians, *in* Hatcher, R. D., Jr., Thomas, W. A., and Viele, G. W., eds., The Appalachian-Ouachita orogen in the United States: Boulder, Colorado, Geological Society of America, The Geology of North America, v. F-2, p. 385–416.

Cousineau, P. A., 1991, The Rivière des Plante ophiolite melange: Tectonic setting and melange formation in the Quebec Appalachians: Journal of Geology, v. 99, p. 81–96.

Cousineau, P. A., and St. Julien, P., 1992, The Saint-Daniel Melange: Evolution for an accretionary complex in the Dunnage terrane of the Quebec Appalachians: Tectonics, v. 11, p. 898–909.

Cousminer, H. L., and Steinkrauss, W. E., 1988, Biostratigraphy of the COST G-2 well (Georges Bank): A record of Late Triassic synrift evaporite deposition; Liassic doming; and mid-Jurassic to Miocene postrift marine sedimentation, *in* Manspeizer, W., ed., Triassic-Jurassic rifting: Amsterdam, Elsevier, p. 167–184.

Crespi, J. M., 1988, Using balanced cross sections to understand early Mesozoic extension faulting, *in* Froelich, A. J., and Robinson, G. R., Jr., eds., Studies of the early Mesozoic basins of the eastern United States: U.S. Geological Survey Bulletin 1776, p. 220–229.

Crough, S. T., 1981, Mesozoic hotspot epeirogeny in eastern North America: Geology, v. 9, p. 2–6.

Culotta, R. C., Pratt, T., and Oliver, J., 1990, A tale of two sutures: COCORP's deep seismic surveys of the Grenville province in the eastern U.S. midconti-

nent: Geology, v. 18, p. 646–649.
Currie, K. L., 1988, The western end of the Avalon zone in southern New Brunswick: Maritime Sediments and Atlantic Geology, v. 24, p. 339–352.
Dallmeyer, R. D., 1975, Incremental $^{40}Ar/^{39}Ar$ ages of biotite and hornblende from retrograded basement gneisses of the southern Blue Ridge: Their bearing on the age of Paleozoic metamorphism: American Journal of Science, v. 275, p. 444–460.
——, 1978, $^{40}Ar/^{39}Ar$ incremental-release ages of hornblende and biotite across the Georgia Inner Piedmont: Their bearing on late Paleozoic–early Mesozoic tectonothermal history: American Journal of Science, v. 278, p. 124–149.
——, 1988, Late Paleozoic tectonothermal evolution of the western Piedmont and eastern Blue Ridge, Georgia: Controls on the chronology of terrane accretion and transport in the southern Appalachian orogen: Geological Society of America Bulletin, v. 100, p. 702–713.
——, ed., 1989a, Terranes in the circum-Atlantic Paleozoic orogens: Geological Society of America Special Paper 230, 277 p.
——, 1989b, Contrasting accreted terranes in the southern Appalachian orogen and Atlantic-Gulf Coastal Plains and their correlations with West African sequences, *in* Dallmeyer, R. D., ed., Terranes in the circum-Atlantic Paleozoic orogens: Geological Society of America Special Paper 230, p. 247–267.
Dallmeyer, R. D., Blackwood, R. F., and Odom, A. L., 1981, Age and origin of the Dover fault: Tectonic boundary between the Gander and Avalon zones of the northeastern Newfoundland Appalachians: Canadian Journal of Earth Sciences, v. 8, p. 1431–1442.
Dallmeyer, R. D., Wright, J. E., Secor, D. T., Jr., and Snoke, A. W., 1986, Character of the Alleghanian orogeny in the southern Appalachians: Part II, Geochronological constraints on the tectonothermal evolution of the eastern Piedmont in South Carolina: Geological Society of America Bulletin, v. 97, p. 1329–1344.
Dallmeyer, R. D., Hermes, O. D., Gil-Ibarguchi, J. I., 1990, $^{40}Ar/^{39}Ar$ mineral ages from the Scituate Granite, Rhode Island: Implications for late Paleozoic tectonothermal activity in New England: Journal of Metamorphic Geology, v. 8, p. 145–157.
Dalziel, I.W.D., 1991, Pacific margins of Laurentia and East Antarctica–Australia as a conjugate rift pair: Evidence and implications for an Eocambrian supercontinent: Geology, v. 19, p. 598–601.
Daniels, D. L., and Leo, G. W., 1985, Geologic interpretation of basement rocks of the Atlantic Coastal Plain: U.S. Geological Survey Open-File Report 85-655, 45 p.
Denison, R. E., Raveling, H. P., and Rouse, J. T., 1967, Age and descriptions of subsurface basement rocks, Pamlico and Albemarle Sound area, North Carolina: American Association of Petroleum Geologists Bulletin, v. 51, p. 268–272.
Denison, R. E., Lidiak, E. G., Bickford, M. E., and Kisvarsanyi, E. B., 1984, Geology and geochronology of Precambrian rocks in the central interior region of the United States: U.S. Geological Survey Professional Paper 1241-C, 33 p.
Dennis, A. J., 1991, Is the central Piedmont suture a low-angle normal fault?: Geology, v. 19, p. 1081–1084.
Dennis, A. J., and Shervais, J. W., 1991, Arc rifting of the Carolina terrane in northwestern South Carolina: Geology, v. 19, p. 226–229.
Dewey, J. F., 1975, Finite plate implications: Some implications for the evolution of rock masses at plate margins, *in* Tectonics and mountain ranges: American Journal of Science, v. 275-A, p. 260–284.
Dillon, W. P., and Popenoe, P., 1988, The Blake Plateau basin and Carolina trough, *in* Sheridan, R. E., and Grow, J. A., eds., The Atlantic continental margin: U.S.: Boulder, Colorado, Geological Society of America, The Geology of North America, v. I-2, p. 291–328.
Dillon, W. P., Klitgord, K. D., and Paull, C. K., 1983, Mesozoic development and structure of the continental margin of South Carolina, *in* Gohn, G. S., ed., Studies related to the Charleston, South Carolina earthquake of 1886—Tectonics and seismicity: U.S. Geological Survey Professional Paper 1313, p. N1–N16.
Diment, W. H., 1968, Gravity anomalies in northwestern New England, *in* Zen, E-an, White, W. S., Hadley, J. B., and Thompson, J. B., Jr., eds., Studies of Appalachian geology: Northern and maritime: New York, Interscience Publishers, p. 399–413.
DiNitto, R. G., Hepburn, J. C., Cardoza, K. D., and Hill, M., 1984, The Marlboro Formation in its type area and associated rocks just west of the Bloody Bluff fault zone, Marlborugh area, Massachusetts, *in* Hanson, L. S., ed., Geology of the coastal lowlands, Boston [Mass.] to Kennebunk, Maine (New England Intercollegiate Geological Conference, annual meeting, 76th): Salem, Massachusetts, Salem State College, p. 271–291.
Dixon, H. R., and Lundgren, L. W., Jr., 1968, Stratigraphy of eastern Connecticut, *in* Zen, E-an, White, W. S., Hadley, J. B., and Thompson, J. B., Jr., eds., Studies of Appalachian geology: Northern and maritime: New York, Interscience Publishers, p. 219–229.
Doll, C. G., Cady, W. M., Thompson, J. B., Jr., and Billings, M. P., compilers, eds., 1961, Centennial geologic map of Vermont: Montpelier, Vermont Geological Survey, scale 1:250,000.
Drake, A. A., Jr., 1985, Tectonic implications of the Indian Run Formation—A newly recognized sedimentary melange in the northern Virginia Piedmont: U.S. Geological Survey Professional Pape 1324, 12 p.
——, 1987, Pre-Taconian deformation in the Piedmont of the Potomac Valley—Penobscottian, Cadomian, or both ?, *in* Contributions to Virginia geology—V: Charlottesville, Virginia Division of Mineral Resources Publication 74, p. 1–18.
——, 1989a, Introduction: Stratigraphy, *in* Hatcher, R. D., Jr., Thomas, W. A., and Viele, G. W., eds., The Appalachian-Ouachita orogen in the United States: Boulder, Colorado, Geological Society of America, The Geology of North America, v. F-2, p. 101–117.
——, 1989b, Tectonics, *in* Hatcher, R. D., Jr., Thomas, W. A., and Viele, G. W., eds., The Appalachian-Ouachita orogen in the United States: Boulder, Colorado, Geological Society of America, The Geology of North America, v. F-2, p. 151–162.
——, 1990, The regional geologic setting of the Franklin-Sterling Hill district, Sussex County, New Jersey, *in* Character and origin of the Franklin-Sterling Hill ore bodies (Symposium proceedings volume): Bethlehem, Pennsylvania, Lehigh University-Franklin-Ogdensburg Mineralogical Society, p. 14–31.
Druhan, R. M., Butler, J. R., Horton, J. W., Jr., and Stoddard, E. F., 1988, Tectonic significance of the Nutbush Creek fault zone, eastern Piedmont of North Carolina and Virginia: Geological Society of America Abstracts with Programs, v. 20, no. 4, p. 261.
Dunning, G. R., and Cousineau, P. A., 1990, U/Pb ages of single zircon from Chain Lakes massif in Quebec: Geological Society of America Abstracts with Programs, v. 22, p. 13.
Edelman, S. H., Liu, A., and Hatcher, R. D., Jr., 1987, The Brevard fault zone in South Carolina and adjacent areas: An Alleghanian orogen-scale dextral shear zone reactivated as a thrust fault: Journal of Geology, v. 95, p. 793–806.
Elliott, D., 1973, Plate tectonics and the problem of too much granitic basement: Geology, v. 1, p. 111.
Eric, J. H., and Dennis, J. G., 1958, Geology of the Concord-Waterford area, Vermont: Vermont Geological Survey Bulletin 11, 66 p.
Espenshade, G. H., Rankin, D. W., Shaw, K. W., and Neuman, R. B., 1975, Geologic map of the east half of the Winston-Salem quadrangle, N.C., Va., Tenn.: U.S. Geological Survey Miscellaneous Geologic Investigations Map I-709B, scale 1:250,000.
Eusden, J. D., Jr., and Barreiro, B., 1988, The timing of peak high-grade metamorphism in central-eastern New England: Maritime Sediments and Atlantic Geology, v. 24, p. 241–255.
Farrar, S. S., 1984, The Goochland granulite terrane: Remobilized Grenville basement in the eastern Virginia Piedmont, *in* Bartholomew, M. J., ed., The Grenville event in the Appalachians and related topics: Geological Society of America Special Paper 194, p. 215–227.
——, 1985, Stratigraphy of the northeastern North Carolina Piedmont: Southeastern Geology, v. 25, p. 159–183.
Feiss, P. G., 1982, Geochemistry and tectonic setting of the volcanics of the

Carolina slate belt: Economic Geology, v. 77, p. 273–293.

Finney, S. C., 1986, Graptolite biofacies and correlation of eustatic subsidence, and tectonic events in the Middle to Upper Ordovician of North America: Palaios, v. 1, p. 435–461.

Flohr, M.J.K., and Pavlides, L., 1986, Thermobarometry of schists from the Quantico Formation and the Ta River Metamorphic Suite, Virginia: Geological Society of America Abstracts with Programs, v. 18, p. 221.

Foland, K. A., and Faul, H., 1977, Ages of the White Mountain intrusives—New Hampshire, Vermont, and Maine, U.S.A.: American Journal of Science, v. 277, p. 888–904.

Frederiksen, N. O., 1984, Stratigraphic, paleoclimatic, and paleobiogeographic significance of Tertiary sporomorphs from Massachusetts: U.S. Geological Survey Professional Paper 1308, 25 p.

Fullagar, P. D., and Butler, J. R., 1979, 325–265 m.y.-old granitic plutons in the Piedmont of the southeastern Appalachians: American Journal of Science, v. 279, p. 161–185.

——, 1980, Radiometric dating in the Sauratown Mountains area, North Carolina, *in* Price, V., Jr., Thayer, P. A., and Ranson, W. A., eds., Geological investigations of Piedmont and Triassic rocks, central North Carolina and Virginia (Carolina Geological Society field trip guidebook): Aiken, South Carolina, E.I. duPont de Neumours & Co., Savannah River Laboratory, p. CGS-80-B-11-1 to CGS-80-B-11-10.

Furcon, A. S., 1939, Geology and mineral resources of the Warrenton quadrangle, Virginia: Virginia Geological Survey Bulletin 54, 94 p.

Fyffe, L. R., and Fricker, A., 1987, Tectonostratigraphic terrane analysis of New Brunswick: Maritime Sediments and Atlantic Geology, v. 23, p. 113–122.

Fyffe, L. R., and Riva, J., 1990, Revised stratigraphy of the Cookson Group of southwestern New Brunswick and adjacent Maine: Atlantic Geology, v. 26, p. 271–275.

Fyffe, L. R., Stewart, D. B., and Ludman, A., 1988, Tectonic significance of black pelites and basalts in the St. Croix terrane, coastal Maine and New Brunswick: Maritime Sediments and Atlantic Geology, v. 24, p. 281–288.

Gastaldo, R. A., and Cook, R. B., 1982, A reinvestigation of the paleobotanical evidence for the age of the Erin Shale, *in* Bearce, D. N., Black, W. W., Kish, S. A., and Tull, J. F., Tectonic studies in the Talladega and Carolina slate belts, southern Appalachian orogen: Geological Society of America Special Paper 191, p. 69–77.

Gates, A. E., 1987, Transpressional dome formation in the southwest Virginia Piedmont: American Journal of Science, v. 287, p. 927–949.

Gates, A. E., and Glover, L., III, 1989, Alleghanian tectono-thermal evolution of the dextral transcurrent Hylas zone, Virginia Piedmont, U.S.A.: Journal of Structural Geology, v. 11, p. 407–419.

Gates, A. E., and Speer, J. A., 1991, Allochemical retrograde metamorphism in shear zones: An example in metapelites, Virginia, USA: Journal of Metamorphic Geology, v. 9, p. 581–604.

Gates, A. E., Simpson, C., and Glover, L., III, 1986, Appalachian Carboniferous dextral/strike-slip faults: An example from Brookneal, Virginia: Tectonics, v. 5, p. 119–133.

Gates, A. E., Speer, J. A., and Pratt, T. L., 1988, The Alleghanian southern Appalachian Piedmont: A transpressional model: Tectonics, v. 7, p. 1307–1324.

Gates, O., 1989, Silurian roundstone conglomerates of coastal Maine and adjacent New Brunswick, *in* Tucker, R. D., and Marvinney, R. G., eds., Structure and stratigraphy: Augusta, Maine Geological Survey, Studies in Maine Geology, v. 2, p. 127–144.

Gates, O., and Moench, R. H., 1981, Bimodal Silurian and Lower Devonian volcanic rock assemblages in the Machias-Eastport area, Maine: U.S. Geological Survey Professional Paper 1184, 32 p.

Gaudette, H. E., Kovach, A., and Hussey, A. M., II, 1982, Ages of some intrusive rocks of southwestern Maine, U.S.A.: Canadian Journal of Earth Sciences, v. 19, p. 1350–1357.

Gee, D. G., 1975, A tectonic model for the central part of the Scandinavian Caledonides, *in* Tectonics and mountain ranges: American Journal of Science, v. 275-A, p. 468–515.

Geiser, P. A., 1989, Description of structures: Central Appalachians, *in* Hatcher, R. D., Jr., Thomas, W. A., and Viele, G. W., eds., The Appalachian-Ouachita orogen in the United States: Boulder, Colorado, Geological Society of America, The Geology of North America, v. F-2, p. 362–375.

Getty, S. R., and Gromet, L. P., 1992a, Geochronological constraints on ductile deformation, crustal extension, and doming about a basement-cover boundary, New England Appalachians: American Journal of Science, v. 292, p. 359–397.

——, 1992b, Evidence for extension of the Willimantic dome, Connecticut: Implications for the late Paleozoic tectonic evolution of the New England Appalachians: American Journal of Science, v. 292, p. 398–420.

Gibson, G. G., Teeter, S. A., and Fedonkin, M. A., 1984, Ediacaran fossils from the Carolina slate belt, Stanley County, North Carolina: Geology, v. 12, p. 387–390.

Glover, L., III, 1989, Tectonics of the Virginia Blue Ridge and Piedmont: Field trip guidebook T 363, 28th International Geological Congress: Washington, D.C., American Geophysical Union, 59 p.

Glover, L., III, and Sinha, A. K., 1973, The Virgilina deformation, a late Precambrian to Early Cambrian(?) orogenic event in the central Piedmont of Virginia and North Carolina: American Journal of Science, The Byron N. Cooper Volume, v. 273-A, p. 234–251.

Glover, L., III, Mose, D. G., Costain, J. K., Poland, F. B., and Reilly, J. M., 1982, Grenville basement in the eastern Piedmont of Virginia: A progress report: Geological Society of America Abstracts with Programs, v. 14, p. 20.

Glover, L., III, Speer, A., Russell, G. S., and Farrar, S. S., 1983, Ages of regional metamorphism and ductile deformation in the central and southern Appalachians: Lithos, v. 16, p. 223–245.

Glover, L., III, and 6 others, 1987, Tectonics and crustal structure in the central Appalachians of Virginia: Reinterpreted from USGS I-64 reflection seismic profile: Geological Society of America Abstracts with Programs, v. 19, p. 86.

Gohn, G. S., 1988, Late Mesozoic and early Cenozoic geology of the Atlantic Coastal Plain: North Carolina to Florida, *in* Sheridan, R. E., and Grow, J. A., eds., The Atlantic continental margin, U.S.: Boulder, Colorado, Geological Society of America, The Geology of North America, v. I-2, p. 107–130.

Gohn, G. S., Houser, B. B., and Schneider, R. R., 1983, Geology of the lower Mesozoic(?) sedimentary rocks in Clubhouse Crossroads test hole #3, near Charleston, South Carolina, *in* Gohn, G. S., ed., Studies related to the Charleston, South Carolina earthquake of 1886—Tectonics and seismicity: U.S. Geological Survey Professional Paper 1313, p. D1–D17.

Goldsmith, R., 1991, Structural and metamorphic history of eastern Massachusetts, *in* Hatch, N. L., Jr., ed., The bedrock geology of Massachusetts: U.S. Geological Survey Professional Paper 1366E-J, p. H1–H63.

Goldsmith, R., and Secor, D. T., Jr., 1993, Proterozoic rocks of accreted terranes in the eastern United States, *in* Reed, J. C., Jr., Bickford, M. E., Jr., Houston, R. S., Link, P. K., Rankin, D. W., Sims, P. K., and Van Schmus, W. R., eds., Precambrian: Conterminous United States: Boulder, Colorado, Geological Society of America, The Geology of North America, v. C-2, p. 422–439.

Goldsmith, R., Milton, D. J., and Horton, J. W., Jr., 1988, Geologic map of the Charlotte 1° × 2° quadrangle, North Carolina and South Carolina: U.S. Geological Survey Miscellaneous Investigations Map I-1251-E, scale 1:250,000.

Goldstein, A. G., 1989, Tectonic significance of multiple motions on terrane-bounding faults in the northern Appalachians: Geological Society of America Bulletin, v. 101, p. 927–938.

Gottfried, D., Annell, C. S., and Byerly, G. R., 1983, Geochemistry and tectonic significance of subsurface basalts near Charleston, South Carolina: Clubhouse Crossroads test holes #2 and #3, *in* Gohn, G. S., ed., Studies related to the Charleston, South Carolina earthquake of 1886—Tectonics and seismicity: U.S. Geological Survey Professional Paper 1313, p. A1–A19.

Gravity Anomaly Map Committee, 1987, Gravity anomaly map of North America: Boulder, Colorado, Geological Society of America Continent-Scale Map 002, scale 1:5,000,000.

Gray, G. G., 1986, Native terranes of the central Klamath Mountains, California:

Tectonics, v. 5, p. 1043–1054.
Green, A. G., and 9 others, 1988, Crustal structure of the Grenville front and adjacent terranes: Geology, v. 16, p. 788–792.
Gromet, L. P., 1989, Avalonian terranes and late Paleozoic tectonics in southeastern New England; Constraints and problems, *in* Dallmeyer, R. D., ed., Terranes in the Circum-Atlantic Paleozoic orogens: Geological Society of America Special Paper 230, p. 193–211.
Gromet, L. P., and Robinson, P., 1990, Isotopic evidence for late Paleozoic gneissic deformation and recrystallization in the core of the Pelham dome, Massachusetts: Geological Society of America Abstracts with Programs, v. 22, no. 2, p. A368.
Grow, J. A., and Sheridan, R. E., 1988, U.S. Atlantic continental margin: A typical Atlantic-type or passive continental margin, *in* Sheridan, R. E., and Grow, J. A., eds., The Atlantic continental margin: U.S.: Boulder, Colorado, Geological Society of America, The Geology of North America, v. I-2, p. 1–7.
Grow, J. A., Klitgord, K. D., and Schlee, J. S., 1988a, Structure and evolution of Baltimore Canyon trough, *in* Sheridan, R. E., and Grow, J. A., eds., The Atlantic continental margin: U.S.: Boulder, Colorado, Geological Society of America, The Geology of North America, v. I-2, p. 269–290.
Grow, J. A., Klitgord, K. D., Schlee, J. S., and Dillon, W. P., 1988b, Representative seismic profiles of the U.S. Atlantic continental margin, *in* Sheridan, R. E., and Grow, J. A., eds., The Atlantic continental margin: U.S.: Boulder, Colorado, Geological Society of America, The Geology of North America, v. I-2, plate 4.
Guidotti, C. U., 1989, Metamorphism in Maine: An overview, *in* Tucker, R. D., and Marvinney, R. G., eds., Igneous and metamorphic geology: Augusta, Maine Geological Survey, Studies in Maine Geology, v. 3, p. 1–17.
Hack, J. T., 1989, Geomorphology of the Appalachian Highlands, *in* Hatcher, R. D., Jr., Thomas, W. A., and Viele, G. W., eds., The Appalachian-Ouachita orogen in the United States: Boulder, Colorado, Geological Society of America, The Geology of North America, v. F-2, p. 459–470.
Hadley, J. B., and Goldsmith, R., 1963, Geology of the eastern Great Smoky Mountains, North Carolina and Tennessee: U.S. Geological Survey Professional Paper 349-B, 118 p.
Hall, L. M., and Robinson, P., 1982, Stratigraphic-tectonic subdivisions of southern New England, *in* St.-Julien, P., and Béland, J., eds., Major structural zones and faults of the northern Appalachians: Geological Association of Canada Special Paper 24, p. 15–41.
Hames, W. E., Tracy, R. J., Ratcliffe, N. M., and Sutter, J. F., 1991, Petrologic, structural, and geochronologic characteristics of the Acadian metamorphic overprint on the Taconian zone in part of southwestern New England: American Journal of Science, v. 291, p. 887–913.
Hanley, T. B., 1986, Petrography and structural geology of Uchee belt rocks in Columbus, Georgia, and Phenix City, Alabama, *in* Neathery, T. L., ed., Southeastern Section of the Geological Society of America: Boulder, Colorado, Geological Society of America, Centennial Field Guide, v. 6, p. 297–300.
Hansen, H. J., 1988, Buried rift basin underlying coastal plain sediments, central Delmarva Peninsula, Maryland: Geology, v. 16, p. 779–782.
Hansen, H. J., and Edwards, J., Jr., 1986, Lithology and distribution of pre-Cretaceous basement rocks beneath the Maryland Coastal Plain: Maryland Geological Survey Report of Investigations 44, 27 p.
Harland, W. B., and Gayer, R. A., 1972, The Arctic Caledonides and earlier oceans: Geologic Magazine, v. 109, p. 289–314.
Harper, S. B., and Fullagar, P. D., 1981, Rb-Sr ages of granitic gneisses of the Inner Piedmont belt of northwestern North Carolina and southwestern South Carolina: Geological Society of America Bulletin, v. 92, Part 1, p. 864–872.
Harris, A. L., and Fettes, D. J., eds., 1988, The Caledonide-Appalachian orogen (Geological Society Special Publication 38): London, Blackwell Scientific Publications, 643 p.
Harris, L. D., Harris, A. G., DeWitt, W., Jr., and Bayer, K. C., 1981, Evaluation of southern eastern overthrust belt beneath Blue Ridge–Piedmont thrust: American Association of Petroleum Geologists Bulletin, v. 65, p. 2497–2505.
Harrison, T. M., Aleinikoff, J. N., and Compston, W., 1987, Observations and controls on the occurrence of inherited zircon in Concord-type granitoids, New Hampshire: Geochimica et Cosmochimica Acta, v. 51, p. 2549–2558.
Hatch, N. L., Jr., 1988, New evidence for faulting along the "Monroe line," eastern Vermont and westernmost New Hampshire: American Journal of Science, v. 288, p. 1–18.
Hatch, N. L., Jr., and Stanley, R. S., 1973, Some suggested stratigraphic relations in part of southwestern New England: U.S. Geological Survey Bulletin 1380, 83 p.
Hatch, N. L., Jr., Moench, R. H., and Lyons, J. B., 1983, Silurian–Lower Devonian stratigraphy of eastern and south-central New Hampshire: Extensions from western Maine: American Journal of Science, v. 283, p. 739–761.
Hatcher, R. D., Jr., 1972, Developmental model for the southern Appalachians: Geological Society of America Bulletin, v. 83, p. 2735–2760.
——, 1978a, Tectonics of the western Piedmont and Blue Ridge, southern Appalachians: Review and speculation: American Journal of Science, v. 278, p. 276–304.
——, 1978b, The Alto allochthon: A major tectonic feature of the Piedmont of northeast Georgia, *in* Short contributions to the geology of Georgia: Georgia Geologic Survey Bulletin 93, p. 83–86.
——, 1987, Tectonics of the southern and central Appalachian internides: Annual Review of Earth and Planetary Sciences, v. 15, p. 337–362.
——, 1989a, Appalachians introduction, *in* Hatcher, R. D., Jr., Thomas, W. A., and Viele, G. W., eds., The Appalachian-Ouachita orogen in the United States: Boulder, Colorado, Geological Society of America, The Geology of North America, v. F-2, p. 1–5.
——, 1989b, Tectonic synthesis of the U.S. Appalachians, *in* Hatcher, R. D., Jr., Thomas, W. A., and Viele, G. W., eds., The Appalachian-Ouachita orogen in the United States: Boulder, Colorado, Geological Society of America, The Geology of North America, v. F-2, p. 511–535.
Hatcher, R. D., Jr., and Odom, A. L., 1980, Timing of thrusting in the southern Appalachians, U.S.A.: Model for orogeny?: Geological Society of London Journal, v. 137, p. 321–327.
Hatcher, R. D., Jr., Howell, D. E., and Talwani, P., 1977, Eastern Piedmont fault system: Speculations on its extent: Geology, v. 5, p. 636–640.
Hatcher, R. D., Jr., Hooper, R. T., Petty, S. M., and Willis, J. D., 1984, Structure and chemical petrology of three southern Appalachian mafic-ultramafic complexes and their bearing upon the tectonics of emplacement and origin of Appalachian ultramafic bodies: American Journal of Science, v. 284, p. 484–506.
Hatcher, R. D., Jr., Thomas, W. A., and Viele, G. W., eds., 1989, The Appalachian-Ouachita orogen in the United States: Boulder, Colorado, Geological Society of America, The Geology of North America, v. F-2, 767 p.
Hatcher, R. D., Jr., Osberg, P. H., Drake, A. A., Jr., Robinson, P., and Thomas, W. A., 1990, Tectonic map of the U.S. Appalachians, *in* Hatcher, R. D., Jr., Thomas, W. A., and Viele, G. W., eds., The Appalachian-Ouachita orogen in the United States: Boulder, Colorado, Geological Society of America, The Geology of North America, v. F-2, plate 1, scale 1:2,105,000.
Hatcher, R. D., Jr., and 11 others, 1994, E-5, Cumberland Plateau (North American craton) to Blake Plateau basin: Boulder, Colorado, Geological Society of America Centennial Continent/Ocean Transect 18, 2 sheets, scale 1:500,000 (in press).
Haworth, R. T., Daniels, D. L., Williams, H., and Zietz, I., 1980, Bouguer gravity map of the Appalachian orogen: Memorial University of Newfoundland Map 3, 2 sheets, scale 1:1,000,000.
Haworth, R. T., Williams, H., and Keen, C. E., 1985, D-1, Northern Appalachian Mountains across Newfoundland to south Labrador Sea: Boulder, Colorado, Geological Society of America Centennial Continent/Ocean Transect 1, 9 p., 2 sheets, scale 1:500,000.
Hayward, J. A., and Gaudette, H. E., 1984, Carboniferous age for the Sebago and Effingham plutons, Maine and New Hampshire: Geological Society of America Abstracts with Programs, v. 16, p. 22.

Heck, F. R., 1989, Mesozoic extension in the southern Appalachians: Geology, v. 17, p. 711–714.
Hepburn, J. C., Hill, M., and Hon, R., 1987, The Avalonian and Nashoba terranes, eastern Massachusetts, U.S.A.: An overview, *in* The Avalon terrane of the Northern Appalachian orogen, Part 1: Maritime Sediments and Atlantic Geology, Special Issue, v. 22, p. 1–12.
Hermes, O. D., and Murray, D. P., 1988, Middle Devonian to Permian plutonism and volcanism in the N. American Appalachians, *in* Harris, A. L., and Fettes, D. J., eds., The Caledonian-Appalachian orogen (Geological Society of London Special Publication 38): London, Blackwell Scientific Publications, p. 559–571.
Hermes, O. D., Ballard, R. D., and Banks, P. O., 1978, Upper Ordovician peralkalic granites from the Gulf of Maine: Geological Society of America Bulletin, v. 89, p. 1761–1774.
Hibbard, J., 1991, The Carolina slate belt–Milton belt contact in north-central North Carolina: Geological Society of America Abstracts with Programs, v. 23, no. 1, p. 44.
Higgins, M. W., 1972, Age, origin, regional relations, and nomenclature of the Glenarm Series, central Appalachian Piedmont: a reinterpretation: Geological Society of America Bulletin, v. 83, p. 989–1062.
Higgins, M. W., and Zietz, I., 1983, Geologic interpretation of geophysical maps of the pre-Cretaceous "basement" beneath the Coastal Plain of the southeastern United States, *in* Hatcher, R. D., Jr., Williams, H., and Zietz, I., eds., Contributions to the tectonics and geophysics of mountain chains: Geological Society of America Memoir 158, p. 125–130.
Higgins, M. W., Sinha, A. K., Zartman, R. E., and Kirk, W. S., 1977, U-Pb zircon dates from the central Appalachian Piedmont—A possible case of inherited radiogenic lead: Geological Society of America Bulletin, v. 88, p. 125–132.
Higgins, M. W., Atkins, R. L., Crawford, T. J., Crawford, R. F., III, and Cook, R. B., 1984, A brief excursion through two thrust stacks that comprise most of the crystalline terrane of Georgia and Alabama (Georgia Geological Society, 19th annual field trip guidebook): Atlanta, Georgia State University, 67 p.
Higgins, M. W., Atkins, R. L., Crawford, T. J., Crawford, R. F., III, Brocks, R., and Cook, R. B., 1988, The structure, stratigraphy, tectonostratigraphy, and evolution of the southernmost part of the Appalachian orogen: U.S. Geological Survey Professional Paper 1475, 173 p.
Higgins, M. W., Crawford, R. F., III, Atkins, R. L., and Crawford, T. J., 1989, The Macon complex: An ancient accretionary complex in the southern Appalachians, *in* Horton, J. W., Jr., and Rast, N., eds., Melanges and olistostromes of the U.S. Appalachians: Geological Society of America Special Paper 228, p. 229–246.
Hildenbrand, T. G., Simpson, R. W., Godson, R. H., and Kane, M. F., 1982, Digital colored residual and regional Bouguer gravity maps of the conterminous United States with cut-off wavelengths of 250 km and 1000 km: U.S. Geological Survey Geophysical Investigations Map GP-953-A, scale 1:7,500,000.
Hill, M. D., Hepburn, J. C., Collins, R. D., and Horn, R., 1984, Igneous rocks of the Nashoba block, eastern Massachusetts, *in* Hanson, L. S., ed., Geology of the coastal lowlands, Boston [Mass.] to Kennebunk, Maine (New England Intercollegiate Geological Conference, annual meeting, 76th): Salem, Massachusetts, Salem State College, p. 61–80.
Hoffman, P. F., 1988, United Plates of America, the birth of a craton: Early Proterozoic assembly and growth of Laurentia: Annual Review of Earth and Planetary Sciences, v. 16, p. 543–603.
Hogan, J. P., and Sinha, A. K., 1989, Compositional variation of plutonism in the coastal Maine magmatic province: Mode of origin and tectonic setting, *in* Tucker, R. D., and Marvinney, R. G., eds., Igneous and metamorphic geology: Augusta, Maine Geological Survey, Studies in Maine Geology, v. 4, p. 1–33.
Holbrook, W. S., and 9 others, 1992, Deep velocity structure of rifted continental crust, U.S. Mid-Atlantic margin, from wide-angle reflection/refraction data: Geophysical Research Letters, v. 19, p. 1699–1702.
Hon, R., and Roy, D. C., 1981, Magmatogenic and stratigraphic constraints on Acadian tectonism in Maine: Geological Society of America Abstracts with Programs, v. 13, p. 138.
Hon, R., and 6 others, 1986, Mid-Paleozoic calc-alkaline igneous rocks of the Nashoba block and Merrimack trough, *in* Newberg, D. W., ed., Guidebook for field trips in southwestern Maine (New England Intercollegiate Geological Conference, annual meeting, 78th): Lewiston, Maine, Bates College, p. 37–52.
Hooper, R. J., Hatcher, R. D., Jr., and Stowell, H., 1991, The Avalon-Piedmont terrane boundary, central Georgia: Geological Society of America Abstracts with Programs, v. 23, no. 1, p. 46.
Hopeck, J. T., 1991, Faulting and related fabrics in the Miramichi and Aroostook-Matapedia tracts, *in* Ludman, A., ed., Geology of the coastal lithotectonic block and neighboring terranes, eastern Maine and southern New Brunswick (New England Intercollegiate Geological Conference, annual meeting 83rd): Flushing, New York, Queens College, City University of New York, p. 294–308.
Hopson, J. L., and Hatcher, R. D., Jr., 1988, Structural and stratigraphic setting of the Alto allochthon, northeast Georgia: Geological Society of America Bulletin, v. 100, p. 339–350.
Horton, J. W., Jr., 1982, Geologic map and mineral resources of the Rosman quadrangle, North Carolina: North Carolina Geological Survey Map GM-185-NE, scale 1:24,000.
——, 1984, Stratigraphic nomenclature in the Kings Mountain belt, North Carolina and South Carolina, *in* Stratigraphic notes, 1983: U.S. Geological Survey Bulletin 1537, p. A59–A67.
Horton, J. W., Jr., and Butler, J. R., 1986, The Brevard fault zone of Rosman, Transylvania County, North Carolina, *in* Neathery, T. L., ed., Southeastern Section of the Geological Society of America: Boulder, Colorado, Geological Society of America, Centennial Field Guide, v. 6, p. 217–222.
Horton, J. W., Jr., and McConnell, K. I., 1991, The western Piedmont, *in* Horton, J. W., Jr., and Zullo, V. A., eds., The geology of the Carolinas: Knoxville, University of Tennessee Press, p. 36–58.
Horton, J. W., Jr., Sutter, J. F., Stern, T. W., and Milton, D. J., 1987, Alleghanian deformation, metamorphism, and granite emplacement in the central Piedmont of the southern Appalachians: American Journal of Science, v. 287, p. 635–660.
Horton, J. W., Jr., Drake, A. A., Jr., and Rankin, D. W., 1989, Tectonostratigraphic terranes and their boundaries in the central and southern Appalachians, *in* Dallmeyer, R. D., ed., Terranes in the circum-Atlantic Paleozoic orogens: Geological Society of America Special Paper 230, p. 213–245.
Horton, J. W., Jr., Drake, A. A., Jr., Rankin, D. W., and Dallmeyer, R. D., 1991, Preliminary tectonostratigraphic terrane map of the central and southern Appalachians: U.S. Geological Survey Miscellaneous Investigations Series Map I-2163, scale 1:2,000,000.
Hubbard, M., and 6 others, 1991, Major dextral strike-slip deformation in the northern Appalachians: The Norumbega fault zone, Maine: Geological Society of America Abstracts with Programs, v. 23, no. 5, p. 311.
Hughes, S., and Luetgert, J. H., 1991, Crustal structure of the western New England Appalachians and Adirondack Mountains: Journal of Geophysical Research, v. 96, p. 16,471–16,494.
Hussey, A. M., II, 1988, Lithotectonic stratigraphy, deformation, plutonism, and metamorphism, greater Casco Bay region, southwestern Maine, *in* Tucker, R. D., and Marvinney, R. G., eds., Structure and stratigraphy: Augusta, Maine Geological Survey, Studies in Maine Geology, v. 1, p. 17–33.
Hussey, A. M., II, and Newberg, D. W., 1978, Major faulting in the Merrimack synclinorium between Hollis, New Hampshire, and Biddeford, Maine: Geological Society of America Abstracts with Programs, v. 10, p. 48.
Hutchinson, D. R., and Klitgord, K. D., 1988a, Evolution of rift basins on the continental margin off southern New England, *in* Manspeizer, W., ed., Triassic-Jurassic rifting: Amsterdam, Elsevier, p. 81–98.
——, 1988b, Deep structure of rift basins from the continental margin around New England, *in* Froelich, A. J., and Robinson, G. R., Jr., eds., Studies of the early Mesozoic basins of the eastern United States: U.S. Geological

Survey Bulletin 1776, p. 211–219.

Hutchinson, D. R., Grow, J. A., and Klitgord, K. D., 1983a, Crustal structure beneath the southern Appalachians: Nonuniqueness of gravity modeling: Geology, v. 11, p. 611–615.

Hutchinson, D. R., Grow, J. A., Klitgord, K. D., and Swift, B. A., 1983b, Deep structure and evolution of the Carolina trough, *in* Watkins, J. S., and Drake, C. L., eds., Studies in continental marine geology: American Association of Petroleum Geologists Memoir 34, p. 129–152.

Hutchinson, D. R., Klitgord, K. D., and Detrick, R. S., 1985, Block Island fault: A Paleozoic crustal boundary on the Long Island platform: Geology, v. 13, p. 875–879.

Hutchinson, D. R., Klitgord, K. D., Lee, M. W., and Trehu, A. M., 1988, U.S. Geological Survey deep seismic reflection profile across the Gulf of Maine: Geological Society of America Bulletin, v. 100, p. 172–184.

Hutson, F. E., and Tollo, R. P., 1992, Mechum River Formation: Evidence for the evolution of a Late Proterozoic rift basin, Blue Ridge anticlinorium, Virginia: Geological Society of America Abstracts with Programs, v. 24, no. 2, p. 22.

Isachsen, Y. W., 1975, Possible evidence for contemporary doming of the Adirondack Mountains, New York, and suggested implications for regional tectonics and seismicity: Tectonophysics, v. 29, p. 169–181.

——, 1976, Contemporary doming of the Adirondack Mountains, New York [abs.]: Eos (Transactions, American Geophysical Union), v. 57, p. 325.

Iverson, W. P., and Smithson, S. B., 1982, Master decollement root zone beneath the southern Appalachians and crustal balance: Geology, v. 10, p. 241–245.

Jacobi, R. D., 1981, Peripheral bulge—A casual mechanism for the Lower/Middle Ordovician unconformity along the western margin of the Northern Appalachians: Earth and Planetary Science Letters, v. 56, p. 245–251.

James, D. E., Smith, T. J., and Steinhart, J. S., 1968, Crustal structure of the Middle Atlantic States: Journal of Geophysical Research, v. 73, p. 1983–2007.

Jonas, A. I., 1928, Geologic map of Virginia: Charlottesville, Virginia Geological Survey, scale 1:500,000.

Karner, G. D., and Watts, A. B., 1983, Gravity anomalies and flexure of the lithosphere at mountain ranges: Journal of Geophysical Research, v. 88, p. 10,449–10,477.

Kaye, C. A., 1982, Bedrock and Quaternary geology of the Boston area, Massachusetts: Geological Society of America Reviews in Engineering Geology, v. 5, p. 25–40.

——, 1983, Discovery of a Late Triassic basin north of Boston and some implications as to post-Paleozoic tectonics in northeastern Massachusetts: American Journal of Science, v. 283, p. 1060–1079.

Kaye, C. A., and Zartman, R. E., 1980, A late Proterozoic Z to Cambrian age for the stratified rocks of the Boston Basin, Massachusetts, U.S.A., *in* Wones, D. R., ed., Proceedings of the Caledonides in the USA, IGCP project 27, Caledonide orogen: Blacksburg, Virginia Polytechnic Institute and State University, Department of Geological Sciences, Memoir 2, p. 257–262.

Keen, C. E., and Haworth, R. T., 1985a, D-2, Transform margin south of Grand Banks: Offshore eastern Canada: Boulder, Colorado, Geological Society of America, Centennial Continent/Ocean Transect no. 2, 6 p., 1 sheet, scale 1:000,000.

——, 1985b, D-3, Rifted continental margin off Nova Scotia: Offshore eastern Canada: Boulder, Colorado, Geological Society of America, Centennial Continent/Ocean Transect no. 3, 7 p., 1 sheet, scale 1:500,000.

Keen, C. E., Stockmal, G. S., Welsink, H., Quinlan, G., and Mudford, B., 1987, Deep crustal structure and evolution of the rifted margin northeast of Newfoundland: Results from LITHOPROBE East: Canadian Journal of Earth Sciences, v. 24, p. 1537–1549.

Keen, M. J., and Williams, G. L., eds., 1990, Geology of the continental margin of eastern Canada: Boulder, Colorado, Geological Society of America, The Geology of North America, v. I-1, 855 p.

Keppie, J. D., 1989, Northern Appalachian terranes and their accretionary history, *in* Dallmeyer, R. D., ed., Terranes in the Circum-Atlantic Paleozoic orogens: Geological Society of America Special Paper 230, p. 159–192.

King, P. B., 1955, A geologic section across the southern Appalachians—An outline of the geology in the segment in Tennessee, North Carolina, and South Carolina, *in* Russell, R. J., ed., Guides to southeastern geology: New York, Geological Society of America, p. 332–373.

——, 1977, The evolution of North America: Princeton, New Jersey, Princeton University Press, 197 p.

Kish, S. A., 1983, A geochronological study of deformation and metamorphism in the Blue Ridge and Piedmont of the Carolinas [Ph.D. thesis]: Chapel Hill, University of North Carolina, 220 p.

——, 1990, Timing of middle Paleozoic (Acadian) metamorphism in the southern Appalachians: K-Ar studies in the Talladega belt, Alabama: Geology, v. 18, p. 650–653.

Kish, S. A., Butler, J. R., and Fullagar, P. D., 1979, The timing of metamorphism and deformation in the central and eastern Piedmont of North Carolina: Geological Society of America Abstracts with Programs, v. 11, p. 184–185.

Kite, L. E., and Stoddard, E. F., 1984, The Halifax County complex: Oceanic lithosphere in the eastern North Carolina Piedmont: Geological Society of America Bulletin, v. 95, p. 422–432.

Kline, S. W., 1991, Provenance of arkosic metasediments in the Virginia Blue Ridge: Constraints on Appalachian suspect terrane models: American Journal of Science, v. 291, p. 189–198.

Klitgord, K. D., and Grow, J. A., 1980, Jurassic seismic stratigraphy and basement structure of western Atlantic magnetic quiet zone: American Association of Petroleum Geologists Bulletin, v. 64, p. 1658–1680.

Klitgord, K. D., and Schouten, H., 1986, Plate kinematics of the central Atlantic, *in* Vogt, P. R., and Tucholke, B. E., eds., The western North Atlantic region: Boulder, Colorado, Geological Society of America, The Geology of North America, v. M, p. 351–378.

Klitgord, K. D., Hutchinson, D. R., and Schouten, H., 1988, U.S. Atlantic continental margin; structural and tectonic framework, *in* Sheridan, R. E., and Grow, J. A., eds., The Atlantic continental margin: U.S.: Boulder, Colorado, Geological Society of America, The Geology of North America, v. I-2, p. 19–55.

Kumarapeli, P. S., Goodacre, A. K., and Thomas, M. D., 1981, Gravity and magnetic anomalies of the Sutton Mountain region, Quebec and Vermont: Expressions of rift volcanics related to the opening of Iapetus: Canadian Journal of Earth Sciences, v. 18, p. 680–692.

Kumarapeli, P. S., Dunning, G. R., Pintson, H., and Shaver, J., 1989, Geochemistry and U-Pb zircon age of comenditic metafelsites of the Tibbit Hill Formation, Quebec Appalachians: Canadian Journal of Earth Sciences, v. 26, p. 1374–1383.

Kunk, M. J., and Sutter, J. F., 1984, $^{40}Ar/^{39}Ar$ age spectrum dating of biotite from Middle Ordovician bentonites eastern North America, *in* Bruton, D. L., ed., Aspects of the Ordovician System: Paleontological Contributions from the University of Oslo, no. 295, p. 11–22.

Laird, J., 1989, Metamorphism, *in* Hatcher, R. D., Jr., Thomas, W. A., and Viele, G. W., eds., The Appalachian-Ouachita orogen in the United States: Boulder, Colorado, Geological Society of America, The Geology of North America, v. F-2, p. 134–151.

Laird, J., and Albee, A. L., 1981, Pressure, temperature, and time indicators in mafic schist: Their application to reconstructing the polymetamorphic history of Vermont: American Journal of Science, v. 281, p. 127–175.

Laird, J., Lanphere, M. A., and Albee, A. L., 1984, Distribution of Ordovician and Devonian metamorphism in mafic and pelitic schists from northern Vermont: American Journal of Science, v. 284, p. 376–413.

Lanphere, M. A., 1983, $^{40}Ar/^{39}Ar$ ages of basalt from Clubhouse Crossroads Test Hole No. 2, near Charleston, South Carolina, *in* Gohn, G. S., ed., Studies related to the Charleston, South Carolina earthquake of 1886—Tectonics and seismicity: U.S. Geological Survey Professional Paper 1313, p. B1–B8.

Laurent, R., and Hébert, R., 1989, The volcanic and intrusive rocks of the Quebec Appalachian ophiolites (Canada) and their island-arc setting: Chemical Geology, v. 77, p. 287–302.

Lefort, J. P., 1984, Mise en évidence d'une virgation carbonifère induite par la dorsale Reguibat (Mauritanie) dans les Appalaches du Sud (U.S.A.), Arguments géophysiques: Bulletin de la Societé Géologique de France, v. 26,

p. 1293–1303.

Lefort, J. P., and Max, M. D., 1991, Is there Archean crust beneath Chesapeake Bay?: Tectonics, v. 10, p. 213–226.

Lemmon, R. E., 1981, An igneous origin for the Henderson augen gneiss, western North Carolina: Evidence from zircon morphology: Southeastern Geology, v. 22, p. 79–90.

Lenk, C., Strother, P. K., Kaye, C. A., and Barghoorn, E. S., 1982, Precambrian age of the Boston Basin: New evidence from microfossils: Science, v. 216, p. 619–620.

Leo, G. W., Barreiro, B., and Mortensen, J. K., 1991, A latest Proterozoic U-Pb age for plutonic basement rocks in southeastern Cape Cod, Massachusetts: Geological Society of America Abstracts with Programs, v. 23, no. 1, p. 58.

Lidiak, E. G., and Hinze, W. J., 1993, Grenville province in the subsurface of eastern United States, *in* Reed, J. C., Jr., Bickford, M. E., Houton, R. S., Link, P. K., Rankin, D. W., Sims, P. K., and Van Schmus, W. R., eds., Precambrian: Conterminous U.S.: Boulder, Colorado, Geological Society of America, The Geology of North America, v. C-2, p. 353–365.

Lipin, B. R., 1984, Chromite from the Blue Ridge province of North Carolina: American Journal of Science, v. 284, p. 507–529.

Long, L. T., 1979, The Carolina slate belt—Evidence of a continental rift zone: Geology, v. 7, p. 180–184.

Luetgert, J. H., and Bottcher, C., 1987, Seismic refraction profiles within the Merrimack synclinorium, central Maine [abs.]: Eos (Transactions, American Geophysical Union, v. 68, p. 347.

Lyons, J. B., 1955, Geology of the Hanover quadrangle, New Hampshire–Vermont: Geological Society of America Bulletin, v. 66, p. 105–146.

Lyons, J. B., Aleinikoff, J. N., and Zartman, R. E., 1986, Uranium-thorium-lead ages of the Highlandcroft Plutonic Suite, northern New England: American Journal of Science, v. 286, p. 489–509.

Lyons, J. B., Bothner, W. A., Moench, R. H., Thompson, J. B., Jr., eds., 1994, Geologic map of New Hampshire: U.S. Geological Survey Map, scale 1:250,000 (in press).

Lyons, P. C., 1979, Biostratigraphy of the Pennsylvanian of Massachusetts and Rhode Island: U.S. Geological Survey Professional Paper 1110, p. A20–A24.

Magnetic Anomaly Map Committee, 1987, Magnetic anomaly map of North America: Boulder, Colorado, Geological Society of America, Continent-Scale Map 003, 4 sheets, scale 1:5,000,000.

Maher, H. D., Jr., 1987, Kinematic history of mylonitic rocks from the Augusta fault zone, South Carolina and Georgia: American Journal of Science, v. 287, p. 795–816.

Maher, H. D., Jr., Dallmeyer, R. D., Sacks, P., and Secor, D. T., Jr., 1990, Age of movement on the Augusta fault zone in the eastern Piedmont of S.C. and Ga., Late Alleghanian extension: Geological Society of America Abstracts with Programs, v. 22, no. 4, p. 23.

Maher, H. D., Jr., Sacks, P. E., and Secor, D. T., Jr., 1991, The eastern Piedmont of South Carolina, *in* Horton, J. W., Jr., and Zullo, V. A., eds., The geology of the Carolinas: Knoxville, University of Tennessee Press, p. 93–108.

Manspeizer, W., 1980, Rift tectonics inferred from volcanic and clastic structures, *in* Manspeizer, W., ed., Field studies, New Jersey geology and guide to field trips (New York Geological Association, annual meeting, 52nd): Newark, New Jersey, Rutgers University, p. 314–350.

Manspeizer, W., and Cousminer, H. L., 1988, Late Triassic–Early Jurassic synrift basins of the U.S. Atlantic margin, *in* Sheridan, R. E., and Grow, J. A., eds., The Atlantic continental margin: U.S.: Boulder, Colorado, Geological Society of America, The Geology of North America, v. I-2, p. 197–216.

Manspeizer, W., and deBoer, J., 1989, Rift basins, *in* Hatcher, R. D., Jr., Thomas, W. A., and Viele, G. W., eds., The Appalachian-Ouachita orogen in the United States: Boulder, Colorado, Geological Society of America, The Geology of North America, v. F-2, p. 319–333.

Manspeizer, W., and 8 others, 1989, Post-Paleozoic activity, *in* Hatcher, R. D., Jr., Thomas, W. A., and Viele, G. W., eds., The Appalachian-Ouachita orogen in the United States: Boulder, Colorado, Geological Society of America, The Geology of North America, v. F-2, p. 319–374.

Marr, J. D., Jr., 1985, Geology of the crystalline portion of the Richmond 1° × 2° quadrangle, a progress report (Seventeenth Annual Virginia Geological Field Conference): Charlottesville, Virginia Division of Mineral Resources, p. 1–22.

Mauger, R. L., Spruill, R. K., Christopher, M. T., and Shafiquallah, M., 1983, Petrology and geochemistry of peralkaline metagranite and metarhyolite dikes, Fountain quarry, Pitt County, North Carolina: Southeastern Geology, v. 24, p. 67–89.

McBride, J. H., and Nelson, K. D., 1988, Comment *on* "Is the Brunswick magnetic anomaly really the Alleghanian suture?": Tectonics, v. 7, p. 343–346.

McBride, J. H., Nelson, K. D., and Brown, L. D., 1989, Evidence and implications of an extensive early Mesozoic rift basin and basalt/diabase sequence beneath the southeast Coastal Plain: Geological Society of America Bulletin, v. 101, p. 512–520.

McConnell, K. I., Hatcher, R. D., Jr., and Heyn, T., 1992, Geology of the Sauratown Mountains window, *in* Dennison, J. M., and Stewart, K. G., eds., Geologic field guides to North Carolina and vicinity: Chapel Hill, Department of Geology, University of North Carolina, Guidebook no. 1, p. 201–212.

McHone, J. G., and Butler, J. R., 1984, Mesozoic igneous provinces of New England and the opening of the North Atlantic: Geological Society of America Bulletin, v. 95, p. 757–765.

McLelland, J., Isachsen, Y., Whitney, P., Chiarenzelli, J., and Hall, L., 1993, Geology of the Adirondack massif, New York, *in* Reed, J. C., Jr., Bickford, M. E., Jr., Houston, R. S., Link, P. K., Rankin, D. W., Sims, P. K., and Van Schmus, W. R., eds., Precambrian: Conterminous United States: Boulder, Colorado, Geological Society of America, The Geology of North America, v. C-2, p. 338–353.

McLeod, M. J., Ruitenberg, A. A., and Johnson, S. C., 1991, Compilation and correlation of southwestern New Brunswick geology: Charlotte, Queen, Kings, Saint John, and Sunbury counties (PROV & C-NBCAMD), *in* Abbott, S. A., ed., Project summaries for 1991, sixteenth annual review of activities: New Brunswick Department of Natural Resources and Energy, Mineral Resources, Information Circular 91-2, p. 128–133.

McMaster, R. L., de Boer, J., and Collins, B. P., 1980, Tectonic development of southern Narragansett Bay and offshore Rhode Island: Geology, v. 8, p. 496–500.

McSween, H. Y., Jr., Speer, J. A., and Fullagar, P. D., 1991, Plutonic rocks, *in* Horton, J. W., Jr., and Zullo, V. A., eds., The geology of the Carolinas: Knoxville, University of Tennessee Press, p. 109–126.

Miller, C. F., Sando, T. W., Fullagar, P. D., and Kish, S. A., 1983, Low potassium plutonism in the Blue Ridge of North Carolina and northern Georgia: Geological Society of America Abstracts with Programs, v. 15, p. 46.

Milton, D. J., 1983, Garnet-biotite geothermometry confirms the premetamorphic age of the Greenbrier fault, Great Smoky Mountains, North Carolina: Geological Society of America Abstracts with Programs, v. 15, p. 90.

Misra, K. C., and Conte, J. A., 1991, Amphibolites of the Ashe and Alligator Back formations, North Carolina: Samples of Late Proterozoic–early Paleozoic oceanic crust: Geological Society of America Bulletin, v. 103, p. 737–750.

Moench, R. H., 1990, The Piermont allochthon, northern Connecticut Valley area, New England—Preliminary description and resource implications, *in* Slack, J. F., ed., Summary results of the Glens Falls CUSMAP project, New York, Vermont, and New Hampshire: U.S. Geological Survey Bulletin 1887, p. J1–J23.

Moench, R. H., and Pankiwskyj, K. A., 1988, Geologic map of western Maine: U.S. Geological Survey Miscellaneous Investigations Series Map I-1692, 21 p, scale 1:250,000.

Moench, R. H., and St. Julien, P., 1989, Northern Appalachian transect: Southeastern Quebec, Canada through western Maine, U.S.A.: Field trip guidebook T358, 28th International Geological Congress: Washington, D.C., American Geophysical Union, 52 p.

Moench, R. H., Bothner, W. A., Marvinney, R. G., and Pollock, S. G., 1992, The Second Lake rift, northern New England: Possible resolution of the Piermont allochthon—Frontenac Formation problems: Geological Society of America Abstracts with Programs, v. 24, no. 3, p. 63–64.

Morgan, B. A., 1972, Metamorphic map of the Appalachians: U.S. Geological Survey Miscellaneous Geologic Investigations Map I-724, scale 1:2,500,000.

Morgan, W. J., 1981, Hotspot tracks and the opening of the Atlantic and Indian oceans, *in* Emiliani, C., ed., The oceanic lithosphere: The sea, Volume 7: New York, Wiley and Sons, p. 443–487.

Moore, J. M., 1986, Introduction: The "Grenville problem" then and now, *in* Moore, J. M., Davidson, A., and Baer, A. J., eds., The Grenville Province: Geological Association of Canada Special Paper 31, p. 1–11.

Mose, D. G., and Nagel, M. S., 1982, Plutonic events in the Piedmont of Virginia: Southeastern Geology, v. 23, p. 25–40.

Mose, D., Cohen, S., and Glover, L., 1980, Rb-Sr whole-rock age of the Occoquan pluton and Fluvanna County pluton in northern Virginia: Geological Society of America Abstracts with Programs, v. 12, p. 202.

Murray, D. P., 1988, Post-Acadian metamorphism in the Appalachians, *in* Harris, A. L., and Fettes, D. J., eds., The Caledonian-Appalachian orogen: (Geological Society Special Publication 38): London, Blackwell Scientific Publications, p. 597–609.

Mussman, W. J., and Read, J. F., 1986, Sedimentology and development of a passive- to convergent-margin unconformity: Middle Ordovician Knox unconformity, Virginia Appalachians: Geological Society of America Bulletin, v. 97, p. 282–295.

Mutter, J. C., 1985, Seaward dipping reflectors and the continent-ocean boundary at passive continental margins: Tectonophysics, v. 114, p. 117–131.

Nance, R. D., 1990, Late Precambrian–early Paleozoic arc-platform transitions in the Avalon terrane of the Northern Appalachians: Review and implications, *in* Socci, A. D., Skehan, J. W., and Smith, G. W., eds., Geology of the composite Avalon terrane of southern New England: Geological Society of America Special Paper 245, p. 1–11.

Naylor, R. S., Boone, G. M., Boudette, E. L., Ashenden, D. D., and Robinson, P., 1973, Pre-Ordovician rocks in the Bronson Hill and Boundary Mountain anticlinoria, New England, U.S.A. [abs.]: Eos (Transactions, American Geophysical Union), v. 50, p. 495.

Nelson, A. E., Horton, J. W., Jr., and Clarke, J. W., 1987, Generalized tectonic map of the Greenville 1° × 2° quadrangle, Georgia, South Carolina, and North Carolina: U.S. Geological Survey Miscellaneous Field Studies Map MF-1898, scale 1:250,000.

——, 1989, Geologic map of the Greenville 1° × 2° quadrangle, Georgia, South Carolina, and North Carolina: U.S. Geological Survey Open-File Report 89-9, scale 1:250,000.

Nelson, K. D., 1992, Are crustal thickness variations in old mountain belts like the Appalachians a consequence of lithospheric deformation?: Geology, v. 20, p. 498–502.

Nelson, K. D., and 8 others, 1985, New COCORP profiling in the southeastern United States. Part I: Late Paleozoic suture and Mesozoic rift basin: Geology, v. 13, p. 714–718.

Nelson, K. D., McBride, J. H., and Arnow, J. A., 1986, Deep reflection character, gravity gradient, and crustal thickness variation in the Appalachian orogen: Relation to Mesozoic extension and igneous activity: Geological Society of America Abstracts with Programs, v. 18, p. 704–705.

Nelson, K. D., Arnow, J. A., and Giguere, M., 1987, Normal-fault boundary of an Appalachian basement massif?: Results of COCORP profiling across the Pine Mountain belt in western Georgia: Geology, v. 15, p. 832–836.

Neuman, R. B., 1967, Bedrock geology of the Shin Pond and Stacyville quadrangles, Penobscot County, Maine: U.S. Geological Survey Professional Paper 524-I, 37 p.

——, 1973, Staurolite zone Caradoc (Middle-Late Ordovician) age, Old World Province brachiopods from Penobscot Bay, Maine: Discussion: Geological Society of America Bulletin, v. 84, p. 1829–1830.

——, 1984, Geology and paleobiology of islands in the Ordovician Iapetus Ocean: Geological Society of America Bulletin, v. 95, p. 1188–1201.

Neuman, R. B., and Harper, D.A.T., 1992, Paleogeographic significance of Arenig–Llanvirn Toquima–Table Head and Celtic brachiopod assemblages, *in* Webby, B. D., and Laurie, J. R., eds., Global perspectives on Ordovician geology: Rotterdam, Balkema, p. 241–254.

Neuman, R. B., and Max, M. D., 1989, Penobscottian-Grampian-Finnmarkian orogenies as indicators of terrane linkages, *in* Dallmeyer, R. D., ed., Terranes in the circum-Atlantic Paleozoic orogens: Geological Society of America Special Paper 230, p. 31–45.

Neuman, R. B., Palmer, A. R., and Dutro, J. T., Jr., 1989, Paleontological contributions to Paleozoic paleogeographic reconstructions of the Appalachians, *in* Hatcher, R. D., Jr., Thomas, W. A., and Viele, G. W., eds., The Appalachian-Ouachita orogen in the United States: Boulder, Colorado, Geological Society of America, The Geology of North America, v. F-2, p. 375–384.

Noel, J. R., Spariosu, D. J., and Dallmeyer, R. D., 1988, Paleomagnetism and $^{40}Ar/^{39}Ar$ ages from the Carolina slate belt, Albermarle, N.C.: Implications for terrane amalgamation with North America: Geology, v. 16, p. 64–68.

O'Brien, B. H., O'Brien, S. J., and Dunning, G. R., 1991, Silurian cover, late Precambrian–Early Ordovician basement, and the chronology of Silurian orogenesis in the Hermitage flexure (Newfoundland Appalachians): American Journal of Science, v. 291, p. 760–799.

O'Brien, S. J., Wardle, R. J., and King, A. F., 1983, The Avalon zone: A pan-African terrane in the Appalachian orogen of Canada: Geological Journal, v. 18, p. 195–222.

Odom, A. L., and Fullagar, P. D., 1973, Geochronologic and tectonic relationships between the Inner Piedmont, Brevard zone, and Blue Ridge belts, North Carolina: American Journal of Science, The Byron N. Cooper Volume, v. 273-A, p. 133–149.

O'Hara, K., and Gromet, L. P., 1985, Two distinct late Precambrian (Avalonian) terranes in southeastern New England and their late Paleozoic juxtaposition: American Journal of Science, v. 285, p. 673–709.

Olsen, P. E., 1989, Stratigraphy, facies, depositional environments, and paleontology of the Newark Supergroup, *in* Hatcher, R. D., Jr., Thomas, W. A., and Viele, G. W., eds., The Appalachian-Ouachita orogen in the United States: Boulder, Colorado, Geological Society of America, The Geology of North America, v. F-2, p. 343–350.

Olszewski, W. J., Jr., 1980, The geochronology of some stratified metamorphic rocks in northeastern Massachusetts: Canadian Journal of Earth Sciences, v. 17, p. 1407–1416.

Olszewski, W. J., Jr., and Gaudette, H. E., 1988, Early Paleozoic thermotectonic history of eastern New England: Cambro-Ordovician metamorphism and plutonism—A distinctive feature of the "Casco" terrane: Geological Society of America Abstracts with Programs, v. 20, no. 1, p. 60.

Osberg, P. H., 1978, Synthesis of the geology of the northeastern Appalachians, U.S.A., *in* Tozer, E. T. and Schenk, P. E., eds., Caledonian-Appalachian orogen of the North Atlantic region: Geological Survey of Canada Paper 78-13, p. 137–147.

——, compiler, 1989, Distribution of structural features and plutons within the Acadian orogen, *in* Hatcher, R. D., Jr., Thomas, W. A., and Viele, G. W., eds., The Appalachian-Ouachita orogen in the United States: Boulder, Colorado, Geological Society of America, The Geology of North America, v. F-2, plate 4.

Osberg, P. H., Hussey, A. M., II, and Boone, G. M., eds., 1985, Bedrock geologic map of Maine: Augusta, Maine Geological Survey, scale 1:500,000.

Osberg, P. H., Tull, J. F., Robinson, P., Hon, R., and Butler, J. R., 1989, The Acadian orogen, *in* Hatcher, R. D., Jr., Thomas, W. A., and Viele, G. W., eds., The Appalachian-Ouachita orogen in the United States: Boulder, Colorado, Geological Society of America, The Geology of North America, v. F-2, p. 179–232.

Pakiser, L. C., and Mooney, W. D., eds., 1989, Geophysical framework of the continental United States: Boulder, Colorado, Geological Society of America Memoir 172, 826 p.

Palmer, A. R., 1971, The Cambrian of the Appalachian and eastern New England regions, eastern United States, *in* Holland, C. H., ed., Cambrian of the New World: New York, Interscience Publishers, p. 169–217.

——, compiler, 1983, The Decade of North American Geology 1983 geologic time scale: Geology, v. 11, p. 503–504.

Panish, P. T., and Hall, L. M., 1985, Geology of the Mt. Prospect region, western Connecticut, *in* Tracey, R. J., ed., Guidebook for field trips in Connecticut and adjacent areas of New York and Rhode Island (New England Intercollegiate Geological Conference, annual meeting, 77th): Connecticut Geological and Natural History Survey Guidebook 6, p. 443–489.

Patterson, J. G., 1989, Deformation in portions of the distal continental margin to ancestral North America: An example from the westernmost internal zone, central and southern Appalachian orogen, Virginia: Tectonics, v. 8, p. 535–554.

Pavlides, L., 1981, The central Virginia volcanic-plutonic belt: An island arc of Cambrian(?) age: U.S. Geological Survey Professional Paper 1231-A, 34 p.

—— , 1989, Early Paleozoic composite melange terrane, central Appalachian Piedmont, Virginia and Maryland; its origin and tectonic history, *in* Horton, J. W., Jr., and Rast, N., eds., Melanges and olistostromes of the U.S. Appalachians: Geological Society of America Special Paper 228, p. 135–193.

Pavlides, L., Stern, T. W., Arth, J. G., Muth, K. G., and Newell, M. F., 1982, Middle and upper Paleozoic granitic rocks in the Piedmont near Fredericksburg, Virginia: Geochronology: U.S. Geological Survey Professional Paper 1231-B, p. B1–B9.

Pease, M. H., Jr., 1982, The Bonemill Brook fault zone, eastern Connecticut, *in* Joeston, R., and Quarrier, S. S., eds., Guidebook for fieldtrips in Connecticut and south-central Massachusetts (New England Intercollegiate Geological Conference, annual meeting, 77th): Connecticut Geological and Natural History Survey Guidebook 5, p. 263–287.

—— , 1989, Correlation of the Oakdale Formation and Paxton Group of central Massachusetts with strata in northeastern Connecticut: U.S. Geological Survey Bulletin 1796, 26 p.

Peper, J. D., and Pease, M. H., Jr., 1976, Summary of stratigraphy in the Brimfield area, Connecticut and Massachusetts, *in* Page, L. R., ed., Contributions to the stratigraphy of New England, Geological Society of America Memoir 148, p. 253–270.

Petersen, T. A., Brown, L. D., Cook, F. A., Kaufman, S., and Oliver, J. E., 1984, Structure of the Riddleville basin from COCORP seismic data and implications for reactivation tectonics: Journal of Geology, v. 92, p. 261–271.

Phinney, R. A., and Roy-Chowdhury, K., 1989, Reflection seismic studies of crustal structure in the eastern United States, *in* Pakiser, L. C., and Mooney, W. D., eds., Geophysical framework of the continental United States: Geological Society of America Memoir 172, p. 613–653.

Pindell, J., and Dewey, J. F., 1982, Permo-Triassic reconstruction of western Pangea and the evolution of the Gulf of Mexico/Caribbean region: Tectonics, v. 1, p. 179–212.

Piper, J.D.A., 1982, The Precambrian paleomagnetic record: The case for the Proterozoic supercontinent: Earth and Planetary Science Letters, v. 59, p. 61–89.

Pitcher, W. S., 1983, Granite type and tectonic environment, *in* Hsü, K. J., ed., Mountain building processes: New York, Academic Press, p. 9–40.

Poag, C. W., and Valentine, P. C., 1988, Mesozoic and Cenozoic stratigraphy of the United States Atlantic continental shelf and slope, *in* Sheridan, R. E., and Grow, J. A., eds., The Atlantic continental margin: U.S.: Boulder, Colorado, Geological Society of America, The Geology of North America, v. I-2, p. 67–85.

Pojeta, J., Jr., Kříž, J., and Berdan, J. M., 1976, Silurian-Devonian pelecypods and Paleozoic stratigraphy of subsurface rocks in Florida and Georgia and related Silurian pelecypods from Bolivia and Turkey: U.S. Geological Survey Professional Paper 879, 32 p.

Pratt, B. R., and Waldron, J.W.F., 1991, A Middle Cambrian trilobite faunule from the Meguma Group of Nova Scotia: Canadian Journal of Earth Sciences, v. 28, p. 1843–1853.

Pratt, T. L., Coruh, C., Costain, J. K., and Glover, L., III, 1988, A geophysical study of the earth's crust in central Virginia: Implications for Appalachian crustal structure: Journal of Geophysical Research, v. 93, p. 6649–6667.

Prowell, D. C., 1989, Cretaceous and Cenozoic tectonism in the Appalachians of the eastern United States, *in* Hatcher, R. D., Jr., Thomas, W. A., and Viele, G. W., eds., The Appalachian-Ouachita orogen in the United States: Boulder, Colorado, Geological Society of America, The Geology of North America, v. F-2, p. 362–366.

Quinlan, G. M., and Beaumont, C., 1984, Appalachian thrusting, lithospheric flexure, and the Paleozoic stratigraphy of the eastern interior of North America: Canadian Journal of Earth Sciences, v. 21, p. 966–973.

Rankin, D. W., 1968, Volcanism related to tectonism in the Piscataquis volcanic belt, an island arc of Early Devonian age in north-central Maine, *in* Zen, E-an, White, W. S., Hadley, J. B., and Thompson, J. B., Jr., eds., Studies of Appalachian geology: Northern and maritime: New York, Interscience Publishers, p. 355–369.

—— , 1975, The continental margin of eastern North America in the southern Appalachians: The opening and closing of the proto-Atlantic Ocean, *in* Tectonics and mountain ranges: American Journal of Science, v. 275-A, p. 298–336.

—— , 1976, Appalachian salients and recesses: Late Precambrian continental breakup and the opening of the Iapetus Ocean: Journal of Geophysical Research, v. 81, p. 5605–5619.

—— , 1977, Introduction and discussion, *in* Rankin, D. W., ed., Studies related to the Charleston, South Carolina, earthquake of 1886—A preliminary report: U.S. Geological Survey Professional Paper 1028, p. 1–15.

—— , 1983, Late Proterozoic rifting of eastern North America: Geological Society of America Abstracts with Programs, v. 15, p. 183.

—— , 1993a, Introduction, *in* Reed, J. C., Jr., Bickford, M. E., Jr., Houston, R. S., Link, P. K., Rankin, D. W., Sims, P. K., and Van Schmus, W. R., eds., Precambrian: Conterminous United States: Boulder, Colorado, Geological Society of America, The Geology of North America, v. C-2, p. 335–338.

—— , compiler, 1993b, Geologic map of Precambrian (Proterozoic) rocks east of the Grenville front and their geologic setting in the United States and Canada, *in* Reed, J. C., Jr., Bickford, M. E., Jr., Houston, R. S., Link, P. K., Rankin, D. W., Sims, P. K., and Van Schmus, W. R., eds., Precambrian: Conterminous United States: Boulder, Colorado, Geological Society of America, The Geology of North America, v. C-2, plate 5A, scale 1:2,000,000.

Rankin, D. W., Espenshade, G. H., and Shaw, K. W., 1973, Stratigraphy and structure of the metamorphic belt in northwestern North Carolina and southwestern Virginia: A study from the Blue Ridge across the Brevard fault zone to the Sauratown Mountains anticlinorium: American Journal of Science, v. 273-A, p. 1–40.

Rankin, D. W., and 6 others, 1988, Plutonism and volcanism related to the pre-Arenig evolution of the Caledonide-Appalachian orogen, *in* Harris, A. L., and Fettes, D. J., eds., The Caledonian-Appalachian orogen (Geological Society Special Publication 38): London, Blackwell, p. 149–183.

Rankin, D. W., Hall, L. M., Drake, A. A., Jr., Goldsmith, R., Ratcliffe, N. M., and Stanley, R. S., 1989, Proterozoic evolution of the rifted margin of Laurentia, *in* Hatcher, R. D., Jr., Thomas, W. A., and Viele, G. W., eds., The Appalachian-Ouachita orogen in the United States: Boulder, Colorado, Geological Society of America, The Geology of North America, v. F-2, p. 10–42.

Rankin, D. W., and 13 others, 1991, E-4, Central Kentucky to Carolina trough: Boulder, Colorado, Geological Society of America, Centennial Continent/Ocean Transect no. 16, 41 p., 2 sheets, scale 1:500,000.

Rankin, D. W., and 7 others, 1993a, Map showing subdivisions of the Grenville orogen and younger tectonostratigraphic terranes in eastern United States and adjacent Canada that may contain Precambrian rocks, *in* Reed, J. C., Jr., Bickford, M. E., Jr., Houston, R. S., Link, P. K., Rankin, D. W., Sims, P. K., and Van Schmus, W. R., eds., Precambrian: Conterminous United States: Boulder, Colorado, Geological Society of America, The Geology of North America, v. C-2, plate 5B, scale 1:5,000,000.

Rankin, D. W., Drake, A. A., Jr., Ratcliffe, N. M., 1993b, Proterozoic North American (Laurentian) rocks of the Appalachian orogen, *in* Reed, J. C., Jr., Bickford, M. E., Jr., Houston, R. S., Link, P. K., Rankin, D. W., Sims, P. K., and Van Schmus, W. R., eds., Precambrian: Conterminous United States: Boulder, Colorado, Geological Society of America, The Geology of North

America, v. C-2, p. 378–422.

Rast, N., and Skehan, J. W., 1983, The evolution of the Avalonian plate: Tectonophysics, v. 100, p. 257–286.

Ratcliffe, N. M., 1971, The Ramapo fracture system in New York and adjacent New Jersey: A case of tectonic heredity: Geological Society of America Bulletin, v. 82, p. 125–142.

—— , 1987, Basaltic rocks in the Rensselaer Plateau and Chatham slices of the Taconic allochthon: Chemistry and tectonic setting: Geological Society of America Bulletin, v. 99, p. 511–528.

—— , 1988, Along-strike variation in tectonics of the Taconian orogen: Geological Society of America Abstracts with Programs, v. 20, no. 1, p. 64.

Ratcliffe, N. M., and Harwood, D. S., 1975, Blastomylonites associated with recumbent folds and overthrusts at the western edge of the Berkshire massif, Connecticut and Massachusetts—A preliminary report, *in* Harwood, D. S., ed., Tectonic studies of the Berkshire massif, western Massachusetts, Connecticut and Vermont: U.S. Geological Survey Professional Paper 888-A, p. 1–19.

Ratcliffe, N. M., Burton, W. C., D'Angelo, R. M., and Costain, J. K., 1986, Low-angle extensional faulting, reactivated mylonites, and seismic reflection geometry of the Newark basin margin in eastern Pennsylvania: Geology, v. 14, p. 766–770.

Ratcliffe, N. M., Burton, W. C., Sutter, J. F., and Mukasa, S. B., 1988, Stratigraphy, structural geology and thermochronology of the northern Berkshire massif and southern Green Mountains, *in* Bothner, W. A., ed., Guidebook for field trips in southwestern New Hampshire, southeastern Vermont, and north-central Massachusetts (New England Intercollegiate Geological Conference, annual meeting, 80th): Durham, University of New Hampshire, Department of Earth Sciences, p. 126–135.

Read, J. F., 1989, Evolution of Cambro-Ordovician passive margin, U.S. Appalachians, *in* Hatcher, R. D., Jr., Thomas, W. A., and Viele, G. W., eds., The Appalachian-Ouachita orogen in the United States: Boulder, Colorado, Geological Society of America, The Geology of North America, v. F-2, p. 42–57.

Reed, J. C., Jr., and Bryant, B., 1964, Evidence for strike-slip faulting along the Brevard zone in North Carolina: Geological Society of America Bulletin, v. 75, p. 1177–1196.

Reed, J. C., Jr., Bickford, M. E., Jr., Houston, R. S., Link, P. K., Rankin, D. W., Sims, P. K., and Van Schmus, W. R., eds., 1993, Precambrian: Conterminous United States: Boulder, Colorado, Geological Society of America, The Geology of North America, v. C-2, p. 658, 7 plates.

Riva, J., and Pickerill, R. K., 1987, The late mid-Ordovician transgressive sequence and the Montmorency fault at Montmorency Falls, Quebec, *in* Roy, D. C., ed., Northeast Section of the Geological Society of America: Boulder, Colorado, Geological Society of America, Centennial Field Guide, v. 5, p. 357–362.

Robinson, P., Thompson, P. J., and Elbert, D. C., 1991, The nappe theory in the Connecticut Valley region: Thirty-five years since Jim Thompson's first proposal: American Mineralogist, v. 76, p. 689–712.

Robinson, P., and 6 others, 1992, The Pelham dome, central Massachusetts: Stratigraphy, geochronology, and Acadian and Pennsylvanian structure and metamorphism, *in* Robinson, P., and Brady, J. B., eds., Guidebook for field trips in the Connecticut Valley region of Massachusetts and adjacent states, Volume 1 (New England Intercollegiate Geological Conference, annual meeting, 84th): Amherst, University of Massachusetts, Department of Geology and Geography Contribution 66, p. 132–169.

Rodgers, John, 1953, Geologic map of east Tennessee with explanatory text: Tennessee Division of Geology Bulletin 58, pt. 2, 168 p.

—— , 1970, The tectonics of the Appalachians: New York, Interscience Publishers, 271 p.

—— , 1971, The Taconic orogeny: Geological Society of America Bulletin, v. 62, p. 1141–1178.

—— , 1972, Latest Precambrian (post-Grenville) rocks of the Appalachian region: American Journal of Science, v. 272, p. 507–520.

—— , 1975, Appalachian salients and recesses: Geological Society of America Abstracts with Programs, v. 7, p. 111–112.

—— , 1981, The Merrimack synclinorium in northeastern Connecticut: American Journal of Science, v. 281, p. 176–186.

—— , 1985, Bedrock geologic map of Connecticut: Hartford, Connecticut Geologic and Natural History Survey, scale 1:125,000.

Rogers, J.J.W., 1982, Criteria for recognizing environments of formation of volcanic suites: Application of these criteria to volcanic suites in the Carolina Slate Belt, *in* Bearce, D. N., Black, W. W., Kish, S. A., and Tull, J. F., eds., Tectonic studies in the Talladega and Carolina slate belts, southern Appalachian orogen: Geological Society of America Special Paper 191, p. 99–107.

Ruitenberg, A. A., Fyffe, L. R., McCutcheon, S. R., St. Peter, C. J., Irrinki, R. R., and Venugopal, D. V., 1977, Evolution of pre-Carboniferous tectonostratigraphic zones in the New Brunswick Appalachians: Geoscience Canada, v. 4, p. 171–181.

Russell, G. S., 1978, U/Pb, Rb-Sr, and K-Ar isotopic studies bearing on the tectonic development of the southernmost Appalachian orogen, Alabama [Ph.D. thesis]: Tallahassee, Florida State University, 197 p.

—— , 1985, Reconnaissance geochronological investigations in the Phenix City Gneiss and Bartletts Ferry mylonite zone, *in* Kish, S. A., Hanley, T. B., and Schamel, S., eds., Geology of the southwestern Piedmont of Georgia: Tallahassee, Florida State University, Department of Geology, v. 9–11.

Russell, G. S., Russell, C. W., Speer, J. A., and Glover, L., III, 1981, Rb-Sr evidence of latest Precambrian to Cambrian and Alleghanian plutonism along the eastern margin of the sub-Coastal Plain Appalachians, North Carolina and Virginia: Geological Society of America Abstracts with Programs, v. 13, p. 543.

Russell, G. S., Russell, C. W., and Golden, B. K., 1984, The Taconic history of the northern Alabama Piedmont: Geological Society of America Abstracts with Programs, v. 16, p. 191.

Russell, G. S., Odom, A. L., and Russell, W., 1987, Uranium-lead and rubidium-strontium isotopic evidence for the age and origin of granitic rocks in the northern Alabama Piedmont, *in* Drummond, M. S., and Green, N. L., eds., Granites of Alabama: Tuscaloosa, Geological Society of Alabama, p. 239–249.

Sacks, P. E., and Secor, D. T., Jr., 1986, Alleghanian down-to-the-north slip along the Modoc zone in the Appalachian Piedmont: Geological Society of America Abstracts with Programs, v. 18, p. 737.

—— , 1990, Delamination in collisional orogens: Geology, v. 18, p. 999–1002.

Sacks, P. E., Maher, H. D., Secor, D. T., Jr., and Shervais, J. W., 1989, Bucks Mountain complex, Kiokee belt, southern Appalachian Piedmont of South Carolina and Georgia, *in* Mittwede, S. K., and Stoddard, E. F., eds., Ultramafic rocks of the Appalachian Piedmont: Geological Society of America Special Paper 231, p. 75–86.

St. Jean, J., 1973, A new trilobite from the Piedmont of North Carolina: American Journal of Science, v. 273-A, p. 196–216.

St. Julien, P., and Hubert, C., 1975, Evolution of the Taconian orogen in the Quebec Appalachians, *in* Tectonics and mountain ranges: American Journal of Science, v. 275-A, p. 337–362.

Samson, S., Palmer, A. R., Robison, R. A., and Secor, D. T., Jr., 1990, Biogeographical significance of Cambrian trilobites from the Carolina slate belt: Geological Society of America Bulletin, v. 102, p. 1459–1470.

Sawyer, D. S., 1988, Thermal evolution, *in* Sheridan, R. E., and Grow, J. A., eds., The Atlantic continental margin: U.S.: Boulder, Colorado, Geological Society of America, The Geology of North America, v. I-2, p. 417–428.

Schamel, S., and Bauer, D., 1980, Remobilized Grenville basement in the Pine Mountain window, *in* Wones, D. R., ed., The Caledonides in the U.S.A., IGCP Project 27: Blacksburg, Virginia Polytechnic Institute and State University, Department of Geological Sciences, Memoir 2, p. 313–316.

Schenk, P. E., 1980, Paleogeographic implications of the Meguma Group, Nova Scotia—A chip of Africa?, *in* Wones, D. R., ed., Proceedings of the Caledonides in the U.S.A., IGCP project 27, Caledonide orogen: Blacksburg, Virginia Polytechnic Institute and State University, Department of Geological Sciences, Memoir 2, p. 27–30.

—— , 1991, Events and sea-level changes on Gondwana's margin: The Meguma zone (Cambrian to Devonian) of Nova Scotia, Canada: Geological Society of America Bulletin, v. 103, p. 512–521.

Schlee, J. S., and Klitgord, K. D., 1988, Georges Bank basin: A regional synthesis, *in* Sheridan, R. E., and Grow, J. A., eds., The Atlantic continental margin: U.S.: Boulder, Colorado, Geological Society of America, The Geology of North America, v. I-2, p. 243–268.

Schroeder, K. E., and Nance, R. D., 1987, Mylonite-hosted gold mineralization at the Howie Mine, Union County, North Carolina: Geological Society of America Abstracts with Programs, v. 19, no. 2, p. 128.

Scotese, C. R., Bambach, R. K., Barton, C., Van der Voo, R., and Ziegler, A. M., 1979, Paleozoic base maps: Journal of Geology, v. 87, p. 217–277.

Secor, D. T., Jr., Samson, S. L., Snoke, A. W., and Palmer, A. R., 1983, Confirmation of the Carolina slate belt as an exotic terrane: Science, v. 221, p. 649–651.

Secor, D. T., Jr., Snoke, A. W., and Dallmeyer, R. D., 1986, Character of the Alleghanian orogeny in the southern Appalachians: Part III, Regional tectonic relations: Geological Society of America Bulletin, v. 97, p. 1345–1353.

Secor, D. T., Jr., Murray, D. P., and Glover, L., III, 1989, Geology of the Avalonian rocks, *in* Hatcher, R. D., Jr., Thomas, W. A., and Viele, G. W., eds., The Appalachian-Ouachita orogen in the United States: Boulder, Colorado, Geological Society of America, The Geology of North America, v. F-2, p. 57–85.

Seiders, V. M., Mixon, R. B., Stern, T. W., Newell, M. F., and Thomas, C. B., Jr., 1975, Age of plutonism and tectonism and a new minimum age limit on the Glenarm Series in the northeast Virginia Piedmont near Occoquan: American Journal of Science, v. 275, p. 481–511.

Shaw, H. F., and Wasserburg, G. J., 1984, Isotopic constraints on the origin of Appalachian mafic complexes: American Journal of Science, v. 284, p. 319–349.

Shelley, S. A., Shervais, J. W., and Secor, D. T., Jr., 1988, Geochemical characterization of metavolcanic rocks of the Carolina slate belt, central South Carolina: Geological Society of America Abstracts with Programs, v. 20, no. 4, p. 314.

Sheridan, R. E., and Grow, J. A., eds., 1988a, The Atlantic continental margin: U.S.: Boulder, Colorado, Geological Society of America, The Geology of North America, v. I-2, 610 p., 8 plates.

——, 1988b, Synthesis and unanswered questions, *in* Sheridan, R. E., and Grow, J. A., eds., The Atlantic continental margin: U.S.: Boulder, Colorado, Geological Society of America, The Geology of North America, v. I-2, p. 595–599.

Sheridan, R. E., Grow, J. A., and Klitgord, K. D., 1988a, Geophysical data, *in* Sheridan, R. E., and Grow, J. A., eds., The Atlantic continental margin: U.S.: Boulder, Colorado, Geological Society of America, The Geology of North America, v. I-1, p. 177–196.

Sheridan, R. E., Mullins, H. T., Austin, J. A., Jr., Ball, M. M., and Ladd, J. W., 1988b, Geology and geophysics of the Bahamas, *in* Sheridan, R. E., and Grow, J. A., eds., The Atlantic continental margin: U.S.: Boulder, Colorado, Geological Society of America, The Geology of North America, v. I-2, p. 329–364.

Sheridan, R. E., Olsson, R. K., and Miller, J. J., 1991, Seismic reflection and gravity study of proposed Taconic suture under the New Jersey Coastal Plain: Implications for continental growth: Geological Society of America Bulletin, v. 103, p. 402–414.

Shride, A. F., 1976, Stratigraphy and correlation of the Newbury Volcanic Complex, northeastern Massachusetts, *in* Page, L. R., Contributions to the stratigraphy of New England: Geological Society of America Memoir 148, p. 147–177.

Simpson, E. L., and Sundberg, F. A., 1987, Early Cambrian age for synrift deposits of the Chilhowee Group of southwestern Virginia: Geology, v. 15, p. 123–126.

Sinha, A. K., 1988, Plutonism in the U.S. Appalachians, *in* Frontiers in petrology: American Journal of Science, v. 288-A, p. IX–XII.

Sinha, A. K., and Glover, L., III, 1978, U-Pb systematics of zircons during dynamic metamorphism—A study from the Brevard fault zone: Contributions to Mineralogy and Petrology, v. 66, p. 305–310.

Sinha, A. K., and Zietz, I., 1982, Geophysical and geochemical evidence for a Hercynian magmatic arc, Maryland to Georgia: Geology, v. 10, p. 593–596.

Sinha, A. K., Hewitt, D. A., and Rimstidt, J. D., 1988, Metamorphic petrology and strontium isotope geochemistry associated with the development of mylonites: An example from the Brevard fault zone, North Carolina, *in* Frontiers in petrology: American Journal of Science, v. 288-A, p. 115–147.

Sinha, A. K., Hund, E. A., and Hogan, J. P., 1989, Paleozoic accretionary history of the North American plate margin (central and southern Appalachians): Constraints from the age, origin, and distribution of granitic rocks, *in* Hillhouse, J. W., ed., Deep structure and past kinematics of accreted terranes: American Geophysical Union Geophysical Monograph 50, p. 219–238.

Slivitzky, A., and St. Julien, P., 1987, Compilation geologique de la région de l'Estrie-Beauce: Québec, Ministere de l'Énergie et des Resources du Québec, MM 85-4, 40 p.

Sloss, L. L., 1963, Sequences in the cratonic interior of North America: Geological Society of America Bulletin, v. 74, p. 93–11.

Smith, A. E., 1903, Carboniferous fossils in Ocoee slates in Alabama: Science, v. 18, p. 244–246.

Smith, G. W., and Socci, A. D., 1990, Late Precambrian sedimentary geology of the Boston basin, *in* Socci, A. D., Skehan, J. W., and Smith, G. W., eds., Geology of the composite Avalon terrane of southern New England: Geological Society of America Special Paper 245, p. 75–84.

Snoke, A. W., and Frost, B. R., 1990, Exhumation of high pressure pelitic schist, Lake Murray spillway, South Carolina: Evidence for crustal extension during Alleghanian strike-slip faulting: American Journal of Science, v. 290, p. 853–881.

Snoke, A. W., and Mosher, S., 1989, The Alleghanian orogeny as manifested in the Appalachian internides, *in* Hatcher, R. D., Jr., Thomas, W. A., and Viele, G. W., eds., The Appalachian-Ouachita orogen in the United States: Boulder, Colorado, Geological Society of America, The Geology of North America, v. F-2, p. 288–307.

Snoke, A. W., Kish, S. A., and Secor, D. T., Jr., 1980, Deformed Hercynian granitic rocks from the Piedmont of South Carolina: American Journal of Science, v. 280, p. 1018–1034.

Society of Exploration Geophysicists, 1982, Gravity anomaly map of the United States (exclusive of Alaska and Hawaii): Tulsa, Oklahoma, Society of Exploration Geophysicists, 2 sheets, scale 1:2,500,000.

Spear, F. S., and Rumble, D., III, 1986, Mineralogy, petrology and P-T evolution of the Orfordville area, west-central New Hampshire and east-central Vermont, *in* Robinson, P., and Elbert, D. C., eds., Regional metamorphism and metamorphic phase relations in northwestern and central New England: Amherst, University of Massachusetts, Department of Geology and Geography, Contribution 59, p. 57–93.

Speed, R. C., compiler, 1991, Tectonic section display: Boulder, Colorado, Geological Society of America, Centennial Continent/Ocean Transects TRA-TSD, 58 p., 2 sheets, scale 1:1,000,000.

Spencer, C., and 7 others, 1989, The extension of Grenville basement beneath the northern Appalachians: Results from the Quebec-Maine seismic reflection and refraction surveys: Tectonics, v. 8, p. 677–696.

Stanley, R. S., and Hatch, N. L., Jr., 1988, The pre-Silurian geology of the Rowe-Hawley zone, *in* Hatch, N. L., Jr., ed., The bedrock geology of Massachusetts: U.S. Geological Survey Professional Paper 1366A, p. A1–A39.

Stanley, R. S., and Ratcliffe, N. M., 1985, Tectonic synthesis of the Taconian orogeny in western New England: Geological Society of America Bulletin, v. 96, p. 1227–1250.

Steckler, M. S., Watts, A. B., and Thorne, J. A., 1988, Subsidence and basin modeling of the U.S. Atlantic passive margin, *in* Sheridan, R. E., and Grow, J. A., eds., The Atlantic continental margin: U.S.: Boulder, Colorado, Geological Society of America, The Geology of North America, v. I-2, p. 399–416.

Stewart, D. B., 1989, Crustal processes in Maine: American Mineralogist, v. 74, p. 698–714.

Stewart, D. B., and Luetgert, J. H., 1990, The formation of the lower crust beneath the Quebec–Maine–Gulf of Maine global geoscience transect: Geological Society of America Abstracts with Programs, v. 22, no. 7, p. A192.

Stewart, D. B., and Lux, D. R., 1988, Lithologies and metamorphic age of the Precambrian rocks of Seven Hundred Acre Island and vicinity, Islesboro, Penobscott Bay, Maine: Geological Society of America Abstracts with Programs, v. 20, p. 73.

Stewart, D. B., and 8 others, 1986, The Quebec–western Maine seismic reflection profile: Setting and first year results, *in* Barazangi, M., and Brown, L. D., eds., Reflection seismology: The continental crust: American Geophysical Union Geodynamics Series, v. 14, p. 189–199.

Stewart, D. B., Wright, B. E., Unger, J. D., Phillips, J. D., and Hutchinson, D. R., compilers, 1991, Global Geoscience Transect 8, Quebec–Maine–Gulf of Maine transect, southeastern Canada, northeastern United States of America: U.S. Geological Survey Open-File Report 91-353, 65 p., 1 plate, scale 1:1,000,000. [Superseded by Stewart, D. B., Wright, B. E., Unger, J. D., Phillips, J. D., and Hutchinson, D. R., compilers, 1993, Global Geoscience Transect 8, Quebec–Maine–Gulf of Maine transect, southeastern Canada, northeastern United States of America: U.S. Geological Survey Miscellaneous Investigations Map I-2329, scale 1:1,000,000.]

Stoddard, E. F., Farrar, S. S., Horton, J. W., Jr., Butler, J. R., and Druhan, R. M., 1991, The eastern Piedmont in North Carolina, *in* Horton, J. W., Jr., and Zullo, V. A., eds., The geology of the Carolinas: Knoxville, University of Tennessee Press, p. 79–92.

Stow, S. H., 1982, Igneous petrology of the Hillabee Greenstone, Northern Alabama Piedmont, *in* Bearce, D. N., Black, W. W., Kish, S. A., and Tull, J. F., eds., Tectonic studies in the Talladega and Carolina slate belts, southern Appalachian orogen: Geological Society of America Special Paper 191, p. 79–92.

Stow, S. H., Neilson, M. J., and Neathery, T. L., 1984, Petrography, geochemistry, and tectonic significance of the amphibolites of the Alabama Piedmont: American Journal of Science, v. 284, p. 414–436.

Supplee, J. A., 1986, Geology of the Kings Mountain gold mine, Kings Mountain, North Carolina [M.S. thesis]: Chapel Hill, University of North Carolina, 140 p.

Sutter, J. F., 1988, Innovative approaches to the dating of igneous events in the early Mesozoic basins, *in* Froelich, A. J., and Robinson, G. R., Jr., eds., Studies of the early Mesozoic basins of the eastern United States: U.S. Geological Survey Bulletin 1776, p. 194–200.

Sutter, J. F., Milton, D. J., and Kunk, M. J., 1983, $^{40}Ar/^{39}Ar$ age spectrum dating of gabbro plutons and surrounding rocks in the Charlotte belt of North Carolina: Geological Society of America Abstracts with Programs, v. 15, p. 110.

Sutter, J. F., Ratcliffe, N. M., and Mukasa, S. B., 1985, $^{40}Ar/^{39}Ar$ and K-Ar data bearing on the metamorphic and tectonic history of western New England: Geological Society of America Bulletin, v. 96, p. 123–136.

Taylor, S. R., 1989, Geophysical framework of the Appalachians and adjacent Grenville Province, *in* Pakiser, L. C., and Mooney, W. D., eds., Geophysical framework of the continental United States: Geological Society of America Memoir 172, p. 317–348.

Thomas, M. D., 1983, Tectonic significance of paired gravity anomalies in the southern and central Appalachians, *in* Hatcher, R. D., Jr., Williams, H., and Zietz, I., eds., Contributions to the tectonics and geophysics of mountain chains: Geological Society of America Memoir 158, p. 113–124.

Thomas, W. A., 1977, Evolution of Appalachian salients and recesses from reentrants and promontories in the continental margin: American Journal of Science, v. 277, p. 1233–1278.

——, 1991, The Appalachian-Ouachita rifted margin of southeastern North America: Geological Society of America Bulletin, v. 103, p. 415–431.

Thomas, W. A., and Schenk, P. E., 1988, Late Paleozoic sedimentation along the Appalachian orogen, *in* Harris, A. L., and Fettes, D. J., eds., The Caledonian-Appalachian orogen (Geological Society of London Special Publication 38): London, Blackwell Scientific Publications, p. 515–530.

Thomas, W. A., Chowns, T. M., Daniels, D. L., Neathery, T. L., Glover, L., III, and Gleason, F. J., 1989, The subsurface Appalachians beneath the Atlantic and Gulf coastal plains, *in* Hatcher, R. D., Jr., Thomas, W. A., and Viele, G. W., eds., The Appalachian-Ouachita orogen in the United States: Boulder, Colorado, Geological Society of America, The Geology of North America, v. F-2, p. 445–458.

Thompson, J. B., Jr., and Norton, S. A., 1968, Paleozoic regional metamorphism in New England and adjacent areas, *in* Zen, E-an, White, W. S., Hadley, J. B., and Thompson, J. B., Jr., eds., Studies of Appalachian geology: Northern and maritime: New York, Interscience Publishers, p. 319–327.

Thompson, J. B., Jr., McLelland, J. M., and Rankin, D. W., 1989, Simplified geologic map of the Glens Falls 1° × 2° quadrangle: New York, Vermont, New Hampshire: U.S. Geological Survey Miscellaneous Field Studies Map MF2073, scale 1:250,000.

Thompson, J. B., Jr., Bothner, W. A., Robinson, P., Isachsen, Y. W., and Klitgord, K. D., 1993, E-1, Adirondacks to Georges Bank: Boulder, Colorado, Geological Society of America Centennial Continent/Ocean Transect no. 17, 55 p., 2 sheets, scale 1:500,000.

Thompson, M. D., and Hermes, O. D., 1990, Ash-flow stratigraphy in the Mattapan Volcanic Complex, greater Boston, Massachusetts, *in* Socci, A. D., Skehan, J. W., and Smith, G. W., eds., Geology of the composite Avalon terrane of southern New England: Geological Society of America Special Paper 245, p. 85–95.

Tobisch, O. T., 1972, Geologic map of the Milton quadrangle, Virginia–North Carolina and adjacent areas of Virginia: U.S. Geological Survey Miscellaneous Geologic Investigations Map I-683, scale 1:62,500.

Tobisch, O. T., and Glover, L., III, 1971, Nappe formation in part of the southern Appalachian Piedmont: Geological Society of America Bulletin, v. 82, p. 2209–2230.

Tollo, R. P., Aleinikoff, J. N., and Gray, K. J., 1991, New U/Pb zircon isotopic data from the Robertson River Igneous Suite, Virginia Blue Ridge: Implications for the duration of Late Proterozoic anorogenic magmatism: Geological Society of America Abstracts with Programs, v. 23, no. 1, p. 139.

Trehu, A. M., Klitgord, K. D., Sawyer, D. S., and Buffler, R. T., 1989, Atlantic and Gulf of Mexico continental margins, *in* Pakiser, L. C., and Mooney, W. D., eds., Geophysical framework of the continental United States: Geological Society of America Memoir 172, p. 349–382.

Trzcienski, W. E., Jr., 1976, Crossitic amphibole and its possible tectonic significance in the Richmond area, southeastern Quebec, Canada: Canadian Journal of Earth Sciences, v. 13, p. 711–714.

Trzcienski, W. E., Jr., Rodgers, J., and Guidotti, C. V., 1992a, Alternative hypothesis for the Chain Lakes "massif," Maine and Quebec: American Journal of Science, v. 292, p. 508–532.

Trzcienski, W. E., Jr., Thompson, J. B., Jr., Rosenfeld, J. L., and Hepburn, J. C., 1992b, The chicken yard line/Whately fault debate: From Springerfield, Vermont to Whately, Massachusetts, *in* Robinson, P., and Brady, J. B., eds., Guidebook for field trips in the Connecticut Valley region of Massachusetts and adjacent states, Volume 2 (New England Intercollegiate Geological Conference, annual meeting, 84th): Amherst, University of Massachusetts, Department of Geology and Geography Contribution 66, p. 291–304.

Tucholke, B. E., 1986, Sediment thickness in the western North Atlantic Ocean, *in* Vogt, P. R., and Tucholke, B. E., eds., The western North Atlantic region: Boulder, Colorado, Geological Society of America, The Geology of North America, v. M, plate 6.

Tucker, R. D., and Robinson, P., 1990, Age and setting of the Bronson Hill magmatic arc: A re-evaluation based on U-Pb zircon ages in southern New England: Geological Society of America Bulletin, v. 102, p. 1404–1419.

Tucker, R. D., Krogh, T. E., Ross, R. J., Jr., and Williams, S. H., 1990, Time-scale calibration by high-precision U-Pb zircon dating of interstratified volcanic ashes in the Ordovician and Lower Silurian stratotypes of Britain: Earth and Planetary Science Letters, v. 100, p. 51–58.

Tull, J. F., and Groszos, M. S., 1990, Nested Paleozoic "successor" basins in the southern Appalachian Blue Ridge: Geology, v. 18, p. 1046–1049.

Tull, J. F., and Stow, S. H., 1980, Structural and stratigraphic settings of arc volcanism in the Talledega slate belt, *in* Frey, R. W., ed., Excursions in southeastern geology, Volume 2: Falls Church, Virginia, American Geologi-

cal Institute, p. 545–581.
Tull, J. F., Harris, A. G., Repetski, J. E., McKinney, F. K., Garrett, C. B., and Bearce, D. N., 1988, New paleontologic evidence constraining the age and paleotectonic setting of the Talladega slate belt, southern Appalachians: Geological Society of America Bulletin, v. 100, p. 1291–1299.
Unrug, R., and Unrug, S., 1990, Paleontological evidence of Paleozoic age for the Walden Creek Group, Ocoee Supergroup, Tennessee: Geology, v. 18, p. 1041–1045.
van der Pluijm, B. A., and van Staal, C. R., 1988, Characteristics and evolution of the Central Mobile Belt, Canadian Appalachians: Journal of Geology, v. 96, p. 535–547.
Van der Voo, R., 1983, A plate-tectonics model for the Paleozoic assembly of Pangea based on paleomagnetic data, *in* Hatcher, R. D., Jr., Williams, H., and Zietz, I., eds., Contributions to the tectonics and geophysics of mountain chains: Geological Society of America Memoir 158, p. 19–32.
Van der Voo, R., and Johnson, R.J.E., 1985, Paleomagnetism of the Dunn Point Formation (Nova Scotia)—High paleolatitude for the Avalon terrane in the Late Ordovician: Geophysical Research Letters, v. 12, p. 337–340.
van Staal, C. R., and Fyffe, L. R., 1991, Dunnage and Gander zones, New Brunswick: Canadian Appalachian region: New Brunswick Department of Natural Resources and Energy, Mineral Resources, Geoscience Report 91-2, 39 p.
van Staal, C. R., and Williams, H., 1991, Dunnage zone–Gander zone relationships in the Canadian Appalachians: Geological Society of America Abstracts with Programs, v. 23, no. 1, p. 143.
van Staal, C. R., Ravenhurst, C. E., Winchester, J. A., Roddick, J. C., and Langton, J. P., 1990, Post-Taconic blueschist suture in the northern Appalachians of northern New Brunswick, Canada: Geology, v. 18, p. 1073–1077.
Vauchez, A., 1987, Brevard fault zone, southern Appalachians: A medium-angle, dextral Alleghanian shear zone: Geology, v. 15, p. 669–672.
Vick, H. K., Channell, J.E.T., and Opdyke, N. D., 1987, Ordovician docking of the Carolina slate belt: Paleomagnetic data: Tectonics, v. 6, p. 573–583.
Vogt, P. R., 1973, Early events in the opening of the North Atlantic, *in* Tarling, D. H., and Runcorn, S. K., eds., Implications of continental drift to the earth sciences: Volume 2: London, Academic Press, p. 693–712.
—— , 1986, Plate kinematics during the last 20 m.y. and the problem of "present" motions, *in* Vogt, P. R., and Tucholke, B. E., eds., The western North Atlantic region: Boulder, Colorado, Geological Society of America, The Geology of North America, v. M, p. 405–425.
Volkert, R. A., and Drake, A. A., Jr., 1991, Tectonostratigraphic terranes beneath the New Jersey Coastal Plain: Geological Society of America Abstracts with Programs, v. 23, no. 1, p. 144.
Vyhnal, C. R., and McSween, H. Y., Jr., 1990, Constraints on Alleghanian vertical displacements in the southern Appalachian Piedmont, based on aluminum-in-hornblende barometry: Geology, v. 18, p. 938–941.
Walcott, C. D., 1891, Correlation papers—Cambrian: U.S. Geological Survey Bulletin 81, 447 p.
Weed, E.G.A., 1991, Mesozoic and Tertiary stratigraphy of Cape Cod and the nearby islands, *in* Hatch, N. L., Jr., ed., The bedrock geology of Massachusetts: U.S. Geological Survey Professional Paper 1366E-J, p. E46–E57.
Weems, R. E., 1985, Field guide to the metamorphic stratigraphy of western Hanover County, Virginia (Seventeenth Annual Virginia Geological Field Conference): Charlottesville, Virginia Division of Mineral Resources, p. 23–52.
Wehr, F., 1985, Stratigraphy of the Lynchburg Group and Swift Run Formation, Late Proterozoic (730–570 Ma), central Virginia: Southeastern Geology, v. 25, p. 225–239.
Wentworth, C. M., and Mergner-Keefer, M., 1983, Regenerate faults of small Cenozoic offset: Probable earthquake sources in the southeastern United States, *in* Gohn, G. S., ed., Studies related to the Charleston, South Carolina earthquake of 1886—Tectonics and seismicity: U.S. Geological Survey Professional Paper 1313, p. S1–S20.
Wernicke, B., and Tilke, P. G., 1989, Extensional tectonic framework of the Central Atlantic passive margin, *in* Tankard, A. J., and Balkwill, H. R., eds., Extensional tectonics and stratigraphy of the North Atlantic margins: American Association of Petroleum Geologists Memoir 46, p. 7–21.
West, D. P., Jr., and Lux, D. R., 1989, Evidence for Late Paleozoic–early Mesozoic extensional tectonism along the Norumbega fault zone, southwestern Maine: Geological Society of America Abstracts with Programs, v. 21, no. 6, p. A141.
West, D. P., Jr., Lux, D. R., and Hussey, A. M., II, 1988, $^{40}Ar/^{39}Ar$ hornblende ages from southwestern Maine: Evidence for late Paleozoic metamorphism: Maritime Sediments and Atlantic Geology, v. 24, p. 225–239.
West, D. P., Jr., Hubbard, M. S., and Lux, D. R., 1992a, Geochronological and structural evidence for differential movement along the Norumbega fault zone, south-central Maine: Geological Association of Canada Program with Abstracts, v. 17, p. A116.
West, D. P., Jr., Ludman, A., and Lux, D. R., 1992b, Silurian age for the Pocomoonshine Gabbro-Diorite, southeastern Maine and its regional implications: American Journal of Science, v. 292, p. 253–273.
White, W. S., and Billings, M. P., 1951, Geology of the Woodsville quadrangle, Vermont–New Hampshire: Geological Society of America Bulletin, v. 62, p. 647–696.
Williams, H., 1978, Tectonic lithofacies map of the Appalachian orogen: St. John's, Memorial University of Newfoundland, Map no. 1, scale 1:1,000,000.
—— , 1979, Appalachian orogen in Canada: Canadian Journal of Earth Sciences, v. 16, p. 792–807.
Williams, H., and Hatcher, R. D., Jr., 1983, Appalachian suspect terranes, *in* Hatcher, R. D., Jr., Williams, H., and Zietz, I., eds., Contributions to the tectonics and geophysics of mountain chains: Geological Society of America Memoir 158, p. 33–53.
Williams, H., and Hiscott, R. N., 1987, Definition of the Iapetus rift-drift transition in western Newfoundland: Geology, v. 15, p. 1044–1047.
Williams, H., and St. Julien, P., 1982, The Baie Verte–Brompton Line: Early Paleozoic continent ocean interface in the Canadian Appalachians, *in* St. Julien, P., and Béland, J., eds., Major structural zones and faults of the northern Appalachians: Geological Association of Canada Special Paper 24, p. 177–207.
Williams, H., Colman-Sadd, S. P., and Swinden, H. S., 1988, Tectonic-stratigraphic subdivisions of central Newfoundland, *in* Current research, Part B: Geological Survey of Canada Paper 88-1B, p. 91–98.
Wilson, J. T., 1966, Did the Atlantic close and then re-open?: Nature, v. 211, p. 676–681.
Wintsch, R. P., 1992, Contrasting P-T-t paths: Thermochronology applied to the identification of terranes and to the history of terrane assembly, southeastern New England, *in* Robinson, P., and Brady, J. B., eds., Guidebook for field trips in the Connecticut Valley region of Massachusetts and adjacent states, Volume 1 (New England Intercollegiate Geological Conference, annual meeting, 84th): Amherst, University of Massachusetts, Department of Geology and Geography Contribution 66, p. 48–66.
Wintsch, R. P., and Aleinikoff, J. N., 1987, U-Pb isotopic and geologic evidence for late Paleozoic anatexis, deformation, and accretion of the Late Proterozoic Avalon terrane, south-central Connecticut: American Journal of Science, v. 287, p. 107–126.
Wintsch, R. P., and Fout, J. S., 1982, Structure and petrology of the Willimantic dome and the Willimantic fault, eastern Connecticut, *in* Joeston, R., and Quarrier, S. S., eds., Guidebook for fieldtrips in Connecticut and south-central Massachusetts (New England Intercollegiate Geological Conference, annual meeting, 74th): Connecticut Geological and Natural History Survey Guidebook 5, p. 465–482.
Wintsch, R. P., Sutter, J. F., Kunk, M. J., Aleinikoff, J. N., and Dorais, M. J., 1992, Contrasting P-T-t paths: Thermochronologic evidence for a late Paleozoic assembly of the Avalon composite terrane in the New England Appalachians: Tectonics, v. 11, p. 672–289.
Wones, D. R., and Goldsmith, R., 1991, Intrusive rocks of eastern Massachusetts, *in* Hatch, N. L., Jr., ed., The bedrock geology of Massachusetts: U.S. Geolog-

ical Survey Professional Paper 1366-E-J, p. I1–I61.

Wones, D. R., and Sinha, A. K., 1988, A brief review of early Ordovician to Devonian plutonism in the North American Caledonides, *in* Harris, A. L., and Fettes, D. J., eds., The Caledonian-Appalachian orogen (Geological Society of London Special Publication 38): London, Blackwell Scientific Publications, p. 381–388.

Wones, D. R., and Thompson, W., 1979, The Norumbega fault zone: A major regional structure in the central eastern Maine: Geological Society of America Abstracts with Programs, v. 11, p. 60.

Woodrow, D. L., and Sevon, W. D., eds., 1985, The Catskill Delta: Geological Society of America Special Paper 201, 254 p.

Woodward, N. B., Gray, D. R., and Spear, D. B., 1986, Including strain data in balanced cross-sections: Journal of Structural Geology, v. 8, p. 313–324.

Woodworth, J. B., and Wigglesworth, E., 1934, Geography and geology of the region including Cape Cod, the Elizabeth Islands, Nantucket, Marthas Vineyard, No Mans Land, and Block Island: Cambridge, Massachusetts, Museum of Comparative Zoology Memoirs, v. LII, 322 p.

Wright, J. E., Sinha, A. K., and Glover, L., III, 1975, Age of zircons from the Petersburg Granite, Virginia: With comments on belts of plutons in the Piedmont: American Journal of Science, v. 275, p. 848–856.

Zartman, R. E., 1988, Three decades of geochronologic studies in the New England Appalachians: Geological Society of America Bulletin, v. 100, p. 1168–1180.

Zartman, R. E., and Gallego, M. D., 1979, *in* Marvin, R. F., and Dobson, S. W., Radiometric ages: Compilation B, U.S. Geological Survey: Isochron/West, no. 26, p. 18.

Zartman, R. E., and Hermes, O. D., 1987, Archean inheritance in zircon from late Paleozoic granites from the Avalon zone of southeastern New England: An African connection: Earth and Planetary Science Letters, v. 82, p. 305–315.

Zartman, R. E., and Naylor, R. S., 1984, Structural implications of some radiometric ages of igneous rocks in southeastern New England: Geological Society of America Bulletin, v. 95, p. 522–539.

Zartman, R. E., Hermes, O. D., and Pease, M. H., Jr., 1988, Zircon crystallization ages, and subsequent isotopic disturbance events, in gneissic rocks of eastern Connecticut and western Rhode Island: American Journal of Science, v. 288, p. 376–402.

Zen, E-an, 1983, Exotic terranes in the New England Appalachians—Limits, candidates, and ages: A speculative essay: Geological Society of America Memoir 158, p. 55–81.

—— , 1989, Tectonostratigraphic terranes in the Northern Appalachians: Their distribution, origin, and age; evidence for their existence (International Geological Congress, 28th, Field trip guidebook T359): Washington, D.C., American Geophysical Union, 69 p.

—— , 1991, Phanerozoic denudation history of the southern New England Appalachians deduced from pressure data: American Journal of Science, v. 291, p. 401–424.

Zen, E-an, and Palmer, A. R., 1981, Did Avalonia form the eastern shore of Iapetus Ocean?: Geological Society of America Abstracts with Programs, v. 13, p. 587.

Zen, E-an, Goldsmith, R., Ratcliffe, N. M., Robinson, P., and Stanley, R. S., compilers, 1983, Bedrock geologic map of Massachusetts: U.S. Geological Survey, 3 sheets, scale 1:250,000.

Zen, E-an, Stewart, D. B., and Fyffe, L. R., 1986, Paleozoic tectonostratigraphic terranes and their boundaries in the mainland Northern Appalachians: Geological Society of America Abstracts with Programs, v. 18, p. 800.

Zietz, I., and Gilbert, F. P., 1980, Aeromagnetic map of part of the southeastern United States: In color: U.S. Geological Survey Geophysical Investigations Map GP-936, scale 1:2,000,000.

Zietz, I., Calver, J. L., Johnson, S. S., and Kirby, J. R., 1977, Aeromagnetic map of Virginia: U.S. Geological Survey Geophysical Investigations Map GP-916, scale 1:1,000,000.

Zoback, M. L., and Zoback, M. D., 1980, State of stress in the conterminous United States: Journal of Geophysical Research, v. 85, p. 6113–6156.

—— , 1989, Tectonic stress field of the conterminous United States, *in* Pakiser, L. C., and Mooney, W. D., eds., Geophysical framework of the continental United States: Geological Society of America Memoir 172, p. 523–539.

Manuscript Accepted by the Society January 6, 1993

Printed in U.S.A.

DNAG Continent-Ocean Transect Volume
Phanerozoic Evolution of North American
Continent-Ocean Transitions
The Geological Society of America, 1994

Chapter 4

Crustal structure and evolution of the southeastern margin of North America and the Gulf of Mexico basin

Richard T. Buffler
Institute for Geophysics, The University of Texas at Austin, 8701 Mopac Boulevard, Austin, Texas 78759
William A. Thomas
Department of Geological Sciences, University of Kentucky, Lexington, Kentucky 40506

INTRODUCTION

The Gulf of Mexico basin on the southeastern margin of North America is one of the most extensively studied basins in the world. Most of this work, however, has been concentrated on the shallower parts of the thick sedimentary section around the periphery of the basin, primarily related to the search for oil and gas and other minerals. Although comparatively little is known about the details of the more deeply buried parts of the sedimentary section and the crust below, enough geological and geophysical data exist now to enable compilation of a reasonable framework for the entire basin and document the evolution of the crust in general terms. The crust in the region contains the record of three successive phases of plate motion:

1. Divergent motion. Late Precambrian to early Paleozoic rifted margin and early Paleozoic south-facing passive margin along the southeastern edge of the North American craton.
2. Convergent motion. Late Paleozoic diachronous Appalachian-Ouachita orogenic belt, including foreland thrust belts, interior belts, and accreted terranes.
3. Divergent motion. Early to middle Mesozoic rifted margin and middle Mesozoic to present passive margin centered around a small ocean basin (present Gulf of Mexico basin).

Mesozoic rifting introduced a new set of extensional faults, including a northeast-trending rift set and a northwest-trending transform set. Some faults used planes of weakness related to late Paleozoic compressional structures, and the Mesozoic transform system essentially duplicated an early Paleozoic transform system. Mesozoic extension resulted in a broad area of stretched or attenuated continental crust (transitional crust), as well as a narrow band of oceanic crust, forming the framework for the Gulf of Mexico basin (Fig. 1). This basin architecture formed during the early breakup of Pangea as Africa/South America separated from North America. Subsidence caused by thermal cooling of the crust and by sediment loading provided the space for deposition of the overlying thick Mesozoic and Cenozoic sedimentary section.

The overall objective of this chapter is to discuss the general evolution of the southeastern margin of North America and the Gulf of Mexico basin. This chapter begins with a discussion of the late Precambrian through Paleozoic crustal structure and tectonic history of the southeastern margin of North America, as interpreted from rocks around the northern periphery of the Gulf basin (plate motion phases 1 and 2). The Paleozoic framework sets the scene for a discussion of the early (Late Triassic through Early Cretaceous) tectonic evolution of the present basin. This latest phase (plate motion phase 3) produced the modern continent-ocean transition. The early evolution of the modern Gulf basin can be discussed in terms of four stages: (1) Late Triassic–Early Jurassic initial rifting, (2) Middle Jurassic attenuation and formation of transitional crust, (3) Late Jurassic generation of oceanic crust in the deep central Gulf, and (4) Late Jurassic through Early Cretaceous cooling and subsidence of the crust and build up of extensive carbonate platforms surrounding the deep basin. This early evolution was followed by the Late Cretaceous to present partial filling of the deep basin by mainly siliciclastic sediments.

LATE PRECAMBRIAN AND PALEOZOIC CRUSTAL STRUCTURE AND TECTONIC HISTORY OF THE SOUTHEASTERN MARGIN OF NORTH AMERICA

Summary of crustal structure

North American Precambrian continental crust that extends southward beneath the Gulf Coastal Plain reflects several episodes of igneous and metamorphic petrogenesis culminating with the Grenville (~1 Ga) event. Late Precambrian to early Paleozoic rifting established the southeastern margin of the North American continent and generated intracratonic faulting that was active

Buffler, R. T., and Thomas, W. A., 1994, Crustal structure and evolution of the southeastern margin of North America and the Gulf of Mexico basin, *in* Speed, R. C., ed., Phanerozoic Evolution of North American Continent-Ocean Transitions: Boulder, Colorado, Geological Society of America, DNAG Continent-Ocean Transect Volume.

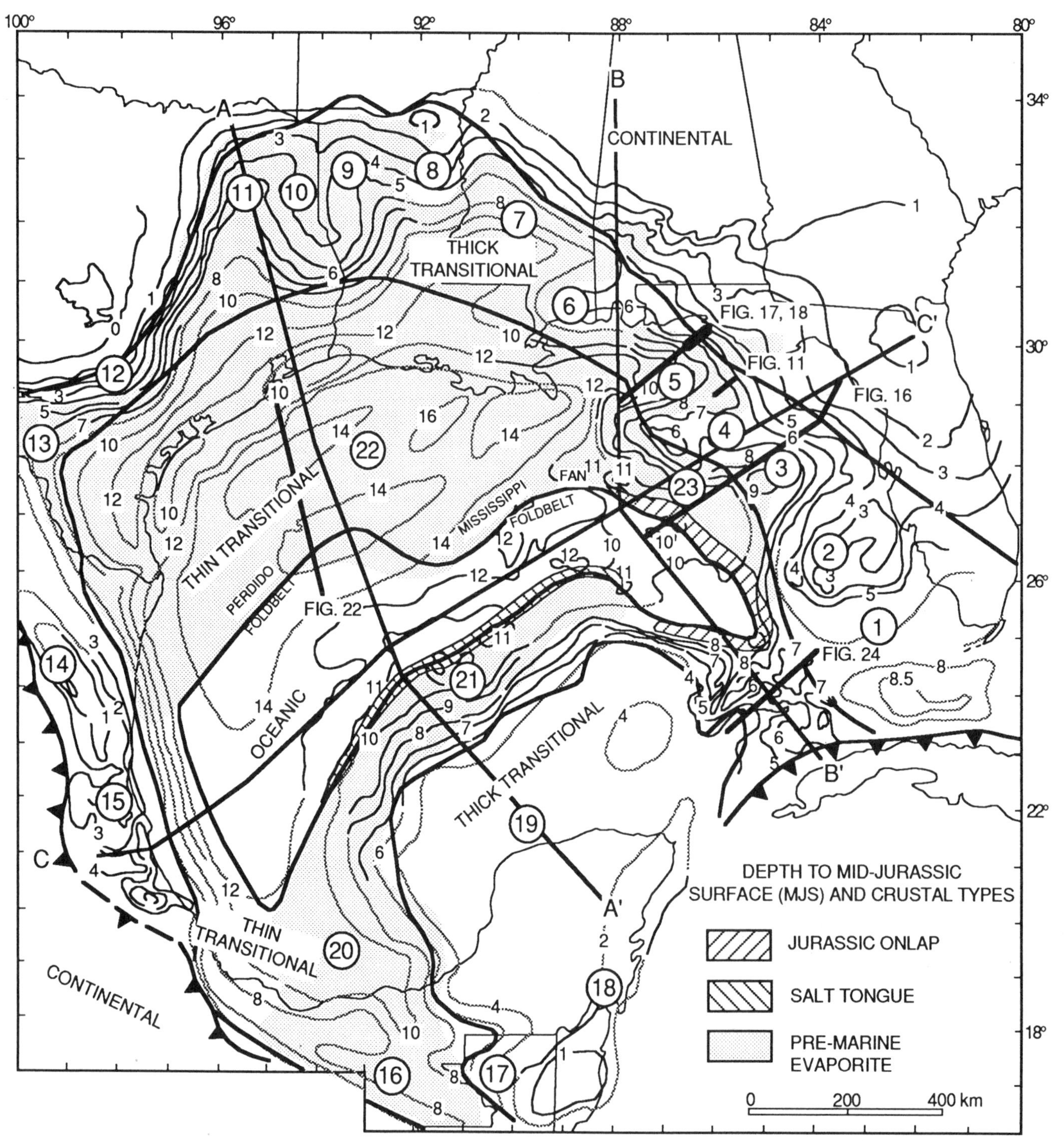

Figure 1. Map of the Gulf of Mexico basin showing generalized depth to mid-Jurassic surface (MJS) in kilometers; distribution of four crustal types—continental, thick transitional, thin transitional, and oceanic crust; and known distribution of mid-Jurassic pre-marine evaporites (Louann Salt and equivalent rocks). Heavy contours indicate areas where the MJS is well constrained by seismic reflection data and well control, whereas light contours indicate areas where the MJS is speculative and based on other geophysical data and extrapolation of trends from adjacent areas. Circled numbers refer to named highs, lows, basins, arches, etc., as follows: 1, South Florida basin; 2, Sarasota arch; 3, Tampa embayment; 4, Middle Ground arch–Southern platform; 5, Northeast Gulf basin–Apalachicola embayment; 6, Wiggins uplift; 7, Mississippi salt basin; 8, Monroe uplift; 9, North Louisiana salt basin; 10, Sabine uplift; 11, East Texas salt basin; 12, San Marcos arch; 13, Rio Grande embayment–Burgos basin; 14, Tamaulipas arch; 15, Tuxpan platform; 16, Macuspana basin; 17, Maya arch; 18, Quintana Roo arch; 19, Yucatan block; 20, Campeche salt basin; 21, Sigsbee salt basin; 22, North Gulf salt basin; and 23, West Florida (salt) basin. Locations of cross sections of Figure 9 and general locations of Figures 11, 16, 17, 18, 22, and 24 are shown.

until Late Cambrian time (Fig. 2; Thomas, 1991). An early to middle Paleozoic passive margin is indicated by an extensive carbonate-shelf succession on continental crust and an off-shelf deep-water mudstone and chert facies on the south.

The primary structural expression of late Paleozoic plate convergence is the sinuous Appalachian-Ouachita orogenic belt (Fig. 3). Structures exposed in the Appalachian Mountains extend southwestward beneath postorogenic Mesozoic-Cenozoic strata of the Gulf Coastal Plain in Alabama. Farther west, except for outcrops in the Ouachita Mountains in Arkansas and Oklahoma and in the Marathon region of southwestern Texas, the late Paleozoic orogenic belt is covered by strata of the Gulf Coastal Plain (Fig. 3; Thomas and others, 1989a). West of central Mississippi, off-shelf-facies rocks were incorporated into an accretionary prism that was thrust over the shelf edge onto shelf-facies rocks and obducted onto the edge of North American continental crust; the orogenic belt is an arc-continent collision orogen of sedimentary, metasedimentary, and volcanic rocks (Fig. 4, O–O′; Viele and Thomas, 1989). In contrast, east of central Mississippi, the thrust belt consists of imbricated shelf-facies rocks, which were tectonically replaced as the cover of North American continental crust by accreted terranes of uncertain affinity (Fig. 4, A–A′). A fragment of African crust (Suwannee terrane) was sutured to the southeastern margin of North American crust (Figs. 3; 4, A–A′) and was left attached to North America during the Mesozoic breakup of Pangea (Thomas and others, 1989b, 1989c). The Suwannee terrane, consisting of late Precambrian volcanic, metamorphic, and plutonic basement and a cover of early and middle Paleozoic sedimentary rocks, underlies the northeastern Gulf and the Gulf Coastal Plain in northern Florida, southern Georgia, and southeastern Alabama (Chowns and Williams, 1983).

North American Precambrian continental crust

Precambrian basement rocks in the south-central part of the North American craton are granite and rhyolite along with minor intrusions of basalt. The eastern part of this region belongs to the Eastern Granite-Rhyolite Province (~1.4 to 1.5 Ga) of Van Schmus and others (1987). These rocks are exposed on the Ozark dome and have been drilled in the subsurface of the Black Warrior and Arkoma basins and Nashville dome (Fig. 3; wells listed in Thomas, 1988a). Locally, in the Mississippi Valley graben, the Precambrian basement is granitic and dioritic gneiss, indicating that the granite-rhyolite province is not totally continuous (Denison, 1984). A possible eastern limit of the province is suggested by other lithologic characteristics, such as low-grade metamorphism of granites in Mississippi (Harrelson and Bicker, 1979), granodiorite composition in eastern Alabama and Georgia, and metamorphic fabric of basement rocks in northwestern Georgia (Neathery and Copeland, 1983). In the western part of the region, younger rocks of the Western Granite-Rhyolite Province (1.34 to 1.4 Ga) extend southward to include granite and gneiss exposed in fault-bounded uplifts in the Arbuckle Mountains (Fig. 3; Ham and others, 1964; Bickford and Lewis, 1979; Denison and others, 1984; Van Schmus and others, 1987; Denison, in Johnson and others, 1988).

South of the Wichita Mountains (Fig. 3) entirely in the subsurface, the Tillman Metasedimentary Group has a probable age of 1.0 to 1.2 Ga (Muehlberger and others, 1967; Coffman and others, 1986). Seismic reflection profiles show a thick succession (7 to 10 km) of layered reflectors that represent the Tillman Group south of the Wichita Mountains, but the layered reflectors terminate abruptly toward the north, suggesting a Precambrian fault between the Tillman Group and rocks of the Western Granite-Rhyolite Province on the north (Brewer and others, 1981, 1983). The southern limit of the Tillman Metasedimentary Group is uncertain, but rocks of the Western Granite-Rhyolite Province are identified farther southwest in western Texas (Van Schmus and others, 1987).

Rocks of the granite-rhyolite provinces are bounded on the east and southeast by the Grenville and Llano Provinces (~1.0 to 1.2 Ga) of metamorphic and plutonic rocks (Van Schmus and others, 1987), but location and structure of the boundary are somewhat uncertain. The Grenville Province is traced from the northern Appalachians to the southwest along the southeastern fringe of North American Precambrian continental crust, and the Llano Province extends southwestward across Texas and possibly into Mexico (Van Schmus and others, 1987; Mosher, 1992). The Grenville-Llano rocks represent the youngest episode of addition of continental crust to southeastern North America during the Precambrian. Later rifting and opening of the Iapetus Ocean produced the North American continental margin along which the late Paleozoic Appalachian-Ouachita orogenic belt subsequently formed.

Late Precambrian–early Paleozoic rifted continental margin

The southern edge of Precambrian North American continental crust generally has an orthogonally zigzag trace that outlines the Alabama promontory, Ouachita embayment, Texas promontory, and Marathon embayment (Fig. 2). The shape of the margin is interpreted to be the result of northeast-striking rift segments offset by northwest-striking transform faults that formed the framework of a south-facing continental margin (Cebull and others, 1976; Thomas, 1976, 1977, 1991). Structural components of the rifted margin are diachronous through the latest Precambrian and earliest Paleozoic (Thomas, 1991).

The location of the southeastern margin of continental crust on the Alabama promontory is interpreted from structural reconstructions and geophysical data. A generally smooth top of Precambrian basement is shown by COCORP seismic reflection profiles (Georgia lines 15, 23, and 24) to dip beneath the foreland thrust belt southeastward at a low angle to a location beneath the present northwest side of the Pine Mountain internal basement massif (Figs. 4, A–A′; 5). Along the southeast side of the Pine Mountain massif, near the northern limit of the Gulf Coastal Plain,

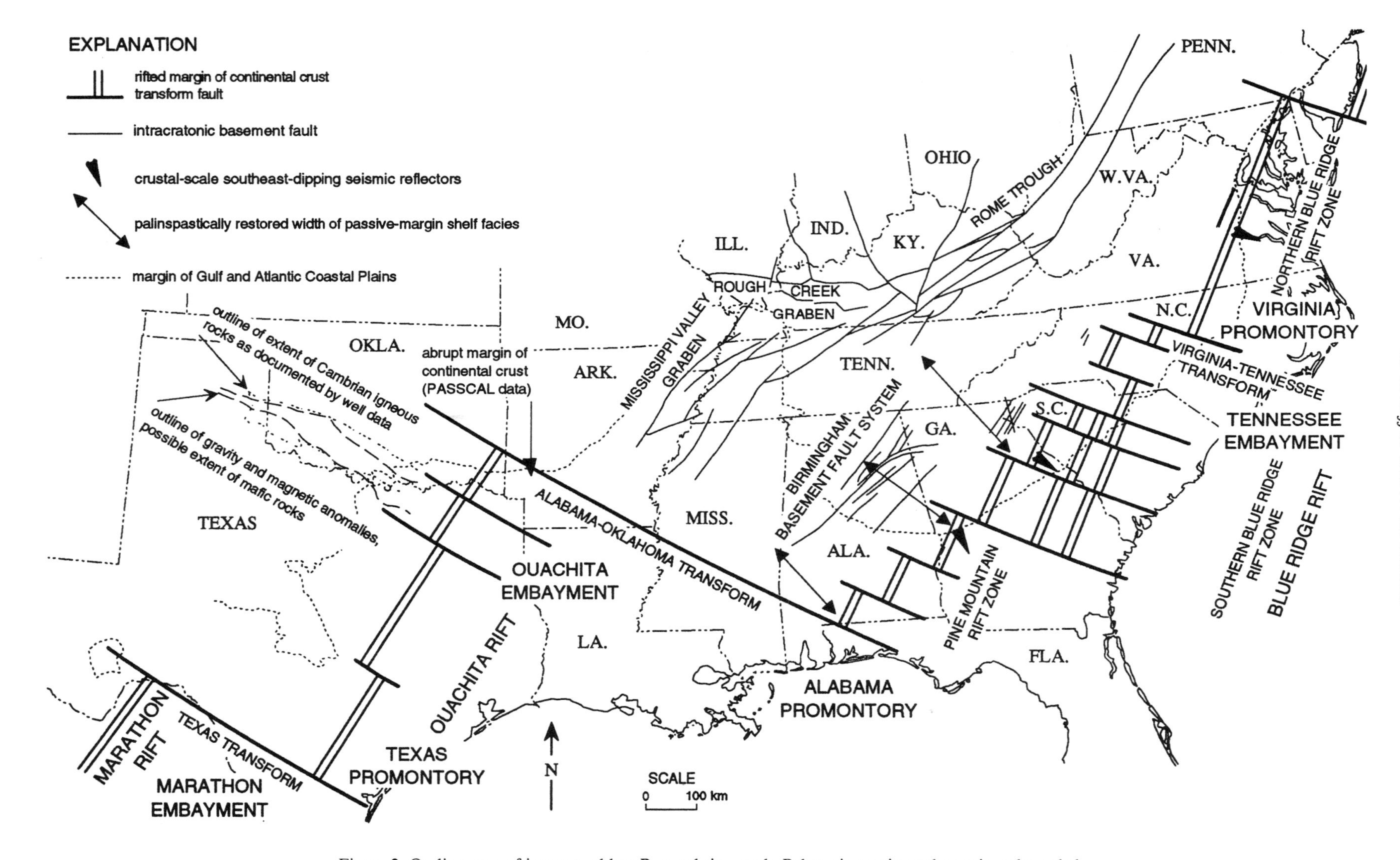

Figure 2. Outline map of interpreted late Precambrian–early Paleozoic continental margin as bounded by rift segments and transform faults (from Thomas, 1991). Map includes locations of observations that provide control for the reconstruction of the continental margin and intracratonic fault systems.

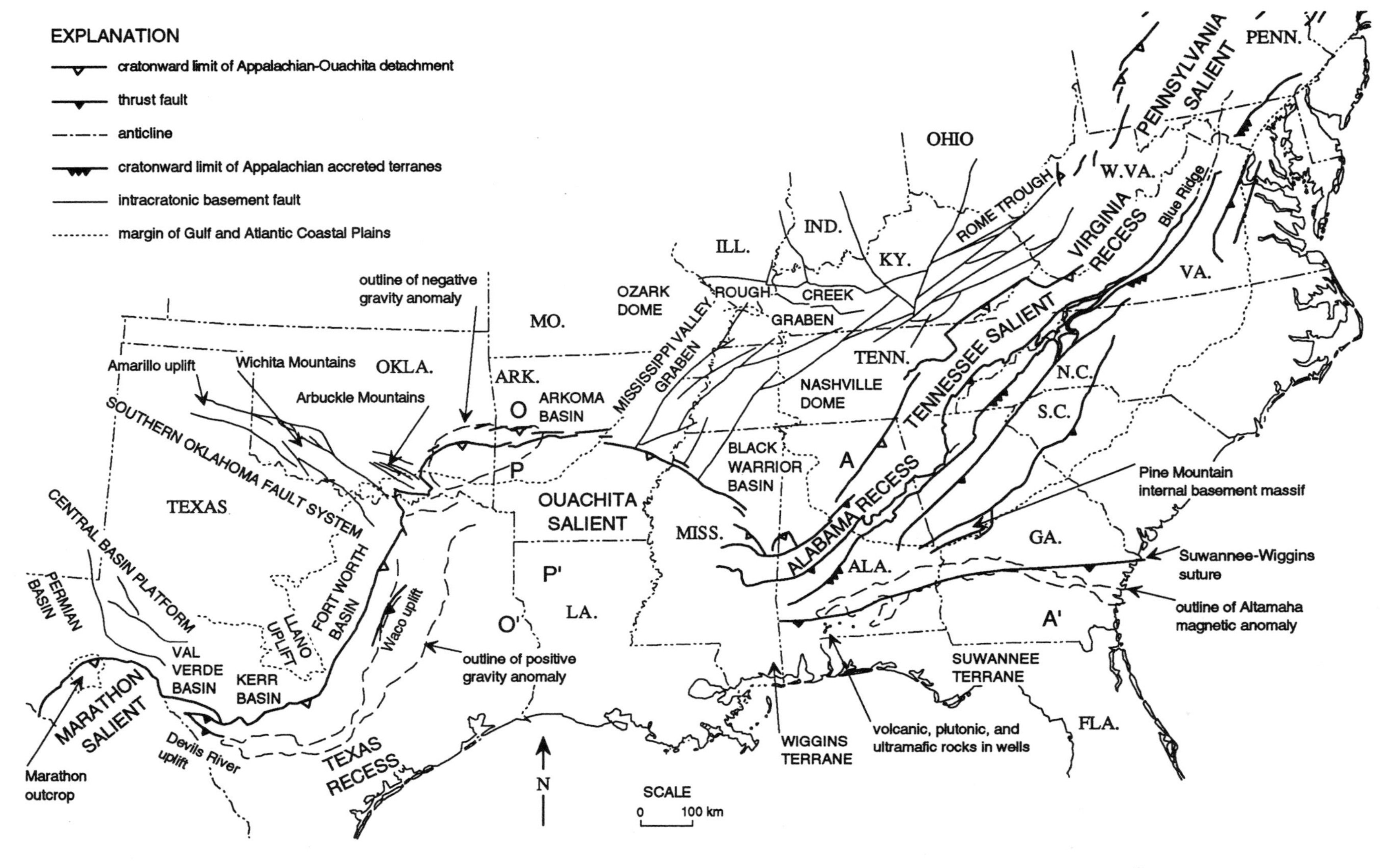

Figure 3. Outline map of late Paleozoic Appalachian-Ouachita orogenic belt, late Paleozoic foreland basins, and early Paleozoic intracratonic fault systems (from Thomas, 1991). Locations of rift-related rocks are shown in present structural position. End points of cross sections of Figure 4 are indicated by letters.

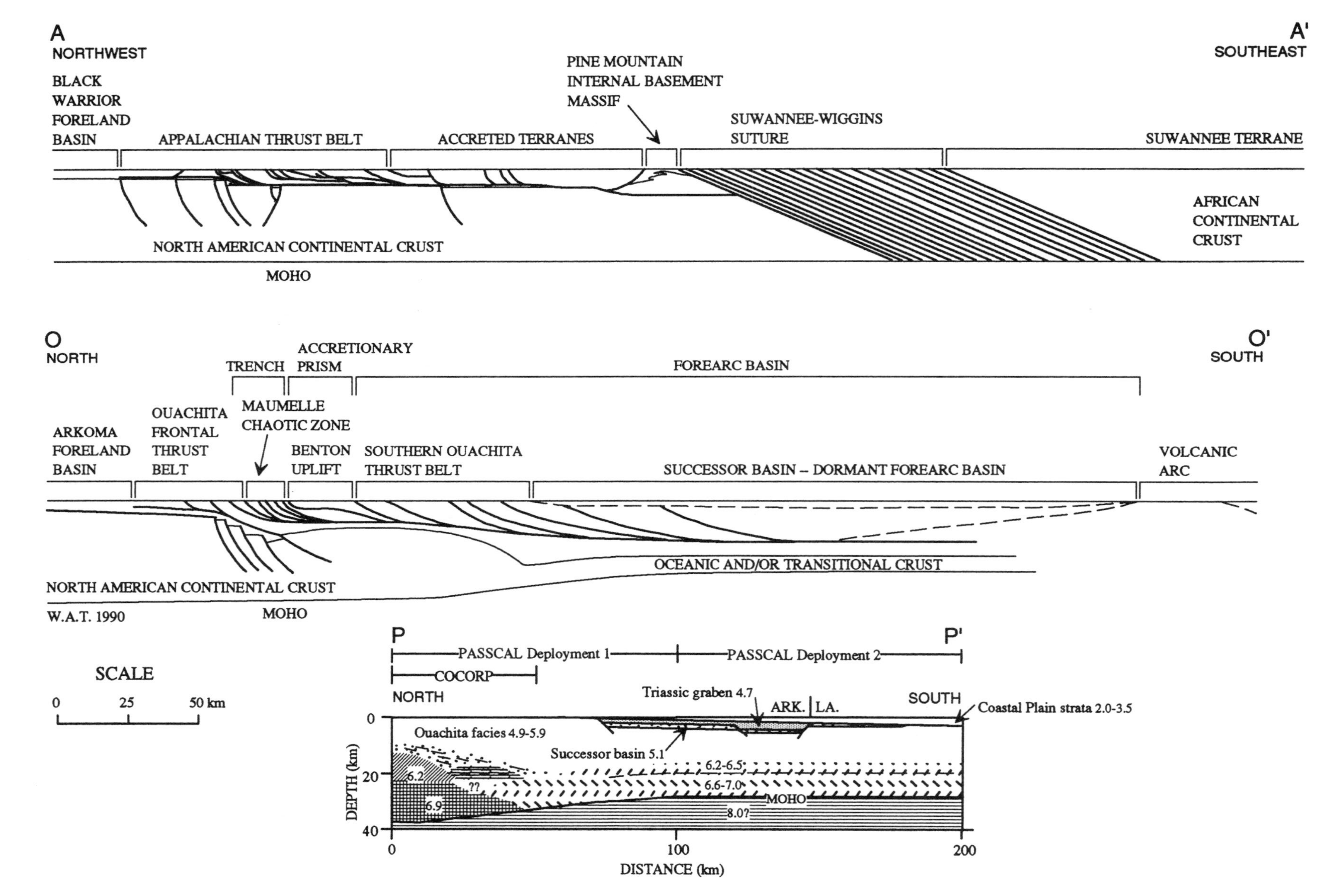

Figure 4. Structural cross section of Appalachian orogen in Alabama, Georgia, and Florida (A–A′) adapted in part from Neathery and Thomas (1983) and from geophysical data presented in Figure 5; and structural cross section of Ouachita orogen in Arkansas, Louisiana, and Texas (O–O′) adapted in part from Viele (1979, 1989) and Lillie and others (1983) and from PASSCAL velocity model (P–P′, from Keller and others, 1989a), which was derived from wide-angle reflection/refraction data. End points of cross sections are shown in Figure 3. Datum of cross sections A–A′ and O–O′ is eroded top of pre-Mesozoic rocks (present erosion surface and sub-Mesozoic unconformity). Numbers on P–P′ indicate average P-wave velocities in km/s.

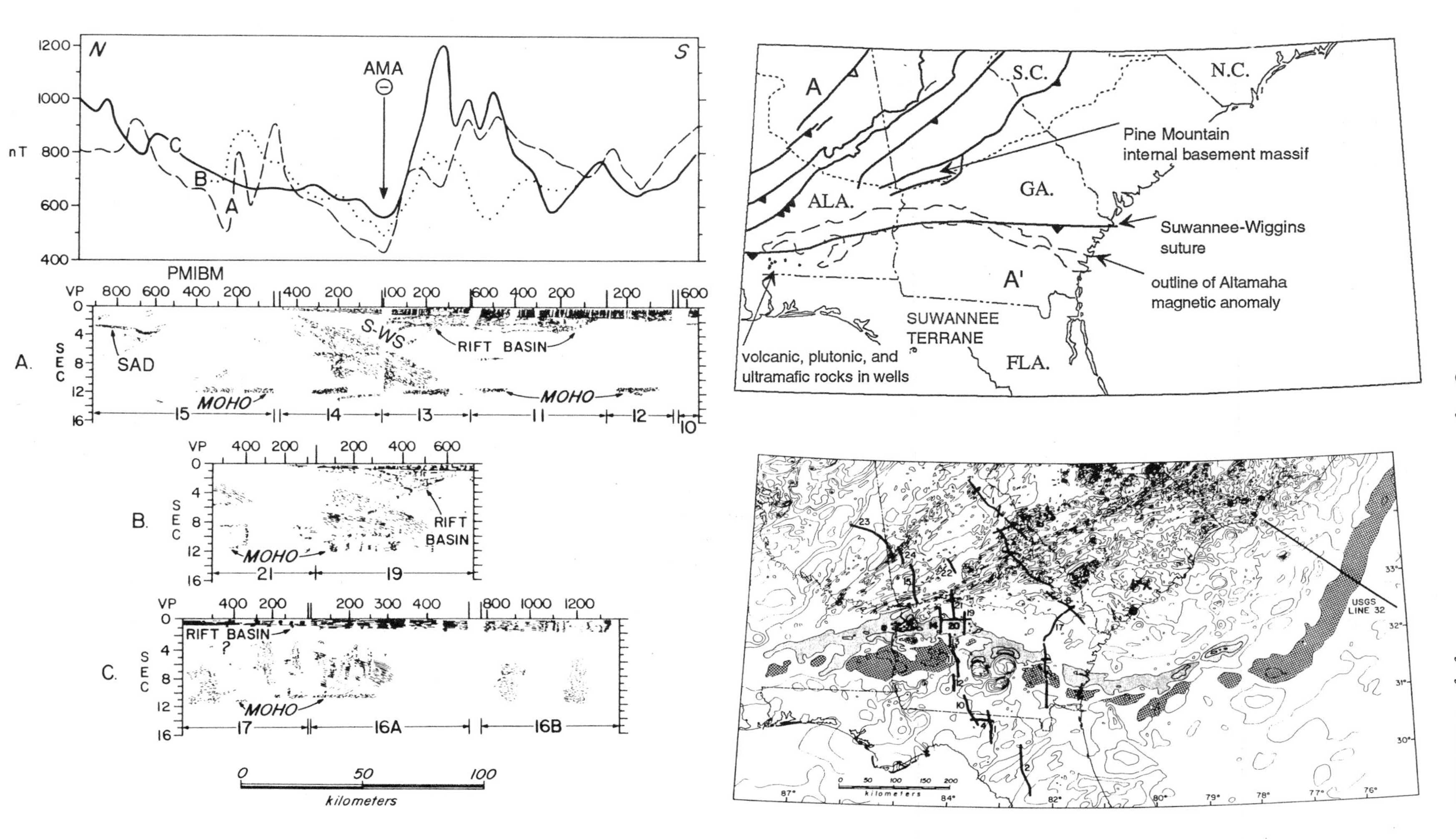

Figure 5. Interpretive line drawings of three COCORP seismic reflection profiles (A, B, C) and three magnetic intensity profiles (from McBride and Nelson, 1988). Locations of seismic profiles shown on magnetic map (Zietz, 1982); excerpt from Figure 3 shown at same scale for comparison. Abbreviations: AMA, Altamaha magnetic anomaly; PMIBM, Pine Mountain internal basement massif; SAD, southern Appalachian décollement; S-WS, Suwannee-Wiggins suture.

continental crustal basement is truncated by reflectors that dip southeastward to the base of the crust, marking the North American continental margin at a boundary with later Paleozoic accreted terranes (Fig. 5; Nelson and others, 1985; Hooper and Hatcher, 1988). In detailed palinspastic reconstructions of balanced structural cross sections, Paleozoic shelf-facies strata now in the Appalachian thrust belt in Alabama extend at least as far southeast as the present position of the Pine Mountain internal basement massif (Figs. 2, 3; Thomas, 1985a; Ferrill, 1989). Both the extent of palinspastically reconstructed shelf strata and the seismic reflection data indicate that the margin of North American Precambrian crust is at least as far southeast as the present position of the Pine Mountain internal basement massif.

The linear Altamaha magnetic anomaly beneath the Gulf Coastal Plain is interpreted to be the signature of the late Paleozoic suture between North American crust and the Suwannee terrane on the south (Figs. 3, 5; Higgins and Zietz, 1983; Horton and others, 1984; Nelson and others, 1985; Tauvers and Muehlberger, 1987; Hooper and Hatcher, 1988; McBride and Nelson, 1988). The magnetic anomaly extends from the southeast side of the Pine Mountain massif to southwestern Alabama, where a cluster of deep wells penetrated volcanic, plutonic, and ultramafic rocks, suggesting an obducted arc and subduction complex, probably near the margin of North American Precambrian continental crust (Figs. 2, 3; Neathery and Thomas, 1975; Thomas and others, 1989b, 1989c). Palinspastic restoration of balanced structural cross sections (Thomas, 1989a) places shelf-facies strata now in the trailing part of the subsurface Appalachian thrust belt at least as far south as the trace of the Altamaha magnetic anomaly in southwestern Alabama, further suggesting the original extent of the continental shelf and the approximate location of the continental margin (Figs. 2, 3).

The southeastern margin of the Alabama promontory is interpreted to be the southwestward continuation of a late Precambrian rifted margin that is well documented in outcrops along the Blue Ridge from Georgia to Pennsylvania (Blue Ridge rift, Figs. 2, 6A; Hatcher, 1972, 1978; Rankin, 1975, 1976; Thomas, 1976, 1977, 1991; Wehr and Glover, 1985; Rast and Kohles, 1986; Schwab, 1986; Simpson and Eriksson, 1989). Inboard (northwest) of the northeast-striking rifted margin, two basement fault systems (Birmingham and Mississippi Valley–Rough Creek–Rome, Fig. 2) strike northeasterly (Thomas, 1985a, 1988a; Ferrill, 1989) and, like the structures of the rifted margin, indicate northwest-southeast extension. Although kinematically consistent, the faults reflect different times of extension. Along the Blue Ridge, post-rift strata within the Unicoi Formation of the Lower Cambrian Chilhowee Group overstep rift boundaries, overlying both syn-rift sedimentary and volcanic rocks and Precambrian crystalline basement rocks, and indicating initiation of a passive margin at the beginning of the Cambrian Period (Fig. 6B; Simpson and Eriksson, 1989; Thomas, 1991). In contrast, downthrown blocks of the inboard fault systems have a clastic sedimentary fill >1 km thick including beds as young as Early Late Cambrian (Woodward, 1961; Webb, 1980; Schwalb, 1982a, 1982b; Howe, 1985; Thomas, 1985b, 1988a, 1991; Weaverling, 1987; Collinson and others, 1988; Ferrill, 1989), and the boundary faults are overstepped by Upper Cambrian carbonate rocks. The difference in age of fault movement documents a westward shift of the locus of extension (Thomas, 1991).

The best documentation for the location of the southern margin of Precambrian continental crust is beneath southward-dipping thrust sheets in the southern part of the Ouachita Mountains of Arkansas near the northern limit of the Gulf Coastal Plain. Wide-angle reflection/refraction seismic data (PASSCAL), interpreted via velocity models, indicate that continental crust thins southward within ~25 km to thin transitional or oceanic crust beneath a thick (>10 km) cover of sedimentary rocks (Figs. 2; 3; 4, P–P′; Keller and others, 1989a). That location of the margin is consistent with COCORP reflection data and with gravity models (Nelson and others, 1982; Lillie and others, 1983; Lillie, 1985; Kruger and Keller, 1986). The narrow zone of transition from continental crust to oceanic crust is typical of a transform fault (Keen, 1982; Scrutton, 1982; Keen and Haworth, 1985), and it contrasts with the broader zone of attenuated continental crust that characterizes rifted margins.

A boundary between regions of contrasting magnetic signatures (Zietz, 1982; Hinze and Braile, 1988) trends northwest-southeast from the seismically defined edge of continental crust beneath the Ouachita thrust belt, indicating the trace of a northwest-striking transform fault (Alabama-Oklahoma transform, Fig. 2). The Alabama-Oklahoma transform is interpreted to connect northeast-striking rift segments, outlining the Alabama promontory and Ouachita embayment (Fig. 2; Thomas, 1976, 1977, 1991).

The southeast side of the Texas promontory beneath the northwestern part of the Gulf Coastal Plain is a northeast-striking segment of the continental margin (Ouachita rift, Fig. 2). The location and trend of the edge of continental crust in eastern Texas are based on models of a northeast-trending linear gravity high (Figs. 2, 3; Kruger and Keller, 1986).

The northwest-striking Southern Oklahoma fault system extends into the continent and is approximately perpendicular to the southeastern margin of the Texas promontory (Figs. 2, 3). The Southern Oklahoma fault system is expressed by prominent gravity and magnetic anomalies and is marked by a narrow zone of igneous rocks, including layered gabbro, basalt-spilite, granite, and rhyolite, which range in age from 577 to 525 ± 25 Ma (Ham and others, 1964; Bowring and Hoppe, 1982; Gilbert, 1983; Lambert and Unruh, 1986). Magnetic signatures (Zietz, 1982) of the Cambrian igneous rocks along the fault system terminate abruptly southeastward, further defining the edge of continental crust (Keller and others, 1989b; Viele and Thomas, 1989).

At the interpreted intersection between the Southern Oklahoma fault system and the Ouachita rift, the northeast-trending gravity high bends abruptly to the east-northeast, crossing the corner of the Ouachita embayment on the outboard side of a deep gravity low, and extending to the margin of continental crust as indicated by the PASSCAL seismic data (Figs. 2, 3). The gravity low suggests a deep sedimentary basin formed during rifting (Kruger and Keller, 1986). The Ouachita rift is not notably offset, although it may bend, at the intersection with the Southern Oklahoma fault system.

The Southern Oklahoma fault system is parallel but not aligned with the Alabama-Oklahoma transform fault (Fig. 2). The orientation of the fault system suggests a transform (or continental transfer) fault that propagated into continental crust. Possibly, the fault system followed a preexisting crustal boundary between the Tillman Metasedimentary Group and the older Precambrian crystalline rocks of the Western Granite-Rhyolite Province on the north (Thomas, 1991).

A northwest-trending segment of the continental margin along the southwestern side of the Texas promontory is defined primarily by gravity models (Keller and others, 1985). The northwest-trending margin of the Texas promontory is interpreted as a transform fault (Texas transform, Fig. 2) that extends beneath Mesozoic strata of the northern limb of the Rio Grande embayment (13, Fig. 1) of the Gulf Coastal Plain toward exposures of Paleozoic rocks in the Marathon thrust belt (Fig. 3). By analogy with the subparallel strikes of the late Paleozoic thrust belt and the early Paleozoic continental margin elsewhere, the continental margin is interpreted to strike southwestward beneath the Marathon thrust belt as a rift segment that intersects the Texas transform fault to frame the Marathon embayment (Fig. 2). Metasedimentary and metavolcanic rocks on the Devils River uplift (a basement massif along the northwest-striking segment of the thrust belt, Fig. 3) are ~850 m thick and include metarhyolites that have ages of 699 ± 26 Ma (Rb-Sr) (Nicholas and Rozendal, 1975; Denison and others, 1977). These possibly rift-related rocks are underlain by more massive metaigneous-metasedimentary basement rocks, which have ages of 1,246 ± 270 to 1,121 ± 244 Ma (Rb-Sr) (Nicholas and Rozendal, 1975; Nicholas and Waddell, 1989), consistent with ages of crustal basement rocks of the Llano Province.

The rifted continental margin westward from the Alabama promontory is deep in the subsurface, and excepting the metasedimentary-metavolcanic rocks on the Devils River uplift, no rift-related rocks have been sampled from either transform or rift segments of the continental margin. Ages of the probable rift-related rocks on the Devils River uplift suggest that the Texas transform may be synkinematic with the Blue Ridge rift in the latest Precambrian (Fig. 6A). Later movement continuing to Early Late Cambrian along the Ouachita rift and Alabama-Oklahoma transform is indicated by the ages of igneous rocks along the Southern Oklahoma fault system (transform) and by the ages of syn-rift sedimentary rocks in the Birmingham and Mississippi Valley–Rough Creek–Rome intracratonic graben systems that reflect extension parallel with the transform direction (Fig. 6B). In contrast to the Early Cambrian age of the base of the transgressive shelf facies of the Sauk sequence adjacent to the Blue Ridge, no shelf-facies strata older than Late Cambrian (or late Middle Cambrian; A. R. Palmer, personal communication, 1989) have been identified in the Ouachita foreland. Syn-rift rocks of the Southern Oklahoma fault system and the Birmingham and Mississippi Valley–Rough Creek–Rome intracratonic graben systems are overlapped by Upper Cambrian strata.

The differences in ages of rift-related rocks and of post-rift sedimentary overstep indicate that the Ouachita margin, as well as the intracratonic fault systems, is younger than the Appalachian margin. The difference in age of rifting suggests a shift in the spreading center at about the beginning of the Cambrian Period from the southwestern part of the Blue Ridge rift to the Ouachita rift (Fig. 6B; Thomas, 1991). The spreading-center shift and initiation of Ouachita rifting were accompanied by initiation of the Alabama-Oklahoma transform fault and the Southern Oklahoma fault system (Figs. 6B, 7A). Most of the extension along the Ouachita rift was transformed along the Alabama-Oklahoma transform fault to the Mid-Iapetus Ridge outboard from the Blue Ridge passive margin (Fig. 6C). A small component of extension propagated northeastward across the Alabama-Oklahoma transform fault into the Mississippi Valley–Rough Creek–Rome and Birmingham intracratonic graben systems. By early in Late Cambrian time, the northeast end of the Ouachita ridge had moved beyond the corner of continental crust on the Alabama promontory, and a passive margin had evolved along the entire rift and transform margin (Fig. 6D).

Early to middle Paleozoic passive margin

The lower part of the Paleozoic sedimentary cover above Precambrian crystalline basement rocks on the southeastern part of the North American craton (along the northwest side of the Blue Ridge and southwestward along the Appalachian thrust belt into Alabama) consists of a basal clastic unit and an overlying thick, regionally extensive, transgressive carbonate unit that represent the Cambrian–Lower Ordovician Sauk sequence of the North American craton (Sloss, 1963, 1988). The basal sandstone (Chilhowee Group) is of Early Cambrian age on the southeast (palinspastically along the Blue Ridge rift and the southeastern part of the Alabama promontory) and is progressively younger toward the northwest onto the craton. The widely extensive transgressive sequence laps over the locally thick graben-fill successions of the Birmingham and Mississippi Valley–Rough Creek–Rome basement fault systems. In the interior of the North American craton, the basal sandstone of the Sauk sequence is of latest Cambrian age (Sloss, 1988).

The passive-margin carbonate facies of the Sauk sequence extends throughout the Appalachian-Ouachita foreland and widely on the North American craton, and a continental-margin shelf edge can be documented or inferred in several places. Shelf-edge facies as old as Early Cambrian have been recognized adjacent to the Blue Ridge (Reinhardt, 1974; Pfeil and Read, 1980), marking the transition to deeper water, off-shelf facies. Seismic reflection profiles document that the passive-margin carbonate facies extends southward beneath the Ouachita thrust belt; however, contemporaneous rocks in the thrust belt are a deep-water off-shelf facies (Viele, 1973; Thomas, 1976; Viele and Thomas,

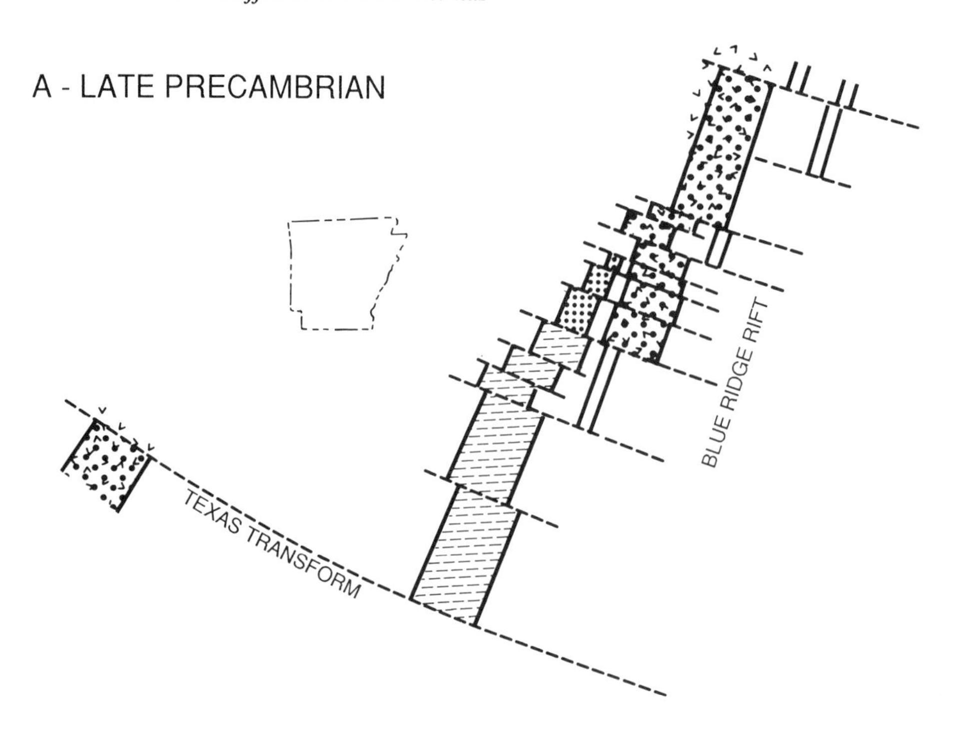

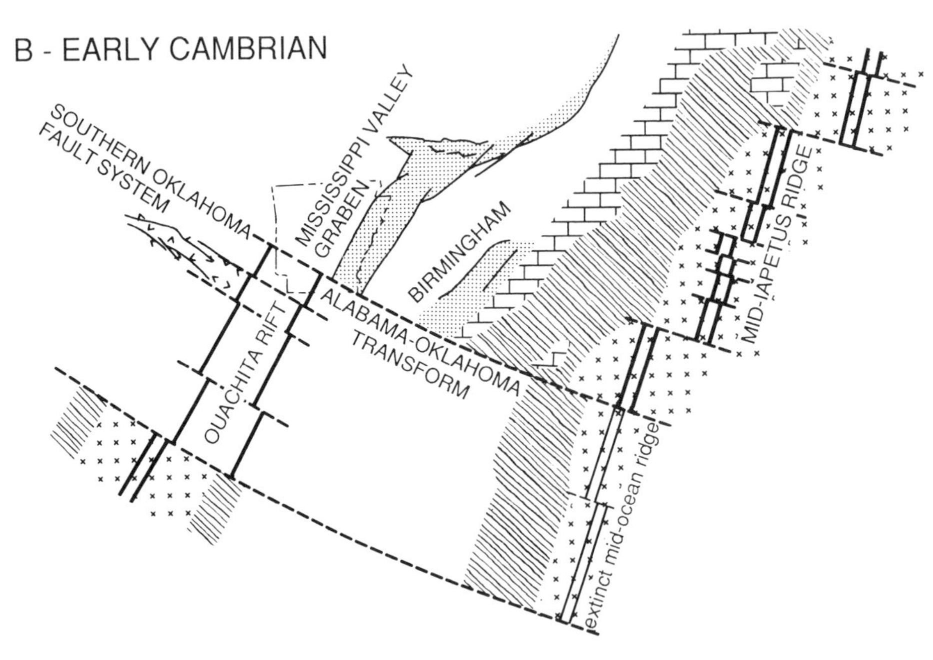

EXPLANATION

- active rift
- transform fault
- rift-fill coarse clastic sedimentary rocks
- rift volcanic and plutonic rocks
- rift-fill sedimentary and volcanic rocks
- inferred thin and/or fine-grained sedimentary rift-fill rocks
- graben-fill sedimentary rocks, intracratonic fault system
- passive-margin shelf facies
- passive-margin off-shelf facies (rifted-margin prism)
- oceanic crust

Figure 6. Sequential diagrammatic maps illustrating interpretation of history of the late Precambrian–early Paleozoic Appalachian-Ouachita rifted margin of southern North America (from Thomas, 1991). Outline of state of Arkansas on each map for consistent location. A, late Precambrian, ~580 Ma; B, Early Cambrian, ~565 Ma; C, Middle Cambrian, ~540 Ma; D, Late Cambrian, ~515 Ma.

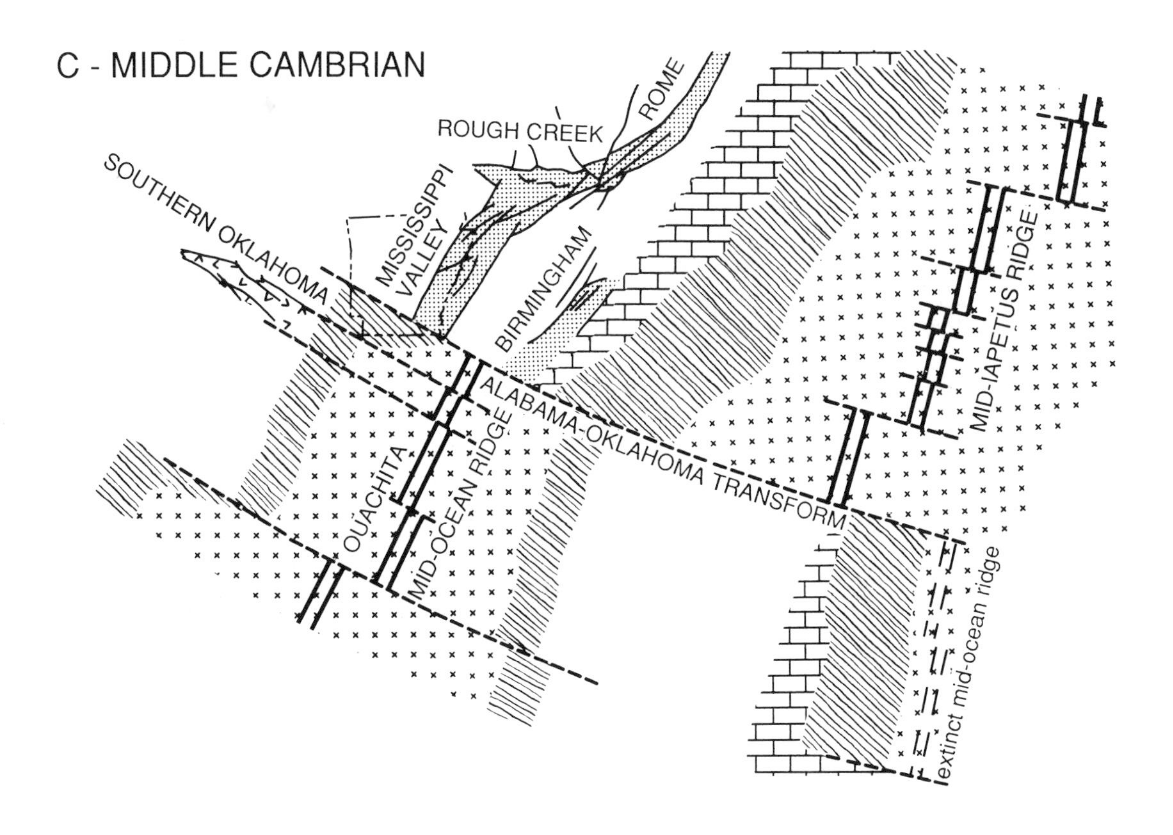
C - MIDDLE CAMBRIAN
ROME
ROUGH CREEK
SOUTHERN OKLAHOMA
MISSISSIPPI VALLEY
BIRMINGHAM
MID-IAPETUS RIDGE
ALABAMA-OKLAHOMA TRANSFORM
OUACHITA MID-OCEAN RIDGE
extinct mid-ocean ridge

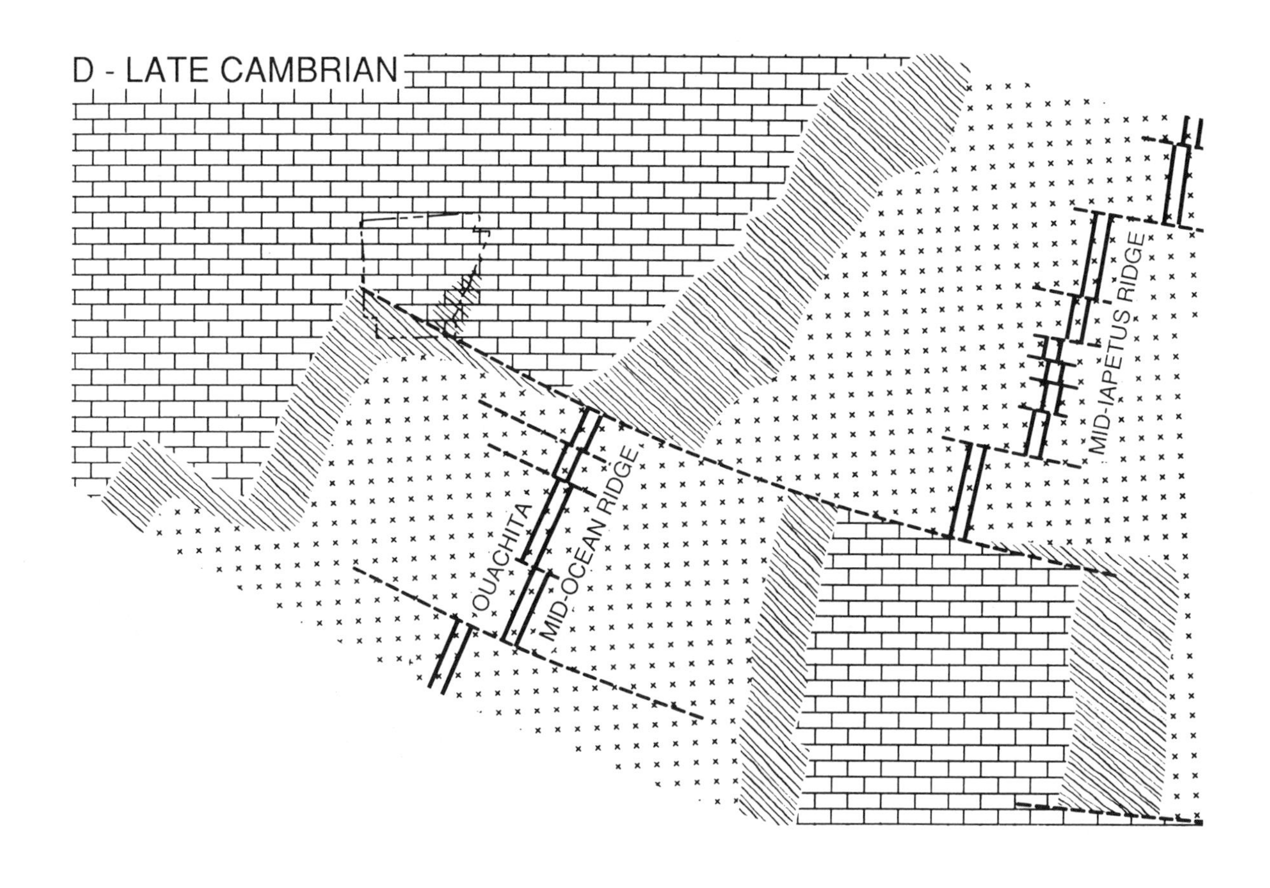
D - LATE CAMBRIAN
MID-IAPETUS RIDGE
OUACHITA MID-OCEAN RIDGE

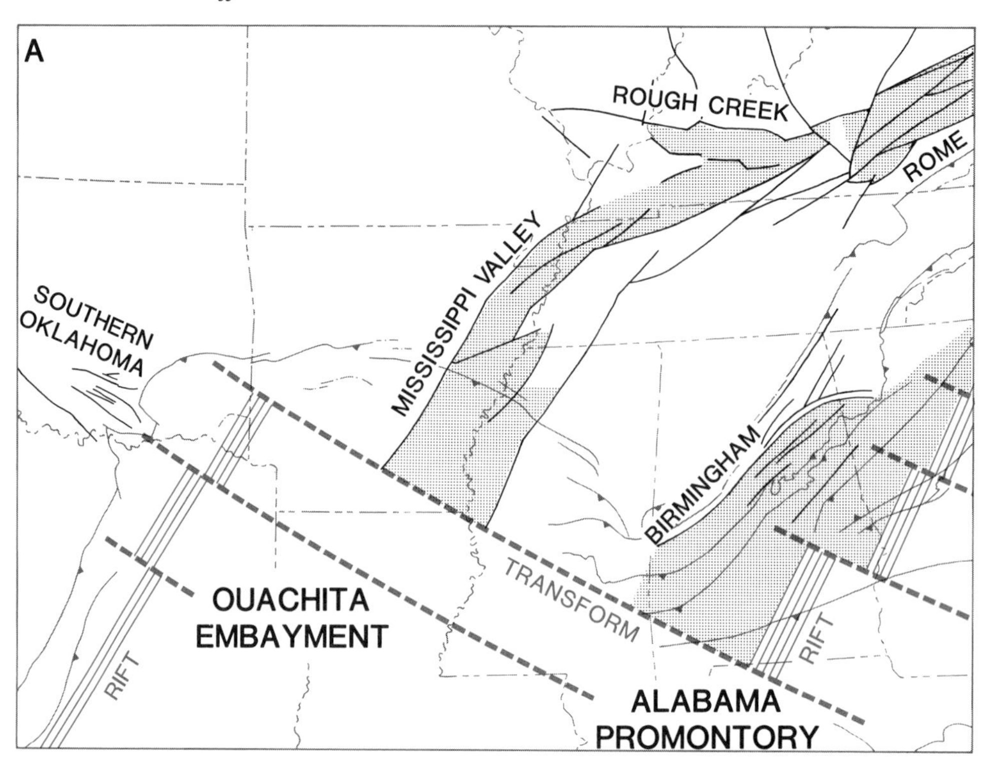

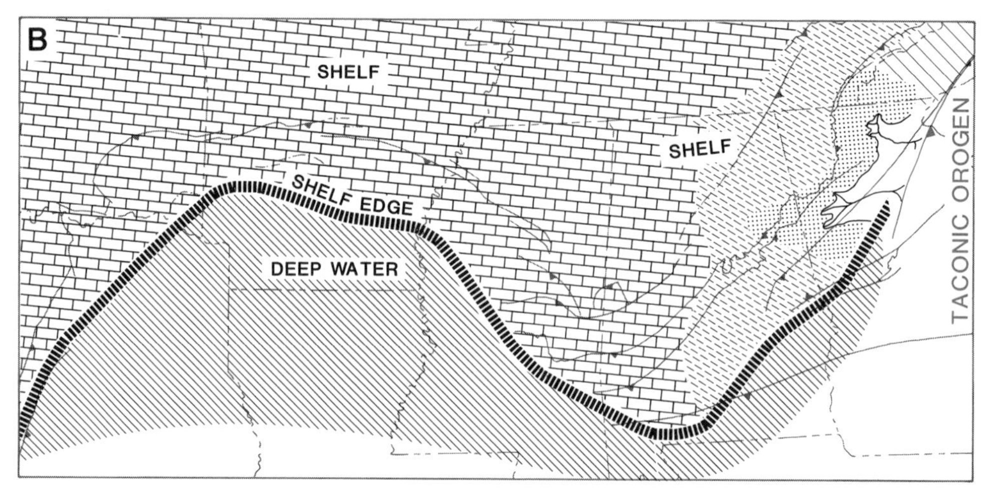

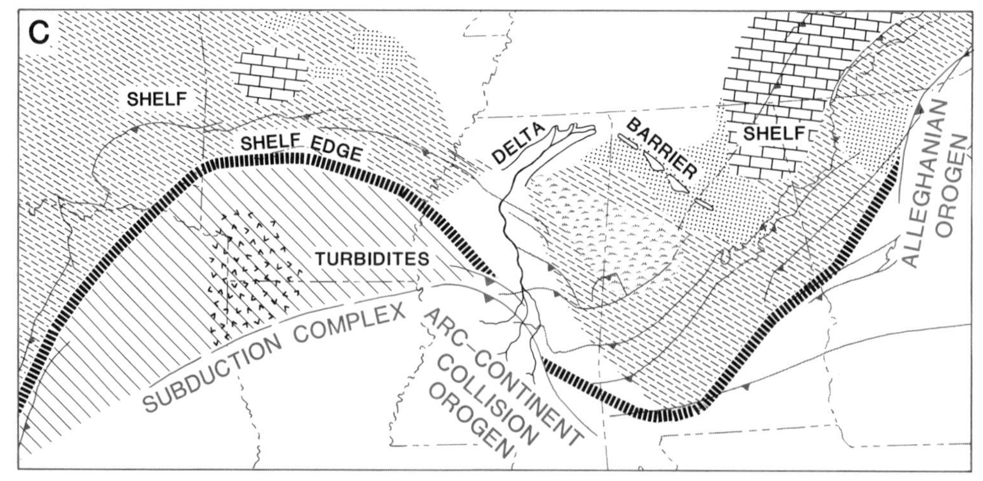

Figure 7. Paleogeographic reconstructions of phases in the tectonic evolution of the Appalachian-Ouachita orogen (from Thomas, 1989b). Base map (gray) shows state boundaries and present structural outlines for location reference. A, Late Precambrian and Early to Middle Cambrian: rifted continental margin; graben-filling sediments on continental crust. B, Middle to Late Ordovician: passive margin around Alabama promontory and Ouachita embayment; carbonate shelf (including deeper or outer shelf in area of Mississippi Valley graben) on continental crust and off-shelf deep-water basin; synorogenic clastic wedge prograding westward from Taconic orogen. C, Late Mississippian: arc-continent collision along southwest side of Alabama promontory; synorogenic clastic wedge prograding northeastward onto shallow shelf in Black Warrior basin and westward into deep basin in Ouachita embayment; shallow-marine shelf and passive margin around Ouachita embayment; separate clastic wedge prograding westward from Alleghanian orogenic source northeast of Alabama promontory. D, Early Pennsylvanian (late Morrowan): continued thrusting and prograding of clastic wedges from southwest (arc-continent collision orogen) and from northeast (Alleghanian orogen) onto Alabama promontory; cratonic delta prograding southward into Arkoma basin; diachronous closing from east to west of remnant ocean basin in Ouachita embayment and prograding of turbidites from both orogenic and cratonic sources into deep remnant ocean basin; initial thrusting and prograding of synorogenic clastic sediment from southeast onto Alabama promontory. E, Middle Pennsylvanian (late Atokan): thrusting along Appalachian-Ouachita orogen; Appalachian-style thrust faults overriding older Ouachita-style thrust faults south of Black Warrior foreland basin on Alabama promontory; synorogenic clastic wedge prograding northward and cratonic delta prograding southward to fill Arkoma foreland basin above older deep-water deposits.

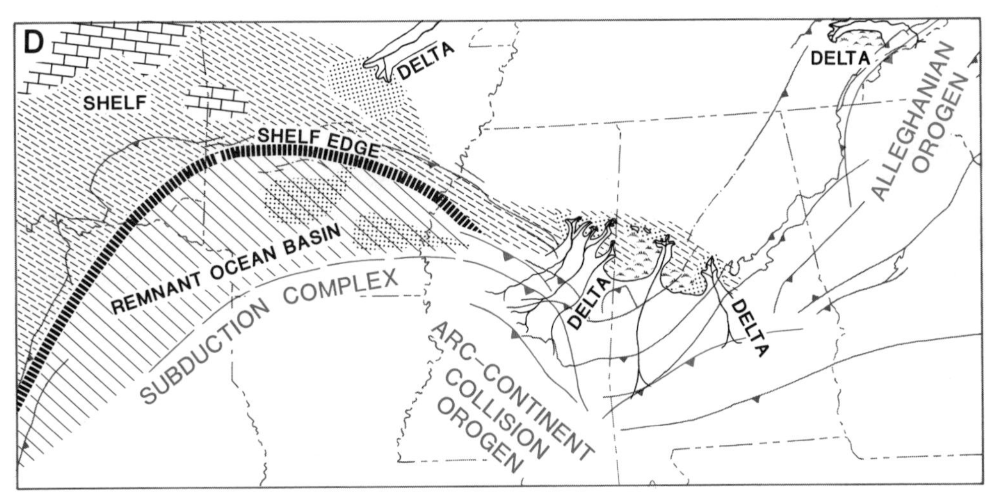
D
SHELF
SHELF EDGE
DELTA
DELTA
REMNANT OCEAN BASIN
SUBDUCTION COMPLEX
ARC-CONTINENT COLLISION OROGEN
DELTA
DELTA
ALLEGHANIAN OROGEN

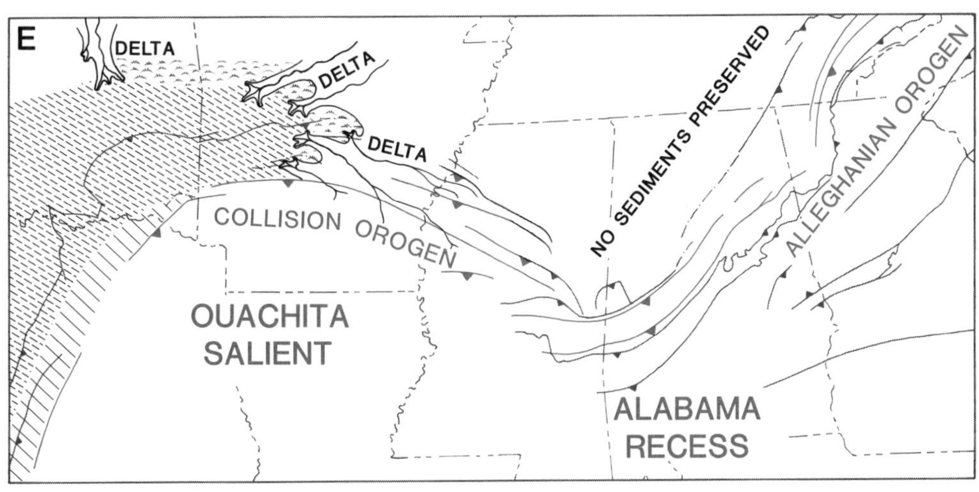
E
DELTA
DELTA
DELTA
COLLISION OROGEN
OUACHITA SALIENT
NO SEDIMENTS PRESERVED
ALLEGHANIAN OROGEN
ALABAMA RECESS

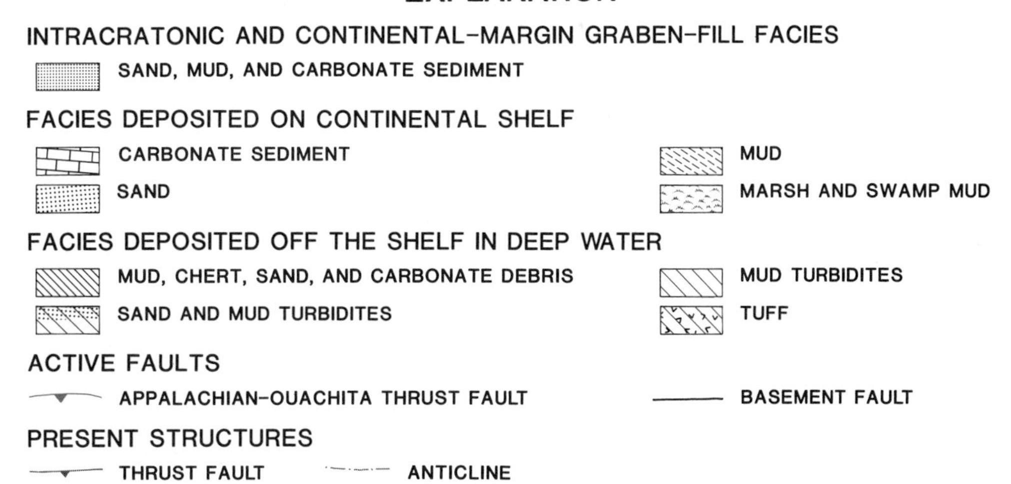
EXPLANATION
INTRACRATONIC AND CONTINENTAL-MARGIN GRABEN-FILL FACIES
SAND, MUD, AND CARBONATE SEDIMENT
FACIES DEPOSITED ON CONTINENTAL SHELF
CARBONATE SEDIMENT
SAND
MUD
MARSH AND SWAMP MUD
FACIES DEPOSITED OFF THE SHELF IN DEEP WATER
MUD, CHERT, SAND, AND CARBONATE DEBRIS
SAND AND MUD TURBIDITES
MUD TURBIDITES
TUFF
ACTIVE FAULTS
APPALACHIAN-OUACHITA THRUST FAULT
BASEMENT FAULT
PRESENT STRUCTURES
THRUST FAULT
ANTICLINE

1989). No shelf-edge rocks have been recognized in the Ouachita region; however, the present structural configuration indicates that off-shelf strata were thrust over the shelf edge onto autochthonous shelf strata during the late Paleozoic Ouachita orogeny. Distribution of the contrasting early Paleozoic carbonate-shelf and deep-water facies indicates a shelf edge approximately coincident with the rifted margin of Precambrian continental crust (Fig. 6D). The oldest known rocks in the off-shelf facies in the Ouachita region are of Late Cambrian age (Ethington and others, 1989), similar to the age of oldest shelf-facies rocks on the southern part of the craton, further suggesting the time of Ouachita rifting. Similar relationships of thrust juxtaposition of shelf and off-shelf facies prevail along the Ouachita orogen southwestward to the Marathon region (King, 1937; Nicholas and Rozendal, 1975; McBride, 1989). Eastward from the Ouachita Mountains, the thrust front cuts diagonally through the facies boundary, and the shelf-carbonate facies is imbricated in the Appalachian allochthon in Alabama, where no shelf-edge facies have been recognized (Thomas, 1989b).

The Upper Cambrian–Lower Ordovician (Knox, Arbuckle-Simpson, Ellenberger) carbonate facies marks maximum transgression during deposition of the Sauk sequence (Sloss, 1988). A cratonwide unconformity at the top of the Sauk sequence decreases in magnitude toward the continental margin. From the southern Appalachian foreland in Alabama westward along the Ouachita-Marathon foreland, the Middle Ordovician through Lower Mississippian succession is dominated by carbonate-shelf facies, signifying passive-margin shelf deposition throughout the southern craton (Fig. 7B). The succession is interrupted by extensive unconformities that reflect eustatic sea-level changes (Vail and others, 1977; Sloss, 1988). On the east, in eastern Alabama and farther northeast along the Appalachian foreland, the carbonate succession contains westward-pinching tongues of synorogenic clastic wedges related to the Taconic (Ordovician-Silurian) and Acadian (Devonian-Mississippian) orogenies in the Appalachians north of the eastern side of the Alabama promontory (Fig. 7B; Thomas, 1977, 1988a), but no record of these events is evident farther west. The passive-margin framework persisted around the Ouachita embayment, Texas promontory, and Marathon embayment from Late Cambrian to Early Mississippian (Figs. 6D, 7B).

Late Paleozoic convergence and the Appalachian-Ouachita orogenic belt

Initial collision at the continental margin is indicated in the Meramecian (middle Mississippian) by northeastward progradation of clastic-wedge sediments onto the shelf facies in the Black Warrior foreland basin on the southwest side of the Alabama promontory (Figs. 3, 7C; Thomas, 1988a, 1989b). A simultaneous abrupt increase in rate of accumulation of deep-water sediment in the Ouachita and Marathon embayments reflects initiation of trench-fill deposition, further indicating plate convergence near the North American continental margin (Fig. 7C; Viele and Thomas, 1989).

Dispersal of clastic sediment toward the northeast and subsidence of the shelf on the southwest indicate that the Black Warrior foreland basin formed in response to a thrust load emplaced on the southwest side of the Alabama promontory and along the Ouachita thrust front (Thomas, 1988a, 1989b). Composition of the sedimentary detritus in the Black Warrior basin documents a sediment source that included a sedimentary thrust belt, a subduction complex, and a volcanic arc (Mack and others, 1983). A comprehensive interpretation of late Paleozoic convergence along the southwest side of the Alabama promontory includes an arc-continent collision orogen initiated in the Meramecian (Fig. 7C).

To the west in the Ouachita embayment, sedimentation rate increased in the deep-water off-shelf setting, but the shelf edge and adjacent shelf remained intact along the northern margin of a remnant ocean basin during the Mississippian (Fig. 7C; Houseknecht, 1986). In the Atokan (Middle Pennsylvanian), rapid subsidence of the shelf was accompanied by down-to-south normal faulting and very rapid sediment accumulation (Houseknecht, 1986). The difference in age of shelf subsidence between the Black Warrior basin and the Arkoma basin indicates that arc-continent collision progressed diachronously from east to west (Figs. 3, 7D), and from Meramecian to Atokan. A very thick succession (>10 km) of deep-water turbidites reflects deposition in the remnant ocean basin, trench, and peripheral foreland basin. The succession grades upward into shallow-marine to deltaic clastic facies as an indication of closing and filling of the foreland basin during the late Atokan (Fig. 7E; Houseknecht, 1986).

Combining outcrop geology of the Ouachita Mountains, deep seismic reflection profiling (COCORP), and wide-angle reflection/refraction seismic surveying (PASSCAL) yields a comprehensive interpretation of Ouachita tectonics (Fig. 4, O–O′; Viele, 1973, 1979; Nelson and others, 1982; Lillie and others, 1983; Keller and others, 1989a; Viele and Thomas, 1989). Lower Paleozoic passive-margin off-shelf rocks exposed in the central (Benton) uplift (Fig. 4, O–O′) of the Ouachitas are deformed by penetrative polyphase fold systems and thrust faults (Viele, 1979). On the foreland (north) side of the Benton uplift, the frontal Ouachita structures are large-scale cratonward-directed thrust faults in Mississippian-Pennsylvanian clastic-wedge rocks (Fig. 4, O–O′; Viele and Thomas, 1989, Fig. 9). The boundary between the Benton uplift and the frontal thrust belt is marked by the Maumelle chaotic zone of Pennsylvanian shales containing blocks of sandstones (Viele, 1979; Thomas and others, 1989a; Viele and Thomas, 1989, Fig. 9). The Maumelle zone is interpreted to be part of a subduction complex, and the Benton uplift to be the top of an accretionary prism (Fig. 4, O–O′; Viele, 1979; Viele and Thomas, 1989). Mississippian-Pennsylvanian clastic-wedge rocks north of the Benton uplift are interpreted to be the fill of a trench and a peripheral foreland basin that formed as the subduction complex overrode the southern edge of North Ameri-

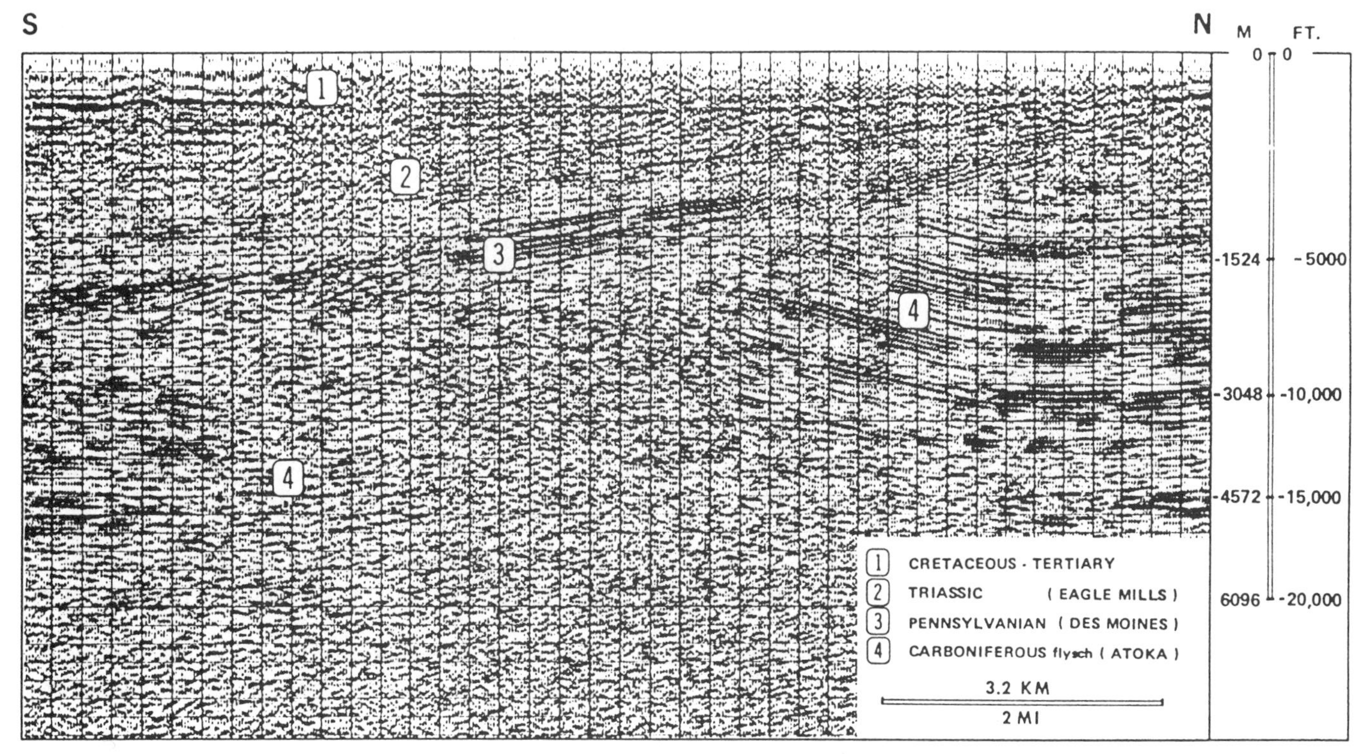

Figure 8. Seismic reflection profile from southwestern Arkansas showing (1) Cretaceous and younger strata of Gulf Coastal Plain; (2) Triassic syn-rift deposits, representing the initial rifting stage of opening of Gulf of Mexico basin; (3) Pennsylvanian successor-basin deposits, the fill of a dormant forearc basin; and (4) deformed Pennsylvanian forearc-basin rocks of the Ouachita southern thrust belt (from Nicholas and Waddell, 1989).

can continental crust (Fig. 4, O–O′). South of the Benton uplift, a very thick accumulation of southward-dipping thrust-imbricated turbidites of Mississippian-Pennsylvanian age is interpreted as forearc-basin deposits (Fig. 4, O–O′; Viele, 1979).

In the Coastal Plain subsurface south of the Ouachita Mountains, thrust-deformed turbidites as young as Atokan are unconformably overlain by nearly undeformed, fluvial to shallow-marine strata of Desmoinesian to Permian age (Fig. 8; Woods and Addington, 1973; Milliken, 1988; Nicholas and Waddell, 1989; Thomas and others, 1989a). These successor-basin deposits are interpreted to be the fill of a dormant forearc basin that has been preserved intact (Fig. 4, O–O′). Preservation of the forearc basin suggests that subduction stopped shortly after the leading edge of the subduction complex rode onto continental crust, and seismic data (Keller and others, 1989a) show that the southern edge of continental crust is only a short distance south of the leading edge of the subduction complex (Fig. 4, P–P′). Preservation of the intact forearc basin south of the Ouachitas contrasts with the thrust collapse of the arc and subduction complex implied by the composition of sedimentary detritus farther east in the Black Warrior basin (Mack and others, 1983). The along-strike change may be a result of oblique collision of the arc with the rifted margin of North American continental crust. The present locations of remnants of the arc are not well documented, but wells in eastern Texas south of the Ouachita outcrops have penetrated volcanic rocks beneath the Desmoinesian successor-basin strata (Nicholas and Waddell, 1989, Fig. 5; Thomas and others, 1989a).

Southwestward from the Arkoma basin, other foreland basins (Fort Worth, Kerr, and Val Verde) along the Ouachita-Marathon thrust belt contain clastic-wedge rocks generally similar to those of the Arkoma basin, and the Marathon thrust belt contains a deep-water clastic wedge similar to that in the Ouachita outcrops. The age of the initial clastic-wedge deposits is constant within the limits of resolution of data from the Black Warrior foreland basin to the Marathon region. Closing and filling of the Fort Worth basin is indicated by Atokan fluvial and deltaic strata (Crosby and Mapel, 1975; Lovick and others, 1982). In contrast, in the Val Verde basin, deposition of deep-water clastic-wedge sediments continued into the Early Permian (King, 1975; Ross, 1986). Movement on the frontal Marathon thrust faults continued into the Early Permian (Ross, 1986; Muehlberger and Tauvers, 1989). The latest episodes of thrusting are diachronous all along the Ouachita-Marathon orogen.

Basement faults of the Southern Oklahoma fault system, as well as other intracratonic fault systems, were reactivated as steep reverse and strike-slip faults during Ouachita compression (Kluth and Coney, 1981; Kluth, 1986; review by Perry, 1989). Thick accumulations of coarse clastic sediment accompanied the fault movement along the Southern Oklahoma fault system (summary by Johnson and others, 1988).

In Alabama, the Appalachian thrust belt strikes northeast nearly perpendicular to strike of the southwest-dipping homocline of the Black Warrior foreland basin and nearly perpendicular to stratigraphic strike of the synorogenic clastic wedge in the Black Warrior basin (Fig. 7C; Thomas, 1988a). Both the structural geometry and the sedimentary fill of the Black Warrior basin indicate that it is a Ouachita foreland basin rather than an Appalachian foreland basin; however, the southeastern end of the Black Warrior basin is truncated by northeast-striking Appalachian thrust faults. In eastern Mississippi, west-striking thrust faults of Appalachian style (shelf facies in the hanging wall) truncate southeast-striking thrust faults of Ouachita style (off-shelf facies in the hanging wall) (Fig. 3; Thomas, 1989b). These relationships indicate that Appalachian thrusting postdates Ouachita thrusting around the edges of the Black Warrior foreland basin and, on a larger scale, around the edges of the Alabama promontory.

The décollement beneath the Appalachian thrust belt in Alabama and Georgia is near the base of the sedimentary cover sequence above Precambrian crystalline basement rocks (Fig. 4, A–A′). The palinspastically restored shelf-facies rocks of the thrust belt and the shallow basement beneath and southeast of the thrust belt are coextensive, indicating that the original sedimentary cover over the Precambrian basement on the southeastern part of the Alabama promontory has been tectonically removed and imbricated (Figs. 2; 4, A–A′). The original shelf-facies cover stratigraphy has been replaced by accreted terranes that are of uncertain affinity and may be of non–North American origin (Williams and Hatcher, 1982; Horton and others, 1989). The Pine Mountain internal basement massif consists of Grenville-age crystalline rocks and a metasedimentary cover sequence; the massif is positioned over the southeast margin of shallow Precambrian basement rocks (Fig. 4, A–A′). Whether the massif is an accreted microcontinent (Thomas, 1977; Neathery and Thomas, 1983; Hooper and Hatcher, 1988) or a fault block from the North American margin (Schamel and others, 1980; Nelson and others, 1987), it is near the autochthonous edge of North American continental crust.

Southeast of the Pine Mountain internal basement massif, south-dipping reflectors extend down to the base of the crust and are interpreted to mark a suture zone (Suwannee-Wiggins suture, Figs. 3; 4, A–A′; 5) between North American crust and the Suwannee terrane to the south (Nelson and others, 1985; Thomas and others, 1989b). In addition to the south-dipping crustal-scale reflectors, the suture is indicated by a contrast in rock types to the north (high-grade gneisses and schists) and south (Suwannee terrane basement and cover) (Chowns and Williams, 1983; Thomas and others, 1989b) and by the trace of the Altamaha magnetic anomaly (Fig. 3; Higgins and Zietz, 1983; Horton and others, 1984; Nelson and others, 1985; Tauvers and Muehlberger, 1987; Hooper and Hatcher, 1988; McBride and Nelson, 1988). Locally in southwestern Alabama, along the south side of the linear Altamaha negative magnetic anomaly, several wells have penetrated volcanic, plutonic, and ultramafic rocks (Fig. 3; Neathery and Thomas, 1975), suggesting the core of a magmatic arc bordered on the north by a south-dipping subduction complex, and further suggesting the westward extent of the Suwannee-Wiggins suture (Thomas and others, 1989b).

South of the Suwannee-Wiggins suture in the subsurface of southeastern Alabama, southern Georgia, and northern Florida, the Suwannee terrane consists of felsic volcanic (~600 Ma) basement and Ordovician-Devonian sedimentary rocks containing faunas of African affinity (Cramer, 1973; Pojeta and others, 1976; Chowns and Williams, 1983). Basement rocks include felsic volcanic rocks, granite, and local metamorphic rocks (Applin, 1951; Milton and Hurst, 1965; Bass, 1969; Milton, 1972; Barnett, 1975; Smith, 1982; Chowns and Williams, 1983). Overlying the basement rocks is a sequence of nearly flat-lying, relatively undeformed sedimentary rocks. The lower unit, which is the most widely extensive and possibly thickest unit of the sequence, consists of sandstones, containing fossils that indicate an Early Ordovician and possibly, in part, Late Cambrian age (Pojeta and others, 1976). Above the Ordovician sandstone, the Paleozoic section is dominated by dark-gray to black shales interbedded with gray fine-grained micaceous sandstones and locally medium- to coarse-grained quartz sandstones (Chowns and Williams, 1983). Fossils document an age range from Middle Ordovician into Middle Devonian (Andress and others, 1969; McLaughlin, 1970; Pojeta and others, 1976). The African affinity of the fossils is confirmed by similarities in the stratigraphic succession of the Suwannee terrane with that in the Bové basin of west Africa (Thomas and others, 1989b).

The westerly striking Suwannee-Wiggins suture crosscuts the northeast-striking structures and terrane boundaries within the Appalachian Piedmont north of the suture (Horton and others, 1989; Thomas and others, 1989b), and the Suwannee terrane is outboard of the other terranes accreted to North America. The Suwannee-Wiggins suture represents the youngest in a succession of episodes of late Paleozoic terrane accretion and reflects continental collision during the Alleghanian orogeny (Thomas and others, 1989b). The relatively undeformed Paleozoic sedimentary cover rocks, however, indicate little deformation as a result of continental collision. The Suwannee terrane apparently includes the complete thickness of African continental crust (Fig. 4, A–A′), and differs from the accreted terranes northwest of the Pine Mountain internal basement massif, where shallow North American crust underlies the accreted rocks.

West of well-documented rocks of the Suwannee terrane, metasedimentary, high-grade metamorphic, and plutonic rocks of the Wiggins terrane also are south of the Suwannee-Wiggins suture (Fig. 3). The Wiggins rocks, however, are of uncertain affinity (Thomas and others, 1989b).

The history of terrane accretion along the southeast side of the Alabama promontory includes tectonic replacement of the North American shelf stratigraphy by accreted terranes emplaced above a shallow basement and also suturing of African continental crust onto the margin of North American crust (Fig. 4, A–A′). The accreted block of African crust (the Suwannee terrane) was left behind attached to North America during the subsequent breakup of Pangea. In contrast, westward from the southwest side of the Alabama promontory along the Ouachita-Marathon orogen, arc-continent collision resulted in the accretion of arc and subduction-complex rocks onto the margin of North America (Fig. 4, O–O′). North American shelf stratigraphy was left in place, and the margin of North American crust was overridden to a limited extent. The accreted rocks are primarily sedimentary rocks of the trench and forearc, as well as some volcanic rocks. These tectonic assemblages constitute much of the continental crust that was extended to form the western part of the Gulf basin during Mesozoic time.

MESOZOIC GEOLOGIC SETTING, TECTONIC FRAMEWORK, AND CRUSTAL STRUCTURE OF THE GULF OF MEXICO BASIN

General setting

The Gulf of Mexico basin is a large, crescent-shaped downwarp of the crust along the southeastern margin of the North American craton. The overall configuration of the basin is illustrated by a generalized contour map (Fig. 1) and a series of three regional cross sections across the basin (Fig. 9; Buffler, 1991; Sawyer and others, 1991). The contour map depicts a prominent *mid-Jurassic surface* (MJS) or unconformity that is recognized throughout the Gulf region. The surface is overlain and onlapped by the extensive Middle Jurassic (Callovian?) pre-marine evaporites (Louann Salt and equivalents) and younger sedimentary rocks that fill the Gulf basin. We use the term "pre-marine evaporites" after Winker and Buffler (1988) for widespread halite-dominated evaporites deposited just prior to the initial extensive marine transgression (Oxfordian) into the Gulf basin. The generalized distribution of these evaporites is shown on Figure 1 (stipple pattern).

The MJS is a prominent unconformity that is distinct in wells and on seismic reflection profiles, and it can be mapped over much of the basin with some confidence (heavy lines, Fig. 1). The MJS truncates all rocks older than the evaporites, which includes the Upper Triassic–Lower Jurassic nonmarine sedimentary and volcanic rocks deposited in rift basins in Mexico, along the northern margin of the Gulf basin, and along the Atlantic margin of North America (La Boca Formation, Eagle Mills Formation, and Newark Group, respectively). These rocks record the earliest rifting during the Mesozoic divergent phase in the evolution of the basin. Also lying below the MJS are the various Precambrian and Paleozoic terranes that record the earlier history of the Gulf region. The MJS mapped on Figure 1 is of mid-Jurassic age, except in the deep central Gulf, where it is the top of Late Jurassic oceanic crust. This crust was emplaced following deposition of the pre-marine evaporites and split the original area of salt deposition.

Examples of the expression of the MJS as identified in wells are shown by three cross sections from northern Louisiana and southern Arkansas (Fig. 10). Here the surface truncates older rocks, including the Upper Triassic–Lower Jurassic Eagle Mills Formation and older Paleozoic sedimentary rocks. A seismic line from the same area shows the same relationships (Fig. 8). Here the MJS is overlapped by Cretaceous strata. Another example of the MJS is identified on a seismic line from the eastern Apalachicola basin in the northeastern Gulf (Fig. 11). Here the unconformable surface truncates an older, dipping sedimentary section. These dipping reflectors probably represent Upper Triassic–Lower Jurassic rocks deposited in rift basins, but they may be Paleozoic sedimentary rocks. Both have been identified in nearby wells. Here the unconformity is overlain by the Louann Salt (pre-marine evaporites) and younger Upper Jurassic strata (Fig. 11).

The MJS is a well-defined acoustic basement on seismic reflection data in the areas of oceanic crust in the deep central Gulf (Fig. 12), lying as deep as 10 to 12 km (Fig. 1). It is also well defined in the southeastern Gulf (Fig. 1), where it is relatively shallow. In some broad areas, however, the surface is less distinct, and the contours there are based on other geophysical data and on extrapolation of trends from adjacent areas of better control (light lines, Fig. 1). For example, under the Texas-Louisiana shelf and slope (22, Fig. 1) and in the Campeche salt area to the south (20, Fig. 1), the surface is very deep and is difficult to image because it is buried beneath a thick section of deformed salt and sedimentary rocks (Fig. 9). The contour values of 14 to 16 km for the area beneath the northern slope are estimated mainly from sparse refraction data, and this depth probably represents maximum crustal subsidence, including the contribution by later loading of a thick prism of Cenozoic rocks (Fig. 9). Under Yucatan (19, Fig. 1), the surface probably is relatively shallow but is poorly defined because of the masking effect of thick carbonate rocks.

The configuration of the MJS shows the overall architecture of the basin, which formed mainly during a Middle Jurassic stage of stretching and attenuation of the crust. This stage of extension resulted in the formation of a broad region of attenuated continental crust or transitional crust (Fig. 1). The configuration of the MJS controlled the distribution of the overlying pre-marine evaporites (Fig. 1), as well as the deposition of the overlying Jurassic and Lower Cretaceous sedimentary rocks (Fig. 9).

The crust of the Gulf basin has been subdivided into four types: continental, thick transitional, thin transitional, and oceanic crust (Figs. 1, 9). The term transitional crust is used to designate crust that has a thickness intermediate between true continental and true oceanic crust. In the Gulf region, the transitional crust represents original Pangean continental crust that has been stretched and attenuated during the breakup of Pangea as Africa/South America separated from North America (e.g., Buffler and Sawyer, 1985; Pindell, 1985; Klitgord and others, 1988).

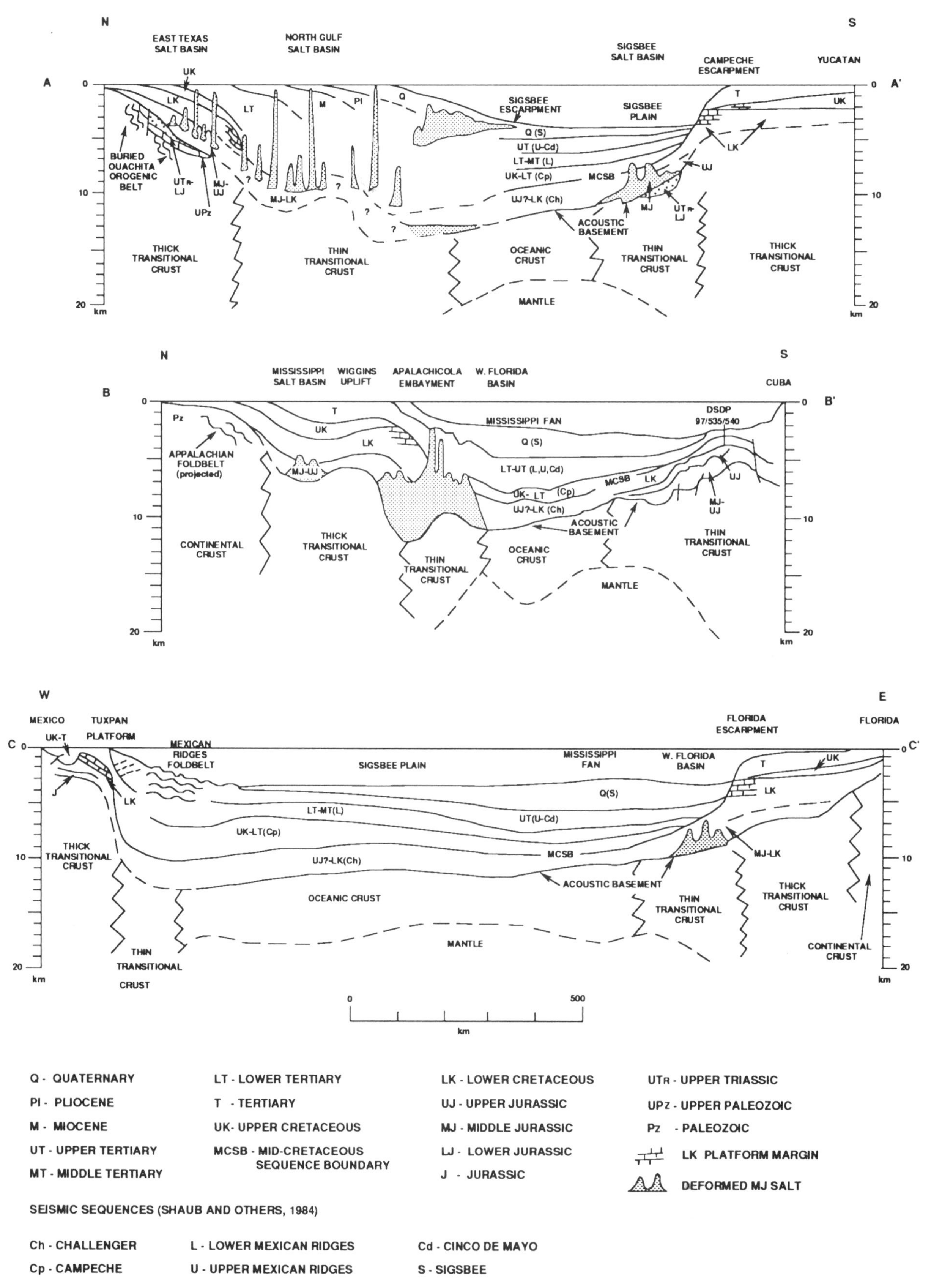

Figure 9. Schematic cross sections of the Gulf of Mexico basin showing crustal types and overlying Mesozoic and Cenozoic sedimentary section. Locations of cross sections shown in Figure 1.

Continental crust comprises the broad area along the southeastern margin of the North American craton characterized by a generally smooth, shallow (2 to 4 km) MJS (Fig. 1). The thick transitional crust consists of Pangean continental crust that was only moderately stretched into a configuration characterized by relatively shallow, broad highs and lows with wave lengths of approximately 300 km. Most of the highs and lows (arches, embayments, etc.) have been given various names (1 to 15, Fig. 1). In contrast, the more deeply subsided area of thin transitional crust consists of a broad area of continental crust that was highly attenuated during the breakup of Pangea (Fig. 1). The widespread pre-marine evaporites overlying the MJS are restricted mainly to the areas of transitional crust (Fig. 1). In the deep central Gulf, a generally east-west–trending band of oceanic crust was emplaced during the final stage of crustal attenuation, splitting the originally contiguous salt basin.

This basic scheme of subdivision of crustal types is based on the depth and overall configuration of the MJS (Fig. 1), as well as the nature of known rocks below the surface, the known crustal structure from seismic refraction data (velocity and thickness or depth to Moho), seismic reflection data, total tectonic subsidence analyses, magnetics, gravity, and the distribution of Jurassic rocks overlying the mid-Jurassic surface. This is purely a descriptive classification and does not imply the specific processes involved in attenuation.

An alternative way of showing the overall crustal type and horizontal crustal extension in the Gulf was presented by Dunbar and Sawyer (1987). Using an earlier version of Figure 1, they constructed a map showing the variation in total tectonic subsidence (TTS) of the basin (Fig. 13). This was computed by correcting the "basement" depth for the loading effect of the sediments. This type of map more closely approximates the original configuration of the surface and is useful in further differentiating the four crustal types. Crustal type and crustal extension (Fig. 14) can be computed from the TTS using a model for the subsidence of passive margins (Dunbar and Sawyer, 1987). Using TTS mapping (Fig. 13) the distribution of oceanic crust corresponds approximately with beta = 4.5 (i.e., where surface area has been extended by a factor of 4.5; see explanation in Fig. 14 caption).

The three regional cross sections across the basin (Fig. 9) show crustal structure, as well as the generalized distribution of the thick sedimentary section deposited in and partially filling the space created by the subsidence of the transitional and oceanic

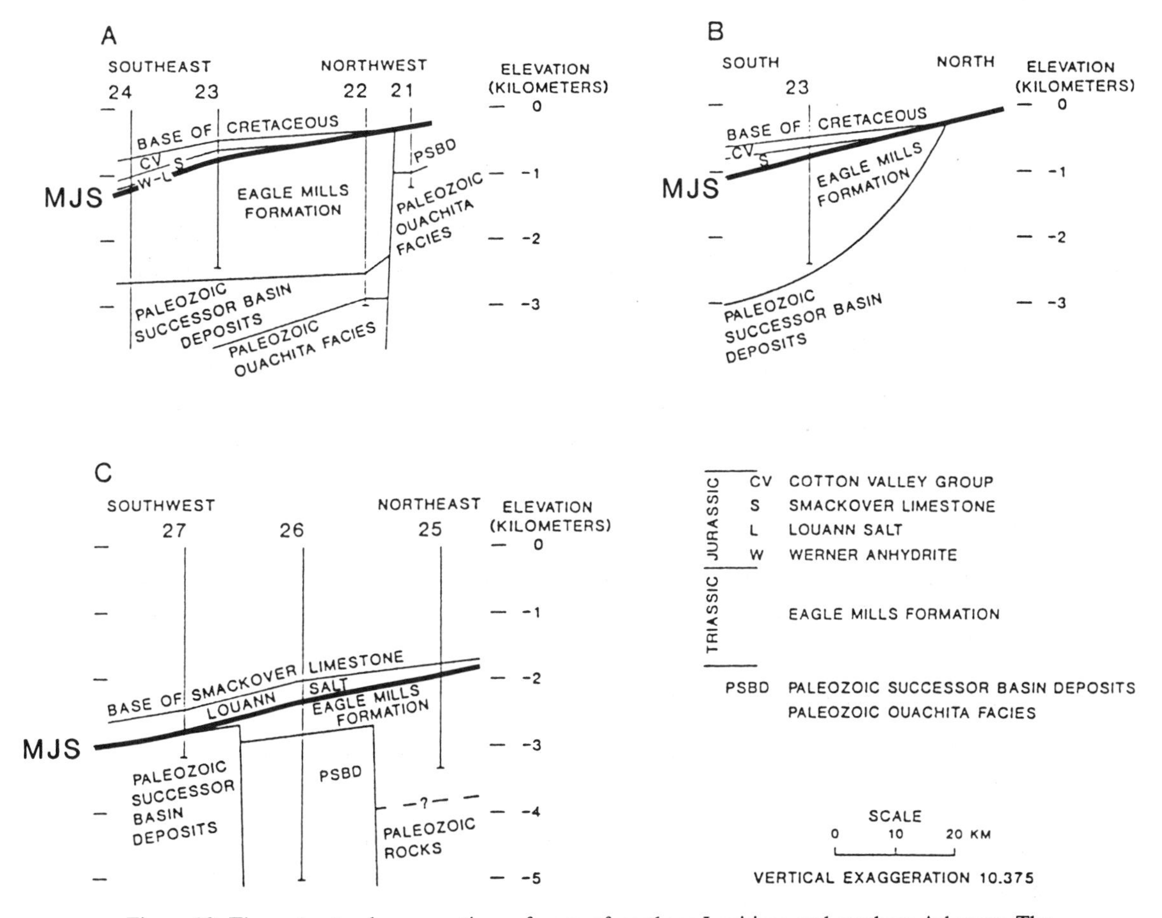

Figure 10. Three structural cross sections of parts of northern Louisiana and southern Arkansas. The mid-Jurassic surface (MJS) truncates older Triassic and Paleozoic rocks and is onlapped and overlain by younger Mesozoic rocks (from Thomas, 1988b). Locations of cross sections shown in Figure 15.

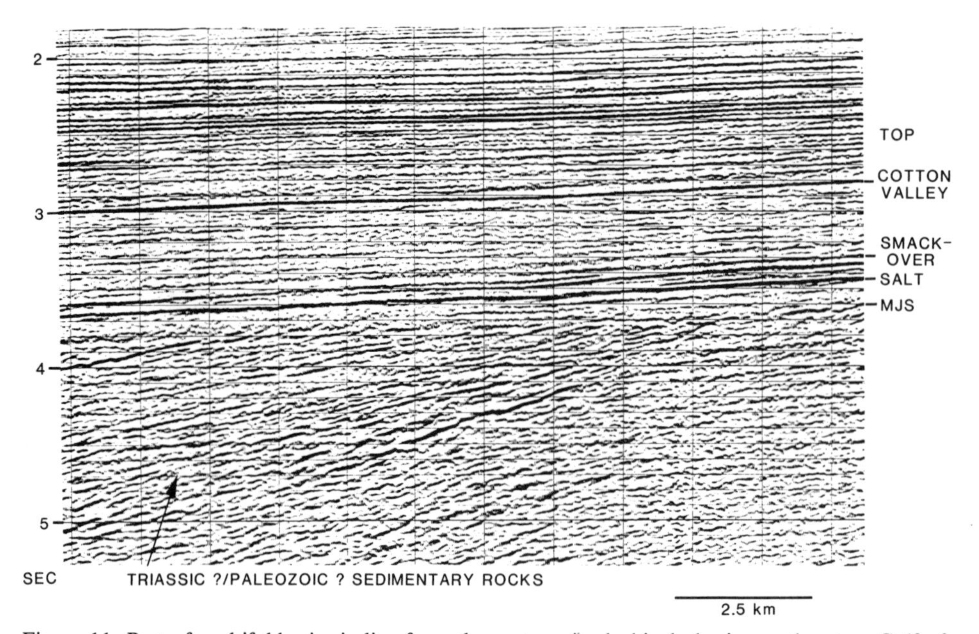

Figure 11. Part of multifold seismic line from the eastern Apalachicola basin, northeastern Gulf of Mexico basin (see Fig. 1 for general location). The mid-Jurassic surface (MJS) is a prominent unconformity that truncates dipping Triassic? or Paleozoic? sedimentary rocks and is overlain by mid-Jurassic, pre-marine evaporites (Louann Salt) and younger Jurassic sedimentary rocks.

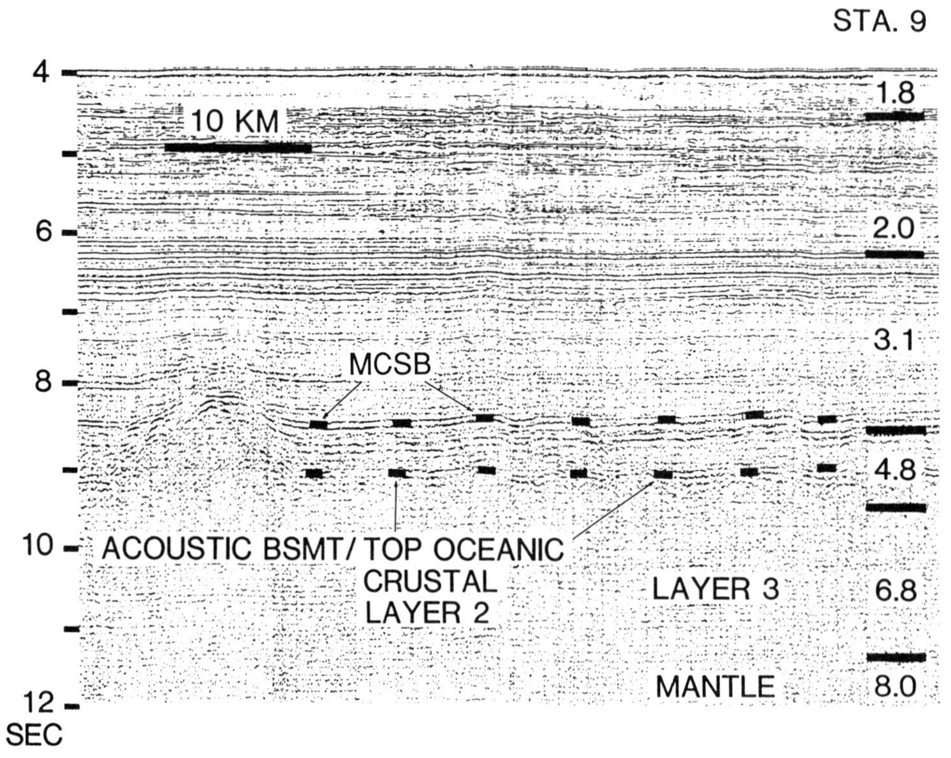

Figure 12. Part of UTIG multifold seismic line in south-central deep Gulf of Mexico basin overlying area of oceanic crust (see Fig. 26 for location). Refraction velocities are from nearby refraction station 9 (modified from Ibrahim and others, 1981).

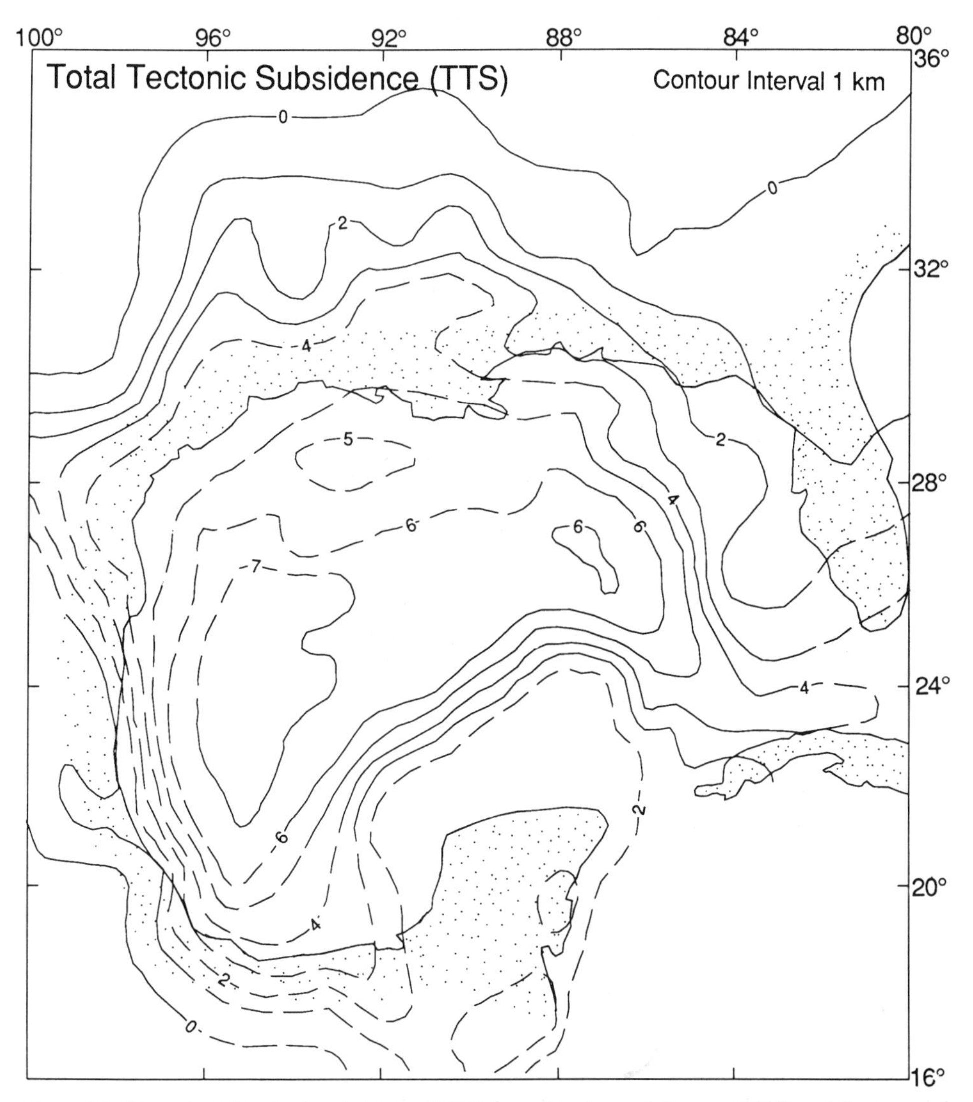

Figure 13. Map of total tectonic subsidence (TTS) from Dunbar and Sawyer (1987) and Sawyer and others (1991). TTS is the observed depth to basement at a point corrected for the loading effect of the sediment. It is calculated using the observed bathymetry, observed basement depth, and a model for the density of the sediments. It is interpreted as the total amount by which basement would have subsided during the formation of the Gulf of Mexico basin if no sediment had been deposited. For this map, the top of basement and the MJS are synonymous.

crust. The sedimentary section is separated into two main sequences by a prominent mid-Cretaceous sequence boundary (MCSB) (Buffler, 1991). The pre-MCSB sequence records the early evolution of the basin (mid-Jurassic through mid-Cretaceous) that was related mainly to thermal or tectonic subsidence of the basin. The lower part of this sequence, which overlies the MJS, contains the extensive pre-marine evaporites in the various salt basins around the Gulf basin (Figs. 1, 9). The later part of this subsidence phase involved the development of widespread carbonate platforms and margins surrounding the entire Gulf basin (Fig. 9). The post-MCSB history consists primarily of filling the deep basin with thick prisms of siliciclastic sediments, first from the west and northwest in Late Cretaceous–early Cenozoic time (Fig. 9, C–C′) and then from the north (Mississippi River drainage) during the late Cenozoic (Fig. 9, A–A′, B–B′). This later stage in the history of the basin is not discussed in this chapter.

Late Triassic–Early Jurassic tectonic framework

Early Mesozoic plate divergence is indicated by a system of sediment-filled, fault-bounded basins that approximately parallels the boundary between the continental crust and thick transitional crust around the northern periphery of the Gulf. The graben fill

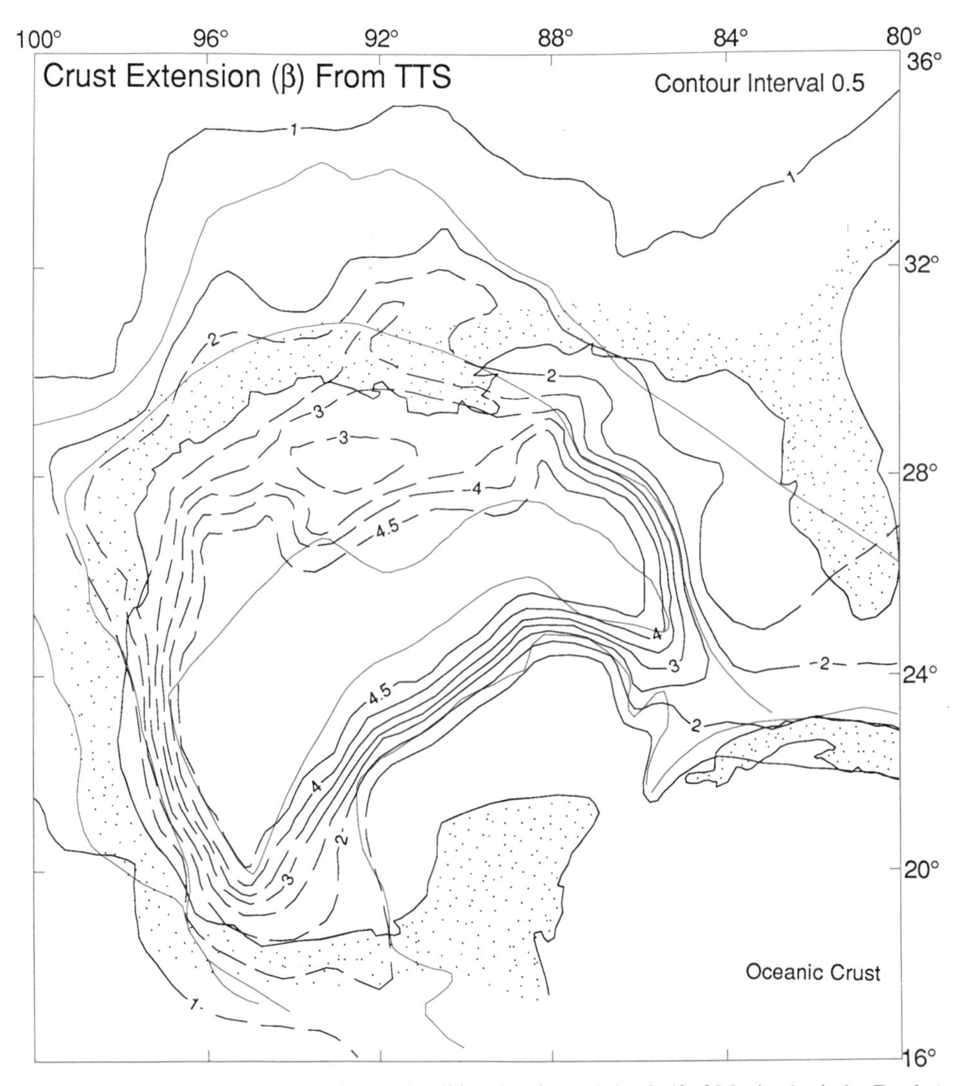

Figure 14. Map of crust extension during the rifting that formed the Gulf of Mexico basin by Dunbar and Sawyer (1987) and Sawyer and others (1991). Extension is described using the extension parameter beta. For example, beta = 2 describes crust extension in which surface area is increased by a factor of 2 and crust is thinned to half the original thickness. Beta = 1 describes unextended crust. Estimates of crust extension are derived from the TTS map in Figure 13. Beta = 4.5 is interpreted to correspond to the boundary between oceanic and transitional crust. Outlined in red are crustal boundaries from Figure 1 for comparison.

consists mainly of a thick section of Upper Triassic–Lower Jurassic (?) nonmarine sedimentary rocks of the Eagle Mills Formation (Figs. 8, 10, 15; Woods and Addington, 1973; Salvador, 1987; Milliken, 1988; Nicholas and Waddell, 1989). Similar rocks are extensive in a broad band across northern Florida and southern Georgia (South Georgia basin) as identified in wells (Fig. 15; Chowns and Williams, 1983) and on seismic reflection profiles (Fig. 5; Nelson and others, 1985; McBride and Nelson, 1988). These rocks also might extend offshore beneath the northeastern Gulf of Mexico, as indicated by seismic reflection data (Fig. 11). Collectively, these rocks are believed to consist of syn-rift nonmarine sediments and minor volcanics that accumulated in asymmetrical rift basins. Everywhere they are covered by younger strata.

The basins generally follow the trend of the underlying late Paleozoic Appalachian-Ouachita orogenic belt (Fig. 15), and therefore, may reflect reactivation of older Paleozoic structural fabrics in the crust that outlined the Alabama promontory, Ouachita embayment, and Texas promontory (Fig. 2). Triassic graben-fill rocks rest unconformably on rocks that range in age from Precambrian to Permian (Thomas and others, 1989b). Across northern Florida and southern Georgia, the basins trend northeasterly and connect with similar basins, both in outcrop and in the subsurface along the Atlantic margin of North America

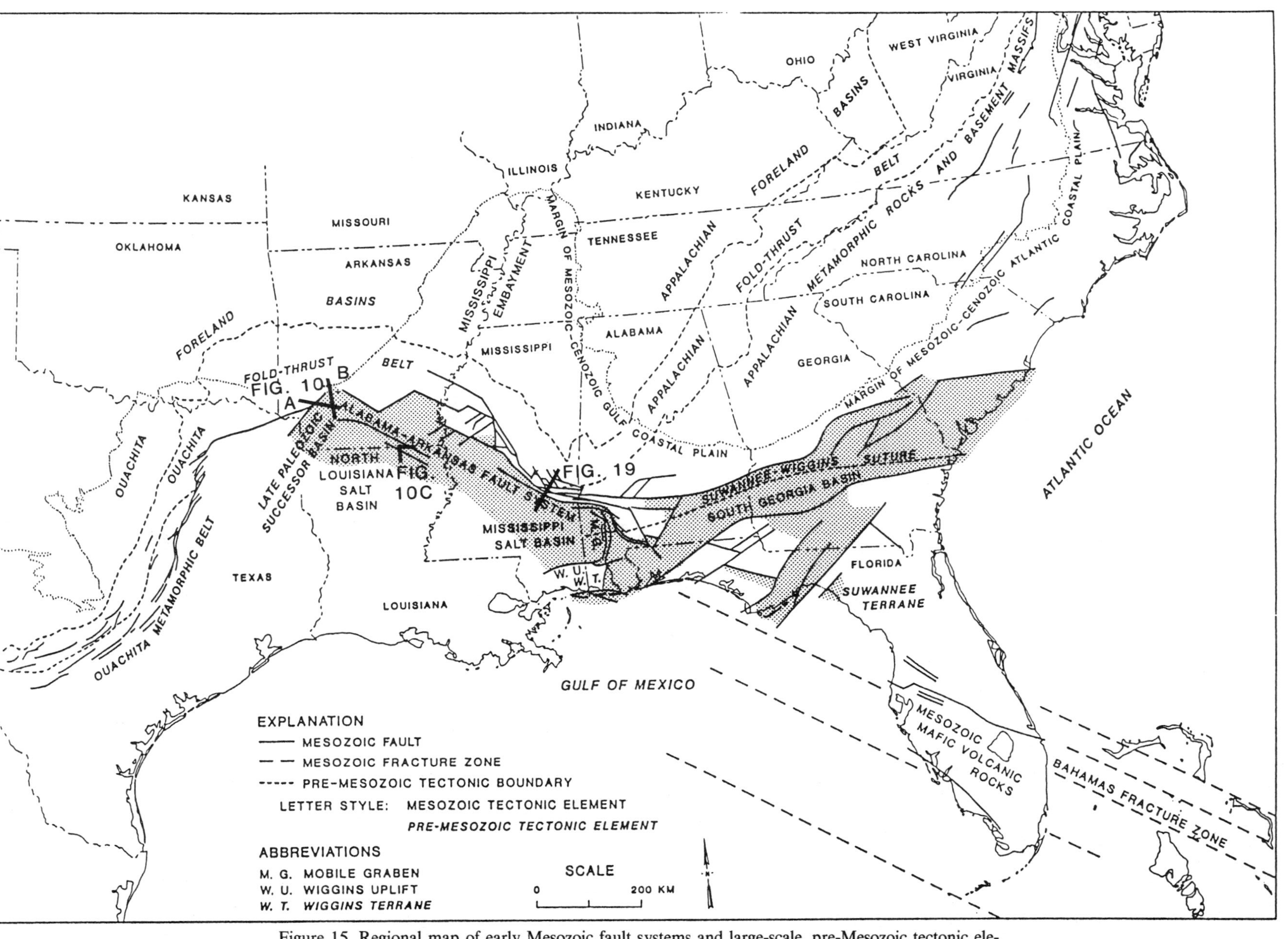

Figure 15. Regional map of early Mesozoic fault systems and large-scale, pre-Mesozoic tectonic elements of southeastern North America (from Thomas, 1988b). Stipple pattern shows distribution of Late Triassic–Early Jurassic graben-fill sedimentary and igneous rocks from Thomas and others (1989c). Shown are locations of Figures 10 and 19.

(Newark Group and associated rocks). The largest is the South Georgia basin. COCORP lines across northern Florida and southern Georgia outline a general structure of the South Georgia basin and reveal as much as 6 km of basin fill (Fig. 5; McBride and Nelson, 1988). From southwestern Alabama to southern Arkansas, rift-related rocks are distributed along a northwestward trend as indicated by scattered wells (Alabama-Arkansas fault system) (Fig. 15; Thomas, 1988b; Thomas and others, 1989c). Abrupt variations in thickness of the rift-related rocks suggest numerous faults bounding relatively small basins.

The extent of Triassic rocks (Eagle Mills Formation) around the northern Gulf margin from northern Louisiana and southern Arkansas to central Texas (Fig. 15) has been recently described and mapped in some detail on the basis of seismic reflection data and wells (Milliken, 1988). Across southern Arkansas and northern Louisiana, the system of basins trends generally east-west, but indidivual basins encompass both northwest and northeast trends (Fig. 15; Milliken, 1988; Thomas, 1988b). The system of basins extends southwestward across eastern Texas, where the distribution of rift-related rocks suggests a relatively simple southeast-dipping homocline (Milliken, 1988), rather than small fault-bounded basins.

The Upper Triassic–Lower Jurassic rocks are truncated over much of the northern Gulf by a prominent unconformity, the MJS (Figs. 1, 8, 10, 11). Thus, we consider these rocks to be part of an early rift stage of initial divergence during the breakup of Pangea. Initial rifting was followed by a mid-Jurassic stage of attenuation that formed the broad area of transitional crust and present structural configuration shown in Figures 1 and 13. These successive stages were apparently part of a more-or-less continuous process of continental rifting and breakup that began in the Triassic and culminated with emplacement of oceanic crust in the Late Jurassic. However, initial rifting apparently was accompanied and followed by erosional truncation of the Upper Triassic–Lower Jurassic rocks over much of the area prior to the mid-Jurassic attenuation. The early rift stage may represent rifting of an initial brittle crust along a narrow area of mantle upwelling, whereas the later phase could represent attenuation of a more ductile crust across a broader area of mantle upwelling. The early rift stage apparently did not involve large amounts of crustal extension or separation of large crustal blocks, in contrast to the mid-Jurassic stage of attenuation.

Mid-Jurassic and later crustal structure

Continental crust. Continental crust comprises a broad area along the southeastern margin of the North American craton (Fig. 1). It is characterized by typical continental crust thicknesses that range from approximately 33 to 48 km. For example, Moho imaged on the COCORP lines across southern Georgia and northern Florida (Fig. 5) is at an approximate depth of 35 km. The MJS has a generally smooth, uniform, basinward-dipping configuration and is relatively shallow, extending from the outcrop to about 2 to 4 km in depth along the boundary with transitional crust (Fig. 1). On the TTS map (Fig. 13), along the boundary between continental crust and transitional crust, the MJS is between 1 and 2 km deep. The continental crust in this region consists of rocks representing (1) the late Precambrian through early Paleozoic rifted continental margin, (2) the early to middle Paleozoic passive margin, (3) the late Paleozoic Appalachian-Ouachita orogenic belt, including various accreted terranes (e.g., Suwanee terrane), and (4) the Late Triassic–Early Jurassic rift. Each of these is discussed in more detail in the previous sections.

Boundary between continental and thick transitional crust. The boundary between continental crust and thick transitional crust (Fig. 1) is based on the distribution of mapped Mesozoic structures. Regionally the boundary corresponds approximately to a major change in character of the MJS from a shallow, relatively smooth, southerly dipping surface to a deeper, more irregular surface characterized by broad highs and lows (Figs. 1, 13). In southern Florida, a major crustal boundary is inferred to correspond with the northwestward projection of the Bahamas fracture zone from the Atlantic (Fig. 15). Identification of this boundary is based primarily on gravity and magnetics data (Klitgord and others, 1984, 1988; Klitgord and Schouten, 1986), but it also corresponds to a zone of Mesozoic (Jurassic) mafic volcanic rocks described from wells in south Florida (Fig. 15; Chowns and Williams, 1983; Mueller and Porch, 1983; Thomas and others, 1989b, 1989c).

The crustal boundary is projected in the northeastern Gulf on the basis of gravity and magnetics data (Klitgord and others, 1984, 1988). Here the boundary corresponds to a regional hinge zone or flexure into the basin of the mid-Jurassic surface as seen on the structure contour maps (Figs. 1, 13) and illustrated on two regional cross sections of the northeastern Gulf (Figs. 16, 17; Dobson, 1990). The southern cross section shows that a flexure of the mid-Jurassic surface is associated with the regional thinning and onlap of the Jurassic sedimentary section from the Tampa embayment to the Middle Ground arch (Fig. 16). The boundary also corresponds to the apparent truncation of the older Triassic-Jurassic rift basins (Fig. 16), as suggested by wells and seismic data in the area (Dobson, 1990). The northern cross section (Fig. 17) shows the regional hinge zone where it corresponds to regional onlap and thinning of the overlying pre-marine evaporites (Louann Salt) in the Apalachicola basin (Dobson, 1990). A seismic profile across the hinge zone shows abrupt thinning of the Jurassic salt (Fig. 18). The area of abrupt thinning of the salt became the locus for later faulting of the overlying Mesozoic section as a result of salt withdrawal. These younger faults are part of a more regional peripheral fault zone around the northern and eastern flank of the Apalachicola basin caused by salt withdrawal (Dobson, 1990; MacRae, 1990).

The onshore northwestward projection of the boundary between continental and transitional crust corresponds to a zone of Mesozoic orthogonal faults mapped from seismic and well data (Fig. 15). This fault zone has been referred to as part of the Gulf Rim fault zone (Martin, 1978; Klitgord and Schouten, 1986) and

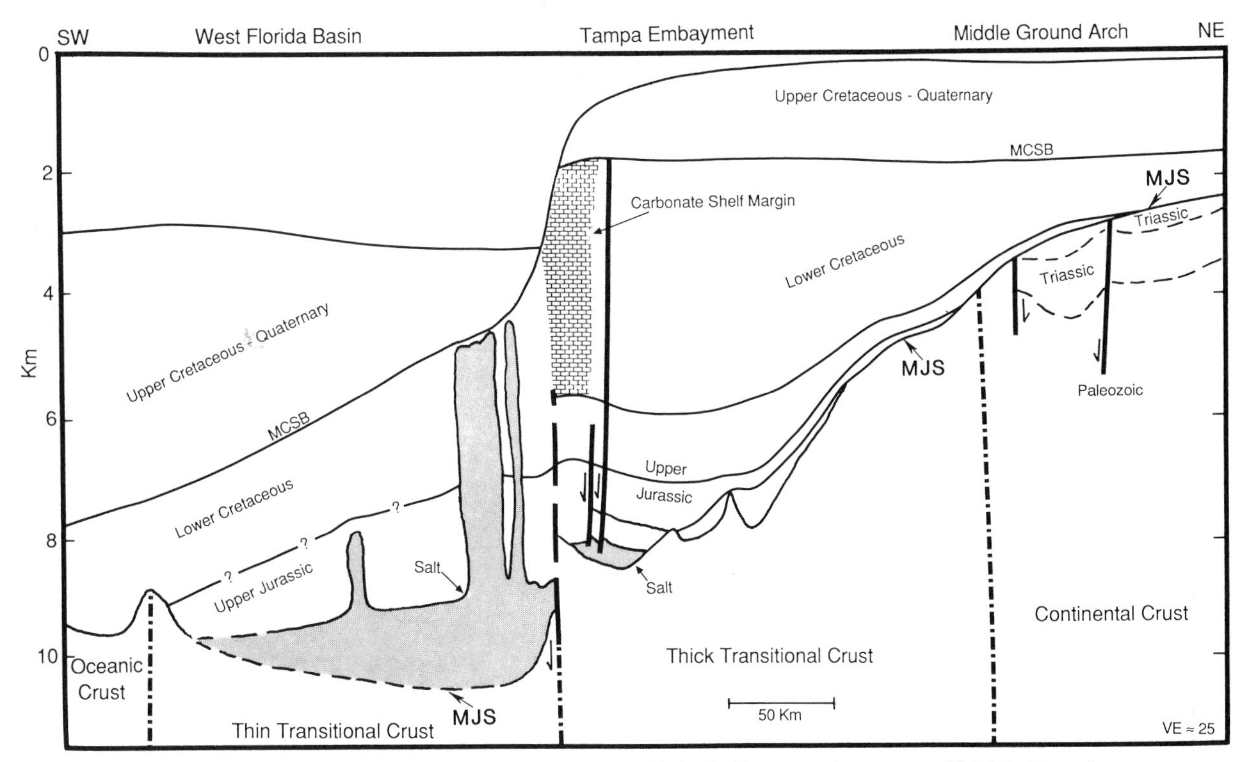

Figure 16. Generalized cross section of the West Florida basin, Tampa embayment, and Middle Ground arch area in the eastern Gulf of Mexico basin by Dobson (1990) showing all four crustal types, the mid-Jurassic surface (MJS), and thinning and onlap of overlying Jurassic section to the east and west (see Fig. 1 for general location).

more recently as the Alabama-Arkansas fault system (Thomas, 1988b). Movement along faults of this zone was greatest from Late Triassic through Late Jurassic (Thomas, 1988b). The unconformity at the MJS is not clearly recognizable at some places in this area, because the later Jurassic faulting probably represents reactivation of Late Triassic–Early Jurassic faults associated with Eagle Mills deposition. In places, the overlying Jurassic clastic rocks are similar to the underlying Eagle Mills. A regional cross section using well data (Fig. 19) defines the mid-Jurassic surface (MJS) truncating Paleozoic rocks and also shows how fault movement influenced deposition of the Jurassic sedimentary section, continuing up into the Cotton Valley Group (Late Jurassic) (Thomas, 1988b). This fault zone controls the regional updip limit of the pre-marine evaporites (Louann Salt) in the region (Figs. 1, 19).

Alignment of the Alabama-Arkansas fault system with the projection of the Bahamas fracture zone (Fig. 15) suggests a large-scale, left-slip system of transform and wrench faults that was active through the Late Triassic–Early Jurassic initial rift phase, as well as the later Middle Jurassic through Late Jurassic opening of the Gulf basin (Thomas, 1988b). The Mesozoic transform boundary coincides in location with an inferred transform fault (the Alabama-Oklahoma transform) along the late Precambrian–early Paleozoic extensional margin of North America (Figs. 2, 6). Both the early Paleozoic and early Mesozoic transform faults framed southeastward projecting promontories of the North American craton (the early Paleozoic Alabama promontory, Figs. 2, 6; and the early Mesozoic Florida promontory; Thomas, 1988b).

The northeastern boundary or hinge zone is similar to a prominent hinge zone along the United States Atlantic continental margin, in that it marks the boundary between continental crust and rifted-continental crust (transitional crust) and the onlap and thinning of the overlying Jurassic sedimentary section (Klitgord and others, 1988). On the Atlantic margin, however, the boundary represents a divergent margin, whereas in the Gulf, the boundary represents a transform or transtensional boundary. The amount of transform movement that might have occurred along the Gulf boundary varies among the models proposed for the opening of the Gulf. The apparent tectonic truncation of the Triassic graben structures in the northeastern Gulf (Fig. 16) suggests some offset along the boundary in this area. Most of the extension and attenuation of the crust apparently took place outboard (southwest) of this boundary and did not affect the area of continental crust to the northeast.

In southwestern Arkansas, the trend of the Mesozoic fault system abruptly turns to the southwest across eastern Texas, and farther south it bends around the San Marcos arch (12, Fig. 1;

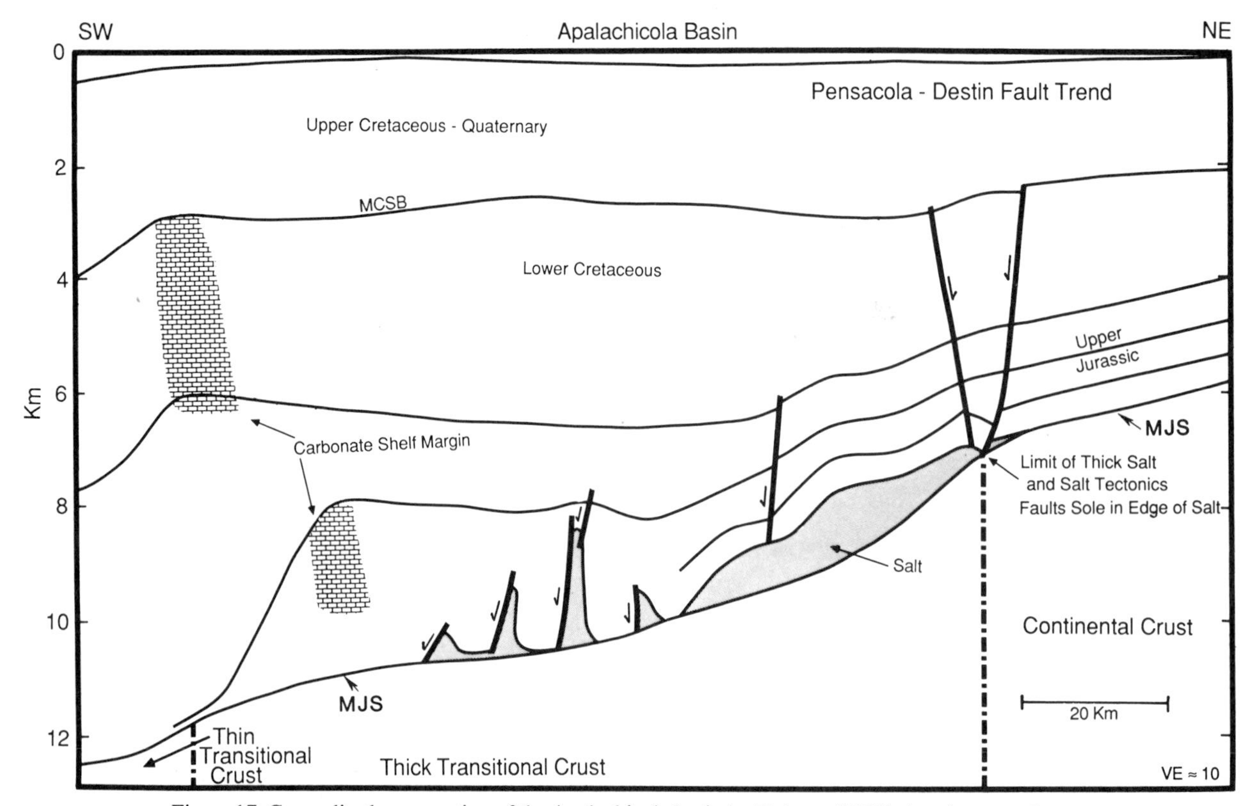

Figure 17. Generalized cross section of the Apalachicola basin by Dobson (1990) showing crustal types, and mid-Jurassic surface (MJS), and overlying Mesozoic section, particularly pinch out of thick salt at inferred crustal boundary (see Fig. 1 for general location; also see Fig. 18). MCSB, mid-Cretaceous sequence boundary.

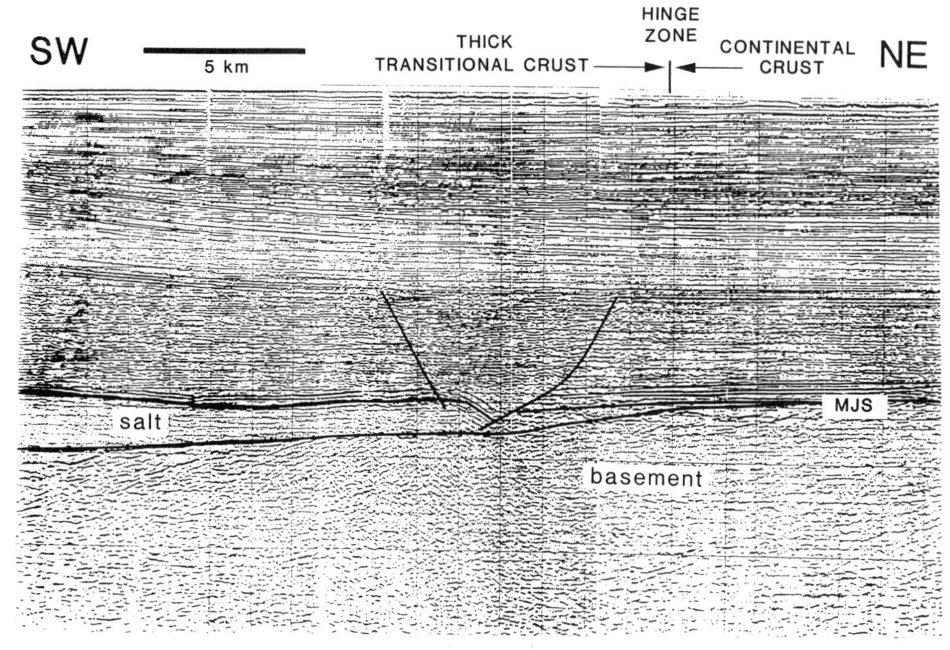

Figure 18. Part of multifold seismic line from northeastern Apalachicola basin crossing hinge zone at boundary between continental and thick transitional crust. This zone is characterized by pinchout of thick Louann Salt and by an associated peripheral fault zone caused by salt withdrawal (also see Fig. 17). See Figure 1 for general location. MJS, mid-Jurassic surface.

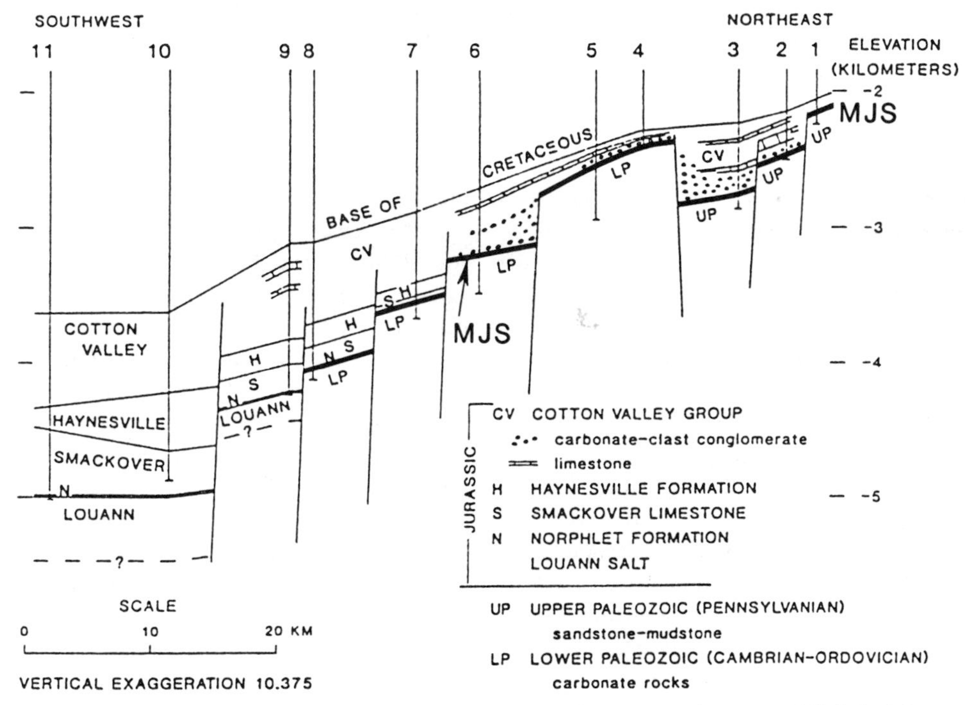

Figure 19. Structural cross section of Alabama-Arkansas fault system in eastern Mississippi from Thomas (1988b). The mid-Jurassic surface (MJS) truncates Paleozoic rocks and is onlapped progressively by Louann Salt and younger Jurassic section. See Figure 15 for location.

compare Fig. 15; Martin, 1978; Thomas, 1988b). Primary movement along the faults was during the Mesozoic, but some activity continued through the Cenozoic. This zone of Mesozoic faults generally follows the trend of the underlying buried Ouachita orogenic belt, reflecting control on extensional structures by older structural fabric.

Thick transitional crust. The thick transitional crust is continental crust that presumably has undergone only moderate attenuation during the Mesozoic breakup of Pangea. Depth to the MJS is greater than on the adjacent continental crust and is characterized by a series of broad highs and lows (arches, uplifts, embayments, etc.) with wavelengths of approximately 300 km (Fig. 1). The highs are relatively shallow (approximately 2 km as restored, Fig. 13) and have been clearly defined by wells and seismic data (heavy lines, Fig. 1), whereas the lows are deeper and more poorly defined (light lines, Fig. 1). Relief between highs and lows generally ranges from 3 to 4 km. The lows appear as broad sag basins; the boundaries with the highs appear to be broad flexures with no large faults. An exception may be the north flank of the Wiggins uplift (6, Fig. 1), where an east-west fault zone is inferred (Fig. 15; Thomas, 1988b).

Each of the basement structures is designated in the literature by one or more names (numbers on Fig. 1). In the northeastern Gulf, the highs generally trend east-west (2, 4, 6, Fig. 1), whereas in the northwest, they trend southeast into the basin (8, 10, 12, Fig. 1). In Mexico, they trend more nearly parallel to the basin (14, Fig. 1). The area underlying the Yucatan block (19, Fig. 1) is poorly defined, but recent gravity and magnetics studies suggest that it also includes large basement highs and lows similar to the Quintana Roo arch (18, Fig. 1) along the southeast margin of the block (Ness and others, 1991).

Potential field data, total tectonic subsidence analyses, and limited refraction data suggest that crustal thickness ranges from 20 to 35 km (Worzel and Watkins, 1973; Klitgord and others, 1984; Nunn and others, 1984, 1989; Kruger and Keller, 1986; Sawyer and others, 1991). Generally, the highs represent blocks of crust with more normal continental crustal thickness (32 to 35 km), whereas the lows are underlain by much thinner crust (20 to 32 km), that is, continental crust that has undergone stretching and moderate attenuation. This relationship is shown by gravity data and tectonic subsidence analysis of the Wiggins uplift–Mississippi salt basin area (6, 7, Fig. 1; Nunn and others, 1984, 1989). Gravity modeling and refraction studies indicate that the Wiggins uplift is underlain by substantially thicker crust (32 km) than the adjacent basins (20 km) (Nunn and others, 1989). However, estimates of crustal thickness variations based on total tectonic subsidence analyses are significantly lower (Nunn and others, 1989). It is not known whether this differential thinning was caused by deformation of an initially uniform crust, as suggested by the constant wavelengths, or was influenced by inherited differences in crustal geology such as older Paleozoic accreted terranes and sutures.

The basement highs where drilled (heavy lines, Fig. 1) consist of a variety of rocks lying below the MJS, including Precambrian and Paleozoic igneous and metamorphic rocks, Paleozoic sedimentary rocks, and the Late Triassic–Early Jurassic rift-related rocks. All of these rocks reflect the earlier history of the basin.

The structural features in the area of thick transitional crust controlled the general distribution (Fig. 1) and thickness of the inferred Callovian salt (pre-marine evaporites), as well as the overlying Jurassic through Lower Cretaceous rocks. The thickest salt is in the broad lows or sag basins (i.e., 5, 7, 9, 11, 13, Fig. 1), whereas the salt is thin or absent over the adjacent highs (Salvador, 1987). In central Mexico (14, 15, Fig. 1), over the Yucatan block (19, Fig. 1), and in the eastern Gulf (1, 2, 3, 4, Fig. 1), salt is generally absent in the area of thick transitional crust. An exception is the western part of the Tampa embayment (3, Fig. 1), where thin salt is indicated by seismic data (Dobson, 1990).

The control of basement structure on the distribution of salt in the northern Gulf indicates that the major phase of stretching and attenuation of the crust to form the basic basin architecture occurred just before and/or during the deposition of the salt, which was mid-Jurassic time, as Africa/South America was separating from North America. These broad structures appear to have had an origin somewhat different from the underlying older Late Triassic–Early Jurassic rift basins. The contrast between the broad downwarps and half-grabens bordered by listric faults suggests that crustal stretching had expanded from more brittle deformation controlled by Paleozoic structural fabrics to widespread attenuation and differential subsidence of a more ductile crust across the entire Gulf region. These two stages are separated over much of the basin by the prominent MJS.

This stretching and attenuation stage may have been caused by an expanding heat source from an upwelling mantle as the continents separated. The final stage of attenuation culminated with emplacement of a band of oceanic crust during the Late Jurassic. Exact mechanisms and processes in stretching of the crust during these two different phases are not clear. Needed are deep crustal reflection and refraction profiles across the basement highs and lows and crustal boundaries to identify crustal thickness, faults, detachment surfaces in the upper crust, and the brittle/ductile boundary.

The area of thick transitional crust forming the northeast margin of the Gulf basin has a unique structural configuration for the Gulf, as well as for continental margins in general. This northwest-trending band is marked by generally east-trending, high-low structural pairs (1 to 7, Fig. 1). The trend of the high-low pairs is at an oblique angle to the overall trend of the margin, which is significantly different, for example, from the adjacent southern Atlantic margin to the southeast (e.g., Trehu and others, 1989). It is postulated that the oblique structures are the signature of transtensional motion. This motion could have included the counterclockwise rotation of smaller individual crustal blocks as part of the attenuation process, thus, forming the high-low pairs. As blocks separated, intervening lows or sag basins formed between them and became the locus for deposition of the overlying Jurassic salt and sedimentary rocks. Alternatively, the orientation may have been controlled by preexisting crustal structures. The scale of the high-low structural pairs suggests that they extend through the crust. The nature and configuration of the boundaries of these structures are critical to understanding the kinematics of opening of the Gulf.

The details of one of these high-low pairs are shown by recent studies of the pre-MJS rocks and the overlying Jurassic and Lower Cretaceous rocks in the Middle Ground arch–Southern platform area and adjacent basins (Apalachicola basin and Tampa embayment). A detailed map of the MJS (Fig. 20) from Dobson (1990) shows the broad east-west Middle Ground arch separated from the Southern platform by a gentle northwest-trending saddle. This saddle is aligned with a northwest-trending linear segment of the Florida escarpment on the southeast, a surface expression of the eroded mid-Cretaceous platform margin in the eastern Gulf (Fig. 16). Here an inferred Mesozoic fault zone possibly offsets basement and controls the location of the overlying Lower Cretaceous platform margin (Fig. 16). This northwest trend from the Florida escarpment to the saddle on the Middle Ground arch may reflect one of the regional transform fault systems that have been postulated for the eastern Gulf (Klitgord and others, 1984, 1988).

This basement configuration controlled the distribution and thickness of the overlying salt (Fig. 21; Dobson, 1990). In the Apalachicola basin to the north, the thickest salt and associated salt structures are in the deepest part of the basin, but thin salt extends beyond into the shallower parts of the basin, as well as onto the continental crust (Figs. 17, 21). The updip limit of thick salt controlled the distribution of later faulting caused by salt withdrawal around the periphery of the Apalachicola basin (Figs. 17, 18). To the south, only thin salt was deposited in the western Tampa embayment, and thick salt was restricted to the adjacent West Florida basin (Figs. 16, 21). Salt is absent over the crest of the Middle Ground arch–Southern platform and the Wiggins uplift–Pensacola arch to the north (Fig. 21).

Differential subsidence of the basins continued throughout the Late Jurassic and Early Cretaceous as indicated by depositional patterns and thicknesses of the overlying units (Figs. 16, 17, 21). For example, Jurassic rocks are much thicker in the Apalachicola basin and the Tampa embayment than over the Middle Ground arch–Southern platform (Figs. 16, 17; Dobson, 1990). In addition, the early Late Jurassic depositional shelf edges appear to follow the MJS contours (Figs. 20, 21; Salvador, 1987; Dobson, 1990). Differential subsidence continued into the Early Cretaceous as outlined in a study by Corso (1987), who mapped three Lower Cretaceous units in the area. Each of these three units is thicker in the axis of the Apalachicola basin than over the adjacent Southern platform. This continued differential subsidence may be, in part, a result of differential sediment loading.

Boundary between thick and thin transitional crust. The basinward boundary of the area of thick transitional crust is

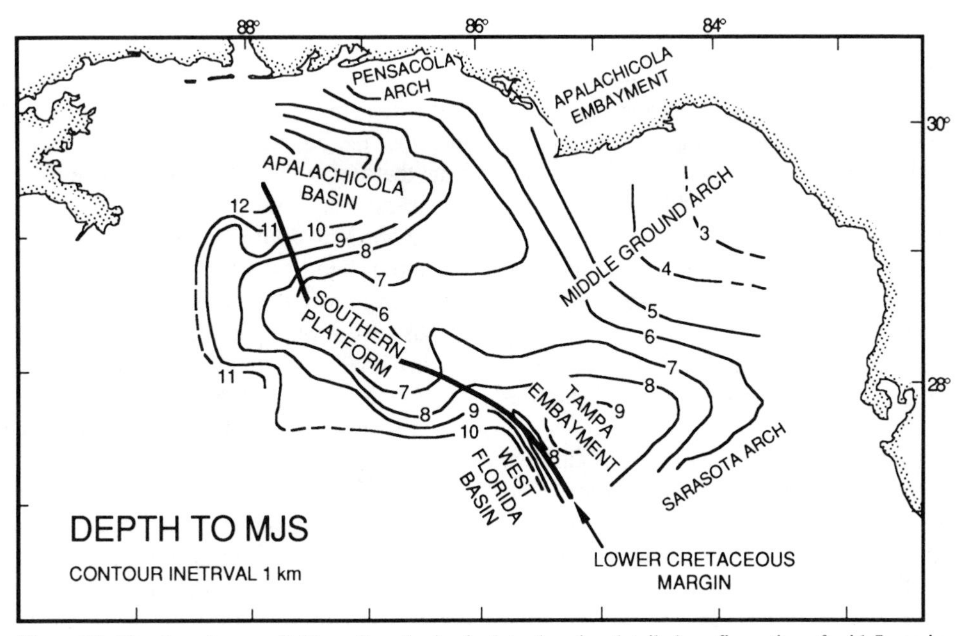

Figure 20. Map based on available well and seismic data showing detailed configuration of mid-Jurassic surface (MJS) in northeastern Gulf of Mexico basin (also see Fig. 1) (from Dobson, 1990). Lower Cretaceous margin along Florida escarpment shown for reference.

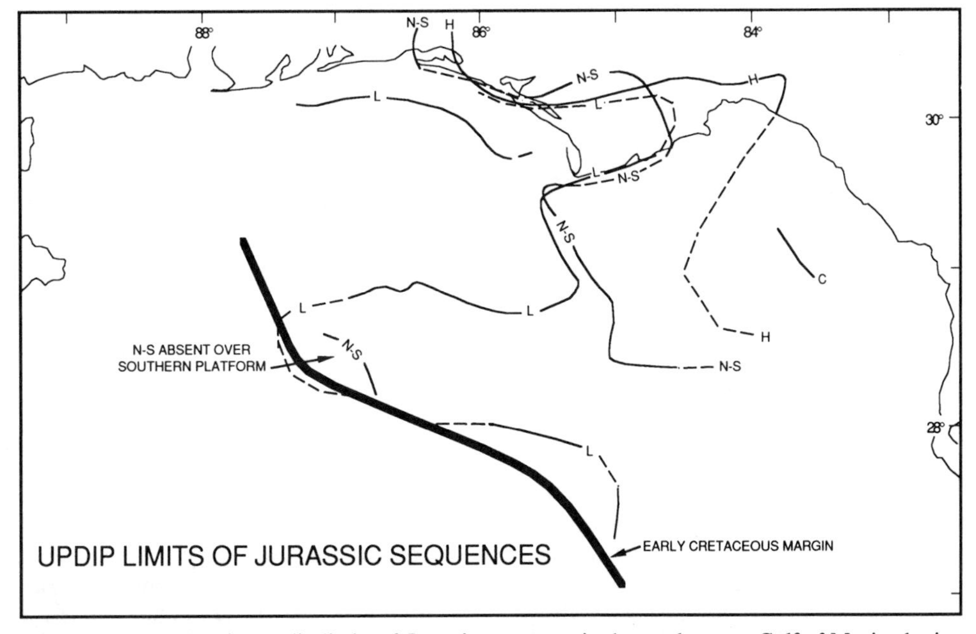

Figure 21. Map showing updip limits of Jurassic sequences in the northeastern Gulf of Mexico basin influenced by basement (mid-Jurassic surface, MJS) configuration (see Fig. 20) (from Dobson, 1990). L, Louann Salt; N-S, Norphlet-Smackover; H, Haynesville; C, Cotton Valley. Early Cretaceous margin along Florida escarpment shown for reference.

characterized by an overall drop in the elevation of the MJS, a major change in the orientation of structural trends and salt subbasins, and an overall decrease in crustal thickness (Figs. 1, 9, 13). For example, in the eastern Gulf the generally basinward-plunging highs and lows of the thick transitional crust are terminated abruptly westward and are bordered by structure that parallels the eastern margin of the area of oceanic crust. The boundary is intepreted to be associated with a circum-Gulf tectonic hinge zone involving both deeper crustal rocks and the overlying Upper Jurassic section (Buffler and Sawyer, 1985; Corso, 1987; Winker and Buffler, 1988). The boundary apparently represents a crustal weakness established during the mid-Jurassic crustal attenuation stage. Differential subsidence and flexuring evidently continued along the boundary into the Late Jurassic or was reactivated in the Late Jurassic, as it influenced deposition of both the Jurassic and overlying Lower Cretaceous rocks.

This boundary is unique to the Gulf basin, and no equivalent boundary has been recognized along the Atlantic margin (Klitgord and others, 1988). A possible exception to this generalization is the area east of the East Coast magnetic anomaly (ECMA) and landward of well-defined oceanic crust (Klitgord and others, 1988). Recently, this area has been postulated by Hall (1990) to represent extremely thinned continental crust, thus making the ECMA equivalent to the thick/thin boundary in this paper.

The boundary between thick and thin transitional crust became the focus for the establishment of the rimmed Lower Cretaceous carbonate platform margin that faced the subsiding deep Gulf basin (Fig. 9; Buffler and Sawyer, 1985; Winker and Buffler, 1988). Here shelf margins were not able to continue to prograde into the deeper part of the basin and began to aggrade vertically as the basin continued to differentially subside. This relationship has been observed at several locations around the Gulf, such as in the south-central Gulf along the Campeche escarpment (Huerta, 1980), and in the southern Texas–northern Mexico area (Smith, 1981). In addition, a flexure along the crustal boundary has been documented in detail in the northeastern Gulf by Corso (1987) and Dobson (1990) using an extensive grid of seismic reflection data. Here the mid-Jurassic surface and overlying Jurassic section increase in dip basinward, and it is where the overlying latest Jurassic–Early Cretaceous carbonate platform margins became established and aggraded vertically (Fig. 17).

Regionally the thick/thin boundary as mapped on Figure 1 and shown on the regional cross sections (Fig. 9) corresponds to the location of the Lower Cretaceous margin. An exception to the thick/thin boundary controlling the linear Lower Cretaceous margin is in the western Gulf basin. Here the margin is broken into a series of platforms and basins controlled by the underlying highs and lows on the MJS. The interpreted crustal boundary corresponds to an inferred northwest-trending linear gradient in the MJS structure (Figs. 1, 9).

In some places there appears to be some faulting associated with the boundary. For example, along the Florida escarpment south of the Southern platform, the Jurassic section is interpreted to be offset by faults (Fig. 16; Dobson, 1990). Here the upthrown block clearly has controlled the location of the overlying Lower Cretaceous platform margin along the Florida escarpment. There also appears to be a large offset in the MJS, that separates an area of thick salt in the West Florida basin from thin salt in the adjacent Tampa embayment (Fig. 16). This tends to support the idea that the boundary was an earlier tectonic feature or hinge zone that was active during the attenuation stage and controlled the distribution of salt.

Thin transitional crust. Seaward of the inferred tectonic hinge zone and Lower Cretaceous margin is an extensive area of thin transitional crust (Fig. 1) characterized by (1) increased depths to the MJS, (2) decreased crustal thickness, and (3) a marked change in orientation of structures. Depths to the MJS range from 5 to 8 km in the southeastern Gulf to 12 to 16 km beneath the northern Gulf shelf/slope. This range in depths mainly reflects differential sedimentary loading; the northern Gulf was loaded by a huge prism of mainly Cenozoic sediments, while the southeastern Gulf remained relatively starved. Depths corrected for sediment loading indicate that the area of thin transitional crust subsided tectonically approximately 4 to 6 km in contrast to 2 to 4 km for the adjacent area of thick transitional crust (Fig. 13; Dunbar and Sawyer, 1987; Sawyer and others, 1991). Crustal thickness ranges from approximately 6 to 20 km on the basis of a few scattered refraction stations, particularly in the southeastern Gulf and the Sigsbee salt basin (21, Fig. 1; Ibrahim and others, 1981; Locker and Chatterjee, 1984).

The strike of major structures appears to be parallel to the boundary of oceanic crust (e.g., the Sigsbee salt basin, 21, and the West Florida basin, 23, Fig. 1). The configuration of the MJS beneath the northern Gulf, however, is speculative (light lines) and is based only on a few refraction stations (Nakamura and others, 1988; Ebeniro and others, 1988; Yosio Nakamura, personal communication, 1991). All of these areas, except for the southeastern Gulf, are overlain by thick salt which generally outlines the original salt basins, particularly in the south and east (20, 21, 23, Fig. 1; Salvador, 1987). Although salt structures mapped in the northern Gulf (Martin, 1980) mainly represent salt now displaced or remobilized by sediment loading, the original distribution must have been somewhat similar (Salvador, 1987). This broad area of salt is known as the North Gulf salt basin (22, Fig. 1). The deep Gulf salt is inferred to be the same age (Callovian) and to have had a similar origin as the salt in the surrounding shallower salt basins (Louann Salt) (Salvador, 1987).

The distribution of thin transitional crust is markedly asymmetric (Fig. 1). The area is relatively narrow in the east, southeast, and south, but is somewhat broader in the southwest. The western part is inferred to be very narrow, corresponding to the steep gradient in the mid-Jurassic surface. Of particular significance, however, is the extremely broad area of stretched crust to the north (as much as 500 km wide). This perhaps is one of the widest areas of thin rifted continental crust in the world.

The thin continental crust is interpreted to represent continental crust that underwent extreme stretching and thinning as South America (including Yucatan) and Africa moved away from North America. Attenuation occurred prior to emplacement of oceanic crust in the deep central Gulf, probably in mid-Jurassic time as indicated by the inferred age of the overlying Callovian(?) salt. Parts of this broad area apparently underwent extreme thinning, particularly in the northwestern Gulf, where a refraction station recorded a shallow Moho and oceanic crustal thicknesses (Fig. 22; Ebeniro and others, 1988). This area is represented on Figure 1 by the inferred narrow basement trough (14 km contour) extending beneath the continental slope. The inferred trough and thin crust may represent a rift that aborted when oceanic crust started to be emplaced to the south (Ebeniro and others, 1988).

Subsidence of this area of stretched crust formed a broad basinal area in which thick evaporites were deposited, probably as a result of the repeated influx of seawater across sills to the west in what is now Mexico (Salvador, 1987). Alternatively, marine waters could have entered the Gulf through the deep southeastern Gulf, as suggested by similar age salt in northern Cuba (Pardo, 1975) and possibly the Straits of Florida as indicated on unpublished seismic data. The lack of any obvious salt in the southeastern Gulf region (Fig. 1), however, tends to support the former idea. The salt probably was deposited very rapidly during a relatively short period, while a delicate balance was maintained between evaporation and the influx of marine waters. The final breakthrough of mantle and emplacement of oceanic crust initiated rapid subsidence of the central basin and allowed for the first marine transgression into the Gulf basin in Oxfordian time (Salvador, 1987), thus terminating evaporite deposition.

In the area of the deep eastern Gulf, the thin transitional crust is characterized by a large asymmetrical salt basin known as the West Florida basin (23, Fig. 1). Thick salt apparently extends along the northeast flank of the basin just west of the Florida escarpment, as suggested by the distribution of shallower salt structures (Fig. 16; Martin, 1980; Lord, 1986, 1987). The salt and inferred overlying Jurassic sedimentary section thins and onlaps the MJS to the west toward the area of oceanic crust (Fig. 16).

In the south-central Gulf, south of the area of oceanic crust is another asymmetrical salt basin known as the Sigsbee salt basin (21, Fig. 1; Lin, 1984). Here the asymmetry is opposite from that of the West Florida basin; the thicker salt is along the northern flank of the basin adjacent to the area of oceanic crust (Fig. 23). The salt and overlying Jurassic section onlaps and pinches out depositionally to the south onto the MJS along the base of the Campeche escarpment (Fig. 23). Here the MJS truncates a widespread layered sequence identified on seismic data. By analogy with similar sections in the northern and northeastern Gulf, this layered sequence is interpreted to represent another Late Triassic–Early Jurassic rift basin.

The deep southeastern Gulf between the Florida and Campeche escarpment is a different area of thin transitional crust. Here the MJS is shallow (5 to 8 km) (Fig. 1) and is overlain by a relatively thin Mesozoic and Cenozoic sedimentary cover (Fig. 9B; Schlager and others, 1984). The crust thickens from north to south, from about 6 km at the ocean crust boundary to approxi-

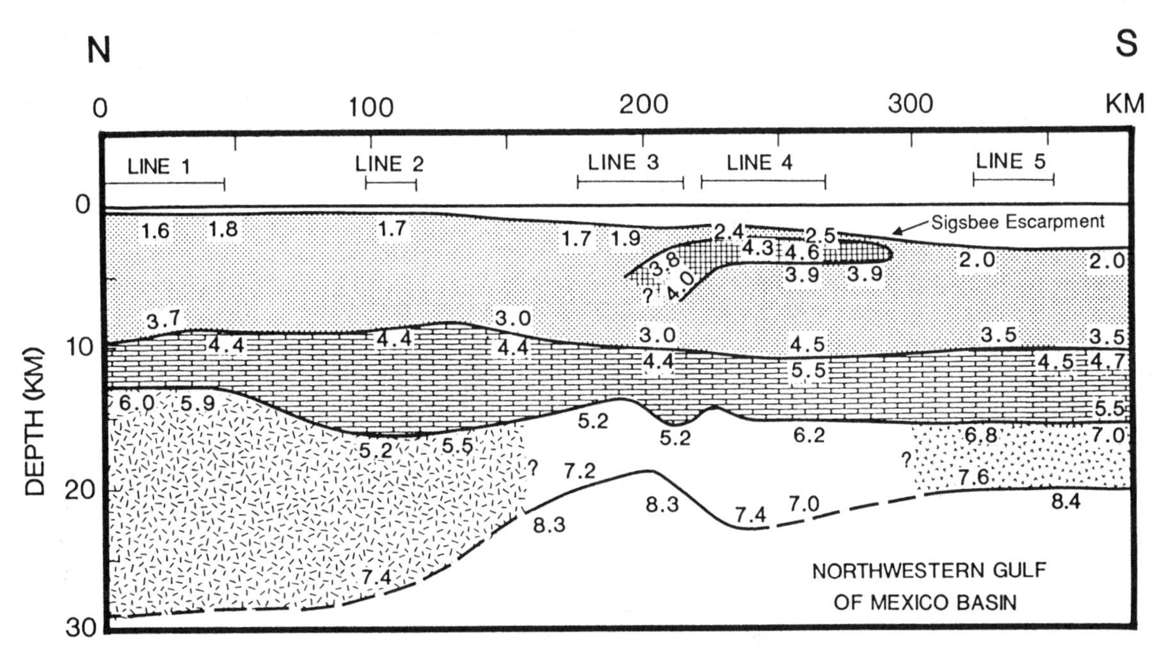

Figure 22. Generalized crustal cross section of the northern Gulf of Mexico (from Ebeniro and others, 1988) based on seismic refraction data. Oceanic crust is indicated by the stippled pattern on the right side of the figure. The velocities and greater thickness of crust on the left are diagnostic of modified continental crust or thin transitional crust. The crustal thickness minimum under line 3 may correspond to an alternative axis of crustal extension that failed to proceed to sea-floor spreading. See Figure 1 for general location.

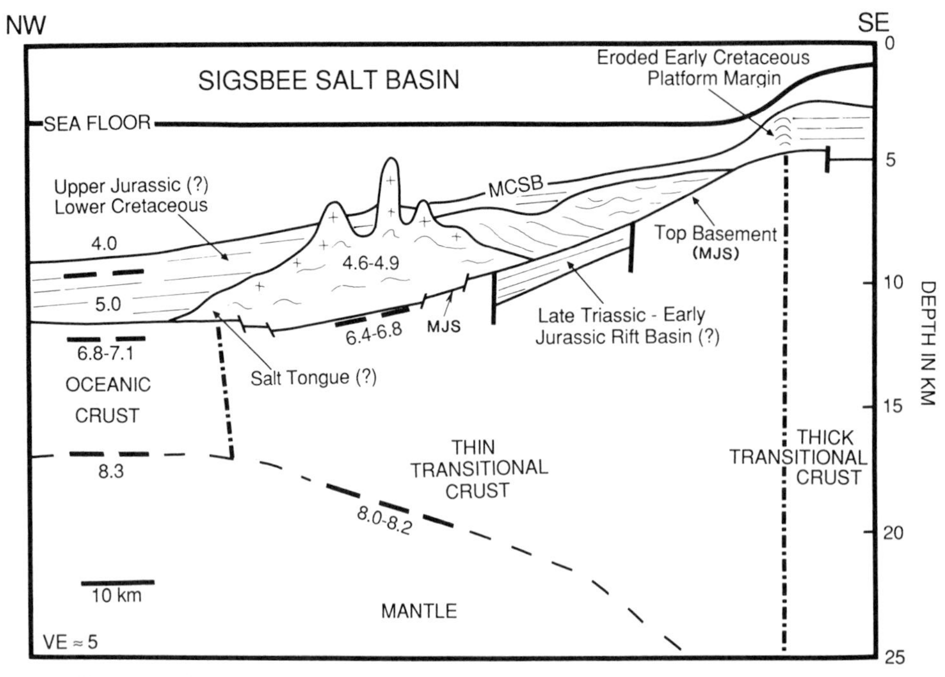

Figure 23. Generalized cross section of the Sigsbee salt basin (21, Fig. 1) showing crustal types and overlying Mesozoic sedimentary section (modified from Lin, 1984; Buffler, 1991; Sawyer and others, 1991).

mately 20 km just north of Cuba (Figs. 9B, 24; Ibrahim and others, 1981; Ebeniro and O'Brien, 1984). The southern part of the area is characterized by broad basement blocks and platforms, whereas the northern part is an extensive area of smaller tilted fault blocks (half-grabens) filled with Jurassic syn-rift deposits (Fig. 25).

DSDP Leg 77 (Buffler and others, 1984) sampled crustal rocks on the top of two of the tilted fault blocks (Sites 537 and 538) and recovered metamorphic rocks with ~500 Ma radiometric ages based on ^{40}Ar-^{39}Ar plateau dates (Dallmeyer, 1984). At Site 538 (Fig. 25) gneissic rocks were intruded by diabase dikes and sills with ages ranging from 190 to 160 Ma. The ages of the metamorphic rocks drilled at Sites 537 and 538 suggests affinities with the Pan-African age rocks in the subsurface of Florida (Dallmeyer, 1984). The diabase ages suggest affinities both with the Late Triassic–Early Jurassic volcanic rift-related rocks in the northern Gulf and with the Jurassic volcanic rocks in the subsurface of southern and central Florida along the projection of Bahamas fracture zone (Fig. 15).

The combination of rocks drilled at Site 538 exemplifies the definition of transitional crust or rifted continental crust. The metamorphic rocks represent original continental crust, and the diabases represent magma injected into the continental crust during Jurassic attenuation. Although this is the only place in the circum-Gulf area where thin transitional crust has been directly sampled, the transitional crust presumably consists everywhere of continental crustal rocks and mafic intrusive rocks similar to those encountered in Leg 77.

The differences in the nature of thin transitional crust between the elongate salt basins in the northern Gulf (21, 22, 23, Fig. 1) and the more faulted area in the southeastern Gulf may reflect different tectonic settings during the Jurassic evolution of the basin. The central Gulf may have been a divergent or pull-apart margin along which elongate salt basins formed, as Yucatan moved out of the northern Gulf. The southeastern Gulf probably was along a transtensional transform margin between Yucatan and Florida and was characterized by block rotation and opening of rhomb grabens.

Oceanic crust. Underlying the deep central Gulf is a band of oceanic crust (Fig. 1), which originally was defined by early seismic refraction studies (Ewing and others, 1960, 1962) and later better defined by combined seismic refraction and reflection studies (Figs. 12, 22, 24; Buffler and others, 1980, 1981; Ibrahim and others, 1981; Buffler and Sawyer, 1985; Ebeniro and others, 1988). The band is more than 300 km wide in the west, but it narrows considerably in the eastern Gulf, where it makes a bend to the southeast (Fig. 1). The oceanic crustal layer is generally 5 to 6 km thick and is characterized by refraction velocities of 6.8 to 7.2 km/s (Figs. 22, 24). This layer in the Gulf probably represents oceanic Layer 3 as defined for oceanic crust in normal ocean basins.

Throughout much of the central and eastern deep basin, an irregular acoustic basement is recognized on reflection data and is interpreted to be the top of oceanic crust (Fig. 12; Buffler and others, 1980, 1981; Ibrahim and others, 1981; Rosenthal, 1987). It is the horizon mapped on Figure 1 in the area of oceanic crust.

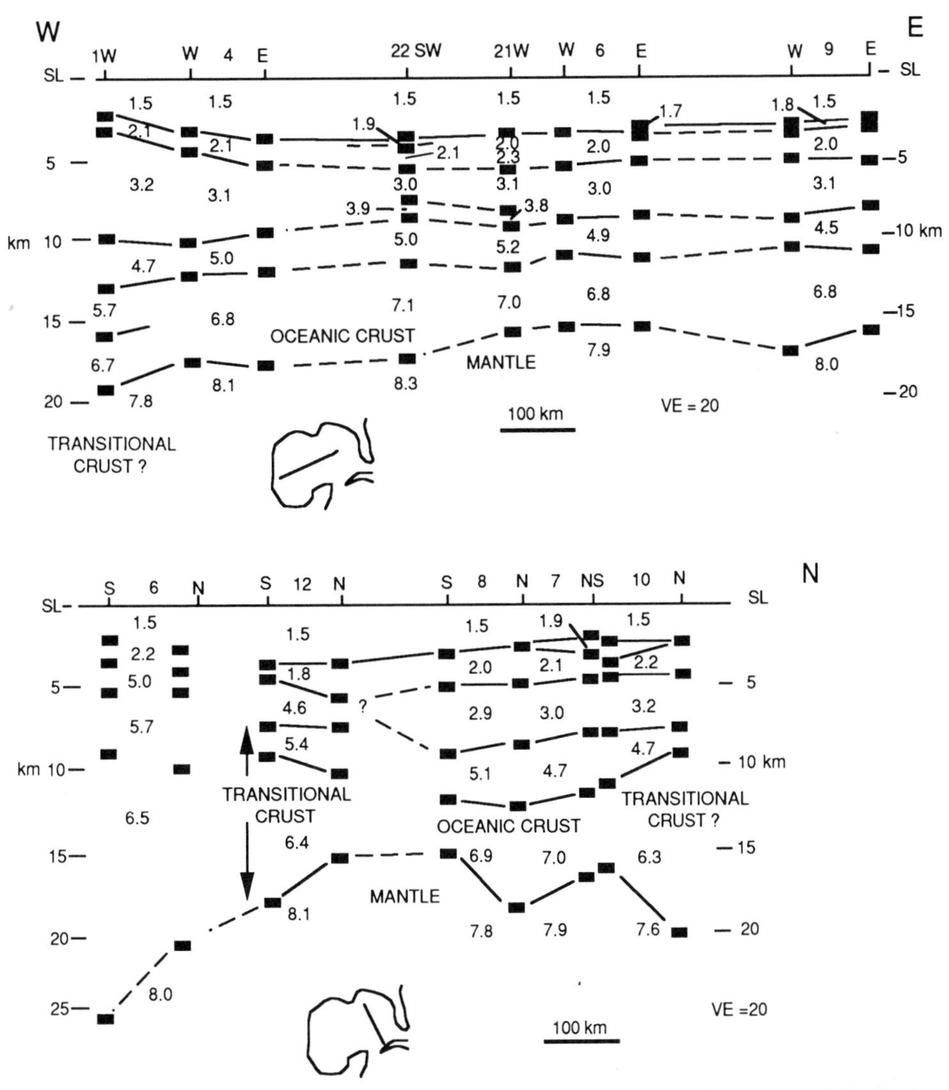

Figure 24. Cross sections across the deep Gulf of Mexico basin showing crustal structure and distribution of crust based on refraction velocities (from Buffler and others, 1981; Ibrahim and others, 1981; Ebeniro and O'Brien, 1984).

This surface does not, however, correspond to the top of the high-velocity crustal layer (6.8 to 7.2 km/s), but it is generally just above it (Fig. 12; Ibrahim and others, 1981). The acoustic basement, therefore, is interpreted to represent the top of the oceanic volcanic layer (Layer 2), of oceanic crust in normal ocean basins. The volcanic layer probably is not expressed as a refraction layer in the Gulf, because it is masked by an overlying high-velocity sedimentary layer (4.8 to 5.2 km/s). This uniformly layered sedimentary sequence onlaps and fills the top of the irregular volcanic layer and is interpreted to be mainly deep-water carbonate rocks and some terrigenous clastic rocks (Fig. 12; Challenger unit of Shaub and others, 1984; Ibrahim and others, 1981; Rosenthal, 1987).

The oceanic crust and boundaries in the eastern deep Gulf were the subject of a recent study by Rosenthal (1987). Using both regional University of Texas seismic lines and a detailed grid of industry lines over part of the area, he was able to identify a series of east-northeast–trending linear highs and lows on the top of oceanic crust (Fig. 26). In one place, these linear features are cut by a prominent north-northwest–trending trough, and they are truncated to the east at an area of flat-lying, smooth oceanic crust (Fig. 26). Rosenthal (1987) suggested that the two different areas of oceanic crust may have had different origins. The linear features are interpreted to be spreading ridges related to a north-northwest opening of the central Gulf, whereas the area of smooth crust may be flood basalts that filled a transtensional basin or leaky transform in the eastern Gulf.

The deep central and eastern Gulf is characterized by low-amplitude magnetic anomalies (Fig. 27; Klitgord and others, 1984; Hall and others, 1984; Sawyer and others, 1991). Although

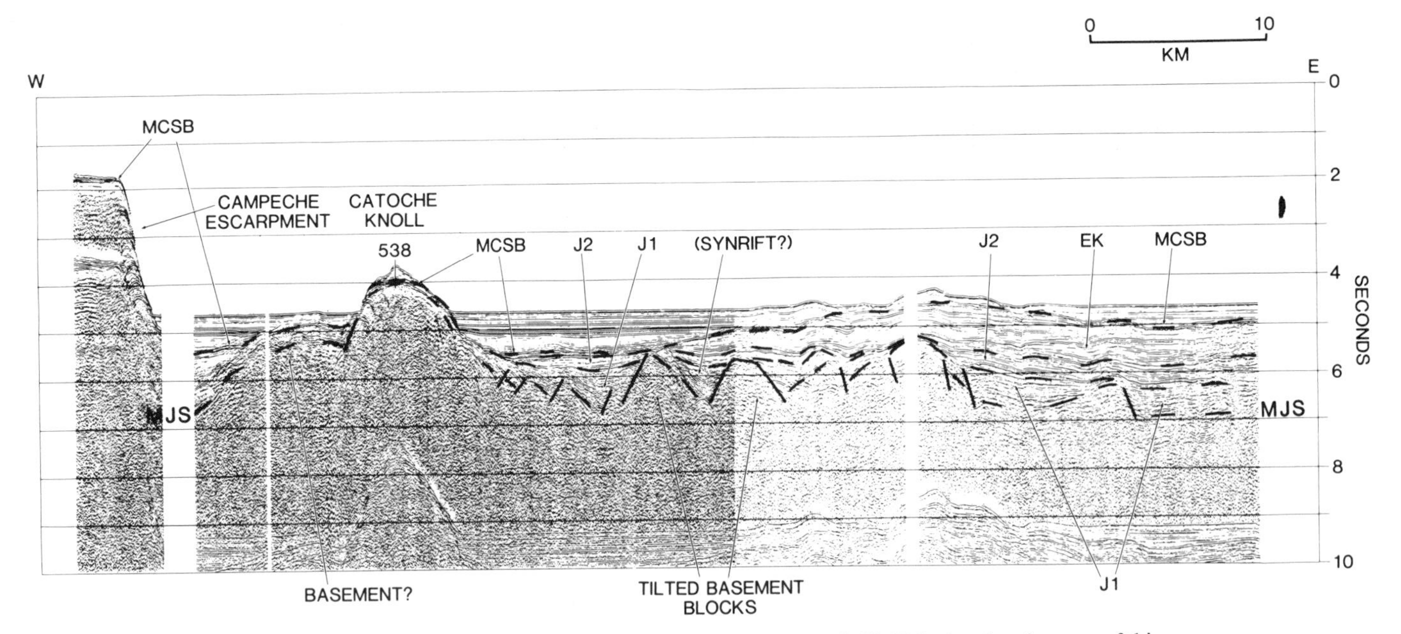

Figure 25. UTIG multifold seismic line SF-2 in deep southeastern Gulf of Mexico showing area of thin transitional crust characterized by block-faulted mid-Jurassic surface (MJS) (from Schlager and others, 1984). Rift basins are filled with Jurassic? syn-rift deposits (J1) and overlain by Upper Jurassic and Lower Cretaceous deep-marine strata (J2 and EK). Deep Sea Drilling Project (DSDP) site 538 drilled on the top of Catoche Knoll encountered Paleozoic metamorphic rocks intruded by Jurassic diabase rocks below the mid-Jurassic surface (MJS), an excellent example of transitional crust. See Figure 1 for general location.

more extensive than the area of oceanic crust mapped by seismic data, the area of the low-amplitude anomalies is similar in shape and probably reflects the overall distribution of oceanic crust (Buffler and Sawyer, 1985; Sawyer and others, 1991). The adjacent areas are characterized by high-amplitude magnetic anomalies, which probably represent a dike-injected transitional crust. No well-defined linear oceanic magnetic anomalies were recognized on the older magnetic data (Fig. 27), perhaps because of poor data modeling (Hall and others, 1984). Preliminary modeling of the anomalies, however, did suggest a tentative correlation with the Mesozoic geomagnetic reversal scale and a Late Jurassic–Early Cretaceous opening (Shepherd, 1983).

More recently, a detailed aeromagnetic survey over the entire central and eastern deep Gulf has identified some generally east-west linear magnetic anomalies in the area previously believed to contain no linear anomalies (Hall and Najmuddin, 1992; Najmuddin, 1992; Najmuddin and Hall, 1992). These anomalies diverge slightly to the west. They were interpreted to be Late Jurassic–Early Cretaceous (151 to 142 Ma) sea-floor spreading anomalies, which suggests a more north-south, counterclockwise opening of the eastern Gulf about a pole located in central Florida. The entire area of low-amplitude anomalies (Fig. 27) was considered to be underlain by oceanic crust, again an area more extensive than that mapped by seismic data (Fig. 1).

Total tectonic subsidence analysis techniques also can be used to estimate the distribution of oceanic crust in the Gulf of Mexico (Buffler and Sawyer, 1985; Dunbar and Sawyer, 1987). The distribution of oceanic crust is indicated by this method to be the area of maximum crustal extension (beta of 4.5, Fig. 14). Although slightly larger, this area corresponds approximately to the outline of oceanic crust interpreted from seismic data (Fig. 14; Buffler and Sawyer, 1985; Dunbar and Sawyer, 1987; Sawyer and others, 1991).

The age of the oceanic crust is believed to be Late Jurassic, but this is not well constrained. It apparently was emplaced after deposition of the Callovian salt and associated deposits, because they onlap and pinch out over transitional crust and are not present over oceanic crust (Phair and Buffler, 1983; Rosenthal, 1987; Salvador, 1987). Tentative correlation of seismic stratigraphic units from DSDP Leg 77 sites in the southeastern Gulf suggests that most of the sedimentary section overlying oceanic crust is Early Cretaceous (Phair and Buffler, 1983; Lord, 1986, 1987), thus bracketing the age of oceanic crust as Late Jurassic–earliest Cretaceous. The interpretation of the recent aeromagnetic survey suggests a latest Jurassic–earliest Cretaceous age (Hall and Najmuddin, 1992; Najmuddin, 1992; Najmuddin and Hall, 1992).

Emplacement of the crust is believed to represent the final rift stage in the evolution of the Gulf basin, as the broad area of crustal attenuation (transitional crust) and mantle upwelling finally broke to the surface along a narrow east-west element of the rift in the central Gulf. Another possible rift zone in the northwest Gulf aborted after the central Gulf rift became established (Fig. 22; Ebeniro and others, 1988). Following formation of oceanic

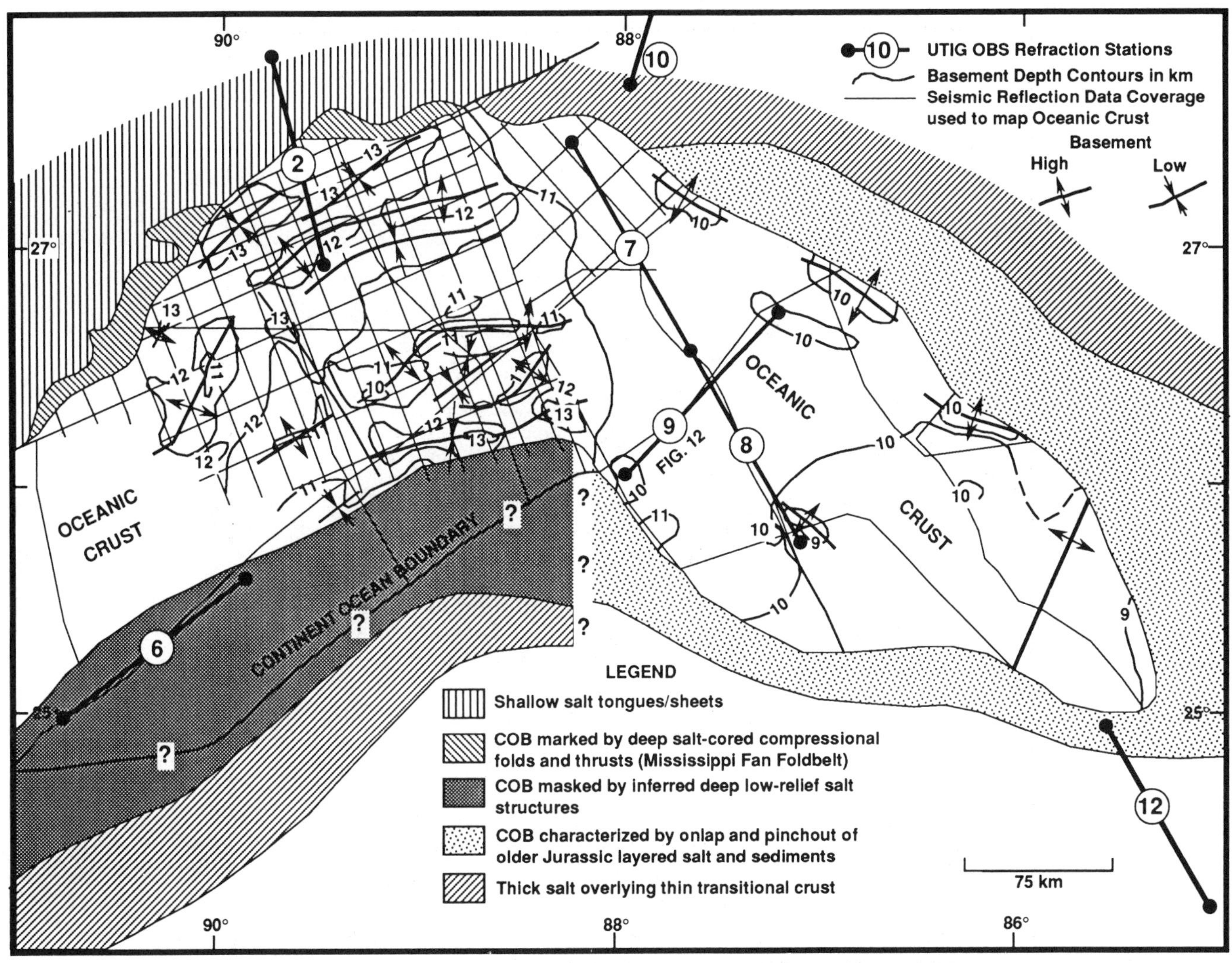

Figure 26. Map of crust boundaries and depth to top of oceanic crust (mid-Jurassic surface, MJS) in the deep eastern Gulf of Mexico basin from Rosenthal (1987) (also see Fig. 1). COB is the continent-ocean boundary.

crust, the deep basin continued to cool and subside through the Early Cretaceous and receive a thick accumulation of mainly deep-water carbonate rocks (Buffler, 1991).

Oceanic/transitional crust boundary. The boundary between oceanic crust and thin transitional crust as mapped on Figure 1 is defined by refraction and reflection data around most of the deep basin. All along the north flank of the Sigsbee salt basin (21, Fig. 1), the exact crustal boundary is not indicated on reflection data, because it is masked by a salt tongue (Figs. 1, 23; Lin, 1984). This salt tongue is onlapped and overlain by the undeformed Upper Jurassic?–Lower Cretaceous deep-marine deposits in the central deep Gulf basin, indicating that it was emplaced during the Late Jurassic evolution of the basin. The boundary is interpreted to lie under the leading edge of the salt tongue on the basis of refraction, magnetic, and gravity data (Fig. 23; Buffler and others, 1980, 1981; Hall and others, 1982; Lin, 1984).

In the eastern Gulf, the distribution of oceanic crust is defined in general terms by refraction data (Figs. 24, 26; Ibrahim and others, 1981). More specifically, the boundary as shown on Figure 1 is defined by the onlap and pinchout of Middle to Upper Jurassic strata (Figs. 9, 16, 26; Phair and Buffler, 1983; Rosenthal, 1987). This older section, which includes Jurassic salt, apparently represents nonmarine to marine rocks deposited on thin transitional crust prior to the emplacement of oceanic crust in Late Jurassic time and prior to rapid subsidence and deposition of the overlying Upper Jurassic?–Lower Cretaceous deep-water sedimentary deposits (Fig. 16).

The boundary of oceanic crust in the northern Gulf is obscured by thick Cenozoic deposits and by shallow, remobilized salt. The location of the boundary is estimated from the inferred original distribution of thick salt (Salvador, 1987). Analogy with the Sigsbee salt basin to the south suggests (1) that salt was deposited on transitional crust and not on oceanic crust, and

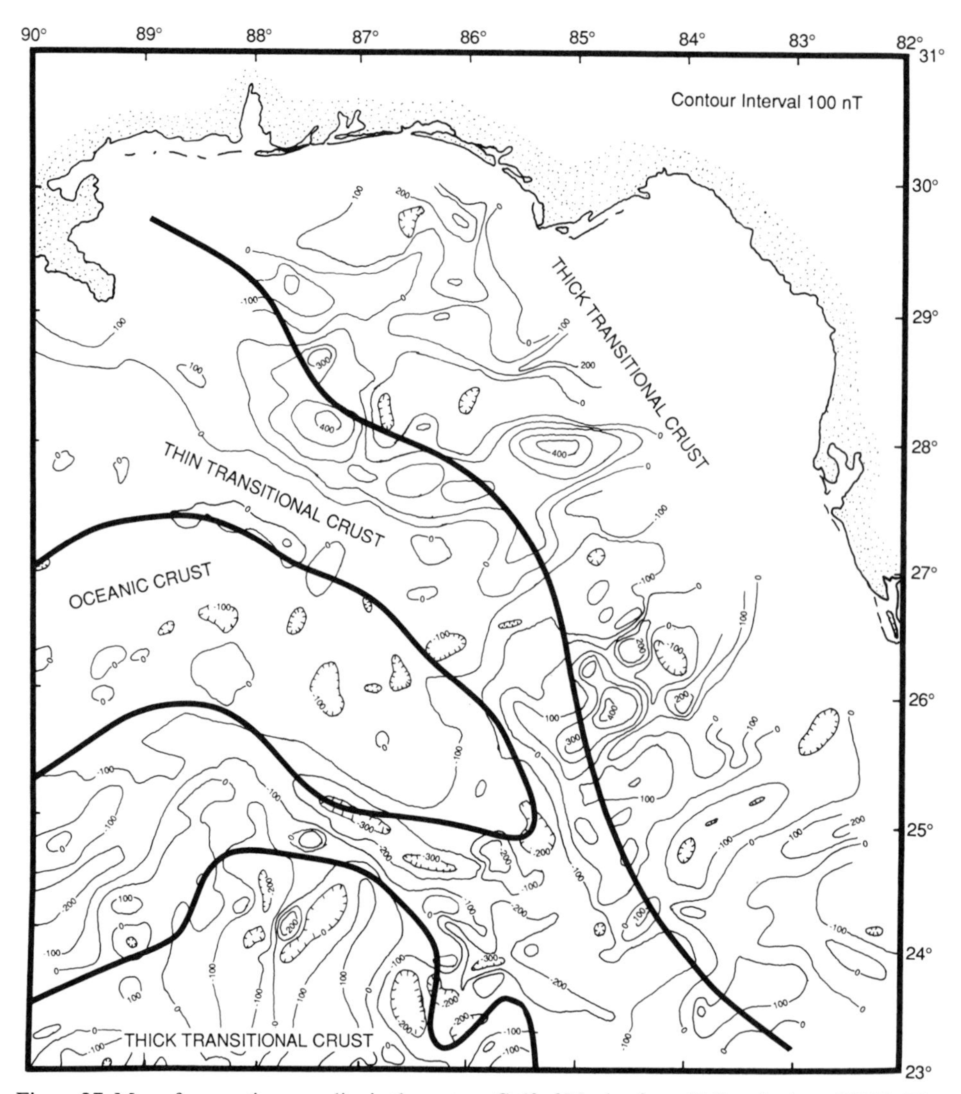

Figure 27. Map of magnetic anomalies in the eastern Gulf of Mexico from Hall and others (1984). The bold lines are the crust-type boundaries from Figure 1. The magnetic anomalies are generally subdued over oceanic crust (modified from Sawyer and others, 1991).

(2) that the present distribution of salt is not greatly different from the original distribution. Most of the northern boundary is more accurately defined by several large compressional (contractional) foldbelts, the Perdido foldbelt in the northwest and Mississippi Fan foldbelt under the Mississippi Fan (Figs. 1, 26, 28; Blickwede and Queffelec, 1988; Worral and Snelson, 1989; Fuqua, 1990; Weimer and Buffler, 1992). Both of these foldbelts are interpreted to be salt-cored (Fig. 28). These foldbelts apparently represent horizontal shortening near the toe of a deep detachment in basal salt caused by sediment loading upslope (Fig. 28). The folds form at the basinward depositional limit of the salt, which presumably was deposited on transitional crust. Here the décollement surface ends, and the transmitted stresses are buttressed against the more rigid deep Gulf sedimentary section that overlies oceanic crust. Thus, the basinward edge of both of these foldbelts is interpreted to approximate the boundary between oceanic crust and transitional crust (Fig. 1). The actual boundary could be somewhat landward of this location, because of some possible early (Jurassic) basinward movement of the salt in the form of a salt tongue similar to that in the Sigsbee salt basin to the south.

These two foldbelts are clearly offset along the lower slope (Fig. 1). The offset is characterized by a zone of deep-seated, northwest-trending salt structures along the lower slope (Lee and others, 1989). Furthermore, this offset is marked by a prominent northwest-trending gravity anomaly, which is even more prominent on terrain-corrected satellite gravity data (Chris Small, personal communication, 1991). This offset, therefore, is interpreted to be a deep-seated structure, possibly a transform boundary related to a northwest-opening direction of the Gulf of Mexico basin.

The western boundary is obscured by thick sediments and the Mexican Ridges foldbelt (Fig. 9), but it can be inferred from a

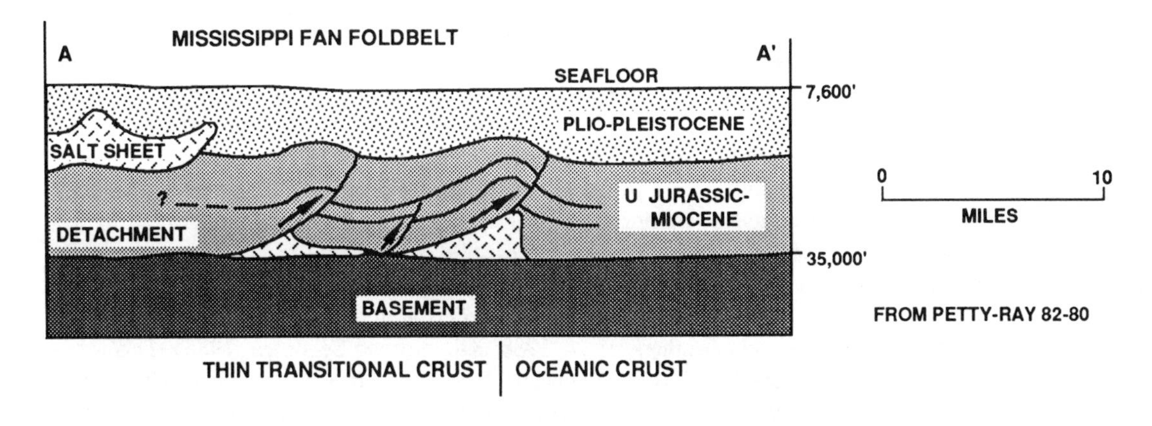

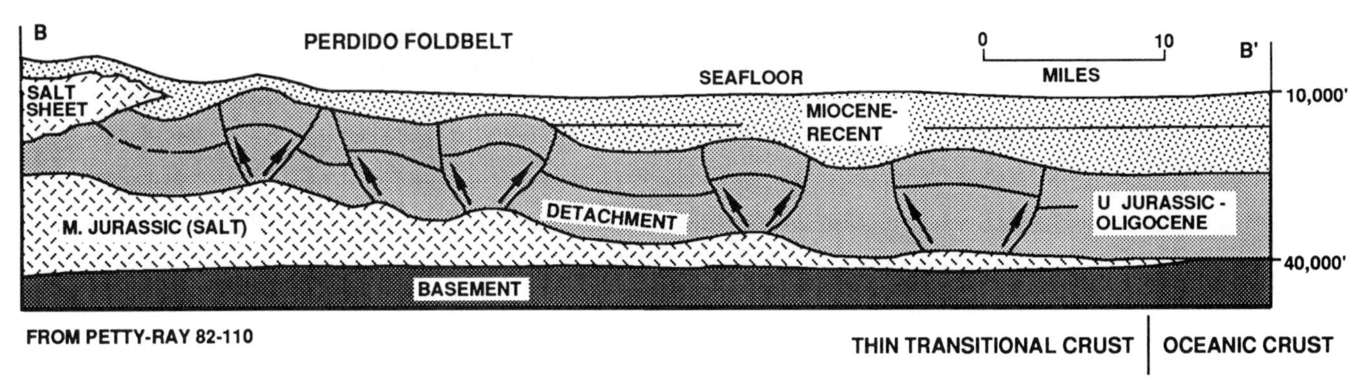

Figure 28. Schematic cross sections across two large foldbelts in the northern deep Gulf of Mexico basin—the Mississippi Fan foldbelt and the Perdido foldbelt (modified from Weimer and Buffler, 1992). The basinward limit of salt overlying the mid-Jurassic surface (MJS; top of basement) and the leading edge of deformation in the foldbelt is interpreted to approximate the boundary between oceanic and thin transitional crust (also see Fig. 1).

few refraction stations (Fig. 24; Swolfs, 1967; Ibrahim and others, 1981) and a postulated abrupt change in relief of the MJS (Fig. 1). This change is not well documented on the seismic reflection data as acoustic basement is deep and is masked by interference from the water-bottom multiple in the western Gulf. The depth (12 to 14 km) can be extrapolated almost to the present shoreline, however, by assuming a constant thickness for the observed overlying Upper Jurassic?–Lower Cretaceous Challenger unit (Fig. 9). Just to the west, the MJS is very shallow (2 to 3 km) beneath the Tuxpan platform (15, Fig. 1), requiring a steep gradient, although the exact orientation of the gradient is speculative (light lines, Fig. 1). This linear gradient has been postulated to represent a broad transform boundary along which the Yucatan block (19, Fig. 1) moved during the opening of the Gulf basin (Pindell, 1985).

Models for the early evolution of the Gulf of Mexico basin. The distribution of crustal types and framework of the Gulf of Mexico basin are interpreted to have formed mainly in the Jurassic as part of the initial breakup of Pangea as Africa/South America separated from North America. Initial rifting in the Gulf area occurred in Late Triassic–Early Jurassic time and was characterized by brittle deformation and formation of half-grabens and basins peripheral to the modern Gulf of Mexico basin. The shape of the rift probably was controlled, at least in part, by older Paleozoic structural fabrics (rifts, transforms, thrust faults, suture zones, etc.) (Thomas, 1988b, 1991). Following a time of erosion, extension continued during the Early to Middle Jurassic with the formation of the large region of transitional crust by stretching of continental crust. This process formed the basic architecture of the present basin (Fig. 1). The overall rifting process culminated with a relatively brief episode of sea-floor spreading and emplacement of oceanic crust in the central Gulf during the Late Jurassic–Early Cretaceous. As opening of the Gulf "ocean" ended, the Gulf component of extension shifted to the Atlantic and proto-Caribbean. Subsequently, cooling and then subsidence through the Early Cretaceous was accompanied by buildup of extensive carbonate platforms surrounding the deep basin.

Various tectonic models have been proposed for the early tectonic evolution of the Gulf basin. A common aspect of all these models is that the formation of the basin involved Mesozoic extension, including rifts and spreading centers that were linked by transform faults and shear zones. Also implied is the formation of broad areas of transitional crust prior to the formation of oceanic crust. Although the number of proposed models is large, nearly all can be classified into one of two basic groups that differ mainly in the original position and subsequent movement of the Yucatan block or terrane (19, Fig. 1).

In one group of models, opening of the Gulf is produced by treating Yucatan as a block that moved in a direction independent of the opening direction of the North Atlantic. In these models Yucatan moved south out of an initial position in the northern Gulf by either a counterclockwise rotation (Pindell and Dewey, 1982; Shepherd, 1983; Buffler and Sawyer, 1985; Pindell, 1985; Dunbar and Sawyer, 1987), a clockwise rotation (Hall and others, 1982; Gose and others, 1982), or more or less directly south with respect to North America (White, 1980; Salvador, 1987). The most detailed and complex model of this group is that of Pindell (1985) (Fig. 29), which not only rotates Yucatan counterclockwise along transforms in the western Gulf, but also moves a second independent block (Florida Straits block) southeastward out the Gulf along transforms generally parallel to the opening of the North Atlantic. The model starts with a late Paleozoic reconstruction that places Yucatan and the Florida Straits blocks in the northern Gulf (Fig. 29A). Subsequent movements first stretch the continental crust creating the large areas of transitional crust around the margins of Gulf before rupturing the continental crust, generating oceanic crust at one or possibly two spreading centers and creating a complex ridge/ ridge/transform triple junction in the eastern Gulf (Fig. 29B).

In the alternative group of models, Yucatan is attached to and forms an integral part of the African/South American plate (e.g., Moore and Castillo, 1974; Van der Voo and others, 1976; Pilger, 1978; Buffler and others, 1980; Walper, 1980; Dickinson and Coney, 1980; Anderson and Schmidt, 1983; Klitgord and others, 1984, 1988; Klitgord and Schouten, 1986). This is best illustrated (Fig. 30) by the models of Klitgord and others (1984, 1988) and Klitgord and Schouten (1986). Sea-floor spreading in the Gulf is simultaneous with that of the central North Atlantic; the Gulf and Atlantic spreading centers are linked by a transform boundary that crosses the Florida platform (Klitgord and others, 1984). Atlantic fracture zones are projected into the Gulf basin, in part, on the basis of gravity and magnetic trends; and thus, opening of the Gulf occurs along flow lines parallel with those of the Atlantic.

The distribution of crust and configuration of basement

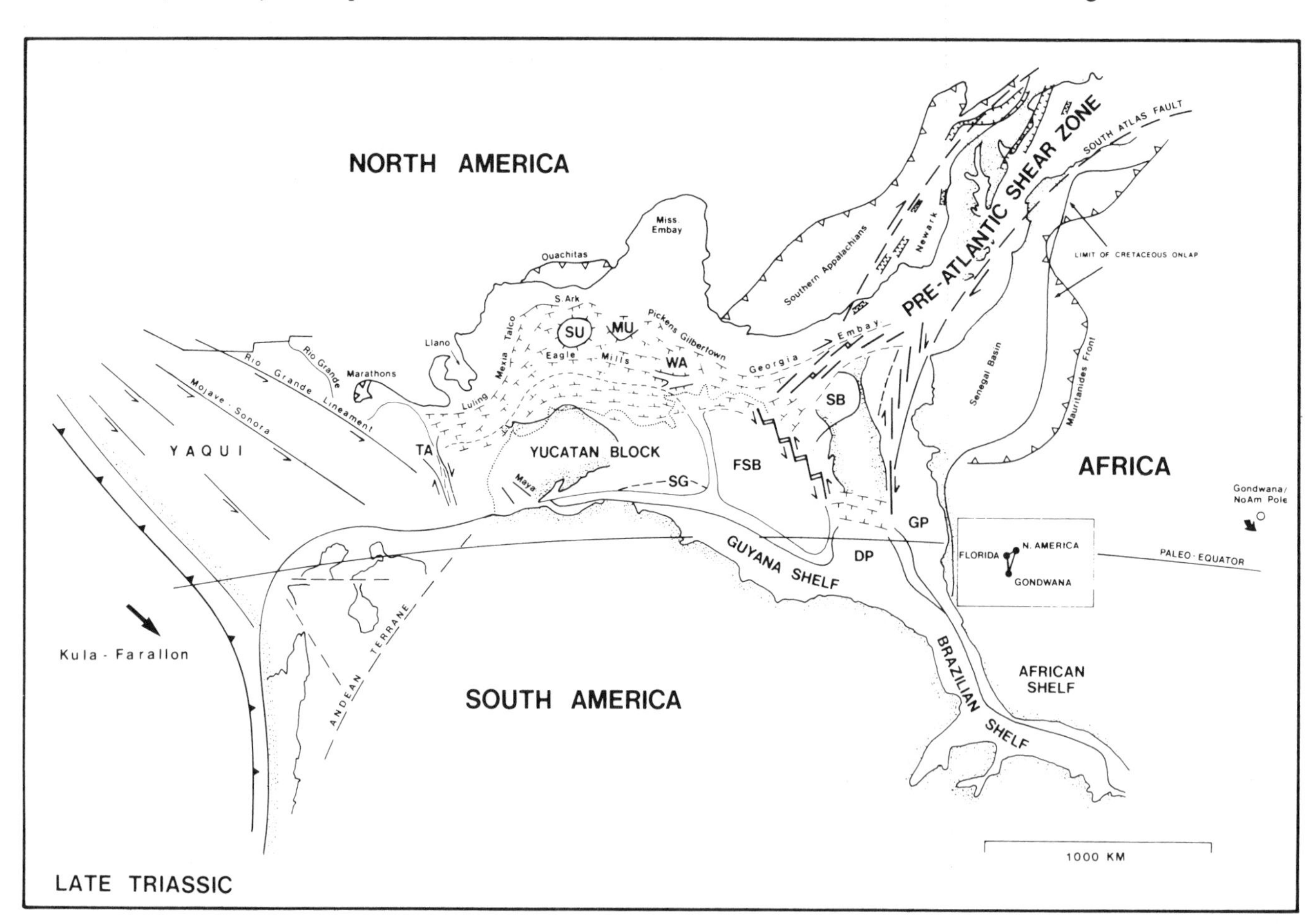

Figure 29 (on this and facing page). Diagrams showing the end members of a model for the early evolution of the Gulf of Mexico basin by Pindell (1985). Opening of the Gulf occurs along two broad transtensional zones but involves the rotation of an independent Yucatan block out of the northern Gulf, as well as the southeastward movement of a Florida Straits block and opening of a proto-Caribbean.

summarized in this chapter, as well as the distribution of overlying salt (Salvador, 1987), generally support the first-described group of models, in which reconstructions place Yucatan in the northernmost Gulf. For example, the eastward narrowing of the area of oceanic and thin transitional crust suggests counterclockwise rotation of Yucatan out of the northern Gulf (Buffler and Sawyer, 1985; Pindell, 1985; Dunbar and Sawyer, 1987). Critical in this evolution scheme is the restoration of the broad area of transitional crust, which accounts for significantly more extension than the emplacement of oceanic crust (Buffler and Sawyer, 1985; Dunbar and Sawyer, 1987). For example, Buffler and Sawyer (1985) calculated 280 km of spreading in forming oceanic crust, and 470 km of spreading in extension of the transitional crust. This model also explains the inferred steep gradient of the MJS along the western Gulf (Fig. 1), which could represent the western transform boundary along which Yucatan rotated. In addition, an independent transform boundary or transtensional zone in the eastern Gulf could explain the area of thick transitional crust characterized by the broad structural highs and lows. The highs, which have an east-west orientation (4, 6, Fig. 1) could represent smaller independent blocks that rotated counterclockwise similar to Yucatan. As they rotated and pulled apart they would have formed the intervening lows or basins in which salt was deposited (5, 7, Fig. 1).

Alternatively, many significant northwest-trending structural features throughout the Gulf basin support a more northwest-southeast opening as in the Klitgord models. These features include the northwest-trending offset of the Perdido and Mississippi Fan foldbelts, the northwest-trending trough in the area of oceanic crust, and inferred transform fault zones through the southeastern Gulf (Buffler and others, 1990). A northwest-trending linear element along the Florida escarpment in the eastern Gulf separates thin from thick transitional crust (Fig. 16; Dobson, 1990). In addition, the entire northern Gulf area of transitional crust is inferred to be characterized by northwest-trending transforms based on the distribution of salt and other geophysical and bathymetric anomalies (Watkins, 1990).

Reconstructions using these observed northwest trends or the more northwesterly trends (Fig. 30) of Klitgord and others (1988), however, present serious space problems. If South America is restored along these northwest trends into the Gulf area using the North America/Africa/South America reconstructions of Klitgord and others (1988) or Dunbar and Sawyer (1987), Yucatan has a significant overlap with South America (Dietmar Muller and Lisa Gahagan, personal communication, 1991). This overlap can be removed by including some counterclockwise rotation of the Yucatan block out of the northeastern Gulf, even though the northwest trends do not appear to reflect this movement. This apparent contradiction is not reconcilable at this time.

In summary, the identification of crustal types and the con-

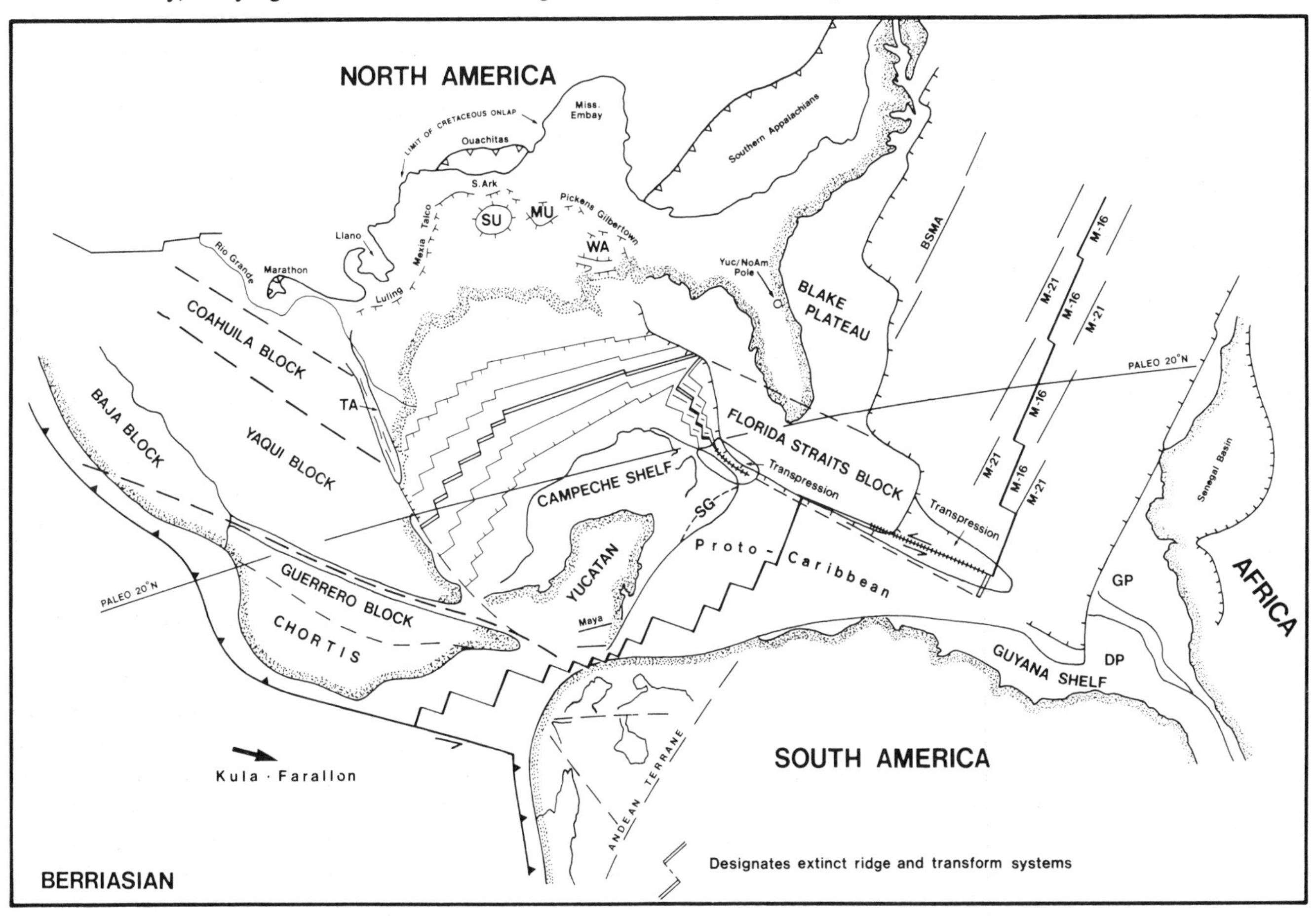

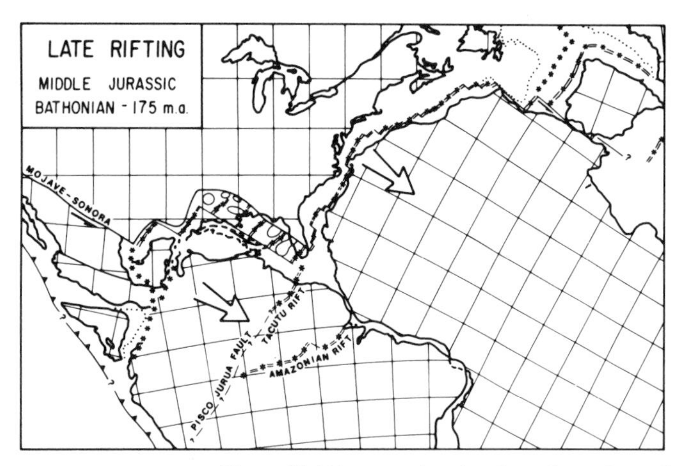

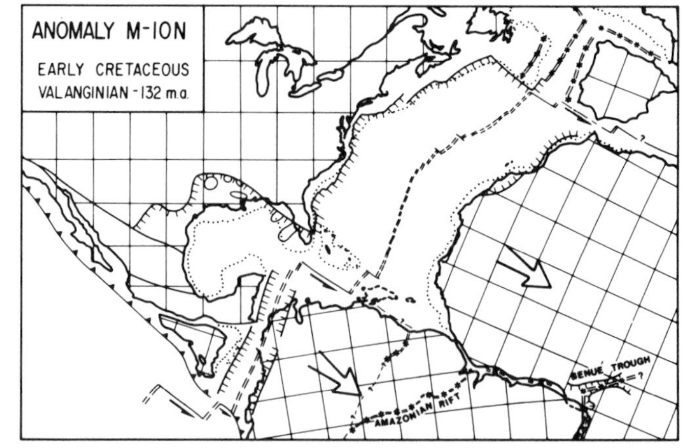

Figure 30. Diagrams showing the end members of a model for the early evolution of the Gulf of Mexico basin by Klitgord and Schouten (1986). Opening of the Gulf occurs along two broad northwest-trending transtensional zones across northern Mexico and Florida that parallel Atlantic fracture zones. Yucatan remains attached to South America.

figuration of the MJS in the Gulf basin is possible using seismic reflection, seismic refraction, gravity, and magnetic data, as well as plate reconstructions, total tectonic subsidence analyses, and geologic information from the periphery of the basins. Although still incomplete, these data provide important constraints on the early opening history of the Gulf region. It is critical, therefore, that interpretations of the early Mesozoic kinematics of the region be consistent with the processes that modify continental crust and generate oceanic crust.

SUMMARY OF EARLY EVOLUTION OF THE GULF OF MEXICO BASIN

A general consensus now exists among many geologists and geophysicists for an interpretation of the early evolution of the Gulf of Mexico basin that includes four main stages in an overall continuous process: (1) a Late Triassic–Early Jurassic stage of early rifting followed by erosion, (2) an Early to Middle Jurassic stage of crustal attenuation and formation of a broad area of transitional crust, (3) a Late Jurassic stage of oceanic crust formation in the deep central Gulf, and (4) a Late Jurassic through Early Cretaceous stage of cooling and subsidence of the crust and buildup of extensive carbonate platforms surrounding the deep basin. The early evolution terminated with the formation of the widespread mid-Cretaceous sequence boundary (MCSB). Each of these stages is outlined briefly here and is shown schematically in Figure 31.

Late Triassic–Early Jurassic rifting

This stage is characterized by the formation of linear zones of rifting of a brittle crust (Fig. 31A). Fault basins are characterized by large and small half-grabens bounded by listric normal faults. They are filled with nonmarine sediments and volcanics. Equivalent rocks are distributed all along the Atlantic Coastal Plain of the United States, across northern Florida and southern Georgia, all along the northern Gulf (Eagle Mills), and into Mexico. Similar rocks possibly underlie parts of the Sigsbee salt basin (21, Fig. 1) in the deep Gulf. All of these rocks represent the initial rift stage in the breakup of Pangea. Rifting apparently occurred, in part, along an older structural fabric inherited from late Precambrian and Paleozoic rifting and the later Paleozoic continental collision that formed Pangea. These rocks are truncated in places by an erosion surface that forms the mid-Jurassic surface (MJS) as defined in this chapter (Fig. 31A). This rift stage formed by relative movements between Africa, South America, and North America and initiated attenuation of the crust throughout the Gulf region.

Middle Jurassic attenuation

During Middle Jurassic, the entire Gulf region underwent attenuation of the crust in association with heating by a broad mantle upwelling (Fig. 31B). This resulted in the formation of the areas of transitional crust and the associated structural highs and lows that form the basic basin architecture of the present basin. During this time the Yucatan block began rotating counterclockwise out of the northern Gulf like a large ball bearing caught between broad transtensional zones across Mexico and Florida. Alternatively, Yucatan moved southeastward out of the northern Gulf along with South America. The periphery of the basin only underwent moderate stretching and the crust remained relatively thick, forming the area of broad arches and basins. The center of the basin, however, underwent considerably more stretching and subsidence to form the large area of thin-transitional crust. Overall, the Gulf basin probably was a large, relatively shallow structural downwarp characterized by broad highs and lows. Thick salt was deposited throughout the broad central area of thin crust, as well as in many of the basins in the adjacent thick crust, as marine water episodically spilled into the basin across sills, maintaining a delicate balance between evaporation and influx. The

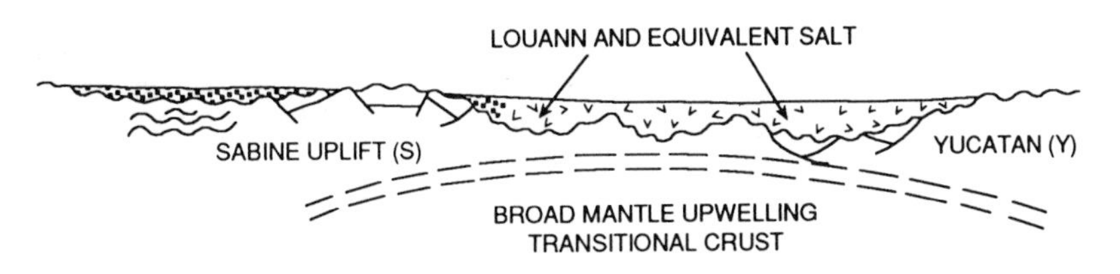

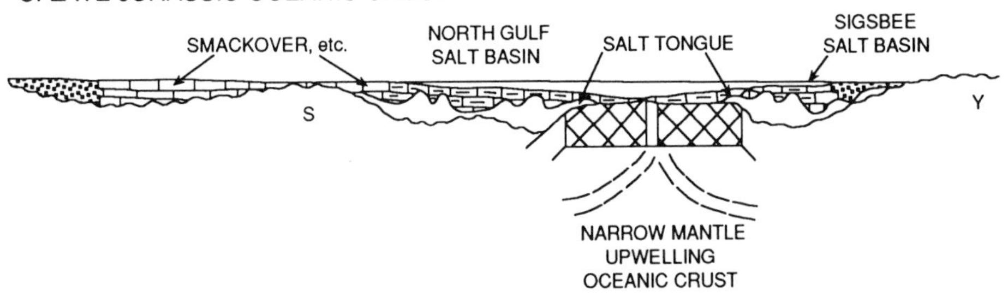

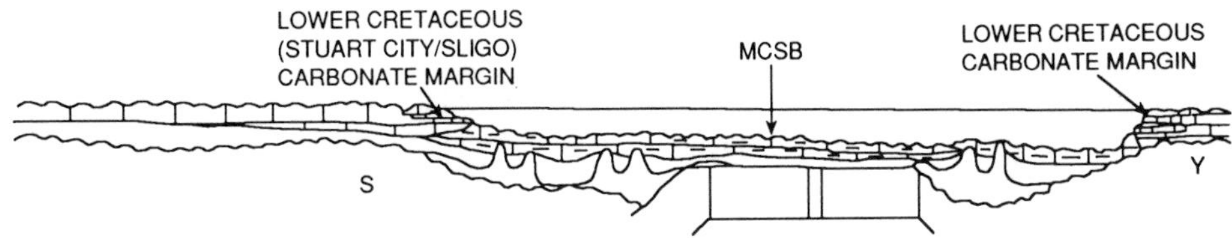

Figure 31. Schematic diagrams showing four stages of the early (Late Triassic through Early Cretaceous) evolution of the Gulf of Mexico basin.

inlet for the marine waters possibly was in central Mexico. Alternatively, water could have entered through the deep southeastern Gulf–Straits of Florida region, as there is approximately contemporaneous Middle Jurassic salt in Cuba. Contemporaneous nonmarine terrigenous clastic sediments were deposited peripherally to the salt basins (Fig. 31B).

Late Jurassic oceanic crust

The Late Jurassic is characterized by the emplacement of oceanic crust along a generally east-west striking rift in the thinning continental crust (Fig. 31C). The area of oceanic crust became wider to the west, possibly as Yucatan continued to rotate in a counterclockwise direction, separating the earlier formed salt basins. The eastern, more narrow area of oceanic crust could represent a leaky transform or transtensional boundary. This phase was associated with a general marine transgression into the Gulf as the crust began to cool and subside and long-term relative sea level rose. Deep-water sediments were deposited over oceanic crust, and broad shallow-to-deep shelf environments were established over the adjacent areas of transitional crust (Smackover and equivalent rocks, Fig. 31C). Gravity and sediment loading caused early basinward flowage of salt tongues along the boundary between oceanic and thin transitional crust. The broad structural highs and lows continued to influence sedimentation around the periphery of the basin, and many of the highs remained emergent throughout most of the Late Jurassic (e.g., Yucatan, Fig. 31C). Extensive rifting and deposition of syn-rift sediments continued in the southeastern Gulf, probably because of interaction between the Yucatan and Florida blocks. Following this rifting, a marine seaway between the Gulf and the Caribbean became well established through the southeastern Gulf, providing a complete link across the Gulf between the Atlantic and Pacific. Thick terrigenous clastic prisms prograded into the basin at places along the margins (Fig. 31C).

Early Cretaceous subsidence

The Gulf basin became locked into its present configuration with respect to North America by Early Cretaceous time (Fig. 31D), as sea-floor spreading aborted in the Gulf and shifted entirely to the Atlantic–proto Caribbean. Broad carbonate platforms with prominent rimmed margins became established along tectonic hinge zones that formed because of differential subsidence at the boundary between thin and thick crust (Figs. 9, 31D). Very abrupt, high-relief, bypass margins had developed by mid-Cretaceous time along the Florida and Campeche escarpments and along the western Gulf in Mexico. Predominantly cyclic, fine-grained carbonates were being deposited in the adjacent deep basin.

Finally, in mid-Cenomanian time, a rapid drop and subsequent short-term rise in sea level superimposed on a long-term rise in sea level and associated environmental changes terminally drowned the outer margins of the carbonate platforms, causing the margins to step back or retreat to more landward positions. Widespread submarine erosion along the base of the Lower Cretaceous margins by submarine currents, subaerial erosion along the platform margins, and an associated change in depositional environment and depositional patterns in the deep central Gulf created the prominent, widespread mid-Cretaceous sequence boundary (MCSB, Fig. 1, 31D). These events and changes in sedimentation patterns represent a major turning point in Gulf history that marks the end of the early tectonic evolution of the deep Gulf basin. It set the scene for the later partial filling of the basin by large clastic prisms that prograded first from the west and northwest in Late Cretaceous–early Cenozoic time and then from the north during the late Cenozoic (Mississippi River drainage) (Fig. 9).

ACKNOWLEDGMENTS

We thank the many geologists and geophysicists with whom we have had the pleasure to work on the Gulf of Mexico basin–southern margin of North America regions over the past many years. They have given freely with their ideas. We particularly thank Rex Pilger, Jeff Nunn, and Bob Speed for critically reviewing the manuscript and providing helpful suggestions. Special thanks go to Bob Speed and Pete Palmer for their patience, understanding, and faith, while still keeping the pressure on us. Ms. Toni Lee Mitchell helped type and edit the manuscript, and Wayne Lloyd helped prepare some of the figures. Partial funding for RTB's part of this project was provided by a grant from the National Science Foundation (OCE 8417771) and Texas Higher Education Coordinating Board Advance Research Program Award 3513. Partial funding for WAT's research was provided by a grant from the National Science Foundation (EAR 8218604). This is University of Texas Institute for Geophysics (UTIG) Contribution Number 933.

REFERENCES CITED

Anderson, T. H., and Schmidt, V. A., 1983, The evolution of Middle America and the Gulf of Mexico–Caribbean Sea region during Mesozoic time: Geological Society of America Bulletin, v. 94, p. 941–966.

Andress, N. E., Cramer, F. H., and Goldstein, R. F., 1969, Ordovician chitinozoans from Florida well samples: Gulf Coast Association of Geological Societies Transactions, v. 19, p. 369–375.

Applin, P. L., 1951, Preliminary report on buried pre-Mesozoic rocks in Florida and adjacent states: U.S. Geological Survey Circular 91, 28 p.

Barnett, R. S., 1975, Basement structure of Florida and its tectonic implications: Gulf Coast Association of Geological Societies Transactions, v. 25, p. 122–142.

Bass, M. N., 1969, Petrography and ages of crystalline basement rocks of Florida; Some extrapolations: American Association of Petroleum Geologists Memoir 11, p. 283–310.

Bickford, M. E., and Lewis, R. D., 1979, U-Pb geochronology of exposed basement rocks in Oklahoma: Geological Society of America Bulletin, v. 90, p. 540–544.

Blickwede, J. F., and Queffelec, T. A., 1988, Perdido foldbelt; A new deep-water frontier in western Gulf of Mexico [abs.]: American Association of Petroleum Geologists Bulletin, v. 72, p. 163.

Bowring, S. A., and Hoppe, W. J., 1982, U/Pb zircon ages from Mount Sheridan Gabbro, Wichita Mountains: Oklahoma Geological Survey Guidebook 21, p. 54–59.

Brewer, J. A., Brown, L. D., Steiner, D., Oliver, J. E., Kaufman, S., and Denison, R. E., 1981, Proterozoic basin in the southern Midcontinent of the United States revealed by COCORP deep seismic reflection profiling: Geology, v. 9, p. 569–575.

Brewer, J. A., Good, R., Oliver, J. E., Brown, L. D., and Kaufman, S., 1983, COCORP profiling across the southern Oklahoma aulacogen; Overthrusting of the Wichita Mountains and compression within the Anadarko basin: Geology, v. 11, p. 109–114.

Buffler, R. T., 1991, Seismic stratigraphy and geologic history of the deep Gulf of Mexico basin, *in* Salvador, A., ed., The Gulf of Mexico basin: Boulder, Colorado, Geological Society of America, The Geology of North America, v. J, p. 353–387.

Buffler, R. T., and Sawyer, D. S., 1985, Distribution of crust and early history, Gulf of Mexico basin: Gulf Coast Association of Geological Societies Transactions, v. 35, p. 333–344.

Buffler, R. T., Watkins, J. S., Worzel, J. L., and Shaub, F. J., 1980, Structure and early geologic history of the deep central Gulf of Mexico, *in* Pilger, R., ed., Proceedings of a Symposium on the Origin of the Gulf of Mexico and the Early Opening of the Central North Atlantic: Louisiana State University, Baton Rouge, Louisiana, March 1980, p. 3–16.

Buffler, R. T., Shaub, F. J., Huerta, R., and Ibrahim, A. K., 1981, A model for the early evolution of the Gulf of Mexico basin: Oceanologica Acta, International Geological Congress, Proceedings, 26th, Geology of Continental Margins Symposium, Paris, July 1980, p. 129–136.

Buffler, R. T., Schlager, W., and Scientific Party, 1984, Initial Reports of the Deep Sea Drilling Project: Washington, D.C., U.S. Government Printing Office, v. 77, 747 p.

Buffler, R. T., Marton, G., Müller, D., Gahagan, L., Sawyer, D., and Rosenthal, D., 1990, Crustal types and northwest trending structural features; Constraints on reconstructing the Gulf of Mexico basin: Geological Society of America Abstracts with Programs, v. 22, no. 7, p. A186.

Cebull, S. E., Shurbet, D. H., Keller, G. R., and Russell, L. R., 1976, Possible role of transform faults in the development of apparent offsets in the Ouachita–southern Appalachian tectonic belt: Journal of Geology, v. 84, p. 107–114.

Chowns, T. M., and Williams, C. T., 1983, Pre-Cretaceous rocks beneath the Georgia Coastal Plain; Regional implications, *in* Gohn, G. S., ed., Studies related to the Charleston, South Carolina, earthquake of 1886; Tectonics and seismicity: U.S. Geological Survey Professional Paper 1313, p. L1–L42.

Coffman, J. D., Gilbert, M. C., and McConnell, D. A., 1986, An interpretation of the crustal structure of the Southern Oklahoma aulacogen satisfying gravity data: Oklahoma Geological Survey Guidebook 23, p. 1–10.

Collinson, C., Sargent, M. L., and Jennings, J. R., 1988, Illinois basin region, *in* Sloss, L. L., ed., Sedimentary cover; North American craton; U.S.: Geological Society of America, The Geology of North America, v. D-2, p. 383–426.

Corso, W., 1987, Development of the Early Cretaceous northwest Florida carbonate platform [Ph.D. dissertation]: University of Texas at Austin, 136 p.

Cramer, F. H., 1973, Middle and Upper Silurian chitinozoan succession in Florida subsurface: Journal of Paleontology, v. 47, p. 279–288.

Crosby, E. J., and Mapel, W. J., 1975, Central and west Texas, *in* McKee, E. D., and Crosby, E. J., coordinators, Paleotectonic investigations of the Pennsylvanian System in the United States: U.S. Geological Survey Professional Paper 853, p. 197–232.

Dalmeyer, R. D., 1984, $^{40}Ar/^{39}Ar$ ages from pre-Mesozoic crystalline basement penetrated at hole 537 and 538A of the Deep Sea Drilling Project Leg 77, Southeastern Gulf of Mexico; Tectonic implications, *in* Buffler, R.T., Schager, W., and others, Initial Reports of the Deep Sea Drilling Project: Washington, D.C., U.S. Government Printing Office, v. 77, p. 497–504.

Denison, R. E., 1984, Basement rocks in northern Arkansas: Arkansas Geological Commission Miscellaneous Publication 18-B, p. 33–49.

Denison, R. E., Burke, W. H., Otto, J. B., and Hetherington, E. A., 1977, Age of igneous and metamorphic activity affecting the Ouachita foldbelt, *in* Stone, C. G., ed., Symposium on the Geology of the Ouachita Mountains, v. 1: Arkansas Geological Commission, p. 25–40.

Denison, R. E., Lidiak, E. G., Bickford, M. E., and Kisvarsanyi, E. B., 1984, Geology and geochronology of Precambrian rocks in the Central Interior region of the United States: U.S. Geological Survey Professional Paper 1241-C, 20 p.

Dickinson, W. R., and Coney, P. J., 1980, Plate tectonic constraints on the origin of the Gulf of Mexico, *in* Pilger, R. H., ed., Proceedings, Symposium on the Origin of the Gulf of Mexico: Louisiana State University, Baton Rouge, p. 27–36.

Dobson, L. M., 1990, Seismic stratigraphy and geologic history of Jurassic rocks, northeastern Gulf of Mexico [M.A. thesis]: The University of Texas at Austin, 115 p.

Dunbar, J., and Sawyer, D. S., 1987, Implications of continental crust extension for plate reconstruction; An example from the Gulf of Mexico: Tectonics, v. 6, p. 739–755.

Ebeniro, J. O., and O'Brien, W. P., 1984, Crustal structure of the South Florida bank derived from Ocean Bottom Seismometers refraction profiles: The University of Texas Institute for Geophysics Technical Report, no. 32, 87 p.

Ebeniro, J. O., Nakamura, Y., Sawyer, D. S., and O'Brien, W. P., Jr., 1988, Sedimentary and crustal structure of the northern Gulf of Mexico: Journal of Geophysical Research, v. 93, p. 9075–9092.

Ethington, R. L., Finney, S. C., and Repetski, J. E., 1989, Biostratigraphy of the Paleozoic rocks of the Ouachita orogen, Arkansas, Oklahoma, west Texas, *in* Hatcher, R. D., Jr., Thomas, W. A., and Viele, G. W., eds., The Appalachian-Ouachita orogen in the United States: Geological Society of America, The Geology of North America, v. F-2, p. 563–574.

Ewing, J., Antoine, J., and Ewing, M., 1960, Geophysical measurements in the western Caribbean Sea and in the Gulf of Mexico: Journal of Geophysical Research, v. 65, p. 4087–4126.

Ewing, J. I., Worzel, J. L., and Ewing, M., 1962, Sediments and oceanic structural history of the Gulf of Mexico: Journal of Geophysical Research, v. 67, p. 2509–2527.

Ferrill, B. A., 1989, Middle Cambrian to Lower Mississippian synsedimentary structures in the Appalachian fold-thrust belt in Alabama and Georgia [Ph.D. thesis]: Tuscaloosa, University of Alabama, 270 p.

Fuqua, A. D., 1990, Seismic structural analysis of the Perdido foldbelt, Alaminos Canyon area, northwestern Gulf of Mexico [M.A. thesis]: The University of Texas at Austin, 89 p.

Gilbert, M. C., 1983, Timing and chemistry of igneous events associated with the Southern Oklahoma aulacogen: Tectonophysics, v. 94, p. 439–455.

Gose, W. A., Belcher, R. C., and Scott, G. R., 1982, Paleomagnetic results from northeastern Mexico; Evidence for large Mesozoic rotations: Geology, v. 10, p. 50–54.

Hall, D. J., 1990, Gulf Coast–East Coast magnetic anomaly I; Root of the main crustal decollement for the Appalachian-Ouachita orogen: Geology, v. 18, p. 862–865.

Hall, D. J., Cavanaugh, T. D., Watkins, J. S., and McMillen, K. J., 1982, The rotational origin of the Gulf of Mexico based on regional gravity data, *in* Watkins, J.,and Drake, C., eds., Studies in continental margin geology: American Association of Petroleum Geologists Memoir 34, p. 115–126.

Hall, S., and Najmuddin, I., 1992, Plate tectonic development of the eastern Gulf of Mexico: Geological Society of America Abstracts with Programs, v. 24, no. 1, p. 14.

Hall, S., Shepherd, A., Titus, M., and Snow, R., 1984, Magnetic total intensity anomalies (east of 90°W long.), *in* Buffler, R., and others, eds., Gulf of Mexico, Atlas 6, Ocean Margin Drilling Program, Regional atlas series: Woods Hole, Massachusetts, Marine Science International, Sheet 3.

Ham, W. E., Denison, R. E., and Meritt, C. A., 1964, Basement rocks and structural evolution of southern Oklahoma: Oklahoma Geological Survey Bulletin 95, 302 p.

Harrelson, D. W., and Bicker, A. R., Jr., 1979, Petrography of some subsurface igneous rocks of Mississippi: Gulf Coast Association of Geological Societies Transactions, v. 29, p. 244–251.

Hatcher, R.D., Jr., 1972, Developmental model for the southern Appalachians: Geological Society of America Bulletin, v. 83, p. 2735–2760.

——, 1978, Tectonics of the western Piedmont and Blue Ridge, southern Appalachians; Review and speculation: American Journal of Science, v. 278, p. 276–304.

Higgins, M. W., and Zietz, I., 1983, Geologic interpretation of geophysical maps of the pre-Cretaceous "basement" beneath the Coastal Plain of the southeastern United States: Geological Society of America Memoir 158, p. 125–130.

Hinze, W. J., and Braile, L.W ., 1988, Geophysical aspects of the craton; U.S., *in* Sloss, L. L., ed., Sedimentary cover; North American craton; U.S.: Geological Society of America, The Geology of North America, v. D-2, p. 5–24.

Hooper, R. J., and Hatcher, R. D., Jr., 1988, Pine Mountain terrane, a complex window in the Georgia and Alabama Piedmont; Evidence from the eastern termination: Geology, v. 16, p. 307–310.

Horton, J. W., Jr., Zietz, I.,and Neathery, T. L., 1984, Truncation of the Appalachian Piedmont beneath the Coastal Plain of Alabama; Evidence from new magnetic data: Geology, v. 12, p. 51–55.

Horton, J. W., Jr., Drake, A. A., Jr., and Rankin, D.W., 1989, Tectonostratigraphic terranes and their Paleozoic boundaries in the central and southern Appalachians: Geological Society of America Special Paper 230, p. 213–245.

Houseknecht, D. W., 1986, Evolution from passive margin to foreland basin; The Atoka Formation of the Arkoma basin, south-central U.S.A.: International Association of Sedimentologists Special Publication 8, p. 327–345.

Howe, J. R., 1985, Tectonics, sedimentation, and hydrocarbon potential of the Reelfoot aulacogen [M.S. thesis]: Norman, University of Oklahoma, 109 p.

Huerta, R., 1980, Seismic stratigraphic and structural analysis of northeast Campeche escarpment, Gulf of Mexico [M.A. thesis]: The University of Texas at

Austin, 107 p.

Ibrahim, A. K., Carye, J., Latham, G., and Buffler, R. T., 1981, Crustal structure in the Gulf of Mexico from OBS refraction and multichannel reflection data: American Association of Petroleum Geologists Bulletin, v. 65, p. 1207–1229.

Johnson, K. S., and 7 others, 1988, Southern Midcontinent region, *in* Sloss, L. L., ed., Sedimentary cover; North American craton; U.S.: Geological Society of America, The Geology of North America, v. D-2, p. 307–359.

Keen, C. E., 1982, The continental margins of eastern Canada; A review: American Geophysical Union Geodynamics Series, v. 6, p. 45–58.

Keen, C. E., and Haworth, R. T., 1985, D-2 Transform margin south of Grand Banks; Offshore eastern Canada: Geological Society of America Centennial Continent/Ocean Transect #2.

Keller, G. R., Smith, R. A., Hinze, W. J., and Aiken, C.L.V., 1985, Regional gravity and magnetic study of west Texas, *in* Hinze, W. J., ed., The utility of regional gravity and magnetic anomaly maps: Society of Exploration Geophysicists, p. 198–212.

Keller, G. R., Braile, L. W., McMechan, G. A., Thomas, W. A., Harder, S. H., Chang, W.-F., and Jardine, W. G., 1989a, Paleozoic continent-ocean transition in the Ouachita Mountains imaged from PASSCAL wide-angle seismic reflection-refraction data: Geology, v. 17, p. 119–122.

Keller, G. R., Kruger, J. M., Smith, K. J., and Voight, W. M., 1989b, The Ouachita system; A geophysical overview, *in* Hatcher, R. D., Jr., Thomas, W. A., and Viele, G. W., eds., The Appalachian-Ouachita orogen in the United States: Geological Society of America, The Geology of North America, v. F-2, p. 689–693.

King, P. B., 1937, Geology of the Marathon region, Texas: U.S. Geological Survey Professional Paper 187, 148 p.

—— , 1975, The Ouachita and Appalachian orogenic belts, *in* Nairn, A.E.M., and Stehli, F. G., eds., The ocean basins and margins, v. 3: New York, Plenum, p. 201–241.

Klitgord, K. D., and Schouten, H., 1986, Plate kinematics of the central Atlantic, *in* Vogt, P. R., and Tucholke, B. E., eds., The western North Atlantic region: Geological Society of America, The Geology of North America, v. M, p. 351–378.

Klitgord, K. D., Popenoe, P., and Schouten, H., 1984, Florida, a Jurassic transform plate boundary: Journal of Geophysical Research, v. 89, p. 7753–7772.

Klitgord, K. D., Hutchinson, D. R., and Schouten, H., 1988, U.S. Atlantic continental margin; Structural and tectonic framework, *in* Sheridan, R. E., and Grow, J. A., eds., The Atlantic continental margin, U.S.: Geological Society of America, The Geology of North America, v. I-2, p. 19–55.

Kluth, C. F., 1986, Plate tectonics of the Ancestral Rocky Mountains: American Association of Petroleum Geologists Memoir 41, p. 353–369.

Kluth, C. F., and Coney, P. J., 1981, Plate tectonics of the Ancestral Rocky Mountains: Geology, v. 9, p. 10–15.

Kruger, J. M., and Keller, G. R., 1986, Interpretation of crustal structure from regional gravity anomalies, Ouachita Mountains area and adjacent Gulf Coastal Plain: American Association of Petroleum Geologists Bulletin, v. 70, p. 667–689.

Lambert, D. D., and Unruh, D. M., 1986, Isotopic constraints on the age and source history of the Glen Mountains layered complex, Wichita Mountains: Oklahoma Geological Survey Guidebook 23, p. 53–59.

Lee, G. H., Bryant, W. R., and Watkins, J. S., 1989, Salt structures and sedimentary basins in the Keathley Canyon area, northwestern Gulf of Mexico; Their development and tectonic implications, *in* Gulf of Mexico salt tectonics, associated processes and exploration potential: Gulf Coast Section Society of Economic Paleontologists and Mineralogists Foundation, Tenth Annual Research Conference, Program and Extended and Illustrated Abstracts, p. 90–93.

Lillie, R. J., 1985, Tectonically buried continent/ocean boundary, Ouachita Mountains, Arkansas: Geology, v. 13, p. 18–21.

Lillie, R. J., and 7 others, 1983, Crustal structure of Ouachita Mountains, Arkansas; A model based on integration of COCORP reflection profiles and regional geophysical data: American Association of Petroleum Geologists Bulletin, v. 67, p. 907–931.

Lin, T., 1984, Seismic stratigraphy and structure of the Sigsbee salt basin, south-central Gulf of Mexico [M.A. thesis]: The University of Texas at Austin, 102 p.

Locker, S. D., and Chatterjee, S. K., 1984, Seismic velocity structure, *in* Buffler, R. T., and others, eds., Gulf of Mexico, Atlas 6, Ocean Margin Drilling Program, Regional atlas series: Woods Hole, Massachusetts, Marine Science International, Sheet 4.

Lord, J., 1986, Seismic stratigraphy and geologic history of the West Florida basin, eastern Gulf of Mexico [M.A. thesis]: Houston, Texas, Rice University, 290 p.

—— , 1987, Seismic stratigraphic investigation of the West Florida basin: Gulf Coast Association of Geological Societies Transactions, v. 37, p. 123–138.

Lovick, G. P., Mazzine, C. G., and Kotila, D. A., 1982, Atokan clastics; Depositional environments in a foreland basin: Oil and Gas Journal, v. 80, p. 181–199.

Mack, G. H., Thomas, W. A., and Horsey, C. A., 1983, Composition of Carboniferous sandstones and tectonic framework of southern Appalachian-Ouachita orogen: Journal of Sedimentary Petrology, v. 53, p. 931–946.

MacRae, G., 1990, Salt tectonism and seismic stratigraphy of the Upper Jurassic in the Destin dome region, northeastern Gulf of Mexico [M.A. thesis]: Texas A&M University, 81 p.

Martin, R. G., 1978, Northern and eastern Gulf of Mexico continental margin; Stratigraphic and structural framework, *in* Bouma, A. H., Moore, G. T., and Coleman, J. M., eds., Framework, facies, and oil-trapping characteristics of the upper continental margin: American Association of Petroleum Geologists Studies in Geology, v. 7, p. 21–42.

—— , 1980, Distribution of salt structures in the Gulf of Mexico; Map and descriptive text: U.S. Geological Survey Map MF-1213, scale 1:2,500,000 at 19°N. latitude.

McBride, E. F., 1989, Stratigraphy and sedimentary history of pre-Permian Paleozoic rocks of the Marathon uplift, *in* Hatcher, R. D., Jr., Thomas, W. A., and Viele, G. W., eds., The Appalachian-Ouachita orogen in the United States: Geological Society of America, The Geology of North America, v. F-2, p. 603–620.

McBride, J. H., and Nelson, K. D., 1988, Integration of COCORP deep reflection and magnetic anomaly analysis in the southeastern United States: Implications for origin of the Brunswick and East Coast magnetic anomalies: Geological Society of America Bulletin, v. 100, p. 436–445.

McLaughlin, R. E., 1970, Palynology of core samples of Paleozoic sediments from beneath the Coastal Plain of Early County, Georgia: Georgia Geological Survey Information Circular 40, 27 p.

Milliken, J. V., 1988, Late Paleozoic and early Mesozoic geologic evolution of the Arklatex area [M.A. thesis]: Houston, Texas, Rice University, 259 p.

Milton, C., 1972, Igneous and metamorphic basement rocks of Florida: Florida Bureau of Geology Bulletin 55, 125 p.

Milton, C., and Hurst, V. J., 1965, Subsurface "basement" rocks of Georgia: Georgia Geological Survey Bulletin 76, 56 p.

Moore, G. W., and del Castillo, L., 1974, Tectonic evolution of the southern Gulf of Mexico: Geological Society of America Bulletin, v. 85, p. 607–618.

Mosher, S., 1992, Western extensions of Grenville age rocks; Texas, *in* Reed, J. C., Jr., and 6 others, eds., The Precambrian; U.S.: Geological Society of America, The Geology of North America, v. C-2, p. 365–378.

Muehlberger, W. R., and Tauvers, P. R., 1989, Marathon fold-thrust belt, west Texas, *in* Hatcher, R. D., Jr., Thomas, W. A., and Viele, G. W., eds., The Appalachian-Ouachita orogen in the United States: Geological Society of America, The Geology of North America, v. F-2, p. 673–680.

Muehlberger, W. R., Denison, R. E., and Lidiak, E. G., 1967, Basement rocks in continental interior of United States: American Association of Petroleum Geologists Bulletin, v. 51, p. 2351–2380.

Mueller, P. A., and Porch, J. W., 1983, Tectonic implications of Paleozoic and Mesozoic igneous rocks in the subsurface of peninsular Florida: Gulf Coast Association of Geological Societies Transactions, v. 33, p. 169–173.

Najmuddin, I. J., 1992, Analysis and interpretation of an aeromagnetic survey over the eastern deep Gulf of Mexico [M.A. thesis]: Houston, Texas, University of Houston, 201 p.

Najmuddin, I. J., and Hall, S. A., 1992, Mesozoic seafloor spreading in the eastern Gulf of Mexico: Geological Society of America Abstracts with Programs, v. 24, no. 1, p. 40.

Nakamura, Y., Sawyer, D. S., Shaub, F. J., MacKenzie, K., and Oberst, J., 1988, Deep crustal structure of the northwestern Gulf of Mexico: Gulf Coast Association of Geological Societies Transactions, v. 38, p. 207–215.

Neathery, T. L., and Copeland, C. W., 1983, New information on the basement and lower Paleozoic stratigraphy of north Alabama: Alabama Geological Survey Open-File Report, 28 p.

Neathery, T. L., and Thomas, W. A., 1975, Pre-Mesozoic basement rocks of the Alabama Coastal Plain: Gulf Coast Association of Geological Societies Transactions, v. 25, p. 86–99.

—— , 1983, Geodynamics transect of the Appalachian orogen in Alabama, *in* Rast, N., and Delany, F. M., eds., Profiles of orogenic belts: American Geophysical Union Geodynamics Series, v. 10, p. 301–307.

Nelson, K. D., and 7 others, 1982, COCORP seismic reflection profiling in the Ouachita Mountains of western Arkansas; Geometry and geologic interpretation: Tectonics, v. 1, p. 413–430.

Nelson, K. D., and 8 others, 1985, New COCORP profiling in the southeastern United States; Part I; Late Paleozoic suture and Mesozoic rift basin: Geology, v. 13, p. 714–718.

Nelson, K. D., Arnow, J. A., Giguere, M., and Schamel, S., 1987, Normal-fault boundary of an Appalachian basement massif?; Results of COCORP profiling across the Pine Mountain belt in western Georgia: Geology, v. 15, p. 832–836.

Ness, G. E., Dauphin, J. P., Garcia-Abdeslem, J., and Alvarado-Omana, M. A., 1991, Bathymetry and gravity and magnetic anomalies of the Yucatan Peninsula and adjacent areas: Geological Society of America, Map and Chart Series MCH073, scale 1:1,565,000 at 21°N.

Nicholas, R. L., and Rozendal, R. A., 1975, Subsurface positive elements within Ouachita foldbelt in Texas and their relation to Paleozoic cratonic margin: American Association of Petroleum Geologists Bulletin, v. 59, p. 193–216.

Nicholas, R. L., and Waddell, D. E., 1989, The Ouachita system in the subsurface of Texas, Arkansas, and Louisiana, *in* Hatcher, R. D., Jr., Thomas, W. A., and Viele, G. W., eds., The Appalachian-Ouachita orogen in the United States: Geological Society of America, The Geology of North America, v. F-2, p. 661–672.

Nunn, J. A., Scardina, A. D., and Pilger, R. H., 1984, Thermal evolution of the north-central Gulf Coast: Tectonics, v. 3, p. 723–740.

Nunn, J. A., Driskill, B. W., and Suh, M., 1989, Thermal and subsidence history of the Wiggins uplift, southern Mississippi [abs.]: EOS Transactions of the American Geophysical Union, v. 70, p. 466.

Pardo, G., 1975, Geology of Cuba, *in* Nairn, A.E.M., and Stehli, F. G., eds., The Gulf of Mexico and Caribbean, ocean basins and margins, v. 3: New York, Plenum Press, p. 553–615.

Perry, W. J., Jr., 1989, Tectonic evolution of the Anadarko basin region, Oklahoma: U.S. Geological Survey Bulletin 1866-A, 19 p.

Pfeil, R. W., and Read, J. F., 1980, Cambrian carbonate platform margin facies, Shady Dolomite, southwestern Virginia, U.S.A.: Journal of Sedimentary Petrology, v. 50, p. 91–116.

Phair, R. L., and Buffler, R. T., 1983, Pre-Middle Cretaceous geologic history of the deep southeastern Gulf of Mexico, *in* Bally, A. W., ed., Seismic expression of structural styles; A picture and work atlas: American Association of Petroleum Geologists Studies in Geology 15, v. 2, p. 2.2.3–141.

Pilger, R. H., 1978, A closed Gulf of Mexico, pre-Atlantic Ocean plate reconstruction and the early rift history of the Gulf and North Atlantic: Gulf Coast Association of Geological Societies Transactions, v. 28, p. 385–393.

Pindell, J. L., 1985, Alleghenian reconstruction and subsequent evolution of the Gulf of Mexico, Bahamas and proto-Caribbean: Tectonics, v. 4, p. 1–39.

Pindell, J., and Dewey, J. D., 1982, Permo-Triassic reconstructions of western Pangea and the evolution of the Gulf of Mexico/Caribbean region: Tectonics, v. 1, p. 179–211.

Pojeta, J., Kriz, J., and Berdan, J. M., 1976, Silurian-Devonian pelecypods and Paleozoic stratigraphy of subsurface rocks in Florida and Georgia and related Silurian pelecypods from Bolivia and Turkey: U.S. Geological Survey Professional Paper 879, 32 p.

Rankin, D. W., 1975, The continental margin of eastern North America in the southern Appalachians: The opening and closing of the proto-Atlantic Ocean: American Journal of Science, v. 275-A, p. 298–336.

—— , 1976, Appalachian salients and recesses; Late Precambrian continental breakup and opening of the Iapetus Ocean: Journal of Geophysical Research, v. 81, p. 5605–5619.

Rast, N., and Kohles, K. M., 1986, The origin of the Ocoee Supergroup: American Journal of Science, v. 286, p. 593–616.

Reinhardt, J., 1974, Stratigraphy, sedimentology and Cambro-Ordovician paleogeography of the Frederick Valley, Maryland: Maryland Geological Survey Report of Investigation, no. 23, 74 p.

Rosenthal, D. B., 1987, Distribution of crust in the deep eastern Gulf of Mexico [M.A. thesis]: The University of Texas at Austin, 149 p.

Ross, C. A., 1986, Paleozoic evolution of southern margin of Permian basin: Geological Society of America Bulletin, v. 97, p. 536–554.

Salvador, A., 1987, Late Triassic–Jurassic paleogeography and origin of Gulf of Mexico basin: American Association of Petroleum Geologists Bulletin, v. 71, p. 419–451.

Sawyer, D. S., Buffler, R. T., and Pilger, R., 1991, The crust under the Gulf of Mexico basin, *in* Salvador, A., ed., The Gulf of Mexico basin: Boulder, Colorado, Geological Society of America, The Geology of North America, v. J, p. 53–72.

Schamel, S., Hanley, T. B., and Sears, J. W., 1980, Geology of the Pine Mountain window and adjacent terranes in the Piedmont Province of Alabama and Georgia: Geological Society of America Southeastern Section, 29th Annual Meeting, Guidebook, 69 p.

Schlager, W., Buffler, R. T., Angstadt, D., and Phair, R. L., 1984, Geologic history of the southeastern Gulf of Mexico, *in* Initial Reports of the Deep Sea Drilling Project: Washington, D.C., U.S. Government Printing Office, v. 77, p. 715–738.

Schwab, F. L., 1986, Latest Precambrian–earliest Paleozoic sedimentation, Appalachian Blue Ridge and adjacent areas; Review and speculation, *in* McDowell, R. C., and Glover, L., III, eds., The Lowry volume; Studies in Appalachian geology: Virginia Tech Department of Geological Sciences Memoir 3, p. 115–137.

Schwalb, H. R., 1982a, Paleozoic geology of the New Madrid area: U.S. Nuclear Regulatory Commission NUREG/CR-2909, 61 p.

—— , 1982b, Geologic-tectonic history of the area surrounding the northern end of the Mississippi Embayment: University of Missouri-Rolla Journal, no. 3, p. 31–42.

Scrutton, R. A., 1982, Crustal structure and development of sheared passive continental margins: American Geophysical Union Geodynamics Series, v. 6, p. 133–140.

Shaub, F. J., Buffler, R. T., and Parsons, J. G., 1984, Seismic stratigraphic framework of the deep Gulf of Mexico basin: American Association of Petroleum Geologists Bulletin, v. 68, p. 1790–1802.

Shepherd, A., 1983, A study of magnetic anomalies in the eastern Gulf of Mexico [M.S. thesis]: University of Houston, 197 p.

Simpson, E. L., and Eriksson, K. A., 1989, Sedimentology of the Unicoi Formation in southern and central Virginia; Evidence for late Proterozoic to Early Cambrian rift-to-passive margin transition: Geological Society of America Bulletin, v. 101, p. 42–54.

Sloss, L. L., 1963, Sequences in the cratonic interior of North America: Geological Society of America Bulletin, v. 74, p. 93–114.

—— , 1988, Tectonic evolution of the craton in Phanerozoic time, *in* Sloss, L. L., ed., Sedimentary cover; North American craton, U.S.: Geological Society of America, The Geology of North America, v. D-2, p. 25–51.

Smith, C. I., 1981, Review of the geologic setting, stratigraphy, and facies distribution of the Lower Cretaceous in northern Mexico, *in* Lower Cretaceous stratigraphy and structure, northern Mexico: West Texas Geological Society

Field Trip Guidebook Publication 81-74, p. 1–27.

Smith, D., 1982, Review of the tectonic history of the Florida basement: Tectonophysics, v. 88, p. 1–22.

Swolfs, H. S., 1967, Seismic refraction studies in the southwestern Gulf of Mexico, [M.S. thesis]: College Station, Texas, Texas A&M University, 42 p.

Tauvers, P. R., and Muehlberger, W. R., 1987, Is the Brunswick magnetic anomaly really the Alleghanian suture?: Tectonics, v. 6, p. 331–342.

Thomas, W. A., 1976, Evolution of Ouachita-Appalachian continental margin: Journal of Geology, v. 84, p. 323–342.

——, 1977, Evolution of Appalachian-Ouachita salients and recesses from reentrants and promontories in the continental margin: American Journal of Science, v. 277, p. 1233–1278.

——, 1985a, Northern Alabama sections, *in* Woodward, N. B., ed., Valley and Ridge thrust belt; Balanced structural sections, Pennsylvania to Alabama (Appalachian Basin Industrial Associates): University of Tennessee Department of Geological Sciences Studies in Geology 12, p. 54–61.

——, 1985b, The Appalachian-Ouachita connection; Paleozoic orogenic belt at the southern margin of North America: Annual Review of Earth and Planetary Sciences, v. 13, p. 175–199.

——, 1988a, The Black Warrior basin, *in* Sloss, L. L., ed., Sedimentary cover; North American craton, U.S.: Geological Society of America, The Geology of North America, v. D-2, p. 471–492.

——, 1988b, Early Mesozoic faults of the northern Gulf Coastal Plain in the context of opening of the Atlantic Ocean, *in* Manspeizer, W., ed., Triassic-Jurassic rifting, continental breakup and the origin of the Atlantic Ocean and passive margins, Developments in geotectonics 22, Part A: New York, Elsevier, p. 463–476.

——, 1989a, Cross sections of Appalachian-Ouachita orogen beneath Gulf Coastal Plain, *in* Hatcher, R. D., Jr., Thomas, W. A., and Viele, G. W., eds., The Appalachian-Ouachita orogen in the United States: Geological Society of America, The Geology of North America, v. F-2, Pl. 9, scale 1:1,000,000.

——, 1989b, The Appalachian-Ouachita orogen beneath the Gulf Coastal Plain between the outcrops in the Appalachian and Ouachita Mountains, *in* Hatcher, R. D., Jr., Thomas, W. A., and Viele, G. W., eds., The Appalachian-Ouachita orogen in the United States: Geological Society of America, The Geology of North America, v. F-2, p. 537–553.

——, 1991, The Appalachian-Ouachita rifted margin of southeastern North America: Geological Society of America Bulletin, v. 103, p. 415–431.

Thomas, W. A., and 6 others, 1989a, Tectonic map of the Ouachita orogen, *in* Hatcher, R. D., Jr., Thomas, W. A., and Viele, G. W., eds., The Appalachian-Ouachita orogen in the United States: Geological Society of America, The Geology of North America, v. F-2, Pl. 9, scale 1:2,500,000.

Thomas, W. A., Chowns, T. M., Daniels, D. L., Neathery, T. L., Glover, L., III, and Gleason, R. J., 1989b, The subsurface Appalachians beneath the Atlantic and Gulf Coastal Plains, *in* Hatcher, R. D., Jr., Thomas, W. A., and Viele, G. W., eds., The Appalachian-Ouachita orogen in the United States: Geological Society of America, The Geology of North America, v. F-2, p. 445–458.

——, 1989c, Pre-Mesozoic paleogeologic map of Appalachian-Ouachita orogen beneath Atlantic and Gulf Coastal Plains, *in* Hatcher, R. D., Jr., Thomas, W. A., and Viele, G. W., eds., The Appalachian-Ouachita orogen in the United States: Geological Society of America, The Geology of North America, v. F-2, Pl. 6, scale 1:2,500,000.

Trehu, A. M., Klitgord, K. D., Sawyer, D. S., and Buffler, R. T., 1989, Atlantic and Gulf of Mexico continental margins, *in* Pakiser, L. C., and Mooney, W. D., eds., Geophysical framework of the continental United States: Boulder, Colorado, Geological Society of America Memoir 172, p. 349–382.

Vail, P. R., Mitchum, R. M., Jr., and Thompson, S., III, 1977, Seismic stratigraphy and global changes of sea level; Part 4; Global cycles of relative changes of sea level: American Association of Petroleum Geologists Memoir 26, p. 83–97.

Van Schmus, W. R., Bickford, M. E., and Zietz, I., 1987, Early and middle Proterozoic provinces in the central United States, *in* Kröner, A., ed., Proterozoic lithospheric evolution: American Geophysical Union Geodynamics Series, v. 17, p. 43–68.

Van der Voo, R., Mauk, F. J., and French, R. B., 1976, Permian-Triassic continental configurations and the origin of the Gulf of Mexico: Geology, v. 4, p. 177–180.

Viele, G. W., 1973, Structure and tectonic history of the Ouachita Mountains, Arkansas, *in* DeJong, K. A., and Scholten, R., eds., Gravity and tectonics: New York, Wiley, p. 361–377.

——, 1979, Geologic map and cross section, eastern Ouachita Mountains, Arkansas: Geological Society of America Map and Chart Series MC-28F, scale 1:250,000.

——, 1989, Ouachita system; Cross sections, C–C′, *in* Hatcher, R. D., Jr., Thomas, W. A., and Viele, G. W., eds., The Appalachian-Ouachita orogen in the United States: Geological Society of America, The Geology of North America, v. F-2, Pl. 11, scale 1:250,000.

Viele, G. W., and Thomas, W. A., 1989, Tectonic synthesis of the Ouachita orogenic belt, *in* Hatcher, R. D., Jr., Thomas, W. A., and Viele, G. W., eds., The Appalachian-Ouachita orogen in the United States: Geological Society of America, The Geology of North America, v. F-2, p. 695–728.

Walper, J. L., 1980, Tectonic evolution of the Gulf of Mexico, *in* Pilger, R. H., ed., The origin of the Gulf of Mexico and the early opening of the central North Atlantic Ocean: Louisiana State University School of Geosciences, Proceedings of a Symposium, p. 87–98.

Watkins, J. S., 1990, Palinspastic restoration of Mesozoic extension in the northern Gulf of Mexico: Geological Society of America Abstracts with Programs, v. 22, no. 7, p. A112.

Weaverling, P. H., 1987, Early Paleozoic tectonic and sedimentary evolution of the Reelfoot–Rough Creek rift system, midcontinent, U.S. [M.S. thesis]: Columbia, University of Missouri, 116 p.

Webb, E. J., 1980, Cambrian sedimentation and structural evolution of the Rome trough in Kentucky [Ph.D. thesis]: Cincinnati, Ohio, University of Cincinnati, 98 p.

Wehr, F., and Glover, L., III, 1985, Stratigraphy and tectonics of the Virginia–North Carolina Blue Ridge: Evolution of a late Proterozoic–early Paleozoic hinge zone: Geological Society of America Bulletin, v. 96, p. 285–295.

Weimer, P., and Buffler, R. T., 1992, Structural geology and evolution of the Mississippi Fan foldbelt: American Association of Petroleum Geologists Bulletin, v. 76, p. 225–251.

White, G. W., 1980, Permian-Triassic continental reconstruction of the Gulf of Mexico–Caribbean area: Nature, v. 283, p. 823–826.

Williams, H., and Hatcher, R. D., Jr., 1982, Suspect terranes and accretionary history of the Appalachian orogen: Geology, v. 10, p. 530–536.

Winker, C. D., and Buffler, R. T., 1988, Paleogeographic evolution of early deep-water Gulf of Mexico and margins, Jurassic to Middle Cretaceous (Comanchean): American Association of Petroleum Geologists Bulletin, v. 72, p. 318–346.

Woods, R. D., and Addington, S. W., 1973, Pre-Jurassic geologic framework, northern Gulf basin: Gulf Coast Association of Geological Societies Transactions, v. 23, p. 92–108.

Woodward, H. P., 1961, Preliminary subsurface study of southeastern Appalachian interior plateau: American Association of Petroleum Geologists Bulletin, v. 45, p. 1634–1655.

Worral, D. M., and Snelson, S., 1989, Evolution of the northern Gulf of Mexico, with emphasis on Cenozoic growth faulting and the role of salt, *in* Bally, A. W., and Palmer, A. R., eds., The Geology of North America: An overview: Boulder, Colorado, Geological Society of America, The Geology of North America, v. A, p. 97–138.

Worzel, J. L., and Watkins, J. S., 1973, Evolution of the northern Gulf Coast deduced from geophysical data: Gulf Coast Association of Geological Societies Transactions, v. 23, p. 84–91.

Zietz, I., compiler, 1982, Composite magnetic anomaly map of the United States: U.S. Geological Survey Geophysical Investigations Map GP-954-A, scale 1:2,500,000.

Manuscript Accepted by the Society June 24, 1992

Printed in U.S.A.

DNAG Continent-Ocean Transect Volume
Phanerozoic Evolution of North American
Continent-Ocean Transitions
The Geological Society of America, 1994

Chapter 5

Phanerozoic tectonic evolution of Mexico

Fernando Ortega-Gutiérrez
Instituto de Geología, Universidad Nacional Autónoma de México, 20, DF, Mexico
Richard L. Sedlock* and Robert C. Speed
Department of Geological Sciences, Northwestern University, Evanston, Illinois 60208

INTRODUCTION

We present a tectonic evolution of Mexico over the past 600 m.y., with focus on the position of southern margin of Proterozoic North America in Mexico, the Phanerozoic events that shaped this margin, and the accretion of terranes that built Mexico out from that margin to its present configuration.

The evolution of Mexico is a peculiar and difficult problem in tectonics. Whereas the kinematic history of the region of Mexico, for example, the end of Proterozoic continental breakup, the collision of North and South America in the late Paleozoic, the drifting apart of those two continents in the Mesozoic, and the motions of plates in the Pacific basin relative to North America in Cretaceous and Cenozoic time, is reasonably well known, the kinematics of terranes now in the region is poorly understood. In large part, this disparity is due to the extraordinary extent and volume of Cenozoic volcanism. Although such volcanism informs us about Cenozoic convergence, it cautions that the crust of Mexico has probably undergone major modification in Cenozoic time. Moreover, the volcanic cover limits greatly the direct access to older rocks and structures necessary to develop the story of an evolution that is well resolved in space and time. A tectonic synthesis of Mexico thus requires more assumptions than for most mountainous regions.

Our model of tectonic evolution of Mexico attempts to satisfy both far-field and sparse near-field kinematic histories and existing geologic and geophysical data. This chapter is a companion to an in-depth synthesis of data, identification and characterization of terranes, and illustration of stages in the accretion history by Sedlock and others (1993). Moreover, our study stems partly from and employs Continent-Ocean Transects H1 (Mitre-Salazar and Roldán-Quintana, 1990) and H3 (Ortega-Gutiérrez, 1990), which are shown in Figure 1.

Our approach differs from previous models of the tectonic evolution of Mexico. Most models divided Mexico into a few poorly defined continental blocks and focused on Mesozoic displacements during the breakup of Pangea and opening of the Gulf of Mexico basin (Walper, 1980; Pindell and Dewey, 1982; Anderson and Schmidt, 1983; Pindell, 1985; Klitgord and Schouten, 1987). De Cserna (1989) summarized the geology of Mexico in the context of morphotectonic provinces, but did not consider terranes or kinematic evolution. Campa-Uranga and Coney (1983) and Coney and Campa-Uranga (1987) divided Mexico into tectonostratigraphic terranes, but did not address their kinematic evolution. Our division of terranes differs from theirs in a number of ways.

We begin with a summary of the modern tectonics of Mexico, follow with a history of ancient regional events that were probably important in Mexico's evolution, discuss briefly North America and terranes in Mexico, and then outline a tectonic model of Mexican evolution since 600 Ma. More complete coverage of all these items can be found in Sedlock and others (1993).

NEOTECTONICS OF MEXICO

Plates

The region of Mexico and its surroundings comprises five plates (Fig. 2): North American, Pacific, Rivera, Cocos, and Caribbean. Mexico is almost entirely on the North American plate. Exceptions are Baja California, which is attached to the Pacific plate, and Mexico's southern tip, which probably moves partly or wholly with the Caribbean plate (Fig. 2). The western edge of mainland Mexico is a right-oblique convergent margin above the subducting oceanic Cocos and Rivera plates and a right-lateral strike-slip margin relative to oceanic lithosphere of the Pacific plate. To the east, Mexico continues geologically into the marine realms of the Gulf of Mexico and Yucatan basin, all of which, including closed basins of oceanic crust, are within the North American plate (Fig. 2). The North American–Caribbean plate boundary, poorly defined west of the Cayman Trough, is apparently a left-lateral strike-slip zone that crosses Mexico between Guatemala and the Gulf of Tehuantepec (Figs. 1 and 2) (Guzmán-Speziale and others, 1989).

The North American–Pacific boundary is a series of short

*Present address: Department of Geology, San Jose State University, San Jose, California 95192-0102.

Ortega-Gutiérrez, F., Sedlock, R. L., and Speed, R. C., 1994, Phanerozoic tectonic evolution of Mexico, *in* Speed, R. C., ed., Phanerozoic Evolution of North American Continent-Ocean Transitions: Boulder, Colorado, Geological Society of America, DNAG Continent-Ocean Transect Volume.

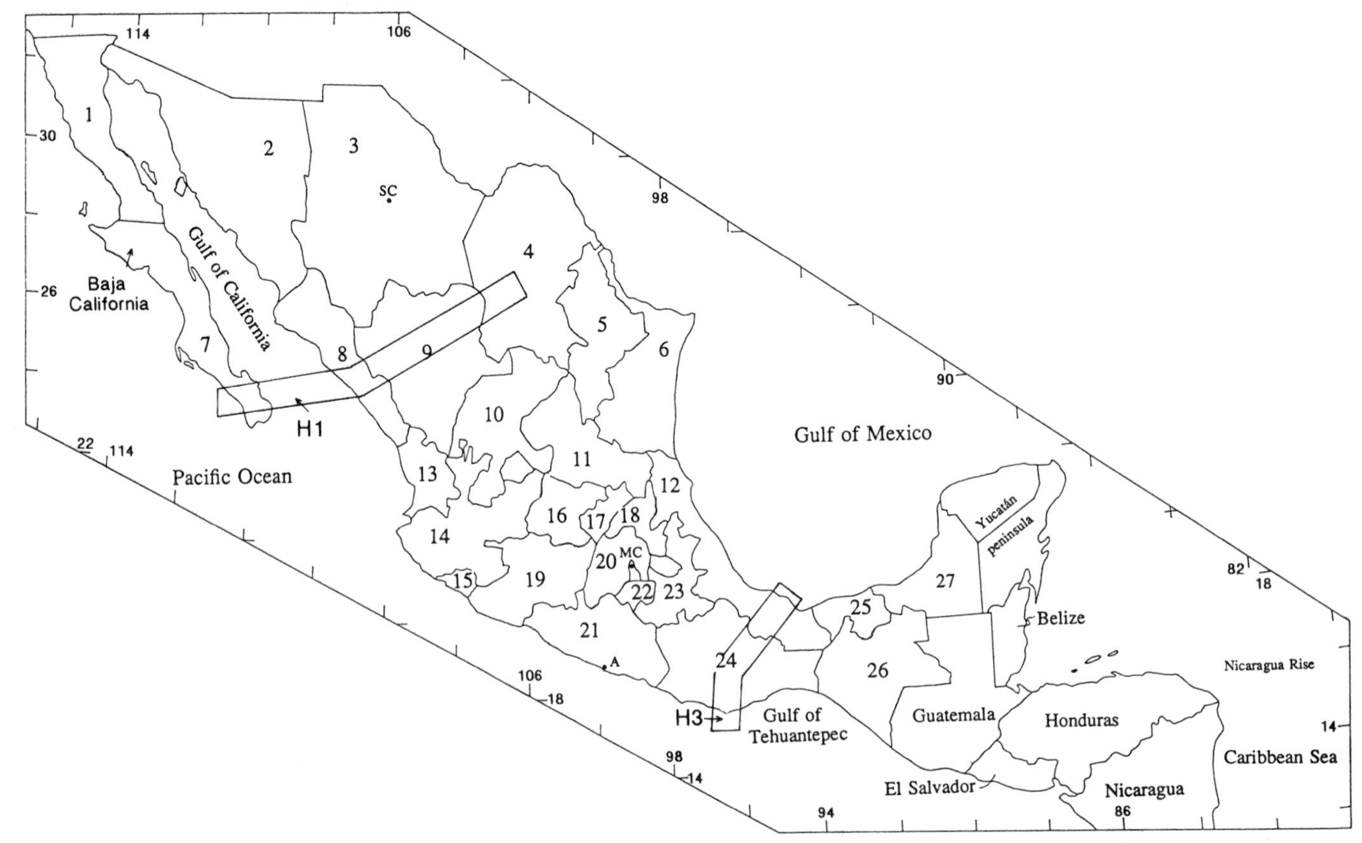

Figure 1. Geographic map of Mexico and part of Central America. States of Mexico are numbered as follows. 1: Baja California Norte; 2: Sonora; 3: Chihuahua; 4: Coahuila; 5: Nuevo León; 6: Tamaulipas; 7: Baja California Sur; 8: Sinaloa; 9: Durango; 10: Zacatecas; 11: San Luis Potosí; 12: Veracruz; 13: Nayarit; 14: Jalisco; 15: Colima; 16: Guanajuato; 17: Querétaro; 18: Hidalgo; 19: Michoacán; 20: México; 21: Guerrero; 22: Morelos; 23: Puebla; 24: Oaxaca; 25: Tabasco; 26: Chiapas; 27: Campeche. SC is Sierra del Cuervo; MC is Mexico City; A is Acapulco. H1 and H3 are transects by Mitre and Roldán (1991) and Ortega (1990), respectively.

segments of the northernmost East Pacific Rise joined by northwest-striking transform faults (Fig. 2). The Gulf of California is a product of continental rifting and sea-floor spreading where the plate boundary apparently stepped inboard of the continental edge of about 6 Ma (Lonsdale, 1989). The relative velocity is 50 mm/yr (S55°E) in the central part of the Gulf of California; this implies a small component of boundary-normal extension (DeMets and others, 1990). Baja California is a fragment formerly attached to continental Mexico before the opening of the Gulf of California and is perhaps one of the world's most evident examples of the kinematics of terrane generation and migration at continental margins.

The North American–Cocos boundary is marked by the Acapulco trench and forearc and by a north-northeast–dipping Benioff zone with maximum dip of 60° and seismicity as deep as 240 km (Burbach and others, 1984; Le Fevre and McNally, 1985). The relative velocity is between 55 and 75 mm/yr (south-southwest), depending on position, and is determined by slip vectors of thrust earthquakes (DeMets and others, 1990). The convergence is slightly counterclockwise to the trench normal (Fig. 2).

The North American–Caribbean plate boundary is best defined at the Cayman Trough (Fig. 2), which is floored by oceanic crust over most its length and contains an active spreading center (Holcombe and others, 1973; Rosencrantz and others, 1988). The Cayman Trough is an elongate pull-apart basin at a step in a left-lateral strike-slip system, defined locally by the Swan Islands and Oriente faults (Fig. 2). The half-spreading rate in the Cayman Trough is about 12 mm/yr (DeMets and others, 1990). West of the Cayman Trough, the North American–Caribbean boundary is poorly demarcated and presumably continues west through Honduras, Guatemala, and southern Mexico (Chiapas) to a triple junction with the Cocos plate in the Gulf of Tehuantepec. The onland portion of this boundary includes parallel left-slip faults with marked topographic expression, the Motagua, Polochic, and Jocotán-Chamelecón faults (Fig. 2), implying that displacements are distributed in a zone of finite width.

The Rivera and Cocos plates are remnants of the Farallon

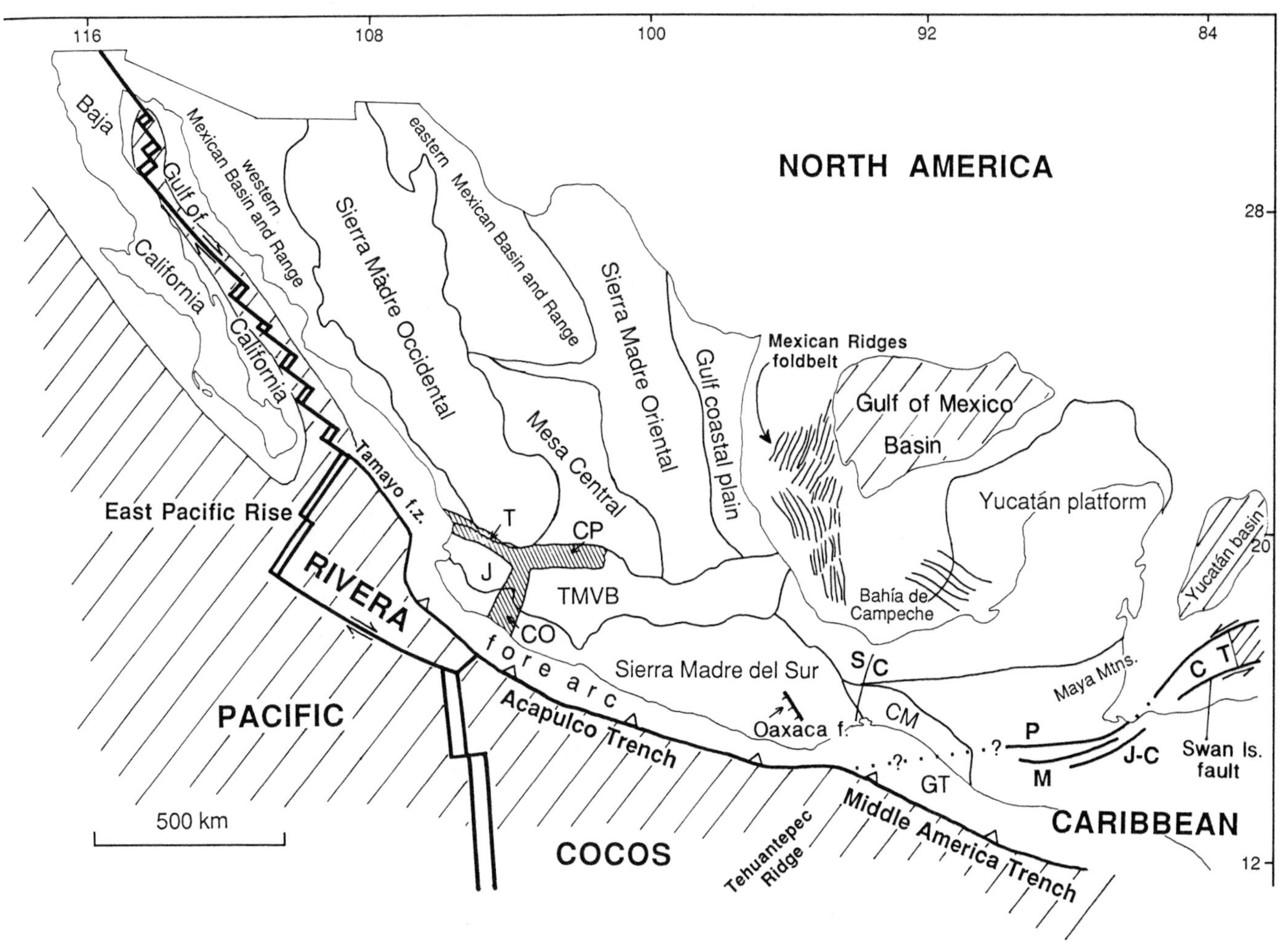

Figure 2. Major neotectonic provinces and structures (light lines) and plate boundary structures (heavy lines) of Mexico. Plate names in capital letters. Lines in Gulf of Mexico are axial traces of folds. Widely spaced line pattern shows oceanic lithosphere. Narrowly spaced lined pattern shows graben system: T, Tepic; CO, Colima; CP, Chapala. J: Jalisco block; TMVB: Trans Mexico volcanic belt; GT: Gulf of Tehuantepec; SC: Salina Cruz fault; CM: Chiapas massif; P: Polochic fault; M: Motagua fault; J-C: Jocotán-Chamelecón fault; CT: Cayman Trough.

plate of oceanic lithosphere that converged with North America all along western Mexico until about 28 Ma (Atwater, 1970, 1989). At that time, the Pacific-Farallon spreading boundary intersected North America in northern Baja California, and a triple junction (North American–Pacific–Farallon [Cocos/Rivera]) has since migrated southeasterly along the Mexican western coast.

Neotectonic provinces

The Basin and Range province of Neogene normal faulting and horst-graben structures occupies a large region of central-northern Mexico (Fig. 3) that includes the margins of the Sierra Madre Occidental province of Paleogene arc magmatism and much of another province to the east of comparatively minor Paleogene magmatism (INEGI, 1980). The onset of Basin and Range tectonism in northern Mexico was about 24 Ma and was concurrent with a change from intermediate to bimodal compositions of volcanic rocks.

The Baja California Peninsula (Fig. 3) is a composite terrane excised from coastal continental Mexico by an inboard jump of the Pacific–North American plate boundary at about 6 Ma. The Gulf of California is the zone of strongly extended continental crust and newly created oceanic crust that borders this new segment of the plate boundary. The locus of drifting in the gulf may follow a Neogene magmatic arc; the remnants of the arc occupy both flanks, but the axis was probably on the Baja side. The zone of rifting of the continent that was related to and preceded opening of the Gulf of California has variable width (Fig. 3) (Stock and Hodges, 1989), perhaps reflecting the width of the preceding arc or/and the confluence with preexisting Basin and Range structures (Fig. 3).

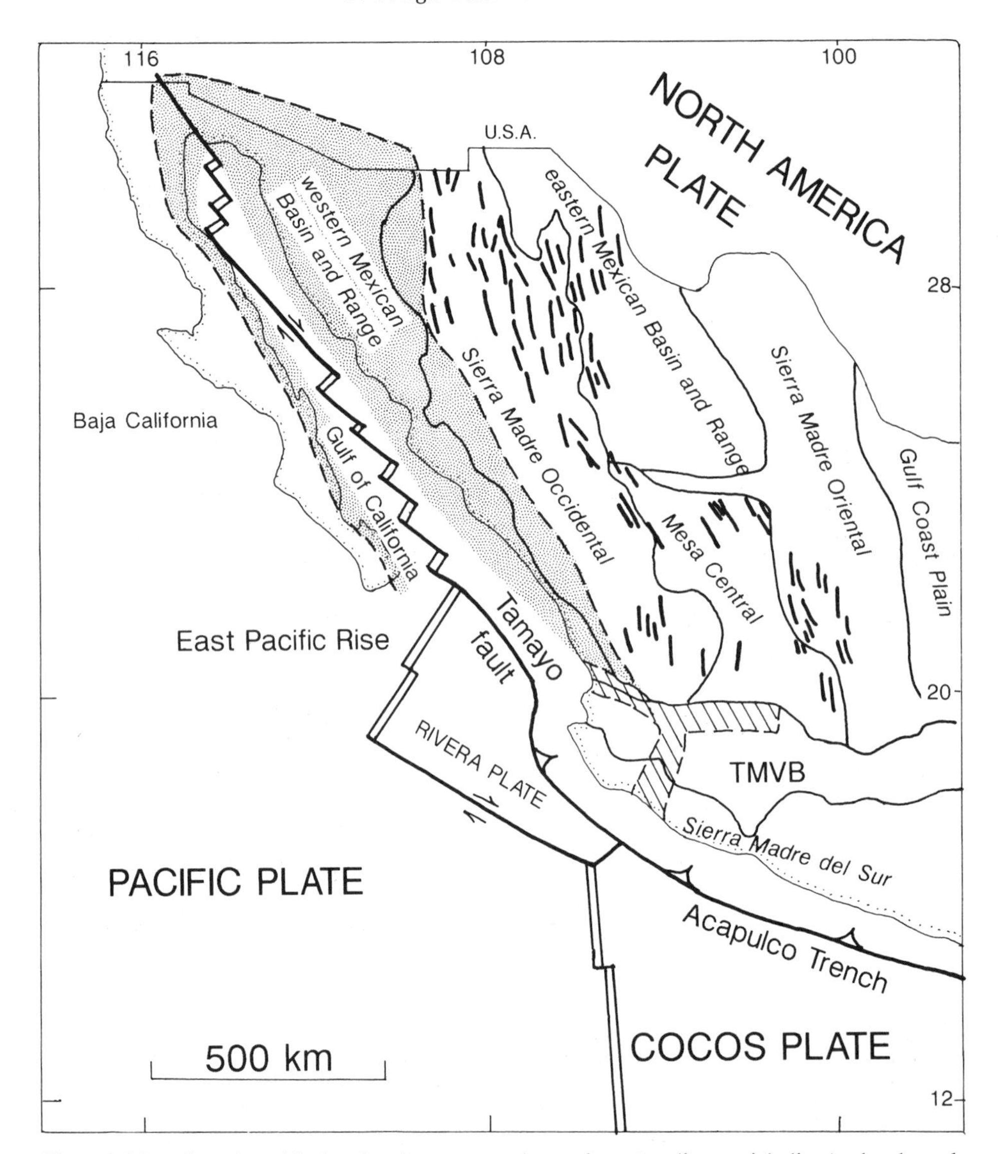

Figure 3. Map of northern Mexico showing neotectonic provinces (medium weight lines), plate boundary faults (heavy continuous and double lines), graben identified on Figure 2 (line pattern), and region of rifting related to opening of the Gulf of California (shaded). Short heavy lines are traces of major Basin-Range faults.

We now address a Neogene evolution of the Gulf of California (Fig. 4), modified from Lonsdale (1989). Magmatism of the continental arc that existed on Baja California until about 12 Ma was driven by the subduction of the Cocos and related small plates. As the Pacific–Cocos–North America triple junction migrated southward, subduction below southern Baja California was supplanted (14 to 12 Ma) by right-lateral transform motion along the outboard Tosco-Abreojos fault zone. From 12 to 5 Ma (Fig. 4), the triple junction migrated to a point south of Baja California, and Pacific–North America strike-slip motion was taken up in a zone of faults that straddled Baja California (Spencer and Normark; 1979; de Cserna, 1989; Sedlock and Hamilton, 1991). Just before 5 Ma (Fig. 4), the triple junction jumped eastward to its present site, initiating the opening of the Gulf of California (Curray and Moore, 1984). As rifting and spreading progressed, most but not all of the plate motion was transferred to the rift-transform system of the gulf (Sedlock and Hamilton, 1991). The Rivera plate detached from the Cocos within the past 5 m.y. (Fig. 4), probably along a former transform, as the triple junction continued southward.

The foot of the Mexican continental slope south of the Gulf of California is the Acapulco trench, the formation of which is related to the subduction of the Cocos and Rivera plates below the North American plate (Fig. 2). The continental slope and shelf above the trench is a forearc caused by the accretion of sediments from the downgoing plates and from the trench floor (Fig. 2) (Watkins and others, 1981; Ladd and Buffler, 1985).

The Sierra Madre Occidental is an elongate plateau (Fig. 2),

Late Cenozoic tectonic evolution of northwestern Mexico

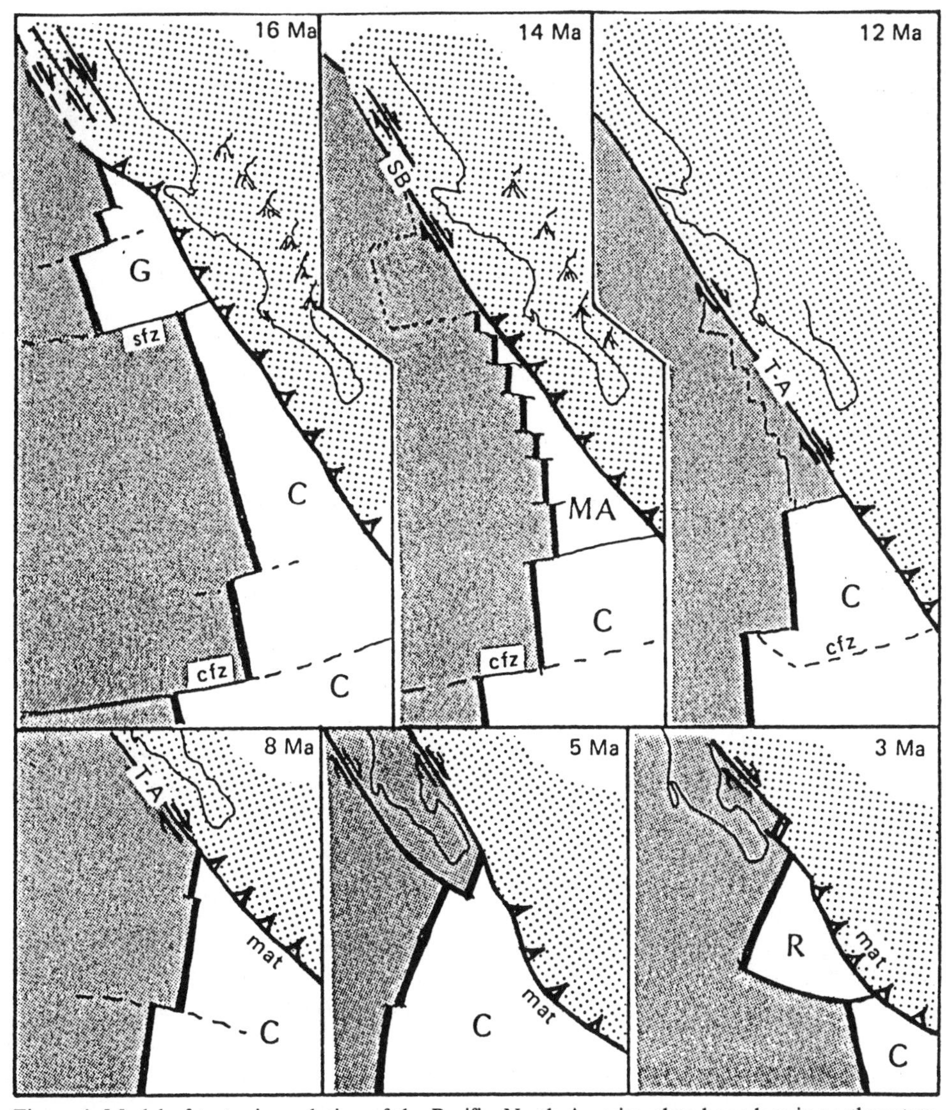

Figure 4. Model of tectonic evolution of the Pacific–North America plate boundary in northwestern Mexico and opening of the Gulf of California, from Sedlock and others (1993), modified from Lonsdale (1989). Dark pattern: Pacific plate; dots: North American plate; G: Guadalupe plate; C: Cocos plate; MA: Magdalena plate; R: Rivera plate; sfz: Shirley fracture zone; cfz: Clarion fracture zone; SB: San Benito fault zone; T-A: Tosca-Abreojos fault zone; mat: Middle America trench.

constructed by copious Cenozoic volcanism that began before 50 Ma and perhaps in the Cretaceous (McDowell and Keizer, 1977). The plateau is a continental arc built above the former Farallon–North America subduction zone. It includes a general eastward decrease in age of magmatism to about 40 Ma that may indicate progressive shallowing of the Benioff zone (Coney and Reynolds, 1977; Damon and others, 1981; Clark and others, 1982). Following the eastward sweep of principally andesitic volcanism, huge calderas with ignimbritic effusions developed widely in the province from about 38 to 27 Ma. The next stage of volcanism was chiefly mafic, and occurred from 32 to 17 Ma, during regional development of basin and range structures in northern Mexico. From 24 to 12 Ma, magmatism shifted westward from the Sierra Madre Occidental to eastern Baja California, which underwent intra-arc extension in middle and late Miocene time that presaged the opening of the Gulf of California (Hausback, 1984; Sawlan and Smith, 1984; Sawlan, 1991).

The Sierra Madre Oriental is also a plateau but with only minor Cenozoic volcanism compared to that of the Sierra Madre Occidental (Fig. 2). The Sierra Madre Oriental is underlain principally by the Late Cretaceous and early Tertiary Laramide fold and thrust belt. This plateau may represent a generally nonextended backarc to the continental arc of the Sierra Madre Occidental.

The Gulf coastal plain of northern Mexico (Fig. 2) is the low-lying region east of the plateaus, presumably representing North America inboard of the provinces affected by Cenozoic convergence of the Farallon plate and its progeny. Within this area, however, are alkaline and ultraalkaline intrusive and extrusive rocks of Oligocene to Quaternary age (Bloomfield and Cepeda, 1973; Cantagrel and Robin, 1979). Such rocks may represent either rifting well inboard of the continental edge, like that of the Rio Grande rift in New Mexico, or the most distal magmatism of subduction of the Pacific margin. Offshore between the Gulf coastal plain and the oceanic crust of the Gulf of Mexico, there is a belt of folded and thrusted Cretaceous and Cenozoic strata called the Mexican Ridges fold belt (Fig. 2) (Buffler and others, 1979) and Cordillera Ordóñez (de Cserna, 1981). The Mexican Ridges fold belt records late Neogene and modern east-west horizontal contraction across the western flank of the Gulf of Mexico basin. The cause of such contraction is not clear, but eastward gravity sliding of the shelf has been suggested (Suter, 1987).

The Trans-Mexican volcanic belt (TMVB), defined by stratovolcanoes, calderas, fields of volcanic rocks of Holocene to Miocene age, and active faults, trends east-west across central Mexico (Fig. 2). The TMVB appears to straddle a zone of terrane boundaries that has undergone episodic reactivation, perhaps to the present (Gastil and Jensky, 1973; Johnson, 1987; Urrutia-Fucugauchi and Bohnel, 1987; DeMets and Stein, 1990; Suter and others, 1991). Magmas of the TMVB may be related to subduction of the Cocos plate (Nixon, 1982).

Cutting through the western TMVB are three grabens (Figs. 2 and 3), the Colima, Tepic, and Chapala, that are associated with active volcanoes. It has been proposed that the graben system represents the breakaway zone of a small continental terrane (Jalisco block) that will become part of the Rivera plate as the Rivera–North American plate boundary jumps inboard (Luhr and others, 1985; Allan, 1986).

The Sierra Madre del Sur is a mountainous province that exposes mainly basement rocks in a complex of structural blocks (Fig. 2). Although it contains widespread Cenozoic volcanic rocks, there is little evidence that this province underwent prolific continental arc volcanism, in spite of its position just landward of the Acapulco Trench. The Sierra Madre del Sur thus differs in this respect from the Sierra Madre Occidental. The Oaxaca fault (Fig. 2) is probably the only intraplate fault of major length (130 km) that is active in southern Mexico (Centeno and others, 1990).

The Chiapas massif is a mountain range whose rocks have geologic affinities with those to the east in Guatemala. The Chiapas massif appears to be in fault contact with terranes at the eastern margin of the Sierra Madre del Sur province. The Yucatán platform is a low-lying plateau, probably a little-stretched microcontinental fragment, between the oceanic basins of the Gulf of Mexico and Yucatán basin (Fig. 2).

SYNOPSIS OF PRE-NEOGENE EVENTS AND MAJOR FEATURES

Proterozoic basement in Mexico and southwestern U.S.

Figure 5 shows sites where the existence and ages of Proterozoic crystalline basement are known in Mexico. Such rocks crop out widely in Sonora and in the southern states of Mexico, but elsewhere in Mexico are known only in a few outcrops and wells and from xenoliths in Quaternary volcanic rocks (Ruiz and others, 1988a, b) at a dozen sites. We surmise that Proterozoic basement probably does not underlie most of west-central mainland Mexico or Baja California.

Three age-range provinces of Proterozoic basement are identified in the United States near the Mexican border (Fig. 6): 1800 to 1700 Ma, 1700–1600 Ma, and Grenvillean (1300 to 1000 Ma) (Bickford and others, 1986). These continue into northern Mexico (Fig. 6), although the sampling on the Grenvillean basement has been sparse and the positions of its boundaries are not well known. The northwest-trending boundary between the 1800 to 1700 and 1700 to 1600 Ma provinces in northwestern Mexico is the Mojave-Sonora megashear (Silver and Anderson, 1974; Anderson and Schmidt, 1983).

In eastern Mexico from Chihuahua south to Oaxaca (Fig. 5), Proterozoic basement is of Grenvillean age. The igneous-metamorphic complexes of the Chiapas massif and the possibly correlative Chuacus Group may include widespread Precambrian rocks, although only a single radiometric date among many (zircon in Chuacus gneiss; 1075 Ma) is that old, and in fact could be from inherited zircons (Gomberg and others, 1968; McBirney and Bass, 1969). Four wells in Yucatán, Guatemala, and Belize (Fig. 5) bottomed in crystalline basement, which is not well dated, but is inferred to be of mid-Paleozoic age (Sedlock and others, 1993).

End of Proterozoic passive margin

In Proterozoic time, today's Proterozoic North America (PNA) was part of a more extensive continent called Rodinia by McMenamin and McMenamin (1990). Rodinia underwent rifting in latest Proterozoic time (800 to 550 Ma) in a far-reaching network (Stewart, 1976; Bond and Komintz, 1984; Bond and others, 1988). The rifting culminated in drifting and the devel-

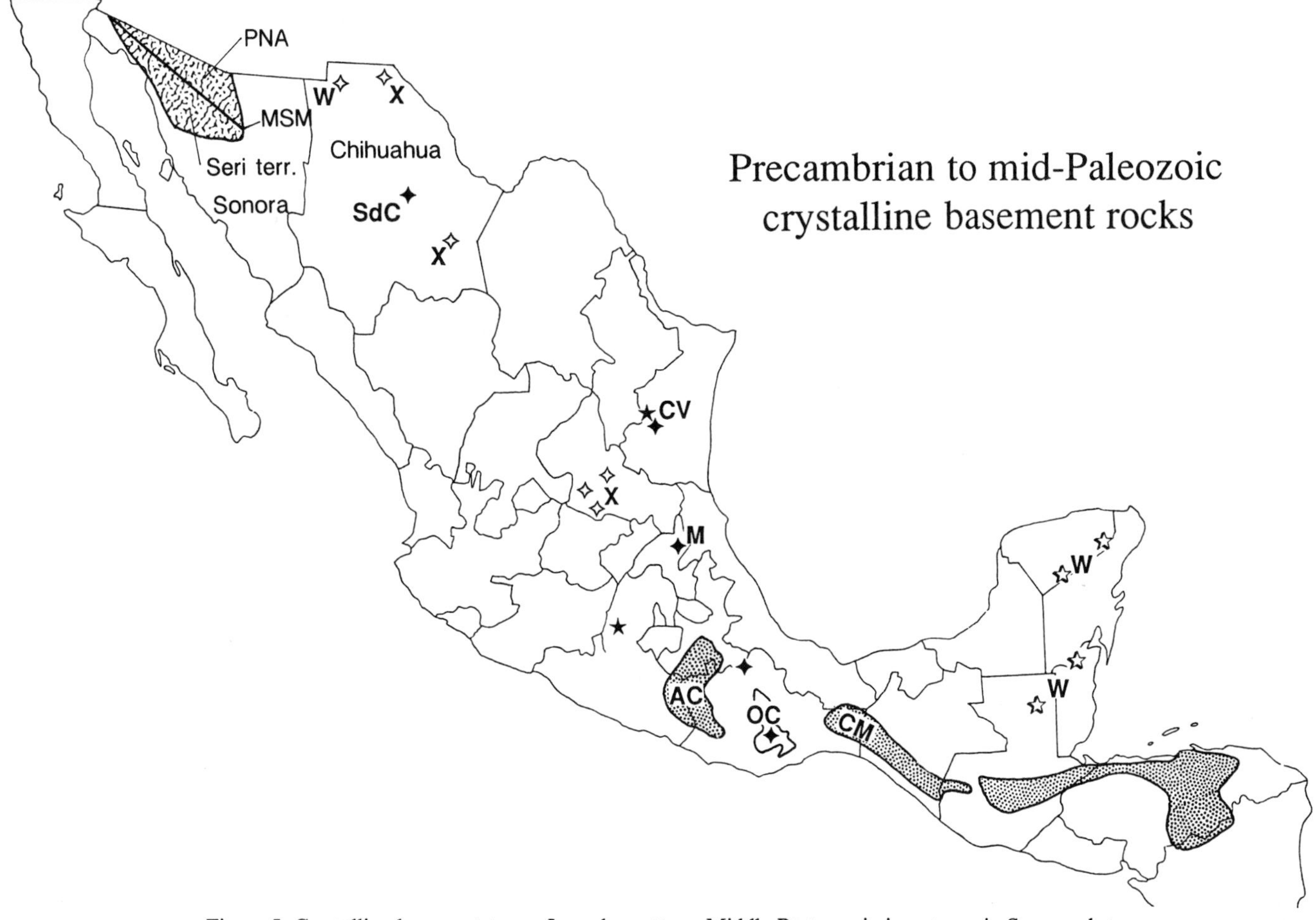

Figure 5. Crystalline basement types. Irregular pattern: Middle Proterozoic in outcrop in Sonora; dot pattern: Late Proterozoic to middle Paleozoic in outcrop, including Acatlán Complex (AC) and Chiapas Massif (CM); filled diamonds: outcrops of Grenvillean rocks, including Oaxacan Complex (OC), Ciudad Victoria (CV), Molango (M), and Sierra del Cuervo (SdC); open diamonds: Grenvillean basement in wells (W) and inferred from xenoliths (X); filled stars: outcrops of basement of probable mid-Paleozoic age; open stars: crystalline basement of possible mid-Paleozoic age. PNA: Proterozoic North America; MSM: Mojave-Sonora megashear.

opment of end of Proterozoic passive margins that shaped PNA more or less in its present configuration. Comparing the modern and end of Proterozoic loci of the edge of PNA, it is clear or probable that the edge has migrated slightly continentward by different mechanisms at different places: by landward thrusting in the Appalachians, by oblique-slip excisions in southern California and the Pacific Northwest, and by rifting and drifting in the Arctic. The fragments removed from the continent during the Phanerozoic presumably have become terranes elsewhere along the margin of North America or of some other continent.

The evolution of the edge of PNA in Mexico in both end of Proterozoic and Phanerozoic time is hard to investigate because of the wide inaccessibility of pre-Cenozoic rock and the likely major Cenozoic crustal transformation (King, 1975). We infer that the present-day edge of PNA in Mexico is the Mojave-Sonora megashear (Fig. 7), and that to the south, Mexico is composed of terranes of mainly Mesozoic ages of attachment. Some questions about the Phanerozoic evolution of the edge are as follows. (1) Did an end of Proterozoic edge to North America develop in or near present-day Mexico? (2) If it did, was it coincident with today's edge of PNA, or was it farther south? (3) Was it in an active or passive margin? (4) If the present and end of Proterozoic edges differ, what were the tectonics of the change? The few constraints to possible answers are the following. (1) The part of Proterozoic North America in modern Mexico evolved as a cratonal platform in early Paleozoic time, not as the subsiding shelf of a rifted passive margin. (2) The Seri terrane (Fig. 7), which was attached to PNA along the Mojave-Sonora megashear by Jurassic or older left slip, contains the only lower Paleozoic passive margin wedge strata in Mexico; other terranes containing Precambrian basement have cover recording platformal conditions in early Paleozoic time. (3) The development of

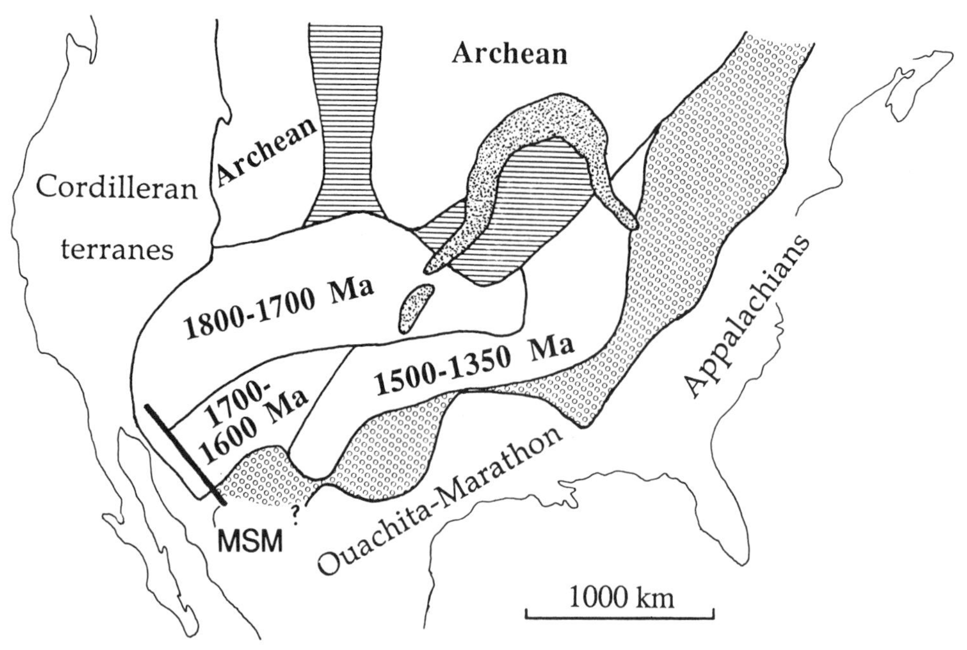

Figure 6. Precambrian domains in southern North America, modified from Anderson and Silver (1981), Bickford and others (1986), and Hoffman (1989). Horizontal lines: 2000–1800 Ma provinces; circle pattern: 1300–1000 Ma Grenville province; irregular dot pattern: 1100 Ma rift. MSM is Mojave-Sonora megashear.

the edges of PNA in regions of the United States that are contiguous to Mexico bear on interpretations of the evolution of PNA in Mexico; these are discussed next.

In the U.S. Gulf Coast (Fig. 8), an edge to PNA almost certainly developed during the end of Proterozoic breakup of Rodinia; there is a long zigzag belt of marked changes in crustal thickness and velocity, and two sites along the belt (Wichita Mountains, Devils uplift) are igneous rocks, both thought to be rift related, that have Cambrian dates (Nicholas and Rosendal, 1975; Keller and others, 1989; Thomas, 1977, 1985; Buffler and Thomas, this volume). This margin is generally obscured by the frontal zone of the late Paleozoic Ouachitan orogen and foreland basins (Fig. 9). The northwest-trending reaches of the belt are marked by platformal Paleozoic cover and by an abrupt transition in crustal thickness. These strands are interpreted as transform segments in the breakup system of Rodinia (Fig. 8b) (Thomas, 1989; Buffler and Thomas, this volume), thereby accounting for the absence of Cambrian rift margin strata and for the narrowness of the margin as at the Southern Grand Banks Transform of Canada (Todd and Keen, 1989; Haworth and others, this volume). Support for this interpretation, which we accept, comes from the similar strike of the many short transform segments in the contiguous rift-dominated zone of the Appalachians (Fig. 8B) (Rankin, this volume). The northeast-trending reaches of the Gulf coast margin are presumably of rift origin, by analogy to the Appalachian system. The complete burial of these segments by the Ouachita orogen (Fig. 9) makes their rift origin difficult to test.

The margin of PNA in the southwestern United States has a markedly different character north and south of Bishop, California (Fig. 8A). To the north, it includes a very broad passive margin stratal wedge (Stewart, 1972; Stewart and Poole, 1974; Bond and Komintz, 1984) that indicates a rift-dominated margin in latest Proterozoic time and the onset of drift at about 570 Ma. Here the edge of PNA parallels the isopachs and facies trends of the passive margin wedge. This may be the only reach of the western United States and Canada where the modern and Proterozoic edges are the same, or alternatively, where the edge has not been substantially eroded tectonically during the Phanerozoic (Speed, 1983). South of Bishop, in contrast, the early Paleozoic passive margin wedge features trend perpendicular to the edge of PNA, and farther south in southern California, platformal Paleozoic cover extends to the edge of PNA (Fig. 8) (Hamilton and Myers, 1966; Stewart and Poole, 1974; Stone and Stevens, 1988; Walker, 1988). The margin south of Bishop is widely regarded as having undergone major excisions during the Phanerozoic (see below).

The Seri terrane (Fig. 7) probably moved at least 500 km southeast from southern California its present position in late Paleozoic and/or in Jurassic time, according to displaced stratal facies and the offset of the boundary of Proterozoic provinces (Fig. 6) (Silver and Anderson, 1974). Younger right-slip motions occurred on a surface outboard of the contact of the Seri terrane and PNA.

Two models of the configuration of PNA in Mexico at the end of Proterozoic time seem most admissible. (1) The edge of

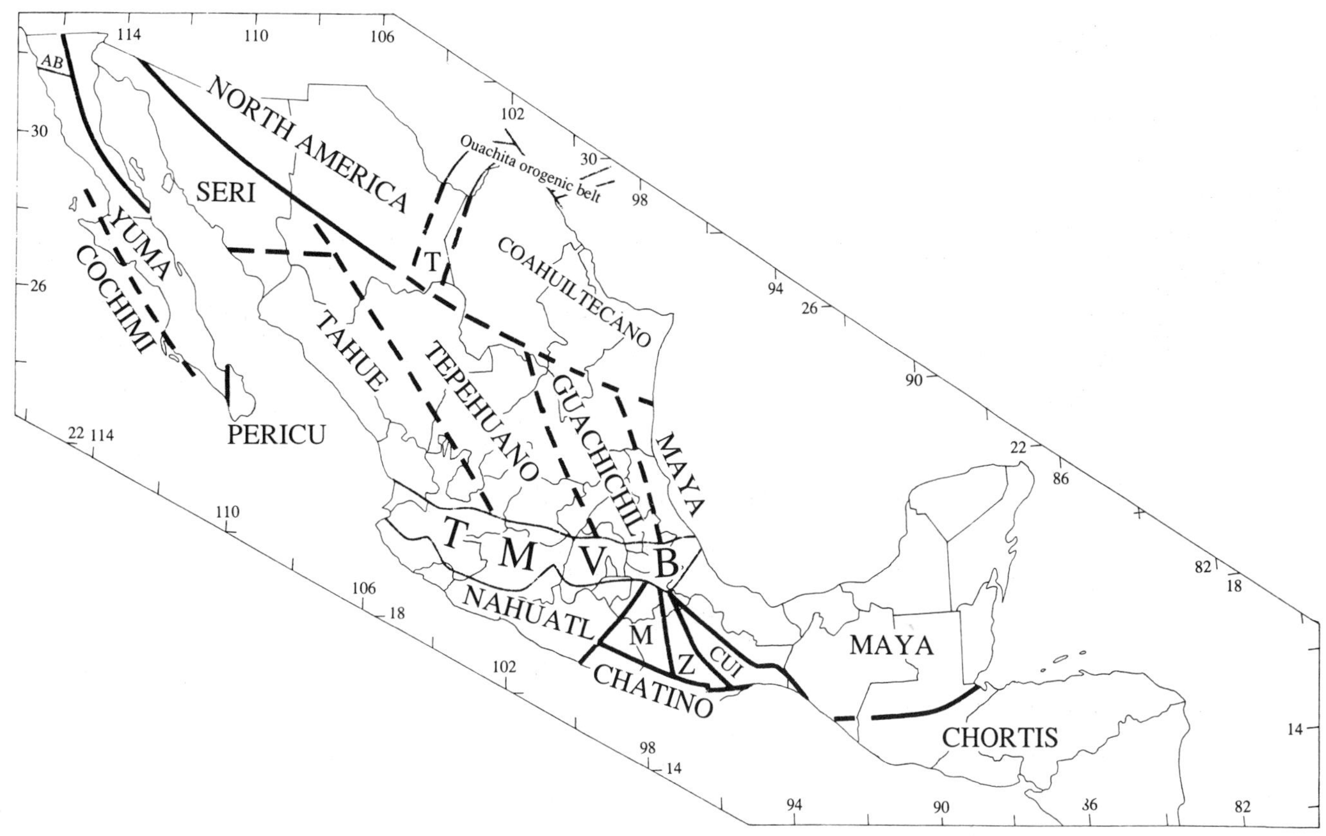

Figure 7. Map of terranes in Mexico and northern Central America; terrane boundaries are very heavy lines. Terranes overlapped by Trans-Mexican volcanic belt (TMVB). CUI: Cuicateco; M: Mixteco; T: Tarahumara; Z: Zapoteco; AB: Agua Blanca fault.

PNA in Mexico was close to that of today (Stewart, 1988, 1990) and was a westerly continuation of the transform-dominated edge of Rodinia in the U.S. Gulf Coast (Fig. 8) (Sedlock and others, 1993); in this case, the Seri terrane may have migrated from a late Paleozoic position in southern California and attached in late Paleozoic or early Mesozoic time to an end of Proterozoic surface in northwestern Mexico. (2) PNA in Mexico and in southern California formed a salient with uncertain margins that extended south and west, perhaps a great distance, from the present edge of PNA (Shurbet and Cebull, 1987); in this model, the end of Proterozoic edge in Mexico was trimmed continentward, probably by Phanerozoic tectonics recognized in southern California.

We can find no evidence with which to deal a deft blow to either hypothesis. We favor the first (Fig. 8), however, for two reasons. First, the westward extrapolation into Mexico of the end of Proterozoic Gulf Coast structure is reasonable, because such structure marks probable abrupt differences in basement. Second, if a large salient of PNA existed in Mexico early in the Phanerozoic but was set adrift and/or thereafter broken up, its relics might be identified as terranes, or a large mass whose paleomagnetic pole path might indicate past proximity to Mexico. We find no support for either expected result. Continental restorations of Scotese and McKerrow (1990) showed no large continental mass near Mexico in the Paleozoic until the collision of South America (see below). It is conceivable, however, that a salient of PNA existed at the time of late Paleozoic collision with South America and was removed as a new addition to South America during Cretaceous drifting. This possibility emerges from the existence of basement rocks that include Grenvillian ages in western northern South America (Kroonenberg, 1982; Rowley and Pindell, 1989).

Paleozoic oceanic crust

The premise of the preceding section is that North America took on an end of Proterozoic configuration in Mexico close to that of PNA today by the development of a southern passive margin dominated by continent-continent transforms at about the end of Proterozoic time. It implies that sea-floor spreading occurred south of the margin and that PNA faced south to early Paleozoic oceanic crust. No remnant of such crust has been identified in Mexico, except for a pre-Mississippian ophiolite in the Acatlán Complex of the Mixteco terrane. We presume that the rest was subducted in the Ouachitan orogeny or during Mesozoic accretion of Mexican terranes.

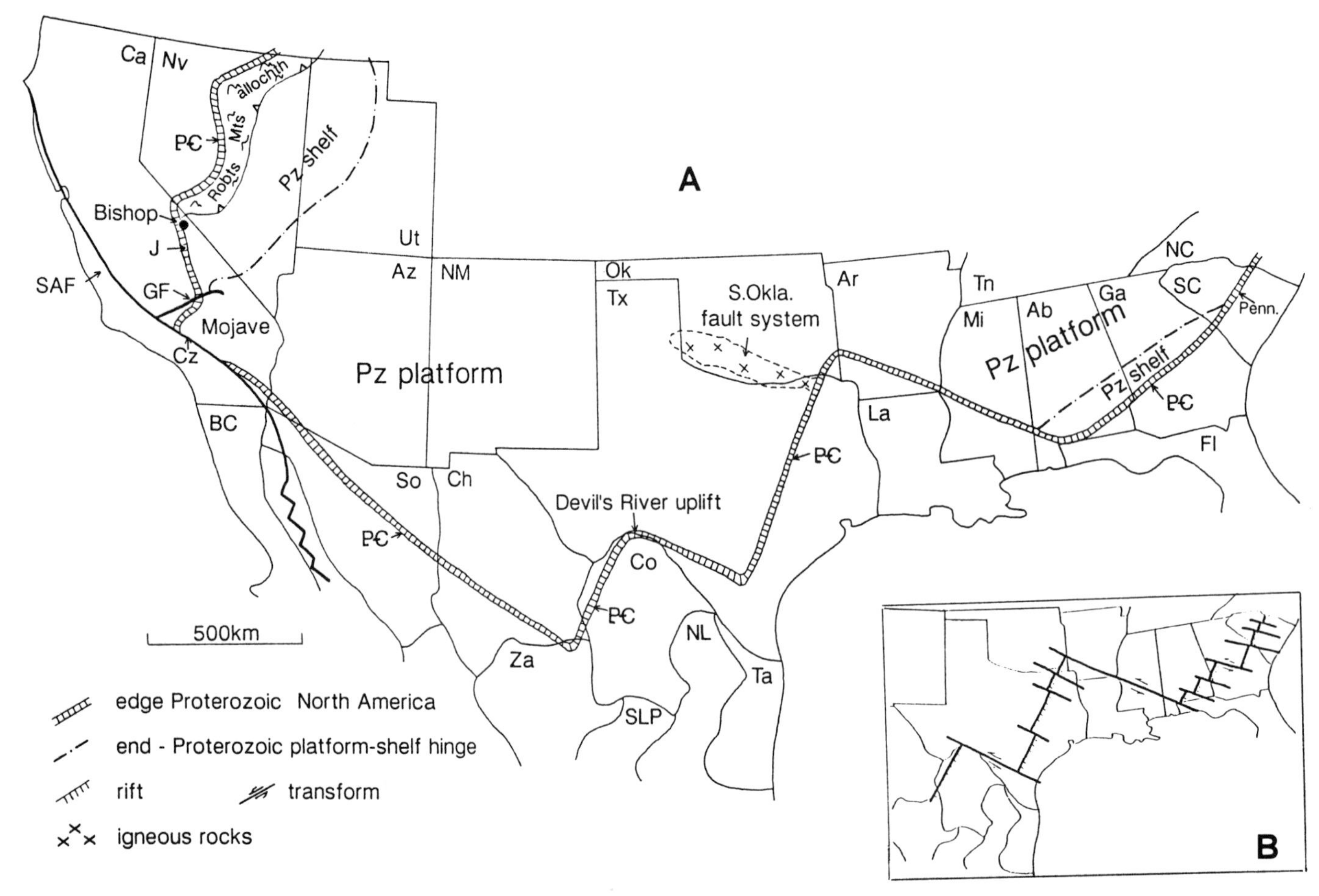

Figure 8. A: Map of northern Mexico and southern United States showing probable locus of modern edge of contiguous Proterozoic North America; edge is almost everywhere deeply buried by sediments and tectonic packets; age symbols attached to segments of the edge are interpreted age of formation of the existing edge; Proterozoic-Cambrian (P-Є): end-Proterozoic drifting of North America; J: Jurassic excision of North America; CZ: Cenozoic excision of North America. SAF: San Andreas fault; GF: Garlock fault; PZ: Paleozoic. B: Interpretation of edge of Proterozoic North America between Coahuila and South Carolina as a ridge-transform system, after Thomas (1989); ridges identified by lines with tick marks.

Ouachitan orogenic belt

The late Paleozoic Ouachitan orogen (Fig. 9) along the southern margin of PNA in the southern United States was the product of diachronous collision of North America and Gondwana (South America and Africa) (King, 1975; Lillie and others, 1983; Flawn and others, 1961; Thomas, 1977, 1985; Handschy and others, 1987; Viele, 1979; Viele and Thomas, 1989; Buffler and Thomas, this volume). Reconstructions based on paleomagnetic and other evidence show that North and South America were in close proximity in the vicinity of Florida to Texas and perhaps west into Mexico (Ross, 1979; Van der Voo and others, 1984; Scotese and McKerrow, 1990). Interpretations of Ouachitan deformation, metamorphism, and synorogenic sedimentation indicate that North America was the subducting plate, the overriding Gondwana sutured at or continentward of the end of Proterozoic edge of PNA, and the collision was right oblique, migrating east to west in Pennsylvanian and Early Permian time (Viele, 1979; Keller and Cebull, 1973; Wickham and others, 1976; Viele and Thomas, 1989; Buffler and Thomas, this volume).

The most westerly exposures of the Ouachitan orogen are in the Marathon region of Texas (Fig. 9). Gravity and lead isotope studies indicate that the belt probably continues southward from Marathon to a poorly defined limit at the northern border of Durango that might coincide with the Mojave-Sonora megashear (Fig. 9) (Handshy and others, 1987; Aiken and others, 1988; James and Henry, 1990). The location of the Ouachitan orogen beyond this point, if any, is unknown.

The structure of the Ouachitan orogen from internal to external (southeast to northwest) zones through Texas is probably as follows (Flawn and others, 1961; King, 1975; Viele, 1979; Nicholas and Rosendal, 1975; Buffler and Thomas, this volume). The most southerly (internal) zone records late Paleozoic high

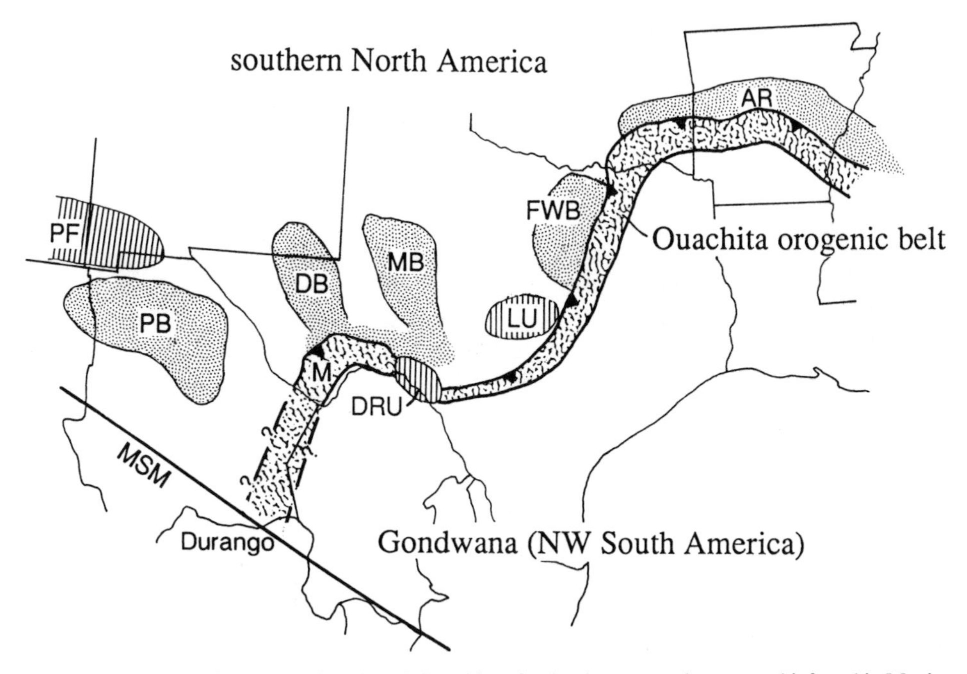

Figure 9. The Ouachitan orogenic belt and Ouachitan foreland structures known and inferred in Mexico and central southern United States foreland basins in dot patterns; uplifts in lined pattern. From Ross (1979), Armin (1987), and Viele and Thomas (1989). DB: Delaware basin; DRU: Devis River uplift; FWB: Fort Worth basin; LU: Llano uplift; M: Marathon; MB: Midland basin; PB: Pedregosa basin; PF: Pedregosa forebulge; AR: Arkoma basin; MSM: Mojave-Sonora megashear.

temperature-pressure metamorphism and continental arc magmatism superposed on siliceous crust of uncertain original age range, but probably including Precambrian. This is covered by deposits of the Upper Permian successor basin. To the north is an inner sedimentary forearc that includes accreted packets of lower Paleozoic ocean-floor sediments, presumably derived in part from the Paleozoic oceanic crust that fronted PNA and perhaps from other oceanic lithospheres. This forearc is covered by late Paleozoic flysch, partly volcanigenic. The most northward (external) zone of the orogen is the outer part of the Ouachitan forearc, composed of accreted late Paleozoic flysch, which probably varied from North American continental rise sediments to outer shelf to piggybacked foreland basin fill. Continentward of the Ouachitan deformation front are many discrete autochthonous Paleozoic foreland basins and uplifts (Fig. 9).

Paleogeography of Pangea

The Ouachitan collision joined South America and western North America in part of Pangea, which existed as a coherent supercontinent from the Early Permian. In Early Jurassic time, large intracontinental stretching began, leading to the opening of the Gulf of Mexico basin. It is unknown whether the North America–South America suture extended westward only to the present limit of the recognized Ouachitan orogenic belt (Fig. 9), or across much or all of present-day northern Mexico. We adopt the first alternative, portrayed in Figure 10, chiefly because a Permian-Triassic continental arc of northerly trend developed in present-day eastern Mexico, and it is difficult to rationalize the existence of a wide expanse of Pangean continental crust rather than subducting oceanic lithosphere to the west of the arc (Sedlock and others, 1993). We infer, therefore, that the collision stopped when the paleogeography (Fig. 10) had been attained, and/or the margin of the colliding continents took on orientations unsuitable for further collision (e.g., the intersection of a north-south western boundary of South America). In support of this interpretation, it is unlikely that a previously more westerly Ouachitan orogenic belt was truncated by the left-lateral Mojave-Sonora megashear, because contemporaneous Paleozoic rocks on opposite walls are not correlatives (Stewart, 1988, 1990).

Rifting and drifting in the Gulf of Mexico

The region of Pangea that is today's eastern Mexico, Yucatán, and the U.S. Gulf Coast began to undergo horizontal extension in Late Triassic time that led to the Jurassic Gulf of Mexico basin (Fig. 2) and to the drifting apart of South America and North America in the Early Cretaceous. The motion of South America relative to North America was S45°E, 30 mm/yr during the Cretaceous (Ladd, 1976; Klitgord and Schouten, 1987; Pindell and others, 1988), but kinematics of the stretching region prior to the onset of drifting are unknown and were probably complex.

Rifting occurred in two stages (Buffler and Sawyer, 1985;

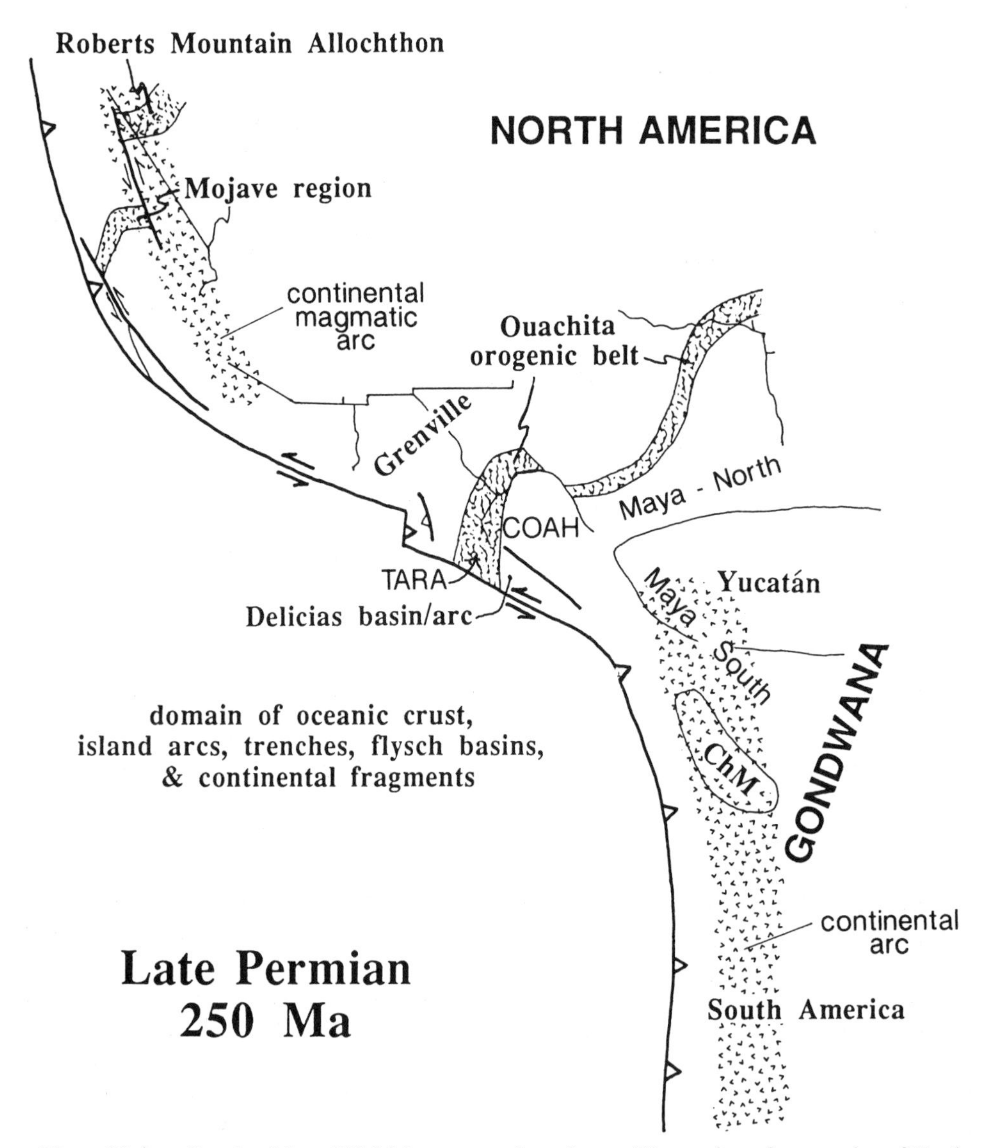

Figure 10. Late Permian (about 250 Ma) reconstruction of part of Pangea just after cessation of North America–Gondwana collision. TARA: Tarahumara terrane; COAH: Coahuiltecano; ChM: Chiapas massif.

Dunbar and Sawyer, 1987; Winker and Buffler, 1988; Buffler and Thomas, this volume; Sawyer, this volume). In the first, Late Triassic and Early Jurassic rifting affected eastern Mexico (Guachachil and Maya terranes) and the U.S. Gulf Coast, forming grabens associated with basaltic volcanism. Total extension and crustal attenuation were small in this stage. The first stage climaxed with the development of a widespread Middle Jurassic unconformity, interpreted as due to a regional rise in the asthenosphere. In the second stage, Middle and Late Jurassic large horizontal stretching took place within the Gulf of Mexico basin, greatly attenuating continental crust around the periphery of the basin and culminating in sea-floor spreading in the deep central basin (Fig. 2). The total horizontal displacement in a southeasterly direction before the Cretaceous was about 450 km by drifting between Yucatán and the Gulf Coast plus about 550 km by stretching of continental lithosphere in Yucatán and the Gulf Coast (Dunbar and Sawyer, 1987).

During rifting stages, the displacements of individual blocks were evidently complex and did not simply follow small circle paths about the North America–South America Euler pole, as is commonly assumed (e.g., Klitgord and Schouten, 1987). Rather, paleomagnetic directions at sites of Jurassic age in northeastern Mexico (Gose and others, 1982) and in Yucatán (Molina-Garza and others, 1990) show substantial counterclockwise rotation relative to Jurassic North America. During the second stage, the locus of rifting was concentrated in the Gulf of Mexico, leaving eastern Mexico only moderately broken up by first-stage extension. A principal difference in the two regions appears to have been that eastern Mexico was above a Permian-Triassic convergent margin shortly before or possibly during the onset of extension, whereas the Gulf of Mexico region was a backarc to that convergent margin. Perhaps the second-stage phenomena in the gulf were related to backarc asthenospheric rise.

A boundary must have developed early in the Jurassic between eastern Mexico and the gulf region where Yucatán continued to move away from North America; this movement apparently culminated in spreading in the Gulf of Mexico basin. For simplicity, we assume that the boundary was a transform fault beneath the continental slope of eastern Mexico (Fig. 2), generally similar to the Tamaulipas–Golden Lane fault of Pindell (1985). The transform occurs within our Maya terrane.

The rifting and drifting of the Gulf of Mexico basin stopped at the end of the Jurassic; structures were then covered by carbonate banks on and around platforms and by hemipelagic sediments in the basins (Salvador, 1987).

The Late Jurassic arrangement of isolated ephemeral islands, shallow-water platforms, and intraplatform deeper-water basins at the western and southern margins of the Gulf of Mexico persisted into the Early Cretaceous (Fig. 11) (Viniegra-O., 1971, 1981; Enos, 1983; Young, 1983; Cantú-Chapa and others, 1985; Winker and Buffler, 1988; de Cserna, 1989). Neocomian transgression transformed Jurassic peninsulas (Coahuila, Tamaulipas) to islands and reduced the area of smaller islands. The Tethyan affinity of reptilian fossils in Neocomian strata in the northern Mixteco terrane indicates exchange of oceanic waters between the Gulf of Mexico basin and the Tethyan realm to the south and east (Ferrusquía-Villafranca and Comas-Rodríguez, 1988). A likely paleogeography shows a nearly closed basin, bounded on the north by the U.S. Gulf Coast, on the west by the Cretaceous arc in western Mexico, and on the south by the southern Maya terrane and Zapoteco-Mixteco terranes, with a single oceanic outlet in the eastern Gulf of Mexico (Fig. 11).

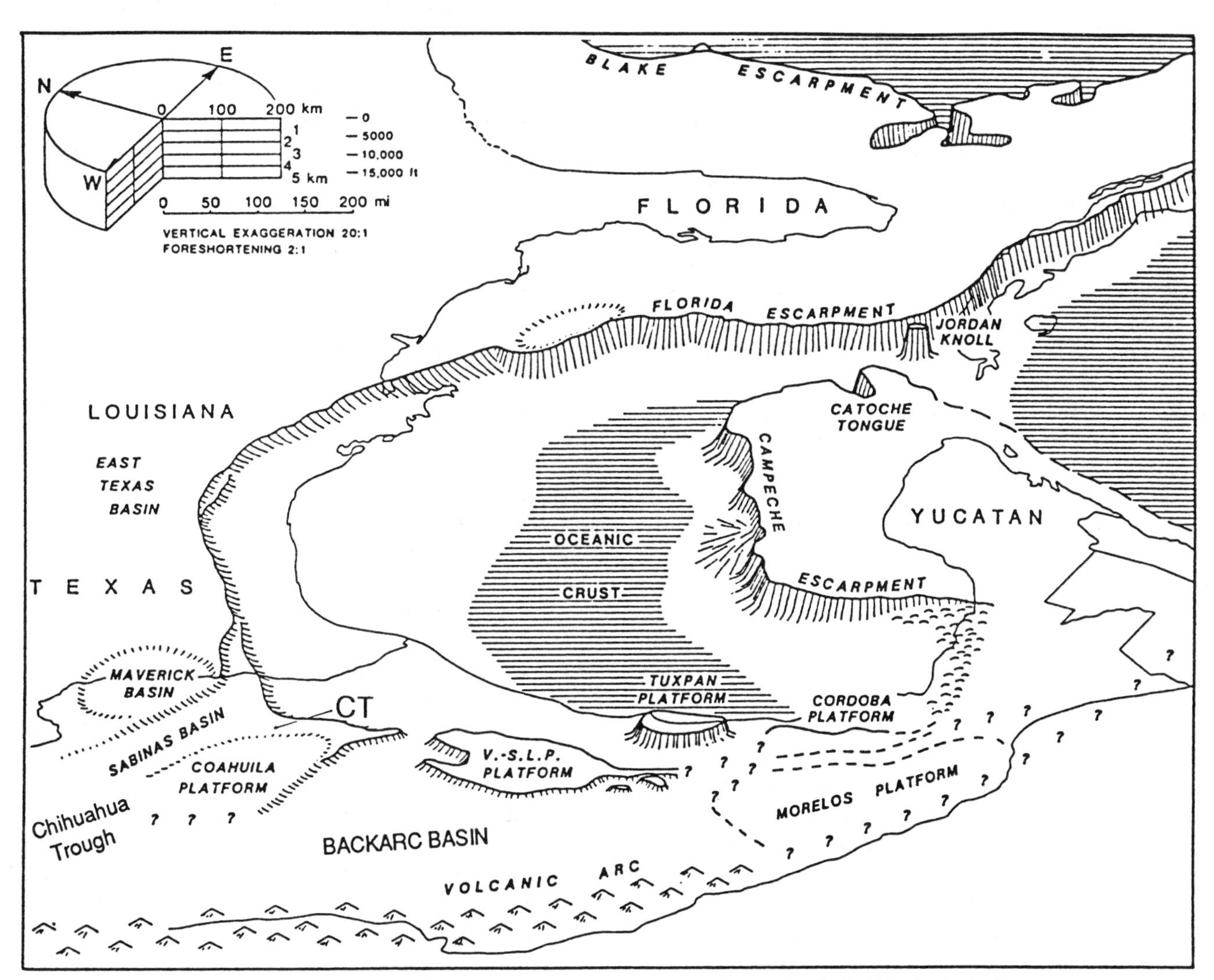

Figure 11. Oblique view of mid-Cretaceous bathymetry of Mexico and the Gulf of Mexico basin, from Winker and Buffler (1988). CT: Coahuiltecano terrane.

After an influx of terrigenous material during late Aptian time, mid-Cretaceous (Albian-Cenomanian) carbonates were deposited during the time of maximum thermal subsidence over a larger part of Mexico than in Neocomian time. Steep-sided, high-relief carbonate-evaporite platforms bounded by reef complexes shed very coarse carbonate detritus (notable as prolific hydrocarbon reservoirs) into adjacent deep-water basins. The decline in the number and size and the step back of carbonate banks around the Gulf of Mexico in Comanchean time (90 to 85 Ma) may have been caused by sea-level rise and changing environments (Winker and Buffler, 1988). Late Cretaceous (post-Turonian) deposition in eastern Mexico was characterized by an increasing terrigeneous component, shallowing of basins, and eastward migration of the shoreline. Throughout the Cretaceous, evaporites and carbonates accumulated on the stable Yucatán platform (Fig. 11). We interpret the similarity of Cretaceous carbonates and clastic rocks in the Chortis terrane to coeval deposits of the open-marine shelf in eastern Mexico to indicate that the regions were contiguous.

Caribbean tectonics

At the end of the Jurassic, western Pangea still existed as a contiguous continent comprising North America and South America. Early in Early Cretaceous time, South America broke away from North America at or near the southern edge of present-day Yucatán (Fig. 2) (Pindell and others, 1988). The drifting, which continued until Albian or Cenomanian time, produced the Caribbean basin. Thus, a large fragment of Paleozoic South America was added to the southern margin of Jurassic North America. The Jurassic edge of North America was probably along the southern edge of Yucatán (Maya terrane, see below) and thence east through north Cuba along the southern edge of the Bahama Banks (Malfait and Dinkleman, 1972; Edgar and others, 1971).

Although the expansion of the Caribbean basin between the drifting continents in the Cretaceous is well established, the processes and events that have filled the gap are not fully understood. At least some Lower Cretaceous oceanic lithosphere almost certainly developed contiguous with the central Atlantic, but relict spreading ridges cannot be recognized within today's Caribbean basin (Donnelly and others, 1973; Malfait and Dinkleman, 1972). Three alternative origins have been conceived for the oceanic lithospheres and marginal arc terranes of the present Caribbean basin. First, the oceanic lithosphere may have had Caribbean birth and may have undergone local breakups and subduction (Freeland and Dietz, 1972; Klitgord and Schouten, 1987). This implies that the left-lateral displacement on the North America–Caribbean plate boundary has not been large, perhaps ≤500 km. Second, the Caribbean oceanic lithosphere may belong wholly or mainly to the Farallon plate, which existed west of and was subducted below Pangea in early Mesozoic time. Such lithosphere may have been inserted into the Caribbean basin as it widened in Early Cretaceous time and, after that, continued moving eastward relative to the continents, and their margins being subducted in the Cretaceous and Paleogene (Malfait and Dinkleman, 1972). Third, a native Lower Cretaceous Caribbean oceanic lithosphere may have been consumed below a salient of the Farallon plate that began eastward migration in Paleogene time and became an independent Caribbean plate by the development of the Middle America Trench (Fig. 2) in mid-Cenozoic time (Sykes and others, 1982; Pindell and Dewey, 1982; Burke and others, 1984; Pindell and Barrett, 1990).

We tentatively employ the third origin because it is most consistent with the controversial evidence of large post–middle Eocene displacement (1100 km; Rosencrantz and others, 1988) at the North America–Caribbean plate boundary at the Cayman Trough (Fig. 2). In one variant of this hypothesis, one or more island arcs collided with southern Mexico (Maya terrane) in latest Cretaceous and were then translated northeastward at the leading edge of the Caribbean plate to their present positions in the Greater Antilles (Pindell and Dewey, 1982). Constraints on the kinematic evolution of the Chortis terrane and Precambrian basement of Cuba (Renne and others, 1989) are best satisfied by this idea. In another variant, the Chortis terrane collided with the Maya terrane at its southern edge in Late Cretaceous time and has had only a few hundred kilometers of displacement along the terrane boundary since then (Donnelly, 1989). This variant best explains the geologic evidence for small cumulative displacement in the Motagua-Polochic-Jocotán-Chameleón fault set in Central America and southern Mexico (Fig. 2), and explains the additional displacement, if any, by stretching in the Chortis terrane to the south.

Permian to Paleogene tectonics of western Mexico

We now address the regional features of Mexico and environs west of the Gulf of Mexico realm that developed after Pangea attained the Late Permian paleogeography shown in Figure 10. Such features are Cordilleran structures in the southwestern United States; the Mojave-Sonora megashear, plates of the proximal Pacific basin and their motions relative to Mexican magmatic arcs; and Laramide foreland deformation.

Mexico north of the Mojave-Sonora megashear in Sonora and Chihuahua is PNA, whereas to the south, it is constituted entirely by terranes that assembled at various times in the late Paleozoic–Paleogene. We employ constraints from the regional features discussed next with properties of terranes, presented later, to frame a tectonic evolution of the terranes.

Evolution of oceanic plates bordering western Mexico

An evolution back to Jurassic time of oceanic plates bordering western Mexico can be reconstructed using plate-hotspot circuits (Engebretson and others, 1985; DeMets and others, 1990) and magnetic anomalies (Stock and Molnar, 1988; Stock and Hodges, 1989). The products of such reconstructions are the normal and tangential components of plate velocities and times of passage of triple junctions as functions of position along an assumed western edge of Mexico (Sedlock and others, 1993). The uncertainties in such reconstructions increase with age and are highly speculative before about 100 Ma.

The essentials of our reconstruction (Sedlock and others, 1993) are as follows: 180 to 145 Ma, rapid convergence of Farallon (Fa) and North America (NA) plates with a small left-slip component; 145 to 119 Ma, slow convergence of Fa and Na with a small left-slip component; 119 to 100 Ma, rapid convergence of Fa and NA with a very small right-slip component; 100 to 85 Ma, moderately rapid convergence with moderate right slip at latitudes north of Mazatlán and 100 to 74 Ma for latitudes south of Mazatlán. At about 85 Ma, spreading began between the Farallon and a new plate (Kula) on the northern side of the Farallon off western Mexico. The Kula plate (Ku) subducted rapidly below NA with large right slip until about 59 Ma (Woods and Davies, 1982; Engebretson and others, 1985). From 74 to 25 Ma, Fa-NA was nearly orthogonally convergent.

Subduction of Fa and Ku below NA was supplanted by highly oblique right-lateral motion at about 25 Ma when the Pacific-Farallon spreading ridge (now Pacific-Cocos) intersected the trench along western Mexico (Fig. 2) (Atwater, 1970, 1989). Since then, the triple junction has migrated southeast along the convergent margin to its current site in the Acapulco Trench (Fig. 2). The right slip of Pacific relative to NA was markedly slower between 20 and 11 Ma than before and after.

Cordilleran tectonics of southwestern United States

Major tectonostratigraphic differences along the western margin of PNA in the southwestern United States were noted earlier; we now expand on these with views toward correlations in Mexico.

The differences in Paleozoic tectonic phenomena along this margin are in the existence of collisional structure, magmatism, and width of excision of continental tracts along the margin. In the Great Basin segment (north of Bishop, Fig. 8), the rift-dominated, end of Proterozoic passive margin (Stewart, 1972) is apparently fully preserved (Speed, 1983) and was overridden in Early Mississippian time by the Roberts Mountains allochthon (Fig. 8), which is constituted by deformed lower Paleozoic sediments and ocean-floor basalt. No Paleozoic magmatism or excision occurred in the Great Basin segment.

In southern California (south of Bishop, Fig. 8), in contrast, the edge of PNA intersects the isopachs and facies changes of Paleozoic passive margin wedge (shelf) deposits and Paleozoic platformal strata of the craton interior. Such intersections are widely regarded as due to truncation of the continental margin at one or more times in the Phanerozoic (Hamilton and Myers, 1966; Anderson and others, 1979; Stewart and others, 1984). There is evidence for strike slip on northwest-striking faults in this region: left lateral in the Pennsylvanian and Permian, right and/or left lateral in the Jurassic, and right lateral in the Cretaceous and Cenozoic (Stone and Stevens, 1988; Walker, 1988; Schweikert and Lahren, 1990; Kistler, 1990; Stevens and others, 1992; Saleeby and others, this volume). Thrusting also occurred during all or part of the Permian and Mesozoic.

In southern California there are tectonic packets of lower Paleozoic ocean-floor strata in the Mojave region (Fig. 8) (Carr and others, 1984), but their structural evolution differs substantially from that of the Roberts Mountains allochthon in the Great Basin. In the Mojave, they are interpreted to have overridden PNA in late Paleozoic or Jurassic time, and they are covered by Late Permian volcanic rocks (Carr and others, 1984; Walker, 1988; Kistler, 1990). Moreover, Permian and Early Triassic plutons occupy a possible margin-parallel belt with PNA in southern California (Fig. 10). Such magmatism, apparently of continental-arc type, implies a component of convergence and subcontinental subduction in the Permian–Early Triassic continental margin tectonics of southern California that was either concomitant with or alternating with left-lateral slip.

The late Paleozoic truncation surface or zone in southern California almost certainly extended southeastward into northern Mexico, either to the locus of the Mojave-Sonora megashear or along a trend not far south of the megashear. Our premise is that the truncation cut southwardly inboard into PNA from Bishop south into southern California and then outward farther south to merge with the end of Proterozoic, transform-dominated passive margin of northern Mexico (Fig. 8).

Mojave-Sonora megashear

The Mojave-Sonora megashear (MSM) is a fault or fault zone in northwestern Mexico whose existence is inferred in Sonora from the abrupt juxtaposition of Precambrian rocks of different ages and cover of lower Paleozoic strata of different facies (Fig. 12) (Silver and Anderson, 1974, 1983; Anderson and Silver, 1979, 1981; Stewart and others, 1984, 1990; Anderson and others, 1990). The MSM is the attachment surface between North America and our Seri terrane. Both the Precambrian basement and lower Paleozoic passive margin sediment wedge of the Seri terrane are like those in the Death Valley region within PNA of southern California (Figs. 8 and 12). It is widely agreed that the Seri terrane was displaced 700 to 800 km southeast from a position in southern California as a result of the continental truncations discussed above. The movement history, however, is controversial.

The MSM is hard to locate because it is covered by younger deposits and thrust sheets (De Jong and others, 1990). It is best defined in Sonora by contrasts in basement and Paleozoic cover on either side of a covered region many kilometers wide. The MSM is more cryptic northwest and southeast of Sonora. There are several northwest-striking faults in eastern Mexico (Fig. 12) that may form a zone of dispersed shear related to the MSM (Sedlock and others, 1993).

The MSM is the approximate locus of Jurassic continental arc intrusive and volcanic rocks in the northern Mexico (Fig. 12). Such rocks extend beyond the MSM region in a nearly continu-

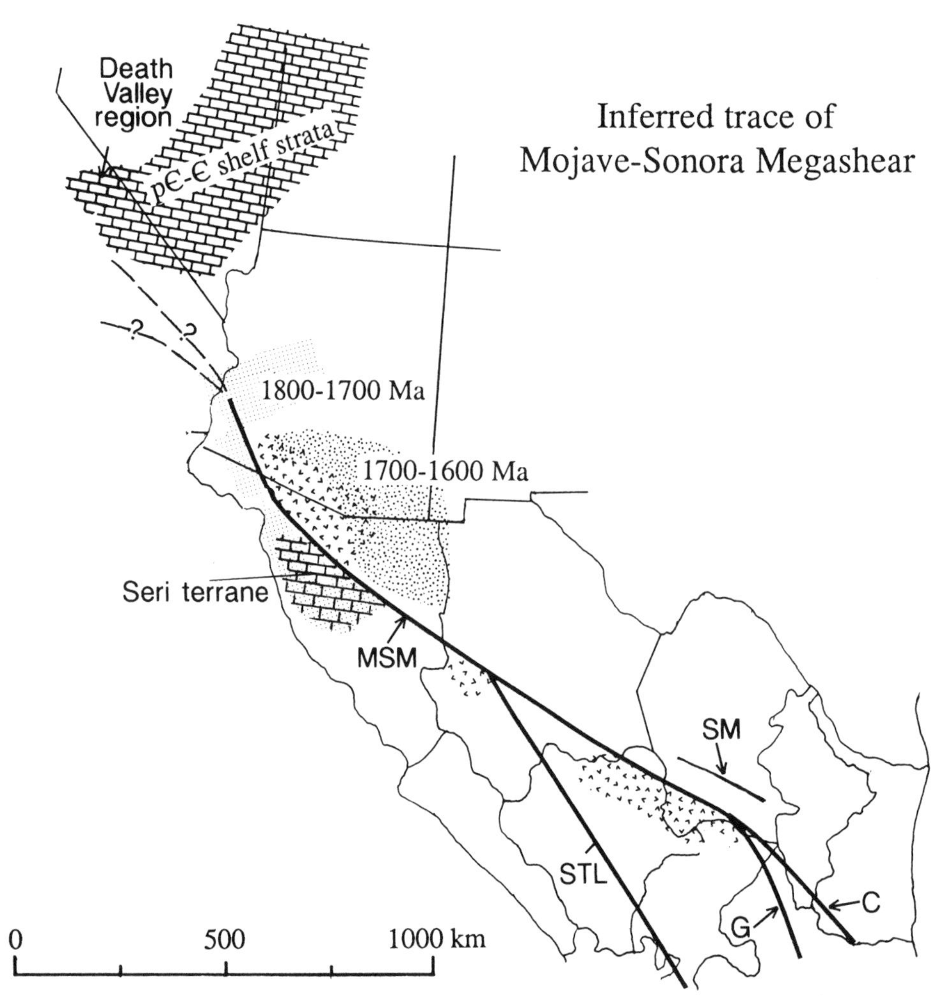

Figure 12. Inferred trace of Mojave-Sonora megashear (MSM) in Sonora and possible branches in eastern Mexico (C: southern boundary of Coahuiltecano terrane; G: a fault separating northern and southern Cuahuiltecano terranes; STL: San Tiburcio lineament of Mitre and Roldán (1991); SM: San Marcos fault). Dot patterns show Proterozoic basement; bricks are Paleozoic passive margin wedge strata; V patterns are Jurassic continental arc volcanic rocks.

ous belt northward into the U.S. Cordillera and southward to southern Mexico. The timing of MSM motions depends on whether the arc of northwestern Mexico is in place or greatly offset. Late Jurassic deformation in Sonora and Zacatecas, near the inferred trace of the MSM, implies displacement of the MSM at least that late (Rodríguez, 1990; Anderson and others, 1991). Stevens and others (1992), however, provided cause to think the offset of the Seri terrane was chiefly in late Paleozoic time. We propose that the MSM underwent left slip in both late Paleozoic and early Mesozoic time. The MSM either coincided with the end of Proterozoic transform-dominated passive margin of PNA in Mexico, or alternatively, the MSM developed slightly inboard of the end of Proterozoic margin by excisions during the late Paleozoic to Jurassic interval in a left-lateral strike-slip plate boundary zone.

Magmatic arcs

Belts of continental arc magmatic rocks in North America and attached terranes of Mexico provide a record of convergence with oceanic plates at times from Permian to the present. The subducting plates almost certainly existed on Mexico's western margin, because continental reconstructions show a proto-Pacific ocean basin on Pangea's western (today's coordinates) side and there is no recognition of convergence in the Gulf of Mexico region since the Ouachitan orogeny (Buffler and Thomas, this volume). In general, the succession of postaccretion continental arcs in Mexico is one of general westward migration, implying that the trench(s) along western Mexico have moved progressively westward.

The earliest continental arc rocks, Early Permian to Middle

Triassic, exist in two belts: easternmost Mexico (López-Ramos, 1985; López-Infanzón, 1986) and the Mojave region of the southwestern United States (Carr and others, 1984; Walker, 1988) (Fig. 10). In Mexico, the magmatic rocks are chiefly plutonic and are known mainly in wells between exposures in the Chiapas batholith in the south and in Coahuila (Las Delicias) in the north. In the United States, they are in outcrops and in both extrusive and intrusive bodies. Both belts are autochthonous, or at least, were not greatly displaced (≤500 km) in post-Pangean time. The two belts evidently reflect the development of convergent margins at places along western Pangea and the subduction of oceanic lithosphere from the proto-Pacific basin. The gap between the two belts in northwestern Mexico is interpreted to reflect a strike-slip segment of the active continental margin in the Permian and Triassic periods and Jurassic offset, probably minor, of the MSM. Similar kinematics were also conceived by Stewart and others (1990).

A second phase of continental arc magmatism occurred from Early to Late Jurassic time in a long belt that probably extended continuously through northern and eastern Mexico in regions that were autochthonous or parautochthonous in the Jurassic. In southern California and in south-eastern Mexico, the second phase of magmatism was coincident with the previous magmatic loci (Kistler, 1974; Dilles and Wright, 1988), but it extended across southern Arizona and northern Mexico where the prior arc had not existed (Fig. 7) (Damon and others, 1984; Asmerom and others, 1990; Tosdal and others, 1990a, b). The locus of this belt is inferred to reflect a change in plate obliquity along western Mexico wherein the former strike-slip segment of the margin became more convergent.

A second province of Late Triassic–Late Jurassic arc magmatism evolved in more far-traveled terranes that now compose parts of central and western Mexico (Cochimí, Pericú, Yuma, Tahue, and Tepehuano) south of the Mojave-Sonora megashear (Fig. 7). Such magmatism, much of it apparently of island arc affinity, implies the consumption of much oceanic lithosphere west of the Mexican margin during transport of terranes.

In Early Cretaceous time a third phase of continental arc magmatism occupied a probably nearly continuous belt along the western margin of northern Mexico that by then was constituted by all but the most outboard of terranes. During the Late Cretaceous, the locus of magmatism in northern Mexico that belonged to North America migrated steadily eastward from Sonora and Sinaloa to Chihuahua, Durango, and Zacatecas (Anderson and Silver, 1969; Clark and others, 1982). The migration was probably caused by progressive shallowing of the slab (Coney and Reynolds, 1977; Clark and others, 1982).

In southwestern Mexico (Guerrero to Jalisco), an Early Cretaceous arc is not evident in terranes (Chatino) that are now at the western margin. A Late Cretaceous arc, however, exists in this region and is continuous with that of northern Mexico. We infer that the Early Cretaceous arc, expected from Farallon–North America convergence, is missing because it was excised from the Mexican margin in mid-Cretaceous time. We infer further that the missing arc segment exists in Baja California, for which paleomagnetic data indicate large northward transport between mid-Cretaceous and Tertiary time.

A fourth phase of continental magmatism in the Paleogene affected Mexico after it had attained its current geography. The Paleogene arc is characterized in northern Mexico by the slow eastward migration of volcanism and the progressive widening of the arc, followed by the comparatively rapid return of the arc to the western margin of Mexico by the end of Paleogene time. Over the same duration southern Mexico records a lower volume of arc volcanism compared to that in the north. A plate tectonic explanation for such geographic differences is hard to come by, unless Paleogene granites in the Chatino terrane are considered the roots of a deeply eroded arc.

Laramide foreland deformation

Only one regional, probably laterally continuous, belt of contractile deformation that developed after terranes were assembled has been identified in Mexico. This is the Laramide or Hidalgoan orogenic belt that evolved in latest Cretaceous through about middle Eocene time from Sonora to Chiapas (Fig. 13) (Guzmán and de Cserna, 1963; de Cserna, 1965; Mossman and Viniegra, 1976; Suter, 1984, 1987; Lovejoy, 1980; Corbitt, 1984; Dyer and others, 1988; Drewes, 1991; Padilla y Sánchez, 1985, 1986; Quintero-Legorreta and Aranda-García, 1985). The Laramide is a thin-skinned thrust and fold belt in which the detachment is mainly at the base of or within the Mesozoic cover in the attached terranes and within or above the Paleozoic platformal strata of PNA. The basement is rarely involved. The region of Laramide deformation was a continental platform behind the active margin of western Mexico in latest Cretaceous and Paleogene time and can be regarded as the foreland in which horizontal contraction due to convergence at the plate boundary surfaced.

Thrusting in the Laramide belt was regionally top to the northeast, but the strikes of thrusts vary locally from east to north, depending on orientations of basement horsts and graben that were inherited from Triassic and Jurassic rifting (Fig. 13). Cumulative horizontal transport varied laterally from 40 to 200 km. The age of onset of deformation appears to decrease toward the foreland, ending in the middle Eocene. Detached sheets locally rotated about vertical axes (Kleist and others, 1984). Laramide deformation was concurrent with Late Cretaceous and Paleogene magmatism, which occurred within the belt west of the thrust front. Such magmatism migrated eastward during this duration as did the deformation front, and the existence and changes of both phenomena may be related to progressive shallowing of subducted oceanic lithosphere, as has been postulated for the Laramide belt of the western United States (Coney and Reynolds, 1977).

Contractile deformations of pre-Laramide Mesozoic age are recognized in regions of Mexico that are in the hinterland of the Laramide belt or west of there. These are hard to date, correlate, and frame in a regional conceptual view. Some can be assumed to

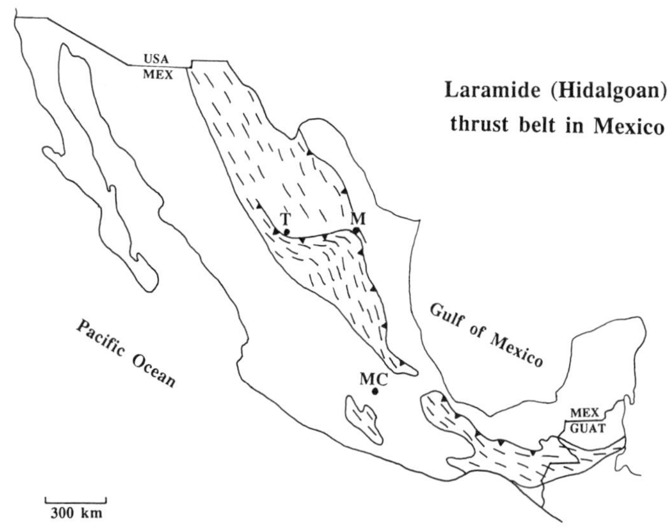

Figure 13. Distribution of foreland thrusting and folding of the Laramide (Hidalgoan) belt in autochthonous Mexico in Late Cretaceous and Paleogene time. MC: Mexico City; T: Torreón; M: Monterrey.

have developed in the arc-forearc zone of the Mexican margin, and others probably within actively displacing terranes. We mention such events in descriptions of individual terranes.

Bakcarc basins of western Mexico

Basins developed on the Cretaceous Mexican continent mainly in back of the continental arc that faced the active margin. Such backarc basins include the Chihuahua Trough (Fig. 11) of Sonora and Chihuahua (Cantú-Chapa and others, 1985; Dickinson and others,1986; Araujo-Mendieta and Estavillo-González, 1987; Salvador, 1987) and poorly defined basins southwest of the Coahuila platform in Durango, Zacatecas, and Guanajuato (Servais and others, 1982).

PROTEROZOIC NORTH AMERICA IN MEXICO

We now address the major Phanerozoic tectonic features and events in Proterozoic North America (PNA) in Mexico. PNA exists in Sonora and Chihuahua and extends with continuity into the southwestern United States (Figs. 1 and 3), as indicated by age and structure of Proterozoic basement and by correlations among sections of Paleozoic cover (Flawn and others, 1961; King, 1975; Dickinson, 1981; Stewart and others, 1990). The position and age of the southern edge of PNA, however, are not firmly established. We proposed above that the edge is the Mojave-Sonora megashear and that its age may be alternatively end of Proterozoic, late Paleozoic, or as young as Jurassic. If the edge is end of Proterozoic, it is a passive margin (here, transform dominated) that fully or mostly blocked out North America as it is today (Fig. 8). If the edge is younger, we propose that it evolved by excision of continental slivers along plate boundaries with mainly left-slip transform motion from a slightly more outboard locus of the end of Proterozoic margin. This southern boundary of PNA has since been a surface of sliding and attachment of terranes that now populate central and southern Mexico.

Below, we present a brief Phanerozoic history of PNA in Mexico.

Event history

The Phanerozoic event history within North America in Mexico is as follows (Fig. 14). Thin cratonal Paleozoic cover was deposited on crystalline basement of Proterozoic age. Near its southeastern edge in Mexico, North America's upper Paleozoic cover is a thick flysch, implying local basin subsidence. North America contains a lacuna of Triassic and Early Jurassic age, implying emergence in that time. From Middle Jurassic through Early Cretaceous time, a continental magmatic arc existed in a belt that lies along the MSM from Arizona at least as far southeast as central Chihuahua (Figs. 1 and 12). Northeast of this magmatic belt in Mexico are Mesozoic basins, apparently of backarc origin. In late Mesozoic and early Paleogene time, North America in Mexico was widely affected by foreland thrusting and folding. The central realm of PNA in Mexico underwent Cenozoic continental arc magmatism of the Sierra Madre Occidental, and Basin and Range extension and basaltic effusion occurred widely.

Crystalline basement

Figure 5 illustrates that basement rocks crop out widely only in northernmost Mexico, whereas in the eastern three-fourths of the country, they are covered (except at Sierra del Cuervo; Fig. 5) and known only in a pair of wells, and as xenoliths in Quaternary lava at two sites (Anderson and Silver, 1977, 1981; Anderson, 1983; Quintero and Guerrero, 1985; Handschy and Dyer, 1987; Blount and others, 1988; Ruiz and others, 1988a, b). U-Pb dates from PNA in northwestern Sonora (Fig. 5) are 1600–1700 Ma for metamorphic rocks and 1410–1460 Ma for granitoids, whereas U-Pb and Sm-Nd dates of metamorphic rocks from Chihuahua (Fig. 5) are of Grenvillean age (900 to 1100 Ma). This geography of dated basement is consistent with the unbroken continuation of the 1600 to 1700 Ma and Grenvillean age domains of Proterozoic rocks south from the United States into northern Mexico (Fig. 6) (Anderson and Silver, 1977a, 1977b, 1981; Bickford and others, 1986; Hoffman, 1989).

Paleozoic Cover

An Ordovician through Lower Permian stratal succession covers the basement unconformably over much of PNA in Mexico, but is widely missing, owing to pre–Upper Jurassic tectonism in Chihuahua and to emergence with or without deformation in western Sonora (Fig. 14). The full Paleozoic section is considered platformal on PNA in Sonora and northern Chihuahua, whereas in the Pedregosa basin of eastern Chihuahua, the platformal sec-

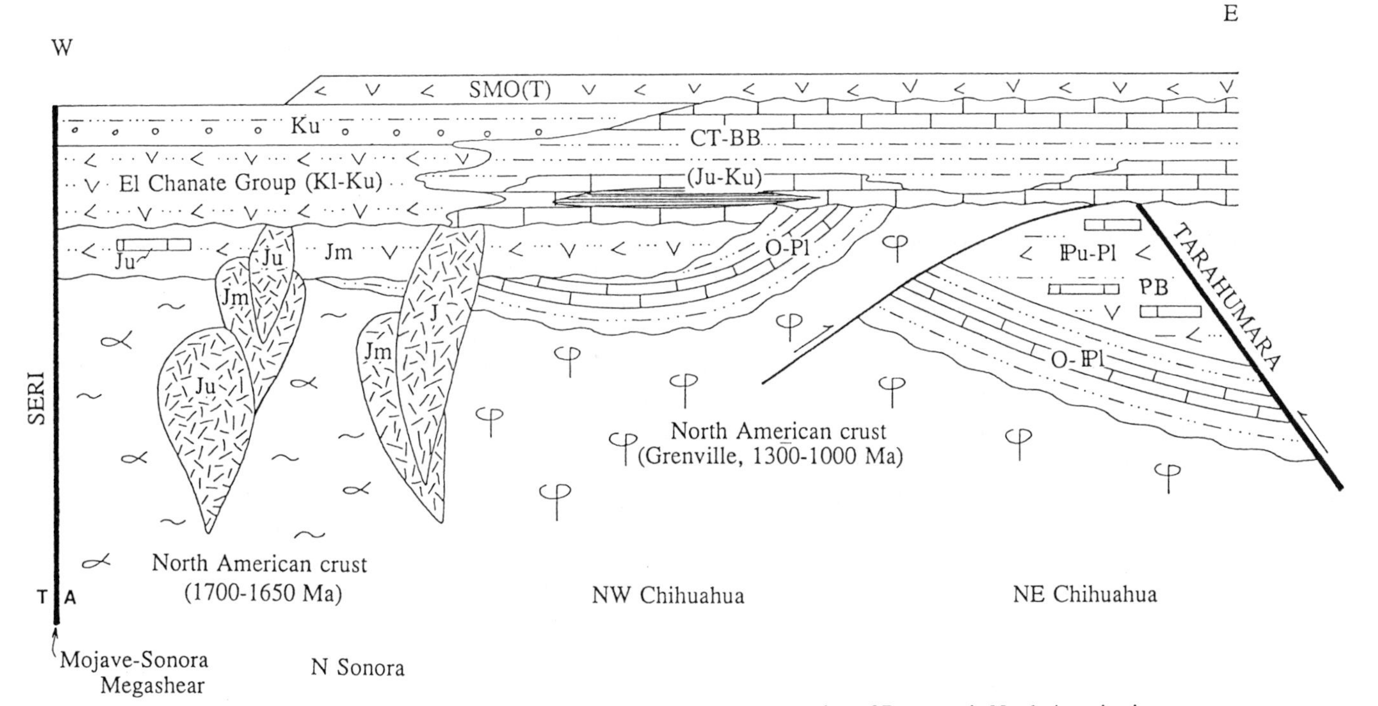

Figure 14. Composite tectonostratigraphy and west-east cross section of Proterozoic North America in northern Mexico. SMO(T): Tertiary volcanic rocks of Sierra Madre Occidental; CT-BB: Chihuahua Trough and Bisbee basin; PB: Pedregosa basin. Effect of post-Jurassic deformation omitted. Upper Paleozoic strata below Tarahumara thrust are in Pedregosa basin.

tion is Ordovician through Lower Pennsylvanian, and this is succeeded by flysch of Late Pennsylvanian and Early Permian ages (Fig. 14) (Imlay, 1936; Ramírez-M. and Acevedo-C., 1957; Bridges, 1964a, b, c; Díaz and Navarro-G., 1964; Navarro and Tovar, 1974; King, 1975; Greenwood and others, 1977; Dyer, 1986; González-León, 1986; Armin, 1987). The flysch contains siliceous tuff and rhyolite clasts, implying proximity of the Pedregosa basin to arc magmatism in Early Permian time.

The tectonic kindred of the Paleozoic succession, aside from that of the Pedregosa basin, is a basic factor in interpretation of the position and origin of the southern edge of North America in Mexico, as yet an unsolved problem. The strata are clearly not part of a typical passive margin wedge by virtue of their thinness and carbonate- and quartz-rich facies. This implies a cratonal setting. The varied platformal facies of this region and the southern fringe of the United States, however, appear to occupy narrow east-west–trending belts and have southerly sediment transport directions; these features have been used to interpret a former south-facing continental edge not far south of today's southernmost outcrops (Stewart, 1988; Stewart and others, 1990; Stevens, 1982; Palmer and others, 1984).

Permian-Triassic deformation

Between Early Permian and Middle Jurassic time, PNA in Mexico underwent emergence and warping, causing erosion of the Paleozoic cover in western Sonora (Fig. 14). Thrust faulting with transport to the southeast brought Grenvillean basement above upper Paleozoic flysch of the Pedregosa basin in eastern Chihuahua (Handschy and Dyer, 1987) (Fig. 14).

Mesozoic arcs and backarc basins

The Proterozoic basement and remnants of Paleozoic cover were the foundation for arc volcanism and plutonism of Middle and Late Jurassic age in a belt that occupies the modern southwestern edge of PNA in Mexico (Fig. 12). The arc extended far to the northwest in the U.S. Cordillera and east in northern Mexico at least as far as the Coahuiltecano terrane (Anderson and others, 1991; Anderson and Schmidt, 1983; Tosdal and others, 1990a). In Sonora, the Jurassic arc is succeeded by Cretaceous magmatic rocks. Marine backarc basins (Chihuahua Trough, Bisbee basin) with axial parallelism to the arc existed on the northeastern flank of the arc from Late Jurassic through Late Cretaceous time.

Mesozoic and Paleogene faulting

North America in northern Mexico and the southwesternmost United States is a zone of northwest-southeast–striking subvertical faults that are thought to have taken up displacements with varied slip at several times in the Mesozoic and perhaps before (Drewes, 1991). One of these, the Texas lineament (Baker, 1935) is thought to be zone of right slip, the movements of which have continued into the late Tertiary. Another is the Mojave-Sonora megashear, the southern boundary of PNA, on which major left slip was Middle Jurassic (Anderson and Silver, 1979) or late Paleozoic (Stevens and others, 1992) or both.

In Late Cretaceous and Paleogene time, virtually the entire region of North America in Mexico was a thin-skinned thrust and fold belt, the Laramide or Hidalgoan orogenic belt (Fig. 13) (Guzmán and de Cserna, 1963; Davis, 1979; Drewes, 1991; Dyer and others, 1988). Moreover, there is evidence for earlier thrusting (mid-Cretaceous, possibly Late Jurassic) in the more internal reach southwest of the thrust and fold belt (Rodríguez, 1988, 1990; Roldán-Quintana, 1982; Rangin, 1978; DeJong and others, 1988; Pubellier and Rangin, 1987). The tectonic transport of the belt was variably to the northeast and east-northeast, with cumulative displacements from 40 to 200 km, and northeasterly migration of its deformation front.

Cenozoic magmatism and extension

The continental arc magmatism of the Sierra Madre Occidental, which began in Paleogene or Cretaceous time, extended to a northern tip in Proterozoic North America in Mexico. The tip is involved in basin and range extension that is principally of Miocene age.

TERRANES

We identified and characterized (Sedlock and others, 1993) 16 tectonostratigraphic terranes in Mexico and northern Central America (Fig. 7). These are discussed below only with regard to their tectonic kindred and displacement history. Each is recognized by different histories of rock formation and deformation and is known or inferred to be fault bounded and to have undergone substantial displacement before attachment at its current site. Some terranes comprise smaller terranes (subterranes) of generally related tectonic kindred and displacement history, and are treated as composites for convenience.

Our terrane division evolved from an earlier one developed at the Institute of Geology, University of Mexico, and used in Continent-Ocean Transects H1 (Mitre-Salazar and Roldán-Quintana, 1990) and H3 (Ortega-Gutiérrez, 1990) (the names of terranes are taken from those of indigenous Precolumbian cultures).

Chatino

The Chatino terrane was a continental arc, now greatly unroofed, in the Jurassic and Early Cretaceous and possibly earlier, before its attachment to the Zapoteco-Mixteco and Náhuatl terranes, probably in middle and Late Cretaceous time. It includes another continental arc of Late Cretaceous–Paleogene age together with an accretionary forearc of Cenozoic age, almost certainly related to nearly orthogonal convergence of the Farallon and Cocos plates to the then-amalgamated western Mexico. The pre-Jurassic evolution of the Chatino terrane is uncertain.

Chortis

The Chortis terrane is currently part of the Caribbean plate (Fig. 7). We infer, however, that Chortis was in outboard western Mexico in the Mesozoic and underwent large left-slip displacement to southern Mexico in the Early Cretaceous, then was translated relatively westward in the Late Cretaceous together with the Baja California superterrane. In early Cenozoic time, Chortis became part of the Cenozoic Caribbean plate and has translated eastward relative to Mexico, as it is still apparently doing. This complex kinematic history is consistent with the obliquity vs. time of convergence of oceanic plates and North America, the paleomagnetic rotation history of Chortis (Gose, 1985), and the motion of other terranes of presumed outboard position (Baja California, the Franciscan rocks of California; Saleeby and others, this volume).

The earlier evolution of the Chortis terrane is unclear. Its continental basement of Paleozoic and Precambrian(?) metamorphic rocks and late Paleozoic intrusions may have come from the Ouachitan collision zone of northern South America, or it may be a fragment exotic to Pangea. From Late Jurassic to Cenozoic time, western Chortis has been a continental magmatic arc, thus presumably confronting oceanic lithosphere of the Pacific basin.

Coahuiltecano

The Coahuiltecano basement, known only from wells and from fragments in sedimentary cover, seems to be continental and to have mid-Paleozoic metamorphic ages. On this basis and from its cover of Pennsylvanian and Permian continental arc–related deposits, the Coahuiltecano terrane is taken to belong to Paleozoic northern South America that collided in the Ouachitan orogen, and to be remnant after the Mesozoic drift of South America. The Coahuiltecano and Maya terranes were probably conjoint before Middle Jurassic time and together underwent Permian-Triassic continental arc magmatism at the western edge of Pangea. Since Middle Jurassic time, the Coahuilatecano has been parautochthonous relative to North America, whereas the Maya terrane has been partly displaced during rifting and drifting of the Gulf of Mexico basin (see below). Extension and large rotations (Gose and others, 1982) occurred within the Coahuiltecano during the gulf breakup and before the Oxfordian onset of drift in the Gulf of Mexico. Laramide foreland contraction and Tertiary Basin and Range extension, both with concurrent magmatism, widely affected this terrane.

Cochimí

This composite terrane comprises several subterranes of early Mesozoic ophiolitic basement of oceanic and/or arc origin. All of these became island arcs or received arc-derived sediment in Middle and Late Jurassic and Early Cretaceous time. The subterranes were blanketed by terrigenous sediments beginning in the Early Cretaceous. The attached subterranes were then underplated by blueschist between mid-Cretaceous and early Tertiary time. The Cochimí terrane underwent large uplift and extension that may have been concurrent with blueschist emplacement.

Paleomagnetic data imply 10°–20° of northward transport of the ophiolitic subterranes between about 25 and 100 Ma (Hagstrum and others, 1985, 1987) and about 25° in the same duration for the blueschist (Hagstrum and Sedlock, 1990). Such northward transport can be inferred for Baja California as a whole from data from the Yuma as well as the Cochimí terrane, and implies an earlier position in present southern Mexico or beyond. The disparity in latitude anomalies among subterranes suggests that the Cochimí may not have fully coalesced until a time during or after northward migration of its parts.

Cuicateco

The basement of the Cuicateco terrane is inferred to be Upper Jurassic and/or Lower Cretaceous oceanic crust on the basis of thrust slices of mafic-ultramafic rocks in its structurally low eastern margin, and on the interpretation that its principal constituent is the sedimentary and volcanic fill of a late Mesozoic rift or pull-apart basin (Carfantan, 1983; Ortega, 1990). The basin fill includes flysch, tuff, greenstone, black slate, and carbonate; Early Cretaceous fossils have been found at a couple of sites (Ortega and González, 1985; Alzaga and Pano, 1989). The fill is strongly deformed but only slightly metamorphosed. Distinctive pebbles in a conglomerate may have been derived from the Zapoteco and Mixteco terranes, which flank the Cuicateco on the west. The westernmost zone of the Cuicateco terrane, below the Zapoteco, is a mylonitic zone that contains relics of many protoliths, including Precambrian gneiss and granitoids of probable pre-Cretaceous age. One relic of a pluton is dated as 130 Ma (Delgado and others, 1986).

The parautochthonous Cuicateco terrane probably filled a rift or left-oblique-slip basin near its present site in southern Mexico. We infer that the basin formed within a broad left-lateral shear zone west of the Maya terrane during drifting in the Gulf of Mexico basin. The Cuicateco basin fill was then contracted and thrust east above the Maya terrane at a time between the mid-Cretaceous and early Tertiary.

Guachichil

The basement, mainly sampled in wells but also from outcrops in two areas in Tamaulipas and Hildago, is Grenvillean gneiss at granulite facies, giving Sm-Nd ages around 900 Ma (Patchett and Ruiz, 1987). It also includes quartzite, anorthosite, amphibolite, and marble. The Guachichil basement appears by all criteria to be related to the Grenville province of eastern PNA (Fig. 3). It is clearly detached and displaced from PNA, because the non-Grenvillean basement of the Coahuiltecano terrane is between them. Moreover, the excision of Guachichil basement from PNA was early, perhaps during end of Proterozoic rifting of southern North America, because the lower Paleozoic strata of the Guachichil terrane that are presumed to have covered the basement differ from those known in the U.S. Gulf Coast (Stewart, 1988).

All of the pre-Permian Paleozoic rocks of the Guachichil occupy fault-bounded blocks, and that their autochthoneity to the basement is unclear. Moreover, such rocks differ in the northern and southern parts of the terrane. In the north, there are turbiditic strata of Middle Silurian age that are succeeded by turbidites and rhyolitic tuffs. These are overlain unconformably by Lower Pennsylvanian platformal strata. There is also a packet of low-grade schist (Granjeno) of basaltic and siliciclastic protoliths plus serpentinite of uncertain age of deposition and of Late Mississippian metamorphism. In the south, Mississippian flysch and Pennsylvanian shallow-marine strata are known besides Grenvillean basement among pre-Permian rocks. A Lower Permian volcanogenic flysch (Guacamaya) exists in north and south Guachichil. The Paleozoic tectonostratigraphy suggests that the terrane may comprise discrete blocks that came together by Permian time.

The Guachichil has extensive Mesozoic cover like that of other terranes in northeastern Mexico. The Upper Triassic through Middle Jurassic strata are continental red beds and volcanic sediments of rift affinity, whereas the Upper Jurassic and Cretaceous strata are carbonate rich and indicative of deposition in postrift environments of decreasing bathymetric differentiation with time. The large paleomagnetic declination anomalies (Gose and others, 1982) and faulting in the rift sequence imply, however, that the Guachichil underwent substantial parautochthonous rearrangement.

The Guachichil was deformed in the Laramide foreland fold and thrust belt in Late Cretaceous and Paleogene time and was intruded by granites early in the Cenozoic, probably as a fringe of the continental arc that swept eastward.

We infer that the Guachichil existed as discrete microcontinental blocks in the ocean south of Paleozoic North America. Such blocks were swept up in late Paleozoic time by encroaching South America and incorporated either in its forearc or at its then-westernmost margin. The Granjeno Schist can be supposed to be a subduction complex in either case, and the Guacamaya flysch links the Guachichil with Permian magmatism, either in frontal northern South America or in an island arc west of South America. The Guachichil, however, was probably not well within the Ouachitan orogen, because it lacks the Late Permian–Triassic continental arc magmatism that appears to have developed along the west-facing edge of Pangea south of the Ouachitan belt (Fig. 10). Therefore, we suppose it either escaped to the west during Ouachitan collision or already existed there. In the early Mesozoic, the Guachichil was probably transferred southeast along coastal Pangea by the left-slip component of motion in the plate boundary zone and was attached in Jurassic time.

Maya

The basement of the Maya terrane (Fig. 7) is known only from outcrops in southern Mexico and Guatemala and from well information in northern and eastern Mexico and in the Yucatán Peninsula. It is constituted by rocks of pre-Pennsylvanian metamorphic ages and locally contains evidence for Precambrian and Carboniferous plutonism and metamorphism. Parts of the Maya basement are overlain by Upper Carboniferous and middle Permian sedimentary rocks that are partly volcanogenic and greatly deformed, but little metamorphosed (Maya Mountains of Belize; Fig. 2). Together, such characteristics permit the interpretation that the Maya basement belonged to Paleozoic South America and that the late Paleozoic magmatism and sedimentation are related to a continental arc that preceded and was concurrent with the collision of the two American continents in the Ouachitan orogeny. The Maya terrane was therefore probably contiguous with the Coahuiltecano in late Paleozoic time in western Pangea.

The western reaches of the Maya terrane contain batholiths of a Late Permian–Triassic continental arc that was probably caused by the eastward subduction below Pangea of oceanic lithosphere of the Pacific basin (Fig. 10).

The Maya terrane underwent heterogeneous rifting related to the Gulf of Mexico basin, beginning in Late Triassic time in the northern reach of the terrane and in Late Jurassic time in southern Mexico, but apparently not at all in the Yucatán platform. The rift phase was succeeded in Early Cretaceous time by carbonate bank deposition at nearly all sites that are onland or shallow marine today. In the offshore deeply submerged realm of the Maya terrane that underwent large second-stage stretching, however, deep-water carbonate accumulated in the Jurassic.

During the Mesozoic drifting related to the development of the Gulf of Mexico basin, much of the Maya terrane (Yucatán and southern Maya) moved south-southeast to southeast relative to PNA, probably from a position in the Ouachitan internal zone east of the Coahuilatecano terrane, which was apparently parautochthonous in the Mesozoic (Fig. 11). The northern part of the Maya terrane may have undergone less southward transport than the southern, and could be parautochthonous with the Coahuilatecano. The dispersal of pieces of the Maya terrane, which accommodated sea-floor spreading in the Gulf of Mexico, probably occurred in a zone of faults that include the Tamaulipas–Golden Lane fault of Pindell (1985). Paleomagnetic data indicate large pre-Cretaceous counterclockwise rotations in the southern Maya terrane (Chiapas massif) in Permian-Jurassic rocks (Molina and others, 1990; Guerrero and others, 1990). This suggests that the dispersal included left-lateral shear. The transport of the component blocks of the Maya terrane ceased at the end of Jurassic time, when drifting jumped to the southern margin of the Maya (south of Yucatán; Fig. 11). That margin then became the northern edge of the Lower Cretaceous Caribbean basin.

The ophiolitic El Tambor subterrane exists above the southern edge of the Maya terrane in Guatemala (Fig. 7) at its contact with the Chortis terrane, or equivalently, the North America–Caribbean plate boundary. The El Tambor was emplaced in the Campanian or Maastrichtian by northward thrusting (Sutter, 1979). It is interpreted to be oceanic crust that was attached to either the Chortis terrane or an unidentified island arc that overrode the Maya terrane in a collision.

Mixteco

Basement rocks of the Mixteco terrane (Acatlán Complex) probably record early to middle Paleozoic subduction and high pressure-temperature metamorphism of clastic sediments of Grenvillean provenance and basaltic rocks below an ophiolite. The resulting complex underwent partial melting, and granitoid bodies were emplaced and deformed in the Devonian near the preserved top of the complex. The melting event was coupled with the collision with and at least partial overthrusting by the Grenvillean crust of the Zapoteco terrane in Devonian time (Ortega, 1979, 1981a, 1990). Sediments apparently of latest Devonian age and of Grenvillean provenance covered the suture between the two terranes. These were subsequently strongly deformed and moderately metamorphosed and overlain unconformably by upper Paleozoic strata as old as Mississippian that are not metamorphosed. Later, the Acatlán Complex and its cover were intruded by a pluton with a 287 Ma date (Ruiz, 1979). The Mixteco terrane thus records an Early Permian tectonic event that involved probable convergence and continental arc magmatism.

The upper Paleozoic cover to the Acatlán Complex consists of marine siliciclastic and carbonate strata of Mississippian, Pennsylvanian, and Permian ages. These are succeeded unconformably by ignimbrite and andesite of apparent Triassic age and are cut by granitic dikes of Triassic to Middle Jurassic age, which suggests that the Mixteco terrane was in the region of eastern Mexico undergoing continental magmatism by the early Mesozoic. Younger strata are epicontinental and marine Jurassic and Cretaceous cover that blanket most of eastern Mexico. Cenozoic volcanogenic rocks represent arc magmatism of the Sierra Madre del Sur.

Náhuatl

This poorly understood probably composite terrane is constituted mainly by upper and lower metamorphic complexes. The lower, the Tierra Caliente Complex (Ortega-Gutiérrez, 1981b), is considered basement, although it is exposed only in the eastern part of the terrane. This basement comprises schists of varied undated flysch-type sedimentary, arc-magmatic, and oceanic plateau protoliths. Protolith ages are probably pre–Late Jurassic and are inferred to be as old as late Paleozoic (de Cserna, 1982). In Michoacán, metamorphosed tholeiitic basalt and sedimentary rocks of Late Triassic age are included (Campa, 1978). Metamorphic facies are prehnite-pumpellyite, greenschist, and lower amphibolite and are products of as many as four phases of metamorphism (Elías, 1987, 1989). The complex is interpreted as an amalgamation of island arc and basinal rocks (Ortega, 1981a; Robinson, 1991). It also has been suggested (Ruiz and others, 1988c) that the Tierra Caliente Complex is correlative with Late Jurassic and Cretaceous rocks of arc and oceanic kindred that occur along the margins of the Caribbean plate.

The upper complex lies stratigraphically above the Tierra Caliente Complex. Conglomerate in the upper complex contains distinctive greenstone clasts that imply a source in the lower complex and autochthoneity of the upper complex.

The upper complex, strongly deformed, is at low metamorphic grades. Its protoliths are of sedimentary and arc-volcanic strata of Cretaceous ages. The upper complex includes andesitic rocks, siliciclastic sediments, and middle and Upper Cretaceous carbonates. Some carbonates are probably correlative with autochthonous carbonates that cover terraces of eastern Mexico. The upper complex probably embraces island-arc lavas and related sediments together with arc-platform and continental-platform carbonate banks. The carbonates suggest that the Náhuatl terrane was attached to its eastern neighbors by the end of the Jurassic. The Náhuatl terrane underwent silicic plutonism and volcanism in Late Cretaceous and early Tertiary time, probably owing to subduction of Mexico by the Farallon and Kula plates.

The terrane was strongly affected by postattachment contraction and easterly thrusting in Late Cretaceous time and by an early phase of Laramide orogenesis. This severely imbricated the upper complex but did not apparently involve basement. The Laramide structures are locally capped by little deformed cover of Maastrichtian and Neogene volcanogenic rocks that are related to convergence on the west side of the Náhuatl terrane.

Pericú

The Pericú terrane probably lay at the Náhuatl-Tahue boundary on mainland Mexico before the detachment of the Baja California block at about 6 Ma. Basement rocks of the Pericú are metasedimentary, of siliciclastic and calcareous protoliths, and granite and gabbro that intrude and metamorphosed the layered rocks to high grade. The plutons could be as old as Late Mississippian (Gastil and others, 1975), and the metamorphism is clearly pre–mid-Cretaceous (Ortega, 1982; Aranda and Pérez, 1989). Such characteristics preclude correlation between basements of the Pericú and Náhuatl terranes because of the predominance of arc-derived protoliths in the Náhuatl. The Pericú basement may, however, be similar to that of the Tahue terrane, which contains late Paleozoic siliciclastic and carbonate strata but few volcanics. If the Pericú-Tahue correlation is valid, both terranes probably originated near the western edge of PNA.

The metamorphic protoliths were invaded by low-K plutons in the Early Cretaceous. The metasedimentary rocks and plutons then underwent penetrative east-west contraction in mid-Cretaceous time. This deformation was followed by Upper Cretaceous plutons, which are undeformed. Both episodes of Cretaceous plutonism can be interpreted as postattachment continental arc magmatism due to subduction of the Farallon and/or Kula plates. The western boundary of the Pericú terrane is the La Paz fault, which is a brittle fault that reactivated an earlier mylonite zone (Aranda and Pérez, 1988)

Seri

The Seri terrane contacts PNA at the Mojave-Sonora megashear (MSM). The terrane is divided into a mainland block and a block in Baja California (Fig. 11). The two blocks apparently have similar tectonostratigraphy but very different displacement histories.

The Seri terrane is distinguished by its pre-Grenvillean crystalline basement, which is correlatable with the 1800 to 1700 Ma basement age province of PNA (Fig. 6) (Silver and Anderson, 1974; Anderson and Silver, 1981) and by its Paleozoic passive margin wedge, apparently the only one in Mexico (Stewart and others, 1984).

The passive margin wedge is constituted by thick shallow-marine strata, the Late Proterozoic and Lower Cambrian strata are rift-facies clastics and basalt, and the Middle Cambrian through Lower Pennsylvanian strata are mainly carbonate and subsiding shelf deposits (Stewart and others, 1984, 1990; Radelli and others, 1987). They are succeeded conformably by Permian clastics that may be foreland basin deposits.

The southern fringe of the passive margin wedge and foreland basin deposits in Sonora is overlain by an allochthon of Paleozoic basinal strata consisting of hemipelagic and minor clastic rocks that have partial continental provenance (Poole and others, 1983; Ketner, 1986). The basinal strata are Ordovician through Permian and include an angular unconformity within the Mississippian that records a phase of moderate deformation. Upper Mississippian turbidites lie above the unconformity. The allochthon was emplaced after Early Permian time, but apparently before the Late Triassic. Although the rocks of the allochthon have some similarities in age and facies with those of the Roberts Mountains allochthon in the Great Basin (Fig. 8), the two allochthons are not correlatable because of important lithic differences (Poole and Madrid, 1986) and different internal structural and emplacement histories.

The Paleozoic rocks are succeeded by Upper Triassic red beds that are probably of rift affinity (Stewart and Roldán, 1991). Such beds include clasts of both the allochthonous and autochthonous Paleozoic rocks noted above. The rifting is appropriate in time to be related to the early breakup of Pangea. Jurassic volcanic and plutonic rocks of the Seri terrane and adjacent North America indicate that a continental magmatic arc, which continued well northward in the U.S. Cordillera, straddled the Mojave-Sonora megashear. It is not yet clear, however, whether the arc developed along a preexisting suture or was displaced greatly by left-slip faulting along its axis. A continental arc continued to exist at least episodically in the Seri terrane in the Cretaceous and Paleogene, giving rise to volcanic rocks and plutons. The Lower Cretaceous volcanics evidently blanket the MSM. Neogene bimodal volcanic rocks erupted during Basin and Range extension in Sonora.

By correlation of Proterozoic age provinces and facies of Paleozoic cover, the Seri terrane probably came from the former western continuation of the broad passive margin wedge of the

Great Basin (Figs. 8 and 9) (Silver and Anderson, 1974; Stewart and others, 1990). The ages of excision and of attachment in Sonora were end of Jurassic at the minimum and late Paleozoic at maximum. Paleomagnetic latitude data from Upper Triassic rocks suggest a Jurassic age of displacement (Cohen and others, 1986), but such data do not provide enough resolution to be decisive. Furthermore, it is unclear whether the MSM was the edge of PNA in Sonora upon Seri attachment, or whether the MSM cut inboard of the earlier edge of PNA. The place at which the allochthon of basinal Paleozoic rocks was attached to the Seri terrane is unknown.

The Baja California block of the Seri terrane (Gastil and others, 1991) was sutured to the Yuma terrane in the Cretaceous. Because the Yuma terrane underwent large transport to the northwest relative to PNA in the Paleogene, the Seri in Baja California must have traveled with the Yuma and been at latitudes of southern Mexico in the Cretaceous. Therefore, we propose that more than one terrane of Seri tectonostratigraphy was generated in southern California in the late Paleozoic, and they have had diverse movement histories down and up the developing western edge of Mexico in Mesozoic and Cenozoic time. The blocks of Seri terrane in Baja California and Sonora are now together, although they have separated again, in part or in whole, on the Gulf of California–San Andreas transform-ridge system.

Tahue

The Tahue terrane is little understood because of nearly complete superposition by young deposits; its tectonostratigraphy is inferred, and its margins are entirely covered (Mitre-Salazar and Roldán-Quintana, 1991). Meager scattered exposures below the Mesozoic arc volcanics consist of hemipelagic and siliciclastic strata that are locally dated Mississippian-Permian (Carrillo-Martínez, 1971; Malpica, 1972) and are lithologically similar to protoliths of nearby outcrops of greenschist and lower amphibolite metamorphic facies. The metamorphic correlatives also include volcanogenic rocks. The depositional basement for these strata is uncertain. A couple of outcrops of gneiss occur in the Tahue terrane, but their relation to the Mississippian-Permian rocks is unclear (Mullan, 1978), and one gneiss may be from faulted post-Mississippian plutons. A suggestion of oceanic crust below the Mississippian-Permian comes from the low initial $^{87}Sr/^{86}Sr$ values in Mesozoic plutons of the Tahue terrane. The Mississippian-Permian rocks are lithologically like the basement rocks of the Pericú terrane, and perhaps evolved from deposits partly of continental provenance deposited on oceanic crust in mid-Paleozoic time. We infer that such sites may have existed near the western margin of PNA and that the terranes traveled southeast in late Paleozoic and/or early Mesozoic time to their present latitudes. The age of attachment of the Tahue is inferred to be in Late Jurassic or Early Cretaceous time by analogy with the Náhuatl terrane.

The Tahue terrane underwent continental arc magmatism from Late Jurassic until mid-Cenozoic time. The rocks are sequences of intermediate and basaltic lavas and rhyolitic tuffs, plutons, and several mafic complexes that may represent arc roots. Such magmatism emerged above convergent margins with the Farallon and succeeding plates. The Cretaceous and older rocks of the arc complex are metamorphosed to as high as greenschist facies. They are overthrust from west to east by nappes, one of mid-Cretaceous sedimentary and arc-volcanogenic rocks, and several of ophiolites (Ortega and others, 1979; Servais and others, 1982, 1986). The ophiolitic nappes are inferred to be the floors of former Mesozoic arc basins, backarc (Ortega and others, 1979) or forearc (Servais and others, 1986).

Much of the Tahue terrane underwent Neogene extension and basalt extrusion; these may be related to transtension related to the Gulf of California and/or to the southern fringe of Basin and Range faulting.

Tarahumara

The lowest exposures are of low-grade metasedimentary rocks that give Pennsylvanian K-Ar and Rb-Sr ages of metamorphic mica (Denison and others, 1969). Their lithology, deformation, and age ally them with the internal zone of Ouachitan orogen in the Marathon region of Texas (Fig. 9) (Flawn and Maxwell, 1958). The terrane boundaries, nowhere exposed, are assigned by the geography of the negative Bouguer gravity anomaly that is associated with the Ouachitan belt in west Texas and continues southward into northeastern Mexico (Handschy and others, 1987). Common lead isotopes in Tertiary igneous rocks of the Tarahumara and adjacent terranes are interpreted to indicate that the buried edge of PNA is about 40 km southeast of the western margin of the Tarahumara terrane (James and Henry, 1990). Moreover, the existence of metasedimentary xenoliths of lower crustal origin in Neogene lava of eastern Chihuahua suggests that PNA extends in the subsurface at least as far south as east-central Chihuahua (Nimz and others, 1986). We infer that the Tarahumara terrane was the accretionary forearc to the late Paleozoic continental arc of northern South America,which is contiguous with Tarahumara in the Coahuiltecano terrane, and that both terranes were on the overriding plate in the collision of the two American continents in the Ouachitan orogen. Moreover, we infer that the southwestern end of the Tarahumara may have been the western end of the Ouachitan orogen and on the Pacific coast of Pangea (Fig. 10).

The metamorphic rocks of the Tarahumara terrane are imbricated in the thin-skinned Laramide fold and thrust belt (Navarro and Tovar, 1974; Padilla y Sanchez, 1986) and are overlain unconformably by Upper Jurrasic and Cretaceous siliciclastic, carbonate, and evaporitic shelfal strata; these are continuous into surrounding terranes. There are no recognized rift facies in the Mesozoic cover.

Tepehuano

This complicated and poorly understood terrane is described here in five highly interpretive tectonostratigraphic units: crystalline basement, early Mesozoic continental arc and backarc basin rocks, Upper Jurassic to mid-Cretaceous shelfal strata, Paleogene continental arc rocks, and Neogene Basin and Range rocks.

The existence of Proterozoic basement below at least the southeastern realm of the terrane can be inferred from xenoliths of gneissic granulite from Quaternary lavas (Fig. 5) (Ruiz and others, 1988a, b). Sm-Nd data indicate a pre-Grenvillean crust that underwent a major infusion of mantle material at about 1000 Ma. Some xenoliths give younger mixing ages, indicating Phanerozoic crustal reworking.

The second unit combines poorly dated, deformed, low-grade metaigneous and metasedimentary rocks from widely separated, sparse outcrops, and these may have originated at highly diverse places and times, probably including the Paleozoic and Mesozoic. Formations included are the Taray, Caopas, Rodeo, Nazas, Zacatecas, Chilitos, Gran Tesoro, and Arperos, which collectively may record the development in Late Triassic and Jurassic time of one or more continental arcs, subduction complexes that include fragments of Paleozoic rocks, and oceanic backarc basin crust and sediment fill. All of these formations underwent large horizontal contraction during magmatism involving imbrication and backarc basin collapse (Servais and others, 1982) on east-west to west-southwest–east-northeast bearings, probably mainly in the Jurassic, but to the west, also in Early Cretaceous time. The deformation is probably related to attachment of the Tepehuano terrane. It is possible that the second unit comprises sequentially emplaced complexes and records a long attachment history from Late Jurassic to mid-Cretaceous time. Without deep insight, however, we assume that the Tepehuano arrived as a whole in the Late Jurassic.

The third unit, Mesozoic cover, includes Upper Jurassic basal strata in the northern and eastern Tepehuano terrane that are continuous with those of the Coahuiltecano and Guachichil terranes, implying attachment of that part of the terrane by the Late Jurassic. To the west, deposition and deformation of the second unit continued into the Cretaceous, and such rocks are covered unconformably by Albian shelfal strata. Thus, the conclusion of deformation and onset of deposition of cover appear to have migrated southwesterly across the Tepehuano terrane.

The Mesozoic cover of unit 3 underwent detachment and imbrication related to north to northeast-vergent Laramide foreland thrusting in Late Cretaceous and Paleogene time (Pantoja-Alor, 1963; Ledesma-Guerrero, 1967; Enciso de la Vega, 1968). Paleocene plutonic rocks, unit 4, invaded the terrane. These are in the eastern realm of the continental arc of the Sierra Madre Occidental that swept eastward, presumably because of shallowing of the slab, in the Paleocene.

The fifth tectonostratigraphic unit comprises volcanic and sedimentary rocks of Neogene age, as old as 24 Ma, that were deposited during Basin and Range extension.

Paleomagnetic data from Upper Triassic rocks in our second unit suggest that such rocks underwent substantial net transport to the south (Nairn, 1976; Cohen and others,1986). This probably took place before Upper Jurassic time, because the shelfal cover, for which paleomagnetic data exist in nearby terranes, has no recognized latitudinal transport. We employ the inferred paleomagnetic transport direction and inferences from Jurassic movements of the Guachichil and Zapoteco terranes, which have Grenvillean basements, to assign a southward component of movement of the Tepehuano before attachment.

Yuma

The Yuma terrane composes most of the Baja California Peninsula and extends into California, where it is called the Guerrero terrane by Howell and others (1985) and the Santa Ana terrane by Howell and Vedder (this volume). It comprises two subterranes; the western one is composed of island arc magmatic rocks of Jurassic and Cretaceous age, and the eastern one is flysch of inferred forearc or intraarc basin origin and Triassic to Cretaceous age (Gastil and others, 1975, 1981, 1986a, b; Todd and others, 1988, 1991).

The flysch of the eastern subterrane is chiefly siliciclastic and includes rare quartzite conglomerate and andesite, indicating mainly continental provenance. It is dated by megafossils of Late Triassic, Jurassic, and Aptian-Albian age. The flysch is very deformed and is intruded by syntectonic S-type granite bodies (Todd and others, 1991) of Early Cretaceous age and by posttectonic granites 105 to 82 Ma in age (Silver and others, 1979). The island arc subterrane contains thick successions of intermediate and siliceous lavas, called the Santiago Peak and Alisitos formations, which are intruded by comagmatic plutons of the western zone of the Peninsular Range batholith, dated 135 to 105 Ma (Silver and others, 1979; Hill and others, 1986; Gromet and Silver, 1987). The tectonic affinity of this arc is based on compositions of initial Sr isotopes, rare earth element patterns, and I-type granites.

Both subterranes and their contact are intruded by the younger (105 to 82 Ma) plutons of the Peninsular Ranges batholith, the compositions of which indicate a continental arc affinity. The two subterranes are joined at a steeply east dipping suture, and the younger volcanics of the arc terrane are locally deposited across it.

The Yuma terrane attached to the Seri terrane between 106 and 97 Ma at one place, according to U-Pb ages of deformed and undeformed plutons in the suture zone (Griffith, 1987; Griffith and Goetz, 1987; Goetz and others, 1988). Postbatholithic strata are uppermost Cretaceous and Paleocene marine clastic rocks of mainly Yuma terrane provenance; Eocene rocks have Seri terrane provenance (Bartling and Abbott, 1983).

Mid-Tertiary volcanic rocks record continental arc volcanism; late Neogene lavas, some of them alkalic, are probably related to rifting leading to formation of the Gulf of California.

The Yuma, Seri (Baja California block), and Cochimí terranes form the Baja California composite terrane, which existed as much as 1500 km south of its present latitude after the mid-Cretaceous and before Oligocene time, according to many paleomagnetic data (Hagstrum and others, 1985; Morris and others,

1986; Filmer and Kirschvink, 1989; Flynn and others, 1989; Lund and others, 1991a, b). If it is assumed that the Seri terrane in Baja California was transported toward present-day southern Mexico during a Jurassic and Early Cretaceous phase of left-lateral boundary-parallel slip, the Yuma terrane was probably also translated southward before the reversal of strike slip in Late Cretaceous time.

Zapoteco

The basement of the Zapoteco terrane is Proterozoic continental crust (Oaxacan Complex) composed of rock types such as charnockite, anorthosite, and calcsilicate, and pelitic gneiss that contains 1100 to 1000 Ma ages of granulite facies metamorphism (Ortega, 1981a, 1981b, 1984). The basement is unconformably overlain by platformal strata, some of which contain Lower Ordovician trilobites like those of southeastern Canada (Avalon terrane), South America, and northwestern Europe (Pantoja-Alor and Robison, 1967). Except for the fossils, such characteristics are almost unequivocal indicators of a Grenville origin. The paleopole of primary magnetization in the Oaxacan complex is >40° from the Grenville loop of the North America polar wander path (Ballard and others, 1989), which can be interpreted to indicate a position of the complex near Quebec during Grenvillean time (Sedlock and others, 1993).

During an interval between the end of Proterozoic breakup of Rodinia and the Devonian, the Zapoteco basement probably drifted south relative to PNA, perhaps during right-lateral transport of terranes in oceans east of the eastern margin of PNA. The Zapoteco collided with and overthrusted the Mixteco terrane in the Devonian, probably in an oceanic realm (Yáñez and others, 1991). Their suture was covered by Upper Devonian(?) clastics of Oaxacan Complex provenance (Ortega, 1990).

Deformation and metamorphism occurred in the Zapoteco-Mixteco (Z-M) composite terrane in latest Devonian time, followed by the development of an Early Mississippian unconformity. This unconformity was covered by shallow-marine siliciclastic strata of Mississippian to Permian age. The Oaxacan Complex was then intruded by granitoids that give Rb-Sr isochron ages from Early Permian to Middle Jurassic (Yáñez and others, 1991). The Permian is succeeded unconformably by Jurassic subaerial and shallow-marine strata and Cretaceous platformal carbonates.

Constraints to interpretation of the late Paleozoic tectonic evolution are: (1) the Early Permian intrusive event; (2) that post-Devonian strata are little deformed and metamorphosed, implying that they did not pass through the Ouachitan collision zone; and (3) paleomagnetic data from the Mixteco that indicate $15° \pm 8°$ of southward translation relative to PNA between about 160 and 110 Ma (Urrutia and others, 1987; Ortega and Urrutia, 1989). We hypothesize that the Z-M terrane was accreted from an oceanic realm south of PNA to the forearc of converging South America in Early Carboniferous time. Its forearc position was well west along northern South America, west of the Tarahumara terrane that was also in the South American forearc. Upon collision with North America of the Tarahumara in Permian time, the Z-M terrane escaped westward and migrated well north along the Permian-Triassic convergent boundary zone of western Pangea (Fig. 10), undergoing episodic continental arc magmatism that gave rise to the Permian granitoid of the Zapoteco and Mixteco terranes and the Triassic calc-alkalic volcanics of the Mixteco. The Z-M terranes then translated southward some 1500 km during the deposition of its Mezozoic cover, presumably in a left-oblique convergent boundary zone that existed along western Mexico in the Jurassic and Early Cretaceous, and had arrived at its current latitude by Albian time. The attachment of the Z-M terranes occurred in the Cretaceous, perhaps upon arrival from the north and certainly by the time of its easterly overthrusting of the Cuicateco terrane, which was no later than early Tertiary.

The Zapoteco terrane has undergone continental arc magmatism throughout the Cenozoic. It has extended east-west in the Neogene, and incipient horst and graben structures have developed (Centeno and others, 1990).

PHANEROZOIC EVOLUTION OF MEXICO

We now employ the data and interpretations of preceding sections to formulate an evolution of Mexico through Phanerozoic time displayed in a time series toward the present.

End Proterozoic

Evidence points to a continuation of the intracontinental rift-transform system across Mexico between the Gulf Coast and Cordillera of the United States that more or less blocked out Proterozoic North America as it is today at the end of Phanerozoic time. This model, agreed upon by some (Palmer and others, 1984; Stewart, 1988), is more probable than the alternative; i.e., that PNA extended greatly toward the southwest in Mexico (in today's coordinates) (Shurbet and Cebull, 1987; Dickinson, 1981) and/or was an active rather than passive margin. Moreover, we surmise that the Proterozoic edge of PNA in Mexico was transform dominated as in the U.S. Gulf Coast, and was the Mojave-Sonora megashear or a surface not far to the south, as shown in Figure 8.

The implications of the model of Figure 8 are that (1) oceanic lithosphere developed south of the margin in early Paleozoic time; (2) the oceanic lithosphere at the continental margin may have had unusual properties (thin crust, great depth to sea floor), using modern characteristics of transform margins (Todd and Keen, 1989; Haworth and others, this volume); (3) the width of the continental transitional zone between craton and ocean floor was narrow (50 km?), possibly little disrupted, and nonsubsiding compared to typical rift margins (Haworth and others, Rankin, and Saleeby and others, this volume); (4) short rift segments may have punctuated and offset the long transform segments; and (5) the intracontinental transform may have existed 100 m.y. or more before separation of PNA from continental relics of Rodinia was complete in Mexico. The cratonal aspect of lower Paleozoic strata of PNA in Mexico implies that the motions of PNA and outboard plates were close to pure strike slip relative to the margin orientation, probably not transtensional, and certainly not transpressional. It can be inferred that the margin may have been sediment starved and that lower Paleozoic sediments that covered the oceanic crust adjacent to PNA were principally noncalcareous hemipelagic and carbonate–quartz-arenitic turbidites.

The Zapoteco terrane was blocked out as a microcontinent of the rift-dominated margin of PNA in eastern Canada. The Guachichil and Tepehuano terranes may also have emerged at the end of Proterozoic time as microcontinents in the Paleozoic ocean, derived from the breakup of Rodinia in the U.S. Gulf Coast or eastern Mexico. The stranding of microcontinents can be attributed to ridge jump.

Early Paleozoic

PNA of northern Mexico in early Paleozoic time had a steeply south facing margin and a probable sediment-starved oceanic basin to the south (Fig. 15). The Zapoteco terrane translated south and west relative to PNA, probably migrating well outboard of eastern PNA by margin-parallel right-slip transport during intermediate and late phases of orogeny of the Appalachians (Rankin, this volume). It underwent platformal sedimentation in the open sea. The Zapoteco and Mixteco terranes were amalgamated in the Devonian at a boundary with a convergent component within the oceanic realm.

Gondwana, the western realm of which was continental South America, had begun to converge on Paleozoic oceanic lithosphere(s) between it and PNA and to develop active margin features at its currently northern edge (Fig. 15). Its Proterozoic basement underwent magmatism that may be as old as Silurian.

The Guachichil terrane received a cover of platformal sediments, then subsided in part in the Devonian(?) and accumulated siliceous volcanogenic material, possibly tuff. Such events may record the encroachment on the Guachichil by Gondwana and ultimately its accretion to the Gondwana forearc.

Convergence between the North American plate and outboard plates at its western edge had begun in the Devonian or earlier time with the growth of the Roberts Mountain allochthon,

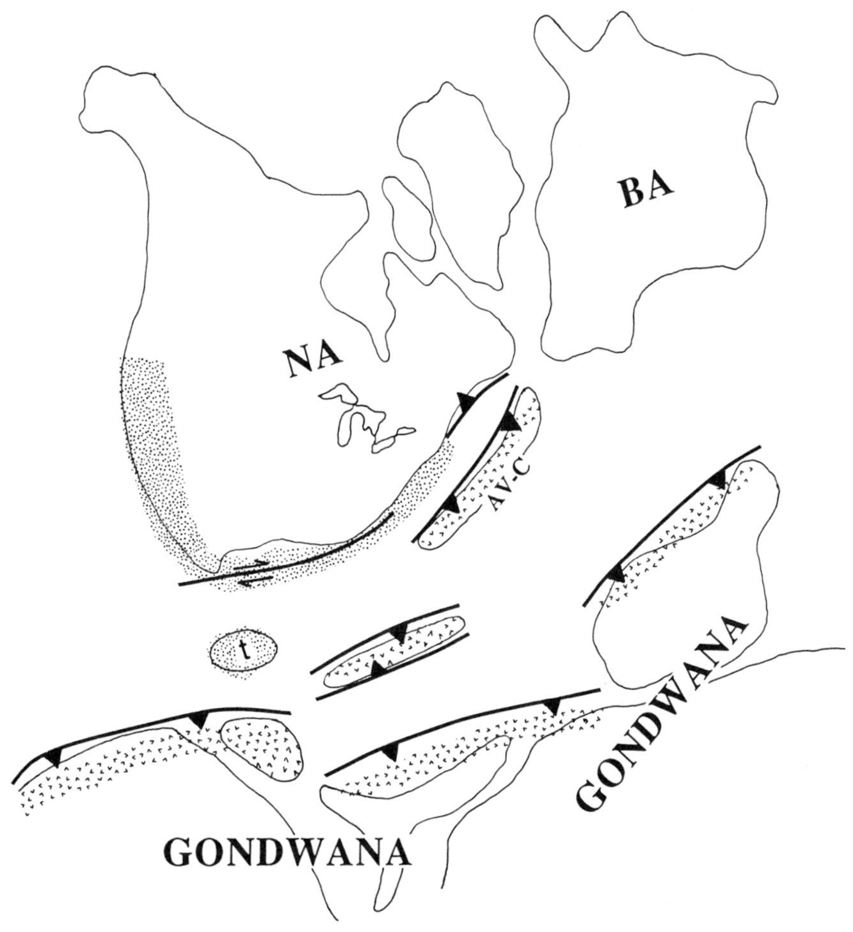

Figure 15. Schematic paleogeography of North America (NA) and Gondwana in early Paleozoic time (Silurian), modified from Bally and others (1989). AV-C: Avalon-Carolina terranes; BA: Baltica; t: terrane of North American origin (Zapoteco, Guachichil) migrating west or southwest relative to Paleozoic North America in modern coordinates within Paleozoic oceanic realm. Dot pattern: platformal or shelfal deposition; V pattern: arc magmatism.

probably as an oceanic accretionary wedge, and the development of lower Paleozoic island arcs west of what is now the Great Basin (Fig. 15). Such convergent boundaries did not continue as far south as Mexico, at least not near the margin of PNA.

Carboniferous and Early Permian

Right-oblique convergence between the North American and Gondwana plates caused diachronous collision of PNA and the Gondwana continents, complete subduction of the formerly intervening Paleozoic oceanic lithosphere, and the formation of Pangea (Fig. 10). Prior to collision, the active margin of Gondwana (Fig. 15) evolved by accretion of strata from the subducting oceanic lithosphere, by continental arc magmatism, and by the deposition of forearc sedimentary cover that probably included trench fill during later stages. The collision migrated east to west along the southern margin of PNA as the Ouachitan orogeny. The encroachment of the overriding Gondwana plate caused downflexing of the southern fringe of PNA accompanied by high-angle faulting, probably of preexisting faults (Buffler and Thomas, this volume). This differentiated subsidence resulted in the series of late Paleozoic foreland basins and local uplifts that are just inboard of the Ouachitan orogen (Fig. 9). The Gondwanan forearc overrode an unknown but probably relatively small width of PNA before cessation of closure (Lillie and others, 1983). The Tarahumara terrane is the sedimentary forearc of the South American part of Gondwana and is equivalent to the external and internal zones of the Ouachitan orogenic belt in the southern United States. The interior region of the South American active margin (continental arc, forearc, and backarc[?] basins) gave rise to the Coahuilatecano terrane, which is currently in its Permian position relative to PNA, and to the Maya and Chortis terrane (see below), both of which have since been displaced with respect to the Coahuilatecano and Permian North America.

The Z-M composite terrane was swept into the Gondwana forearc, deformed and metamorphosed very late in the Devonian. Its position was well west within the Gondwana forearc, perhaps on the west-facing margin of Gondwana. At the time of continental collision, the Z-M terrane may have faced the proto-Pacific oceanic lithosphere, west of the Tarahumara terrane in the Ouachitan belt (Fig. 10), and then may have translated northwest as a microplate relative to the margin of Permian PNA (by then, Pangea). Alternatively, the Z-M terrane could have been a piece of the Gondwana forearc that escaped westward across a free boundary to the proto-Pacific during the continental collision.

The Guachichil terrane had a late Paleozoic history similar to that of the Z-M terrane in being incorporated in the Gondwana forearc. To account for the differences in the two terranes, however, we postulate that Guachichil was maintained at a shallow level in the forearc, permitting deposition of tuff from the continental arc and preventing metamorphism and large deformation, whereas the Z-M was initially deeper. The Granjeno schist of the Guachichil terrane may have been an accretionary packet of sediments and basalt that followed an initially deep track through the Gondwana forearc and a later shallowing track, which juxtaposed it against the Grenvillean gneissic block during progressive deformation of the forearc. Like the Z-M terrane, the Guachichil terrane either was west of the Tarahumara belt or escaped westward from the forearc, perhaps to continue northwestward migration relative to Pangea in Pennsylvanian and or Early Permian time.

The Tepehuano terrane is assigned a late Paleozoic history similar to that of the Guachichil, by virtue of its probable content of Grenvillean basement.

Most of the Seri terrane was blocked out in the southern Cordillera of the United States, south of Bishop (Fig. 8), by the onset of strike slip on northwest-trending faults that cut the margin of PNA beginning in Pennsylvanian time. The Proterozoic basement and rift-dominated lower Paleozoic passive margin wedge probably translated southeast relative to PNA, most likely only partly but conceivably all the way to Sonora in the Permian. The allochthonous lower Paleozoic basinal strata of the Seri terrane and the Mojave region of California were probably deposited on the Paleozoic proto-Pacific ocean floor that fronted PNA after the end of Proterozoic breakup of Rodinia. They probably evolved as accretionary prisms in Mississippian and earlier periods, as did the Roberts Mountains allochthon to their north (Fig. 8), but, in contrast to the Roberts Mountains allochthon, which collided with western PNA in early Carboniferous time, they apparently stayed outboard of the continental margin during the late Paleozoic. These strata were probably translated southeast on the left slip faults that affected the basement of the Seri terrane.

The basements of the Tahue and Pericú terranes are inferred, with considerable uncertainty, to have evolved from sedimentary protoliths of Mississippian age that were deposited on oceanic basement. Because these terranes arrived in the mid-Mesozoic, relatively late in the evolution of the post-Paleozoic active margin of western Mexico, they probably underwent large left-lateral transport relative to PNA. Therefore, we assume that such basements emerged in the proto-Pacific, west of late Paleozoic North America, and were perhaps related to island arcs and accretionary forearcs; the Sonomia and Golconda allochthon terranes are equivalents in the U.S. Cordillera (Speed, 1979). If true, the translation paths of the Pericú and Tahue terranes in the early Mesozoic and/or late Paleozoic may have been similar to, but outboard of, that of the Seri terrane.

Early Permian–Late Triassic

Calc-alkalic magmatism was initiated in a continental magmatic arc at the western margin of Pangea in the Early to Late Permian in eastern Mexico and northwestern South America and in the Late Permian to Early Triassic in the southwestern United States (south of Bishop, Fig. 8; Figs. 10 and 16). The onset of magmatism records eastward subduction of oceanic lithosphere of one or more plates in the proto-Pacific basin beneath western Pangea. The onset of magmatism was roughly coeval with the mid-Permian cessation of Gondwana–North America convergence, and probably reflects a major reorganization of relative plate motions. Late Permian convergence was probably left oblique, based on the sense of slip inferred in the southwestern United States (Avé Lallemant and Oldow, 1988). Sinistral displacement on the boundary probably contributed to the progressive southeastward displacement of the Seri terrane and more outboard Mexican terranes. The gap in the Permian-Triassic arc in northern Mexico is explained by plate motion nearly parallel to the northwesterly-trending segment of the Pangean margin (Stewart and others, 1990).

In the U.S. Cordillera north of Bishop (Fig. 8), the Late Permian–Early Triassic motion of proto-Pacific plates was left-oblique convergent relative to PNA, but the subduction zone probably faced west (Speed, 1979, 1983). Late Paleozoic terranes composed of island arcs and accretionary complexes collided with and overrode the passive margin of PNA early in the Triassic, but without continental magmatism. There, subduction switched to east facing in Middle Triassic time, and continental arc magmatism and an active margin began in the central U.S. Cordillera (Fig. 16).

The Triassic magmatism of the Z-M terrane and perhaps the Tepehuano terrane may have occurred above the east-facing subduction zone at the PNA margin in southern California (Fig. 16). Furthermore, the emplacement in Late Permian or Triassic time

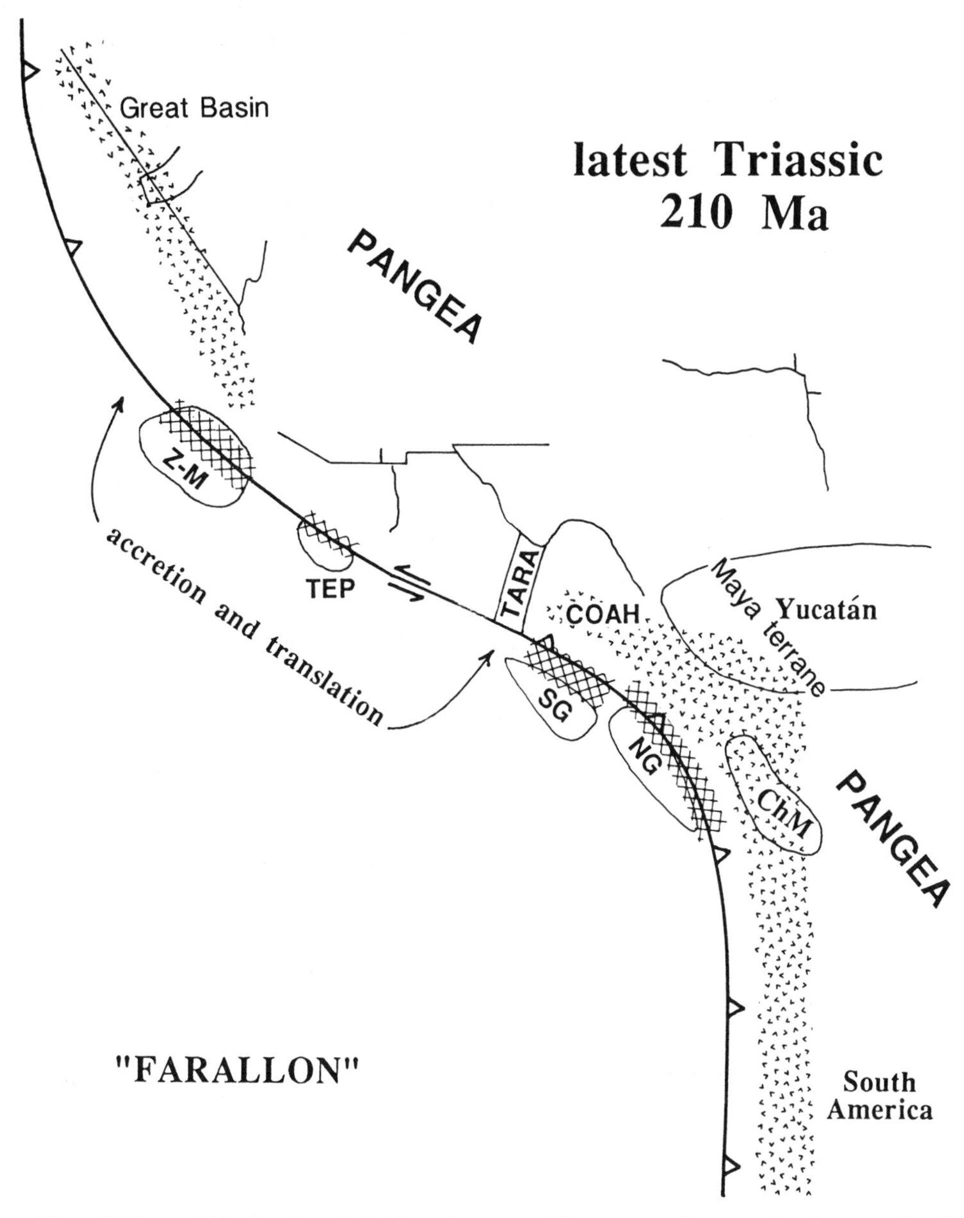

Figure 16. Latest Triassic (about 210 Ma) paleogeographic reconstruction; cross-hatch: zones of collision, obduction, and accretion; Vs: continental arc magmatism. Seri terrane may exist somewhere between California and Sonora. Z-M: Zapoteco-Mixteco composite terrane; TEP: Tepehuano; SG and NG: southern and northern parts of Guachichil; other abbreviations as in Figure 10.

of the allochthonous lower and upper Paleozoic ocean-floor rocks on the Seri terrane and in the Mojave region may have been related to such left-oblique convergence.

The Guachichil terrane, in contrast, did not undergo continental magmatism during this interval. We suppose that it migrated southeast along the mainly strike-slip boundary of northern Mexico (Fig. 16). The thrusting within PNA in Chihuahua of Grenvillean crust over Lower Permian flysch (Fig. 14) may have occurred by Late Permian time (Handschy and Dyer, 1987); if so, it must have occurred along the transform-dominated segment of northern Mexico, perhaps due to local convergence at a short rift increment (Fig. 10).

Late Triassic–Jurassic

Continental Mexico may have developed most markedly toward its present form during this interval. There were two principal kinematic events; the breakup of Pangea and onset of relatively southeasterly drift of South America from the central or southern locus of the Ouachitan orogen, which extended west across Pangea (Fig. 9), and the left-oblique convergence of oceanic lithosphere (Farallón and/or other plates) of the proto-Pacific basin with the evolving western margin of Mexico (Figs. 16–18).

The Pangean breakup began in the Late Triassic with the development of rift basins and horsts in the Maya, Guachichil, Coahuiltecano, and Tepehuano terranes and perhaps, west into PNA and the Seri terrane. This involved small total stretch. In the Middle Jurassic (Fig. 17), the stretching became concentrated in and east of the Maya terrane in the Gulf of Mexico basin, causing further breakup of the Maya terrane, large crustal extension leading to spreading in the northern Maya, and the southeasterly drifting of the little-extended Yucatan block of the Maya terrane (Fig. 17). The concentration of stretching and drifting, producing

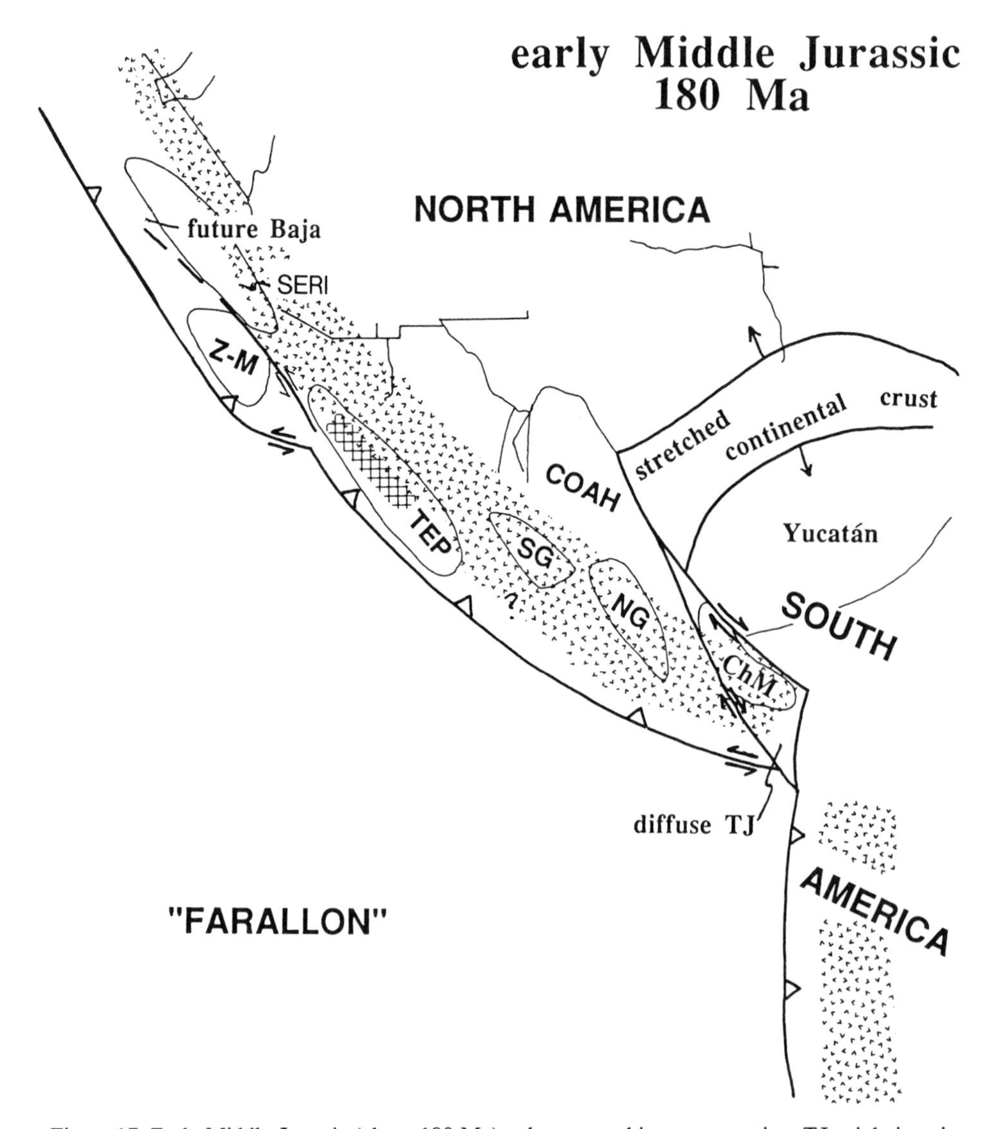

Figure 17. Early Middle Jurassic (about 180 Ma) paleogeographic reconstruction; TJ: triple junction; abbreviations as in Figure 16.

some 1000 km of Jurassic displacement, was accompanied by a rise in the asthenosphere, perhaps caused in part by backarc mantle motions. The right-lateral boundary between little- and greatly-stretched Mexico in the Jurassic is along or below the present-day continental slope of eastern Mexico (Fig. 17). This boundary was in fact probably a broad zone that included a number of blocks within the Maya terrane (such as the Chiapas massif; Fig. 17) (Molina-Garza and others, 1990).

The kinematic connection between the extension of eastern Mexico and convergence in western Mexico is poorly controlled and controversial. We argue that the displacement of Yucatán and South America relative to PNA did not occur along the Mojave-Sonora megashear and did not cause migration of terranes relative to PNA (Fig. 17). Rather, the inboard terranes (Guachichil, Tepehuano) migrated southeasterly in the western plate boundary and attached at various times. This boundary zone included an array of margin-parallel left-slip faults that operated episodically and repositioned the inboard terranes during the Jurassic (Figs. 17–19). The Mojave-Sonora megashear, perhaps the inherited Proterozoic edge of PNA, is the most inboard of left-slip faults within the zone. The sense of obliquity at the convergent margin comes from offsets on discrete strike-slip faults of this age and from paleomagnetic data, which indicate the southward migration and counterclockwise rotation of terranes relative to PNA (Urrutia and others, 1987). The Middle Jurassic convergence probably continued from the preceding duration, but was more orthogonal to the margin, at least with respect to the northwest-striking edge of PNA in northern Mexico (Figs. 16 and 17). The oblique convergence caused the growth of a continuous continental magmatic arc, initially along the margin of PNA from the southwestern United States to western South America through the Ouachitan suture. It also caused the margin-parallel migration of terranes from northwest to southeast, the attachment of terranes, and the southwestward growth of the Mexican mar-

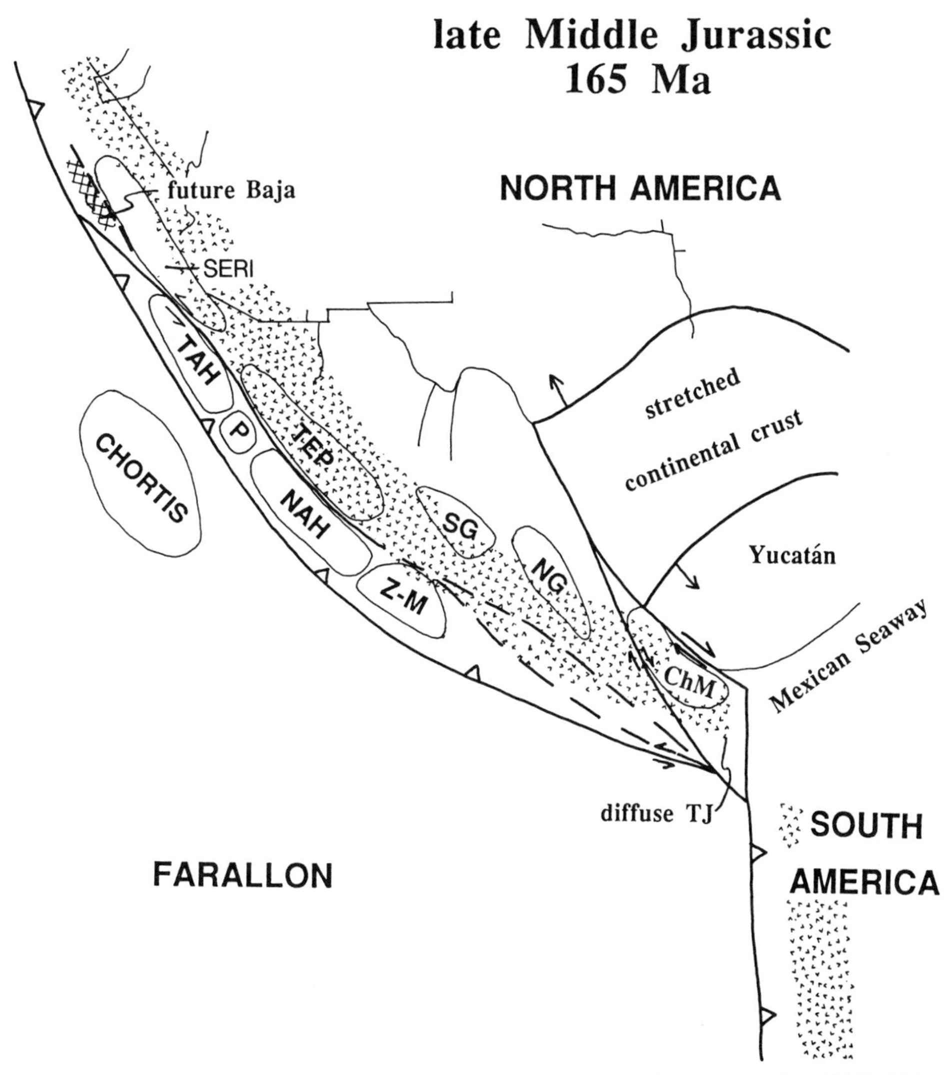

Figure 18. Late Middle Jurassic (about 165 Ma) paleogeographic reconstruction; TAH: Tahue; P: Pericú; NAH: Náhuatl; other abbreviations as in Figures 16 and 17.

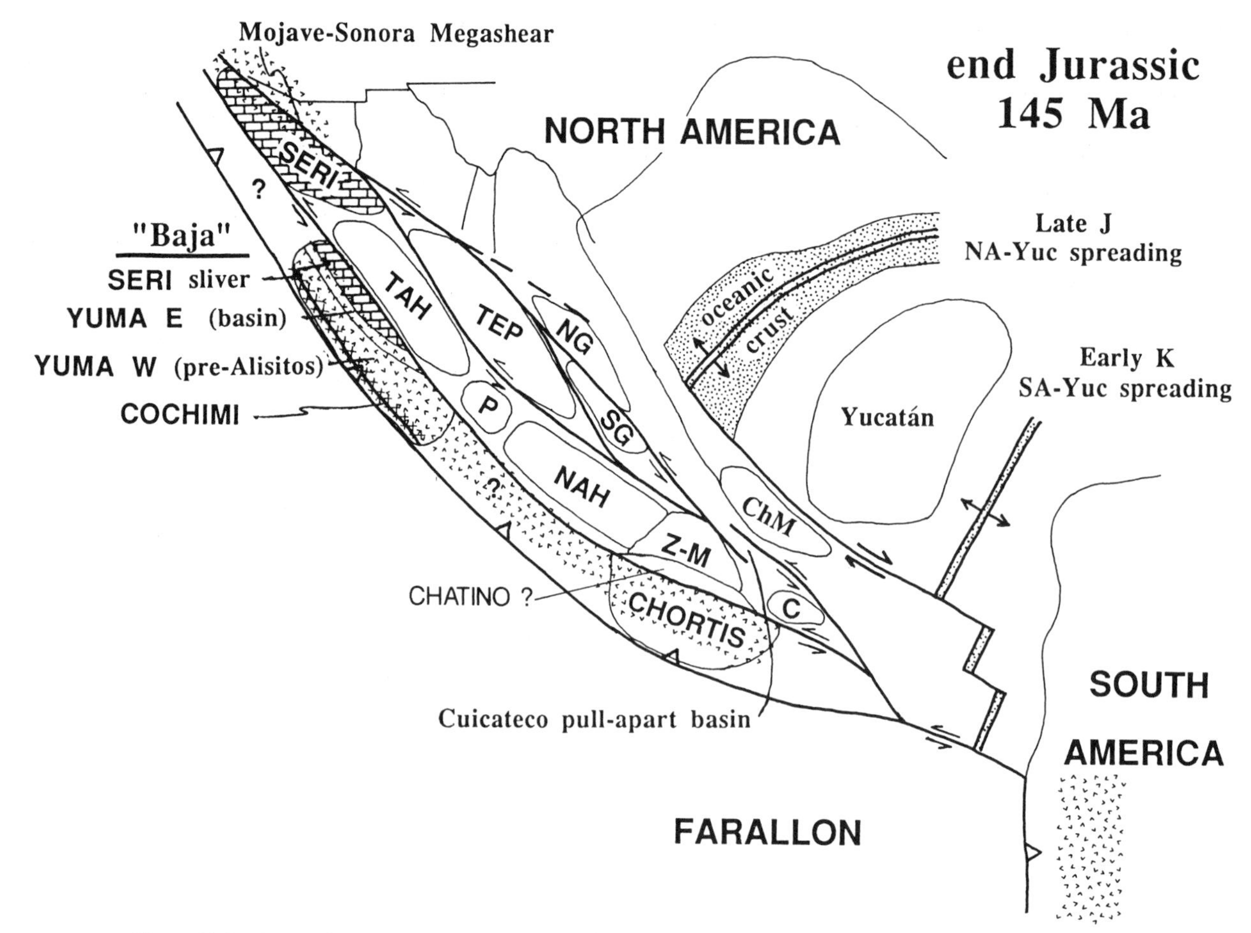

Figure 19. End Jurassic (about 145 Ma) paleographic reconstruction; C: Cuba; other abbreviations as in Figures 16–18.

gin. Such growth produced a generally westerly migration of the central reach of the continental arc and the evolution of a second belt or zone of island arcs that coalesced during terrane attachment in western Mexico.

The Guachichil terrane, perhaps as two or more discrete blocks, probably migrated only a small distance along the boundary before attaching early in this duration. The continental magmatic arc then jumped west from the former western edge of Triassic North America to the Guachichil early in the Jurassic (Figs. 17 and 18). The Tepehuano terrane probably migrated to a position west of the accreted Guachichil and incorporated a piece of accretionary forearc (Taray Formation) that fronted the Guachichil. The Middle Jurassic accretion of the Tepehuano caused further outstepping of the continental arc (Fig. 18).

The Seri terrane probably completed its longshore migration from southern California along the inboard of the active margin zone in Middle Jurassic time or earlier (Fig. 18). The amalgamated Zapoteco-Mixteco composite terrane migrated southeast in a more outboard realm of the zone of oblique convergence in the Jurassic from a possible initial position at the margin in the southwestern United States (Figs. 17 and 18).

In the Middle Jurassic the Tahue and Pericú terranes probably migrated southeast along the active continental margin in outboard positions from initial intraoceanic sites off western PNA, perhaps north of southern California (Fig. 18). The Náhuatl is interpreted to have arisen as an island magmatic-forearc complex in the Jurassic within a southeast-migrating terrane assembly (Fig. 18). The Tahue, Pericú, and Náhuatl terranes probably attached to continental Mexico late in the Jurassic (Fig. 19).

The position of the Chatino terrane in the Jurassic is unknown. We suggest, however, that the Chatino is a fragment of the Chortis terrane that was transferred to the Zapoteco-Mixteco composite terrane during left-lateral translation of the Chortis at the end of Jurassic time (Figs. 19 and 20).

The Grenvillean basement and Paleozoic cover that is part of Cuba today may be a fragment of the Zapoteco terrane that broke away during or after the southeastward translation of the composite Z-M terrane (Fig. 19). The fragment presumably stayed in the outboard left-lateral flow along the active margin and arrived in the region of Caribbean sea-floor spreading in Early Cretaceous time (Fig. 20).

At or shortly after the end of Jurassic time, drift began between South America and North America along the southeastern margin of the Yucatán block of the Maya terrane (Fig. 19).

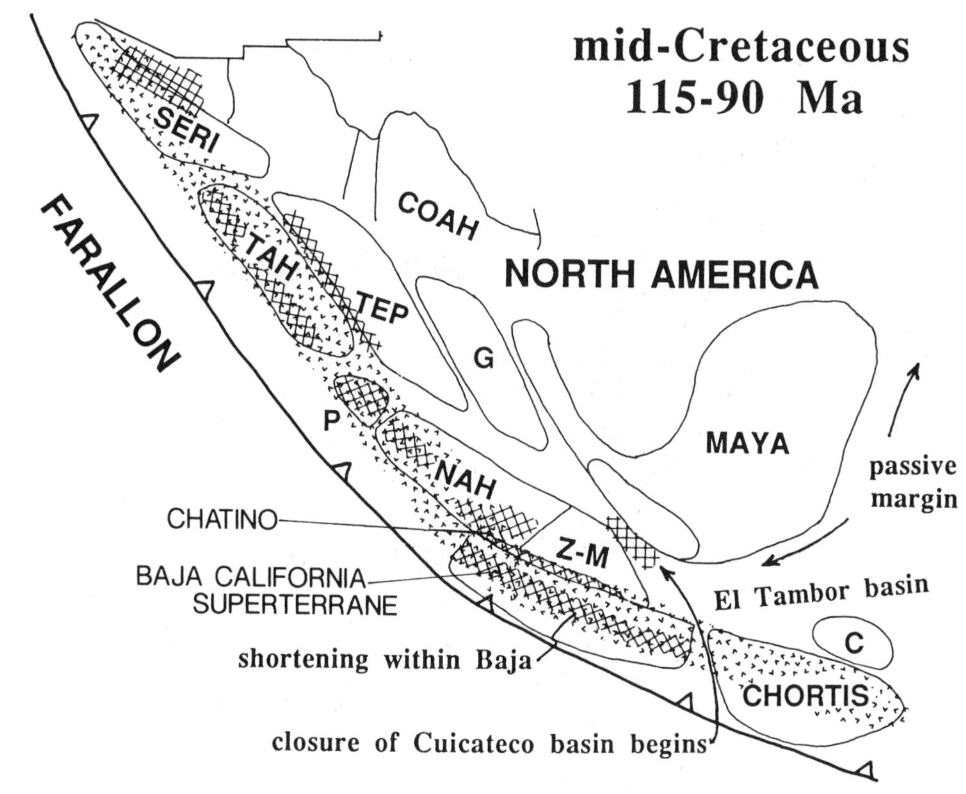

Figure 20. Mid-Cretaceous (115–90 Ma) reconstruction; G: Guachichil; abbreviations as in Figures 16–19.

This began the opening of the Caribbean basin, which provided the first deep seaway between Atlantic and Pacific realms.

We propose that Late Jurassic continental Mexico was composed of all terranes except for Chortis, Cochimí, and part of the Yuma (Fig. 19), which lay within the zone of major plate-boundary motions. In the Late Jurassic, however, a broad zone of left-oblique slip developed by the reactivation of earlier faults, mainly terrane boundaries and, especially, the Mojave-Sonora megashear (Seri-PNA boundary), the Coahuiltecano-Maya, and the Guachichil-Tepehuano, faults between the northern and southern Guachichil blocks, and the San Marcos fault. Left-oblique slip at the southern end of this zone caused the opening of the basin and spreading that led to the Cuicateco terrane (Fig. 19).

Before or during the Late Jurassic, a piece was excised from the outboard flank of the Seri terrane, which overthrusted flysch of the Yuma terrane (Yuma East) that was probably in a forearc at or near the principal trench of the active margin (Fig. 19). This composite terrane then migrated southeast within the frontal zone of the active margin and ultimately collided with one or more island arcs that are the Yuma West terrane at the leading edge of the continent (Fig. 19).

The Chortis terrane is notably hard to track before the Cretaceous. We suppose that it originated from Pangea in western South America in the early Mesozoic and got to an outboard position in the active margin of southern Mexico by the end of Jurassic time (Fig. 19).

Cretaceous

During the Cretaceous, continental Mexico was nearly fully assembled, except for the Baja California composite terrane and the Chortis terrane. It was fronted to the southwest by an active margin, which included a continental arc, a forearc that probably included the still-migrating terranes together with accretionary complexes and one or a chain of backarc basins east of the continental arc (Figs. 20 and 21). The active margin rode above the oceanic Farallon and Kula plates. A major switch in the sense of the boundary-parallel component of plate motion, from left to right slip, occurred at about the middle of the Cretaceous (Fig. 21).

The final attachment of the Zapoteco-Mixteco, Chatino, Náhuatl, Pericú, and Tahue terranes may have been complete just before or during Early Cretaceous time. All of these together with the Tepehuano, Cuicateco, Seri, and PNA underwent at least local pre-Laramide deformation in the Cretaceous with generally continent-verging tectonic transport (Fig. 20). The cause of such contraction may have been an increase in boundary-normal convergence of the Farallon and North American plates during Early Cretaceous time, or perhaps the final attachment of the terranes named above. The continental arc probably extended continuously along the Mexican margin throughout the Cretaceous.

The Baja California composite terrane (Yuma east plus a piece of Seri) migrated toward southern Mexico through Early Cretaceous time, arriving by the middle of the period and under-

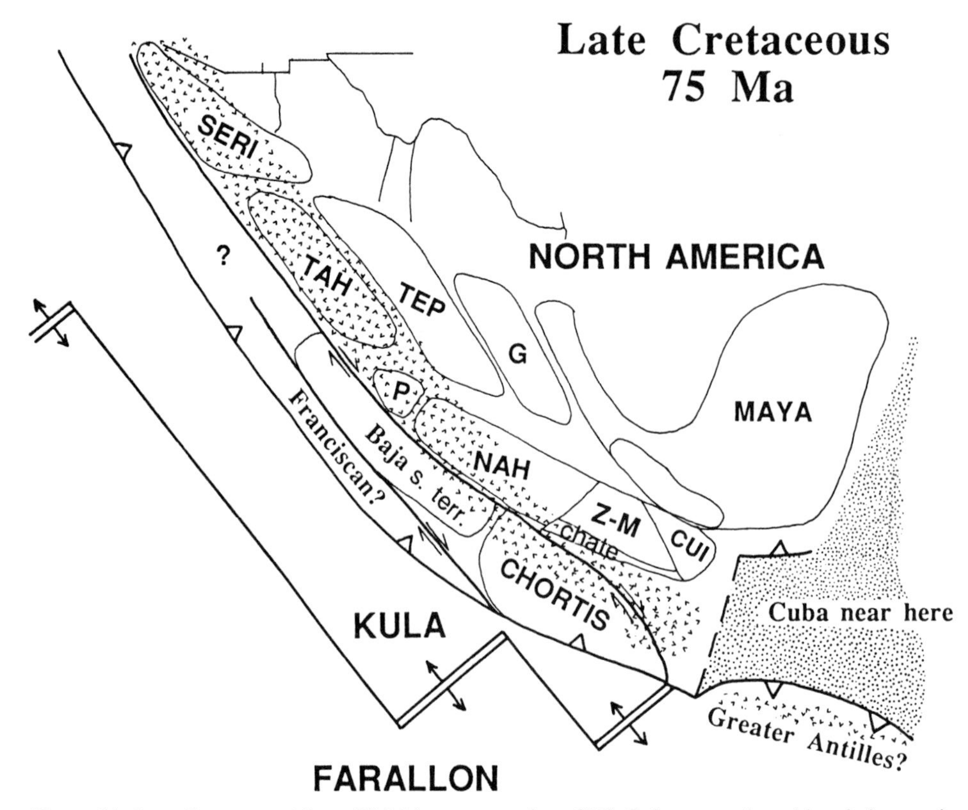

Figure 21. Late Cretaceous (about 75 Ma) reconstruction; CUI: Cuicateco; other abbreviations as in Figures 16, 17, 18, and 20.

going continent arc magmatism at the front of the active margin. Perhaps together with Chortis, Baja California then migrated northwesterly along the active margin in Late Cretaceous time. Baja California probably continued to evolve as sedimentary forearc material accreted to and underplated its outboard side; such material was derived from the adjacent oceanic plate, whose boundary-parallel velocity component, relative to North America, exceeded that of Baja California (Fig. 21).

The Chortis terrane may have begun the Cretaceous Period within southern Mexico's active margin (Fig. 19), and may have migrated east-southeast along the margin with the Baja California composite terrane in the Early Cretaceous (Fig. 20). In the Late Cretaceous, Chortis may have moved northward with Baja California (Fig. 21), although, alternatively, it may have stayed south of the Maya terrane in its Early Cretaceous position, as advocated by Donnelly (1989). The oceanic El Tambor basin formed south of the Maya terrane in mid-Cretaceous time (Fig. 20), probably as part of the opening of the Caribbean basin. Local convergence between the El Tambor basin and North America caused both subduction and obduction of oceanic El Tambor crust in the Late Cretaceous (Fig. 21). The convergence may have involved the collision of an unidentified island arc or the collision of the Chortis and Maya terranes (Donnelly, 1989).

The Laramide foreland thrusting and folding began in almost all of northern and eastern Mexico near the end of Cretaceous time (Fig. 13). The onset of right slip on the Trans-Mexican volcanic belt that culminated in the Paleogene probably began in Late Cretaceous time.

Paleogene

Subduction of the Farallon plate at Mexico's active western margin continued through the Paleogene (Fig. 22). The Baja California composite terrane probably arrived near its present position by middle Eocene time (Lund and others, 1991b). The Chortis terrane plus much of southern Mexico may have moved as much as 400 km northwest relative to northern Mexico on the right-lateral Trans-Mexican volcanic belt (Fig. 22). The Laramide orogeny continued in the continental foreland until about the middle Eocene. The continental arc migrated eastward in northern Mexico during the Paleogene (Fig. 22). Both the arc migration and the foreland deformation may have been related to progressive shallowing of the Farallon slab.

In late Eocene time, the left-lateral North America–Caribbean plate boundary developed south of Maya terrane and extended west to a triple junction (trench-transform-trench) in the Gulf of Tehuantepec (Fig. 23). If it is assumed that Chortis existed in western Mexico in the middle Eocene (Fig. 22), then it has since translated about 1000 km eastward relative to Mexico as a part of the Caribbean plate.

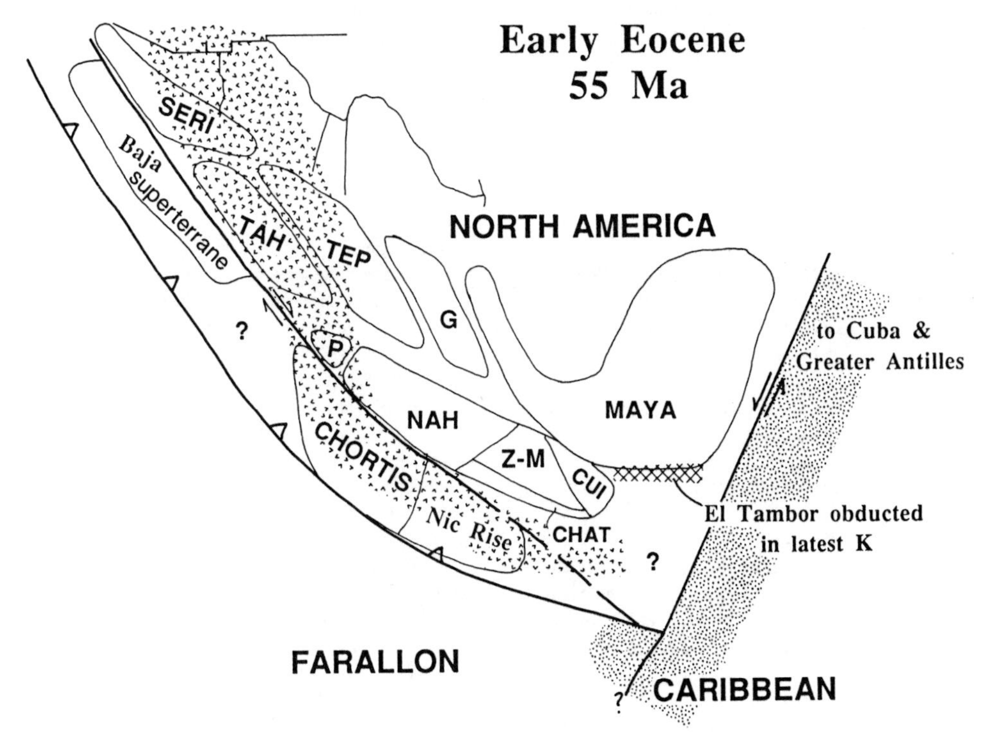

Figure 22. Early Eocene (about 55 Ma) reconstruction; Nic Rise: Nicaragua Rise that was or became eastern part of Chortis terrane; CUI: Cuicateco; CHAT: Chatino; other abbreviations as in Figures 16, 17, and 20.

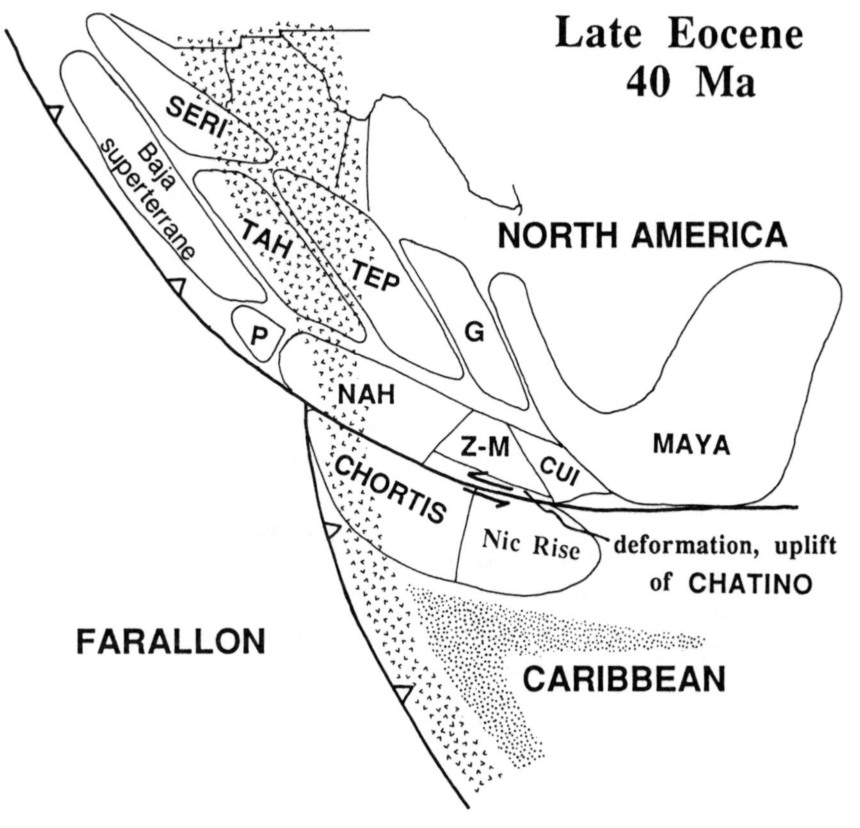

Figure 23. Late Eocene (about 40 Ma) reconstruction; abbreviations as in Figure 22.

ACKNOWLEDGMENTS

We are indebted to J. H. Stewart and R. T. Buffler for thorough reviews of our manuscript and to Judith Gendlin Jones for research assistance.

REFERENCES CITED

Aiken, C. L., Schellhorn, R. W., and de la Fuente, M. F., 1988, Gravity of northern Mexico, *in* Clark, K. F., Goodell, P. C., and Hoffer, J. M., eds., Stratigraphy, tectonics, and resources of parts of Sierra Madre Occidental province, Mexico: El Paso Geological Society Field Conference Guidebook, p. 119–133.

Allan, J. F., 1986, Geology of the northern Colima and Zacoalco grabens, southwest Mexico: Late Cenozoic rifting in the Mexican volcanic belt: Geological Society of America Bulletin, v. 97, p. 473–485.

Alzaga, H., and Pano, A., 1989, Origen de la Formacion Chivillas y presencia del Jurásico Tardío en la región de Tehuacán, Puebla, México: Instituto Mexicano del Petróleo, Revista, v. 21, p. 476–484.

Anderson, J. L., 1983, Proterozoic anorogenic granite plutonism of North America, *in* Medaris, L. G., Jr., Byers, C. W., Mickelson, D. M., and Shanks, W. C., eds., Proterozoic geology; selected papers from an international Proterozoic symposium: Geological Society of America Memoir 161, p. 133–154.

Anderson, J. L., and 9 others, 1990, San Gabriel (Tujunga) terrane—Coming home to Mojave: Geological Society of America Abstracts with Programs, v. 22, p. A303.

Anderson, T. H., and Silver, L. T., 1977a, Geochronometric and stratigraphic outlines of the Precambrian rocks of northwestern Mexico: Geological Society of America Abstracts with Programs, v. 9, no. 7, p. 880.

——, 1977b, U-Pb isotope ages of granitic plutons near Cananea, Sonora: Economic Geology, v. 72, p. 827–836.

——, 1979, The role of the Mojave-Sonora Megashear in the tectonic evolution of northern Sonora, *in* Anderson, T. H., and Roldán-Quintana, J., eds., Geology of northern Sonora: University of Pittsburgh, Guidebook, Field Trip 27, p. 59–68.

——, 1981, An overview of Precambrian rocks in Sonora, Mexico: Universidad Nacional Autónoma de México, Instituto de Geología, Revista, v. 5, p. 131–139.

Anderson, T. H., and Schmidt, V. A., 1983, The evolution of Middle America and the Gulf of Mexico–Caribbean Sea region during Mesozoic time: Geological Society of America Bulletin, v. 94, p. 941–966.

Anderson, T. H., Eells, J. L., and Silver, L. T., 1970, Precambrian and Paleozoic rocks of the Caborca region, Sonora, México, *in* Anderson, T. H., and Roldán-Quintana, J., eds., Geology of northern Sonora (Geological Society of America Section meeting, field trip guidebook): Pittsburgh, Pa., University of Pittsburgh, p. 1–22.

Anderson, T. H., McKee, J. W., and Jones, N. W., 1990, Jurassic(?) melange in north-central Mexico: Geological Society of America Abstracts with Programs, v. 22, no. 3, p. 3.

——, 1991, A northwest-trending, Jurassic fold nappe, northernmost Zacatecas, México: Tectonics, v. 10, p. 383–401.

Aranda, J. J., and Pérez, J. A., 1988, Estudio geológico de Punta Coyotes, Baja California Sur: Universidad Nacional Autónoma de México, Instituto de Geología, Revista, v. 7, p. 1–21.

——, 1989, Estratigrafía del complejo cristalino de la región de Todos Santos, Estado de Baja California Sur: Universidad Nacional Autónoma de México, Instituto de Geología, Revista, v. 8, p. 149–170.

Araujo-Mendieta, J., and Estavillo-González, C. F., 1987, Evolución tectónica sedimentaria del Jurasico Superior y Cretácico Inferior en el NE de Sonora, México: Revista del Instituto Mexicano del Petróleo, v. 19, p. 4–67.

Armin, R. A., 1987, Sedimentology and tectonic significance of Wolfcampian (Lower Permian) conglomerates in the Pedregosa basin: Southeastern Arizona, southwestern New Mexico, and northern México: Geological Society of America Bulletin, v. 99, p. 42–65.

Asmerom, Y., Snow, J. K., Holm, D. K., Jacobsen, S. B., Wernicke, B. P., and Lux, D. R., 1990a, Rapid uplift and crustal growth in extensional environments: An isotopic study from the Death Valley region, California: Geology, v. 18, p. 223–226.

Asmerom, Y., Zartman, R. E., Damon, P.E., and Shafiqullah, M., 1990b, Zircon U-Th-Pb and whole-rock Rb-Sr age patterns of lower Mesozoic igneous rocks in the Santa Rita Mountains, southeast Arizona: Implications for Mesozoic magmatism and tectonics in the southern Cordillera: Geological Society of America Bulletin, v. 102, p. 961–968.

Atwater, T., 1970, Implications of plate tectonics for the Cenozoic tectonic evolution of western North America: Geological Society of America Bulletin, v. 81, p. 3513–3536.

——, 1989, Plate tectonic history of the northeast Pacific and western North America, *in* Winterer, E. L., Hussong, D. M., and Decker, R. W., eds., The eastern Pacific region: Boulder, Colorado, Geological Society of America, The Geology of North America, v. N, p. 21–72.

Avé Lallement, H. G., and Oldow, J. S., 1988, Early Mesozoic southward migration of Cordilleran transpressional terranes: Tectonics, v. 7, p. 1057–1075.

Baker, C. L., 1935, Major structural features in Trans-Pecos, Texas: Texas University Bulletin 3401, p. 137–214.

Ballard, M. M., van der Voo, R., and Urrutia-Fucugauchi, J., 1989, Paleomagnetic results from the Grenvillian-aged rocks from Oaxaca, México: Evidence for a displaced terrane: Precambrian Research, v. 42, p. 343–352.

Bartling, W. A., and Abbott, P. L., 1983, Upper Cretaceous sedimentation and tectonics with reference to the Eocene, San Miguel Island and San Diego area, California, *in* Larue, D. K., and Steel, R. J., eds., Cenozoic marine sedimentation, Pacific margin, USA: Los Angeles, Pacific Section, Society of Economic Mineralogists and Paleontologists, p. 133–150.

Bickford, M. E., Van Schmus, W. R., and Zietz, I., 1986, Proterozoic history of the mid-continent region of North America: Geology, v. 14, p. 492–496.

Bloomfield, K., and Cepeda, L., 1973, Oligocene alkaline igneous activity in NE Mexico: Geological Magazine, v. 110, p. 551–555.

Blount, J. G., Walker, N. W., and Carlson, W. D., 1988, Geochemistry and U-Pb zircon ages of mid-Proterozoic metaigneous rocks from Chihuahua, México: Geological Society of America Abstracts with Programs, v. 20, p. A205.

Bond, G. C., and Kominz, M. A., 1984, Construction of tectonic subsidence curves for the early Paleozoic miogeocline, southern Canadian Rocky Mountains: Implications for subsidence mechanisms, age of breakup, and crustal thinning: Geological Society of America Bulletin, v. 95, p. 155–173.

Bond, G. C., Komintz, M. A., and Grotzinger, J. P., 1988, Cambro-Ordovician eustasy: Evidence from geophysical modeling of subsidence in Cordilleran and Appalachian passive margins, *in* Kleinspen, K. L., and Paola, C., eds., New perspectives in basin analysis: New York, Springer Verlag, p. 129–160.

Bridges, L. W., 1964a, Stratigraphy of Mina Plomosas–Placer de Guadalupe area, *in* Geology of Mina Plomosas–Placer de Guadalupe area, Chihuahua, México (Field trip guidebook): Midland, West Texas Geological Society Publication 64-50, p. 50–59.

——, 1964b, Structure of Mina Plomosas–Placer de Guadalupe area, *in* Geology of Mina Plomosas–Placer de Guadalupe area, Chihuahua, Mexico (Field trip guidebook): Midland, West Texas Geological Society Publication 64-50, p. 60–61.

——, 1964c, Regional speculations in northern México, *in* Geology of Mina Plomosas–Placer de Guadalupe area, Chihuahua, México (Field trip guidebook): Midland, West Texas Geological Society Publication 64-50, p. 93–98.

Buffler, R. T., and Sawyer, D. S., 1985, Distribution of crust and early history, Gulf of Mexico basin: Gulf Coast Geological Society Transactions, v. 35, p. 333–344.

Buffler, R. T., Shaub, F. J., Watkins, J. S., and Worzel, J. L., 1979, Anatomy of the Mexican Ridges, southwestern Gulf of Mexico, *in* Watkins, J. S., Montadert, L., and Dickerson, P. W., eds., Geological and geophysical investigations of continental margins: American Association of Petroleum Geologists Memoir 29, p. 319–327.

Burbach, G. V., Frohlich, C., Pennington, W. D., and Matumoto, T., 1984,

Seismicity and tectonics of the subducted Cocos plate: Journal of Geophysical Research, v. 89, p. 7719–7735.

Burke, K., Cooper, C., Dewey, J. F., Mann, P., and Pindell, J. L., 1984, Caribbean tectonics and relative plate motion, *in* Bonini, W. E., Hargraves, R. B., and Shagam, R., eds., The Caribbean–South American plate boundary and regional tectonics: Geological Society of America Memoir 162, p. 31–63.

Burkhart, B., Deaton, B. C., Dengo, G., and Moreno, G., 1987, Tectonic wedges and offset Laramide structures along the Polochic fault of Guatemala and Chiapas, Mexico: Tectonics, v. 6, p. 411–422.

Campa, M. F., 1978, La evolución tectónica de Tierra Caliente, Guerrero: Sociedad Geológica Mexicana Boletín, v. 39, p. 52–64.

Campa-Uranga, M. F., and Coney, P. J., 1983, Tectono-stratigraphic terranes and mineral resources distributions of México: Canadian Journal of Earth Sciences, v. 20, p. 1040–1051.

Cantagrel, J. M., and Robin, C., 1979, K-Ar dating on eastern Mexican volcanic rocks—Relations between the andesitic and alkaline provinces: Journal of Volcanology and Geothermal Research, v. 5, p. 99–114.

Cantú-Chapa, C. M., Sandoval-Silva, R., and Arenas-Partida, R., 1985, Evolución sedimentaria del Cretácico Inferior en el norte de México: Revista del Instituto Mexicano del Petróleo, v. 17, p. 14–37.

Carfantan, J. C., 1983, Les ensembles géologiques du Mexique meridional. Evolution géodynamique durant le Mésozoïque et le Cénozoïque: Geofísica Internacional, v. 22, p. 9–37.

Carr, M. D., Christiansen, R. L., and Poole, F. G., 1984, Pre-Cenozoic geology of the El Paso Mountains, southwestern Great Basin, California: A summary, *in* Lintz, J., Jr., ed., Western geological excursions, Volume 4: Reno, Nevada, Mackey School of Mines, p. 84–93.

Carrillo-Martínez, M., 1971, Geología de la Hoja San José de Gracia, Sinaloa (Tesis Profesional): Ciudad de México, Universidad Nacional Autónoma de México, Faculty Ingeniería, 154 p.

Centeno, E., Ortega, F., and Corona, R., 1990, Oaxaca fault—Cenozoic reactivation of the suture between the Zapoteco and Cuicateco terranes, southern Mexico: Geological Society of America Abstracts with Programs, v. 22, no. 3, p. 13.

Clark, K. F., Foster, C. T., and Damon, P. E., 1982, Cenozoic mineral deposits and subduction-related magmatic arcs in México: Geological Society of America Bulletin, v. 93, p. 533–544.

Cohen, K. K., Anderson, T. H., and Schmidt, V. A., 1986, A paleomagnetic test of the proposed Mojave-Sonora megashear in northwestern México: Tectonophysics, v. 131, p. 23–51.

Coney, P. J., and Campa-Uranga, M. F., 1987, Lithotectonic terrane map of Mexico (west of the 91st meridian): U.S. Geological Survey Miscellaneous Field Studies Map MF-1874-D, scale 1:2,500,000.

Coney, P. J., and Reynolds, S. J., 1977, Cordilleran Benioff zones: Nature, v. 270, p. 403–406.

Corbitt, L. L., 1984, Tectonics of fold and thrust belt of northwestern Chihuahua, *in* Geology and petroleum potential of Chihuahua, México (Field trip guidebook): West Texas Geological Society Publication 84-80, p. 174–180.

Curray, J. R., and Moore, D. G., 1984, Geologic history of the Gulf of California, *in* Crouch, J. K., and Bachman, S. B., eds., Tectonics and sedimentation along the California margin: Los Angeles, Pacific Section, Society of Economic Paleontologists and Mineralogists, p. 17–35.

Damon, P. E., Shafiqullah, M., and Clark, K. F., 1981, Age trends of igneous activity in relation to metallogenesis in the southern Cordillera, *in* Dickinson, W. R., and Payne, W. D., eds., Relations of tectonics to ore deposits in the southern Cordillera: Arizona Geological Society Digest, v. 14, p. 137–154.

Damon, P. E., Shafiqullah, M., and Roldán, J., 1984, The Cordilleran Jurassic arc from Chiapas (southern Mexico) to Arizona: Geological Society of America Abstracts with Programs, v. 16, no. 6, p. 482.

Davis, G. H., 1979, Laramide faulting and folding in southeastern Arizona: American Journal of Science, v. 279, p. 543–569.

de Cserna, Z., 1965, Reconocimiento geológico de la Sierra Madre del Sur de México, entre Chilpancingo y Acapulco, Estado de Guerrero: Boletín del Instituto de Geología, Universidad Nacional Autónoma de México, v. 62, p. 1–77.

——, 1981, Margen continental de colisión activo en la parte suroccidental del Golfo de México: Universidad Nacional Autónoma de México, Instituto de Geología, Revista, v. 5, p. 255–261.

——, 1982, Hoja Tejupilco 14Q-g(9), con resumen de la geología de la Hoja Tejupilco, Estados de Guerrero, México y Michoacán: Instituto de Geología, Universidad Nacional Autónoma de México, Carta Geológica de México, serie de 1:100,000.

——, 1989, An outline of the geology of México, *in* Bally, A. W., and Palmer, A. R., eds., The geology of North America—An overview: Boulder, Colorado, Geological Society of America, The Geology of North America, v. A, p. 233–264.

DeJong, K. A., Escárcega-Escárcega, J. A., and Damon, P. E., 1988, Eastward thrusting, southwestward folding, and westward backsliding in the Sierra La Víbora, Sonora, México: Geology, v. 16, p. 904–907.

DeJong, K. A., García y Barragán, J. C., Damon, P. E., Miranda, M., Jacques-Ayala, C., and Almazán-Vázquez, E., 1990, Untangling the tectonic knot of Mesozoic Cordilleran orogeny in northern Sonora, NW México: Geological Society of America Abstracts with Programs, v. 22, p. A327.

Delgado, L. A., Rubinovich, R., and Gasca, A., 1986, Descripción preliminar de la geología y mecánica de emplazamiento del complejo ultrabásico del Cretácico de Loma Baya, Guerrero, México: Geofísica International, v. 25, p. 537–558.

DeMets, C., and Stein, S., 1990, Present-day kinematics of the Rivera plate and implications for tectonics of southwestern Mexico: Journal of Geophysical Research, v. 95, p. 21,931–21,948.

DeMets, C., Gordon, R. G., Argus, D. F., and Stein, S., 1990, Current plate motions: Geophysical Journal International, v. 101, p. 425–478.

Denison, R. E., Kenny, G. S., Burke, W. H., Jr., and Hetherington, E. A., Jr., 1969, Isotopic ages of igneous and metamorphic boulders from the Haymond Formation (Pennsylvanian), Marathon basin, Texas, and their significance: Geological Society of America Bulletin, v. 80, p. 245–256.

Díaz, T., and Navarro-G., A., 1964, Lithology and stratigraphic correlation of the upper Paleozoic in the region of Palomas, Chihuahua, *in* Geology of Mina Plomosas–Placer de Guadalupe area, Chihuahua, Mexico (Field trip guidebook): Midland, West Texas Geological Society, Publication 64-50, p. 65–84.

Dickinson, W. R., 1981, Plate tectonic evolution of the southern Cordillera: Arizona Geological Society Digest, v. 14, p. 1–35.

Dickinson, W. R., Klute, M. A., and Swift, P. N., 1986, The Bisbee Basin and its bearing on late Mesozoic paleogeographic and paleotectonic relations between the Cordilleran and Caribbean regions, *in* Abbott, P. L., ed., Cretaceous stratigraphy of western North America: Los Angeles, Pacific Section, Society of Economic Paleontologists and Mineralogists, p. 51–62.

Dilles, J. H., and Wright, J. E., 1988, The chronology of early Mesozoic arc magmatism in the Yerington district of western Nevada and its regional implications: Geological Society of America Bulletin, v. 100, p. 644–652.

Donnelly, T. W., 1989, Geologic history of the Caribbean and Central America, *in* Bally, A. W., and Palmer, A. R., eds., The geology of North America—An overview: Boulder, Colorado, Geological Society of America, The Geology of North America, v. A, p. 299–321.

Donnelly, T. W., Melson, W., Kay, R., and Rogers, J. W., 1973, Basalts and dolerites of late Cretaceous age from the central Caribbean, *in* Edger, N. T., and Saunders, J. B., eds., Initial Reports of the Deep Sea Drilling project, v. 15: Washington, D.C., U.S. Government Printing Office, p. 989–1012.

Drewes, H., 1991, Description and development of the Cordilleran orogenic belt in the southwestern United States and northern Mexico: U.S. Geological Survey Professional Paper 1512, p. 92.

Dunbar, J. A., and Sawyer, D. S., 1987, Implications of continental extension for plate reconstruction: An example from the Gulf of Mexico: Tectonics, v. 6, p. 739–755.

Dyer, R., 1986, Precambrian and Paleozoic rocks of Sierra el Carrizalillo, Chihuahua, México—A preliminary report: Geological Society of America Abstracts with Programs, v. 18, p. 353.

Dyer, R., Chávez-Quirarte, R., and Guthrie, R. S., 1988, Cordilleran orogenic belt of northern Chihuahua, México: American Association of Petroleum Geologists Bulletin, v. 72, p. 99.

Edgar, N. T., Ewing, J. I., and Hennion, J., 1971, Seismic refraction and reflection in the Caribbean Sea: American Association of Petroleum Geologists Bulletin, v. 55, p. 833–870.

Elías, H., 1987, Metamorphic geology of the Tierra Caliente Complex, Tejupilco region, State of Mexico: Geological Society of America Abstracts with Programs, v. 19, p. 654.

——, 1989, Geología metamórfica del área de San Lucas de Maíz, Estado de México: Universidad Nacional Autónoma de México, Instituto de Geología, Boletín, v. 105, 79 p.

Enciso de la Vega, S., 1968, Hoja Cuencamé 13R-l(7), con resumen de la geología de la Hoja Cuencamé, Estado de Durango: Instituto de Geología, Universidad Nacional Autónoma de México, Carta Geológica de México, serie de 1:100,000.

Engebretson, D. C., Cox, A., and Gordon, R. G., 1985, Relative motions between oceanic and continental plates in the Pacific basin: Geological Society of America Special Paper 206, 59 p.

Enos, P., 1983, Late Mesozoic paleogeography of México, *in* Reynolds, M. W., and Dolly, E. D., eds., Mesozoic paleogeography of the west-central United States (Rocky Mountain Paleogeography Symposium 2): Denver, Colorado, Rocky Mountain Section, Society of Economic Paleontologists and Mineralogists, p. 133–157.

Ferrusquía-Villafranca, I., and Comas-Rodríguez, O., 1988, Reptiles marinos mesozoicos en el sureste de México y su significación geológico-paleontológica entre Sierra de Gamón y Laguna de Santiaguillo, Estado de Durango: Instituto de Geología, Universidad Nacional Autónoma de México, Revista, v. 7, p. 136–147.

Filmer, P. E., and Kirschvink, J. L., 1989, A paleomagnetic constraint on the Late Cretaceous paleoposition of northwestern Baja California, México: Journal of Geophysical Research, v. 94, p. 7332–7342.

Flawn, P. T., and Maxwell, R. A., 1958, Metamorphic rocks in the Sierra del Carmen, Coahuila, México: American Association of Petroleum Geologists Bulletin, v. 42, p. 2245–2249.

Flawn, P. T., Goldstein, A., Jr., King, P. B., and Weaver, C. E., 1961, The Ouachitan System: Austin, University of Texas Publication 6120, 401 p.

Flynn, J. J., Cipolletti, R. M., and Novacek, M. J., 1989, Chronology of early Eocene marine and terrestrial strata, Baja California, México: Geological Society of America Bulletin, v. 101, p. 1182–1196.

Freeland, G. L., and Dietz, R. S., 1972, Plate tectonic evolution of the Caribbean–Gulf of Mexico region: Nature, v. 232, p. 20–23.

Gastil, R. G., and Jensky, W. A., II, 1973, Evidence for strike-slip displacement beneath the Trans-Mexican volcanic belt, *in* Kovach, R. L., and Nur, A., eds., Proceedings of Conference on Tectonic Problems of San Andreas Fault System: Stanford University Publications in Geological Sciences, v. 13, p. 171–180.

Gastil, R. G., Phillips, R. P., and Allison, E. C., 1975, Reconnaissance geology of the state of Baja California: Geological Society of America Memoir 140, 170 p.

Gastil, R. G., Morgan, G. J., and Krummenacher, D., 1981, The tectonic history of peninsular California and adjacent México, *in* Ernst, W. G., ed., The geotectonic development of California (Rubey Volume I): Englewood Cliffs, New Jersey, Prentice-Hall, p. 284–306.

Gastil, R. G., Diamond, J., and Knaack, C., 1986a, The magnetite ilmenite line in peninsular California: Geological Society of America Abstracts with Programs, v. 18, p. 109.

Gastil, R. G., Miller, R. H., and Campa-Uranga, M., 1986b, The Cretaceous paleogeography of peninsular California and adjacent México, *in* Abbott, P. L., ed., Cretaceous stratigraphy of western North America: Los Angeles, Pacific Section, Society of Economic Paleontologists and Mineralogists, p. 41–50.

Gastil, R. G., and 10 others, 1991, The relation between the Paleozoic strata on opposite sides of the Gulf of California, *in* Jacques, C., and Pérez, E., eds., Studies of Sonoran Geology: Geological Society of America Special Paper 254, p. 7–18.

Goetz, C. W., Girty, G. H., and Gastil, R. G., 1988, East over west ductile thrusting along a terrane boundary in the Peninsular Ranges: Rancho El Rosarito, Baja California, México: Geological Society of America Abstracts with Programs, v. 20, no. 3, p. 164.

Gomberg, D. N., Banks, P. D., and McBirney, A. R., 1968, Preliminary zircon ages from the Central Cordillera: Science, v. 162, p. 121–122.

González-León, C., 1986, Estratigrafía del Paleozoico de la Sierra del Tule, noreste de Sonora: Instituto de Geología, Universidad Nacional Autónoma de México, Revista, p. 117–135.

Gose, W. A., 1985, Paleomagnetic results from Honduras and their bearing on Caribbean tectonics: Tectonics, v. 4, p. 565–585.

Gose, W. A., Belcher, R. C., and Scott, G. R., 1982, Paleomagnetic results from northeastern México: Evidence for large Mesozoic rotations: Geology, v. 10, p. 50–54.

Greenwood, E., Kottlowski, F. E., and Thompson, S., III, 1977, Petroleum potential and stratigraphy of Pedregosa Basin: Comparison with Permian and Orogrande basins: American Association of Petroleum Geologists Bulletin, v. 61, p. 1448–1469.

Griffith, R. C., 1987, Geology of the southern Sierra Calamajué area; structural and stratigraphic evidence for latest Albian compression along a terrane boundary, Baja California, México [M.S. thesis]: San Diego, California, San Diego State University, 115 p.

Griffith, R. C., and Goetz, C. W., 1987, Structural and geochronological evidence for mid-Cretaceous compressional tectonics along a terrane boundary in the Peninsular Ranges: Geological Society of America Abstracts with Programs, v. 19, p. 384.

Gromet, L. P., and Silver, L. T., 1987, REE variations across the Peninsular Ranges batholith: Implications for batholithic petrogenesis and crustal growth in magmatic arcs: Journal of Petrology, v. 28, p. 75–125.

Guerrero, J. C., Herrero-Bervera, E., and Helsley, C. E., 1990, Paleomagnetic evidence for post-Jurassic stability of southeastern Mexico: Maya terrane: Journal of Geophysical Research, v. 95, p. 7091–7100.

Guzmán, E. J., and de Cserna, Z., 1963, Tectonic history of Mexico: American Association of Petroleum Geologists Memoir 2, p. 113–129.

Guzmán-Speziale, M., Pennington, W., and Matumoto, T., 1989, The triple junction of the North America, Cocos, and Caribbean plates: Seismicity and tectonics: Tectonics, v. 8, p. 981–997.

Hagstrum, J. T., and Sedlock, R., 1990, Remagnetization and northward translation of Mesozoic red chert from Cedros Island and the San Benito Islands, Baja California, Mexico: Geological Society of America Bulletin, v. 102, p. 983–991.

Hagstrum, J. T., McWilliams, M., Howell, D. G., and Gromme, C. S., 1985, Mesozoic paleomagnetism and northward translation of the Baja California peninsula: Geological Society of America Bulletin, v. 96, p. 221–225.

Hagstrum, J. T., Sawlan, M. G., Hausback, B. P., Smith, J. G., and Gromme, C. S., 1987, Miocene paleomagnetism and tectonic setting of the Baja California peninsula, México: Journal of Geophysical Research, v. 92, p. 2627–2640.

Hamilton, W., and Myers, W. B., 1966, Cenozoic tectonics of the western United States: Reviews of Geophysics, v. 4, p. 509–549.

Handschy, J. W., and Dyer, R., 1987, Polyphase deformation in Sierra del Cuervo, Chihuahua, Mexico—Evidence for ancestral Rocky Mountain tectonics in the Ouachita foreland of northern Mexico: Geological Society of America Bulletin, v. 99, p. 618–632.

Handschy, J. W., Keller, G. R., and Smith, K. J., 1987, The Ouachita system in northern México: Tectonics, v. 6, p. 323–330.

Hausback, B. P., 1984, Cenozoic volcanic and tectonic evolution of Baja California Sur, México, *in* Frizzell, V. A., Jr., ed., Geology of the Baja California Peninsula, Volume 39: Los Angeles, Pacific Section, Society of Economic Paleontologists and Mineralogists, p. 219–236.

Hill, R. I., Silver, L. T., and Taylor, H. P., Jr., 1986, Coupled Sr-O isotope variations as an indicator of source heterogeneity for the northern Peninsular

Ranges batholith: Contributions to Mineralogy and Petrology, v. 92, p. 351–361.

Hoffman, P. F., 1989, Precambrian geology and tectonic history of North America, *in* Bally, A. W., and Palmer, A. R., eds., The geology of North America—An overview: Boulder, Colorado, Geological Society of America, The Geology of North America, v. A, p. 447–512.

Holcombe, T. L., Vogt, P. R., Matthews, J. E., and Murchison, R. R., 1973, Evidence for seafloor spreading in the Cayman Trough: Earth and Planetary Science Letters, v. 20, p. 357–371.

Howell, D. G., and 7 others, 1985, C-3, Pacific abyssal plain to the Rio Grande Rift: Boulder, Colorado, Geological Society of America, Continent/Ocean transect no. 5, 3 sheets with text, scale 1:500,000.

Imlay, R. W., 1936, Evolution of the Coahuila Peninsula, Part IV, Geology of the western part of the Sierra de Parras: Geological Society of America Bulletin, v. 47, p. 1091–1152.

INEGI, 1980, Carta geológica de México: Instituto Nacional de Estadística, Geografía e Informática, scale 1:1,000,000, 8 sheets.

James, E. W., and Henry, C. D., 1990, Accreted Paleozoic terrane in Trans-Pecos Texas: Isotopic mapping of the buried Ouachita front: Geological Society of America Abstracts with Programs, v. 22, p. A329.

Johnson, C. A., 1987, Regional tectonics in central México: Active rifting and transtension within the Mexican volcanic belt: Eos (Transactions, American Geophysical Union), v. 68, p. 423.

Keller, G. R., and Cebull, S. E., 1973, Plate tectonics and the Ouachita System in Texas, Oklahoma and Arkansas: Geological Society of America Bulletin, v. 83, p. 1659–1666.

Keller, G. R., Braile, L. W., McMechan, G. A., Thomas, W. A., Harder, S. H., Chang, W. F., and Jardine, W. G., 1989, Paleozoic continent-ocean transition in the Ouachita Mountains imaged from PASSCAL wide-angle seismic reflection-refraction data: Geology, v. 17, p. 119–122.

Ketner, K. B., 1986, Eureka Quartzite in México?—Tectonic implications: Geology, v. 14, p. 1027–1030.

King, P. B., 1975, The Ouachita and Appalachian orogenic belts, *in* Nairn, A.E.M., and Stehli, F. G., eds., The ocean basins and margins: Volume 3: The Gulf of Mexico and the Caribbean: New York, Plenum Press, p. 201–241.

Kistler, R. W., 1974, Phanerozoic batholiths in western North America: Annual Review of Earth and Planetary Sciences, v. 2, p. 403–418.

—— , 1990, Two different lithosphere types in the Sierra Nevada, California, *in* Anderson, J. L., ed., The nature and origin of cordilleran magmatism: Geological Society of America Memoir 174, p. 271–281.

Kleist, R., Hall, S. A., and Evans, I., 1984, A paleomagnetic study of the Lower Cretaceous Cupido Formation, northeast México: Evidence for local rotation within the Sierra Madre Oriental: Geological Society of America Bulletin, v. 95, p. 55–60.

Klitgord, K. D., and Schouten, H., 1987, Plate kinematics of the Central Atlantic, *in* Tucholke, B. E., and Vogt, P. R., eds., The western Atlantic region: Boulder, Colorado, Geological Society of America, The Geology of North America, v. M, p. 351–378.

Kroonenberg, S. B., 1982, A Grenvillian granulitic belt in the Colombian Andes and its relation to the Guiana Shield: Geologie en Mijnbouw, v. 61, p. 325–333.

Ladd, J. W., 1976, Relative motion of South America with respect to North America and Caribbean tectonics: Geological Society of America Bulletin, v. 87, p. 969–976.

Ladd, J. W., and Buffler, R. T., eds., 1985, Middle America Trench off western Central America: Woods Hole, Massachusetts, Marine Science International, Ocean Basins and Margins Drilling Program Atlas Series, No. 7, 12 p.

Ledezma-Guerrero, O., 1967, Hoja Parras 13R-l(6), con resumen de la geología de la Hoja Parras, Estados de Coahuila, Durango y Zacatecas: Instituto de Geología, Universidad Nacional Autónoma de México, Carta Geológica de México, serie de 1:100,000.

LeFevre, L. V., and McNally, K. C., 1985, Stress distribution and subduction of aseismic ridges in the Middle America subduction zone: Journal of Geophysical Research, v. 90, p. 4495–4510.

Lillie, R. J., Nelson, K. D., de Voogd, B., Brewer, J., Oliver, J. E., Brown, L. D., Kaufman, S., and Viele, G. W., 1983, Crustal structure of the Ouachita Mountains, Arkansas; a model based on integration of COCORP reflection profiles and regional geophysical data: American Association of Petroleum Geologists Bulletin, v. 67, p. 907–931.

Londsdale, P., 1989, Geology and tectonic history of the Gulf of California, *in* Winterer, E. L., Hussong, D. M., and Decker, R. W., eds., The eastern Pacific region: Boulder, Colorado, Geological Society of America, The Geology of North America, v. N, p. 499–521.

López-Infanzón, M., 1986, Petrología y radiometría de rocas ígneas y metamórficas de México: Asociación Mexicana de Geólogos Petroleros Boletín, v. 38, p. 59–98.

López-Ramos, E., 1985, Geología de México, Tomo II: Mexico City, edición 3a (Privately printed), 453 p.

Lovejoy, E.M.P., ed., 1980, Sierra de Juárez, Chihuahua, Mexico: Structure and stratigraphy: El Paso, Texas, El Paso Geological Society, 59 p.

Luhr, J. F., Nelson, S. A., Allan, J. F., and Carmichael, I.S.E., 1985, Active rifting in southwestern Mexico: Manifestations of an incipient eastward spreading-ridge jump: Geology, v. 13, p. 54–57.

Lund, S. P., Bottjer, D. J., Whidden, K. J., and Powers, J. E., 1991a, Paleomagnetic evidence for the timing of accretion of the Peninsular Ranges terrane in southern California: Geological Society of America Abstracts with Programs, v. 23, no. 2, p. 74.

Lund, S. P., Bottjer, D. J., Whidden, K. J., Powers, J. E., and Steele, M. C., 1991b, Paleomagnetic evidence for Paleogene terrane displacements and accretion in southern California, *in* Abbott, P. L., and May, J. A., eds., Eocene geologic history, San Diego region, Volume 68: Los Angeles, Pacific Section, Society of Economic Paleontologists and Mineralogists, p. 99–106.

Malfait, B. T., and Dinkelman, M. G., 1972, Circum-Caribbean tectonic and igneous activity and the evolution of the Caribbean plate: Geological Society of America Bulletin, v. 83, p. 251–272.

Malpica, C. R., 1972, Rocas marinas del Paleozoico tardío en el área de San José de Gracia, Sinaloa: Sociedad Geológica Mexicana, II Convención Nacional, Libro de Resúmenes, p. 174–175.

McBirney, A. R., and Bass, M. N., 1969, Structural relations of pre-Mesozoic rocks of northern Central America, *in* Tectonic relations of northern Central America and the western Caribbean: American Association of Petroleum Geologists, Memoir 11, p. 269–280.

McDowell, F. W., and Keizer, R. P., 1977, Timing of mid-Tertiary volcanism in the Sierra Madre Occidental between Durango City and Mazatlán, México: Geological Society of America Bulletin, v. 88, p. 1479–1487.

McMenamin, M.A.S., and McMenamin, D.L.S., 1990, The emergence of animals; the Cambrian breakthrough: New York, Columbia University Press, 217 p.

Mitre-Salazar, L-M., and Roldán-Quintana, J., 1990, H-1 La Paz to Saltillo, Northwestern and Northern Mexico: Boulder, Colorado, Geological Society of America, Continent/Ocean Transect no. 14, 2 sheets with text, scale 1:500,000.

Molina-Garza, R. S., Van der Voo, R., and Urrutia-Fucugauchi, J., 1990, Paleomagnetic data from Chiapas, southern México: Implications for tectonic evolution of the Gulf of Mexico: Eos (Transactions, American Geophysical Union), v. 71, p. 491.

Morris, L. K., Lund, S. P., and Bottjer, D. J., 1986, Paleolatitude drift history of displaced terranes in southern and Baja California: Nature, v. 321, p. 844–847.

Mossman, R. W., and Viniegra, F., 1976, Complex fault structures in Veracruz province of México: American Association of Petroleum Geologists Bulletin, v. 60, p. 379–388.

Mullan, H. S., 1978, Evolution of the Nevadan orogen in northwestern Mexico: Geological Society of America Bulletin, v. 89, p. 1175–1188.

Nairn, A.E.M., 1976, A paleomagnetic study of certain Mesozoic formations in northern México: Physics of the Earth and Planetary Interiors, v. 13, p. 47–56.

Navarro, A., and Tovar, J., 1974, Stratigraphy and tectonics of the state of Chihuahua, *in* Geologic Field Trip Guidebook through the states of Chihuahua and Sonora: West Texas Geological Society Publication 74-63, p. 87–91.

Nicholas, R. L., and Rosendal, R. A., 1975, Subsurface positive elements within the Ouachita footbelt in Texas and their relationship to the Paleozoic cratonic margin: American Association of Petroleum Geologists Bulletin, v. 59, p. 193–216.

Nimz, G. J., Cameron, K. L., Cameron, M., and Morris, J., 1986, Petrology of the lower crust and upper mantle beneath southeastern Chihuahua: Geofísica Internacional, v. 25, p. 85–116.

Nixon, G. T., 1982, The relationship between Quaternary volcanism in central México and the seismicity and structure of subducted oceanic lithosphere: Geological Society of America Bulletin, v. 93, p. 514–523.

Ortega, F., 1979, The tectonothermic evolution of the Paleozoic Acatlán Complex of southern Mexico: Geological Society of America Abstracts with Programs, v. 11, p. 490.

—— , 1981a, Metamorphic belts of southern Mexico and their tectonic significance: Geofísica Internacional, v. 20, p. 177–202.

—— , 1981b, La evolución tectónica premisisípica del sur de México: Instituto de Geología, Universidad Nacional Autónoma de México, Revista, v. 5, p. 140–157.

—— , 1982, Evolución magmática y metamórfica del complejo cristalino de La Paz, Baja California Sur: Sociedad Geológica Mexicana, VI Convención Nacional, Libro de Resúmenes, p. 90.

—— , 1984, Evidence of Precambrian evaporites in the Oaxacan granulite complex of southern Mexico: Precambrian Research, v. 23, p. 377–393.

—— , 1990, H-3, Acapulco Trench to the Gulf of Mexico across southern Mexico: Boulder, Colorado, Geological Society of America, Continent/Ocean Transect no. 13, one sheet with text, scale 1:500,000.

Ortega, F., and González, C., 1985, Una edad cretácica de las rocas sedimentarias deformadas de la Sierra de Juárez, Oaxaca: Universidad Nacional Autónoma de México, Instituto de Geología, Revista, v. 6, p. 100–101.

Ortega, F., and Urrutia, J., 1989, Paleogeography and tectonics of the Mixteca terrane, southern Mexico, during the interval of drifting between North and South America and their Gulf of Mexico rifting (abstract): EOS, Transactions of the American Geophysical Union, v. 70, p. 1314.

Ortega, F., Prieto, R., Zúñiga, Y., and Flores, S., 1979, Una secuencia volcánico-plutónico-sedimentaria cretácica en el norte de Sinaloa—un complejo ofiolítico?: Universidad Nacional Autónoma de México, Instituto de Geología, Revista, v. 3, p. 1–8.

Padilla y Sánchez, R. J., 1985, Las estructuras de la Curvatura de Monterrey, Estados de Coahuila, Nuevo León, Zacatecas y San Luis Potosí: Instituto de Geología, Universidad Nacional Autónoma de México, Revista, v. 6, p. 1–20.

—— , 1986, Post-Paleozoic tectonics of northeast Mexico and its role in the evolution of the Gulf of Mexico: Geofísica Internacional, v. 25, p. 157–206.

Palmer, A. R., DeMis, W. D., Muehlberger, W. R., and Robison, R. A., 1984, Geologic implications of Middle Cambrian boulders from the Haymond Formation (Pennsylvanian) in the Marathon basin, west Texas: Geology, v. 12, p. 91–94.

Pantoja-Alor, J., 1963, Hoja San Pedro del Gallo 13R-k(3), con resumen de la geología de la Hoja San Pedro del Gallo, Estado de Durango: Carta Geológica de México, serie de 1:100,000.

Pantoja-Alor, J., and Robison, R., 1967, Paleozoic sedimentary rocks in Oaxaca, México: Science, v. 157, p. 1033–1035.

Patchett, P. J., and Ruiz, J., 1987, Nd isotopic ages of crust formation and metamorphism in the Precambrian of eastern and southern México: Contributions to Mineralogy and Petrology, v. 96, p. 523–528.

Pindell, J., 1985, Alleghenian reconstruction and subsequent evolution of the Gulf of Mexico, Bahamas, and proto-Caribbean: Tectonics, v. 4, p. 1–39.

Pindell, J. L., and Barrett, S. F., 1990, Geological evolution of the Caribbean region; a plate-tectonic perspective, *in* Dengo, G., and Case, J. E., eds., The Caribbean region: Boulder, Colorado, Geological Society of America, The Geology of North America, v. H, p. 405–432.

Pindell, J. L. and Dewey, J. F., 1982, Permo-Triassic reconstruction of western Pangea and the evolution of the Gulf of Mexico/Caribbean region: Tectonics, v. 1, p. 179–212.

Pindell, J. L., Cande, S. C., Pitman, W. C., III, Rowley, D. B., Dewey, J. F., Labrecque, J., and Haxby, W., 1988, A plate-kinematic framework for models of Caribbean evolution: Tectonophysics, v. 155, p. 121–138.

Poole, F. G., and Madrid, R. J., 1986, Paleozoic rocks in Sonora (México) and their relation to the southwestern continental margin of North America: Geological Society of America Abstracts with Programs, v. 18, p. 720–721.

Poole, F. G., Murchey, B. L., and Jones, D. L., 1983, Bedded barite deposits of middle and late Paleozoic age in central Sonora, México: Geological Society of America Abstracts with Programs, v. 15, p. 299.

Pubellier, M., and Rangin, C., 1987, Mise en évidence d'une phase cénomano-turonienne en Sonora central (Mexique)—Conséquences et domaine téthysien: Comptes Rendus Academie Science Paris, Serie II, v. 305, p. 1093–1098.

Quintero-Legorreta, O., and Aranda-García, M., 1985, Relaciones estructurales entre el Anticlinorio de Parras y el Anticlinorio de Arteaga (Sierra Madre Oriental), en la región de Agua Nueva, Coahuila: Instituto de Geología, Universidad Nacional Autónoma de México, Revista, v. 6, p. 21–36.

Quintero-Legorreta, O., and Guerrero, J. C., 1985, Una nueva localidad del basamento Precámbrico de Chihuahua en el área de Carrizalillo: Instituto de Geología, Universidad Nacional Autónoma de México, Revista, v. 6, p. 98–99.

Radelli, L., and 12 others, 1987, Allochthonous Paleozoic bodies of central Sonora: Departamento de Geología, University of Sonora Boletín, p. 1–15.

Ramírez-M., J. C., and Acevedo-C., F., 1957, Notas sobre la geología de Chihuahua: Asociación Mexicana de Geólogos Petroleros Boletín, v. 9, p. 583–770.

Rangin, C., 1978, Speculative model of Mesozoic geodynamics, central Baja California to northwestern Sonora, Mexico, *in* Mesozoic Paleogeography of the western United States: Los Angeles, Pacific Section, Society of Economic Paleontologists and Mineralogists, p. 85–106.

Renne, P. R., and 6 others, 1989, $^{40}Ar/^{39}Ar$ and U-Pb evidence for Late Proterozoic (Grenville age) continental crust in north-central Cuba and regional tectonic implications: Precambrian Research, v. 42, p. 325–341.

Robinson, K. L., 1991, U-Pb zircon geochronology of basement terranes and the tectonic evolution of southwestern mainland Mexico [M.S. thesis]: San Diego State University, 190 p.

Rodríguez, J. L., 1990, Relaciones estructurales en la parte centroseptentrional del Estado de Sonora: Universidad Nacional Autónoma de México, Instituto de Geología, Revista, v. 9, p. 51–61.

Rodríguez-Castañeda, J. L., 1988, Estratigrafía de la región de Tuape, Sonora: Instituto de Geología, Universidad Nacional Autónoma de México, Revista, v. 7, p. 52–66.

Roldán-Quintana, J., 1982, Evolución tectónica del Estado de Sonora: Instituto de Geología, Universidad Nacional Autónoma de México, Revista, v. 5, p. 178–185.

Rosencrantz, E., Ross, M., and Sclater, J., 1988, Age and spreading history of the Cayman Trough as determined from depth, heat flow, and magnetic anomalies: Journal of Geophysical Research, v. 93, p. 2141–2157.

Ross, C. A., 1979, Late Paleozoic collision of North and South America: Geology, v. 7, p. 41–44.

Rowley, D. B., and Pindell, J. L., 1989, End-Paleozoic and early Mesozoic western Pangea reconstruction and its implications for the distributions of Precambrian and Paleozoic rocks around meso-America: Precambrian Research, v. 42, p. 411–444.

Ruiz, J., Patchett, P. J., and Arculus, R. J., 1988a, Nd-Sr isotopic composition of lower crustal xenoliths—Evidence for the origin of mid-Tertiary felsic volcanics in México: Contributions to Mineralogy and Petrology, v. 99, p. 36–43.

—— , 1988b, Reply *to* Comment *on* "Nd-Sr isotopic composition of lower crustal xenoliths—Evidence for the origin of mid-Tertiary felsic volcanics in México": Contributions to Mineralogy and Petrology, v. 104, p. 615–618.

Ruiz, J., Patchett, P. J., and Ortega-Gutiérrez, F., 1988c, Proterozoic and Phanerozoic basement terranes of México from Nd isotopic studies: Geological Society of America Bulletin, v. 100, p. 274–281.
Ruiz-Castellanos, M., 1979, Rubidium-strontium geochronology of the Oaxaca and Acatlán metamorphic areas of southern Mexico [Ph.D. thesis]: Dallas, University of Texas, 188 p.
Salvador, A., 1987, Late Triassic–Jurassic paleogeography and origin of Gulf of Mexico basin: American Association of Petroleum Geologists Bulletin, v. 71, p. 419–451.
Sawlan, M. G., 1991, Magmatic evolution of the Gulf of California rift: *in* Dauphin, J. P., and Simoneit, B.R.T., eds., The Gulf and Peninsular Province of the Californias: American Association of Petroleum Geologists Memoirs 47, p. 301–369.
Sawlan, M. G., and Smith, J. G., 1984, Petrologic characteristics, age and tectonic setting of Neogene volcanic rocks in northern Baja California Sur, México, *in* Frizzell, V. A., Jr., ed., Geology of the Baja California Peninsula, Volume 39: Los Angeles, Pacific Section, Society of Economic Paleontologists and Mineralogists, p. 237–251.
Schweikert, R. A., and Lahren, M. M., 1990, Speculative reconstruction of the Mojave–Snow Lake fault: Implications for Paleozoic and Mesozoic orogenesis in the western United States: Tectonics, v. 9, p. 1609–1629.
Scotese, C. R., and McKerrow, W. S., 1990, Revised world maps and introduction, *in* Mckerrow, W. S., and Scotese, C. R., eds., Paleozoic Paleogeography and Biogeography: Geological Society of London Memoir 12, p. 1–21.
Sedlock, R., and Hamilton, D. H., 1991, Late Cenozoic tectonic evolution of southwestern California: Journal of Geophysical Research, v. 96, p. 2325–2352.
Sedlock, R. L., Ortega-Gutiérrez, F., and Speed, R. C., 1993, Tectonostratigraphic terranes and tectonic evolution of Mexico: Geological Society of America Special Paper 278, 146 p.
Servais, M., Rojo-Yaniz, R., and Colorado-Liévano, D., 1982, Estudio de las rocas básicas y ultrabásicas de Sinaloa y Guanajuato: Postulación de un paleogolfo de Baja California y de una digitación Tethysiana en México central: Geomimet, v. 3, p. 53–71.
Servais, M., Cuevas-Pérez, E., and Monod, O., 1986, Une section de Sinaloa à San Luis Potosi: nouvelle approche de l'évolution du Mexique nord-occidental: Bulletin de la Société Géologique de France, ser. 8, v. 2, p. 1033–1047.
Shurbet, D. H., and Cebull, S. E., 1987, Tectonic interpretation of the westernmost part of the Ouachita-Marathon (Hercynian) orogenic belt, west Texas-Mexico: Geology, v. 15, p. 4358–4361.
Silver, L. T., and Anderson, T. H., 1974, Possible left-lateral early to middle Mesozoic disruption of the southwestern North American craton margin: Geological Society of America Abstracts with Programs, v. 6, p. 956.
—— , 1983, Further evidence and analysis of the role of the Mojave-Sonora megashear(s) in Mesozoic Cordilleran tectonics: Geological Society of America Abstracts with Programs, v. 15, p. 273.
Silver, L. T., Taylor, H. P., Jr., and Chappell, B., 1979, Peninsular Ranges Batholith, San Diego and Imperial Counties, California, Mesozoic crystalline rocks: San Diego State University, Geological Society of America, Cordilleran Section, Field Trip Guidebook, p. 83–110.
Speed, R. C., 1979, Collided Paleozoic microplate in the western United States: Journal of Geology, v. 87, p. 279–292.
—— , 1983, Evolution of the sialic margin in the central-western United States *in* Watkins, J. S., and Drake, C. L., eds., Studies in Continental Margin Geology: American Association of Petroleum Geologists Memoir 34, p. 457–470.
Spencer, J. E., and Normark, W. R., 1979, Tosco-Abreojos fault zone: A Neogene transform plate boundary within the Pacific margin of southern Baja California, México: Geology, v. 7, p. 554–557.
Stevens, C.H ., 1982, The Early Permian *Thysanophyllum* coral belt: Another clue to Permian plate-tectonic reconstructions: Geological Society of America Bulletin, v. 93, p. 798–803.
Stevens, C. H., Stone, P., and Kistler, R. W., 1992, A speculative reconstruction of the middle Paleozoic continental margin of southwestern North America: Tectonics, v. 11, p. 405–419.
Stewart, J. H., 1972, Initial deposits in the Cordilleran geosyncline: Evidence of a late Precambrian (<850 m.y.) continental separation: Geological Society of America Bulletin, v. 83, p. 1345–1360.
—— , 1976, Late Precambrian evolution of North America: Plate tectonics implications: Geology, v. 4, p. 11–15.
—— , 1988a, Tectonics of the Walker Lane belt, western Great Basin: Mesozoic and Cenozoic deformation in a zone of shear, *in* Ernst, W. G., ed., Metamorphism and crustal evolution of the western United States (Rubey Volume VII): Englewood Cliffs, New Jersey, Prentice-Hall, p. 683–713.
—— , 1988b, Latest Proterozoic and Paleozoic southern margin of North America and the accretion of Mexico: Geology, v. 16, p. 186–189.
—— , 1992, Late Proterozoic and Paleozoic southern margin of North America in northern Mexico, *in* Clark, K. F., Roldán-Quintana, J., and Schmidt, R. H., eds., Geology and mineral resources of northern Sierra Madre Occidental Province, Mexico: El Paso Geological Society, Guidebook for 1992 field conference, p. 291–300.
Stewart, J. H., and Roldán, J., 1991, Upper Triassic Barranca Group—nonmarine and shallow marine rift basin deposits of northwestern Mexico, *in* Jacques, C., and Pérez, E., eds., Studies of Sonoran Geology: Geological Society of America Special Paper 254, p. 19–36.
Stewart, J. H., and Poole, F. G., 1974, Lower Paleozoic and uppermost Precambrian miogeocline, Great Basin, western United States, *in* Dickinson, W. R., ed., Tectonics and sedimentation: Society of Economic Paleontologists and Mineralogists Special Publication 22, p. 28–57.
Stewart, J. H., McMenamin, M.A.S., and Morales-Ramírez, J. M., 1984, Upper Proterozoic and Cambrian rocks in the Caborca region, Sonora, México—Physical stratigraphy, biostratigraphy, paleocurrent studies, and regional relations: U.S. Geological Survey Professional Paper 1309, 36 p.
Stewart, J. H., Poole, F. G., Ketner, K. B., Madrid, R. J., Roldán-Quintana, J., and Amaya-Martínez, R., 1990, Tectonics and stratigraphy of the Paleozoic and Triassic southern margin of North America, Sonora, México, *in* Gehrels, G. E., and Spencer, J. E., eds., Geologic excursions through the Sonoran Desert region, Arizona and Sonora: Arizona Geological Survey Special Paper 7, p. 183–202.
Stock, J. M., and Hodges, K. V., 1989, Pre-Pliocene extension around the Gulf of California, and the transfer of Baja California to the Pacific plate: Tectonics, v. 8, p. 99–116.
Stock, J., and Molnar, P., 1988, Uncertainties and implications of the Late Cretaceous and Tertiary position of North America relative to the Farallón, Kula, and Pacific plates: Tectonics, v. 7, p. 1339–1384.
Suter, M., 1984, Cordilleran deformation along the eastern edge of the Valles–San Luis Potosí carbonate platform, Sierra Madre Oriental fold-thrust belt, east-central México: Geological Society of America Bulletin, v. 95, p. 1387–1397.
—— , 1987, Structural traverse across the Sierra Madre Oriental fold-thrust belt in east-central México: Geological Society of America Bulletin, v. 98, p. 249–264.
—— , 1991, State of stress and active deformation in Mexico and western Central America, *in* Slemmons, D. B., Engdahl, E. R., Zoback, M. D., and Blackwell, D. D., eds., Neotectonics of North America: Boulder, Geological Society of America, p. 401–421.
Suter, M., Quintero-Legarreta, O., and Johnson, C. A., 1992, Active faults and state of stress in the central part of the Trans-Mexican volcanic belt, Mexico; part 1, The Venta de Bravo fault: Journal of Geophysical Research, v. 97, no. B8, p. 11983–11993.
Sutter, J. F., 1979, Late Cretaceous collisional tectonics along the Motagua fault zone, Guatemala: Geological Society of America Abstracts with Programs, v. 11, no. 7, p. 525.
Sykes, L. R., McCann, W. R., and Kafka, A. L., 1982, Motion of Caribbean plate during last 7 million years and implications for earlier Cenozoic movements: Journal of Geophysical Research, v. 87, p. 10,656–10,676.
Thomas, W. A., 1977, Evolution of the Appalachian-Ouachita salients and recesses from reentrants and promontories in the continental margin: Ameri-

can Journal of Science, v. 277, p. 1233–1278.

—— , 1985, The Appalachian-Ouachitan connection; Paleozoic orogenic belt at the southern margin of North America: Annual Review of Earth and Planetary Sciences, v. 13, p. 175–199.

—— , 1989, The Appalachian-Ouachitan belt beneath the Gulf Coastal Plain between the outcrops in the Appalachian and Ouachitan Mountains, *in* Hatcher, R. D., Thomas, W. A., and Viele, G. W., eds., The Appalachian-Ouachitan orogen in the United States: Boulder, Colorado, Geological Society of America, The Geology of North America, v. F-2, p. 537–553.

Todd, B. J., and Keen, C. E., 1989, Temperature effects and their geological consequences at transform margins: Canadian Journal of Earth Sciences, v. 26, p. 2591–2603.

Todd, V. R., Erskine, B. G., and Morton, D. M., 1988, Metamorphic and tectonic evolution of the northern Peninsular Ranges batholith, southern California, *in* Ernst, W. G., ed., Metamorphism and crustal evolution of the western United States (Rubey Volume VII): Englewood Cliffs, New Jersey, Prentice Hall, p. 894–937.

Todd, V. R., Shaw, S. E., Girty, G. H., and Jachens, R. C., 1991, A probable Jurassic plutonic arc of continental affinity in the Peninsular Ranges batholith, southern California: Tectonic implications: Geological Society of America Abstracts with Programs, v. 23, no. 2, p. 104.

Tosdal, R. M., Haxel, G. B., Anderson, T. H., Connors, C. D., May, D. J., and Wright, J. E., 1990a, Highlights of Jurassic, Cretaceous to early Tertiary, and mid-Tertiary tectonics, south-central Arizona and north-central Sonora, *in* Gehrels, G. E., and Spencer, J. E., eds., Geologic excursions through the Sonoran Desert region, Arizona and Sonora: Arizona Geological Survey Special Paper 7, p. 76–89.

Tosdal, R. M., Haxel, G. B., and Wright, J. E., 1990b, Jurassic geology of the Sonoran Desert region, southern Arizona, southeastern California, and northernmost Sonora, *in* Jenny, J. P., and Reynolds, S. J., eds., Geologic evolution of Arizona: Arizona Geological Society Digest, v. 17, p. 397–434.

Urrutia, J., Morán, D. J., and Cabral, E., 1987, Paleomagnetism and tectonics of Mexico: Geofísica Internacional, v. 26, p. 429–458.

Urrutia-Fucugauchi, J., and Bohnel, H. N., 1987, Discussion *of* "Tectonic interpretation of the Trans-Mexican volcanic belt": Tectonophysics, v. 138, p. 319–323.

Van der Voo, R., Peinado, J., and Scotese, C. R., 1984, A paleomagnetic reevaluation of Pangea reconstructions, *in* Van der Voo, R., Scotese, C. R., and Bonhommet, N., eds., Plate reconstructions from Paleozoic paleomagnetism: American Geophysical Union Geodynamics Series, v. 12, p. 11–26.

Viele, G. W., 1979, Geological map and cross section, eastern Ouachitan Mountains, Arkansas; Map Summary: Geological Society of America Bulletin, v. 90, p. 1096–1099.

Viele, G. W., and Thomas, W. A., 1989, Tectonic synthesis of the Ouachita orogenic belt, *in* Hatcher, R. D., Thomas, W. A., and Viele, G. W., eds., The Appalachian-Ouachitan orogen in the United States: Boulder, Colorado, Geological Society of America, The Geology of North America, v. F-2, p. 695–728.

Viniegra-O., F., 1971, Age and evolution of salt basins of southeastern Mexico: American Association of Petroleum Geologists, Bulletin, v. 55, p. 478–494.

—— , 1981, Great carbonate bank of Yucatan, southern Mexico: Journal of Petroleum Geology, v. 3, p. 247–278.

Walker, J. D., 1988, Permian and Triassic rocks of the Mojave Desert and their implications for timing and mechanisms of continental truncation: Tectonics, v. 7, p. 685–709.

Walker, N. W., 1988, U-Pb zircon evidence for 1305–1231 Ma crust in the Llano uplift, central Texas: Geological Society of America Abstracts with Programs, v. 20, p. A205.

Walper, J. L., 1980, Tectonic evolution of the Gulf of Mexico, *in* Pilger, R. H., Jr., ed., The origin of the Gulf of Mexico and the early opening of the central north Atlantic Ocean: Baton Rouge, Louisiana State University, p. 87–98.

Watkins, J. S., McMillen, K. J., Bachman, S. B., Shipley, T. H., Moore, J. C., and Angevine, C., 1981, Tectonic synthesis, leg 66—Transect and vicinity and geodynamics: Washington, D.C., U.S. Government Printing Office, Initial reports of the Deep Sea Drilling project, v. 66, p. 837–849.

Wickham, J., Roeder, D. R., and Briggs, G., 1976, Plate tectonic models for the Ouachita folded belt: Geology, v. 4, p. 173–176.

Winker, C. D., and Buffler, R. T., 1988, Paleogeographic evolution of early deep-water Gulf of Mexico and margins, Jurassic to Middle Cretaceous (Comanchean): American Association of Petroleum Geologists Bulletin, v. 72, p. 318–346.

Woods, M. T., and Davies, G. F., 1982, Late Cretaceous genesis of the Kula plate: Earth and Planetary Science Letters, v. 58, p. 161–166.

Yáñez, P., Ruiz, J., Patchett, P. J., Ortega, F., and Gehrels, G., 1991, Isotopic studies of the Acatlán complex, southern Mexico—Implications for Paleozoic North American tectonics: Geological Society of America Bulletin, v. 103, p. 817–828.

Young, K., 1983, Mexico, The Phanerozoic geology of the world, II: The Mesozoic B: Amsterdam, Elsevier, p. 61–88.

Manuscript Accepted by the Society December 30, 1992

Printed in U.S.A.

DNAG Continent-Ocean Transect Volume
Phanerozoic Evolution of North American
Continent-Ocean Transitions
The Geological Society of America, 1994

Chapter 6

Synopsis of geologic history portrayed along Corridor C-3: Southern California borderland–Rio Grande rift

D. G. Howell and J. G. Vedder
U.S. Geological Survey, 345 Middlefield Road, Menlo Park, California 94025

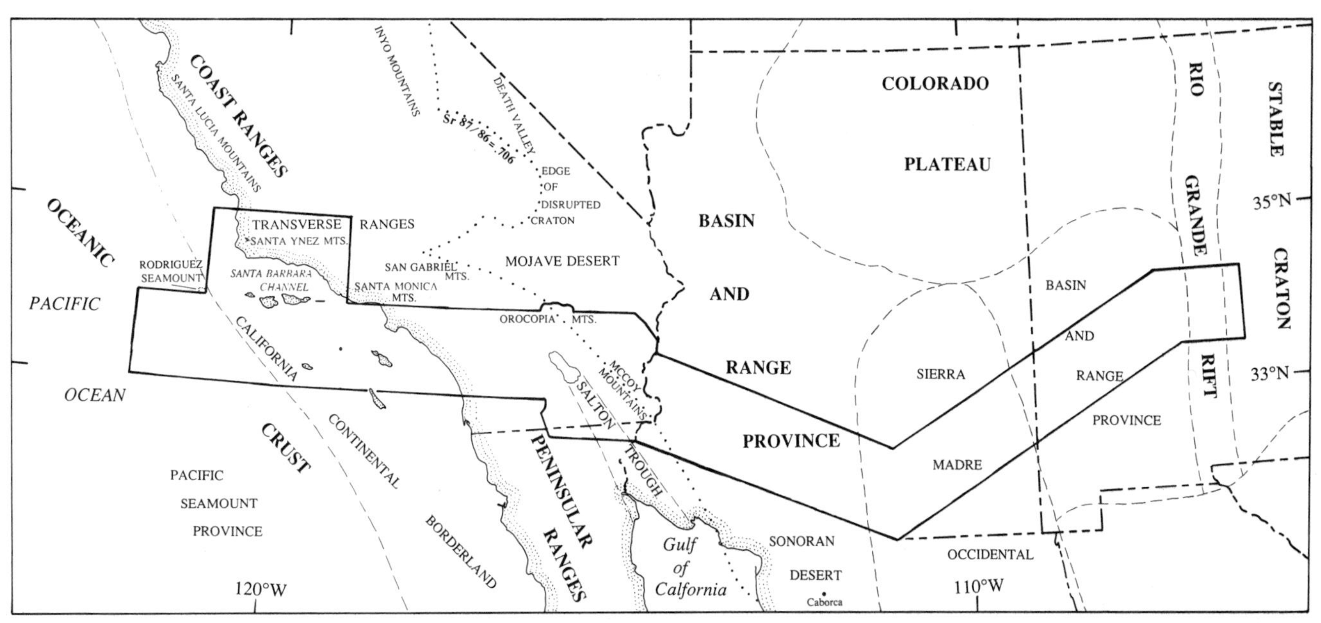

Figure 1. Location of the C-3 transect.

INTRODUCTION

Corridor C-3 (Fig. 1) traverses the transition from oceanic crust to cratonal rocks of North America along a 1,500-km transect from offshore southern California to central New Mexico (Howell and others, 1985; Gibson and others, 1985). The western end of the transect is offshore in an oceanic seamount province of the Pacific plate. Eastward, it crosses the southern California borderland and emerges onshore in the batholith-dominated Peninsular Ranges. C-3 spans the Salton trough, a transform rift floored by nascent oceanic crust, and the Basin and Range structural province, formed by extension and disruption of cratonal rocks. The eastern end of C-3 is on the stable craton.

Neogene tectonism has severely disrupted the older structural grain between the oceanic lithosphere and the craton. Much of this young tectonism is attributed to interactions along the plate edges after the spreading ridge system separating the Pacific and Farallon plates inpinged upon the North American plate in late Oligocene time. Two triple junctions—one migrating northward, the other southward relative to the Pacific plate—formed as the spreading system moved eastward relative to the North American plate. The San Andreas fault system began to develop within a broad zone of right shear as the triple junctions moved apart. The main strand of the San Andreas is but one of many northwest-trending right-lateral faults in the system. The boundary between the Pacific and North American plates is best perceived as a broad zone of right shear rather than a specific fault trace.

The continental crust of the western half of C-3 is composed

Howell, D. G., and Vedder, J. G., 1994, Synopsis of geologic history portrayed along Corridor C-3: Southern California borderland–Rio Grande rift, *in* Speed, R. C., ed., Phanerozoic Evolution of North American Continent-Ocean Transitions: Boulder, Colorado, Geological Society of America, DNAG Continent-Ocean Transect Volume.

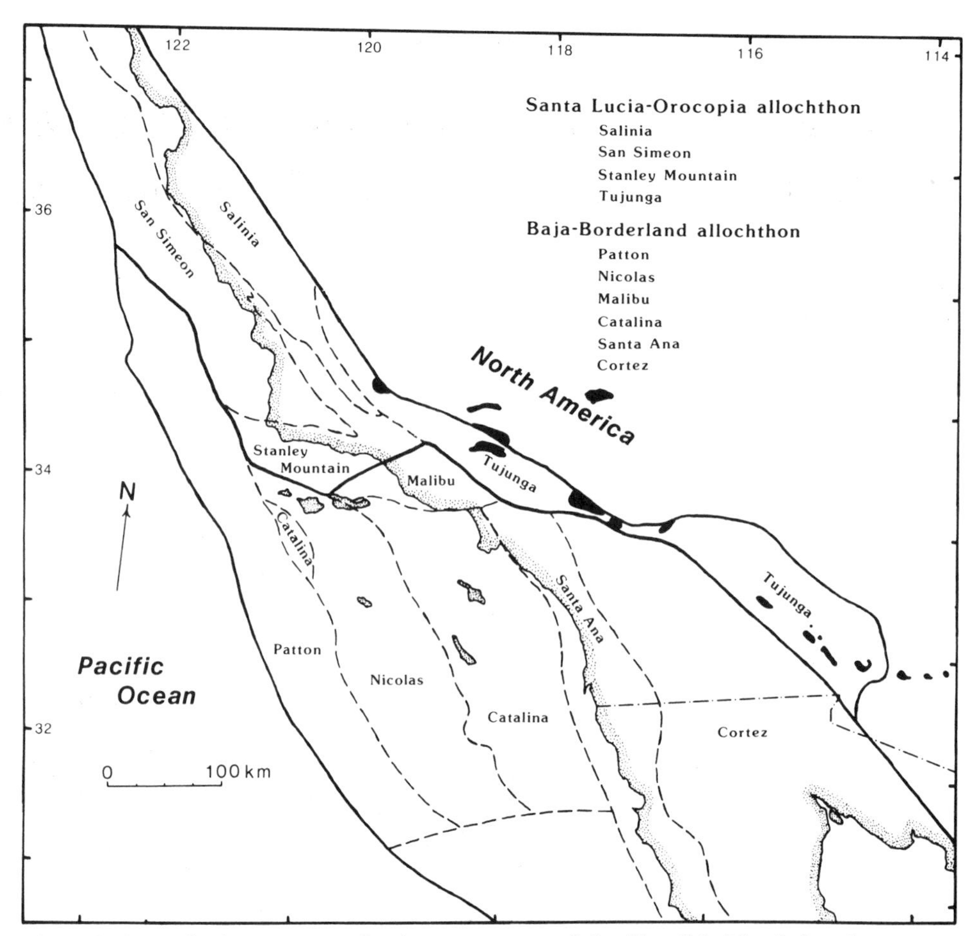

Figure 2. Generalized terrane map for the western part of Corridor C-3. The darkened areas are exposures of schistose rock composing the Baldy terrane; its overall distribution is problematic (see Fig. 5).

of displaced tectonostratigraphic terranes (Fig. 2). This chapter is framed in terms that conform to the definition of a tectonostratigraphic terrane as a fault-bounded entity that is characterized by a specific stratigraphy and a distinct structural history. The stratigraphy of overlapping sedimentary sequences, together with the distribution and emplacement ages of stitching plutons, chronicle the timing of terrane assembly (amalgamation), accretion, and dispersion. The inferred boundaries and rock sequences of each terrane are depicted on the Corridor C-3 of Howell and others (1985), and Figure 4 of this chapter depicts the general stratigraphy of each terrane and chronicles the amalgamation/accretion history among all the terranes. We should caution the reader that this tectonic scenario is still in the realm of hypothesis; alternate opinions abound in the literature, and we encourage interested students to explore these contrasting views.

TECTONIC HISTORY

Corridor C-3 crosses the North American continental margin between two orogenic belts, the Cordilleran to the north and the Ouachitan to the southeast. Each belt began as a rifted margin: the Cordilleran rift occurred approximately 550 Ma, and the Ouachitan about 525 Ma. After these rifting events, a spur-like remnant of Precambrian crust projected southwestward through parts of New Mexico, Arizona, and southeastern California. During the Paleozoic, this old crustal remnant was part of the cratonal platform, upon which a thin veneer of sediment was deposited.

Paleozoic tectonic events that affected regions adjoining Corridor C-3 include the Mississippian Antler orogeny, the Carboniferous and Permian Ouachitan orogeny, and the Triassic Sonoma orogeny. These orogenies apparently resulted in some crustal accretion to the craton, though evidence for accretion of allochthons during the Antler orogeny is equivocal. Tectonically induced sea-level fluctuations controlled Late Devonian sedimentation in southern Arizona and New Mexico. Intraplate deformation accompanying the Ouachitan orogeny created the ancestral Rocky Mountains and their associated depositional basins, such as the Pedrogosa basin. Although the Mojave Desert region was involved in a Permian orogenic disturbance (Davis and others, 1978), the effects of this event apparently did not extend into the Corridor C-3 area.

Transform faulting along the Mojave-Sonora megashear

Throughout Paleozoic and early Mesozoic time the area beneath the eastern end of Corridor C-3 was autochthonous craton interior to a passive margin and was unaffected by any kind of margin-related tectonism. The so-called miogeoclinal and eugeoclinal belts of the Cordillera in Nevada are absent in Corridor C-3. Their equivalent position in C-3 is taken by accreted terranes and batholiths of Mesozoic and Cenozoic age. These missing exterior regions of the passive margin may now lie to the south, in Mexico, because we now believe that at approximately 160 Ma, older Cordilleran structural and depositional trends were truncated in southeastern California, southwestern Arizona, and northern Sonora. Across this region, a northwest-trending shear zone—tens of kilometers wide at places—separates two different provinces of Precambrian strata. This shear zone, named the Mojave-Sonora megashear by Silver and Anderson (1974), is believed to be an ancient transform fault that connected nascent spreading centers in the Gulf of Mexico to a possible subduction zone bordering the California margin. Similarities between stratigraphic sequences in the Inyo Mountains–Death Valley region in California and the Caborca region in Mexico suggest as much as 700 to 800 km of cumulative left slip along the megashear. A Middle Jurassic age for slip along the megashear is presumed because all strata older than this are offset, and a Middle Jurassic arc seemingly stitches the suture in southern Arizona and Mexico; additionally, a Middle Jurassic age of remanent magnetization is inferred for the basal strata of the McCoy Mountains graben, a transtensional structure that may have formed along the megashear (Harding and others, 1983).

Late Jurassic deformation

In the western part of Corridor C-3, forearc-basin strata on Santa Cruz Island and in the Santa Monica Mountains underwent low-grade metamorphism at the same time that the late Jurassic Nevadan orogeny occurred. Part of the sequence in the Santa Monica Mountains contains late Oxfordian and/or Kimmeridgian fossils, and the schistose rocks on Santa Cruz Island are intruded by undeformed 140-Ma tonalite. By analogy with Late Jurassic ophiolitic and arc sequences in the Sierra Nevada and Klamath Mountains, the coeval strata in the Santa Monica Mountains may have been deposited on arc-related ophiolite, but the nature and significance of such an orogenic pulse remain mysterious.

Laramide Orogeny

North of Corridor C-3, the Cordilleran fold-and-thrust belt extends from Canada to the Mojave Desert (Burchfiel and Davis, 1975) and involves thick miogeoclinal strata. South of the transect, an east-directed fold-and-thrust belt, also in miogeoclinal strata, forms the Sierra Madre Oriental across the eastern continental margin of Mexico (Davis, 1979). Along the C-3 transect, the thick Paleozoic miogeoclinal strata are absent, and the aforementioned fold-and-thrust belt is seemingly interrupted across a 500-km-wide region.

Thrust faulting, greenschist-facies metamorphism, and peraluminous granitic intrusions occurred in southern Arizona and New Mexico during the latest Cretaceous and early Tertiary (Haxel and others, 1984). Thrust faults in this region involve crystalline Precambrian basement as well as Mesozoic plutonic rocks. Instead of thrusts following layer detachments, which characterize thrust belts to the north and south, shortening within C-3 occurred along reactivated Precambrian and Jurassic faults by shear strain and slip.

Terrane amalgamation and accretion

Within the limits of Corridor C-3, thirteen allochthonous terranes lie outboard of disrupted Precambrian cratonal North America in southwestern Arizona and southeastern California (Fig. 2). The terranes are assembled into two composite assemblages (Fig. 3), the Santa Lucia–Orocopia allochthon and the Baja-Borderland allochthon (Howell and others, 1987). The Santa Lucia–Orocopia allochthon is a five-terrane assemblage that moved northward along the western edge of North America in Late Cretaceous and early Tertiary time and accreted to the craton at approximately 55 Ma (Fig. 4). The Baja-Borderland allochthon is a six-part assemblage of terranes that moved northward since the Eocene and before middle Miocene time (Howell and others, 1987 (Fig. 5). Using the nomenclature of Blake and others (1982), the events that indicate the timing and nature of terrane amalgamation are briefly described below.

Cretaceous and Tertiary accretionary events

Salinia and Tujunga terranes. Cretaceous plutons in the Tujunga terrane are similar in Rb/Sr chemistry, petrology, and time of emplacement (105 to 80 Ma) to plutons in the Salinia terrane (Ehlig and Joseph, 1977; Silver, 1971). Felsic compositions predominate; granodiorite, tonalite, and quartz monzonite form the majority of the intrusive bodies. Isolated occurrences of the so-called polka-dot granite are distributed from the Orocopia Mountains to the La Panza Range (Joseph, 1981). From these relations, we infer that the Salinia and Tujunga terranes must have been amalgamated by middle Cretaceous time.

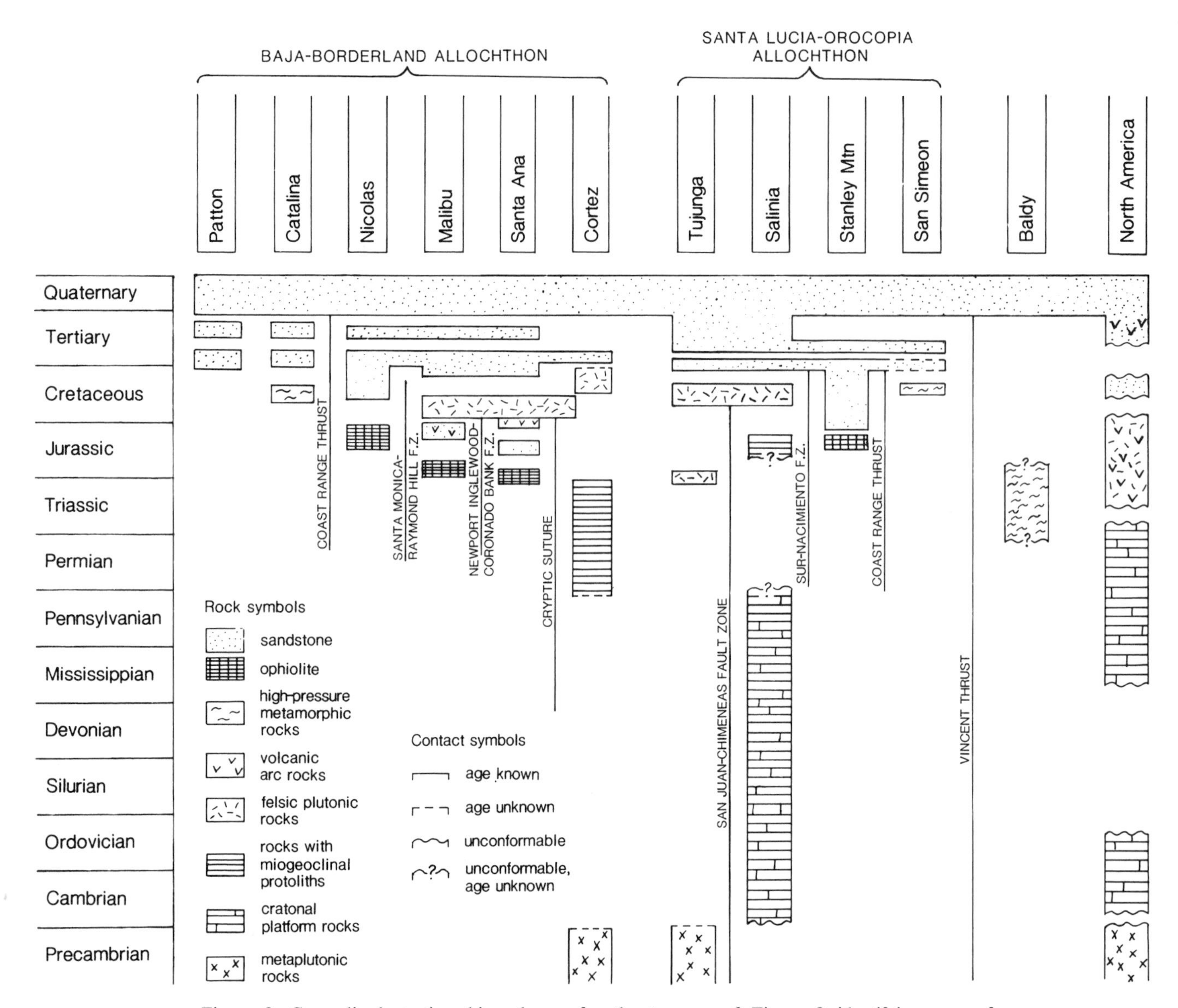

Figure 3. Generalized stratigraphic columns for the terranes of Figure 2 identifying ages of amalgamation and accretion as well as the major faults that separate the terranes. Note the packaging of terranes into two super terranes or composite allochthons.

Santa Lucia–Orocopia allochthon. Campanian and Maastrichtian nonmarine lithofacies are present in both the Stanley Mountain and Salinia terranes. These correlative strata have similar petrography and sediment dispersal patterns. Marine equivalents crop out in each terrane, and similar marine strata lie on the mélange of the San Simeon terrane (Fig. 4). No strata older than early Eocene, however, physically lap across these terranes. A Late Cretaceous and early Tertiary borderland-like setting is inferred for these terranes to explain the heterogeneity of lithofacies (Howell and Vedder, 1978; Nilsen, 1978). The borderland setting formed within a strike-slip system during the Late Cretaceous amalgamation that resulted in the terrane amalgam that we call the Santa Lucia–Orocopia allochthon.

Accretion of this allochthon is inferred to have occurred in the late Paleocene (about 55 Ma) along the Vincent-Orocopia thrust fault (Fig. 5). This emplacement was accompanied by prograde metamorphism of the underlying Baldy terrane (the Vincent-Orocopia-Pelona-Rand Schist) and retrograde metamorphism of the overlying Tujunga terrane (Haxel and Dillon, 1978). This allochthon, possibly together with the underlying Baldy ter-

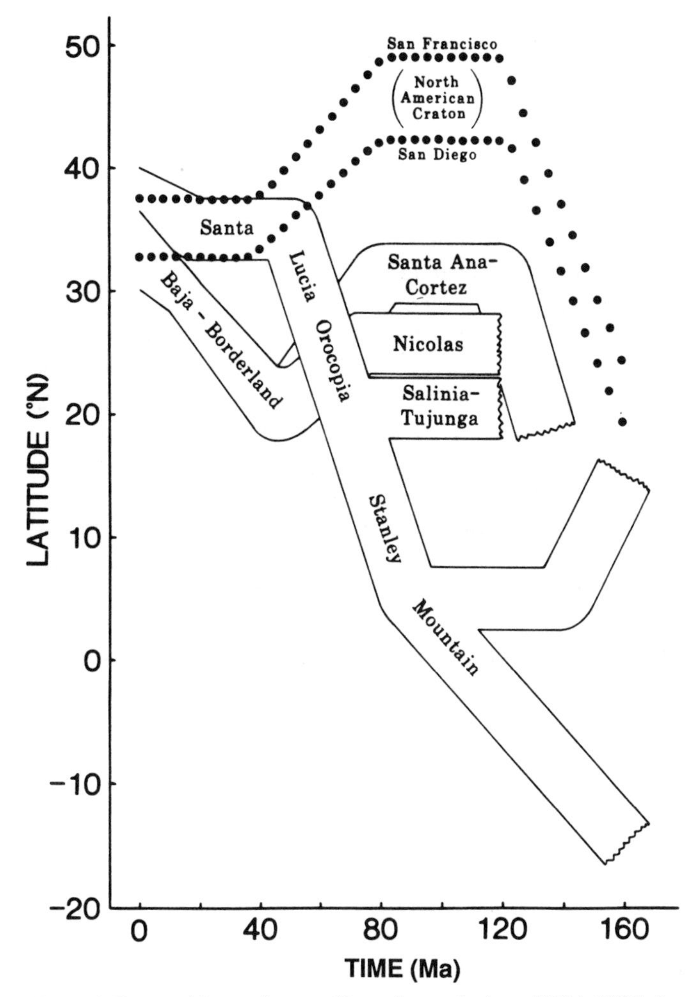

Figure 4. Terrane kinematics; see Champion and others (1984, 1986) for further details regarding the paleomagnetic data. Note that, with the exception of the ophiolitic basement and its chert carapace of the Stanley Mountain terrane and possibly some exotic components within the mélange of the San Simeon terranes, these terrane trajectories are along the west margin of continental North America.

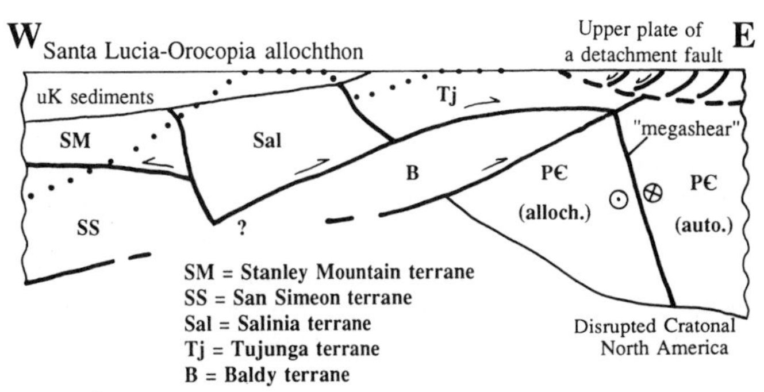

Figure 5. Schematic cross section depicting the structural relations among the terranes of the Santa Lucia–Orocopia allochthon and the tectonic relations involving the allochthon, Baldy terrane, and cratonal North America. The bold dotted line shows the level of erosion following accretion (ca. 55 Ma). Circled dot indicates relative southeastward displacement, circled "x" indicates relative northwestward displacement.

rane, was thrust over the leading edge of North American cratonal rocks and has since been sliced into northwest-trending slivers and distended to the north by the Neogene San Andreas fault system. The actual thrust surface that separates the allochthon from cratonal rocks has not been identified. We wonder if it is not concealed by the upper plate of a post-Miocene detachment fault (see Fig. 5).

The relations between the Baldy terrane and Santa Lucia–Orocopia allochthon prior to the late Paleocene accretion event are not known. The original site of deposition of the graywacke protolith of the Baldy terrane is also not known, but a portion of the protolith was derived from sialic crust. Thus, all the terranes lie along some part of western North America, but the intercalated chert and basalt of the Baldy terrane may indicate a depositional environment on oceanic crust. Miocene plutons that intrude the schistose rocks of the Baldy terrane have initial Sr 87/86 rations of 0.706 or higher, evidence suggesting that the Baldy terrane and the structurally higher Santa Lucia–Orocopia allochthon were already emplaced above the North American craton. A regional unconformity of late Paleocene to early Eocene age is present in these terranes as well as above the McCoy Mountains Formation. Development of this unconformity presumably accompanied emplacement of the allochthon. A phase of metamorphism of the Pelona-type schist, which forms the main constituent of the Baldy terrane, is thought to have been associated with this suturing event.

Baja-Borderland allochthon. Early Cretaceous plutons of the Peninsular Range batholith intruded dissimilar basement rocks (Silver and others, 1979) of the Malibu, Santa Ana, and Cortez terranes. The former two are volcanic arcs on oceanic basement, while the Cortez terrane has an inferred continental basement. The plutons from the western region of the batholith are part of a stationary magmatic arc that persisted from about 135 to 105 Ma. After 105 Ma, the locus of arc volcanism migrated systematically toward the east. Popenoe (1973) correlated overlapping Turonian strata in the Santa Monica Mountains of the Malibu terrane with similar strata in the Santa Ana Mountains of the Santa Ana terrane. Thus, amalgamation of these terranes by Late Cretaceous time is confirmed by pluton-stitching and overlapping sedimentary sequences.

Upper Cretaceous strata that apparently were deposited in a west-facing forearc setting cover large parts of the Nicolas, Malibu, and Santa Ana terranes. Compatible lithofacies trends suggest original continuity between sedimentary sequences of the Nicolas terrane forearc strata on ophiolitic basement, and coeval strata of the Malibu and Santa Ana terranes. A Late Cretaceous amalgamation of the Nicolas terrane with the already assembled Malibu–Santa Ana–Cortez terranes is therefore inferred. Paleomagnetic data (Howell and others, 1987) indicate that this terrane amalgam lay at least 1,000 km to the south in Late Cretaceous time. Because Miocene volcaniclastic strata overlie these terranes and occur in regions of Mexico east of the Baja Peninsula (Hausback, 1984), movement of the terrane assembly to a position west of the Gulf of California must have occurred by 25 Ma; strata of this age seem to lap onto the peninsula and the Sierra Madre Occidental of mainland Mexico (Hausback, 1984).

Accretionary events of uncertain age

Rodriguez terrane. The Rodriguez terrane, a basaltic seamount, was accreted to the continental borderland of southern California during the early Neogene. This terrane separated from the Pacific plate during the transition from subduction to transform regimes in the mid-Tertiary. Faults that bound the Rodriguez terrane probably have been reactivated since the terrane accreted. The seamount is not dated, but the adjacent oceanic crust is Miocene.

Patton terrane. Low-grade metasedimentary and metaultramafic rocks form the known basement of the Patton terrane. Rock types include altered basalt, laumontitic graywacke and argillite, serpentinite, pyroxenite, and albite-epidote rock (Howell and Vedder, 1981). These rocks are similar to the Franciscan complex of the Coastal belt of northern California (Blake and Jones, 1981) and may represent subduction-related accretion along the interface of the Farallon and North American plates.

Catalina terrane. Blueschist and related high-pressure metamorphic rocks constitute much of the Catalina terrane. Blueschist-facies metagraywacke and metavolcanic rocks occur on Santa Catalina Island, in the Palos Verdes Hills, and on a number of borderland banks and ridges (Howell and Vedder, 1981). On Santa Catalina Island, blueschist-facies rocks are structurally overlain by glaucophane-bearing greenschist that is, in turn, overlain by amphibolite and serpentinite (Platt, 1975). The age of metamorphism of the schistose rocks probably is middle Cretaceous (ca. 100 Ma; Suppe and Armstrong, 1972).

Blueschist detritus first appears in overlapping strata on the adjoining Nicolas and Santa Ana terranes in early and middle Miocene time, respectively. The Catalina terrane crops out between the Nicolas and Santa Ana terranes, possibly reflecting some kind of complicated strike-slip faulting or, more likely, an extensional episode during which detachment faulting unroofed the Catalina terrane as lower plate core rocks (e.g., Platt, 1986).

TERRANE DISPERSION

Late Tertiary interaction of the Farallon-Pacific spreading ridge with the North American margin, together with initiation of the San Andreas transform and simultaneous rifting, displaced parts of all the terranes crossed by the transect. Normal and strike-slip displacements occurred along numerous faults from offshore southern California to central New Mexico, disrupting a 1,400-km-wide belt of the North American continent and continental margin. Regions of extension include (1) California Continental Borderland; (2) Salton Trough/San Andreas system; (3) southern Basin and Range, including metamorphic core complexes; and (4) Rio Grande rift.

During the late Cenozoic, a continental borderland developed just east of the Pacific plate across the terranes that earlier accreted to North America. Northward translation of the ancestral parts of the borderland began at approximately 25 Ma. Paleomagnetic and paleocurrent data indicate 90 degrees or more of clockwise rotation for the northwesternmost part of the borderland in the last 15 m.y. (Kamerling and Luyendyk, 1981).

The San Andreas fault is part of the system of structures that resulted from Pacific plate motion past the North American plate. The displacement history of the southern segment of the San Andreas fault presumably originated as recently as 5 Ma with the opening of the Gulf of California, but other fault segments in the system may have begun moving by 12 to 15 Ma (Atwater and Molnar, 1973). The kinematic history of the San Andreas fault has been discussed by many authors, including Hill and Dibblee (1953), Crowell (1962), Ehlig (1968), Powell (1981), and Matti and Frizzell (1986).

EXTENSIONAL FAULTING AND CORE COMPLEXES

Two styles of late Cenozoic extension characterize the region between the Salton Trough and the Rio Grande: (1) low-angle detachment surfaces and metamorphic core complexes, and (2) basin-and-range extension by high-angle normal faults. Basin-and-range extension and low-angle detachment faulting began about 30 to 25 Ma in the Sonoran Desert and was approximately synchronous with initiation of transform faulting along the western continental margin of North America. This phase of extension also was contemporaneous with the cessation of voluminous mid-Tertiary intra-arc, calc-alkaline, basaltic andesite eruptions as well as high-silica rhyolite flows and tuffs. These intra-arc events were followed by less extensive tholeiitic, alkaline-olivine basaltic magmatism. Whereas the principal phase of extension ceased by 18 Ma, transform faulting along the coast and continental rifting to the east along the Rio Grande have persisted to the present.

In the Basin and Range Province, north-trending horsts and grabens are interpreted to have formed as a result of fragmentation of a thin upper crustal slab over a ductilely extending lower crust. Fragmentation was accompanied by tilting produced by asymmetric graben development, by rotation of large blocks along downward-flattening faults, or by isostatic rotation of large buoyant blocks "floating" on a viscous lower crust. Basin-and-range faulting is similar to low-angle detachment faulting in that both models have (1) a ductile lower crust, (2) a 12- to 15-km-thick fragmented upper crust as well as strata rotated by movement along curviplanar faults, and (3) new high-angle faults that cut older faults rotated to low angles. The high-angle normal faults are more widely spaced in basin-and-range extension than in low-angle detachment faulting. Regional estimates of the amount of extension across the Basin and Range Province generally range from 10 to 50 percent. At places, extension has been more than 100 percent.

SUMMARY

The geodynamic history of the continental margin traversed by Corridor C-3 can be encapsulated into seven principal episodes or events. (1) During the latest Precambrian (ca. 570 Ma), rifting produced a passive continental margin that prevailed for the remainder of the Paleozoic and early Mesozoic. Rocks of late Precambrian to early Mesozoic ages in the New Mexico, Arizona, and southeastern California portions of C-3 reflect only platform facies of the passive margin. The continental margin facies were displaced to the southeast as a result of the second major event. (2) The mid-Jurassic Mojave-Sonora megashear most probably crossed Corridor C-3 in the region of eastern California. The left-slip fault system connected a spreading ridge in the Gulf of Mexico with a newly formed subduction zone somewhere off the west coast. This subduction introduced the third significant episode. (3) From ca. 140 to 105 Ma, arc volcanism and the construction of major batholiths characterized the coastal settings. (4) Beginning after 105 Ma, the loci of arc volcanism systematically migrated eastward. Although the 105 to 80 Ma sites remained relatively close to the older coastal settings, the 80 to 35 Ma loci advanced rapidly to the east, culminating with arc volcanism in New Mexico. The present latitudinal position of major portions of the batholiths, however, is the result of coastwise translation from south to north. (5) The Laramide orogeny (80 to 40 Ma), coincident with the eastward shift of volcanism, represented a variety of phenomena and is interpreted in different ways. The main events of significance for Corridor C-3 were the basement-deforming folding and thrusting that disrupted the region from eastern California to New Mexico and the accretion of composite terranes in the coastal region. The causes and effects of these events are not well understood. These compressional and collisional episodes were followed by the sixth episode. (6) From roughly 35 to 18 Ma, much of the region crossed by Corridor C-3 experienced crustal attenuation. Arc volcanism retrogressed to the coast, and crustal uplift is expressed by core-complexes, detachment, and block faulting that created the physiography of the Basin and Range. Pull-apart basins began to form in the coastal region. The locus of crustal stretching shifted for the seventh episode. (7) By 18 Ma, all basin-and-range extension ceased, with the exception of the Rio Grande rift. Along the coast, transform right-slip between the Pacific and North American plates dominated tectonism in that area. Both of these tectonic styles characterize the neotectonic setting of Corridor C-3.

REFERENCES CITED

Atwater, T., and Molnar, P., 1973, Relative motion of the Pacific and North American Plates deduced from sea-floor spreading in the Atlantic, Indian, and South Pacific Oceans: Stanford University Publications in Geological Science, v. 13, p. 136–148.

Blake, M. C., Jr., and Jones, D. L., 1981, The Franciscan Assemblage and related rocks in northern California; A reinterpretation, *in* Ernst, W. G., ed., The geotectonic development of California, Rubey v. 1: Englewood Cliffs, New Jersey, Prentice-Hall, p. 306–328.

Blake, M. C., Jr., Howell, D. G., and Jones, D. L., 1982, Preliminary tectonostratigraphic terrane map of California: U.S. Geological Survey Open-File Report 82-593, 10 p.

Burchfiel, B. C., and Davis, G. A., 1975, Nature and control of Cordilleran orogenesis, western United States; Extensions of an earlier synthesis: American Journal of Science, v. 272, p. 97–118.

Champion, D. E., Howell, D. G., and Gromme, C. S., 1984, Paleomagnetic and geologic data indicating 2,500 km of northward displacement for the Salinian and related terranes, California: Journal of Geophysical Research, v. 89, p. 7736–7752.

Champion, D. E., Howell, D. G., and Marshall, M., 1986, Paleomagnetism of Cretaceous and Eocene strata, San Miguel Island, California borderland and the northward translation of Baja California: Journal of Geophysical Research, v. 91, p. 11557–11570.

Crowell, J. C., 1962, Displacement along the San Andreas fault, California: Geological Society of America Special Paper 71, 61 p.

Davis, G. A., Monger, J.W.H., and Burchfiel, B. C., 1978, Mesozoic construction of the Cordilleran "collage," central British Columbia to central California, *in* Howell, D. G., and McDougall, K. A., eds., Mesozoic paleogeography of the western United States: Society of Economic Paleontologists and Mineralogists, Pacific Section, Pacific Coast Paleogeography Symposium 2, p. 1–32.

Davis, G. H., 1979, Laramide folding and faulting in southeastern Arizona: American Journal of Science, v. 297, p. 543–569.

Ehlig, P. L., 1968, Causes of distribution of Pelona, Rand, and Orocopia Schists along the San Andreas and Garlock faults, *in* Dickinson, W. R., and Grantz, A., eds., Proceedings of conference on geologic problems of San Andreas fault system: Stanford University Publications, Geological Sciences, v. 11, p. 294–306.

Ehlig, P. L., and Joseph, S. E., 1977, Polka Dot Granite and correlation of La Panza Quartz Monzonite with Cretaceous batholithic rocks north of Salton Trough, *in* Howell, D. G., Vedder, J. G., and McDougall, K. A., eds., Cretaceous Geology of the California Coast Ranges, west of the San Andreas fault: Society of Economic Paleontologists and Mineralogists, Pacific Section, Pacific Coast Paleogeography Field Guide 2, p. 91–96.

Gibson, J. D., Howell, D. G., Fuis, G. S., and Vedder, J. G., 1985, Pacific abyssal plain to the Rio Grande Rift: Geological Society of America, North America continent-ocean transect C–3 text, 23 p.

Harding, L. E., Butler, R. F., and Coney, P. J., 1983, Paleomagnetic evidence for Jurassic deformation of the McCoy Mountains Formation, southeastern California and southwestern Arizona: Earth and Planetary Science Letters, v. 62, p. 104–114.

Hausback, B. P., 1984, Cenozoic volcanic and tectonic evolution of Baja California Sur, Mexico, *in* Frizzell, V. A., Jr., ed., Geology of the Baja California Peninsula: Society of Economic Paleontologists and Mineralogists, Pacific Section, v. 39, p. 219–236.

Haxel, G. B., and Dillon, J., 1978, The Pelona-Orocopia Schist and Vincent–Chocolate Mountains thrust system, southern California, *in* Howell, D. G., and McDougall, K. A., eds., Mesozoic paleogeography of the western United

States: Society of Economic Paleontologists and Mineralogists, Pacific Section, Pacific Coast Paleogeography Symposium 2, p. 453–469.

Haxel, G. B., Tosdal, R. M., May, D. J., and Wright, J. E., 1984, Latest Cretaceous and early Tertiary orogenesis in south-central Arizona; Thrust faulting, regional metamorphism, and granitic plutonism: Geological Society of America Bulletin, v. 95, p. 631–653.

Hill, M. L., and Dibblee, T. W., Jr., 1953, San Andreas, Garlock, and Big Pine faults, California; A study of the character, history, and tectonic significance of their displacement: Geological Society of America Bulletin, v. 64, p. 443–458.

Howell, D. G., and Vedder, J. G., 1978, Late Cretaceous paleogeography of the Salinian block, California, *in* Howell, D. G., and McDougall, K. A., eds., Mesozoic paleogeography of the western United States: Society of Economic Paleontologists and Mineralogists, Pacific Section, Pacific Coast Paleogeography Symposium 2, p. 107–116.

—— , 1981, Structural implications of stratigraphic discontinuities across the southern California borderland, *in* Ernst, W. G., ed., The geotectonic development of California, Rubey v. 1: Englewood Cliffs, New Jersey, Prentice-Hall, p. 535–558.

Howell, D. G., Gibson, J. D., Fuis, G. S., Knapp, J. H., Haxel, G. B., Keller, B. R., Silver, L. T., and Vedder, J. G., 1985, C-3 Pacific abyssal plain to the Rio Grande rift: Geological Society of America, Centennial continent/ocean transect #5.

Howell, D. G., Champion, D. E., and Vedder, J. G., 1987, Terrane accretion, crustal kinematics, and basin evolution, southern California, *in* Ingersoll, R., and Ernst, G., eds., Cenozoic basin development of coastal California, Rubey v. 6: Englewood Cliffs, New Jersey, Prentice-Hall, p. 242–259.

Joseph, S. E., 1981, Isotopic correlation of the La Panza Range granitic rocks with similar rocks in the central and eastern Transverse Ranges [M.S. thesis]: Los Angeles, California State University, 77 p.

Kamerling, M. J., and Luyendyk, B. P., 1981, Paleomagnetism and tectonics of the islands of the southern California continental borderland [abs.]: EOS Transactions of the American Geophysical Union, v. 62, p. 855.

Matti, J. C., and Frizzell, V. A., 1986, Distinctive Triassic megaporphyritic monzogranite displaced 160 ± 10 km by the San Andreas fault, southern California; A new constraint for palinspastic reconstructions: Geological Society of America Abstracts with Programs, v. 18, no. 2, p. 154.

Nilsen, T. H., 1978, Late Cretaceous geology of California and the problem of the proto-San Andreas fault, *in* Howell, D. G., and McDougall, K. A., eds., Mesozoic paleogeography of the western United States: Society of Economic Paleontologists and Mineralogists, Pacific Section, Pacific Coast Paleogeography Symposium 2, p. 559–573.

Platt, J. P., 1975, Metamorphic and deformational processes in the Franciscan Complex, California; Some insights from the Catalina Schist terrane: Geological Society of America Bulletin, v. 86, p. 1337–1347.

—— , 1986, Dynamics of orogenic wedges and uplift of high-pressure metamorphic rocks: Geological Society of America Bulletin, v. 97, p. 1037–1053.

Popenoe, W. P., 1973, Southern California Cretaceous formations and faunas with special reference to the Simi Hills and Santa Monica Mountains, *in* Cretaceous stratigraphy of the Santa Monica Mountains and Simi Hills, Pacific Section, Society of Economic Paleontologists and Mineralogists, Fall Field Trip Guidebook.

Powell, R. E., 1981, Geology of the crystalline basement complex, eastern Transverse Ranges, southern California; Constraints on regional tectonic interpretation [Ph.D. thesis]: Pasadena, California Institute of Technology, 441 p.

Silver, L. T., 1971, Problems of crystalline rocks of the Transverse Ranges: Geological Society of America Abstracts with Programs, v. 3, no. 2, p. 193–194.

Silver, L. T., and Anderson, T. H., 1974, Possible left-lateral early to middle Mesozoic disruption of the southwestern North American craton margin: Geological Society of America Abstracts with Programs, v. 6, no. 7, p. 955–956.

Silver, L. T., Taylor, H. P., and Chappell, B., 1979, Some petrological, geochemical, and geochronological observations of the Peninsular Ranges batholith near the international border of the U.S.A. and Mexico, *in* Abbott, P. L., and Todd, V. R., eds., Mesozoic crystalline rocks, Peninsular Ranges Batholith and pegmatites, Point Sal Ophiolite: San Diego State University, Department of Geological Sciences, p. 83–110.

Suppe, J., and Armstrong, R. L., 1972, Potassium-argon dating of the Franciscan metamorphic rocks: American Journal of Science, v. 272, p. 217–233.

Manuscript Accepted by the Society November 18, 1987

NOTES ADDED IN PROOF

Generally, one hopes that the schedule of scientific publications stays ahead of the acquisition of new data and the consequent refinement of ideas. The latter, of course, is the normal progression of scientific thought. In this case, the concepts captured in 1986, but now scheduled for publication in 1994, need some comment in order to bring the reader up to date. The principal area of interest involves paleomagnetic data and how these data constrain the amount and timing of coast-wise translations of some of the terranes in southern California. Quite candidly, much of the new data are just as preliminary as the data referred to in this text. Newer results, however, bring into question large-scale, Paleocene displacements of Cretaceous and older parts of the Salinia and Tujunga terranes. Most readers of the literature must realize the suspect nature of both Salinia and Tujunga. Most of these problems remain, e.g., the fate of the "lost" western part of Salinia and the geodynamics of emplacement of Tujunga terrane over Baldy terrane, to name just two. But it now seems that all the bonafide exposures of Cretaceous granite in these two terranes resided somewhere in southern California or northern Mexico in Late Cretaceous time. Nonetheless, the putative Cretaceous allochthoneity is still valid for the remaining terranes of the Santa Lucia–Orocopia allochthon, including the Pigeon Point part of "Salinia," and all of the Baja-Borderland allochthon.

Printed in U.S.A.

DNAG Continent-Ocean Transect Volume
Phanerozoic Evolution of North American
Continent-Ocean Transitions
The Geological Society of America, 1994

Chapter 7

Tectonic evolution of the central U.S. Cordillera: A synthesis of the C1 and C2 Continent-Ocean Transects

J. B. Saleeby
Department of Geological and Planetary Sciences, California Institute of Technology, Pasadena, California 91125
R. C. Speed
Department of Geological Sciences, Northwestern University, Evanston, Illinois 60201
M. C. Blake
U.S. Geological Survey, 345 Middlefield Road, Menlo Park, California 94025

INTRODUCTION

The evolution of the North American continent and adjacent ocean basins in the central Cordillera of the western United States in Phanerozoic time was governed by three sequential tectonic regimes. The first included the creation of a passive margin during the latest Proterozoic to Early Cambrian (Stewart, 1976) and the removal of an unknown amount of sialic crust from the western margin of the continent. The second regime maintained a passive continental margin of western North America from Middle Cambrian to Triassic time, but permitted collisions of outboard terranes with the sialic margin in Mississippian and Permian-Triassic time (Speed, 1982; Dickinson and others, 1983). Since Triassic time, western North America, adjacent oceanic plates, and intervening microplates and other tectonic packets have existed in a regime of active margin tectonics (Hamilton, 1969; Coney and others, 1980; Saleeby, 1983; Saleeby and Busby-Spera, 1992) driven mainly by eastward subduction of oceanic lithosphere. This third and currently operating regime has been marked by diverse phenomena including subduction of oceanic lithosphere below the continent and phases of highly oblique convergence and suture-zone or intra-arc spreading, ridge-trench collision, growth of a continental arc, major foreland contraction and extension, and the accretion of displaced terranes to the sialic edge.

Corridors C1 and C2 of the Ocean-Continent Transect Program traverse all essential elements of the transition from Pacific plate oceanic crust to cratonal North America that have resulted from these three tectonic regimes. Figure 1 shows the locations of the corridors in relation to the major tectonic elements and Figure 2 shows the corridors in relation to regional morphotectonic domains. Corridor C1 begins offshore at the Mendocino triple junction; it crosses accretionary terranes of the California Coast Ranges, southern Klamath Mountains–northern Sierra Nevada and northwestern Nevada, traverses the deformed foreland, including the Sevier and Idaho-Wyoming foreland thrust belts, and ends in the central U.S. craton. Corridor C2 begins in the Pacific plate; it crosses east through an inactive Paleogene–early Neogene trench complex, the San Andreas fault, accretionary terranes similar to those of C1, and the deformed foreland, and ends in disrupted craton of the Colorado plateau. Subduction-related continental volcanic-plutonic arcs of the modern Cascades and Mesozoic Sierra Nevada are traversed by the two corridors as well as the modern Basin and Range extensional province and its precursor, the mid-Tertiary extensional belt. Corridor C1 also crosses the southernmost edge of the late Cenozoic high lava plains of southern Oregon-Idaho and northernmost Nevada-California.

The objective of this chapter is an overview of critical phenomena and products of Phanerozoic ocean-continent interactions in the regions traversed by transects C1 and C2. Several alternative interpretations to those advocated along various segments of the C1 and C2 displays are also reviewed. Systematic coverage of the transects and more extensive reference lists are provided with the graphic displays (Saleeby and others, 1986; Blake and others, 1987).

PRESENT PLATE REGIME AND NEOTECTONICS

The present regime of generally convergent plate tectonics between western North America and mainly oceanic plates to the west began in the Triassic at least as early as ~240 Ma. The record of this protracted convergent regime is held in the rocks and structures of the western continent, the relative and absolute plate motions of North America, and the structures and magnetic stripes of the Pacific, Juan De Fuca, and Cocos plates. The main

Saleeby, J. B., Speed, R. C., and Blake, M. C., 1994, Tectonic evolution of the central U.S. Cordillera: A synthesis of the C1 and C2 Continent-Ocean Transects, *in* Speed, R. C., ed., Phanerozoic Evolution of North American Continent-Ocean Transitions: Boulder, Colorado, Geological Society of America, DNAG Continent-Ocean Transect Volume.

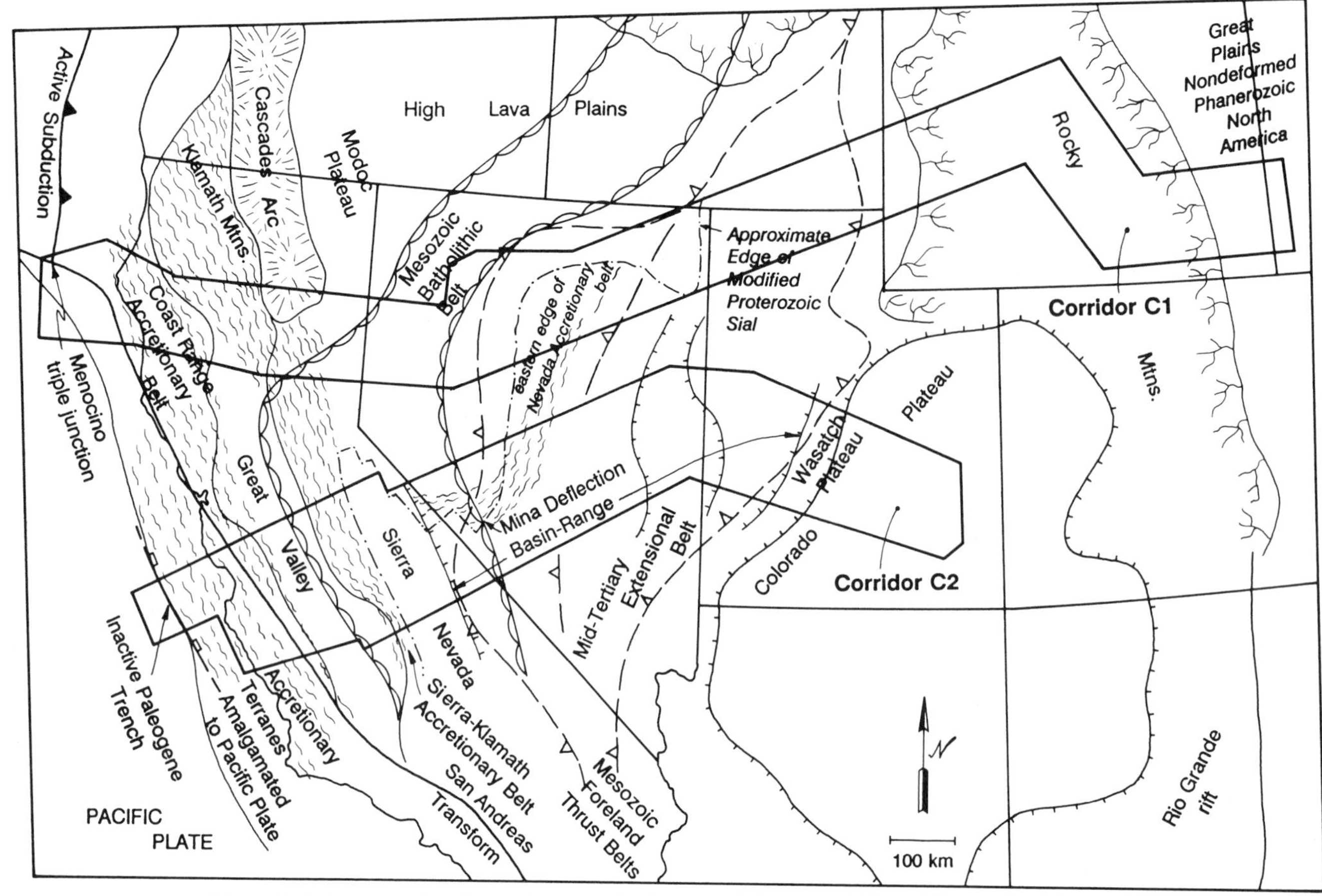

Figure 1. Index map showing major tectonic features of the central United States Cordillera and locations of corridors C1 and C2.

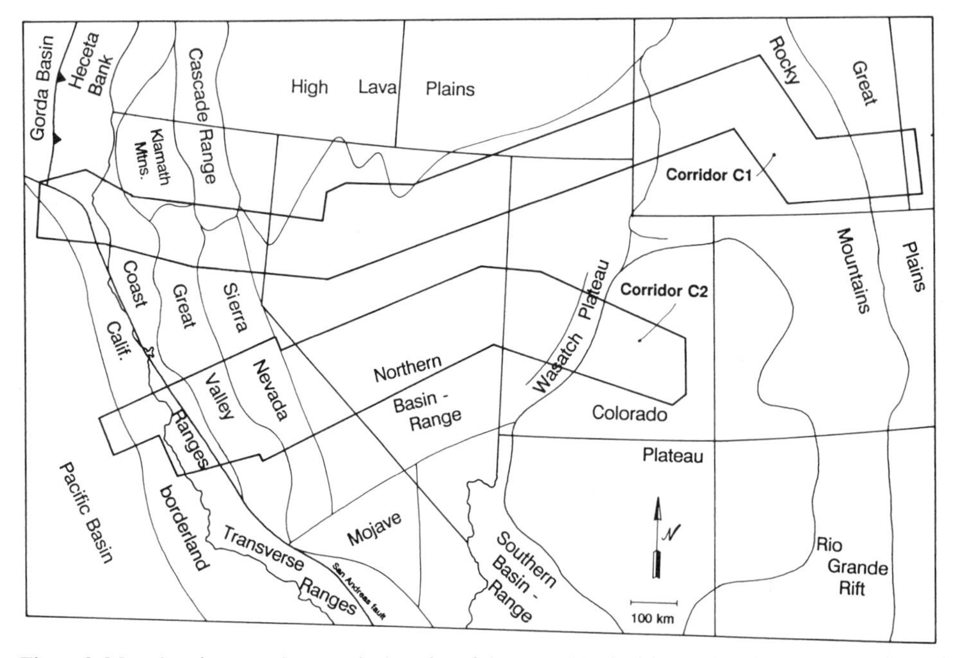

Figure 2. Map showing morphotectonic domains of the central United States Cordillera and locations of corridors C1 and C2.

alliances throughout this phase are the generally persistent closure between North America and adjacent oceanic lithospheres to the west (Coney, 1978; Engebretson and others, 1985) and the great width of western North America, over which deformational and thermal features of presumed convergent plate tectonic origin have been generated.

Today's active margin features within continental western United States occupy a 2000-km-wide zone between the Great Plains and oceanic lithosphere of the Pacific and Juan de Fuca plates. The main morphotectonic divisions are the Rocky Mountains, the Rio Grande rift system, the Colorado Plateau, the Basin and Range province, the Snake River Plain, the high lava plains, the Sierra Nevada, the Great Valley, the Klamath province, the Cascade arc and related sediment-filled subduction trench, including the Heceta bank, the Coast Ranges east of the San Andreas fault, the San Andreas plate boundary, the Mendocino triple junction, accretionary terranes west of the San Andreas fault joined to the Pacific plate, the continental slope and slope basins, and the abyssal crust of the Pacific plate.

Morphotectonic regions east of the Great Valley share the common characteristic only of high mean elevation (>1 km), and differ in other properties such as heat flow, style and intensity of neotectonic phenomena, and crustal structure. Moreover, the ages at which the regions began to develop in their present form differ: the Rocky Mountains, 40 Ma; Rio Grande rift, 30 Ma (Chapin, 1979); the Colorado Plateau and Sierra Nevada, 5–10 Ma (Christiansen, 1966; Hamblin, 1965; Thompson and Zoback, 1979); the high lava plains and the Snake River Plain, 12 Ma (Christiansen and McKee, 1978); and the Basin and Range province, between 5 and 10 Ma (Stewart, 1978).

The present plate juncture system consisting of the San Andreas transform, the Mendocino triple junction, and the Juan De Fuca–North America subduction zone began its evolution at ~30 Ma (Atwater, 1970). Disruption of an early Tertiary subduction zone, which bordered the entire California margin, resulted from the oblique collision of a segment of the actively spreading East Pacific Rise. The spreading rise produced the Farallon oceanic plate off its eastern flank and Pacific plate off its western flank. The progressive consumption of Rise segments beneath the California margin left only small remnants of the subducted Farallon oceanic plate along the northeast Pacific edge, including the Juan De Fuca Plate in C1. The present configuration of the Mendocino triple junction and the survival of the Juan De Fuca plate appear to have resulted from a large transform offset (Mendocino fracture zone) in the East Pacific Rise near its segments, which initially impacted the California margin. Consumption of the Rise segments south of the fracture zone resulted in the transform-trench-transform Mendocino triple junction. The triple junction is migrating northward at a rate of ~50 mm/yr with respect to North America, leaving in its wake the progressively lengthening San Andreas transform. Subduction of the Juan De Fuca plate beneath North America is also occurring at ~50 mm/yr and is strongly oblique with a northward dextral component. Transpression between the Pacific plate and North America just south of the triple junction since about 3 Ma accounts for active uplift and thrusting near Cape Mendocino, including obduction of the King Range terrane discussed below (Fig. 3) (McLaughlin and others, 1982). Northward triple junction migration corresponds roughly in time and latitude to the extinction of Cenozoic arc volcanism in the Sierra Nevada and southern California, reflecting the cessation of subduction-related mantle magma production.

Subduction of East Pacific Rise segments has juxtaposed Pacific plate oceanic crust against much of the California margin. In C2, a well-defined seismic reflector from the top of oceanic basement indicates that oceanic crust dips eastward beneath an inactive accretionary prism. The remnants of magnetic anomaly 7 (Atwater and Menard, 1970) trend north-south, intersecting the axial region of the inactive trench with an obliquity of ~45°. Anomaly 7 corresponds to ~26 Ma (La Brecque and others, 1977), the presumed age of Pacific crust last consumed along the C2 segment of the margin prior to establishment of the transform juncture. Marine seismic data show a normal thickness of oceanic crust; an upper mantle velocity anisotropy direction is consistent with generally east-west spreading along the consumed segments of the East Pacific Rise (Raitt and others, 1969; Shor and others, 1971).

In C1 the fingerprints of east-west ocean-floor spreading are shown by the north-northeast trends of the magnetic anomaly patterns (Raff and Mason, 1961) that generally parallel the Gorda Rise located just off the western edge of the corridor. Anomaly 3A (~5 Ma, La Brecque and others, 1977) is directly outboard of the subduction zone. The Juan De Fuca plate exhibits a departure from rigid plate tectonics by internal deformation, discordant strikes of the Mendocino and Blanco fracture zones, and convergence across the Mendocino fracture zone. Much of the Juan De Fuca plate and bounding fracture zone is heavily sedimented, and thus extensive basement sampling by dredging is not feasible. However, the Mendocino fracture zone is a major morphotectonic feature, having a ridge and trough bathymetric profile. First motion studies suggest right-lateral motion along the fracture and a small southward-underthrusting component (Bolt and others, 1968; Seeber and others, 1970). Reflection profiling across the fracture shows deformation and erosion of Pleistocene strata (Silver, 1971), that, in conjunction with the dredging of rounded basaltic cobbles from the fracture ridge, suggests emergence in the recent past (Krause and others, 1964). A system of steeply dipping faults striking parallel to the rift of the Gorda Rise is actively deforming the Juan De Fuca plate and its overlying sediments (Silver, 1971). Analysis of marine geological data and first motion studies (McEvilly, 1968; Tobin and Sykes, 1968) led Silver (1971) to conclude that this fault system has a predominant left-slip movement pattern. The young oceanic crust of the Juan De Fuca plate is thus being fragmented prior to its subduction.

Marine and onland seismic data and field mapping show landward-dipping thrust faults within the accretionary wedge above the Juan De Fuca plate. Offshore to the south of the triple junction, where transform faulting of the San Andreas is now the dominant tectonic process, there is a thick blanket of flat-lying,

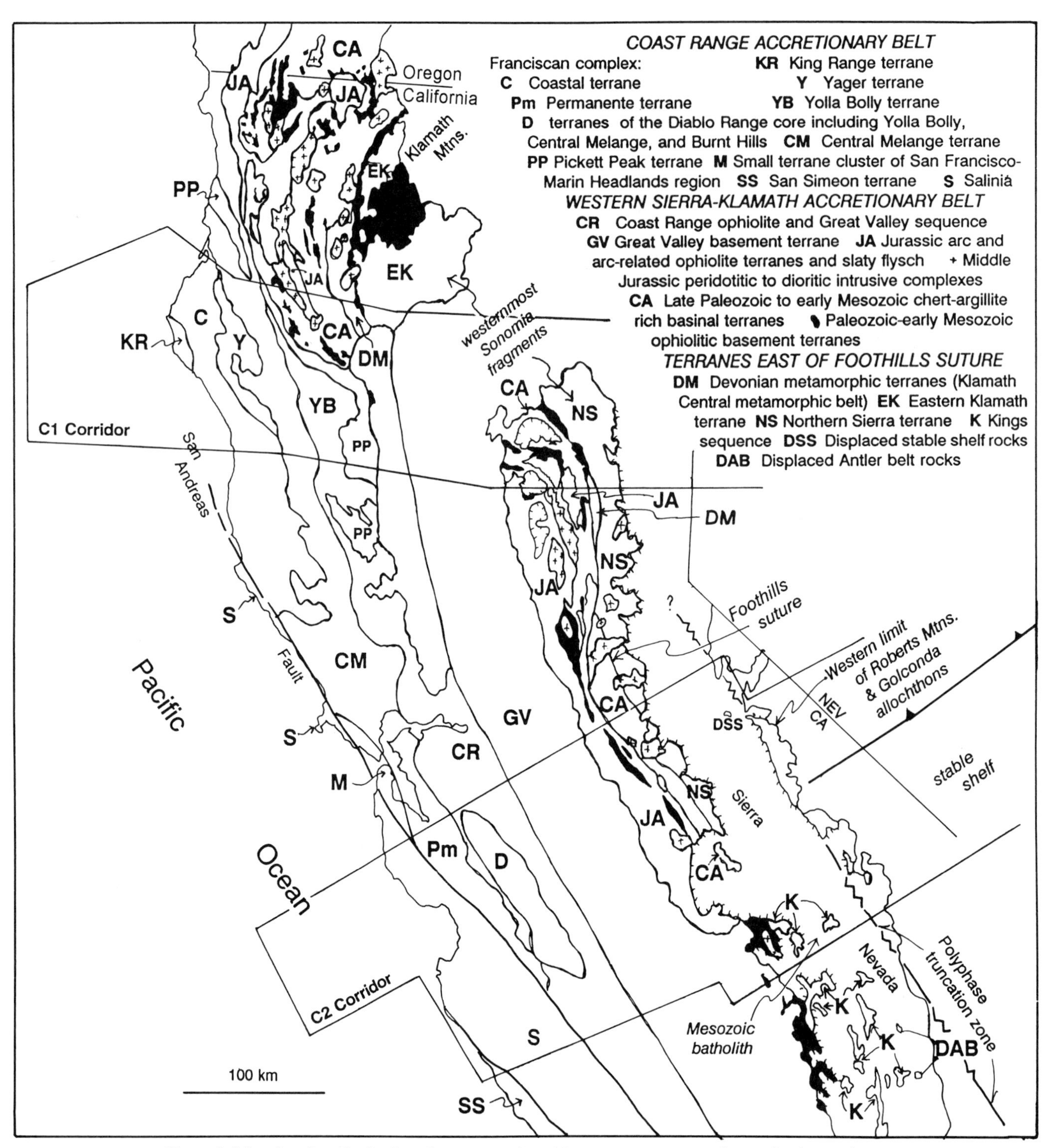

Figure 3. Map showing distribution of tectonostratigraphic terranes or terrane clusters for the northern and central California regions.

late Cenozoic sedimentary rocks, presumably overlying the oceanic plate and an accretionary complex formed prior to the propagation of the San Andreas fault system into the region. On land, south of the triple junction and east of the San Andreas fault, Quaternary northeastward-directed thrusting resulted from northeast-southwest convergence between the Pacific and North American plates (Griscom, 1980; McLaughlin and others, 1982). Seismicity and gravity anomalies (Cockerham, 1984) indicate that the Juan De Fuca plate dips shallowly beneath the Cape Mendocino region extending at least as far east as the Coast Range thrust near the northwest end of the Great Valley. In the immediate region of the triple junction, the San Andreas fault system appears to splay northward into thrusts over a broad region.

The effects of the underthrusting Juan De Fuca plate have significance for tectonic features elsewhere in the western portion of C1. Major uplift in the eastern Coast Ranges has been dated at no earlier than 3.1 Ma based on the first appearance of Franciscan detritus in the late Cenozoic deposits of the Great Valley (Blake and others, 1984). This uplift appears to have been accommodated on the eastern side of the Coast Range largely by reactivated high-angle faulting on the Coast Range thrust. Seismicity on the Coast Range thrust, near the northern end of the Great Valley (California Department of Water Resources, unpublished data), supports this possibility. Farther east within the Cascade Range, underthrusting of the Juan De Fuca plate is recorded in the calc-alkaline volcanoes, including Mount Lassen, which was last active in A.D. 1916.

The San Andreas transform system extends ~1000 km south from the Mendocino triple junction to the Salton Trough of southeastern California, where it serves as a complex plate boundary along the eastern edge of the Salton Trough. The trough is the northernmost point to which oblique spreading of the Gulf of California has propagated. The gulf spreading system consists of right-stepping short ridge segments with long transform offsets which are roughly on trend with the main San Andreas system. The gulf ridge-transform system continues into the actively spreading East Pacific Rise at the mouth of the Gulf of California.

The dextral San Andreas transform system consists of the San Andreas fault zone and its numerous subsidiary branches. In C2 the system consists of (from west to east) the San Gregorio–Hosgri fault, the San Andreas fault, the Calaveras fault, which ties into the Hayward fault near the north edge of C2, and several small dextral faults to the east. Taken together the current active system is ~90 km wide. The San Andreas and Calaveras faults constitute the main branches of the transform system in C2. Major structural consequences of the system in C2 are: (1) juxtaposition of totally different terranes, such as Salinia and the Franciscan Complex along the San Andreas zone; and (2) fragmentation, juxtaposition, refragmentation, and dispersal of earlier accreted terranes, as seen for the Calaveras fault within the Franciscan complex. The offshore San Gregorio–Hosgri fault may also be an important branch of the San Andreas system (Graham, 1979). Graham and Dickinson (1978) suggested 115 km of post–early Miocene dextral slip along this fault.

The first-order geologic consequence of the San Andreas transform system is the juxtaposition of the Salinian sialic crystalline terrane of the Coast Ranges outboard of Franciscan oceanic subduction complex rocks. This requires between 500 and 600 km of dextral slip. However, only 310 km of slip is documented for the San Andreas fault itself. Graham (1979) suggested that the discrepancy between 310 km and 500 to 600 km may be accounted for if slip histories of all branches of the transform system are taken into consideration. On the basis of plate kinematic models, Atwater (1970) suggested up to 1800 km post–30 Ma dextral slip between the Pacific and North American plates. Geologic correlations across the San Andreas fault (Dickinson, 1972; Crowell, 1979; Graham, 1979; Nilsen, 1984) support Atwater's idea that the San Andreas fault took up most of the relative motion, but lead to the much more limited Neogene offset. However, paleomagnetic investigations suggesting northward transport of Salinia of >1000 km (Champion and others, 1984) argue for a significant "proto-San Andreas" movement history. Stratigraphic relations discussed in Nilsen (1984) indicate that any pre–30 Ma northward transport occurred prior to Eocene time and thus cannot alleviate the discrepancy between plate and fault displacements. More recent plate kinematic analysis (Atwater, 1989) based on Stock and Molnar's (1988) global circuit data suggests only 800 ± 100 km plate displacement in the past 20 m.y. during San Andreas history. Thus if the total geologic offset of the San Andreas system is considered, and the possibility of ~100 km of dextral sense spreading in the Basin and Range as well as other increments of distributed finite strain are considered, the discrepancy disappears.

Major dextral displacements of Coast Range terranes began perhaps as early as the Late Cretaceous during northward migration of the intersection between the Kula-Farallon ridge and the California margin (Crowell, 1979). Reconstructions of Pacific basin plate geometry and motions suggest a very rapid (>100 mm/yr) northward motion for the Kula plate during latest Cretaceous–earliest Cenozoic time (Engebretson and others, 1986). Perhaps large displacements of the "proto-San Andreas system" occurred then. Once the Farallon plate was in contact with the central California region, significant northward displacement may have ceased until the establishment of the modern plate juncture system.

Establishment of the San Andreas transform system accounts for the abandonment of the Tertiary accretionary prism of C2. The truncation of magnetic anomaly 7 by the filled Tertiary trench suggests that the East Pacific Rise and the equivalent of ~10 m.y. of the Pacific plate were subducted prior to the establishment of the transform system or, if not subducted, that undetected major strands of the transform system along the trench axis removed fragments of the Rise and Pacific plate from the C2 area. The new transform system developed through older accretionary terranes of the Coast Ranges rather than along the Tertiary subducting trench. The Tertiary accretionary prism has thus migrated northward as part of the Pacific plate along the San Andreas transform. An important question is whether dextral slip along the Tertiary trench, or in faults not discovered east of the inactive trench, can account for the large post–30 Ma offset suggested by Atwater (1970). The major north-directed thrust zone shown on the C2 strip map within the accretionary prism suggests a significant component of northward oblique subduction during growth of the prism.

The Pacific–North American relative plate velocity in the area of the C2 corridor, based on global circuits, is N36 ± 2W at 48 ± 1 mm/yr (De Mets and others, 1990). This value diverges from the N40W strike of the San Andreas fault along its straight segment in and to the south of C2, and from the 36 ± 5 mm/yr Quaternary slip rate along that segment of the fault determined by Sieh and Jahns (1984). Oblique extension distributed across the Basin and Range represented by movement of the Sierra Nevada block relative to stable North America is N36 ± 3W at 11 ± 1 mm/yr (Argus and Gordon, 1991). Using these data, Argus and Gordon (1991) calculated a discrepancy vector of N20 ± 17W of 6 ± 2 mm/yr between the San Andreas slip rate plus Sierra Nevada block motion and the global plate circuit model. This resolves into discrepancies of 5 ± 3 mm/yr parallel to and 2 ± 2 mm/yr normal to the San Andreas fault. This analysis suggests that parallel strands of the San Andreas transform system in the offshore area and Coast Ranges of the C2 region do not have large active slip rates, and that transpressive shortening is occurring in the region, as discussed in Aydin and Page (1984).

BASIN AND RANGE PROVINCE

The Basin and Range province is a broad region of horizontal Cenozoic extension between the Sierra Nevada and Colorado Plateau, identified by normal faulting and high seismicity (Hamilton and Myers, 1966; Thompson, 1966; Stewart, 1971, 1978; Thompson and Burke, 1973; Smith, 1978; Zoback and others, 1981; Anderson and others, 1983; Klemperer and others, 1986; Zoback, 1989). The province may be viewed as an intracontinental rift zone that, at the latitude of C1 and C2, is the widest in the world. Extension began widely in the Oligocene (Zoback and others, 1981; Gans and Miller, 1983), perhaps locally in Eocene time (Fouch, 1979), and continues to the present.

Relative to Cenozoic tectonics at the western edge of the North American plate, the development of the Basin and Range can be traced in three stages: (1) from 40 to 25 Ma, the province was intra-arc and above a flat slab of the subducting Farallon plate (Lipman and others, 1972; Engebretson and others, 1986); (2) from 25 to 10 Ma, it was in back of a narrow, partly extinct southern Cascades magmatic arc, following the detachment or sinking of the Farallon slab (Christiansen and Lipman, 1972; Snyder and others, 1976; Eaton, 1982); (3) from 10 Ma to present, it has been landward of the San Andreas transform system and may be part of a wide zone of deformation along the boundary between the North American and Pacific plates (Atwater, 1970). Basin and Range extension in the first two stages may have arisen from the rise and fall of a flat slab of the Farallon plate (Coney and Reynolds, 1977). In the last stage, the maximum principal elongation from fault-slip data and focal mechanisms is thought to be approximately east-west in the western and eastern marginal zones of the Basin and Range province, but to be northwest to west-northwest in the interior (Zoback, 1989). The ongoing northwesterly transport of the Sierra Nevada block relative to cratonal North America (Argus and Gordon, 1991) suggests that right-lateral shear on northwest-striking planes occurs within the Basin and Range province because there is little evident strike-slip faulting along the Sierra Nevada–Basin and Range boundary. The orientation of maximum principal elongations suggests that such shear is concentrated in the marginal zones of the Basin and Range province. The cause of the strain orientation in the interior region is unclear.

Extensional phenomena of the Basin and Range at the surface and near surface in C1 and C2 vary substantially with position and depth, and it is debated whether they have changed with time. Below we divide discussions of phenomena between eastern and western halves. The division is specific to latitudes between and adjacent to the C1 and C2 corridors. To the south the Basin and Range province narrows and many of the features of the eastern and western halves merge into the highly extended region, including the Death Valley area and the Owens Valley–high eastern Sierra fault-scarp system (Wernicke and others, 1988).

Eastern Basin and Range

The eastern half of the Basin and Range province, approximately between Carlin and Eureka, Nevada, on the west and the Colorado Plateau on the east, contains extensional structures of three general styles: brittle thick skinned, brittle thin skinned, and ductile.

Brittle thick-skinned structures are the classic broad horsts and grabens and half grabens that make up the main ranges and basins and are bounded by high-angle planar or listric normal faults. Such faults continue down to maximum seismogenic depths (12–15 km), presumably the present brittle-ductile transition (Anderson and others, 1983; Smith and Bruhn, 1984). The eastern margin (Wasatch Plateau) and western margin of the eastern Basin and Range contain thick-skinned structures; the interior part may not, although this is debated.

The brittle thin-skinned style has master low-angle normal detachment faults that can be traced from the surface by seismic reflection profiling downdip as far as 120 km and to depths up to 20 km (McDonald, 1976; Allmendinger and others, 1983, 1986; Smith and Bruhn, 1984; Gans and others, 1985). The hanging walls of such detachment faults are broken by planar and listric high-angle normal faults into successions of horst and grabens and/or half grabens. The spacing between high-angle faults in the hanging walls is proportional to depth to the detachment fault (Allmendinger and others, 1986; Wernicke, 1985), such that thin- and thick-skinned structures are gradational. The footwalls of the detachment faults are thought to be nondeformed (Allmendinger and others, 1986; Wernicke, 1985). Master detachment faults have developed sequentially at given places and have westerly and easterly dips. Sequential detachment faults in the Schell Creek Range caused imbrication, rotation, and inversion of early faults and bedding in the hanging walls (Gans and Miller, 1983).

Ductile zones in the eastern half contain Cenozoic mylonites that were derived from amphibolite and greenschist-grade metamorphic and intrusive rocks and represent strongly stretched mid-level crust exhumed 10–15 km. The ductile zones are exposed in several nodes of maximum uplift, called core complexes, that are in the Albion–Grouse Creek Mountains (Compton and others, 1977; Miller, 1980), Pilot Range (Miller, 1984); Ruby Mountains (Howard, 1980; Snoke, 1980; Dallmeyer and others, 1986; Snoke and Miller, 1987), and Snake Range (Miller and others, 1983). The mylonites are the footwalls to detachment faults with unmetamorphosed, brittlely faulted hanging walls. Strain in the mylonites diminishes downward, over about 1 km in the Ruby Mountains, to nil. Maximum stretch in the mylonites is parallel to that in the suprajacent hanging wall (Snoke, 1980; Miller and others, 1983).

The direction of maximum principal elongation in the eastern Basin and Range is east-west $\pm$ 20° (Miller and others, 1983; Zoback, 1989; Snoke and Miller, 1987). The magnitude of elongation in the eastern half is generally large but heterogeneous. Gans (1987) estimated from balanced sections 80% for the whole eastern half in C2, 250% across a 200-km-wide region containing the core complex of the Snake Range (Gans and Miller, 1983), and as much as 500% within the core complex mylonite (Miller and others, 1983).

From seismic and other data, many workers believe the detachment faults of the eastern Basin and Range to be low-angle normal faults with displacements of 30–60 km that cut subplanarly through much of the crust and across the brittle-ductile transition (Wernicke, 1981, 1985; Allmendinger and others, 1983, 1986; Bartly and Wernicke, 1984; Snoke and Miller, 1987). In this model, the mylonites are zones of large simple shear as well as the footwall of the downdip continuation of the detachment faults into the ductile realm. The exhumation of the mylonites is due to extreme denudation of footwall mid-crustal rocks, extreme attenuation of the hanging-wall upper crustal rocks, and to isostatic uplift of greatly extended crust. An opposing view is that detachment faults developed at the brittle-ductile transition and that the brittle and ductile zones extended longitudinally together with only minor displacement at their mutual boundary (Miller and others, 1983; Gans and others, 1985).

The most conspicuous surface manifestations of extension in the eastern Basin and Range are the horsts, grabens, and half grabens, which are probably less than 10 m.y. old (Stewart, 1978). Zoback and others (1981) suggested that these represent a later phase of extension superposed on the earlier phase of predominantly low angle detachment faulting, the change probably due to cooling and thickening of the brittle zone. Allmendinger and others (1986) and Wernicke (1985), however, argued that the high-angle fault structures occur only in the hanging walls of detachment faults and that no change in fault style took place. Their evidence is that the high-angle faults do not appear to cut the low-angle faults in seismic sections and that detachment faults are known to have been active in the past 1 m.y.

Western Basin and Range

In contrast to the eastern half, the extensional style of the western Basin and Range is almost entirely thick-skinned brittle; the thin-skinned brittle style is recognized at only two places, and no evidence for the ductile style exists at the surface. Within northern Nevada in the western half, faults that are range bounding and those internal to ranges are high angle, and those still active apparently continue steeply to maximum seismogenic depths (15 km) (Ryall and Malone, 1971; Ryall and Priestley, 1975; Okaya and Thompson, 1985; Anderson and others, 1983). These are mainly north to northeast-striking and record east-west extension that may have been supplanted by northwest-southeast extension in the past 10 m.y. (Zoback, 1989). A local zone, the Walker lane, which lies along and parallel to the California-Nevada border, also includes northwest-striking right-slip and east-northeast–striking left-slip faults. It is unclear whether the Walker lane takes up nonrotational deformation with east-west extension on conjugate shears or whether there is net simple shear on one of the strike-slip sets.

Thin-skinned brittle extension of Cenozoic age is demonstrated at Yerington (Proffett, 1977) and north of Austin (Smith, 1992). At other places in the western half, detachment faults that penetrate the basement to Tertiary cover have not been recognized at the surface or in sparse seismic sections (Anderson and others, 1983). Ductile extension of Cenozoic age is not documented in the western half, although exposed low-grade metamorphic rocks in the Toiyabe Range north of Austin (Smith, 1992) and near Silver Peak (Kirsch, 1971) could represent deep levels of hanging walls above core complexes.

The western Basin and Range has undergone meager Cenozoic unroofing compared to the eastern, implying substantially less extension than the eastern. A maximum of 20% elongation is estimated for the western half (Speed and others, 1987), a value that corresponds to extension only by the thick-skinned style. Thus, the Basin and Range as a whole appears to have undergone heterogeneous extension with a maximum in eastern Nevada along the zone of ductile features (core complexes) (Fig. 1).

Corridor C1 also includes the southern terminus of the Cascade Range and Modoc Plateau. Recent gravity and seismic refraction data from northeastern California (Zucca and others, 1984; Fuis and others, 1987) show crustal structure beneath these two provinces similar to that of the Sierra Nevada. In addition, both provinces are characterized by northwest-trending block faults of Quaternary age that appear to mark the western margin of the Basin and Range faulting. The crustal structure beneath the Modoc Plateau is not consistent with highly attenuated or rifted crust. The structural pattern is more suggestive of a block-faulted boundary between the Basin and Range province and continental arc crust thickened by Mesozoic and Cenozoic arc magmatism.

Deep crustal structure

Seismic data show that the crust of the Basin and Range contains a wealth of reflectors compared to that of the Colorado Plateau and other cratonal regions (McDonald, 1976; Allmendinger and others, 1983; Smith and Bruhn, 1984; Gans and others, 1985; Klemperer and others, 1986). Sets of shallowly dipping reflections emerge from varied depths as great as the refraction Moho, below which no reflections are generated. In the eastern Basin and Range, some extensive reflectors can be connected with the surface traces of low-angle detachment faults. One of these, the Sevier Desert detachment, can be traced on seismic profiles subplanarly downdip over 120 km to depths of about 20 km (Allmendinger and others, 1983). Deeper reflection sets in the lower crust may be due to subhorizontal bodies of magma and mafic igneous rock and/or to major ductile shear zones (Klemperer and others, 1986).

The reflection Moho below the Basin and Range is defined as the base of the zone of crustal reflections and the top of a seismically opaque region. It appears to be a smooth, continuous surface that is discordant to crustal reflectors and is disharmonic to major Cenozoic crustal structures that record more than 10 km of throw. This implies that the Moho of the Basin and Range is a young, probably late Neogene, feature. Its origin is plausibly a mantle differentiation surface across which basalt passes up into the lower crust, and below which crystal residues of ultramafic composition accumulate within ultramafic residue host rock (Lachenbruch and Sass, 1978). The convective heat transfer implied by such an origin of the Moho is supported by other phenomena: high surface heat flow (averaging 80 mW/m^2); thin lithosphere (45 to 65 km) as measured by surface-wave dispersion (Priestly and Brune, 1978; Priestly and others, 1982); low headwave velocities in the upper mantle (7.6–8.0 km/s) compared to craton values (Eaton, 1963; Prohdehl, 1979); and crustal low-velocity zones in at least the eastern Basin and Range (Braille and others, 1974).

The C2 section shows a pronounced crustal thickening along the western edge of the Basin and Range province corresponding to the crustal root of the Sierra Nevada (Eaton, 1966; Oliver, 1977; Pakiser and Brune, 1980). Reanalysis of the geophysical data used to construct the Sierran crustal root model in conjunction with recent teleseismic and regional P_n arrival studies has brought this model into question (Jones and others, 1990). The alternative working model proposed by Jones and others (1990) is contrasted with the C2 model in Figure 4. The alternative model shows a relatively small crustal root for the Sierra Nevada with a ~35 km depth to the Moho and an asthenosphere upwelling beneath the region. In this hypothesis, horizontal flow of quartz-bearing middle to deep crustal rocks smooths out the Moho beneath areas of strong topographic and structural relief (Block and Royden, 1990; Wernicke, 1990). This is not consistent with observations of deep-level batholithic exposures of the southern Sierra, which reveal a predominance of quartz-bearing rocks (mafic tonalite) to depths of at least 30 km and evidence for these rocks to have flowed plastically for a prolonged time period under a wide range of metamorphic conditions (Saleeby and others, 1986; Pickett and Saleeby, 1989, 1993; Saleeby, 1990b). The relative roles of deep crustal flow of quartz-bearing rocks versus the mantle magmatic differentiation mechanism discussed above for producing a smooth Moho under the Basin and Range province are unknown.

CRYPTIC EDGE OF PRECAMBRIAN NORTH AMERICAN SIAL

The present edge of contiguous sialic Precambrian North America in the central western United States is approximated in Figure 1 and shown with respect to surface geology on the strip maps of C1 and C2. The sialic edge is a basement feature and is generally cryptic due to burial by younger rocks and nappes or obscuration by magmatism and metamorphism. The margin's surface trace is estimated by outermost outcrops of autochthonous continental platform or shelf facies (Roberts and others, 1958; Kay and Crawford, 1964; Matti and McKee, 1977), by ratios of initial Sr, Nd, and Pb isotopes and the petrochemistry of autochthonous Phanerozoic magmatic rocks (Kistler and Peterman, 1973; Doe and Delevaux, 1973; Zartman, 1974; Armstrong and others, 1977; Miller and Bradfish, 1980; Farmer and DePaolo, 1983), and locally (south of 41°N) by maxima in gravity gradients that indicate rapid lateral changes in crustal thickness and/or composition (Speed, 1982).

The cryptic sialic edge evolved as a Late Proterozoic passive margin that developed along the entire length of western North America. The existence of the passive margin was interpreted from autochthonous and parautochthonous upper Precambrian and lower Paleozoic shelf strata that crop out between the platform-shelf hinge on the east and the region of accreted terranes on the west (Stewart, 1972, 1976; Stewart and Suczek, 1977). Lower strata of this sequence are coarse clastics with basalt (Windermere Group) that represent various rifting phases between about 600 and 900 Ma. The final rifting followed by postrift subsidence and continental separation is marked by a terrigenous, marginward-thickening clastic wedge. The onset of postrift subsidence is estimated by thermal models to have been at about 600 Ma, near the end of Proterozoic time (Stewart and Suczek, 1977; Armen and Mayer, 1983). The clastic prism is succeeded by rocks of shallow-marine subsiding shelf environments (Kay and Crawford, 1964; Matti and McKee, 1977), as shown in the C1 and C2 sections of northern and central Nevada.

The present sialic crust in the region affected by passive margin tectonics, west of the platform shelf hinge in Utah (see Fig. 7), is between 25 and 35 km thick except perhaps where the Mesozoic continental arc is superposed on it, where it is interpreted to be as much as 55 km thick. The prevalent thicknesses are less than those east of the hingeline, possibly reflecting an

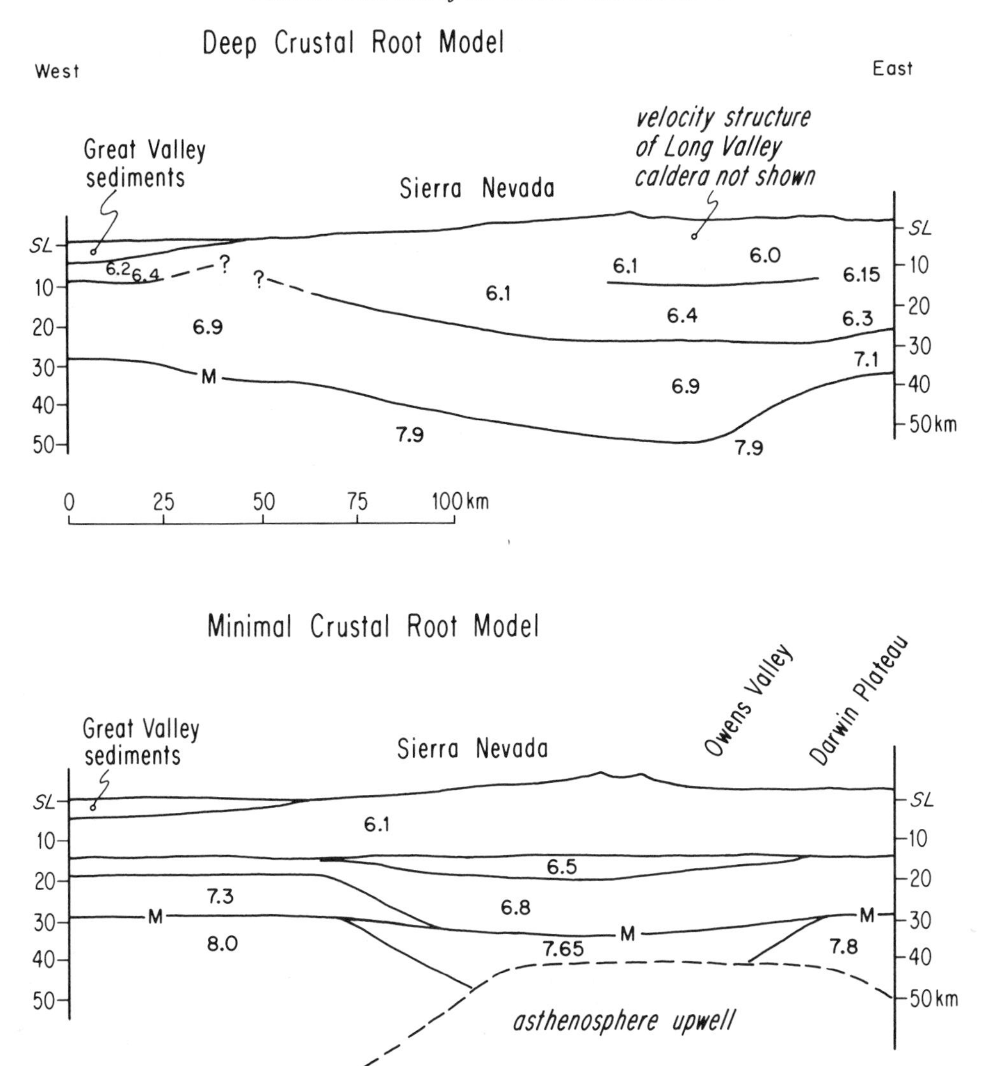

Figure 4. Generalized seismic P-wave crustal structure model for the central Sierra Nevada as shown in C2 synthesis and alternative model suggested by Jones and others (1990), which replaces Sierran root with asthenosphere upwell in central to southern Sierra Nevada region. M = Moho.

unusually broad zone of rifting and attenuation during passive margin formation. The magnitude of Precambrian attenuation cannot be treated quantitatively because of probable changes in crustal thickness due to Mesozoic contraction and Cenozoic attenuation and magmatism.

Evidence that the Late Proterozoic passive margin may be preserved in Nevada comes from differences among autochthonous and parautochthonous lower Paleozoic strata that can be interpreted as inner and outer shelf facies (Kay, 1960; Matti and McKee, 1977; Rowell and others, 1979). At sites in western North America outside Nevada, however, it is inferred that an outer zone of continental rock, including the Proterozoic sialic edge, was removed in post-Paleozoic time. Evidence is the absence of the pre-Jurassic tectonostratigraphic units that are sutured against or thrust on the sialic continent in Nevada (Speed, 1982) and from the truncation of the Paleozoic hingeline and projected Paleozoic structures by the present sialic edge (Hamilton and Myers, 1966). Such a truncation zone is hypothesized along the axial region of the central and southern Sierra Nevada in the C2 region. The processes of tectonic attrition of the sialic edge are considered below.

The approximate trace of the outer edge of cryptic Precambrian sial is shown to change from a north-south trend to an east-west trend in west-central Nevada (Fig. 1 and C2 strip map), which is called the Mina deflection. The deflection may be a product of late Mesozoic right-lateral shear on northwest-striking planes (Speed, 1978; Saleeby, 1981), or an earlier salient of the sialic edge, perhaps formed during Proterozoic rifting (Oldow, 1984). If the deflection is due to shear, the right-lateral displacement and/or shear strain are greatly accentuated along the axis of the Sierra Nevada, where a northwest-trending prong of possible cryptic sial is hypothesized (Fig. 1; Kistler and Peterman, 1973,

1978). The western edge of the prong coincides roughly with the western Sierra Nevada suture zone, a complex accretionary zone floored by Paleozoic ophiolite (Saleeby, 1983, 1990a, 1992). The northwest-trending prong of hypothetical cryptic sialic rocks is based primarily on initial Sr isotopic data on Mesozoic batholithic rocks of the Sierra Nevada (Kistler and Peterman, 1973, 1978) and corresponds mainly to belts of Paleozoic shelf-slope-rise and Early Mesozoic, shelfal, metamorphic-framework remnants. The metasedimentary stratigraphic units may have rested depositionally and/or tectonically above Precambrian sial prior to the generation of the Sierra Nevada batholith. Alternatively, as recent isotopic and petrologic studies suggest, the Precambrian sialic component within the axial Sierran batholithic magmas was incorporated from detrital rocks derived from sialic basement sources (DePaolo, 1981; Saleeby and others, 1987; Kistler, 1990; Saleeby, 1990b). Thus the axial Sierran prong may represent in part the general trace of a major North American basement-derived sediment wedge instead of the actual Mesozoic bounds on such basement.

TECTONIC ACCRETION AND DISPLACED TERRANES

Accreted terranes of Nevada and California are grouped in three belts (Fig. 3); from east to west and older to younger, these are Nevada, western Sierra–Klamath, and Coast Range accretionary belts. Terranes of the Nevada belt accreted against and above the passive margin of western North America in Mississippian and Permian-Triassic times. Terranes of the western Sierra–Klamath belt accreted at an active margin that was probably truncated during late Paleozoic to early Mesozoic time. The western Sierra–Klamath terranes were brought into contact with both Precambrian North America and the earlier terranes of the Nevada belt and were accompanied by continental arc magmatism that cut across the former passive margin and its boundary with the Nevada belt (Speed, 1978; Saleeby, 1981). Coast Range accretion began in the Early Cretaceous and is active today. Arc magmatism related to Coast Range accretion was and is superimposed on the two inner accretionary belts.

The Nevada accretionary belt consists of three exposed terranes: the Roberts Mountains allochthon, Golconda allochthon, and Sonomia (Fig. 1). Sonomia may be continuous beneath cover between the Golconda allochthon and the western Sierra–Klamath belt and include at its western edge the eastern Klamath terrane and the northern Sierra terrane (Fig. 3) (Speed, 1979). Alternatively, the latter two terranes may be distinct from Sonomia. A cryptic fourth terrane of the Nevada belt, Antleria, is postulated to be a mid-Paleozoic arc that lies tectonically below Sonomia (Speed, 1979). A contrasting view of the terrane composition of the Nevada accretionary belt is provided by Silberling and others (1987), who divided the region shown here as Sonomia into many smaller terranes. Their division is based on lithotectonic differences of Mesozoic layered rocks, all of which we regard as parautochthonous and originally deposited on Sonomian basement. The tectonic transport in Mesozoic time among pieces of the Sonomia terrane may have indeed been substantial, perhaps a few hundred kilometers.

The Roberts Mountains allochthon consists of a tectonic assemblage of pelagic, hemipelagic, turbiditic, and volcanic rocks of early Paleozoic age. It laps over lower Paleozoic strata of the North American continental shelf at least 130 km from the sialic edge and was almost certainly emplaced from the west during the Antler orogeny early in Mississippian time (Roberts and others, 1958; Smith and Ketner, 1968; Stewart and Poole, 1974; Speed and Sleep, 1982; Dickinson and others, 1983). The Golconda allochthon possesses rocks and gross structure similar to those of the Roberts Mountains, except that the rocks are of Mississippian to Permian age (Silberling, 1973; Stewart and others, 1977; Speed, 1977, 1979; Miller and others, 1981, 1983; Breuckner and Snyder, 1985). They were emplaced in Early Triassic time at least 100 km inboard of the sialic edge and above the earlier Roberts Mountains allochthon and its late Paleozoic and earliest Triassic cover (Speed, 1979).

Sonomia is a lithospheric fragment of sequential magmatic arc–related tectonostratigraphic units surmounted by a Permian magmatic arc (Speed, 1979); the fragment was accreted to North America early in the Triassic. Its main exposures are at the terrane margins, where late Mesozoic deformation transported rocks of Sonomia to the surface by imbricate thrusting as in Nevada, and by steep tilting and flattening as in the northern Sierra Nevada. The central regions of Sonomia are deeply buried below thick Triassic flysch and continental arc volcanics that succeeded Sonomia (Speed, 1978). Sonomia is the earliest of exposed displaced crystalline terranes to have accreted to southwestern North America. There is some question whether Sonomia was a contiguous mass upon initial collision or whether it includes sequentially accreted fragments. Paleomagnetic data show no significant anomalies in Sonomia with respect to North America in Cretaceous and possibly Late Jurassic time (Gromme and others, 1967; Hannah and Verosub, 1980; Russell and others, 1982; Frei and others, 1984; Bogen and Schweickert, 1986), thus permitting Sonomia between the Sierra-Klamath belt and Golconda allochthon to have been parautochthonous since at least that time.

Because of the closeness in minimum ages of rocks in the Golconda allochthon and Sonomia, their long reach of probable contact, and the occurrence of copious Early Triassic magmatic arc–derived debris in a foreland basin of the Golconda allochthon at Candelaria (C2 strip map), the emplacement of the two terranes was probably linked and constituted the Sonoma orogeny of Silberling and Roberts (1962). Sonomia is modeled as an island arc system migrating with closing and lateral components along coastal North America and the Golconda allochthon as its accretionary forearc (Speed, 1977, 1979). During at least the later stages of its history, the arc surmounted a subduction zone in which the downgoing slab was oceanic lithosphere attached to the passive margin of North America. Closure ceased when continental lithosphere started down below Sonomia, by which time the forearc, the Golconda allochthon, was almost fully emplaced on the outer continental shelf. If the oceanic slab attached to Paleozoic North America broke off and sank upon cessation of closure, a finite but probably minor width of sialic continental edge may have been taken into the mantle with it. Two migration paths for Sonomia have been envisioned. One is along the western continental margin with possible large transport distance (Speed, 1977; Breuckner and Snyder, 1985). The other is normal to the margin, involving an arc reversal and consumption of backarc basin lithosphere (Burchfiel and Davis, 1972; Miller and others, 1983). Upon welding to North America, a new ocean-facing convergent zone developed somewhere west of Sonomia, and Sonomia underwent thermal contraction and subsidence due to loss of subduction-related heating. Subsequently, a deep-water Triassic successor basin developed approximately coextensive with Sonomia.

The Mississippian emplacement of the Roberts Mountains allochthon as part of the Antler orogeny has similar manifestations to that of the Golconda: transport from an oceanic region as a predeformed tectonic mass, absence of related continental magmatism and metamorphism, and lack of pervasive shortening of the continent shelf below the allochthon. Thus, this allochthon is also thought to be an accretionary prism (Dickinson, 1977; Speed and Sleep, 1982) that was underrun by the continental shelf before final collision between the continent and a migrating island arc system. The postulated magmatic arc terrane is Antleria, no longer visible because of subduction below Sonomia. Alternatively, Antleria could constitute basement of Sonomia, having collided with North America during the Antler orogeny and subsequently rifted off prior to the Sonomia phase of arc volcanism. This interpretation was discussed by Miller (1987) and Rubin and others (1990); their "McCloud arc" corresponds generally to the Sonomia-Antleria composite.

A terrane map of the central and southern California region is given in Figure 3. Major groupings consist of the North American autochthon, the western Sierra–Klamath accretionary belt, and the Coast Range accretionary belt. As discussed below, Sonomia may include the oldest and easternmost terranes of the western Sierra–Klamath belt. Small vestiges of the Nevada accretionary belt are also present in the east-central Sierra Nevada (Schweickert and Lahren, 1987). To the west of these small vestiges, the constitution of terranes and their structure contrast strongly to those of the Nevada accretionary belt. The boundary is a polyphase truncation zone which, starting in late Paleozoic time, truncated the stable shelf and Antler belt rocks and possibly constituted a transform tear system in the emplacement of Sonoman thrust plates. The transform displacement was sinistral and may have been as much as 1000 km (Saleeby, 1981, 1992; Walker, 1988; Stevens and others, 1992). Terranes of the Sierra Nevada metamorphic framework were emplaced and deposited adjacent to the late Paleozoic truncation zone and underwent subsequent phases of trench-linked transform faulting in Mesozoic time consisting of both dextral and sinistral senses.

Another important feature of the Figure 3 terrane map is the inclusion of the Great Valley basement and the Coast Range ophiolite in the western Sierra–Klamath belt; this is because the Great Valley–Coast Range ophiolitic basement rocks were formed and accreted in close conjunction with similar rocks of the western Sierra–Klamath belt. The Coast Range accretionary belt consists of terranes of the Franciscan complex as well as the Salinia terrane, a major displaced fragment of Mesozoic sialic crystalline rock.

The accretionary terranes shown in Figure 3 in general become younger from east to west. We will begin our discussion of the terranes with the older (eastern) assemblages.

In the eastern Klamath Mountains major thrust sheets of mafic and mica-rich schists with Devonian minimum metamorphic ages (Lanphere and others, 1968; Irwin, 1977), Ordovician ophiolitic rocks (Mattinson and Hopson, 1972; Lindsley-Griffin, 1977; Quick, 1981; Jacobsen and others, 1984), and Devonian volcanic arc rocks (Irwin, 1977) constitute the composite eastern Klamath terrane (Fig. 3). The northern Sierra terrane consists of the Shoo Fly complex, which comprises Ordovician (and possibly older) to earliest Devonian(?) imbricated rise and slope deposits and fragments of oceanic crust. Devonian and Permian volcanic arc successions lie unconformably on the complex (D'Allura and others, 1977; Bond and DeVay, 1980; Schweickert and Snyder, 1981; Varga and Moores, 1981; Harwood, 1983). Ophiolitic melange within the Shoo Fly complex contains a 600 Ma plagiogranite block with oceanic crustal isotopic signatures (Saleeby, 1990a). The ophiolitic melange may represent oceanic basement for the Shoo Fly rise-slope sequence and possibly the only known remnants of oceanic lithosphere generated during late Precambrian rifting along the Cordilleran margin. Metaclastic rocks of the Shoo Fly complex extend ~300 km southward of C1, where they are cut out by the Sierra Nevada batholith (C2 strip map). Ordovician ophiolite remnants are seen again along the western Sierra near the southern margin of C2 (Shaw and others, 1987). Thus the Paleozoic record along the eastern margin of the western Sierra–Klamath accretionary belt is variable and fragmented.

In the southern Sierra Nevada there are two contrasting terranes that make up the prebatholithic metamorphic framework. In the southeast there are sinistrally displaced Antler belt rocks (Walker, 1988; Dunne and Suczek, 1991). Small pendants on strike to the north of these rocks may include displaced stable shelf strata in addition to Antler belt rocks (Lahren and others, 1990), suggesting that the Roberts Mountains thrust may have been displaced along the eastern Sierra as well. To the west along the axial and western southern Sierra Nevada the Kings sequence constitutes the main prebatholithic assemblage (Saleeby and others, 1978; Saleeby and Busby, 1993). The Kings sequence consists of an eastern shelfal facies with interbedded shallow-margin ignimbrites and a western basinal facies, which includes mafic flows and sits depositionally on Paleozoic ophiolite (Saleeby and others, 1978). Late Triassic–Early Jurassic fossils have been recovered from both facies, and comparable volcanic and hypabyssal rock U/Pb zircon ages have been determined from both facies (Saleeby and Busby-Spera, 1992; Saleeby and Busby, 1993). The Kings sequence represents an intra-arc shelf to forearc basin depositional system that was established in early Mesozoic time along the truncated stable shelf edge and its veneer of displaced Antler belt rocks. The western facies lapped across Paleozoic ophiolite, which was previously accreted along the truncation zone in late Paleozoic time as well (Saleeby, 1992). The eastern facies may also have been deposited in part on accreted Paleozoic ophiolite (Kistler, 1990), or alternatively, on disrupted stable shelf and displaced Antler belt rocks (Saleeby and Busby-Spera, 1992; Saleeby and Busby, 1993). The western facies of the Kings sequence extends northward as the Calaveras complex of the Sierra Foothills metamorphic belt; part of its eastern facies may correlate to the Sailor Canyon Formation, which lies in depositional sequence above the Shoo Fly complex (Saleeby and Busby, 1993).

From the southern Sierra to the Klamath Mountains region, the truncated stable shelf and its veneer of displaced Antler belt rocks, the northern Sierra terrane, and the eastern Klamath terrane in series formed a tectonic buttress against which terranes of oceanic basement and basinal sedimentary and island arc sequences of the western Sierra–Klamath accretionary belt were accreted in Triassic and Jurassic time. The contact zone between the outer terranes and the regional "buttress" is called the Foothills suture and is postulated to have had a complex movement history of eastward underthrusting and transform motion (Saleeby, 1981, 1982, 1990a, 1992). Our concept of the Foothills suture is fundamentally different from that of Moores (1970), Schweickert and Cowan (1975), Schweickert (1981), and Day and others (1985), who put the suture within Jurassic assemblages of the Foothills belt (cf. Fig. 5A). In our concept the Foothills suture marks the locus of accretion of equatorial proto-Pacific oceanic basement, seamount, and pelagic sedimentary sequences along a west-facing subducting margin in Triassic to Early Jurassic time (Davis and others, 1978; Saleeby, 1981, 1983, 1990a, 1992; Wright, 1981, 1982). All Jurassic assemblages of the Foothills belt, as well as their Paleozoic ophiolitic basement, are west of the suture. Arc-related volcanic and sedimentary strata of the Foothills belt as well as the Kings sequence formed unconformably above the accreted oceanic assemblages west of the suture. Continental arc magmatism related to eastward subduction along and to the west of the suture was widespread in eastern California and Nevada. The superposition of Jurassic arc plutonic and volcanic sequences with time across the accreted oceanic sequences (Snoke and others, 1981; Sharp, 1988; Saleeby, 1990a; Saleeby and Busby-Spera, 1992) may mark forearc magmatism and/or migration of the magmatic arc across its former forearc. The most intense phase of thrusting and dynamic metamorphism in the western Sierra–Klamath accretionary belt occurred during Middle Jurassic time and is called the Siskiyou event (Coleman and others, 1988; Saleeby and Busby-Spera, 1992). Siskiyou deformation was superimposed on accretionary and postaccretionary extensional structures of the western Sierra–Klamath belt. Deformation may have been driven by a grazing collision of a major Pacific northwest terrane such as Wrangellia and the Alexander terrane assemblage or by strongly coupled subduction.

A major reorganization within the continental arc framework at ~165 to 160 Ma is indicated by major extension and silicic ignimbrite activity (Saleeby and Busby-Spera, 1992). The forearc underwent oblique rifting with the generation of interarc and intra-arc ophiolites (Saleeby and others, 1979, 1982; Hopson and others, 1981; Saleeby, 1981, 1982, 1983, 1990a, 1992). Within the western Klamath Mountains, a complete oceanic crust and upper mantle sequence (Josephine ophiolite, Fig. 5B) was generated during the rifting event (Harper, 1980, 1984). In the northern Sierra the Smartville and related ophiolitic rocks just south of C1 were generated by intra-arc extensional magmatism within the Paleozoic ophiolitic basement framework (Saleeby, 1982, 1983, 1990a; Saleeby and others, 1989a). The Coast Range ophiolite represents a major track of juvenile oceanic crust created during the 165 to 160 Ma rifting event (Fig. 5, C and D). Earlier phases of extension are recorded for Late Triassic–Early Jurassic time within the western Sierra–Klamath accretionary belt, but remnants of complete oceanic crust sequences were not generated or did not survive subsequent tectonic events (Saleeby, 1990a, 1992; Saleeby and Busby-Spera, 1992).

Between 160 and 150 Ma, the arc-interarc basin framework of the western Sierra–Klamath accretionary belt collapsed with the Nevadan culmination of accretion and related deformation. This event was called the Nevadan orogeny by Blackwelder (1914). Young interarc basin crust (Josephine ophiolite) was thrust eastward beneath the Klamath Mountains, resulting in the collision of an outer rifted arc segment with earlier accreted terranes to the east which were surmounted by a Middle Jurassic remnant arc (Fig. 5B) (Snoke, 1977; Saleeby and others, 1982; Harper and Wright, 1984). In contrast, young interarc basin and rifted arc rocks of the northern Sierra, Great Valley, and Coast Range ophiolite belt were at least locally thrust eastward over earlier terranes of the western Sierra–Klamath accretionary belt (C1 and C2 cross sections). Steeply dipping reverse faults with westward heaves of the Foothills fault system were at least locally superposed on the east-directed thrusts in the western Sierra (Day and others, 1985). The Great Valley basement is identified as the

Figure 5. Alternative plate tectonic models for Late Jurassic Nevadan orogeny shown in general west to east cross-sectional views. A: Collision of exotic arc in western Sierra Foothills (after Moores, 1970; Schweickert and Cowan, 1975; Day and others, 1985). B: Opening of Josephine ophiolite interarc basin followed by basin closure and thrusting (after Snoke, 1977; Saleeby and others, 1982; Harper and Wright, 1984). C: Opening of Great Valley interarc basin followed by arc reversal and basin closure along axial Great Valley to western Sierra Foothills area (after C2 tectonic section and explanatory text). D: Transtensional interarc basin spreading along Great Valley–Coast Range ophiolite followed by transpressional intraoceanic thrusting without complete basin closure (after Saleeby, 1981, 1992; Saleeby and Busby-Spera, 1992).

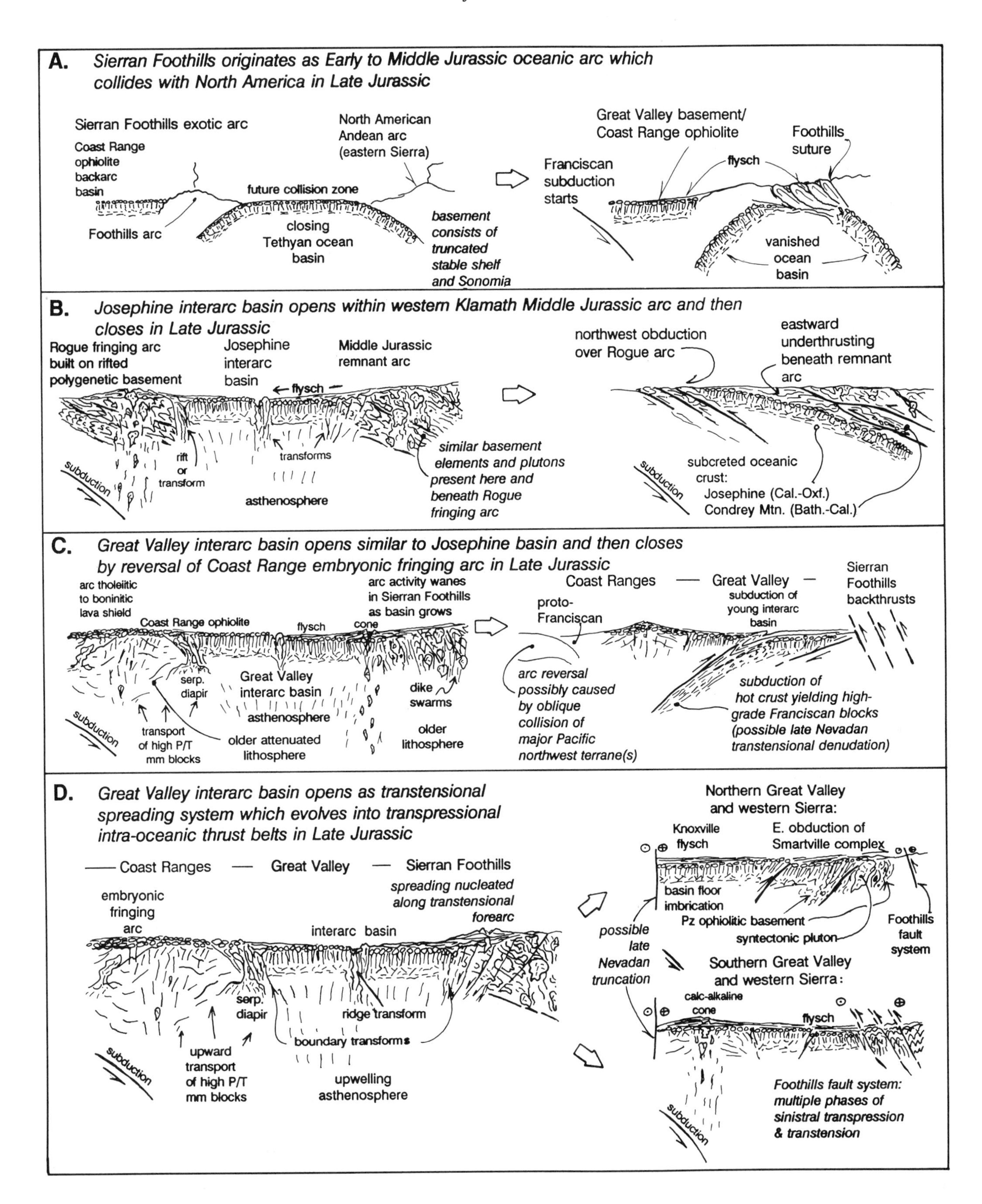
A. Sierran Foothills originates as Early to Middle Jurassic oceanic arc which collides with North America in Late Jurassic
Sierran Foothills exotic arc
Coast Range ophiolite backarc basin
North American Andean arc (eastern Sierra)
future collision zone
Foothills arc
closing Tethyan ocean basin
basement consists of truncated stable shelf and Sonomia
Great Valley basement/ Coast Range ophiolite
Foothills suture
Franciscan subduction starts
flysch
vanished ocean basin
B. Josephine interarc basin opens within western Klamath Middle Jurassic arc and then closes in Late Jurassic
Rogue fringing arc built on rifted polygenetic basement
Josephine interarc basin
Middle Jurassic remnant arc
flysch
rift or transform
transforms
asthenosphere
subduction
similar basement elements and plutons present here and beneath Rogue fringing arc
northwest obduction over Rogue arc
eastward underthrusting beneath remnant arc
subcreted oceanic crust:
Josephine (Cal.-Oxf.)
Condrey Mtn. (Bath.-Cal.)
C. Great Valley interarc basin opens similar to Josephine basin and then closes by reversal of Coast Range embryonic fringing arc in Late Jurassic
arc tholeiitic to boninitic lava shield
Coast Range ophiolite
flysch
arc activity wanes in Sierran Foothills as basin grows
cone
serp. diapir
Great Valley interarc basin
asthenosphere
dike swarms
transport of high P/T mm blocks
older attenuated lithosphere
older lithosphere
Coast Ranges — Great Valley —
Sierran Foothills backthrusts
proto-Franciscan
subduction of young interarc basin
arc reversal possibly caused by oblique collision of major Pacific northwest terrane(s)
subduction of hot crust yielding high-grade Franciscan blocks (possible late Nevadan transtensional denudation)
D. Great Valley interarc basin opens as transtensional spreading system which evolves into transpressional intra-oceanic thrust belts in Late Jurassic
— Coast Ranges — Great Valley — Sierran Foothills
embryonic fringing arc
spreading nucleated along transtensional forearc
interarc basin
serp. diapir
ridge transform
boundary transforms
upward transport of high P/T mm blocks
upwelling asthenosphere
subduction
Northern Great Valley and western Sierra:
Knoxville flysch
E. obduction of Smartville complex
basin floor imbrication
Pz ophiolitic basement
syntectonic pluton
Foothills fault system
possible late Nevadan truncation
Southern Great Valley and western Sierra:
calc-alkaline cone
flysch
Foothills fault system: multiple phases of sinistral transpression & transtension

west-dipping magnetic slab that underlies most of the Great Valley (C1 and C2 sections). Possible related thrusts of the Nevadan culmination within the Great Valley basement are interpreted as west-dipping mid-crustal reflectors (C2 tectonic section). Basement cores and field relations in the region of the seismic line and a closer examination of basement-rock map relations along the path of the seismic line suggests, however, that the reflectors could represent Cretaceous structures related to the Sierra Nevada batholith. This alternative interpretation is presented in Figure 6 and is discussed below in conjunction with batholithic structure and tectonics.

Various plate tectonic geometries for the Nevadan culmination were presented by Schweickert and Cowan (1975), Snoke (1977), Harper (1980), Saleeby (1981, 1982), Harper and Wright (1984), Day and others (1985), Sharp (1988), and Saleeby and Busby-Spera (1992). General aspects of these models are shown in Figure 5. The critical differences between these models lie in the interpretation of the arc-related ophiolites and Callovian to Oxfordian arc-related sequences along the western edge of the western Sierra–Klamath accretionary belt. Schweickert and Cowan (1975) and Day and others (1985) emphasized thrust faults along the structural bases of these sequences, related the thrusts to west-dipping subduction, and considered the sequences to be an exotic east-facing island arc that collided with the North American continental arc during the Nevadan orogeny (Fig. 5A and C1 tectonic section). The west-dipping magnetic slab of the Great Valley is considered a direct expression of the exotic arc and west-dipping subduction zone.

In contrast, similar basement elements in the hypothetical exotic arc and in pre-Nevadan terranes of the western Sierra–Klamath accretionary belt lying in the continental forearc, in addition to a crosscutting belt of Jurassic plutons scattered throughout the belt, were cited by Saleeby (1981) and Harper and Wright (1984) as evidence for a nonexotic origin for the western arc and ophiolitic rocks. Alternative models offered by these workers as well as by Snoke (1977), Harper (1980), and Saleeby and others (1982), call on interarc basin formation and/or forearc rifting for the formation of the arc-related ophiolites within an existing accretionary framework at the leading edge of the North American plate. Changes in relative plate motions and/or a grazing collision by major northward-migrating terranes of the Pacific northwest (Wrangellia-Alexander) resulted in the Nevadan collapse of the arc-basin system. Both eastward and westward vergences of Nevadan structures are envisaged as direct expressions of this deformational regime, not the collision of an exotic arc. Figure 5B shows a well-delimited model for the western Klamath region, where rifting and interarc basin formation of the Josephine ophiolite was followed by west-vergent thrust imbrication of the resultant arc-interarc basin-remnant arc system (Snoke, 1977; Harper, 1980; Saleeby and others, 1982; Harper and Wright, 1984). Figure 5C shows an alternative to the Figure 5A model for the Great Valley–western Sierra region (Saleeby, 1992, and C2 tectonic section). Here interarc basin formation was followed by an arc reversal, which resulted in eastward imbrication of the western Sierra and Great Valley basinal and arc terranes. Figure 5D is an alternative model after Saleeby (1981, 1992). In this model, oblique spreading within the interarc basin system was controlled mainly by large margin-parallel transforms. The Nevadan culminations resulted from a major Late Jurassic change in plate motion patterns (May and Butler, 1986), whereby transform and rift edge systems evolved into oblique convergence zones forming intra-oceanic thrust belts. The west-dipping structural fabric of the western Sierra and Great Valley is partly related to eastward overthrusting along basin floor and boundary transforms, but is also related strongly to Cretaceous low-angle extensional faulting off the western margin of the Sierra Nevada batholith (Fig. 6).

The later views expressed for the Nevadan orogeny (Fig. 5, B, C, and D) consider the Foothills suture as the eastern margin of the western Sierra–Klamath accretionary belt and the contact between oceanic terranes accreted in Triassic to Late Jurassic time to rocks of the Nevada accretionary belt and truncated stable shelf rocks to the south. The earlier views consider the main suture as lying along Nevadan-age thrust faults within the western Sierra–Klamath accretionary belt bounding the hypothetical exotic arc rocks.

Terranes here shown in the western Sierra–Klamath accretionary belt (Fig. 3) are in fact groupings of smaller "terranes." These include Paleozoic ophiolite, which forms the basement for a number of the smaller constituent early Mesozoic terranes; late Paleozoic to early Mesozoic basinal sedimentary and volcanic arc sequences; and Middle to Late Jurassic island arc sequences, arc-related ophiolites, and flysch. Due to structural complexity and depositional environments with rapid and abundant facies changes, a number of workers have split the terranes delineated in Figure 3 into numerous terranes and have suggested that links between terranes of the Sierra Nevada and Klamath Mountains existed no earlier than Late Jurassic time (i.e., C1 explanatory pamphlet). However, there are remarkable similarities between terranes of the Sierras and Klamaths that point clearly to them as being part of the same major paleogeographic and tectonic accretionary belt, from perhaps early Paleozoic and certainly from Late Triassic time. The similarities include (1) early Paleozoic quartz-rich clastic rocks in the more eastern terranes; (2) Ordovician ophiolite remnants; (3) Devonian amphibolite facies metamorphic terranes derived from oceanic crustal rocks; (4) early Mesozoic (pre–Middle Jurassic) high pressure-temperature (P/T) metamorphic rocks; (5) Permian to Lower Jurassic chert-rich basinal terranes that locally contain carbonate clasts and lenses derived from paleobiogeographically diverse sources; (6) Carboniferous ophiolitic rocks; (7) Late Triassic ophiolitic rocks; (8) an initiation of island arc activity under marine basinal conditions in western terranes during the Late Triassic; (9) sequences of Early to Middle Jurassic quartz-rich marine strata locally interstratified with the aforementioned chert-rich basinal strata; (10) a distinct suite of Middle Jurassic (165–170 Ma) peridotitic to dioritic intrusive complexes, which cut nearly every constituent terrane but are cut by Late Jurassic (Nevadan) structures; (11) a distinct

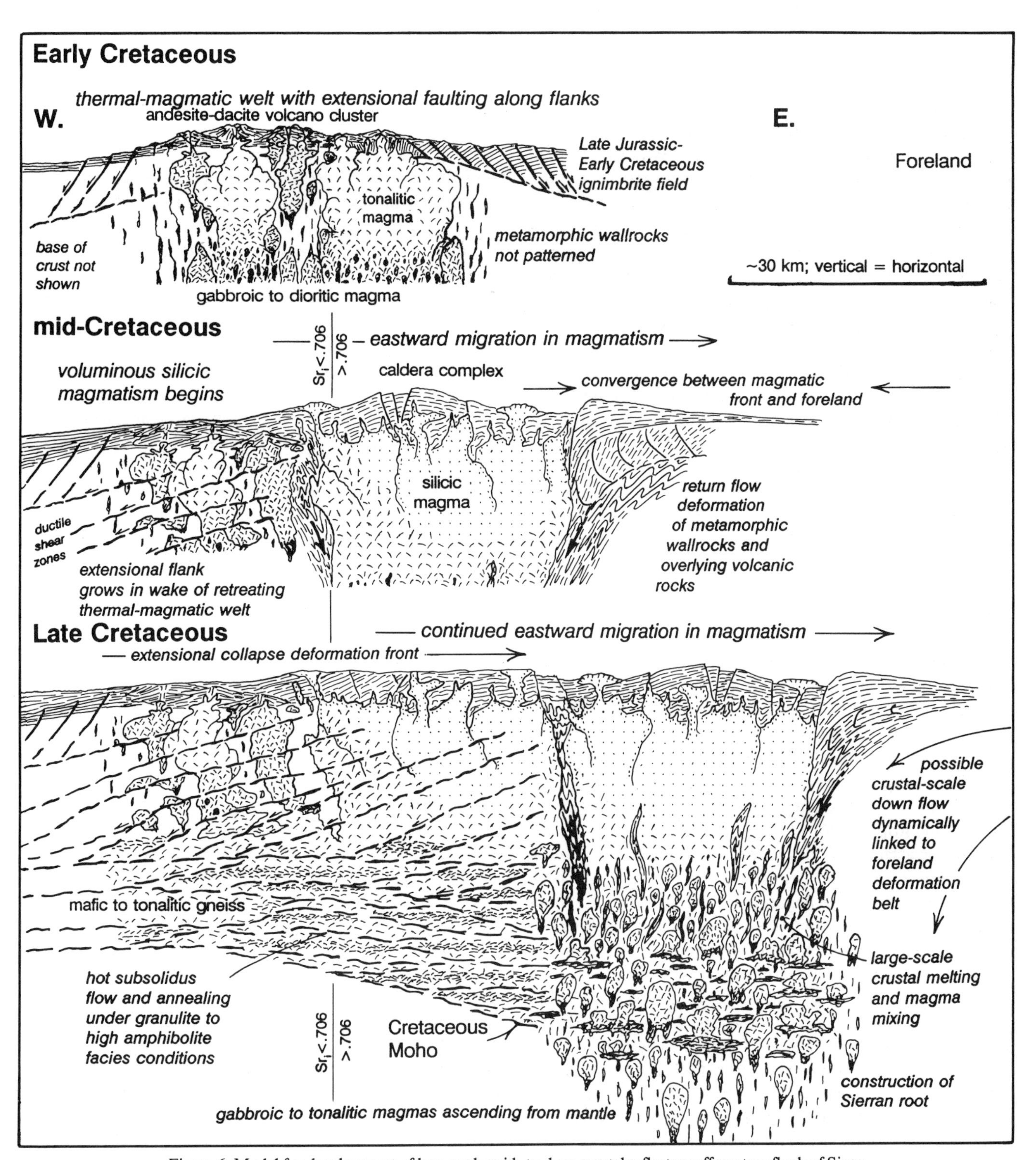

Figure 6. Model for development of low-angle mid- to deep crustal reflectors off western flank of Sierra Nevada batholith (alternative to that shown in C2 tectonic section). Reflection geometry here results from the eastward migration of an extensional collapse front in the wake of the eastward migrating thermal-magmatic welt of the Sierra Nevada batholith. Uppermost crustal extension is accommodated regionally by upper batholith spreading and caldera formation (model modified from Saleeby, 1990b).

pulse of Callovian to Oxfordian island arc volcanism in western terranes; and (12) deposition of chert detritus–rich flysch in the early Kimmeridgean and slaty cleavage formation prior to the end of the Kimmeridgean (Nevadan orogeny). Furthermore, Sierran and Klamath segments of the western Sierra–Klamath accretionary belt both underwent at least intermittent deformation and metamorphism in the Late Triassic–Early Jurassic, and major deformation and metamorphism in the Middle Jurassic that culminated in the Late Jurassic Nevadan orogeny. Such correspondence demonstrates the belt as one major paleogeographic-tectonic system. The only noteworthy differences are in the late Paleozoic volcanic stratigraphies in the Sonomia-related eastern terranes. However, these differences can be explained by expected facies variations within the Sonomia volcanic arc (Miller and others, 1988). In summary, terranes of the western Sierra–Klamath accretionary belt represent a family of fringing island arc, related forearc and interarc basin, and accretionary prism–like assemblages, which were accreted in a series of events beginning with the Permian-Triassic Sonoma orogeny and ending with the Late Jurassic Nevadan orogeny.

Plutons of the continental arc as old as ~140 Ma crosscut the youngest accretion-related structures of the western Sierra–Klamath accretionary belt (Saleeby and Sharp, 1980; Saleeby, 1983; Saleeby and others, 1988). Such plutons indicate a minimum age of accretion along this belt, and they are concurrent with the onset of eastward subduction, which led to the development of the Coast Range accretionary belt. In contrast to the earlier widely held views of the Coast Range belt (Franciscan complex) representing a westward-stepping accretionary prism complex, the C1 and C2 syntheses show clearly that the Coast Range belt was constructed by the accretion of distinct and mainly unrelated terranes. Together these terranes constitute the Franciscan complex. The different terranes of the Franciscan complex can be readily differentiated on the tectonostratigraphic diagrams and strip maps of C1 and C2 as well as in Figure 3. The Coast Range accretionary belt as used here includes the Franciscan complex of ensimatic terranes as well as Salinia, a large Mesozoic sialic crystalline fragment. The Coast Range accretionary belt formed with major involvement of strike-slip faulting and eastward overthrusting, in addition to east-dipping subduction of Pacific basin plates beneath North America. The hanging wall of the east-dipping subduction zone is widely considered to be the Coast Range ophiolite, which separates the western Sierra–Klamath accretionary belt and the overlying Cretaceous-Tertiary Great Valley forearc basin sequence from the Franciscan complex of the Coast Range accretionary belt. The Coast Range ophiolite forms the basement for the Great Valley forearc succession.

The tectonic boundary between the Franciscan complex and the Coast Range ophiolite is defined as the Coast Range thrust, the trace of which extends along the western margin of the Great Valley (Fig. 2) (Bailey and others, 1970). This fault is inferred from (1) the high *P-T* metamorphism and tectonic disruption of the Franciscan complex and absence of these features in the overlying, coeval Great Valley sequence; (2) plate tectonic models requiring that the Franciscan rocks formed in the east-dipping subduction zone; and (3) local low-angle fault contact of the Coast Range ophiolite on the Franciscan (e.g., the Del Puerto Canyon area of C2). Throughout C1, however, where the fault contact is well exposed, it is steeply dipping, usually toward the west (Blake and others, 1984). In these places the faults are Cenozoic and probably still active. Seismic profiling near C2 (Wentworth and others, 1984; Zoback and Wentworth, 1986) suggests that the Coast Range thrust intersects the top of the west-dipping mafic crustal layer of the western Sierra–Klamath accretionary belt, which is inferred to underlie both the Great Valley sequence and perhaps part of the Franciscan complex so that the Franciscan rocks form a tectonic wedge between the underlying Great Valley basement and the overlying Coast Range ophiolite and Great Valley sequence (C1 and C2 sections).

The implications of this cross-sectional view are that some of the older and inner Franciscan terranes with high *P-T* metamorphic mineral assemblages (Pickett Peak and Yolla Bolly) were metamorphosed in a subduction zone outboard of their present position and were subsequently driven eastward and upward onto the Great Valley basement. At the same time the overlying Coast Range ophiolite and Great Valley sequence were delaminated and backthrust as the highest sheets. The timing of this event is not well determined, but must postdate the mid-Cretaceous metamorphic ages (~92 Ma) of the Yolla Bolly terrane (exposed in both C1 and C2) and predate the Late Cretaceous–early Cenozoic strike-slip faulting described below. In C2 the thrust sequence and the Coast Range roof thrust were deformed into a broad late Cenozoic antiform which forms the axis of the Diablo Range. Here much of the Coast Range thrust has been folded to steep dips. The high-angle fault below the Coast Range ophiolite–Great Valley sequence and western Klamaths in C1 is probably an active reverse fault formed in response to subduction of the Gorda plate.

West of the Coast Range thrust are a number of imbricate low-angle thrust faults within the Pickett Peak and Yolla Bolly terranes that emplace more highly deformed and metamorphosed rocks on less-deformed and lower-grade rocks. These thrusts are probably part of the late Mesozoic(?) Coast Range thrust system, which was partly responsible for bringing high *P-T* metamorphic rocks to the surface. Upward transport of the high *P-T* metamorphic rocks was also related to episodes of postaccretional extensional and strike-slip faulting (Platt, 1986; Jayko and others, 1987; McLaughlin and others, 1988). The western margin of the Yolla Bolly and Pickett Peak terranes and associated thrust faults in C1 are truncated by a system of north-northwest–trending high-angle faults that is marked by a broad zone of tectonic melange containing blocks and slabs of Yolla Bolly and Pickett Peak rocks as well as Coast Range ophiolite and Great Valley sequence. Much or all of the melange mixing resulted from pre-San Andreas right-lateral shear. In addition, within this broad zone of right-lateral, strike-slip faulting are numerous blocks and slabs of Middle Jurassic to mid-Cretaceous oceanic seamount volcanic rocks and overlying pelagic radiolarian chert or limestone, believed on the basis of paleomagnetic and paleontological data to have been transported from equatorial latitudes (Alvarez and others, 1980; Sliter, 1984). In the area of C2 a semicoherent sequence of equatorial seamount and pelagic beds of the Permanente terrane lies in a tectonic position analogous to Central melange belt of C1. The Permanente terrane and smaller blocks to the north thus appear to have been accreted and perhaps simultaneously dispersed along a transpressive trench. The trench kinematics are considered transpressive because of the shearing out and dispersal of Permanente terrane and related rocks along the transform melange zone. The subduction zone megathrust may correspond to a deep, east-dipping low-angle reflector observed in C2 to project up to the main exposures of the Permanente terrane and downward beneath the Great Valley.

Plate motion reconstructions (Engebretson, 1982; Page and Engebretson, 1984) suggest that the Franciscan Central melange and Permanente transpressive accretion zone may be a manifestation of the passage of the Kula-Farallon ridge that occurred after 88 Ma, the age of the youngest equatorial fossils in the melange, and before 50 Ma, the age of overlap deposits across these structures in southwest Oregon (Blake and others, 1985).

The western margin of the Central melange in the Coast Range accretionary belt in C1 is a zone of east-dipping thrust faults that juxtaposes the melange above young coherent terranes (Coastal and Yager). These thrust faults appear to mark a change to more normal convergence, probably a return to Farallon–North American plate interaction. The mid-Cenozoic thrusting is delimited by the youngest underthrust rocks (~38 Ma) and late Cenozoic overlap deposits (~24 Ma). The Coastal terrane contains fragments of Late Cretaceous oceanic seamounts and local blocks of high *P-T* metamorphic rock (Coleman and Lee, 1962; Sliter, 1984), whereas the Yager terrane represents a trench-slope basin of Paleocene to Eocene age (Underwood, 1983).

The most recent tectonic event in the Coast Ranges of the C1 area is the eastward thrusting of the King Range terrane onto the Coastal terrane (Griscom, 1980; McLaughlin and others, 1982), probably as a result of compression between the Pacific and North American plates during the passage of the Mendocino triple junction (~2 Ma). The King Range terrane consists of trench turbidites and pelagic deposits as young as middle Miocene (McLaughlin and others, 1982). This and the Yager terrane may be part of the early Tertiary accretionary prism that lies on the eastern edge of the Pacific plate in C2. Migrating with the accretionary prism on the Pacific plate along the San Andreas transform system are Late Cretaceous accretionary prism-type rocks of the San Simeon terrane (Page, 1981) and the Cretaceous sialic crystalline rocks of the Salinia terrane (Ross, 1978). Neither the parentland nor the nature of their common boundary is known for these highly contrasting terranes. The early Tertiary accretionary prism of the Pacific plate is overlapped by Miocene and younger marine basinal strata, which include the active Monterey submarine-fan sequence.

Tectonic accretion is undoubtedly occurring along the inner wall of the Juan De Fuca plate–North American plate junction. Heavy sedimentation has hidden the trench and built out the Heceta bank. Deep-penetration marine seismic studies have not been published for this area.

RESPONSES OF THE NORTH AMERICAN PLATE TO ACCRETIONARY TECTONICS

Accretionary tectonics along the central U.S. Cordilleran margin occurred in two major regimes: (1) an active continental margin, wherein accretion was driven by subduction of Pacific basin crust beneath North America, and (2) a passive continental margin, wherein accretion resulted from the collision between and overriding of the continent by island arc terranes. The Mississippian Antler and Permian-Triassic Sonoma orogenies resulted from collisions of magmatic arcs with the passive margin and the overriding of sedimentary prisms across the outer continental shelf. The effects on the continent of such collisions differ markedly from those associated with active margin tectonics (Speed and Sleep, 1982). There were no thermal effects, magmatism, or metamorphism at shallow levels within the continent during those orogenies. There was no pervasive crustal shortening or mountain building within the continental crust. Deformation within the overriding terranes preceded emplacement, and that of the overridden continental shelf consists of local shear strain and/or thrust imbrication in thin zones below the respective allochtons (Kay and Crawford, 1964; Gilluly and Gates, 1965).

A major effect of the Mississippian Antler orogeny was the generation of an asymmetric foreland basin with an amplitude of ~3.5 km that rimmed the continentward edge of the Roberts Mountain allochthon (Poole, 1974). The basin probably reflects elastic or elasticoviscous downflexing of strong continental lithosphere during vertical loading by the allochthon and broadening during progressive sedimentation, which widened the region of loading (Speed and Sleep, 1982). Certain stratigraphic features below the foreland basin fill suggest that an upbulge migrated eastward ahead of the downwarp, as predicted by the theory of flexure.

In contrast to the extensive foreland basin developed during the Mississippian continental margin event, foreland basin deposits associated with the Triassic Golconda allochthon are recognized only adjacent to the southern third of that allochthon. This may be because the northern Golconda allochthon was too small to cause significant amplitude of flexure, because the allochthon stayed submerged and provided no orogenic sediment to record such a basin, or because the allochthon was later thrust over related foreland basin strata.

The principal effects of the accretion of Sonomia to the passive sialic margin were the addition of a large width to the morphologic continent and the generation of a deep-water basin west of the Sonomian suture and the edge of sialic North America. The Triassic basin was successor to and approximately coextensive with Sonomia and accumulated early pelagic and hemipelagic sediments, and then great thicknesses of Late Triassic terrigenous turbidite (Speed, 1978). The subsidence of the successor basin is hypothesized to have been due to thermal contraction of the lithosphere of Sonomia upon loss of subduction-related heating (Speed, 1977, 1979). Alternatively, if the emplacement of the Golconda allochthon represents a relatively minor backarc submarine thrust belt (e.g., Burchfiel and Davis, 1972), then the Triassic basin may have been largely inherited from the Havallah basin as well as Sonomia bathymetry (Saleeby and Busby-Spera, 1992).

An active margin developed on western North America in Middle Triassic time following the Sonoma orogeny. In addition to terrane accretion, the main effects of active margin tectonism on the continent can be divided into four categories: (1) Mesozoic continental magmatic arc; (2) thin-skinned foreland contraction of Jurassic-Paleocene age; (3) Laramide compressive basement uplifts of early Cenozoic age; and (4) Cenozoic extension and thermal effects. Categories 1–3 are discussed below; category 4 was addressed previously.

Mesozoic continental arc

A continental magmatic arc (Hamilton, 1969) emerged in eastern California and western Nevada (Fig. 1 and C1-C2 strip maps) in Middle to Late Triassic time, probably between 225 and 240 Ma according to oldest dates of silicic volcanic and plutonic rocks (Kistler, 1966, 1990; Brook, 1977; Dilles and Wright, 1988; Saleeby and Busby-Spera, 1992). The arc crossed the suture between accreted terranes of the Nevada accretionary belt and sialic North America at about 38°N (Fig. 1). Manifestations of the arc are silicic and intermediate volcanogenic (ash-flow tuff, breccia, lava, sediments) and plutonic (tonalitic to granitic and local gabbroic) rocks that are as young as ~80 Ma. The arc region is defined by the existence of Mesozoic volcanic rocks and by a high proportion of plutonic rocks (at least half) among pre-Tertiary exposures (Speed, 1982).

The general locus of voluminous Mesozoic continental arc magmatism in eastern California and western Nevada is shown in Figure 1. Magmatic activity was nearly continuous from Late Triassic to Cretaceous time; minima in activity occurred between 190 and 175 Ma and between 160 and 140 Ma (Evernden and Kistler, 1970; Saleeby and Sharp, 1980; Stern and others, 1981; Chen and Moore, 1982; Saleeby and others, 1989a, 1989b). The second magmatic minima has been widely referred to as a gap; no gap exists, however, because plutons, dike swarms, and ignimbrites of this age are widespread but not voluminous as compared to other parts of the Mesozoic (Saleeby and Busby-Spera, 1992). The main locus of Triassic and earliest Jurassic magmatism is poorly known due to tectonic disruption and the invasion of Jurassic and Cretaceous magma systems. Middle Jurassic magmatism appears to be in a belt trending N40°W in present coordinates, extending from the southeast margin of the Sierra Nevada and adjacent Basin and Range to the northwest Sierra and the Klamath Mountains. The most voluminous Cretaceous magmatic locus clearly cuts the Middle Jurassic belt, trends about N20°W along the Sierra Nevada, and then appears to turn northeastward into western Nevada (Kistler, 1974). There is a distinct eastward migration pattern in voluminous magmatism across the batholithic belt throughout Cretaceous time (Evernden and Kistler, 1970; Stern and others, 1981; Chen and Moore, 1982).

The main effects of arc development on the continent were crustal thickening and intra-arc deformation. Such deformation included phases of major extension as well as contraction. Where continental arc rocks invade the sialic continent, crustal thickness is 35 to possibly 55 km, perhaps 5 to 15 km greater than that of normal craton, such as the presumably undeformed crust of the Colorado Plateau (Eaton, 1963; Johnson, 1968; Prohdehl, 1979; Smith, 1978; C2 sections). Moreover, the crust below Sonomia thickens substantially below the Mesozoic continental arc (Eaton, 1963; Prohdehl, 1979; Stauber, 1980; Speed, 1982). Chemical and isotopic studies imply that major magma additions to the crust were mantle derived, but reflective of the different lithospheres in which they were generated and/or contaminated (Hamilton, 1969; Kistler and Peterman, 1973, 1978; Miller and Bradfish, 1980; DePaolo, 1981).

Crustal structure along the axis of the main Cretaceous continental arc in the Sierra Nevada is shown in a three-layer model based on seismic P-wave velocities (C2 section). An upper 6.1 km/s layer is thought to be primarily granitic batholithic material with subordinate steeply dipping screens (pendants) of metamorphic wall rocks. A deeper 6.4 km/s layer is depicted as high amphibolite grade wall rock, probably resulting from crustal anatexis, and recrystallized intermediate to mafic rock that was either magmatically emplaced from greater depth or gravitationally fractionated from the higher level silicic plutons. Sillimanite gneiss and diorite amphibolite xenoliths in late Cenozoic lavas from the C2 area are direct samples of this layer (Domenick and others, 1983). The 6.9 km/s deepest layer of the batholith is depicted as a mixture of granulitic residues from crustal anatexis and recrystallized mafic and ultramafic cumulates contributed from mantle-derived magmas. Garnet granulite and plagioclase-garnet pyroxenite xenoliths are direct samples of this layer (Domenick and others, 1983). Similar features are exposed in deep batholithic rocks of the southern Sierra Nevada, where they occur in association with garnet-bearing mafic tonalites (Ross, 1980, 1983; Sharry, 1981; Sams and others, 1983; Saleeby and others, 1986, 1987; Saleeby, 1990b; Pickett and Saleeby, 1993). Observations in this region indicate abundant mixing between mantle-derived magma and crustal anatectic melts during batholith generation. Nd and Sr isotopic patterns within the central batholith suggest subequal contributions from mantle (subduction zone) sources and recycled sialic material (DePaolo, 1981), although direct studies of deeper-level rocks to the south suggest a higher proportion of mantle-derived material, perhaps ~80% (Saleeby and others, 1987; Pickett and Saleeby, 1989). Fundamental questions about the Sierran batholith are whether the recycled sialic material was Proterozoic crystalline rock or sedimentary derivatives of such materials, and/or the relative proportions of these two. The C2 model for Sierran batholithic structure along with direct observations of deep-level batholithic rocks at the southern end of the range suggest that the middle to lower crust was completely reconstituted during mid-Cretaceous batholithic activity (Saleeby and others, 1986; Saleeby, 1990b).

As discussed above, there is research underway that questions whether a substantial crust root exists under the Sierra Nevada batholith, as shown in the C2 sections (Jones and others, 1990). Tectonic modification of an original root may have taken place in a series of steps, most recently involving lower crustal flow during Basin and Range extension, and early-stage lateral flow and horizontal extension during and immediately following batholith genesis may also have helped attenuate the batholithic crust. The key question is how deep quartz-rich rocks occurred in abundance within the batholith. Recent work on deep-level exposures of the southern Sierra Nevada batholith revealed an abundance of quartz-rich rocks (mafic tonalite gneiss) that were at least 30 km deep within the original batholithic crust. Thus the "root" levels of the batholith may have been highly susceptible to tectonic modification by plastic flow over a very broad thermal range.

Recent gravity and seismic refraction data across northeastern California suggest that the Modoc Plateau and southern Cascade Range in the vicinity of Lassen Peak are underlain by relatively thick, 38 to 45 km, crust (Fuis and others, 1987). These data contradict earlier ideas (Blake and Jones, 1977; Hamilton, 1978) that the area was underlain by rift-related oceanic-type crust. The relative amount of Mesozoic versus Cenozoic crustal thickening by continental arc activity in this region is unknown.

Deformation related to the continental arc consists of protracted shortening strains normal to the arc, arc-parallel shear, and at least local arc-vergent thrusting at its backside. Progressive east-northeast subhorizontal coaxial shortening of Jurassic and mid-Cretaceous age in arc-related strata, a maximum of about 80% in older rocks, was demonstrated by Tobisch and others (1978) and Tobisch and Fiske (1982). The flattening plane parallels axial-planar fabric orientations throughout the pre-Tertiary layered rocks of the arc region between 37° and 40°N (Kistler, 1966; Speed, 1978; Saleeby and Sharp, 1980; Schweickert, 1981; Nokelberg and Kistler, 1980). Such intra-arc strain is thought to represent deformation related to local voluminous magma injection rather than whole arc contraction. Arguments for net crustal extension were given by Tobisch and others (1986), and, along with Saleeby and others (1986) and Saleeby (1990b), they discussed the intra-arc shortening strains (± down-dip constrictional fabrics) in the context of both pluton shouldering strains and downward return flow in the metamorphic septa related to the rise of the silicic magma bodies.

Field and laboratory studies in the deep-level batholithic rocks of the southern Sierra Nevada (Pickett and Saleeby, 1990, 1993; Saleeby, 1990) and deep seismic reflection profiling along the western margin of the central (high-level) Sierra Nevada batholith (Zoback and others, 1985; Geophysical Systems Corporation, 1982, unpublished data; R. Clayton, unpublished data) suggest that the deep levels of the composite batholith may tectonically delaminate during and shortly after igneous emplacement. A family of low, west-dipping middle to deep crustal reflectors along the west-central Sierra were originally interpreted as Nevadan-age (prebatholithic) east-vergent thrusts (Zoback and others, 1985; C2 tectonic section). However, the west-dipping reflectors were imaged to the east past the limits of the western metamorphic belt and into the Sierra Nevada batholith, well beyond the possible extent of Nevadan structures. Field, geochronologic, petrographic, and thermobarometric data from the deep-level batholithic rocks of the southern Sierra Nevada show that at depths of at least 30 km, the predominate quartz-bearing rocks (mainly mafic tonalite) deform penetratively with concentrated strain along anastomosing ductile shear zones during solidus and hot subsolidus conditions. This deformation may accommodate substantial horizontal differential movement within and between slowly cooling plutonic rocks and magma. Figure 6 combines these deep-level observations with the seismic reflection geometry, the high-level batholithic extensional and inflation-return flow strain patterns discussed above, and the migration pattern in Cretaceous magmatism in a regional dynamic model for the composite batholith. The model offers an alternative view for the low-dipping reflectors observed along the western margin of the batholith; i.e., the reflectors originated as a family of low-angle extensional shear zones that developed off the western flank of the eastward-migrating thermal-magmatic axis. These structures sole out into subhorizontal flow structures in the lower crust that were also ductile over an extended time period. Foreland contractile deformation occurred simultaneously to the east, or front edge, of the migrating thermal-magmatic axis.

Mesozoic plutonic rocks occur east of the continental arc as defined earlier, in the region of foreland deformation. These igneous bodies differ from those of the arc by their small diameters at present exposure levels, small proportion among pre-Tertiary rocks, and compositional differences, mainly S-type versus I-type arc plutons (Miller and Bradfish, 1980; Kistler and others, 1981). In eastern Nevada and western Utah, K-Ar uplift dates of plutons are strongly bimodal at 80 and 160 Ma, implying two short-lived pulses of magmatism (Armstrong and Suppe, 1973; Allmendinger and Jordan, 1981). This magmatism may not be subduction-related because of the constancy in position and time of the continental arc in eastern California and western Nevada. Heating of the sialic crust and basal strata of the 10–12-km-thick cover may have been due to thickening by thrust imbrication and/or to convective circulation in a shallow backarc asthenosphere, causing anatexis in the foreland region (Armstrong, 1968; Best and others, 1974; Kistler and others, 1981; Snoke and Lush, 1984; Snoke and Miller, 1987; Speed and others, 1987).

Foreland deformation

The region east of the continental arc in westernmost Nevada and eastern California to the front of the Rocky Mountains in Wyoming and Colorado was the foreland of western North America that underwent mainly contractile deformation in late Mesozoic and early Cenozoic time. Such contraction must have been related to plate boundary tectonics at the western edge of North America, then in and west of the Sierran foothills, and/or to motions of the North American plate relative to subjacent tracts. Foreland deformation occurred in two main regions, discussed below, over different durations. The first extends from the continental arc east to and including the Sevier thrust belt in Utah and southern Nevada and its northerly prolongation, commonly called the Idaho-Wyming thrust belt. Its deformation is mainly thin or moderately thin skinned, and occurred from Jurassic to Paleocene time. The second, the Laramide deformation, occurs east of the first in Wyoming, Utah, and Colorado. Its deformation is mainly thick skinned and basement involved, and occurred in Paleocene and Eocene time.

Western foreland region

In Nevada and Utah, foreland deformation of the western region was heterogeneous in surface distribution and depth and occurred over a remarkably broad region compared to others of the world's foreland thrust belts (Armstrong, 1968; Speed, 1982). The Paleozoic and lower Mesozoic strata that were deposited on the North American shelf record thin-skinned tectonics in a braided map pattern (Fig. 7). In this region, the three main thrust belts in which horizontal displacements ramped to surface are the Sevier belt (Armstrong, 1968; Bruhn and Beck, 1981; Allmendinger and others, 1983, 1986), the Eureka belt (Speed, 1982), and the Winnemucca belt (Speed, 1978, 1982; Oldow, 1981, 1984; Speed and others, 1987). Between the Sevier and Eureka belts, there is a broad tectonic enclave of little shallow deformation (Armstrong, 1972; Speed, 1982) but substantial deformation and metamorphism at depths below 5–10 km (Miller, 1980; Allmendinger and Jordan, 1981, 1984; Snoke and Miller, 1987). Between the Eureka and Winnemucca belts there is a belt of moderate unroofing of low-grade metamorphic rocks (Speed and others, 1987).

Foreland thrusting of the western region continues north into the Idaho-Wyoming thrust belt (Armstrong and Oriel, 1965; Royse and others, 1975; Dixon, 1982) (Fig. 7) only from the Sevier thrust belt. The more westerly belts are either missing to the north, due perhaps to tectonic attrition and rafting away in late Mesozoic time (Speed, 1982; Saleeby and Busby-Spera, 1992), or they are unrecognized or engulfed in the Late Cretaceous Idaho batholith (Fig. 7).

Both the Sevier and Idaho-Wyoming thrust belts are zones in which faults ramp up to the free surface with maximum throw of ~7 km, syntectonic sedimentation, eastward vergence, and little involvement of crystalline basement reflecting emplacement of nappes from deep levels. The Sevier belt lies above the Paleozoic craton-shelf hinge and rapidly eastward-thinning cover section. The few crystalline slices within the Sevier belt are probably due to intersection of the frontal footwall ramp with the top of the west-facing late Precambrian hinge in the sialic crust. The Sevier belt and regions immediately west of it record cratonward piggyback propagation of thrust sheets from Late Jurassic through Paleocene time; the total displacement is about 80 km (Armstrong, 1968; Royse and others, 1975; Allmendinger and Jordan, 1981; Heller and others, 1986). The timing, propagation, and total displacement of the Eureka belt are less well known, but the belt includes Lower Cretaceous(?) sedimentary rocks (Newark Canyon Group) that are probably syntectonic.

The tectonic enclave (Fig. 7) is defined by an upper interval of cover strata of variable thickness from the Triassic down to Mississippian or deeper that underwent little or no deformation or stripping in Mesozoic time (Armstrong, 1972; Gans and Miller, 1983). Below this upper interval, however, Mesozoic deformation and metamorphism are known in deeper cover strata, mainly upper Precambrian and Cambrian, that are now exposed in areas unroofed in Cenozoic core complexes (Howard, 1971, 1980; Armstrong, 1972; Snoke, 1980; Miller, 1982; Allmendinger and Jordan, 1984; Snoke and Miller, 1987). The Mesozoic structures in the lower stratal interval include recumbent fold nappes, thrusts, and low-angle normal faults. Such relations indicate that Mesozoic displacements were probably continuous between the Sevier and Eureka belts on an extensive detachment through the lower interval of the tectonic enclave (Speed, 1982). The detachment was a thrust flat, whereas the two thrust belts were ramp zones. The upper interval in the enclave was thus a coherent thrust sheet of unusual dimensions. It was either decoupled from the more deformable lower interval that underwent horizontal contraction or, more likely, was the top layer in a system of horizontal heterogeneous simple shear. The preservation of the Paleozoic-Triassic sedimentary cover across the enclave into mid-Cenozoic time (Armstrong, 1972) indicates that large uplift and tectonic or erosional stripping at the surface did not occur in the Mesozoic, and that horizontal simple shear was more significant than contraction in the displacements through the lower interval. The Sevier and Eureka belts both lie above major declivities in crustal layering (Speed, 1982). The Sevier is above the west-facing craton-shelf hinge, which has an amplitude of 10 km on the top of crystalline basement. The Eureka is at the toe of the Robert Mountains allochthon, which caused a pronounced west-facing declivity in the Mississippian, as recorded by the related foreland basin. In contrast, layering in the tectonic enclave may have been more plane parallel, thus without cause for displacement climb.

The northern half of the Winnemucca thrust belt contains three regional structures (Fig. 7): the Fencemaker allochthon, the Willow Creek allochthon, and an intervening autochthon that lies above or near the Precambrian edge of sialic North America (Speed and others, 1987). The Fencemaker allochthon is constituted mainly of the Triassic and Jurassic basinal strata that are so widespread in northwestern Nevada. Contraction of the basin in a west-northwest–east-southeast direction at times in the Jurassic and Cretaceous caused the allochthon to develop and ride relatively west above the continental arc at its western side and relatively southeast above the western flank of the Winnemucca autochthon on its eastern margin (the Fencemaker thrust). The Willow Creek allochthon rode west at least 30 km over the

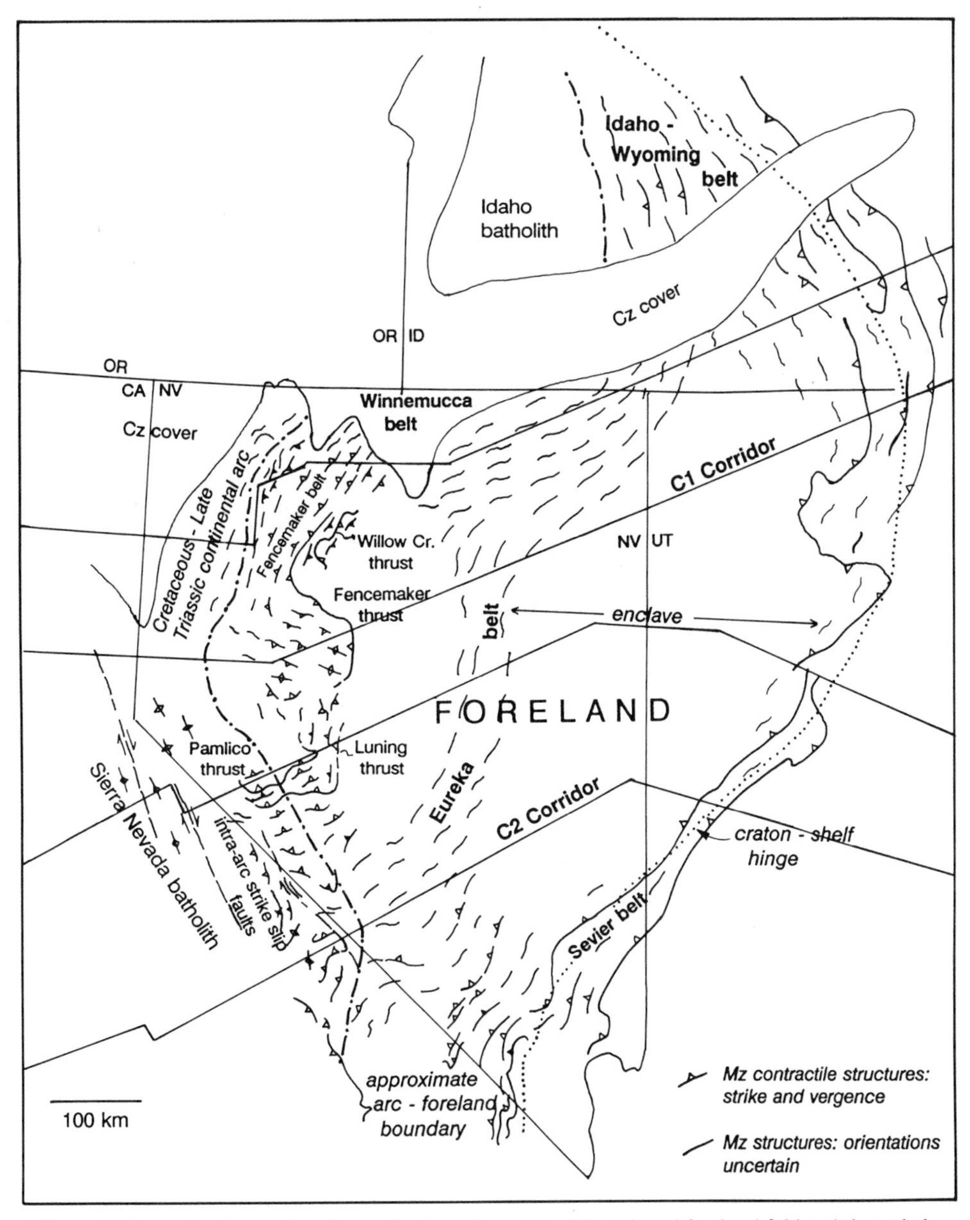

Figure 7. Generalized map showing major branches of the thin-skinned foreland fold and thrust belt.

eastern flank of the Winnemucca autochthon in Late Jurassic time and perhaps later, after eastward emplacement of the eastern edge of the Fencemaker allochthon (Elison and Speed, 1989; Speed and others, 1987; Heck and Speed, 1987). The Willow Creek allochthon may have been due to gravity spreading westward from the unroofed region between the Winnemucca and Eureka thrust belts. The unroofed region may represent upward volume transfer from a horizontally contracting zone of ductile lower crust, and the Eureka belt may have been conjugate to the Winnemucca on the east side of the unroofed zone.

The southern part of the Winnemucca thrust belt contains the Luning, Pamlico, and unnamed thrusts of Jurassic and Cretaceous age (Fig. 7) (Speed, 1978; Speed and Kistler, 1980; Oldow, 1981, 1984). These had Late Jurassic–Early Cretaceous southeasterly, continentward transport that may have been between 50 and 500 km (Oldow, 1984; Oldow and others, 1984; Maher and Kirschvink, 1988). The tectonic transport is thought to include substantial left-lateral shear on northwest-striking planes, displacement increasing toward the southwest (Silberling and John, 1989). This transport pattern is modeled as relative motion of the northwest Nevada basinal terrane against the northeast-trending tapered edge of Proterozoic North America and its allochthons in response to North America's abrupt northwest acceleration in Late Jurassic–earliest Cretaceous time

(May and Butler, 1986; Saleeby and Busby-Spera, 1992; Saleeby and Busby, 1993).

The detachment fault and/or horizontal shear zone postulated to have connected displacements of the Sevier and Eureka belts below the tectonic enclave may have extended west to the Winnemucca thrust belt (Speed, 1982). A more recent model implies that this detachment and another below the western Winnemucca thrust belt merge in the zone of ductile basement below the unroofed Winnemucca autochthon between the Winnemucca and Eureka thrust belts (Speed and others, 1987). The entire region of the Winnemucca belt to at least the western Sevier belt was probably kinematically connected in the Jurassic, and the motions propagated eastward in the Sevier belt in the Cretaceous. Assuming that crustal sections are pinned in the unroofed region on the west and at the frontal thrust of the Sevier belt, balanced-section analysis suggests an average of 18% shortening across the foreland region between the pinlines (Speed and others, 1987). This analysis does not consider the tangential component of motion that appears to have been important on the Luning system (Oldow, 1984; Oldow and others, 1984).

Laramide foreland region

The 400-km-wide Laramide belt developed on cratonal North America where the thickness of the cover above Archean basement in Wyoming and much of Utah in C1 and C2 did not exceed 1 km (Lowell, 1974; Smithson and others, 1978; Blackstone, 1979; Brewer and others, 1982). The Laramide motions were Paleocene to late Eocene, apparently overlapping those of the Sevier belt to the west. The Laramide belt is notable for its position far inland from the contemporaneous margin, its large structural relief, and its development at about the time the continental arc became inactive.

The Laramide belt consists mainly of fault-bounded blocks of older Precambrian crystalline basement. Faults include major plane thrusts that have been traced by reflection profiling as deep as 24 km and record significant crustal contraction (Smithson and others, 1978; Brewer and others, 1982). Vergences on major thrusts are outward from the boundaries of the belt, and regional contraction was between east-west and northeast-southwest. Basement blocks are also bounded by steeply dipping reverse and normal faults connected by transfer structures and are overlain by drape-folded cover.

Plate tectonic explanations for the Laramide deformation have appealed to marked acceleration and related shallowing of the dip of the subcontinental slab at the end of the Cretaceous, thus causing compression and uplift of this far-inland region (Lowell, 1974; Dickinson and Snyder, 1978). This explanation also accounts for the extinction of the continental arc and rapid eastward sweeping of magmatism in early Cenozoic time, at least in the southern Cordillera (Coney and Reynolds, 1977). In this view, subsequent mid-Tertiary extension in the eastern Basin and Range province to the west of the Laramide uplifts resulted from a pulling effect related to deceleration and steepening of the subcontinental slab. Subsequent to this phase of tectonics the modern plate regime was established, as discussed earlier.

TECTONIC ATTRITION, RIFTING, AND TRANSFORM MODIFICATIONS

An outstanding feature of the southwest Cordilleran margin in the region of the C1 and C2 transects is the evidence for substantial tectonic removal of Proterozoic North America and accreted terranes. The initial margin as well as its modern plate juncture configuration exhibit such removal events, and there is much evidence to suggest removal events during intermediate phases of development. The earliest removal event is clearly marked by the formation of the passive margin. Later events appear to have involved transform faulting in conjunction with subduction or rifting. There is no evidence that subduction alone resulted in the loss of sialic North American crust.

Passive margin formation

The greatest modification to western North America was the formation of a passive margin in late Precambrian time (Stewart, 1972, 1976). The event modified North America by the rifting and drifting away to the west of an unknown but possibly large tract of continent and the growth of an ocean basin against the rifted margin. The strike of the new margin was apparently at high angles to structures of earlier layered rocks, suggestive that Proterozoic rifting was due to a major plate rearrangement. The Proterozoic passive margin in western North America is also noteworthy for the unusually wide zone of subsided, presumably attenuated sialic crust that is covered by the clastic wedge. This zone is as wide as 600 km, although changes in width due to superposed Phanerozoic contraction and extension are difficult to evaluate.

Sierran region continental truncation

The axis of the Sierra Nevada batholith and its metamorphic wall-rock terranes appear to trend obliquely across facies trends in the Paleozoic stable shelf sequence to the east as well as the general trend of the Nevada accretionary belt (Fig. 1). These relations were cited by Hamilton (1969), Burchfiel and Davis (1972), and Saleeby and others (1978) as evidence for an early Mesozoic (post-Sonoman) continental margin truncation event along the Sierran trend. Structures produced by the hypothetical truncation are masked by the continental arc, related structures, intra-arc northwest-striking dextral shears including the Mina deflection, and by foreland thrust-belt structures which bend on the east side of the Sierra Nevada into the presumed truncation zone. In the southern Sierra Nevada, a belt of pendants consisting of displaced lower Paleozoic stable shelf strata and displaced Antler belt rocks lies outboard of the truncation zone (Walker, 1988; Lahren and Schweickert, 1989; Lahren and others, 1990; Dunne and Suczek, 1991). The only known Paleozoic rocks to the west of this belt of pendants are ophiolitic melange, which locally lies unconformably beneath Upper Triassic to Lower Jurassic basinal strata of the Kings sequence (Saleeby and others, 1978; Saleeby and Busby-Spera, 1992). The locus of juxtaposition between Paleozoic ophiolitic basement and rocks to the east is the Foothills suture. To the north the Foothills suture joins similar Paleozoic ophiolite and overlying early Mesozoic ensimatic terranes to the Sonomia-related northern Sierra and eastern Klamath terranes (Saleeby, 1981, 1982, 1990a, 1992). The southwestern end of Sonomia appears to have been truncated along this segment of the suture prior to or during juxtaposition of the ophiolite-based terranes.

The occurrence of transform fracture-zone ophiolitic assemblages along the Foothills suture (Saleeby, 1977, 1979, 1982, 1990a, 1992) suggests that the suture was originally a plate boundary transform system. The boundary structure is in the proper tectonic position and of the right age to be the hypothetical truncation zone discussed above. The transform system apparently cut obliquely across Sonomia as well as Precambrian sialic North America and its stable shelf, and presumably caused the tectonic removal of fragments of each (Speed, 1977, 1979; Saleeby, 1977, 1981, 1992). The main removal event was probably as early as Permian-Carboniferous time (Walker, 1988; Saleeby, 1981; Stevens and others, 1992), superposed trench-linked transform displacement having occurred possibly in Triassic–Early Jurassic time, Late Jurassic–earliest Cretaceous time, and in Late Cretaceous time (Saleeby and Busby-Spera, 1992; see discussions below on intra-arc strike-slip faulting).

Transform fault-zone assemblages with igneous ages ranging from ~300 Ma to ~210 Ma occur along the Foothills suture zone (Saleeby and Sharp, 1980; Saleeby, 1982, 1990; Shaw and others, 1987; Saleeby and others, 1989a). Saleeby (1981) proposed an original sinistral transform truncation system with the southward offset of Antler belt rocks to the northern Mojave region in Sonoman or immediate post-Sonoman time. Walker (1988) indicated sinistral offset of the Antler belt rocks in Permian-Carboniferous time, and suggested a late Paleozoic age for the truncation event. This is in apparent agreement with an ~300 Ma clustering of igneous ages in the Sierran foothills transform ophiolite assemblages. The implications of these time relations are that the truncation event may have been active during the left-oblique encroachment of Sonomia onto the North American margin (Speed, 1979; Saleeby, 1992).

In addition to the fragments of displaced Antler belt rocks in the northern Mojave and southeastern Sierra regions, other candidates for the displaced fragments include the Caborca-Hermosillio block of northwest Mexico (Silver and Anderson, 1974; Walker, 1988; Saleeby, 1992; Stevens and others, 1992) and Stikine and the Yukon-Tanana terranes of the northwest Cordillera (see Monger, 1975, 1977; Tempelman-Kuit, 1979; Coney and others, 1980). The former constitutes a major sialic crust-stable shelf fragment, the latter, Sonomia and possible early Paleozoic North America outer shelf and continental rise fragments. The latter would require major dextral offset during or after the truncation event in order to emplace these terranes into their current Pacific Northwest settings. A reasonable alternative to the truncation model is that the apparent truncation zone was inherited from a bight in the late Precambrian rifted margin (Kistler, 1978; Oldow, 1984) and that intra-arc shear of a lesser magnitude has given the apparent truncation zone a younger (Phanerozoic) appearance. The concentration of batholithic plutonism along the truncation zone greatly complicates its interpretation.

Tectonic attrition of the sialic edge in Oregon and Idaho

A large fragment of sialic North America and possibly some early accreted terranes (Sonomia and possibly others) is thought to have been removed from the region of southeastern Oregon, southwestern Idaho, and possibly northern Nevada and California just north of C1 (Hamilton, 1976; Speed, 1982; Saleeby and Busby-Spera, 1992; Saleeby, 1992). Evidence is that (1) as discussed above, the late Precambrian edge and pre-Jurassic accreted terranes and shelf cover that exist in Nevada are mainly missing from the region to the north; (2) the present edge of Proterozoic North America in Idaho intersects the Paleozoic platform-shelf hinge, an unlikely primary arrangement; and (3) displaced terranes with large mid-Mesozoic rotations (Hillhouse and others, 1982), that are absent in central Nevada, sutured in the Cretaceous against the sialic edge in Idaho. The removal of the continental fragment in Oregon and Idaho may have resulted from a trench-linked transform system that propagated into the inboard of the sialic margin during highly oblique convergence. Continued oblique convergence replaced the missing continental fragment(s) with the accreted ensimatic terranes along the truncation locus (Idaho suture zone). The earliest phases of continental truncation along the Idaho suture may have begun as early as the late Paleozoic truncation event, which affected the Sierra Nevada region (Eisbacher, 1983; Saleeby, 1992). Like the Sierra Nevada truncation zone, the Idaho–eastern Oregon zone most clearly displays younger Mesozoic structures, and its crosscutting batholithic complex greatly complicates its interpretation.

Reconfiguration of the continental edge by intra-arc strike-slip motions

The continental arc was the locus of significant margin-parallel dextral shear during Cretaceous time. Within the main axis of arc activity, displacements have been difficult to quantify due to the arcs superpositioning along longitudinally oriented accreted terranes and earlier strike slip as well as contractile structures. Lahren and others (1990) suggested that there was as much as 400 km dextral offset of lower Paleozoic stable shelf strata in a metamorphic pendant of the east-central Sierra, although the hypothetical fault responsible for the offset has been obliterated by the batholith.

Large components of longitudinal translation within the

composite batholith appear to have occurred as the crust distended obliquely in increments along successive chains of plutons. A number of broad structural relations within and between plutons suggest a distinct ~100 km scale right-stepping pattern in the loci of magmatism repeated in time as well as space. Furthermore, a number of substantial steeply dipping ductile shear zones appear to bound the margins of, and extend laterally beyond, some of the largest elongate plutons. Some of the shear zones are synbatholithic dextral faults (Saleeby and Busby-Spera, 1986; Segall and others, 1990; Tikoff and Teyssier, 1992). The shear zones could represent transform-like shear zones that linked en echelon batholithic spreading centers. The distinct northward extending prong in the initial $^{87}Sr/^{86}Sr = 0.706$ isopleth along the Sierra Nevada batholith may reflect in part successive episodes of dextral displacement in sialic domains of the batholith as the batholith formed.

To the east of the main locus of Cretaceous arc magmatism the Proterozoic sialic edge, Sonomia, and the parautochthonous Triassic and Jurassic continental arc and foreland cover are apparently in a late Mesozoic orocline, the Mina deflection (Fig. 1 and C1 strip map). Although the deformation paths and timing of the deflection are uncertain, at least some of the displacements are probably related to north-northwest–striking right-slip faults and/or shear zones within and paralleling the axis of the Mesozoic arc (Speed, 1978, 1982; Kistler and others, 1971, 1980; Albers, 1967; Stewart and others, 1968). If it is entirely deformational, the Mina deflection probably reflects eastward-diminishing right-slip offsets and shear strain with an eastern boundary ~100 km into western Nevada. Paleomagnetic evidence indicates that if significant rotations occurred in the Mina deflection, they occurred before Late Cretaceous time (Geissman and others, 1984). Kistler (1978) discounted the existence of a shear zone and explained the Mina deflection as a transform offset in the Precambrian passive margin. Oldow (1984) argued that both left- and right-lateral shear occurred in this region at times in the Mesozoic, but that the Mina deflection is a pre-Mesozoic feature and unrelated to Mesozoic tectonics. Saleeby and Busby-Spera (1992) interpreted the Mina deflection as a Middle Jurassic feature that formed along the back edge of the continental arc during Siskiyou dextral transpression.

Strike-slip faulting within accretionary prisms

As discussed earlier, our structural synthesis of the Franciscan complex shows the widely believed idea that it is an autochthonous accretionary prism to be a misleading oversimplification. In fact, the Franciscan complex comprises a number of displaced accretionary prism fragments, as well as other tectonic elements, that have been assembled by major strike-slip faulting and eastward overthrusting, as well as by east-dipping subduction. In addition, the main Central melange zone of the Franciscan complex is viewed here as a major trench-linked dextral transform zone. We suggest that major fragments of accreted sediments and oceanic crustal rocks were removed from varied forearc positions and underwent major displacements by such transform motion. Examples of dispersed accretionary prism complexes include the Paleogene–early Neogene prism and the Cretaceous San Simeon terrane that compose part of the eastern edge of the Pacific plate. The fundamental question concerning transform removal events throughout the Coast Range accretionary belt is whether the transforms represent major plate junctures, such as the modern San Andreas boundary transform, or whether they formed within active accretionary prism-forearc areas by trench-linked transform faulting during oblique subduction phases. Another important question concerns the possibility that the eastward overthrust events in the Coast Range belt were driven by major transform episodes. Modern eastward overthrusting in the Coast Ranges is rooted into the San Andreas fault (C2 strip map and sections) and in the C1 area is related to the passage of the Mendocino triple junction (C1 strip map). The tectonic removal phenomena of the modern plate juncture system were discussed earlier. The problem of total offset along the San Andreas fault and the possibility of a proto–San Andreas fault relate directly to the broader issue of Franciscan transform tectonics.

TECTONIC INHERITANCE AND REMOBILIZATION

East of the Sierra Nevada, there is abundant evidence for the influence of older features on the form and position of younger structures, but reactivation of sutures in this region to new major displacement zones or continental boundaries does not seem to have occurred.

The edge of Proterozoic North America in Nevada has probably been maintained through Phanerozoic time as a distinct but covered boundary. This ancient ocean-continent transition influenced later tectonics by forming a buttress to which island-arc related terranes sutured in Mississippian and Permian-Triassic time. This transition is thought to reflect large differences in temperature gradients and degree of subsequent thermal contraction across the suture between the sialic continent and the arc terrane of Sonomia, which became inactive upon collision (Speed, 1979). It approximately underlies a major Triassic transition between a carbonate shelf to the east and deep-water basin that was successor to Sonomia to the west. The ancient sialic edge also influenced the positions of zones of imbrication and surface breakthrough of displacements during Jurassic and Cretaceous foreland contraction. The imbrication may reflect a ramp on the basement surface across the sutured edge of the continent (Speed, 1982).

Other evident inheritances occur in the foreland deformation belts. The Sevier belt is coincident over 1000 km length with the hingeline between continental platform and subsiding shelf. The hingeline thus defines a west-facing ramp on the top of the crystalline basement and also in horizons in suprajacent layered rocks. Another ramp effect is the localization of the Eureka thrust belt at the eastern edge of the Roberts Mountains allochthon, a locus of maximum rate of change of downbuckling of horizons in the autochthon due to loading by the allochthon (Speed and Sleep, 1982).

The compressional block faulting of early Cenozoic Laramide foreland deformation probably reflects the influence and/or reactivation of Precambrian basement structures. The rectangular plan form of Laramide uplifts and basins with bounding reverse and normal faults connected by strike-slip faults is interpreted to reflect failure partly controlled by the ancient grain of the crystalline basement (Matthews and Work, 1978).

The zonation of Neogene faulting in the western Basin and Range province between the Walker lane, which contains strike-slip and dip-slip faults, and the classic mainly dip-slip faults in zones to the east may be influenced by older structure. The Walker lane includes at least part of the region in which arc-parallel strike-slip faulting occurred in Mesozoic time. The most conspicuous set of Neogene strike-slip faults parallels the Mesozoic faults and may represent reactivations of the older faults. It is not clear whether the north-northwest–trending right-slip faults of the Walker lane take up some of the Pacific–North American relative plate motion or whether they are simply faults of a conjugate set that take up pure shear with east-west extension. If the first hypothesis is correct, the very wide zone of Neogene plate boundary deformation may be attributable to inherited structures of appropriate orientation.

There is an evident spatial correspondence in the extent of deformation in western sialic North America during three discrete pulses that represent greatly different tectonic settings: (1) a region of vertically attenuated, presumably rifted sialic crust in late Precambrian passive margin formation, (2) a region of thin-skinned contraction during late Mesozoic foreland deformations, and (3) a region of late Cenozoic extensional basin and range faulting. The cause of repetitive deformation over such a broad zone is not clear, but could be related to pre–Late Proterozoic lithospheric structure.

Tectonic inheritance and remobilization are manifest in several ways along the axial Sierra Nevada. (1) As noted above, Paleozoic–early Mesozoic ophiolite assemblages of the western Sierra–Klamath accretionary belt with transform fracture-zone features and a zone of apparent continental truncation coincide with the Foothills suture. Such relations suggest nucleation of suture-zone thrust tectonics along a transform boundary (Saleeby, 1981, 1992). (2) Intra-arc dextral shears parallel the truncation and suture zone as well as the long dimension of large batholithic structures. The largest and generally youngest batholithic plutons step spatially in a dextral fashion, suggesting that the entire crust of the Sierra Nevada extended in a regional dextral shear field during Late Cretaceous (i.e., 85–90 Ma) magma emplacement events (Saleeby, 1981). (3) At the southern end of the Smartville intra-arc ophiolite, its Middle Jurassic (~162 Ma) sheeted dike swarms and tabular gabbroidal plutons intruded along a preexisting melange fabric in Paleozoic ophiolitic basement of the Jurassic arc (Saleeby, 1982). (4) Late-phase Nevadan high-angle reverse faulting of the eastward-directed Nevadan nappe sequence follows in strike and dip direction earlier east-dipping structures of the Foothills suture zone.

Between the northern Sierra Nevada and southern Klamath Mountains there is evidence for ~100 km of Cretaceous left slip in an east-west direction (Jones and Irwin, 1971). The opposing vergence of Nevadan structures between the Klamath Mountains (west vergence) and the northern Sierra (east vergence) requires a Late Jurassic left-lateral tear in the Sierra-Klamath accretionary belt between the two mountain ranges. The Cretaceous offset between the ranges apparently followed this trend.

In considering processes in tectonic inheritance and remobilization, the C2 transect shows clearly an unpredictable element. The modern San Andreas transform system did not propagate along the preexisting trench axis, but left intact a Paleogene accretionary prism; in addition, the system not only broke the North American plate inboard of the prism, but also liberated a large fragment (Salinia) of the batholithic material from the Mesozoic continental arc in doing so. This seemingly unpredictable behavior may stem from major earlier transform breaks within the Coast Range accretionary belt and in the Mesozoic continental arc which were remobilized during the Neogene plate margin regime.

MAJOR BASINS AND THEIR SEDIMENTARY ACCUMULATIONS

A wide variety of basins developed in conjunction with the evolution of the Phanerozoic southwest Cordillera. Major basinal regimes are discussed below and consist of oceanic, transitional, and continental.

Oceanic basins

The oceanic basins include the Pacific, Farallon, and Kula plates that have interacted with the Cordilleran margin since mid-Mesozoic time as well as plates of Panthalassa, the Pacific basin predecessor. The development of a passive margin late in Proterozoic time presumably generated Panthalassa oceanic lithosphere not far west of the present sialic edge. The length of the Late Proterozoic oceanic plate(s) along the newly generated margin was at least 2000 km (Stewart, 1976); its maximum width is unknown but was probably substantial. The clastic wedge that was shed west from the continent is recognized only above the region of attenuated sialic continental crust, and the whereabouts of the large sedimentary rise that presumably lay above oceanic crust is uncertain. In central Nevada, Silurian and Devonian shelf-edge carbonates exist near to or above the sialic edge. If former extensive rise strata of late Precambrian and Early Cambrian age existed, they may since have been subducted below collided terranes or obducted above the North American shelf during such collisions, for example, the Roberts Mountains allochthon. Quartz-rich clastic strata of the early Paleozoic Shoo Fly complex may also represent vestiges of such rise strata (Schweickert and Snyder, 1981), and if so, interleaved ophiolitic melange with 600 Ma igneous blocks (Saleeby, 1990a) may represent a rare vestige of the eastern Panthalassa plate. Such a view would imply incorporation of rise strata and its oceanic basement into the Sonomia arc basement.

The early Paleozoic paleogeography west of the sialic edge during the passive margin regime is difficult to reconstruct. Alternatives are that open oceans with arcs migrating toward or along North America existed for most of the Paleozoic (Moores, 1970; Speed and Sleep, 1982) or that marginal basins developed between the sialic edge and fringing island arcs of long-term residence (Burchfiel and Davis, 1972). The alternatives govern interpretations of sources of Paleozoic sedimentary rocks in the Roberts Mountains and Golconda allochthons. The open ocean case holds that both allochthons contain turbiditic strata of North American provenance together with pelagic beds swept up by migrating arcs over a long reach of the North American margin. The marginal basin case argues for a more local provenance.

Important remnants of Panthalassa lithosphere are preserved in the Trinity peridotite massif of the eastern Klamath Mountains (C1 strip map; Mattinson and Hopson, 1972; Lindsley-Griffin, 1977; Quick, 1981; Jacobsen and others, 1984) and the Kings River ophiolite of the southwest Sierra Nevada (Saleeby, 1978; Shaw and others, 1987). The Trinity massif consists primarily of voluminous upper mantle tectonites with little crustal section. The Kings River ophiolite consists of a complete upper mantle and mafic crustal section, although pelagic sediments are not present. The critical feature of both the Trinity massif and the Kings River ophiolite is their highly depleted parent mantle source as recorded in their initial Sr, Nd, and Pb isotopic systematics. The isotopic signatures of the source appear to require a mantle melting and magma extraction at ~2.6 Ga, and isolation of the mantle regime for ~2 b.y. prior to the generation of the Trinity and Kings River lithospheric sequences (Shaw and others, 1987). It is interesting that the only two Ordovician oceanic lithosphere remnants thus far recognized in the North American Cordillera reflect such a geochemical evolution, suggesting that a sizable body of such Paleozoic Panthalassa mantle once existed. Rare samples of modern or young mid-ocean ridge basalt (MORB) displaying extremely depleted Nd, Sr, and Pb isotopic compositions may represent extractions from remnants of such mantle. The existence of such mantle also suggests the possibility of mantle refluxing with sialic materials as a mechanism for yielding the relatively less depleted MORB associations of the modern ocean basins.

Remnants of late Paleozoic equatorial Panthalassa crust occur as basaltic seamount fragments with shallow-water carbonate buildups and pelagic-hemipelagic sediments in the Calaveras belts of the western Sierra and North Fork terrane of the central Klamaths. The distinct feature of these oceanic supracrustal assemblages is a Permian-Carboniferous fauna of "Tethyan" (paleoequatorial oceanic) affinity in the carbonate buildups (Douglass, 1967; Davis and others, 1978; Saleeby, 1979; Ando and others, 1983). Ophiolitic igneous sequences with ages clustering around 300 Ma are coextensive with the seamounts and in a number of locations are closely related to such supracrustal oceanic rocks (Saleeby and Sharp, 1980; Saleeby, 1982, 1990; Ando and others, 1983; Saleeby and others, 1989a). A number of these sequences contain transform fracture-zone assemblages. The late Paleozoic equatorial Panthalassa remnants were accreted to North America during Triassic and Jurassic active margin tectonism in the western Sierra–Klamath accretionary belt.

Small remnants of Cretaceous oceanic crust from the conjugate to that in the modern western Pacific basin and overlying pelagic sediments are preserved in the Coast Range accretionary belt as a semicoherent terrane in C2 (Permanente), and in C1 as tectonic fragments in the Central melange and in melange units of accretionary prism complexes to the west. Such remnants consist of basaltic pillow lava of seamount origin with overlying Upper Cretaceous radiolarian chert and pelagic foraminiferal limestone. Paleomagnetic and paleobiogeographic data indicate that these sequences originated in equatorial paleolatitudes (Alvarez and others, 1980; Sliter, 1984). Stratigraphic and paleontological data from Deep Sea Drilling Project sites 310, 464, 465, and 466 (Vallier and others, 1983; Sliter, 1984) on the Hess Rise of the central Pacific indicate that these upper oceanic crustal remnants could have originated as part of that rise system. This suggests northeast plate transport of thousands of kilometers for the Coast Range remnants of the rise system following spreading separation from Pacific plate counterparts. Fundamental questions remain on whether the Coast Range remnants of the rise system traveled on the Farallon, Kula, or some unrecognized Pacific basin plate, and about the relation between accretion to North America and the passage of the Farallon-Kula triple junction.

Structural and stratigraphic features of the Pacific and Juan De Fuca plates underlying the western ends of C1 and C2 were discussed above. Neogene deposits and unconformities above these plates represent basinal and emergent regimes evolving laterally in heavily sedimented trench, migrating triple junction, and transform fault environments. Sedimentological studies of marine sedimentary rocks in the vicinity of Cape Mendocino demonstrate that the Neogene forearc basin began to rise about 3 Ma as the subducting Juan De Fuca plate began to interact with the northward-propagating Mendocino triple junction. Today, this uplift has resulted in the stripping off of the onland portion of the basin that once extended at least 60 km inland from the present shoreline, based on downfaulted remnants. In C2, late Miocene to Holocene sediments of the active Monterey submarine fan have completely concealed the Paleogene accretionary prism and have built depositional lobes for nearly 300 km onto Pacific plate oceanic crust. The Monterey fan is fed by two submarine canyons, the Monterey and the smaller Ascension (C1 strip map; Normark and Hess, 1980). Fan growth is for the most part within a stable tectonic environment, except for its most proximal end, which is cut by the active San Gregorio–Hosgri branch of the San Andreas transform system.

Transitional basins

A number of different types of basins developed within the framework of accreted terranes following and during their attachment to sialic North America. Most notable are successor, rift, and forearc basins, each representing very different processes in marginal tectonism and sedimentation. A successor basin developed at and west of the North American sialic edge in Nevada after the accretion of Sonomia in the Permian-Triassic (Speed, 1978). Triassic strata above the arc volcanic rocks of Sonomia indicate a starved basin environment above carbonate compensation depth, and later, rapid filling with turbidites of mainly terrigenous provenance. The approximate correspondence of the Triassic basin-shelf boundary to the slightly older Sonomia–North America suture indicates that the Triassic basin was a successor to the collided terrane. The extent to which the early Mesozoic basin was inherited from Sonomia bathymetry is unclear. Significant basin subsidence may have occurred due to thermal contraction of the Sonomian arc lithosphere during and after suturing. It has also been proposed that the Triassic basin was created by spreading in back of the continental arc (Dickinson, 1976). Backarc spreading is an unlikely origin for the Triassic basin because (1) the basin existed before magmatism of the continental arc began in northwestern Nevada, according to stratigraphic relations (Russell, 1984), and (2) Triassic basinal strata east of the arc contain no volcanogenic sediment or mafic dikes, which might be expected if spreading and sedimentation were synchronous.

Remnants of Late Triassic to Early and locally Middle Jurassic basinal assemblages are preserved along the western Sierra–Klamath accretionary belt and continue as highly metamorphosed screens or pendants in the southern Sierra Nevada batholith (Irwin, 1972; Clark, 1976; Behrman, 1978; Davis and others, 1978; Irwin and others, 1978; Saleeby and others, 1978; Wright, 1981; Saleeby and Busby-Spera, 1992). The basinal assemblages developed primarily on disrupted Paleozoic–early Mesozoic ophiolitic assemblages as well as accreted "Tethyan"-affinity oceanic sedimentary sequences. They consist primarily of radiolarian chert with various admixtures of argillite, water-laid tuff, and continent-derived turbidites. Local facies representing outer pyroclastic apron deposits from adjacent coeval arc-volcanic centers are also recognized. The early Mesozoic basinal assemblages are accumulations within forearc, intra-arc and/or interarc basin frameworks. In the southern Sierra Nevada, possible facies relations between such deposits and more shelfal quartz-rich clastic and carbonate deposits of the Jurassic-Triassic Kings sequence are poorly preserved but nevertheless present in high-grade contact-metamorphic rocks (Saleeby and others, 1978). The change from shelfal to basinal facies of Mesozoic age is across the southern Foothills suture. Several large early Mesozoic caldera complexes occur within the shelfal "sialic" regime (Saleeby and others, 1978; Busby-Spera, 1984). These belong to the early Mesozoic continental arc. Possible relations between the Jurassic-Triassic basinal assemblages of the western Sierra–Klamath accretionary belt and the Sonomia successor basin are obscured by the intervening igneous rocks of the Cretaceous continental arc and the modern Cascade arc and high lava plains.

As discussed above, the Middle Jurassic arc and related basinal framework was rifted at ~160 to 165 Ma, causing generation of new oceanic crust in interarc and intra-arc basins (Josephine, Coast Range, and Smartville ophiolites). The rifting and spreading may have accompanied major transform motions along the Jurassic margin, leading to dispersal of earlier accreted terranes and the introduction of new terranes (Saleeby, 1981, 1990b, 1992; Saleeby and Busby-Spera, 1992). The Late Jurassic basinal system received distal and/or proximal volcanic-arc deposits, local pelagic deposits, and subsequently Late Jurassic flysch of the Mariposa-Galice slate belt and lowermost Great Valley sequence. The Late Jurassic flysch represents synorogenic deposits shed from uplifted continental and accretionary terrane sources which began to rise during the Middle Jurassic Siskiyou event. Segments of the Late Jurassic basinal system that formed along the western Sierra–Klamath belt closed by convergence during the Nevadan culmination, resulting in conversion of the respective Nevadan flysch deposits into slate.

Remnants of the Late Jurassic rift (and Nevadan flysch) basin system persisted throughout Cretaceous and Cenozoic time as the Great Valley in California. In Cretaceous time the Great Valley was an active forearc basin lying between the Franciscan subduction regime of the Coast Ranges and the Sierra Nevada batholith (Ernst, 1970). An Early Cretaceous change in sandstone compositions in the lower Great Valley sequence from orogenic to magmatic arc source regimes marks the end of Nevadan flysch and the beginning of forearc basin sedimentation (Dickinson and Rich, 1972; Ingersoll, 1983). This provenance change marks the end of accretion-related uplift and the continental arc superposed across the Nevadan orogenic zone. A significant Late Cretaceous unconformity within the forearc basin sequence (Bartow and others, 1984) probably corresponds to the eastward overthrusting of a Franciscan wedge into the oceanward reaches of the forearc basin as discussed earlier. Up to 5 km of forearc basin strata are shown in the apparent depositional axis of the Great Valley basin, which lies well to the west of the geographic centerline (C1 and C2 sections).

Throughout the development of the Coast Range accretionary belt, numerous trench slope basin and trench axis deposits accumulated and accreted. However, remnants of such deposits are fragmented to the point that a coherent picture of trench-related basinal sedimentation cannot be developed. The fragmented history of trench tectonics is due to important strike-slip and eastward overthrust tectonic phases as well as to accretionary prism growth phases related to eastward subduction. Only by detailed studies combining biostratigraphy, sedimentary petrology, and isotopic dating of detrital zircons will it be possible to unravel the complex history of the many accreted fragments that make up the Franciscan complex.

Continental Basins

Sedimentary basins developed across Precambrian North American sialic basement during both passive and active margin phases in the Phanerozoic. Basins locally developed in subtidal to supratidal carbonate realms on the continental shelf during subsidence in early and late Paleozoic time. They are defined by trough-shaped isopachs of shallow-marine carbonate facies

(Pennsylvanian-Permian Arcturus Group; Stevens, 1977), very thick accumulations of resedimented carbonates in possibly originally steep-walled basins (Pennsylvanian Oquirrh Group; Jordan and Douglas, 1980), and carbonate half-basins seaward of the edge of shoalwater carbonate buildups (Devonian "reef" complex of Roberts Mountains and vicinity; Matti and Mckee, 1977; Cook and Taylor, 1977). The tectonic cause of most of these shelf basins is not clear. In general the timings and positions of basins and buildups within the Paleozoic carbonate shelf province have no clear correspondence to tectonic events or processes at the margin. One possible correspondence is that late Paleozoic transform truncation tectonics along the axial Sierra Nevada region partly disrupted the shelf province just as they offset previously accreted oceanic rocks of the Antler orogeny.

In contrast to the carbonate basins, troughs with mainly siliciclastic debris on the continental foreland are in most cases easily related to contemporaneous tectonics. Such foreland basins (exogeosynclines or foredeeps) are associated with allochthons on their oceanward flank. Well-known examples are the Mississippian Antler foreland basin flysch and molasse (Poole, 1974; Dickinson and others, 1983) and Late Jurassic and Cretaceous clastic wedges associated with the Sevier thrust belt (Royse and others, 1975; Wiltschko and Dorr, 1983; Heller and others, 1986). Less evident but of similar tectonic significance is the Cretaceous Newark Canyon Formation (Nolan and others, 1956), possibly related to the Eureka thrust belt in central Nevada (Speed, 1982), and the Early Triassic Candelaria Formation, related to the Golconda allochthon (Speed, 1977).

The foreland basins were asymmetric, thickening toward the allochthon, which probably provided their outboard flanks. Sediment sources were mainly the allochthons, and the coarse clastic fill prograded away from the source with time. Such basins were almost certainly generated by elastic downflexure of the continental lithosphere due to vertical loading of the adjacent tract by the allochthons (Jordan, 1981; Speed and Sleep, 1982). Observed basin shapes, widths, and amplitudes can be modeled successfully with reasonable values of allochthon shape, thickness, and density, and lithospheric flexure parameters.

The Antler foreland basin is particularly interesting for two reasons. First, stratigraphic features below the Mississippian clastic wedge (Poole and Sandberg, 1977) can be interpreted to record the passage of a migrating upbulge 5 to 10 m.y. before the arrival of the basin axis (Speed and Sleep, 1982). The first features are an unconformity above prebulge Late Devonian carbonates that is succeeded by shoalwater carbonates, followed by increasingly basinal carbonate and calcturbidite of eastern provenance; these are overlain by a Mississippian siliciclastic wedge of westerly source. The upbulge is predicted by the theory of elastic flexure. Second, the Roberts Mountains allochthon overrode the early basinal (hemipelagic) sediments but not the clastic wedge during Mississippian emplacement (Jansma and Speed, 1990). The implication is that the allochthon wedged into the foreland basin fill, causing offscraping of the clastic wedge and accretion of it to the toe of the allochthon.

In contrast, all but the youngest foreland basin strata of the Sevier thrust belt are allochthonous imbricated clastic wedge sediments between nappes of Paleozoic carbonates. Here, at least some clastic wedges were overthrust by the allochthons that generated their basins. Moreover, each sequential foreland basin deposit was transferred from the autochthon to the thrust sheet complex as the eastward-propagating detachment fault below the basins successively ramped to the surface. Only the easternmost of the clastic wedges remains autochthonous. The different tectonic styles of the Antler and Sevier foreland basins reflect the different level of allochthon detachment: the depositional and detachment surfaces were approximately the same in the Antler, whereas the detachment was at depth below the depositional surface in the Sevier.

THERMAL EVOLUTION IN RELATION TO PLUTONISM, METAMORPHISM, AND TECTONISM

The Phanerozoic thermal history of the ocean-continent transition along the southwest Cordillera comprises two main phases, distinguished mainly in the continent. The first, or passive margin, phase, late Precambrian to Middle Triassic, included only apparently minor thermal phenomena within the sialic continent but the probable accretion of hot arc terranes to its edge. The succeeding active margin phase has seen nearly continuous but unevenly distributed thermal activity within a wide zone of the continent and its armor of accretionary terranes. Accretionary terranes throughout this phase arrived at the margin in both cold and hot states at shallow levels.

During Late Proterozoic passive margin formation, basalt invaded at least part of the attenuating crust. Present exposures permit recognition of such volcanics in shelf strata only near the craton-shelf hinge. Thus, it is difficult to assess the extent of volcanism and heating by horizontal stretching in the wide zone of presumably rifted sialic crust.

Collisions of island-arc systems with the passive margin in Mississippian and Permian-Triassic time seem to have caused no magmatism or metamorphism within currently exposed levels of the sialic continent. Metamorphism and thermally activated ductile deformation presumably occurred in rocks of the lower continental slope and rise and perhaps deep below the outer shelf during attempted subduction below the colliding island arcs. Such metamorphic rocks, if they exist, have not been returned to the surface. The collided island arc systems (Sonomia and the postulated Antleria) were active upon initial attachment but are supposed to have lost their subduction zone heat source thereafter, accounting for contraction upon accretion (Speed, 1977; Speed and Sleep, 1982).

The phase of active margin tectonics produced a complex of thermal phenomena within the morphologic continent during its accretionary growth from the Middle Triassic edge in the western Sierra–Klamath accretionary belt to its present locus within and along the edge of the Pacific–Juan De Fuca oceanic plates.

The continental arc whose relics comprise batholiths and

volcanic rocks of the Sierra Nevada and westernmost Nevada seems to have persisted from at least Middle Triassic to Late Cretaceous time. The distribution of magmatism in time and position within this arc before the Cretaceous is not well known, but, as described earlier, regional age and map patterns suggest a lull and reorientation from Late Jurassic to earliest Cretaceous time. The generation of the arc magmas from subcrustal sources was almost certainly related to persistent subduction along the California margin.

Metamorphism along the Cretaceous axis of the continental arc ranges up to granulite facies primarily as a function of depth (Sharry, 1981; Ross, 1985; Saleeby and others, 1986, 1987; Sams and Saleeby, 1986; Saleeby, 1990b): upper amphibolite to granulitic rocks are exposed at the extreme southern end of the Sierra Nevada, where unroofing has been ~30 km in ~100 m.y. Within this high-grade zone most preexisting crustal rocks were partly to completely melted in the upper amphibolite facies. Rarely did such rocks undergo subsolidus equilibration in granulite facies. Most granulitic rocks in this region are mid-Cretaceous orthogneisses that partly reequilibrated following descent through the solidus (Saleeby and others, 1987; Saleeby, 1990b; Pickett and Saleeby, 1993). Throughout most of the Cretaceous batholith, wall-rock complexes exhibit amphibolite and/or hornblende hornfels facies and lesser greenschist and/or albite-epidote hornfels facies contact assemblages. Schistose fabrics predominate over hornfelsic or granofelsic fabrics in the contact-metamorphic rocks on a regional scale. Hornfelsic textures are most common at highest structural levels such as the eastern Sierra in the C2 region, where the remnants of a 100 Ma caldera and resurgent pluton are emplaced within older metamorphic framework rocks. Along the axial pendants hornfelsic textures overprint schistose textures adjacent to major plutonic contacts, reflecting late-stage annealing adjacent to the major heat sources. The predominance of schistose textures in association with flattening and downdip constrictional fabrics represents deformation of wall-rock complexes during batholith emplacement by both shouldering and return-flow kinematic patterns (Tobisch and others, 1986; Saleeby and others, 1986; Sams and Saleeby, 1986; Saleeby, 1990b).

The western Sierra–Klamath accretionary belt exhibits a polyphase metamorphic history reflecting numerous episodes of preaccretion, synaccretion, and postaccretion tectonism prior to the superposing of the continental arc. Most notable is the widespread occurrence of amphibolite facies mafic schists within and adjacent to Paleozoic ophiolite complexes. Such tectonites appear to have formed in at least three distinct phases, including the Devonian (Lanphere and others, 1968; Bohlke and McKee, 1984; Saleeby and others, 1988), Permian-Carboniferous (Behrman, 1978; Saleeby, 1978, 1982; Shaw and others, 1987), and Triassic–Early Jurassic (Morgan, 1976; Weisenberg and Avé Lallemant, 1977; Behrman, 1978; Saleeby and Sharp, 1980; Sharp and Leighton, 1987; Saleeby and others, 1987). Devonian tectonites formed during major thrusting events involving early Paleozoic oceanic assemblages. Permian-Carboniferous tectonites formed in conjunction with ophiolitic igneous activity in a high strain environment, which may have been a transform fracture zone (Saleeby, 1978, 1979, 1990a, 1992). The Triassic to Early Jurassic tectonites formed in association with both early Mesozoic fracture-zone assemblages and major thrust complexes. Middle Jurassic high greenschist to amphibolite facies metamorphism was widespread in conjunction with Siskiyou contractile deformation (Coleman and others, 1988; Sharp, 1988). At the southern termination of the Smartville complex, amphibolite facies metamorphic tectonites formed in Paleozoic ophiolitic melange host rocks as ~164 Ma sheeted dikes and layered gabbroic intrusions were folded under solidus and hot subsolidus conditions (Saleeby, 1982). This may represent Siskiyou or early Nevadan deformation. All of the above tectonites clearly predate low-grade (prehnite-pumpellyite to low greenschist) Nevadan metamorphism. Late Nevadan (latest Jurassic–earliest Cretaceous) metamorphism locally attained high greenschist to amphibolite facies conditions in the southern Foothills belt in conjunction with dike and pluton emplacement (Tobisch and others, 1989; Saleeby and others, 1989b; Wolf and Saleeby, 1990).

Local vestiges of an early Mesozoic high *P-T* metamorphic belt are preserved in the central Klamath Mountains and northern Sierra Nevada (Hotz and others, 1977; Schweickert and others, 1980). The central Klamath rocks consist of volcanic and sedimentary protoliths with ~220 Ma metamorphic ages and are believed to be part of a regional Triassic belt that extends northward through much of British Columbia (Davis and others, 1978). The northern Sierra rocks are primarily metaclastic and have been remobilized and disrupted along a late Nevadan high-angle fault. Minimum ages within the Middle Jurassic were reported by Schweickert and others (1980), and considering regional disturbance patterns in zircon populations from adjacent metamorphic rocks (Saleeby and others, 1988), these ages should be considered highly disturbed. The northern Sierra blueschists are likely to be the southern continuation of the regional Triassic belt. Reasons for the distinct clustering of ages within the Triassic along this belt, rather than a range of ages reflecting the duration of active margin tectonism, have not been offered.

The Franciscan complex contains a number of distinct high *P-T* metamorphic terranes that range in age from Middle Jurassic to Late Cretaceous. The oldest is represented by high-grade blocks of eclogite, garnet amphibolite, and glaucophane schist derived primarily from mafic protoliths (Coleman and Lanphere, 1971). The blocks occur in scaly clay and serpentinite melanges of the Franciscan Central belt melange as well as in serpentinite melanges of the Coast Range ophiolite, and in protrusive and sedimentary serpentinite bodies of the lower Great Valley sequence. The serpentinite melanges of the ophiolite belt and the protrusive serpentinites within the lower Great Valley sequence demonstrate that upwardly mobile serpentinite penetrated upper mantle, igneous crust, and sedimentary veneer levels of the Great Valley interarc-forearc basin and, in doing so, sampled a deep cryptic metamorphic terrane. An important aspect of the high-grade blocks with respect to the Coast Range ophiolite and Great

Valley basement is that the main cluster of ophiolite igneous ages is identical to the upper range of the high-pressure metamorphic ages (Coleman and Lanphere, 1971; Suppe and Armstrong, 1972; Suppe and Foland, 1978; Ross and Sharp, 1988).

Several models are considered here to account for the regional distribution and age relations of the high-grade blocks. Model 1 is depicted in the C2 tectonic section and is based on the plate tectonic geometry envisaged by Schweickert and Cowan (1975) and Day and others (1985) for the Nevadan orogeny in the western Sierra Nevada. Figure 5A shows the western Foothills belt as an exotic east-facing arc and the proto–Great Valley as a trailing backarc basin. The Nevadan orogeny in this scenario occurred when the arc collided with North America, along the western Sierra Foothills. To account for a number of difficulties in this model, most important of which are common precollisional histories of the collided entities, the C2 tectonic section depicts the Great Valley basin as forming over an east-dipping subduction zone, the Sierran segment of the Nevadan orogeny resulting from a subsequent short-lived arc-reversal event (Fig. 5C). This resulted in the westward subduction of young, hot interarc basin crust beneath the fringing arc as it impinged on the western Foothills. High-pressure mafic tectonites formed by near synchronicity of interarc basin spreading and subduction, as in the current subduction of the actively spreading Woodlark basin beneath the recently reversed Solomons arc (Doutch and others, 1981). At the close of the Nevadan orogeny, the deep-seated metamorphic terrane was disrupted and fragments of the terrane were dispersed to their current structural positions. Possible disruption mechanisms include eastward underthrusting of an outer collisional mass, which drove the arc reversal as in the Ontong-Java plateaus collision with the Solomons arc, or alternatively, late to post-Nevadan transform faulting and/or transtensional denudation.

Model 2 (Fig. 5D) asserts that distinct phases in forearc extension and transtension may disrupt the underlying high-pressure metamorphic environment and drive rapid denudation of the derivative metamorphic blocks (Saleeby, 1992). Widespread suprasubduction-zone ophiolite formation suggests a lithosphere-scale dynamic pattern that could conceivably affect the subduction-zone metamorphic environment. The key question is whether the high-pressure phases could survive the possible thermal effects of nearby mantle diapirism. That tholeiitic and boninitic lavas that are widespread in the Late Jurassic ophiolites approximate upper mantle liquids derived from such diapirism offers the best possible situation for such a mechanism to work, because magma ascent should be relatively rapid and large upper mantle–lower crustal magma chambers need not be present. The extensional denudation model in Figure 5D asserts that, as the forearc region evolved into an extensional/transtensional tectonic regime and spreading commenced, the deep metamorphic environment was drawn up and rapidly denuded, terminating the preexisting subduction-zone metamorphic conditions and thereby setting the isotopic age systems. Survival of high-pressure metamorphic blocks was restricted to zones of preexisting lithosphere that were rifted apart as subarc asthenosphere upwelled and created new ophiolitic lithosphere beneath interarc basins. The isotopic systems within the metamorphic blocks, for the most part, record the initial denudation, which coincides with the igneous production of forearc and interarc basin ophiolites. Subsequent serpentinite diapirism and tectonic transport patterns further dispersed the metamorphic blocks into their current structural positions.

Semi-intact regional blueschist facies metamorphic rocks occur in the Franciscan complex of the Coast Ranges. These include the Pickett Peak, Yolla Bolly, and Burnt Hills terranes. The Pickett Peak terrane extends from the C1 area to southern Oregon, the Yolla Bolly terrane extends nearly the entire distance of the Coast Ranges, and the Burnt Hills terrane is restricted to the C2 region within the core of the Diablo antiform. The Pickett Peak terrane consists of metavolcanic and turbiditic protoliths that underwent blueschist facies metamorphism between 110 and 140 Ma (Blake and others, 1988; Suppe, 1973). The Yolla Bolly terrane was derived primarily from metaclastic rocks that underwent blueschist facies metamorphism at ~92 Ma (Mattinson and Echeverria, 1980). As discussed earlier, both of these terranes were transported as semi-intact nappes out of their metamorphic environments eastward onto Sierran (Great Valley) basement in the Late Cretaceous. The Burnt Hills terrane also consists primarily of blueschist facies metaclastic rocks, but with a notably younger protolith age (mid-Cretaceous) as compared to the Yolla Bolly terrane (Late Jurassic–Early Cretaceous). Burnt Hills metamorphism occurred along an east-dipping subduction zone now exposed as the core of the Diablo antiform. This latest Cretaceous event represents the only Coast Range high *P-T* environment which can easily be related to a paired metamorphic belt with a west-facing subduction polarity. Pickett Peak, Yolla Bolly, and Nevadan metamorphic terranes all require more complicated geometries calling for major accretion, collision, reversal in vergence, and even extension events in addition to subduction tectonics.

Metamorphism to the east of the Mesozoic continental arc was in general coextensive with apparent low-volume igneous activity. Magmatism and metamorphism occurred at deep crustal levels within the continental foreland between the arc and the Paleozoic hingeline. The timing of such phenomena is not precisely known but seems to have peaked in both the early Late Jurassic and the Late Cretaceous (Lee and others, 1981; Allmendinger and others, 1984; Snoke and Miller, 1987). Thermal phenomena of the foreland differ from those of the continental arc by (1) the restriction to relatively deep levels and the apparent absence of volcanism (except for one site, the Late Jurassic Pony Trail Group; Muffler, 1964), (2) the smallness and discreteness of individual plutons, and (3) the compositions of foreland plutons indicating a significant crustal or sedimentary precursor. Mesozoic metamorphism and deformation of Precambrian and lower Paleozoic strata below 5–10 km within the continental shelf succession occurred without corresponding effects on shallow upper Paleozoic cover (Howard, 1971, 1980; Snoke, 1980; Miller, 1980; Snoke and Miller, 1987). Such relations are recognized in Cenozoic core complexes. Mesozoic magma generation and metamorphism in the foreland are likely to have been related processes. Metamorphic minerals indicate depths as great as ~15 km (Snoke and Miller, 1987), somewhat deeper than the undisturbed, expected stratigraphic depth of the metamorphosed strata. Thus, metamorphism occurred either in strata of anomalous depositional thickness or where strata were thickened tectonically. In the second case, thickening may reflect horizontal shortening, perhaps by duplexing at deep levels of the foreland during the Mesozoic (Armstrong, 1972; Snoke and Miller, 1987).

A fundamental question is whether the elevation of the geotherms at depth in the Mesozoic foreland was directly related to subduction or whether it was the product of motions within the continental lithosphere and/or subjacent asthenosphere. An appeal to foreland heating by the same slab that caused magmatism in the continental arc would require the unlikely condition that melting occurred across a zone >500 km wide in two apparently short pulses, in Late Jurassic and Late Cretaceous time. Anatexis of the sialic crust and lowermost sediments is the more probable origin and was a result of tectonic thickening, as first postulated by Armstrong (1972). Mesozoic plutonism and metamorphism of the southwest U.S. Cordilleran foreland strongly resemble similar age activity in the Omineca crystalline belt of British Columbia, and the two belts may be continuous (Saleeby and Busby-Spera, 1992).

In contrast to Mesozoic time, the time-space distribution of Cenozoic magmatism can be reasonably accounted for by excursions of dip and velocity of the downgoing oceanic slab. The early Cenozoic lull in magmatism indicates a change from the preceding protracted generation of magma below the continental arc (Snyder and others, 1976; Stewart and Carlson, 1976; Coney and Reynolds, 1977). Volcanism first reappeared in the Eocene at its most continentward limit in association with the Laramide compressive uplifts. This has been thought to represent a marked shallowing of the easterly dip of the slab from the Cretaceous to Eocene positions of magmatism. The shallowing may have been consequent to the large increase of relative velocity between the North American and subducting Farallon plates at ~60 Ma (Coney, 1978). After the Eocene, magmatism involving copious intrusion and extrusion of calc-alkaline magmas was concentrated between 20 and 40 Ma in the roughly central zone of the Basin and Range province, and there was possible southward migration during that time (Stewart and Carlson, 1976). The change of locus from Eocene to Oligocene is thought to represent the result of asthenosphere upwelling through a break in the shallow slab from which the eastern part of slab was detached and the western part sank (Snyder and others, 1976).

The core complexes of the eastern Basin and Range province include much 20–40 Ma silicic extrusive rock; eruption was contemporaneous with major intrusions at mid-crustal levels (Miller and others, 1983; Snoke and Miller, 1987). It is unclear whether such intrusion was ubiquitous across the Basin and Range province or whether it occurred in proportion to the magnitude of extension, i.e., mainly in the zone of exhumed ductilely stretched structures.

The final, Neogene magmatic phase of the Basin and Range is marked by minor bimodal basalt-rhyolite volcanism and uplift and extensional rifting in the province and minor calc-alkaline volcanism along both boundaries of the province. The generally high heat flow and the shallow asthenosphere of the Basin and Range province indicates continuing thermal activity following the regime outlined above, but it is not certain whether such activity is increasing or declining relative to the mid-Cenozoic levels. Heat flow across the eastern and western margins of the Basin and Range averages ~80 mW/m^2; local data points are in excess of 200 mW/m^2 (C1 and C2 sections). In contrast, mean heat flow is ~60 mW/m^2 in the adjacent Colorado plateau and ~30 mW/m^2 across the Sierra Nevada. The limited high heat flow values for the Modoc Plateau (C1 display) suggest that this province is probably undergoing the same extension seen in the adjacent Basin and Range. The heat-flow values of the southern Cascade province, however, range widely (46–231 mW/m^2), and probably reflect local circulation of ground water. An east-west profile drawn across the Cascade Range farther north is consistent with the high values seen in other magmatic arcs. Extensional tectonism of the active Basin and Range province and the demise of copious calc-alkaline volcanism have been correlated with passage of the triple junction between the North American–Farallon–Pacific plates north along the margin in California (Atwater, 1970; Atwater and Molnar, 1973). In this plate kinematic model, the Modoc Plateau lies in the southern Cascades backarc–Basin and Range transition zone.

The sharp gradient in heat flow across the western edge of the Basin and Range province, Sierra Nevada, and Great Valley (C2 section and profile) reflects a juxtaposition in the Basin and Range of Neogene high heat flow and the anomalously low regional pattern inherited from Mesozoic tectonism. The juxtaposition may reflect the recent upwelling of asthenosphere along the eastern Sierra region (Fig. 4, alternative crustal structure model). The extremely low heat flow (25–35 mW/m^2) in the Great Valley–western Sierra Nevada region reflects a combination of downward heat loss to cool underthrust oceanic slabs in addition to upward fluxing of heat-producing elements from crustal rocks during batholith generation followed by erosion. A major consequence is a steep westward deepening across the Sierra Nevada of both the modern brittle-ductile transition and the seismogenic zone as seen in microearthquake focal depths (C2 sections; Wong and Savage, 1983; Vetter and Ryall, 1983; Smith and Bruhn, 1984). Heat flow across the Coast Ranges averages ~80 mW/m^2; there is greater scatter westward across the San Andreas fault (C2 section and profile). Heat-flow values over the northern Coast Range portion of C1 are uniformly low (28–76 mW/m^2), and there is no apparent difference north and south of the triple junction, probably reflecting the fact that the subducting Juan De Fuca plate underlies the entire area.

PRESENT UNDERSTANDING OF PALINSPASTIC RESTORATIONS AND PALEOTECTONIC MODELS

Palinspastic restorations along active continental margins must be considered from two distinct perspectives. (1) In regions of the foreland and craton transition, amounts of shortening and/or extensional crustal deformation may be approached quantitatively from the construction of balanced cross sections, and strike-slip displacements may be approached quantitatively by restoration of piercing points. (2) Within the voluminous batholithic belt and the zone of accretionary terranes, strain in displacement zones and in individual terranes may be approached quantitatively, but many major fault displacements can only be considered qualitatively through the integration of regional structural-stratigraphic data, paleobiogeographic data, and gross and often ambiguous paleomagnetic and plate kinematic data. Accordingly, the treatment of palinspastic restorations is divided between continent and accreted terranes.

Restorations within Precambrian North America in the areas of C1 and C2 have been attempted using balanced sections for Basin and Range extension and for Mesozoic foreland contraction. Gans (1987) estimated maximum principal elongation for the eastern half of the northern (40°N) Basin and Range at 80% and 250% for a 200-km-wide segment in the eastern half that includes thin-skinned brittle and ductile stretching (Schell Creek Range and Snake Range). Wernicke and others (1982) calculated 80%–100% elongation for the whole width of the Basin and Range at about 38°N, using estimates of individual fault displacements. Bogen and Schweickert (1986) used paleomagnetics to indicate that a north-south gradient in east-west displacement was small or nil. Speed and others (1987) used all the above data to estimate between 0% and 20% elongation in the western half of the northern Basin and Range province.

A number of paleomagnetic studies in the Basin and Range province, cited earlier, indicate that latitudinal transport relative to the craton has been nil, at least since mid-Cretaceous time. The sensitivity of such data is about 800 km, however, and displacements up to that magnitude would not be recognized. Similar studies for Mesozoic igneous rocks of the Sierra Nevada indicate latitude anomalies relative to North America that are either not significant statistically, or marginally so. Frei and others (1984) concluded that the Sierra Nevada range may have moved as much as 1000 km north since 80 Ma.

Restorations using balanced sections with good seismic control were made over a 150 km length across the Sevier thrust belt in C2 by Allmendinger and others (1986). They found about 40 km of Basin and Range extension in this area of thin-skinned stretching and about 110 km of Cretaceous shortening. Restorations suggest the lowest, youngest thrusts root in crystalline basement. They do not show whether the Cenozoic Sevier Desert detachment fault reactivated a Mesozoic thrust or not. Royse and others (1975) made restorations of the Idaho-Wyoming part of the Sevier thrust belt (Fig. 3) and found a total of 80 km of contraction over a present width of about 80 km.

Displacement of the present leading edge of Roberts Mountains and Golconda allochthons relative to the edge of sialic North America are about 135 and 100 km, respectively, without correction for late Mesozoic and Cenozoic motions. Additional major uncertainties here are the precise position of the buried sialic edge and the maximum original landward extents of the allochthons.

Within the region of accreted terranes, there are at present few palinspastic restoration and coherent paleogeographic synthesis data. Paleobiogeographic analyses of several accreted terranes provide useful information. For example, Ross and Ross (1983) presented a synthesis of late Paleozoic reef-forming fusulinids. They showed reefal accumulations on Sonomia as being distinct from coeval North American shelfal accumulations. In their model, Sonomia evolved in temperate east Pacific regions isolated from the continental shelf and possibly south of the equator. Moreover, late Paleozoic reefal accumulations on seamount fragments dispersed along terranes of the western Sierra–Klamath accretionary belt are distinct from both Sonomia and North American shelf forms and are strongly akin to Tethyan and/or equatorial Panthalassa forms. The depositional and tectonic immersion of the oceanic seamount and reef sequences in younger pelagic sediments presumably records seamount subsidence and burial during trans-Panthalassa migration en route to continental margin accretion. These paleobiogeographic analyses are also critical for subsequent events in the western Sierra–Klamath accretionary belt. As discussed by Saleeby (1983), the various Jurassic arc terranes that underwent final phases of accretion during the Nevadan orogeny contained basement rocks and/or mixed sedimentary-tectonic clasts containing the Sonomian and equatorial Panthalassa fauna. Implications of such relations are that the arc terranes formed within a continental margin framework built from Sonomia and Paleozoic Panthalassa ocean-floor remnants.

Paleobiogeographic and paleomagnetic data from the Permanente terrane of the C2 Franciscan complex and related melange blocks dispersed northward to the C1 region suggest derivation in the mid-Pacific as part of the Hess Rise system (Vallier and others, 1983; Sliter, 1984). This analysis represents one of the few attempts to relate an accreted terrane to an existing remote parent mass.

Debate exists whether Salinia sialic crystalline terrane of the Coast Ranges was derived from the southern termination of the Sierra Nevada or from thousands of kilometers to the south. Paleomagnetic data from Late Cretaceous strata of Salinia suggest northward displacement of ~2500 km (Champion and others, 1984). In contrast, a synthesis of field, petrologic, and isotopic geochemical data (James, 1986; Silver and Mattinson, 1986) ties the Salinia fragment to the southernmost Sierra Nevada–western Mojave Desert region. Such data would delimit Salinia offset to ~310 km along the San Andreas system (see neotectonics section).

Topical discussions herein were integrated with current and in some cases competing ideas in plate tectonic models. Only the broadest generalizations can be made for ancient plate geometries and the derivative continental margin tectonic phenomena. It is clear that the Cordilleran stable shelf originated off a passive margin, and that the processes of terrane accretion and Mesozoic continental arc generation represent subduction of oceanic crust beneath North America. However, many details, such as polarities of accreted arcs, recognition of particular thrust structures as subduction zones, and detailed relations between foreland deformation and accretionary events, remain obscure. Major difficulties in the construction of such detailed models arise from incomplete geologic records, particularly those that have disruptions in sedimentary-tectonic facies relations. The loss of such records is caused by multistage processes that include attrition during accretion by uplift and erosion of high-level nappes, tectonic burial by imbricate underthrusting and perhaps subduction loss, and further attrition by synaccretion and postaccretion strike-slip faulting and rifting. Remobilization during subsequent accretionary events also causes these record losses. Furthermore, synaccretionary and postaccretionary metamorphic and plutonic events have often obscured much of the remaining record. Such problems in reading the paleotectonic record may be amplified along the C1 and C2 region of the North American Cordillera. This arises from the apparent operation of first-order tectonic attritionary processes in series and in parallel with accretionry processes throughout most or all of the history of the region. A cursory view of the *Tectonic Map of North America* (King, 1966) readily offers such an impression by comparison of accreted terrane volume and mean terrane size between the C1-C2 region and regions to the north in the Cordillera, as well as the Appalachian orogen. Except for the largely covered and variably extended Sonomia terrane, the southern Cordillera offers roughly an order of magnitude less accretionary material compared to the latter areas. The limitations in detailed plate tectonic analysis and model construction lead to the question of outstanding problems to be addressed by future work.

OUTSTANDING PROBLEMS

The C1-C2 transect synthesis presents data and ideas on the ocean-continent transition current to the mid-1980s. The major problems posed by the synthesis can be broken into two groups. (1) What is the physical and geological constitution of the crust of the Earth across the ocean-continent transition? (2) How did it evolve to its current state? The question of evolution relates to geologic history and the construction of plate tectonic models. Advancement in the second broad field of problems should be slow and dependent on advancement in the first group of problems. This arises from the nature of the geologic record, or its incompleteness, and by the fact that a first-order awareness of major accretionary terranes, plutonic-metamorphic belts, intact facies systems, and major surface structures already exists, as shown in the C1 and C2 displays.

A multidisciplinary approach to the accurate assessment of the crust is essential, and will require deep crustal geophysical and drill-hole data in conjunction with careful surface mapping and petrographic-geochemical analyses. A fundamental problem is the nature of the Moho and whether, in passing from one tectonic domain to another, it is the same age. The origin of deep, flat-lying seismic wave reflectors is clearly related to this problem and to problems related to structures and terrane domains. Are flat-lying structures a primary feature in the crust created by igneous processes, and do very early phase boundaries form in conjunction with the igneous processes? If such layering is primary, what influence is imparted during subsequent nappe detachment, foreland contraction, or regional extensional tectonism? The depth to which quartz-bearing rocks occur in abundance and the thermal evolution of the crust are critical aspects of these problems.

The integration of geophysical parameters and isotopic geochemical data is essential for better definition of the current geographic distribution of the ancient sialic edge and possible displaced fragments of such sial. The question of juvenile mantle additives and recycled sialic materials in the production of new sialic crust must be addressed by integrated field, petrologic, and isotopic geochemical data. Advanced techniques in isotopic geochemistry will also facilitate correlation of major protolith assemblages into and within complex metamorphic terranes.

Careful acquisition and compilation of surface geologic data will be critical for the interpretation of remote data sets and will refine the interpretation of the geologic and paleotectonic history. The concepts of terrane analysis as set forth by Jones and others (1983) should be followed as a means of objectively analyzing fragments parts of the continental margin assemblages. However, terrane differentiation should be considered not as an end in itself, but rather as a tool for objective tectonic analysis: such terrane analyses for major tectonic units of the C1 and C2 areas are presented in the tectonostratigraphic evolution diagrams of each display.

REFERENCES CITED

Albers, J. P., 1967, Belt of sigmoidal bending and right-lateral faulting in the western Great Basin: Geological Society of America Bulletin, v. 78, p. 143–156.

Allmendinger, R. W., and Jordan, T. E., 1981, Mesozoic evolution, hinterland of the Sevier orogenic belt: Geology, v. 9, p. 309–314.

—— , 1984, Mesozoic structure of the Newfoundland Mountains, Utah: Horizontal shortening and subsequent extension in the hinterland of the Sevier belt: Geological Society of America Bulletin, v. 95, p. 1280–1292.

Allmendinger, R. W., Sharp, J. W., Von Tish, D., Serpa, L., Brown, L., Oliver, J., Kaufman, S., and Smith, R. B., 1983, Cenozoic and Mesozoic structure of the eastern Basin and Range province, Utah, from COCORP seismic reflection data: Geology, v. 11, p. 532–536.

Allmendinger, R. W., Miller, D. M., and Jordan, T. E., 1984, Known and inferred Mesozoic deformation in the hinterland of the Sevier Belt, northwest Utah: Utah Geological Association Publication 13, p. 21–34.

Allmendinger, R. W., Farmer, H., Hauser, E., Sharp, J., Vontish, E., Oliver, J., and Kaufman, S., 1986, Phanerozoic tectonics of the Basin and Range Colorado Plateau transition from COCORP data and geologic data: A review: American Geophysical Union Geodynamics Series, v. 14,

p. 257–268.
Alvarez, W., Kent, D. V., Silva, I. P., Schweickert, R. A., and Larson, R. A., 1980, Franciscan Complex limestone deposited at 17° south paleolatitude: Geological Society of America Bulletin, v. 91, p. 476–484.
Anderson, R. E., Zoback, M. L., and Thompson, G., 1983, Implications of selected subsurface data on the structural form and evolution of some basins in the northern Basin-Range provinces: Geological Society of America Bulletin, v. 94, p. 1055–1072.
Ando, C. J., Irwin, W. P., Jones, D. L., and Saleeby, J. B., 1983, The ophiolitic North Fork terrane in the Salmon River region, central Klamath Mountains, California: Geological Society of America Bulletin, v. 94, p. 236–252.
Argus, D. F. and Gordon, R. G., 1991, Current Sierra Nevada North American motion from very long basement interferometry; implications for the kinematics of the western United States: Geology, v. 19, p. 1085–1088.
Armen, R., and Mayer, L., 1983, Subsidence analyses of the Cordilleran miogeocline: Implications for timing of late Proterozoic rifting and amount of extension: Geology, v. 11, p. 702–705.
Armstrong, F. C., and Oriel, S. S., 1965, Tectonic development of the Idaho-Wyoming thrust belt: American Association of Petroleum Geologists Bulletin, v. 49, p. 1847–1866.
Armstrong, R. L., 1968, Sevier orogenic belt in Nevada and Utah: Geological Society of America Bulletin, v. 79, p. 429–458.
—— , 1972, Low-angle (denudation) faults, hinterland of the Sevier orogenic belt, eastern Nevada and western Utah: Geological Society of America Bulletin, v. 83, p. 1729–1754.
Armstrong, R. L., and Suppe, J., 1973, Potassium-argon geochronometry of Mesozoic igneous rocks in Nevada, Utah, and southern California: Geological Society of America Bulletin, v. 84, p. 1375–1392.
Armstrong, R. L., Taubeneck, W. H., and Hales, P. O., 1977, Rb-Sr and K-Ar geochronometry of Mesozoic granitic rocks and their Sr isotopic compositions, Oregon, Washington, and Idaho: Geological Society of America Bulletin, v. 88, p. 397–411.
Atwater, T., 1970, Implications of plate tectonics for the evolution of western North America: Geological Society of America Bulletin, v. 81, p. 3513–3536.
Atwater, T., and Menard, H. W., 1970, Magnetic lineations Pacific: Earth and Planetary Science Letters, v. 7, p. 442–454.
Atwater, T., and Molner, P., 1973, Relative motion of the American plates deduced from sea-floor spreading in the Atlantic, Indian, and south Pacific oceans, *in* Kovach, R. L., and Nur, A., eds., Proceedings of the conference on tectonic problems of the San Andreas fault system: Stanford, California, Stanford University Publishing, Geological Sciences, v. f 13, p. 136–148.
Aydin, A., and Page, B. M., 1984, Diverse Pliocene-Quaternary tectonics in a transform environment, San Francisco Bay region, California: Geological Society of America Bulletin, v. 95, p. 1303–1317.
Bailey, E. H., Blake, M. C., and Jones, D. L., 1970, On-land Mesozoic oceanic crust in California Coast Ranges: U.S. Geological Survey Professional Paper 700-C, p. C70–C81.
Bartley, J. M., and Wernicke, B. P., 1984, The Snake Range decollement interpreted as a major extensional shear zone: Tectonics, v. 3, p. 647–657.
Bartow, J. A., Lettis, W. R., Sonneman, H. S., and Switzer, J. R., Jr., 1984, Geologic map of the eastern flank of the Diablo Range from Hospital Creek to Poverty Flat, San Joaquin, Stanislaus, and Merced counties, California: U.S. Geological Survey Miscellaneous Geologic Investigations Series Map I-1496, scale 1:125,000.
Behrman, P. G., 1978, Pre-Callovian rocks west of the Melones Fault Zone, central Sierra Nevada foothills, *in* Howell, D. E., and McDougall, K. A., eds., Mesozoic paleogeography of the western United States: Los Angeles, California, Pacific Section, Society of Economic Paleontologists and Mineralogists, Pacific Coast Paleogeography Symposium 2, p. 337–348.
Best, M. G., Armstrong, R. L., Graustein, W. C., Embree, G. F., and Ahlborn, R. C., 1974, Mica granites in the Kern Mountains pluton, White Pine Co., NV: Remobilized basement of the Cordilleran miogeosyncline?: Geological Society of America Bulletin, v. 85, p. 1277–1286.
Blackstone, D. L., 1979, The overthrust belt salient of the Cordilleran fold belt, western Wyoming: Wyoming Geological Association, 29th Annual Field Conference Guidebook, p. 367–384.
Blackwelder, E. A., 1914, A summary of the orogenic epochs in the geologic history of North America: Journal of Geology, v. 22, p. 633–654.
Blake, M. C., and Jones, D. L., 1977, Plate tectonic history of the Yolla Bolly Junction, northern California (Geological Society of America Cordilleran Section Meeting Guidebook): Sacramento, California State University, 14 p.
Blake, M. C., Jayko, A., Chaves, V., Saleeby, J. B., and Seel, K., 1984, Tectonostratigraphic terranes of Magdalena Island, Baja, California Sur, *in* Frizzell, V. A., Jr., ed., Geology of Baja California: Los Angeles, Society of Economic Paleontologists and Mineralogists, p. 183–191.
Blake, M. C., Engebretson, D. C., Jayko, A. S., and Jones, D. L., 1985, Tectonostratigraphic terranes in southwest Oregon, *in* Tectonostratigraphic terranes of the Circum-Pacific region: Circum-Pacific Council for Energy and Mineral Resources Earth Science Series, p. 147–157.
Blake, M. C., and 17 others, compilers, 1987, C-1 Mendocino triple junction to North American craton: Boulder, Colorado, Geological Society of America, Centennial Continent/Ocean Transect no. 12, 3 sheets with text, scale 1:500,000.
Blake, M. C., Jayko, A. S., McLaughlin, R. J., and Underwood, M. B., 1988, Metamorphic and tectonic evolution of the Franciscan complex, northern California, *in* Ernst, W. G., ed., Metamorphism and crustal evolution of the western United States (Rubey Volume VII): Englewood Cliffs, New Jersey, Prentice-Hall, p. 1035–1060.
Block, L., and Royden, L. H., 1990, Core complex geometries and regional scale flow in the lower crust: Tectonics, v. 9, p. 557–567.
Bogen, N. L., and Schweickert, R. A., 1986, Magnitude of crustal extension across the northern Basin and Range province: Constraints from paleomagnetism: Earth and Planetary Science Letters, v. 75, p. 93–100.
Bohlke, J. K., and McKee, E. H., 1984, K/Ar ages relating to metamorphism, plutonism, and gold-quartz vein mineralization near Allegheny, Sierra County, California: Isochron/West, v. 39, p. 3–7.
Bolt, B. A., Lomnitz, C., and McEvilly, T. V., 1968, Seismological evidence on the tectonics of central and northern California and the Mendocino Escarpment: Seismological Society of America Bulletin, v. 58, p. 1725–1767.
Bond, G. D., and DeVay, J. C., 1980, Pre-Upper Devonian quartzose sandstone in the Shoo Fly Formation, northern California—Petrology, provenance, and implications for regional tectonics: Journal of Geology, v. 88, p. 285–308.
Braille, L., Smith, R. C., Keller, G. R., Welch, R., and Meyer, R. A., 1974, Crustal structure across the Wasatch Front from detailed seismic refraction studies: Journal of Geophysical Research, v. 79, p. 1295–1317.
Breuckner, H. K., and Snyder, W. S., 1985, Structure of the Havallah sequence, Golconda allochthon; evidence for prolonged evolution in an accretionary prism: Geological Society of America Bulletin, v. 96, p. 1113–1130.
Brewer, J. A., Allmendinger, R. W., Brown, L. D., Oliver, J. E., and Kaufman, S., 1982, COCORP profiling across the northern Rocky Mountain Front in southern Wyoming, Part I, Laramide structure: Geological Society of America Bulletin, v. 93, p. 1242–1252.
Brook, C. A., 1977, Stratigraphy and structure of the Saddlebag Lake roof pendant, Sierra Nevada: Geological Society of America Bulletin, v. 88, p. 321–331.
Bruhn, R. L., and Beck, S. L., 1981, Mechanics of thrust faulting in crystalline basement, Sevier orogenic belt: Geology, v. 9, p. 200–204.
Burchfiel, B. C., and Davis, G. A., 1972, Structural framework and evolution of the southern part of the Cordilleran orogen, western United States: American Journal of Science, v. 272, p. 97–118.
Busby-Spera, C. J., 1984, Large-volume rhyolite ash-flow eruptions and submarine caldera collapse in the lower Mesozoic Sierra Nevada, California: Journal of Geophysical Research, v. 89, p. 8417–8427.
Champion, D. E., Howell, D. G. and Gromme, C. S., 1984, Paleomagnetic and geologic data indicating 2500 km of northward displacement for the Salinian and related terranes: Journal of Geophysical Research, v. 89, p. 7736–7752.

Chapin, C. E., 1979, Evolution of the Rio Grande rift—A summary, *in* Reicker, R. E., ed., Rio Grande rift—Tectonics and magmatism: Washington, D.C., American Geophysical Union, p. 1–5.

Chen, J. H., and Moore, J. G., 1982, Uranium-lead isotopic studies of the southern Sierra Nevada batholith, California: Geological Society of America Abstracts with Programs, v. 10, p. 99–100.

Christiansen, M. N., 1966, Late Cenozoic crustal movements in the Sierra Nevada of California: Geological Society of America Bulletin, v. 77, p. 163–182.

Christiansen, R. L., and Lipman, P. W., 1972, Tectonic evolution of the western United States—Part II, Late Cenozoic: Royal Society of London Philosophical Transactions, ser. A., v. 271, p. 249–284.

Christiansen, R. L., and McKee, E. H., 1978, Late Cenozoic volcanic and tectonic evolution of the Great Basin and Columbia intermontane regions, *in* Smith, R. B. and Eaton, G. P., eds., Cenozoic Tectonics and Regional Geophysics of the Western Cordillera: Geological Society of America Memoir 152, p. 283–311.

Clark, L. D., 1976, Stratigraphy of the north half of the western Sierra Nevada metamorphic belt, California: U.S. Geological Survey Professional Paper 923, 26 p.

Cockerham, R. S., 1984, Evidence for a 180-km-long subducted slab beneath northern California: Seismological Society of America Bulletin, v. 74, p. 569–576.

Coleman, R. G., and Lanphere, M. A., 1971, Distribution and age of high-grade blueschists, associated eclogites, and amphibolites from Oregon and California: Geological Society of America Bulletin, v. 82, p. 2397–2412.

Coleman, R. G., and Lee, D. E., 1962, Metamorphic aragonite in the glaucophane schists of Cazadero, California: American Journal of Science, v. 260, p. 577–595.

Coleman, R. G., Manning, C. E., Mortimer, N., Donato, M. M., and Hill, L. B., 1988, Tectonic and regional metamorphic framework of the Klamath Mountains and adjacent Coast Ranges, California and Oregon, *in* Ernst, W. G., ed., Metamorphism and crustal evolution of the western United States (Rubey Volume VII): Englewood Cliffs, New Jersey, Prentice-Hall, p. 1061–1097.

Compton, R. R., Todd, V. R., Zartman, R. E., and Naeser, C. W., 1977, Oligocene and Miocene metamorphism, folding, and low-angle faulting in northwestern Utah: Geological Society of America Bulletin, v. 88, p. 1237–1250.

Coney, P. J., 1978, Mesozoic-Cenozoic Cordilleran plate tectonics *in* Smith, R. B. and Eaton, G. P., eds., Cenozoic Tectonics and Regional Geophysics of the Western Cordillera: Geological Society of America Memoir 152, p. 33–50.

Coney, P. J., and Reynolds, S. J., 1977, Cordilleran Benioff zones: Nature, v. 270, p. 403–406.

Coney, P. J., Jones, D. L., and Monger, J.W.H., 1980, Cordilleran suspect terranes: Nature, v. 288, p. 329–333.

Cook, H. E., and Taylor, M. E., 1977, Comparison of continental slope and shelf environments in the upper Cambrian and lowest Ordovician of Nevada: Society of Economic Paleontologists and Mineralogists Special Publication 25, p. 51–81.

Crowell, J. C., 1979, The San Andreas fault system through time: Geological Society of London Journal, v. 136, p. 293–302.

Dallmeyer, R. D., Snoke, A. W., and McKee, E. H., 1986, The Mesozoic-Cenozoic tectonothermal evolution of the Ruby Mountains, East Humboldt Range, Nevada: A Cordilleran metamorphic core complex: Tectonics, v. 5, p. 931–945.

D'Allura, J. A., Moores, E. M., and Robinson, L., 1977, Paleozoic rocks of the northern Sierra Nevada: Their structural and paleogeographic significance, *in* Stewart, J. H., Stevens, C. H., and Fritsche, A. F., eds., Paleozoic paleogeography of the western United States: Los Angeles, California, Pacific Section, Society of Economic Paleontologists and Mineralogists, Pacific Coast Paleogeography Symposium, p. 394–408.

Davis, G. A., Monger, J.W.H., and Burchfiel, B. C., 1978, Mesozoic construction of the Cordilleran 'collage,' central British Columbia to central California, *in* Howell, D. E., and McDougall, K. A., eds., Mesozoic paleogeography of the western United States: Los Angeles, California, Pacific Section, Society of Economic Paleontologists and Mineralogists, Pacific Coast Paleogeography Symposium 2, p. 1–32.

Day, J. W., Moores, E. M., and Tuminas, A. C., 1985, Structure and tectonics of the northern Sierra Nevada: Geological Society of America Bulletin, v. 96, p. 436–450.

DeMets, C., Gordon, R. G., Argus, D. F., and Stein, S., 1990, Current plate motions: Geophysical Journal International, v. 101, p. 425–478.

DePaolo, D. J., 1981, A neodymium and strontium isotopic study of the Mesozoic calc-alkaline granitic batholiths of the Sierra Nevada and Peninsular Ranges, California: Journal of Geoophysical Research, v. 86, p. 10470–10488.

Dickinson, W. R., 1972, Evidence for plate-tectonic regimes in the rock record: American Journal of Science, v. 272, p. 551–576.

——, 1976, Sedimentary basins developed during evolution of Mesozoic-Cenozoic arc-trench system in western North America: Canadian Journal of Science, v. 13, p. 1268–1287.

——, 1977, Paleozoic plate tectonics and the evolution of the Cordilleran continental margin, *in* Stewart, J. H., Stevens, C. H., and Fritsche, A. F., eds., Paleozoic paleogeography of the western United States: Los Angeles, California, Pacific Section, Society of Economic Paleontologists and Mineralogists, Pacific Coast Paleogeography Symposium 1, p. 137–157.

Dickinson, W. R., and Rich, E. I., 1972, Petrologic intervals and petrofacies in the Great Valley Sequence, Sacramento Valley, California: Geological Society of America Bulletin, v. 83, p. 3007–3024.

Dickinson, W. R., and Snyder, W. S., 1978, Plate tectonics of the Laramide orogeny *in* Matthews III, V., ed., Laramide folding associated with basement block faulting in the western United States: Geological Society of America Memoir 151, 370 p.

Dickinson, W. R., Harbaugh, D. W., Antler, A. H., Heller, P. L., and Snyder, W. S., 1983, Detrital modes of upper Paleozoic sandstones derived from the Antler orogen in Nevada: American Journal of Science, v. 283, p. 481–509.

Dilles, J. H., and Wright, J. E., 1988, The chronology of early Mesozoic arc magmatism in the Yerington district of western Nevada and its regional implications: Geological Society of America Bulletin, v. 100, p. 644–652.

Dixon, J. S., 1982, Regional structural synthesis, Wyoming salient of the western overthrust belt: American Association of Petroleum Geologists Bulletin, v. 66, p. 1560–1580.

Doe, B. R., and Delevaux, M. H., 1973, Variations in lead isotopic compositions in Mesozoic granitic rocks of California; a preliminary investigation: Geological Society of America Bulletin, v. 84, p. 3513–3526.

Domenick, M. A., Kistler, R. W., Dodge, F.C.W., and Tatsumoto, M., 1983, Nd and Sr isotopic study of crustal and mantle inclusions from the Sierra Nevada xenoliths and implications for batholith petrogenesis: Geological Society of America Bulletin, v. 94, p. 713–719.

Douglass, R. C., 1967, Permian Tethyan fusulinids from California: U.S. Geological Survey Professional Paper 583-A, p. 7–43.

Doutch, H. F., and eight others, 1981, Plate tectonic map of the circum-Pacific region, southwest quadrant: Tulsa, Oklahoma, American Association of Petroleum Geologists, scale 1:10,000,000.

Dunne, G. C., and Suczek, C. A., 1991, Early Paleozoic eugeoclinal strata in the Kern Plateau pendants, southern Sierra Nevada, California, *in* Cooper, J., and Stevens, C., eds., Paleozoic paleogeography of the western United States, II: Society of Economic Paleontologists and Mineralogists Bulletin, v. 67, p. 677–692.

Eaton, J. P., 1963, Crustal structure from San Francisco, California, to Eureka, Nevada, from seismic refraction measurements: Journal of Geophysical Research, v. 68, p. 5789–5806.

——, 1966, Crustal structure in southern and central California from seismic evidence, *in* Bailey, E. H., ed., Geology of northern California: California Division of Mines and Geology Bulletin, v. 190, p. 419–426.

——, 1982, The Basin and Range province; origin and tectonic significance: Annual Review of Earth and Planetary Sciences, v. 10, p. 409–440.

Eisbacher, G. H., 1983, Devonian-Mississippian sinistral transcurrent faulting along the cratonic margin of western North America: A hypothesis: Geology, v. 11, p. 7–10.

Elison, M. W., and Speed, R. C., 1989, Structural development during flysch basin collapse, the Fencemaker allochthon, East Range, Nevada: Journal of Structural Geology, v. 11, p. 523–538.

Engebretson, D. C., 1982, Relative motions between oceanic and continental plates in the Pacific basin [Ph.D. thesis]: Stanford, California, Stanford University, 211 p.

Engebretson, D. C., Cox, A., and Gordon, R. G., 1986, Relative motions between oceanic and continental plates in the Pacific Basin: Geological Society of America Special Paper 206, 59 p.

Ernst, W. G., 1970, Tectonic contact between the Franciscan melange and the Great Valley sequence—Crustal expression of a late Mesozoic Benioff zone: Journal of Geophysical Research, v. 75, p. 886–901.

Evernden, J. F., and Kistler, R. W., 1970, Chronology of emplacement of Mesozoic batholithic complexes in California and western Nevada: U.S. Geological Survey Professional Paper 623, 67 p.

Farmer, G. L., and DePaolo, D. J., 1983, Origin of Mesozoic and Tertiary granite in the western United States and implications for pre-Mesozoic crustal structure—1. Nd and Sr isotopic studies in the geocline of the northern Great Basin: Journal of Geophysical Research, v. 88, p. 3379–3401.

Fouch, T. D., 1979, Character and paleogeographic distribution of Upper Cretaceous and Paleocene nonmarine sedimentary rocks in east-central Nevada, *in* Armentrout, J. M., Cole, M. R., and Terbest, H., Jr., eds., Cenozoic paleogeography of the western United States: Los Angeles, California, Pacific Section, Society of Economic Paleontologists and Mineralogists, Pacific Coast Paleogeography Symposium 3, p. 97–111.

Frei, L. S., Magill, J. R., and Cox, A., 1984, Paleomagnetic results from the central Sierra Nevada: Constraints on reconstructions of the western United States: Tectonics, v. 3, p. 157–177.

Fuis, G. S., Zucca, J. J., Mooney, W. D., and Milkereit, B., 1987, A geologic interpretation of seismic-refraction results in northeastern California: Geological Society of America Bulletin, v. 98, p. 53–65.

Gans, P. B., 1987, An open system, two layer crustal stretching model from the eastern Great Basin: Tectonics, v. 6, p. 1–12.

Gans, P. B., and Miller, E. L., 1983, Style of mid-Tertiary extension in east-central Nevada, *in* Gurgel, K. D., ed., Geologic excursions in the overthrust belt and metamorphic core complexes of the Intermountain region: Utah Geological and Mineral Survey Special Studies 59, p. 108–160.

Gans, P. B., Miller, E. L., McCarthy, J., and Ouldott, M. L., 1985, Tertiary extension faulting and evolving brittle-ductile transition zones in the northern Snake Range and vicinity—New insights from seismic data: Geology, v. 13, p. 189–193.

Geissman, J. W., Callian, J. T., Oldow, J. S., and Humphries, S. E., 1984, Paleomagnetic assessment of oroflexural deformation in west-central Nevada and significance for emplacement of allochthonous assemblages: Tectonics, v. 3, p. 179–200.

Gilluly, J., and Gates, O., 1965, Tectonic and igneous geology of the northern Shoshone Range: U.S. Geological Survey Professional Paper 465, 153 p.

Graham, S. A., 1979, Tertiary paleotectonics and paleogeographics of the Salinian block, *in* Armentrout, J. M., Cole, M. R., and Terbest, H., Jr., eds., Cenozoic paleogeography of the western United States: Pacific Section, Society of Economic Paleontologists and Mineralogists, Pacific Coast Paleogeography Symposium 3, p. 45–51.

Graham, S. A., and Dickinson, W. R., 1978, Apparent features across the San Gregorio–Hosgri fault trend: Science, v. 199, p. 179–181.

Griscom, A., 1980, Aeromagnetic and interpretation maps of the King Range and Chamise Mountain instant study areas, northern California: U.S. Geological Survey Miscellaneous Field Studies Map MF-1196B, scale 1:24,000.

Gromme, C. S., Merrill, R. T., and Verhoogen, J., 1967, Paleomagnetism of Jurassic and Cretaceous rocks of the Sierra Nevada and its significance for polar wandering and continental drift: Journal of Geophysical Research, v. 72, p. 5661–5684.

Hamblin, W. K., 1965, Origin of reverse drag on the downthrown side of normal faults: Geological Society of America Bulletin, v. 76, p. 1145–1164.

Hamilton, W., 1969, Mesozoic California and the underflow of the Pacific mantle: Geological Society of America Bulletin, v. 80, p. 2409–2430.

—— , 1976, Tectonic history of the Riggins region, western Idaho: Geological Society of America Abstracts with Programs, v. 8, p. 378.

—— , 1978, Mesozoic tectonics of the western United States, *in* Howell, D. G., and McDougall, K. A., eds., Mesozoic paleogeography of the western United States: Los Angeles, California, Pacific Section, Society of Economic Paleontologists and Mineralogists, Pacific Coast Paleogeography Symposium 2, p. 33–70.

—— , 1979, Tectonics of the Indonesian region: U.S. Geological Survey Professional Paper 1078, 345 p.

Hamilton, W., and Meyers, W. B., 1966, Cenozoic tectonics of the western United States: Review of Geophysics, v. 4, p. 509–549.

Hannah, J. L., and Verosub, K. L., 1980, Tectonic upper Paleozoic strata of the northern Sierra Nevada: Geology, v. 8, p. 520–524.

Harper, G. D., 1980, Structure and petrology of the Josephine ophiolite and overlying metasedimentary rocks, northwestern California [Ph.D. thesis]: Berkeley, University of California, 260 p.

—— , 1984, The Josephine ophiolite, northwestern California: Geological Society of America Bulletin, v. 95, p. 1009–1026.

Harper, G. D., and Wright, J. E., 1984, Middle to Late Jurassic tectonic evolution of the Klamath Mountains, California-Oregon: Tectonics, v. 3, p. 759–772.

Harwood, D. S., 1983, Stratigraphy of upper Paleozoic volcanic rocks and regional unconformities in part of the northern Sierra terrane, California: Geological Society of America Bulletin, v. 94, p. 413–422.

Heck, F. R., and Speed, R. C., 1987, Triassic olistostrome and shelf-basin transition in the western Great Basin: Paleogeographic implications: Geological Society of America Bulletin, v. 99, p. 539–551.

Heller, P. L., Bowdler, S. S., Chambers, H. P., Coogan, J. C., Hagen, E. S., Chuster, M. W., and Winslow, N. S., 1986, Time of initial thrusting in the Sevier orogenic belt, Idaho-Wyoming and Utah: Geology, v. 14, p. 388–391.

Hillhouse, J. W., Gromme, C. S., and Vallier, T. L., 1982, Paleomagnetism and Mesozoic tectonics of the Seven Devils arc, Oregon: Journal Geophysical Research, v. 87, p. 3777–3794.

Hopson, C. A., Mattinson, J. M., and Peassagno, E. A., Jr., 1981, Coast Range ophiolite, western California, *in* Ernst, W. G., ed., The geotectonic development of California (Rubey Volume 1): Englewood Cliffs, New Jersey, Prentice-Hall, p. 307–328.

Hotz, P. E., Lanphere, M. A., and Swanson, D. A., 1977, Triassic blueschist from northern California and north-central Oregon: Geology, v. 5, p. 659–663.

Howard, K. A., 1971, Paleozoic metasediments in the northern Ruby Mountains, Nevada: Geological Society of America Bulletin, v. 82, p. 259–264.

—— , 1980, Metamorphic infrastructures in the northern Ruby Mountains, Nevada, *in* Crittenden, M. D., Jr., Coney, P. J., and Davis, G. H., Cordilleran metamorphic core complexes: Geological Society of America Memoir 153, p. 335–347.

Ingersoll, R. V., 1983, Petrofacies and provenance of late Mesozoic forearc basin, northern and central California: American Association of Petroleum Geologists Bulletin, v. 67, p. 1125–1142.

Irwin, W. P., 1972, Terranes of the western Paleozoic and Triassic belt in the southern Klamath Mountains, California: U.S. Geological Survey Professional Paper 800-C, p. 103–111.

—— , 1977, Review of Paleozoic rocks of the Klamath Mountains, *in* Stewart, J. H., Stevens, C. H., and Fritsche, A. F., eds., Paleozoic paleogeography of the western United States: Los Angeles, California, Pacific Section, Society of Economic Paleontologists and Mineralogists, Pacific Coast Paleogeography Symposium 1, p. 441–454.

Irwin, W. P., Jones, D. L., and Kaplan, T. A., 1978, Radiolarians from pre-Nevadan rocks of the Klamath Mountains, California and Oregon, *in* Howell, D. E., and McDougall, K. A., eds., Mesozoic paleogeography of the western United States: Los Angeles, California, Pacific Section, Society of Economic Paleontologists and Mineralogists, Pacific Coast Paleogeography Symposium 2, p. 303–310.

Jacobsen, S. B., Quick, J. E., and Wasserbury, G. J., 1984, A Nd and Sr evolution: Earth and Planetary Science Letters, v. 68, p. 361–378.

James, E. W., 1986, Pre-Tertiary paleogeography along with northern San Andreas fault: (Eos (Transactions, American Geophysical Union), v. 67, p. 1215.

Jansma, P. E., and Speed, R. C., 1990, Omissional faulting during regional Mesozoic contraction at Carlin Canyon, NV: Geological Society of America Bulletin, v. 102, p. 417–427.

Jayko, A. S., Blake, M. C., and Harms, T., 1987, Attenuation of the Coast Range ophiolite by extensional faulting, and nature of the Coast Range "thrust," California: Tectonics, v. 6, p. 475–488.

Johnson, L. R., 1968, Crustal structure between Lake Mead, Nevada and Mono Lake, CA: Journal of Geophysical Research, v. 70, p. 2863–2872.

Jones, C. H., Kanamori, H., and Roecker, S. W., 1990, The missing root: Teleseismic and regional P_n arrival observed in the southern Sierra Nevada, California: EOS (Transactions, American Geophysical Union), v. 71, p. 1558.

Jones, D. L., and Irwin, W. P., 1971, Structural implications of an offset Early Cretaceous shoreline in northern California: Geological Society of America Bulletin, v. 82, p. 815–822.

Jones, D. L., Howell, D. G., and Coney, P. J., 1983, Recognition, character, and analysis of tectonostratigraphic terranes in western North America, *in* Hashimoto, M., and Uyeda, S., eds., Accretion tectonics in the circum-Pacific regions: Boston, D. Reidel Publishing Co., p. 21–36.

Jordan, T. E., 1981, Thrust loads and foreland basin evolution, Cretaceous western United States: American Association of Petroleum Geologists Bulletin, v. 65, p. 2506–2520.

Jordan, T. E., and Douglass, R. C., 1980, Paleogeographic and structural development of the Late Pennsylvanian to Early Permian Oquirrh Basin, northwestern Nevada, *in* Fouch, T. D., and Magathan, E. R., eds., Paleozoic paleogeography of the west-central U.S.: Los Angeles, California, Society of Economic Paleontologists and Mineralogists Symposium 1, p. 217–238.

Kay, M., 1960, Paleozoic continental margin in central Nevada, western U.S.: Copenhagen, International Geological Congress, 21st, Report, Part 12, p. 94–103.

Kay, M., and Crawford, J. P., 1964, Paleozoic facies to eugeosynclinal belt in thrust slices, central Nevada: Geological Society of America Bulletin, v. 75, p. 425–454.

King, P. B., 1966, The tectonic map of North America: U.S. Geological Survey, scale 1:5,000,000.

Kirsch, S. A., 1971, Chaos structure and turtle back dome, Mineral Ridge, Esmeralda County, Nevada: Geological Society of America Bulletin, v. 82, p. 3169–3176.

Kistler, R. W., 1966, Structure and metamorphism in the Mono Craters quadrangle, Sierra Nevada, California: U.S. Geological Survey Bulletin, v. 1221-E, p. E1–E53.

—— , 1974, Phanerozoic batholiths in western North America: A summary of some recent work on variations in time, space chemistry, and isotopic compositions: Annual Review of Earth and Planetary Sciences, v. 2, p. 403–418.

—— , 1978, Mesozoic paleogeography of California, *in* Howell, D. E., and McDougall, K. A., eds., Mesozoic paleogeography of the western United States: Los Angeles, California, Pacific Section, Society of Economic Paleontologists and Mineralogists, Pacific Coast Paleogeography Symposium 2, p. 75–85.

—— , 1990, Two different lithosphere types in the Sierra Nevada, California, *in* Anderson, J. L., ed., The nature and origin of Cordilleran magmatism: Geological Society of America Memoir 174, p. 271–281.

Kistler, R. W., and Peterman, Z. E., 1973, Variations in Sr, Rb, K, Na, and initial $^{87}Sr/^{86}Sr$ in Mesozoic granitic rocks and intruded wall rocks in central California: Geological Society of America Bulletin, v. 84, p. 3489–3512.

—— , 1978, Reconstruction of crustal blocks of California on the basis of initial strontium isotopic compositions of Mesozoic granitic rocks: U.S. Geological Survey Professional Paper 1071, 17 p.

Kistler, R. W., Evernden, J. F., and Shaw, H. R., 1971, Sierra Nevada plutonic cycle: Part I, Origin of composite granitic batholiths: Geological Society of America Bulletin, v. 82, p. 853–868.

Kistler, R. W., Robinson, A. C., and Fleck, R. W., 1980, Mesozoic right-lateral fault in eastern California: Geological Society of America Abstracts with Programs, v. 12, p. 115.

Kistler, R. W., Ghent, E. D., and O'Neil, J. R., 1981, Petrogenesis of garnet two-mica granites in the Ruby Mountains, Nevada: Journal of Geophysical Research, v. 86, p. 10591–10606.

Klemperer, S. L., Hauge, T. A., Hauser, E. C., Oliver, T. E., and Potter, C. J., 1986, The Moho in the northern Basin and Range province, NV, along the COCORP 40° seismic reflection transect: Geological Society of America Bulletin, v. 97, p. 603–618.

Krause, D. C., Plenard, H. W., and Smith, S. M., 1964, Topography and lithology of the Mendocino Ridge: Journal of Marine Research, v. 22, p. 236–249.

La Brecque, J. L., Kent, D. V., and Cande, S. C., 1977, Revised magnetic polarity time scale for Late Cretaceous and Cenozoic time: Geology, v. 5, p. 330–335.

Lachenbruch, A. H., and Sass, J. H., 1978, Models of an extending lithosphere and heat flow in the Basin and Range province, *in* Smith, R. B., and Eaton, G. P., eds., Cenozoic Tectonics and Regional Geophysics of the Western Cordillera: Geological Society of America Memoir 152, p. 209–250.

Lahren, M. M., and Schweickert, R. A., 1989, Proterozoic and Lower Cambrian miogeoclinal rocks of Snow Lake pendant, Yosemite-Emigrant Wilderness, Sierra Nevada, California: Evidence for major Early Cretaceous dextral translation: Geology, v. 17, p. 156–160.

Lahren, M. M., Schweickert, R. A., Mattinson, J. M., and Walker, J. D., 1990, Evidence of uppermost Proterozoic to Lower Cambrian miogeoclinal rocks and the Mojave–Snow Lake fault: Snow Lake pendant, central Sierra Nevada, California: Tectonics, v. 9, p. 1585–1608.

Lanphere, M. A., Irwin, W. P., and Hotz, P. E., 1968, Isotopic age of the Nevadan orogeny and older plutonic and metamorphic events in the Klamath Mountains, California: Geological Society of America Bulletin, v. 79, p. 1027–1052.

Lee, D. E., Kistler, R. W., Friedman, I., and van Loenen, R. E., 1981, Two-mica granites of northeastern Nevada: Journal of Geophysical Research, v. 86, p. 10607–10616.

Lindsley-Griffen, N., 1977, The Trinity ophiolite, Klamath Mountains, California: State of Oregon Department of Geology and Mining Industries Bulletin, v. 95, p. 107–120.

Lipman, P. W., Protoka, H. J., and Christiansen, R. L., 1972, Cenozoic volcanism and plate tectonic evolution of the United States: Part I, Early and middle Cenozoic: Royal Society of London Philosophical Transactions, v. 271, p. 217–248.

Lowell, J. D., 1974, Plate tectonics and foreland basin deformation: Geology, v. 2, p. 275–278.

Maher, K. A., and Kirschvink, J., 1988, Paleomagnetism and tectonics of the Jackson Mountains, NW Nevada: Eos (Transactions, American Geophysical Union), v. 15.

Matthews, V., III, and Work, D. F., 1978, Laramide folding associated with basement block faulting along the northeastern flank of the Front Range, Colorado, *in* Matthews, III, V., ed., Laramide folding associated with basement block faulting in the western United States: Geological Society of America Memoir 151, p. 101–124.

Matti, J. C., and McKee, E. H., 1977, Silurian and Lower Devonian paleogeography of the outer continental shelf of the Cordilleran miogeocline, central Nevada, *in* Stewart, J. H., Stevens, C. H., and Fritsche, A. F., eds., Paleozoic paleogeography of the western United States: Los Angeles, California, Pacific Section, Society of Economical Paleontologists and Mineralogists, Pacific Coast Paleogeography Symposium 1, p. 181–215.

Mattinson, J. M., and Echeverria, L. M., 1980, Ortigalita Peak gabbro, Franciscan Complex: U-Pb dates of intrusion and high-pressure low temperature metamorphism: Geology, v. 8, p. 589–593.

Mattinson, J. M., and Hopson, C. A., 1972, Paleozoic ophiolitic complex in Washington and northern California: Carnegie Institution Yearbook 71,

p. 578–583.
May, S. R., and Butler, R. F., 1986, North American Jurassic apparent polar wander—Implications for plate motions, paleogeography and Cordilleran tectonics: Journal of Geophysical Research, v. 19, p. 1519–1544.
McDonald, R. E., 1976, Tertiary tectonics and sedimentary rocks along the transition, Basin and Range province to plateau and thrust belt province, Utah: Denver, Colorado, Rocky Mountain Association of Geologists, Symposium, p. 281–317.
McEvilly, T. V., 1968, Sea-floor mechanics north of Cape Mendocino, California: Nature, v. 220, p. 901–903.
McLaughlin, R. J., Kling, S. A., Poore, R. Z., McDougal, K., and Beutner, E. C., 1982, Post-middle Miocene accretion of Franciscan rocks, northwestern California: Geological Society of America Bulletin, v. 93, p. 595–605.
McLaughlin, R. J., Blake, M. C., Jr., and Griscom, A., 1988, Tectonics of formation, translation, and dispersal of the Coast Range ophiolite of California: Tectonics, v. 7, p. 1033–1056.
Miller, C. F., and Bradfish, L. J., 1980, An inner Cordilleran belt of muscovite-bearing plutons: Geology, v. 8, p. 412–416.
Miller, D. M., 1980, Structural geology of the northern Albion Mountains, south-central Idaho, *in* Crittenden, Max D., Jr., Coney, P. J., and Davis, G. H., Cordilleran metamorphic core complexes: Geological Society of America Memoir 153, p. 399–423.
——, 1984, Sedimentary and igneous rocks of the Pilot Range, Utah and Nevada: Utah Geological Association Publication, v. 13, p. 45–63.
Miller, E. L., Bateson, J., Dinter, D., Dyer, J. D., Harbaugh, D., and Jones, D. L., 1981, Thrust emplacement of the Schoonover sequence, northern Independence Range, NV: Geological Society of America Bulletin, v. 92, p. 730–737.
Miller, E. L., Holdsworth, B. K., Whiteford, W. B., and Rodgers, D., 1983, Stratigraphy and structure of the Schoonover complex, northeastern Nevada: Geological Society of America Bulletin, v. 95, p. 1063–1076.
Miller, E. L., Gans, P. B., Wright, J. E., and Sutter, J. F., 1988, Metamorphic history of the east-central Basin and Range province: Tectonic setting and relationship to magmatism, *in* Ernst, W. G., ed., Metamorphism and crustal evolution of the western United States (Rubey Volume VII): Englewood Cliffs, New Jersey, Prentice-Hall, p. 649–682.
Miller, M. M., 1987, Dispersed remnants of a northeast Pacific fringing arc: Upper Paleozoic terranes of Permian McCloud faunal affinity, western U.S.: Tectonics, v. 6, p. 807–830.
Miller, R. B., 1982, Geology of the Rimrock Lake pre-Tertiary inlier, southern Washington Cascades: Geological Society of America Bulletin, v. 14, p. 217.
Monger, J.W.H., 1975, Correlation of eugeosynclinal tectono-stratigraphic belts in the North American Cordillera: Geoscience Canada, v. 2, p. 4–9.
——, 1977, Upper Paleozoic rocks of the Western Canadian Cordillera and their bearing on Cordilleran evolution: Canadian Journal of Earth Sciences, v. 14, p. 1832–1859.
Moores, E. M., 1970, Ultramafics and orogeny, with models of the U.S. Cordillera and the Tethys: Nature, v. 228, p. 837–842.
Morgan, B. A., 1976, Geology of Chinese camp and Moccasin quadrangles, Tuolumne County, California: U.S. Geological Survey Miscellaneous Field Studies Map MF-840, scale 1:24,000.
Muffler, L.J.P., 1964, Geology of the Frenchie Creek quadrangle, Nevada: Geological Society of America Bulletin, v. 47, p. 241–251.
Nilsen, T. H., 1984, Offset along the San Andreas fault of Eocene strata from the San Juan Bautista area and western San Emidgio Mountains, California: Geological Society of America Bulletin, v. 95, p. 599–609.
Nokleberg, W. J., and Kistler, R. W., 1980, Paleozoic and Mesozoic deformations in the central Sierra Nevada, California: U.S. Geological Survey Professional Paper 1145, 24 p.
Nolan, T. B., Merriam, C. W., and Williams, J. S., 1956, The stratigraphic section in the vicinity of Eureka, Nevada: U.S. Geological Survey Professional Paper 276, 77 p.
Normark, W. R., and Hess, G. R., 1980, Quaternary growth patterns of California submarine fans, *in* Field, M., Bouma, A., Colburn, I., Douglas, R. G., and Ingle, J. C., eds., Quaternary depositional environments of the Pacific Coast: Los Angeles, California, Pacific Section, Society of Economic Paleontologists and Mineralogists, Pacific Coast Paleogeography Symposium 4, p. 201–210.
Okaya, D. A., and Thompson, G. A., 1985, Geometry of Cenozoic extensional faulting, Dixie Valley, Nevada: Tectonics, v. 4, p. 107–125.
Oldow, J. S., 1981, Kinematics of late Mesozoic thrusting, Pilot Mountains, Nevada: Journal of Structural Geology, v. 3, p. 39–51.
——, 1984, Spatial variability in the structure of the Roberts Mountains allochthon, western Nevada: Geological Society of America Bulletin, v. 95, p. 174–185.
Oldow, J. S., Avé Lallemant, H. G., and Schmidt, W. J., 1984, Kinematics of plate convergence deduced from Mesozoic structures in the western Cordillera: Tectonics, v. 3, p. 210–217.
Oliver, H. W., 1977, Gravity and magnetic investigations of the Sierra Nevada batholith, California: Geological Society of America Bulletin, v. 88, p. 445–461.
Page, B. M., 1981, The Southern Coast Ranges, *in* Ernst, W. G., ed., The geotectonic development of California (Rubey Volume 1): Englewood Cliffs, New Jersey, Prentice-Hall, p. 329–417.
Page, B. M., and Engebretson, D. C., 1984, Correlation between the geologic record and computed plate motions for California: Tectonics, v. 3, p. 133–155.
Pakiser, L. C., and Brune, J. N., 1980, Seismic models of the root of the Sierra Nevada: Science, v. 210, p. 1088–1094.
Pickett, D. A., and Saleeby, J. B., 1989, Petrogenetic and isotopic studies of mid-crustal batholithic rocks of the Tehachapi Mountains, Sierra Nevada, California: Geological Society of America Abstracts with Programs, v. 21, A197.
——, 1990, *P-T-X* conditions in the Tehachapi Mts., Sierra Nevada, CA—Metamorphism and tectonic disruption in the deep Cretaceous batholith: Geological Society of America Abstracts with Programs, v. 22, p. A30.
——, 1993, Thermobarometric constraints on the depth of exposure and conditions of plutonism and metamorphism at deep levels of the Sierra Nevada Batholith, Tehachapi Mountains, California: Journal of Geophysical Research, v. 98, p. 609–629.
Platt, J. P., 1986, Dynamics of orogenic wedges and the uplift of high-pressure metamorphic rocks: Geological Society of America Bulletin, v. 97, p. 1037–1053.
Poole, F. G., 1974, Flysch deposits of the Antler United States, *in* Dickinson, W. R., ed., Tectonics and sedimentation: Society of Economic Paleontologists and Mineralogists Special Publication 22, p. 58–82.
Poole, F. G., and Sandberg, C. A., 1977, Mississippian paleogeography and tectonics of the western United States, *in* Stewart, J. H., Stevens, C. H., and Fritsche, A. F., eds., Paleozoic paleogeography of the western United States: Los Angeles, California, Pacific Section, Society of Economic Paleontologists and Mineralogists, Pacific Coast Paleogeography Symposium 1, p. 67–85.
Priestley, K., and Brune, J., 1978, Surface waves and the structure of the Great Basin of Nevada and western Utah: Journal of Geophysical Research, v. 83, p. 2265–2272.
Priestley, K., Ryall, A., and Fezie, G. S., 1982, Crustal and upper mantle structure in the northwest Basin and Range province: Seismological Society of America Bulletin, v. 72, p. 911–923.
Proffett, J. M., Jr., 1977, Cenozoic geology of the Yerington district, Nevada, and implications for the nature and origin of Basin and Range faulting: Geological Society of America Bulletin, v. 88, p. 247–266.
Prohdehl, C., 1979, Crustal structure of the western U.S.: U.S. Geological Survey Professional Paper 1034, 74 p.
Quick, J. E., 1981, Petrology and petrogenesis of the Trinity peridotite, an upper mantle diapir in the eastern Klamath Mountains, northern California: Contributions to Mineralogy and Petrology, v. 78, p. 413–422.
Raff, A. D., and Mason, R. G., 1961, Magnetic survey off the west coast of North America, 40°N latitude to 52°N latitude: Geological Society of America Bulletin, v. 72, p. 1267–1270.

Raitt, R. W., Shor, G. G., Francis, T.J.G., and Morris, G. B., 1969, Anisotropy of the Pacific upper mantle: Journal of Geophysical Research, v. 74, p. 3095–3109.

Roberts, R. J., Hotz, P. E., Gilluly, J., and Fergusen, H. G., 1958, Paleozoic rocks of north-central Nevada: American Association of Petroleum Geologists Bulletin, v. 74, p. 2813–2857.

Ross, D. C., 1978, The Salinian block—A Mesozoic granitic orphan in the California Coast Ranges, *in* Howell, D. E., and McDougall, K. A., eds., Mesozoic paleogeography of the western United States: Los Angeles, California, Pacific Section, Society of Economical Paleontologists and Mineralogists, Pacific Coast Paleography Symposium 2, p. 509–522.

——, 1980, Reconnaissance geologic map of basement rocks of the southernmost Sierra Nevada (North to 35°30′N): U.S. Geological Survey Open-File Report 80-307, 22 p.

——, 1983, Hornblende-rich, high-grade metamorphic terranes in the southernmost Sierra Nevada, California, and implications for crustal depths and batholith roots: U.S. Geological Survey Open-File Report 83-465, 51 p.

——, 1985, Mafic gneiss complex (batholithic root?) in the southernmost Sierra Nevada, California: Geology, v. 13, p. 288–291.

Ross, C. A., and Ross, R. P., 1983, Late Paleozoic accreted terranes of western North America, *in* Stevens, C. H., ed., Pre-Jurassic rocks in western North America suspect terranes: Los Angeles, California, Society of Economic Paleontologists and Mineralogists, p. 7–22.

Ross, J. A., and Sharp, W. D., 1988, The effects of sub-blocking temperature metamorphism on the K/Ar systematics of hornblendes: $^{40}Ar/^{39}Ar$ dating of polymetamorphic garnet amphibolite from the Franciscan Complex, California: Contributions to Mineralogy and Petrology, v. 100, p. 213–221.

Rowell, A. J., Rees, M. N., and Suczek, C. A., 1979, Margin of the North American continent in Nevada during Late Cambrian time: American Journal of Science, v. 279, p. 1–18.

Royse, F., Warner, M. A., and Resse, D. L., 1975, Thrust belt structural geometry and related stratigraphic problems Wyoming–Idaho–northern Utah: Denver, Colorado, Rocky Mountain Association of Geologists, p. 41–54.

Rubin, C. M., Saleeby, J. B., Cowan, D. S., Brandon, M. T., and McGroder, M. F., 1990, Regionally extensive mid-Cretaceous west-vergent thrust system in the northwestern Cordillera: Implications for continent-margin tectonism: Geology, v. 18, p. 276–280.

Russell, B. J., 1984, Mesozoic geology of the Jackson Mountains, northwestern Nevada: Geological Society of America Bulletin, v. 95, p. 313–323.

Russell, B. J., Beck, M. E., Burmester, R. F., and Speed, R., 1982, Cretaceous magnetizations in northwestern Nevada and tectonic implications: Geology, v. 10, p. 243–250.

Ryall, A., and Malone, S. D., 1971, Earthquake distribution and mechanism of faulting in the Rainbow Mountain–Dixie Valley–Fairview Peak area, NV: Journal of Geophysical Research, v. 76, p. 7241–7248.

Ryall, A., and Priestley, K. F., 1975, Seismicity, secular strain, and maximum magnitude in the Excelsior Mountains area, NV and CA: Geological Society of America Bulletin, v. 56, p. 1105–1135.

Saleeby, J. B., 1977, Fracture zone tectonics, continental margin fragmentation, and emplacement of the Kings-Kaweah ophiolite belt, southwest Sierra Nevada, California, *in* Coleman, R. G., and Irwin, W. P., eds., International geological correlation program, North American ophiolite volume: Oregon State University, Geology Department, p. 123–140.

——, 1978, Kings River ophiolite, Southwest Sierra Nevada Foothills, California: Geological Society of America Bulletin, v. 89, p. 617–636.

——, 1979, Kaweah serpentine melange, southwest Sierra Nevada Foothills, California: Geological Society of America Bulletin, v. 90, p. 29–46.

——, 1981, Ocean floor accretion and volcano-plutonic arc evolution of the Mesozoic Sierra Nevada, California, *in* Ernst, W. G., ed., The geotectonic development of California (Rubey Volume I): Englewood Cliffs, New Jersey, Prentice-Hall, Inc., p. 132–181.

——, 1982, Polygenetic ophiolite belt of the California Sierra Nevada, geochronological and tectonostratigraphic development: Journal of Geophysical Research, v. 87, p. 1802–1824.

——, 1983, Accretionary tectonics of the North American Cordillera: Annual Review of Earth and Planetary Sciences, v. 15, p. 45–73.

——, 1990a, Geochronological and tectonostratigraphic framework of Sierran-Klamath ophiolitic assemblages, *in* Howard, D. S., and Miller, M. M., eds., Paleozoic and early Mesozoic paleogeographic relations in the Klamath Mountains, Sierra Nevada, and related terranes: Geological Society of America Special Paper 255, p. 93–114.

——, 1990b, Progress in tectonic and petrogenetic studies in an exposed cross section of young (~100 Ma) continental crust, southern Sierra Nevada, California, *in* Salisbury, M. H., and Fountain, D. M., eds., Exposed cross sections of the continental crust: Dordrecht, Holland, D. Reidel Publishing Co., p. 137–158.

——, 1992, Petrotectonic and paleogeographic settings of U.S. Cordillera ophiolites, *in* Burchfiel, B. C., Zoback, M. L., and Lipman, P. W., eds., The Cordilleran orogen: Conterminous U.S.: Boulder, Colorado, Geological Society of America, The Geology of North America, v. G-3, p. 653–682.

Saleeby, J. B., and Busby-Spera, C. V., 1986, Fieldtrip guide to the metamorphic framework rocks of the Lake Isabella area, southern Sierra Nevada, California, *in* Mesozoic and Cenozoic structural evolution of selected areas, east-central California (Geological Society of America Cordilleran Section field trip guidebook): Los Angeles, Geology Department, California State University, p. 81–94.

——, 1992, Early Mesozoic tectonic evolution of the U.S. Cordilleran orogen, *in* Burchfiel, B. C., Zoback, M. L., and Lipman, P. W., eds., The Cordilleran orogen: Conterminous U.S.: Boulder, Colorado, Geological Society of America, The Geology of North America, v. G-3, p. 107–168.

Saleeby, J. B., and Busby, C. V., 1993, Paleogeographic and tectonic setting of axial and western metamorphic framework rocks of the southern Sierra Nevada, California, *in* Dunne, G. C., and McDougall, K. A., eds., Mesozoic volume on the paleogeographic development of the western United States: Los Angeles, Pacific Section, Society of Economic Paleontologists and Mineralogists, Book 71, p. 197–226.

Saleeby, J. B., and Sharp, W. D., 1980, Chronology of the structural and petrologic development of the southwest Sierra Nevada Foothills, California: Geological Society of America Bulletin, part I, v. 91, p. 317–320.

Saleeby, J. B., Busby-Spera, C., Goodin, W. D., and Sharp, W. D., 1978, Early Mesozoic paleotectonic-paleogeographic reconstruction of the southern Sierra Nevada region, California, *in* Howell, D. E., and McDougall, K. A., eds., Mesozoic paleogeography of the western United States: Los Angeles, California, Pacific Section, Society of Economical Paleontologists and Mineralogists, Pacific Coast Paleography Symposium 2, p. 311–336.

Saleeby, J. B., Mattinson, J. M., and Wright, J. E., 1979, Regional ophiolite terranes of California: Vestiges of two complex ocean floor assemblages: Geological Society of America Abstracts with Programs, v. 11, p. 509.

Saleeby, J. B., Harper, G. D., Sharp, W. D., and Snoke, A. W., 1982, Time relations and structural-stratigraphic patterns in ophiolite accretion, west-central Klamath Mountains, California: Journal of Geophysical Research, v. 87, p. 3831–3848.

Saleeby, J. B., and 12 others, 1986, C-2, Monterey Bay offshore to the Colorado Plateau: Boulder, Colorado Geological Society of America Centennial Continent-Ocean Transect no. 10, 2 sheets, scale 1:500,000, 87 p.

Saleeby, J. B., Hannah, J. L., and Varga, R. J., 1987, Isotopic age constraints on middle Paleozoic deformation in the northern Sierra Nevada, California: Geology, v. 15, p. 757–760.

Saleeby, J. B., Longiaru, S., Kistler, R. W., Moore, J. G., and Nokleberg, W. J., 1988, Middle Cretaceous silicic metavolcanic rocks in the Kings Canyon area, central Sierra Nevada, California: Geological Society of America Abstracts with Program, v. 20, no. 3, p. 226.

Saleeby, J. B., Shaw, H. F., Niemeyer, S., Edelman, S. H., and Moores, E. M., 1989a, U/Pb, Sm/Nd and Rb/Sr geochronological and isotopic study of northern Sierra Nevada ophiolitic assemblages, California: Contributions to Mineralogy and Petrology, v. 102, p. 205–220.

Saleeby, J. B., Geary, E. E., Paterson, S. R., and Tobisch, O. T., 1989b, Isotopic systematics of Pb/U (zircon) and $^{40}Ar/^{39}Ar$ (biotite/hornblende) from rocks of the Central Foothills terrane, Sierra Nevada, California: Geological Society of America Bulletin, v. 101, p. 1481–1492.

Sams, D. B., and Saleeby, J. B., 1986, U/Pb zircon ages, and implications on the structural and petrologic development of the southernmost Sierra Nevada: Geological Society of America Abstracts with Programs, v. 18, p. 180.

Sams, D. B., Saleeby, J. B., Ross, D. C., and Kistler, R. W., 1983, Cretaceous igneous, metamorphic and deformational events of the southernmost Sierra Nevada, California: Geological Society of America Abstracts with Programs, v. 15, p. 294.

Schweickert, R. A., Armstrong, R. L., and Harakal, J. E., 1981, Tectonic evolution of the Sierra Nevada Range, *in* Ernst, W. G., ed., The geotectonic development of California of California (Rubey Volume I): Englewood Cliffs, New Jersey, Prentice-Hall, Inc., p. 87–131.

Schweickert, R. A., and Cowan, D. S., 1975, Early Mesozoic tectonic evolution of the western Sierra Nevada, California: Geological Society of America Bulletin, v. 86, p. 1329–1336.

Schwieckert, R. A., and Lahren, M. M., 1987, Continuation of Antler and Sonoma orogenic belts to the eastern Sierra Nevada, California, and Late Triassic thrusting in a compressional arc: Geology, v. 15, p. 270–273.

Schweickert, R. A., and Snyder, W., 1981, Paleozoic plate tectonics of the Sierra Nevada and adjacent regions, *in* Ernst, W. G., ed., The geotectonic development of California (Rubey Volume I): Englewood Cliffs, New Jersey, Prentice-Hall, Inc., p. 182–202.

Schweickert, R. A., and others, 1980, Lawsonite blueschist in the northern Sierra Nevada, California: Geology, v. 8, p. 27–31.

Seeber, L., Barazangi, M., and Nowroozi, A. A., 1970, Microearthquake seismicity and tectonics of coastal northern California: Seismological Society of America Bulletin, v. 60, p. 1669–1699.

Segall, P., McKee, E. H., Martel, S. J., and Turrin, B. D., 1990, Late Cretaceous age of fractures in the Sierra Nevada batholith, California: Geology, v. 18, p. 1248–1251.

Sharp, W. D., 1988, Pre-Cretaceous crustal evolution in the Sierra Nevada region, California, *in* Ernst, W. G., ed., Metamorphism and crustal evolution of the western United States (Rubey Volume VII): Englewood Cliffs, New Jersey, Prentice-Hall, Inc., p. 823–864.

Sharp, W. D., and Leighton, C. W., 1987, Accretion of the Foothills ophiolite, western Sierra Nevada Foothills, California: Geological Society of America Abstracts with Programs, v. 19, no. 6, p. 450.

Sharry, J., 1981, Geology of the western Tehachapi Mountains, California [Ph.D. thesis]: Cambridge, Massachusetts, Massachusetts Institute of Technology, 215 p.

Shaw, H. F., Chen, J. H., Saleeby, J. B., and Wasserburg, G. J., 1987, Nd-Sr-Pb systematics and age of the Kings River Ophiolite, California: Contributions to Mineralogy and Petrology, v. 96, p. 281–290.

Shor, G. G., Jr., Menard, H. W., and Raitt, R. W., 1971, Structure of the Pacific Basin, *in* Plaxwell, A. E., ed., The sea, Volume 4, Part II: New York, Wiley and Sons, p. 3–27.

Sieh, K. E., and Jahns, R. H., 1984, Holocene activity of the San Andreas fault at Wallace Creek, California: Geological Society of America Bulletin, v. 95, p. 883–896.

Silberling, N. J., 1973, Geologic events during Permian-Triassic time along the Pacific margin of the United States, *in* Logan, A. and Hills, L. V., eds., The Permian and Triassic systems and their mutual boundary (International Permian-Triassic Conference, Calgary, Canada): Canadian Society of Petrologists and Mineralogists, p. 345–362.

Silberling, N. J., and John, D. A., 1989, Geologic map of pre-Tertiary rocks of the Paradise Range and south Lodi Hills, west-central Nevada: U.S. Geological Survey Miscellaneous Field Studies Map MF 2062, scale 1:62,500.

Silberling, N. J., and Roberts, R. J., 1962, Pre-Tertiary stratigraphy and structure of northwestern Nevada: Geological Society of America Special Paper 72, 58 p.

Silberling, N. J., Jones, D. L., Blake, M. C., Jr., and Howell, D. G., 1987, Lithotectonic terrane map of the western conterminous United States: U.S. Geological Survey Miscellaneous Field Studies Map MF 1874C, scale 1:62,500.

Silver, E. A., 1971, Tectonics of the Mendocino triple junction: Geological Society of America Bulletin, v. 82, p. 2965–2978.

Silver, L. T., and Anderson, T. H., 1974, Possible left-lateral early to middle Mesozoic disruption of the southwestern North American craton margin: Geological Society of America Abstracts with Programs, v. 6, p. 955–956.

Silver, L. T., and Mattinson, J. M., 1986, "Orphan Salinia" has a home: Eos (Transactions, American Geophysical Union), v. 67, p. 1215.

Sliter, W. V., 1984, Foraminifers from Cretaceous limestone of the Franciscan Complex, northern California, *in* Blake, M. C., Jr., ed., Franciscan geology of northern California: Los Angeles, California, Pacific Section, Society of Economic Paleontologists and Mineralogists, p. 149–162.

Smith, D. L., 1992, History and kinematics of Cenozoic extension in the northern Toiyabe Range, Lander County, Nevada: Geological Society of America Bulletin, v. 104, p. 789–801.

Smith, J. F., and Ketner, K. B., 1968, Devonian and Mississippian rocks and the date of the Roberts Mountains thrust in the Carlin-Pinon Range area, Nevada: U.S. Geological Survey Bulletin, v. 1251-I, p. 11–118.

Smith, R. B., 1978, Seismicity, crustal structure, and intraplate tectonics of the interior of the western Cordillera, *in* Smith, R. B. and Eaton, G. P., eds., Cenozoic Tectonics and Regional Geophysics of the Western Cordillera: Geological Society of America Memoir 152, p. 111–144.

Smith, R. B., and Bruhn, R. L., 1984, Intraplate extensional tectonics of the eastern Basin-Range: Inferences on structural style from seismic reflection data, regional tectonics and thermal-mechanical models of brittle-ductile deformation: Journal of Geophysical Research, v. 89, p. 5733–5762.

Smithson, S. B., Brewer, J. A., Kaufman, S., Oliver, J., and Hurich, C., 1978, Nature of the Wind River thrust, Wyoming, from COCORP deep reflection data and from gravity data: Geology, v. 6, p. 648–652.

Snoke, A. W., 1977, A thrust plate of ophiolitic rocks in the Preston Peak area, Klamath Mountains, California: Geological Society America Bulletin, v. 88, p. 1641–1659.

——, 1980, Transition from infrastructure to suprastructure in the northern Ruby Mountains, Nevada, *in* Crittenden, M. D., Jr., Coney, P. J., and Davis, G. H., eds., Cordilleran metamorphic core complexes: Geological Society of America Memoir 153, p. 287–333.

Snoke, A. W., and Lush, A. P., 1984, Polyphase Mesozoic-Cenozoic deformational history of the northern Ruby Mountains–East Humboldt Range, Nevada, *in* Lintz, J., Jr., ed., Western geological excursions, Volume 4 (Geological Society of America Annual Meeting Guidebook): Reno, Nevada, Mackay School of Mines, p. 232–260.

Snoke, A. W., and Miller, D. M., 1987, Metamorphic and tectonic history of the northeastern Great Basin, *in* Ernst, W. G., ed., Metamorphism and crustal evolution of the western United States (Rubey Volume VII): Englewood Cliffs, New Jersey, Prentice-Hall, Inc., p. 606–648.

Snoke, A. W., Quick, J. E., and Bowman, H. R., 1981, Bear Mountain igneous complex, Klamath Mountains, California: An ultrabasic to silicic calc-alkaline suite: Journal of Petrology, v. 22, p. 501–552.

Snyder, W. S., Dickinson, W. R., and Silberman, M. L., 1976, Tectonic implications of space-time patterns of Cenozoic magmatism in the western United States: Earth and Planetary Science Letters, v. 32, p. 91–106.

Speed, R. C., 1977, Island arc and other paleogeographic terranes of late Paleozoic age in the western Great Basin, *in* Stewart, J. H., Stevens, C. H., and Fritsche, A. F., eds., Paleozoic paleogeography of the western United States: Los Angeles, California, Pacific Section, Society of Economic Paleontologists and Mineralogists, Pacific Coast Paleogeography Symposium 1, p. 349–362.

——, 1978, Paleogeographic and plate tectonic evolution of the early Mesozoic marine providence of the western Great Basin, *in* Howell, D. E., and McDougall, K. A., eds., Mesozoic paleogeography of the western United States: Los Angeles, California, Pacific Section, Society of Economic Paleontologists and Mineralogists, Pacific Coast Paleography Symposium 2, p. 253–270.

——, 1979, Collided Paleozoic microplate in the western United States: Journal of Geology, v. 87, p. 279–292.

—— , 1982, Evolution of the sialic margin in the central-western United States, *in* Watkins, J. and Drake, C., eds., Geology of continental margins: American Association of Petroleum Geologists Memoir 34, p. 457–468.

Speed, R. C., and Kistler, K. W., 1980, Cretaceous volcanism, Excelsior Mountains, Nevada: Geological Society of America Bulletin, v. 91, p. 392–398.

Speed, R. C., and Sleep, N. H., 1982, Antler orogeny, a model: Geological Society of America Bulletin, v. 93, p. 815–828.

Speed, R. C., Elison, M. W., and Heck, F. R., 1987, Phanerozoic tectonic history of the Great Basin, *in* Ernst, W. G., ed., Metamorphism and crustal evolution of the western United States (Rubey Volume VII): Englewood Cliffs, New Jersey, Prentice-Hall, Inc., p. 572–605.

Stauber, D. A., 1980, Crustal structure in the Battle Mountain heat flow high in northern Nevada from seismic refraction profiles and Rayleigh wave phase [Ph.D. thesis]: Stanford, California, Stanford University, 316 p.

Stern, T. W., Bateman, P. C., Morgan, B. A., Newell, M. F., and Peck, D. L., 1981, Isotopic U-Pb ages of zircon from granitoids of the central Sierra Nevada, CA: U.S. Geological Survey Professional Paper 1185, 17 p.

Stevens, C. H., 1977, Permian depositional provinces and tectonics, western United States, *in* Stewart, J. H., Stevens, C. H., and Fritsche, A. F., eds., Paleozoic paleogeography of the western United States: Los Angeles, California, Pacific Section, Society of Economic Paleontologists and Mineralogists, Pacific Coast Paleogeography Symposium 1, p. 113–135.

Stevens, C. H., Stone, P., and Kistler, R. W., 1992, A speculative reconstruction of the middle Paleozoic continental margin of southwestern North America: Tectonics, v. 11, p. 405–419.

Stewart, J. H., 1971, Basin-Range structure—A system of horsts and graben produced by deep-seated extension: Geological Society of America Bulletin, v. 82, p. 1019–1044.

—— , 1972, Initial deposits in the Cordilleran geosyncline: Evidence of a late Precambrian (<850 m.y.) continental separation: Geological Society of America Bulletin, v. 83, p. 1345–1360.

—— , 1976, Late Precambrian evolution of North America: Plate tectonics implication: Geology, v. 4, p. 11–15.

—— , 1978, Basin and Range structure in western North America, a review, *in* Smith, R. B. and Eaton, G. P., eds., Cenozoic Tectonics and Regional Geophysics of the western Cordillera: Geological Society of America Memoir 152, p. 1–31.

Stewart, J. H., and Carlson, J. E., 1976, Cenozoic rocks of Nevada—Four maps and a brief description of distribution, lithology, age, and centers of volcanism: Nevada Bureau of Mines and Geology Map 52, scale 1:1,000,000.

Stewart, J. H., and Poole, F. G., 1974, Lower Precambrian Cordilleran miogeocline, Great Basin, western United States, *in* Dickinson, W. R., ed., Tectonics and sedimentation: Society of Economic Paleontologists and Mineralogists Special Publication 22, p. 28–57.

Stewart, J. H., and Suczek, C. A., 1977, Cambrian and paleogeography and tectonics in the western United States, *in* Stewart, J. H., Stevens, C. H., and Fritsche, A. F., eds., Paleozoic paleogeography of the western United States: Los Angeles, California, Pacific Section, Society of Economic Paleontologists and Mineralogists, Pacific Coast Paleogeography Symposium 1, 502 p.

Stewart, J. H., Albers, J. P., and Poole, F. G., 1968, Summary of regional evidence for right-lateral displacement in the western Great Basin: Geological Society of America Bulletin, v. 79, p. 1407–1414.

Stewart, J. H., MacMillan, J. R., Nichols, K. M., and Stevens, C. H., 1977, Deep-water upper Paleozoic rocks in north-central Nevada—A study of the type area of the Havallah Formation, *in* Stewart, J. H., Stevens, C. H., and Fritsche, A. F., eds., Paleozoic paleogeography of the western United States: Los Angeles, California, Pacific Section, Society of Economic Paleontologists and Mineralogists, Pacific Coast Paleogeography Symposium 1, p. 337–347.

Stock, J., and Molnar, P., 1988, Uncertainties and implications of the Late Cretaceous and Tertiary position of North America relative to the Farallon, Kula, and Pacific plates: Tectonics, v. 7, p. 1339–1384.

Suppe, J., 1973, Geology of the Leech Lake Mountain—Ball Mountain region, California: Berkeley, University of California Publications in the Geological Sciences, v. 107, p. 1–81.

Suppe, J., and Armstrong, R. L., 1972, Potassium-argon dating of Franciscan metamorphic rocks: American Journal of Science, v. 272, p. 217–233.

Suppe, J., and Foland, K. A., 1978, The Goat Mountain schists and Pacific Ridge complex: A redeformed but still intact late Mesozoic schupper complex, *in* Howell, D. E., and McDougall, K. A., eds., Mesozoic paleogeography of the western United States: Los Angeles, California, Pacific Section, Society of Economic Paleontologists and Mineralogists, Pacific Coast Paleogeography Symposium 2, p. 431–451.

Tempelmann-Kluit, D. J., 1979, Transported cataclastite, ophiolite and granodiorite in Yukon, evidence of arc-continent collision: Geological Survey of Canada Paper 79-14, 27 p.

Thompson, G. A., 1966, The rift system of the western United States: Geological Survey of Canada Department of Mines Technical Survey Paper 66-14, p. 280–290.

Thompson, G. A., and Burke, D. B., 1973, Rate and direction of spreading in Dixie Valley, Basin and Range province, Nevada: Geological Society of America Bulletin, v. 84, p. 627–632.

Thompson, G. A., and Zoback, M. L., 1979, Regional geophysics of the Colorado Plateau: Tectonophysics, v. 61, p. 149–181.

Tikoff, B., and Teyssier, C., 1992, Crustal-scale, en echelon "P-shear" tensional bridges: A possible solution to the batholithic room problem: Geology, v. 20, p. 927–930.

Tobin, D. G., and Sykes, L. R., 1968, Seismicity and tectonics of the northeast Pacific Ocean: Journal of Geophysical Research, v. 73, p. 3821–3849.

Tobisch, O. T., and Fiske, R. S., 1982, Repeated parallel deformation in part of the eastern Sierra Nevada, California, and its implications for dating structural events: Journal of Structural Geology, v. 4, p. 177–195.

Tobisch, O. T., Fiske, R. S., Sacks, S., and Taniguchi, D., 1978, Strain in metamorphosed volcaniclastic rocks and its bearing on the evolution of orogenic belts: Geological Society of America Bulletin, v. 88, p. 23–40.

Tobisch, O. T., Saleeby, J. B., and Fiske, R. S., 1986, Structural history of continental volcanic arc rocks along part of the eastern Sierra Nevada, California: A case for extensional tectonics: Tectonics, v. 5, p. 65–94.

Tobisch, O. T., Paterson, S., Saleeby, J. B., and Geary, E. E., 1989, Nature and timing of deformation in the Foothills terrane, central Sierra Nevada, California: Its bearing on orogenesis: Geological Society of America Bulletin, v. 101, p. 401–413.

Underwood, M. B., 1983, Depositional setting of the Paleogene Yager Formation, northern Coast Ranges of California, *in* Larue, D. K., and Steel, R. J., eds., Cenozoic marine sedimentation, Pacific margin, U.S.A. (Symposium Volume): Los Angeles, California, Pacific Section, Society of Economic Paleontologists and Mineralogists, p. 81–101.

Vallier, T. L., Dean, W. E., Rea, D. K. and Thiede, J., 1983, Geologic evolution of Hess Rise, central North Pacific Ocean: Geological Society of America Bulletin, v. 94, p. 1289–1307.

Varga, R. J., and Moores, E. M., 1981, Age, origin, and significance of an unconformity that predates island-arc volcanism in the northern Sierra Nevada: Geology, v. 9, p. 512–518.

Vetter, U. R., and Ryall, A. S., 1983, Systematic change of focal mechanism with depth in the western Great Basin: Journal of Geophysical Research, v. 88, p. 8237–8250.

Walker, J. D., 1985, 1988, Permian and Triassic rocks of the Mojave Desert and their implications for timing and mechanisms of continental truncation: Tectonics, v. 7, p. 685–709.

Weisenberg, C. W., and Avé Lallemant, H., 1977, Permo-Triassic emplacement of the Feather River ultramafic body, northern Sierra Nevada Mountains, California: Geological Society of America Abstracts with Programs, v. 9, p. 525.

Wentworth, C. M., Blake, M. C., Jr., Jones, D. L., Walter, A. W., and Zoback, M. D., 1984, Tectonic wedging associated with emplacement of the Franciscan assemblage, California Coast Range, *in* Blake, M. C., Jr., ed., Franciscan geology of northern California: Los Angeles, California, Pacific

Section, Society of Economic Paleontologists and Mineralogists, p. 163–173.
Wernicke, B., 1981, Low-angle normal faults in the Basin and Range Province: Nappe tectonics in an extending orogen: Nature, v. 291, p. 645–648.
—— , 1985, Uniform sense normal simple shear of the continental lithosphere: Canadian Journal of Earth Sciences, v. 22, p. 108–125.
—— , 1990, The fluid crustal layer and its implications for continental dynamics, *in* Salisbury, M. H., and Fountain, D. M., eds., Exposed cross-sections of the continental crust: Dordrecht, Holland, D. Reidel Publishing Co., p. 509–544.
Wernicke, B., Spencer, J. E., Burchfiel, B. C., and Guth, P. L., 1982, Magnitude of crustal extension in the southern Great Basin: Geology, v. 10, p. 499–502.
Wernicke, B., Axen, G. J., and Snow, J. K., 1988, Basin and Range extensional tectonics at the latitude of Las Vegas, Nevada: Geological Society of America Bulletin, v. 100, p. 1738–1757.
Wiltschko, D. V., and Dorr, J. A., 1983, Timing of deformation in overthrust belt and foreland of Idaho, Wyoming, and Utah: American Association Petroleum Geologists Bulletin, v. 67, p. 1304–1322.
Wolf, M. B., and Saleeby, J. B., 1990, Crustal extension during the Nevadan orogeny in the southwestern Sierra Nevada Foothills terrane (FT), CA: Geological Society of America Bulletin, v. 100, p. 859–876.
Wong, I. G., and Savage, W. U., 1983, Geologic and geophysical features of calderas: Eugene, Oregon, University of Oregon Center for Volcanology, 87 p.
Wright, J. E., 1981, Geology and U-Pb geochronology of the western Paleozoic and Triassic subprovince, Klamath Mountains, CA [Ph.D. thesis]: Santa Barbara, California State University, 350 p.
—— , 1982, Permo-Triassic accretionary subduction complex, southwestern Klamath Mountains, northern California: Journal of Geophysical Research, v. 87, p. 3805–3818.
Zartman, R. E., 1974, Lead isotope provinces in the Cordillera of the western United States and their geologic significance: Economic Geology, v. 69, p. 792–805.
Zoback, M. D., and Wentworth, C. M., 1986, Crustal studies in central California using an 800-channel seismic reflection recording system, *in* Barazangi, M. and Brown, L., eds., Reflection seismology: A global perspective: American Geophysical Union Geodynamics Series, v. 13, p. 183–196.
Zoback, M. D., Wentworth, C. M., and Moore, J. G., 1985, Central California seismic reflection transect: II—The eastern Great Valley and western Sierra Nevada: Eos (Transactions, American Geophysical Union), v. 65, p. 1102.
Zoback, M. L., 1989, State of stress and modern deformation of the northern Basin and Range province: Journal of Geophysical Research, v. 94, p. 7105–7128.
Zoback, M. L., Anderson, R. E., and Thompson, G. A., 1981, Cenozoic evolution of state of stress and style of tectonism of the Basin and Range province of western United States: Royal Society of London Philosophical Transactions, v. 300, p. 407–434.
Zucca, J. J., Fuis, G. S., Milkereit, B., Mooney, W. D., and Catchings, R. D., 1984, Crustal structure of northeastern California from seismic-refraction data: Journal of Geophysical Research, v. 91, p. 7359–7382.

MANUSCRIPT ACCEPTED BY THE SOCIETY DECEMBER 16, 1992

Printed in U.S.A.

DNAG Continent-Ocean Transect Volume
Phanerozoic Evolution of North American
Continent-Ocean Transitions
The Geological Society of America, 1994

Chapter 8

Continent-ocean transitions in western North America between latitudes 46 and 56 degrees: Transects B1, B2, B3

J.W.H. Monger
Geological Survey of Canada, 100 West Pender St., Vancouver, British Columbia V6B 1R8, Canada
R. M. Clowes
Department of Geological Sciences, University of British Columbia, Vancouver, British Columbia V6T 2B4, Canada
D. S. Cowan
Department of Geological Sciences, University of Washington, Seattle, Washington 98195
C. J. Potter
Department of Geological Sciences, Cornell University, Ithaca, New York 14853
R. A. Price
Geological Survey of Canada, 601 Booth St., Ottawa, Ontario K1A 0E8, Canada
C. J. Yorath
Pacific Geoscience Centre, Sidney, British Columbia, Canada

INTRODUCTION

This chapter is intended to supplement rather than reproduce material presented in Transects B–1, B–2, and B–3 and their accompanying texts, although some overlap is unavoidable, particularly in summary descriptions of major lithological and structural packages that form the crust of this segment of the Cordillera. The chapter discusses results of recent studies by COCORP within the east-central part of the corridor of transect B–3A, and by LITHOPROBE within the corridors of B–2 and B–3A on and west of Vancouver Island and between B–2 and B–3A in the east-central Cordillera (Fig. 1). The new results confirm the general principles on which the transects were constructed but constrain depths and attitudes of major lithological packages and structures and provide insights into processes within deeper parts of the crust.

Transects B–1, B–2, and B–3 lie in the segment of the Cordillera that includes the southern half of British Columbia and Alberta in Canada, and Washington, northern Idaho, and Montana in the United States (Fig. 1). Each transect emphasizes a somewhat different aspect of Cordilleran geology. Transect B–1 (Yorath and others, 1985c) extends through a region between latitudes 51°N and 56°N and longitudes 126°W to 133°W. It exploits relatively abundant data in the offshore region to illustrate complex interactions near the plate junction between the Explorer (northern subplate of the Juan de Fuca Plate), Pacific, and North American plates and extends eastward across the granitic and high-grade metamorphic terrane of the Coast Mountains to terminate in the volcanogenic and sedimentary terranes of the central Canadian Cordillera. Transect B–2 (Monger and others, 1985) lies within a corridor between latitudes 48°N and 52°N and longitudes 114°W and 127°W; the corridor runs east-northeastward from the Juan de Fuca Plate across the entire Cordillera to terminate in the Alberta Plains east of the limit of Cordilleran deformation. The transect crosses what is geologically and geophysically the best known part of the Canadian Cordillera and includes all five morphostructural belts characteristic of the Canadian segment (Fig. 2A). Transect B–3 (Cowan and Potter, 1986) involves two corridors. The northern one, B–3A, lies between latitudes 48°N and 49°N and longitudes 113°W and 127°W. It overlaps B–2 at its western end and then diverges from it to extend across the entire Cordillera just south of the boundary between Canada and the United States. B–3A is comparable to B–2, but differs in two ways. First, it crosses the North Cascade Mountains, which are, at least in part, the southern extension of the (Canadian) Coast Mountains but with a geological record far less obscured by the extensive granitic intrusions present farther north. Second, in the eastern Cordillera, B–3A crosses the thick accumulation of mid-Proterozoic Belt-Purcell strata that are absent at surface along strike at the latitude of B–2. Transect B–3B

Monger, J.W.H., Clowes, R. M., Cowan, D. S., Potter, C. J., Price, R. A., and Yorath, C. J., 1994, Continent-ocean transitions in western North America between latitudes 46 and 56 degrees: Transects B1, B2, B3, *in* Speed, R. C., ed., Phanerozoic Evolution of North American Continent-Ocean Transitions: Boulder, Colorado, Geological Society of America, DNAG Continent-Ocean Transect Volume.

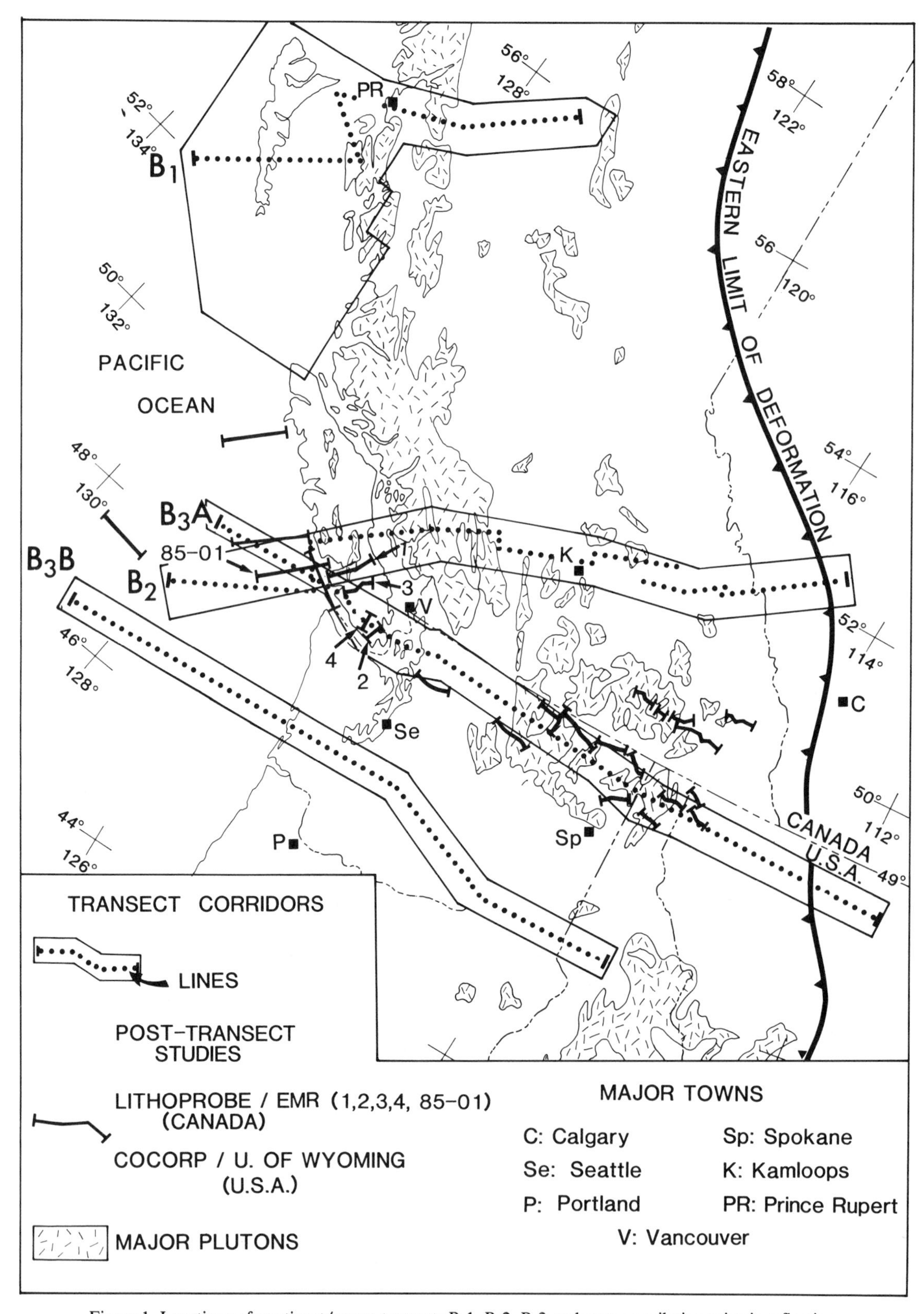

Figure 1. Locations of continent/ocean transects B-1, B-2, B-3 and postcompilation seismic reflection studies, mainly by COCORP and LITHOPROBE.

lies between latitudes 46°N and 47°N and longitudes 116°W and 130°W. It extends from the Juan de Fuca Ridge across the extensive Tertiary and Quaternary volcanics of the High Cascades and Columbia Plateau to the Idaho Batholith. In this discussion we emphasize Transects B–2 and B–3A, which together contain a particularly complete record of Cordilleran and flanking Pacific crustal evolution.

The authors had the following responsibilities in this paper. Clowes provided part of the Neogene and Recent section, in particular the discussion of new, mainly LITHOPROBE, geophysical results from the vicinity of the present continental margin. Cowan contributed material for the southern part of the Neogene and Recent section, and his work in northwestern Washington strongly influenced the North Cascades discussion. Monger was responsible for the general organization of the paper, for attempting to "blend" the different contributions, and for sections on accreted terranes and structures of the Intermontane and Coast belts. Potter provided geophysical and geological information and interpretation, particularly on central and eastern parts of Transect B–3A, based in large part on his COCORP studies. Price contributed data and concepts on the Omineca and Foreland belts. Yorath provided material for the Neogene and Recent section and on the Insular Belt, based in part on his involvement with LITHOPROBE. Other contributors to Transects B–1 and B–2, whose data and ideas were drawn upon for this chapter but who may not be directly referenced, are R. R. Currie, T. Høy, R. D. Hyndman, V. Preto, R. P. Riddihough, P. S. Simony, P. D. Snavely, and G. J. Woodsworth.

Geological and geophysical bases for transect construction

The regions containing the transect corridors are covered by 1:250,000-scale geological maps and/or maps on scales of 1:50,000, 1:63,360, or smaller, and by regional gravity and offshore/onland magnetic maps. Gravity and (for B–2) seismic models defined the thickness of the crust in the transects and the general crustal structure. The only seismic reflection studies that were available at the time of transect compilation (1983–84) provided details of subsurface structure in the offshore region and in the Foreland Belt. Details of crustal composition and structure for most parts of the transects were based on interpretation of the primary structural relationships between major tectonostratigraphic units seen at surface and were guided by petrochemical indications of the nature of the deeper crust. Since compilation of the transects, COCORP and LITHOPROBE studies have defined the geometry of deeper structures across the continent/ocean transition and in parts of the central and eastern Cordillera (Fig. 1).

Geological Overview

Geological record. The crust of this segment of the Cordillera appears to be the product of the following tectonic regimes and deformational episodes (Figs. 2, 3):

(1) A Late Archean/Early Proterozoic (2.8/1.8 Ga) crystalline basement, the southwestern extension of rocks formed in the Canadian Shield, has a strong northeasterly structural grain at the latitudes of Transects B–2 and B–3 and underlies much of the eastern Foreland Belt (Fig. 2A).

(2) This basement was rifted during at least three periods of extension, in Middle and Late Proterozoic and early Paleozoic times, and thick sequences of sediments were deposited during rifting episodes on at least partly attenuated Precambrian crust.

(3) In the eastern Cordillera, between Early Cambrian and Late Jurassic times, platform-shelf-slope-rise or miogeoclinal, mainly carbonate-shale deposits accumulated along the passive western margin of ancestral North America. In the western Cordillera, there is a record of plate convergence between oceanic and arc terranes that had uncertain paleogeographic positions relative to the eastern passive margin.

(4) The mid-Mesozoic to Recent active continental margin evolved in three stages:

(4a) In Middle Jurassic (ca. 170 Ma) to Paleocene (ca. 60 Ma) time, extensive terranes (Insular, Intermontane Superterranes of Figs. 2C, 3) composed of intra-oceanic arc and ocean floor rocks of Paleozoic and younger ages were accreted to the ancient continental margin, a process accompanied and followed by crustal thickening by stacking of thrust sheets, mainly eastward-verging in the eastern Cordillera and westward-verging in the western Cordillera, and by dextral transpression, calc-alkaline magmatism, regional metamorphism (Fig. 2B), and subduction of Pacific Ocean crust.

(4b) In Eocene (ca. 60 to 40 Ma) time, crustal extension and transtension in the central part of the orogen were associated with widespread calc-alkaline and alkaline magmatic arc activity; in late Eocene time there was local accretion of oceanic basalt and sediment near the present continental margin.

(4c) In late Tertiary (40 Ma) to Recent time, magmatic arc and back-arc activity, areally more restricted than that of the early Tertiary interval, was associated with subduction of the downgoing Juan de Fuca Plate.

The Middle Jurassic to early Tertiary interval is clearly the time when most specifically "Cordilleran crust" was created by (1) accretion to the ancient continental margin of large amounts of ensimatic material; (2) crustal thickening involving stacking in thrust sheets of ensimatic material, reworked Precambrian basement, and continental margin deposits; and (3) addition of volumetrically unknown amounts of magma to the crust. At the latitudes of B–1, B–2, and B–3, the proportion of ensimatic material in Cordilleran crust probably slightly exceeds that of cratonic and craton-derived material.

Regional structural pattern. Areally, the large-scale crustal structure of this segment of the Cordillera is most clearly delineated by reference to a metamorphic map (Fig. 2B). The Cordilleran infrastructure of high-grade metamorphic and granitic rock is exposed in two tectonic welts (Omineca, Coast-Cascade Belts) separating three regions of deformed but low-grade metamorphic and unmetamorphosed rock (Foreland,

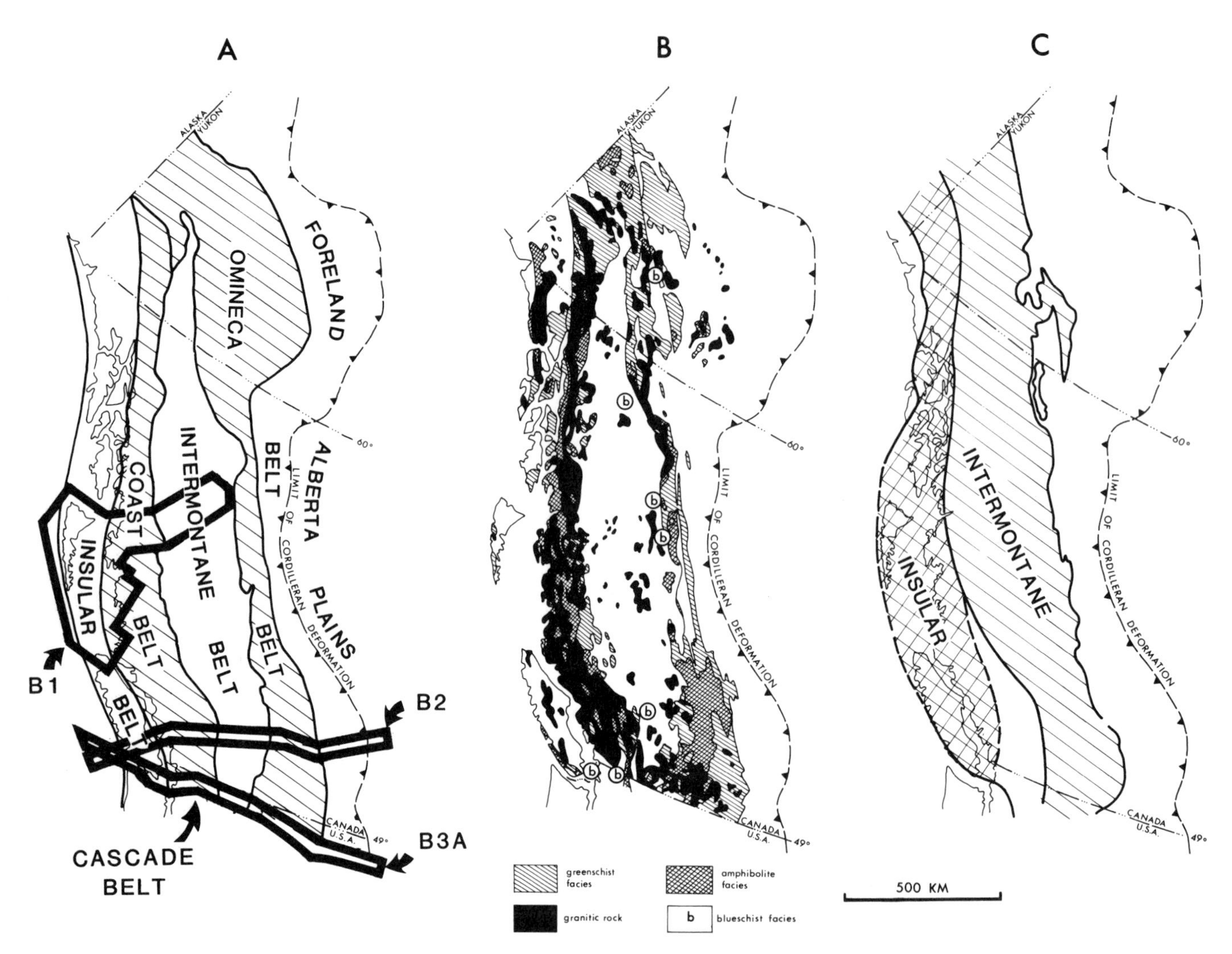

Figure 2. Areal distribution of different aspects of the geology involving Transects B-1, B-2, B-3A. A: Morphostructural belts with transect locations. Foreland Belt: Northeasterly thinning wedge of mid-Proterozoic to Upper Jurassic miogeoclinal and platformal carbonate and craton-derived clastic rock and overlying Jurassic to Paleogene clastic rock; horizontally compressed and displaced for up to 200 km northeastward onto craton in Late Jurassic to Paleogene time. Omineca Belt: Archean/Proterozoic crystalline basement, mid-Proterozoic to mid-Paleozoic sedimentary and minor volcanic and granitic rocks, upper Paleozoic and lower Mesozoic volcanic and sedimentary rock, highly deformed and metamorphosed in mid-Jurassic to early Tertiary time, and intruded by Jurassic and Cretaceous plutons. Intermontane Belt: Upper Paleozoic to Tertiary volcanic and comagmatic plutonic rock and sedimentary rock; deformed at various times from early Mesozoic to Paleogene. Coast Belt: Jurassic, mainly Cretaceous and Tertiary granitic rock, engulfing Paleozoic, locally Precambrian(?), through Tertiary volcanic and sedimentary rocks that are variably metamorphosed up to high-metamorphic grades. Insular Belt: Sedimentary and volcanic strata of Paleozoic, locally Proterozoic, through Tertiary ages, with comagmatic intrusive rocks; deformed at various times from Paleozoic to Recent. B: Metamorphic map of the Canadian Cordillera: unpatterned area denotes subgreenschist and/or unmetamorphosed rocks. C: Distribution of superterranes, composed of smaller terranes amalgamated prior to accretion to the continental margin. Note that the boundaries of these terranes roughly coincide with the Omineca and Coast Belts of Figure 2A.

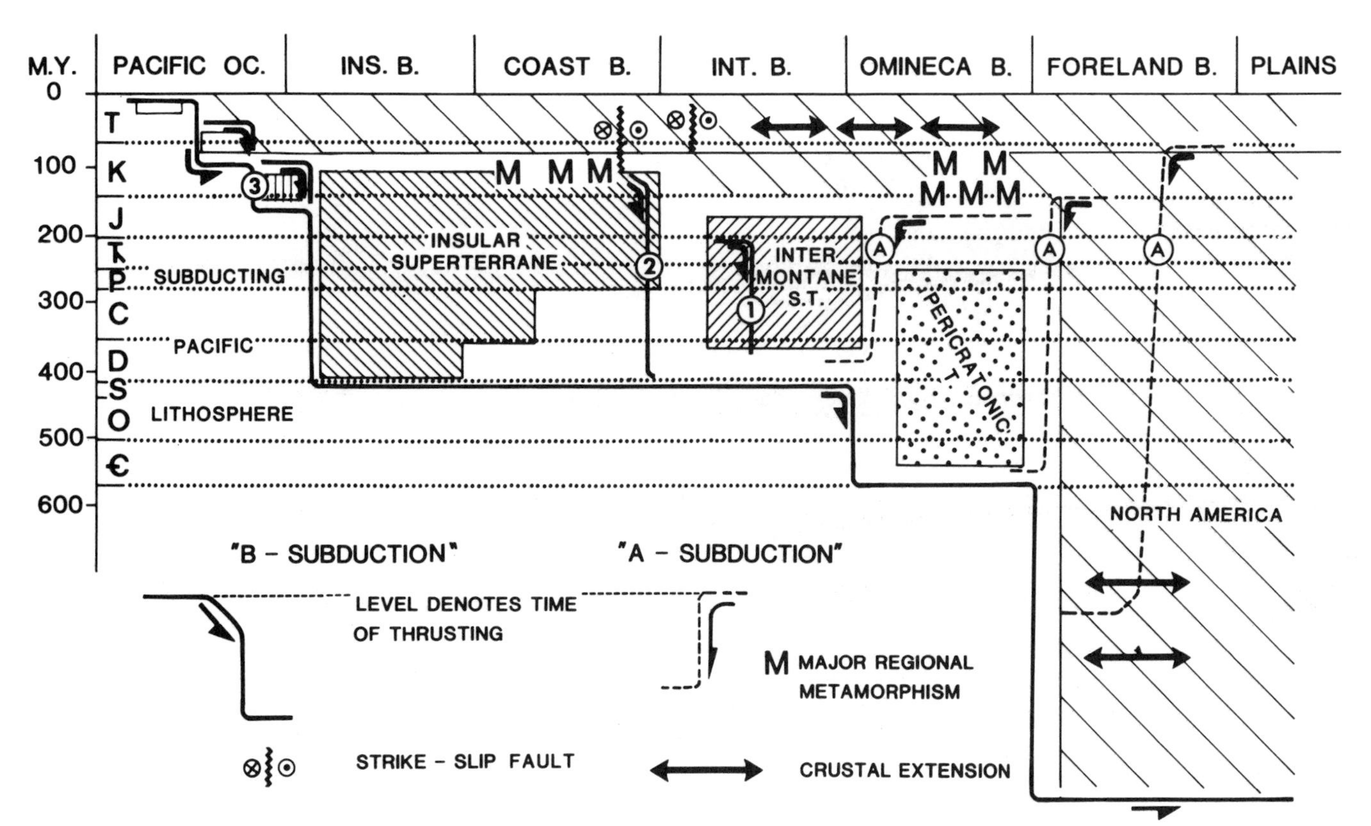

Figure 3. A generalized diagram for the region along the corridor of Transect B-2, showing stratigraphic, structural, and metamorphic relationships within space (morphostructural belts of Fig. 2A) and time coordinates. "A-subduction" refers to a thrust system with continental or transitional crust in the lower plate; "B-subduction" refers to a thrust system with oceanic crust in the lower plate. Distribution of the superterranes is shown in Figure 2C; "pericratonic terrane" refers to rocks that appear to have been deposited along the continental margin but that cannot be definitely linked with it; the dots and heavy hachured patterns distinguish the major terranes; the light, left-side-up hachured pattern denotes the incremental westward building of the continental margin by mainly Mesozoic accretions; numbers (1), (2), and (3) and letter (A) denote thrusts that are part of (1) an early Mesozoic thrust stack, (2) a Cretaceous thrust stack, (3) a late Tertiary to Recent thrust stack, and (A) Middle Jurassic to Paleocene thrust stack.

Intermontane, and Insular Belts) that make up the Cordilleran suprastructure (see Fig. 2A for summary descriptions of belts).

Interpretations of crustal structure as shown on Transects B-2 and B-3A, define the Cordilleran structural grain in vertical section as broadly "two-sided," with east-verging structures on the east side of the Cordillera and west-verging structures on the west side (Fig. 4). These interpretations reflect the apparent dominance of the effects of crustal convergence in forming the structural fabric of the Cordillera. In section, as interpreted from surface structural relationships, much of the crust appears to consist of enormous, downward-younging stacks of thrust sheets that are soled by basal thrusts or décollements (Figs. 3, 5). Conventionally, imbricated wedges that are situated at convergent plate margins are termed "accretionary prisms," while those on land are termed "fold and thrust belts" or simply "thrust and nappe systems." In this chapter we use the neutral, if inelegant, term "thrust stack" to include both offshore and on-land examples, in recognition of their overall geometric, structural, and dynamic similarities (Davis and others, 1983).

Much of the eastern Cordilleran crust, exclusive of Precambrian cratonic basement below the basal thrust, consists of an eastward-verging, eastward-younging thrust stack, largely composed of sedimentary rocks scraped off the cratonic margin, in which deformation ranges in age from Middle Jurassic to Paleocene. Much of the western Cordilleran crust appears to be composed of at least four west-verging, temporally and spatially discrete thrust stacks. The easternmost and oldest thrust stack, which consists of the Cache Creek Terrane of the Intermontane Belt (Fig. 6), is formed from late Paleozoic and early Mesozsoic ocean floor and overlying deposits that were probably off-scraped

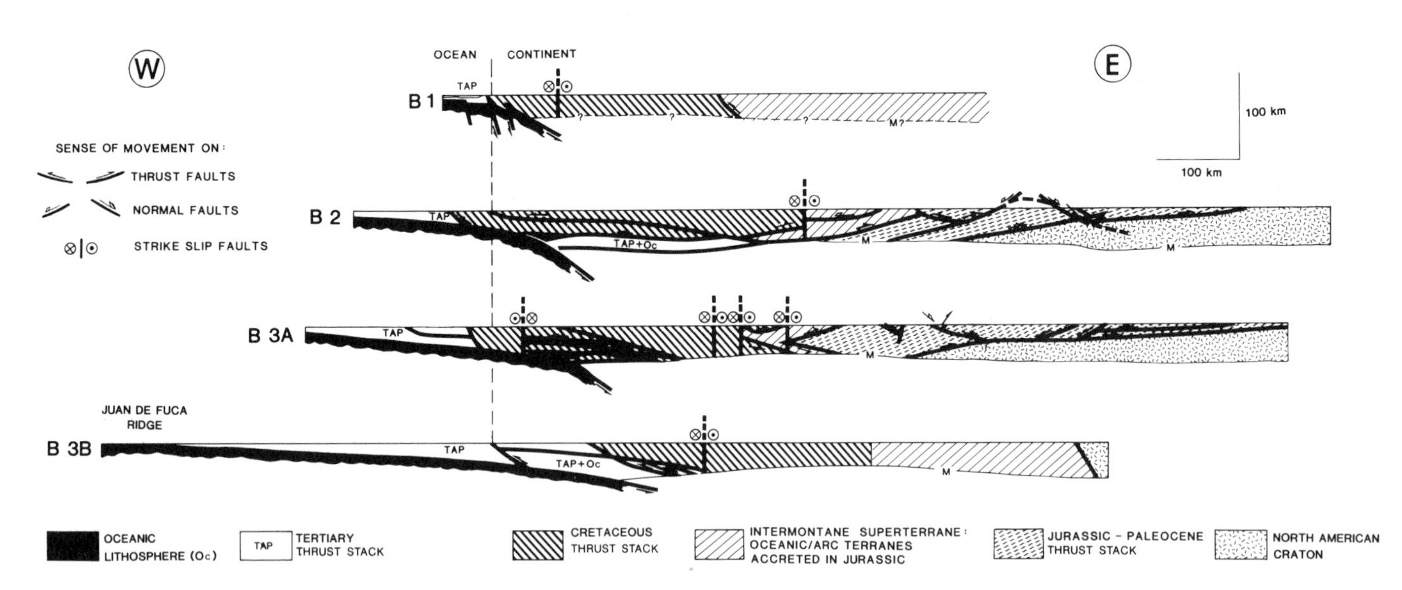

Figure 4. Comparative generalized cross sections, taken directly from Transects B–1, B–2, B–3A, B–3B, showing major crustal elements couched in terms of thrust stacks. The "lid" of both the west-verging Cretaceous thrust stack (which includes Insular Superterrane) and the east-verging Jurassic through Paleocene thrust stack is the Intermontane Superterrane. Note strike-slip and normal faults that disrupt these thrust stacks.

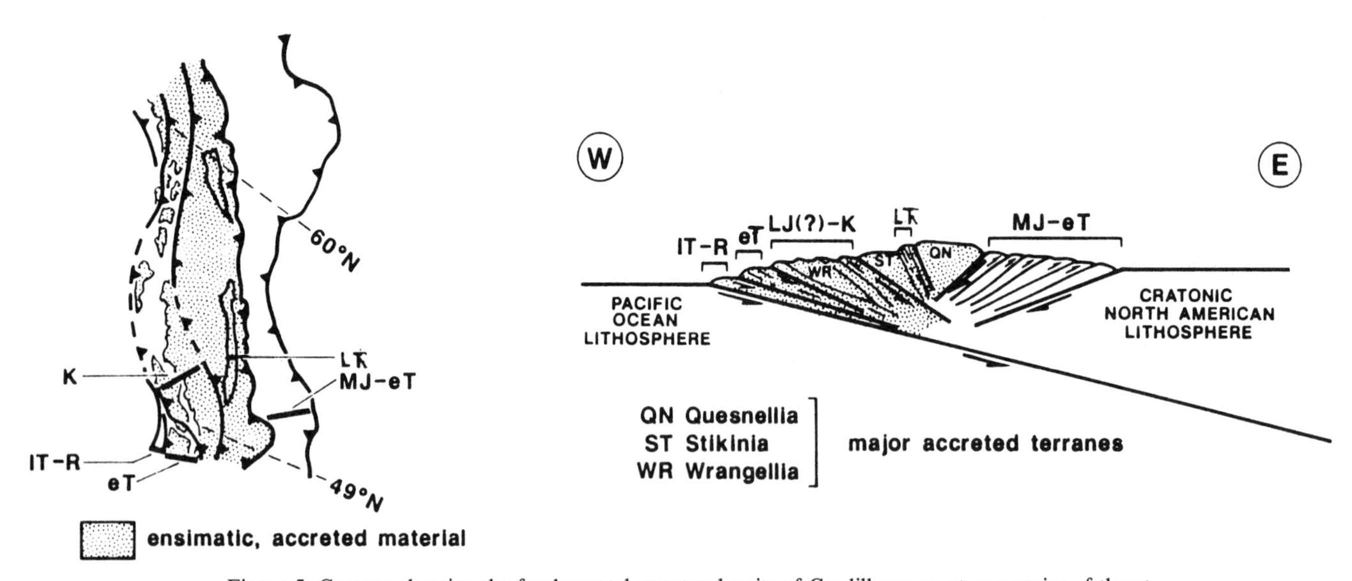

Figure 5. Cartoon showing the fundamental structural units of Cordilleran crust as a series of thrust stacks; terranes are merely parts of these thrust stacks, detached from their lower crust and mantle and incorporated as nappes or "flakes." Tilted 'Ŧ' symbol denotes Triassic.

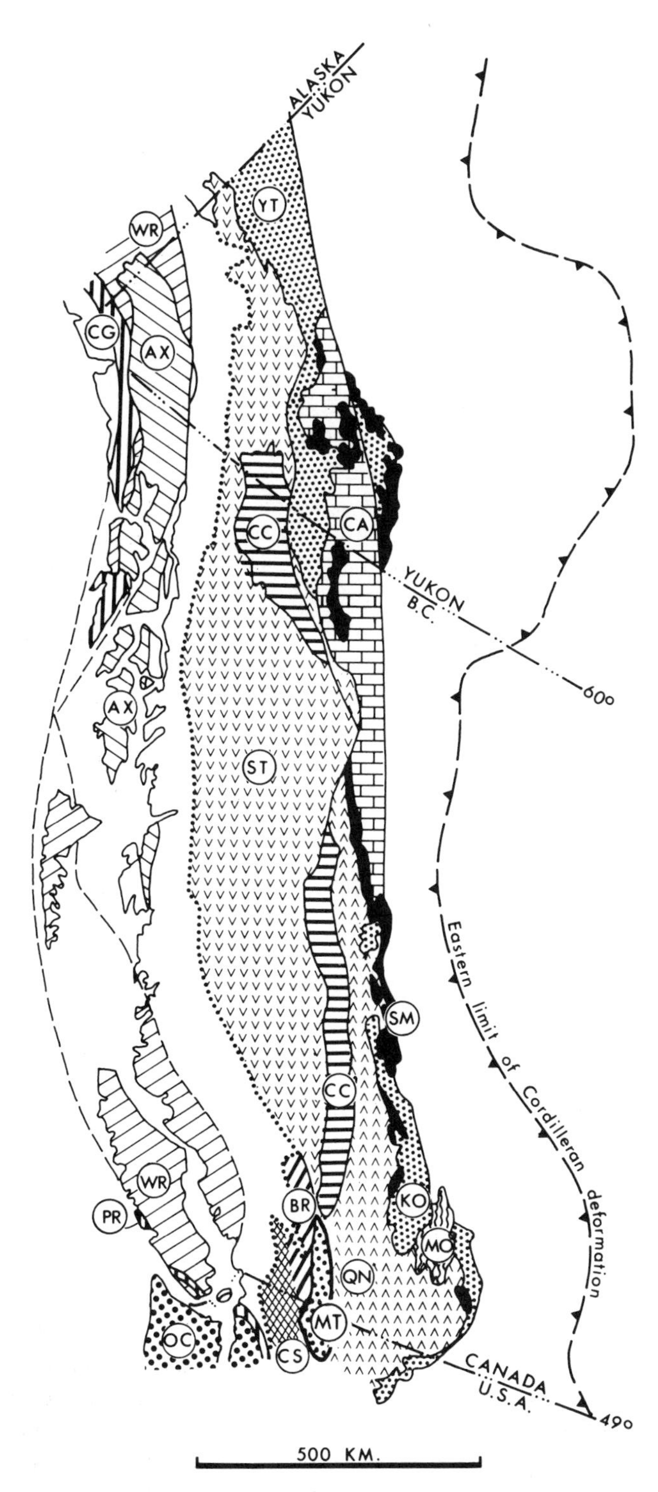

Figure 6. Terrane map of the Canadian segment of the Cordillera; compare with Figure 2C; terranes shown are: AX, Alexander; BR, Bridge River; CA, Cassiar; CC, Cache Creek; CG, Chugach; CS, "Cascade"; KO, Kootenay; MT, Methow-Tyaughton; OC, Olympic; PR, Pacific Rim; QN, Quesnellia; SM, Slide Mountain; ST, Stikine; WR, Wrangellia; YT, Yukon-Tanana. The inner terranes, SM, QN, CC, and ST apparently were associated prior to accretion to form the Intermontane Superterrane; the outer terranes, AX and WR, were together to form the Insular Superterrane prior to accretion. Terrane descriptions in text.

during early Mesozoic east-dipping subduction. As shown on Transect B–3A, a largely Cretaceous thrust stack comprises the North Cascades in Washington and consists of paleogeographically suspect terranes of diverse lithologies and ages. It extends northward into the Coast Belt of British Columbia, where its fundamental structural character is obscured by voluminous granitic rocks. Beneath it—and exposed on southernmost Vancouver Island, in the Olympic Peninsula, and on the Oregon coast—is a thrust stack of oceanic and continental margin rocks that was underplated beneath the Cretaceous stack in early Tertiary time. Westernmost and youngest is a late Tertiary and Recent thrust stack composed of offscraped submarine fan and abyssal plain turbidites, floored by the currently subducting Juan de Fuca Plate. Thus, it is possible to interpret specifically Cordilleran crust as largely consisting of these thrust stacks (Fig. 5).

In addition, dextral transcurrent faults are particularly common in this segment of the Cordillera, and their presence has long been interpreted as a response within the leading edge of the North American Plate to strong northward motions of various contiguous Pacific plates in Late Cretaceous to Recent time (e.g., Engebretson and others, 1985). The faults are at least in part contemporaneous with east-dipping subduction. The contribution to the crustal fabric of the Cordillera of such faults is extremely difficult to assess at present. One early Tertiary fault system, the Fraser River–Straight Creek Fault System (Figs. 28, 30), can be linked to reverse and normal faults in a classic wrench-fault pattern, so that the transcurrent structures presumably flatten out at some unknown depth within the crust and become indistinguishable from other contraction and extension structures. In Transects B–1, B–2, and B–3 this problem was avoided by showing major strike-slip faults cutting vertically through the entire crust (Fig. 4).

Extensional structures of early Tertiary age clearly make a major contribution to the crustal fabric in the central parts of the Cordillera along the lines of Transects B–2 and B–3A. They are prominent on LITHOPROBE and COCORP reflection profiles (Figs. 19 and 29), where they offset earlier contractional structures. Middle Proterozoic to Paleozoic extension faults, inferred to have formed during early rifting of the cratonic margin, are obscured by the eastern, Jurassic through Tertiary thrust faults.

Relationship of structures to accreted terranes. The accreted terranes in this segment of the Cordillera include both large terranes with coherent stratigraphic records (e.g., Stikinia, Wrangellia, Alexander) and disrupted terranes that appear to be former accretionary prisms (e.g., Cache Creek, some terranes in the North Cascades; Fig. 6). The times of juxtaposition of the contrasting terranes, as deduced from overlap assemblages, crosscutting plutons, etc., indicate that most terranes of the western Cordillera were amalgamated into two large composite or superterranes prior to accretion to the continental margin (Figs. 2C, 3; Monger and others, 1982). The eastern, Intermontane Superterrane (Composite Terrane I of Monger and others, 1982; Stikinian Superterrane of Saleeby, 1983) underlies the Intermontane Belt and comprises at least four smaller terranes that probably were together by the end of the Triassic and were thrust eastward over North American rocks in Middle Jurassic time. The western, Insular Superterrane (Composite Terrane II of Monger and others, 1982; Wrangellian Superterrane of Saleeby, 1983) underlies the Insular Belt and comprises two terranes that were together by Late Jurassic time and probably were thrust beneath terranes to the east in Cretaceous time. The boundaries of the superterranes roughly coincide with the two high-grade metamorphic and plutonic welts (Fig. 2B), and the times at which they were accreted—Middle Jurassic in the east and Early Cretaceous in the west—correspond with times of initiation of widespread Mesozoic metamorphism and crustal shortening in these welts (Fig. 3). From these observations it is reasonable to assume that the welts were initiated by accretion of the superterranes, although there is little doubt that most of the deformation and metamorphism in them postdated initial accretions and continued for at least 110 million years (ca. 170 to 60 Ma) in the eastern welt and 80 million years (ca. 120 to 40 Ma) in the western welt.

New geophysical constraints

Since compilation and publication of Transects B–1, B–2, and B–3, new geophysical data have been obtained that modify and constrain the earlier interpretations of crustal structure in the transects, although the principles on which they were constructed and the general pattern shown in them still hold. The most important of these data in Canada are from LITHOPROBE multichannel reflection lines on Vancouver Island (Fig. 1; Yorath and others, 1985a, b; Green and others, 1986; Clowes and others, 1987a), related lines offshore Vancouver Island (Clowes and others, 1987b), and LITHOPROBE reflection studies in the southeastern Canadian Cordillera (Fig. 1; Cook and others, 1987). In the United States the data are from COCORP lines in central and eastern parts of the U.S. Cordillera (Fig. 1; Potter and others, 1986; Potter and others, in preparation; Yoos and others, in preparation). In addition, LITHOPROBE magnetotelluric data (Kurtz and others, 1986), more seismicity data, an interpretation of heat-flow data, and a revised interpretation of refraction data have helped refine the structural model for the Vancouver Island region of Transects B–2 and B–3A.

Seismicity. Seismicity in the Puget Sound–Vancouver Island region has been reviewed by Crossan (1983) and Rogers (1983). They show (e.g., Crossan, 1983, Fig. 8) division between a shallow and deep suite of earthquakes. The deep suite dips west-northwesterly between depths of 40 and 70 km and may be subduction related.

A recently expanded network of seismic stations in the southwestern part of British Columbia reveals a suite of earthquakes extending from the continental shelf to Georgia Strait (between Vancouver Island and the mainland) with an eastward dip of about 15° (G. C. Rogers, 1986, personal communication).

Electrical conductivity. Gough (1986) presented a tectonic model based on Cordilleran geomagnetic studies and other geological and geophysical considerations (Fig. 7). Three principal structures are associated with high electrical conductivity in the crust and upper mantle. The most pervasive of these is the Canadian Cordillera Regional conductor (CCR) that extends throughout the Intermontane and Omineca Belts. Its western extent is poorly defined, but its eastern limit coincides approximately with the southern Rocky Mountain Trench north of latitude 51°N and with the Kootenay Arc to the south. Estimates of depth to CCR conductor are poorly constrained but place its upper surface in the range of 50 to 30 km with a thickness of between 30 and 70 km. The second major conductive structure is the Northern Rockies conductor that lies parallel with the regional strike of the western Rocky Mountains (Foreland Belt) between latitudes 52°N and 54°N (Bingham and others, 1985). It is interpreted as a conductive ridge within 10 km of the surface, formed by thickening of the CCR near its eastern limit, and as such can be considered a part of the regional conductor. Gough's (1986) model proposes mantle upflow beneath the Intermontane and Omineca Belts with associated extensional stress in these belts and compression in the craton to the east (Fig. 7). The third structure is an elongate conductive body in the Precambrian crust or upper mantle that strikes northeast-southwest across southern Alberta into southeastern British Columbia. Gough (1986) states that it is aligned with a seismically defined structure in the Precambrian cratonic basement (Fig. 14; Kanasewich and others, 1969) and suggests that the North American Craton extends at least as far west as the Kootenay Arc.

PRESENT TECTONIC REGIME: THE CONTINENT-OCEAN TRANSITION

Features of the current tectonic regime, which was established 40 m.y. ago, will be reviewed first, followed by those of three earlier tectonic regimes (the intervals 1500 to 170 Ma, 170 to 60 Ma, 60 to 40 Ma) to illustrate the successive, partly superimposed contributions of each regime to the development of the crust of this segment of the Cordillera.

The present tectonic regime appears to have persisted for the past 40 m.y. and succeeds late Eocene accretion of mainly basalt in the Olympic Peninsula, southern Vancouver Island, and adjacent offshore regions (Crescent terrane of Transects B–2 and B–3; Clowes and others, 1987a). During this interval, the continental margin at latitudes encompassing Transects B–1, B–2, and B–3 appears to have been the locus of convergent plate motion along much of its length, but with transform plate motion in the north (Fig. 8). North of the North America–Pacific–Juan de Fuca triple junction that lies between the Queen Charlotte Islands and Vancouver Island, the present continental margin corresponds with the Queen Charlotte dextral transform fault that separates the North American and Pacific Plates (Fig. 8). South of the triple junction, the Juan de Fuca Plate and small Explorer Plate are components of a convergent regime that extends southward to the Mendicino Fracture Zone off northern California. Atwater (1970), Engebretson and others (1985), and Gordon and Jurdy (1986) concluded that the triple junction has maintained a rela-

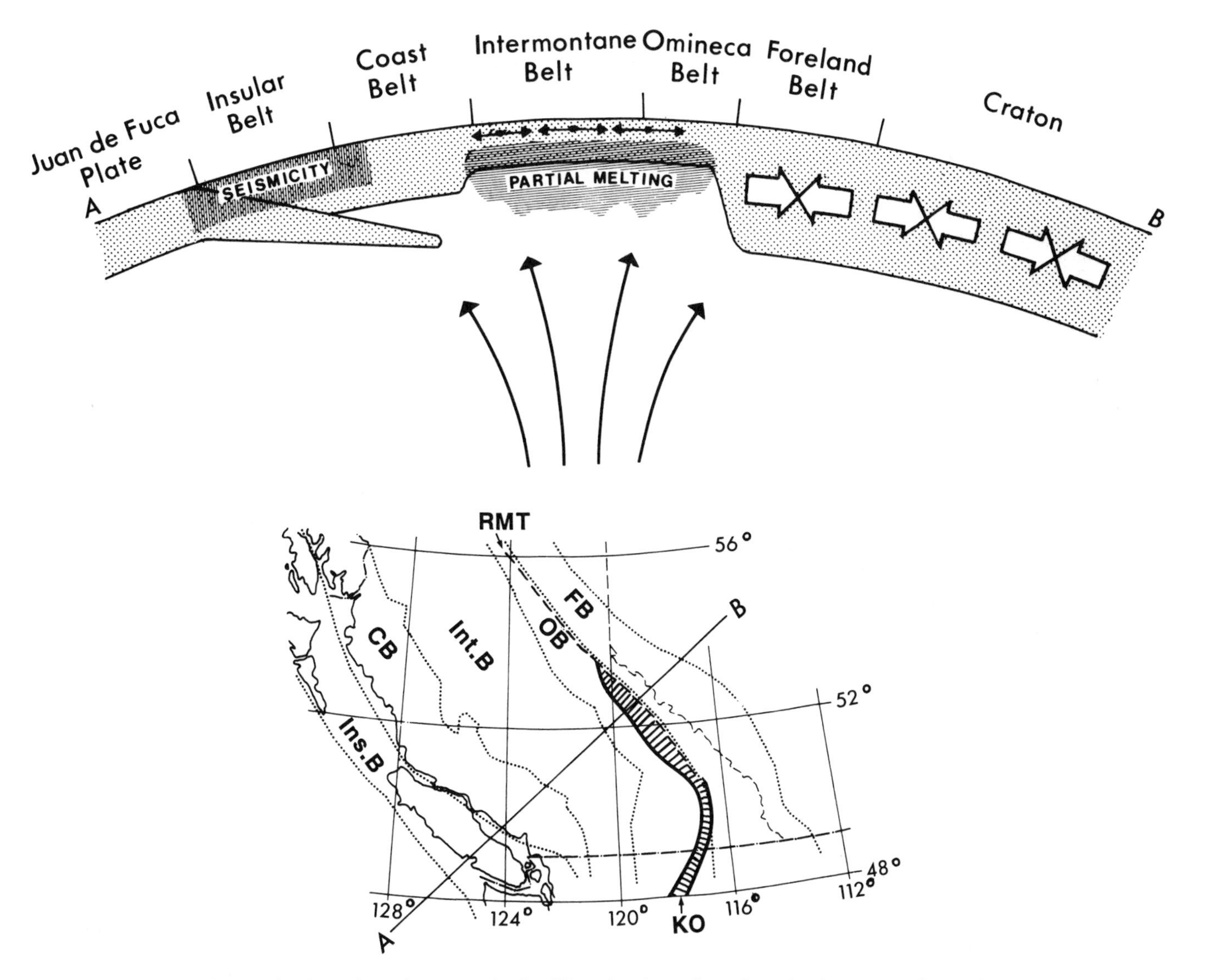

Figure 7. A schematic section across the Cordillera showing upflow of mantle, slow convergence at the continental edge, a short and thin subducted plate, extensional stress in the Intermontane and Omineca Belts, compression in the craton, and no deep Benioff zone north of latitude 49°N (after Gough, 1986). Map shows section line and boundaries of morphostructural belts of Figure 2A. KO: location of Kootenay Arc.

tively constant position west of northern Vancouver Island since 40 Ma. Onshore, the High Cascade magmatic arc, initiated nearly 40 m.y. ago, lies about 200 km east of the convergent plate boundary, and its northern and southern limits correspond closely with the limits of the Juan de Fuca Plate. Motions of Pacific, North American, and Juan de Fuca Plates, with reference to the global hot-spot reference frame, are shown in Fig. 9 (Riddihough and Hyndman, 1992).

The convergent part of the modern continent/ocean transition zone comprises, from west to east, the following throughgoing morphostructural elements (Fig. 8): (1) Juan de Fuca Ridge; (2) the oceanic Juan de Fuca Plate; (3) the continental shelf and slope underlain by the submerged Neogene and Recent accretionary prism of clastic sediments; (4) the coastal topographic high that includes the diverse terranes of Vancouver Island, the Olympic Peninsula, and the Oregon coastal ranges; (5) the topographic depression encompassing Willamette Valley, Puget Sound, and Georgia Strait; (6) the topographically elevated High Cascade–North Cascade–Coast Belt, coincident with the locus of Neogene calc-alkaline magmatism; and (7) the lower Columbia River–Interior Plateau region with extensive tholeiitic to alkaline basalt flows.

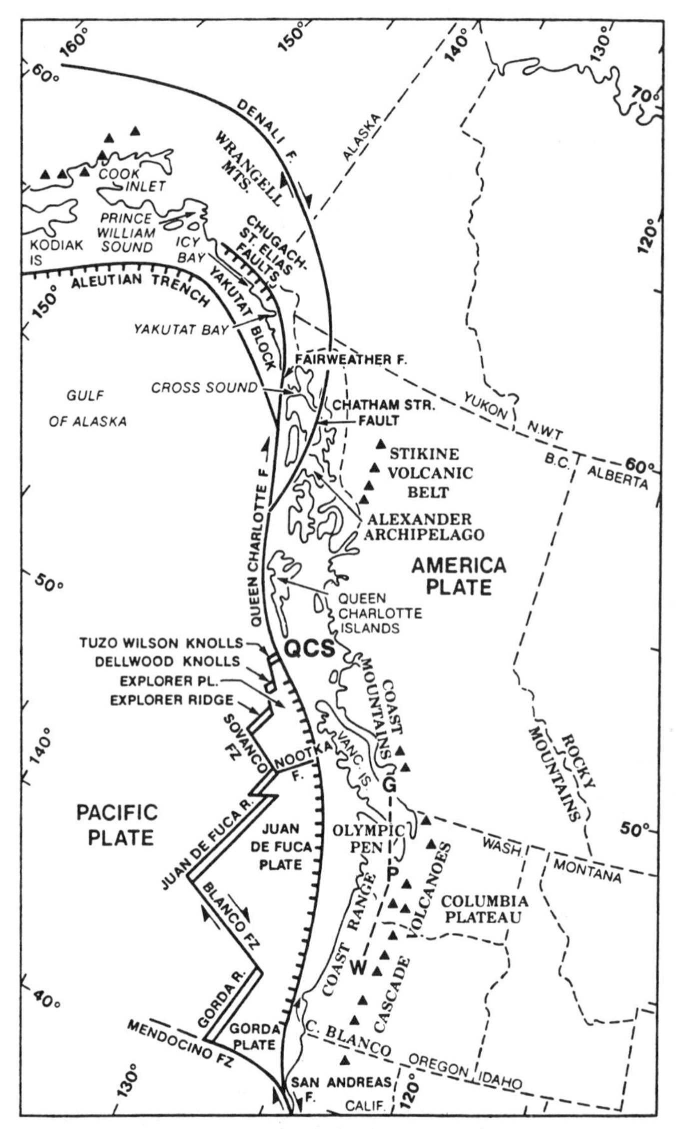

Figure 8. Recent/Neogene features of the northwestern Cordillera and Pacific Ocean margin (modified from Riddihough and Hyndman, 1992) (QCS: Queen Charlotte Strait; W-P-G: Willamette-Puget-Georgia depression).

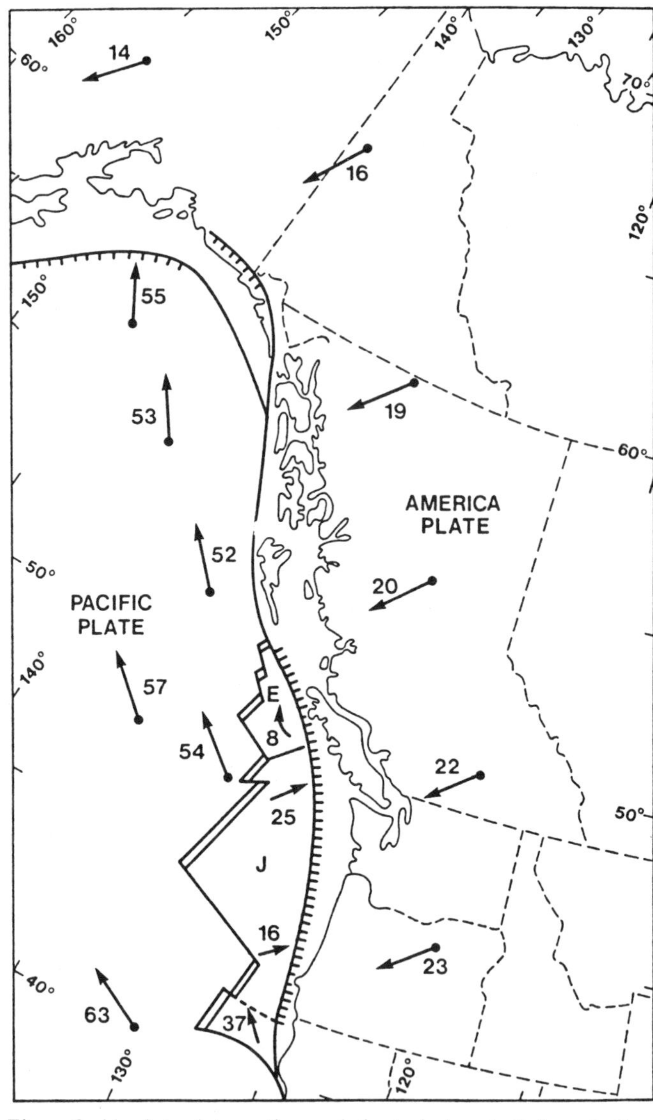

Figure 9. Absolute plate motions, relative to hot spots. J: Juan de Fuca Plate; E: Explorer Plate (from Riddihough and Hyndman, 1992).

Juan de Fuca Ridge

New oceanic crust is forming on the fragmented and complex Juan de Fuca Ridge at rates between 40 and 60 mm/a as determined from magnetic anomaly patterns. The well-established morphology of the ridge, shown on the recent compilation of SEA BEAM bathymetry and SEAMARC acoustic image data (Davis and others, 1985), is an en echelon series of ridge segments with offsetting transform faults. Cowan and others (1986) suggested that the morphology of the Sovanco Fracture Zone, which connects Juan de Fuca and Explorer Ridges (Fig. 8), is that of a series of rhomboid rotated crustal blocks, so that dextral shear on the Sovanco Fracture Zone is distributed through a 15- to 20-km-wide zone rather than concentrated on one fault. The ridge segments have highly variable topography that is believed to result from adjustments to variable plate motions and variations in magma supply, both along the ridge and in time (Riddihough and Hyndman, 1992). Sub-bottom structure associated with the ridge, and the presence of magma chambers, have not been established because of a lack of deep sounding data. However, a recent multichannel crustal reflection profile across the ridge at latitude 48°N, where hydrothermal vents and sulfide chimneys have been observed, shows a prominent narrow reflection ("bright spot") that could be associated with a magma chamber (Rohr, in prepa-

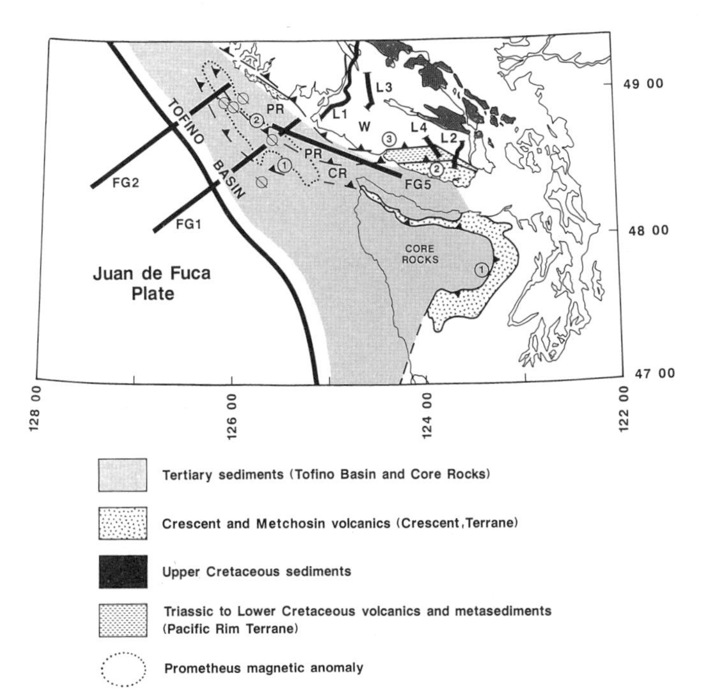

Figure 10. Principal tectonic units of the southern part of the Canadian Pacific continental margin and adjacent parts of the U.S.A., with location of Geological Survey of Canada "Frontier Geoscience" (FG1, 2, 5) and LITHOPROBE (L1, 2, 3, 4) seismic reflection profiles. W = Wrangellia; PR = Pacific Rim Terrane; CR = Crescent Terrane; (1) Hurricane Ridge–Crescent Fault; (2) Leech River–Tofino Fault; (3) San Juan–West Coast Fault.

ration). The profile also shows an oceanic Moho and thus shows that an identifiable oceanic crust has evolved within 5 km of the central valley of the ridge.

Juan de Fuca Plate

The Juan de Fuca Plate was isolated from its Farallon parent about 40 Ma (Engebretson and others, 1985). Its oldest magnetic anomalies are dated at 8 to 9 Ma. In the last 3 to 5 m.y., the northern end of the Juan de Fuca Plate broke off to form the Explorer Plate (Fig. 8; Riddihough and Hyndman, 1992). Whereas the Juan de Fuca Plate is presently subducting beneath the overriding North American Plate, the small Explorer Plate may have ceased active subduction so that it is being passively overridden by and/or colliding with the North American Plate.

In the region at the western end of Transects B–2 and B–3A, the sub-bottom structure is defined by seismic reflection and refraction data, the former obtained by Geological Survey of Canada and LITHOPROBE programs (Fig. 10; Clowes and others, 1987b; Yorath and others, 1985b). The interpretations of these multichannel seismic reflection data are supported by a wide range of other geophysical and geological studies, including magnetics, gravity, magnetotellurics, seismic activity, seismic refraction, and heat flow experiments (Riddihough and Hyndman, 1992). In addition, the extensive offshore program conducted by Shell Canada Resources Ltd. in the 1960s resulted in the drilling of six wells, data from which were reported by Shouldice (1971). On reflection line 85-01 (Fig. 1; = FG1 of Fig. 10; Fig. 11a), a strong reflector delineates the top of the Juan de Fuca Plate beneath Cascadia Basin and the lower continental slope. The basaltic layer of the oceanic crust dips eastward at 1°. Refraction data have defined the subsedimentary structure of the Juan de Fuca Plate (Waldron, 1982; Au and Clowes, 1982; Spence and others, 1985) and show that in general a standard crustal structure with layers 2A, 2B, 3A, and 3B is present. Oceanic crustal thickness is 6 to 7 km, and the underlying upper mantle velocity ranges from 7.9 to 8.2 km/s, although significant lateral variations in structure occur along and across the plate. Above this basement, at the seaward end of the profile (Fig. 11a), the sediments in Cascadia Basin comprise a pre-Pleistocene relatively seismically transparent unit up to nearly 1 km thick and an overlying, rapidly deposited sequence of Pleistocene turbidites up to 2 km thick. Total sediment thickness ranges from negligible near the ridge to less than 3 km at the deformation front near the base of the slope, which marks the seaward limit of folding and thrusting related to the offscraping of sediments from the descending plate.

The reflection from the top of the descending plate is lost intermittently beneath much of the continental slope but reappears below the innermost slope and can be followed for 70 km beneath the shelf.

Accretionary prism

A Tertiary and Recent accretionary prism extends eastward from the base of the continental slope to the coast of Vancouver Island. There is no bathymetric trench, only landward thickening of the sediments. On the 1985 multichannel line 85-01 (FG1 of Fig. 10), landward-dipping thrust faults are clearly visible just east of the deformation front. Associated with these thrusts are rotated sediments in the hanging walls that form anticlines near the surface. The thrusts penetrate the turbidite section and probably extend deeper to intersect the top of the oceanic basement surface. However, landward-thickening of the sub-Pleistocene section suggests shallower detachment perhaps associated with local duplexing. These reflection data have not yet been analyzed in detail, but line 85-01 (FG1) shows that beneath the continental shelf, folded Tertiary sediments occur in the hanging walls and footwalls of thrust sheets that contain basaltic rocks of the Crescent Terrane (Eocene Metchosin Formation) and metasediments and volcanic rocks of the Pacific Rim Terrane. The latter have been tectonically thickened by imbricate thrust faults. Specialized processing of parts of 85-01 (FG1) has revealed the presence of other thrust faults that may intersect the surface of the upper continental slope. These also intersect the oceanic basement sur-

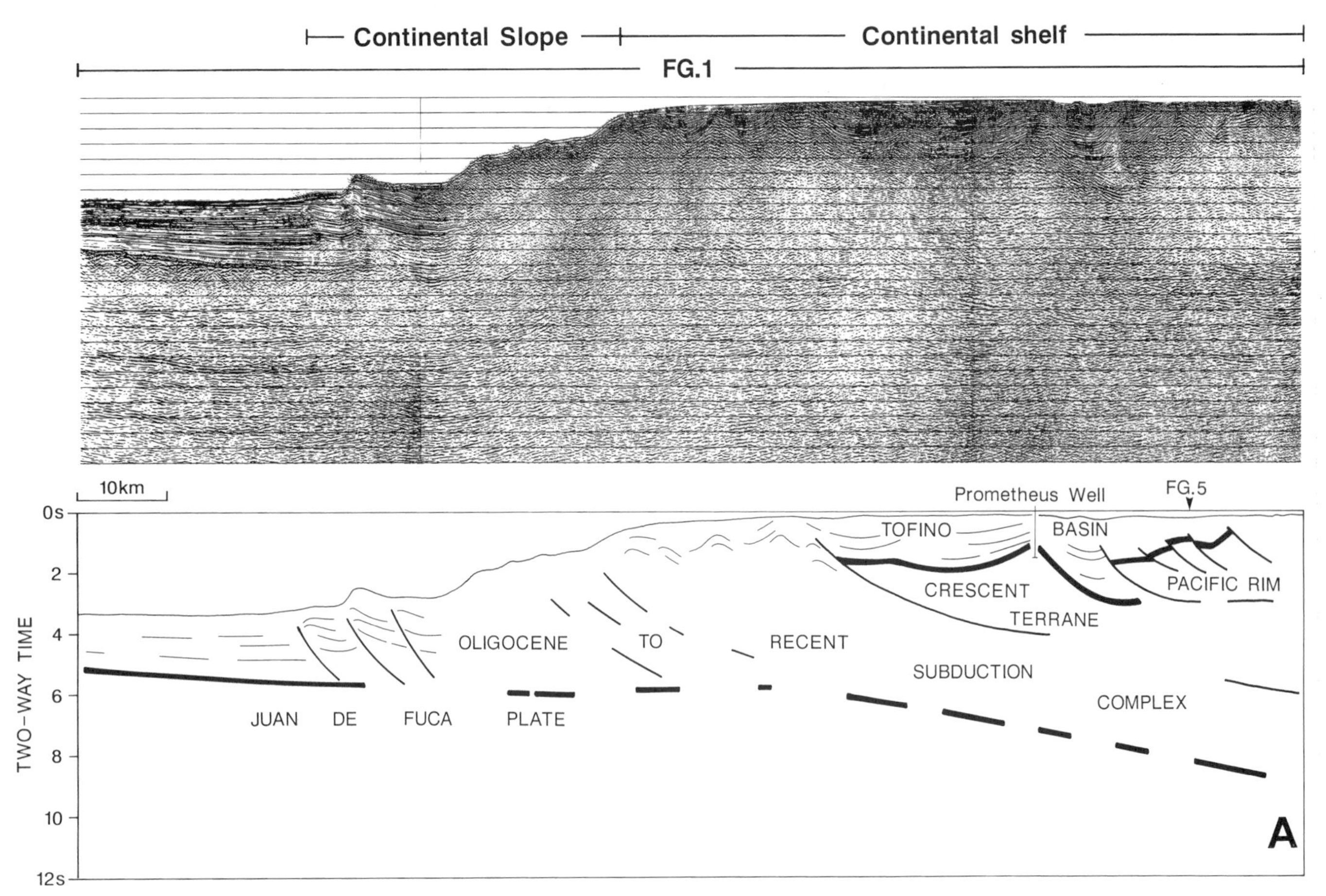

Figure 11a. Migrated record section (above) and partial interpretation (below) of multichannel seismic reflection line FG1 (85-01 of Fig. 1) across the continental margin of Vancouver Island.

face and, together with those at the deformation front, indicate that all of the oceanic sediments are being scraped off the surface of the descending plate and are accreting to the continental margin. Beneath the west coast of Vancouver Island the combined thickness of the Crescent and Pacific Rim Terranes, together with subducted and underplated sediments of the current subduction complex above the descending Juan de Fuca Plate, is about 24 km.

Coastal topographic high

The core of the Olympic Mountains consists of deformed Cenozoic marine clastic sedimentary rocks that occur in the footwall of the Hurricane Ridge (= Crescent) Thrust Fault (Fig. 10). These strata make up part of the current subduction complex imaged on 85-01 (FG1) and LITHOPROBE seismic reflection profiles across Vancouver Island (L1 of Fig. 10). In the hanging wall, Eocene basalt of the Crescent Formation rims the Olympic Mountains and is correlative with the Metchosin Complex of southern Vancouver Island. Together, the Crescent and Metchosin rocks make up the Crescent Terrane of Transect B–2, which on southern Vancouver Island occurs in the footwall of the Leech River Fault, beneath Jura-Cretaceous metasediments of the Leech River Formation. The Leech River in turn occurs in the footwall of the San Juan Fault, which is thought to be coplanar with the Westcoast Fault along the west coast of Vancouver Island (Fig. 10). There, equivalent rocks are represented by the Pacific Rim Complex, which together with the Leech River rocks makes up the Pacific Rim Terrane of Transect B–2. The hanging wall of the San Juan–Westcoast Faults is Wrangellia (Fig. 10), which includes Paleozoic through Jurassic volcanic and sedimentary strata and associated intrusions.

In Fig. 11b, line L1 illustrates the geometry of subsurface reflectors across the width of Vancouver Island and within the corridors of Transects B–2 and B–3A. The gap separating the eastern end of line 85-01 (FG1; Fig. 11a) from the western end of L1 (Fig. 11b) is about 17 km. As described in Yorath and others (1985b) and Clowes and others (1987a), the reflective interval

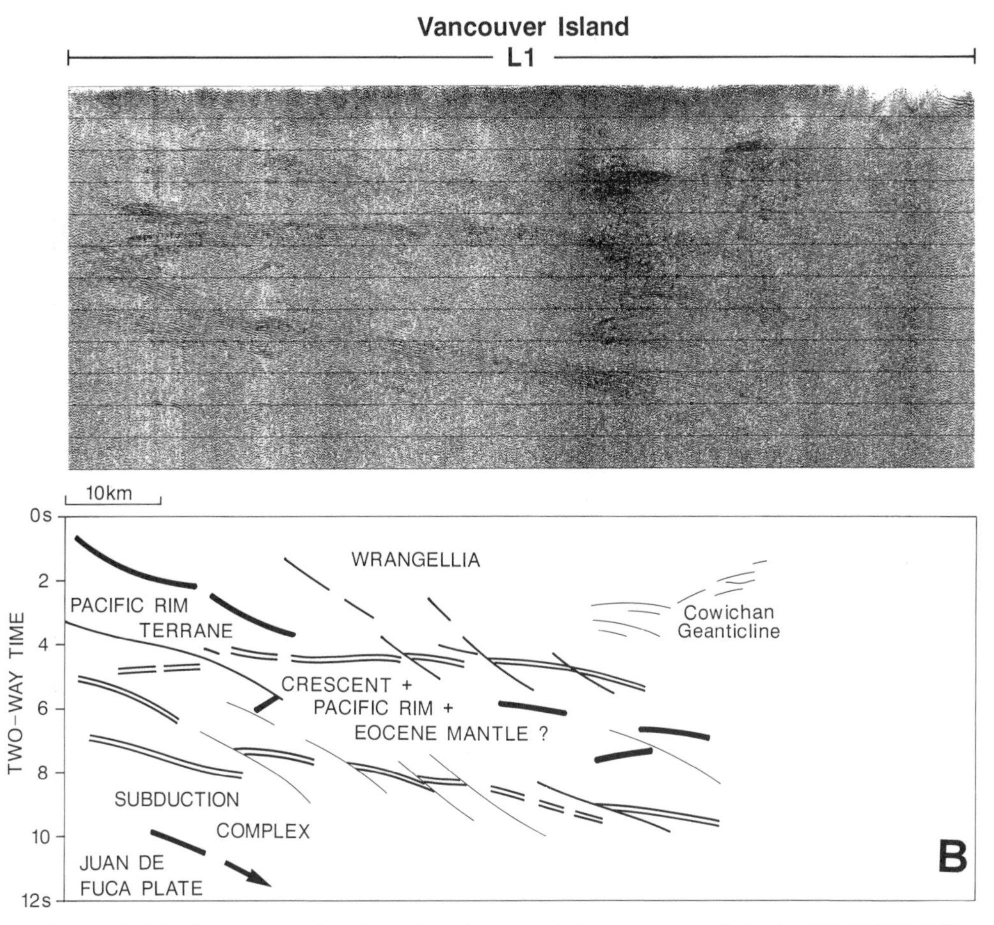

Figure 11b. Unmigrated record section (above) and partial interpretation (below) of LITHOPROBE seismic reflection line L1 across south-central Vancouver Island (location on Fig. 10).

D1–D2 is thought to represent a décollement zone at the base of the 15- to 20-km-thick Wrangellia Terrane. At the western end of L1, the reflector B is believed to be the Westcoast Fault, beneath which the Jura-Cretaceous mélange (including uppermost Triassic arc-volcanics) of the Pacific Rim Terrane, together with the Crescent Terrane, was emplaced beneath Wrangellia during the late Eocene (42 to 38 Ma) and prior to latest Eocene deposition of the overlapping Carmanah Group (Fairchild and Cowan, 1982; Rusmore and Cowan, 1985). This interpretation differs from that presented in Yorath and others (1985b) and Clowes and others (1987a), wherein reflector B was identified as the Tofino Fault beneath which the Crescent Terrane was thought to occur. Based on subsequently acquired offshore reflection, gravity, and magnetic data, the Crescent Terrane is now thought to be part of the underplated zone beneath the Pacific Rim Terrane and Wrangellia (Fig. 12).

The lateral continuity of the D1–D2 interval of line L1 (Fig. 11b) is interrupted by several breaks and easterly dipping events, some of which are listric within the zone. Most of these are interpreted as faults that arise from the décollement zone and project to the surface, where they coincide with mapped thrust faults or plutonic contacts. Insofar as these faults are probably related to late Eocene underplating of Wrangellia by the Pacific Rim and Crescent Terranes (Fig. 12), they help to explain the uplift of central and western Vancouver Island, which resulted in exposure of the plutons and their plutonic/metamorphic root zone along the west coast. At the eastern end of L1 the high-level faults are not clearly imaged, but surface geological studies (Sutherland Brown and Yorath, 1985) indicate that these are broadly transpressive with individual thrust displacements of up to one kilometer. These faults are probably Campanian in age.

Beneath Wrangellia and the Pacific Rim Terrane, a thick layer of underplated material is represented by the interval between D1–D2 and E1–E2 (Figs. 11b, 12). This zone of low reflectivity corresponds to an interval of high velocity (7.7 km/s), as determined from regional refraction studies (Spence and others, 1985), and has been interpreted as either accreted oceanic crust (Green and others, 1986) or imbricated metamorphic rocks

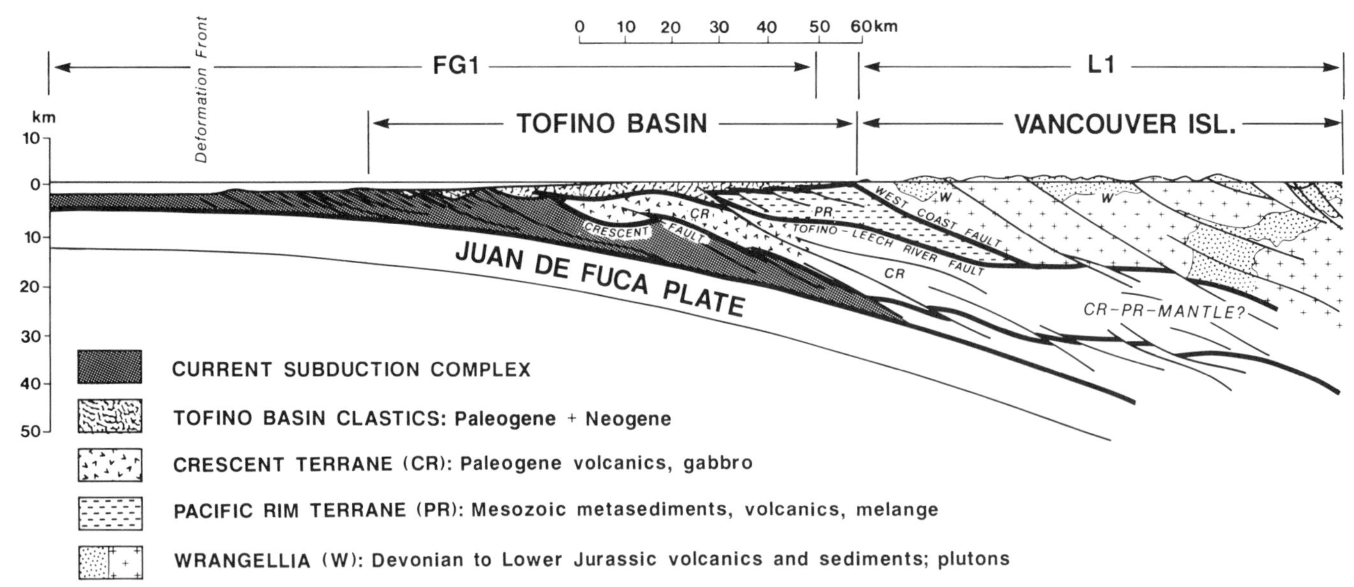

Figure 12. Summary interpretation of the crustal structure of the convergent margin of the southern Insular Belt incorporating reflection lines FG1 and L1 and other geophysical data.

derived from the offscraped sediments of the subducting plate (Clowes and others, 1987b). The seismic signature of the underplated zone in the offshore record of 85-01 (FG1) is unclear. However, at the position of the Prometheus well, volcanics chemically identical with those of the Mechosin Complex and Crescent Formation were penetrated at a depth of approximately 1800 m. These volcanics are the source of the "Prometheus magnetic anomaly" (Fig. 10; Mcleod and others, 1977). Landward of the positive magnetic anomaly, a magnetic low with similar orientation is clearly due to the offshore continuation of the Pacific Rim Terrane (Leech River Formation). Line FG5 (Fig. 10) clearly shows the Crescent Terrane as underlying the Pacific Rim Terrane and beyond, where it intersects line 85-01 (FG1). Thus it appears that the "underplated zone" of Yorath and others (1985) consists at least in part of Crescent Terrane. The Tofino–Leech River Fault (Figs. 10, 12) separates the Crescent Terrane from the overlying Pacific Rim Terrane and in Figure 12 is shown to merge with the décollement zone. Other components of the underplated zone may be additional parts of the Pacific Rim Terrane that were decoupled from the overlying rocks during the accretion process, as well as possible Eocene mantle.

The possibility that part of the underplated zone is Eocene mantle is supported by magnetotelluric and geomagnetic depth sounding studies, which identified a conductive layer some 6 km thick with a high free-water porosity of between 1 and 4 percent (Kurtz and others, 1986) coincident with the top of the zone of decoupling between the North American and Juan de Fuca Plates. Given that this part of the crust appears to be isothermal at about 450°C (Lewis and others, 1988), it is possible that Eocene mantle could have sustained retrograde metamorphism close to the greenschist-amphibolite boundary, through the agency of dehydration reactions in and above the subducting Juan de Fuca Plate. Such reactions would account for the high seismic velocities necessary to define the crustal structure across Vancouver Island.

In the offshore section (85-01 [FG1]; Fig. 11a), basalt of the Crescent Terrane is overlain by bathyal sediments of the Carmanah Group (Cameron, 1980) that were uplifted following underplating and the establishment of the subduction zone at its present location. The Carmanah Group of Eocene and Oligocene age, together with overlying Neogene marine clastics, is equivalent to rocks in the core of the Olympic Mountains. In profile 85-01 (FG1), these occur in the footwall of the extrapolated Hurricane Ridge–Crescent Faults and make up part of the current subduction complex.

In summary, the coastal topographic high is a consequence of subduction-accretion of oceanic materials beneath the edge of the North American Plate. The accretion and underplating of the Pacific Rim and Crescent Terranes, together with the sedimentary section of the current subduction complex, have led to uplift of western Vancouver Island and Campanian subsidence of the Georgia depression, within which up to 2,700 m of nonmarine sediments accumulated (Yorath and others, 1992).

Willamette-Puget-Georgia Depression

The topographic depression extending north from the Wilamette Valley in Oregon, through Puget Sound to Georgia

Strait, can be described from its position west of the Cascade-Coast arc as a forearc depression (Fig. 8). Rogers (1983) suggested that this depression is related to phase changes in the underlying, downgoing Juan de Fuca plate. Another possibility, originally proposed by Yorath and Hyndman (1983) for the area east of the Queen Charlotte Islands and noted above, is that the depression is related to lithospheric flexing; uplift caused by underthrusting and underplating beneath the westernmost continental margin depresses the area to the east.

Cascade–Coast arc

In Washington, the Cascade magmatic arc (Fig. 8) was initiated at 37 Ma. In the southern High Cascades (Transect B–3B), the arc is represented by a section of andesitic to dacitic volcaniclastic rocks, silicic ash-flow deposits, and andesitic lavas and pyroclastic rocks. The sequence has an aggregate thickness of at least 5 km and has yielded dates ranging from 36 to 17 Ma. In the Northern Cascades (B–3A) this volcanic cover has been eroded, but coeval, calc-alkaline plutons are intermittently exposed from near Mount Rainier (lat. 48°N) to north of the International Boundary and range from early Oligocene to middle Miocene. In places these rocks are cut and capped by Pleistocene and Recent volcanic cones.

In Canada, there are two Late Tertiary and Quaternary magmatic belts (Souther, 1992). The older Pemberton Belt, which trends northwesterly, includes granitic intrusions (composite Chilliwack batholith; 30? to 14 Ma) and volcanics (e.g., Coquihalla volcanics; 23 Ma). The younger Garibaldi Volcanic Belt consists of north-trending calc-alkaline centers that range in age from 3.8 Ma to 1.34 ka.

Columbia Plateau

Extensive areas of basalt lie east of, and in a back-arc position relative to, the Coast-Cascade arc (Fig. 8). In southwestern Washington and northeastern Oregon, tholeiitic flood basalts of the Columbia River Basalt Group were erupted from swarms of north-northwest–trending dykes in Miocene time. These basalts reach a maximum thickness of about 3 km in the Pasco Basin (Transect B–3B). Most of the sequence was erupted between 16 to 14 Ma, but the entire section ranges from 17 to 6 Ma. The sequence east of the Columbia River is flat lying, but to the west it has been deformed into a series of widely spaced, predominantly east-west trending anticlines and associated north- and south-verging thrusts. East northeast–west southwest extension in the middle Miocene is recorded by the dyke swarms; most of the north-south contraction recorded by the folds and thrusts has occurred in the last 10 m.y. In Canada, the Chilcotin Group comprises alkaline and tholeiitic basalts that yield ages of 15 to 2 Ma and occur in a similar back-arc setting (Bevier, 1983). Local olivine basalts that yield ages of ca. 23 Ma may be related to the older Pemberton belt and lie with locally strong angular unconformity on faulted Eocene strata (Monger, 1985a).

Tectonic regime near and north of the triple junction

The morphology of the continental margin near the triple junction is very different from that to the south, and the coastal topographic high is interrupted by Queen Charlotte Sound between the northern tip of Vancouver Island and the Queen Charlotte Islands (Fig. 8). In latest Oligocene–Miocene time, western Wrangellia was disrupted by rifting and crustal extension centered on Queen Charlotte Sound; subsidence accompanied by clastic sedimentation followed (Yorath and Chase, 1981; Yorath and Hyndman, 1983). Underthrusting along the continental margin, which began at about 6 Ma, caused depression in the crust east of the Queen Charlotte Islands. The continental margin west of the Queen Charlotte Islands is delineated by the predominantly transform Queen Charlotte Fault Zone (Transect B–1). Across the fault, a narrow belt of material accreted during oblique subduction was deformed during transform faulting to produce narrow upthrust slivers that are bounded by mainly vertical faults, in a manner similar to that observed in on-land wrench fault systems (Dehler and Clowes, in preparation).

OLDER TECTONIC REGIMES

As outlined in the regional overview, the origins and evolution of Cordilleran crust can be discussed within the context of three general tectonic regimes: the mid-Proterozoic to Middle Jurassic (1500 to 170 Ma) largely passive cratonic margin, dominated by crustal extension; the Middle Jurassic to Paleocene (170 to 60 Ma) active margin in which the dominant features of Cordilleran crust were established in a contractional and transpressive regime; and the Eocene (60 to 40 Ma) active margin, which features extension and transtension within a broad magmatic arc regime at the latitudes of Transects B–2 and B–3A.

Middle Proterozoic to Middle Jurassic cratonic margin

Archean and (?) Early Proterozoic crystalline basement, which is the southwestern extension of the Canadian shield ("little modified continental crust" of Transects B–2 and B–3A), can be traced geophysically in the subsurface as far west as the major valleys called the Rocky Mountain Trench in the north and almost to the Purcell Trench in the south. Structurally above and west of the basement are three packages of rocks that can be distinguished on the strengths of their primary linkages with the craton (Fig. 13).

The first package includes rocks that were deposited either on the craton or along its margin and that can be clearly linked to the craton on grounds of stratigraphic and structural continuity. These are the Middle Proterozoic mainly clastic deposits of the Belt and Purcell Supergroups, the Late Proterozoic clastic deposits of the Windermere Supergroup, and Cambrian through Jurassic platform-shelf-slope-rise deposits.

The second package includes Early Proterozoic rocks and their younger cover, with a strong Mesozoic structural fabric;

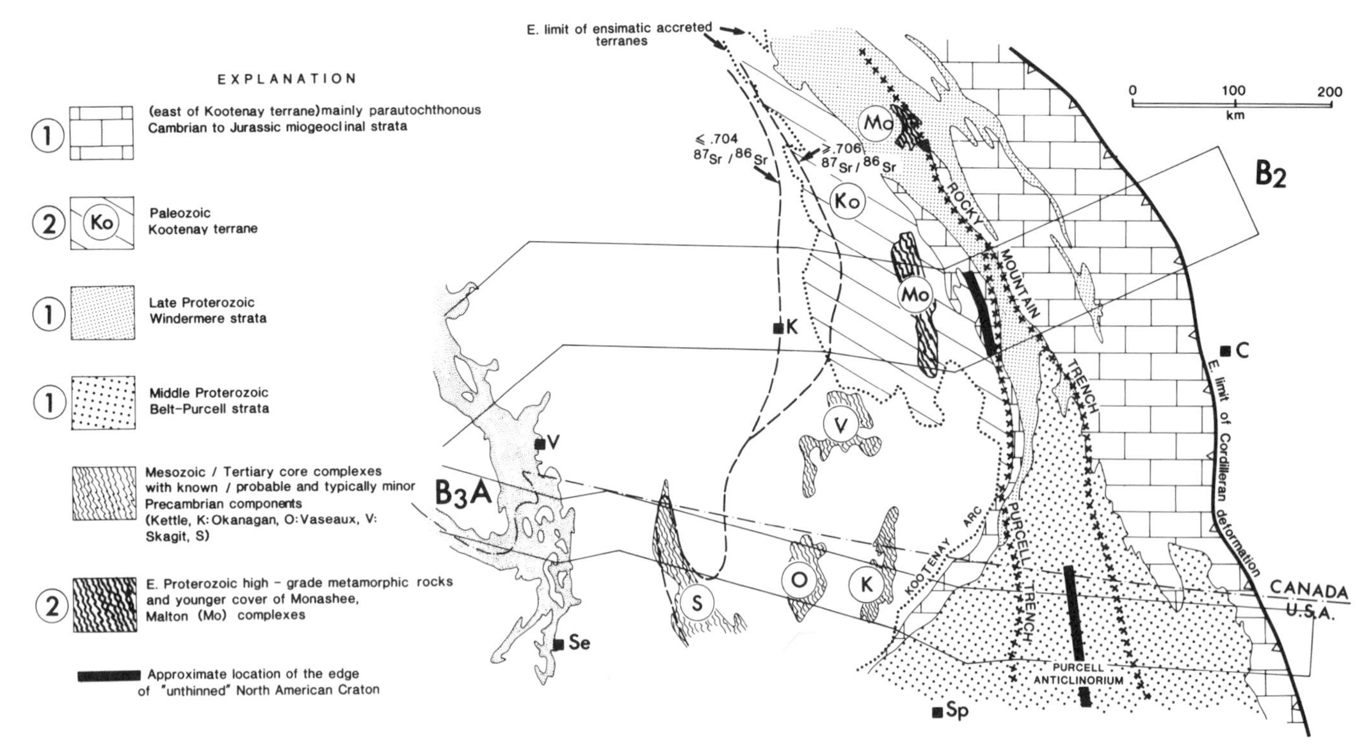

Figure 13. Cratonic margin elements: (1) rocks (Belt, Windermere, Cambrian through Jurassic strata) that can be linked with the craton; (2) rocks (Kootenay Terrane, Monashee-Malton Gneisses) with possible links; (3) rocks with links that are tenuous. The map shows location of edge of "unthinned" North American Craton and location of initial strontium ratio lines, which indicate the possible western limits of old crust. C: Calgary; K: Kamloops; Se: Seattle; Sp: Spokane; V: Vancouver.

these rocks traditionally have been regarded as western-facies equivalents of rocks to the east. However, the rocks of the second package cannot be unequivocally linked to the eastern rocks because the package is isolated and its geological record differs from that of the rocks to the east. The Monashee and Malton Metamorphic Complexes ("tectonized continental crust" of Monashee Terrane of Transect B–2) and variably metamorphosed and deformed, clastic and volcanic strata and granitic intrusions of Paleozoic age (Kootenay terrane of Transect B–2) constitute this package.

The third package includes rocks in the gneissic and granitic "core complexes" that are exposed farther west in uplifts surrounded by rocks of the accreted terranes. Their links with the craton are tenuous at best and are based largely on the isotopic identification of a Precambrian component.

Precambrian Crystalline Basement. As discussed below, Precambrian basement identified seismically and in deep wells east of the deformation front can be traced westward beneath the Foreland Belt by using geophysics. This basement, with its typical northeast-trending structural grain, has long been regarded as Hudsonian (1.8 Ga) on the basis of K-Ar, Rb-Sr, and U-Pb dates. New work on eight basement cores from wells in a zone extending 400 km eastward of the Cordilleran deformation front, and lying just north of latitude 49°, gives Sm-Nd crustal residence ages averaging 2.8 Ga (Frost and Burwash, 1986). This means either that the 1.8-Ga numbers reflect complete resetting of older, 2.8-Ga crust in Hudsonian time or that mantle material was combined with older than 2.8-Ga crust to produce the younger crustal residence age. The 2.8-Ga ages are similar to ones from the western Churchill Province of the Canadian Shield and from the Wyoming Province; whether 2.8-Ga crust is continuous between the two areas is not known.

Rocks deposited on and along the ancient cratonic margin. At the latitude of Transect B–3, the first episode in the evolution of the Cordilleran miogeocline was the creation of a deep basin in which as much as 15 km of Middle Proterozoic fine-grained clastic and carbonate rocks of the Belt-Purcell Supergroup accumulated (Harrison, 1972; McMechan, 1981) during the interval between 1500 and 1300 to 1350 Ma (McMechan and Price, 1982). Block faulting accompanied the filling of the basin, and an aulacogen may have extended northeastward from the basin (McMechan, 1981) into the Southern Alberta Rift of Kanasewich and others (1969), which shows up as a conspicuous trough and ridge on a contour map of the Moho (Fig. 14). The

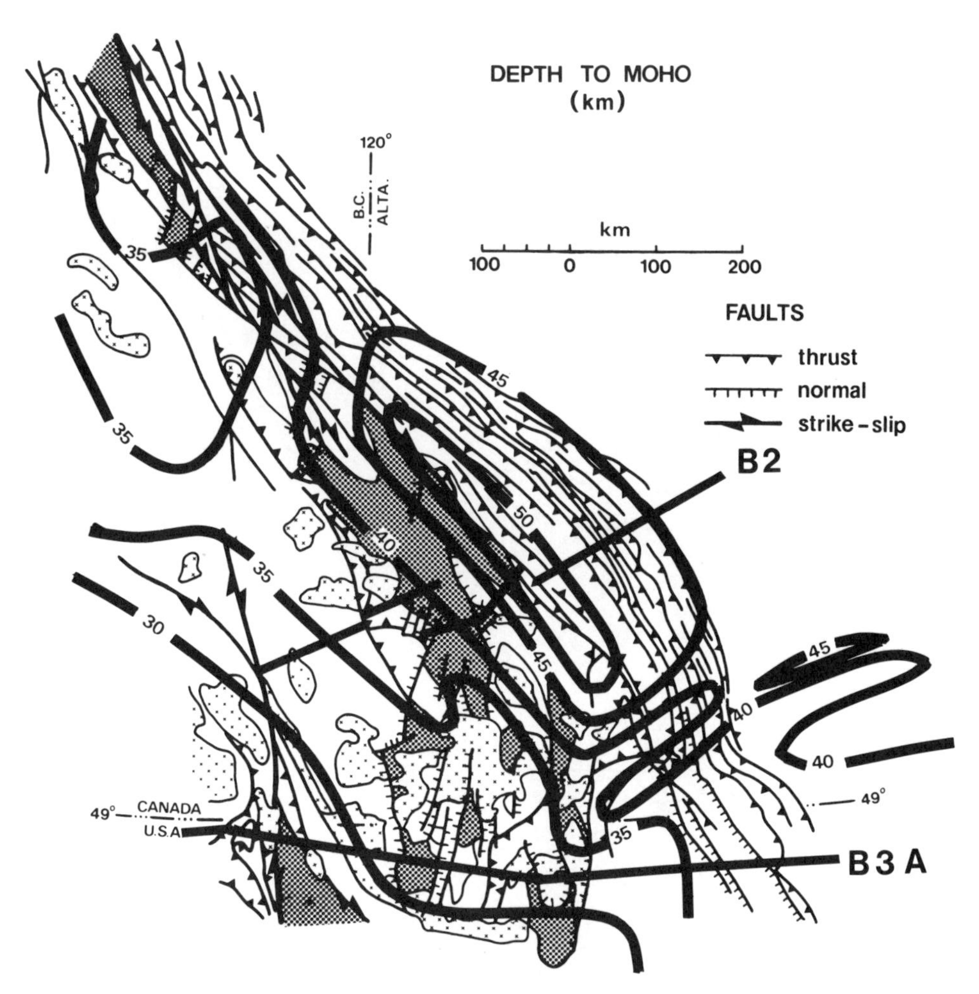

Figure 14. Crustal thickness map showing contours of depth to Moho. Close stipple: high-grade metamorphic rocks; open stipple: granitic rocks. (Compiled by R. A. Price, 1979, from Berry and Forsyth, 1975; Kanasewitch and others, 1969; Mereu and others, 1977; Bennett and others, 1975; Chandra and Cumming, 1972; Spence and others, 1977.)

East Kootenay Orogeny, manifested by local folding, metamorphism, and granitic intrusion, terminated this episode of Cordilleran tectonic evolution about 1300 to 1350 Ma (McMechan and Price, 1982). At the latitude of Transect B–2, Belt-Purcell deposits are absent at the surface but are inferred on the transect to be present in the subsurface.

A second episode of rift-controlled clastic deposition in Late Proterozoic time was marked by local but deep (4 km) erosion into uplifted blocks of the Belt-Purcell Supergroup and by the accumulation of about 8 km of clastic sediment of the Windermere Supergroup (Lis and Price, 1976). Western and northern parts of the Belt-Purcell basin may have been rifted from North America at this time and transported to another part of the globe (Sears and Price, 1978).

Thick wedges of early Paleozoic quartz sandstones unconformably overlap both Windermere and Belt-Purcell strata and locally contain intercalated basic volcanic rocks. They record a third interval of extension, separation, and possible removal of parts of the Late Proterozoic continental terrace wedge. Analyses of rates of basin subsidence in the eastern margin of the miogeocline suggest that an episode of rifting, culminating in sea-floor spreading, occurred in latest Proterozoic or earliest Paleozoic time and initiated an episode of passive margin evolution (Bond and Kominz, 1984) that seems to have persisted from Cambrian to Middle Jurassic time.

Passive margin sedimentation was disrupted during Cambrian and Middle Devonian time by block faulting and differential erosion and subsidence within the continental terrace wedge, near the hinge zone between the platform and the miogeocline (Figs. 15, 16). North of latitude 50°N, a shale basin developed outboard of the Middle Cambrian to Devonian carbonate bank margin that was localized by differential subsidence and block

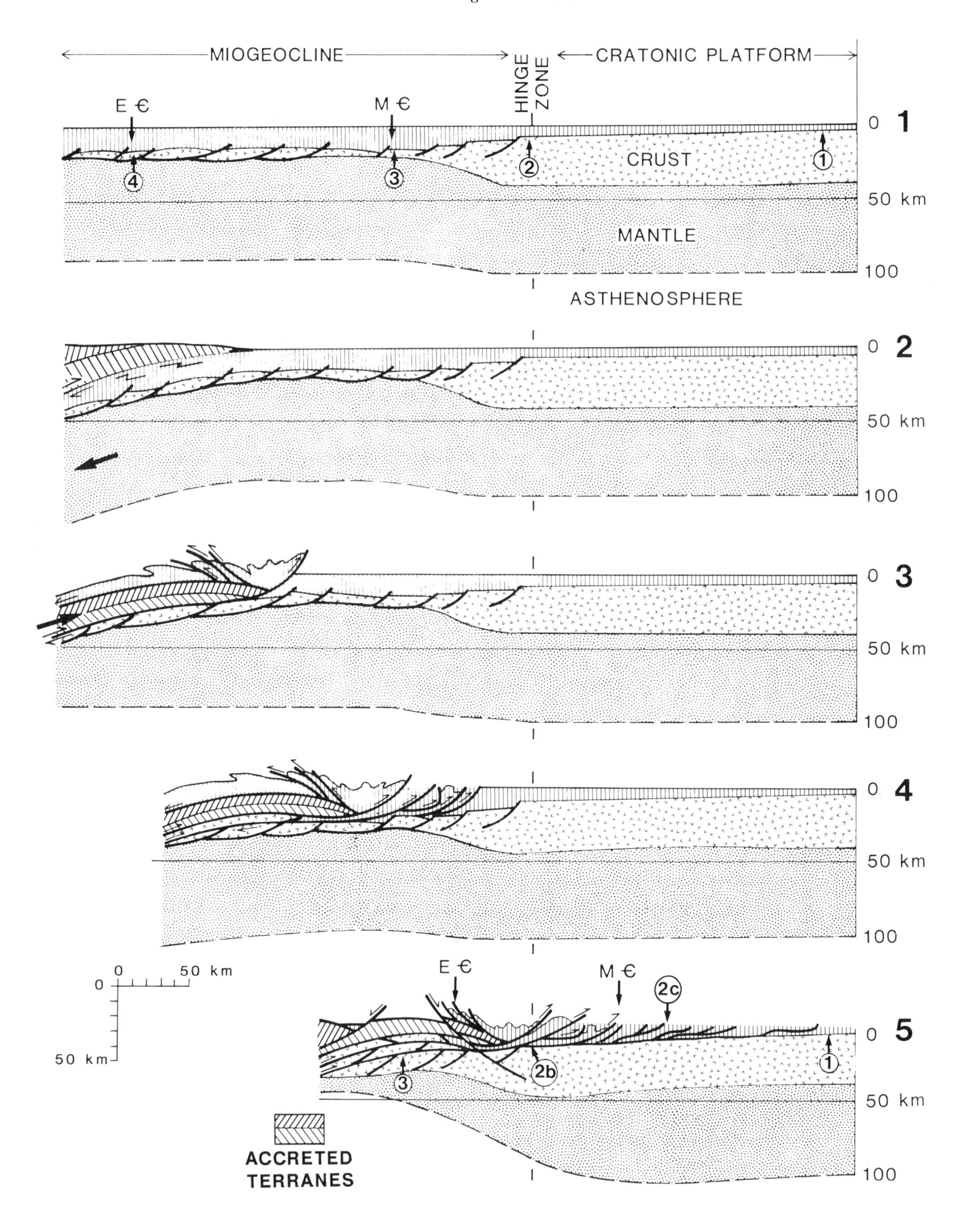

MIOGEOCLINE
HINGE ZONE
CRATONIC PLATFORM
E €
M €
1
CRUST
MANTLE
ASTHENOSPHERE
0
50 km
100
2
3
4
5
2c
2b
0 50 km
50 km
ACCRETED TERRANES

faulting. At about 50°N latitude the locus of block faulting extended across the continental shelf to the zone marked by transverse northeast-trending reverse faults (Fig. 17). Farther south, the area underlain by Belt-Purcell strata (Montania) behaved as a cratonic platform throughout Paleozoic time; this situation further accentuated the contrasts between segments of the Cordillera traversed by Transects B–2 and B–3A.

Monashee (and Malton) Metamorphic Complexes. Metamorphic rocks shown on Transect B–2 as stacked slices of tectonized continental crust of the Monashee terrane include hornblende biotite paragneiss and orthogneiss of 2.1 to 2.5+ Ga ages and mantling semipelitic, psammitic, and calc-silicate gneiss, marble, and quartzite, cut by orthogneiss at least as old as 1.5 Ga. Since these ages do not correspond closely with those known from the craton to the east, and since no direct linkages can be made with the rocks deposited on or along the craton, these rocks are treated in Transect B–2 as a suspect terrane, although one that may be parautochthonous.

Kootenay Terrane. Kootenay Terrane (Ko, Figs. 5, 13) comprises variably deformed Cambro-Ordovician and younger clastics of the Lardeau Group, minor carbonate, and basic and acidic volcanics, which are in part of Devonian age and coeval with intrusive rocks, and are stratigraphically overlain by less deformed, Permo-Carboniferous clastic and volcanic rock of the Milford Group. This sequence has been regarded traditionally as a western "eugeosynclinal," distal, offshelf equivalent of rocks to the east that were deposited on and along the craton. However, it is shown as a suspect terrane on Transect B–2 to emphasize the pre-Milford record of Paleozoic deformation, metamorphism, and intrusion that is not recognized in rocks to the east. The base is possibly a major bedding-plane fault marked by small ultramafic bodies, which lies near the base of the Lardeau Group (J. O. Wheeler, personal communication, 1983). Carboniferous

Figure 15. Stages in the evolution of the eastern Cordillera, based on palinspastic reconstruction of the eastern part of Transect B-2 (after Price and others, 1985). Symbols mark the initial and final positions of the Middle Cambrian and Early Cambrian carbonate bank margins. The numbers identify reference points on the top of the crystalline basement, the subsidence history of which is illustrated in Figure 16. Section 1: pre-collision palinspastic reconstruction. Section 2: partial westward subduction of the outboard part of the miogeocline beneath Intermontane Superterrane in Middle Jurassic time; initiation of east-verging accretionary prism. Section 3: development of west-verging structures above suspect and allochthonous terranes by tectonic wedging and delamination of the outboard part of the miogeocline and its basement (creating the asymmetric but two-sided structural fabric of the Omineca Belt). Section 4: Late Jurassic–Early Cretaceous continuing horizontal compression and eastward translation of the outer miogeocline by convergence between Intermontane Superterrane and North America. Section 5. Eocene tectonic denudation of part of the metamorphic infrastructure following Late Cretaceous–Paleocene convergence that resulted in horizontal compression and eastward displacement of the miogeocline over the edge of the craton; 2b, 2c are final positions of, respectively, basement and cover rocks from the hinge zone.

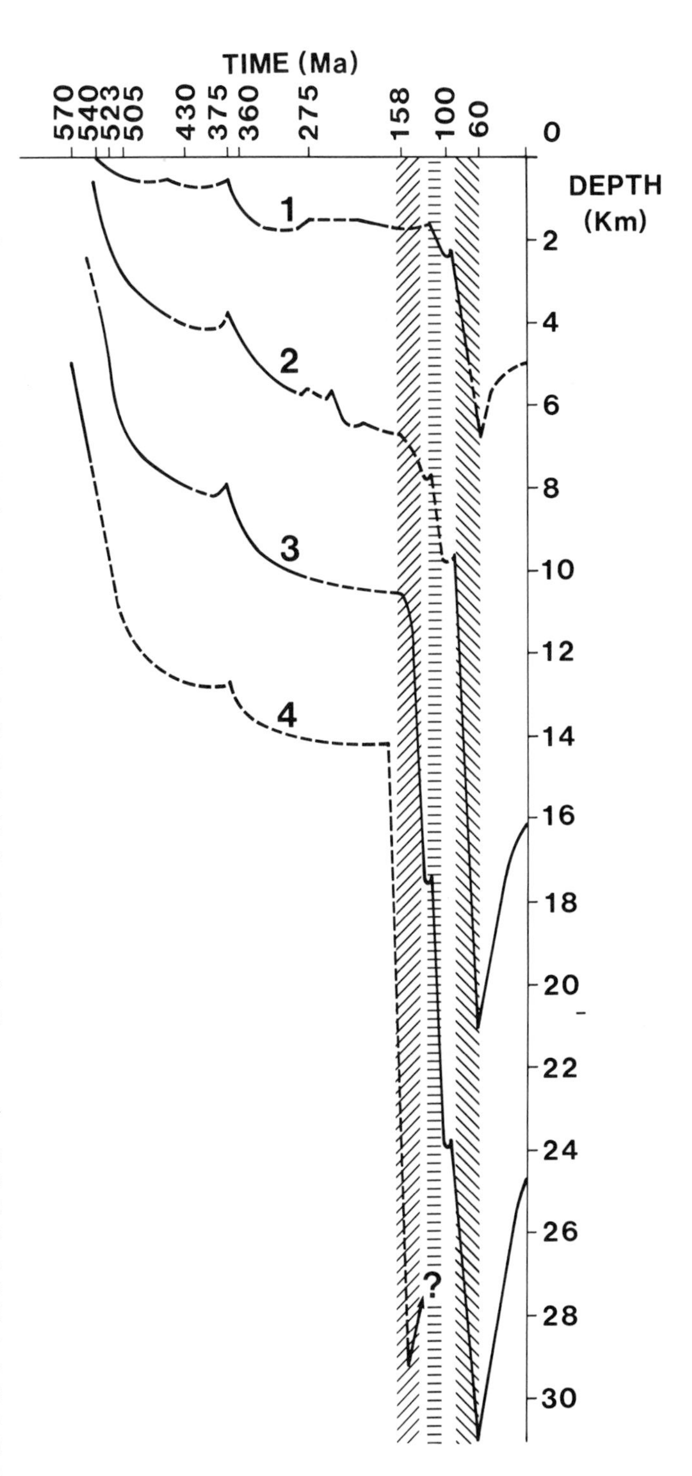

Figure 16. Subsidence history of the Precambrian basement surface (after Price and others, 1985); numbers refer to reference points in Figure 15; line patterns show time spans of Jurassic- through Paleocene-dated convergent events.

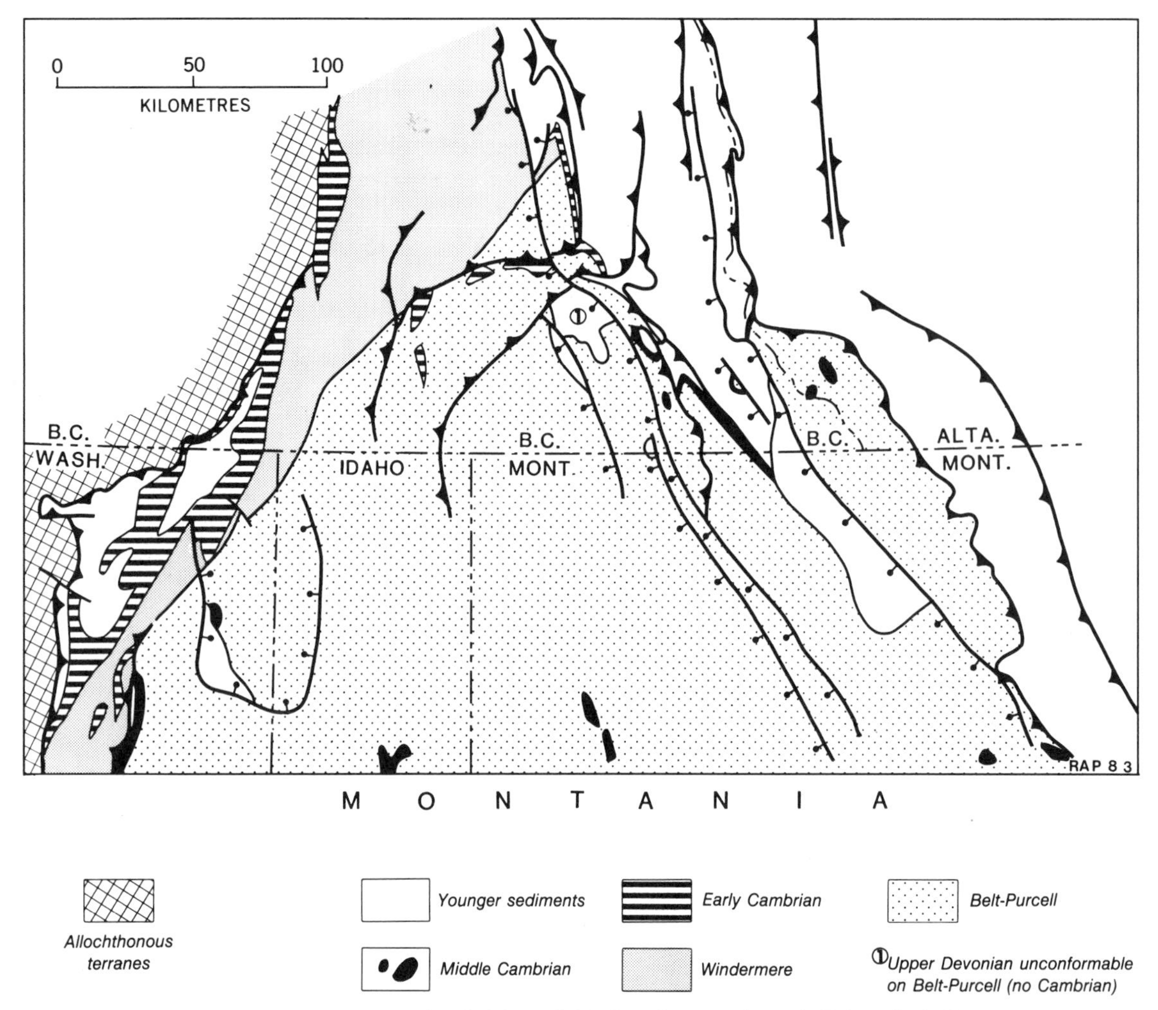

Figure 17. The northwestern margin of "Montania," a Late Proterozoic–early Paleozoic shallow water "cratonic" platform. Cambrian subcrop reconstruction with no palinspastic restoration. (Compiled and interpreted by R. A. Price from various published maps of the Geological Survey of Canada and the U.S. Geological Survey.)

volcanics in the upper part have been correlated by Klepacki and Wheeler (1985) with those of the ensimatic Slide Mountain Group to the west.

Rocks of core complexes. High grade metamorphic rocks, mainly orthogneisses, occur in several of the Mesozoic/Tertiary core complexes (Fig. 13). Many rocks in these appear to be of late Mesozoic–earliest Tertiary ages but contain isotopic evidence for at least some Precambrian component (Mattinson, 1972; R. L. Armstrong, personal communication; R. Parrish, personal communication).

Western limits of the mid-Proterozoic to mid-Jurassic craton. The Middle Proterozoic to mid-Jurassic "fossil" sialic edge of the North American craton is concealed beneath displaced supracrustal rocks and obscured by the effects of mid-Jurassic to Paleocene compressional deformation and Eocene extension. Elucidation of the location and nature of the sialic edge relies on three different types of indirect evidence: (1) geophysical imaging and modelling of the present deep structure of the crust; (2) mapping of initial $^{87}Sr/^{86}Sr$ values of igneous rocks to establish the presence or absence of continental crust at the time they formed and in their place of formation; and (3) palinspastic restorations of the fold and thrust belt to establish the location and nature of the continental margin basins in which rocks of the continental terrace wedges accumulated prior to mid-Jurassic and younger compression. Geophysical, geochemical, and geological evidence suggests that little modified Precambrian crust extends to slightly west of the Rocky Mountain Trench, near Transect B–2, and nearly to the Kootenay Arc near latitude 49°N

(Fig. 13). West of this, all older rocks have a strong Mesozoic "Cordilleran" fabric. As noted above, some rocks, such as those of the Monashee and Malton Metamorphic Complexes, may well represent the western, reworked extensions of the Precambrian shield, possibly pulled away during extension in the Proterozoic and then pushed back during Mesozoic compression. Rocks in the core complexes contain unknown amounts of Precambrian material, possibly merely Precambrian detrital minerals incorporated in Mesozoic gneissic rocks.

Geophysical evidence. Extensive seismic reflection profiling in the southern Canadian Rocky Mountains by the petroleum exploration industry has demonstrated that the relatively flat platformal cover rocks of the Archean and (?) Early Proterozoic crystalline basement slope southwestward from the Churchill Province of the Canadian Shield, under the Rocky Mountains, and at least as far west as the Rocky Mountain Trench, where they are at a depth of about 10 km below sea level (Bally and others, 1966). Local seismic reflection profiles west of the Trench (Cook, 1985), including some recent LITHOPROBE profiles that cross the Rocky Mountain Trench and Purcell Mountains (Figs. 1, 29; Cook and others, 1987), show that the platformal cover on the peneplained basement extends about 30 km west of the Trench to at least as far as the central Purcell Mountains, where it is at a depth of about 15 km. In northwestern Montana, COCORP data from 50 km east of the Purcell Trench show thrust faults above a basal detachment at about 21 km depth (Potter and others, in preparation; Yoos and others, in preparation). The ARCO-Marathon Paul Gibbs 1 well, drilled to a depth of nearly 6,000 m in the Purcell Anticlinorium (Fig. 13), showed that prominent reflections on industry seismic data can be correlated with Proterozoic sills (Harrison and others, 1985), and it is unlikely that a significant thickness of Paleozoic sediments, such as that shown in Transect B–3A, exists beneath the Purcell Anticlinorium. Instead, the 21-km-thick imbricated stack probably consists almost entirely of the Belt-Purcell Supergroup, with limited basement involvement (Yoos and others, in preparation). The crust below the basal detachment at this place is nonreflective. This is the most westerly point to which the peneplained surface of the crystalline basement rocks can be traced.

By contrast, on the COCORP seismic profile west of the Purcell Trench and across the Kootenay Arc, dipping reflectors are seen throughout the crust above a prominent set of relatively flat reflectors (Figs. 18, 19; Potter and others, 1986). The latter reflectors occur at 10.5 to 11.0 s two-way travel times, corresponding to depths of 33 to 35 km, which are close to refraction-determined depths to Moho in this region (Hill, 1972).

The Bouguer regional gravity anomaly map (Fig. 20) shows an elongate negative anomaly of more than 200 milligals over the Rocky Mountain Trench and western Rocky Mountains. The east side of this negative anomaly is clearly a consequence of crustal thickening due to the stacking of thrust sheets above the undeformed but isostatically flexed crystalline continental crust (Price, 1973, 1981). Formerly, the west side of the Bouguer anomaly was interpreted by Price (1981) as marking the edge of the little-modified craton, west of which continental crust had been thinned by rifting during Proterozoic and earliest(?) Paleozoic times, when the "fossil" sialic western edge of North America was created. However, the recognition of large-scale Eocene extension in the Omineca and Intermontane Belts (Price, 1979; Ewing, 1980; Parrish and others, 1985; Tempelman-Kluit and Parkinson, 1986) raises the question of whether the gradient on the west side of the Bouguer anomaly may result from Eocene crustal thinning.

Regional variations in crustal thickness in the southern part of the Canadian Cordillera, as outlined by seismic refraction, are shown on Transect B–2 and summarized in Figures 14 and 21. Under the Intermontane and western Omineca Belts, crustal thickness ranges from 30 to 40 km. In the Rocky Mountains east of longitude 120°W, the crust is generally more than 45 km thick. In the narrow zone situated over the Rocky Mountain Trench and the western Rocky Mountains, where the Bouguer gravity anomaly is less than 200 milligals, the crust is more than 50 km thick. The variation in depth to the Moho outlined by seismic refraction experiments follows the broad outlines of the Bouguer gravity anomaly map.

A notable feature of the crustal thickness map (Fig. 14) is the transverse, northeast-trending ridge and trough that extend from the cratonic platform east of the Foreland Belt at about latitude 50°N to the west side of the Purcell Anticlinorium at about latitude 49°N. This transverse structure was recognized as a deep rift in the Precambrian basement by Kanasewich and others, 1969) on the basis of seismic refraction and reflection and the Bouguer gravity and magnetic anomalies. It is aligned with the electrical conductivity anomaly discussed by Gough (1986) and with the northeast-trending faults in the Purcell Anticlinorium. Conspicuous changes in stratigraphic relationships involving early Paleozoic and Middle and Late Proterozoic rocks (Fig. 17) show that these Late Jurassic and/or Early Cretaceous transverse faults follow the locus of older structures across which there was large-scale vertical displacement during Late Proterozoic and early Paleozoic time (Lis and Price, 1976; Benvenuto and Price, 1979). Although the supracrustal rocks in which the transverse faults occur were displaced about 200 km northeastward from the region in which they were deposited in late Mesozoic time (Price, 1981), the rift structure in the Precambrian basement of the craton apparently extended beyond the craton, into the continental margin basin in which Late Proterozoic and early Paleozoic strata accumulated, and was active and affected patterns of erosion and deposition while they were accumulating.

The regional magnetic anomaly field in the southeastern Canadian Cordillera provides additional information on the deep geology of the region and supports conclusions drawn from other lines of geophysical evidence. Northeast-trending, large amplitude, large wavelength magnetic anomalies that are aligned with the tectonic grain of the Churchill Province of the Canadian Shield extend under the Foreland Belt at least as far west as the Rocky Mountain Trench; in southern British Columbia the anomalies may extend well beyond the Trench (Fig. 22). The

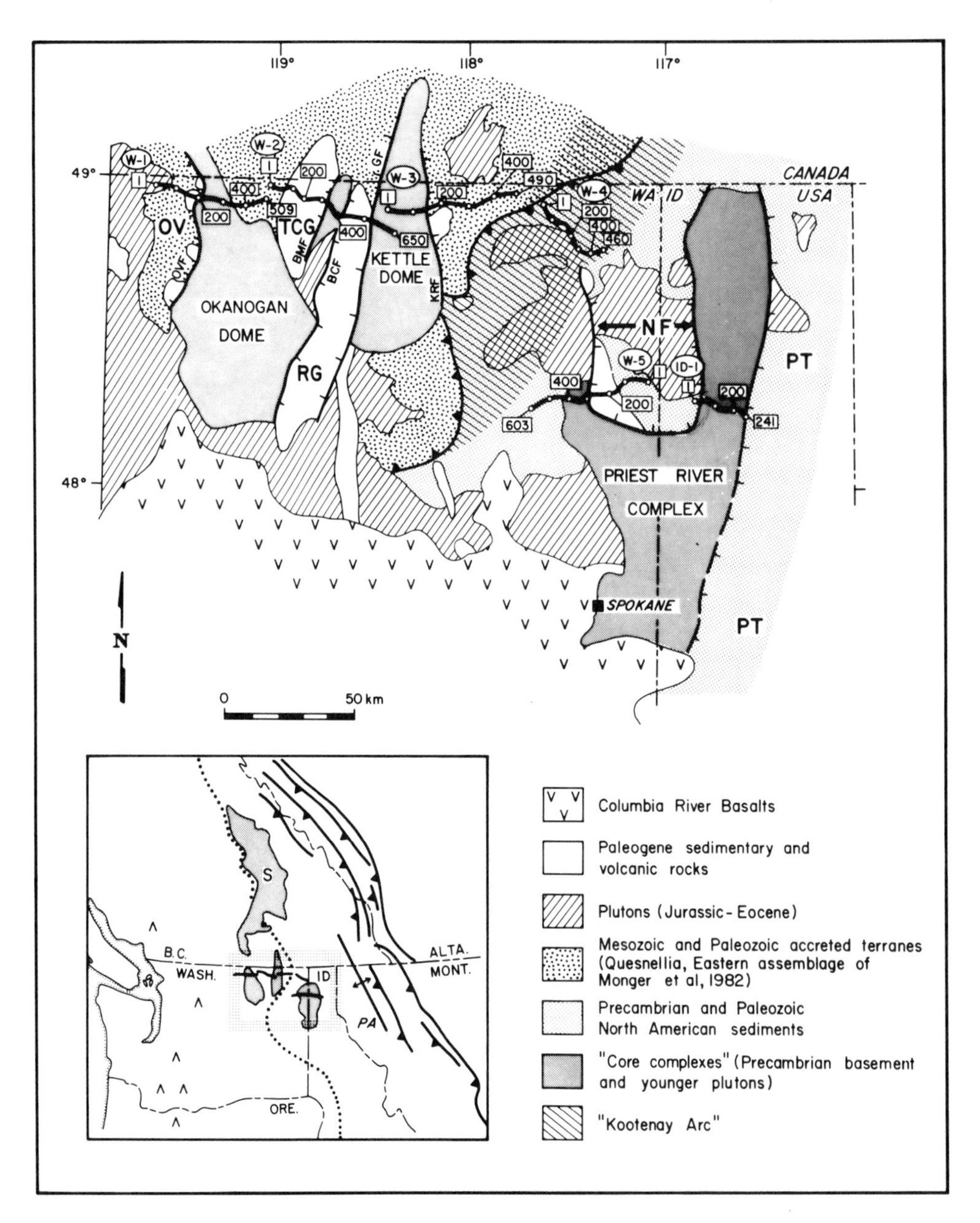

Figure 18. Location and geologic setting of the COCORP deep seismic reflection lines in parts of northeastern Washington and Idaho. W-1, W-2, W-3, W-4, and W-5 refer to COCORP Washington lines 1–5; ID-1 refers to COCORP Idaho line 1. Numbered boxes refer to station numbers (vibration points) on COCORP lines. OV, Okanagan Valley; OVF, Okanagan Valley Fault; TCG, Toroda Creek Graben; RG, Republic Graben; KRF, Kettle River Fault; NF, Newport Fault; PT, Purcell Trench; BCF, Bacon Creek Fault; GF, Granby Fault; BMF, Bodie Mountain Fault. Inset is index map showing the location of the seismic lines with respect to major features of the Cordillera; on inset, stippled box is the area shown on the detailed location map. S, Shuswap Complex; PA, Purcell Anticlinorium; dotted line represents the approximate eastern limit of accreted terranes mapped at the surface.

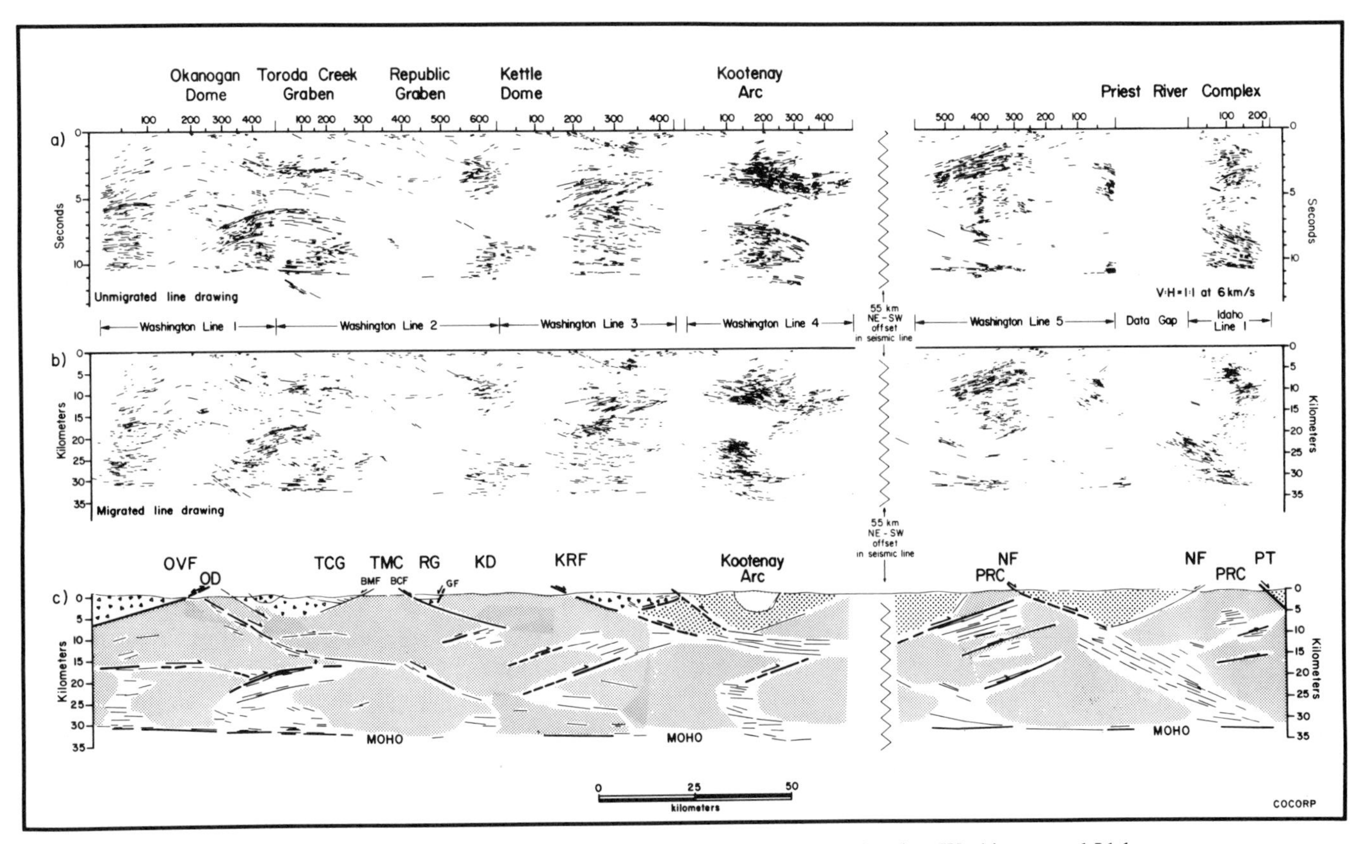

Figure 19. Summary of COCORP data collected in the Cordilleran interior, Washington and Idaho, during the summer of 1984. (a) Line drawings of all seismic sections. Data are unmigrated, and drawings are scaled so that H-V for a velocity of 6.0 km/s. (b) Migrated line drawings, produced by applying a migration routine to the same sections. The migration velocity used was 6.0 km/s. (c) Geologic interpretation of the seismic sections. This section incorporates the reflection data and surface geology into a model for crustal structure. Major features include west-dipping Mesozoic thrust structures, traceable to depths of 23 km beneath the Okanagan Dome and shallowing to 5 km beneath the Priest River Complex; Eocene normal fault zones traceable to midcrustal depths and interpreted to feed into broad lower crustal dipping deformation zones that sole into the Moho; and a relatively flat Moho geometry that is probably a product of Eocene extension and magmatism. Refer to Potter and others (1986) for discussion of rationale for the interpretations shown. Stipple pattern, sialic basement; dot pattern, Proterozoic and Paleozoic North American supracrustal rocks; random triangle pattern, Intermontane Accreted Terrane; diagonal ruled pattern, Spirit pluton; BCF, Bacon Creek Fault; BMF, Bodie Mountain Fault; GF, Granby Fault; KD, Kettle Dome; KRF, Kettle River Fault; NF, Newport Fault; OD, Okanagan Dome; OVF, Okanagan Valley Fault; PRC, Priest River Complex; PT, Purcell Trench; RG, Republic Graben; TCG, Toroda Creek Graben; TMC, Tenas Mary Complex.

Omineca Belt is "magnetically quiet," but in the Intermontane Belt the magnetic anomalies trend north-northwest, parallel with the Cordilleran tectonic grain imposed in Middle Jurassic to Paleocene time.

Geochemical evidence. Some constraint is provided on the western limit of ancient continental crust by the use of $^{87}Sr/^{86}Sr$ initial ratios. Low ratios ($\leqslant$0.704) occur in mantle-derived, mainly ensimatic, rocks such as those of the accreted terranes, whereas higher ratios ($\geqslant$0.706) reflect either derivation of magmatic rocks from Precambrian rocks or the contamination of magmas rising through old crust. Thus, the initial $^{87}Sr/^{86}Sr$ value for an igneous rock indicates whether it was underlain by continental crust at the time at which it was created and in the place in which it was created. Most igneous rocks in the Cordillera have been structurally displaced laterally relative to underlying rocks since they were intruded or extruded. Therefore, the use of the $^{87}Sr/^{86}Sr$ initial ratios greater or less than 0.705 to discriminate between the presence or absence of underlying continental crust may lead to large errors if displacement of the igneous rocks subsequent to their crystallization cannot be ruled out. Further-

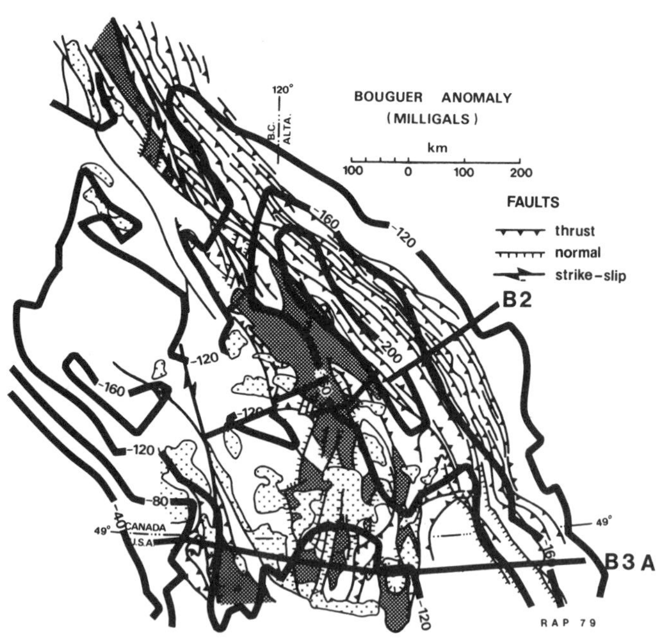

Figure 20. Bouguer regional gravity anomaly map for region of Transects B-2, B-3A. (After Stacey and others, 1973).

more, continental crust that occurs beneath the allochthonous (or parautochthonous) rocks may also be allochthonous, and accordingly, even if the presence of continental crust can definitely be established beneath the allochthonous rocks, this continental crust itself may be allochthonous and must be considered "suspect." With these reservations in mind, the line separating igneous rocks with initial $^{87}Sr/^{86}Sr$ values greater than 0.705 from those with values less than 0.705 indicates that parautochthonous North American continental crust may extend as far west as the central Intermontane Belt in Transect B–2 and the metamorphic core of the North Cascades in Transect B–3A (Fig. 13; Armstrong and others, 1977; Armstrong, 1988; Jung, 1986).

Geological evidence. Palinspastic restoration of the Foreland Belt, utilizing the principle of tectonic inversion, is a powerful tool for reconstructing the nature and location of the "fossil" continental margin. Simply stated, the principle of tectonic inversion is that if the contents of a sedimentary basin are detached from their basement and displaced over an adjacent cratonic platform, the structural relief of the basin is inverted in the displaced mass. The greater thicknesses of sediment, which come from the deeper parts of the basin, form higher parts of the structural culminations in the displaced mass. Furthermore, the margins of the basin, from which the sedimentary fill has been displaced by thrust faulting, form footwall ramps over which other displaced rocks are draped in regional fault-bend folds (Price, 1981). The principle is illustrated in Figure 23 by the Purcell Anticlinorium, which is the tectonically inverted counterpart to the continental margin basin in which the Purcell Supergroup was deposited. Furthermore, the Kootenay Arc (Fig. 13) is a crustal scale monoclinal flexure of about 20 km amplitude that can be interpreted as a fault-bend fold in which the entire parautochthonous supracrustal sequence of this part of the Cordillera is draped over a footwall ramp marking the "fossil" edge of the North American Craton (Price, 1981). In Transect B–2 the "fossil" edge of the continental craton is inferred to lie beneath the structural culmination of the eastern Selkirk Mountains (long. 117°30′W) and is indicated schematically by the inferred listric, west-dipping normal faults in the autochthonous basement that have Late Proterozoic Miette (Windermere) strata preserved along their hanging wall. The offset parautochthonous counterparts of these autochthonous strata occur in the Bourgeau thrust sheet, about 200 km to the northeast (Transect 2B). In Transect B–3A the "fossil" edge of the cratonic platform lies between the Rocky Mountain and Purcell Trenches (Fig. 13), where the Bouguer gravity anomaly increases westward from −160 milligals to between −160 and −120 milligals (Fig. 20). This anomaly increase marks the eastern edge of the basin in which the Middle Proterozoic Belt–Purcell Supergroup was deposited. The strata deposited in this part of the basin now lie in the leading edge of the Lewis thrust sheet, about 150 km to the northeast. Using the principle of tectonic inversion and the location within the Foreland Belt of the hinge zone between the thinner platform cover sequence and the thicker miogeoclinal sequence, it can be shown that the locus of the Middle Proterozoic to Middle Jurassic "fossil" edge of the continental craton lies just west of the Rocky Mountain Trench from the latitude of Transect B–3A southward to about 50°N latitude. There the locus shifts westward to the western part of the Purcell anticlinorium, which it follows southward, beyond Transect B–2 (Figs. 13, 15).

The distal "fossil" sialic edge of the ancient North American continent is far more difficult to delineate than the margin of the craton. The rifting and crustal stretching in Middle Proterozoic, Late Proterozoic, and (?) Early Cambrian time that produced the "fossil" continental margin and the basins in which the Belt/Purcell Supergroup, the Windermere Supergroup, and the early Paleozoic continental margin (miogeoclinal) rocks accumulated probably does not result in a sharp rupture separating continental from oceanic crust (Price, 1981; and Figs. 15, 23). Much or all of the miogeoclinal continental terrace wedge probably was floored by tectonically attenuated continental crust, some of which was probably incorporated in the Jurassic-Paleogene thrust stack during collisions with the accreted terranes. In Transect B–2, Early Proterozoic and older crystalline basement rocks (Malton Gneiss) have been projected into the section under the Selkirk Mountains from the region to the north where they are exposed (Simony and others, 1980). They may represent parautochthonous North American crystalline basement that was stretched during Late Proterozoic and/or Early Cambrian rifting and detached from its underlying mantle and displaced over the North American Craton during Middle Jurassic to Paleocene crustal convergence. Likewise, the sheets of basement rock that are inferred to occur

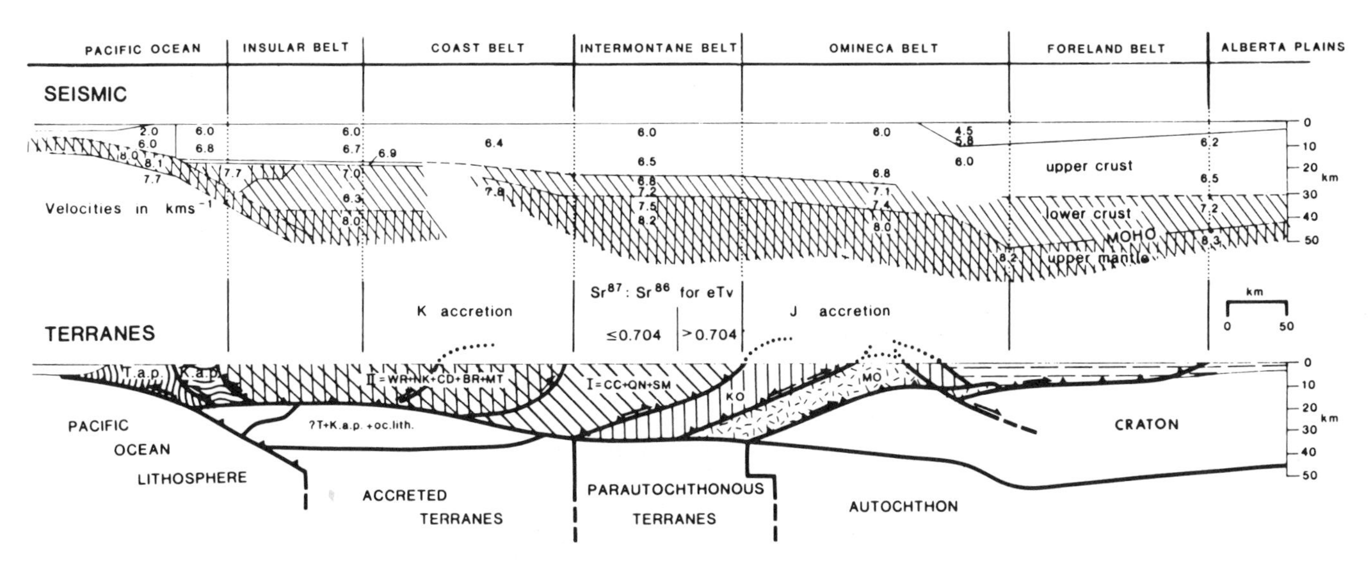

Figure 21. Seismic refraction model of Transect B-2 compared with transect model derived largely from surface geology to show interpreted vertical distribution of terranes. I = Intermontane Superterrane; II = Insular Superterrane; T.a.p. = Tertiary accretionary prism; K.a.p. = Cretaceous accretionary prism exclusive of Wrangellia and other "Cascade" terranes; KO, Kootenay Terrane and MO, Monashee Terrane—together these constitute the "parautochthonous" or "pericratonic" terranes of Figure 3 (from Monger and others, 1985a). Other terrane abbreviations as in Figure 6.

below this sheet, and to form a duplex structure that is partly responsible for the structural culmination in which the Monashee Complex is exposed, also may represent slices of parautochthonous North American crystalline basement with the same history. This is the interpretation that was adopted in Transect B–2 and in Figure 15. An equally plausible but alternative interpretation is that these Precambrian crystalline basement rocks are allochthonous and have moved into place against North America along with the other suspect terranes. Until some compelling evidence to the contrary is presented, these crystalline basement rocks must be treated as suspect terranes that may be allochthonous with respect to adjacent parts of North America.

Middle Jurassic to Paleocene active margin

As noted earlier in the "Geological Overview" section, the Middle Jurassic to Paleocene interval was the time of formation of most uniquely "Cordilleran crust." Major structural elements and tectonostratigraphic terranes are shown in Figure 24. Notable are the two superterranes composed of smaller terranes that were amalgamated prior to accretion. The rocks of the small terranes are of intra-oceanic origin, and some carry faunal and/or paleomagnetic records to indicate that they are displaced with respect to the craton. The terranes appear to consist mainly of nappes or "flakes" of supracrustal rock that were delaminated from their underlying lithosphere during the process of subduction/accretion, in a manner analogous to the removal of the original lower crust and mantle of Wrangellia on Vancouver Island discussed

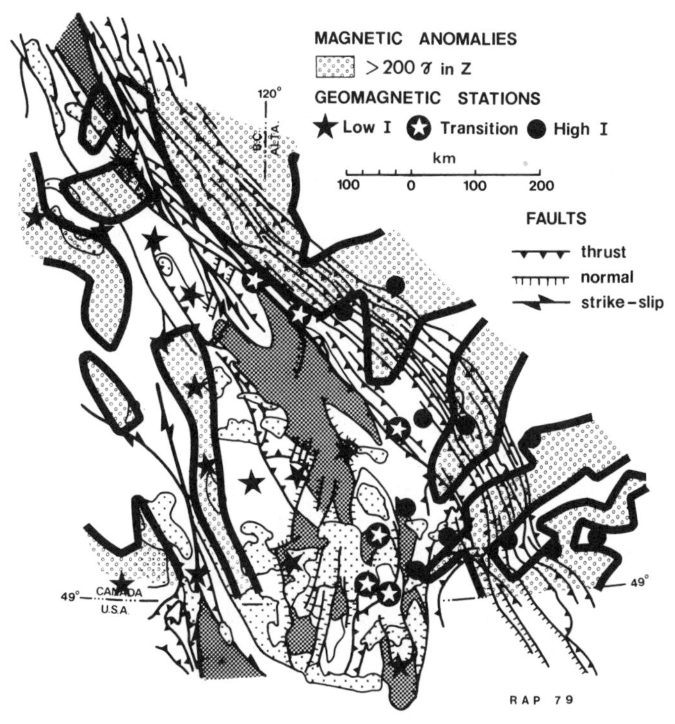

Figure 22. Magnetic anomaly map and geomagnetic depth-sounding stations for region of Transects B-2 and B-3A. Low I indicates high electrical conductance (after Dragert and Clarke, 1977, and Lajoie and Caner, 1970; see also Gough, 1986). Magnetic anomalies after Coles and others (1976) and Kanasewich and others (1969).

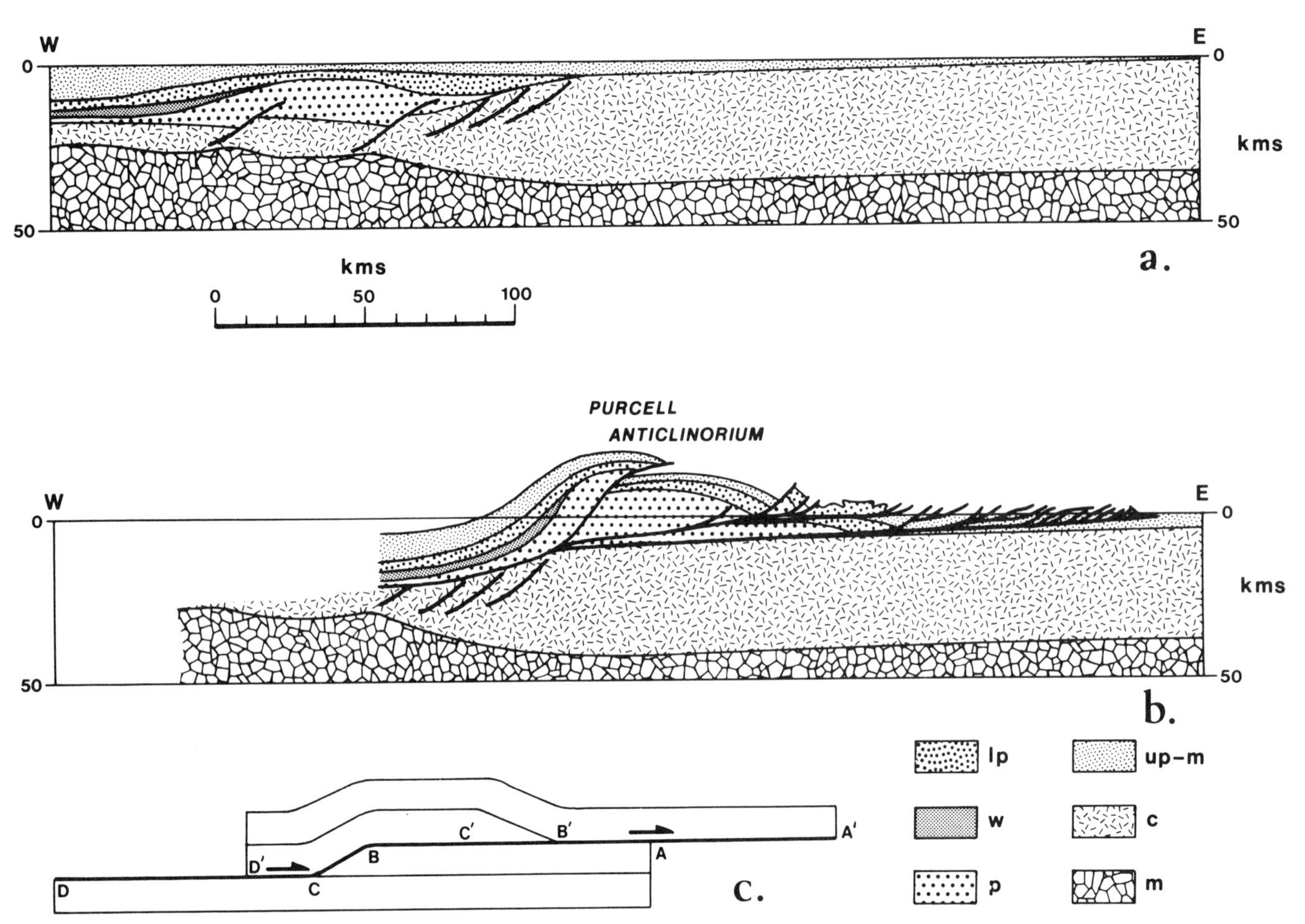

Figure 23. Model for the origin of the Purcell Anticlinorium (after Price, 1981). a. Cordilleran miogeocline (continental margin sedimentary basin) before thrusting. b. Inversion of the continental margin sedimentary basin due to thrust faulting and juxtaposition over the flat surface of the continental craton. c. Schematic representation of the inversion. m, mantle; c, continental crust; p, Belt-Purcell; w, Windermere; lp, lower Paleozoic; up-m upper mantle Paleozoic-Mesozoic.

earlier. The eastern Intermontane Superterrane was accreted to the ancient continental margin in the Jurassic, and the Insular Superterrane to it in the Cretaceous. Both accretions took place within the general framework of subducting Pacific Ocean crust and concomitant upper plate magmatism.

The eastern boundary of the Intermontane Superterrane, herein called the Omineca Suture (Fig. 24), is interpreted in Transect B–2 as the uppermost thrust fault of an enormous stack of nappes. This thrust stack, which constitutes much of the Omineca Belt and the entire Foreland Belt, prograded eastward between Middle Jurassic and Paleocene time to lie on the craton with its thin veneer of Phanerozoic sedimentary rocks.

At the latitudes of Transects B–2, B–3A, the western boundary of the Intermontane Superterrane is delineated by the Coast-Cascade Suture that lies on the east side of the Coast-Cascade Belts. The boundary is interpreted in Transect B–3A to be the uppermost thrust of an imbricated stack that prograded westward in Cretaceous time to form a thrust stack that underlies the North Cascades and probably the southern Coast Belt and the Insular Belt (Fig. 24). At the latitude of Transect B–1, the uppermost thrust lies on the west side of the Coast Belt. Beneath this thrust stack are younger rocks accreted in Tertiary and Recent time.

Below, we briefly describe the accreted terranes, then the evidence for amalgamation of most of them into the two superterranes, the sutures, and finally the geometry of the two accretionary prisms.

Intermontane Superterrane. The Intermontane Superterrane comprises the following four smaller terranes that were amalgamated by the end of Triassic time (Figs. 2C, 5, 24).

Slide Mountain Terrane. The Slide Mountain Terrane appears to consist of the deposits and floor of a marginal oceanic basin. Near the latitude of Transect B–2, Slide Mountain Terrane (Fennell Formation and Kaslo Group) comprises predominantly

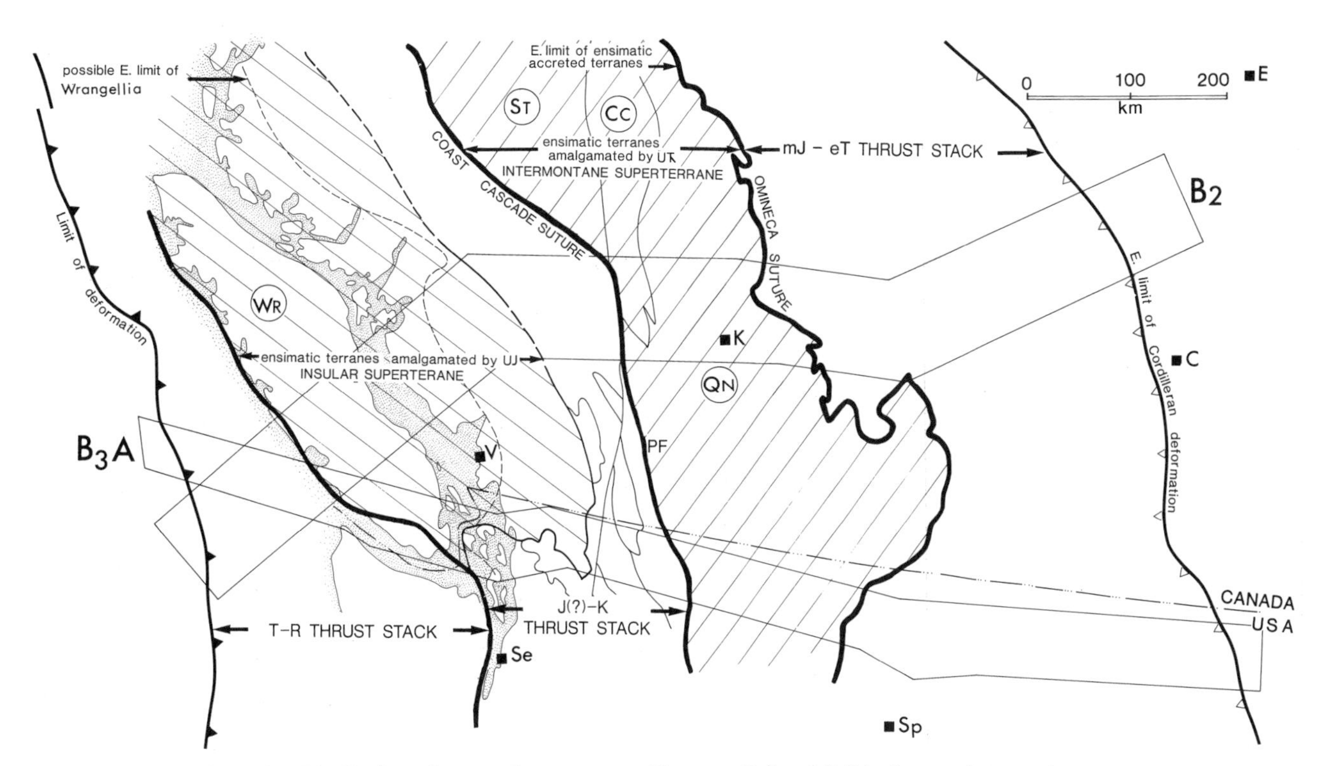

Figure 24. Distribution of accreted terranes near Transects B-2 and B-3A; the area between terranes accreted in Jurassic and those accreted in Cretaceous time comprises several smaller terranes described in text. Names of major terranes: CC, Cache Creek; QN, Quesnellia; ST, Stikinia; WR, Wrangellia. Distribution of small terranes in Cretaceous accretionary prism is shown in Figure 28. C, Calgary; E, Edmonton; K, Kamloops; Se, Seattle; Sp, Spokane; V, Vancouver.

tholeiitic pillow basalt with minor ultramafic rock, gabbro and diorite, and fine grained clastic rock and chert that range in age from mid-Carboniferous to Permo-Triassic. The terrane is locally overlain by Middle (?) and Upper Triassic clastic rocks (Slocan Group, "black phyllite") that grade westward into arc volcanics of Quesnellia.

In places where detailed structural and paleontological studies have been made on this terrane, as in the Cariboo and Cassiar Mountains (lats. 54°N and 59°N; Struik and Orchard, 1985; Harms, 1986), the terrane consists entirely of thin, imbricated thrust sheets. There is a local record of Permian deformation within the terrane. In northern British Columbia, a thrust involving Early Permian rocks of this terrane is cut by a Permian pluton (Harms, 1986). Just north of latitude 49°N, the Kaslo Group of the Slide Mountain Terrane consists of two parts separated by a thrust fault. The age of rocks in the lower plate is Early Permian, and the thrust is cut by Permian diorite (Klepacki, 1983; Klepacki and Wheeler, 1985). In addition, Klepacki and Wheeler (1985) suggest that volcanics in the Milford Group (the upper unit of Kootenay Terrane) are correlative with the Kaslo Group; if so, Slide Mountain and Kootenay Terranes were contiguous in the late Paleozoic.

Quesnellia (MzPz3 of Transect B–3A). Quesnellia comprises lower Mesozoic volcanic and sedimentary rocks and associated granites of arc affinity, together with stratigraphically underlying latest Devonian through Permian volcanic and sedimentary strata, some of which are probably correlative with Slide Mountain Terrane. The cumulative stratigraphic thickness is between 5 and 10 km. This terrane extends into Washington and disappears southward beneath Neogene Columbia River basalts. The great width (300 km) of the terrane between Transects B–2 and B–3A is probably due largely to Eocene extension (discussed below), because farther north (latitude 53°N) it is a narrow sliver (Fig. 5).

Late Paleozoic rocks of Quesnellia include two distinct lithologic assemblages for which the original stratigraphic relationships are unknown. One comprises clastic and volcaniclastic rocks and carbonate (Harper Ranch Group, Anarchist Group, Mount Roberts Formation, Mission Argillite). The other includes argillite, chert, basic volcanics, and minor ultramafic rock (Kobau and Chapperon Groups) and is lithologically similar to and partly coeval with Slide Mountain Terrane. Permian Harper Ranch faunas (and Slide Mountain faunas farther north in British Columbia) are unlike those of coeval cratonic faunas presently at the same latitude but resemble faunas in Stikinia, the eastern Klamaths of California, and northwestern Nevada. Ross and

Ross (1983) locate these faunas (and their hosting terranes) at low latitudes in the southern hemisphere in eastern Panthalassa.

Unconformably overlying both assemblages are Middle and Upper Triassic and earliest Jurassic volcanic rocks (Nicola, Rossland Groups) and their eastern sedimentary facies (Slocan Group). Both volcanics and associated coeval intrusions have a west-to-east, calcalkaline-to-alkaline trend and accordingly have been interpreted as a west-facing volcanic arc (Monger, 1985a; Mortimer, 1986). The low initial strontium ratios of the magmatic rocks ($\leqslant$0.704) suggest their intra-oceanic or mantle origin (Armstrong, 1988). A pre-211 Ma, metamorphic and plutonic complex derived in part from Triassic rocks, exposed along the western margin of part of Quesnellia (Mount Lytton Complex), possibly represents deeper crust of this arc (Monger, 1985a). Unconformably above the earliest Mesozoic rocks are Lower and Middle Jurassic sedimentary strata, including conglomerates that are largely derived from the underlying granitic and volcanic rocks but that locally include, probably high in the section, quartzite clasts for which the only known source is the miogeoclinal succession to the east. Triassic and Jurassic faunas in the Mesozoic strata are purported to have lived in eastern Panthalassa, at lower latitudes than their present positions relative to the continent would suggest (Tipper, 1981; Tozer, 1982).

Cache Creek Terrane. The Cache Creek Terrane is interpreted as an early Mesozoic accretionary prism that lay west of the early Mesozoic Quesnellia Arc. It consists, in order of decreasing abundance, of radiolarian chert and argillite, carbonate, basalt, and alpine-type ultramafic rock, with minor graywacke and volcaniclastic rock. At the latitude of Transect B-2, it comprises a mélange consisting of blocks of chert, carbonate, basalt, volcaniclastics, and serpentinite in a matrix of highly disrupted chert and carbonaceous argillite, together with large (10 by 20 km) bodies of shallow-water carbonate. Fossiliferous clasts in the mélange range in age from Late Pennsylvanian to Late Permian; the matrix ranges from possibly Early, definitely Late, Permian to Late Triassic. Some clasts within the mélange, as well as fault-bounded slivers of volcaniclastic and ignimbritic material within the terrane, resemble lithologies in the western part of the Upper Triassic Nicola Group of Quesnellia and lend credence to the idea that the Cache Creek represents the subduction complex related to the Nicola Arc. Fusulinids and conodonts of mid to early Late Permian age are similar to Permian Tethyan or western Panthalassan forms (Ross and Ross, 1983). Juxtaposition of this fauna against the different but coeval fauna of Quesnellia implies post-Permian, pre-Upper Triassic displacements of great magnitude for at least part of the Cache Creek Terrane.

Stikinia. At the latitude of Transect B-1 (54°N; Fig. 5), Stikinia displays its typical succession of late Paleozoic, Triassic, and Lower to Middle Jurassic arc-related rocks. The lowest unit is Permian limestone with associated clastic and volcanic rocks. It is overlain, in places disconformably and elsewhere with angular unconformity, by Upper Triassic volcanic and sedimentary rocks of the Takla Group. Uppermost are Lower to Middle Jurassic volcanic rocks and sedimentary rocks of the Hazelton Group and comagmatic intrusives. The sequence probably has a cumulative stratigraphic thickness in excess of 10 km. Stikinian Permian, Triassic, and Jurassic faunas are in general similar to those of Quesnellia and indicate derivation from more southerly latitudes.

Just north of Transect B-2, the southernmost exposures shown as Stikinia on Figure 24 comprise Upper Triassic clastic and carbonate rocks and possible Jurassic volcanics. They are separated from the Cache Creek Terrane by the Tertiary Fraser River dextral strike-slip fault (Fig. 28).

Evidence for Amalgamation. The four terranes were associated by the end of Triassic time to form the Intermontane Superterrane. Quesnellia and the Slide Mountain Terrane were together by Middle(?) and Late Triassic time, because overlapping sedimentary rocks are common to both terranes (Monger and others, 1982; Monger, 1984). The Cache Creek Terrane, which east of Transect B-1 at lat. 55°N contains blueschist dated as Late Triassic (Paterson and Harakal, 1974), appears to be the early Mesozoic accretionary complex complementary to the Nicola (Quesnellia) Arc. Parts of the Cache Creek, at least, were inactive by Early Jurassic time, for it is locally cross-cut by 200-Ma plutons (Armstrong, 1988). The primary structural relationship between arc and accretionary complex has been disturbed, for as demonstrated by Travers (1982) and shown on Transect B-2, Cache Creek rocks were thrust eastward over Quesnellia in Jurassic-Cretaceous time. The Stikine and Cache Creek Terranes in northern British Columbia were probably associated by Late Triassic time, and definitely by mid-Jurassic time, when Cache Creek rocks were thrust westward over Stikinia and shed detritus onto it (Monger, 1984).

As noted above, and according to Klepacki and Wheeler (1985), the Slide Mountain Terrane in southern British Columbia is a partial facies equivalent of the Carboniferous Milford Group, which forms the upper part of Kootenay Terrane (Transect B-2; Klepacki and Wheeler, 1985). The relationship between Kootenay Terrane and rocks definitely deposited along the ancient cratonic margin is a possible thrust fault. Clasts probably derived from the miogeocline occur in probably Middle Jurassic conglomerates on Quesnellia.

Analogies have been drawn between the tectonic setting of the western North American/eastern Panthallassan region in early Mesozoic time and the present-day western/southwestern Pacific (Saleeby, 1983; Silver and Smith, 1983; Monger, 1984). A western accretionary complex (Cache Creek Terrane) was flanked by offshore arcs (Quesnellia, Stikinia) built on older arc and oceanic/marginal basin rocks. The arc/accretionary complex was separated from the continental margin by an early Mesozoic sedimentary basin of unknown width that was floored partly by ocean/marginal basin rocks and (probably) partly by attenuated continental crust. The accretion of this material to the continental margin is discussed following description of rocks of the Insular Superterrane.

Insular Superterrane. The Insular Superterrane is separated from the Intermontane Superterrane by a number of small terranes at the latitudes of Transects B-2 and B-3A (Figs. 5, 27) and juxtaposed against it at the latitude of Transect B–1. It includes the Wrangellia and Alexander Terranes at the latitude of Transect B–1, but only Wrangellia is recognized at the latitudes of B–2 and B–3A. These terranes were together by the end of the Jurassic.

Alexander. The Alexander Terrane has a well-preserved stratigraphic record in southeastern Alaska and extends south-southeastward in the subsurface beneath the northwestern part of Queen Charlotte Islands and Hecate Strait and for an unknown distance into the granitic and metamorphic Coast Belt (Fig. 26; Transect B–1; Yorath and Chase, 1981; Woodsworth and Orchard, 1985). In Alaska, the Alexander Terrane preserves a remarkably complete Cambrian—and possibly Precambrian—through Triassic record (Gehrels and Berg, 1984; Gehrels and Saleeby, 1984). Uppermost Precambrian through Silurian volcanic and calc-alkaline plutonic rocks, clastic, and carbonate rocks appear to record a succession of magmatic arcs, separated by deformational and metamorphic episodes in pre-Ordovician and Siluro-Devonian time. Middle Devonian through Triassic sedimentary and subordinate volcanic rocks record deposition in tectonically stable, perhaps rift-related settings. Gehrels and Saleeby (1984; G. E. Gehrels, personal communication) speculate that in the early Paleozoic the Alexander Terrane had affinities with rocks now in eastern Australia, but that by Triassic time it was probably on the eastern side of Panthallassa but south of the paleoequator.

Wrangellia. Wrangellia comprises a sequence of early Paleozoic through Jurassic volcanic and associated plutonic rocks, clastic sedimentary rock, and carbonate, with a remarkably similar stratigraphy on Vancouver Island, in the Queen Charlotte Islands, and in southern Alaska. The oldest rocks (Sicker Group) include arc-related volcanic, plutonic, and sedimentary rocks of Late Silurian–Early Devonian ages, overlain by late Paleozoic, mainly sedimentary rocks (Brandon and others, 1986). Above this is Upper Triassic tholeiitic basalt (Karmutsen Formation), up to 6 km thick, and overlying sediments (Muller, 1977). Uppermost are Early Jurassic calc-alkaline volcanics (Bonanza, Yakoun groups) and associated plutonic rocks, the roots of which probably form the Westcoast Metamorphic/Plutonic Complex (Isachsen and others, 1985), which are the structurally lowest Wrangellian rocks known. Paleomagnetic results from Triassic basalts are in excellent agreement from different parts of Wrangellia and suggest that Wrangellia formerly had less latitudinal spread than it does today and was located at more southerly latitudes in Triassic time, relative to the North American Craton (e.g., Yole and Irving, 1980).

Evidence for amalgamation. In Alaska, rocks of both Alexander Terrane and Wrangellia are overlapped by the Upper Jurassic and Lower Cretaceous (Oxfordian and younger) sedimentary and volcanic Gravina-Nutzotin Belt (Berg and others, 1972). Although the boundary between the Alexander Terrane and Wrangellia is concealed in the Queen Charlotte Islands, Yorath and Chase (1981) cited the presence there of an Early Cretaceous clastic wedge and Jura-Cretaceous intrusions as evidence that the two terranes collided at this time. Farther south, the eastern margin of Wrangellia was intruded and obscured by Jura-Cretaceous granitic rocks of the western Coast Belt, but it probably can be traced in septa for at least 40 km east of the western margin of the Coast Belt (Roddick and Woodsworth, 1977). Near latitude 50°N, a latest Jurassic–Early Cretaceous clastic and volcanic sequence (Gambier and Fire Lake Groups; Peninsula and Brokenback Hill Formations; Fig. 27) lies on probable Wrangellian rocks on the west side of the Coast Belt and on Triassic-Jurassic rocks of the Nooksack Terrane, west of Harrison Lake. The latter can be linked with strata on the west side of the North Cascades.

Omineca Suture and eastern thrust stack. The suture separating North American rocks from the ensimatic terranes accreted to them lies along the Omineca Crystalline Belt (Figs. 2A, C; 24). The Foreland Belt, together with those rocks in the Omineca Belt that lie east of the suture, can be considered as a single complex thrust stack composed of mainly northeast-verging imbricate thrust slices of supracrustal rocks and rare slices of continental crystalline basement rock that were scraped off attenuated and normal continental lithosphere during Middle Jurassic to Paleocene convergence between North America and the accreted terranes.

Two contrasting suites of supracrustal rocks occur in the eastern thrust stack. Each formed in a distinctly different tectonic environment. The thick (10 to 20 km) northeast-tapering miogeoclinal prism of Middle Proterozoic to Middle Jurassic rocks of North American provenance accumulated on and adjacent to the margin of the North American Craton, apparently as a westward-prograded continental shelf and slope wedge. An overlying northeast-tapering wedge of sediments, of Late Jurassic to Paleocene age, older parts of which are imbricated with the accretionary prism, consists of detrital outwash from the growing stack of thrusts.

The Omineca Suture typically is marked by initial emplacement on thrust faults of rocks of the accreted terranes over rocks to the east, followed by backfolding of the thrust-faulted rocks to the west (e.g., Brown and Read, 1983; Schiarizza and Preto, 1984; Brown and others, 1986), by metamorphism up to high grades, and by granitic intrusion (e.g., Archibald and others, 1983, 1984). A hypothesis showing stages in the evolution of the structure of the Omineca and Foreland Belts is illustrated in Fig. 15. As discussed above, the miogeocline is assumed to be underlain by thinned continental crust. Geobarometry suggests that in Middle to Late Jurassic time, the outboard part of the miogeocline was carried to depths of about 25 km beneath the allochthonous terranes (Ghent and others, 1982). During continuing contraction from Middle Jurassic to (?)Early Cretaceous time, the accreted terranes were driven as tectonic wedges between the North American supracrustal rocks and their basement to produce major southwest verging folds and thrust faults (Archibald and others, 1983; Price, 1986), so that in places the suture is an east-dipping fault system with North American rocks in the hanging wall and accreted terranes in the footwall. During Late Jurassic and Early Cretaceous time, the thrust stack prograded eastward almost as far as the hinge zone between the platform and miogeocline, and the earliest of the two clastic wedges accumulated in the eastward migrating foreland basin. During Late Cretaceous and Paleocene time, the accretionary prism prograded over the continental craton, and the foreland basin expanded across the craton.

The time of suturing is not precisely controlled but based on the following considerations was probably in the Middle Jurassic (ca. 180 Ma). Earliest Jurassic (ca. 200 Ma) volcanic rocks (Rossland Group), which are probably the easternmost alkaline part of the early Mesozoic Quesnellia Arc, have low initial strontium isotope ratios (0.703 to 0.704; Beddoe-Stephens and Lambert, 1981), which supports their intraoceanic origin. They are overlain by marine clastic rocks as young as Toarcian (187 to 193 Ma; Tipper, 1984), which are the youngest marine rocks in the region. Granitic rocks (Kuskanax Batholith) emplaced in both Slide Mountain Terrane and Kootenay Terrane rocks yield slightly discordant U-Pb ages of 173 Ma and contain traces of inherited (2 Ga) zircon, implying either the existence of a Precambrian basement at depth at that time or the availability of old, continent-derived, detrital zircon (Parrish and Wheeler, 1983). Subsequent crustal thickening and uplift are shown by the following features. Geochronometry and geobarometry of granitic intrusions in the Kootenay Arc suggest that between 166 and 156 Ma (Middle-Late Jurassic boundary), both Windermere Supergroup and Kootenay Terrane were carried to depths of 20 to 24 km and intruded by granitic rocks whose higher-level equivalents were being emplaced at the same time into the overlying allochthonous terrane (Archibald and others, 1983). Probable Middle Jurassic conglomerates (Ashcroft Formation) on Quesnellia locally contain quartzite clasts, for which the only known source is uplifted miogeoclinal strata (Monger, 1985a). The first clastic rocks shed from the west into the miogeoclinal Triassic-Jurassic Fernie basin of the Foreland Belt are Oxfordian to Kimmeridgian (163 to 152 Ma); these are followed by thick and coarse grained clastics of the earliest foreland basin in latest Jurassic (ca. 150 Ma) time.

Major vertical displacements in the evolution of this part of Transect B–2 are illustrated in Figure 16. During Paleozoic and early Mesozoic accumulation of the continental terrace wedge or miogeocline, the basement surface subsided to depths of between 2 km (eastern cratonic platform) and 14 km (thickest part of miogeocline). During obduction of the accreted terranes and the initial growth of the thrust stack, the oldest strata in the western part of the miogeocline were buried tectonically to depths of 30 km or more, the fossil edge of the craton (the hinge zone) was buried to depths of more than 20 km beneath the accretionary prism, and the cratonic basement surface was buried to a depth of more than 6 km beneath the sedimentary fill of the foreland basin.

Isostatic subsidence of the continental craton under the weight of the displaced supracrustal rocks involved flexure of the lithosphere and the formation of a moat—the foreland basin—in which the synorogenic detrital outwash from the expanding and emerging accretionary prism was trapped (Fig. 25) to form the synorogenic, northeast-tapering clastic wedges that filled the foreland basin (Price, 1973). The synorogenic clastic wedge deposits in the western part of the foreland basin were cannibalized by the prograding accretionary prism as the thrust and fold belt continued to expand and to incorporate more of the supracrustal cover of the continental craton.

The tectonic accretion that formed the two major composite allochthonous terranes and welded them to North America was a process that involved the disappearance (subduction) of large volumes of lithosphere but the preservation of much of the supracrustal cover from that lithosphere. The supracrustal rocks that occur in the Foreland Belt and the eastern part of the Omineca Belt are stacked in east-verging imbricate thrust sheets above a strip of North American lithosphere that is about 300 km narrower near its surface than the strip of lithosphere they covered before the collisional orogeny began (see Fig. 15).

The very dominant easterly vergence of the mid-Jurassic through Paleocene structures in the eastern accretionary prism making up the Foreland Belt and the eastern part of the Omineca Belt suggests that the subduction zone was west-dipping, which is anomalously opposite in vergence to the east-dipping subduction of oceanic crust in the present-day, Tertiary, Cretaceous, and (probably) early Mesozoic arcs. This difference may be due to a major change of North American "absolute" plate movements between 200 and 180 Ma. In Late Triassic and earliest Jurassic time there appears to have been a passive cratonic margin with an offshore arc (Quesnellia) above an east-dipping subduction zone. After this time, there was an active continental margin in which the westernmost, attenuated edge of the North American Plate was initially driven into the back of the old off-shore arc, which triggered development of the east-verging accretionary prism. This prism continued to grow in a back-arc setting until the end of Paleocene time, during continuing plate convergence, as younger arcs were built on the edge of the North American Plate above east-dipping subduction zones, and ceased to grow in the Eocene extensional regime discussed below.

Coast-Cascade suture and accretionary prism. The present western margin of North America, from Mexico to Alaska, is underlain by a vast collage that accreted from Late Jurassic to Recent times, with the largest volume probably added in the Cretaceous (Fig. 26). The collage is dominated in the south by intra-oceanic and continent-derived sediments (Franciscan Complex) and in the north by extensive intra-oceanic accreted terranes. No one mode of accretion can be ascribed to the collage as a whole, and at present it is impossible to say whether it accumulated by predominantly convergent or by predominantly dextral transcurrent movements, although clearly both modes were involved.

With this caveat, and following Transect B–3A, the fundamental structure of the Coast-Cascade Belt is interpreted herein as a stack of west-vergent thrust faults, although it should be noted that other interpretations place considerable emphasis on transcurrent movements during accretion (e.g., Davis and others,

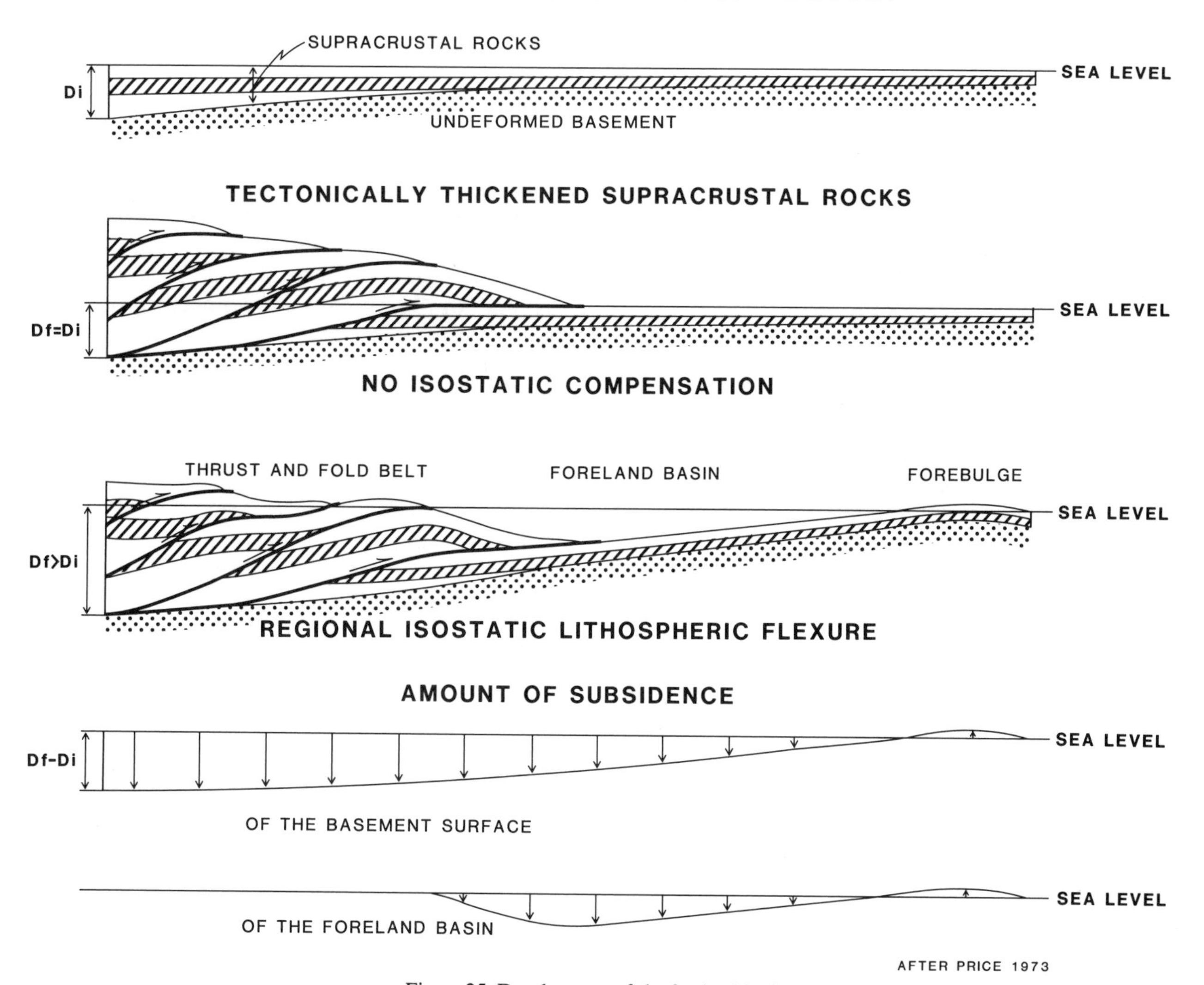

Figure 25. Development of the foreland basin.

1978; Brandon and others, 1988; Brown, 1987). The nature of this structural complex has been most thoroughly investigated in the North Cascade region (lat. 48°N), although no single view of its evolution currently prevails there. Farther north, the primary structural character is obscured by voluminous granitic intrusions and by high-grade metamorphism.

In the North Cascades, at the latitude of the southern limit of Transect B–3A (latitude 48°N) where no Insular Superterrane rocks are present, a zone 150 km wide comprises at least nine small terranes separated by generally east-dipping thrust faults. This zone forms the North Cascades structural province, which lies between and structurally above Tertiary accretionary prisms to the west and the previously accreted Intermontane Superterrane to the east (Fig. 27; Transect 3–A). The primary structural and stratigraphic relationships observed in the Cascades are used herein as a guide to the nature of the suture at and north of the latitude of Transect B–2 (50°N). Near Transect B–2, voluminous, mainly Cretaceous, granitic intrusions of the Coast Belt obscure many primary structural relationships. At this latitude, the Insular Superterrane, which can be thought of as the largest and nearly structurally lowest block in the Cretaceous thrust stack, extends eastward for at least 40 km to underlie the western part of the Coast Belt.

The eastern one-third of the Coast Belt, a zone about 75 km wide, comprises several small terranes that are the compressed northerly continuations of several terranes to the south in the western North Cascades (Fig. 27). Between this and the insular superterrane, is a region with metamorphic rocks of unknown

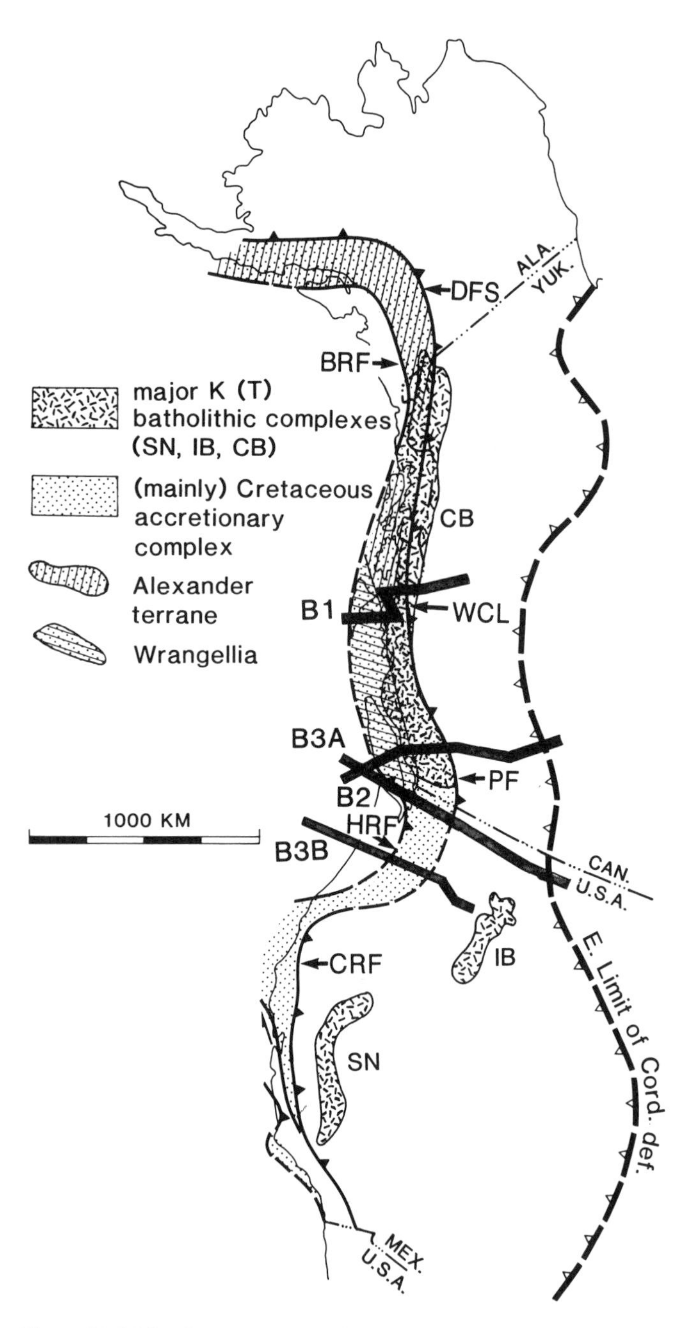

Figure 26. Major Cretaceous tectonic elements along the western margin of North America. The map shows the distribution of the (mainly) Cretaceous collage (interpreted herein as a thrust stack with major terranes within it) in the northern Cordillera, and the crosscutting (mainly) Cretaceous and Tertiary granitic Coast Belt. BRF, Border Ranges Fault; CRF, Coast Range Fault; DFS, Denali Fault System; HRF, Hurricane Ridge Fault; PF, Pasayten Fault; WCL, Work Channel Lineament; CB, Coast Belt; IB, Idaho Batholith; SN, Sierra Nevada.

stratigraphic affinity, preserved as septa within granites, and Triassic to Cretaceous strata of the Nooksack terrane (Fig. 27). In the north, at the latitude (54°N) of Transect B1, the Insular superterrane (shown as Alexander terrane on Transect B1) abuts the Intermontane Superterrane (shown as Stikinia on B1) on the west side of the Coast Belt. The boundary between the western terranes and the Intermontane Superterrane follows the Pasayten Fault north from Washington into the complexly faulted eastern boundary of the Coast Belt near latitude 52°N and then crosses almost to the west side of the Coast Belt near latitude 54°N to lie along the Work Channel Lineament (Fig. 26; Transect B1). Following, we first discuss the North Cascades region, as our hypothesis concerning its evolution appears to be critical to understanding the nature of the whole suture zone, and follow it by discussion of the Coast Belt farther north. The North Cascades include, from east to west and probably from structurally highest to lowest, at least the following nine small terranes (Fig. 27; Monger, 1985b; Brandon and others, 1988): (1) *Methow Terrane,* with a basement of Triassic (MORB) basalt, overlain by Jurassic and Cretaceous clastic and minor volcanic rocks; (2) *Bridge River Terrane,* composed of disrupted Permian to Middle Jurassic chert, argillite, basalt, an ultramafics; (3) *Cadwallader Terrane,* which contains Triassic basalt of probable arc affinity, clastics, and carbonate, overlain by Jurassic and Cretaceous clastics; (4) *Shuksan Terrane,* containing Jurassic and (?)Triassic basalt and hemipelagic clastic rocks that underwent blueschist metamorphism in Jura-Cretaceous (130 to 140 Ma) time (the Settler Schist in Canada has been tentatively correlated with this terrane by Monger, 1985a); (5) *Chilliwack Terrane,* which includes Devonian through early Jurassic clastics and local arc-related volcanic rock and carbonate; (6) *Nooksack Terrane,* which comprises Middle Triassic to Lower Cretaceous clastic and volcanic rocks and may be linked to the Chilliwack Terrane by Early Jurassic time by the recent discovery within it (Arthur, 1986) of Toarcian conglomerate with clasts containing Permian Chilliwack-type fossils; (7) *Decatur Terrane,* which comprises Jurassic and Cretaceous clastics on a Jurassic ophiolite basement; (8) *Constitution Terrane,* with Permian and Jurassic volcanics and cherts overlain by Jura-Cretaceous clastics; and (9) *Turtleback Terrane,* consisting of Paleozoic arc volcanic and plutonic rocks (Monger, 1985b, 1986; Brandon and others, 1988). In addition, small slices of metamorphic rock yielding Paleozoic and Precambrian ages occur along several of the terrane-bounding faults.

As noted by Danner (1977) and Davis and others (1978), lithologically and faunally the Chilliwack Terrane has some similarity with the Harper Ranch Group in Quesnellia, and the Constitution Terrane to the Cache Creek Terrane. The 140 to 130 Ma dates from blueschists in the Shuksan Terrane are similar to those from the eastern belt of the Franciscan Complex in California. Thus this region contains some elements similar to those accreted earlier in the Intermontane Belt to the east, Jura-Cretaceous presumably subduction-related metamorphic rocks, truly exotic material represented by Wrangellia, and partly overlying Jura-Cretaceous overlap assemblages.

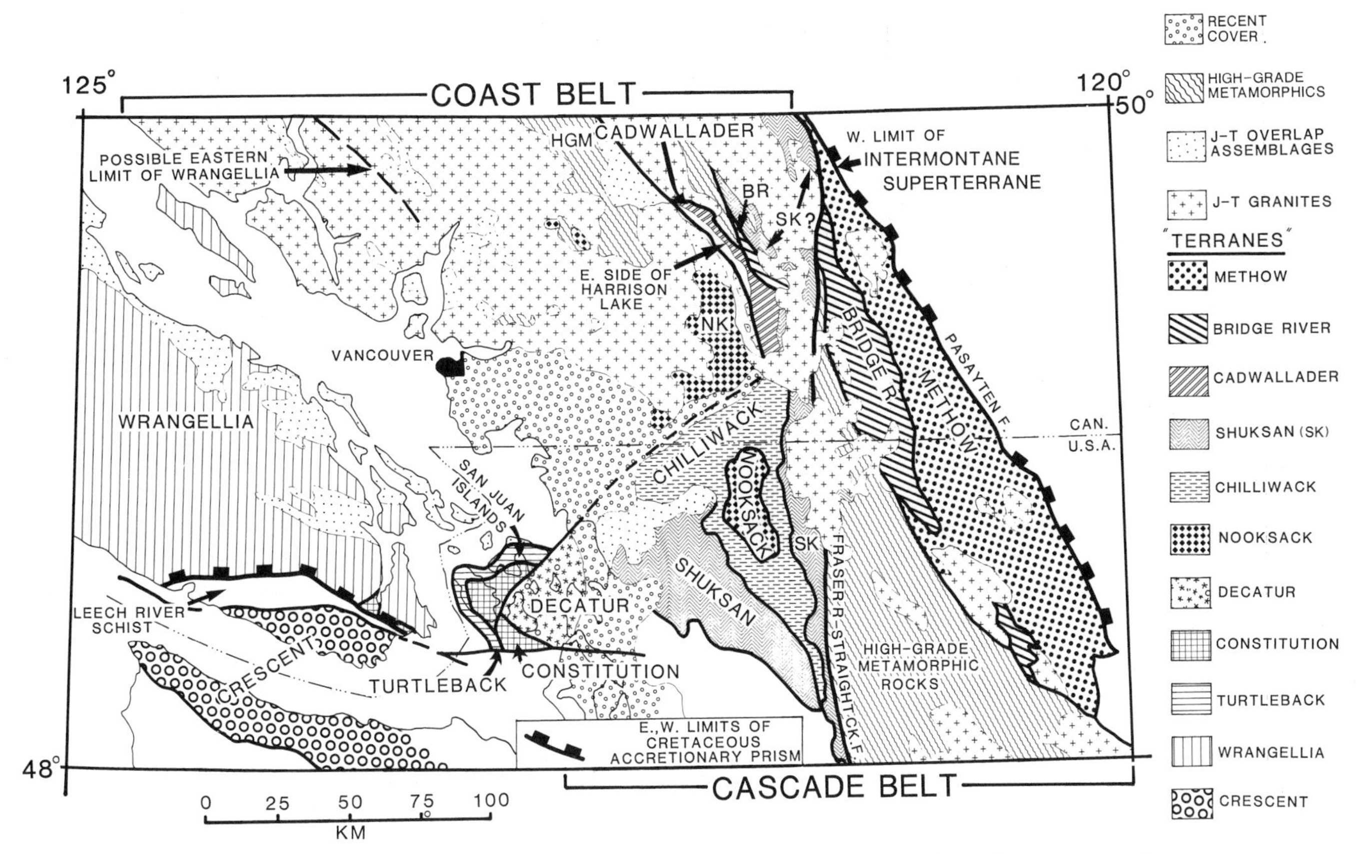

Figure 27. The boundary between the southern Coast Belt and Cascade Belt. Figure shows eastern and western limits of Cretaceous thrust stack, location of small terranes (mainly included in "Cascade" Terrane of Figure 5), and overlapping rock units.

In general, and excepting the Methow Terrane and parts of the Bridge River Terrane, metamorphism of the terranes increases from west to east. The metamorphic core of the Cascades contains gneisses and extensive kyanite and sillimanite schists. The Shuksan Terrane contains regional blueschist and greenschist. The Chilliwack, Decatur, Constitution, and Turtleback and probably Nooksack (M. Brandon, personal communication) Terranes have been affected by regional low-grade, high-pressure lawsonite-albite metamorphism that was apparently related to structural burial and rapid uplift during thrusting.

The deformation and metamorphism in these rocks appear to range in age from early Early Cretaceous in the east to early Late Cretaceous in the west, although small slices of older, high-grade metamorphic rocks occur along faults bounding the terranes. West of the high-grade metamorphic core, the Shuksan Terrane underwent blueschist metamorphism at 130 to 140 Ma (Brown and others, 1982). This terrane can possibly be traced north into Canada and into the high-grade core of the Cascades, east of Harrison Lake (Fig. 27; Monger, 1986), where it is juxtaposed against other terranes (Bridge River, Cadwallader) by thrust faults that are cross-cut by plutonic rocks that yield concordant K-Ar ages as old as 120 Ma (R. Parrish, personal communication), although typically they are slightly younger (110 to 80 Ma). Abundant age dates (e.g., Mattinson, 1972) show that high-grade metamorphism in the Cascade core typically is of late Early Cretaceous–early Late Cretaceous ages (100 to 80 Ma). Stratified rocks and syn-deformational granites in the high-grade core were buried as much as 27 km (Bartholomew, 1977) and presumably rapidly uplifted, as U-Pb ages of 110 Ma (R. L. Armstrong, personal communication) are not very different from K-Ar dates from the same intrusions (120 to 90 Ma). By contrast, in the San Juan Islands in the west (Fig. 27), the main structural and metamorphic event can be dated as early Late Cretaceous (between 100 and 84 Ma; Brandon and others, 1988).

A simple general hypothesis for the evolution of this area is that all oceanic/marginal basins (represented by the mainly Triassic Bridge River Terrane and by the mainly Jurassic Shuksan Terrane) were closed by earliest Cretaceous time. Eastern terranes in this collage were stacked by thrusting, buried to depths of 27 km, metamorphosed, and intruded by mid–Early Cretaceous time. To the west, mainly westward-directed thrusting continued into early Late Cretaceous time, within previously amalgamated terranes.

North of the Cascades, the predominantly Cretaceous granites of the Coast Plutonic Complex obscure much of the record. In Transect B–2, the Insular Superterrane occupies a large part of the region between the Intermontane Superterrane and Tertiary accretionary prisms to the west, with the eastern one-third of the Coast Belt containing terranes that can be tracked north from the Cascades. Near the latitude of Transect B–1 (54°) Crawford and Hollister (1984) and Hollister and Crawford (1986) propose that the crust was doubled in thickness in Cretaceous time by thrusting of the Insular Superterrane (Alexander) beneath Intermontane Superterrane (Stikinia) as shown on Transect B–1, by magmatic intrusion, and by underplating, and was rapidly uplifted in early Tertiary time. The juxtaposition of these two major terranes suggests that, at this latitude, suturing has been so complete that any intervening material has been completely eliminated at the surface.

The eastern margin of the Intermontane Belt and the eastern part of the Coast Belt feature west-dipping structures. Dating of these structures at the latitude of Transect B–2 appears to be post–Middle Jurassic, from the youngest rocks involved, and pre-mid-Cretaceous, from overlapping strata (Travers, 1982; Monger, 1985a). A U-Pb date of 137 Ma from gneissic rocks from this structural zone, which is coeval with dates from Shuksan blueschists to the west, may suggest an episode of back-folding above a predominantly east-dipping subduction zone, analogous but with reversed vergence to Jurassic back-folding (associated with the Jurassic through Paleocene, west-dipping thrust stack) on the east side of the Intermontane Belt.

However, the kinematic picture is not one of simple orthogonal convergence, for there is widespread evidence for late Mesozoic and early Tertiary dextral transcurrent faulting across this segment of the Cordillera. In the Cascade and southeastern Coast Belts, north northwest–trending faults and shear zones with subhorizontal stretching lineations have been used by Brown and Engebretson (1985) as evidence that subduction/accretion took place during oblique convergence between Farallon and North American Plates in the interval 150 to 100 Ma. In Canada, such shear zones locally cut pre–120 Ma thrust faults that juxtapose terranes (Monger, 1986). In addition, the case for transcurrent movements is supported by a notable magmatic lull in this segment of the Cordillera between 70 and 60 Ma (Armstrong, 1988), which may be due to the western Cordillera becoming coupled with the strongly northward-moving Kula Plate (Vance, 1977).

There is evidence for dextral transcurrent displacements farther inboard at about latest Mesozoic–earliest Tertiary time. Price and Carmichael (1986) have shown that the Tintina–Northern Rocky Mountain Trench Fault Zone and the (Eocene) Fraser River–Straight Creek Fault System (Fig. 28) both follow concentric small-circle arcs, as is required for transform faults that are kinematically linked en echelon. However, the southern part of the Tintina–Northern Rocky Mountain Trench Fault Zone, which deviates near latitude 54°N toward the North American Craton from the small-circle locus of the rest of the fault, must be a zone of right-hand transpressive deformation. They suggest that there the transformation of strike-slip into oblique convergence provides a simple explanation for the otherwise enigmatic observation that the Foreland Belt is wider and topographically higher in the south and involves twice as much horizontal shortening there (⩾200 km) as in the Northern Rocky Mountains (⩽100 km).

There is disagreement about the magnitude of the right-hand shear between cratonic North America and the outboard part of the miogeocline and terranes that had been accreted to it in Cretaceous time. On the basis of paleomagnetic data from mid-Cretaceous plutonic rocks in the high-grade metamorphic core of the Cascades in Washington and British Columbia, Beck and Nosun (1972) and Irving and others (1985) conclude that more than 2,000 km of northward, post-mid-Cretaceous displacement is required. By contrast, Price and Carmichael (1986) conclude, on the basis of analysis of structure, that since mid-Cretaceous time the total north and northwest displacement of the accreted allochthonous and suspect terranes has been less than 500 km. Most of this displacement was transformed southward into right-hand oblique convergence that produced large-scale thrusting and folding in the Canadian Rockies in Late Cretaceous and Paleocene time, but part of the displacement was transformed southward during early and middle Eocene time, via a large area of distributed shear and crustal stretching, to the en echelon Fraser River–Straight Creek Fault Zone.

Although a narrow belt of Late Cretaceous (85 to 64 Ma) magmatic rocks runs for much of the length of the Canadian Cordillera (Armstrong, 1988), there is no known corresponding accretionary prism to the west in Canada. Jura-Cretaceous rocks of the Chugach Terrane in southern Alaska perhaps were removed from a more southerly location by transcurrent movements along the continental margin at the end of the Eocene, as suggested by Cowan (1982).

Eocene active margin

Although an active margin persisted through Eocene time and a broad magmatic arc that is subalkaline on the west and alkaline on the east existed across much of the Cordillera (Armstrong, 1988), there was a dramatic change in tectonic style at about 60 Ma. The contractional tectonics of the eastern Cordillera, which had persisted since Middle Jurassic time, were replaced in the interior part of the Cordillera by widespread crustal extension and transtension (Fig. 28; Price, 1979; Ewing, 1980; Tempelman-Kluit and Parkinson, 1986; Parrish and others, 1988). Coney and Harms (1984) speculated that this was a result of collapse and lateral spreading of overthickened crust that had formed from Middle Jurassic to Paleogene time, aided by widespread magmatism that reduced the viscosity of the crust and by a reduction of stress caused by changing plate movements.

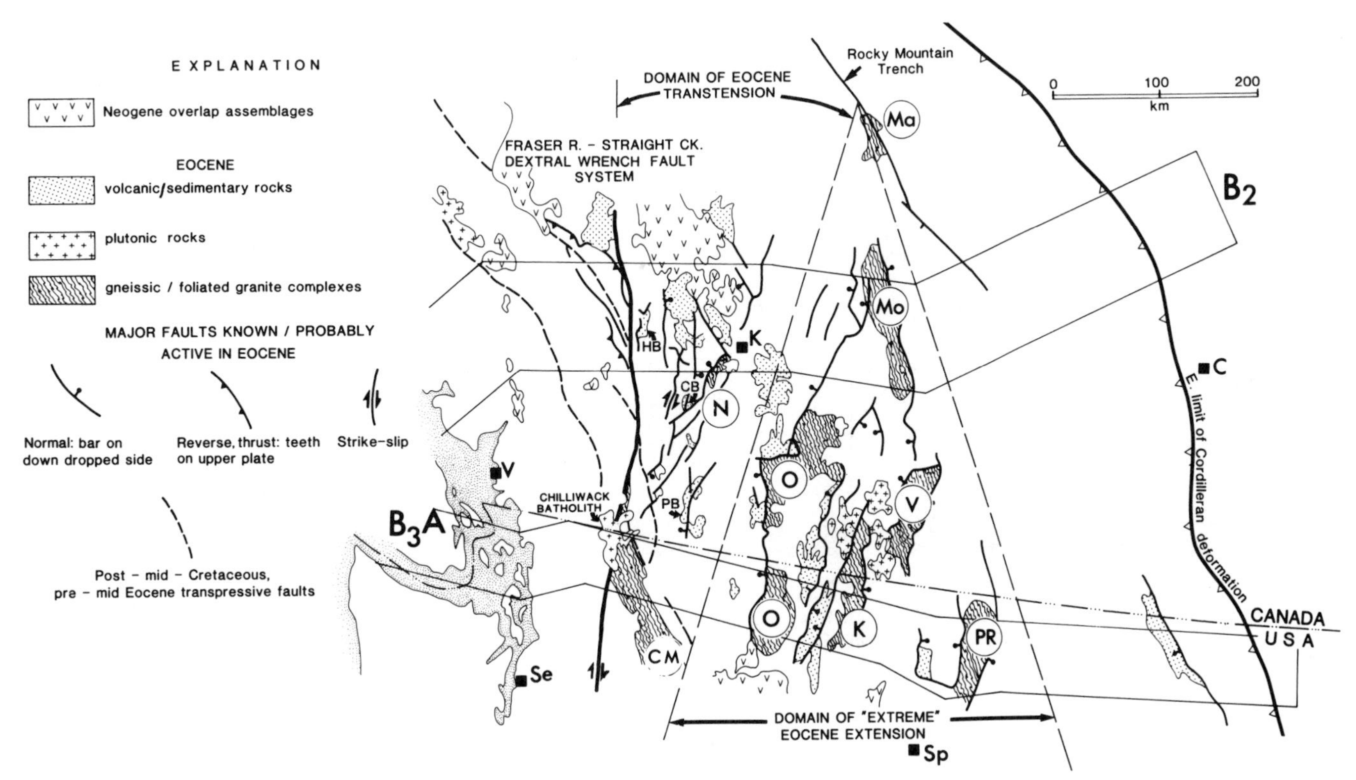

Figure 28. Distribution of major Tertiary structural elements along the corridors of Transects B-2 and B-3A. Circled letters indicate location of the core complexes: CM, Cascade metamorphic core; K, Kettle Dome; Ma, Malton Gneiss; Mo, Monashee Complex; N, Nicola Batholith; V, Valhalla Complex. Basins: CB, Coldwater Basin; HB, Hat Creek Basin; PB, Princeton Basin. C, Calgary; K, Kamloops; Se, Seattle; Sp, Spokane; V, Vancouver.

At the end of this interval, at about 40 Ma, the oceanic Crescent Terrane was accreted in coastal Oregon, in northwestern Washington, and beneath southern Vancouver Island.

Domain of Eocene extension. The style of Eocene structures in the interior part of the Cordillera suggests that there are two structural domains that are probably gradational with one another, an eastern one dominated by extension and a western one dominated by transtension (Fig. 28). The eastern domain is a triangular region whose apex lies near the Malton Gneiss in the Rocky Mountain Trench and whose base extends from near the Jurassic suture near Kootenay Arc on the east to at least the Cretaceous suture (Pasayten Fault) on the west. Within the triangular area are several fault-bounded core-complexes composed of crystalline rocks of amphibolite metamorphic grade that underwent rapid uplift from mid-crustal levels and tectonic denudation in Eocene time (Price, 1979; Harms and Price, 1983), as postulated from regionally consistent patterns of K-Ar ages from hornblende and biotite (Miller and Engels, 1975; Fox and others, 1977; Mathews, 1981, 1983). The core complexes (Armstrong, 1982) comprise a crystalline lower plate containing broad, gently dipping, strongly lineated mylonite zones, cut by brittle detachment zones with low to moderate dips, and flanked by an upper plate of less metamorphosed and unmetamorphosed rocks (Rhodes and Cheney, 1981; Tempelman-Kluit and Parkinson, 1986). Protoliths of rocks in the core complexes include early Proterozoic (Monashee Complex), mid-Paleozoic (Priest River Complex), Mesozoic, and early Cenozoic (Figs. 13; 28; Okanagan, Kettle, Valhalla Complexes) (Cheney, 1980; Read, 1980; Orr, 1985; R. Parrish, personal communication, 1986).

Palinspastic restoration of the Rocky Mountain Thrust Belt (e.g., Fig. 15; Price, 1981; Transect B–2) requires that Mesozoic thrust faults be represented within the core complexes, and such thrusts have been identified in places within the Monashee, Valhalla, Kettle, and Priest River complexes (Brown and others, 1986; Parrish and others, 1985, 1988; Orr, 1985; Rhodes and Hyndman, 1984). Above the core complexes, the less-metamorphosed upper plate rocks include Proterozoic through lower Paleozoic sedimentary sequences with North American affinities (above the Priest River complex), Quesnellia/Slide Mountain rocks (above the Okanagan and Kettle Domes), and thick se-

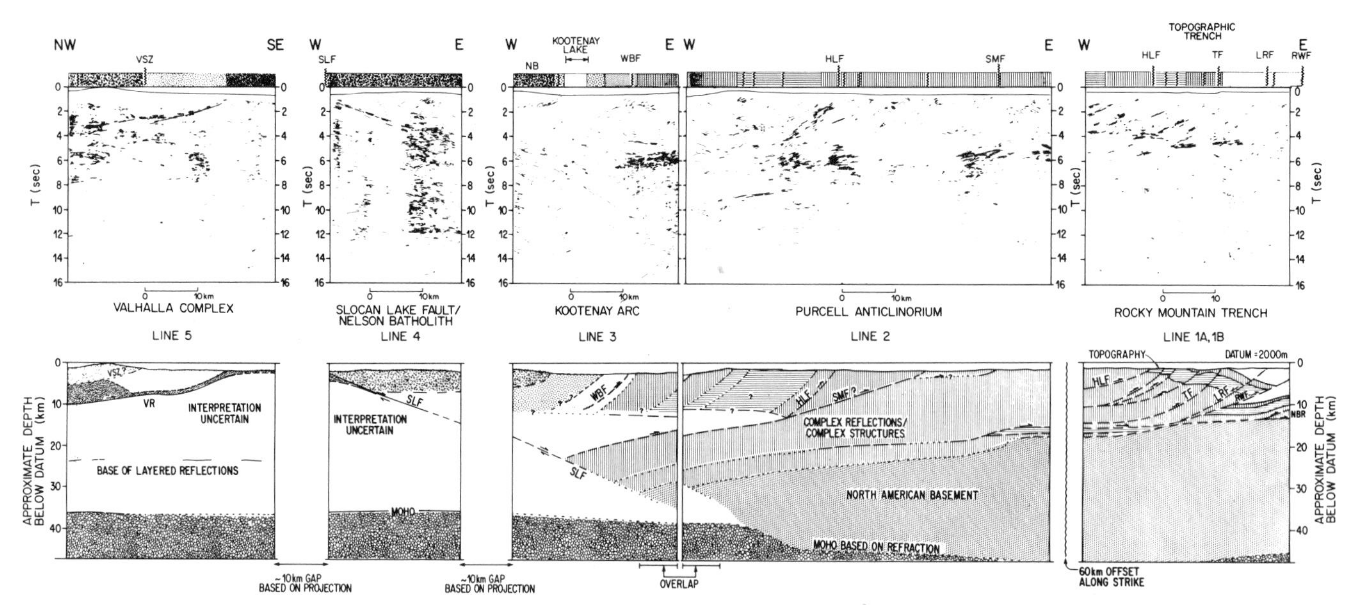

Figure 29. LITHOPROBE seismic reflection line diagrams (above) and interpretation (below) from the eastern Cordillera, showing Tertiary structures crosscutting older fabrics formed during contractional tectonics (from Cook and others, 1987). Location of lines in eastern Cordillera between Transects B-1 and B-2, indicated in Figure 1. HLF, Hall Lake Fault; LRF, Lussier River Fault; NB, Nelson Batholith; NBR, near basement reflector; RWF, Redwall Fault; SLF, Slocan Lake Fault; SMF, St. Mary's Fault; TF, Torrent Fault; VR, Valhalla reflector; VSZ, Valkyr shear zone; WBF, West Bernard Fault.

quences of Eocene sediments and volcanic rocks that were deposited contemporaneously with faulting and that are themselves complexly faulted (Monger, 1968; Church, 1973; Pearson and Obradovich, 1977).

The possible relationships between the Tertiary and older fabrics of the core complexes is shown in the structural interpretation of the Monashee Terrane on Transect B–2 as a west-dipping imbricated stack of thrust slices (a "crustal duplex") of Proterozoic crystalline and sedimentary rocks, with thrust faults that are Jurassic near the top and Paleocene at the base. The stack is truncated and bounded by normal faults. The eastern of these, shown on Transect B–2 as terminating within the crust, may well extend to the base of the crust where its extension would coincide with the step in the Moho beneath the Rocky Mountain Trench (Purcell Fault) as determined from seismic and gravity data (Fig. 21). Seismic reflection studies, carried out by COCORP and LITHOPROBE since the transects were drawn (Figs. 19, 29), show structures with opposing dips in mid to lower crust; east-dipping structures (beneath Kettle Dome, Priest River Complex) cut west-dipping ones. Much crustal fabric for this region consists of gently dipping reflectors, cut by steeper structures that at surface are known Tertiary faults (Figs. 19, 29; Potter and others, 1986; Cook and others, 1987).

Domain of Eocene transtension. Northwest of the region dominated by Eocene extension is an area that features transtension and is linked to the west to the Fraser River–Straight Creek dextral wrench-fault system (Figs. 28, 30; Price, 1979; Ewing, 1980; McMechan, 1983; Monger, 1985a). The one core complex recognized in this area is the Nicola Batholith (Ewing, 1980), and the repetitive Tertiary normal faulting seen to the southeast is not as strongly developed, so that large areas of Eocene rocks are little deformed. There are, however, a few deep basins (Coldwater, Princeton, Hat Creek basins; Fig. 28) with Eocene clastic and volcanic infillings; the bounding faults of these basins may have moved both normally and with a dextral transcurrent sense (McMechan, 1983).

The Fraser River (in British Columbia)–Straight Creek (in Washington) Fault is a dextral strike-slip fault linked to flanking and coeval northeast-trending normal faults and with northwest-trending reverse faults in the distinctive wrench-fault pattern (Monger, 1985a). Recent workers (Kleinspehn, 1985; Mathews and Rouse, 1984; Monger, 1985a; J. A. Vance, personal communication) estimate displacements on the Fraser River–Straight Fault System ranging from 70 to 110 km, based on offsets of older stratigraphic and structural elements across it (Fig. 30). The time of movement is given by the following features directly associated with the fault: (1) mid-Cretaceous (and [?] younger) structures and rock units are offset by the fault; (2) Eocene rocks are downdropped along the fault; (3) two separate northwest-trending reverse and thrust faults splaying off the main fault are post–Middle Eocene (e.g., Hungry Valley F. on Fig. 30; Mathews and Rouse, 1984; Monger, 1985a), and the main fault is cut near latitude 49°N by the Oligocene (35 Ma and younger) Chilliwack Batholith. Transects B–2 and B–3A show the Fraser River–Straight Creek Fault as penetrating the entire crust; there is no evidence that it does this, but equally no evidence that it flattens and detaches at some level within the crust.

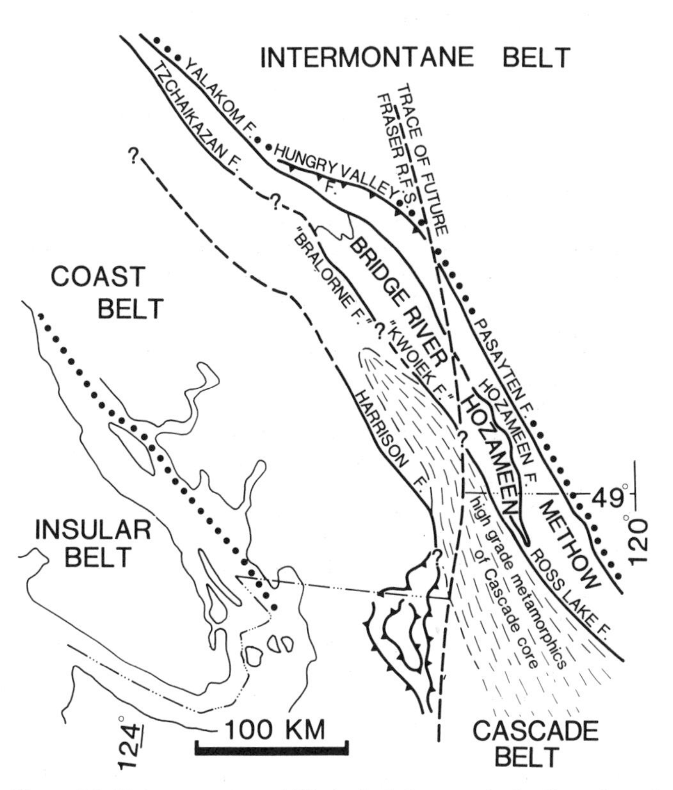

Figure 30. Major tectonic and lithological elements in the Cascade and Eastern Coast Belts, restored to their pre-Eocene positions by removal of 80 to 100 km of right-lateral displacement on the Fraser River–Straight Creek Fault Zone.

CONCLUSIONS

Perhaps the most important conclusion reached here is that the major crust-forming structures of this segment of the Cordillera can be interpreted as enormous east- and west-dipping stacks of imbricated thrust sheets, with thrusts that decrease in age from top to bottom and toward marginal parts of the orogen. From the evidence of the seismic profile across Vancouver Island, even large terranes (e.g., Wrangellia) become detached from their lower crust and mantle and are incorporated as large nappes or "flakes" within the thrust stacks.

There are four east-dipping thrust stacks, all of which appear to involve subducdtion of oceanic crust. The youngest is forming today beneath the continental shelf and slope off Vancouver Island and Washington. East of this and structurally higher, in the Olympic Mountains, southernmost Vancouver Island, and offshore Vancouver Island, is a small late Eocene thrust slice. The structure of the North Cascades and probably much of the Coast and Insular Belts is herein interpreted as a Cretaceous thrust stack (although many of the major faults may record transcurrent movements), within which the extensive Insular Superterrane (Wrangellia and Alexander Terranes) may be considered as a large nappe. A fragmented, Upper Triassic accretionary prism forms the Cache Creek Terrane of the Intermontane Belt. As an approximation,and based simply on the asymmetry of the magnetic stripes in the Pacific Ocean (UNESCO, 1976), the Cretaceous to Recent east-dipping prisms represent crustal shortening of at least 5,000 km since mid-Cretaceous time.

A west-dipping thrust stack underlies the Foreland Belt and much of the Omineca Belt. The thrust stack formed from Middle Jurassic through Paleocene time and apparently involved subduction of (probably) attenuated and normal continental crust in a back-arc position relative to subduction-related magmatic rocks farther west. Crustal shortening in this stack in excess of 300 km, as determined from palinspastic reconstructions, is at least an order of magnitude less than that recorded by the east-dipping prisms.

LITHOPROBE geophysical data and interpretation from the Vancouver Island area imply that the crust of the upper plate there is thickening from beneath, by underplating or subcretion, by a mechanism analogous to that determined from the structural geometry and age relationships of rocks in the Foreland Belt.

The geophysical results from Vancouver Island also show that Wrangellia is less than 20 km thick, which implies that it was a thin slice when emplaced or, more likely, that its lower lithosphere has been removed by subduction erosion.

"Backfolding" near the suture, or upper bounding fault of the thrust stack, took place early in the history of both the west-dipping Jurassic to Tertiary stack and the Cretaceous east-dipping stack.

The Cretaceous and older thrust stacks have been disrupted by partly coeval Cretaceous and Tertiary transcurrent faults that probably formed as the result of the interaction of northward-moving Pacific plates with the western margin of the North American Plate, in a manner analogous to that seen today at the western end of Transect B–1.

Widespread Eocene extensional/transtensional faults in the interior of the Cordillera have disrupted older, flatter-lying "accretionary" structures, with results that appear very prominently on new COCORP and LITHOPROBE lines in those areas.

The crustal pattern shown on Transects B–1, B–2, and B–3 has held up, so far, in the light of the new geophysical information. Some details, such as depths and attitudes of lithological units and structures, are constrained by the new data. Other details, such as the physical and chemical nature of the lower crust and the vertical distribution through the crust of various magmatic rocks that have been added to the crust or remelted from it, remains to be resolved.

REFERENCES CITED

Archibald, D. A., Glover, J. K., Price, R. A., Farrar, E., and Carmichael, D. M., 1983, Geochronology and tectonic implications of magmatism and metamorphism, southern Kootenay Arc and neighbouring regions, southeastern British Columbia; Part 1, Jurassic to mid-Cretaceous: Canadian Journal of Earth Sciences, v. 20, p. 1891–1913.

Archibald, D. A., Krogh, T. E., Armstrong, R. L., and Farrar, E., 1984, Geochronology and tectonic implications of magmatism and metamorphism, southern Kootenay Arc and neighboring regions, southeastern British Columbia; Part 2, Mid-Cretaceous to Eocene: Canadian Journal of Earth Sciences, v. 21, p. 567–584.

Armstrong, R. L., 1982, Cordilleran metamorphic core complexes from Arizona to Southern Canada: Annual Review, Earth and Planetary Sciences, v. 10, p. 129–154.

——, 1988, Mesozoic and Early Cenozoic magmatic evolution of the Canadian Cordillera; Rogers Symposium Volume: Geological Society of America Special Paper 218, p. 55–91.

Armstrong, R. L., Taubeneck, W. H., and Hales, P. O., 1977, Rb-Sr and K-Ar geochronometry of Mesozoic granitic rocks and their Sr isotopic composition, Oregon, Washington, and Idaho: Geological Society of America Bulletin, v. 88, p. 398–411.

Arthur, A., 1986, Stratigraphy along the west side of Harrison Lake, southwestern British Columbia: Current Research, Part B, Geological Survey of Canada Paper 86–1B, report 75, p. 715–720.

Atwater, T., 1970, Implications of plate tectonics for the Cenozoic tectonic evolution of western North America: Geological Society of America Bulletin, v. 81, p. 3513–3536.

Au, D., and Clowes, R. M., 1982, Crustal structure from an OBS survey of the Nootka fault zone off western Canada: Geophysical Journal of the Royal Astronomical Society, v. 68, p. 27–47.

Bally, A. W., Gordy, P. L., and Stewart, G. A., 1966, Structure, seismic data, and orogenic evolution of southern Canadian Rockies: Bulletin of Canadian Petroleum Geology, v. 14, p. 337–381.

Bartholemew, P. R., 1977, Geology and metamorphism of the Yale Creek area, B.C. [M.Sc. thesis]: Vancouver, University of British Columbia, 105 p.

Beck, M. E., Jr., and Nosun, L., 1972, Anomalous paleolatitudes in Cretaceous granitic rocks: Nature, v. 235, p. 11–13.

Beddoe-Stephens, B., and Lambert, R. St.J., 1981, Geochemical, mineralogical, and isotopic data relating to the origin and tectonic setting of the Rossland volcanic rocks, southern British Columbia: Canadian Journal of Earth Sciences, v. 18, p. 858–868.

Bennett, G. T., Clowes, R. M., and Ellis, R. M., 1975, A seismic refraction survey along the southern Rocky Mountain Trench, Canada: Bulletin of the Seismological Society of America, v. 65, p. 37–54.

Benvenuto, G. L., and Price, R. A., 1979, Structural evolution of Hosmer Thrust Sheet, Southeastern British Columbia: Bulletin of Canadian Petroleum Geology, v. 27, p. 360–394.

Berg, H. C., Jones, D. L., and Richter, D. H., 1972, Gravina-Nutzotin Belt, Tectonic significance of an upper Mesozoic sedimentary and volcanic sequence in southern and southeastern Alaska: U.S. Geological Survey Professional Paper 800–D, p. D1–D24.

Berry, M. J., and Forsyth, D. A., 1975, Structure of the Canadian Cordillera from seismic refraction and other data: Canadian Journal of Earth Sciences, v. 12, p. 182–208.

Bevier, M. L., 1983, Regional stratigraphy and age of Chilcotin Group basalts, south-central British Columbia: Canadian Journal of Earth Sciences, v. 20, p. 515–524.

Bingham, D. K., Gough, D. I., and Ingham, M. R., 1985, Conductive structures under the Canadian Rocky Mountains: Canadian Journal of Earth Sciences, v. 22, p. 384–398.

Bond, G. C., and Kominz, M. A., 1984, Construction of tectonic subsidence curves for the early Paleozoic miogeocline, southern Canadian Rocky Mountains; Indications for subsidence mechanisms, age of breakup, and crustal thinning: Geological Society of America Bulletin, v. 95, p. 135–175.

Brandon, M. T., Orchard, M. J., Parrish, R. R., Sutherland Brown, A., and Yorath, C. J., 1986, Fossil ages and isotopic dates from the Paleozoic Sicker Group and associated intrusive rocks, Vancouver Island, British Columbia: Current Research, Part A, Geological Survey of Canada Paper 86–1A, p. 683–696.

Brandon, M. T., Cowan, D. S., and Vance, J. A., 1988, Stratigraphic and structural framework of the Late Cretaceous San Juan Islands and Cascades thrust system: Geological Society of America Special Paper 221, 81 p.

Brown, E. H., 1987, Structural geology and accretionary history of the northwest Cascades system: Geological Society of America Bulletin, v. 99, p. 201–214.

Brown, E. H., and Engebretson, D. C., 1985, Structural history and plate-tectonic interpretation of the Easton-Shuksan blueschist, North Cascades, Washington: Geological Society of America Abstracts with Programs, v. 17, p. 344.

Brown, E. H., Wilson, D. L., Armstrong, R. L., and Harakal, J. E., 1982, Petrologic, structural, and age relations of serpentinite, amphibolite, and blueschist in the Shuksan Suite of the Iron Mountain–Gee Point area, North Cascades, Washington: Geological Society of America Bulletin, v. 93, p. 1087–1098.

Brown, R. L., and Read, P. B., 1983, Shuswap terrane of British Columbia; A Mesozoic "core complex": Geology, v. 11, p. 164–168.

Brown, R. L., Journeay, J. M., Lane, L. S., Murphy, D. C., and Rees, C. J., 1986, Obduction, backfolding, and piggyback thrusting in the metamorphic hinterland of the southeastern Canadian Cordillera: Journal of Structural Geology, v. 8, no. 3/4, p. 255–268.

Cameron, B.E.B., 1980, Biostratigraphy and depositional environment of the Escalante and Hesquiat Formations (Early Tertiary) of Nootka Sound area, Vancouver Island, British Columbia: Geological Survey of Canada Paper 78–9, 28 p.

Chandra, N. M., and Cumming, G. L., 1972, Seismic refraction studies in western Canada: Canadian Journal of Earth Sciences, v. 9, p. 1099–1109.

Cheney, E. S., 1980, Kettle dome and related structures of northeastern Washington: Geological Society of America Memoir 153, p. 463–482.

Church, B. N., 1973, Geology of the White Lake Basin: British Columbia Ministry of Energy, Mines, and Petroleum Resources Bulletin 61, 120 p.

Clowes, R. M., Brandon, M. T., Green, A. G., Yorath, C. J., Sutherland Brown, A., Kanasewich, E. R., and Spencer, C., 1987a, Lithoprobe–southern Vancouver Island; Cenozoic subduction complex imaged by deep seismic reflections: Canadian Journal of Earth Sciences, v. 24, p. 31–51.

Clowes, R. M., Yorath, C. J., and Hyndman, R. D., 1987b, Reflection mapping across the convergent margin of Western Canada: Geophysical Journal of the Royal Astronomical Society, v. 89, p. 79–84.

Coles, R. L., Haines, G. V., and Hannaford, W., 1976, Large scale magnetic anomalies over western Canada and the Arctic; A discussion: Canadian Journal of Earth Sciences, v. 13, p. 790–802.

Coney, P. J., and Harms, T. A., 1984, Cordilleran metamorphic core complexes; Cenozoic extensional relics of Mesozoic compression: Geology, v. 12, p. 550–554.

Cook, F. A., 1985, Deep basement seismic reflection profiling of the Purcell Anticlinorium using a land air gun source: Journal of Geophysical Research, v. 90, no. 1B), p. 651–662.

Cook, F. A., and 9 others, 1987, LITHOPROBE Southern Canadian Cordilleran Transect; Rocky Mountain Thrust Belt to Valhalla Gneiss Complex: Geophysical Journal of the Royal Astronomical Society, v. 89, p. 91–98.

Cowan, D. S., 1982, Geological evidence for post-40 m.y. B.P. large-scale northwestward displacement of part of southeastern Alaska: Geology, v. 10, p. 309–313.

Cowan, D. S., and Potter, C. J., principal compilers, 1986, Juan de Fuca spreading ridge to Montana Thrust belt: Geological Society of America Continent-ocean Transect B3.

Cowan, D. S., Botros, M., and Johnson, H. D., 1986, Bookshelf tectonics; Rotated crustal blocks within the Sovanco Fracture Zone: Geophysical Research Letters, v. 13, no. 10, p. 995–998.

Crawford, M. L., and Hollister, L. S., 1984, Thermal and tectonic processes in an orogenic belt: Geological Society of America Abstracts with Programs, v. 16, p. 478.

Crossan, R. S., 1983, Review of seismicity in the Puget Sound region from 1970 through 1978: U.S. Geological Survey Open-File Report 83–19, p. 6–18.

Danner, W. R., 1977, Paleozoic rocks of northwestern Washington and adjacent parts of British Columbia, *in* Stewart, J. H., Stevens, C. H., and Fritsche, A. E., eds., Paleozoic paleogeography of the western United States: Society of Economic Paleontologists and Mineralogists, Pacific Section, p. 481–502.

Davis, D., Suppe, J., and Dahlen, F. A., 1983, Mechanics of fold and thrust belts and accretionary wedges: Journal of Geophysical Research, v. 88, p. 1153–1172.

Davis, E., Currie, R., Sawyer, B., Riddihough, R., and Holmes, M., 1985, Juan de Fuca Ridge atlas; Regional Seamarc II acoustic image mosaics and Seabeam bathymetry: Geological Survey of Canada Open File 1144, Scale 1:250,000.

Davis, G. A., Monger, J.W.H., and Burchfiel, B. C., 1978, Mesozoic construction of the Cordilleran "collage," central British Columbia to central California, *in* Howell, D. G., and McDougall, K. A., eds., Mesozoic paleogeography of the western United States: Society of Economic Paleontologists and Mineralogists Pacific Coast Paleogeography Symposium, p. 1–32.

Dragert, H., and Clarke, G.K.C., 1977, A detailed investigation of the Canadian Cordillera transition anomaly: Journal of Geophysics, v. 42, p. 373–390.

Engebretson, D. C., Cox, A., and Gordon, R. G., 1985, Relative motions between oceanic and continental plates in the Pacific Basin: Geological Society of America Special Paper 206, 59 p.

Ewing, T. E., 1980, Paleogene tectonic evolution of the Pacific Northwest: Journal of Geology, v. 88, p. 619–638.

Fairchild, L. H., and Cowan, D. S., 1982, Structure, petrology, and tectonic significance of the Leech River Complex northwest of Victoria, Vancouver Island: Canadian Journal of Earth Sciences, v. 19, p. 1817–1835.

Fox, K. F., Jr., Rinehart, C. D., and Engels, J. C., 1977, Plutonism and orogeny in north-central Washington: U.S. Geological Survey Professional Paper 989, 27 p.

Frost, C. D., and Burwash, R. A., 1986, Nd evidence for extensive Archean basement in western Churchill Province, Canada: Canadian Journal of Earth Science, v. 23, p. 1433–1437.

Gehrels, G. E., and Berg, H. C., 1984, Geologic map of southeastern Alaska: U.S. Geological Survey Open-File Report 84-0886, p. 28, scale 1:600,000.

Gehrels, G. E., and Saleeby, J. B., 1984, Paleozoic geologic history of the Alexander Terrane in S.E. Alaska, and comparisons with other orogenic belts: Geological Society of America Abstracts with Programs, v. 16, p. 516.

Ghent, E. D., Knitter, C. C., Raeside, R. P., and Stout, M. Z., 1982, Geothermometry and geobarometry of pelitic rocks, upper kyanite and sillimanite zones, Mica Creek, British Columbia: Canadian Mineralogy, v. 20, p. 295–305.

Gordon, R. G., and Jurdy, D. M., 1986, Cenozoic Global Plate Motions: Journal of Geophysical Research, v. 91, no. B12, p. 12389–12406.

Gough, D. I., 1986, Mantle upflow tectonics in the Canadian Cordillera: Journal of Geophysical Research, v. 91, p. 1909–1919.

Green, A. G., Berry, M. J., Spencer, C. P., Kanasewich, E. R., Chiu, S., Clowes, R. M., Yorath, C. J., Stewart, D. B., Unger, J. D., and Poole, W. H.,1986, Recent seismic reflection studies in Canada, *in* Barazangi, M., and Brown, L., eds., Deep structure of the continental crust; Results from reflection seismology; American Geophysical Union Geodynamics Series, v. 13, p. 85–97.

Harms, T. A., 1986, Structural and tectonic analysis of the Sylvester Allochthon, northern British Columbia; Implications for paleogeography and accretion [Ph.D. thesis]: Tucson, University of Arizona, 80 p.

Harms, T. A., and Price, R. A., 1983, The Newport Fault; Eocene crustal stretching, rocking, and listric normal faulting in NE Washington and Idaho: Geological Society of America Abstracts with Programs, v. 15, p. 309.

Harrison, J. E., 1972, Precambrian belt basin of the northwestern United States; Its geometry, sedimentation, and copper occurrences; Geological Society of America Bulletin, v. 83, p. 1215–1240.

Harrison, J. E., Cressman, E. R., and Kleinkopf, M. D., 1985, Regional structure, the Atlantic-Richfield-Marathon Oil No. 1 Gibbs borehole, and hydrocarbon resource potential west of the Rocky Mountain trench in northwestern Montana: U.S. Geological Survey Open-File Report 85-249, 9 p.

Hill, D. P., 1972, Crust and upper mantle structure of the Columbia Plateau from long range seismic refraction experiments: Geological Society of America Bulletin, v. 83, p. 1639–1648.

Hollister, L. S., and Crawford, M. L., 1986, Melt-enhanced deformation; A major tectonic process: Geology, v. 14, p. 558–561.

Irving, E., Woodsworth, G. J., Wynne, P. J., and Morrison, A., 1985, Paleomagnetic evidence for displacement from the south of the Coast Plutonic Complex, British Columbia: Canadian Journal of Earth Sciences, v. 22, p. 584–598.

Isachsen, C., Armstrong, R. L., and Parrish, R. R., 1985, U-Pb, Rb-Sr, and K-Ar geochronometry of Vancouver Island igneous rocks: Geological Association of Canada, Victoria Section, Symposium on deep structure of Vancouver Island, Program and Abstracts, p. 21–22.

Jung, A., 1986, Geochronometry and geochemistry of the Thuya, Takomkane, Raft, and Baldy batholiths, west of Shuswap Metamorphic Complex, south-central British Columbia [B.Sc. thesis]: Vancouver, University of British Columbia, 140 p.

Kanasewich, E. R., Clowes, R. M., and McCloughan, C. H., 1969, A buried Precambrian rift in western Canada: Tectonophysics, v. 8, p. 513–527.

Kleinspehn, K. L., 1985, Cretaceous sedimentation and tectonics, Tyaughton-Methow Basin, southwestern British Columbia: Canadian Journal of Earth Science, v. 22, p. 154–174.

Klepacki, D. W., 1983, Stratigraphic and structural relations of the Milford, Kaslo, and Slocan groups, Roseberry Quadrangle, Lardeau map area, British Columbia: Geological Survey of Canada Paper 83-1A, p. 229–233.

Klepacki, D. W., and Wheeler, J. O., 1985, Stratigraphic and structural relations of the Milford, Kaslo, and Slocan Groups, Goat Range, Lardeau and Nelson map areas, British Columbia: Geological Survey of Canada Paper 85-1A, p. 277–286.

Kurtz, R. D., DeLaurier, J. M., and Gupta, J. C., 1986, A magnetotelluric sounding across Vancouver Island detects the subducting Juan de Fuca Plate: Nature, v. 321, p. 596–599.

Lajoie, J. J., and Caner, B., 1970, Geomagnetic induction anomaly near Kootenay Lake; A strike-slip feature in the lower crust?: Canadian Journal of Earth Sciences, v. 7, p. 1568–1579.

Lewis, T. J., Bentkowski, W. H., Davis, E. E., Hyndman, R. D., Souther, J. G., and Wright, J. A., 1988, Subduction of the Juan de Fuca Plate; Thermal consequences: Journal of Geophysical Research, v. 93, p. 15, 207–15, 225.

Lis, M. G., and Price, R. A., 1976, Large-scale block faulting during deposition of the Windermere Supergroup (Hadrynian) in south-eastern British Columbia: Geological Survey of Canada Paper 76-1A, p. 135–136.

McLeod, N. S., Tiffin, D. L., Snavely, P. D., Jr., and Currie, R. G., 1977, Geologic interpretation of magnetic and gravity anomalies in the Strait of Juan de Fuca, U.S.-Canada: Canadian Journal of Earth Sciences, v. 14, p. 223–238.

McMechan, M. E., 1981, The middle Proterozoic Purcell Supergroup in the southwestern Rocky and southeastern Purcell Mountains, British Columbia, and the initiation of the Cordilleran miogeocline, southern Canada and adjacent United States; Bulletin of Canadian Petroleum Geology, v. 29, p. 583–621.

McMechan, M. E., and Price, R. A., 1982, Superimposed low-grade metamorphism in the Mount Fisher area, southeastern British Columbia; Implication for the East Kootenay orogeny: Canadian Journal of Earth Sciences, v. 19, p. 476–489.

McMechan, R. D., 1983, Geology of the Princeton Basin: British Columbia Ministry of Energy, Mines, and Petroleum Resources Paper 1983-3, 52 p.

Mathews, W. H., 1981, Early Cenozoic resetting of potassium-argon dates and geothermal history of north Okanagan area, British Columbia: Canadian Journal of Earth Sciences, v. 18, p. 1310–1319.

——, 1983, Early Tertiary resetting of potassium-argon dates in the Kootenay Arc, southeastern British Columbia: Canadian Journal of Earth Sciences,

v. 20, p. 867–872.

Mathews, W. H., and Rouse, G. E., 1984, The Gang Ranch–Big Bar area, south-central British Columbia; Stratigraphy, geochronology, and palynology of the Tertiary beds and their relationships to the Fraser Fault: Canadian Journal of Earth Sciences, v. 21, p. 1132–1144.

Mattinson, J. M., 1972, Ages of zircons from the Northern Cascade Mountains, Washington: Geological Society of America Bulletin, v. 83, p. 3769–3784.

Mereu, R. F., Majumdar, S. C., and White, R. E., 1977, The structure of the crust and upper mantle under the highest ranges of the Canadian Rockies from a seismic refraction survey: Canadian Journal of Earth Sciences, v. 14, p. 196–208.

Miller, F. K., and Engels, J. C., 1975, Distribution and trends of discordant ages of the plutonic rocks of northeastern Washington and northern Idaho: Geological Society of America Bulletin, v. 86, p. 517–528.

Monger, J.W.H., 1968, Early Tertiary stratified rocks, Greenwood map area (82E/2), British Columbia: Geological Survey of Canada Paper 67-42, 39 p.

——, 1984, Cordilleran tectonics; A Canadian perspective: France Geological Society Bulletin, v. 26, p. 255–278.

——, 1985a, Structural evolution of the southwestern Intermontane Belt, Ashcroft and Hope map areas, British Columbia: Current Research, Geological Survey of Canada Paper 85-1A, p. 349–358.

——, 1985b, Terranes in the southeastern Coast Plutonic Complex and Cascade Fold Belt: Geological Society of America Abstracts with Programs, v. 17, p. 371.

——, 1986, Geology between Harrison Lake and Fraser River, Hope map area, southwestern British Columbia: Current Research, Part B, Geological Survey of Canada Paper 86-1B, p. 699–706.

Monger, J.W.H., Price, R. A., and Tempelman-Kluit, D. J., 1982, Tectonic accretion and the origin of the two major metamorphic and plutonic welts in the Canadian Cordillera: Geology, v. 10, p. 70–75.

Monger, J.W.H., Clowes, R. M., Price, R. A., Simony, P. S., Riddihough, R. P., and Woodsworth, G. J., Juan de Fuca plate to Alberta Plains: Geological Society of America Continent-ocean transect B2.

Mortimer, N., 1986, Early Mesozoic, subduction-related, high and low K magmatism of western Quesnellia, British Columbia: Geological Society of America Abstracts with Programs, v. 18, p. 161.

Muller, J. E., 1977, Evolution of the Pacific Margin, Vancouver Island, and adjacent regions: Canadian Journal of Earth Sciences, v. 14, p. 2062–2085.

Orr, K. E., 1985, Structural features along the margins of Okanogan and Kettle Domes, northeastern Washington and southern British Columbia [Ph.D. thesis]: Seattle, University of Washington, 109 p.

Parrish, R. R., and Wheeler, J. O., 1983, A U-Pb zircon age from the Kuskanax batholith, southeastern British Columbia: Canadian Journal of Earth Sciences, v. 20, p. 1751–1756.

Parrish, R. R., Carr, S. D., and Brown, R. L., 1985, Valhalla gneiss complex, southeast British Columbia; 1984 Fieldwork: Current Research, Part A, Geological Survey of Canada Paper 85-1A, p. 81–87.

Parrish, R. R., Carr, S. D., and Parkinson, D. L., 1988, Eocene extensional tectonics and geochronometry of the southern Omineca Belt, British Columbia and Washington: Tectonics, v. 7, p. 181–212.

Paterson, I. A., and Harakal, J. E., 1974, Potassium-argon dating of blueschists from Pinchi Lake, central British Columbia: Canadian Journal of Earth Sciences, v. 11, p. 1007–1011.

Pearson, R. C., and Obradovich, J. D., 1977, Eocene rocks in northeast Washington—radiometric ages and correlation: U.S. Geological Survey Bulletin 1433, 41 p.

Potter, C. J., Sanford, W. E., Yoos, T. R., Prussen, E. I., Keach, R. W. II, Oliver, J. E., Kaufman, S., and Brown, L. D., 1986, COCORP deep seismic reflection traverse of the interior of the North American Cordillera, Washington and Idaho; Implications for orogenic evolution: Tectonics, v. 5, p. 1007–1025.

Price, R. A., 1973, Large-scale gravitational flow of supracrustal rocks, southern Canadian Rockies, *in* De Jong, K. A., and Scholten, R., eds., Gravity of Tectonics: New York, Wiley, p. 491–502.

——, 1979, Intracontinental ductile crustal spreading linking the Fraser River and northern Rocky Mountain Trench transform fault zones, south-central British Columbia and northeast Washington: Geological Society of America Abstracts with Programs, v. 11, p. 499.

——, 1981, The Cordilleran foreland thrust and fold belt in the southern Canadian Rocky Mountains, *in* Price, N. J., and McClay, K., eds., Thrust and Nappe Tectonics: Geological Society of London Special Publications 9, p. 427–448.

——, 1986, The southeastern Canadian Cordillera; Thrust faulting, tectonic wedging, and delamination of the lithosphere: Journal of Structural Geology, v. 8, nos. 3/4, p. 239–254.

Price, R. A., and Carmichael, D. M., 1986, Geometric test for Late Cretaceous–Paleoogene intracontinental transform faulting in the Canadian Cordillera: Geology, v. 14, p. 468–471.

Price, R. A., Monger, J.W.H., and Roddick, J. A., 1985, Cordilleran cross-section; Calgary to Vancouver, *in* Tempelman-Kluit, D., ed., Field guides to geology and mineral deposits in the Southern Canadian Cordillera, Geological Society of America Cordilleran Section Annual Meeting, Vancouver, B.C., May 1985, p. 3-1–3-85.

Read, P. B., 1980, Stratigraphy and structure; Thor-Odin to Frenchman Cap "domes," Vernon east-half map area, southern British Columbia: Current Research, Part A, Geological Survey of Canada Paper 80-1A, p. 19–25.

Rhodes, B. P., and Cheney, E. C., 1981, Low-angle faulting and origin of the Kettle Dome; A metamorphic core complex in northeastern Washington: Geology, v. 9, p. 366–369.

Rhodes, B. P., and Hyndman, D. W., 1984, Kinematics of mylonites in the Priest River "metamorphic core complex," northern Idaho and northeastern Washington: Canadian Journal of Earth Sciences, v. 21, p. 1161–1170.

Riddihough, R. P., and Hyndman, R. D., 1992, The modern plate tectonic regime of the continental margin of western Canada, *in* Gabrielse, H., and Yorath, C. J., eds., Geology of the Cordilleran orogen in Canada: Ottawa, Geological Survey of Canada, Geology of Canada series, v. 4 (Geological Society of America, The Geology of North America, v. G-2), p. 435–455.

Roddick, J. A., and Woodsworth, G. J., 1977, Coast Mountains Project: Geological Survey of Canada Paper 77-1A, p. 271–272.

Rogers, G. C., 1983, Some comments on the seismicity of the northern Puget Sound–southern Vancouver Island region: U.S. Geological Survey Open-File Report 83-19, p. 19–39.

Ross, C. A., and Ross, J.R.P., 1983, Late Paleozoic accreted terranes of western North America, *in* Stevens, C. H., ed., Pre-Jurassic rocks in western North American suspect terranes: Society of Economic Geologists and Paleontologists, Pacific Section, p. 7–22.

Rusmore, M. E., and Cowan, D. S., 1985, Jurassic-Cretaceous rock units along the southern edge of Wrangellia terrane on Vancouver Island: Canadian Journal of Earth Sciences, v. 41, p. 1341–1354.

Saleeby, J. B., 1983, Accretionary tectonics of the North American Cordillera: Annual Review of Earth and Planetary Sciences, v. 15, p. 45–73.

Schiarizza, P., and Preto, V. A., 1984, Geology of the Adams Plateau–Clearwater Area: British Columbia Ministry of Energy, Mines, and Petroleum Resources, Preliminary Map 56, scale 1:100,000.

Sears, J. W., and Price, R.A., 1978, The Siberian Connection; A case for Precambrian separation of the North American and Siberian cratons: Geology, v. 6, p. 267–270.

Shouldice, D. H., 1971, Geology of the western Canadian continental shelf: Bulletin of Canadian Petroleum Geology, v. 19, p. 405–436.

Silver, E. A., and Smith, R. B., 1983, Comparison of terrane accretion in modern Southeast Asia and Mesozoic North American Cordillera: Geology, v. 11, p. 198–202.

Simony, P. S., Ghent, E. D., Craw, D., Mitchell, W., and Robbins, D. B., 1980, Structural and metamorphic evolution of the northeast flank of Shuswap Complex in Southern Canoe River Area, British Columbia; Cordilleran core complexes: Geological Society of America Memoir 153, p. 445–462.

Souther, J. G., 1992, Volcanic regimes, *in* Gabrielse, H., and Yorath, C. J., eds., Geology of the Cordilleran orogen in Canada: Geological Survey of Canada,

Geology of Canada, no. 4 (also Geological Society of America, The Geology of North America, v. G-2), p. 459–492.

Spence, G. D., Ellis, R. M., and Clowes, R. M., 1977, Depth limits on the M-discontinuity in the southern Rocky Mountain Trench: Bulletin of the Seismological Society of America, v. 67, p. 543–546.

Spence, G. D., Clowes, R. M., and Ellis, R. M., 1985, Seismic structure across the active subduction zone of western Canada: Journal of Geophysical Research, v. 90, p. 6754–6772.

Stacey, R. A., Boyd, J. B., Stephens, L. C., and Burke, W.G.F., 1973, Gravity measurements in British Columbia: Earth Physics Branch, Department of Energy, Mines and Resources, Gravity Map Series 152-153, scale 1:1,000,000.

Struik, L. C., and Orchard, M. J., 1985, Late Paleozoic conodonts from ribbon chert delineate imbricate thrusts within the Antler Formation of the Slide Mountain terrane, central British Columbia: Geology, v. 13, p. 794–798.

Sutherland Brown, A., and Yorath, C. J., 1985, LITHOPROBE profile across southern Vancouver Island; Geology and tectonics: Geological Society of America Cordilleran Section Annual Meeting Field Guidebook 8, p. 8-1–8-23.

Tempelman-Kluit, D. J., and Parkinson, D., 1986, Extension across the Eocene Okanogan crustal shear in southern British Columbia: Geology,v. 14, p. 318–321.

Tipper, H. W., 1981, Offset of an Upper Pliensbachian geographic zonation in the North American Cordillera by transcurrent movement: Canadian Journal of Earth Sciences, v. 18, p. 1788–1792.

—— , 1984, The age of the Jurassic Rossland Group of southeastern British Columbia: Geological Survey of Canada Paper 84-1A, p. 631–632.

Tozer, E. T., 1982, Marine Triassic faunas of North America: Geologische Rundschau, v. 71, p. 1077–1104.

Travers, W. B., 1982, Possible large-scale overthrusting near Ashcroft, British Columbia; Implications for petroleum prospecting: Canadian Petroleum Geology Bulletin, v. 30, no. 1, p. 1–8.

United Nations Educational, Scientific, and Cultural Organization (UNESCO), 1976, Choubert, G., and Faure-Muret, A., general coordinators, Geological world atlas, Pacific Ocean: UNESCO, sheet 20, scale 1:10,000,000.

Vance, J. A., 1977, The stratigraphy and structure of Orcas Island, San Juan Islands, *in* Brown, E. H., and Ellis, R. C., eds., Geological excursions in the Pacific Northwest: Geological Society of America Annual Meeting Field-guide, Western Washington University, Bellingham, p. 170–203.

Waldron, D. A., 1982, Structural characteristics of a subducting oceanic plate off Western Canada [M.Sc. thesis]: Vancouver, University of British Columbia, 121 p.

Woodsworth, G. J., and Orchard, M. J., 1985, Upper Paleozoic to lower Mesozoic strata and their conodonts, western Coast Plutonic Complex, British Columbia: Canadian Journal of Earth Sciences, v. 22, p. 1329–1344.

Yole, R. W., and Irving, E., 1980, Displacement of Vancouver Island; Paleomagnetic evidence from the Karmutsen Formation: Canadian Journal of Earth Sciences, v. 17, p. 1210–1228.

Yorath, C. J., and Chase, R. L., 1981, Tectonic history of the Queen Charlotte Islands and adjacent areas; A model: Canadian Journal of Earth Sciences, v. 18, no. 11, p. 1717–1739.

Yorath, C. J., and Hyndman, R. D., 1983, Subsidence and thermal history of Queen Charlotte Basin: Canadian Journal of Earth Sciences, v. 20, p. 135–159.

Yorath, C. J., Clowes, R. M., Green, A. G., Sutherland Brown, A., Brandon, M. T., Massey, N.W.D., Spencer, C., Kanasewich, E. R., and Hyndman, R. D., 1985a, LITHOPROBE, Phase 1, southern Vancouver Island; Preliminary analyses of reflection seismic profiles and surface geological studies: Geological Survey of Canada Paper 85-1A, p. 543–554.

Yorath, C. J., Green, A. G., Clowes, R. M., Sutherland Brown, A., Brandon, M. T., Kanasewich, E. R., Hyndman, R. D., and Spencer, C., 1985b, LITHOPROBE, southern Vancouver Island; Seismic reflection sees through Wrangellia to the Juan de Fuca plate: Geology, v. 13, p. 759–762.

Yorath, C. J., Woodsworth, G. J., Riddihough, R. P., Currie, R. G., Hyndman, R. D., Rogers, G. C., Seeman, D. A., and Collins, A. D., 1985c, Intermontane Belt (Skeena Mountains) to Insular Belt (Queen Charlotte Islands): Geological Society of America Continent-ocean transect B1.

Yorath, C. J., Sutherland Brown, A., Campbell, R. B., and Dodds, C. J., 1992, The Insular Belt; Upper Jurassic to Paleogene Assemblages, *in* Gabrielse, H., and Yorath, C. J., eds., Geology of the Cordilleran Orogen in Canada: Geological Survey of Canada, Geology of Canada, no. 4; (also Geological Society of America, The Geology of North America, v. G-2), p. 354–360.

Manuscript Accepted by the Society November 18, 1987
Geological Survey of Canada Contribution No. 49486

Printed in U.S.A.

DNAG Continent-Ocean Transect Volume
Phanerozoic Evolution of North American
Continent-Ocean Transitions
The Geological Society of America, 1994

Chapter 9

Continent-ocean transition in Alaska: The tectonic assembly of eastern Denalia

Thomas E. Moore and Arthur Grantz
U.S. Geological Survey, 345 Middlefield Road, MS 904, Menlo Park, California 94025
Sarah M. Roeske
Department of Geology, University of California, Davis, California 95616

INTRODUCTION

Alaska is the eastern, subaerial part of a large subcontinent of distinctive tectonic character that serves as an isthmus between nuclear North America, with its fringing belt of allochthonous terranes, and the accreted terranes and volcanic belts that constitute northeastern Russia. Physiographically, this subcontinent, which we name Denalia, is a bulge in the continental platform in the vicinity of Alaska, the Chukotsk Peninsula, and the broad continental shelf of the Bering Sea. The bulge is convex to the south and is bounded on the east and west by constrictions in the width of the continental platform and on the north and south by the edge of the continental shelf (Fig. 1). Tectonically, Denalia is characterized by geologic youthfulness and complexity, an abundance of convergent and transcurrent faults, and absence of autochthonous cratonic rocks. It contains a profusion of lithotectonic terranes of diverse origin and age that were emplaced in late Mesozoic and Cenozoic time. In addition, it includes the superimposed Cenozoic Aleutian arc and subduction zone and the Queen Charlotte–Fairweather transform fault system. Parts of Denalia were created by pre-middle Mesozoic tectonic events, but these took place elsewhere, before the affected rocks were tectonically transported and incorporated into the landmass of Denalia. Except for a small area in the Porcupine Plateau region along the Alaska-Yukon boundary, the only Precambrian rocks that have been recognized in the subcontinent are in tectonically emplaced fragments, the largest of which is the Arctic Alaska terrane in the Brooks Range, Arctic Foothills, and Arctic Coastal Plain of northern Alaska (Fig. 2). Thus, there is no cratonic core against which the constituent terranes of Denalia accreted; rather, most of Denalia was assembled by processes that operated at Mesozoic and Paleogene continental margins. Study of the continent-ocean transition in Alaska thus requires a consideration of the geology of the entire state from the Pacific rim to the Arctic Ocean basin.

In this paper we present a somewhat speculative synopsis of the evolution of the continent-ocean transition in eastern Denalia from the principal perspective of the crust along the Continent-Ocean Transect A-3: Gulf of Alaska to Arctic Ocean (Grantz and others, 1991) (Fig. 2). We extend our view east and west of the transect to include similar features across the central part of the state. The record is still incomplete, and in many instances our discussion is based on speculative models derived from meager data. In spite of these obstacles, we attempt a unified discussion of the tectonic assembly and evolution of the continent-ocean transition of eastern Denalia in central Alaska.

We begin our discussion by outlining the present tectonic status of Alaska—its physiography, active faults, seismicity, lithotectonic terranes, paleomagnetic domains, and present continental margins. Following this outline is a proposed history of the continent-ocean evolution in Alaska that draws heavily on geologic relations near the A-3 transect corridor. This history is inextricably intermeshed with the assembly, or aggregation, of Alaska by tectonic processes in Mesozoic and Cenozoic time. We conclude our discussion with generalizations about the tectonic character of the continent-ocean transition in Alaska, which may provide useful insights into the tectonic character and origin of other areas where continental geology is dominated by the products of continental-margin structural processes. A discussion of the geologic and geophysical data upon which the A-3 transect is based is presented in the explanatory pamphlet that accompanies the transect. Many of the features and hypotheses discussed below are illustrated by maps and cross-sectional sketches taken from the A-3 transect, and it may be helpful to consult the transect. Throughout the text, we use the geologic time scale of Palmer (1983).

The lithotectonic terrane concept of Coney and others (1980) and Jones and others (1983) is used throughout this report to facilitate the presentation and discussion of the geology and tectonics of Alaska. This concept, which holds that Alaska

Moore, T. E., Grantz, A., and Roeske, S. M., 1994, Continent-ocean transition in Alaska: The tectonic assembly of eastern Denalia, *in* Speed, R. C., ed., Phanerozoic Evolution of North American Continent-Ocean Transitions: Boulder, Colorado, Geological Society of America, DNAG Continent-Ocean Transect Volume.

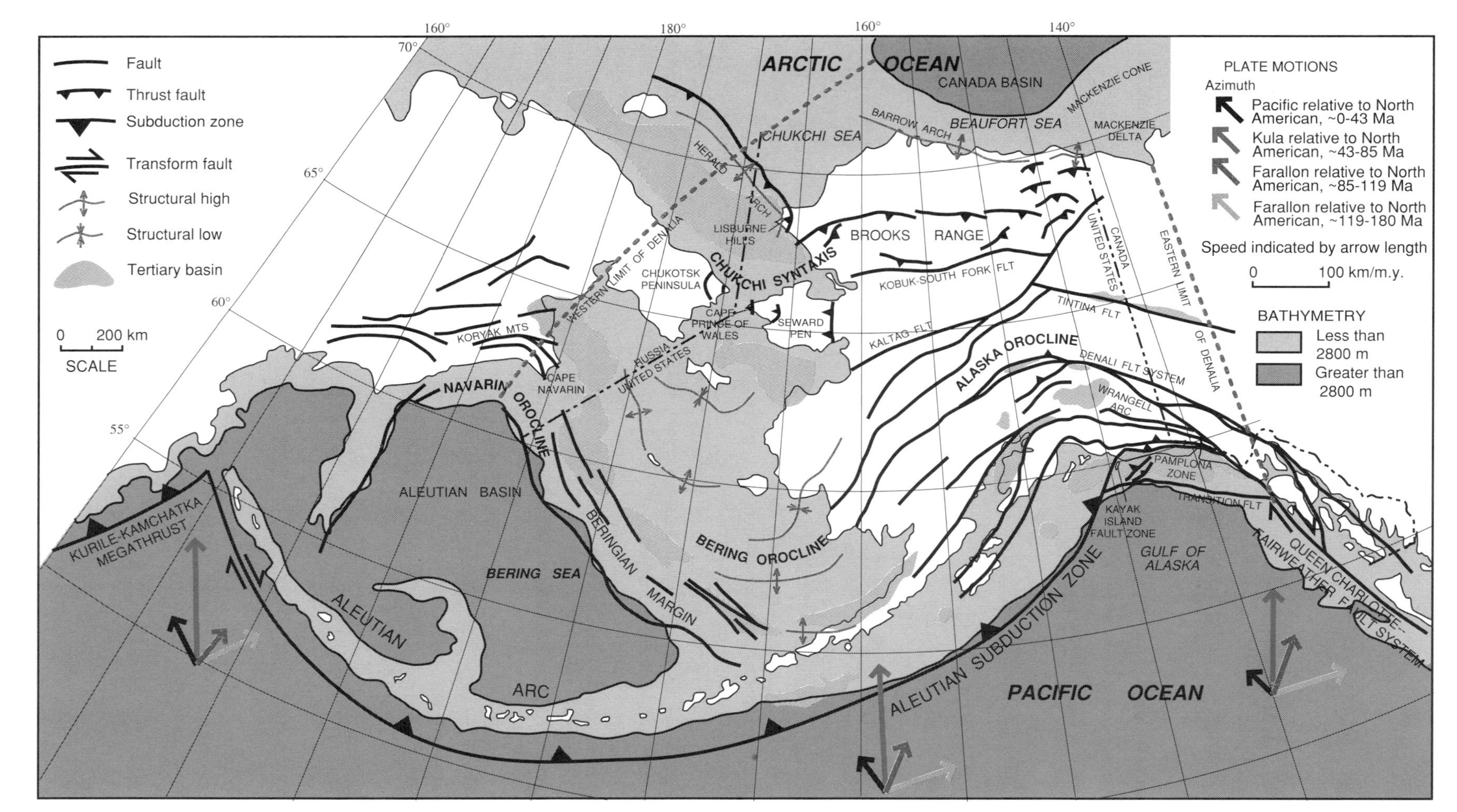

Figure 1. Major tectonic features of the subcontinent Denalia. Eastern and western limits of Denalia are denoted by dashed red lines; the northern limit, by the northern continental margin of Alaska; and the southern limit, by the southern continental margin of southern Alaska and the Beringian margin in the Bering Sea.

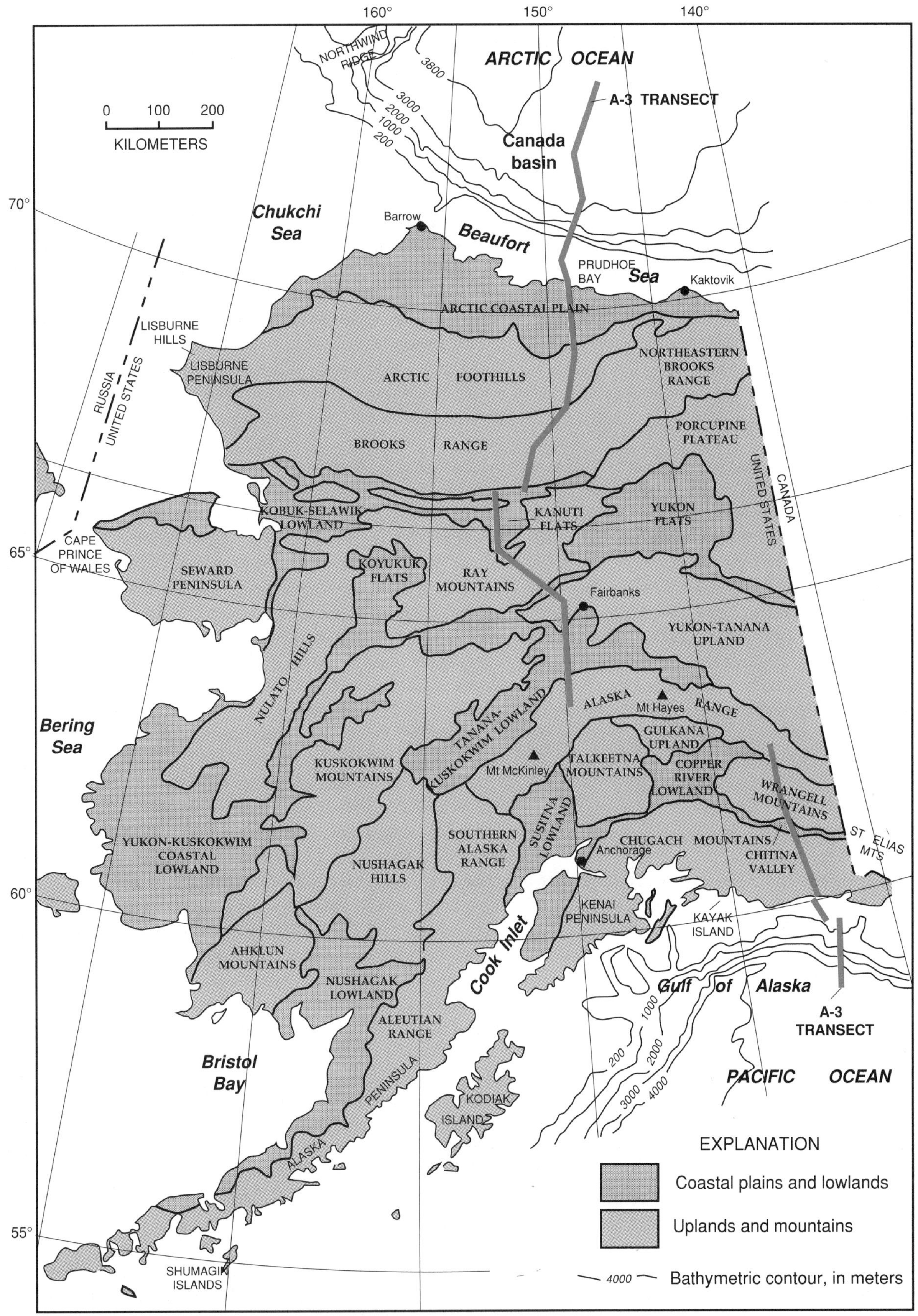

Figure 2. Physiographic provinces of Alaska (after Wahrhaftig, 1965, Plate 1). Heavy red lines mark trace of Continent-Ocean Transect A-3 of Grantz and others (1991).

comprises a large number of geologically unique lithotectonic terranes separated by structural boundaries, has been applied to the entire state by Jones and others (1987). Many of these terranes have clearly defined structural boundaries along which significant tectonic transport has occurred, but others do not. This latter group of terranes is considered suspect because, even though significant tectonic transport has not been documented along their structural boundaries, these terranes differ distinctly in stratigraphy from their neighbors and may be structurally isolated from them. The amount of tectonic transport required to create a lithotectonic terrane is only that which places a terrane out of stratigraphic context with its neighbors. The source of lithotectonic terranes and the amount of tectonic transport the terranes have undergone are important, but are not central to the identification or usefulness of terranes as mappable crustal entities. For understanding the evolution of ancient continental margins, terranes that have been displaced only tens or hundreds of kilometers are just as significant as those that may have been displaced thousands of kilometers and which might therefore be called "exotic." Although the terrane concept is controversial (Hudson, 1987; Dover, 1990a; Sengor and Dewey, 1990), and many proposed terranes are inadequately defined, the terrane concept facilitates understanding of the bewildering array of rock units, which have been recognized and mapped across Alaska. Our analysis of the geology of Alaska for the A-3 transect was based on our own assessment of the rocks that constitute the terranes and not solely on descriptions by previous workers. The stratigraphic composition of the lithotectonic terranes referred to in this report is given in an appendix in the explanatory pamphlet that accompanies the A-3 transect (Grantz and others, 1991).

CURRENT TECTONIC STATUS OF DENALIA IN ALASKA

Denalia was amalgamated and consolidated into a continental platform by tectonic processes that have operated since the Jurassic, and today some of these processes continue to shape its landscape and modify its geology. Three of the most important of these processes responsible for the assembly of Denalia have been (1) middle Cretaceous to Holocene dextral transpression between western North America and the Pacific and associated plates, as outlined by Engebretson and others (1985); (2) middle Cretaceous translational rotation of Arctic Alaska from the Canadian Arctic Islands to produce the Amerasia basin of the Arctic Ocean (Grantz and May, 1983; Grantz and others, 1990b); and (3) oroclinal deformation of Denalia as a result of Laramide (late Late Cretaceous and Paleogene) latitudinal convergence of North America and Eurasia (Grantz, 1966; Coe and others, 1985). Whereas many have elaborated on these processes and their consequences for the tectonic evolution of Alaska, these ideas can be traced to two pioneering papers. Carey (1958) was the first to propose the Alaska orocline theory and sphenochasmic rifting for the Amerasia basin. St. Amand (1957) noted that the major faults of southern Alaska are parallel to the Pacific coast and are an extension of the San Andreas and Queen Charlotte fault systems. To provide further insight into the tectonic processes that have shaped Denalia, we begin with a discussion of its bathymetry, seismicity, and neotectonic record.

Denalia and its boundaries

Denalia is bounded on the south by a compound arcuate convergent margin composed of an extinct segment in the Bering Sea and an active segment, the Aleutian subduction zone, in the north Pacific Ocean (Fig. 1). The extinct Bering Sea (Beringian) margin, between long 168°W and 175°E (Fig. 1), was active in latest Cretaceous to Eocene time and has since been buried by post-Eocene sedimentary fill of the deep Aleutian basin of the Bering Sea (Cooper and others, 1987). In the Eocene, the Aleutian subduction zone and associated Aleutian arc were superimposed obliquely across the Beringian margin onto oceanic crust of the north Pacific basin, intersecting the Beringian margin near long 168°W. The active Aleutian subduction zone extends northeastward from long 168°W as an unbroken arc to the Kayak Island area in the northwestern Gulf of Alaska (Fig. 1). Near Kayak Island, the Aleutian subduction zone enters the continent, whereupon plate convergence is dissipated in a series of thrust faults and associated folds. The southeastern limit of Denalia (Fig. 1) is at the intersection of the subduction zone with the northwest-trending continental margin of southeastern Alaska. The southwestern limit is marked by the intersection of the Beringian margin with the northeast-trending Siberian margin near Cape Navarin (Fig. 1). These intersections lie at large northward embayments in the North Pacific rim.

On the north, Denalia is bounded by a rifted continental margin of Hauterivian (middle Early Cretaceous) age that underlies the outer shelf and slope of the Chukchi and Beaufort seas of the Arctic Ocean (Fig. 1). Hauterivian to Holocene marine sedimentary deposits in the Mackenzie cone, derived principally from the Mackenzie River basin and the glacial drainage of Amundsen Gulf, have buried this margin to depths from 6 km beneath the abyssal plain of the Canada basin to more than 12 km near the Mackenzie delta. The embayment of the Canada basin at the Mackenzie delta anchors the eastern limit of Denalia; the western limit is anchored at the intersection of the continental margin of the Chukchi Sea shelf with the southern end of the Northwind Ridge of the Arctic Ocean basin (Figs. 1 and 2). Although the eastern and western boundaries of Denalia are not associated with any specific preexisting geologic feature, they delimit much of the region between nuclear North America on the east and nuclear Eurasia on the west.

Physiography and active faults

The highest and most vigorously eroded mountains in the world are found where rates of late Cenozoic crustal deformation, especially convergence, are greatest. In Alaska the effect is shown dramatically by the close association of active (displaced during the Quaternary) convergent and transcurrent faults with the highest and most extensive mountain systems (Figs. 1 and 2). The tectonic processes that can be inferred from the location of the highest mountains in present-day Alaska offer insight into the tectonic processes that shaped Tertiary Alaska.

The rugged and extensive Chugach–St. Elias mountain system of coastal southern Alaska (Fig. 2) occupies the region where the Queen Charlotte dextral transcurrent fault system of the Pacific–North American plate boundary (Fig. 1) enters the continent as the Fairweather fault and is partly transformed into the Chugach–St. Elias and associated right transpressive faults (Fig. 3). These mountains have also been affected by uplift related to subduction at the Wrangell Mountains volcanic arc segment of the Aleutian arc (Fig. 1). Convergence at the Aleutian subduction zone (Fig. 1) is redistributed among other structures and dissipated into uplift where it enters continental southern Alaska. Vectors of relative plate boundary motions that produce this deformation and uplift in southern Alaska are taken from Engebretson and others (1985) and are shown in Figure 1.

The highstanding Mount McKinley (6194 m) and Mount Hayes (4216 m) massifs of the central Alaska Range (Fig. 2) are also the product of transpression related to Pacific–North American plate motion along the Queen Charlotte fault system (Fig. 1). These massifs lie adjacent to the Denali dextral transcurrent fault system west of its junction with the Totschunda fault, which receives its dextral transcurrent motion from the master Queen Charlotte transform fault system via both the Shakwak and the Fairweather fault trends (Fig. 3). This augmented transcurrent displacement is transformed into dextral transpression in the axial region of the Alaska orocline (Fig. 1), where, in addition to the Mount Hayes uplift, it accounts for the rise of the Mount McKinley massif to the highest elevation in North America.

North of the Alaska Range, the highest and most rugged mountains are in the northeastern Brooks Range (2700 m maximum elevation) (Fig. 2). These mountains overlie a convergent (thrust fault) tectonic system of Eocene to Holocene age (Grantz and others, 1990a; Wallace and Hanks, 1990). The convergence is visible in thrust folds in the northern foothills and coastal plain of the northeastern Brooks Range and the continental shelf and slope to the north and dies out beneath the continental rise 170 to 200 km north of Kaktovich (Fig. 1). These young folds and faults die out to the west along the western front of the northeastern Brooks Range (the sinistral, northeast-trending Canning displacement zone of Grantz and May, 1983) (Fig. 4). To the east they die out in a northerly trending zone of earthquakes and extensional and dextral faults west of the lower Mackenzie River (Basham and others, 1977). A local seismic network (Biswas and others, 1986) shows that seismicity in northern Alaska is limited to the highstanding northeastern Brooks Range: the Arctic Coastal Plain west of the northeastern Brooks Range is aseismic. Regional epicenter maps (Meyers, 1976, Figs. 1 and 2; Grantz and others, 1990c) show that the lower Mackenzie River valley east of the Richardson Mountains, the extension of the northeastern Brooks Range into northern Canada, is also aseismic. In addition, the convergence has lifted the slope and rise of the adjacent continental margin as much as 1 km with respect to the continental margin to the west (Grantz and others, 1990a, Fig. 15).

Seismicity

Historic earthquakes in Alaska provide a record of Quaternary tectonic activity that is in part more detailed, but less complete, than that offered by active faults (Fig. 4). The faults provide an integrated record of activity over a period of thousands to millions of years. The historical seismic record in Alaska goes back only to 1786 (Meyers, 1976) and is shorter than the recurrence interval of earthquakes associated with active faults in many areas. Accordingly, the record of Alaskan seismicity in Figure 4, showing the distribution of earthquakes in central Alaska from 1971 to 1984 larger than magnitude 3 to 4.5, provides an invaluable, but incomplete, record of the distribution of tectonic seismicity and structures beneath Alaska. A minimum-magnitude criterion was not applied to the earthquakes shown in Figure 4 because the figure was designed to show where, and at what depths, earthquakes occur in central Alaska, and not to compare relative levels of seismicity across the state. The procedures by which the data were thinned, and some probable errors, were discussed in Grantz and others (1991). The large active faults of Alaska, seismically determined contours on the Wadati-Benioff zones beneath Alaska, and the orientation of inferred principal stress axes and earthquake mechanisms are also plotted in Figure 4 to show their relation to the seismic zones. The following discussion of seismicity is modified from the transect text of the Continent-Ocean Transect A-3 of Grantz and others (1991, p. 6–7).

The dominant seismogenic feature in Alaska is the northeast-trending zone of earthquakes that defines the northwest-dipping Aleutian Wadati-Benioff zone beneath the eastern Aleutian arc (Fig. 4). These earthquakes are characterized by thrust-fault focal mechanisms that yield northwesterly plunging maximum compressive stress axes (*P*-axes). In the area of the Denali fault system in central Alaska, the density of earthquakes diminishes sharply, and the overall distribution of intermediate-depth earthquakes changes from a north-south to northeast-southwest trend. These changes mark the eastward shift from the Aleutian Wadati-Benioff zone to the Wrangell Wadati-Benioff zone, which extends eastward from Prince William Sound along the Wrangell Mountains and the continental margin in the Gulf of Alaska. The Wrangell Wadati-Benioff zone is seismically less active than the Aleutian Wadati-Benioff zone, but the Wrangell zone has a northerly dip and is defined by earthquakes yielding thrust-fault focal mechanisms with northwesterly plunging *P*-axes (Stephens and others, 1984; Biswas and others, 1986). The orientation of the stress axes indicates that convergence is oblique to the trend of the Wrangell Wadati-Benioff zone but subparallel to Pacific– North American plate convergence (Fig. 4). Thus, the Aleutian and Wrangell Wadati-Benioff zones are distinctive seg-

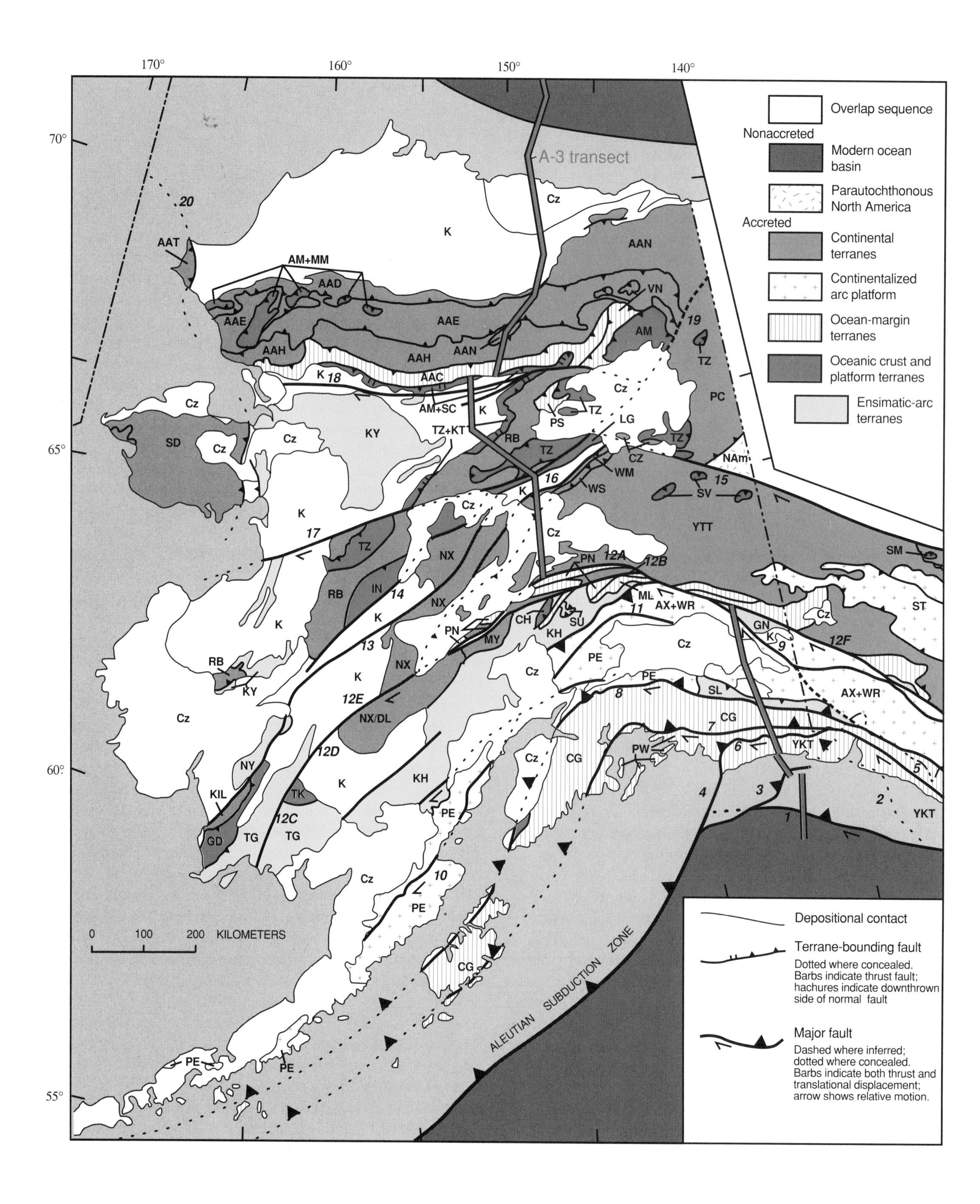
170°
160°
150°
140°
70°
65°
60°
55°
A-3 transect
Overlap sequence
Nonaccreted
Modern ocean basin
Parautochthonous North America
Accreted
Continental terranes
Continentalized arc platform
Ocean-margin terranes
Oceanic crust and platform terranes
Ensimatic-arc terranes
Depositional contact
Terrane-bounding fault
Dotted where concealed. Barbs indicate thrust fault; hachures indicate downthrown side of normal fault
Major fault
Dashed where inferred; dotted where concealed. Barbs indicate both thrust and translational displacement; arrow shows relative motion.
0 100 200 KILOMETERS
ALEUTIAN SUBDUCTION ZONE

ments of a single subduction zone that extends along the northern and western margins of the Gulf of Alaska. Clusters of epicenters also are present over the Transition fault zone and its western projection in the Gulf of Alaska, and along the Denali fault zone of eastern interior Alaska (Fig. 4).

North of the Denali fault system, the belt of interior Alaska seismicity bifurcates into west-northwest– and north-trending branches. The west-northwest–trending branch of seismicity extends to the Seward Peninsula and the southern Chukchi Sea and consists of crustal earthquakes characterized by normal-fault mechanisms and northeast-trending minimum compressive stress axes (*T*-axes). The geologic association(s) of this zone of shallow earthquakes is not known. The seismicity is subparallel to the trend of the Tintina and eastern Denali dextral transcurrent faults, and could possibly result from related intracrustal transcurrent faults that are not well expressed in the surface geology. The north-trending branch of seismicity projects northward from interior Alaska to the eastern Beaufort Sea along the trend of the Aleutian Wadati-Benioff zone. Near its northern end, this belt of dispersed earthquakes marks the western limit of seismicity and significant Cenozoic contractional deformation and uplift in northeastern Alaska. The seismicity dies out to the north in the western part of the Cenozoic thrust-fold province on the Beaufort Sea shelf and slope. These north-vergent Cenozoic thrusts and thrust folds of the northeastern Brooks Range and Beaufort Sea appear to be related to Pacific–North American plate convergence transmitted through the crust across Alaska from the Wrangell Mountains segment of the Aleutian subduction zone to the continental margin in the Beaufort Sea. Thus, the thrusts and folds may be the bow wave of Pacific–North American plate convergence in northwestern North America. A possible position of this hypothetical thrust displacement beneath the Brooks Range and Koyukuk basin is shown by open circles along mainly older faults in cross-section III on sheets 2 and 3 of Grantz and others (1991).

Figure 3. Terranes and major faults of Alaska (modified from Jones and others, 1987). Selected terranes: AM, Angayucham; AX+WR, Alexander-Wrangellia; CG, Chugach; CH, Chulitna; CZ, Crazy Mountains, DL, Dillinger; GD, Goodnews; GN, Gravina-Nutzotin; IN, Innoko; KH, Kahiltna; KIL, Kilbuk; KT, Kanuti; KY, Koyukuk; LG, Livengood; ML, MacClaren, MM, Misheguk Mountain; MY, Mystic; NX, Nixon Fork; NY, Nyac; PC, Porcupine; PE, Peninsular; PN, Pingston; PS, Prospect Creek; PW, Prince William; RB, Ruby; SC, Slate Creek; SD, Seward; SM, Slide Mountain; SL, Strelna; ST, Stikine; SU, Susitna; SV, Seventymile; TG, Togiak; TK, Tikchik; TZ, Tozitna; VN, Venetie; WM, White Mountains; WS, Wickersham; YKT, Yakutat; YTT, Yukon-Tanana. Selected subterranes of Arctic Alaska terrane: AAC, Coldfoot; AAD, DeLong Mountains; AAE, Endicott Mountains; AAH, Hammond; AAN, North Slope; AAT, Tigara. Other important tectonic units: Cz, Cenozoic overlap sequence; K, Cretaceous overlap sequence; NAm, parautochthonous North America. Selected faults: 1, Transition fault; 2, Dangerous River zone; 3, Pamplona zone; 4, Kayak Island fault zone; 5, Fairweather fault; 6, Chugach–St. Elias fault; 7, Contact fault; 8, Border Ranges fault; 9, Totschunda fault; 10, Bruin Bay fault; 11, Talkeetna thrust fault; 12, Denali fault system (12A, Hines Creek strand; 12B, McKinley strand; 12C, Togiak-Tikchik fault; 12D, Holitna fault; 12E, Farewell fault; 12F, Shakwak fault); 13, Nixon Fork fault; 14, Susulatna lineament; 15, Tintina fault; 16, Beaver Creek fault; 17, Kaltag fault; 18, Kobuk-South Fork fault; 19, Porcupine lineament; 20, Herald thrust fault. Heavy red lines mark trace of Continent-Ocean Transect A-3 of Grantz and others (1991).

First-order deformational features

The most distinctive structural feature of Alaska is the great arcuate system of northward-convex transcurrent faults that sweeps across the state (Fig. 1). This arcuate pattern contrasts strikingly with the generally linear structural trends in most of the Canadian Cordillera to the southeast. Between long 145°W and 150°W the major lithologic boundaries and structural features in Alaska south of the Brooks Range change from the predominant southeast-striking structural grain of the Canadian Cordillera to the southwest-striking structural grain of southwestern and western Alaska (Fig. 3). Carey (1958) named this change in trend the Alaska orocline, and proposed that the bend was caused by pivoting of the western two-thirds of Alaska about a pole in the Alaska Range. Geologic data (Grantz, 1966) suggest that the main development of the Alaska orocline occurred in the early Tertiary, and that the deformation was accomplished by rotation of a series of elongate prisms of northeast-southwest trend that were bounded by right-lateral strike-slip faults of similar trend in the active western limb of the orocline (Fig. 1); the eastern limb of the Alaska orocline, the Canadian Cordillera, remained passive. Although specific data are lacking from the continental shelf of the Bering Sea, outcrops near Cape Navarin in northernmost Russia (Popeko, 1985) suggest that the axis of a second orocline, the Navarin orocline, lies about 90 km west of Cape Navarin (Fig. 1). The strike of the axis of the Navarin orocline is northerly, as is that of the Alaska orocline to the east, and both axes project into the reentrants in the continental margin that define the southern physiographic margin of Denalia. If the Navarin orocline is indeed complementary to the Alaska orocline, its western limb, which is well exposed in northeastern Russia, should be passive. Its eastern limb, however, covered by the Bering Sea, may be underlain by northwest-trending strike-slip faults with left-lateral displacement.

Between the Alaska and Navarin oroclines is the opposing Bering orocline, which encompasses a broad belt of arcuate, southward-convex structures (Fig. 1). This large orocline constitutes the southern two-thirds of Denalia. The eastern limb of the Bering orocline is the active western limb of the Alaska orocline; the western limb of the Bering orocline is the postulated active eastern limb of the Navarin orocline. Like the Alaska orocline, the Bering orocline was formed during the Late Cretaceous and/or Paleogene.

Oroclinal folds also affect the rocks and structures of the Brooks Range orogen. The southward bend in the Brooks Range orogen in westernmost Alaska and the Chukotsk Peninsula, termed the Chukchi syntaxis by Tailleur and Brosgé (1970), is

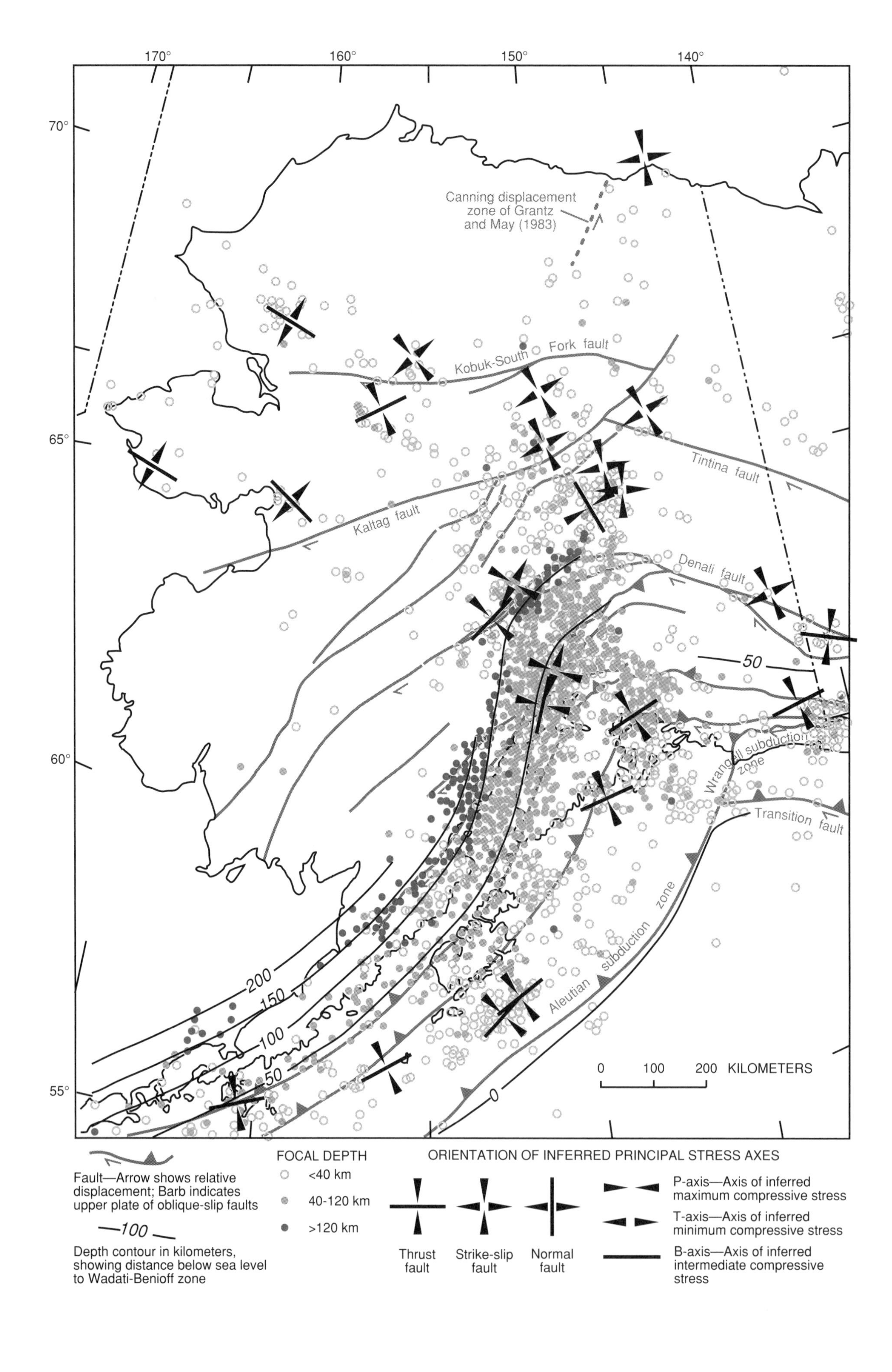

170°
160°
150°
140°
70°
65°
60°
55°
Canning displacement zone of Grantz and May (1983)
Kobuk-South Fork fault
Tintina fault
Kaltag fault
Denali fault
Wrangell subduction zone
Transition fault
Aleutian subduction zone
200
150
100
50
0
0 100 200 KILOMETERS
Fault—Arrow shows relative displacement; Barb indicates upper plate of oblique-slip faults
—100—
Depth contour in kilometers, showing distance below sea level to Wadati-Benioff zone
FOCAL DEPTH
<40 km
40-120 km
>120 km
ORIENTATION OF INFERRED PRINCIPAL STRESS AXES
Thrust fault
Strike-slip fault
Normal fault
P-axis—Axis of inferred maximum compressive stress
T-axis—Axis of inferred minimum compressive stress
B-axis—Axis of inferred intermediate compressive stress

roughly coaxial and concordant with the Bering orocline and is interpreted to lie at its core (Fig. 1). The junction of the physiographically and structurally north-trending Lisburne Hills with the southwest-trending western Brooks Range (Fig. 1) also resembles an oroclinal fold (Patton and Tailleur, 1977). This feature (Grantz, 1966), however, represents crosscutting of the northwest-verging western Brooks Range orogen of Middle Jurassic to Late Cretaceous or early Tertiary age by the east-verging Herald thrust fault zone of post-Albian or Cenomanian and pre-Eocene age in the central Chukchi Sea. The Herald fault zone possibly correlates with a thrust fault that places carbonate rocks (Lisburne Group?) over continental rocks of North American affinity (Seward terrane) at Cape Prince of Wales, on the westernmost tip of the Seward Peninsula. Convergence on the Herald thrust-fault zone may be related to the east-west shortening that created the Bering orocline in Laramide time, and constitutes the paleostructural boundary between Eurasia and North America.

Genetic classification and distribution of lithotectonic terranes in Alaska

Denalia consists almost entirely of lithotectonic terranes that were tectonically assembled against the ancient margin of western North America (Jones and others, 1987) and northeastern Eurasia (Fujita and Newberry, 1982, 1983). We consider only the Alaska terranes in this paper.

Some of the terranes of Alaska are of a non–North American or "exotic" origin, whereas others have lithologic affinities with North America or its continental margin. The terranes in Alaska and their known or inferred structural boundaries are shown in Figure 3 and are listed in the appendix *Tables of Lithotectonic Terranes of Alaska and Vicinity* in Grantz and others (1991). The terranes shown are modified from Jones and others (1987) to incorporate the results of more recent investigations and to emphasize significant geologic relations. These revisions are discussed in Grantz and others (1991, p. 10–11).

To understand the evolution of the continent-ocean boundary in Alaska, an interpretation of the crustal structure of each terrane boundary along the A-3 transect is required, as well as an examination of the character of the terranes. Accordingly, much attention is directed to the nature of the terrane-bounding structures, which in many areas is less well understood than the character and origin of the terranes. Where data on these structures were contradictory or lacking, we resorted to analogy with better known areas to produce a coherent tectonic interpretation for the entire crust across central Alaska.

To further aid our analysis, we classified the Alaska lithotectonic terranes according to a system similar to that of Grantz and others (1991), and shown in Figure 3. The classification is based on lithology but is assigned genetic designators to amplify the significance of its categories in the tectonic assembly of Alaska and in the evolution of the region's multiple continent-ocean transitions. Some of the terranes contain stratal sequences assignable to more than one terrane class as a result of structuring. Where such complexities were reported, we classified the terrane by the oldest constituent sequence because the oldest sequence most likely reflects the fundamental tectonic character of the terrane. We made no further attempt to subdivide the complex terranes. Terranes classified as "accreted," designate terranes that were tectonically assembled onto the margin of North America by any of a variety of processes (that is, in the sense of King, 1977); we do not mean to imply that these terranes were emplaced solely by subduction accretion. Figure 3 shows that, despite their great number and structural complexity, once classified the terranes reveal a degree of tectonic order. As knowledge of the character and origin of terranes increases, a more sophisticated classification will reveal even higher degrees of tectonic order.

Nonaccreted terranes. Two classes of Alaska's lithotectonic terranes formed more or less in place and are grouped as nonaccreted terranes in Figure 3. The first class consists of a small area of parautochthonous Precambrian and younger rocks near the Yukon River at the Alaska-Yukon border. These rocks were assigned to cratonic North America by Grantz and others (1991) to emphasize their close association with undisplaced rocks of the North American craton, although strictly speaking they are allochthonous. The second class, modern ocean basins, consists of the oceanic crust of the modern Pacific and Arctic oceans.

Overlap sequences. Overlap sequences are widespread in Alaska. These comprise sedimentary successions unconformably deposited in part or completely upon the older, tectonically accreted terranes, and therefore have mainly stratigraphic rather than structural boundaries.

Accreted terranes. The accreted terranes of Alaska, those that have been assembled by tectonic processes, are herein grouped into four main classes. Although a terrane of one class may be distributed among those of a different class, each class tends to be localized within a specific area of Alaska (Fig. 3). Each of the terranes is observed or inferred to be bounded by faults along which the terrane was transported by one or more of the first-order regional tectonic stress fields prior to burial beneath an overlap sequence.

Terranes that have been displaced the least resemble parts of the western margin of the North American craton and are classified as continental terranes of North American affinity under the general heading of continental terranes (Fig. 3). Some, such as the Ruby and Yukon-Tanana terranes, consist of crystalline rock of continental character inferred to have been transported northward and incorporated into the body of Denalia by convergence on major transcurrent and thrust-fault systems of the continental margin in late Mesozoic and Paleogene time. Others, such as the

Figure 4. Seismicity (larger than magnitude 3 to 4.5) and inferred regional stress of central Alaska, 1971–1984 (after Biswas and others, 1986). See Grantz and others (1991) for selection parameters of earthquake epicenters and principal stress axes.

Hammond, Endicott, Coldfoot, and De Long subterranes of the Arctic Alaska terrane, consist of continental shelf and slope strata that were tectonically delaminated from their crystalline basement and now form nappes in the Brooks Range orogen. All of the Arctic Alaska teranes are fragments of the western Canadian shield. The terranes were separated from the shield during the rotational rifting that formed the Amerasia basin of the Arctic Ocean in middle Cretaceous time.

A second subclass of continental terranes, continental terranes of uncertain affinity, is characterized by continental rock types that have no obvious counterparts with continental rocks of western North America. The provenance, origin, and displacement history are unknown for these terranes. From currently available knowledge, this class of terranes may be of Eurasian or North American origin. The Kilbuck, Nixon Fork–Dillinger, and Mystic terranes are assigned to this class. They lie outboard (more distal from cratonic North America) of the continental terranes of North American affinity.

The second class of accreted terranes, continentalized arc platform (Fig. 3), consists of island-arc plutonic and volcanic rocks of late Paleozoic and Early Jurassic age that have been extensively deformed and intruded by younger arc rocks. These superimposed arcs include the Chisana arc of Albian age and the Wrangell arc (Fig. 1) of late Cenozoic age within the Alexander-Wrangellia terrane. Plutonism and deformation, accompanied by regional dynamothermal metamorphism, have thickened the crust of the continentalized terranes so that the terranes now exhibit continental characteristics, such as coherence and tectonic stability. The crustal thickening and stability are expressed as gravity lows, moderately deep erosion, including well-developed surfaces of marine planation, and thick postvolcanic sedimentary prisms derived from the continentalized and uplifted arc rocks. In the upper parts of these terranes, gently dipping unconformities of Early or Middle Jurassic age are overlain by moderately deformed Middle Jurassic to Upper Cretaceous continental-terrace deposits. Examples of continentalized arc-platform terranes are the Alexander-Wrangellia and the Peninsular terranes.

Ocean-margin terranes constitute the third class of Alaska terranes (Fig. 3). These terranes were deposited on oceanic or transitional crust adjacent to ancient continental margins of North America. In northern Alaska, they include Paleozoic to Lower Jurassic terrigenous clastic deposits (Slate Creek, Venetie, and Prospect Creek terranes) derived from the adjacent Arctic Alaska and Ruby continental terranes of North American affinity. The northern ocean-margin terranes originated at the margins of the paleo-Pacific (Angayucham) Ocean in the Paleozoic and early Mesozoic. In southern Alaska, they include Upper Cretaceous and Cenozoic terrigenous clastic deposits (Chugach, Prince William, and Yakutat terranes) derived from British Columbia and the Pacific Northwest during the Cretaceous and Paleogene and from southern Alaska in the late Cenozoic. These southern ocean-margin terranes originated at the margins of the Pacific and related oceanic plates.

The fourth class of Alaska terranes comprises oceanic crust and platform terranes. The terranes of this class are numerous and lithologically diverse, comprise Paleozoic to middle Cretaceous rocks, and lie outboard of terranes with continental affinities. To clarify their character and distribution, the oceanic crust and platform terranes can be divided into four subclasses.

The largest subclass (grouped under oceanic crust and platform terranes in Fig. 3) is characterized by diabase, pillow basalt, and chert with variable, but typically subordinate, amounts of tuff, argillite, graywacke, and limestone. Terranes in this subclass consist mostly of Devonian to Upper Triassic or Lower Jurassic rocks, but age control is sketchy. Typical examples of this subclass include the Angayucham, Innoko, Seventymile, and Tozitna terranes. The rocks constituting these terranes formed in the paleo–Pacific Ocean marginal to western North America. Because evidence in Alaska for the existence of a pre-Jurassic Pacific Ocean is contained mainly in the Angayucham and related terranes, we herein refer to this inferred ocean, constituting all or part of the paleo-Pacific Ocean, as the "Angayucham Ocean." Terranes of this subclass are accordingly grouped under upper oceanic crust terranes of the Angayucham Ocean. Similar rocks with Cambrian to Devonian fossils in the Goodnews and Livengood terranes are thought to be remnants of an earlier oceanic tract or possible tectonically emplaced exotic terranes. These comparable terranes are classified as upper oceanic crust terranes of uncertain affinity (oceanic crust and platform terranes in Fig. 3).

A third subclass of oceanic crust and platform terranes is formed by ensimatic island-arc volcanic terranes (distinguished in Fig. 3), some of which have large epiclastic volcanic aprons and others of which have both epiclastic volcanic aprons and intercalated or enveloping sedimentary prisms of terrigenous flysch and hemipelagic deposits. Many of these terranes also contain genetically related plutons. Examples of ensimatic-arc terranes in which andesitic or basaltic island-arc rocks are dominant are the Koyukuk, Nyac, and Togiak terranes. Examples in which epiclastic or enveloping sedimentary aprons are dominant are the Gravina-Nutzotin and Kahiltna terranes of southern Alaska. Both types are interspersed with the upper oceanic crust terranes of the Angayucham Ocean and, in Alaska, the ensimatic-arc terranes lie mainly north of the continentalized arc-platform terranes. Most of the ensimatic-arc terranes are thought to have developed on oceanic rocks of the Angayucham Ocean, but some, such as the Gravina-Nutzotin and Strelna terranes, may rest in part on the Alexander-Wrangellia continentalized-arc terrane. Most of the ensimatic arcs are Jurassic to middle Cretaceous in age, indicating a major episode of convergence and crustal consumption at multiple sites in the Angayucham Ocean during that time. Older crustal consumption events in the Angayucham Ocean are recorded in the Upper Triassic Clearwater and Susitna terranes and in the Pennsylvanian to Lower Permian(?) Strelna terrane.

The fourth subclass of oceanic crust and platform terranes, ophiolites of the Angayucham Ocean (grouped with oceanic crust and platform terranes in Fig. 3), consists of ultramafic rocks, gabbro, diabase, serpentinite, and in some cases accompanying basaltic dikes and flows, and radiolarian chert. The ophiolitic terranes comprise the Lower Jurassic Kanuti and Misheguk Mountain terranes and part of the Mississippian to Triassic Seventymile terrane. All three of these terranes were emplaced as klippen on the continental terranes of North American affinity, but originated in the lower crust and mantle beneath the Angayucham Ocean. The Kanuti and Misheguk Mountains terranes yield Middle Jurassic crystallization ages and Middle to Late Jurassic emplacement ages, which makes their uplift and obduction generally coeval with the initiation of arc volcanism in the ensimatic-arc terranes. Partial ophiolites are also minor constituents of several other oceanic terranes of the Angayucham Ocean, such as the Chulitna and Goodnews terranes, but these terranes consist mainly of upper oceanic crust and therefore are classified as upper oceanic crust terranes of the Angayucham Ocean.

Distribution patterns of accreted terranes. The terrane map (Fig. 3) and paleomagnetic data (Fig. 5) suggest that in general the lithotectonic terranes of Alaska become increasingly allochthonous and oceanic toward the south. For example, parautochthonous North American rocks and the continental terranes of North American affinity, such as the Arctic Alaska, Ruby, and Yukon-Tanana terranes, lie in northern and central Alaska. Structurally admixed with these terranes, but mainly adjacent to them on the south, lie belts of continental terranes of uncertain affinity, such as the White Mountains, Nixon Fork, and Dillinger terranes. Oceanic terranes associated with the paleo–Pacific (Angayucham) Ocean are intermingled with, or lie south of, the terranes with continental affinity. In south-central Alaska, the oceanic terranes are bounded to the south by a belt of far-traveled continentalized-arc terranes (Peninsular and Alexander-Wrangellia). The ensimatic-arc terranes represent the closing of the Angayucham Ocean mainly in Late Triassic to middle Cretaceous time. South of the continentalized-arc terranes lies the ocean-margin terranes of the Pacific rim, emplaced in Cenozoic time, and the Pacific and related oceanic plates. These areal relations between lithologically distinct classes of terranes were an important consideration in constructing the cross section along the A-3 transect. The significance of these relations for the evolution of the complex continent-ocean transition in Alaska is explained below.

PLATE TECTONIC SETTING

Denalia was assembled and deformed in a succession of plate tectonic environments that have been inferred from seafloor magnetic anomalies. Models based on the assumption of relatively fixed hotspots (Engebretson and others, 1985; Debiche and others, 1987; Harbert, 1990) and summing plate motions along globe-circling paths (Stock and Molnar, 1987) have established (with an accuracy and reliability that decreases with time) the orientation, velocity, character, and age of the successive Late Jurassic to modern plate motions that created these tectonic environments (Fig. 1). A simplified history of these settings is presented below. In general, the geologically documented Late Jurassic to modern structural events in Denalia can be correlated with the succession of plate tectonic settings for Denalia, the North Pacific, and the Arctic regions. Earlier relative plate motions, inferred from Denalia's geologic record, predate the preserved paleomagnetic record of oceanic-plate movements.

Relative motion of North America and plates of the Pacific realm

Of fundamental importance to an understanding of the construction of Denalia are the relative plate motions of the North American plate and the oceanic plates known and inferred along the western margin of North America. The models of relative plate motions of Engebretson and others (1985) are used in this paper and, in the following discussion, are related to the tectonic processes affecting Denalia.

The model of Engebretson and others (1985) predicts that during the Jurassic and Neocomian, large areas of the northern and eastern Pacific basin were occupied by the Farallon plate. The eastern part of the Farallon plate, which was located far east of its generative spreading center, must have contained the oldest oceanic crust in the Farallon plate. A record of mainly ensimatic-arc rocks of Jurassic and Neocomian age in the terranes of southern Denalia indicates that this region contained subduction zones that lay outboard of North America. Thus, the history of plate motion along the margin of North America for Jurassic and earliest Cretaceous time may have been far more complex than shown in Engebretson and others (1985). On this basis and the evidence given below, we hypothesize that in Jurassic and Neocomian time southern Denalia was populated with multiple arcs of related, but unknown, polarity and geometry. A modern analog is the present western Pacific basin, where older crust of the Pacific plate is being subducted along a series of complexly related island arcs.

From the beginning of the paleomagnetic record at 180 Ma to the time of separation of the Kula plate at 85 Ma, the North American plate was bounded to the west by the Farallon plate. The motion of the Farallon plate relative to the North American plate from 180 to about 115 Ma was sinistral oblique, but changed to near orthogonal between 115 Ma and about 85 Ma (Fig. 1). Thus, as currently reconstructed, terranes situated in southern Denalia could not have been derived from North American sources to the south prior to 115 Ma. The Farallon plate, however, could have carried terranes from low latitudes to dock along the margin of northwestern North America in the middle Cretaceous (Debiche and others, 1987) if they were derived from non–North American sources. It is significant that during the time of nearly orthogonal convergence and presumed subduction of the Farallon plate toward North America from about 115 to 100 Ma, major batholithic belts were emplaced along the margin of North America and in the continental terranes of the central part of Alaska.

Between its birth at 85 Ma and its death at 43 Ma, the Kula plate lay outboard of northwest North America and con-

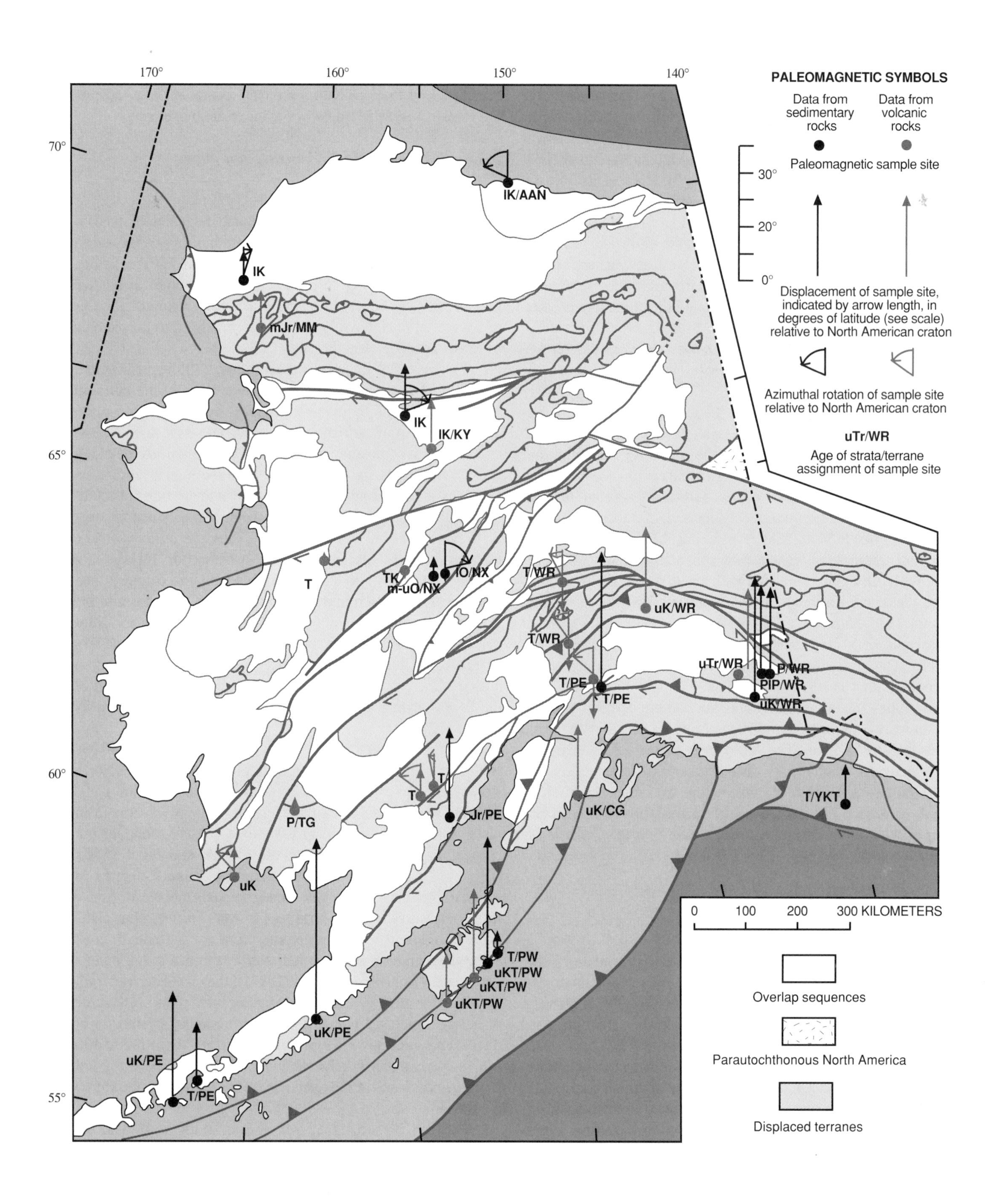
170°
160°
150°
140°
70°
65°
60°
55°
PALEOMAGNETIC SYMBOLS
Data from sedimentary rocks
Data from volcanic rocks
Paleomagnetic sample site
30°
20°
0°
Displacement of sample site, indicated by arrow length, in degrees of latitude (see scale) relative to North American craton
Azimuthal rotation of sample site relative to North American craton
uTr/WR
Age of strata/terrane assignment of sample site
0 100 200 300 KILOMETERS
Overlap sequences
Parautochthonous North America
Displaced terranes
lK/AAN
lK
mJr/MM
lK
lK/KY
T
TK
m-uO/NX
lO/NX
T/WR
uK/WR
T/WR
T/PE
T/PE
uTr/WR
P/WR
PIP/WR
uK/WR
T
T
Jr/PE
uK/CG
T/YKT
P/TG
uK
T/PW
uKT/PW
uKT/PW
uKT/PW
uK/PE
uK/PE
T/PE

stituted the oceanic crust of the entire northeastern Pacific basin. The motion of the Kula plate relative to northwestern North America was oblique dextral convergence from 85 to 55 Ma and similar, but more highly oblique, convergence between 55 and 43 Ma (Lonsdale, 1988) (Fig. 1). A major plate reorganization heralded by the onset of rapid spreading on the Kula-Pacific ridge resulted in a dramatic increase in the relative velocity of the Kula plate with respect to North America from 56 to 43 Ma. The high relative velocities between the Kula and North American plates and their increasing obliquity with time explain the large-scale northward displacement of North American terranes in southern Denalia.

Since 43 Ma, when the Kula plate became fused to the Pacific plate, the oceanic Pacific plate has been moving northwestward subparallel to the northwestern margin of North America (Fig. 1). This period of relative movement is recorded in right-slip transform motion along the San Andreas and Queen Charlotte–Fairweather fault systems and in related subduction and magmatism along the northern rim of the Pacific basin at the Aleutian arc, which extends from the Wrangell Mountains westward almost 4000 km along the Aleutian island chain.

Relative motion of North America and Eurasia

The construction of Denalia was also affected by the relative motions of the North American and Eurasian plates (Harbert and others, 1990), which are currently juxtaposed in western Denalia. During the Late Cretaceous, relative motion between these two plates was obliquely convergent. A period of strong convergence from about 70 to about 50 Ma correlates with evidence for contraction in the Bering and Chukchi seas (see below). After about 50 Ma, relative motion was transtensional, which may provide an explanation for the observed subsidence of the Bering shelf and the formation of pull-apart basins along the Beringian margin. Of key importance are the poorly known relative motions of North America and Eurasia prior to Neocomian time. Comparison of apparent polar wander paths from North America and Eurasia suggests that an ocean basin may have existed between these two plates in the late Paleozoic and early Mesozoic (Harbert and others, 1990) that was apparently consumed by the Late Cretaceous.

PALEOMAGNETIC DATA AND THE TECTONIC ASSEMBLY OF ALASKA

Paleomagnetic studies have played a fundamental role in the development of the lithotectonic terrane concept and consequently in our understanding of the tectonic assembly of Alaska. Results from studies in the 1970s and early 1980s provided the first quantitative evidence that many terranes of southern Alaska have traveled thousands of kilometers with respect to nuclear North America before reaching their present positions (Fig. 5). The paleomagnetic data from northern Alaska are more equivocal. Recent studies have refined the timing and relative displacement of some terranes and have demonstrated that large sections of Alaska have rotated about vertical axes. The major tectonic implications of the paleomagnetic studies for the tectonic assembly of Denalia in Alaska are discussed further, but see summaries by Stone and others (1982), Coe and others (1985, 1990), Hillhouse (1987), Stone and McWilliams (1989), and Harbert (1990) for thorough reviews of the data.

Tectonic assembly

Southern Alaska. Paleomagnetic studies of the Wrangellia terrane in southern Alaska (Hillhouse, 1977) show that this terrane originated far south of its present position. Results from the Alexander terrane in southeastern Alaska, however, suggest that the Alexander terrane has been in place with respect to North America since the Late Triassic (Hillhouse and Grommé, 1980). These results contradict geologic evidence that these two terranes were amalgamated into the composite Alexander-Wrangellia terrane by the Late Pennsylvanian (Gardner and others, 1988). Reexamination of the Late Triassic data from the Alexander terrane (Haeussler and Coe, 1987) indicates that the magnetization thought to be Triassic is probably secondary and that higher temperature magnetization records a paleolatitude similar to that found in the Wrangellia terrane. These studies show that the combined Alexander-Wrangellia terrane has moved a minimum of $28° \pm 6°$ (about 3000 km) northward since the Late Triassic. Furthermore, paleomagnetic data from Permian rocks in the Wrangellia terrane (Panuska and Stone, 1981) and the Alexander terrane (Panuska and Stone, 1985) permit the two terranes to have had the same paleolatitude since Permian time. The Peninsular terrane also formed in the general proximity of Wrangellia, because the two terranes contain similar Norian bivalve assemblages (Newton, 1983). Stratigraphic successions of the Peninsular and Alexander-Wrangellia terranes indicate that they were amalgamated into the southern Alaska superterrane (SAS) by the Middle Jurassic (Jones and others, 1982).

Estimates for the timing and location of the collision of the SAS with North America range from the Middle Jurassic to the middle Cretaceous, but both geologic and paleomagnetic data now demonstrate that significant northward movement of the SAS took place after its collision with North America. Some of the latitudinal displacement of the SAS occurred along the Tintina fault system of east-central Alaska and the Canadian Cordillera (Fig. 3). Paleomagnetic measurements require displacements on the Tintina of >900 km and of 1500 ± 950 km (Marquis and Globerman, 1988). Displacement on the Tintina since the Late Cretaceous is commonly estimated to be about 450 km, but more

Figure 5. Relative latitudinal displacement and rotation of terranes in Alaska compared with cratonal North America, as determined from paleomagnetic data. See Grantz and others (1991) for sources of paleomagnetic data and Figure 3 for explanation of abbreviations.

than 900 km of total displacement has been suggested on the basis of Paleozoic stratigraphic relations (Gabrielse, 1985). Paleomagnetic data from volcanic flows in the Paleocene Cantwell Formation of the north-central Alaska Range indicate that the northern part of the SAS arrived at its present latitude no later than 59 to 54 Ma (Hillhouse and Grommé, 1982; Hillhouse and others, 1985), whereas paleomagnetic data from the Arkose Ridge Formation in the southern part of the Peninsular terrane indicate that the southern part of the SAS has moved a minimum of 400 km northward with respect to the North American craton since deposition in the Paleocene and Eocene (Stamatakos and others, 1989). These conflicting results indicate displacement has occurred on a currently unidentified fault between the southern and northern SAS, perhaps the McKinley strand of the Denali fault system. These data also indicate that the SAS was assembled in its present position over a period of time by cumulative displacement along a number of faults both continentward of, and within, the SAS.

Paleomagnetic data from Upper Cretaceous and lower Tertiary rocks of the terranes of southern Alaska also indicate hundreds to thousands of kilometers of postdepositional latitudinal displacement of these terranes with respect to nuclear North America (Fig. 5). Lower Paleocene volcanic flows in the Ghost Rocks Formation of the Chugach terrane have been displaced $25° \pm 7°$ (about 2700 km) northward since their deposition (Plumley and others, 1983), and lower Eocene (57 ± 1 Ma) volcanic rocks in the Orca Group of the Prince William terrane have been displaced $13° \pm 9°$ (1400 km) northward since deposition (Coe and others, 1990). At least part of these major displacements occurred along the Border Ranges and Contact faults, where evidence for dextral strike-slip displacements of several hundreds of kilometers has been recognized (Roeske and others, 1990; Bol and Roeske, 1993), and unknown amounts of displacement may have occurred on numerous small faults within the Prince William and Chugach terranes. Paleomagnetism of cores from wells in the Yakutat terrane, the most outboard terrane of southern Alaska, indicates that the Yakutat terrane was displaced northward 13° (about 1400 km) since the middle Eocene.

Thus, paleomagnetic data indicate that some geologic units in southern Alaska have undergone hundreds to thousands of kilometers of postdepositional latitudinal displacement. These data support geologic inferences that southern Alaska has been built by a process of terrane assembly along transcurrent splays of interregional fault systems of the Pacific rim, although the specific faults responsible for some of the slip are not currently known. This process of assembly continues today with the emplacement of the Yakutat terrane in the eastern Gulf of Alaska by northward translation along the Queen Charlotte–Fairweather fault system and its splays (Bruns, 1985; Plafker, 1987).

Central and northern Alaska. Models for the tectonic assembly of central and northern Alaska would appear to be amenable to testing by paleomagnetic methods because the models call for large-scale crustal rotations or for latitudinal displacement along strike-slip faults. Unfortunately, most paleomagnetic studies in this region have involved rocks that have been chemically remagnetized, apparently in the Late Cretaceous or early Tertiary.

Among the rocks of central Alaska that have not undergone Cretaceous remagnetization is the thick Ordovician to Upper Devonian carbonate sequence of the Nixon Fork terrane. This terrane preserves a paleomagnetic signature that indicates that it was colatitudinal with nuclear North America in the Ordovician (Plumley, 1984). These results corroborate stratigraphic and paleontologic evidence that suggests that the Nixon Fork terrane is a fragment of North America displaced from an original position in the Canadian Cordillera (Blodgett, 1983).

Paleomagnetic results from rocks of northern Alaska address two topics key to tectonic reconstructions of Denalia: (1) the site of origin and mechanism of emplacement of the Arctic Alaska terrane and (2) the original latitudinal position of a terrane of the Angayucham (paleo-Pacific) Ocean. Models for the emplacement of the Arctic Alaska terrane and opening of the Amerasia basin of the Arctic Ocean are constrained by paleomagnetic data that suggest that the Arctic Alaska terrane has rotated about 70° counterclockwise since the deposition of its Hauterivian (124 to 131 Ma) strata (Halgedahl and Jarrard, 1987). Restoring this amount of rotation would place the Arctic coast of Alaska against the Queen Elizabeth Islands of Arctic Canada, closing the Amerasia basin. This model supports geologic data that indicate that the Amerasia basin began opening in the late Hauterivian (Grantz and May, 1983) and that the strata of the Arctic Alaska terrane were once contiguous with those of the Sverdrup basin of the Canadian Arctic Islands (Grantz and others, 1979; Embry, 1990). Paleomagnetic data from sheeted dikes of the Misheguk Mountain terrane, a Middle Jurassic ophiolite terrane of the Angayucham Ocean now at the top of the Brooks Range nappe sequence in western Alaska, indicate that it was formed at a paleolatitude of 32°N (Plumley and others, 1990). Relative to the present position of the terrane as part of the North American continent, the Misheguk Mountain terrane was displaced a minimum of 12° (about 1500 km) northward between the time of its crystallization at 170 Ma and emplacement on top of the Arctic Alaska terrane during the Late Jurassic to Neocomian. However, the amount of northward translation of the Misheguk Mountain terrane may be considerably larger because post-Neocomian counterclockwise rotation of the Misheguk Mountain terrane with the Arctic Alaska terrane is not accounted for by the paleomagnetic data. The preemplacement northward translation of the Misheguk Mountain terrane is probably related to convergence within the large Devonian to Jurassic Angayucham Ocean basin between the ensimatic arcs within the basin and the continental margin of North America represented by the Arctic Alaska terrane.

Late Cretaceous and Tertiary crustal-scale rotations

Recent paleomagnetic studies indicate that the western two-thirds of Alaska were rotated counterclockwise $44° \pm 11°$ between 65 and 35 Ma (Harbert and others, 1987; Coe and others, 1990). These paleomagnetic data are in agreement with the amount and sense of rotation required to account for the Alaska orocline (Fig. 1) and with recent detailed structural data collected in the hinge of the bend along the Contact fault and Border Ranges fault systems that suggest that the oroclinal bending occurred in the early to middle Eocene (Plafker and others, 1986; Bol and Coe, 1987; Little and Naeser, 1989). The mechanism for the bending is still debated, but the geologic, plate tectonic, and paleomagnetic data suggest that Grantz's (1966) hypothesis of block rotation in southern Alaska is reasonable (Fig. 6). The block-rotation model relates the oroclinal folding to compression between Eurasia and North America.

EVOLUTION OF THE CONTINENT-OCEAN TRANSITION IN ALASKA AND ITS TECTONIC ASSEMBLY IN MESOZOIC AND CENOZOIC TIME

Before Denalia—Our premise

Terranes in Alaska for which there is significant evidence for North American affinities include the Arctic Alaska, Ruby, Yukon-Tanana, Seward, Porcupine, Crazy Mountains, and Wickersham (Fig. 3). Continental terranes of uncertain origin, possibly derived from North America, include the Dillinger, Nixon Fork, and White Mountains (Fig. 3). All of these allochthonous terranes, while unique in detail, contain crystalline rocks and Proterozoic to lower Mesozoic sedimentary sequences. Although Precambrian crystalline rocks have not been mapped in many of these terranes, isotopic evidence suggests that they may occur at depth beneath some of the terranes.

It is our premise that some or all of the terranes of continental affinity are fragments of the pre-Jurassic continental margin of North America. This origin is indicated by the following lines of evidence, which have led numerous workers to correlate many Alaskan terranes with North American miogeoclinal rocks in the Canadian Cordillera (Blodgett, 1983; Tempelman-Kluitt, 1984; Dover, 1990b; Weber, 1990; Lane, 1991).

1. Deformed metasedimentary rocks—quartzose pelitic schist and quartzite and associated metabasite, marble, felsic metavolcanic rocks, and orthogneiss—of the Arctic Alaska, Yukon-Tanana, Seward, and Ruby terranes are lithologically similar.

2. Metamorphosed orthogneiss and associated felsic volcanic rocks of the continental terranes yield similar U-Pb ages of 391 to 343 Ma; Sm-Nd isotopic-model ages of about 2 Ga; high initial ^{87}Sr-^{86}Sr ratios of 0.707 to 0.714; and U-Pb detrital-zircon ages as old as 2.2 Ga (Aleinikoff and others, 1986, 1993; Patton and others, 1987; Hemming and others, 1989; Nelson and others, 1989; Rubin and others, 1990). These data are generally similar to isotopic data from continental rocks of North American affinity in the Canadian Cordillera (Hoffman, 1989) and suggest that the Devonian to Mississippian granitic rocks of the continental terranes constituted the northern part of a major arc developed along the western margin of North America in Devonian time (Rubin and others, 1990).

3. Each of the continental terranes displays a clockwise path of metamorphism that began with high-pressure metamorphism in the Jurassic and/or Early Cretaceous and that was overprinted by greenschist facies metamorphism at lower pressures in the middle Cretaceous (Dusel-Bacon and others, 1989; Dusel-Bacon and Hansen, 1992; Roeske and others, 1992). The younger retrograde metamorphism was coeval with regional extensional tectonism that affected all the continental terranes (Pavlis, 1989; Miller and Hudson, 1991).

4. In general, sedimentary sequences of the continental terranes were deposited on stable continental platforms throughout much of the Paleozoic to early Mesozoic.

5. Devonian and younger fossils of the continental terranes appear to be of North American affinity.

A relation of great importance to understanding the middle Mesozoic history of terranes in Alaska is the superposition of mafic and ultramafic rocks of the Angayucham, Misheguk Mountain, Tozitna, Kanuti, Seventymile, and Innoko terranes on continental rocks of the Arctic Alaska, Ruby, and Yukon-Tanana terranes (Fig. 3). The mafic and ultramafic terranes comprise enormous thrust sheets that were emplaced at high structural levels in Jurassic time (Roeder and Mull, 1978; Jones and others, 1986). Similar mafic and ultramafic rocks (Slide Mountain terrane) were emplaced on continental rocks of the North American miogeocline along much of the length of the Canadian Cordillera during the Jurassic (Monger and others, 1982; Harms and others, 1984). We contend that these tectonic events are correlative (although possibly diachronous) and represent obduction of oceanic rocks along much of the northwestern continental margin of North America. Thus, structural superposition of the mafic and ultramafic terranes constrains the Arctic Alaska, Ruby, and Yukon-Tanana terranes to a position adjacent to the North American continental margin by the time of the Jurassic obduction event. If the direction of thrusting of each of the mafic and ultramafic terranes could be determined, it might be possible to gain a better understanding of how the underlying continental terranes have been displaced kinematically since Jurassic time. Kinematic analyses of geologic contacts reflect later extensional displacement between the terranes rather than the earlier emplacement by thrusting (Gottschalk and Oldow, 1988; Harris, 1988; Moore and Murphy, 1989; Miller and Hudson, 1991; Pavlis and others, 1993).

Some have taken the view that the continental terranes of Alaska are essentially autochthonous and reflect an intricate continental margin (Patton and others, 1977; Box, 1985). Jones and others (1986), however, have pointed out that substantial displacement of the Ruby terrane with respect to the Arctic Alaska terrane is required to accommodate the emplacement of some of the allochthonous mafic and ultramafic rocks in the eastern part of the Arctic Alaska terrane. The Yukon-Tanana terrane has been displaced relative to the Ruby terrane a minimum of 450 km, and possibly much more, along the Tintina fault. Because the amount

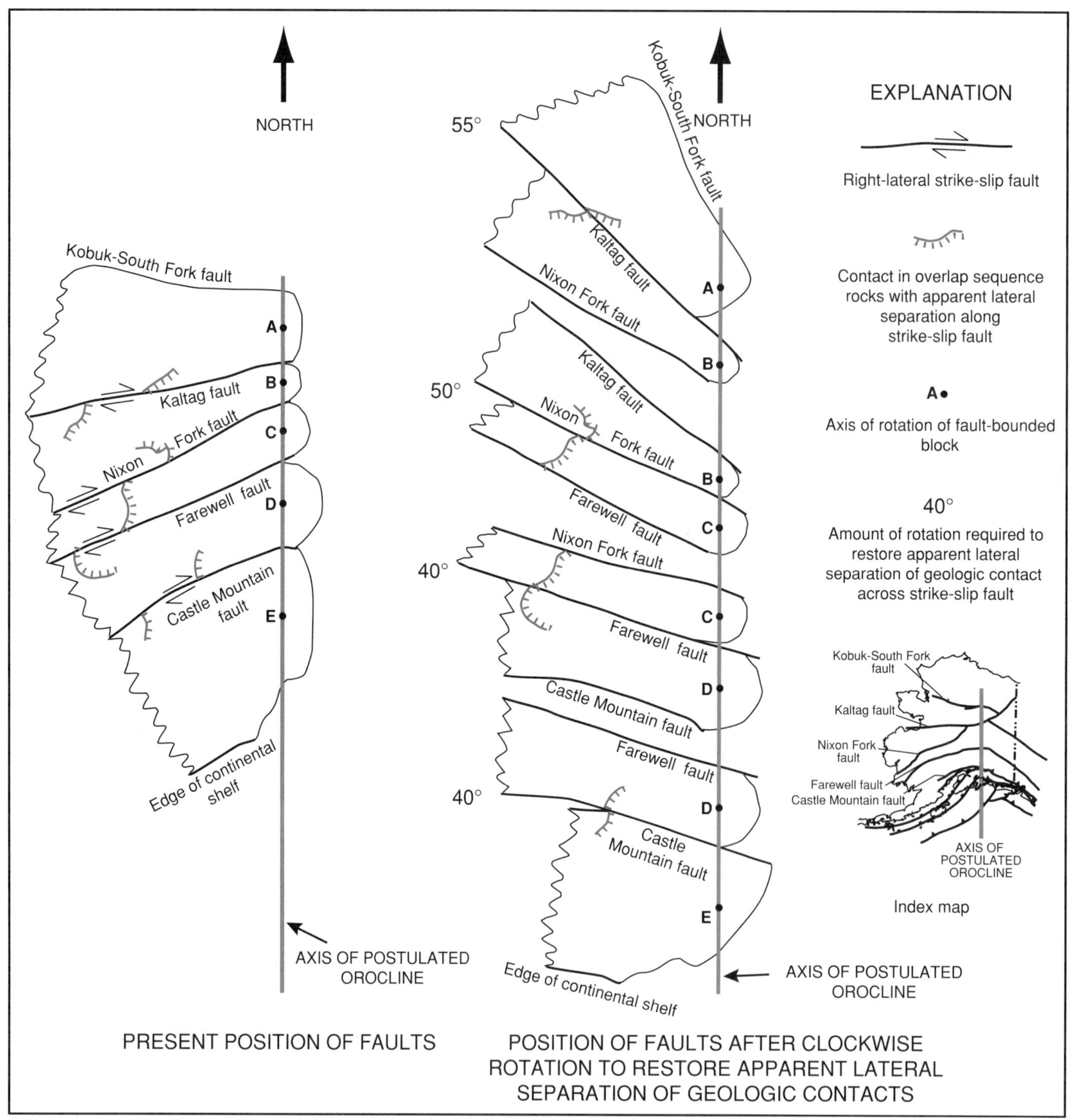

Figure 6. Block-rotation model of Grantz (1966) for restoring displacement on Late Cretaceous and Tertiary faults of the western limb of the Alaska orocline (see Fig. 1). Only faults that terminate eastward at axis of postulated orocline were considered for model. Amount of rotation across faults estimated from restoration of apparent lateral separation in Late Cretaceous and Tertiary rocks of overlap sequence. Model recently revised to account for 100 to 115 km separation across Castle Mountain fault (A. Grantz, 1993, unpublished data).

and timing of these displacements are poorly constrained, and because large-scale rotation of some of the terranes is probable (Tailleur, 1980; Plumley, 1984), we assume that the northwestern margin of the North American continent was originally linear to broadly sinuous, similar in profile to that of modern North America and South America. This assumption is supported by the Devonian arc assemblages discussed above, which probably formed as a sublinear feature along the continental margin. Furthermore, a more or less linear reconstruction of the Paleozoic and early Mesozoic continental margin of North America would require more conservative amounts of convergence between allochthonous mafic and ultramafic terranes and the continental rocks of the North American margin than that required along an intricate margin. In addition, most modern continental margins of the world are sublinear to broadly sinuous. Later oblique convergence and rifting structurally complicated this margin by imposing large transcurrent displacements, doubling-up, overthrusting, and rotation of continental fragments.

Paleozoic and early Mesozoic northwestern margin of the North American continent

We summarize below the geology of the Arctic Alaska terrane as an example of the pre-middle Cretaceous history of the northwestern margin of North America. The Arctic Alaska terrane is the largest continental terrane in Alaska and contains both subcrustal crystalline and supracrustal sedimentary rocks. Because Mesozoic foredeep strata are preserved in this terrane and because the terrane has largely escaped the late Mesozoic and Cenozoic transpressional deformation and magmatism that affected the continental terranes of central Alaska and the Canadian Cordillera, the Arctic Alaska terrane preserves the most detailed record of the northwestern margin of North America.

The Arctic Alaska terrane is divided generally into three primary stratigraphic units that reflect its tectonic history. In ascending order, these are (1) structurally and stratigraphically complex Proterozoic to middle Paleozoic assemblages (Franklinian sequence of Lerand, 1973); (2) a laterally extensive uppermost Devonian and Lower Mississippian to Lower Cretaceous nonmarine to marine succession deposited on a south-facing passive margin (Ellesmerian sequence of Lerand, 1973); and (3) Jurassic to Holocene siliciclastic foredeep deposits (Brookian sequence of Lerand, 1973). The two lower rock units record early Paleozoic convergence followed by rifting and tectonic quiescence, whereas the uppermost rock unit consists of sediments shed northward from the Brookian orogen, a collisional belt marked by obducted ophiolites of the Angayucham Ocean in northern Alaska. The Brookian orogen was formed during the Mesozoic and Cenozoic assembly of Denalia (see below).

Early Paleozoic convergence. The pre-Mississippian rocks of the Arctic Alaska terrane consist of metasedimentary schist, orthogneiss, and marble; phyllite along the length of the southern Brooks Range; locally schistose metasedimentary and granitic rocks in the northeastern salient of the Brooks Range; and clastic sedimentary rocks in the Lisburne Peninsula. Along the main axis of the Brooks Range, pre-Mississippian rocks typically constitute stratigraphic assemblages of Proterozoic to Devonian carbonate-platform rocks, lower Paleozoic clastic shelfal rocks and marble, and deformed Cambrian and Ordovician deep-basinal successions of chert, argillite, and mafic volcanic rocks (Moore and others, 1992). In addition, an Ordovician volcanic-arc assemblage is present in the central Brooks Range (Moore, 1987; Julian, 1989). Wells drilled into pre-Mississippian rocks in the North Slope indicate that much of that region is underlain by deformed lower Paleozoic argillite and fine-grained quartzose clastic rocks (Carter and Laufield, 1975), but to the south magnetic anomalies suggest that the pre-Mississippian succession contains igneous rocks. In the southern Brooks Range, pre-Mississippian rocks were mildly (nonpenetrative) deformed, but in the northeastern Brooks Range and North Slope they were strongly (penetrative) deformed and metamorphosed in the Late Silurian and Devonian (Oldow and others, 1987a, b). Widespread, large granitic batholiths that yield discordant Early Devonian U-Pb crystallization ages intrude the pre-Mississippian rocks in the southern and northeastern Brooks Range and North Slope subsurface (Dillon, 1989; Aleinikoff and others, 1993). We previously interpreted metamorphosed Devonian silicic volcanic rocks associated with Late Devonian or younger metabasites in the southern Brooks Range as bimodal volcanic rocks (Grantz and others, 1991), but on the basis of reconnaissance geochemical (Rubin and others, 1990) and geochronologic (Aleinikoff and others, 1993) data, we provisionally reinterpret the silicic volcanic rocks as the extrusive equivalents of the Early Devonian batholithic rocks.

Although largely obscured and disrupted by younger Brookian deformation and metamorphism, we suggest that the Lower Devonian and older rocks of the Arctic Alaska terrane record deposition along a passive margin to North America in the Late Proterozoic and early Paleozoic. Evidence of this margin is provided by the depositional environments of the Lower Devonian and older rocks, which range from carbonate-platform to deep-basinal settings and commonly contain North American fauna. However, Ordovician volcanic-arc rocks in the central Brooks Range and a prominent Early Devonian magmatic belt in the southern Brooks Range indicate that basin-closing events affected northern Alaska in the Ordovician and Devonian. Additional evidence of basin closing in the early Paleozoic may be provided by Silurian and/or Devonian deformation in the northeastern Brooks Range and North Slope subsurface and by the presence of a trilobite fauna of insular Siberian affinity (Dutro and others, 1984; Palmer and others, 1984) in some stratigraphic successions in the central Brooks Range. We do not know whether this closure resulted from convergence in a back-arc basin, from accretion of exotic terranes by subduction or by transpression, or from other tectonic mechanisms. However, the evolved isotopic and compositional characteristics of the Devonian magmatic belt indicate that the belt intruded a continental platform package that had been assembled by the Early Devonian. The inferred suturing faults are shown schematically (and hypothetically) in the A-3 transect as juxtaposing paleo-peri-Siberia and paleo-North America at upper mid-crustal levels beneath the Brooks Range, although the position of such a suture has not been identified on existing geologic maps.

Middle Devonian to Early Jurassic rifting and tectonic quiescence. Tectonic relaxation followed early Paleozoic arc magmatism in the Arctic Alaska terrane and resulted in deposition of clastic strata in local extensional basins in the Middle to Late Devonian. Extensional tectonism in middle Paleozoic time is supported by pre-Mississippian graben-filling sequences observed in seismic data and well cores beneath the North Slope (Kirschner and others, 1983) and by an exposed sequence of coarse-grained marine to nonmarine clastic rocks in the northeastern Brooks Range (Anderson and Wallace, 1990).

In the northern part of the Arctic Alaska terrane, the graben-filling strata are overlain by the Ellesmerian sequence on a regional sub-Mississippian unconformity. Mississippian strata consist of compositionally mature, locally derived nonmarine rocks that are overlain by shallow-marine shale (Endicott Group) succeeded by Mississippian and Pennsylvanian platform limestone (Lisburne Group). Except for northerly derived outpourings of clastic sediments that produced localized coarse-grained coastal-plain deposits along the northern margin of the terrane in the Permian and Triassic (Sadlerochit Group), Permian to Lower Cretaceous rocks consist mainly of shallow-shelf to shelf-basin shale rich in organic materials (Shublik Formation and Kingak Shale). Well and seismic data show that the Ellesmerian sequence onlaps the deformed pre-Mississippian rocks toward the north and that lithofacies within the Ellesmerian sequence generally become younger and coarsen northward. Facies relations indicate that regional northward transgressions occurred in the Mississippian, Permian, and Late Triassic.

To the south in the Brooks Range orogenic belt, strata correlative with the Ellesmerian sequence compose the Endicott Mountains and structurally higher De Long Mountains subterranes (Martin, 1970; Mayfield and others, 1988; Moore and others, 1992). These subterranes are nappes that were thrust northward onto the North Slope subterrane in the Late Jurassic and/or Early Cretaceous. Structurally simple reconstructions suggest that the De Long Mountains subterrane traveled farther than the Endicott Mountains subterrane and that both the Endicott Mountains and De Long Mountains subterranes originated in a more distal position relative to the North Slope subterrane (Mayfield and others, 1988). Stratigraphic successions of the Endicott Mountains and De Long Mountains subterranes fine upward from Devonian and Carboniferous nonmarine and shallow-marine deposits into fine-grained Permian to Jurassic deposits that are similar to those of the North Slope subterrane, but that display more distal and deeper water facies than the North Slope subterrane. In addition, the base of the Endicott Mountains allochthon consists of a thick, coarse-grained fluvial-deltaic sequence (Endicott Group) that began prograding southward across the rifted margin of the Arctic Alaska terrane in the Late Devonian. Stratigraphic relations among the North Slope, Endicott Mountains, and De Long Mountains subterranes, coupled with the presence of coeval oceanic rocks of the Angayucham and related terranes south of the Arctic Alaska terrane, indicate that the Arctic Alaska terrane was a fragment of a slowly subsiding, south-facing passive continental margin that persisted from the latest Devonian to Early Cretaceous.

In summary, the presence of extensional-basin deposits beneath the Ellesmerian sequence of the North Slope subterrane suggests that rifting began in the Middle Devonian and resulted in sea-floor spreading by the Late Devonian. Subsidence and tectonic quiescence following rifting resulted in deposition of a Late Devonian to Early Cretaceous passive-margin succession under gradually deepening conditions and waning input of clastic sediment.

Mesozoic assembly of proto-Denalia

First crust—Devonian to Early Cretaceous Angayucham Ocean. The oceanic tract postulated to have been beneath and seaward of the paleo-Pacific-facing North American continental margin constitutes the Angayucham Ocean. This ancient ocean is represented by ocean-margin, ensimatic-arc, ophiolite, and basalt-chert terranes that now lie structurally upon, or outboard of, the continental terranes of North American affinity in Alaska (Fig. 7). Although modified by structural attenuation, erosion, and younger deformation, the oceanic terranes are arranged in a structural succession of thrust sheets; the ocean-margin terranes at the structural base are succeeded upward, in turn, by the basalt-chert, ophiolite, and ensimatic-arc terranes. Reconstruction of these terranes is fundamental to understanding the initial phase of processes that led to the formation of Denalia.

The structurally lowest terranes are ocean-margin terranes (for example, Slate Creek, Venetie, Prospect Creek, and the Seventymile [lower part] terranes), which consist in part of quartzose flysch. Rare palynomorph and plant fossils indicate that the flysch is Devonian in age. Murphy and Patton (1988) interpreted the Slate Creek and Venetie terranes as slope and rise deposits correlative with the thick Upper Devonian fluvial-deltaic deposits of the Endicott Mountains subterrane because of compatible age, composition, and lithofacies.

The basalt-chert terranes include the Angayucham, Tozitna, Innoko, Goodnews, Tikchik, and Seventymile (middle part) terranes, which are composed of fault slivers, typically imbricated parallel to dip, of pillow basalt, diabase, gabbro, volcanogenic graywacke, chert, argillite, and minor carbonate rocks. Radiolarians in chert beds indicate that the fault slices range from Devonian at the base to Jurassic at the top (Hoare and Jones, 1981; Jones and others, 1988; Pallister and others, 1989). Shallow-marine carbonate rocks, which are present locally in the basalt at the base of the assemblage, contain Devonian and Carboniferous fossils (Patton and others, 1994). Geochemical data and lithofacies indicate that most of the mafic rocks were extruded on seamounts in an ocean basin (Barker and others, 1988; Pallister and others, 1989), although basalts of mid-ocean-ridge affinity are also present (S. E. Box, 1991, oral commun.). In addition, Triassic and Jurassic island-arc rocks have been identified within the Angayucham and Innoko terranes (Karl, 1991; Patton and others, 1994). High-pressure metamorphic assemblages in the Innoko, Tozitna, Goodnews, and Seventymile terranes (Foster and others, 1987; Roeske and others, 1992; Patton and others, 1994) indicate that these terranes were probably deformed and assembled in a subduction zone.

Structurally overlying the basalt-chert terranes are the ophiolitic Misheguk Mountain, Kanuti, and Seventymile (upper part) terranes. In most places, only the lower ultramafic and gabbroic parts of the ophiolite suite are present in these terranes; the upper diabase and basalt parts have been largely eroded or covered by younger sedimentary rocks. Although the age of the ophiolitic terranes is not well known, plagiogranite from the Misheguk Mountain terrane has yielded a U-Pb crystallization age of 170 Ma (Moore and others, 1993) and hornblende from the Kanuti terrane has yielded K-Ar ages that range from 172 to 155 Ma (Patton and others, 1977, 1994). Harris (1988) interpreted crystallization sequences and mineral chemistries from rocks of this subterrane as indicative of crystallization in an arc rather than at a mid-ocean ridge. Amphibolite facies metamorphic rocks are present locally at the bases of the ophiolitic terranes and are inferred to be fragments of basalt-chert terranes that were metamorphosed beneath hot ophiolitic terranes and underplated in a newly formed subduction zone (Zimmerman and Frank, 1982; Boak and others, 1987; Loney and Himmelberg, 1989). $^{40}Ar-^{39}Ar$ ages indicate metamorphism of the amphibolite facies rocks at 170 to 163 Ma, less than 10 m.y. after crystallization of the ophiolitic sequences.

Regional map relations indicate that the Koyukuk, Nyac, Togiak, and Stikine volcanic-arc terranes and the Kahiltna and Gravina-Nutzotin volcanogenic-flysch terranes are the structurally highest of the Angayucham Ocean (Grantz and others, 1991;

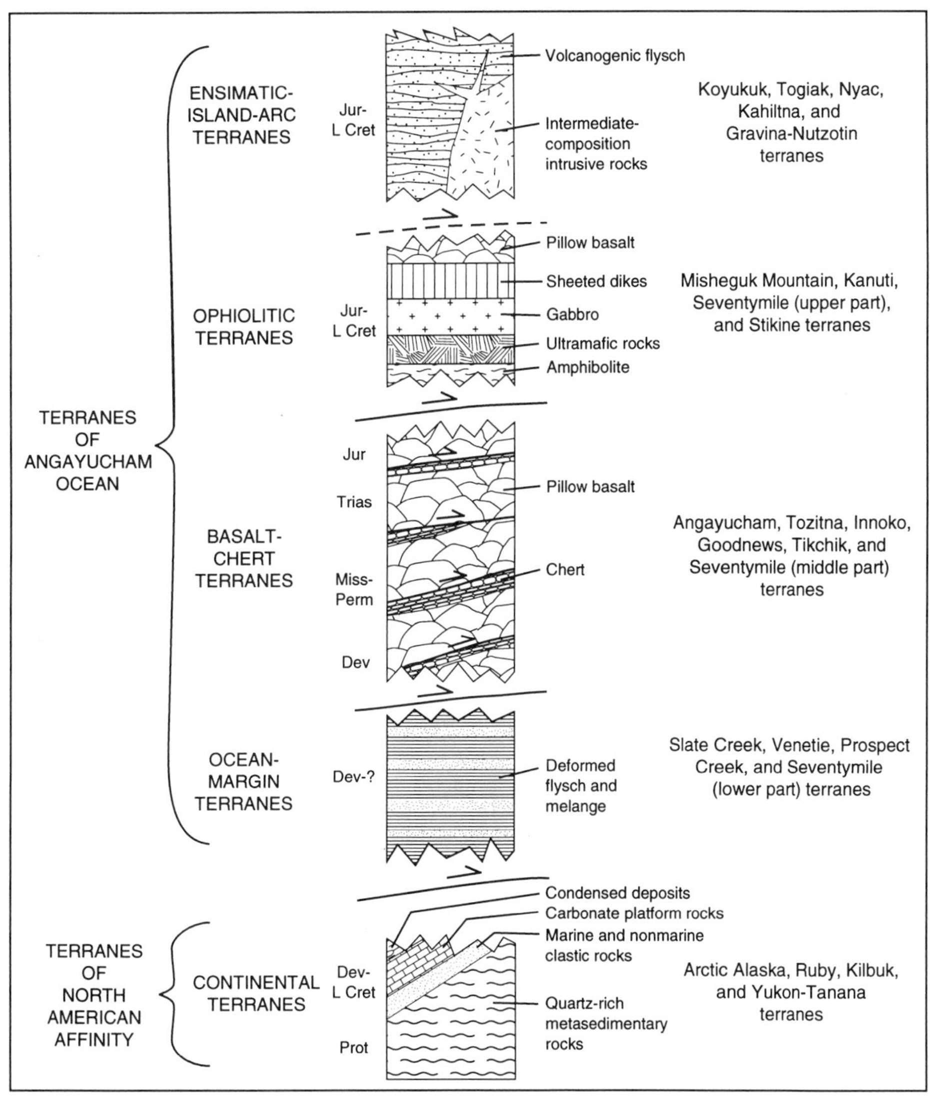

Figure 7. Schematic stratigraphic column depicting structural stacking of Angayucham Ocean terranes.

Patton and others, 1994). The arc terranes are composed of variable proportions of basalt, and lesser andesite, volcaniclastic rocks, plutonic rocks, chert, argillite, and clinopyroxene-rich graywacke turbidites, all of which show close geochemical affinity with modern ensimatic volcanic arcs (Box and Patton, 1989; Wallace and others, 1989). The arc terranes form enormous, structurally complex assemblages comprising arc edifices and associated sedimentary prisms. Continentally derived detritus is not present in most of these terranes, although quartz-rich debris of possible North American origin is present in the Gravina-Nutzotin terrane (Nokleberg and others, 1985). Most of the arc terranes are poorly dated but yield Jurassic and Early Cretaceous, and locally Triassic, isotopic and fossil ages (S. E. Box, 1991, oral commun.). We suggest that these arc terranes are fragments of a number of mainly ensimatic arcs of variable ages, but we cannot eliminate the possibility that the terranes are fragments of a single long-lived arc.

The fossils, lithofacies, and present structural position of the oceanic terranes attest to the existence of an ocean basin of great complexity off the northwest margin of North America from the Devonian to the Early Cretaceous (about 250 m.y.). Because arc-volcanic rocks of Paleozoic age are absent in the terranes of the Angayucham Ocean, we suggest that the Angayucham Ocean was in a constructional phase before the Triassic. Convergent processes were initiated within the basin in the Triassic, which resulted in the consumption of its ocean floor by subduction and in the erection of intraoceanic arcs. The polarity and paleogeographic position of these arcs is unknown. By analogy with the modern western Pacific realm, where a large number of ensimatic arcs have been active for 10 to 50 m.y. and have evolved in complex ways, we postulate that a number of arcs, possibly of diverse polarity and configuration, populated the Angayucham Ocean basin from the Triassic to the Early Cretaceous.

First assembly—Jurassic and Early Cretaceous convergence between the Angayucham Ocean and the North American continental margin. The Angayucham Ocean was consumed in Triassic to Early Cretaceous time by convergence at local centers of subduction at an unknown number of volcanic arcs that include the older parts of the Nyac, Togiak, Koyukuk, and Kahiltna terranes. Subduction assemblages were constructed in front of these, and probably other, arcs, and incorporated imbricated parts of oceanic platforms, seamounts, older arc rocks, and their overlying sedimentary cover. With time, subduction resulted in the accretion of progressively more proximal parts of the Angayucham basin, on the basis of the upward decrease in age of superimposed fault slivers within the ocean-margin and basalt-chert terranes, and the upward progression in lithofacies from (1) slope and rise prisms of continentally derived Devonian sedimentary rocks, to (2) Devonian and Carboniferous basaltic rocks and associated carbonate bodies, which may have been generated in an extensional environment during initial opening of the Angayucham basin, to (3) mixtures of Permian, Triassic, Jurassic, and Lower Cretaceous radiolarian chert, seamount basalt, and mid-ocean-ridge basalt.

Local fault slivers of tuffaceous graywacke of Middle Jurassic age (Patton and Tailleur, 1964) and arc detritus in Brookian sequence foredeep deposits in the Brooks Range (Mayfield and others, 1978) corroborate our hypothesis that volcanic arcs were emplaced at a structurally high level along the margin of the Pacific Ocean, but are now largely eroded away. On the basis of age, chemical signatures, and structural position, we suggest that the ophiolitic terranes represent suprasubduction-zone ophiolites that floored the forearc (or possibly interarc basins) of the volcanic arcs. The Jurassic and Early Cretaceous ensimatic-arc terranes of southern and western Alaska (for example, Koyukuk, Nyac, and Togiak terranes) may be the preserved remnants of these volcanic arcs, but are in uncertain relation with the ophiolitic terranes because of separation by later high-angle faults (for example, the Kobuk–South Fork fault in Fig. 1). The total displacement along these faults is unknown, but may include a large amount of strike-slip displacement (Grantz, 1966). Thus, a direct correlation between obducted ophiolite terranes (for example, the Misheguk Mountain terrane) and adjacent arc terranes (for example, the Koyukuk terrane) may be misleading because of the possibility of significant lateral displacement on later faults.

Amphibolite facies metamorphism of the basalt-chert (subduction complex) terranes as a result of inferred underthrusting beneath hot suprasubduction-zone ophiolite in the Middle Jurassic indicates that subduction was underway at that time. Subduction may have been initiated outboard of the North American continent in the Angayucham Ocean, because no continental rocks are involved in the metamorphism. The earliest involvement of continental rocks in the subduction is indicated by the oldest (Upper Jurassic) foredeep deposits in the Brooks Range. This sedimentary detritus suggests that subduction in the Angayucham Ocean began to emplace terranes onto the North American continental margin by the Late Jurassic. The involvement of continental rocks in subduction produced high-pressure metamorphism, deformation, and thrusting along the southern margin of the continental terranes. In the Arctic Alaska terrane, this tectonic event was the early part of the Brookian orogeny. Because isotopic dating of the high-pressure metamorphic rocks is complicated by the presence of metastable mineral phases, the timing of subduction of continental rocks has not been unequivocally dated, but is thought to have occurred during the Jurassic and/or Early Cretaceous (Armstrong and others, 1986; Miller and Hudson, 1991).

The emplacement of the terranes of the Angayucham Ocean onto the Arctic Alaska terrane resembles the obduction of the oceanic Slide Mountain terrane onto terranes of North American affinity in the Canadian Cordillera. Similarly, the Stikine terrane, which consists of ensimatic-arc rocks of late Paleozoic and Jurassic age (Samson and others, 1991), structurally overlies high-pressure metamorphic rocks of the Yukon-Tanana terrane in east-central Alaska (Dusel-Bacon and Hansen, 1992). These relations support our hypothesis that much of the continental margin of northwestern North America was subducted and metamorphosed beneath volcanic arcs of the Angayucham

Ocean during the middle Mesozoic (also see Coney, 1989; Gehrels and others, 1990).

The collision of volcanic arcs of the Angayucham Ocean with the northwestern margin of North America in the Late Jurassic and Early Cretaceous resulted in the emplacement of the subduction assemblages, ophiolite, and probably the arc terranes onto the margin of North America. The presence of high-pressure prograde metamorphic assemblages in the continental terranes (Arctic Alaska, Ruby, Yukon-Tanana terranes) (Armstrong and others, 1986; Dusel-Bacon and Hansen, 1992; Roeske and others, 1992), the absence of Jurassic and Neocomian magmatic belts in these continental terranes, and the presence of contractional structures in the Arctic Alaska terrane that verge away from the oceanic rocks are evidence that the margin of North America was subducted oceanward beneath one or more ensimatic arcs (Fig. 8). The collision and subduction of the northwestern margin of North America by an arc of the Angayucham Ocean in the Jurassic and Early Cretaceous may be analogous to modern subduction of Australia beneath the Banda arc in Indonesia and of Eurasia beneath the Luzon arc in Taiwan (Box, 1985).

Alternatively, the relation between terranes of North American affinity and those of the Angayucham Ocean can be explained by a tectonic model that calls for oceanward-facing arcs that fringed the northwestern margin of North America in the Jurassic, but were separated from the continental margin by marginal basins (Fig. 9) (Miller and Hudson, 1991). In this model, the marginal basins constitute the Anagayucham Ocean. Northeastward subduction of oceanic crust beneath the arcs may have driven backarc thrusting that collapsed marginal basins of the Angayucham Ocean and emplaced them as imbricate thrust sheets onto the North American margin in the Early Cretaceous (Fig. 9). Although driven by subduction of opposite polarity from that of the arc-continent collision model, the direction of structural transport and the stacking order of terranes onto the margin of North America would be similar in both scenarios. The long span (Devonian to Jurassic) of the Angayucham terrane and the deep-marine character of its sedimentary deposits, however, suggest that the Angayucham basin was located away from a continental margin and was older than arcs to which it could be linked if formed as a backarc basin. The backarc-basin model might satisfy the known relations, however, if the backarc region were a marginal sea floored by Paleozoic ocean crust that was starved of continental sediment from the Mississippian to Jurassic.

A third model, favored by one of us, proposes that subduction of oceanic crust of the Angayucham basin occurred along an oceanward-facing subduction zone bordering the western margin of the North American continent. This model has the same subduction polarity as that in Figure 9, but lacks a fringing arc and marginal basin. In this scenario, the Angayucham terrane was assembled as an accretionary prism along the margin of North America above eastward-subducting oceanic crust of the Angayucham Ocean. The accretionary prism was later collapsed along back thrusts that partly obducted and partly underplated the prism onto the outboard margin of North America, similar to the "flake tectonic" model of Oxburgh (1972). The oceanward-facing subduction model would be confirmed if there were a Jurassic and Early Cretaceous magmatic belt in the Arctic Alaska terrane. Arc-volcanic rocks, however, are restricted to the sparse fault slivers of Middle Jurassic tuff preserved near the structural top of the Arctic Alaska terrane (Grantz and others, 1991; C. G. Mull, 1993, oral commun.), and plutonic rocks of this age have not been found in the Arctic Alaska terrane. This model may be valid, nonetheless, if subduction were amagmatic, because the subducted plate descended at a low angle, as in part of the modern Andean arc.

Although the arc-continental collisional model for the Brookian orogen (Fig. 8) best explains the history of high-pressure metamorphism, thrusting, and absence of Jurassic and Cretaceous magmatism in the Arctic Alaska terrane and is assumed throughout the remainder of this paper, it fails to explain the northern position of the deep crustal root of the Brooks Range required by the Bouguer gravity field (Grantz and others, 1991, sheet 3) and recent seismic data (Levander and others, 1991). It also does not account satisfactorily for the present position of the original basement of the continental nappes (De Long and Endicott Mountains subterranes) of the Brooks Range. In addition, it requires a large amount (in excess of 500 km, the amount of shortening estimated within the Brooks Range) of subduction of relatively light continental crust beneath the thinner and heavier oceanic crust. If the overriding plate is assumed to be the Koyukuk terrane to the south of Arctic Alaska, further difficulties are (1) the age of the Koyukuk (arc) terrane, which formed in the Early Cretaceous, well after the Misheguk Mountain terrane was emplaced as the highest nappes of the Brooks Range in the late Middle Jurassic, and (2) the relatively short strike length along which the Brooks Range faces the Koyukuk terrane. For these reasons, we conclude that one or more arcs, larger and older than the Koyukuk arc, were associated with the subduction, underplating, and obduction that produced the Brooks Range orogen, and that the Koyukuk arc represents only a late and relatively minor phase of the orogeny.

We have adopted a structural model that provides for tectonic wedging in the subducted continental plate and hence northward underthrusting beneath the Arctic Alaska terrane, as well as over it (Grantz and others, 1991, sheet 3). Such continentward underthrusting, shown schematically in Figure 9, resulted in oceanic and lower continental crust being underplated beneath the thick continental crust of the Arctic Alaska terrane in the Brooks Range. This underplating provides for the thickening of the lower crust beneath the present-day Brooks Range. Seismic and magnetotelluric data may support a similar scenario for the southern margin of the Yukon-Tanana terrane (see below). Although such underplating would be most easily explained by back thrusting in an oceanward-facing subduction zone with or without a marginal basin, underplating may also have resulted from subduction beneath continentward-facing arcs if subduction were caused by ridge-push forces.

Stratigraphic evidence and isotopic dating indicate that the

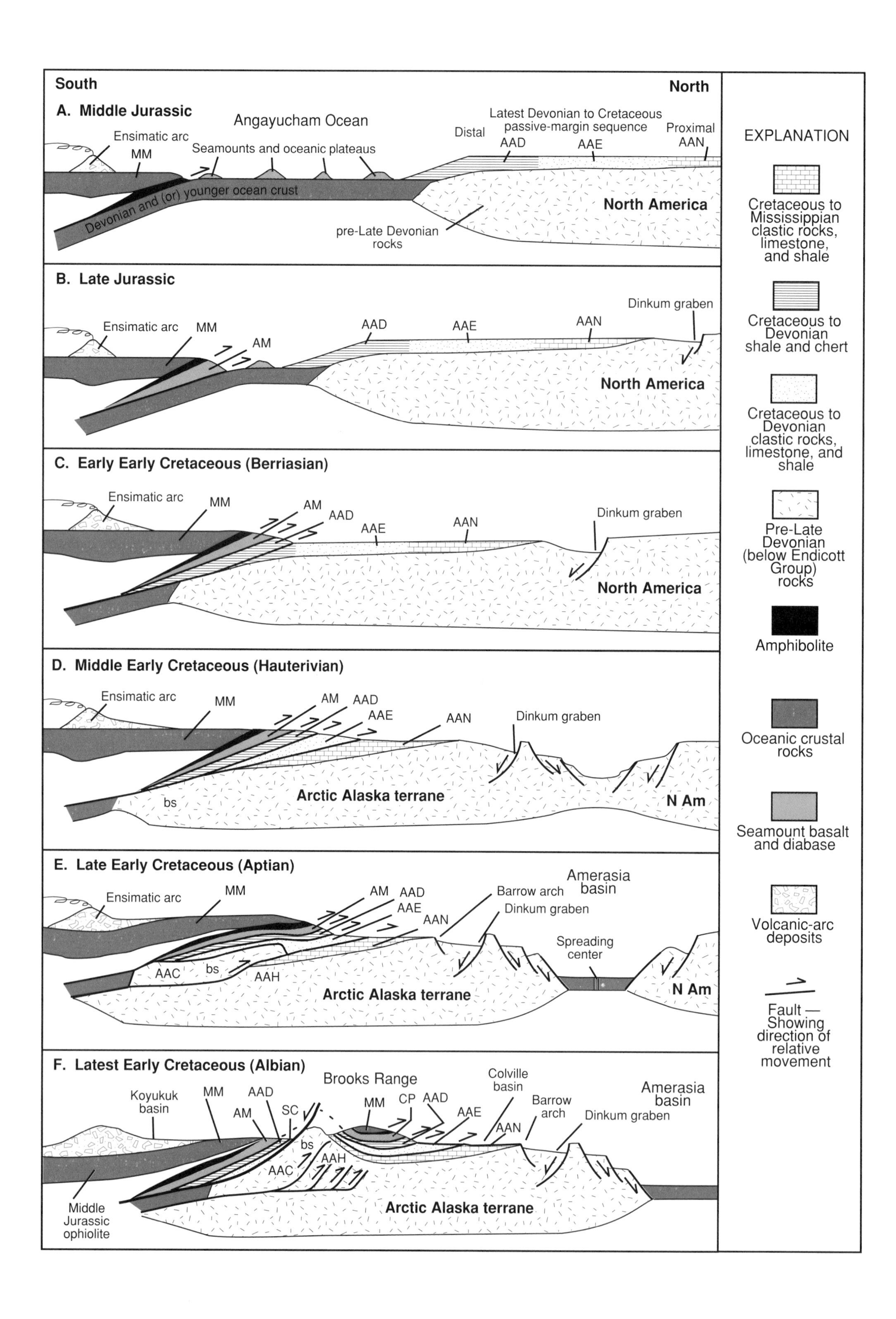
South
North
A. Middle Jurassic
Angayucham Ocean
Ensimatic arc
MM
Seamounts and oceanic plateaus
Distal
Latest Devonian to Cretaceous passive-margin sequence
Proximal
AAD
AAE
AAN
Devonian and (or) younger ocean crust
North America
pre-Late Devonian rocks
B. Late Jurassic
Ensimatic arc
MM
AM
AAD
AAE
AAN
Dinkum graben
North America
C. Early Early Cretaceous (Berriasian)
Ensimatic arc
MM
AM
AAD
AAE
AAN
Dinkum graben
North America
D. Middle Early Cretaceous (Hauterivian)
Ensimatic arc
MM
AM
AAD
AAE
AAN
Dinkum graben
bs
Arctic Alaska terrane
N Am
E. Late Early Cretaceous (Aptian)
Ensimatic arc
MM
AM
AAD
AAE
AAN
Barrow arch
Amerasia basin
Dinkum graben
Spreading center
AAC
bs
AAH
Arctic Alaska terrane
N Am
F. Latest Early Cretaceous (Albian)
Brooks Range
Koyukuk basin
MM
AAD
AM
SC
MM
CP
AAD
AAE
Colville basin
Barrow arch
Amerasia basin
Dinkum graben
AAN
bs
AAH
AAC
Middle Jurassic ophiolite
Arctic Alaska terrane
EXPLANATION
Cretaceous to Mississippian clastic rocks, limestone, and shale
Cretaceous to Devonian shale and chert
Cretaceous to Devonian clastic rocks, limestone, and shale
Pre-Late Devonian (below Endicott Group) rocks
Amphibolite
Oceanic crustal rocks
Seamount basalt and diabase
Volcanic-arc deposits
Fault — Showing direction of relative movement

regional phase of convergent deformation and high-pressure metamorphism in the Brooks Range during the Late Jurassic and Early Cretaceous was succeeded in the middle Cretaceous by widespread uplift, resetting of K-Ar and Ar-Ar ages, ductile deformation, and overprinting, or in some instances, total replacement, of high-pressure mineral assemblages by lower pressure assemblages (Dusel-Bacon and others, 1989). This uplift and deformational event was broadly coeval with filling of the adjacent Colville (north) and Koyukuk (south) sedimentary basins and with the onset of normal faulting along the southern margin of the Brooks Range (Gottschalk and Oldow, 1988; Miller and Hudson, 1991). The overprinting metamorphism, normal faulting, uplift, and erosional unroofing during the middle Cretaceous reflects gravity extension of crust that had been tectonically thickened during the earlier, convergent phase of Brookian deformation (Fig. 8).

Similarly, the Ruby and Yukon-Tanana terranes record metamorphism in the Jurassic followed by unroofing and resetting of K-Ar ages in the middle Cretaceous (Dusel-Bacon and others, 1989). The structurally lower parts of the Yukon-Tanana terrane record ductile deformation and normal faulting approximately coeval with analogous structures in the Arctic Alaska terrane (Pavlis and others, 1993). In addition, the Ruby and Yukon-Tanana terranes (1) were intruded by middle Cretaceous granitic rocks that are highly evolved and (2) locally underwent higher grade metamorphism than rocks of the Arctic Alaska terrane. The Ruby and Yukon-Tanana terranes may have undergone more extension than the Arctic Alaska terrane because the overlying ophiolitic terranes (Tozitna, Seventymile terranes) directly overlie crystalline, rather than sedimentary, rocks. We therefore conclude that the middle Cretaceous extension documented in the Arctic Alaska terrane occurred elsewhere along the northwestern margin of North America as a consequence of arc-continent collision, and that lateral dispersal of these terranes did not occur until after the initiation of extensional deformation and related metamorphism and plutonism.

Figure 8. Arc-continent subduction model for Brookian orogen (modified from Moore and others, 1992). A: Southerly directed subduction (present coordinates) of late Paleozoic and early Mesozoic oceanic crust of Angayucham Ocean in Middle Jurassic; amphibolitic metabasite generated by underthrusting of oceanic crust and platforms beneath young, hot suprasubduction-zone oceanic crust (MM). B: Subduction of oceanic crust nearly complete; Angayucham Ocean basin mostly closed. Subduction complex consists mostly of seamount and oceanic plateau fragments scraped from subducted oceanic crust (AM). Failed Jurassic rifting in Dinkum graben prior to opening of Amerasia basin is shown at right. C: Subduction of seaward part of Arctic Alaska terrane in latest Jurassic or earliest Cretaceous; outboard sedimentary cover (AAD) detached from basement and imbricated. D: Subduction of continental substructure to depth sufficient to produce blueschist facies metamorphism (bs); medial sedimentary cover (AAE) detached from basement, imbricated, and underthrust by inboard sedimentary cover and basement (AAN); rifting and continental breakup occurs along northern margin of Arctic Alaska terrane. E: Continental-substructure rocks containing blueschist facies assemblages uplifted and emplaced at higher structural levels by large-scale out-of-sequence faulting, imbrication, duplexing, and crustal thickening of lower crust (AAC); Amerasia basin opened and Barrow arch formed. F: Continued contraction thickened middle crustal rocks (AAH); coeval or later crustal attenuation thinned imbricated crustal superstructure along down-to-south normal faults, particularly in southern Brooks Range. Resulting uplift and erosional unroofing produced huge volumes of clastic detritus shed southward into Koyukuk basin and northward into Colville basin. Terranes of Angayucham Ocean: AM, Angayucham terrane; CP, Copter Peak terrane; MM, Misheguk Mountain terrane; SC, Slate Creek terrane. Subterranes of Arctic Alaska terrane: AAC, Coldfoot subterrane; AAD, De Long Mountains subterrane; AAE, Endicott Mountains subterrane; AAH, Hammond subterrane; AAN, North Slope subterrane.

Separation of northern Denalia from North America by middle Cretaceous rotational rifting in the Amerasia basin of the Arctic Ocean. Depositional facies and onlap relations demonstrate that a large continental highland, named Barrovia by Tailleur (1973), bordered the northern margin of the Arctic Alaska terrane and formed its source area during the late Paleozoic to late Mesozoic. The Amerasia basin now occupies the site of this highland, indicating that the Arctic Alaska terrane was severed from its northern source by the creation of this basin. Consequently, the evolution of the Amerasia basin, particularly the Canada basin (Fig. 1), is critical for understanding the evolution of northern Denalia.

The Canada basin, the part of the Amerasia basin south of the Alpha-Mendeleev Ridge and adjoining northern Alaska, northern Canada, and northeastern Russia, is the deepest part of the Arctic Ocean. It is floored by an abyssal plain that lies at a depth of about 4000 m and is underlain by more than 6 km of sedimentary deposits and a lower crustal basement 6 to 8 km thick (Grantz and others, 1990b), a thickness typical of old oceanic crust. Using seismic refraction methods, Mair and Lyons (1981) identified layers 2 and 3 within the lower crustal rocks of the basin and estimated their seismic velocities at 6.6 and 7.5 km/s, respectively; these layers rest on upper mantle that has a median velocity of 8.3 km/s. These data, although modified by us somewhat (May and Grantz, 1990; Grantz and others, 1991), are similar to those from other modern ocean basins and are accepted by most workers as strong evidence that the Canada basin is underlain by oceanic crust. However, magnetic anomaly patterns in the basin (Taylor and others, 1981; Vogt and others, 1982), which might provide direct evidence of sea-floor spreading, are poorly defined because of either thick sediment cover or an origin during a magnetically quiet interval. Seismic-reflection data calibrated by well data from the basin margins provide stratigraphic control on Albian and younger reflectors but do not clearly identify the age of the oldest sedimentary deposits in the basin. Consequently, the age of the oceanic crust underlying the basin can be dated only as pre-Albian (late Early Cretaceous). An Early Cretaceous age for formation of the basin has been estimated from Mesozoic stratigraphy of the North Slope of Alaska, Alaska Beaufort Sea shelf-bathymetry and heat-flow data, and comparison with other ocean basins (Grantz and May, 1983). Onshore

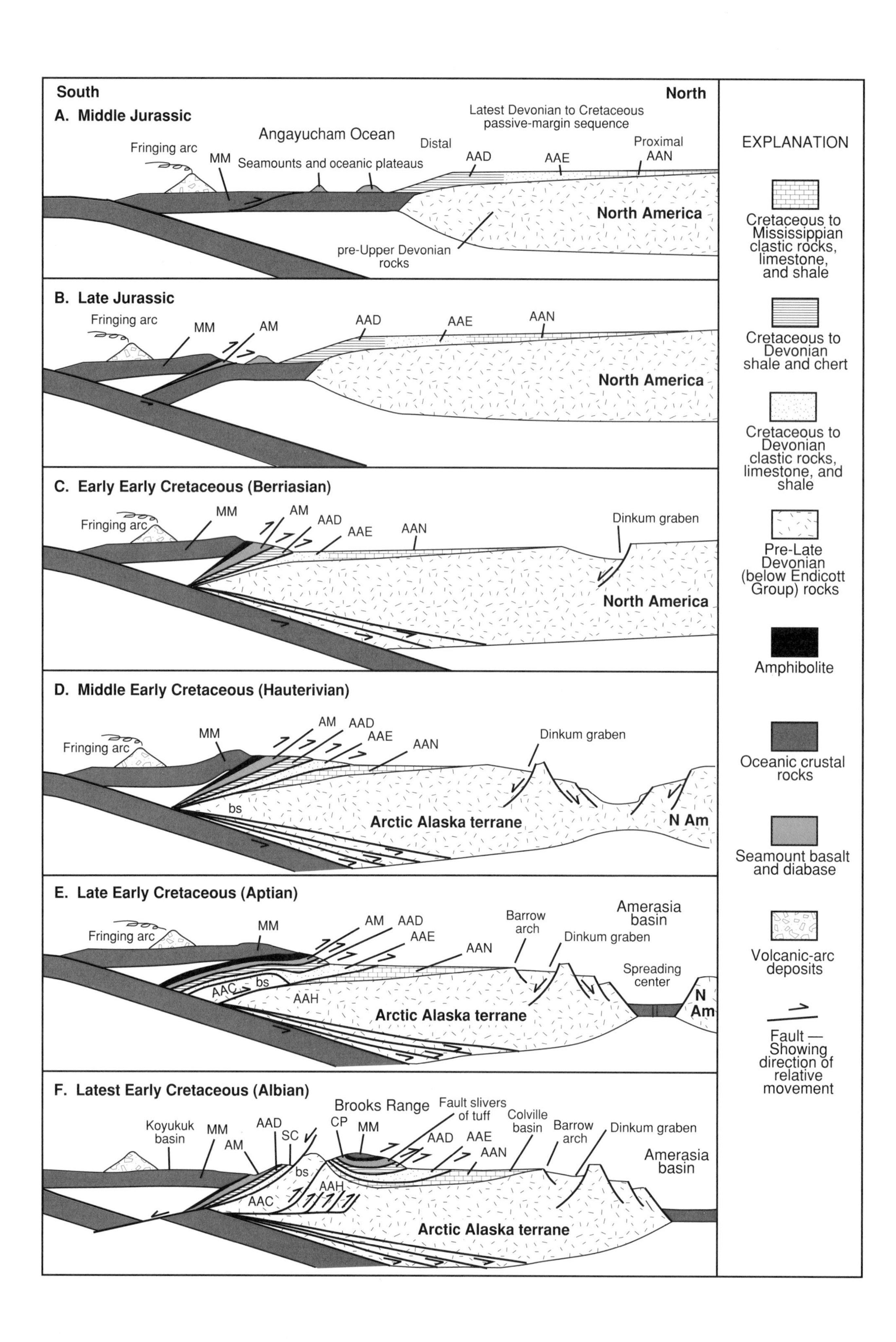
South
North
A. Middle Jurassic
Latest Devonian to Cretaceous passive-margin sequence
Angayucham Ocean
Fringing arc
MM
Seamounts and oceanic plateaus
Distal
AAD
AAE
Proximal AAN
North America
pre-Upper Devonian rocks
B. Late Jurassic
Fringing arc
MM
AM
AAD
AAE
AAN
North America
C. Early Early Cretaceous (Berriasian)
Fringing arc
MM
AM
AAD
AAE
AAN
Dinkum graben
North America
D. Middle Early Cretaceous (Hauterivian)
Fringing arc
MM
AM
AAD
AAE
AAN
Dinkum graben
bs
Arctic Alaska terrane
N Am
E. Late Early Cretaceous (Aptian)
Fringing arc
MM
AM
AAD
AAE
AAN
Barrow arch
Amerasia basin
Dinkum graben
Spreading center
AAC
bs
AAH
Arctic Alaska terrane
N Am
F. Latest Early Cretaceous (Albian)
Koyukuk basin
MM
AM
AAD
SC
Brooks Range
CP
MM
Fault slivers of tuff
AAD
AAE
AAN
Colville basin
Barrow arch
Dinkum graben
Amerasia basin
bs
AAH
AAC
Arctic Alaska terrane
EXPLANATION
Cretaceous to Mississippian clastic rocks, limestone, and shale
Cretaceous to Devonian shale and chert
Cretaceous to Devonian clastic rocks, limestone, and shale
Pre-Late Devonian (below Endicott Group) rocks
Amphibolite
Oceanic crustal rocks
Seamount basalt and diabase
Volcanic-arc deposits
Fault — Showing direction of relative movement

well and seismic-reflection data from the basin margin along the coast of northern Alaska record uplift, erosion, and extensional faulting into the Valanginian, followed by subsidence and onlap in late Hauterivian and younger time. The resulting Lower Cretaceous unconformity has been interpreted as a breakup unconformity that dates the initiation of sea-floor spreading in the Amerasia basin (Grantz and May, 1983).

On the basis of regional stratigraphic and tectonic relations, Rickwood (1970) suggested that the modern continental margin of northern Alaska is an Atlantic (passive) type and was formed at the end of deposition of the northerly derived upper Ellesmerian sequence. An Atlantic-type continental margin was confirmed by Grantz and May (1983), who presented seismic-reflection profiles across this margin and identified Lower Cretaceous to Quaternary unconformities, subsidence hinge lines, oceanward (northward) progradational sedimentary prisms, down-to-basin normal faults, large-scale slumps, growth faults, and other features typical of passive margins. The seismic reflection profiles also revealed the 10- to 40-km-wide and 170-km-long Dinkum graben beneath the Beaufort Sea shelf that Grantz and May (1983) interpreted as a fossil rift valley containing Jurassic and Cretaceous sedimentary rocks. The Dinkum graben and stratigraphic relations along the basin margin suggest that rifting began in the Middle Jurassic and that drift was initiated in the Hauterivian (131 to 124 Ma). Calculations based on heat-flow measurements and the depth of unloaded oceanic crust suggest that spreading continued until late Albian or Cenomanian time (about 105 to 95 Ma) (Langseth and others, 1990).

Because the precise age and spreading history of the Amerasia basin is unknown, most speculation about the position of the Arctic Alaska terrane prior to opening of the Amerasia basin in the Early Cretaceous has evolved into two general models: (1) the opening of the Amerasia basin by a 67° counterclockwise rotation about a pole in the Mackenzie delta and the restoration of northern Alaska to the northern Canadian continental margin (Grantz and others, 1979, 1990b) or (2) the nonrotational opening of the Amerasia basin along a ridge oriented normal to the northern Canadian continental margin and the consequent left slip of the Arctic Alaska terrane along a transform fault from a position against the Lomonosov ridge, now lying in the central Arctic, and the Eurasian Barents shelf (Nilsen, 1981; Hubbard and others, 1987; Oldow and others, 1987a, 1989). The opening of the Amerasia basin by 67° of counterclockwise rotation is supported by matching 1000 m isobaths and Paleozoic and Mesozoic stratigraphy between the northern Alaska and northern Canadian Arctic margins (Grantz and others, 1979, 1990b) and by paleomagnetic data from oriented drill cores in northern Alaska (Halgedahl and Jarrard, 1987). Conversely, left-slip transform displacement along the northern Canadian continental margin is not supported by seismic-reflection data offshore (Dixon and Dietrich, 1990), and the stratigraphies of northern Greenland, Spitsbergen, and northern Alaska, as well as geologic relations onshore in the Yukon and interior Alaska, are difficult to reconcile with the required 1200 km of displacement (Lane, 1992). Moreover, large-scale left-slip displacement in the Late Jurassic or Early Cretaceous would place the Arctic Alaska, Angayucham, and Misheguk Mountains terranes in positions near northern Greenland, far from their kindred Ruby, Yukon-Tanana, Tozitna, and Seventymile terranes of the northern Alaska Cordillera. Rotational opening of the Amerasia basin, in contrast, would restore the Arctic Alaska and Angayucham terranes to a position at the northernmost end of the Pacific margin of North America and close to the restored positions of other similar terranes in the Cordillera.

We conclude that rifting of the northern margin of the Arctic Alaska terrane was achieved by a 67° counterclockwise rotation of Arctic Alaska about a pole near lat 68°30′N, long 135°W in the lower Mackenzie River valley. The rotational model suggests that northern Alaska was attached to Arctic Canada in the pre-middle Cretaceous and that it is a fragment of the North American craton. On the basis of geophysical calculations, the opening of the Amerasia basin was completed by the Cenomanian. The rotation of the Arctic Alaska terrane and the structurally overlying Angayucham terrane about the pole in the Mackenzie delta produced a hinge in the sublinear northern margin of northwestern North America and resulted in a westward salient of North American rocks surrounded by oceanic regions. The emplacement of this salient of North America, the Arctic Alaska terrane, amid the middle Cretaceous oceanic rocks began the assembly of the subcontinent of Denalia.

Figure 9. Oceanward-facing subduction model for Brookian orogen. A: Ocean-facing, northerly directed (present coordinates) subduction of oceanic crust in Pacific basin beneath an ensimatic arc that fringed the northwestern margin of North America; formation of ophiolite (MM), back-arc thrusting, and amphibolite facies metamorphism initiated in back-arc basin in Angayucham Ocean. B: Collapse of marginal basin nearly complete and Angayucham basin mostly closed; rifting that lead to creation of the Amerasia basin begins. C: Outboard sedimentary cover (AAD) detached from basement and imbricated beneath upper plate of marginal-basin rocks; basement rocks to AAD detached and tectonically underplated. D: High-pressure metamorphism (bs) in underplated continental rocks; medial sedimentary cover (AAD) detached from basement, imbricated, and underthrust by inboard sedimentary cover and basement (AAN); rifting and continental breakup occurs along northern margin of Arctic Alaska terrane. E: Continental rocks with high-pressure assemblages uplifted and emplaced at higher structural levels by large-scale out-of-sequence faulting, imbrication, duplexing, and crustal thickening of lower crust (AAC); Amerasia basin opened and Barrow arch formed. F: Continued contraction thickened middle and lower crustal rocks; coeval or later crustal attenuation thinned imbricated crustal superstructure along down-to-south normal faults, particularly in southern Brooks Range. Resulting uplift and erosional unroofing produced huge volumes of clastic detritus shed southward into Koyukuk basin and northward into Colville basin. A generally similar oceanward-facing subduction model that depicts an Andean arc rather than a fringing arc and marginal basin is proposed in the text. Evidence of an Andean arc might be provided by fault slivers of tuffaceous volcanic-arc rocks in F. See Figure 8 for explanation of abbreviations.

Assembly of central and southern Denalia during oblique convergence beginning in middle Cretaceous time

In contrast to northern Denalia, which consists of a fragment of North America emplaced by creation of part of the Arctic Ocean, central and southern Denalia are constructed of terranes assembled along the paleo-Pacific margin of western North America. The origin of the terranes of central and southern Alaska from southerly paleolatitudes is indicated by a number of lines of information, including (1) paleomagnetic evidence, reviewed above, showing that important terranes originated at paleolatitudes 28° or more south of their present positions; (2) stratigraphic evidence indicating that some southern Alaska terranes are correlative with units in the Canadian Cordillera to the south and demonstrate large dextral slip on major faults, such as the Denali and Tintina faults (Eisbacher, 1976; Churkin and others, 1982; Gabrielse, 1985; Nokleberg and others, 1985; Plafker, 1987); and (3) paleontologic evidence suggesting that faunas in some southern Alaskan terranes were deposited at warmer paleolatitudes (Jones and others, 1977; Taylor and others, 1984; Silberling, 1985). These independent lines of evidence for northward displacement of major Alaska terranes are in agreement with implications of modern seismicity and fault motions along the margins of northwestern North America and with motions of North American and Pacific plates over the past 100 m.y. as determined by Engebretson and others (1985) and Debiche and others (1987) from hotspot tracks. Controversies, however, have arisen over the amount of displacement, kinematic history, sites of origin and collision, and paleogeographic reconstructions of the terranes.

The Jurassic to modern record of paleomagnetic data in the Pacific provides two models for emplacement of the southern Alaska terranes: (1) northward motion of non–North American terranes as part of the oceanic plates of the paleo-Pacific and docking at their present sites at any time since the Early Jurassic; and (2) northwestward propulsion of terranes by right-oblique subduction along the margin of the North American continent. The first model allows in excess of 6000 km of northward translation of terranes, but requires terrane origin at a non–North American location. The second model permits the origin of terranes along the margin of North America, but the amount of possible northward translation is limited to the northward component of northward-oblique plate convergence along the Pacific margin. Plate motions show that northward-oblique plate convergence along the North American continental margin occurred after 100 Ma (Avé Lallemant and Oldow, 1988); older northerly directed plate vectors along the North American margin are not known. Nonetheless, the component of plate motion tangential to the margin of North America since 100 Ma would produce about 3000 km of northward latitudinal displacement, sufficient to explain the northward displacement of the Alaska terranes required by paleomagnetic data. The two models are not mutually exclusive; some terranes may have traveled with oceanic plates, whereas others may have been propelled northward by oblique convergence along the margin of North America. Alternatively, terranes may have traveled with oceanic plates until being accreted along the margin of North America by collisional suturing or other processes at southerly paleolatitudes prior to 100 Ma, and subsequently been translated northward.

Although we suspect that the terranes of central and southern Alaska had a variety of origins and displacement histories, many of the most important terranes can be linked to North America by their composition, stratigraphy, fauna, or other lines of evidence, or contain Jurassic and Cretaceous overlap sequences that indicate residence along the margin of North America by the Cretaceous. We therefore conclude that the formation of the southern part of Denalia was in large part the product of translational tectonism since 100 Ma.

Continental and associated terranes of central Denalia. Along the A-3 transect, between the west-trending Kobuk and South Fork faults on the north and the Denali fault system on the south, interior Alaska consists principally of several large terranes of displaced continental rocks (Ruby, Yukon-Tanana, and Nixon Fork terranes) overlain by oceanic terranes (Tozitna, Innoko, and Seventymile terranes) and by a number of smaller terranes along the Tintina and Denali faults (Livengood, Manley, White Mountains, Wickersham, Pingston, and McKinley terranes) (Grantz and others, 1991, sheet 1). These terranes are overlapped by deformed middle and Upper Cretaceous flysch basins (Koyukuk and Kuskokwim basins). To the west and south of these terranes are widespread ensimatic island-arc terranes (Togiak, Nyac, and Koyukuk terranes) and middle Cretaceous volcanogenic-flysch terranes (Kahiltna and Gravina-Nutzotin terranes).

On the basis of similar age, lithology, structural and metamorphic history, and isotopic characteristics, the southern Arctic Alaska terrane (Coldfoot subterrane), the Ruby terrane, and the Yukon-Tanana terrane must have once formed a roughly linear belt along the margin of North America (Grantz and others, 1991). By the Early Cretaceous, these rocks had been partly subducted and metamorphosed under high-pressure conditions and overridden by oceanic terranes from the floor of the adjacent Angayucham Ocean. If the Arctic Alaska terrane was emplaced by rotational rifting in the Amerasia basin, as discussed above, then the correlative Ruby and Yukon-Tanana terranes, together with their structurally overlying ocean-floor assemblages, probably originated well to the south in the present Canadian Cordillera.

Assuming the Arctic Alaska, Ruby, and Yukon-Tanana terranes once formed a linear belt, the present configuration of these terranes provides a clue to the assembly of the interior Alaska terranes. The Arctic Alaska terrane and the Yukon-Tanana terrane are now elongate along the northwest to west trend of the Cordillera. The Ruby terrane, however, lies between these two terranes and has a northeasterly strike. Thus, these three terranes are arranged in an oroclinal Z-fold, the northeast-striking Ruby terrane forming the short central limb of the fold (Tailleur, 1980) (Fig. 10). The geometry of this apparent fold suggest right-lateral displacement in a northwest-trending shear couple. Apparent

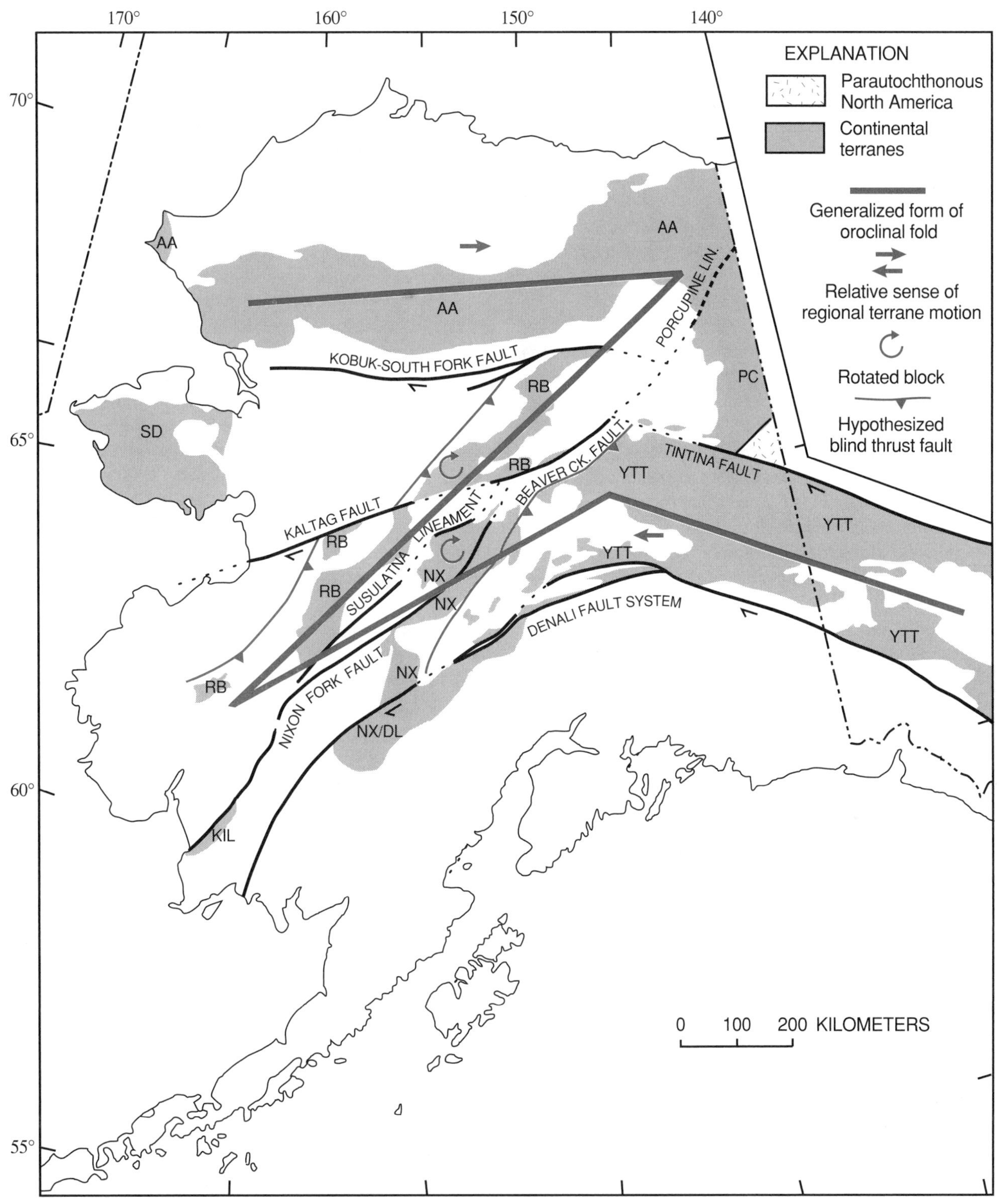

Figure 10. Apparent mega-Z fold of Alaska continental terranes and associated structures (after Tailleur, 1980). See Figure 3 for explanation of abbreviations.

right-slip displacement across the shear couple may be about 500 km. Actual displacement, however, may be significantly greater because the hinges of the oroclinal fold are truncated by the Kobuk and associated faults on the north and by the Susulatna lineament and the Beaver Creek and Kaltag faults on the south. Right-slip displacement would be at least 900 km if the continental terranes of southwest Alaska (such as Kilbuk) originated along the continental margin of North America.

If the Arctic Alaska, Ruby, and Yukon-Tanana terranes were emplaced in a right-slip shear couple as a crude mega-Z-fold, rotation of the short limb would be expected. Unfortunately, deformation and metamorphism has destroyed evidence bearing directly on the Z-fold in the Ruby terrane. Paleomagnetic data have been obtained, however, from the Nixon Fork terrane, which lies between the Ruby and Yukon-Tanana terranes (Fig. 10). These data indicate clockwise rotation of the Nixon Fork terrane of as much as 130° since the Paleozoic (Plumley, 1984). The paleomagnetic data also suggest little latitudinal displacement for Ordovician strata of the Nixon Fork terrane with respect to the present North American field. These data are supported by stratigraphic and faunal data, which are of North American affinity and suggest an origin of the terrane in the Canadian Cordillera (Blodgett, 1983). Because the Nixon Fork terrane is not structurally overlain by oceanic rocks (for example, the Angayucham, Tozitna, Innoko, or Seventymile terranes) and consists of mostly undeformed Paleozoic sedimentary cover resting on Proterozoic basement (Dillon and others, 1985), its original position along the margin of North America must have been relatively inboard of the metamorphic Ruby and Yukon-Tanana terranes.

Tectonic rotation of the crystalline Ruby and Nixon Fork terranes creates a space problem in the lithosphere. Rotation of these large terranes as rigid blocks in the upper crust would be accompanied by plastic deformation in subjacent lower crustal rocks. Therefore, we hypothesize the existence of a decollement under the Ruby terrane that verges toward the leading edge of the rotating blocks and terminates as blind thrusts in the Lower and middle Cretaceous flysch of the Koyukuk terrane, which was positioned originally along the continental margin of North America (Grantz and others, 1991, sheet 2). Consequently, the Ruby terrane is shown as a crustal prism that was thrust onto Lower and middle Cretaceous flysch of the Koyukuk terrane. Similarly, the Yukon-Tanana terrane is shown to be thrust northwestward onto Lower and middle Cretaceous flysch of the Manley terrane on a blind thrust beneath the Wickersham and White Mountains terranes (Grantz and others, 1991, sheet 2).

Island-arc and ocean-margin terranes of southern Denalia. South of the Denali fault, southern Alaska consists principally of accretionary prism, flysch, and continentalized arc-platform terranes that contain a great variety of paleomagnetic, paleobiogeographic, and other evidence of origin south of their present latitudes. Chief among these are the Alexander, Peninsular, and Wrangellia terranes, or the SAS (Fig. 3). Although the Alexander and Wrangellia terranes have distinct stratigraphic aspects, reconnaissance studies suggest that the contact between them was stitched by Pennsylvanian granitic rocks, thus indicating amalgamation by Pennsylvanian time (Gardner and others, 1988; Plafker and others, 1989). Although the contact between the Peninsular and Wrangellia terranes is faulted, coarse-grained Middle Jurassic volcanogenic rocks of arc derivation in both terranes indicate that amalgamation occurred by the early Middle Jurassic. The Strelna terrane of Grantz and others (1991), which is thrust onto the southern margin of the Wrangellia terrane (Fig. 3), may be a metamorphosed equivalent of the basal part of the unmetamorphosed Wrangellia terrane (Plafker and others, 1989) and is included herein in the SAS. In eastern Alaska, the Alexander terrane appears to plunge gently westward beneath the Wrangellia terrane to form its stratigraphic basement. This interpretation is supported by the thick crustal section beneath the SAS required by Bouguer gravity (Grantz and others, 1991) and refraction seismology (Goodwin and others, 1989).

The terranes of the SAS preserve an upper Proterozoic to Lower Cretaceous geologic record that is dissimilar to other successions of western North America. The SAS stratigraphy records (1) Late Proterozoic to Silurian arc magmatism followed by Silurian and earliest Devonian orogenesis; (2) a stable Middle Devonian and Carboniferous marine platform; (3) Carboniferous and Permian arc magmatism; (4) Late Triassic rift or plume volcanism and platform deposition; and (5) Middle Jurassic and then Early Cretaceous arc magmatism and shallow-marine to basinal deposition (Gehrels and Saleeby, 1987; Plafker and others, 1989; Nokleberg and others, 1994). Triassic and Jurassic faunas of the SAS, although fundamentally of North American origin (Savage, 1988), indicate marine depositional environments warmer than those found in North American successions at the same latitude (Tipper, 1981; Tozer, 1982; Silberling, 1985; Jones and others, 1986). The paleoclimatic observations corroborate the paleomagnetic data that indicate that the SAS has moved 28° northward since the Late Triassic, such that part arrived by the Paleocene or early Eocene. Middle Cretaceous deformation of the volcanogenic-flysch terranes (Gravina-Nutzotin and Kahiltna terranes) that intervene between the Yukon-Tanana terrane and the SAS has been cited as evidence that collision of the SAS with North America occurred in the middle Cretaceous (Coney and others, 1980; Csejtey and others, 1982; Nokleberg and others, 1985). However, cessation of arc magmatism in the Peninsular terrane in the Late Jurassic, stratigraphic relations of the Alexander terrane in southeastern Alaska, and terrane linkages based on correlations of magmatic belts in the Canadian Cordillera suggest that collision of the SAS with North America occurred in the Jurassic (Wallace and others, 1989; Rubin and Saleeby, 1992; Van der Heyden, 1992). Widespread middle Cretaceous granitic rocks in the continental terranes of interior Alaska and approximately coeval arc volcanism and plutonism in the SAS indicate that continentward subduction outboard of the SAS had begun by the middle Cretaceous. We conclude that the SAS was joined to the North American margin in the Jurassic about 28° south of its present location and was then translated coastwise along with other terranes to its present location during the Late

Cretaceous and Cenozoic. Deformation of the intervening flysch-basin terranes were caused in part by convergence during translation in the middle and Late Cretaceous (Jones and others, 1986; Plafker and others, 1989; Wallace and others, 1989) (see below).

The SAS is bordered on the south by the Chugach terrane at the Border Ranges fault (Fig. 3). The Chugach terrane consists, from north to south, of (1) blueschist and greenschist; (2) a melange of Triassic, Jurassic, and Lower Cretaceous mafic volcanic rocks, chert, limestone, and graywacke; and (3) deformed Upper Cretaceous volcanogenic flysch and oceanic volcanic rocks. These rocks record intermittent tectonism and deformation in the Jurassic and Cretaceous. Because it contains depositional debris that could only have been derived from the SAS, the Chugach terrane is regarded as a subduction complex that corresponds to the Jurassic and younger arc rocks of the SAS (Plafker and others, 1989). The Border Ranges fault is therefore a subduction-zone roof thrust, but because of its great length (about 2000 km) and the truncation of geologic structures in the SAS, this fault probably also accommodated younger strike-slip displacement of several hundred kilometers or more (Plafker and others, 1989; Roeske and others, 1990).

The Chugach terrane is extensively intruded and metamorphosed by early Tertiary intermediate-composition plutons that record partial melting of an underplated accretionary complex (Hudson and Plafker, 1982; Sisson and others, 1989). High-temperature metamorphism and igneous rocks of intermediate composition are unusual in active accretionary complexes. In this instance, these thermal effects may reflect subduction of hot oceanic crust of the Kula-Farallon ridge beneath the Chugach terrane (James and others, 1989; Plafker and others, 1989; Grantz and others, 1991, sheet 1). This explanation is supported by the timing of the plutonism and metamorphism, which was nearly coeval with the termination of accretion in the Chugach terrane.

The Chugach terrane is bordered on the south by the Prince William terrane at the Contact fault (Fig. 3). This terrane, consisting of Paleocene and Eocene flysch, oceanic volcanic rocks, and pelagic sedimentary deposits, was accreted to the Chugach terrane by 51 Ma (Plafker and others, 1989). Paleomagnetic data indicate 13° to 25° of northward latitudinal displacement of the Prince William terrane since the Paleocene. The Yakutat terrane, which is to the south of the Prince William terrane (Fig. 3), is an accretionary prism and fossil subduction zone displaced about 13° from its initial position in the Canadian Cordillera since the Eocene. The Yakutat terrane is currently being deformed against southern Alaska at the Pamplona fold and thrust belt (Plafker, 1987; Grantz and others, 1991, sheet 1) (Fig. 1; see below).

Flysch basins at inboard margin of SAS. The inboard limit of the SAS and the southern limit of terranes of continental origin correspond to the general position of the Denali fault and its splays (Fig. 3). This region is also occupied by thick, widespread, and highly deformed volcanogenic-flysch deposits assigned to the Kahiltna terrane in southwestern Alaska and the Gravina-Nutzotin terrane in south-central Alaska (Coney and Jones, 1985; Jones and others, 1986). These terranes form the northern part of a belt of flysch terranes that extends discontinuously for 800 km through southeastern Alaska.

The flysch terranes consist of Upper Jurassic and Lower Cretaceous strata derived mainly from volcanic-arc sources, but locally from the SAS and minor continental sources (Nokleberg and others, 1985, 1994). The basement rocks of the flysch terranes are mostly unknown, although in places the flysch rests on rocks of the SAS (Nokleberg and others, 1985; Wallace and others, 1989). In south-central Alaska, small rootless terranes of ophiolitic, volcanic-arc, and continental affinity (for example, the Chulitna and McClaren terranes) are engulfed in the Kahiltna terrane (Jones and others, 1986; Nokleberg and others, 1994), but evidence of high-pressure metamorphism is absent. The flysch terranes have been interpreted as an accretionary prism on the leading, inboard (northwestern) edge of the SAS, which, as discussed above, may have been deformed during the proposed collision between the SAS and North America in the middle Cretaceous (Coney and others, 1980; Csejtey and others, 1982; Nokleberg and others, 1985). North-dipping deep-crustal reflections along the southern margin of the Yukon-Tanana terrane (Fisher and others, 1989) and magnetotelluric data suggesting Mesozoic flysch beneath the Yukon-Tanana terrane (Stanley and others, 1990) may support this model.

Our model (Grantz and others, 1991, sheets 1 and 2), put forward above, accounts for paleomagnetic, stratigraphic, and paleontologic data by suggesting that the SAS collided with North America in the Jurassic, prior to subsequent north latitudinal displacement of 28°. The collisional relation of the SAS to North America is obscured by younger deformation, but may be related to the Jurassic obduction of ophiolitic rocks of the Seventymile terrane and volcanic-arc rocks of the Stikinia terrane onto the Yukon-Tanana terrane, and the consequent high-pressure metamorphism of the underlying Yukon-Tanana terrane. The collision of the SAS with North America may thus be a southern extension of the tectonism that led to the obduction of ophiolites in the Brooks Range in the Late Jurassic and Early Cretaceous. As with the orogenesis of the Brooks Range, we conclude that collision of the SAS with North America may have resulted from either (1) subduction of the margin of North America beneath a continentward-facing arc (Fig. 8), or (2) continentward backarc thrusting inboard of an oceanward-facing arc that fringed North America (Figs. 8 and 9). In the north, the polarity of subduction became ocean-facing subsequent to ophiolite emplacement and may have resulted in northward underplating of continental and oceanic rocks beneath the Yukon-Tanana terrane in the middle Cretaceous (shown for convenience as the Gravina-Nutzotin terrane in the lower crust of Grantz and others, 1991, sheet 2). Thus, we view the lower crustal structure of the southern margin of the Yukon-Tanana terrane as a crustal wedge, like that proposed for the Arctic Alaska terrane (see above).

In the Late Cretaceous and Cenozoic, northward translation of the SAS trapped parts of the Kahiltna terrane inboard of the SAS. Coeval north-dipping subduction along the Pacific rim outboard of the SAS resulted in further deformation of the Kahiltna and Gravina-Nutzotin flysch basins and emplacement of plutonic-arc rocks within the SAS and adjoining terranes. This model implies that (1) flysch basins of somewhat different origins are juxtaposed in the Kahiltna and, possibly, the Gravina-Nutzotin terranes and (2) there was significant northward translational displacement with respect to nuclear North America along an ancestral Denali fault, perhaps in excess of the nearly 400 km of slip attributed to the modern-day Denali fault by many (for example, Nokleberg and others, 1985, 1994).

Oroclinal deformation of Denalia by latitudinal convergence between North America and Eurasia

The remarkable southward-concave bend in regional stratigraphic, structural, and topographic trends of the Alaska orocline has been attributed to disparity in the geometric accumulation of terranes (Stone, 1980) or to preexisting protrusions of the North American continent (Box, 1985). However, abundant paleomagnetic data, presented above, indicate that southwestern Alaska rotated about 40° counterclockwise on a vertical axis from about 65 to 50 Ma (Coe and others, 1985). These data are in agreement with oroclinal bending models (for example, Grantz, 1966) (Fig. 6). Restoration of this rotation restores the structural trends in southwestern Alaska to a west or northwest strike, which is compatible with our postulated northwest-trending, sublinear margin for North America in pre-Cretaceous time. Rotation of the southwest-trending leg of the Alaska orocline in the latest Cretaceous and/or Paleogene implies that the development of the Bering and Navarin oroclines was coeval with that of the Alaska orocline and was probably part of the same plate-scale tectonic deformation.

Formation of the oroclines of southern Denalia in latest Cretaceous and/or Paleogene time is nearly coincident with (1) east-vergent thrusting along the Herald Arch, Lisburne Peninsula, and Seward Peninsula (Herald thrust zone) in northwestern Alaska (Grantz and others, 1981), (2) plate reorganization in the Bering Sea (Cooper and others, 1987), and (3) rapid northward translation of terranes along the Cordilleran margin (Engebretson and others, 1985; Harbert, 1990). In addition, relative plate motions between Eurasia and North America were strongly convergent between about 70 to 50 Ma (Engebretson and others, 1985; Harbert and others, 1990). We therefore relate oroclinal deformation in Denalia to convergence between North America and Eurasia, as well as northward migration of terranes along the Pacific margin.

In northern Denalia, convergence between North America and Eurasia produced basement-involved, east-vergent thrusting in the Herald thrust system, which extends southward from Wrangel Island to at least St. Lawrence Island. In southern Denalia, convergence between Eurasia and the terranes of southeastern Denalia, which were being propelled rapidly northward by the obliquely underthrusting Kula plate, resulted in oroclinal bending of southern Denalia and thrusting in the Koryak Mountains in the Russian Far East. The Kobuk–South Fork and Kaltag fault systems, which lie between these structural domains, must have accommodated significant relative displacements between the domains. The basement-involved thrusting north of the Kobuk–South Fork fault system may reflect the presence of thick and old continental crust at depth, whereas oroclinal folding south of the Kaltag fault may have been favored by the thinner and more heterogeneous crust of southern Denalia.

Cretaceous and Cenozoic dextral-slip fault systems and transpressional assembly of southern Alaska

According to our tectonic model for assembly of southern Denalia, transpressional emplacement of some of the constituent terranes requires orogen-parallel, dextral transcurrent displacements of 3000 km or more relative to nuclear North America. This displacement is transmitted via faults that enter southern and interior Alaska, where they form many terrane boundaries, truncate and displace parts of other terranes, and offset thrust faults that emplaced the Jurassic and Lower Cretaceous nappes of interior Alaska. There is substantial disagreement, however, on the timing and displacement of these faults. We summarize below our analysis of transcurrent displacements in southeastern Denalia by arranging the faults into three general systems: the Tintina fault system, the Denali fault system, and faults outboard of the SAS.

Tintina fault system. The Tintina fault system is a major tectonic boundary that juxtaposes the continental terranes of central Denalia against parautochthonous North America (Fig. 3). Most of the transcurrent displacement of east-central Alaska and the Canadian Cordillera was on the Tintina fault, but displacement in central and western Alaska was distributed among a series of splay faults, including the Kobuk–South Fork, Kaltag, Beaver Creek, Nixon Fork, and Susulatna faults. Although the Tintina is mainly a Late Cretaceous and Tertiary fault, seismic activity indicates that the Tintina fault system is active today (Estabrook and others, 1988).

About 450 km of post–Late Cretaceous right slip was suggested for the Tintina fault by Roddick (1967), and nearly 1000 km of displacement is suggested by offset Paleozoic facies trends in the Canadian Cordillera (Gabrielse, 1985). Some of the discrepancy may be accounted for by additional displacement on the fault in middle or Late Cretaceous time. A minimum of 900 km of displacement would be required to restore the continental terranes of possible North American affinity along the Tintina fault system in southwest Alaska (for example, Kilbuk and Ruby terranes) to a position along a sublinear margin of North America. Even greater displacement may be required if these continental terranes extend farther southwestward beneath the Bering Sea shelf. On this basis, we suggest a cumulative minimum displacement of at least 1000 km for the Tintina fault system. Furthermore, displacement of this magnitude would be required to accommodate the large-scale rotation of the Ruby and Nixon Fork terranes, as suggested by Grantz and others (1991) (see above).

Denali fault system. The arcuate Denali fault system, along which the A-3 transect is offset, is more than 2000 km long and forms one of the major tectonic boundaries of North America (Fig. 1). In south-central Alaska, the Denali system roughly separates relatively less displaced continental terranes to the north from relatively more displaced arc and ocean-margin terranes to the south (Fig. 3). In east-central Alaska, the Denali system consists of the Shakwak fault and, in south-central and southwestern Alaska, the Hines Creek and McKinley strands and the Togiak-Tikchik, Holitna, and Farewell faults (Fig. 3). On the basis of terrane correlations, Eisbacher (1976) and Nokleberg and others (1985) suggested 300 to 400 km of Late Cretaceous and younger right-lateral displacement across the fault. The McKinley strand has had as much as 6.5 km of Quaternary right slip (Stout and others, 1973), and Quaternary rates of movement on the eastern part of the Denali system average 1 to 2 mm/yr (Plafker and other, 1977).

The largest displacements on the Denali fault occurred in the Late Cretaceous and early Tertiary when the Shakwak segment and Hines Creek strand were active. Lesser Cenozoic displacement is transferred northward across the SAS from the Queen Charlotte fault system to the Denali fault system via the Fairweather and Totschunda faults (Figs. 1 and 3), although the southern continuation of the Totschunda has not been located precisely. About 600 km of late Cenozoic displacement is estimated for the Fairweather fault, but a large fraction of this displacement has been transferred to the Transition and Chugach–St. Elias faults (Fig. 3) and thrust faults in the eastern Chugach Mountains (Fig. 2). Although displacement across the Totschunda fault was estimated at less than a kilometer by Lahr and Plafker (1980), correlation of the southern margin of the flysch of the Kahiltna and Gravina-Nutzotin terranes across the fault suggests that the SAS has been displaced about 200 km to the northwest since the middle Cretaceous. Some of the Cenozoic right-slip displacement on the Denali fault may be taken up in northwest-vergent thrust faults in the southern Alaska Range (Lahr and Plafker, 1980; Plafker and others, 1992).

Faults outboard of SAS. The Border Ranges fault is a north-dipping fault along which the Chugach terrane was accreted to the SAS (Fig. 3). Although the Border Ranges fault is primarily a Mesozoic subduction-zone thrust fault, it has a complex kinematic history (Little and Naeser, 1989). Evidence for middle Cretaceous ductile deformation was reported by Pavlis and Crouse (1989), and Pavlis (1982) suggested Late Cretaceous and/or early Tertiary right-slip displacement. More than several hundred kilometers of right-slip displacement is possible for the fault (Roeske and others, 1991), but large-scale lateral translation on the fault may be mostly pre-Paleocene (Little and Naeser, 1989).

The north-dipping Contact fault system consists of thrust faults that place the Chugach terrane over the Prince William terrane (Fig. 3). These faults are well expressed with seismic refraction data (Fuis and others, 1991), and probably constitute an early Tertiary fault system with both thrust and right-slip displacement (Plafker and others, 1986; Bol and Roeske, 1993).

Following thrusting and right slip in the Cretaceous and early Tertiary, the Border Ranges and Contact faults became the sites of uplift and normal faulting (Little and Naeser, 1989). The Cenozoic uplift may have resulted from isostatic adjustment to the underplating of oceanic-crust and ocean-margin terranes beneath the Chugach terrane (Fuis and others, 1991). The uplift of thick sedimentary prisms on the south side of the faults imposed steep dips on the faults, which may have originally dipped northward at moderate angles.

Displacement history for southern Denalia. Geologic data indicate that the Tintina, Denali, Border Ranges, and Contact fault systems were active primarily in the Late Cretaceous and early Tertiary, coincident with rapid, oblique subduction of the Kula plate beneath the northwestern margin of North America 85 to 43 Ma (Engebretson and others, 1985). These relations suggest that a wide region along the outboard margin of North America was partially coupled to the Kula plate and propelled northward as a consequence of oblique subduction. About 1500 km of displacement between the SAS and North America can be accounted for on known faults. Significant additional right slip on faults outboard of the SAS, including the Border Ranges and Contact faults, is likely, but displacement on these faults may be only a fraction of the total displacement outboard of North America if oblique slip is assumed for all faults in the accretionary prism of southern Alaska (for example, the Pamplona zone, Chugach–St. Elias fault) (Grantz and others, 1991, sheet 1).

The cumulative right-slip displacement on the Denali and Tintina fault systems accounts for half of that required to explain the 28° of latitudinal shift of the SAS with respect to nuclear North America. We suggest that some of this displacement may have occurred on unrecognized middle Cretaceous faults in the Kahiltna and Gravina-Nutzotin (flysch) terranes, or may be shrouded in mylonite zones in the metamorphic rocks of the Yukon-Tanana and Ruby terranes. Zones of displacement in the Yukon-Tanana and Ruby terranes would be expected if they were subject to large-scale rotation, as we propose (Grantz and others, 1991). In addition, if the right-slip faults of southern Alaska flatten into decollements at depth, as proposed by Oldow and others (1989), some northward displacement may be represented by uplifted flat or shallow-dipping faults.

Little evidence exists for transform displacement between the middle Tertiary and the Pliocene inboard of the continental margin of Alaska. Plate reconstructions indicate that relative motion between the Pacific plate and North America during this interval was slow and nearly parallel, or slightly transtensional, to the regional trend of the continental margin. As a result, plate-boundary faulting during this interval may have been concentrated at the Queen Charlotte–Fairweather fault system (Fig. 1).

A slight change to the northwest in the relative motion of the Pacific plate relative to the North American plate at about 5 Ma caused transform displacement to resume inboard of the continental margin in the late Cenozoic (Plafker and others, 1992). This change of relative motion reactivated parts of the Denali fault and other fault systems in southern Alaska.

Superposition of the Aleutian arc on southeastern Denali in the Cenozoic

Magmatism in the Aleutian arc began at about 55 to 50 Ma. This coincides with capture of a sector of the Kula plate in the Bering Sea and the resultant change of relative plate motion of this plate with respect to both the Pacific and North American plates. The eastern half of the Aleutian magmatic arc is developed mainly on the SAS in the Alaska Peninsula and in south-central Alaska; the western half is developed on oceanic crust between the North Pacific Ocean and the deep Aleutian basin of the Bering Sea (Fig. 1). Magmatism has been active from the early Eocene to early Oligocene and from the middle Miocene to the present (Worrall, 1991). Since the late Eocene, arc volcanism along the Aleutian arc has been a product of subduction at the Aleutian trench of the Pacific plate.

The west-trending Wrangell Mountains arc segment of the Aleutian arc lies on the SAS in south-central Alaska (Fig. 1) between the Alaska Peninsula segment of the Aleutian arc on the west and the Fairweather-Totschunda fault system on the east. Volcanism in the Wrangell Mountains has been active since 26 Ma. A volcanic gap of about 400 km separates the westernmost volcano, Mt. Drum, of the Wrangell Mountains segment from the Alaska Peninsula segment of the Aleutian arc (Richter and others, 1990). The Wrangell Mountains arc is thought to be the eastward extension of the Aleutian arc because of its tectonic location, similar age and composition, and because of the weak seismicity that defines the Wadati-Benioff zone 100 km below the volcanoes of the Wrangell Mountains arc (Page and others, 1989; Grantz and others, 1991, sheet 1).

Because the Wrangell Mountains arc lies on the hanging wall of the collisional zone along the northern margin of the Yakutat terrane (see below) and because seismic-reflection data do not support active subduction of the Pacific plate beneath the Yakutat terrane in the eastern Gulf of Alaska, we interpret the Wrangell Mountains arc as the product of subduction of oceanic basement of the western part of the Yakutat terrane beneath terranes of south-central Alaska at the Chugach–St. Elias fault (Grantz and others, 1991, sheet 1) (Fig. 3). The western margin of the Yakutat terrane is being subducted simultaneously at the Aleutian subduction zone and its northeastward extension, the convergent Pamplona zone, which lies on the continental shelf and coastal plains in the western Gulf of Alaska (Fig. 3).

Collision and accretion of the Yakutat terrane and transmission of Pacific–North American plate convergence from the Pacific rim to the Arctic basin in northeastern Alaska

Regional seismicity and seismic refraction, magnetic, and gravity data indicate that west of the A-3 transect the Chugach terrane of south-central Alaska is underlain by oceanic crust that dips gently north above the Wrangell Mountains Wadati-Benioff zone (Fuis and others, 1991; Fuis and Plafkker, 1991; Grantz and others, 1991, sheet 1). The mafic crust probably represents an underplated section of the Kula plate that was accreted to the base of the Chugach and Prince William terranes in the Eocene. Subduction velocity and direction below and outboard of the Chugach and Prince William terranes may have regenerated after either (1) subduction of the Kula plate ridge at about 50 Ma or (2) a marked change of relative plate vectors and speed of the Kula plate ending at its capture by the Pacific plate at 43 Ma. The ongoing collisional accretion of the Yakuktat terrane, the most outboard terrane in southern Alaska, is a consequence of the late Cenozoic continuation of subduction at the modern Wadati-Benioff zone.

The basal part of the Yakutat terrane is a displaced Cretaceous and Cenozoic composite ocean margin that is colliding with the Pacific rim in southern Alaska (Plafker, 1987). To the west, the displaced margin consists of Tertiary sedimentary material deposited on Paleocene or Eocene ocean crust; to the east, it composes a Paleogene accretionary prism. The transition between the east and west parts of the terrane is at the Dangerous River zone (Fig. 3), the west-facing front of the Paleogene accretionary prism (Grantz and others, 1991). Thus, the Yakutat terrane preserves a fossil Paleogene subduction complex that has been transported northward along the Queen Charlotte–Fairweather fault system in the Cenozoic. Provenance studies, regional geology, and paleomagnetic data suggest that the basal units of the Yakutat terrane were assembled along coastal North America between Chatham Strait and Puget Sound, about 500 to 1500 km southeast of the present position of the Yakutat terrane (Van Alstine and others, 1985; Plafker, 1987). The amalgamation of the basal units probably took place as they moved northward in response to oblique underthrusting of the Kula plate relative to the North American plate.

Collision of the western part of the Yakutat terrane in south-central Alaska at the Aleutian subduction zone has resulted in subduction of its oceanic basement and offscraping and accretion of its flat-lying sedimentary cover. The accretion of the sedimentary cover occurs seaward of the Kayak Island fault zone and the Chugach–St. Elias fault, and is expressed as the seismically active southeast-facing Pamplona fold and thrust zone, which arcs eastward past the 141st meridian (Fig. 1). The Pamplona zone thus marks the surface position of the northeastern continuation of the Aleutian subduction zone. This geometry, together with the apparent absence of seismicity outboard of the Pamplona zone in the Yakutat terrane, indicates that the colliding Yakutat terrane is currently being subducted with the Pacific plate, rather than being fully accreted to the Prince William and Chugach terranes above a zone of active underthrusting (Plafker, 1987).

The collision of the structurally thickened, eastern part of the Yakutat terrane in the Neogene may have had tectonic effects across the width of Alaska. The northeasterly striking zone of compressional earthquake epicenters that extends northward from the intersection of the Aleutian and Wrangell Mountains segments of the Aleutian Wadati-Benioff zone in central Alaska to northeastern Alaska and the Beaufort Sea shelf (Fig. 4) may reflect strain transmitted into the interior of Alaska. In northeastern Alaska, seismic-reflection surveys, geologic mapping, geochronology, and seismicity indicate that north-vergent thrust faults having Eocene to Holocene displacement underlie the Beaufort Sea shelf and the topographically high northeastern salient of the Brooks Range (Grantz and others, 1983, 1987; Moore and others, 1985; Wallace and Hanks, 1990). In the absence of any active plate boundaries in northeastern Alaska and its surrounding region, the north-trending zone of seismicity across Alaska and the coincidence in timing of deformation suggest that the young thrusting and associated folding in northern Alaska may be a distal effect of Pacific–North American plate convergence at the Aleutian subduction zone. This hypothesis requires an active decollement that extends from the Aleutian Wadati-Benioff zone northward under all of Alaska at the Moho and in the lower crust, and that is exposed at the continental rise of the Beaufort Sea shelf (Grantz and others, 1991, sheets 1 and 2). The apparent continuity, but divergent dips and contrasting seismicity rates, led Page and others (1989) to conclude that the Aleutian and Wrangell Mountains segments of the Wadati-Benioff zone define a north-dipping buckle, or cuspidal ridge, in the surface of the subducted lithosphere beneath southern Alaska. The position of the buckle projects northward toward the northeasterly trend of young contractional deformation and seismicity that crosses Alaska. We suggest that partial coupling of the North American plate to the subducted lithosphere where it is buckled allows transmission of contractional strain across a broad swath of eastern Denalia.

Modern continent-ocean transition at Alaska's continental margin

Southern Alaska continental margin. The modern continental margin of southern Alaska is an active plate boundary caused by the northwestward movement of the Pacific plate relative to the North American plate which, in the Gulf of Alaska, travels northwestward at about 63 mm/yr (Engebretson and others, 1985). Along the Alaska-Aleutian border of the North Pacific, the Pacific plate consists of Paleogene oceanic crust that is 6 to 7 km thick and a sedimentary cover that is 0.2 to nearly 4 km thick. In the Gulf of Alaska, relative plate motion is convergent and nearly orthogonal to the regional strike of the modern continental margin. The convergence results in a fast rate of subduction, accretion of ocean-floor deposits at the Aleutian Trench, and andesitic volcanism in the Aleutian and Wrangell Mountains arcs. A forearc basin extending from Cook Inlet to the Shumagin Islands has formed in front of the Aleutian arc. In southeastern Alaska, the relative plate motion is right slip and follows the continental margin along the Queen Charlotte–Fairweather transform fault system (Fig. 1).

Between this transform segment and the convergent Aleutian subduction zone, the transform motion of the Pacific–North American plate boundary is being partitioned northward via the Totschunda fault to the Denali fault system and being dissipated on transpressional thrust faults, including the Pamplona zone, Chugach–St. Elias fault, and Border Ranges fault (Fig. 3). The transpressional faulting results in active uplift and thrusting in a wide area of south-central Alaska, principally in the Chugach–St. Elias Mountains and in the Alaska Range (Fig. 2). Right-oblique seismicity on the Transition fault (Lahr and Plafker, 1980), the southern boundary of the Yakutat terrane, indicates that the Transition fault is also currently accommodating part of the transform motion on the Queen Charlotte–Fairweather fault system. On the basis of gravity data, about 50 km of north-over-south thrust displacement is proposed across the fault (Grantz and others, 1991). We therefore conclude that the Transition fault is a transpressional fault and that it is the southernmost splay of the Queen Charlotte–Fairweather fault system in southern Alaska. Thus, the modern Pacific–North American plate motion between the Fairweather–Queen Charlotte fault system and the Aleutian Trench is accommodated in a diffuse transpressional zone that extends northward across the Yakutat terrane to the Denali fault.

Northern Alaska continental margin. The gently arcuate modern continental margin of northern Alaska is a passive margin of the nonvolcanic, or cold, rift type that is covered by progradational continental-terrace deposits and marked by synrift and postrift extensional faults (Grantz and May, 1983; Grantz and others, 1990a). The Canada basin to the north is underlain by 6 to 8 km of oceanic crust of late Hauterivian to middle(?) Late Cretaceous age, and is overlain by 6 to >12 km of sedimentary cover of late Neocomian and younger age (Grantz and others, 1990b). The crustal boundary between the continental rocks of northern Alaska (Arctic Alaska terrane) and the oceanic rocks of the Canada basin is inferred from gravity data to lie beneath the slope-rise transition in the Beaufort Sea. Seismic-reflection profiles across the Alaska Beaufort Sea shelf and slope and the subsurface stratigraphy of coastal northern Alaska indicate that rifting led to continental breakup and drift and initiated formation of the oceanic Canada basin in late Neocomian (Hauterivian) time, at about 130 Ma.

Clastic deposition was initiated across the newly created continental margin in the late Neocomian and, by Aptian time, shifted to deposition of thick progradational clastic sequences. A submarine canyon at least 1.2 km deep and perhaps as deep as 1.8 km was cut into the progradational sedimentary prism by the late Cenomanian or Turonian (Kirschner and Rycerski, 1988), demonstrating that the progradational prism was building into deep water early in its history. The clastic deposition has continued into the Cenozoic, resulting in a post-Neocomian continental slope-rise sedimentary prism that in places is thicker than 12 km. The postrift sedimentary prism, which overstepped older rift-related structures, is extensively deformed by listric normal faults, some of which are growth faults. Along the eastern part of the continental margin, the continental-terrace deposits are deformed by large north-dipping, north-vergent thrust faults and detachment folds of Cenozoic age. These compressional structures are the distal effects of convergence between the Pacific and North American plates, described above.

SUMMARY AND IMPLICATIONS OF THE TECTONIC EVOLUTION OF THE CONTINENT-OCEAN TRANSITION IN ALASKA

During the Mesozoic and Cenozoic, the northwestern margin of North America expanded oceanward 600 to 1000 km by the assembly, or accretion (King, 1977), of a collage of terranes against the parautochthonous and autochthonous core of North America. These assembled terranes constitute most of the eastern part of a major subcontinental region here called Denalia. Through terrane analysis, we have arrived at a speculative evolution and kinematic history for the growth of eastern Denalia. The evolution of Denalia is complex in detail, but can be ascribed to a few major tectonic processes, which are described below and shown schematically in Figure 11. Our analysis is based on the premise that in late Paleozoic and early Mesozoic time the northwestern margin of North America was a west-facing, sublinear passive margin (Fig. 11A), part of which is preserved in the Arctic Alaska terrane.

(1) Ophiolite obduction and high-pressure metamorphism of continental terranes. Continental terranes marginal to North America, including the Arctic Alaska, Ruby, and Yukon-Tanana terranes, contain evidence of Jurassic to Early Cretaceous high-pressure metamorphism, continent-directed folding and thrusting, and overthrusting of oceanic rocks. The overthrust oceanic rocks, which include suprasubduction-zone ophiolites, are now represented by the Angayucham, Misheguk Mountain, Tozitna, Kanuti, Innoko, and Seventymile terranes (Fig. 11B). Ophiolite obduction may have resulted from (1) subduction of the margin of North America beneath a continentward-facing arc and forearc; (2) continentward backarc thrusting of a marginal basin inboard of an oceanward-facing arc that fringed North America; or (3) wedge tectonics generated by continentward thrusting above a continentward-dipping subduction zone. Because the obducted terranes are intruded by middle Cretaceous and younger plutonic-arc rocks and are truncated by large-scale right-slip faults, terrane emplacement occurred prior to the onset of Pacific ocean-floor subduction and coastwise right slip along the northwestern margin of North America in middle Cretaceous time and later.

Arc collision and ophiolite obduction created a major orogenic welt along the Pacific margin of North America in Jurassic and Early Cretaceous time that underwent attenuation and extensional collapse in the middle Cretaceous (Fig. 11C). Erosional unroofing, overprinting of high-pressure metamorphic assemblages by lower pressure assemblages, ductile deformation, and setting of middle Cretaceous K-Ar isotopic ages occurred in the affected rocks in the Arctic Alaska, Ruby, and Yukon-Tanana terranes. The distribution of the overprinted assemblage suggests that extensional collapse occurred prior to, or accompanied, the large-scale coastwise dextral translation that displaced parts of the orogenic welt in the middle Cretaceous and later. In the Ruby and Yukon-Tanana terranes, the extensional phase was accompanied, or soon followed, by emplacement of granitic rocks that are either (1) partial melts of lower crustal rocks produced by mantle upwelling or (2) highly evolved magmatic-arc intrusive rocks. If the granitic rocks in the Ruby and Yukon-Tanana terranes were generated from arc magmatism along the Pacific margin of North America, then their presence may indicate that plutonism followed large-scale coastwise translation of part of the orogenic welt, so that the Arctic Alaska terrane was tectonically removed or isolated from the magmatism.

(2) Rifting and sea-floor spreading. Late Neocomian rifting and middle Cretaceous sea-floor spreading about a pole in the Mackenzie delta (Fig. 11B) separated Arctic Alaska from the Canadian Arctic Islands of northwestern North America and rotated it 67° counterclockwise to its present position relative to North America. This separation and rotation formed the Arctic Alaska terrane, the northern margin of Denalia, and the Amerasia basin of the Arctic Ocean. Rotation of the Arctic Alaska terrane was partly coeval with the onset of large-scale coastwise dextral translation along the Pacific margin of North America (Fig. 11C).

(3) Large-scale right slip. Large-scale right slip shaped the Pacific margin from middle Cretaceous time to the present (Fig. 11, C and D). The translation was parallel to the trend of the Cordilleran orogen of North America and was driven by partial coupling of hanging-wall continental-margin terranes to obliquely underthrusted oceanic crust beneath western North America. Known displacements on the Denali and Tintina fault systems account for about half of the translation required along the Pacific coast fault system by paleomagnetic data from southern Alaska; the remainder may be distributed on unknown or poorly documented displacement zones. Distributed shear in the orogen-parallel translational fault system dislodged inboard continental terranes, such as the Ruby, Nixon Fork, and Yukon-Tanana terranes, and caused large-scale rotations about vertical axes of the Ruby and Nixon Fork terranes (Fig. 11, C and D). Some of the translation occurred outboard of the SAS within active and inactive accretionary-prism assemblages along faults, such as the Border Ranges, Contact, and Chugach–St. Elias faults.

(4) Subduction-zone accretion. Following Jurassic to Early Cretaceous subduction, arc collision, and ophiolite obduction, underthrusting of Pacific basin oceanic crust was initiated along the northwestern margin of North America, and continues today along the Aleutian-Wrangell Mountains subduction zone (Fig. 11, C, D, and E). Subduction resulted in emplacement of orogen-parallel magmatic belts within Alaska and the construction of accretionary prisms of slices of oceanic crust, minor oceanic sediment, and trench-turbidite sequences. The subduction involved oblique convergence of the Kula, Pacific, and possibly the Farallon plates against the northwestern margin of North America, and was coeval with large-scale orogen-parallel translation of terranes along the Cordilleran orogen.

(5) Eurasian–North American convergence. Convergence between the North American and Eurasian plates generated the basement-involved, east-directed Herald thrust system in northern Alaska and the linked Alaska, Bering, and Navarin oroclines in southwestern Alaska and the Bering Sea region (Fig. 11D). About 40° of counterclockwise rotation has been recorded in the western limb of the Alaska orocline in southwestern Alaska by paleomagnetic data (Fig. 11E) and similar amounts of rotation (but opposite sense) are likely in the eastern limb of the Navarin orocline. Oroclinal rotation and the construction of a southward-concave bend in southern Denalia are also partly the products of northward terrane migration along the western margin of North America and collision against a Eurasian backstop. Zones of dislocation along the Kobuk–South Fork and Kaltag fault systems may have accommodated strain between the domain of oroclinal folding in southern Denalia and the region of basement-involved thrusting in northern Denalia.

(6) Terrane collision and transpression. Northward translation and offscraping of ocean-margin deposits of the Yakutat terrane at the eastern, or Wrangell Mountains, segment of the Aleutian subduction zone formed a southeast-facing fold-thrust belt (the Pamplona zone), and renewed strike-slip faulting along the Denali fault system and transpressional displacement on numerous faults between the Transition fault in the Gulf of Alaska and the Denali fault system in the Alaska Range (Fig. 11F). A distal effect of collision of the Yakutat terrane may be north-vergent contraction in the northeastern Brooks Range and adjoining continental margin.

The following are some implications of the speculative tectonic model for the evolution of eastern Denalia.

1. Many terranes in Alaska, in particular the continental terranes north of the Denali fault, were probably marginal to North America by middle Mesozoic time. Other terranes may have been part of one or more fringing arc-trench–marginal-basin systems, although their specific point of origin may never be known.

2. Relatively more distal ocean-margin and ocean-floor terranes are typically found as thin thrust sheets resting on more inboard continental terranes and modified by younger extensional structures. Basement is usually absent in the structurally higher terranes and may have been delaminated and tectonically underplated beneath the continental terranes.

3. Northward displacement of paleomagnetic poles in the southern Alaska superterrane approximately equals the maximum northward displacement along the Pacific margin predicted by plate-motion studies of Engebretson and others (1985). This near equivalence since the middle Cretaceous indicates a high degree of efficiency in the transmission of translational strain from subducted oceanic crust to hanging-wall terranes of the Pacific margin.

4. The accretion of some terranes, such as the Yakutat terrane, may produce tectonic effects hundreds of kilometers from the site of accretion.

5. Denalia was assembled by a variety of dynamic forces that overlap in time and space and therefore cannot be identified as discrete orogenic events.

6. Many of the processes recorded in the Pacific margin can be linked qualitatively with relative plate motions between the North American plate and the plates of the Pacific Ocean, as calculated by Engebretson and others (1985).

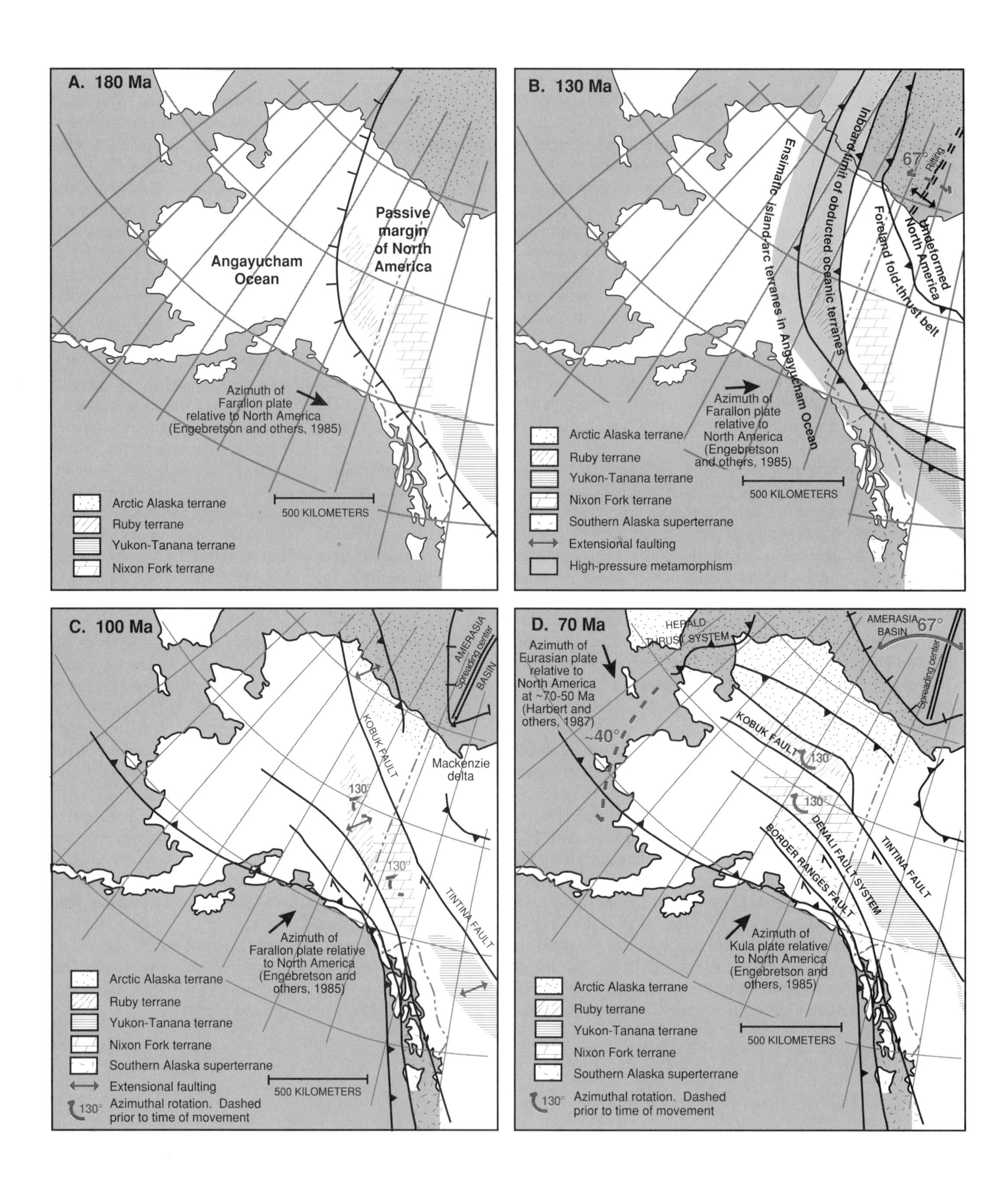
A. 180 Ma
Angayucham Ocean
Passive margin of North America
Azimuth of Farallon plate relative to North America (Engebretson and others, 1985)
500 KILOMETERS
Arctic Alaska terrane
Ruby terrane
Yukon-Tanana terrane
Nixon Fork terrane
B. 130 Ma
Ensimatic island-arc terranes in Angayucham Ocean
Inboard limit of obducted oceanic terranes
Foreland fold-thrust belt
Undeformed North America
Rifting
67°
Azimuth of Farallon plate relative to North America (Engebretson and others, 1985)
500 KILOMETERS
Arctic Alaska terrane
Ruby terrane
Yukon-Tanana terrane
Nixon Fork terrane
Southern Alaska superterrane
Extensional faulting
High-pressure metamorphism
C. 100 Ma
AMERASIA BASIN
Spreading center
KOBUK FAULT
Mackenzie delta
130°
TINTINA FAULT
Azimuth of Farallon plate relative to North America (Engebretson and others, 1985)
Arctic Alaska terrane
Ruby terrane
Yukon-Tanana terrane
Nixon Fork terrane
Southern Alaska superterrane
Extensional faulting
130° Azimuthal rotation. Dashed prior to time of movement
500 KILOMETERS
D. 70 Ma
HERALD THRUST SYSTEM
AMERASIA BASIN
67°
Spreading center
Azimuth of Eurasian plate relative to North America at ~70-50 Ma (Harbert and others, 1987)
~40°
KOBUK FAULT
130°
DENALI FAULT SYSTEM
BORDER RANGES FAULT
TINTINA FAULT
Azimuth of Kula plate relative to North America (Engebretson and others, 1985)
Arctic Alaska terrane
Ruby terrane
Yukon-Tanana terrane
Nixon Fork terrane
Southern Alaska superterrane
500 KILOMETERS
130° Azimuthal rotation. Dashed prior to time of movement

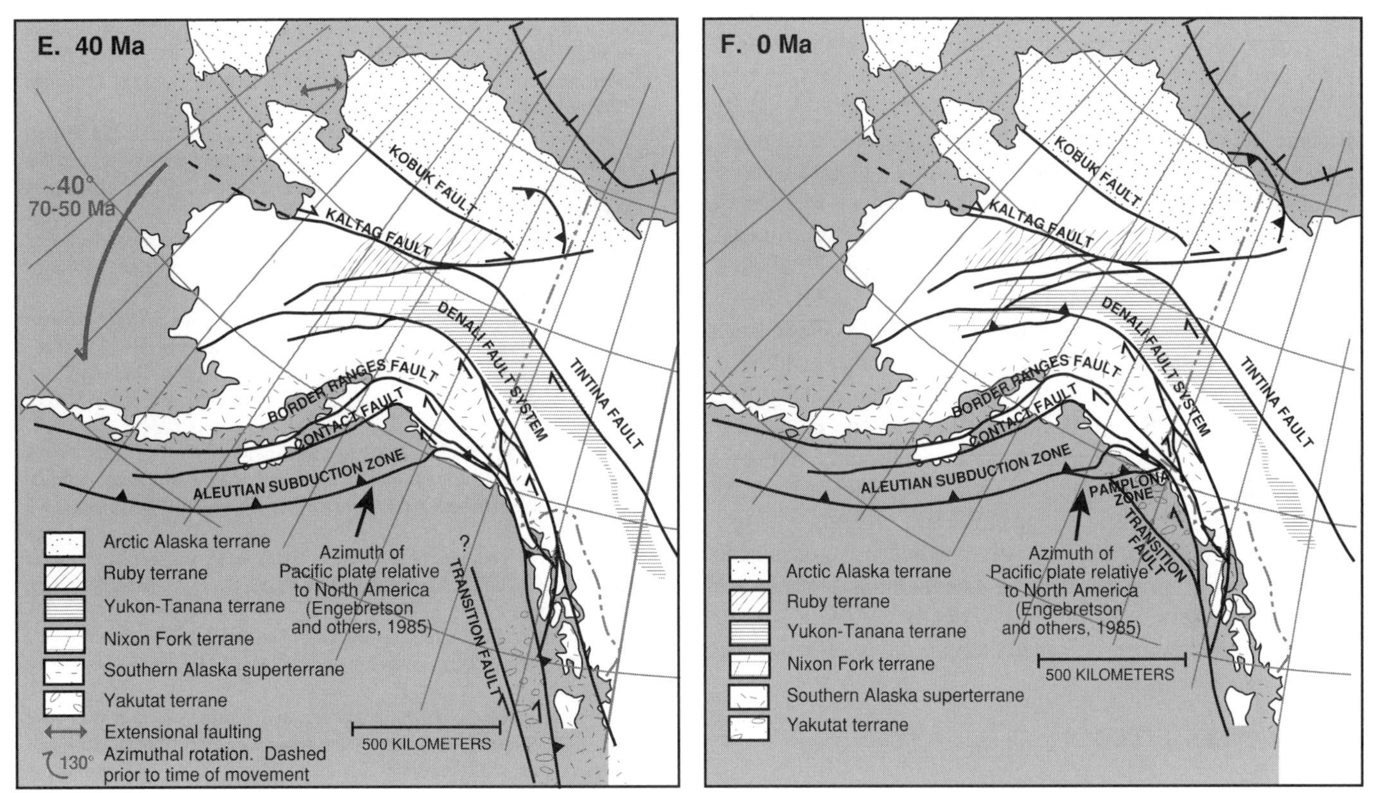

Figure 11. Conceptual model for tectonic evolution of Alaska used in constructing Continent-Ocean Transect A-3 of Grantz and others (1991). A: At 180 Ma (Middle Jurassic), the northwestern margin of North America was a passive margin that had existed for about 180 m.y. B: By 130 Ma (Early Cretaceous), subduction of oceanic lithosphere was initiated in the Pacific basin. In this model, continentward-facing subduction culminated in descent of the North American margin, as indicated by the distribution of high-pressure metamorphism and the absence of arc magmatism in continental terranes. Relative plate motions suggest that subduction may have been accompanied by left-lateral faulting. C: By 100 Ma, arc-continent subduction had ceased and oceanward-facing subduction was initiated along the Pacific margin. Subduction was slightly to moderately oblique northward, initiating the period of northward terrane migration that continues today. Onset of oceanward-facing subduction along the Pacific margin coincides with possible rotations of continental terranes, such as the Ruby and Nixon Fork terranes, and attenuation and extension in regions of the earlier collisional orogen. To the north, continental breakup and drift in the Amerasia basin had occurred, and the Arctic Alaska terrane was rotating about a pivot in the Mackenzie delta region. D: At 70 Ma, outboard terranes of the Pacific margin were arriving in southern Denalia, propelled rapidly northward by the accelerated oblique subduction of the Kula plate. In northern Denalia where continental crust was old and thick, Eurasian–North American plate convergence produced the Herald thrust system; in southern Denalia, where crust was thinner and consisted largely of oceanic and ocean-margin terranes, accumulation of terranes against a Eurasian plate backstop resulted in oroclinal folding. The northern and southern structural domains may have been separated by the Kaltag and Kobuk faults. E: In the early Tertiary, oroclinal rotation of the Alaska orocline was completed, subduction was established at the Aleutian subduction zone, and the Aleutian arc was constructed astride the continental margin. The Yakutat terrane, part of an older subduction complex formed at mid-latitudes, was detached from the North American margin and began riding with the Pacific plate. F: The Yakutat terrane is currently colliding with the Aleutian subduction zone; underthrusting of the terrane has resulted in development of a diffuse zone of transpressional deformation in south-central Alaska and the possible transmission of contractional strain across the width of Alaska to the continental rise in the Beaufort Sea.

ACKNOWLEDGMENTS

This paper and the A-3 transect have benefited from the data, ideas, and/or observations of John N. Aleinikoff, Joseph G. Arth, Fred Barker, David B. Barnes, Robert B. Blodgett, Stephen E. Box, William P. Brosgé, Terry R. Bruns, John W. Cady, Richard B. Campbell, Robert M. Chapman, Michael Churkin, Jr., Robert S. Coe, Grant W. Cushing, John T. Dillon, Chris J. Dodds, James H. Dover, Cynthia Dusel-Bacon, Michael A. Fisher, Helen L. Foster, Gary S. Fuis, William Harbert, Anita G. Harris, David L. Jones, John Lahr, Steven D. May, Elizabeth L. Miller, Thomas P. Miller, James W.H. Monger, Charles G. Mull, Bonita L. Murchey, Warren J. Nokleberg, Robert A. Page, Bruce C. Panuska, William W. Patton, Jr., Terry L. Pavlis, Ellen N. Penner, Garnett H. Pessel, George Plafker, David B. Stone, Irvin L. Tailleur, Alison B. Till, Wesley K. Wallace, and Florence R. Weber. Our ideas and writing were significantly improved through the thoughtful reviews of David W. Scholl and David G. Howell and the invaluable editing of Karen E. Adams.

We take this opportunity to point out that the aeromagnetic-anomaly profile on sheet 2 of Continent-Ocean Transect A-3 (Grantz and others, 1991) was inadvertently shifted during construction of the transect. We regret any inconvenience this may have caused users of the transect.

REFERENCES CITED

Aleinikoff, J. N., Dusel-Bacon, C., and Foster, H. L., 1986, Geochronology of augen gneiss and related rocks, Yukon-Tanana terrane, east-central Alaska: Geological Society of America Bulletin, v. 97, p. 626–637.

Aleinikoff, J. N., Moore, T. E., Nokleberg, W. J., and Walter, M., 1993, U-Pb ages of zircon, monazite and sphene from Devonian metagranites and metafelsites, eastern Brooks Range, *in* Dusel-Bacon, C., and Till, A. B., eds., Geologic studies in Alaska by the U.S. Geological Survey, 1991: U.S. Geological Survey Bulletin 2068, p. 59–70.

Anderson, A. V., and Wallace, W. K., 1990, Middle Devonian to Lower Mississippian clastic depositional cycles southwest of Bathrub Ridge, northeastern Brooks Range, Alaska: Geological Association of Canada/Mineralogical Association of Canada Program with Abstracts, Annual Meeting, v. 15, p. A2.

Armstrong, R. L., Harakal, J. E., Forbes, R. B., Evans, B. W., and Thurston, S. P., 1986, Rb-Sr and K-Ar study of metamorphic rocks of the Seward Peninsula and southern Brooks Range, *in* Evans, B. W., and Brown, E. H., eds., Blueschists and eclogites: Geological Society of America Memoir 164, p. 185–203.

Avé Lallemant, H. G., and Oldow, J. S., 1988, Early Mesozoic southward migration of Cordilleran transpressional terranes: Tectonics, v. 7, p. 1057–1088.

Barker, F., Jones, D. L., Budahn, J. R., and Coney, P. J., 1988, Ocean plateau–seamount origin of basaltic rocks, Angayucham terrane, central Alaska: Journal of Geology, v. 96, p. 368–374.

Basham, P. W., Forsyth, D. A., and Wetmiller, R. J., 1977, The seismicity of northern Canada: Canadian Journal of Earth Sciences, v. 14, p. 1646–1667.

Biswas, N. N., Aki, K., Pulpan, H., and Tytgat, G., 1986, Characteristics of regional stresses in Alaska and neighboring areas: Geophysical Research Letters, v. 13, p. 177–180.

Blodgett, R. B., 1983, Paleobiogeographic affinities of Devonian fossils from the Nixon Fork terrane, southwestern Alaska, *in* Stevens, C. H., ed., Pre-Jurassic rocks in western North American suspect terranes: Los Angeles, California, Pacific Section, Society of Economic Paleontologists and Mineralogists, p. 125–130.

Boak, J. M., Turner, D. L., Henry, D. J., Moore, T. E., and Wallace, W. K., 1987, Petrology and K-Ar ages of the Misheguk igneous sequence—an allochthonous mafic and ultramafic complex—and its metamorphic aureole, western Brooks Range, Alaska, *in* Tailleur, I., and Weimer, P., eds., Alaska North Slope geology: Bakersfield, California, Pacific Section, Society of Economic Paleontologists and Mineralogists, Book 50, p. 737–745.

Bol, A. J., and Coe, R. S., 1987, Paleomagnetism of pillow basalt and sheeted dikes, Resurrection Peninsula, Alaska: Confirmation of large northward displacement in S. Alaska since early Tertiary: Geological Society of America Abstracts with Programs, v. 19, p. 594.

Bol, A. J., and Roeske, S. M., 1993, Strike-slip faulting and block rotation along the Contact fault system, eastern Prince William Sound, Alaska: Tectonics, v. 12, p. 49–62.

Box, S. E., 1985, Early Cretaceous orogenic belt in northeastern Alaska: Internal organization, lateral extent, and tectonic interpretation, *in* Howell, D. G., ed., Tectonostratigraphic terranes of the circum-Pacific region: Houston, Texas, Circum-Pacific Council for Energy and Mineral Resources Earth Science Series no. 1, p. 137–145.

Box, S. E., and Patton, W. W., 1989, Igneous history of the Koyukuk terrane, western Alaska: Constraints on the origin, evolution, and ultimate collision of an accreted island-arc terrane: Journal of Geophysical Research, v. 94, p. 15,843–15,867.

Bruns, T. R., 1985, Tectonics of the Yakutat block, an allochthonous terrane in the northern Gulf of Alaska: U.S. Geological Survey Open-File Report 85-13, 112 p.

Carey, S. W., 1958, The tectonic approach to continental drift, *in* Carey, S. W., ed., Continental drift, a symposium: Hobart, Tasmania University, p. 177–355.

Carter, C., and Laufeld, S., 1975, Ordovician and Silurian fossils in well cores from North Slope of Alaska: American Association of Petroleum Geologists Bulletin, v. 59, p. 457–464.

Churkin, M., Jr., Foster, H. L., Chapman, R. M., and Weber, F. R., 1982, Terranes and suture zones in east-central Alaska: Journal of Geophysical Research, v. 87, p. 3718–3730.

Coe, R. S., Globerman, B. R., Plumley, P. W., and Thrupp, G. A., 1985, Paleomagnetic results from Alaska and their tectonic implications, *in* Howell, D. G., ed., Tectonostratigraphic terranes of the circum-Pacific region: Houston, Texas, Circum-Pacific Council for Energy and Mineral Resources Earth Science Series no. 1, p. 85–108.

Coe, R. S., Bol, A. J., Gromme, C. S., and Hillhouse, J. W., 1990, Early Eocene location and orientation of the Kula-Farallon Ridge: Paleomagnetic results from the Resurrection Peninsula, Alaska [abs.]: Eos (Transactions, American Geophysical Union), v. 71, p. 1589.

Coney, P. J., 1989, Structural aspects of suspect terranes and accretionary tectonics in western North America: Journal of Structural Geology, v. 11, p. 107–125.

Coney, P. J., and Jones, D. L., 1985, Accretion tectonic and crustal structure in Alaska: Tectonophysics, v. 119, p. 265–283.

Coney, P. J., Jones, D. L., and Monger, J.W.H., 1980, Cordilleran suspect terranes: Nature, v. 288, p. 329–333.

Cooper, A. K., Marlow, M. S., and Scholl, D. W., 1987, Geologic framework of the Bering Sea crust, *in* Scholl, D. W., Grantz, A., and Vedder, J. G., eds., Geology and resource potential of the continental margin of western North America and adjacent ocean basins—Beaufort Sea to Baja California: Houston, Texas, Circum-Pacific Council for Energy and Mineral Resources Earth Science Series no. 6, p. 73–102.

Csejtey, B., Jr., Cox, D. P., Evarts, R. C., Stricker, D. G., and Foster, H. L., 1982, The Cenozoic Denali fault system and the Cretaceous accretionary development of southern Alaska: Journal of Geophysical Research, v. 87, p. 3741–3754.

Debiche, M. G., Cox, A., and Engebretson, D. C., 1987, The motion of allochthonous terranes across the North Pacific basin: Geological Society of America Special Paper 207, 49 p.

Dillon, J. T., 1989, Structure and stratigraphy of the southern Brooks Range and northern Koyukuk basin near the Dalton Highway, *in* Mull, C. G., and Adams, K. E., eds., Dalton Highway, Yukon River to Prudhoe Bay, Alaska: Alaska Division of Geological and Geophysical Surveys Guidebook 7, v. 2, p. 157–187.

Dillon, J. T., Patton, W. W., Jr., Mukasa, S. B., Tilton, G. R., Blum, J., and Moll, E. J., 1985, New radiometric evidence for the age and thermal history of the metamorphic rocks of the Ruby and Nixon Fork terranes, *in* Bartsch-Winkler, S., and Reed, K. M., eds., The United States Geological Survey in Alaska: Accomplishments during 1983: U.S. Geological Survey Circular 945, p. 13–18.

Dixon, J., and Dietrich, J. R., 1990, Canadian Beaufort Sea and adjacent land areas, *in* Grantz, A., Johnson, L., and Sweeney, J. F., eds., The Arctic Ocean region: Boulder, Colorado, Geological Society of America, The Geology of North America, v. L, p. 239–256.

Dover, J. H., 1990a, Problems of terrane terminology—Causes and effects: Geology, v. 18, p. 485–580.

——, 1990b, Geology of east-central Alaska: U.S. Geological Survey Open-File Report 90-289, 91 p.

Dusel-Bacon, C., and Hansen, V. L., 1992, High-pressure amphibolite-facies metamorphism and deformation within the Yukon-Tanana and Taylor Mountain terranes, eastern Alaska, *in* Bradley, D. C., and Dusel-Bacon, C., eds., Geologic studies in Alaska by the U.S. Geological Survey, 1991: U.S. Geological Survey Bulletin 2041, p. 140–159.

Dusel-Bacon, C., Brosgé, W. P., Till, A. B., Doyle, E. O., Mayfield, C. F., Reiser, H. N., and Miller, T. P., 1989, Distribution, facies, ages, and proposed tectonic associations of regionally metamorphosed rocks in northern Alaska: U.S. Geological Survey Professional Paper 1497-A, 44 p., 2 sheets, scale 1:1,000,000.

Dutro, J. T., Jr., Palmer, A. R., Repetski, J. E., and Brosgé, W. P., 1984, Middle Cambrian fossils from the Doonerak anticlinorium, central Brooks Range, Alaska: Journal of Paleontology, v. 58, p. 1364–1371.

Eisbacher, G. H., 1976, Sedimentology of the Dezadeash flysch and its implications for strike-slip faulting along the Denali fault, Yukon Territory and Alaska: Canadian Journal of Earth Sciences, v. 13, p. 1495–1513.

Embry, A. F., 1990, Geologic and geophysical evidence in support of the hypothesis of anticlockwise rotation of northern Alaska: Marine Geology, v. 93, p. 317–329.

Engebretson, D. C., Cox, A., and Gordon, R. G., 1985, Relative motions between oceanic and continental plates in the Pacific Basin: Geological Society of America Special Paper 206, 59 p.

Estabrook, D. H., Stone, D. B., and Davies, J. N., 1988, Seismotectonics of northern Alaska: Journal of Geophysical Research, v. 93, p. 12,026–12,040.

Fisher, M. A., Brocher, T. M., Nokleberg, W. J., Plafker, G., and Smith, G. L., 1989, Seismic reflection images of the crust of the northern part of the Chugach terrane, Alaska: Results of a survey for the Trans-Alaska Crustal Transect (TACT): Journal of Geophysical Research, v. 94, p. 4424–4440.

Foster, H. L., Keith, T.E.C., and Menzie, W. D., 1987, Geology of east-central Alaska: U.S. Geological Survey Open-File Report 87-188, 59 p.

Fuis, G. S., and Plafker, G., 1991, Evolution of deep structure along the Trans-Alaska crustal transect, Chugach Mountains and Copper River basin, southern Alaska: Journal of Geophysical Research, v. 96, p. 4229–5253.

Fuis, G. S., Ambos, E. L., Mooney, W. D., Christensen, N. J., and Geist, E., 1991, Crustal structure of accreted terranes in southern Alaska, Chugach Mountains and Copper River basin, from seismic refraction results: Journal of Geophysical Research, v. 96, p. 4187–4227.

Fujita, K., and Newberry, J. T., 1982, Tectonic evolution of northeastern Siberia and adjacent regions: Tectonophysics, v. 89, p. 337–357.

——, 1983, Accretionary terranes and tectonic evolution of northeast Siberia, *in* Hashimoto, M., and Uyeda, S., eds., Accretion tectonics in the circum-Pacific regions: Tokyo, Terra Scientific Publishing Company, p. 43–57.

Gabrielse, H., 1985, Major dextral transcurrent displacements along the northern Rocky Mountain Trench and related lineaments in north-central British Columbia: Geological Society of America Bulletin, v. 96, p. 1–14.

Gardner, M. C., Bergman, S. C., MacKevett, E. M., Jr., Plafker, G., Campbell, R. C., Cushing, G. W., Dodds, C. J., and McClelland, W. D., 1988, Middle Pennsylvanian pluton stitching of Wrangellia and the Alexander terrane, Wrangell Mountains, Alaska: Geology, v. 16, p. 967–971.

Gehrels, G. E., and Saleeby, J. B., 1987, Geologic framework, tectonic evolution, and displacement history of the Alexander terrane: Tectonics, v. 6, p. 151–174.

Gahrels, G. E., McClelland, W. C., Samson, S. D., Patchett, P. J., and Jackson, J. L., 1990, Ancient continental margin assemblage in northern Coast Mountains, southeast Alaska and northwest Canada: Geology, v. 18, p. 208–211.

Goodwin, E. B., Fuis, G. S., Nokleberg, W. J., and Ambos, E. L., 1989, The crustal structure of the Wrangellia terrane along the east Glenn Highway, eastern-southern Alaska: Journal of Geophysical Research, v. 94, p. 16,037–16,058.

Gottschalk, R. R., Jr., and Oldow, J. S., 1988, Low-angle normal faults in the south-central Brooks Range fold and thrust belt, Alaska: Geology, v. 16, p. 395–399.

Grantz, A., 1966, Strike-slip faults in Alaska: U.S. Geological Survey Open-File Report, 82 p.

Grantz, A., and May, S. D., 1983, Rifting history and structural development of the continental margin north of Alaska, *in* Watkins, J. S., and Drake, C. L., eds., Studies in continental margin geology: American Association of Petroleum Geologists Memoir 34, p. 77–100.

Grantz, A., Eittreim, S., and Dinter, D. A., 1979, Geology and tectonic development of the continental margin north of Alaska: Tectonophysics, v. 59, p. 263–291.

Grantz, A., Eittreim, S., and Whitney, O. T., 1981, Geology and physiography of the continental margin north of Alaska and implications for the origin of the Canada basin, *in* Nairn, A.E.M., Churkin, M., and Stehli, F. G., eds., The ocean basins and margins: Volume 5: New York, Plenum Publishing Corporation, v. 5, p. 439–492.

Grantz, A., Dinter, D. A., and Biswas, N. N., 1983, Map, cross sections, and chart showing late Quaternary faults, folds, and earthquake epicenters on the Alaskan Beaufort Shelf: U.S. Geological Survey Miscellaneous Investigations Series Map I-11B2-C, scale 1:500,000.

Grantz, A. Dinter, D. A., and Culotta, R. C., 1987, Geology of the continental shelf north of the Arctic National Wildlife Refuge, *in* Tailleur, I. L., and Weimer, P., eds., Alaskan North Slope geology: Los Angeles, California, Pacific Section, Society of Economic Paleontologists and Mineralogists, and Anchorage, Alaska Geological Society, Book 50, p. 759–762.

Grantz, A., May, S. D., and Hart, P. E., 1990a, Geology of the Arctic continental margin of Alaska, *in* Grantz, A., Johnson, L., and Sweeney, J. F., eds., The Arctic Ocean region: Boulder, Colorado, Geological Society of America, The Geology of North America, v. L, p. 257–288.

Grantz, A., May, S. D., Taylor, P. T., and Lawver, L. A., 1990b, Canada basin, *in* Grantz, A., Johnson, L., and Sweeney, J. F., eds., The Arctic Ocean region: Boulder, Colorado, Geological Society of America, The Geology of North America, v. L, p. 379–402.

Grantz, A., Green, A. R., Smith, D. G., Lahr, J. C., and Fujita, K., 1990c, Major Phanerozoic tectonic features of the Arctic Ocean region, *in* Grantz, A., Johnson, G. S., and Sweeney, J. F., eds., The Arctic Ocean region: Boulder, Colorado, Geological Society of America, The Geology of North America, v. L, plate 11, scale 1:6,000,000.

Grantz, A., Moore, T. E., and Roeske, S. M., 1991, A-3 Gulf of Alaska to Arctic Ocean: Boulder, Colorado, Geological Society of America, Centennial Continent/Ocean Transect no. 15, 72 p., 3 sheets, scale 1:500,000.

Haeussler, P. J., and Coe, R. S., 1987, New paleomagnetic results from the Hound Island volcanics and implications for the Alexander terrane: Geological Society of America Abstracts with Programs, v. 19, p. 385.

Halgedahl, S., and Jarrard, R., 1987, Paleomagnetism of the Kuparuk River Formation from oriented drill core: Evidence for rotation of the North Slope block, *in* Tailleur, I. L., and Weimer, P., eds., Alaskan North Slope geology: Los Angeles, California, Pacific Section, Society of Economic Paleontologists and Mineralogists, and Anchorage, Alaska Geological Society, Book 50, p. 581–617.

Harbert, W., 1990, Paleomagnetic data from Alaska: Reliability, interpretation and terrane trajectories: Tectonophysics, v. 184, p. 111–135.

Harbert, W., Frei, L. S., Cox, A., and Engebretson, D. C., 1987, Relative motions between Eurasia and North America in the Bering Sea region: Tectonophysics, v. 134, p. 239–261.

Harbert, W., Frei, L., Jarrard, R., Halgedahl, S., and Engebretson, D., 1990, Paleomagnetic and plate-tectonic constraints on the evolution of the Alaskan–eastern Siberian Arctic, *in* Grantz, A., Johnson, L., and Sweeney, J. F., eds., The Arctic Ocean region: Boulder, Colorado, Geological Society of America, The Geology of North America, v. L, p. 567–592.

Harms, T. A., Coney, P. J., and Jones, D. L., 1984, The Sylvester allochthon, Slide Mountain terrane, British Columbia–Yukon: A correlative of oceanic terranes of northern Alaska: Geological Society of America Abstracts with Programs, v. 16, p. 288.

Harris, R. A., 1988, Origin, emplacement and attenuation of the Misheguk Mountain allochthon, western Brooks Range, Alaska: Geological Society of America Abstracts with Programs, v. 20, p. A112.

Hemming, S., Sharp, W. D., Moore, T. E., and Mezger, K., 1989, Pb/U dating of detrital zircons from the Carboniferous Nuka Formation, Brooks Range, Alaska: Evidence for a 2.07 Ga provenance: Geological Society of America Abstracts with Programs, v. 21, no. 6, p. A190.

Hillhouse, J. W., 1977, Paleomagnetism of the Triassic Nikokai Greenstone, McCarthy quadrangle, Alaska: Canadian Journal of Earth Sciences, v. 14, p. 2578–2592.

—— , 1987, Accretion in southern Alaska: Tectonophysics, v. 139, p. 107–122.

Hillhouse, J. W., and Grommé, C. S., 1980, Paleomagnetism of the Triassic Hound Island volcanics, Alexander terrane, southeastern Alaska: Journal of Geophysical Research, v. 85, p. 2594–2602.

—— , 1982, Limits to northward drift of the Paleocene Cantwell Formation, central Alaska: Geology, v. 10, p. 552–556.

Hillhouse, J. W., Grommé, C. S., and Csejtey, B., Jr., 1985, Tectonic implications of paleomagnetic poles from early Tertiary volcanic rocks, south-central Alaska: Journal of Geophysical Research, v. 90, p. 12,523–12,535.

Hoare, J. M., and Jones, D. L., 1981, Lower Paleozoic radiolarian chert and associated rocks in the Tikchik Lakes area, southwestern Alaska, *in* Albert, N.R.D., and Hudson, T., ed., The United States Geological Survey in Alaska: Accomplishments during 1979: U.S. Geological Survey Circular 823-B, p. 44–45.

Hoffman, P. F., 1989, Precambrian geology and tectonic history of North America, *in* Bally, A. W., and Palmer, A. R., eds., The Geology of North America—An overview: Boulder, Colorado, Geophysical Society of America, The Geology of North America, v. A, p. 447–512.

Hubbard, R. J., Edrich, S. P., and Rattey, R. P., 1987, Geologic evolution and hydrocarbon habitat of the Arctic Alaska microplate, *in* Tailleur, I. L., and Weimer, P., eds., Alaskan North Slope geology: Bakersfield, California, Pacific Section, Society of Economic Paleontologists and Mineralogists, v. 50, p. 797–830.

Hudson, T. L., 1987, Suspect philosophy?: Geology, v. 15, p. 1177.

Hudson, T., and Plafker, G., 1982, Paleogene metamorphism of an accretionary flysch terrane, eastern Gulf of Alaska: Geological Society of America Bulletin, v. 93, p. 1280–1290.

James, T. S., Hollister, L. S., and Morgan, W. J., 1989, Thermal modeling of the Chugach metamorphic complex: Journal of Geophysical Research, v. 94, p. 4411–4423.

Jones, D. L., Silberling, N. J., and Hillhouse, J., 1977, Wrangellia—A displaced terrane in northwestern North America: Canadian Journal of Earth Sciences, v. 14, p. 2565–2577.

Jones, D. L., Silberling, N. J., Gilbert, W., and Coney, P., 1982, Character, distribution, and tectonic significance of accretionary terranes in the central Alaska Range: Journal of Geophysical Research, v. 87, p. 3709–3717.

Jones, D. L., Howell, D. G., Coney, P. J., and Monger, J.W.H., 1983, Recognition, character, and analysis of tectonostratigraphic terranes in western North America, *in* Hashimoto, M., and Uyeda, S., eds., Accretion tectonics in the Circum-Pacific regions: Tokyo, Terra Scientific Publishing Company, p. 21–35.

Jones, D. L., Silberling, N. J., and Coney, P. J., 1986, Collision tectonics in the Cordillera of western North America: Examples from Alaska, *in* Coward, M. P., and Ries, A. C., eds., Collision tectonics: Geological Society of London Special Publication 19, p. 367–387.

Jones, D. L., Silberling, N. J., Coney, P. J., and Plafker, G., 1987, Lithotectonic terrane map of Alaska (west of the 141st meridian): U.S. Geological Survey Miscellaneous Field Studies Map MF-1874-A, 1 sheet, scale 1:2,500,000.

Jones, D. L., Coney, P. J., Harms, T. A., and Dillon, J. T., 1988, Interpretive geologic map and supporting radiolarian data from the Angayucham terrane, Coldfoot area, southern Brooks Range, Alaska: U.S. Geological Survey Miscellaneous Field Studies Map MF-1993, 1 sheet, scale 1:63,360.

Julian, F. E., 1989, Structure and stratigraphy of lower Paleozoic rocks, Doonerak window, central Brooks Range, Alaska [Ph.D. thesis]: Houston, Texas, Rice University, 127 p., 1 sheet, scale 1:31,680.

Karl, S. M., 1991, Arc and extensional basin geochemical and tectonic affinities for the Maiyumerak basalts in the western Brooks Range, *in* Bradley, D. C., and Ford, A. B., eds., Geologic studies in Alaska by the U.S. Geological Survey, 1990: U.S. Geological Survey Bulletin 1999, p. 141–155.

King, P. B., 1977, The evolution of North America (revised edition): Princeton, New Jersey, Princeton University Press, 197 p.

Kirschner, C. E., Gryc, G., and Molenaar, C., 1983, Regional seismic lines in the National Petroleum Reserve in Alaska, *in* Bally, A. W., ed., Seismic expression of structural styles, Volume 1: American Association of Petroleum Geologists Studies in Geology, no. 15, p. 1.2.5-1–1.2.5-14.

Kirschner, D. E., and Rycerski, B. A., 1988, Petroleum potential of representative stratigraphic and structural elements in the National Petroleum Reserve in Alaska, *in* Gryc, G., ed., Geology and exploration of the National Petroleum Reserve in Alaska, 1974 to 1982: U.S. Geological Survey Professional Paper 1399, p. 191–208.

Lahr, J. C., and Plafker, G., 1980, Holocene Pacific–North American plate interaction in southern Alaska: Implications for the Yakataga seismic gap: Geology, v. 8, p. 483–486.

Lane, L. S., 1991, The pre-Mississippian "Neruokpuk Formation," northeastern Alaska and northwestern Yukon: Review and new regional correlation: Canadian Journal of Earth Sciences, v. 28, p. 1521–1533.

—— , 1992, Kaltag fault, northern Yukon, Canada: Constraints on evolution of Arctic Alaska: Geology, v. 20, p. 653–656.

Langseth, M. G., Lachenbruch, A. H., and Marshall, B. V., 1990, Geothermal observations in the Arctic region, *in* Grantz, A., Johnson, L., and Sweeney, J. F., eds., The Arctic Ocean region: Boulder, Colorado, Geological Society of America, The Geology of North America, v. L, p. 133–151.

Lerand, M., 1973, Beaufort Sea, *in* McCrossam, R. G., ed., The future petroleum provinces of Canada—Their geology and potential: Canadian Society of Petroleum Geology Memoir 1, p. 315–386.

Levander, A. R., Wissinger, E. S., Fuis, G. S., and Lutter, W. J., 1991, The 1990 Brooks Range seismic experiment: Near vertical incidence reflection images [abs.]: EOS (Transactions, American Geophysical Union) v. 72, no. 44 supplement, p. 296.

Little, T. A., and Naeser, S. W., 1989, Tertiary tectonics of the Border Ranges fault system, Chugach Mountains, Alaska: Deformation and uplift in a forearc setting: Journal Geophysical Research, v. 94, p. 4333–4360.

Loney, R. A., and Himmelberg, G. R., 1989, The Kanuti ophiolite, Alaska: Journal of Geophysical Research, v. 94, p. 15,869–15,900.

Lonsdale, P., 1988, Paleogene history of the Kula plate: Offshore evidence and onshore implications: Geological Society of America Bulletin, v. 100, p. 755–766.

Mair, J. A., and Lyons, J. A., 1981, Crustal structure and velocity anistrophy beneath the Beaufort Sea: Canadian Journal of Earth Sciences, v. 18, p. 724–741.

Marquis, G., and Globerman, B. R., 1988, Northward motion of the Whitehouse Trough: Paleomagnetic evidence from the Upper Cretaceous Carmacks

Group: Canadian Journal of Earth Sciences, v. 25, p. 2005–2016.

Martin, A. J., 1970, Structure and tectonic history of the western Brooks Range, De Long Mountains, and Lisburne Hills, northern Alaska: Geological Society of America Bulletin, v. 81, p. 3605–3622.

May, S. D., and Grantz, A., 1990, Sediment thickness in the southern Canada Basin: Marine Geology, v. 93, p. 331–347.

Mayfield, C. F., Tailleur, I. L., Mull, C. G., and Silberman, M. L., 1978, Granitic clasts from Upper Cretaceous conglomerate in the northwestern Brooks Range, *in* Johnson, K. M., ed., The United States Geological Survey in Alaska: Accomplishments during 1977: U.S. Geological Survey Circular 772-B, p. B11–B13.

Mayfield, C. F., Tailleur, I. L., and Ellersieck, I., 1988, Stratigraphy, structure, and palinspastic synthesis of the western Brooks Range, northwestern Alaska, *in* Gryc, G., ed., Geology and exploration of the National Petroleum Reserve in Alaska, 1974 to 1982: U.S. Geological Survey Professional Paper 1399, p. 143–186.

Meyers, H., 1976, A historical summary of earthquake epicenters in and near Alaska: Boulder, Colorado, National Oceanic and Atmospheric Administration, NOAA technical Memorandum EDS NGSDC-1, 57 p.

Miller, E. L., and Hudson, T. L., 1991, Mid-Cretaceous extensional fragmentation of a Jurassic–Early Cretaceous compressional orogen, Alaska: Tectonics, v. 10, p. 781–796.

Monger, J.W.H., Price, R. A., and Tempelman-Kluit, D. J., 1982, Tectonic accretion and the origin of the two major metamorphic and plutonic welts in the Canadian Cordillera: Geology, v. 10, p. 70–75.

Moore, T. E., 1987, Geochemistry and tectonic setting of some volcanic rocks of the Franklinian Assemblage, central and eastern Brooks Range, *in* Tailleur, I., and Weimer, P., eds., Alaskan North Slope geology: Bakersfield, California, Pacific Section, Society of Economic Paleontologists and Mineralogists, and Alaska Geological Society Book 50, p. 691–710.

Moore, T. E., and Murphy, J. M., 1989, Nature of the basal contact of the Tozitna terrane along the Dalton Highway, northeast Tanana quadrangle, Alaska, *in* Dover, J. H., and Galloway, J. P., Geologic studies in Alaska by the U.S. Geological Survey, 1988: U.S. Geological Survey Bulletin 1903, p. 46–53.

Moore, T. E., Whitney, J. W., and Wallace, W. K., 1985, Cenozoic north-vergent tectonism in northeastern Alaska [abs.]: Eos (Transactions, American Geophysical Union), v. 66, p. 862.

Moore, T. E., Wallace, W. K., Bird, K. J., Karl, S. M., Mull, C. G., and Dillon, J. T., 1992, Stratigraphy, structure, and geologic synthesis of northern Alaska: U.S. Geological Survey Open-File Report OF 92-330, 183 p., 1 map sheet, scale 1:2,500,000.

Moore, T. E., Aleinikoff, J. N., and Walter, M., 1993, Middle Jurassic U-Pb crystallization age for Siniktanneyak Mountain ophiolite, Brooks Range, Alaska: Geological Society of America Abstracts with Programs, v. 25, no. 5, p. 124.

Murphy, J. M., and Patton, W. W., Jr., 1988, Geologic setting and petrography of the phyllite and metagraywacke thrust panel, north-central Alaska, *in* Galloway, J. P., and Hamilton, T. D., eds., Geologic studies in Alaska by the U.S. Geological Survey during 1987: U.S. Geological Survey Circular 1016, p. 104–108.

Nelson, B. K., Nelson, S. W., and Till, A. B., 1989, Isotopic evidence of an Early Proterozoic crustal source for granites of the Brooks Range, northern Alaska: Geological Society of America Abstracts with Programs, v. 21, p. A105.

Newton, C. R., 1983, Paleozoogeographic affinities of Norian bivalves from the Wrangellian, Peninsular, and Alexander terranes, western North America, *in* Stevens, C. H., ed., Pre-Jurassic rocks in western North American suspect terranes: Los Angeles, California, Pacific Section, Society of Economic Paleontologists and Mineralogists, p. 37–48.

Nilsen, T. H., 1981, Upper Devonian and Lower Mississippian redbeds, Brooks Range, Alaska, *in* Miall, A. D., ed., Sedimentation and tectonics in alluvial basins: Geological Association of Canada Special Paper 23, p. 187–219.

Nokleberg, W. J., and Aleinikoff, J. N., 1985, Summary of stratigraphy, structure, and metamorphism of Devonian igneous-arc terranes, northeastern Mount Hayes quadrangle, eastern Alaska Range, *in* Bartsch-Winkler, S., ed., The United States Geological Survey in Alaska—Accomplishments during 1984: U.S. Geological Survey Circular 967, p. 66–71.

Nokleberg, W. J., Jones, D. L., and Silberling, N. J., 1985, Origin and tectonic evolution of the Maclaren and Wrangellia terranes, eastern Alaska Range, Alaska: Geological Society of America Bulletin, v. 96, p. 1251–1270.

Nokleberg, W. J., Plafker, G., and Wilson, F. H., 1994, Geology of south-central Alaska, *in* Plafker, G., and Berg, H. C., eds., The Cordilleran orogen: Alaska: Boulder, Colorado, Geological Society of America, The Geology of North America, v. G-1 (in press).

Oldow, J. S., Avé Lallemant, H.G.A., Julian, F. E., and Seidensticker, C. M., 1987a, Ellesmerian(?) and Brookian deformation in the Franklin Mountains, northeastern Brooks Range, Alaska, and its bearing on the origin of the Canada basin: Geology, v. 15, p. 37–41.

Oldow, J. S., Seidensticker, C. M., Phelps, J. C., Julian, F. E., Gottschalk, R. R., Boler, K. W., Handschy, J. W., and Ave Lallemant, H. G., 1987b, Balanced cross sections through the central Brooks Range and North Slope, Arctic Alaska: Tulsa, Oklahoma, American Association of Petroleum Geologists Special Pupblication 19 p., 8 plates, scale 1:200,000.

Oldow, J. S., Bally, A. W., Avé Lallemant, H. G., and Leeman, W. P., 1989, Phanerozoic evolution of the North American Cordillera; United States and Canada, *in* Bally, A. W., and Palmer, A. R., eds., The Geology of North America—An overview: Boulder, Colorado, Geological Society of America, The Geology of North America, v. A, p. 139–232.

Oxburgh, E. R., 1972, Flake tectonics and continental collision: Nature, v. 239, p. 202–204.

Packer, D. R., and Stone, D. B., 1974, Paleomagnetism of Jurassic rocks from southern Alaska, and the tectonic implications: Canadian Journal of Earth Sciences, v. 11, p. 976–997.

Page, R. A., Stephens, C. D., and Lahr, J. C., 1989, Seismicity of the Wrangell and Aleutian Wadati-Benioff zones and the North American plate along the Trans-Alaska Crustal Transect, Chugach Mountains and Copper River basin, southern Alaska: Journal of Geophysical Research, v. 94, p. 16,059–16,082.

Pallister, J. S., Budahn, J. R., and Murchey, B. L., 1989, Pillow basalts of the Angayucham terrane: Oceanic-plateau and island crust accreted to the Brooks Range: Journal of Geophysical Research, v. 94, p. 15,901–15,923.

Palmer, A. R., 1983, The Decade of North American Geology, 1983 time scale: Geology, v. 11, p. 503–504.

Palmer, A. R., Dillon, J. T., and Dutro, J. T., Jr., 1984, Middle Cambrian trilobites with Siberian affinities from the central Brooks Range, northern Alaska: Geological Society of America Abstracts with Programs, v. 16, p. 327.

Panuska, B. C., and Stone, D. B., 1981, Late Paleozoic paleomagnetic data for Wrangellia: Resolution of the polarity ambiguity: Nature, v. 293, p. 561–563.

—— , 1985, Latitudinal motions of the Wrangellia and Alexander terranes and the southern Alaska superterrane, *in* Howell, D. B., ed., Tectonostratigraphic terranes of the circum-Pacific region: Houston, Texas, Circum-Pacific Council for Energy and Mineral Resources Earth Science Series no. 1, p. 109–120.

Patton, W. W., and Tailleur, I. L., 1964, Geology of the Killik-Itkillik region, Alaska: U.S. Geological Survey Professional Paper 303G, p. 409–500.

—— , 1977, Evidence in the Bering Strait region for differential movement between North America and Eurasia: Geological Society of America Bulletin, v. 88, p. 1298–1304.

Patton, W. W., Tailleur, I. L., Brosgé, W. P., and Lanphere, M. A., 1977, Preliminary report on the ophiolites of northern and western Alaska: Oregon Department of Geology and Mineral Industries Bulletin 95, p. 51–57.

Patton, W. W., Jr., Stern, T. W., Arth, J. G., and Carlson, C., 1987, New U-Pb ages from granite and granite gneiss in the Ruby geanticline and southern Brooks Range, Alaska: Journal of Geology, v. 95, p. 118–126.

Patton, W. W., Jr., Box, S. E., Moll-Stalcup, E. J., and Miller, T. P., 1994, Geology of western-central Alaska, *in* Plafker, G., and Berg, H. C., eds., The Cordilleran orogen: Alaska: Boulder, Colorado, Geological Society of Amer-

ica, The Geology of North America, v. G-1 (in press).
Pavlis, T. L., 1982, Origin and age of the Border Ranges fault of southern Alaska and its bearing on the late Mesozoic tectonic evolution of Alaska: Tectonics, v. 1, p. 343–368.
——, 1989, Middle Cretaceous orogenesis in the northern Cordillera; a Mediterranean analog of collision-related extensional tectonics: Geology, v. 17, p. 947–950.
Pavlis, T. L., and Crouse, G. W., 1989, Late Mesozoic strike-slip movement on the Border Ranges fault system in the eastern Chugach Mountains, southern Alaska: Journal of Geophysical Research, v. 94, p. 4321–4332.
Pavlis, T. L., Sisson, V. B., Foster, H. L., Nokleberg, W. J., and Plafker, G., 1993, Mid-Cretaceous extensional tectonics of the Yukon-Tanana terrane, Trans-Alaska crustal transect (TACT), east-central Alaska: Tectonics, v. 12, p. 103–122.
Plafker, G., 1987, Regional geology and petroleum potential of the northern Gulf of Alaska continental margin, *in* Scholl, D. W., Grantz, A., and Vedder, J. G., eds., Geology and resource potential of the continental margin of western North America and adjacent ocean basins—Beaufort Sea to Baja California: Houston, Texas, Circum-Pacific Council for Energy and Mineral Resources Earth Science Series no. 6, p. 229–268.
Plafker, G., Hudson, T. L., and Richter, D. H., 1977, Preliminary observations on late Cenozoic displacements along the Totschunda and Denali fault systems: U.S. Geological Survey Circular 751-B, p. B67-B69.
Plafker, G., Nokleberg, W. J., Lull, J. S., Roeske, S. M., and Winkler, G. R., 1986, Nature and timing of deformation along the Contact fault system in the Cordova, Bering Glacier, and Valdez quadrangles, *in* Bartsch-Winkler, S., and Reed, K. M., eds., Geologic studies in Alaska by the U.S. Geological Survey during 1985: U.S. Geological Survey Circular 978, p. 74–77.
Plafker, G., Nokleberg, W. J., and Lull, J. S., 1989, Bedrock geology and tectonic evolution of the Wrangellia, Peninsular, and Chugach terranes along the Trans-Alaska Crustal Transect in the Chugach Mountains and southern Copper River basin, Alaska: Journal of Geophysical Research, v. 94, p. 4255–4295.
Plafker, G., Naeser, C. W., Zimmerman, R. A., Lull, J. S., and Hudson, T., 1992, Cenozoic uplift history of the Mount McKinley area in the central Alaska Range based on fission-track dating, *in* Bradley, D. C., and Dusel-Bacon, C., eds., Geologic studies in Alaska by the U.S. Geological Survey, 1991: U.S. Geological Survey Bulletin 2041, p. 202–212.
Plumley, P. W., 1984, A paleomagnetic study of the Prince William terrane and Nixon Fork terrane, Alaska [Ph.D. thesis]: Santa Cruz, University of California 190 p.
Plumley, P. W., Coe, R. S., and Byrne, T., 1983, Paleomagnetism of the Paleocene Ghost Rocks Formation, Prince William terrane, Alaska: Tectonics, v. 2, p. 295–314.
Plumley, P. W., Wirth, K. R., Bird, J. M., and Harding, D. J., 1990, Paleomagnetic results from sheeted dike sequence, western Brooks Range Ophiolite, Alaska [abs.]: Eos (Transactions, American Geophysical Union), v. 71, p. 1295.
Popeko, V. A., compiler, 1985, Geological map of the northern part of the Koryak upland: Leningrad, Academy of Science on the USSR, Far Eastern Scientific Center, Institute of Tectonics and Geophysics, 1 sheet, scale 1:1,000,000.
Richter, D. H., Smith, J. G., Lanphere, M. A., Dalrymple, G. B., Reed, B. L., and Shew, N., 1990, Age and progression of volcanism, Wrangell volcanic field, Alaska: Bulletin of Volcanology, v. 53, p. 29–44.
Rickwood, F. K., 1970, The Prudhoe Bay field, *in* Adkison, W. L., and Brosgé, M. M., eds., Proceedings of the geological seminar on the North Slope of Alaska: Los Angeles, California, Pacific Section, American Association of Petroleum Geologists, p. L1–L11.
Roddick, J. A., 1967, Tintina trench: Journal of Geology, v. 75, p. 23–32.
Roeder, D. H., and Mull, C. G., 1978, Tectonics of Brooks Range ophiolites (Alaska): American Association of Petroleum Geologists Bulletin, v. 62, p. 1696–1702.
Roeske, S. M., Pavlis, T. L., Sisson, V. B., and Smart, K., 1990, Cretaceous strike slip along the Border Ranges fault system in eastern and southeastern Alaska: Geological Association of Canada Abstracts with Programs, v. 15, p. A113.
Roeske, S. M., Snee, L. W., and Pavlis, T. L., 1991, Strike-slip and accretion events along the southern Alaska plate margin in the Cretaceous and early Tertiary: Geological Society of America Abstracts with Programs, v. 23, p. A428.
Roeske, S. M., Bradshaw, J. Y., Snee, L. W., 1992, High P/low T metamorphism of oceanic and continental crust in the Ruby geanticline, north-central Alaska: Geological Society of America Abstracts with Programs, v. 24, no. 5, p. 79.
Rubin, C. M., and Saleeby, J. B., 1992, Tectonic history of the eastern edge of the Alexander terrane, southeast Alaska: Tectonics, v. 11, p. 586–602.
Rubin, C. M., Miller, M. M., and Smith, G. E., 1990, Tectonic development of Cordilleran mid-Paleozoic volcanoplutonic complexes: Evidence for convergent margin tectonism, *in* Harwood, D. S., and Miller, M. M., eds., Paleozoic and early Mesozoic paleogeographic relations; Sierra Nevada, Klamath Mountains, and related rocks: Geological Society America Special Paper 225, p. 1–16.
Samson, S. D., Patchett, P. J., McClelland, W. C., and Gehrels, G., 1991, Nd isotopic characterization of metamorphic rocks in the Coast Mountains, Alaskan and Canadian Cordillera: Ancient crust bounded by juvenile terranes: Tectonics, v. 10, p. 770–780.
Savage, N. M., 1988, Devonian faunas and major depositional events in the southern Alaska terrane, southeastern Alaska, *in* McMillan, N. J., Embry, A. T., and Glass, D. J., eds., Devonian of the world, Volume III: Paleontology, paleoecology, and biostratigraphy: Canadian Society of Petroleum Geologists Memoir 14, p. 257–264.
Sengor, A.M.C., and Dewey, F.R.S., 1990, Terranology: Vice or virtue?: Royal Society of London Philosophical Transactions, ser. A, v. 331, p. 457–477.
Silberling, N. J., 1985, Biogeographic significance of the Upper Triassic bivalve Monotris in Circum-Pacific region, *in* Howell, D. G., ed., Tectonostratigraphic terranes of the circum-Pacific region: Houston, Texas, Circum-Pacific Council for Energy and Mineral Resources Earth Science Series no. 1, p. 63–70.
Sisson, V. B., Hollister, L. S., and Onstott, T. C., 1989, Petrologic and age constraints on the origin of a low pressure/high temperature metamorphic complex, southern Alaska: Journal of Geophysical Research, v. 94, p. 4392–4410.
St. Amand, P., 1957, Geological and geophysical synthesis of the tectonics of portions of British Columbia, the Yukon Territory and Alaska: Geological Society of America Bulletin, v. 68, p. 1343–1370.
Stamatakos, J. A., Kodama, K. P., Vittorio, L. F., and Pavlis, T. L., 1989, Paleomagnetism of Cretaceous and Paleocene sedimentary rocks across the Castle Mountain Fault, south-central Alaska, *in* Hillhouse, J. W., ed., Deep structure and past kinematics of accreted terranes: Washington, D.C., American Geophysical Union, Geophysical Monograph, v. 50, p. 151–178.
Stanley, W. D., Labson, V. F., Nokleberg, W. J., Csejtey, B., Jr., and Fisher, M. A., 1990, The Denali fault system and Alaska Range of Alaska: Evidence for suturing and thin-skinned tectonics from magnetotellurics: Geological Society of America Bulletin, v. 102, p. 160–173.
Stephens, C. D., Fogleman, K. A., Lahr, J. C., and Page, R. A., 1984, Wrangell Benioff zone, southern Alaska: Geology, v. 12, p. 373–376.
Stock, J., and Molnar, P., 1987, Revised history of early Tertiary plate motion in the southwest Pacific: Nature, 325, p. 495–499.
Stone, D. B., 1980, The Alaskan orocline, the paleomagnetism and paleogeography of Alaska: Tectonophysics, v. 63, p. 63–73.
Stone, D. B., and McWilliams, M. O., 1989, Paleomagnetic evidence for relative motion in western North America, *in* Ben-Avraham, Z., ed., The evolution of the Pacific Ocean margins: Oxford, England, Oxford University Press, p. 53–72.
Stone, D. B., Panuska, B. C., and Packer, D. R., 1982, Paleolatitude versus time for southern Alaska: Journal of Geophysical Research, v. 87, p. 3697–3707.
Stout, J. H., Brady, J. B., Weber, F. R., and Page, R. A., 1973, Evidence for Quaternary movement on the McKinley strand of the Denali fault in the

Delta River area, Alaska: Geological Society of America Bulletin, v. 84, p. 939–947.

Tailleur, I. L., 1973, Probable rift origin of Canada basin, *in* Pitcher, M. G., ed., Arctic geology: American Association of Petroleum Geologists Memoir 19, p. 526–535.

—— , 1980, Rationalization of Koyukuk "crunch," northern and central Alaska [abs.]: American Association of Petroleum Geologists Bulletin, v. 64, p. 792.

Tailleur, I. L., and Brosge, W. P., 1970, Tectonic history of northern Alaska, *in* Adkinson, W. L., and Brosge, M. M., eds., Proceedings of the geological seminar on the North Slope of Alaska: Los Angeles, Pacific section, American Association of Petroleum Geologists, p. E1-E-19.

Taylor, D. G., Callomon, J. H., Smith, R., Tipper, H. W., and Westermann, G.E.G., 1984, Jurassic ammonite biogeography and paleogeography of North America: Geological Association of Canada Special Paper 27-13, 27 p.

Taylor, P. T., Kovacs, L. C., Vogt, P. R., and Johnston, G. L., 1981, Detailed aeromagnetic investigation of the Arctic Basin, 2: Journal of Geophysical Research, v. 86, p. 6323–6333.

Tempelman-Kluit, D. J., 1984, Counterparts of Alaska's terranes in Yukon, *in* Symposium on Cordilleran geology and mineral exploration: Status and future trends: Vancouver, Cordilleran Section, Geological Association of Canada, p. 41–44.

Tipper, H. W., 1981, Offset of an upper Pliensbachian geographic zonation in the North American Cordillera by transcurrent faulting: Canadian Journal of Earth Science, v. 18, p. 1788–1792.

Tozier, E. T., 1982, Marine Triassic faunas of North America: Their significance in assessing plate and terrane movements: Geologische Rundschau, v. 17, p. 1077–1104.

Van Alstine, D. R., Bazard, D. R., and Whitney, J. W., 1985, Paleomagnetism of cores from the Yakutat well, Gulf of Alaska: Eos (Transactions, American Geophysical Union), v. 66, no. 46, p. 865.

Van der Heyden, P., 1992, A Middle Jurassic to early Tertiary Andean-Sierran arc model for the Coast Belt of British Columbia: Tectonics, v. 11, p. 82–97.

Vogt, P. R., Taylor, P. T., Kovacs, L. C., and Johnson, G. L., 1982, The Canada Basin: Aeromagnetic constraints on structure and evolution: Tectonophysics, v. 89, p. 295–336.

Wahrhaftig, C., 1965, Physiographic divisions of Alaska: U.S. Geological Survey Professional Paper 482, 52 p.

Wallace, W. K., and Hanks, C. L., 1990, Structural provinces of the northeastern Brooks Range, Arctic National Wildlife Refuge, Alaska: American Association of Petroleum Geologists Bulletin, v. 74, p. 1100–1118.

Wallace, W. K., Hanks, C. L., and Rogers, J. F., 1989, The southern Kahiltna terrane: Implications for the tectonic evolution of southwestern Alaska: Geological Society of America Bulletin, v. 101, p. 1389–1407.

Weber, F. R., 1990, Correlations across the western part of the Tintina fault system and their implications for displacement history: Geological Association of Canada Program with Abstracts, v. 15, p. 138.

Worral, D. M., 1991, Tectonic history of the Bering Sea and the evolution of Tertiary strike-slip basins of the Bering shelf: Geological Society of America Special Paper 257, 120 p.

Zimmerman, J., and Frank, C. O., 1982, Possible obduction-related metamorphic rocks at the base of the ultramafic zone, Avan Hills complex, DeLong Mountains, *in* Coonrad, W. L., ed., The United States Geological Survey in Alaska: Accomplishments during 1980: U.S. Geological Survey Circular 844, p. 27–28.

MANUSCRIPT ACCEPTED BY THE SOCIETY FEBRUARY 9, 1994

Printed in U.S.A.

DNAG Continent-Ocean Transect Volume
Phanerozoic Evolution of North American
Continent-Ocean Transitions
The Geological Society of America, 1994

Chapter 10

The polar continent-ocean transition in Canada (Corridor G)

J. F. Sweeney
Pacific Geoscience Centre, Box 6000, Sidney, British Columbia V8L 4B2, Canada
U. Mayr
Institute of Sedimentary and Petroleum Geology, 3303 33rd Street NW, Calgary, Alberta T2L 2A7, Canada
L. W. Sobczak
Earth Physics Branch, Department of Energy, Mines, and Resources, Ottawa, Ontario K1A 0Y3, Canada
H. R. Balkwill
Petro-Canada, Box 2844, Calgary, Alberta T2P 2M7, Canada

INTRODUCTION

The Canadian Arctic transect corridor (Corridor G) extends across a wide range of geologic settings. In the south at Somerset Island, Archean crystalline rocks of the craton are exposed (Fig. 1). Northward across Cornwallis and Devon Islands the crystalline rocks are covered by Proterozoic and early Paleozoic platform deposits that were uplifted, eroded, and deformed in Silurian through Early Carboniferous time. Overlying these rocks between Devon and Ellef Ringnes Islands are late Paleozoic and Mesozoic strata of the Sverdrup Basin. These strata were locally deformed by evaporite diapirism and repeated episodes of mafic igneous intrusion and, east of Corridor G, were regionally deformed in latest Cretaceous and Paleogene time. Northwest of Ellef Ringnes Island the corridor traverses the latest Cretaceous and younger clastic terrace wedge that covers the continental margin of the Canada Basin. The corridor terminates at abyssal depths over the continental rise.

The corridor was selected to incorporate deep crustal seismic results between Cornwallis and northern Ellef Ringnes Islands (Sander and Overton,1965; Hobson and Overton, 1967) and offshore refraction and reflection work along the east coast of Ellef Ringnes Island (Sobczak, 1982; Sobczak and Overton, 1984). Phanerozoic strata are almost completely represented along the corridor, and reflection seismic data (courtesy of Panarctic Oils Ltd., Calgary) locate stratigraphic horizons at several points. Borehole data are available for Cornwallis, Bathurst, Amund Ringnes, and Ellef Ringnes Islands, and there is regional coverage by gravity (Sobczak, 1978) and aeromagnetic (Coles and others, 1976) data, including detailed measurements in the Sverdrup Basin and northward.

Control is poor in offshore areas because of a lack of boreholes, and north of Ellef Ringnes Island, seismic information is also lacking. Consequently, the stratigraphy and tectonic history of the outer shelf and slope remain largely unknown.

With emphasis on rocks within Corridor G, this report will focus on four main topics: the geophysical character and structure of the continent-ocean transition; the mode of origin and age of the continent-ocean transition; the Phanerozoic structural and tectonic history of the Canadian Arctic archipelago; and recommendations for future study of the polar continental boundary.

THE CONTINENT-OCEAN TRANSITION

Geophysical character and geological structure

The crust below the southern Canada Basin has long been considered oceanic in character (Oliver and others, 1955). Seismic refraction data across the polar margin are available only from Prince Patrick Island and indicate that the crust thins significantly, from about 35 km beneath the island to about 12 km below the continental slope 300 km to the northwest (Berry and Barr, 1971). Similarly, from the continental rise north of Alaska, the crust thins seaward to about 10 km beneath Canada Basin (Mair and Lyons, 1981; Baggeroer and Falconer, 1982). Moho depths below the central part of the abyssal plain have not been measured.

Along Corridor G the polar shelf extends about 175 km from the shoreline to the shelf-slope break at about 600 m below the sea surface (Fig. 2), yielding an average bathymetric gradient of about 3.4 m/km. The gradient of the continental slope averages about 9.4 m/km.

Aeromagnetic data over the Canadian Arctic margin have been collected at line spacings of 20 km or more at altitudes of

Sweeney, J. F., Mayr, U., Sobczak, L. W., and Balkwill, H. R., 1994, The polar continent-ocean transition in Canada (Corridor G), *in* Speed, R. C., ed., Phanerozoic Evolution of North American Continent-Ocean Transitions: Boulder, Colorado, Geological Society of America, DNAG Continent-Ocean Transect Volume.

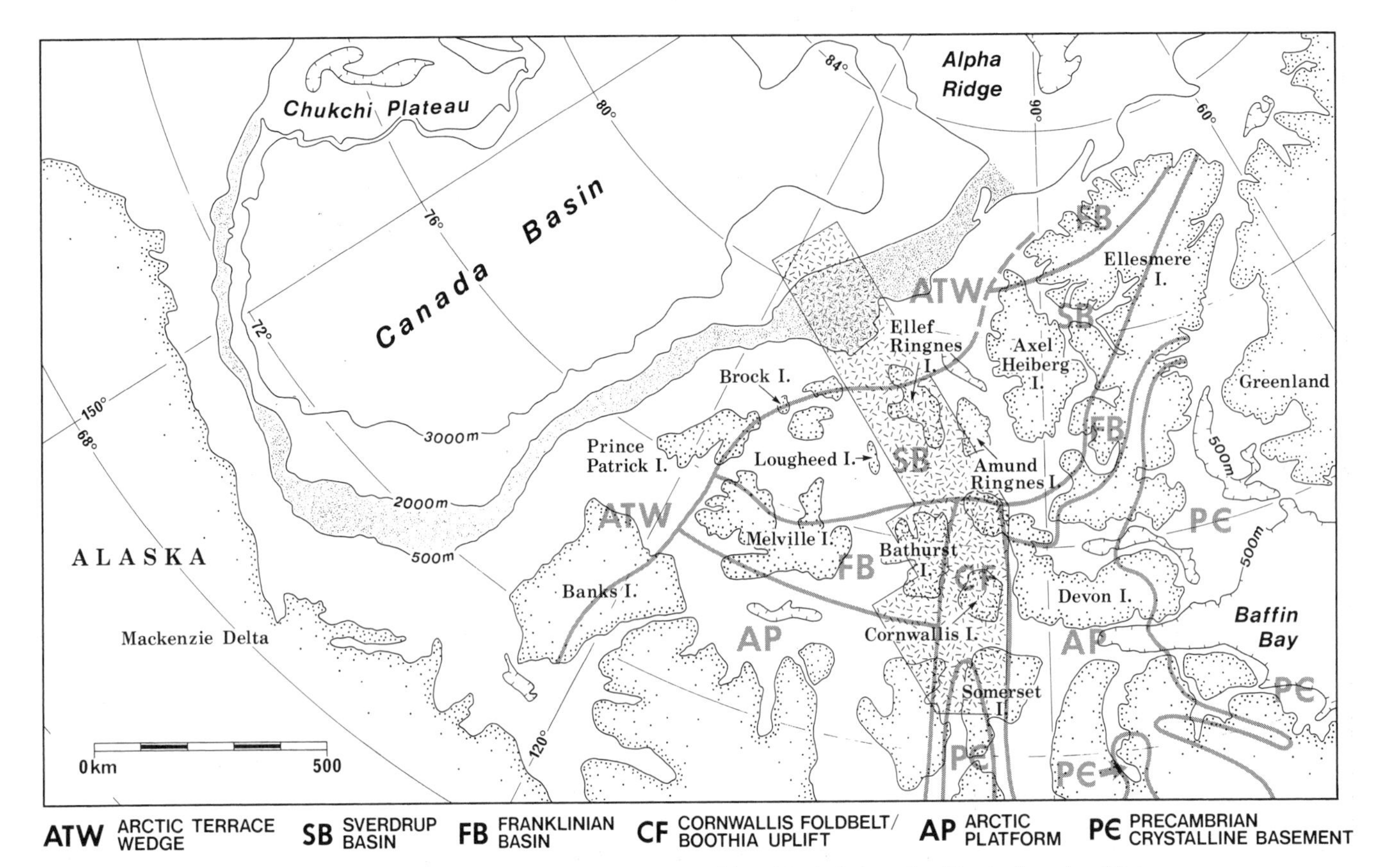

Figure 1. Nomenclature and continent–ocean transition (shaded) in polar North America. Major geologic provinces outlined. Corridor G area hatchured.

3.5 to 5.0 km (see Coles and others, 1978 for details). Within the corridor, there is also closely spaced (2 to 3 km) low-level (300-m) coverage from Amund Ringnes Island to the Canada Basin (Bhattacharyya, 1968). A poorly defined, weak (200 nT) magnetic low lies over much of the polar margin and separates a landward, relatively featureless anomaly field from moderately intense short-wavelength anomalies over the deep ocean basin (Fig. 3; Riddihough and others, 1973; Hood and Bower, 1976; Coles and others, 1976). North of Ellef Ringnes Island the shoreward flank of this low is bounded by a weak (200 nT) magnetic high along the outer shelf (Bhattacharyya, 1968).

The axis of the magnetic low transects the polar margin at low angle, extending from the base of the continental slope north of Prince Patrick Island to exposed rocks of the Sverdrup and Franklinian Basins on northern Ellesmere Island (Riddihough and others, 1973). The axis could represent a zone of widespread faulting and thick sedimentation, possibly related to the development of the present continental margin (Bhattacharyya, 1968). Coles and others (1978) inferred that magnetic lows over the continental boundary indicate deep burial of magnetic basement. In Corridor G, Bhattacharyya (1968) calculated that the magnetic low represents 19 km of weakly magnetic material, probably sediments, present on the continental slope northwest of Ellef Ringnes Island.

Free-air gravity over the North American margin of the Canada Basin includes a sinuous belt of elliptical highs that closely follow the trend of the continental shelf break (Fig. 4) and mark the transition to thinner crust and deeper water of the ocean basin. In Corridor G the anomaly high, greater than 100 mGal, is flanked landward and seaward by parallel but more diffuse negative anomaly belts with –20 to –50 mGal amplitudes. A major part of the free-air anomalies can be accounted for by assuming crustal depth and density variations, such as basement arches or belts of dense rock beneath the outer shelf (Weber, 1963; Wold and others, 1970), and/or regional downflexure of the lithosphere loaded by prograding shelf sediments (Walcott, 1972; Sobczak, 1975).

Arches or large density contrasts within the basement are not evident in P-wave velocity profiles across the margin north of Alaska (Grantz and others, 1979). Such features may well be present elsewhere along the continental boundary but available magnetic and seismic data indicate only the presence of thick

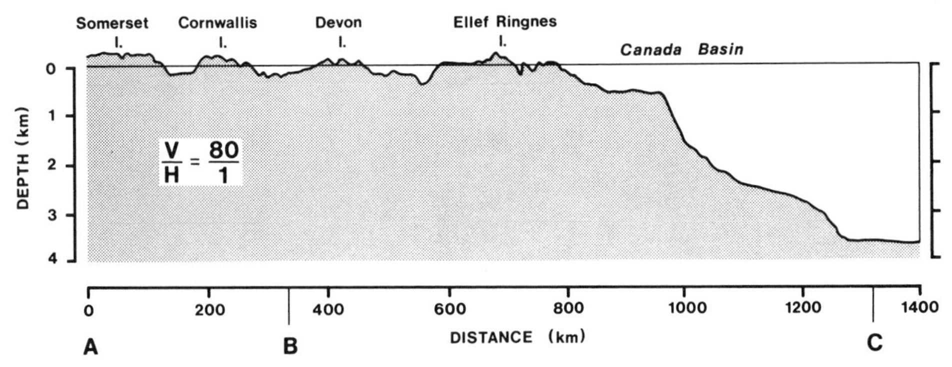

Figure 2. Topography and bathymetry along Corridor G centerline (see Figs. 3 and 4).

sediments (Berry and Barr, 1971; Dixon, 1982; Willumsen and Cote, 1982). Cenozoic clastics alone are over 3 km thick at several points along the margin (Balkwill, 1978; Grantz and others, 1979; Norris and Yorath, 1981). In Corridor G, northwest of Ellef Ringnes Island, refraction results (Hobson and Overton, 1967) indicate a seaward-thickening sediment wedge at least 3 km thick across the inner shelf.

Earthquakes occur mainly within the crust along the polar continental margin and rarely exceed magnitude 5 (Basham and others, 1977; Wetmiller and Forsyth, 1978). Most epicenters lie along the steep gradients of the major gravity highs (Figs. 4 and 5). The thick sedimentary loads thought to produce the gravity highs and magnetic lows may also create the deviatoric stress responsible for the contemporary seismicity (Basham and others, 1977).

Alternatively, the release of stresses associated with deglaciation may produce much of the observed seismicity. This appears to be the case along the passive margin of eastern Canada where over 150 m of post-glacial rebound has already taken place (Stein and others, 1979). However, stresses produced by glacial unloading may not be important along the polar margin between Axel Heiberg Island and the Mackenzie Delta because less than 20 m of post-Pleistocene uplift has occurred there (Farrand and Gajda, 1962; Paterson, 1977).

Shelf-edge gravity highs and prograded clastic wedges are typical features of passive margins. Seismicity is not. The polar margin may have a relatively reduced threshold for seismic activity because: (1) focal mechanism solutions suggest that the horizontal component of deviatoric tensional stress within the crust is nearly normal to the polar margin between Ellef Ringnes Island and the Mackenzie Delta (Hasegawa, 1977; Hasegawa and others, 1979), and (2) oceanic crust beneath Canada Basin is downflexed under an unusually thick sediment cover (2 to 5 km: Hall, 1973; Grantz and others, 1979; Mair and Lyons, 1981) and this may induce great stress in the adjacent continental margin, which is coupled to the oceanic crust. The polar margin, therefore, may require less additional stress input than typical passive margins to become a locus of seismic activity.

Mode of origin and age

Discussion of the geophysics and geology of the Arctic margin of North America can be found in Sobczak and Weber (1973), Sobczak (1975), Grantz and others (1979), Churkin and Trexler (1981), Dutro (1981), Norris and Yorath (1981), and Sweeney (1982a). The directions and times of plate motions that opened the Canada Basin are uncertain (see Clark, 1981, p. 619–623 for a review). Several workers following Carey (1955) have argued that the Canada Basin was produced when a continental block containing northern Alaska rotated away from Arctic Canada about a pivot near the Mackenzie Delta. The argument is based on comparisons of upper Paleozoic and Mesozoic geology of the Alaskan and Canadian margins (Rickwood, 1970; Tailleur, 1973; Grantz and others, 1979). Paleomagnetic studies in northernmost Alaska have yielded results consistent with the rotation model, but the data are as yet too few to assess the regional extent of the apparent rotation (Hillhouse and Grommé, 1983).

Passive margins appear to evolve through successive phases (Bally and Snelson, 1980; Scrutton, 1982): (a) at the outset of continental breakup, tensional faulting occurs, and rocks are often uplifted and eroded near the newly forming continental edge; (b) as the continental blocks separate, their opposing margins begin to subside and receive sediments. The time of margin formation can therefore be bracketed by: (a) the youngest rocks to undergo faulting and erosion during the breakup phase and (b) the oldest rocks deposited during the subsidence phase. Dating of these phases for the Canadian Arctic margin is made difficult by the scarcity of geophysical data and almost complete lack of geological control north of the modern coastline.

Along the inner shelf between Brock and Axel Heiberg Islands, the oldest recognized subsidence phase sediments, basal

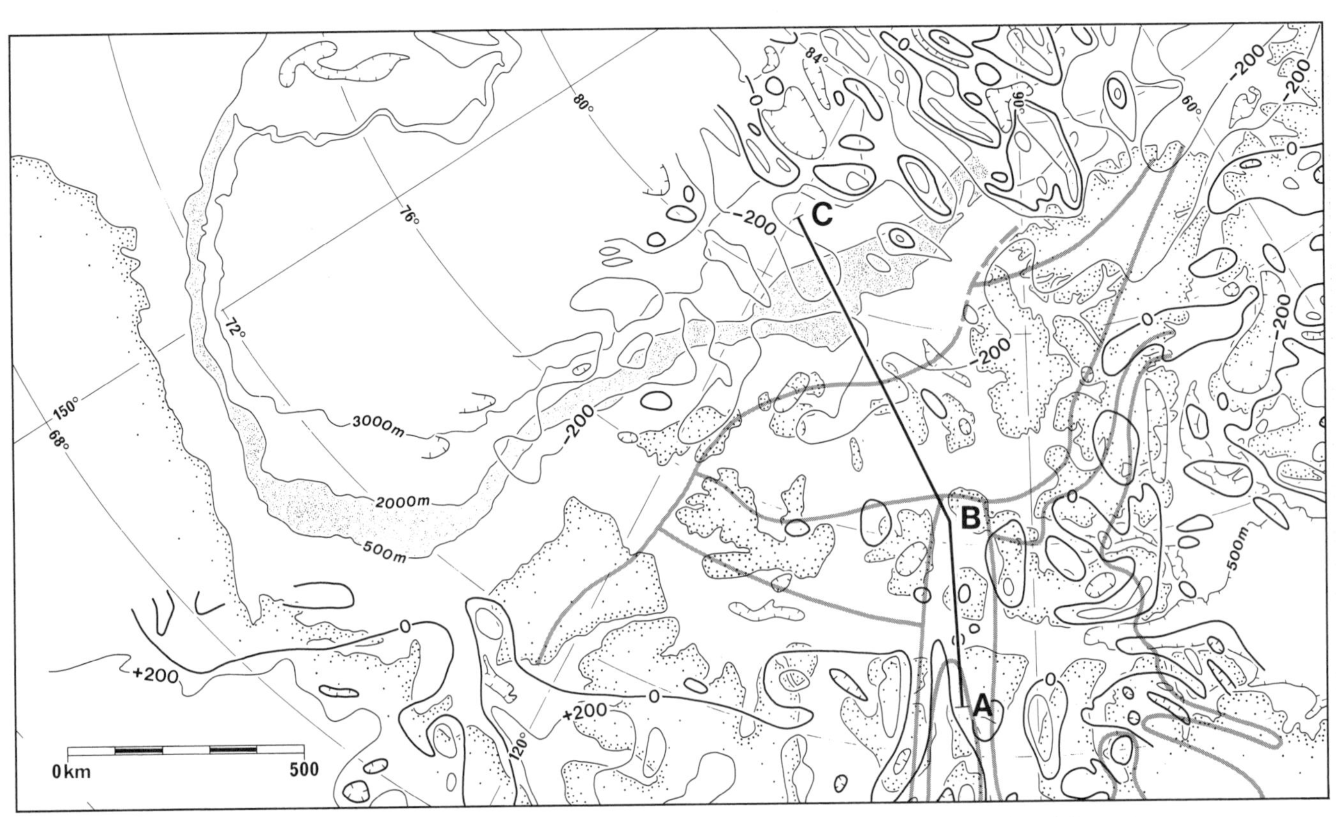

Figure 3. Vertical component magnetic field residuals, relative to IGRF (modified after Coles and others, 1976). Corridor G centerline indicated.

beds of the Arctic Terrace Wedge, are Maastrichtian in age (Meneley and others, 1975; Balkwill, 1978). They lie unconformably on rocks that are downfaulted to the north and are not known to be younger than Late Jurassic–earliest Cretaceous (Meneley and others, 1975). This segment of the polar continental margin was formed, therefore, sometime between about 75 and 144 Ma, using the time scale of Harland and others (1982).

Marine Oxfordian/Kimmeridgian shales overstep older Sverdrup Basin rocks southwestward to Banks Island where the shales are locally in the subsurface (Miall, 1975). They are very tuffaceous, implying nearby volcanism. There are no known Late Jurassic basalt flows in the Sverdrup Basin, but there are some hypabyssal mafic sills of Late Jurassic age (Balkwill, 1978). The mafic rocks could indicate that rifting of the Arctic margin began in the early Late Jurassic, and that the Oxfordian/Kimmeridgian shales are the basal beds overlying the breakup unconformity.

Several lines of evidence, however, suggest that the main phase of rifting took place during the Cretaceous period:

- a great volume of hypabyssal and extrusive tholeiites is interbedded between Valanginian and Santonian rocks (131 to 83 Ma) in the north-central part of the Sverdrup Basin, adjacent to the Arctic margin (Thorsteinsson and Tozer, 1970; Balkwill, 1978; A. F. Embry, personal communication, 1984). Furthermore, Sverdrup Basin underwent vigorous subsidence and sediment infilling during Hauterivian through Albian time (131 to 98 Ma).
- application of depth–age relations proposed by Parsons and Sclater (1977) have resulted in estimates between 125 and 90 Ma for the age of oceanic crust beneath the southern Canada Basin (Eittreim and Grantz, 1979; Baggeroer and Falconer, 1982; Lawver and Baggeroer, 1983).
- application of age–heat flow relations proposed by Parsons and Sclater (1977) indicate an age of 110 to 80 Ma for the crust below central Canada Basin (Lawver and Baggeroer, 1983).

In the southern Canada Basin, Taylor and others (1981) determined that marine magnetic anomalies have low amplitude and are poorly lineated, possibly indicating oceanic crustal generation by other than normal sea-floor spreading processes (Taylor, 1983). In view of the crustal ages estimated from seismic and heat-flow measurements, it is possible that most of this sea floor was created during the interval between 118 and 83 Ma when the Earth's magnetic field did not change polarity (Harland and others, 1982). The observed magnetic variations would thereby reflect topography or structure of sea-floor basement.

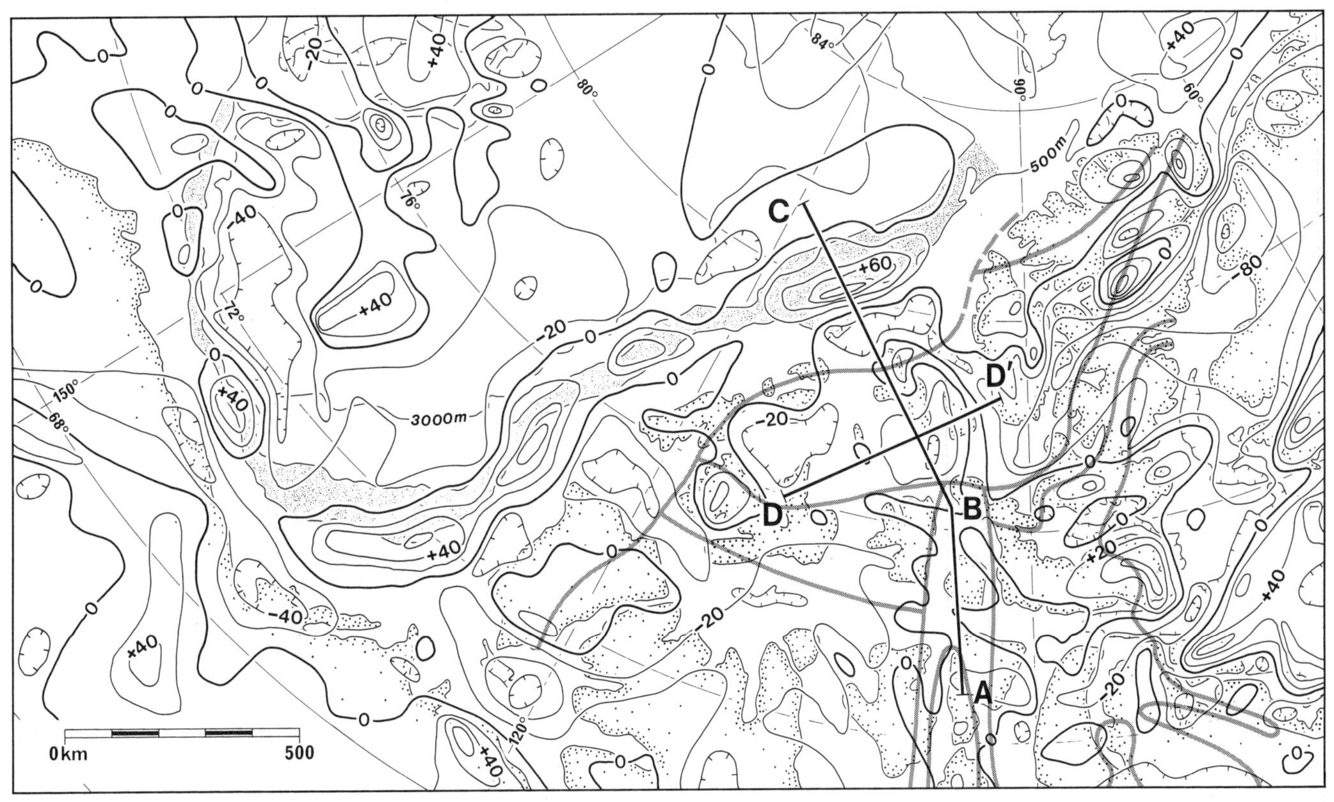

Figure 4. Mean free air gravity (½° latitude × 2° longitude) in 20 mGal contours (from Sobczak, 1978). Figure 6 line of section (DD′) and Corridor G centerline indicated.

PHANEROZOIC HISTORY OF THE POLAR MARGIN AND CORRIDOR G REGION

Devonian and older Paleozoic rocks outcrop in northern Greenland (Dawes, 1976), northern Ellesmere and Axel Heiberg Islands (Trettin and others, 1972), and northern Alaska (Dutro, 1981) and are in the subsurface at several points along the polar margin in North America (Meneley and others, 1975; Miall, 1976; Norris and Yorath, 1981; Churkin and Trexler, 1981). Across northern Greenland and adjacent parts of Canada, shallow-dipping Cambrian platform rocks extend from the craton northward into dominantly shelf facies, which in turn grade northward into a dominantly clastic trough facies that on Ellesmere Island was bounded on the north mainly by shallow marine and volcanic rocks (Dawes, 1976; Trettin and Balkwill, 1979). Trough sediments on Ellesmere Island were derived from northern as well as southern sources, although the former declined steadily during the Cambrian (Trettin and Balkwill, 1979).

On northernmost Ellesmere Island, felsic to basic volcanism occurred in Early and Middle Ordovician time (Trettin and Balkwill, 1979) and was accompanied by the emplacement of granitic and ophiolitic rocks (Trettin and others, 1982). This was followed by local deformation and rapid deposition of flysch derived in part from terrestrial Cambrian or older, low-grade, metamorphic assemblages within a major new or rejuvenated source terrane to the north (Trettin and others, 1972).

Preliminary paleomagnetic reconstructions (Morel and Irving, 1978; Scotese and others, 1979) show that during Ordovician time, no major continental blocks made contact with polar North America. The source terrane, therefore, may have been an isolated continental fragment that was sutured to what is now northern Ellesmere Island in early Middle Ordovician time, but the sequence of geologic events is far from clear (Trettin and others, 1982). The area is structurally and stratigraphically complex; geological and geochronological studies are ongoing.

Clastic deposition advanced southward on Ellesmere Island during the Silurian and Early Devonian periods. Both the volume and clast size of deposited material increases to the northeast (Trettin and others, 1972; Trettin and Balkwill, 1979). Rocks on northern Ellesmere and Axel Heiberg Islands were uplifted and folded during the Early Devonian, contemporary with and possibly related to major Caledonian or orogenic events to the east in Greenland.

In Corridor G, Middle Cambrian to Silurian carbonate,

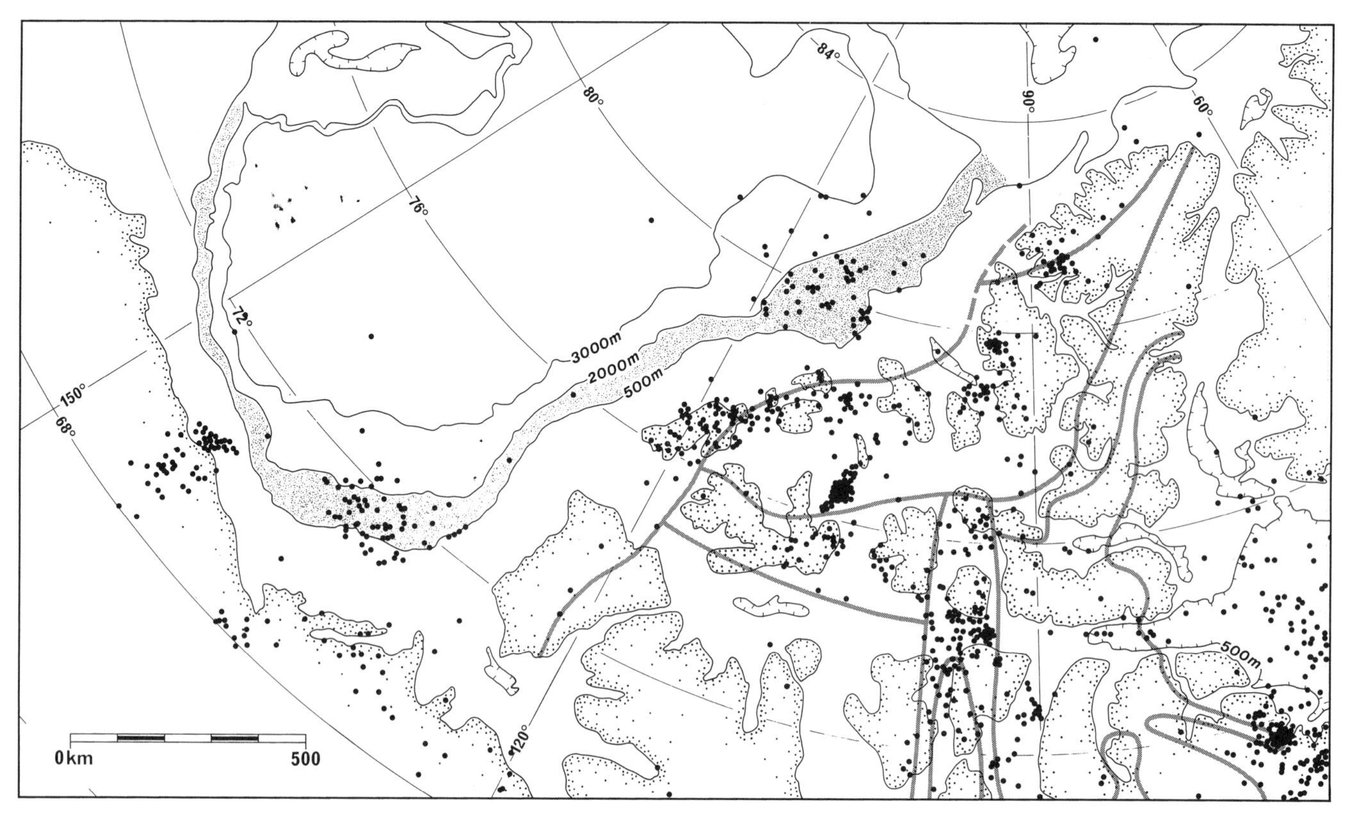

Figure 5. Major geologic provinces in polar Canada and seismicity for the period 1908–1983.

shale, and gypsum-anhydrite sequences lie unconformably on Precambrian sedimentary and crystalline rocks along the axis of the north-trending Boothia uplift (Kerr and Christie, 1965). The sedimentary units thicken northward and were deposited in shallow shelf environments that remained stable until Early Devonian time (Trettin and others, 1972).

The Boothia structure, a broad north-plunging horst, was formed in several pulses from Proterozoic through Devonian time (Kerr, 1981). The major episode of uplift and erosion, called the Cornwallis Disturbance by Kerr (1977), produced the Cornwallis Foldbelt in the Early Devonian. During the Devonian period, at least 5 km of vertical displacement took place along steep basement faults that splay outward near the surface to become the high-angle reverse faults and overturned folds of the Cornwallis Foldbelt (Kerr, 1981).

Connections between the Cornwallis Disturbance and major Caledonian(?) uplift and erosion to the north are unclear. Shortly afterward, from Late Devonian to Early Carboniferous time, major tectonism (Ellesmerian orogeny) affected rocks throughout the Franklinian Basin (Fig. 1). Local intrusion and metamorphism plus regional folding and faulting took place. Tectonic strike was inherited partly from basement structure and partly from depositional trends established during the earlier Paleozoic (Trettin and Balkwill, 1979). On Ellesmere Island the trend of dominant folds and faults varies, north to south, from northeasterly to northerly and then turns east-west across the southern margin of the orogen, best observed in Corridor G on Bathurst Island. The latter trend is interrupted by north-striking structures of the Cornwallis Foldbelt and Boothia Horst, which were positive features immediately before and during the Ellesmerian Orogeny (Kerr, 1981).

The orogeny produced a thick southward-tapering prism of clastic sediments that paleocurrent studies suggest were transported mainly from northeast to southwest across the Canadian Arctic Islands (Embry and Klovan, 1976). Uplift along what is now the polar continental margin included northern Ellesmere and Axel Heiberg Islands in early Middle Devonian time (Embry and Klovan, 1976, p. 597). By the onset of the Early Carboniferous the uplifted zone had expanded southwest to northern Prince Patrick Island (Embry and Klovan, 1976, p. 601).

Plate motions responsible for the Ellesmerian Orogeny remain uncertain. Present paleomagnetic constraints indicate that during the early Paleozoic the Siberian Shield approached polar North America from what is now the east and that intervening sea floor was probably eliminated in Devonian time as the two cratons apparently slid past each other (Morel and Irving, 1978; Scotese and others, 1979). The Ellesmerian event may, thereby, be a product of collision and/or left-lateral shearing between Siberia and North America (Sweeney, 1982b).

Unconformably overlying deformed rocks of the Franklinian Basin are the Early Carboniferous to latest Cretaceous stratigraphic sequences of the Sverdrup Basin (Fig. 1). The succession is as much as 13 km thick and is composed of marine and nonmarine terrigenous clastic sediment, carbonate and evaporite strata, locally abundant gabbro dikes and sills, and some basalt flows. Rocks are mainly concordant in the axial part of the basin, but there are unconformities within the marginal facies.

The depth-sections along Corridor G suggest that Sverdrup Basin geology is continuous from island to island. The inter-island waterways likely represent a drowned Paleogene fluvial system that has been smoothed and fluted by later glaciation (Pelletier, 1964; Balkwill, 1978).

Basin subsidence may have been initiated by thermal contraction of the lithosphere following the metamorphic and intrusive events of the Ellesmerian orogeny. Although mid-Paleozoic thermal features are documented only on northern Ellesmere and Axel Heiberg Islands, well north of the Sverdrup Basin depoaxis, tensional effects, possibly produced by lithospheric cooling, exist within Sverdrup Basin sequences at three distinct times: Carboniferous–Early Permian, Late Jurassic, and late Early Cretaceous.

The late Paleozoic tensional features are half grabens linked by en echelon northwest-trending faults, and by mafic volcanism and intrusion. Mesozoic extension within the basin caused rapid increases in basin subsidence rates, produced northeast-striking listric normal faults, and facilitated injection of gabbroic dikes and sills. As indicated earlier, these later periods of extension within the Sverdrup Basin may be related to the opening of the Canada Basin.

The geologic history of the Sverdrup Basin is, thereby, divided into three cycles. Each cycle begins with a period of incipient rifting, basin foundering, igneous intrusion, and marine transgression. This is followed by the gradual filling of the newly formed depression, mainly by clastic sediments. Subaerial deposits characterize the end of each cycle. Almost all of the terrigenous clastic fill delivered to the Sverdrup Basin is mature to supermature quartz arenite derived from platform and cratonic source regions to the south and southeast.

Large oval diapirs with halite cores (Davies, 1975) extend along a broad northeast-trending zone from the northeastern tip of Melville Island to west-central Ellesmere Island. In Corridor G, diapirs are emplaced on Ellef Ringnes Island. Faunal and lithostratigraphic evidence indicates a Carboniferous age for deposition of the evaporite. Halokinetic migration of deeply buried evaporite and its intrusion into overlying rocks probably surged during phases of rapid loading during late Middle to early Late Triassic and Albian time (Balkwill, 1978).

Sverdrup Basin subsidence terminated in Campanian-Maastrichtian time with the Eurekan orogeny, which began with uplift and erosion of basin fill and older rocks on eastern Ellesmere Island. In late Maastrichtian–early Paleocene time, this was followed by folding, uplift of broad northwest-trending arches, and syntectonic deposition of clastics in intervening depositional basins within the central and eastern Arctic Islands. Along Corridor G, these structures are shown on Ellef Ringnes Island and adjacent regions offshore. In the southern and eastern parts of the archipelago, large grabens developed during this interval, some of which were later filled with several kilometres of clastic sediment (Daae and Rutgers, 1975).

A compressional regime produced folds, reverse faults, and thrusts in the eastern archipelago between middle Eocene and early Miocene time. The orientation of the fold/fault system indicates that regional compression was directed along a northwest-southeast axis. The surface expression of the compressional belt extends westward to the large subtle folds in the Ringnes islands. At least 60 km of crustal shortening may have taken place (Balkwill and Bustin, 1980). A P-wave velocity/crustal density model across the Arctic islands (Fig. 6) suggests that between Lougheed and Axel Heiberg Islands the crust thickens eastward by about 8 km and the Moho appears to be buckled (Sobzcak, 1982; Sobczak and Overton, 1984). This could be evidence that the Tertiary compressional belt extends at depth somewhat farther southwest than is apparent at the surface. The time of suggested Moho buckling is unknown, and Early Cretaceous dike/fault features plus present seismicity in the Ellef Ringnes–Melville Island area (Hasegawa, 1977; Forsyth and others, 1979) indicate a long history of northeast-southwest-directed compression, normal to the Tertiary axis but subparallel to the line of section DD′ (Figs. 4 and 6).

Clastic detritus from uplifts produced by Eurekan tectonism migrated mainly northwestward from Campanian time onward, forming the thick deposits of the Arctic Terrace Wedge that cover the present continent–ocean transition. In Corridor G, these sediments outcrop on northernmost Ellef Ringnes Island and are interpreted from gravity measurements to thicken seaward to about 11 km over the outer shelf and slope.

Tectonic connections between the Eurekan orogeny in the eastern Arctic Islands and the broadly contemporary opening of both the Labrador Sea–Baffin Bay, and the North Atlantic–eastern Arctic Basin are not well established. Discussion centers around the motion of Greenland relative to polar Canada during the Campanian-Eocene interval (see Soper and others, 1982, for a review). It has long been suggested that the opening of Baffin Bay was achieved by simple left-lateral shearing along Nares Strait (Taylor, 1910; Wilson, 1963). The magnitude of the offset is unclear, and estimates range from 0 to 25 km, based on geological arguments (Kerr, 1967; Christie and others, 1981), to 250 km based on aeromagnetic data (Kovacs, 1982). It has recently been proposed that Baffin Bay genesis could have been accommodated in northern Greenland and Ellesmere Island by nonrigid plate behavior in the form of crustal shortening and/or sinistral motion distributed across a sequence of en echelon faults (H. R. Jackson, personal communication, 1984; Hugon, 1983).

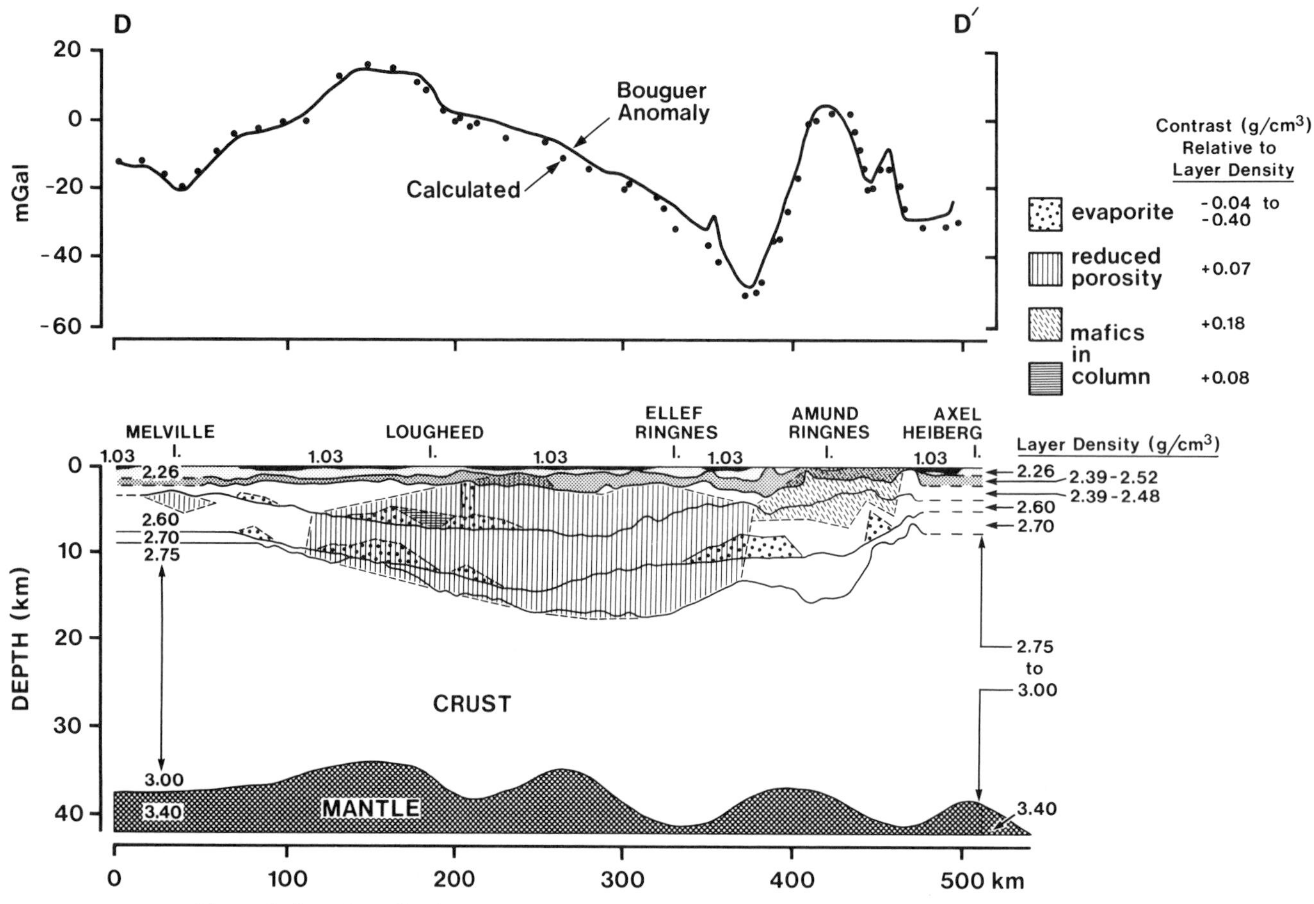

Figure 6. Crustal profile along Sverdrup Basin depoaxis (DD′ in Fig. 4). Crustal horizons and Moho from detailed refraction profiling and borehole control along line of section (Forsyth and others, 1979; Sobczak, 1982). Local masses, layer densities and density contrasts interpreted from refraction seismic and gravity data (Sobczak and Overton, 1984).

FUTURE RESEARCH

Major geoscientific problems facing Arctic workers, along with priorities for future directions of study, have been summarized (Gudmandsen, 1980; Sweeney, 1981). The goals are better understanding of the geodynamic history of the Arctic as a region and improved control in placing north polar tectonic development in a global context by extending geoscientific knowledge of the north to deeper crustal levels on both a local and a regional scale. Much needed fundamental information about the Arctic continent–ocean transition will come from integrated studies of the structure and physical properties of the Canadian polar shelf and slope that are underway and planned to continue year-round from an ice island base slowly drifting from Ellesmere Island toward the Mackenzie Delta (Weber and others, 1984). It is expected that the new data will identify major structures within the deeper crust seaward of the inner shelf and indicate the thickness and distribution of the overlying sediments.

REFERENCES CITED

Baggeroer, A. B., and Falconer, R., 1982, Array refraction profiles and crustal models of the Canada Basin: Journal of Geophysical Research, v. 87, p. 5461–5476.

Balkwill, H. R., 1978, Evolution of Sverdrup Basin, Arctic Canada: American Association of Petroleum Geologists Bulletin, v. 62, p. 1004–1028.

Balkwill, H. R., and Bustin, R. M., 1980, Late Phanerozoic structures, Canadian Arctic Archipelago: Palaeogeography, Palaeoclimatology, Palaeocology, v. 30, p. 219–227.

Bally, A. W., and Snelson, S., 1980, Realms of subsidence, *in* Mull, A., ed., Facts and Principles of World Petroleum Occurrence: Canadian Society of Petroleum Geologists Memoir 6, p. 9–94.

Basham, P. W., Forsyth, D. A., and Wetmiller, R. J., 1977, Seismicity of northern Canada: Canadian Journal of Earth Sciences, v. 14, p. 1646–1667.

Bhattacharyya, B. K., 1968, Analysis of aeromagnetic data over the Arctic Islands and continental shelf of Canada: Geological Survey of Canada Paper 68–44, p. 1–14.

Berry, M. J., and Barr, K. G., 1971, A seismic refraction profile across the polar continental shelf of the Queen Elizabeth Islands: Canadian Journal of Earth Sciences, v. 8, p. 347–360.

Carey, S. W., 1955, The orocline concept in geotectonics: Royal Society of Tasmania Papers and Proceedings, v. 89, p. 255–288.

Christie, R. L., Dawes, P. R., Frisch, T., Higgins, A. K., Hurst, J. M., Kerr, J. W., and Peel, J. S., 1981, Geological evidence against major displacement in Nares Strait: Nature, v. 291, p. 478–480.

Churkin, M., Jr., and Trexler, J. H., 1981, Continental plates and accreted oceanic terranes in the Arctic, *in* Nairn, A.E.M., Churkin, M., Jr., and Stehli, F. G., eds., The Ocean Basins and Margins; v. 5, The Arctic Ocean: New York, Plenum Press, p. 1–20.

Clark, D. L., 1981, Geology and geophysics of the Amerasia Basin, *in* Nairn, A.E.M., Churkin, M., Jr., and Stehli, F. G., eds., The Ocean Basins and Margins; v. 5, The Arctic Ocean: New York, Plenum Press, p. 599–634.

Coles, R. L., Haines, G. V., and Hannaford, W., 1976, Large scale magnetic anomalies over western Canada and the Arctic; A discussion: Canadian Journal of Earth Sciences, v. 13, p. 790–802.

Coles, R. L., Hannaford, W., and Haines, G. V., 1978, Magnetic anomalies and the evolution of the Arctic, *in* Sweeney, J. F., ed., Arctic Geophysical Review: Ottawa, Publications of the Earth Physics Branch, v. 45, p. 51–66.

Daae, H. D., and Rutgers, A.T.C., 1975, Geological history of the Northwest Passage: Bulletin of Canadian Petroleum Geology, v. 23, p. 84–108.

Davies, G. R., 1975, Hoodoo L–41—diapiric halite facies of the Otto Fiord Formation in the Sverdrup Basin, Arctic Archipelago: Geological Survey of Canada Paper 75–1C, p. 23–29.

Dawes, P. R., 1976, Precambrian to Tertiary of northern Greenland, *in* Escher, A., and Watts, W. S., eds., Geology of Greenland: Copenhagen, Grønlands Geologiske Undersogelse, p. 248–303.

Dixon, J., 1982, Upper Oxfordian to Albian geology, Mackenzie Delta, arctic Canada, *in* Embry, A. F. and Balkwill, H. R., eds., Arctic geology and geophysics: Canadian Society of Petroleum Geologists Memoir 8, p. 29–42.

Dutro, J. T., Jr., 1981, Geology of Alaska bordering the Arctic Ocean, *in* Nairn, A.E.M., Churkin, M., Jr., and Stehli, F. G., eds., The Ocean Basins and Margins; v. 5, The Arctic Ocean: New York, Plenum Press, p. 21–36.

Eittreim, S., and Grantz, A., 1979, CDP seismic sections of the western Beaufort continental margin, *In:* C. E. Keen, ed., Crustal Properties across Passive Margins, Tectonophysics, 59, p. 251–262.

Embry, A. F., and Klovan, J. E., 1976, The Middle-Upper Devonian clastic wedge of the Franklinian geosyncline: Bulletin of Canadian Petroleum Geology, v. 24, p. 485–639.

Farrand, W. R., and Gajda, R. T., 1962, Isobases on the Wisconsin marine limit in Canada: Ottawa, Department of Mines and Technical Surveys, Geographic Bulletin 17, p. 5–22.

Forsyth, D. A., Mair, J. A., and Fraser, I., 1979, Crustal structure of the central Sverdrup Basin: Canadian Journal of Earth Sciences, v. 16, p. 1581–1598.

Grantz, A., Eittreim, S., and Dinter, D. A., 1979, Geology and tectonic development of the continental margin north of Alaska, *in* Keen, C. E., ed., Crustal Properties across Passive Margins: Tectonophysics, v. 59, p. 263–291.

Gudmandsen, P., ed., 1980, Marine Science Problems in the Eastern Arctic: Copenhagen, Commission for Scientific Work in Greenland, p.1–135.

Hall, J. K., 1973, Geophysical evidence for ancient seafloor spreading from Alpha Cordillera and Mendeleev Ridge, American Association of Petroleum Geologists Memoir 19, p. 542–561.

Harland, W. B., Cox, A. V., Llewellyn, P. G., Pickton, C.A.G., Smith, A. G., and Walters, R., 1982, A Geologic Time Scale: Cambridge University Press, 131 p.

Hasewaga, H. S., 1977, Focal parameters of four Sverdrup Basin, Arctic Canada, earthquakes in November and December of 1972: Canadian Journal of Earth Sciences, v. 14, p. 2481–2494.

Hasegawa, H. S., Chou, C. W., and Basham, P. W., 1979, Seismotectonics of the Beaufort Sea: Canadian Journal of Earth Sciences, v. 16, p. 816–830.

Hillhouse, J. W., and Grommé, C. S., 1983, Paleomagnetic studies and the hypothetical rotation of Arctic Alaska, *in* Mull, C. G., and Reed, K. M., eds., The Origin of the Arctic Ocean (Canada Basin): Journal of the Alaska Geological Society, v. 2, p. 27–39.

Hobson, G. D., and Overton, A., 1967, A seismic section of the Sverdrup Basin, Canadian Arctic Islands, *in* Musgrave, A. W., ed., Seismic Refraction Prospecting: Tulsa, Society of Exploration Geophysists, p. 550–562.

Hood, P. J., and Bower, M. E., 1976, Arctic Ocean; Low-level aeromagnetic profiles obtained in 1975: Geological Survey of Canada Paper 76–1A, p. 421–424.

Hugon, H., 1983, Ellesmere–Greenland fold belt; Structural evidence for left-lateral shearing: Tectonophysics, v. 100, p. 215–225.

Kerr, J. W., 1967, Nares submarine rift valley and the relative rotation of northern Greenland: Bulletin of Canadian Petroleum Geology, v. 15, p. 483–520.

——, 1977, Cornwallis foldbelt and the mechanism of basement uplift: Canadian Journal of Earth Sciences, v. 14, p. 1374–1401.

——, 1981, Evolution of the Canadian Arctic Islands; A Transition between the Atlantic and Arctic oceans, *in* Nairn, A.E.M., Churkin, M., and Stehli, F. G., eds., The Ocean Basins and Margins; v. 5, The Arctic Ocean: New York, Plenum Press, p. 105–199.

Kerr, J. W., and Christie, R. L., 1965, Tectonic history of Boothia Uplift and Cornwallis foldbelt, Arctic Canada: American Association of Petroleum Geologists Bulletin, v. 49, p. 905–926.

Kovacs, L. C., 1982, Motion along Nares Strait recorded in the Lincoln Sea; Aeromagnetic evidence, *in* Dawes, P. R., and Kerr, J. W., eds., Nares Strait and the Drift of Greenland; a Conflict in Plate Tectonics: Meddelelser om Grønland, v. 8, p. 275–290.

Lawver, L. A., and Baggeroer, A., 1983, A note on the age of the Canada Basin, *in* Mull, C. G., and Reed, K. M., eds., The Origin of the Arctic Ocean (Canada Basin): Journal of the Alaska Geological Society, v. 2, p. 57–66.

Mair, J. A., and Lyons, J. A., 1981, Crustal structure and velocity anisotrophy beneath the Beaufort Sea: Canadian Journal of Earth Sciences, v. 18, p. 724–741.

Meneley, R. A., Henao, D., and Merritt, R. K., 1975, The northwest margin of the Sverdrup Basin, *in* Yorath, C. J., Parker, E. R., and Glass, D. J., eds., Canada's Continental Margins: Canadian Society of Petroleum Geologists Memoir 4, p. 531–544.

Miall, A. D., 1975, Post-Paleozoic geology of Banks, Prince Patrick, and Eglinton islands, *in* Yorath, C. J., Parker, E. R., and Glass, D. J., eds., Canada's Continental Margins: Canadian Society of Petroleum Geologists Memoir 4, p. 557–588.

——, 1976, Devonian geology of Banks Island, Arctic Canada, and its bearing on the tectonic development of the circum-Arctic region: Geological Society of America Bulletin, v. 87, p. 1599–1608.

Morel, P., and Irving, E., 1978, Tentative paleocontinental maps for the early Phanerozoic and Proterozoic: Journal of Geology, v. 86, p. 535–561.

Norris, D. K., and Yorath, C. J., 1981, The North American plate from the Arctic Archipelago to the Romanzof Mountains, *in* Nairn, A.E.M., Churkin, M., Jr., and Stehli, F. G., eds., The Ocean Basins and Margins; v. 5, The Arctic Ocean: New York, Plenum Press, p. 37–104.

Oliver, J., Ewing, M., and Press, F., 1955, Crustal structures of the Arctic regions from L_g phase: Geological Society of America Bulletin, v. 66, p. 1063–1074.

Parsons, B., and Sclater, J. G., 1977, An analysis of the variation of ocean floor bathymetry and heat flow with age: Journal of Geophysical Research, v. 82, p. 803–827.

Paterson, W.S.B., 1977, Extent of the Late-Wisconsin glaciation in northwest Greenland and northern Ellesmere Island; A review of the glaciological and geological evidence: Quaternary Research, v. 8, p. 180–190.

Pelletier, B. R., 1964, Development of submarine physiography in the Canadian Arctic and its relation to crustal movements: Dartmouth, Nova Scotia, Bedford Institute of Oceanography Report 64–16, p. 1–45.

Rickwood, F. K., 1970, The Prudhoe Bay field, *in* Adkison, W. L., and Brosgé, M. M., eds., Proceedings of the Geological Seminar on the North Slope of Alaska: Los Angeles, Pacific Section of the American Association of Petroleum Geologists, p. L1–L21.

Riddihough, R. P., Haines, G. V., and Hannaford, W., 1973, Regional magnetic anomalies of the Canadian Arctic: Canadian Journal of Earth Sciences, v. 10, p.147–163.

Sander, G. W., and Overton, A., 1965, Deep seismic refraction investigations in the Canadian Arctic Archipelago: Geophysics, v. 30, p. 87–96.

Scotese, C., Bambach, R. K., Barton, C., Van der Voo, R., and Ziegler, A. M., 1979, Paleozoic base maps: Journal of Geology, v. 87, p. 217–277.

Scrutton, R. A., 1982, Passive continental margins: a review of observations and mechanisms *in* Scrutton, R. A., ed., Dynamics of Passive Margins, Geodynamics Series Volume 6, American Geophysical Union, p. 5–11.

Sobczak, L. W., 1975, Gravity and deep structure of the continental margin of Banks Island and the Mackenzie Delta: Canadian Journal of Earth Sciences, v. 12, p. 378–394.

——, 1978, Gravity from 60°N to the North Pole, *in* Sweeney, J. F., ed., Arctic Geophysical Review: Ottawa, Publications of the Earth Physics Branch, v. 45, p. 67–74.

——, 1982, Fragmentation of the Canadian Arctic Archipelago, Greenland and surrounding oceans, *in* Dawes, P. R., and Kerr, J. W., eds., Nares Strait and the Drift of Greenland; a Conflict in Plate Tectonics: Meddelelser om Grønland, v. 8, p. 221–236.

Sobczak, L. W., and Overton, A., 1984, Shallow and deep crustal structure of the western Sverdrop Basin, Arctic Canada: Canadian Journal of Earth Sciences, v. 21, p. 902–919.

Sobczak, L. W., and Weber, J. R., 1973, Crustal structure of the Queen Elizabeth islands and polar continental margin, American Association of Petroleum Geologists Memoir 19, p. 517–525.

Soper, N. J., Dawes, P. R., and Higgins, A. K., 1982, Cretaceous-Tertiary magmatic and tectonic events in North Greenland and the history of adjacent ocean basins, *in* Dawes, P. R., and Kerr, J. W., eds., Nares Strait and the Drift of Greenland; a Conflict in Plate Tectonics: Meddelelser om Grønland, v. 8, p. 205–220.

Stein, S., Sleep, N. H., Geller, R. J., Wang, S-C., and Kroeger, G. C., 1979, Earthquakes along the passive margin of eastern Canada: Geophysical Research Letters, v. 6, p. 537–540.

Sweeney, J. F., 1981, Arctic research for the 1980s; An expose of problems: Geoscience Canada, v. 8, p. 162–166.

——, 1982a, Structure and development of the polar margin of North America, *in* Scrutton, R. A., ed., Dynamics of Passive margins: American Geophysical Union and Geological Society of America Geodynamics Series, v. 6, p. 17–29.

——, 1982b, Mid-Paleozoic travels of Arctic-Alaska: Nature, v. 298, p. 647–649.

Tailleur, I. L., 1973, Probable rift origin of Canada Basin, Arctic Ocean, *in* Pitcher, M. G., ed., Arctic Geology: American Association of Petroleum Geologists Memoir 19, p. 526–535.

Taylor, F. B., 1910, Bearing of the Tertiary mountain belt on the origin of the Earth's plan: Geological Society of America Bulletin, v. 21, p. 179–226.

Taylor, P. T., 1983, Nature of the Canada Basin; Implications from satellite-derived magnetic anomaly data, *in* Mull, C. G., and Reed, K. M., eds., The Origin of the Arctic Ocean (Canada Basin): Journal of the Alaska Geological Society, v. 2, p. 1–8.

Taylor, P. T., Kovacs, L. C., Vogt, P. R., and Johnson, G. L., 1981, Detailed aeromagnetic investigation of the Arctic Basin, 2: Journal of Geophysical Research, v. 86, p. 6323–6333.

Thorsteinsson, R., and Tozer, E. T., 1970, Geology of the Arctic Archipelago, *in* Douglas, R.J.W., ed., Geology and Economic Minerals of Canada, edition 5: Ottawa, Geological Survey of Canada Economic Geology Report 1, p. 548–590.

Trettin, H. P., and Balkwill, H. R., 1979, Contributions to the tectonic history of the Innuitian Province, Arctic Canada: Canadian Journal of Earth Sciences, v. 16, p. 748–769.

Trettin, H. P., Frisch, T. O., Sobczak, L. W., Weber, J. R., Niblett, E. R., Law, L. K., DeLaurier, J., and Whitham, K., 1972, The Innuitian Province, *in* Price, R. A., and Douglas, R.J.W., eds., Variations in Tectonic Styles in Canada: Geological Association of Canada Special Paper 11, p. 83–179.

Trettin, H. P., Loverdige, W. D., and Sullivan, R. W., 1982, U-Pb ages on zircon from the M'Clintock West massif and the Markham Fiord pluton, northernmost Ellesmere Island: Geological Survey of Canada Paper 82–1C, p. 161–166.

Walcott, R. I., 1972, Gravity, flexure and the growth of sedimentary basins at a continental edge: Geological Society of America Bulletin, v. 83, p. 1845–1848.

Weber, J. R., 1963, Gravity anomalies over the polar continental shelf: Contributions from the Dominion Observatory, v. 5, no. 17, p. 3–10.

Weber, J. R., Forsyth, D. A., Judge, A. S., and Jackson, H. R., 1984, A geoscience program for the Canadian ice island research project: Ottawa, Department of Energy, Mines, and Resources, 51 p.

Wetmiller, R. J., and Forsyth, D. A., 1978, Seismicity of the Arctic, 1908–1975, *in* Sweeney, J. F., ed., Arctic Geophysical Review: Ottawa, Publications of the Earth Physics Branch, v. 45, p. 15–24.

Willumsen, P. S., and Cote, R. P., 1982, Tertiary sedimentation in southern Beaufort Sea, Canada, *in* Embry, A. F. and Balkwill, H. R., eds., Arctic Geology and Geophysics: Canadian Society of Petroleum Geologists Memoir 8, p. 43–53.

Wilson, J. T., 1963, Hypothesis of the Earth's behavior: Nature, v. 198, p. 925–929.

Wold, R. J., Woodzick, T. L. and Ostenso, N. A., 1970, Structure of the Beaufort Sea continental margin: Geophysics, v. 35, p. 849–861.

MANUSCRIPT ACCEPTED BY THE SOCIETY JANUARY 6, 1988
CONTRIBUTION FROM THE EARTH PHYSICS BRANCH NO. 1175

ACKNOWLEDGMENTS

Panarctic Oils Ltd., Calgary, provided unpublished seismic reflection data. R. J. Wetmiller (Earth Physics Branch, Ottawa) and P. H. McGrath (Geological Survey of Canada, Ottawa) contributed current maps of seismicity (RJW) and total magnetic field (PHM), as well as a magnetic profile (PHM). An early draft of the manuscript was critically reviewed by A. F. Embry, A. V. Okulitch, R. C. Speed, H. P. Trettin, and C. J. Yorath.

DNAG Continent-Ocean Transect Volume
Phanerozoic Evolution of North American
Continent-Ocean Transitions
The Geological Society of America, 1994

Chapter 11

Rifted continental margins of North America

Dale S. Sawyer
Department of Geology and Geophysics, Rice University, P.O. Box 1892, Houston, Texas 77251

INTRODUCTION

During the late 1970s, the study of rifted margins was changed fundamentally by the recognition that simple models of thinning of continental lithosphere could account for many features of the subsidence observed on rifted or passive continental margins (McKenzie, 1978). The 1980s have produced at least three major new foci of research in the study of such margins and continental rifting in general. These new directions have emerged partly as a result of new data acquisition, primarily deep penetration seismic reflection profiling and complementary long offset seismic reflection-refraction experiments on rifted margins worldwide, and partly as a result of the application of land geologic observations of continental rifts to the study of margins.

The first of these new directions was the recognition of the importance of low-angle normal faulting in continental rifting. This avenue of research has changed our expectations of the large-scale patterns of rifting and rifted margin formation. We had come to expect that most rifts should be symmetric and that, upon plate reconstruction, conjugate rifted margins would also show symmetry at the time of initiation of sea-floor spreading. A number of investigators (Wernicke, 1981, 1985; Lister and others, 1986) observed asymmetry in the shallow structures in continental rifts, and extended this idea to the whole lithosphere scale. The new model, usually referred to as the simple shear or delamination model, concentrates shearing along narrow zones at low angle to the surface and through the continental lithosphere. Most of the strain during continental extension takes place within this shear zone, while relatively intact hanging-wall and footwall blocks separated. This model predicts a significant asymmetry in the characteristics of the upper plate or the hanging-wall block and the lower plate or the footwall block. Most investigators now anticipate this type of asymmetry when examining a new margin. Symmetry has become the oddity rather than the norm.

The second new research focus of the 1980s was the role of magmatism in the formation of rifted margins. At the beginning of the decade, most margins were classed as nonvolcanic (the name should probably be nonmagmatic) rifted margins, and only a few margins were thought to contain significant accumulations of igneous rocks. New high-quality seismic reflection and refraction data and drilling (Legs 80 and 81, Deep Sea Drilling Project, and Leg 104, Ocean Drilling Program) have shown that igneous rocks are present on most margins that have been studied. In many cases, volcanic rocks are part of seaward-dipping reflector sequences that are thought to have been extruded subaerially during rifting (Mutter and others, 1982; Mutter, 1985). In addition, magma seems to have ponded and cooled at the base of rifting continental crust. At the rate these studies are progressing, there soon may not be any margins interpreted to be "nonvolcanic," and the study of magmatism on margins will move away from the question of whether magmatism happened to questions of how much, when, and where magma was emplaced. With the realization that magmatism is nearly ubiquitous on rifted margins, there are new tools for incorporating the study of magma genesis into quantitative models. McKenzie and Bickle (1988) devised a scheme that uses the results of hundreds of peridotite melting experiments to predict the amount and major element chemistry of basaltic magma that would be produced by any pressure-temperature path during decompression melting of the asthenosphere. White and McKenzie (1989) used this tool to show that elevated temperature in the asthenosphere, such as would be found in the vicinity of a mantle plume and/or hot spot, prior to rifting can lead to tremendously increased melt production during rifting. This increased melt production, likely the result of mantle plume activity, can explain many of the variations in amount of volcanism on margins.

The third new research direction during the 1980s was the increasing use of the yield stress envelope model of continental lithosphere strength in dynamic, as opposed to kinematic, models of continental rifting and the formation of rifted continental margins. At the beginning of the decade, the picture of the continental lithosphere was a strong elastic plate overlying a weaker, viscous mantle. Different investigators had very different ideas about the depth of the boundary between these layers. Physical properties were often assumed to be uniform within each layer. This simple conception of lithospheric strength did not allow the prediction of the diversity of observed continental rift and rifted margin styles. Hence, these diverse rifts were often categorized as distinct phe-

Sawyer, D. S., 1994, Rifted continental margins of North America, *in* Speed, R. C., ed., Phanerozoic Evolution of North American Continent-Ocean Transitions: Boulder, Colorado, Geological Society of America, DNAG Continent-Ocean Transect Volume.

nomena. Goetze and Evans (1979) and Brace and Kohlstedt (1980) used laboratory observations of rock deformation to conceive of a new way to parameterize the variation of strength of the continental lithosphere. Their model predicted that the upper continental crust (modeled using laboratory results for quartz) would undergo brittle failure during extension, while the lower continental crust would undergo ductile flow. The strongest part of the continental crust would be in the middle, near the transition in failure mechanism known as the brittle-ductile transition. Similarly, a thin layer at the top of the mantle (modeled using laboratory results for olivine) might undergo brittle failure during extension, while the mantle below would undergo ductile flow. The strongest part of the mantle would lie at the brittle-ductile transition, or if the entire mantle was in the ductile regime, just below the Moho. Within regions of brittle failure, both in the crust and mantle, the strength of the rock would increase with depth because confining pressure increases with depth. Within regions of ductile flow, both in the crust and mantle, the strength of the rock would decrease with depth because temperature increases with depth and the strength is independent of pressure. This model predicts that there are two strong layers in the continental lithosphere; the stronger of the two probably lies in the upper mantle, and the "weaker" strong zone lies in the middle crust. The upper part of the crust is weak and brittle, and the lower part of the crust is weak and ductile. The mantle below the uppermost strong part is weak and ductile.

This characterization of continental lithosphere as, from top to bottom, weak-strong-weak-strong-weak has had a profound effect on thinking about the dynamics of continental extension. It leads to the view of continental lithosphere as a "composite" material. Well-known composite materials such as fiberglass have strength and failure characteristics that are different from the characteristics of either of their components taken separately. By analogy, we should expect that continental lithosphere will exhibit properties quite different from either a brittle layer or a ductile layer alone. Explanations for much of the incredible diversity of continental rifts and rifted margins will likely emerge from this conceptual analog.

The objective for this chapter is to describe briefly the diversity of rifted continental margins of North America and then apply to them some of the progress that has been made in these three areas of research emphasis during the 1980s. I place the most emphasis on the progress in the area of modeling the dynamics of continental extension, and conclude by highlighting some unanswered questions regarding the formation of rifted continental margins.

OVERVIEW OF RIFTED MARGINS OF NORTH AMERICA

Highly diverse rifted continental margins bound a significant proportion of the North American continent. Rifted continental margins occur on both sides of the Gulf of Mexico basin, up the Atlantic coasts of the United States and Canada, the Labrador Sea coast of Canada, from the Canadian Polar islands to the MacKenzie Delta, and the Alaskan polar margin. I attempt below a summary of some of the most significant features of those margins.

The Gulf of Mexico basin was formed by continental rifting between the Yucatan block of the Maya terrane (Ortega, this volume) and the North American craton (Salvador, 1991; Buffler and Thomas, this volume) (Fig. 1). Seismic and magnetic data indicate that oceanic crust forms the basement of the central Gulf of Mexico. The timing, duration, and direction of sea-floor spreading are poorly constrained. No unequivocal or dateable magnetic lineations are observed in the Gulf of Mexico, and the oldest sediments are deeply buried. Estimates of opening time range from about 175 Ma, contemporaneous with opening in the adjacent Atlantic Ocean, to as young as Late Jurassic. The motions of terranes during stretching and spreading are also poorly constrained, but the bulk of the evidence points to a counterclockwise rotation of Yucatan out of an original position snug against Louisiana, Alabama, Mississippi, and the panhandle of Florida (Pindell and Dewey, 1982; Pindell, 1985; Dunbar and Sawyer, 1987). The pole for this rotation was probably offshore northeastern Florida. This means that the margins of Alabama, Mississippi, Louisiana, eastern Texas, and the northern and western margins of the Yucatan are rifted continental margins. The western and eastern parts of the Gulf of Mexico basin are probably transform margins formed at the edges of the rotating Yucatan block. The zone underlain by rifted, thinned continental crust is much wider on the North American rifted margin than on the Yucatan margin (400–600 km vs. 50–100 km). One explanation for this asymmetry is that the northern margin includes a block of relatively less extended, more coherent, continental crust. Another explanation, currently favored for many asymmetric rifted margins and suggested but not yet demonstrated for the Gulf of Mexico, is that the rifting process was controlled by whole lithosphere simple shear (Marton and Buffler, unpublished). In the later explanation, the North American margin is interpreted to be the footwall block of the simple shear system (Marton and Buffler, unpublished). The crust-type distribution of the North American margin is difficult to determine because of thick salt deposited during rifting when the Gulf of Mexico was a semiclosed basin, and mobilized by later sediment loading. The salt makes seismic observations of the crust difficult or impossible. There is little or no evidence of significant volcanic activity associated with the rifting that formed the Gulf of Mexico basin, but if volcanic rocks exist, they would now be deeply buried under sediment and difficult to identify. It seems likely that the rifting that formed the Gulf of Mexico basin was localized by preexisting weakness trends associated with the Ouachita Mountain belt.

The history of rifting that formed the Atlantic and Labrador Ocean margins of North America is summarized in Figure 2, taken from Srivastava and Verhoef (1992). The sea-floor spreading events designated A through L in Figure 2 are referred to in the following paragraphs. The North Atlantic Ocean was formed

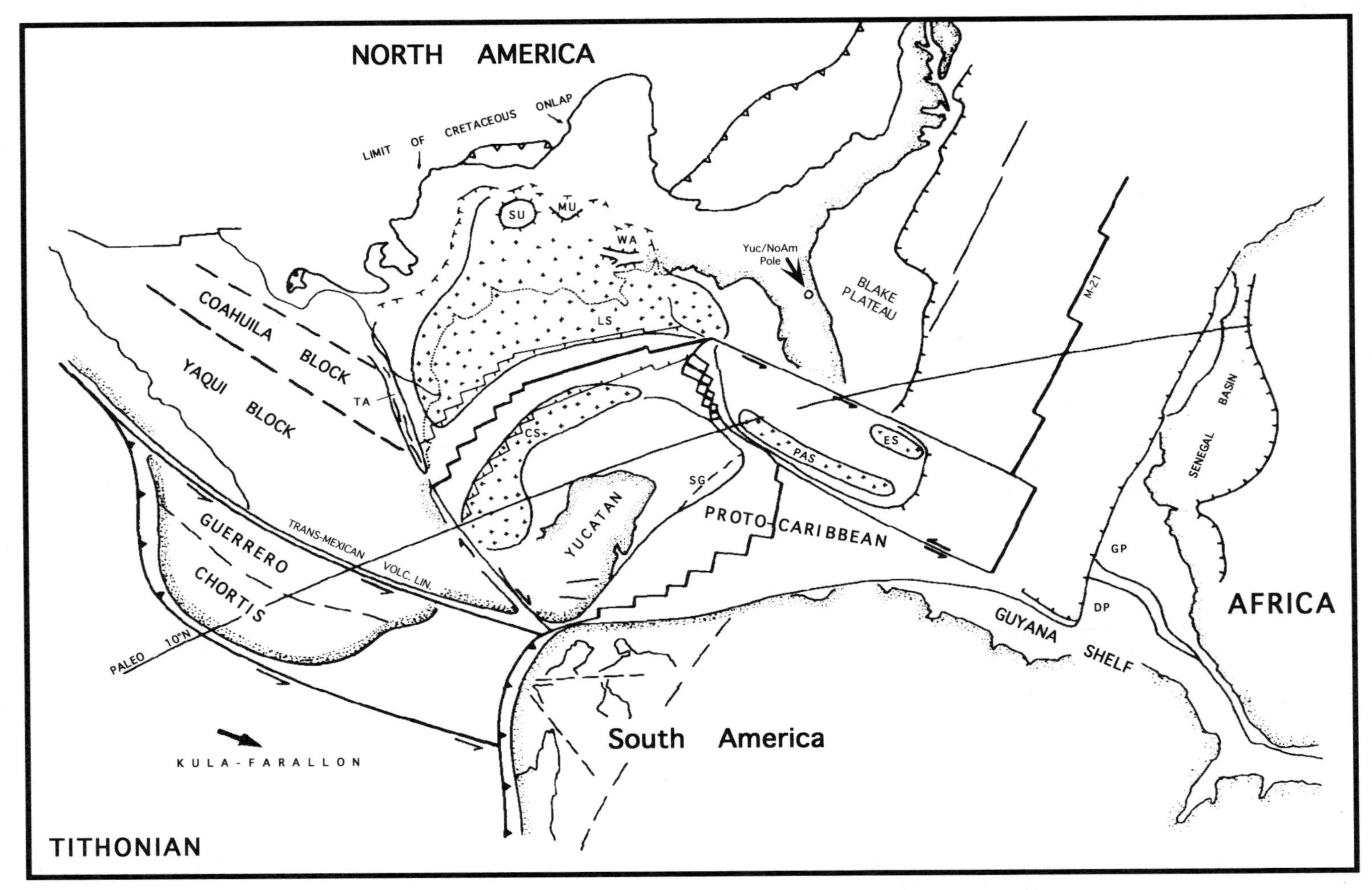

Figure 1. Reconstruction of the Gulf of Mexico basin about 150 Ma, according to Pindell (1985). At this time, salt deposition has ceased and sea-floor spreading has begun. The Yucatan Peninsula is rotating counterclockwise away from North America. CS, Campeche salt; DP, Demarara Plateau; ES, Exuma Sound salt; GP, Guinea Plateau; LS, Louann salt; MU, Monroe uplift; PAS, Punta Alegre salt; SG, Sierra Guaniguanico of present-day western Cuba; SU, Sabine uplift; TA, Tampa arch; WA, Wiggins arch. The plus sign pattern is salt.

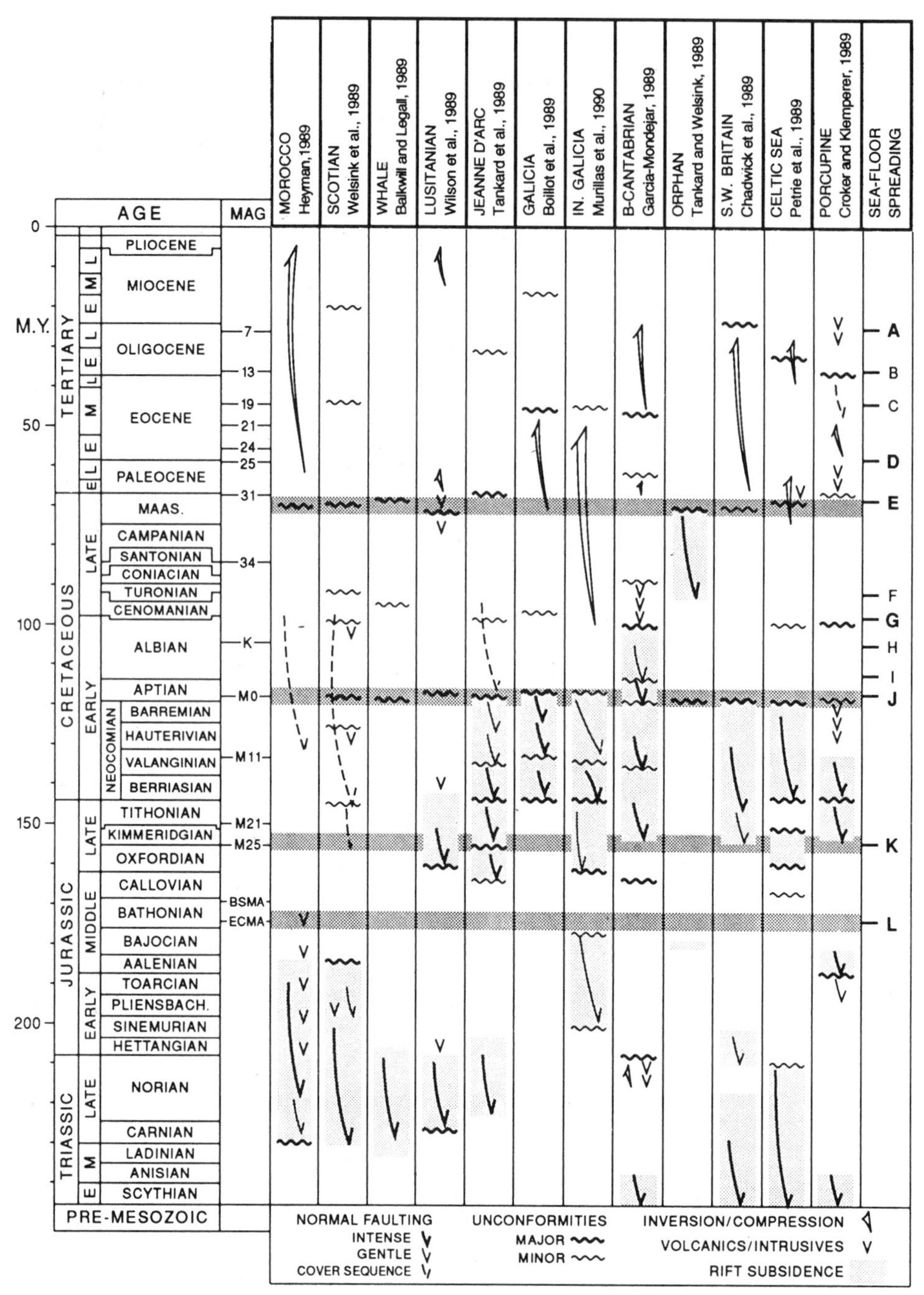

Figure 2. Tectono-stratigraphy of the sedimentary basins around the central North Atlantic and sea-floor spreading events (A to L) in the North Atlantic (this is Table 1 of Srivastava and Verhoef, 1992). A: Iberia moving with Eurasia (Srivastava and others, 1990); B: sea-floor spreading in the Labrador Sea ceases and Greenland starts to move with North America (Roest and Srivastava, 1989); C: Iberia moving as an independent plate (Roest and Srivastava, 1991); D: separation of Greenland from Eurasia and volcanism in North America (Srivastava and Tapscott, 1986); E: change of direction of motion between Eurasia and North America (Srivastava and others, 1988); F: active sea-floor spreading starting in the Labrador Sea (Roest and Srivastava, 1989); G: complete separation of Eurasia from North America (Srivastava and others, 1988); H: separation of Goban Spur from Flemish Cap (De Graciansky and others, 1985); I: separation of Galicia Bank from Flemish Cap (Boillot and others, 1988); J: formation of volcanic ridges (known as "J" anomaly ridge) between Iberia, North America, and Africa and excessive volcanism along the Newfoundland Fracture Zone; K: possible separation of southern part of Iberia from North America (Mauffret and others, 1989; Srivastava and others, 1990); L: separation of Africa from North America (Klitgord and Schouten, 1986).

by sea-floor spreading that propagated from the south to the north between the Late Triassic (prior to the beginning of sea-floor spreading between Africa and North America at event L) and the Oligocene (event B). Although the sea-floor spreading propagated northward, Figure 2 shows that most of the basins along the margin were affected by the same rifting and subsidence events before, during, and after the local initiation of sea-floor spreading.

The Atlantic margin of the United States (Klitgord and others, 1988) and Nova Scotia (Wade and MacLean, 1990) is composed of a series of offshore basins: the Blake Plateau basin, the Carolina trough, the Baltimore Canyon trough, the Georges Bank basin offshore of the United States (Fig. 3), and the Scotian basin off Nova Scotia. The basins and troughs are separated by arches, where the basement is shallower and the sediment is thinner. The reason for the formation of alternating basins and arches along the strike of a margin is poorly understood. The arches may correspond to transfer zones (Bosworth, 1985, 1987) where the simple shear polarity of the margin reverses along strike. The polarity reversals, if present, may affect the sense of shear in the whole lithosphere or, if the shear zones bottom in a lower crustal decollement, only the upper crust. Alternatively, the arches and basins may be upper plate and lower plate margin segments, respectively.

Klitgord and others (1988) presented a clear synopsis of the tectonic development of the United States Atlantic rifted margin, which I summarize here. Construction of the megacontinent Pangea by the collision of a number of terranes, microcontinents, and continents was complete by the late Paleozoic. Many large east-dipping thrust faults were generated during the collision phase. These included thin-skinned thrust systems over the North American craton and thick-skinned thrust systems largely within the accreted terranes. These faults represented potential weaknesses that could be reactivated by later extension. Regional extension began in the Late Permian, and may have caused some uplift, explaining the absence of sediment deposition. Roughly

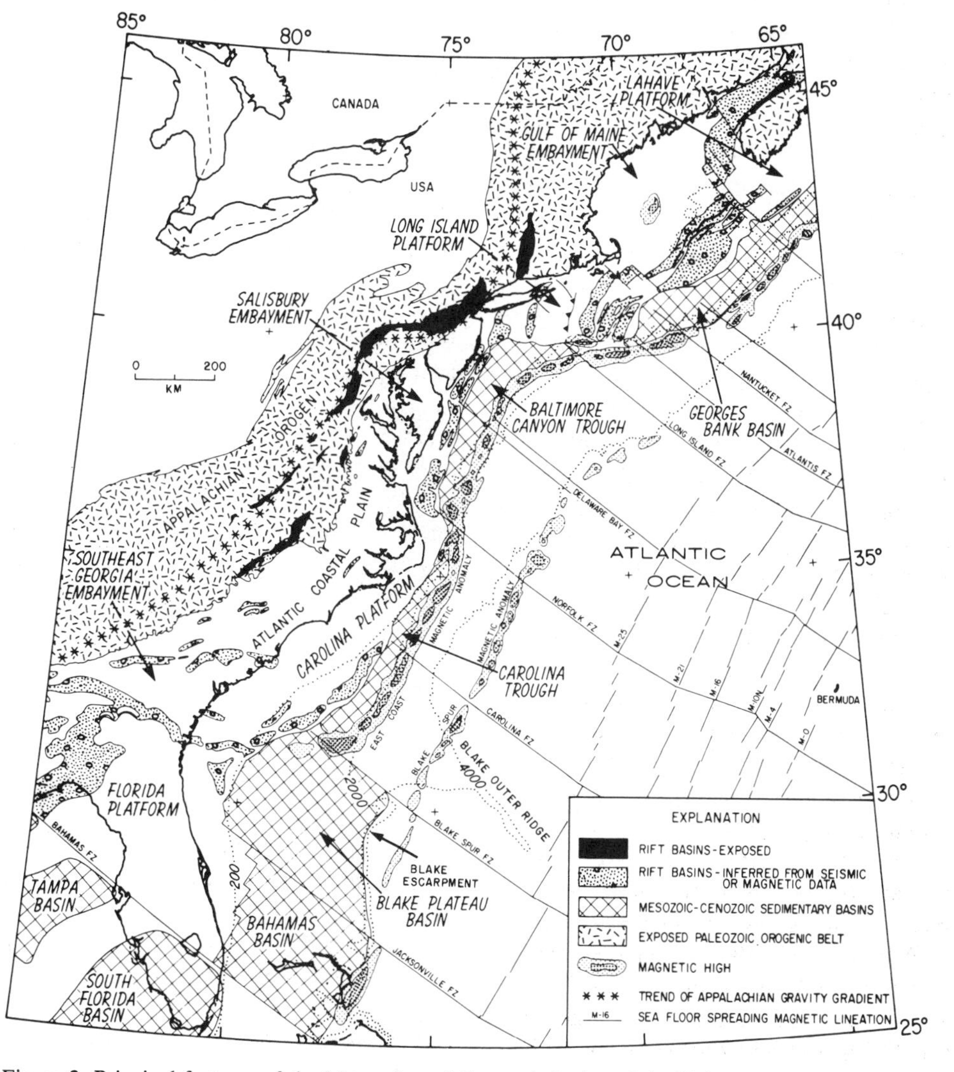

Figure 3. Principal features of the Mesozoic and Cenozoic basins of the United States Atlantic continental margin (Grow and Sheridan, 1988).

east-west–directed rifting began in the Late Triassic in a wide (300 km) band from the Grand Banks to the Gulf of Mexico (Fig. 4). This rifting continued for about 25–35 m.y., reactivating thrust faults from the late Paleozoic and forming a series of half-graben basins, of which the Newark basin (the large exposed rift basin in northern New Jersey; Fig. 3) is a prominent example. Basaltic magmatism affected many of these basins late during this phase of the rifting. A second phase of rifting, narrower than the first (in some places as little as 100 km), but of similar extent along the margin, began in the Late Jurassic. Because the second phase was narrower than the first, some of the Triassic rift basins were not affected by the later rifting, and were stranded on one side of the future Atlantic Ocean or the other. The landward edge of the region affected by the later stage of rifting is the hinge zone (Fig. 4), an inflection of the basement surface with shallow dip on the landward side and steeper dip on the seaward side. This second rifting phase may have included some rift flank uplift, which manifests itself in erosion of basement blocks landward of the hinge zone. The second rift phase was 20–25 m.y. in duration and ended with the initiation of sea-floor spreading at about 175 Ma (event L) in the Baltimore Canyon trough and Carolina trough segments in the margin. Sea-floor spreading began somewhat later to the north and south in the Blake Plateau basin and Georges Bank basin. Since the Klitgord and others (1988) synthesis, it has become clear that significant magma was emplaced at the base and erupted at the surface of the extended continental crust during the second rifting phase (Austin and others, 1990; Sheridan and others, 1993; Holbrook and others, unpublished). The basalt that was erupted at the end of the first rifting phase, the underplated and erupted magma of the second rifting phase, and the formation of oceanic crust during sea-floor spreading, are probably associated and represent a continuum of magmatic activity that began significantly after the initiation of rifting, and then increasingly dominated the rifting process. Rifting ceased at the initiation of sea-floor spreading in each margin segment. Regions affected by the second rifting stage underwent large snyrift subsidence and even greater regional postrift (presumably thermally controlled) subsidence. Regions affected by only the first phase of rifting underwent significant local (half-graben controlled) synrifting subsidence, but little postrifting subsidence.

There are significant discontinuities in the width of extended continental crust at the southern boundary of the Georges Bank basin and the northern boundary of the Blake Plateau basin. These may represent boundaries of margin segments at which the strength of prerift continental lithosphere changed, or they may mark transfer zones at which the polarity of the whole lithosphere simple shear system reversed. On the basis of inrepretations of fault-block tilt direction, and regional synrifting and postrifting subsidence history, Wernicke and Tilke (1989) interpreted the George Bank basin to be an upper plate margin and the Baltimore Canyon trough to be a lower plate margin. In the nomenclature of Courtillot (1982), the Blake Plateau and Georges Bank basins would correspond to "locked zones" in the rifting process, and the Baltimore Canyon and Carolina troughs would be "loci of rift initiation."

The basins along the United States and Nova Scotian Atlantic margins all contain 12–18 km of sediment in their deepest parts, making seismic imaging of the transition zone difficult or impossible. This margin was once thought to be a so-called "nonvolcanic" rifted margin. That designation has been shown to be grossly incorrect by recent seismic studies (Austin and others, 1990; Oh and others, 1991; Sheridan and others, 1993), which show seaward-dipping reflector sequences (Fig. 5). These seaward-dipping reflectors are presumably subaerial extruded lava, and the associated lower crust of velocity 7.1 to 7.5 km/s is formed by the addition of basic magma to the lower crust by either intrusion or underplating (LASE Study Group, 1986; Trehu and others, 1989; Holbrook and Kelemen, 1993). The East Coast magnetic anomaly parallels the margin and was until recently thought to be caused by the boundary between magnetized oceanic crust and unmagnetized continental crust (Keen, 1969; Klitgord and Behrendt, 1979). As it has become clearer that there is a significant component of extrusive and intrusive volcanic material in the transitional crust of the Atlantic margin, the ECMA is now thought to be associated with the boundaries of the bodies of synrifting magma (Austin and others, 1990; Talwani and others, 1992).

The Grand Banks and Newfoundland basin (Tankard and Welsink, 1989; Grant and McAlpine, 1990) were formed as the Iberian Peninsula rifted from North America about 110–120 Ma (Fig. 2, events I and J, and Fig. 6). Rifting here is characterized by a broad zone of half-graben basins, possibly similar in mechanism of formation to the Triassic basins in the United States, forming the Grand Banks. The individual basins, such as the Jeanne D'Arc, Whale, and Horseshoe, contain up to 14 km of sediment mostly deposited during and shortly after rifting. The average depth of the basement between these half grabens is 2–4 km, the expected total subsidence for only mildly thinned continental crust. The basins and arches of the Grand Banks were truncated by the prominent Avalon unconformity, associated with broad Avalon uplift in the late Mesozoic. East of the Grand Banks is the Newfoundland basin, which is interpreted to be underlain by oceanic crust by some (de Voogd and Keen, 1987), and highly extended continental crust by others. In seismic reflection profiles, Tucholke and others (1989) saw faulted and eroded basement blocks that they interpreted to be rifted continental crust. This crust is conjugate to highly extended continental or oceanic crust under the Iberia abyssal plain off Portugal (Whitmarsh and others, 1990). A transect across these conjugate margins will be a focus of Ocean Drilling Program activity beginning with Leg 149 in 1993.

The Orphan Knoll basin and northeast Newfoundland shelf margins were formed in the Early Cretaceous, about 85–95 Ma (event G), when Europe rifted from North America (Srivastava and Verhoef, 1992) (Fig. 6). Extension in this region is very complicated because there was also rifting associated with the

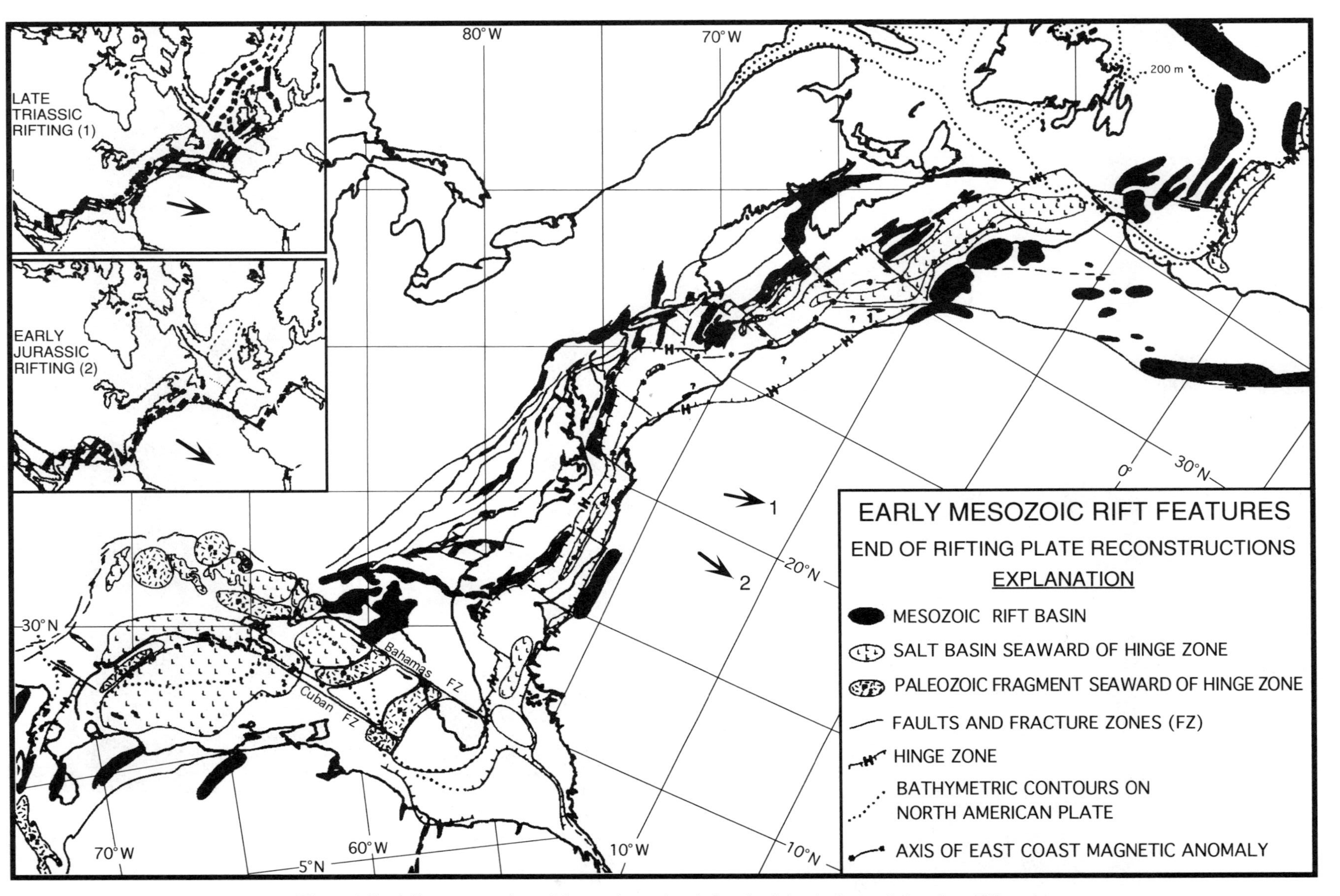

Figure 4. Predrift reconstruction of the continents bordering the Atlantic Ocean (taken from Klitgord and others [1988]). Plate motions and major fault motion trends are indicated for the early synrift (1) and late synrift (2) phases. Insets show Late Triassic rift configuration (1) with the North Atlantic–Arctic system and the Early Jurassic rift configuration (2) with the Tethys system.

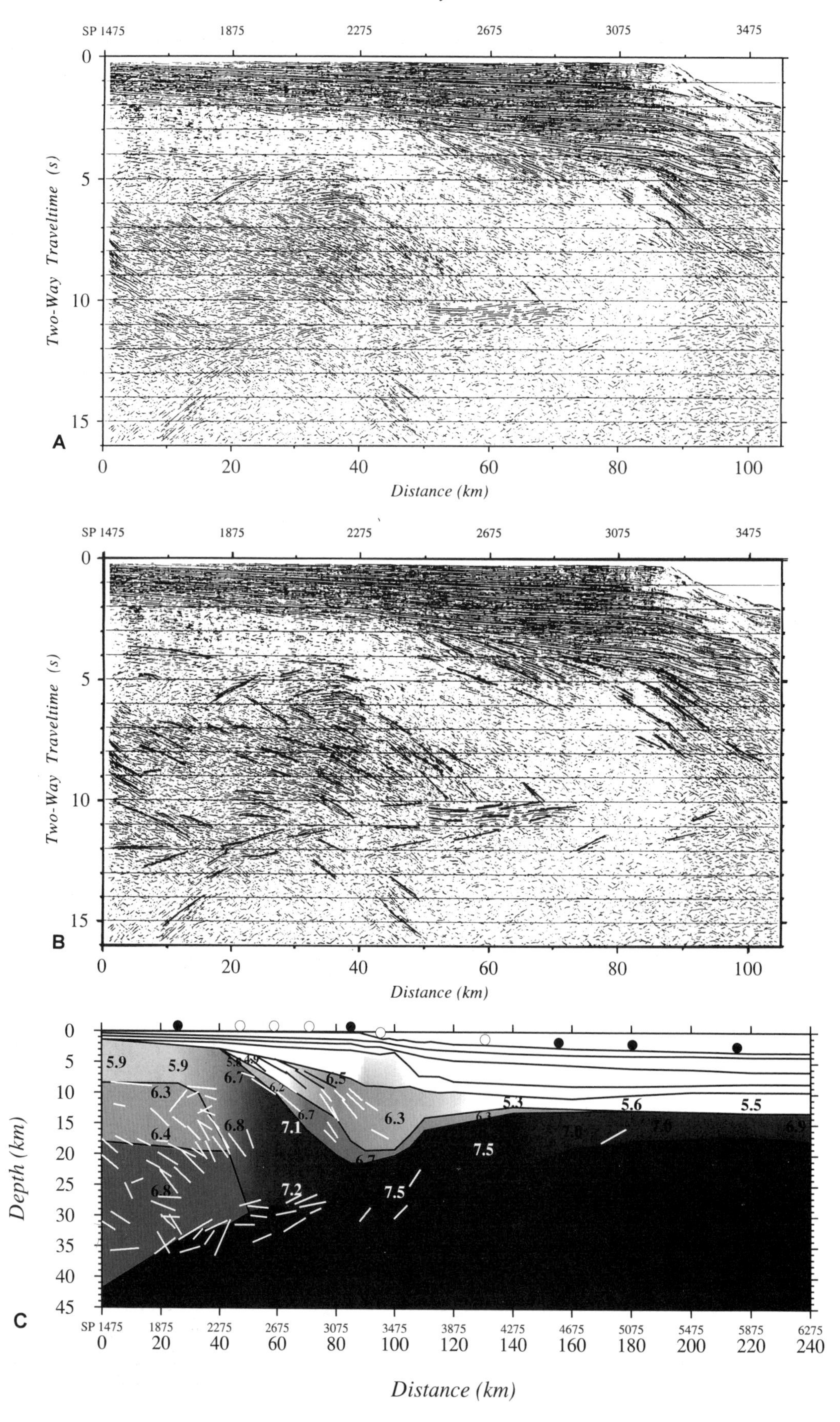
SP 1475
1875
2275
2675
3075
3475
Two-Way Traveltime (s)
A
Distance (km)
B
Depth (km)
C
5.9
5.9
6.3
6.4
6.8
6.7
6.2
6.5
6.7
6.8
7.1
6.3
6.7
7.2
7.5
7.5
5.3
5.6
5.5
7.0
6.9
Distance (km)

Figure 5. Seismic reflection (Sheridan and others, 1993) and refraction (Holbrook and Kelemen, 1993) velocity sections of the Virginia continental margin along seismic reflection-refraction profile MA-801. This margin was only recently recognized as a volcanic rifted margin. The seaward-dipping reflector sequence is seen in A and B between 50 and 95 km distance and 3 and 8 s two-way traveltime. The velocity model (C) was obtained using ray-tracing to explain arrivals from 10 ocean-bottom seismograph data sets acquired at the locations indicated by the open and filled circles along the sea floor. The white reflector segments are the same as the interpreted reflectors in B. There is a seaward increase in the velocity of the lower transitional crust at a distance of about 45 km. Higher velocities are interpreted to indicate intrusion or underplating of basaltic magma into or under the extended continental crust.

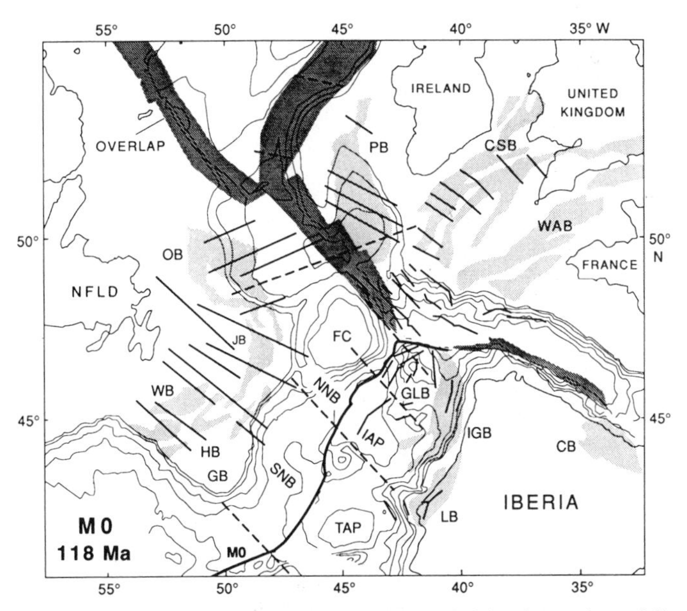

Figure 6. Reconstruction of the north Central Atlantic at chron M0 (event J) showing a simplified bathymetry on each plate, outlines of the sedimentary basins (shaded regions) and their tectonic features (taken from Fig. 2 of Srivastava and Verhoef, 1992). Also shown are the directions of plate motion (dashed lines) and resulting overlap between plate boundaries (dark stippled regions). The region of overlap reflects extension of the continental crust during rifting. JB, Jeanne d'Arc basin; WB, Whale basin; HB, Horseshoe basin; OB, Orphan basin; FC, Flemish Cap; CSB, Celtic Sea basin; WAB, Western Approaches basin; PB, Porcupine basin; GLB, Galicia Bank; IGB, Inner Galicia basin; LB, Lusitanian basin; CB, Cantabrian basin; NNB, North Newfoundland basin; SNB, South Newfoundland basin; IAP, Iberia Abyssal Plain; TAP, Tagus Abyssal Plain. NFLD is Newfoundland.

earlier breakup to the south. There is a substantially broader zone of extended crust here (200–300 km) than to the northwest in the Labrador Sea margins (50–100 km). Along Labrador, breakup follows the northwest-trending structural grain of the 2 b.y. old suture between the Nain and Rae provinces, and the rifting process could easily exploit conveniently oriented weaknesses (Dunbar and Sawyer, 1989a). Northeast of Newfoundland where the rift had to cut across the northeast-trending structural grain of the Grenville and Appalachian belts, there were no conveniently oriented weaknesses and the rift was much wider. Dunbar and Sawyer (1989a) argued that where rifting can use conveniently oriented weaknesses, less total extension of continental crust preceded the initiation of sea-floor spreading.

The Labrador shelf margin (Balkwill and McMillan, 1990) was formed when Greenland rifted from North America with extension beginning in the Early Cretaceous. Sea-floor spreading probably began in the Late Cretaceous (Srivastava and Verhoef, 1992) (Fig. 2, event F). The Avalon unconformity and a delay in the beginning of thermal subsidence relative to the onset of spreading (Royden and Keen, 1980) of both the Labrador shelf and the Orphan Knoll basin–Northeast Newfoundland basin are evidence for a modest amount of uplift prior to the initiation of sea-floor spreading.

It is not clear whether Baffin Bay (MacLean and others, 1990) is underlain by oceanic or continental crust. Rifting took place in the latest Cretaceous or Paleocene (Fig. 2, events D and E), a time of a major change in the relative velocity of Greenland with respect to North America. Rifting and/or sea-floor spreading almost certainly ended in the late Eocene (event B), when Greenland ceased moving with respect to North America and began moving with Eurasia. There is a clear record of postrifting uplift in the Baffin Bay region; some of the earliest rift phase sedimentary rocks are currenly elevated by up to 600 m above sea level. This is an unusual, and difficult to explain, feature for a rifted continental margin.

The postrifting subsidence history of most of the basins that make up the United States and Canadian Atlantic margin follow one of two major patterns: (1) postrifting uplift or at least an absence of significant postrifting subsidence, or (2) significant postrifting thermal subsidence that is roughly in proportion to thinning of the lithosphere during rifting. In the second case, postrifting subsidence is often a factor of two greater than the synrifting subsidence. The first pattern seems to apply to basins of the Grand Banks, shelf basins of the Scotian margin and Triassic basins of the United States margin. The second pattern seems to by typical for basins formed adjacent to the site of eventual sea-floor spreading along the margin, such as the Baltimore Canyon Trough and the Carolina Trough.

There are a number of proposed plate reconstructions for the Canada Basin and the adjacent North Slope of Alaska and Arctic Canadian margins (Lawver and Scotese, 1990). In the "rotational model," the continental margin between Ellsemere Island, Canada, and Point Barrow, Alaska, is thought to have formed by counterclockwise rotation of Alaska away from the Canadian Polar islands about a pole located in the MacKenzie Delta region between about 120 and 80 Ma (Fig. 7). In this model, both the North Slope of Alaska and the Arctic Canadian margins are rifted continental margins. There were at least two rifting events prior to formation of the Canada basin, which is underlain by oceanic crust. Other classes of models, which treat

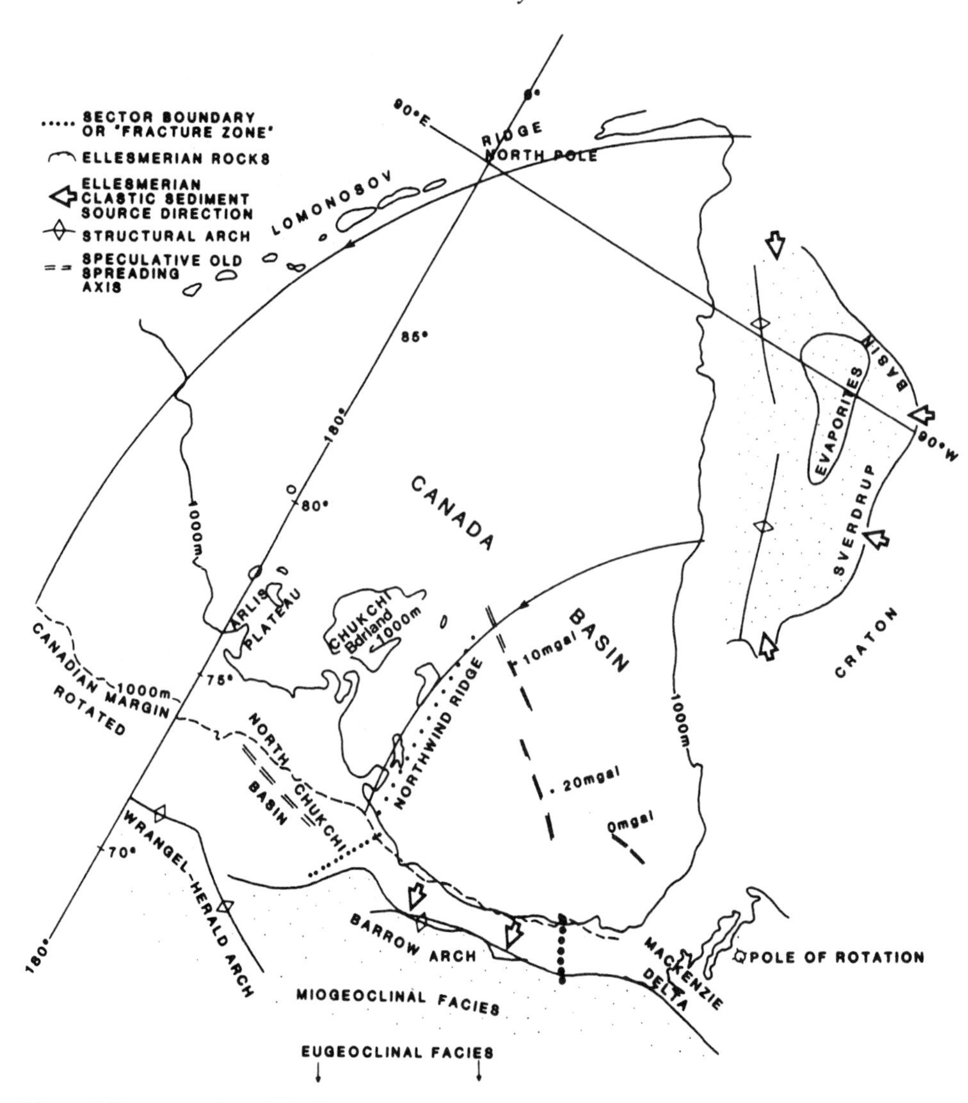

Figure 7. Reconstruction of the Canada basin using the "rotational model" (Grantz and others, 1979). The 1000 m bathymetric contour from the Canadian side has been rotated about a pole in the MacKenzie Delta. The Lomonosov Ridge and the Northwind Escarpment lie nearly on a small circle about the pole of rotation.

either the North Slope margin or the Arctic Canadian margin as a strike-slip margin, have also been proposed. Where the North Slope margin is treated as strike slip, the Arctic Canadian margin is interpreted to have rifted from either northeastern Siberia or the Chuckchi Plateau. Where the Arctic Canadian margin is treated as strike slip, the North Slope of Alaska is interpreted to have rifted from either the Lomonosov Ridge or the Alpha-Mendeleev ridges. The range of possible models reflects the relative lack of knowledge of the crustal structure in this region because it is frequently ice covered, and few seismic investigations have been possible.

KINEMATICS OF RIFTED MARGIN FORMATION

At the beginning of the decade of the 1980s, it was widely accepted that continental lithosphere was extended and thinned during continental rifting and prior to the beginning of sea-floor spreading. The most common model of uniform crust and lithosphere thinning McKenzie (1978) predicted that there would be subsidence during extension due to the replacement, between the surface and the depth of isostatic equilibrium, of light continental crust with denser upwelling asthenosphere. Superimposed on this synrifting extension and continuing after rifting stopped and sea-floor spreading began, the model predicted subsidence due to cooling and contraction of rocks brought closer to the surface during rifting. The model predicts a particular ratio of synrifting to postrifting subsidence that varies with time since rifting, but not with amount of extension. This model was tested using subsidence data from wells and found to describe generally the subsidence amounts and history (Steckler and Watts, 1978, 1980; Royden and Keen, 1980). The model said nothing about how extension and thinning of the lithosphere would be distributed laterally, either along margin or across margin.

More extensive comparison of McKenzie's (1978) model with subsidence data identified numerous cases in which the ratio of synrift and postrift subsidence deviated significantly from its predictions. In some cases there even appeared to have been uplift during rifting, followed by postrift subsidence. The model was therefore modified to allow the amount of extension and thinning of the crust to be different from that of the subcrustal lithosphere (Royden and Keen, 1980). This model could fit most of the known subsidence data, although it was not clear what rifting kinematics could lead to this vertically varying extension. It seemed clear that if the principal extension was integrated in each particular layer between unextended continental interiors, total extension would have to be equal in all layers. This lead to models in which total extension in the subcrustal lithosphere was distributed over a larger area than extension in the crust (Rowley and Sahagian, 1986). This pattern predicts local subsidence of thinning crust, and local uplift of unthinned crust adjacent to thinning crust, followed by regional postrift subsidence. Even this fairly general constraint on vertical and lateral distribution of extension was breached when it was shown that subcrustal lithosphere thinning (and all its subsidence or uplift implications) could probably occur without extension by adding heat to the base of the lithosphere (Royden and others, 1982). This is true because the "base of the lithosphere" in these models is a thermal, rather than a chemical, boundary.

On a whole lithosphere scale, most rifts were assumed to form in a generally symmetric way (Fig. 8). Hower, there was clearly asymmetry in surface manifestation of rifting, typically half-graben systems that periodically changed orientation along the strike of a rift zone (Bally, 1982; Bosworth, 1987). This asymmetry was not typically thought to penetrate deeper than the probably mid-crustal depth of the bounding fault decollement surface, until Wernicke (1981, 1985) suggested that the pattern of shallowly dipping faults might penetrate not only to the mid-crust, but perhaps through the Moho and the upper mantle (Fig. 8). Extension of surface faults might be a zone of intense shear strain, rather than a fault, but on a lithosphere scale it would act like a fault, allowing relatively intact lithospheric blocks to move relative to one another. The orientation of the fault would impose a profound asymmetry on the development of the rift and the resulting rifted margins.

In Wernicke's simple shear model, the two rifted margins, formed on the hanging-wall (upper plate) and footwall (lower plate) blocks, show distinct subsidence histories (Fig. 8). The footwall block subsidies during both synrift and postrift periods, and exposes mid-crustal or deeper rocks as the basement surface. The denser upper mantle is being thinned under the hanging-wall block, which subsides less than the footwall block. In certain geometries there may even be uplift of the hanging-wall block during rifting. The hanging-wall block then subsides during the postrift period, and no unusual crustal rocks are exposed. The long-term equilibrium subsidence of the crust on the hanging-wall block is less than that on the footwall block because the crust is predominantly lighter upper crust in the first case and predominantly heavier lower crust in the second case. There is also a predicted asymmetry in temperatures in overlying sediments; hotter basins are over exposed mid- to lower crustal rocks of the footwall block. Note that the simple shear model, in contrast to those mentioned earlier, predicts a systematic lateral distribution of crustal extension and subcrustal extension. This pattern can be tested, and the testing continues as complementary data are collected on conjugate rifted margins. Critical data include seismic reflection and refraction data showing the distribution of crust thickness, crust reflectivity, and sediment; drilling data constrain ages and depositional environments of sedimentary horizons, and identify the nature of the basement rocks.

When the term "pure shear" is applied to a margin, it is usually because the rift is observed to be symmetric. Lateral distribution of extension may vary in the crust and upper mantle. This effect is seen in Figure 8, where the upper mantle has extended over a wider area than the crust. The result is rift flank uplift and erosion. There is actually nothing about the pure shear mechanism that requires the resulting rift to be symmetric.

The term "simple shear" is frequently assigned to rifts or pairs of conjugate margins that are asymmetric. The essential element for a simple shear rift is that the locus of maximum extension in the upper mantle be laterally offset from that in the crust. In Figure 8 the mechanism of this offset is a fault or narrow shear zone. This has lead to the notion that simple shear in rifting is a process that, like fracture in rock-mechanics experiments, should occur at some prescribed angle in homogeneous rocks. Simple shear is just a large-scale description for the process by which laterally offset weaknesses in the strong layers of the lithosphere are exploited during rifting. This explanation suggests that there should be no prescribed "proper" dip for simple shear zones. Further, there is no reason to believe that the shear zone has to be planar. It may cross the upper crust at one point, flatten in the weaken lower crust, and then cross into the upper mantle elsewhere. A dynamic model of this type is described more fully below.

MAGMATISM DURING RIFTED MARGIN FORMATION

At the beginning of the 1980s, most investigators thought that the primary processes of rifting were extension and thinning of continental lithosphere. Magmatism, sometimes voluminous, was known to occur in rifting, but it was not understood how they were related in space or time. Synrift igneous rocks were not often discussed as important components of rifted margins because much of the crust at margins is deeply buried under sediments. As such, it was largely inaccessible to the drill and difficult to image seismically.

Modern seismic reflection profiling data, acquired with increasingly large sources and longer receiver arrays, are now commonly showing seaward-dipping reflector sequences forming the top of the transitional crust at rifted margins (Fig. 9). Mutter and others (1982, 1984) and Mutter (1985) noted the following characteristic features of seaward-dipping reflector sequences (Fig. 9). (1) Reflectors have ubiquitous seaward dip, perhaps flattening to near horizontal at the continental end. (2) Individual reflectors have arcuate, convex-upward shapes. The curvature is highly variable. There is typically an overall seaward offlap or outbuilding aspect to the sequence. (3) The seaward limit of the sequence is rarely well developed. The reflectors generally lose strength and the length of individual reflectors decreases until they are no longer identifiable. (4) Sequences often do not show a distinct basal reflector. Exceptions are the Outer Voring Plateau (Talwani and others, 1983; Mutter and others, 1984), offshore Southwest Africa (Austin and Uchupi, 1982), Antarctica (Hinz, 1981), and the Carolina trough (Austin and others, 1990). (5) Where it has been possible to define, the seaward-dipping reflector sequence lies just continentward of the oldest mapped sea-floor spreading magnetic anomaly.

Palmason (1973, 1980) proposed an elegant model for formation of dipping lava flows in Iceland that is often cited as an explanation for seaward-dipping reflector sequences. He considered intrusion of dikes and extension of flows into and over preexisting oceanic crust at a subaerial spreading center. Each new flow was thickest at the spreading center and thinner away from the center. The loading of one flow on another created an arcuate shape, deflecting it downward toward the spreading center. If this process operated during rifting of continental lithosphere, it would be expected to leave behind a seaward-dipping reflector sequence on each rifted margin. Mutter and others (1982) drew a cartoon of this process (Fig. 10). If this process created seaward-dipping reflector sequences, then the lava flows had to have been formed at or above sea level. This apparent requirement of subaerial magma extrusion posed problems for existing extensional models of rifting, because they predicted synrifting subsidence for most reasonable choices of crust parameters. The most likely mechanisms for keeping rifting lithosphere elevated to or above sea level include excess heating and a large magma supply. Alternatively, a spreading ridge could be surrounded by crustal barriers, depressing the local sea level.

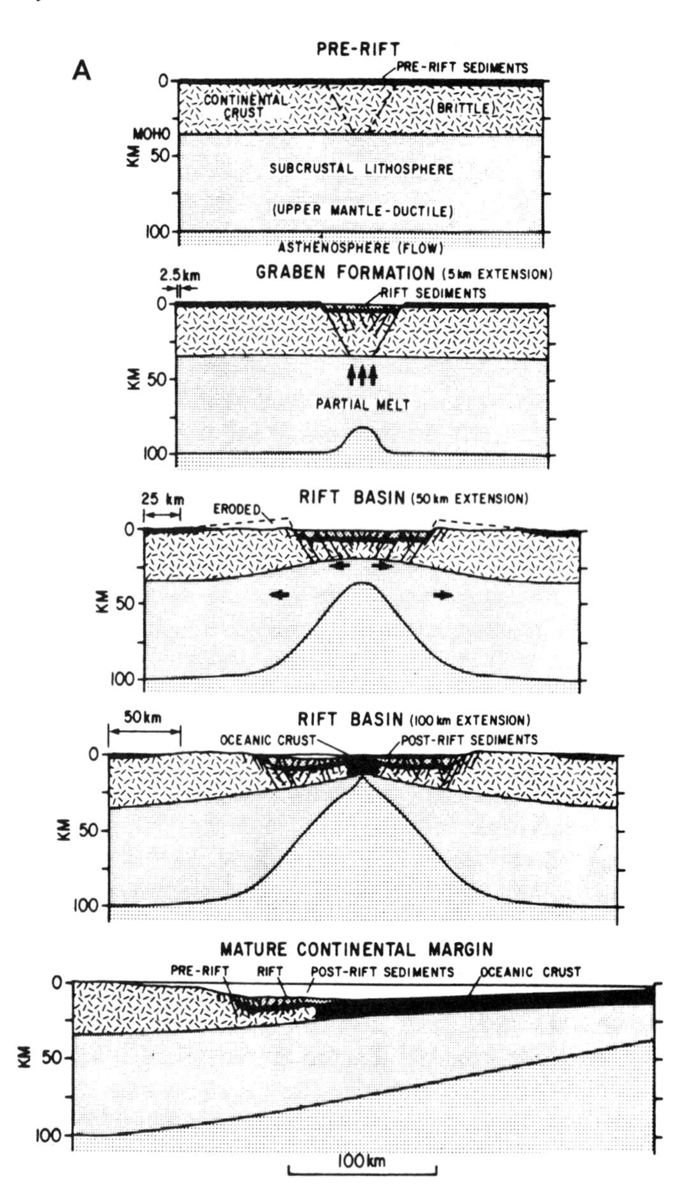

Deep penetration seismic reflection-refraction experiments at rifted margins have frequently identified anomalously high-velocity (7.1 to 7.5 km/s) lower transitional crust (Fig. 5). This is sometimes modeled as a basaltic layer underplated to the base of thinned, prerift continental crust. The basaltic magma is like that forming oceanic crust, generated by decompression melting in the upwelling asthenosphere and ponded at the base of the crust during rifting. Some of the magma is erupted to the surface to form a seaward-dipping reflector sequence. The magma not erupted to the surface is frozen in place at the base of the crust.

White and McKenzie (1989) attributed observed along-strike variations in the amount of volcanism at rifted margins to proximity to an active hot spot or mantle plume. Plume activity is likely to increase temperature of the asthenosphere, and they showed that the amount of magma produced during decompression melting is very sensitive to asthenosphere temperature. Thus, rifted margins formed near hot spots should be highly volcanic, and those farther away from hot spots are unlikely to undergo as much volcanism. These two general margin types are most often described as "volcanic" margins and "non-volcanic" margins, although, in light of the many "nonvolcanic" margins recently identified as having seaward-dipping reflector sequences and high-velocity lower crust, it is now more appropriate to call the latter "less volcanic" margins.

DYNAMIC MODELS OF RIFTED MARGIN FORMATION

During the 1980s models of the formation of rifted margins progressed from kinematic to dynamic. Kinematic models are those that specify a strain-distribution history for the lithosphere during rifting and then use physical principles to predict the ob-

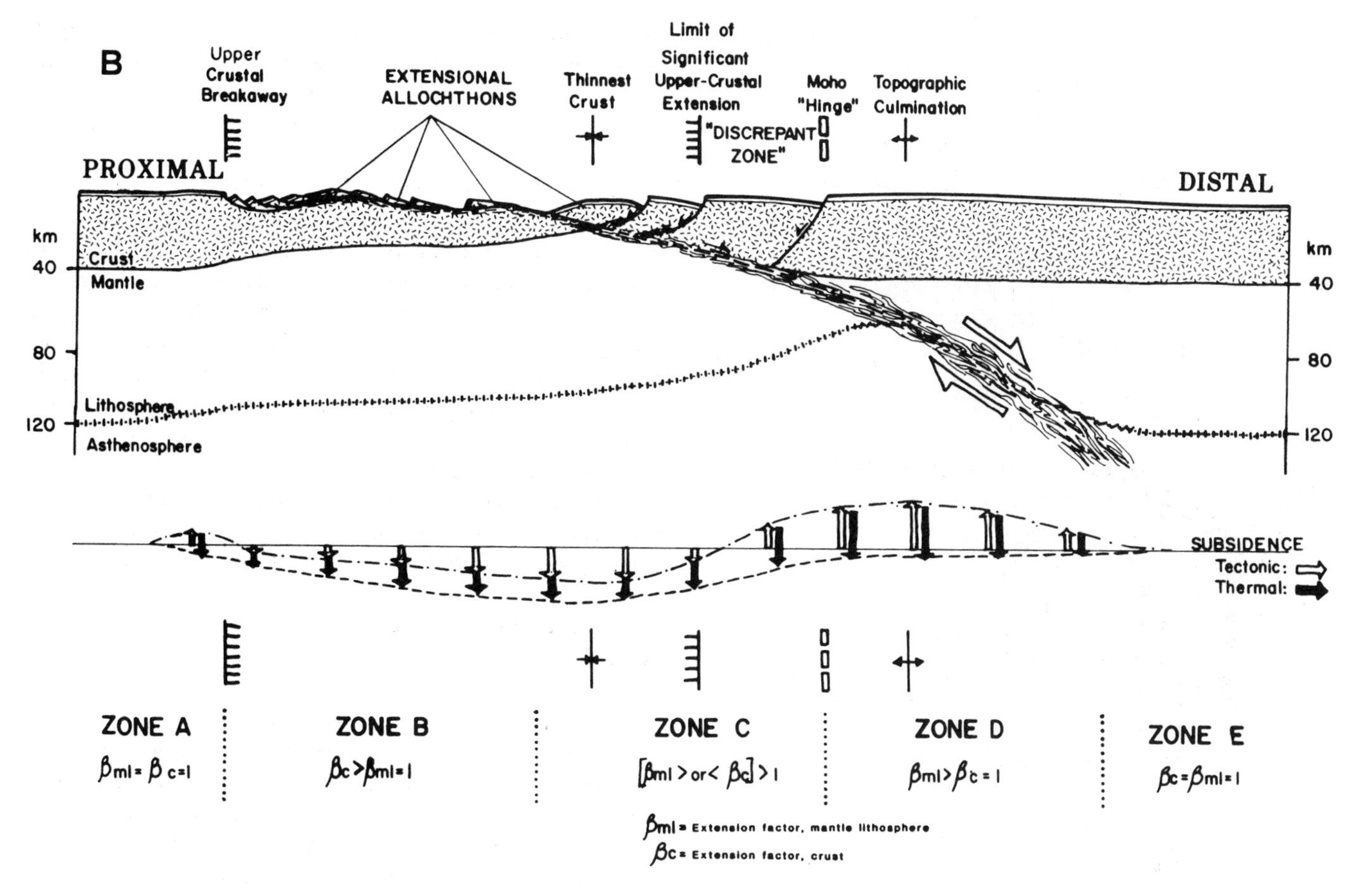

Figure 8 (on this and facing page). Cartoons of "pure" (A) and "simple" (B) shear style of rifted continental margin development (Hellinger and Sclater, 1983; Wernicke, 1985).

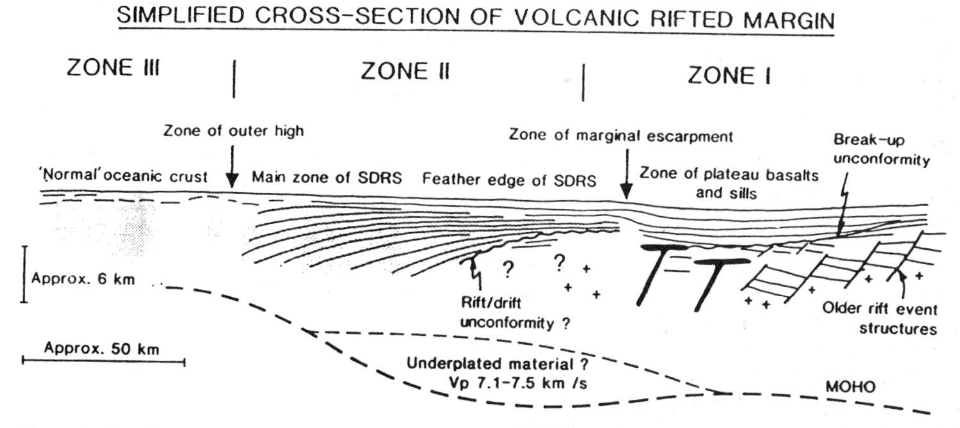

Figure 9. This idealized cross section of a volcanic rifted margin from Larsen and Sawyer (1991). Zone I is the relatively unmodified continental lithosphere adjacent to the seaward-dipping reflector sequence (SDRS). The continental lithosphere sometimes shows signs of faulting, but more typically is not obviously deformed during rifting. Zone II encompasses the SDRS crust with a pod of possibly underplated material at its base. Zone III encompasses submarine oceanic crust that may or may not be normal oceanic crust in a structural, thermal, or subsidence sense. Vp is compressional wave velocity.

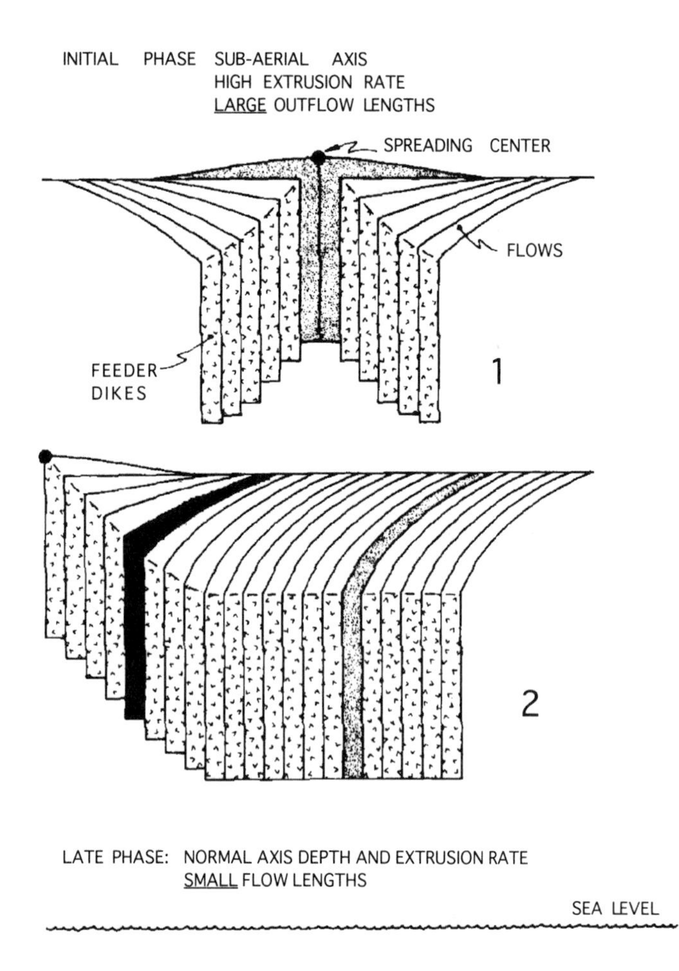

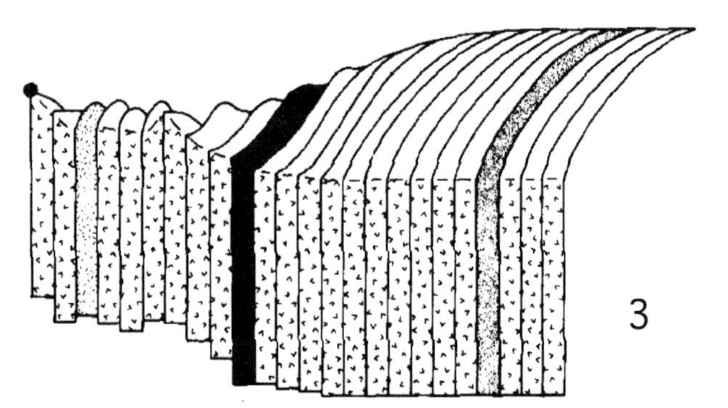

Figure 10. Mutter and others (1982) cartoon of formation of a seaward-dipping reflector sequence (SDRS) by subaerial sea-floor spreading. An SDRS is initiated when magma supply is adequate to build the spreading center to sea level. Lava flows then have significant length, and when intercalated with sediments, form good seismic reflectors. The flows take on a dip toward the spreading center as they are progressively loaded with subsequent flows (1). The SDRS can continue to form as long as the spreading center remains above sea level (2). When spreading center submerges, lava can no longer flow and the emplacement of normal submarine oceanic crust begins (3). Several feeder dike and/or dipping layer pairs are shaded differently so that they can be traced from panel to panel. This is the sequence presumed for many volcanic rifted margins.

servable characteristics of the resulting continental margins. McKenzie's (1978) model was a one-dimensional kinematic model. Wernicke's (1985) model was a two-dimensional kinematic model. Kinematic models have the drawback that the strain-distribution history is not coupled to the physical properties of the lithosphere or to the distributions of heterogeneities in that lithosphere. Because of this, one cannot know if the kinematics are even vaguely realistic. Dynamic models are those that specify the state of the lithosphere prior to rifting and the temperature and stress boundary conditions that are applied to the model lithosphere during rifting. The strain-distribution history emerges naturally from a dynamic model, along with the suite of predictions made using kinematic models. The yield stress envelope model of lithosphere strength has been the key to making dynamic modeling of rifting possible. In the following sections, I describe this model and how it is applied to determine the rheological behavior of continental lithosphere, and then show how it is incorporated into dynamic models of rifted margin formation.

Yield stress envelope model of continental lithospheric strength

The yield stress envelope model (Fig. 11) defines stress required to cause rocks to fail under compression or tension as a function of depth in the lithosphere (Brace and Kohlstedt, 1980). It is constructed from several stress-strain relations, each of which will control failure under certain conditions of temperature, pressure, and lithology. The two principal failure regimes incorporated in these models are the ductile regime and the brittle regime. The ductile regime is discussed first here because most investigators actually handle deformation in the brittle regime as a modification of the ductile flow equations. These stress-strain relations are combined with lithosphere structure to produce yield stress envelope curves that are the key element of current continental rifting models.

Ductile Regime

Deformation of rocks at higher temperatures is thought to be controlled by diffusion processes that can be modeled as power law fluid flow. In such a fluid, the relation between principal stresses σ_H and σ_V, in the horizontal and vertical directions, and the strain rate in the horizontal direction, $\dot{\epsilon}_H$, takes the form:

$$\dot{\epsilon}_H = A\,(\sigma_H - \sigma_V)^n \exp\left[-\frac{Q}{RT}\right] \qquad (1)$$

where A, Q, and n are material properties, T is temperature, and R is the universal gas constant. All stresses in this paper will be taken to be positive when tensional. For olivine and quartz, often used to represent the mantle and crust, respectively, the power law exponent, n, is ≈ 3. Q is an activation energy that controls temperature sensitivity of the flow. As temperature increases, strain rate increases exponentially (Fig. 11). Strain rate

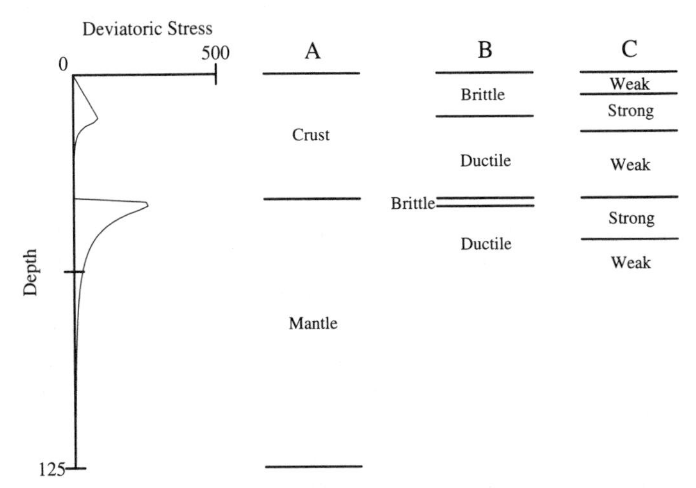

Figure 11. Stress required to produce strain at a geologically significant rate (10^{-15} s^{-1}) as a function of depth in the lithosphere. Depth is given in kilometers. The horizontal component of the deviatoric stress is given in MPa. Pore pressure is assumed to be hydrostatic. The crust is composed of wet quartz diorite and the mantle is composed of wet dunite (Carter and Tsenn, 1987). A normal continental geotherm is used (Sclater and others, 1980). At each depth, both the stress required to produce brittle failure and the stress required to produce ductile failure at the prescribed rate are calculated. The process that requires the lowest stress is presumed to operate at that depth. This leads to the qualitative description of the model, based on deformation mechanism, shown in column B. In this plot, stress may also be interpreted as strength, and the integral of stress vs. depth represents total strength of the lithosphere. This leads to a qualitative description of the vertical zonation of the lithosphere based on strength, shown in column C.

increases as about the cube of differential stress. There is no pressure term in the strain-rate function and hence no direct depth dependence. Variation of strain rate with depth comes from variation of temperature with depth in the Earth. To determine the yield stress, equation 1 can be rearranged to give:

$$\sigma_H - \sigma_V = \left(\frac{\dot{\epsilon}_H}{A}\right)^{\frac{1}{n}} \exp\left[\frac{Q}{nRT}\right] \quad (2)$$

There is a unique value of stress that will cause a material to deform at any given strain rate. For extensional provinces, the strain rate selected usually lies in the range 10^{-16} to 10^{-14} s^{-1}.

Power-law fluids, particularly those with high exponents, tend to deform heterogeneously, lots in some regions but little in others. Power-law fluids with lower power-law exponents tend to deform more uniformly. Regions of stress concentrations are very susceptible to localized deformation in power law fluids.

Brittle Regime

Goetze and Evans (1979) proposed that the dominant deformation mechanism in the upper part of the continental crust is fictional sliding on preexisting faults. They treated rocks as having a large number of small, randomly oriented fractures, any of which could slide to release stress. Each fracture would be expected to slide, or fail, when the shear stress across it exceeded the friction on the fracture surface. The friction is controlled by the effective pressure, which is the lithostatic pressure less the pore-fluid pressure, and the coefficient of friction (Fig. 11). The ratio of the pore-fluid pressure to the lithostatic pressure is usually called λ. When λ is near 1, very little stress is required to cause brittle failure. As λ decreases, the brittle strength of the rock increases: λ is often chosen to be ≈ 0.42, which corresponds to hydrostatic pore pressure, or 0.0 which corresponds to dry rocks. The governing equations (modified from Brace and Kohlstedt [1980] by Houseman and England [1986]), in terms of the horizontal component of the deviatoric stress, τ_{xx}, and the vertical principal stress, σ_v, are:

$$\tau_{xx} = -0.39\,(1-\lambda)\sigma_v \qquad |\sigma_v| < 972 \text{ MPa} \quad (3)$$

$$\tau_{xx} = 34.0 \text{ MPa} - 0.34\,(1-\lambda)\sigma_v \qquad |\sigma_v| > 972 \text{ MPa} \quad (4)$$

The following discussion treats only the extensional situation. The vertical principal stress component is a function of depth, z, the gravitational acceleration, g, and the average density of the overburden, ρ.

$$\sigma_v = -\rho g z \quad (5)$$

The presence of randomly oriented fractures is important because it ensures that there will always be a conveniently oriented fracture for a given stress field, and this allows direction to be removed from the mathematics. Byerlee's Law, which governs this frictional sliding, has been shown to be largely insensitive to strain rate, temperature, and rock type (Byerlee, 1968). Thus, the principal control on the stress required to produce failure in brittle rocks is the effective pressure, through its simple association with depth of burial and pore pressure (Fig. 11).

Putting it together for continental lithosphere

In early yield stress envelope studies, continental crust was modeled as quartz and the mantle was modeled as olivine, because flow-law parameters for quartz and olivine had been measured (Goetze and Evans, 1979). Parameters were unavailable for other rocks that certainly better represent the bulk lithology of the continental crust. Of the two, the use of flow-law parameters for olivine in the mantle is more likely to be a reasonable choice. Ductile flow data for many more rock types are now available (Fig. 11).

At a given depth, temperature, pore-fluid pressure, and lithology, the stress required to deform rocks by brittle failure and by ductile flow at a particular strain rate can be calculated. The dominant mechanism is presumed to be the one that produces failure at the lowest stress (Fig. 11). In the shallower crust and sometimes in the upper mantle, the dominant failure mechanism is brittle failure. In the lower crust and certainly in all but the upper few kilometers of the mantle, the dominant failure mechanism is ductile flow.

The boundary between regions of brittle failure and ductile flow in a single material is called the brittle-ductile transition (Fig. 11). It occurs at the depth where the yield stress for brittle failure equals the stress required to produce ductile flow at the chosen strain rate. Because the stress required for ductile flow is dependent on the choice of strain rate that defines failure of the rock, then the depth of the brittle-ductile transition is also dependent on that choice. The brittle-ductile transition depth for a process that is deforming the lithosphere quite rapidly would be much deeper than for a slower deformation process.

The simple yield stress envelope model does a poor job of predicting stress required to produce deformation near the brittle-ductile transition depth. By assuming that exactly one mechanism is operative at each depth, the yield stress envelope model does not take into account that at least two roughly equal deformation mechanisms are operating near this transition. The depth to the brittle-ductile transition and the exact deformation mechanisms operating there are of most direct concern to us when we use earthquake seismic hypocenter data to constrain the deformation mechanism in tectonically active areas. Earthquakes are believed to occur in rocks that are deforming by frictional sliding and/or brittle failure. Chen and Molnar (1983) used a large data base of earthquake hypocenters to show that earthquakes in continental crust are most common in the upper crust and are rare in the lower crust. Earthquakes sometimes occur in the upper mantle. Chen and Molnar (1983) interpreted their results to support the yield stress envelope model and to indicate that the upper mantle can sometimes deform in the brittle domain.

For modeling continental extension, the processes near the brittle-ductile transition are not very important. The contrasts between srong (high-yield stress) and weak (low-yield stress) layers are more important than the contrasts between brittle and ductile layers. The important vertical zonation of the quartz-olivine yield stress envelope of the continental lithosphere is, from the top down, weak, strong, weak, strong, weak (Fig. 11). The upper weak layer is quartz deforming brittlely at low yield stress. The upper strong layer is composed of quartz deforming in a combination of mechanisms at high-yield stress near the brittle-ductile transition. The middle weak layer is quartz deforming by ductile flow at very low stress at the base of the crust. The lower strong layer is olivine deforming by ductile flow and perhaps brittle failure at very high stress in the upper mantle. The lower weak layer is olivine flowing at low stress in the mantle below the upper few kilometers.

TWO-DIMENSIONAL FINITE ELEMENT MODELS

A recent and important use of the yield stress envelope method of characterizing strength of lithosphere has been in construction of dynamic models of the formation of rifted continental margins. This group of models considers evolution during rifting of a cross section of a rift or conjugate pair of rifted margins. The models are based on the assumption that the structures extend infinitely and without variation along the strike of the rift. Rifting is localized by emplacement of some type of heterogeneity, usually a weakness such as weaker rock or hotter rock, in otherwise uniform continental lithosphere. Extensional stresses are applied either by body forces within the model or by constant stress or velocity boundary conditions at the edge of the model. Models are run to simulate some appropriate period of geologic time. Results are presented in a variety of forms, including deformed meshes, nodal velocities, strain-rate contours, stress contours, temperature contours, elevation history, and extension history. Some of the modeling studies represent attempts to match timing and rifting geometry of a particular rift or rifted margin; however, most illustrate generic rift features (Braun and Beaumont, 1987, 1989a, 1989b, 1989c; Dunbar and Sawyer, 1988, 1989b; Chery and others, 1990; Keen and Boutilier, 1990; Sawyer and Harry, 1991; Harry and Sawyer, 1992a, 1992b; Harry and others, 1993).

In dynamic models of continental extension, brittle failure should be modeled as perfect plasticity. This approach is valid if the scale of the individual faults is small with respect to the thickness of the brittle layer. Perfect plasticity is difficult to model because once the yield stress is reached or exceeded, the amount and rate of strain are not defined. This problem is avoided by approximating the behavior of plastic materials with high n power-law fluids. The location of individual faults is not predicted with this method, although it is possible to predict the amount of total extension, which we expect will appear as faulting, that should be observable in a region.

Isostasy is incorporated in all these models through a combination of gravitational body forces and forces on the edges of the model (Fig. 12). Forces on the edges of the model are required because the model covers only the region of interest. Forces on the side keep the material from collapsing outward (like a dam keeps water in a lake). They are most often applied as zero or constant-velocity horizontal displacements of the side boundary nodes. Edge forces on the bottom boundary of the model are required to simulate the buoyancy effect of the lithosphere floating on the asthenosphere.

Keeping track of temperature in these models is vital because it couples strongly with viscosity (through the exponential term involving temperature in equation 1) in the ductile flow law rheology. Temperature changes during the execution of these models due to advection and conduction of heat and the generation of heat by radioactive materials. Advection of heat is handled automatically by deformation of the finite element method grid in each of these models. Conduction of heat and heat generation are usually handled by a separate finite element method code executed between executions of the mechanical model steps.

Temperature boundary conditions common to most of the models are 0 °C on the top and no lateral heat flow, or insulating boundaries, on the sides (Fig. 12). The bottom boundary condition can be constant temperature or constant vertical heat flow. The constant vertical heat-flow boundary condition is probably preferable. The constant temperature bottom boundary condition leads to problems when the bottom of the model is elevated

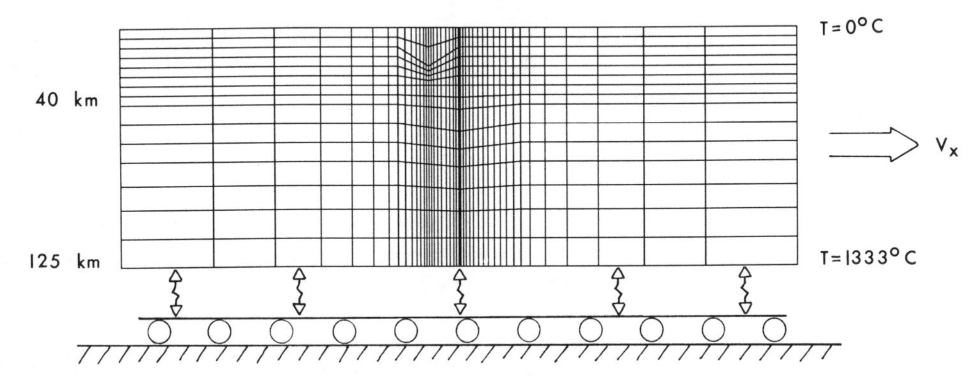

Figure 12. Finite element grid with boundary conditions. this particular grid is taken from Harry and Sawyer (1992b) and relates to the model shown in Figure 15. Individual finite elements can behave as either a plastic (an approximation to brittle failure on a fine scale) or as a power-law fluid, depending on local conditions of temperature, pressure, strain rate, and lithology. The lines bounding the elements move with the straining rocks (except when the finite element method (FEM) grid has been highly sheared and needs to be remeshed). The symbols along the bottom boundary indicate that the FEM grid is sitting on an elastic foundation (simulates isostasy), and that the bottom elements are free to move side to side.

significantly above its original depth. In this case, the heat flow into the bottom of the model is unrealistically high. However, if there is significant convection in the uplifted asthenosphere, as suggested by Mutter and others (1988), then the constant temperature bottom boundary condition may be most appropriate.

Braun and Beaumont (1989a) categorized the sources of stress available for driving lithosphere extension (Fig. 12). "Distant pull," by analogy to "slab pull," is an in-plane stress that represents a state of horizontal deviatoric tensile stress. Only near the surface does it represent an absolute tensile stress. "Local gravity push" results from gravitational potential diffrences between a lithospheric column and a reference column. The reference column is usually taken to be the mid-ocean ridge, but can also be unweakened continental lithosphere. Local gravity push has been invoked to explain the extension of the Basin and Range province, where the potential anomaly is the result of overthickening of the continental crust. A mantle plume impinging on the bottom of continental lithosphere can also elevate the lithosphere and cause tensional stresses due to local gravity push.

In the models discussed here, both distant pull and local gravity push are considered. Distant pull can be invoked in two ways. The most common is specifying a constant, or time varying, horizontal velocity on one side of a model while keeping the other side fixed in the horizontal direction. The other method involves applying a constant, or time varying, stress to one boundary while keeping the other fixed. The constant stress method seems more intuitive, but the constant velocity is easier to constrain with geologic observations. Presumably, plate motions are due to a near balance between large forces, and therefore plate velocity is buffered. Local gravity push is usually invoked as part of the finite element method and is implicitly included in all the models discussed. When a constant velocity side boundary condition is applied to a model, the model lithosphere must yield at that rate somewhere inside the model in each layer. In the ductilely weak layers this happens by ductile flow. In the cooler, ductilely stronger layers, it must happen by plastic (simulating brittle) failure at stress exceeding the yield stress. This type of failure tends to be strongly localized by weaknesses and lateral strength variations in the model lithosphere.

In most of the models of this type, the strength of the prerift lithosphere is laterally heterogeneous. The heterogeneity usually takes the form of a weakened area surrounded by unweakened lithosphere. A variety of different weakness-producing mechanisms has been considered and found to lead to very different rifting styles. Rifting style includes lateral distribution and timing of subsidence or uplift, faulting, and magmatism, and the location and timing of initiation of sea-floor spreading.

Unweakened lithosphere is modeled as a one- or two-lithology crust overlying a single-lithology mantle (Fig. 13). Typical choices of mantle lithology are wet olivine and wet dunite. Single-lithology crust models have used wet quartz and quartz diorite. Dual-lithology crust models include wet quartz over wet feldspar, wet granite over quartz diorite, and wet quartz over plagioclase. Where two lithologies are used in the crust, the models exhibit three strong zones, with intervening weak zones, in the continental lithosphere. Where one lithology is used in the crust, there are only two strong zones in the lithosphere. In most cases, the strongest portion of the continental lithosphere is located in the upper mantle.

Rifting in these models manifests itself in the distribution of plastic extension of the elements along the Earth's surface and in the subsidence of the surface. These models do not yet predict the distribution of individual faults that will accompany rifting. A surface element that has been plastically extended in the model would correspond to a region in which there would be some faulting observed. The orientation of that faulting would likely be controlled by small-scale (below the resolution of the finite element method models) heterogeneities in the near-surface rocks.

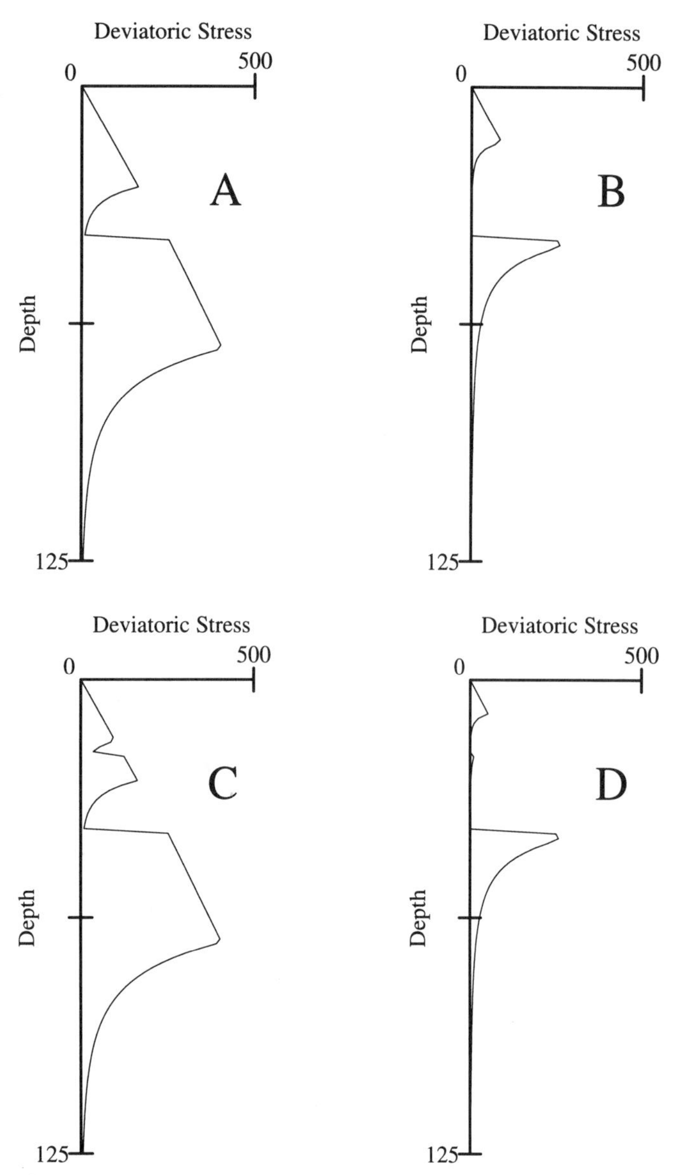

Figure 13. Several models for unweakened continental lithosphere with their yield strength vs. depth. A strain rate of 10^{-15} s^{-1} is used. Depth is given in kilometers. The horizontal component of the deviatoric stress is given in MPa. Pore pressure is assumed to be hydrostatic. Rheologic parameters were given by Carter and Tsenn (1987). The models are as follows. A: Quartz-diorite crust over dunite mantle with cold continental geotherm; B: Quartz-diorite crust over dunite mantle with normal continental geotherm; C: Westerly granite upper crust over quartz-diorite lower crust over dunite mantle with cold continental geotherm; D: Westerly granite upper crust over quartz-diorite lower crust over dunite mantle with normal continental geotherm.

The form of the strength variation introduced into the model lithosphere can have at least as big an impact on the rift that will result as the forces that cause the rifting (Fig. 14). The strength variation introduced often takes the form of emplacing weaknesses at some positions and depths in the model. Extension is typically localized at or near those weak spots. Even emplacing a strong spot in an otherwise uniform lithosphere will serve to nucleate localized extension. This can occur because stress is concentrated at the edge of the strong spot, and, in power-law viscous flow, the apparent viscosity of the material is inversely related to stress. Thus, the edges of the strong spots become loci of extension.

Dunbar and Sawyer (1988, 1989b) argued that if the lithosphere is composed of strong and weak layers, then rifting should largely be controlled by the presence of weaknesses in the strong layers (Figs. 13 and 14). Weaknesses in the weaker layers, although undoubtedly present, should have relatively less impact on the location or style of rifting. This approach, and the single-lithology crustal model they adopted, suggested that lithospheric weaknesses could be classed as "upper mantle" weaknesses or "middle crustal" weaknesses (this scheme can be generalized to a multiple-layer crust if desired). Further, they suggested that there might be differences in rifting style that should result from extending the two types of weaknesses. They emplaced a crustal-scale granite batholith into an otherwise uniform continental lithosphere to simulate a middle crustal weak region. It was a weakness because the granite was weaker than the quartz diorite of the adjacent crust. In a companion model, Dunbar and Sawyer locally thickened the continental crust by 10 km to simulate an upper mantle weakness. The area with thicker crust was weaker because they replaced relatively strong upper mantle rocks with weaker lower crustal rocks. The middle crustal strong zone is unaffected by this change in crust thickness, while the upper mantle strong zone is weakened. Dunbar and Sawyer's model results showed that there were distinct differences in rifting style that resulted from the difference in weakness type. Rifts at crustal weaknesses were characterized by rapid initial subsidence and surface faulting, late-stage volcanism, and long rifting times. Rifts at mantle weaknesses were characterized by initial doming, early volcanism, late-stage surface faulting and subsidence, and relatively shorter rifting times. Note that these are some of the characteristics often ascribed to "passive" and "active" rifts, although the models were all constructed to be "passive" rifts. These types of weaknesses could also be combined to produce rifting with a range of intermediate styles, and, if laterally offset, can produce another suite of interesting rift characteristics, explored further below.

The importance of two- (and perhaps three-) dimensional finite-element modeling becomes particularly clear in models of laterally offset crust and mantle weaknesses. This type of problem was addressed by Dunbar and Sawyer (1988, 1989b), Braun and Beaumont (1989a, 1989b), Sawyer and Harry (1991), and Harry and Sawyer (1992a, 1992b). Dunbar and Sawyer (1988, 1989b) emplaced mid-crustal and upper mantle weaknesses with their centers 100 km apart (Figs. 13 and 14). This arrangement might correspond to weakness distribution at an old continental suture. As Dunbar and Sawyer stretched the model, they found that early strain in the crust was concentrated at the crustal weakness and early strain in the mantle was concentrated at the mantle weakness. Between the two weaknesses, there was shearing in the weak lower crust to accommodate the offset loci of deformation.

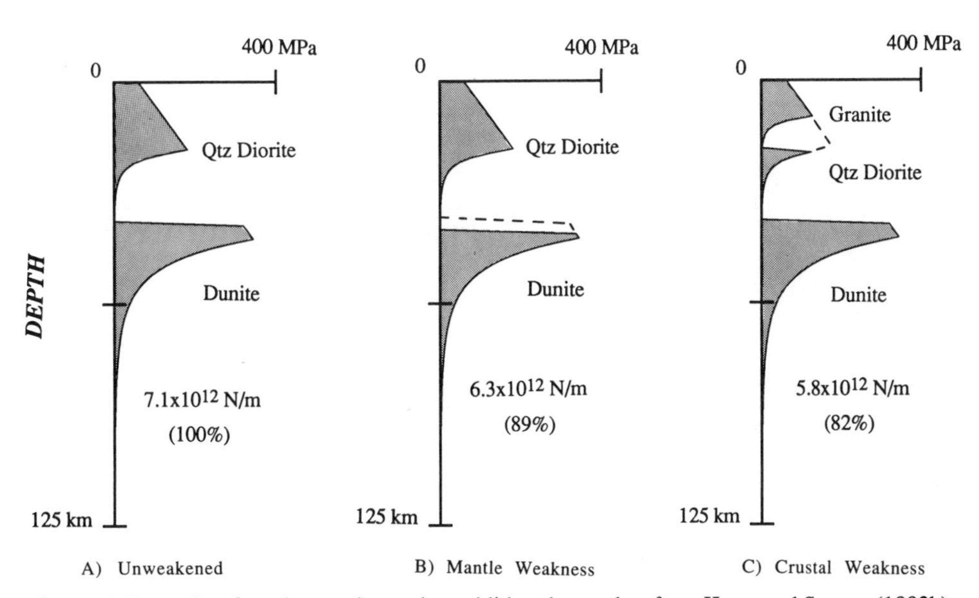

Figure 14. Examples of weaknesses in continental lithosphere, taken from Harry and Sawyer (1992b). A is unweakened continental lithosphere and its total integrated strength is represented as 100%. B is for lithosphere with the crust thickened by 5 km. Note that the extra crust is very weak and replaces the normally strong upper mantle. This configuration weakens the lithosphere by about 20%, the weakening coming in the upper mantle strong zone. C is for lithosphere in which part of the quartz diorite in the upper crust has been replaced with granite. Because granite is much weaker in ductile deformation than quartz diorite, the crustal strong zone is weakened and the total lithosphere strength is reduced by about 20%. Note that the upper mantle is not weakened in this case. Harry and Sawyer (1992b) suggested that many other types of lithosphere weaknesses may be classified in terms of whether they weaken the upper mantle and/or middle crustal strong layers.

This geometry resembles large-scale simple shear of the lithosphere. The simple shear system emerges not from any inherent material property of the lithosphere, but from the distribution of prerifting weakness. If the continental lithosphere had an inherent tendency to form low-angle detachments, then all such systems should form at a preferred dip. Dunbar and Sawyer suggest that detachments tend to form to connect laterally offset middle crust and upper mantle weaknesses, and as such, detachments should be able to form at any angle. As rifting progresses, however, the pattern changes substantially. The heat flow into the base of the crust is elevated at the position of the mantle deformation (thinning). This heat gradually weakens the overlying crust until the newly created thermal weakness is more profound than the original crustal weakness. When this occurs, the locus of rifting in the continental crust jumps to the new position and abandons the site of the original crustal weakness. With the loci of crustal extension and mantle extension now aligned vertically, extension becomes increasingly localized, leading to the generation of new oceanic lithosphere. When extension in the two layers is aligned, there is no longer a need for simple shear in the lower crust, so the shear zone shuts down and the extension looks like large-scale pure shear. Thus the models suggest that we might expect to see a progression from large-scale simple shear to pure shear during the rifting process. This may explain why many marginal basins have proven hard to classify as solely pure or simple shear.

Whereas all the finite element method models discussed above have been generic in nature, Sawyer and Harry (1991) and Harry and Sawyer (1992a) took the offset weakness model and used it to explain some aspects of the geology of the Baltimore Canyon trough. In the model (Fig. 15) they varied the width, position, and magnitude of the crust and mantle weaknesses to reproduce the general timing and lateral distribution of extension along the margin transect when the model was extended at the correct relative plate velocity. The model develops as described in the previous paragraph. The abandoned rift basin over the crust weakness is analogous to the Triassic rift basins that parallel the margin. These basins were active for about the first half of the rifting interval, and then appear to have shut down as the locus of rifting jumped to the east. The hinge zone, observed all along the United States Atlantic margin, has as its model analog the boundary between the crust that is affected by both the early and late phases of rifting and the crust that affected only by the early phase.

The third facet of the geology of the Baltimore Canyon trough area that the model may explain is the short-lived basaltic volcanism in the Triassic basins near the end of their formation. There is not enough lithosphere thinning directly under these

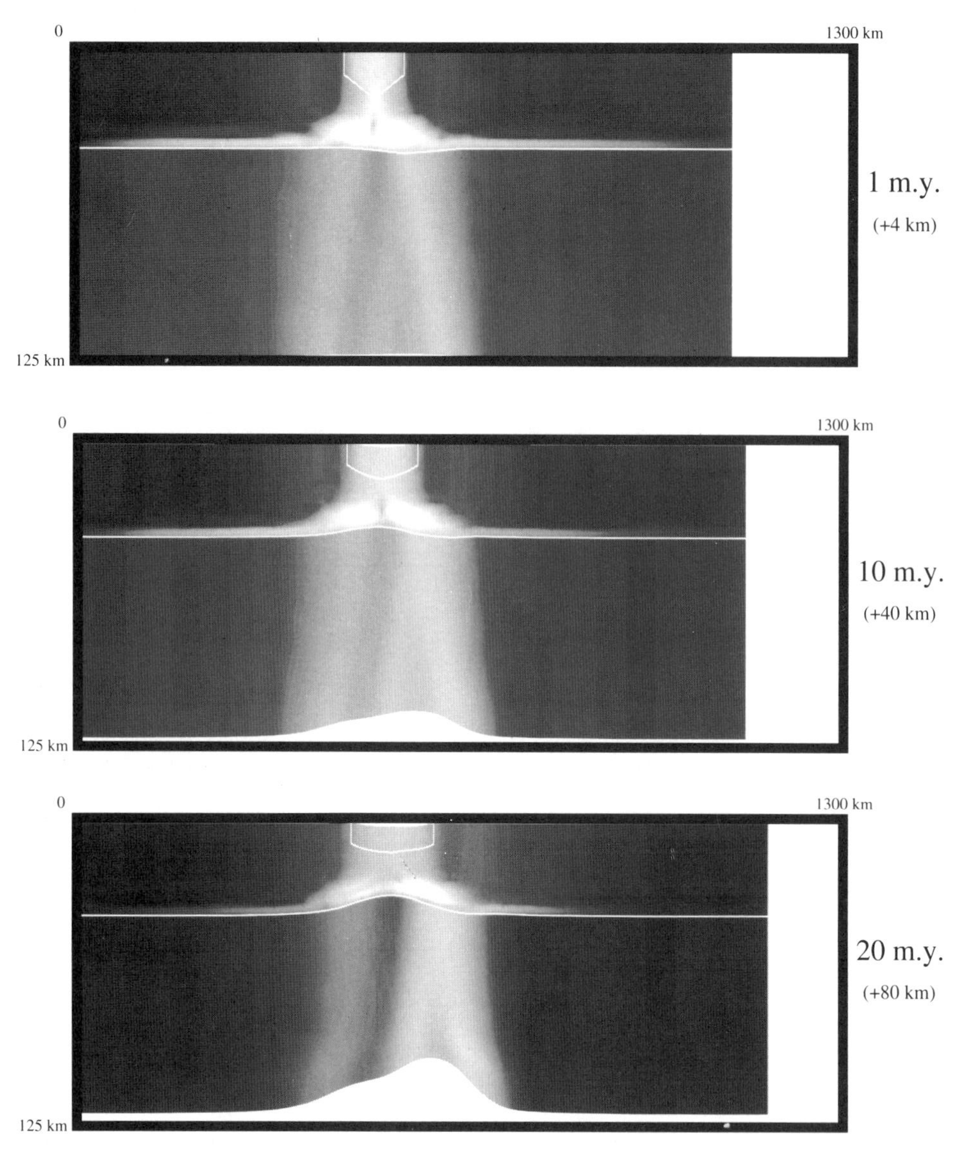

Figure 15 (on this and facing page). Deformation resulting from stretching lithosphere with laterally offset crust and mantle weaknesses (Harry and Sawyer, 1992b). Model parameters were chosen to approximate the formation of the Newark basin and Baltimore Canyon Trough. During the first 25 m.y. of the model run, the deformation in the crust is localized where the crust was initially weak and deformation in the upper mantle is localized where the upper mantle was weak. A simple shear zone develops in the lower crust to connect the localized deformation in the two strong layers. After about 25 m.y., the heat flow from the thinning upper mantle into the crust weakens the crust and in the representation for 30 m.y., thinning in the crust is occurring both in the original weak zone and over the mantle weakness where the crust was not initially weakened. The growing thermal weakness takes over by 40 m.y. and the thinning in the region of original crustal weakness ceases. At this stage, the deformation is more like what is commonly called pure shear. The model progresses to the point where sea-floor spreading would begin at 50 m.y. Note the pronounced asymmetry of the rifted margins that are produced. The Newark and other Triassic basins form in the region analogous to the original weakened crust. The Baltimore Canyon Trough forms over the region of original mantle weakness and where the late-stage crustal thinning occurred. The hinge zone corresponds to the landward edge of the late-stage crustal thinning.

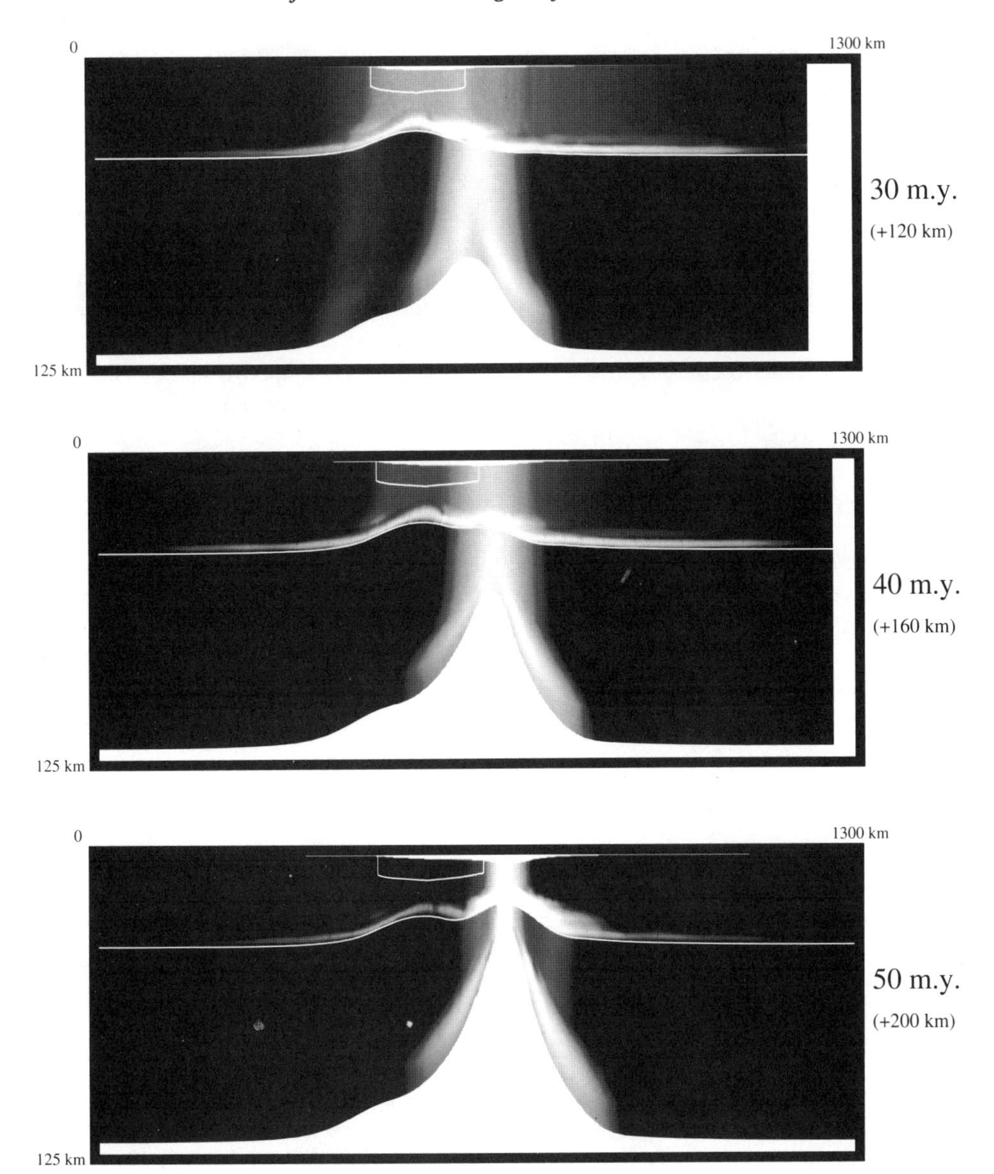

basins to argue that the magma could be produced by decompression melting of the asthenosphere. The model suggests, however, that there is likely to have been decompression melting of the asthenosphere under the mantle weakness at about the correct time for the onset of this volcanism. What remains to be explained is how the magma can be moved laterally by about 100 km to the site of the Triassic basins. Careful examination of the model shows that there is a substantial lateral pressure gradient in the lower crust that draws weak lower crustal rocks toward the position of the crust weakness (the Triassic basins). This pressure is the result of the strong part of the crust thinning over a full-strength mantle strong zone. The mantle does not bend easily in isostatic response to the crust thinning. It is mechanically easier to suck in weak lower crustal rocks to fill the space than it is to bend the mantle up. Harry and Sawyer (1992b) suggested that magma might be drawn laterally from where it is made at the mantle weakness to where it needs to erupt at the crustal weakness even more easily than relatively stronger lower crustal rock. If this hypothesis is correct, then the eruption of the magma in the Triassic basins would be expected to stop when the extension in the basins stopped, because the mechanism of lateral transport would be gone. This is generally consistent with the observations. Although speculative, this is a possible mechanism for the emplacement of some very enigmatic basalts.

These dynamic models are computationally expensive. Examples of two-dimensional models have taken hours on modern supercomputers, and days on widely available work stations. Three-dimensional models, which will be required to solve important problems involving rift propagation along margins, may take days on supercomputers or months on workstations. Faster computers, better algorithms, and insight to the important physical processes in continental rifting will be required to make these tools most useful.

RIFTED MARGIN RESEARCH IN THE 1990S

Our understanding of the process of rifting and the formation of rifted continental margins is far from complete. Most of the current models of vertical crustal motions during and after rifting are not able to explain the many regional unconformities seen in basins. They also have trouble explaining the formation of seaward-dipping reflector sequences at sea level late in the rifting process. Mechanisms of magma production and emplacement in rifted margins are only beginning to be explored. Indications of subsurface temperature and pressure both before and during rifting can be expected to emerge with studies of the chemistry of rift volcanic rocks. Very little is known about how basaltic magma moves through the mantle and crust, although this process is key to understanding the association of underplated crust and seaward-dipping reflector sequences. Modern deep-penetration seismic reflection and refraction methods have yet to be applied to most of the rifted margins of North America, or anywhere else. Better coordination of onshore and offshore seismic experiments is required to produce better subsurface images of rifted margins beneath the shoreline. Three-dimensional, or at least finer grids of two-dimensional, seismic reflection data will be required to unravel complex rift geometry. Drilling of conjugate margins will be required to determine the age and depositional environment of sediments imaged seismically, and to identify basement rock types. The Ocean Drilling Program is about to embark on a study of both volcanic and non-volcanic rifted margins in the North Atlantic by drilling on the conjugate margins off Newfoundland and Iberia, with the objectives of sampling basement rocks and penetrating complete sedimentary sections, and drilling at several locations off southeast Greenland to explore a seaward-dipping reflector sequence formed at various distances from the Iceland hotspot. Modeling a variety of well-observed rifted margins may tell us what physical factors are most important in controlling rifting style, as well as constraining our intuition about Earth processes.

REFERENCES CITED

Austin, J. A., Jr., and Uchupi, E., 1982, Continental-oceanic crustal transition off Southwest Africa: American Association of Petroleum Geologists Bulletin, v. 66, p. 1328–1347.

Austin, J. A., Jr., Stoffa, P. L., Phillips, J. D., Oh, J., Sawyer, D. S., Purdy, G. M., Reiter, E., and Makris, J., 1990, Crustal structure of the Southeast Georgia embayment–Carolina trough: Preliminary results of a composite seismic image of a continental suture(?) and a volcanic passive margin: Geology, v. 18, p. 1023–1027.

Balkwill, H. R., and Legall, F. D., 1989, Whale basin, offshore Newfoundland: Extension and salt diapirism, *in* Tankard, A. J., and Balkwill, H. R., eds., Extensional tectonics and stratigraphy of the North Atlantic margins: American Association of Petroleum Geologists Memoir 46, p. 233–246.

Balkwill, H. R., and McMillan, N. J., 1990, Mesozoic-Cenozoic geology of the Labrador Shelf, *in* Keen, M. J., and Williams, G. L., eds., Geology of the continental margin of Eastern Canada: Boulder, Colorado, Geological Society of America, Geology of North America, v. I-1, p. 295–324.

Bally, A. W., 1982, Musings over sedimentary basin evolution: Royal Society of London Philosophical Transactions, ser. A, v. 305, p. 323–338.

Boillot, G., Winterer, E. L., and Meyer, A. W., 1988, Proceedings of the Ocean Drilling Program, Scientific results, Volume 103: College Station, Texas, Ocean Drilling Program, 858 p.

Boillot, G., Mougenot, D., Girardeau, J., and Winterer, E. L., 1989, Rifting processes on west Galicia margin, Spain, *in* Tankard, A. J., and Balkwill, H. R., eds., Extensional tectonics and stratigraphy of the North Atlantic margins: American Association of Petroleum Geologists Memoir 46, p. 363–378.

Bosworth, W., 1985, Geometry of propagating continental rifts: Nature, v. 316, p. 625–627.

Bosworth, W. R., 1987, Off-axis volcanism in the Gregory rift, east Africa: Implications for models of continental rifting: Geology, v. 15, p. 397–401.

Brace, W. F., and Kohlstedt, D. L., 1980, Limits on lithospheric stress imposed by laboratory experiments: Journal of Geophysical Research, v. 85, p. 6248–6252.

Braun, J., and Beaumont, C., 1987, Styles of continental rifting: Results from dynamic models of lithospheric extension, *in* Beaumont, C., and Tankard, A. J., eds., Sedimentary basins and basin forming mechanisms: Canadian Society of Petroleum Geologists, Memoir 12, p. 241–258.

——, 1989a, Contrasting styles of lithospheric extension: Implications for differences between the Basin and Range province and rifted continental margins, *in* Tankard, A. J., and Balkwill, H. R., eds., Extensional tectonics and stratigraphy of the North Atlantic margins: American Association of Petroleum Geologists, Memoir 46, p. 53–79.

——, 1989b, Dynamical models of the role of crustal shear zones in asymmetric continental extension: Earth and Planetary Science Letters, v. 93, p. 405–423.

——, 1989c, A physical explanation of the relation between flank uplifts and the breakup unconformity at rifted continental margins: Geology, v. 17, p. 760–764.

Byerlee, J. D., 1968, Brittle-ductile transition in rocks: Journal of Geophysical Research, v. 73, p. 4741–4750.

Carter, N. L., and Tsenn, M. C., 1987, Flow properties of continental lithosphere: Tectonophysics, v. 136, p. 27–63.

Chadwick, R. A., Livermore, R. A., and Penn, I. A., 1989, Continental extension in southern Britain and surrounding areas and its relationship to the opening of the North Atlantic, *in* Tankard, A. J., and Balkwill, H. R., eds., Extensional tectonics and stratigraphy of the North Atlantic margins: American Association of Petroleum Geologists, Memoir 46, p. 411–424.

Chen, W. P., and Molnar, P., 1983, Focal depths of intracontinental and intraplate earthquakes and their implications for the thermal and mechanical properties of the lithosphere: Journal of Geophysical Research, v. 88, p. 4183–4214.

Chery, J., Lucazeau, F., Daignieres, M., and Vilotte, J. P., 1990, The deformation of continental crust in extensional zones: A numerical approach, *in* Pinet, B., and Bois, C., eds., The potential for deep seismic profiling for hydrocarbon exploration: Paris, Institut Francais de Petrole, Edition Technip, p. 35–44.

Courtillot, V., 1982, Propagating rifts and continental breakup: Tectonics, v. 1, p. 239–250.

Croker, P. F., and Klemperer, S. L., 1989, Structure and stratigraphy of the Porcupine Basin: Relationships to deep crustal structure and the opening of the North Atlantic, *in* Tankard, A. J., and Balkwill, H. R., eds., Extensional tectonics and stratigraphy of the North Atlantic margins: American Association of Petroleum Geologists, Memoir 46, p. 445–460.

De Graciansky, P. C., and 14 others, 1985, The Goban Spur transect: Geological evolution of a sediment-starved passive continental margin: Geological Society of America Bulletin, v. 96, p. 58–76.

de Voogd, B., and Keen, C. E., 1987, Lithoprobe East: Results from reflection profiling of the continental margin: Grand Banks region: Royal Astronomical Society Geophysical Journal, v. 89, p. 195–200.

Dunbar, J. A., and Sawyer, D. S., 1987, Implications of continental crust extension for plate reconstruction: An example from the Gulf of Mexico: Tectonics, v. 6, p. 739–755.

—— , 1988, Continental rifting at pre-existing lithospheric weaknesses: Nature, v. 333, p. 450–452.

—— , 1989a, Effects of continental heterogeneity on the distribution of extension and shape of rifted continental margins: Tectonics, v. 8, p. 1059–1078.

—— , 1989b, How preexisting weaknesses control the style of continental breakup: Journal of Geophysical Research, v. 94, p. 7278–7292.

Garcia-Mondejar, J., 1989, Strike-slip subsidence of the Basque-Cantabrian Basin of northern Spain and its relationship to Aptian-Albian opening of Bay of Biscay, *in* Tankard, A. J., and Balkwill, H. R., eds., Extensional tectonics and stratigraphy of the North Atlantic margins: American Association of Petroleum Geologists, Memoir 46, p. 395–410.

Goetze, C., and Evans, B., 1979, Stress and temperature in the bending lithosphere as constrained by experimental rock mechanics: Royal Astronomical Society Geophysical Journal, v. 59, p. 463–478.

Grant, A. C., and McAlpine, K. D., 1990, The continental margin around Newfoundland, *in* Keen, M. J., and Williams, G. L., eds., Geology of the continental margin of eastern Canada: Boulder, Colorado, Geological Society of America, Geology of North America, v. I-1, p. 241–292.

Grantz, A., Eittreim, S., and Dinter, D. A., 1979, Geology and tectonic development of the continental margin north of Alaska: Tectonophysics, v. 59, p. 263–291.

Grow, J. A., and Sheridan, R. E., 1988, U.S. Atlantic continental margin: A typical Atlantic-type or passive continental margin, *in* Sheridan, R. E., and Grow, J. A., eds., The Atlantic continental margin: U.S.: Boulder, Colorado, Geological Society of America, Geology of North America, v. I-2, p. 1–7.

Harry, D. L., and Sawyer, D. S., 1992a, Basaltic volcanism, mantle plumes, and the mechanics of rifting: The Paraná flood basalt province of South America: Geology, v. 20, p. 207–210.

—— , 1992b, A dynamic model of lithospheric extension in the Baltimore Canyon Trough region: Tectonics, v. 11, p. 420–436.

Harry, D. L., Sawyer, D. S., and Leeman, W. P., 1993, The mechanics of continental extension in Western North America: Implications for the magmatic and structural evolution of the Great Basin: Earth and Planetary Science Letters, v. 117, p. 59–71.

Hellinger, S. J., and Sclater, J. G., 1983, Some comments on two-layer extensional models for the evoluton of sedimentary basins: Journal of Geophysical Research, v. 88, p. 8251–8269.

Heyman, M.A.W., 1989, Tectonic and depositional history of the Moroccan continental margin, *in* Tankard, A. J., and Balkwill, H. R., eds., Extensional tectonics and stratigraphy of the North Atlantic margins: Memoir 46, p. 323–340.

Hinz, K., 1981, A hypothesis on terrestrial catastrophes: Wedges of very thick oceanward dipping layers beneath passive continental margins—Thin origin and paleoenvironmental significance: Geologisches Jahrbuch Reihe E, Geophysik, H., v. 22, p. 3–28.

Holbrook, W. S., and Kelemen, P. B., 1993, Large igneous province on the U.S. Atlantic margin and implications for magmatism during continental breakup: Nature, v. 364, p. 433–436.

Houseman, G., and England, P., 1986, A dynamical model of lithosphere extension and sedimentary basin formation: Journal of Geophysical Research, v. 91, p. 719–729.

Keen, C. E., and Boutilier, R., 1990, Geodynamic modelling of rift basins: The synrift evolution of a simple half-graben, *in* Pinet, B., and Bois, C., eds., The potential of deep seismic profiling for hydrocarbon exploration: Paris, Institut Francais de Petrole, Edition Technip, p. 23–33.

Keen, M. J., 1969, Possible edge effect to explain magnetic anomalies off the eastern seaboard of the U.S.: Nature, v. 222, p. 72–74.

Klitgord, K. D., and Behrendt, J. C., 1979, Basin structure of the U.S. Atlantic margin, *in* Watkins, J. S., Montadert, L., and Dickerson, P. W., eds., Geological and Geophysical Investigations of Continental Margins: American Association of Petroleum Geologists Memoir 29, p. 85–112.

Klitgord, K. D., and Schouten, H., 1986, Plate kinematics of the central Atlantic, *in* Vogt, P. R., and Tucholke, B. E., eds., The Western North Atlantic region: Boulder, Colorado, Geological Society of America, Geology of North America, p. 351–378.

Klitgord, K. D., Hutchinson, D. R., and Schouten, H., 1988, U.S. Atlantic continental margin: Structural and tectonic framework, *in* Sheridan, R. E., and Grow, J. A., eds., The Atlantic continental margin: Boulder, Colorado, Geological Society of America, Geology of North America, v. I-2, p. 19–55.

Larsen, H. C., and Sawyer, D. S., 1991, North Atlantic rifted continental margins: College Station, Texas, Ocean Drilling Program, 96 p.

LASE Study Group, 1986, Deep structure of the U.S. east coast passive margin from large aperture seismic experiments (LASE): Marine and Petroleum Geology, v. 3, p. 234–242.

Lawver, L. A., and Scotese, C. R., 1990, A review of tectonic models for the evolution of the Canada Basin, *in* Grantz, A., Johnson, L., and Sweeney, J. F., eds., The Arctic Ocean region: Boulder, Colorado, Geological Society of America, Geology of North America, v. L, p. 593–618.

Lister, G. S., Etheridge, M. A., and Symonds, P. A., 1986, Detachment faulting and the evolution of passive continental margins: Geology, v. 14, p. 245–250.

MacLean, B., Williams, G. L., and Srivastava, S. P., 1990, Geology of Baffin Bay and Davis Strait, *in* Keen, M. J., and Williams, G. L., eds., Geology of the continental margin of Eastern Canada: Boulder, Colorado, Geological Society of America, Geology of North America, v. I-1, p. 325–348.

Mauffret, A., Mougenot, D., Miles, P. R., and Malod, J. A., 1989, Results from multichannel reflection profiling on the Tagus Plain (Portugal)—Comparison with the Canadian margin, *in* Tankard, A. J., and Balkwill, H. R., eds., Extensional tectonics and stratigraphy of the North Atlantic margins: American Association of Petroleum Geologists, Memoir 46, p. 379–393.

McKenzie, D., 1978, Some remarks on the development of sedimentary basins: Earth and Planetary Science Letters, v. 40, p. 25–32.

McKenzie, D., and Bickle, M. J., 1988, The volume of melt generated by extension of the lithosphere: Journal of Petrology, v. 29, p. 625–679.

Murillas, J., Mougenot, D., Boillot, G., Comas, M. C., Banda, E., and Mauffret, A., 1990, Structure and evolution of the Galicia interior basin (Atlantic western Iberian continental margin): Tectonophysics, v. 184, p. 297–319.

Mutter, J. C., 1985, Seaward dipping reflectors and the continent-ocean boundary at passive continental margins: Tectonophysics, v. 114, p. 117–131.

Mutter, J. C., Talwani, M., and Stoffa, P. L., 1982, Origin of seaward-dipping reflectors in oceanic crust off the Norwegian margin by "subaerial sea-floor spreading": Geology, v. 10, p. 353–357.

—— , 1984, Evidence for a thick oceanic crust adjacent to the Norwegian margin: Journal of Geophysical Research, v. 89, p. 483–502.

Mutter, J. C., Buck, W. R., and Zehnder, C. M., 1988, Convective partial melting 1. A model for the formation of thick basaltic sequences during the initiation of spreading: Journal of Geophysical Research, v. 93, p. 1031–1048.

Oh, J., Phillips, J. D., Austin, J. A., Jr., and Stoffa, P. L., 1991, Deep penetration reflection profiling across the southeastern United States continental margin, *in* Meissner, R. O., and 5 others, eds., Continental lithosphere: Deep seismic reflections: American Geophysical Union, Geodynamics Series, v. 22, p. 225–240.

Palmason, G., 1973, Kinematics and heat flow in a volcanic rift zone, with application to Iceland: Royal Astronomical Society Geophysical Journal, v. 33, p. 451–481.

—— , 1980, A continuum model of crustal generation in Iceland; kinematic aspects: Journal of Geophysics, v. 47, p. 7–18.

Petrie, S. H., Brown, J. R., Granger, R. J., and Lovell, J.P.B., 1989, Mesozoic history of the Celtic Sea basins, *in* Tankard, A. J., and Balkwill, H. R., eds., Extensional tectonics and stratigraphy of the North Atlantic margins: American Association of Petroleum Geologists, Memoir 46, p. 433–444.

Pindell, J. L., 1985, Alleghenian reconstruction and subsequent evolution of the Gulf of Mexico, Bahamas, and proto-Caribbean: Tectonics, v. 4, p. 1–39.

Pindell, J. L., and Dewey, J. D., 1982, Permo-Triassic reconstructions of western Pangea and the evolution of the Gulf of Mexico/Caribbean region: Tectonophysics, v. 1, p. 179–211.

Roest, W. R., and Srivastava, S. P., 1989, Sea-floor spreading in the Labrador Sea: A new reconstruction: Geology, v. 17, p. 1000–1004.

—— , 1991, Kinematics of the plate boundaries between Eurasia, Iberia, and Africa in the North Atlantic from the Late Cretaceous to present: Geology, v. 19, p. 613–616.

Rowley, D. B., and Sahagian, D., 1986, Depth-dependent stretching—A different approach: Geology, v. 14, p. 32–35.

Royden, L., and Keen, C. E., 1980, Rifting process and thermal evolution of continental margin of eastern Canada determined from subsidence curves: Earth and Planetary Science Letters, v. 51, p. 343–361.

Royden, L. H., Horvath, F., and Burchfiel, B. C., 1982, Transform faulting, extension, and subduction in the Carpathian Pannonian region: Geological Society of America Bulletin, v. 93, p. 717–725.

Salvador, A., 1991, Origin and development of the Gulf of Mexico basin, *in* Salvador, A., ed., The Gulf of Mexico basin: Boulder, Colorado, Geological Society of America, Geology of North America, v. J, p. 389–444.

Sawyer, D. S., and Harry, D. L., 1991, Dynamic modeling of divergent margin formation: Application to the U.S. Atlantic margin: Marine Geology, v. 102, p. 29–42.

Sclater, J. G., Jaupart, C., and Galson, D., 1980, The heat flow through oceanic and continental crust and the heat loss of the Earth: Reviews of Geophysics and Space Physics, v. 18, p. 269–311.

Sheridan, R. E., and eight others, 1993, Deep seismic refelction data of EDGE U.S. mid-Atlantic continental-margn experiment: Implications for Appalachian sutures and Mesozoic rifting and magmatic underplating: Geology, v. 21, p. 563–567.

Srivastava, S. P., and Tapscott, C. R., 1986, Plate kinematics of the North Atlantic, *in* Vogt, P. R., and Tucholke, B. E., eds., The western North Atlantic region: Boulder, Colorado, Geological Society of America, Geology of North America, v. M, p. 379–404.

Srivastava, S. P., and Verhoef, J., 1992, Evolution of Mesozoic sedimentary basins around the North Central Atlantic: A preliminary plate kinematic solution, *in* Parnell, J., ed., Basins on the Atlantic seaboard: Petroleum geology, sedimentology and basin evolution: Geological Society (London), Special Publication 62, p. 397–420.

Srivastava, S. P., Verhoef, J., and MacNab, R., 1988, Results from a detailed aeromagnetic survey across the northeast Newfoundland margin, Part II: Early opening of the North Atlantic between the British Isles and Newfoundland: Marine and Petroleum Geology, v. 5, p. 324–337.

Srivastava, S. P., Roest, W. R., Kovacs, L. C., Levesque, S., Verhoef, J., and MacNab, R., 1990, Motion of Iberia since the Late Jurassic: Results from detailed aeromagnetic measurements in the Newfoundland Basin: Tectonophysics, v. 184, p. 229–260.

Steckler, M. S., and Watts, A. B., 1978, Subsidence of the Atlantic-type continental margin off New York: Earth and Planetary Science Letters, v. 41, p. 1–13.

—— , 1980, The Gulf of Lion: Subsidence of a young continental margin: Nature, v. 287, p. 425–429.

Talwani, M., Mutter, J. C., and Hinz, K., 1983, Ocean continent boundary under the Norwegian continental margin, *in* Bott, M.H.P., Saxov, S., Talwani, M., and Thiede, J., eds., Structure and development of the Greenland-Scotland Ridge: New methods and concepts: New York, Plenum Press, p. 121–132.

Talwani, M., Ewing, J., Sheridan, R. E., Musser, D. L., Glover, L., Holbrook, S., and Purdy, M., 1992, EDGE lines of the U.S. Mid-Atlantic margin and the East Coast Magnetic Anomaly: Eos (Transactions, American Geophysical Union), v. 73, p. 490.

Tankard, A. J., and Welsink, H. J., 1989, Mesozoic extension and styles of basin formation in Atlantic Canada, *in* Tankard, A. J., and Balkwill, H. R., eds., Extensional tectonics and stratigraphy of the North Atlantic margins: American Association of Petroleum Geologists, Memoir 46, p. 175–195.

Tankard, A. J., Balkwill, H. R., and Jenkins, W.A.M., 1989, Structure styles and stratigraphy of the Jeanne d'Arc basin, Grand Banks of Newfoundland, *in* Tankard, A. J., and Balkwill, H. R., eds., Extensional tectonics and stratigraphy of the North Atlantic margins: American Association of Petroleum Geologists, Memoir 46, p. 265–282.

Trehu, A. M., Ballard, A., Dorman, L. M., Gettrust, J. F., Klitgord, K. D., and Schreiner, A., 1989, Structure of the lower crust beneath the Carolina trough, U.S. Atlantic continental margin: Journal of Geophysical Research, v. 94, p. 585–600.

Tucholke, B. E., Austin, J. A., Jr., and Uchupi, E., 1989, Crustal structure and rift-drift evolution of the Newfoundland basin, *in* Tankard, A. J., and Balkwill, H. R., eds., Extensional tectonics and stratigraphy of the North Atlantic margins: American Association of Petroleum Geologists, Memoir 46, p. 247–263.

Wade, J. A., and MacLean, B. C., 1990, The geology of the southeastern margn of Canada, *in* Keen, M. J., and Williams, G. L., eds., Geology of the continental margin of eastern Canada: Boulder, Colorado, Geological Society of America, Geology of North America, v. I-1, p. 169–238.

Welsink, H. J., Dwyer, J. D., and Knight, R. J., 1989, Tectono-stratigraphy of passive margin off Nova Scotia, *in* Tankard, A. J., and Balkwill, H. R., eds., Extensional tectonics and stratigraphy of the North Atlantic margins: American Association of Petroleum Geologists, Memoir 46, p. 215–232.

Wernicke, B., 1981, Low-angle normal faults in the Basin and Range province: Nappe tectonics in an extending orogen: Nature, v. 291, p. 645–648.

—— , 1985, Uniform-sense normal simple shear of the continental lithosphere: Canadian Journal of Earth Sciences, v. 22, p. 108–125.

Wernicke, B., and Tilke, P. G., 1989, Extensional tectonic framework of the U.S. central Atlantic passive margin, *in* Tankard, A. J., and Balkwill, H. R., eds., Extensional tectonics and stratigraphy of the North Atlantic margins: American Association of Petroleum Geologists, Memoir 46, p. 7–21.

White, R., and McKenzie, D., 1989, Magmatism at rift zones: The generation of volcanic rifted margins and flood basalts: Journal of Geophysical Research, v. 94, p. 7685–7730.

Whitmarsh, R. B., Miles, P. R., and Mauffret, A., 1990, The ocean-continent boundary off the western continental margin of Iberia 1. Crustal structure at 40°30′N: Geophysical Journal, v. 103, p. 509–531.

Wilson, R.C.L., Hiscott, R. N., Willis, M. G., and Gradstein, F. M., 1989, The Lusitanian basin of west-central Portugal: Mesozoic and Tertiary tectonic stratigraphy and subsidence history, *in* Tankard, A. J., and Balkwill, H. R., eds., Extensional tectonics and stratigraphy of the North Atlantic margins: American Association of Petroleum Geologists, Memoir 46, p. 341–362.

MANUSCRIPT ACCEPTED BY THE SOCIETY JULY 28, 1993

Printed in U.S.A.

DNAG Continent-Ocean Transect Volume
Phanerozoic Evolution of North American
Continent-Ocean Transitions
The Geological Society of America, 1994

Chapter 12

Modern active oceanic margins of North America

Roland von Huene*
U.S. Geological Survey, 345 Middlefield Road, Menlo Park, California 94025
Robin Riddihough
Geological Survey of Canada, 601 Booth St., Ottawa, Ontario K1A 0E8, Canada

INTRODUCTION

Modern convergent and strike-slip margins are crossed by six continent/ocean transects (Fig. 1). Along these margins, continental and oceanic crust are separated by a seismically active tectonic contact, which is the boundary between the North American Plate and the Pacific or Cocos Plates. Four transects cross convergent margins where oceanic crust underthrusts continental crust (A-2, B-2, B-3, and H-2[1]). One transect (B-1) is across the Queen Charlotte Transform Fault, a part of the Pacific–North American plate boundary, which is as long as the better-known San Andreas Fault. Another transect (A-3) crosses a collision zone and the controversial transition fault zone, which is interpreted by some as a highly oblique thrust and by others as inactive during the past 5 m.y. or more, and only recently an active fault.

We summarize here the tectonic interpretations of these boundaries made by the transect teams, and results from work completed after the transects were assembled. After transect compilation, the multichannel seismic records across active margins in A-2 and H-2 were reprocessed, and a new record was acquired in the B-2 transect area.

MODERN ACTIVE MARGIN DATA

Within the North American continent/ocean transects, the rocks making up convergent margins, from the tectonically active areas of the lower slope to inactive segments landward, are diverse rather than a gradual progression of increasingly older accreted rocks. Accretion in the tectonically active lower continental slopes at the margins was accompanied by subcrustal accretion of subducted sediment along the subduction zone below the forearc, and periods of tectonic erosion left rocks with vastly different ages juxtaposed. Structural differences between the four convergent margins were probably linked to local sediment supply, rates and direction of convergence, or incoming sea-floor topography along each zone of convergence. The following descriptions give an overview of structure along each active-margin part of the transects.

Kodiak to Kushkokwim Alaska, Transect A-2

The A2 transect (von Huene and others, 1985) crosses the Aleutian Trench where the convergence rate between the North American and Pacific Plates is 64 km/m.y. along a vector normal to the trend of the trench axis (Engebretson and others, 1985). At DSDP (Deep Sea Drilling Project) Site 178 north of the seaward end of the transect, the igneous rock of the Pacific Plate is overlain first by thin (20 to 30 m) Paleogene pelagic sediment and then by a 1- to 2-km-thick sequence of upper Miocene through Quaternary mud and silt turbidites of the Surveyor deep-sea fan. Much of the sediment of the fan sampled at DSDP Site 178 is younger than 5 Ma (von Huene and Kulm, 1973). Only minor faulting accompanies down-bending of the Pacific plate into the Aleutian Trench, and in seismic records most faults are masked by the thick sediment. Magnetic anomaly 21 (47 to 50 Ma) has been identified over ocean crust near the seaward end of the transect (Naugler and Wageman, 1973).

The trench axis has received a turbidite sequence consisting principally of mud with silt turbidites at DSDP site 180, about 300 km northeast of the transect (von Huene and Kulm, 1973). Because the transect is farther from the mainland Alaska source area than site 180, much of the trench fill is probably of a fine-grained facies. However, input from Surveyor Channel, which crosses the Gulf of Alaska and empties into the trench just upstream of the transect, and a local glacial channel crossing the adjacent shelf may have contributed coarse material. In the transect area, trench fill is about 1 km thick. The sediments entering the subduction zone, consisting of 1 to 2 km of Surveyor fan strata and the 1 km of trench-fill strata, are thus about 3 km thick. The trench is sediment-flooded.

The oceanic crust can be followed in seismic reflection and

*Present address: GEOMAR, Research Center, Marine Geosciences, Division of Oceanic Geodynamics, Christian-Albrechts University, Wischhofstrasse 1–3, 2300 Kiel 14, Germany.

[1]Transect H-2 is unpublished.

von Huene, R., and Riddihough, R., 1994, Modern active oceanic margins of North America, *in* Speed, R. C., ed., Phanerozoic Evolution of North American Continent-Ocean Transitions: Boulder, Colorado, Geological Society of America, DNAG Continent-Ocean Transect Volume.

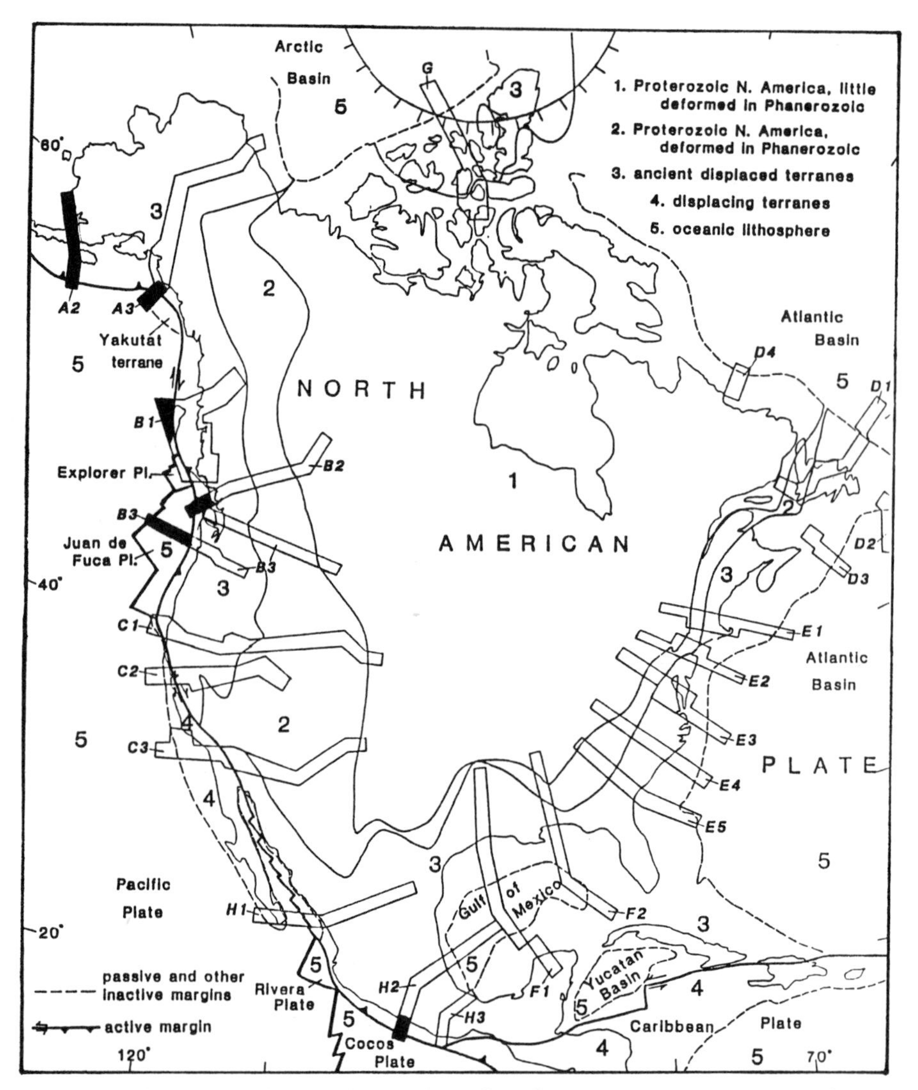

Figure 1. Map of North America showing locations of continent/ocean transects. Sectors of transects that cross active oceanic margins are shown in black. These transects are: A-2, Kodiak to Kuskokwim, Alaska (Von Huene and others, 1985); A-3, Gulf of Alaska to Beaufort Sea (Grantz and others, 1988); B-1, Intermontane Belt (Skeena Mountains) to Insular Belt (Queen Charlotte Islands) (Yorath and others, 1985); B-2, Juan de Fuca Plate to Alberta Plains (Monger and others, 1985); Juan de Fuca Spreading Ridge to Montana Thrust Belt (Cowan and others, 1986); and H-2, Acapulco Trench to Gulf of Mexico.

refraction data to beneath Kodiak Island (von Huene and others, 1985). Initially, it dips only 4°. Accretion is first recognized where the upper half of the sediment stack is folded and faulted above a décollement and the lower half is subducted below it (Fig. 2). In the first 20 km of the subduction zone, the subducted strata undergo no significant disruption. The lack of disruption indicates a reduction of friction at the décollement. Low friction is possible if pore fluids become overpressured; quantitative estimates of pore-fluid pressure indicate that at the front of the subduction zone the décollement is at 95 percent of lithostatic pressure (Davis and von Huene, 1987).

The accreted sediment above the décollement is tectonically thickened by folding, tilting, and thrust faulting and covered by slope sediment, as illustrated in the seismic section (Fig. 2). Thus, the original 1.5-km-thick section above the décollement is thickened upon accretion by compressional deformation and sedimentation to perhaps 4 km. The seismic data do not illustrate how thickening proceeds landward of the lower slope but does show little active folding and faulting there. The continued uplift landward is attributed to underplating (Fisher and von Huene, 1980).

The mid-slope structural boundary between the areas of rapid and slow deformation is also the inferred boundary between the Pilocene(?) accretion complex and an Eocene accreted sequence. This is also where the slope steepens from 3° or 5° to 15°, indicating a major structural break. The interval of undated rocks between the lower-slope Pliocene and the upper-slope Eocene strata is insufficient to contain a continuous progression of materials representing 35 m.y. of accretion from Oligocene to Pliocene time, and these sediments are presumed to be missing. This interpretation is consistent with similar but better-displayed structural breaks along the Shumagin and central segments of the Aleutian Trench (Bruns and others, 1987; Scholl and others, 1987; McCarthy and Scholl, 1985; von Huene and others, 1986a,b). Since the area close to the Alaskan mainland probably received considerable sediment in Oligocene and Miocene time, tectonic erosion is assumed. The missing material was probably subducted, because convergence was never highly oblique during that period (Engebretson and others, 1985). Some of the subducted material may have been underplated, which would help explain the uplift and erosion of the adjacent shelf during early and middle Miocene time (von Huene and others, 1988).

The Gulf of Alaska to the Beaufort Sea, Transect A-3

The continental margin of the central Gulf of Alaska, where transected by corridor A3, makes the transition from the Pacific Ocean basin to continental crust across the Yakutat Terrane. The Yakutat Terrane is currently subducting and colliding with North America in the northern Gulf of Alaska (Fig. 3). The terrane is bordered on the northeast by the Fairweather and Queen Charlotte Faults, on the north by the Chugach–St. Elias Thrust Fault system, on the west by the Kayak Zone, and on the south by the Transition Fault (Bruns, 1983a, 1985; Bruns and others, 1987; Plafker, 1983, 1987; Perez and Jacob, 1979). Convergence is roughly normal to the plate boundary along the Kayak Zone, a 60- to 70-km-wide zone of compressional deformation across the continental shelf and slope (Bruns, 1985) that is an extension of the Aleutian Trench (Fig. 3). The direction of convergence across the Chugach–St. Elias Thrust-Fault system is oblique. The southern boundary is the Transition Fault, which separates the terrane from the Pacific Plate. Associated with the Transition Fault is a magnetic anomaly that also extends 160 km west of the Aleutian Trench and marks the subducted part of the Transition Fault and thus the subducted southern edge of the Yakutat Terrane (Bruns, 1985; Schwab and others, 1980).

In the past 5 m.y., displacement along the Transition Fault has produced relatively small structures such as those imaged in seismic records (Fig. 3). The age of the downward flexure of the igneous oceanic crust against the fault occurred prior to 5 Ma and is similar to the structure along the Queen Charlotte Islands Fault. On the landward side of the Transition Fault, the locally imaged deformation is considerably less than at the base of the continental slope along the Queen Charlotte Island Fault. At the eastern end of the Yakutat Terrane, where the Transition Fault crosses the continental slope and shelf to join the Fairweather–Queen Charlotte Fault System (Fig. 3), no deformation was detected in seismic records that might indicate 5 m.y. of Pacific–North American Plate motion along the Transition Fault (von Huene and others, 1979; Bruns, 1983a, 1985). However, the sea floor has a nearly 2-km-elevation difference across the southern boundary of the Yakutat Terrane, and the crust at this boundary thickens landward and has a continental velocity structure (Bayer and others, 1977; von Huene and others, 1979). The strike-slip Queen Charlotte Fault separates crusts with similar differences, whereas the Aleutian Trench is associated with a more gradual thickening of the continental crust (von Huene and others, 1979; see A-2 transect, Fig. 1). The character of the deformation along the Transition Fault has greater similarity to the adjacent strike-slip Queen Charlotte margin than to the Aleutian Trench subduction zone. From seismic reflection records, the Transition Fault appears to have been locked, at least locally, after an early period of activity; it may have only recently become active again.

Two proposed models for the late Cenozoic tectonic evolution of the northern Gulf of Alaska and for the displacement history along the Transition Fault differ principally in the migration history of the Yakutat Terrane. In one model, the Transition Fault has been locked, and the Yakutat Terrane has traveled as a part of the Pacific Plate during at least Pliocene to Quaternary time and possibly since about Eocene time (Bruns, 1983a, 1985). The model is supported by the seismic reflection records showing little deformation along the fault except locally (von Huene and others, 1979; Bruns, 1983a, 1985). Paleoclimates inferred by a comparison of microfossil assemblages from the terrane and from the stable North American continent indicate migration of the terrane past California, Oregon, and Washington (Keller and others, 1984). The other model is based on inferred source areas for Yakutat Terrane rocks from British Columbia (Plafker, 1983, 1987). In this model, the Yakutat Terrane migrated from about the latitude of Dixon Entrance in British Columbia, northwest along Alaska, between two transform faults, rotated counterclockwise 20°, and has been in place for about the last 15 m.y. The Pacific Plate would then have been subducted obliquely at the Transition Fault since emplacement of the terrane. Recent seismic activity along the Transition Fault includes two earthquakes that have thrust focal mechanisms and a magnitude of six and greater (Gawthrop and others, 1973).

A paleomagnetic observation on rocks from an industry well in the terrane indicates northward transport from the latitude of Seattle, which falls between the points of origin suggested in the two models (van Alstine and others, 1985).

Trying to resolve the apparent conflict between the models requires reevaluation of the seismic reflection records indicating inactivity across the Transition Fault since at least Pliocene time, as well as direct correlation of rocks between the Yakutat Terrane and the Pacific Northwest. The thrust focal mechanism along the Transition Fault and the areas of local deformation can be explained by transfer of deformation from the inboard Chugach–St. Elias fault zone to the outboard Transition Fault. As collision proceeds along the Chugach–St. Elias thrust zone, the resistance to faulting by the thickening crust may have exceeded that at an existing boundary at the base of the continental slope. If such a

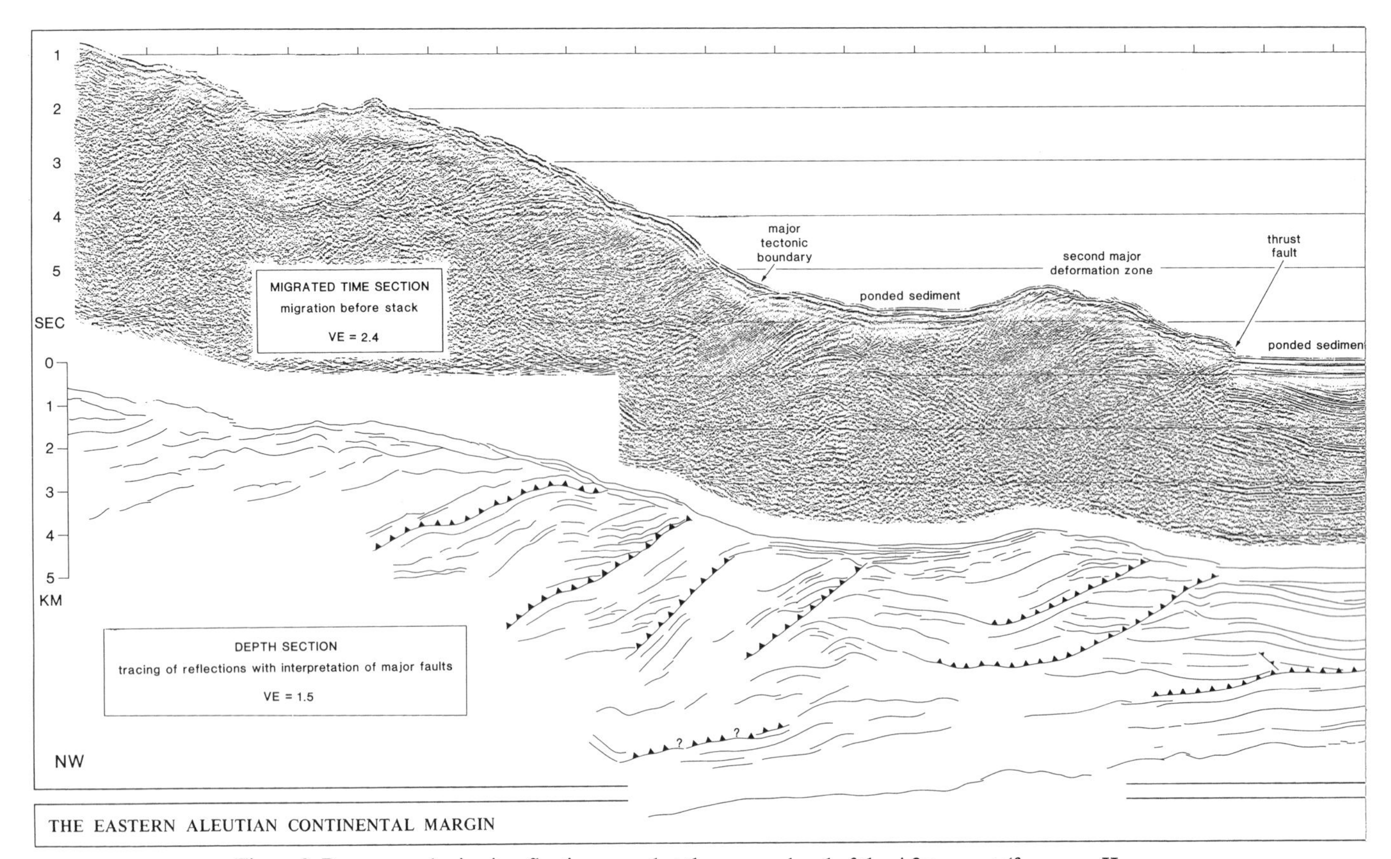

Figure 2. Reprocessed seismic reflection record at the seaward end of the A2 transect (from von Huene and others, 1986b). Upper display is a migrated time secton. Lower display is a line drawing of the depth section showing major reflectors and interpretation of major faults.

jump is presently in progress, it could accommodate both the terrane history of migration with the Pacific Plate and the modern indications of thrusting. The paleomagnetic and paleoclimatic evidence suggest an area of origin for the terrane between northern California and northern Washington. The active plate boundary in the A-3 corridor may have been inland until very recently, and the Transition Fault may have been inactive or principally a strike-slip boundary to which that oblique convergent motion has recently been transferred.

The Intermontane Belt (Skeena Mountains) to the Insular Belt (Queen Charlotte Islands), Transect B-1

A relatively narrow transform system (the Queen Charlotte Fault) marks the boundary between the Pacific Plate and the North American continent in the eastern Gulf of Alaska. Motion along the Queen Charlotte Fault is primarily right-lateral strike slip at between 50 and 60 km/m.y. but may contain a 10 km/m.y. component of convergence off the southern Queen Charlotte Islands. The proportion of strike slip to convergence, given by the relative plate motion vector, as estimated from global plate motions (Minster and Jordan, 1978) depends on the orientation of the fault trace. Convergence may be absent north of the Queen Charlotte Islands. Along the southern Queen Charlotte Islands margin, the fault may have at least two geologically active strands, with a principal inner fault about 15 km offshore and an outer fault at the base of the slope about 35 km offshore. Earthquake foci detected by ocean-bottom seismometers (Hyndman and Ellis, 1981) indicate that the inner fault is the more active and undergoes brittle failures to a depth of over 20 km. Between these two faults is a midslope area that contains en echelon folds and compressional features. Seismic refraction, gravity, and magnetic interpretation suggest the terrace is composed of higher velocity, probably lithified, sediments that are underlain by oceanic crust. Seismic profiling across the margin is limited to single-channel data. A trough at the base of the slope seems to contain normal faults. However, sediments on the Pacific Ocean crust to the southwest are tilted northeastward toward the trough, a tilt that has been interpreted as due either to lithospheric flexure associated with convergence or to the influence of the offshore Kodiak-Bowie Seamount Chain (Oshawa Rise).

Crustal thickness estimates shown in Transect B-1 suggest that if oceanic crust underthrusts the Queen Charlotte Islands and the Wrangellia Terrane, the latter is no more than 10 to 15 km thick. The present convergent regime may have been initiated at only 6 Ma (Yorath and Hyndman, 1983); alternatively, excess

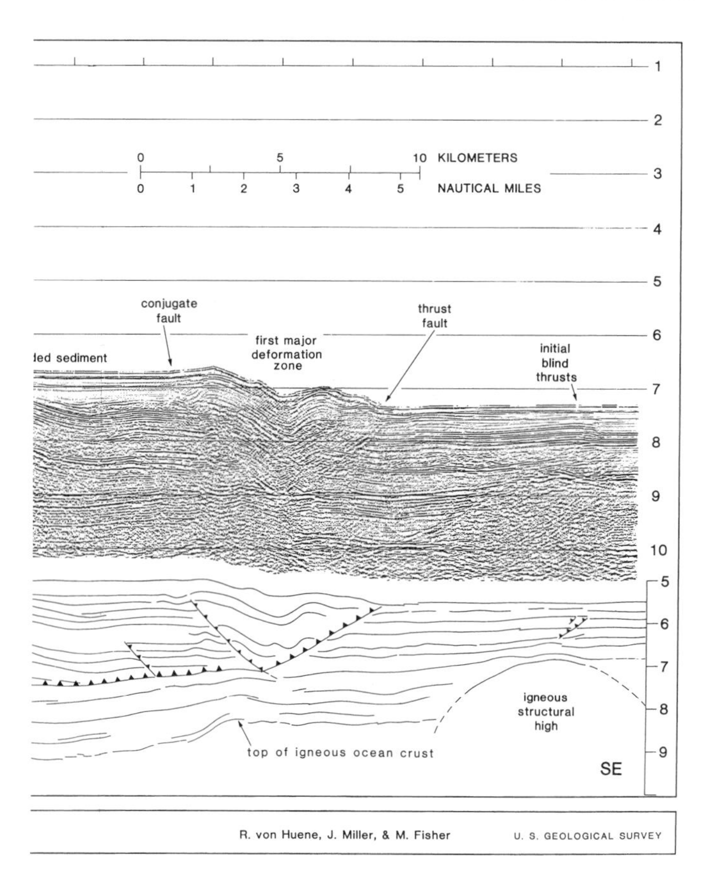

material may be removed and transported laterally along the Queen Charlotte Islands Fault, as has occurred in the past (Riddihough and Hyndman, 1989).

The Juan de Fuca Plate to the Alberta Plains, Transect B-2

The B2 corridor crosses the convergent margin off southern Vancouver Island. Here the 6- to 7-Ma Juan de Fuca Plate is orthogonally underthrusting the terranes that make up the Insular Belt of British Columbia at a rate of about 45 km/m.y. Plate motion calculations based on magnetic anomalies show that this rate has decreased from 60 km/m.y. during the last 10 m.y. The sediment thickness on the incoming plate is approximately 2 km.

From detailed bathymetry, SEAMARC II data, and seismic profiles, it is evident that deformation of the incoming sediments begins with a series of anticlinal ridges of 5 to 800 m height, which occur at the base of the continental slope about 100 km offshore. The ridges were shown by Barnard (1978) to contain deformed upper Pleistocene sediments and are therefore relatively young. Detailed multichannel seismic profiles (Fig. 4; Yorath and Brandon, 1986) show a series of ramped anticlines with underlying landward-dipping thrust faults. A major décollement zone has been identified, beneath which sediments are subducted with the downgoing plate. Below the décollement, the sediment may thicken landward and be highly deformed. Offshore, the décollement is tentatively identified as the boundary between a basal hemipelagic sequence and the overlying Cascadia Fan turbidites observed in DSDP 174 (Kulm and others, 1973).

Offshore seismic reflection records are continuous with a deep onland crustal seismic-reflection record profile, which has imaged the downgoing Juan de Fuca Plate (Clowes and others, 1987). That record shows major underplating at the base of the Wrangellia Terrane and other older terranes, which occurred during the past 40 m.y. The underplated sediment may be partially correlative with the turbidites in the core of the Olympic Peninsula to the southeast. In both areas the turbidite underthrusts a slice of Eocene oceanic crust, the Metchosin Formation.

The Juan de Fuca Spreading Ridge to the Montana Thrust Belt, Transect B-3

The tectonic framework of the convergence zone of this transect is almost identical to that of B-2 to the north. The young (8 Ma) Juan de Fuca Plate converges with the North American continent at a slightly slower rate of 40 to 42 km/m.y., but carries a similar section of Cascadia Fan sediments of 1.8 to 2 km thickness. Because the margin trends north-south, the northeast convergence is right-oblique. The geometry of the descending plate and the overlying underplated oceanic sediments and basalts is similar to that of southern Vancouver Island (B-2).

At the toe of the slope, anticlinal ridges similar to those of the B-2 transect are also observed. However, from single-channel seismic records, the frontal accretionary complex has folds verging landward rather than seaward (Snaveley, 1987; Seely, 1977). The landward-verging folds were interpreted as hanging-wall or ramp anticlines that developed above oceanward-dipping thrusts.

This structure is in direct contrast to the landward-dipping thrusts seen in B-2 and along the Oregon margin to the south. To test the vergence, a deep-tow seismic reflection experiment was carried out (Lewis, 1984), which confirmed the seaward-dipping thrust fault interpretation. The contrast in accretionary style within a short lateral distance in an almost identical tectonic setting provides a target for more detailed multichannel seismic and swath-mapping investigations.

ACAPULCO TRENCH TO THE GULF OF MEXICO, TRANSECT H-2

The H-2 corridor transects the Middle America Trench off the Pacific coast of Mexico. The continental shelf is narrow, and a steep landward trench slope ends at the 5-km-deep trench axis. A large submarine canyon cuts the upper slope and channels much sediment into the trench axis in the vicinity of the multichannel seismic records that are generally used to represent this area (Fig. 5). The abundant trench sediment near the canyon mouth has been accreted into a large wedge of landward-dipping beds and thrust faults, and seismic records farther from the relatively thick trench fill show smaller complexes (Shipley, 1982). Because of its well-imaged accretionary complex in a grid of multichannel seismic reflection records, this area was extensively drilled during DSDP Leg 66 (Watkins and others, 1982).

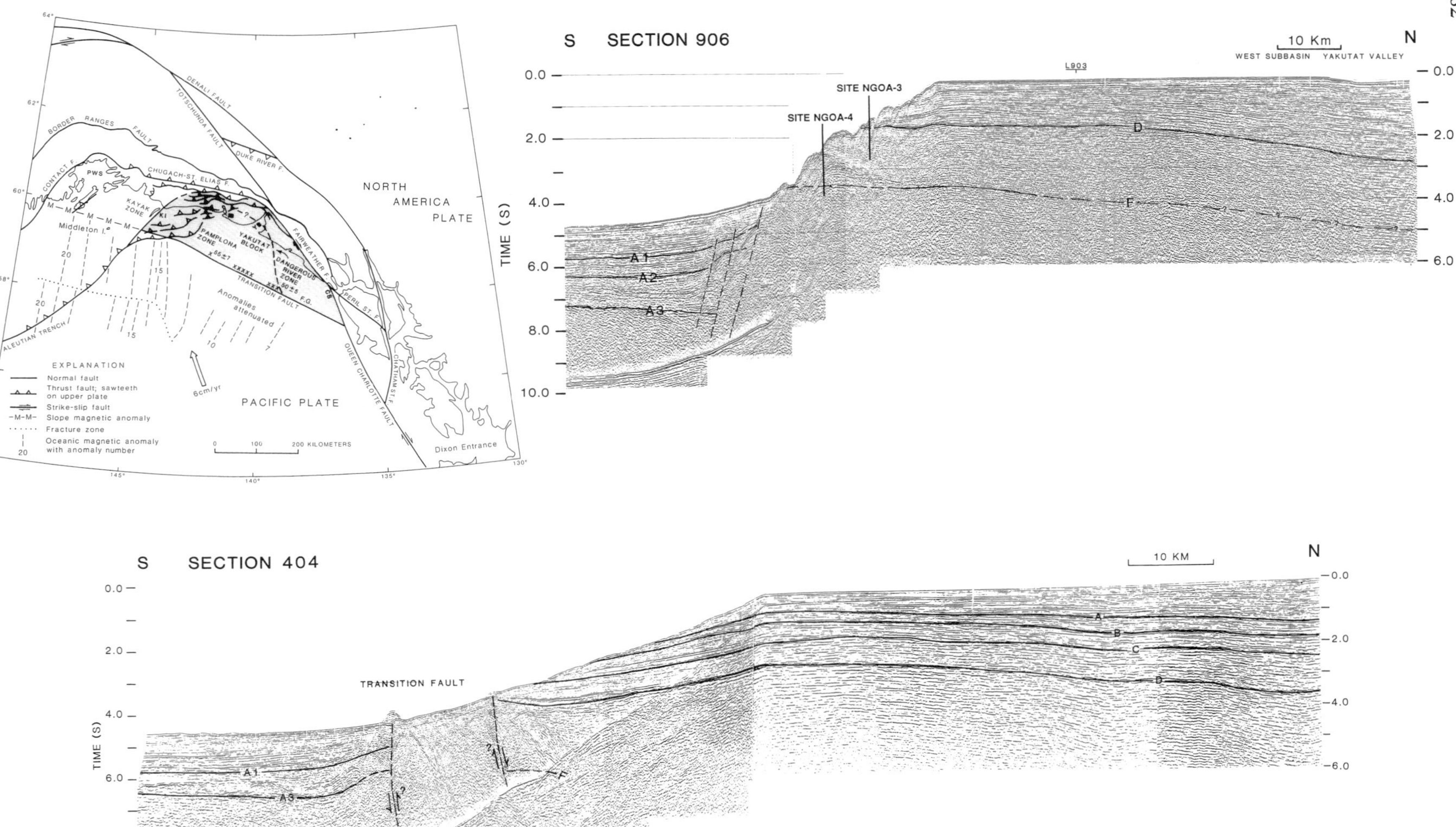

Figure 3. Map showing the Yakutat Terrane in the northern Gulf of Alaska. Two seismic records show the structure of the Transition Fault (from Bruns, 1983a). Letters on highlighted horizons correspond to ages as follows: D, base of upper Miocene; F, lower Eocene; A1, base of Pleistocene; A2, base of Pliocene; A3, basement. A. B. and C are prominent reflectors of unspecified ages.

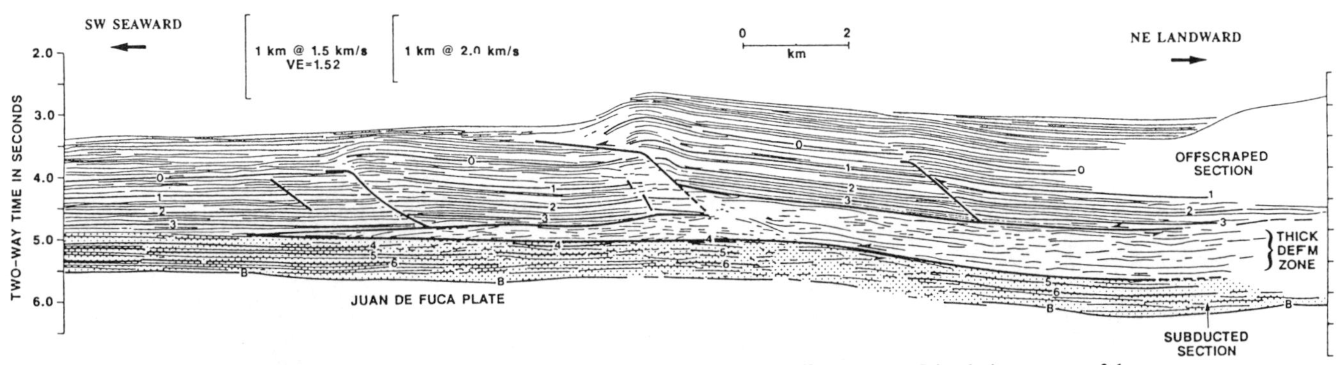

Figure 4. Preliminary line drawing of a seismic reflection record off Vancouver Island, shot as part of the Canadian LITHOPROBE program. Letters and numbers are prominent reflectors of unspecified ages (from M. T. Brandon, written communication, 1986).

The underthrusting Cocos Plate converges relative to the overriding North American Plate at about 70 km/m.y. Leg 66 drilling results suggest the continental margin in the area was truncated between 35 and 10 Ma, by transcurrent faulting or subduction erosion (Moore and others, 1982). An uplift of the coastal mountains, possibly accompanying transcurrent faulting, indicates truncation across the H-2 corridor about 20 to 30 Ma. The accreted complex was subsequently developed from offscraping sediment in addition to underplating as inferred below. No forearc basin such as those on many trench slopes is evident in the seismic lines, and sediment is transported directly from the adjacent continent down the slope.

The results of studies of a network of seismic reflection records (Shipley, 1982), drilling (Moore and others, 1982), and a reprocessed seismic record (Collins and Watkins, 1986) provide much information about this active margin. Sediments about 800 m thick overlie oceanic crust in the trench axis. The seaward third of the seismic section (Fig. 5) resolves the top of subducting igneous oceanic crust. Sediments about 800 m thick overlie oceanic crust in the trench axis. A wedge-shaped packet of sediment at the base of the trench slope has been scraped off the subducting oceanic crust and is the youngest of a sequence of packets forming the trench slope. The numerous landward-dipping reflections below the trench slope reflect the progressive offscraping, uplift, and tilting of sediments initially deposited in the trench axis. Leg 66 drilling results document the increasing age of thrust packets landward. Multiple thrust faults are inferred in the accreted sediment, and they probably occur along bedding planes. No thrust faults were documented biostratigraphically during Leg 66, probably because of the low angles of faulting and high rates of sedimentation, which minimize the age differences across faults.

The accreted complex is thickened at the front of the margin by thrust faulting and tilting of beds. The maximum angle of tilting is achieved about 5 km landward of the trench axis; farther landward, the lower slope accretionary mass responds to tectonism without further tilting. Nonetheless, the structural geometry and the paleobathymetric history from studies of benthic foraminiferal assemblages in the slope cover document continued uplift of the slope and thus continued crustal thickening. To explain continued uplift, underplating is proposed (Moore and others, 1982). About one-half of the accretionary complex is thought to have been underplated. In the reprocessed record, a layer of subducting sediment is visible (Collins and Watkins, 1986). Some of the subducting sediment must therefore detach from the ocean plate beneath the landward part of the accretionary complex.

The accretionary complex is stacked against the truncated continental crust of Mexico. The lithologic and paleobathymetric data from DSDP cores indicate subaerial erosion until 22 Ma of a platform cut into metamorphic rocks equivalent to those exposed on shore, depression of the erosion surface during 3.5 m.y. to about 3 km, and subsequent uplift since 18 Ma during accretion at the front of the margin (Moore and others, 1982). The present accretionary environment was preceded by an erosional one, for which the tectonic conditions are not clear.

DISCUSSION

Information from the six modern active margins included in the continent/ocean transect ranges from model-dependent interpretation based on little more than single-channel seismic records to extensive data bases including reprocessed multichannel data and DSDP drilling. Reprocessing of seismic reflection data, including careful migration, has been shown to greatly enhance the structural information in the convergent margin tectonic setting. However, the greater depth of penetration in a multichannel record that has been processed without migration lacks the picture of clear structural geometry that constrains an interpretation of tectonic processes beneath the lower slope where modern tectonism is affecting the continent most actively. The combination of SEABEAM bathymetry, reprocessed and well-migrated seismic reflection records, and DSDP drill holes as acquired in the H-2 corridor reveals a complex history and changing tectonic pro-

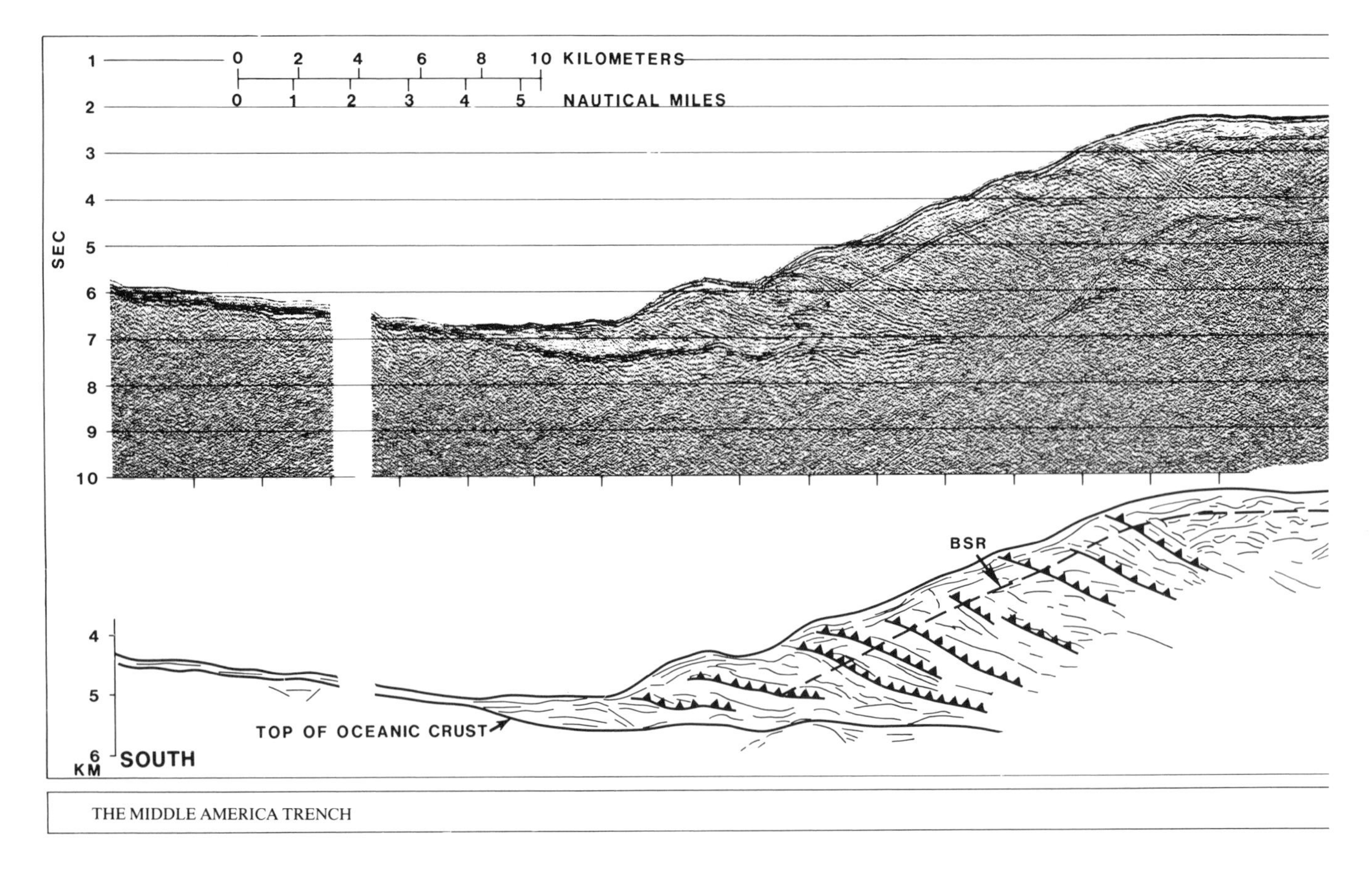

Figure 5. Reprocessed seismic reflection record at the seaward end of the H-2 transect (from Collins and Watkins, 1986). Upper and lower displays are migrated time and depth sections as in Figure 2. BSR, bottom-simulating reflector.

cesses. Here the Mexican conntinental crust, consisting of high-grade metamorphic Paleozoic and Mesozoic rocks intruded by multiple plutonic bodies as young as Cenozoic in age, was subareally eroded to the deep levels where such metamorphism originates. The margin then subsided to about 3 km below sea level, probably simultaneously with tectonic erosion. The erosion may have involved both strike-slip transport as well as subduction erosion because of the approximately 10-m.y. gap in time between proposed strike-slip truncation and the onset of accretion. A major drainage provided the locus for sediment transported to the trench, which locally flooded the margin and promoted accretion since about 12 Ma. Mass-balance calculations indicate the division of the sediment into three equal parts; one part is attached to the continent by frontal accretion, a second by underplating, and the third is transported by subduction well landward of the front of the margin. Thus, at various times in a 22-m.y. period, a broad spectrum of tectonic processes was active in this one area.

Similar tectonic histories may be revealed by data along the other transects once the structural resolution of modern seismic reflection techniques and swath-mapped bathymetric data can be combined with temporal control from studies of drill cores. A general overview of most transects suggests there were dominant tectonic processes through periods of 5 to 20 m.y. The eastern Aleutian margin has been dominated by accretion since the late Miocene and may have been erosional for the preceding 15 to 20 m.y. (von Huene and others, 1987; Bruns and Carlson, 1987). Although addition of material must be the dominant process to maintain the continents, the local geologic record may be broken by gaps in which material was removed and transported elsewhere. One of the better understood tectonic processes is the rifting of a continental area and drifting away of outboard fragments on another plate. This has occurred along the Queen Charlotte margin. Another important process is the subduction of sediment, a process observed along most convergent margins of North America. Less well documented is the tectonic erosion and subduction of parts of the front of convergent margins, a process required by the missing continental borders that have not been found elsewhere as far-travelled fragments. Across southern Vancouver Island, it is clear that if the crustal thickness of the Wrangellia Terrane was ever greater than 10 to 15 km, major subcrustal erosion occurred during the last 40 m.y. Where are the

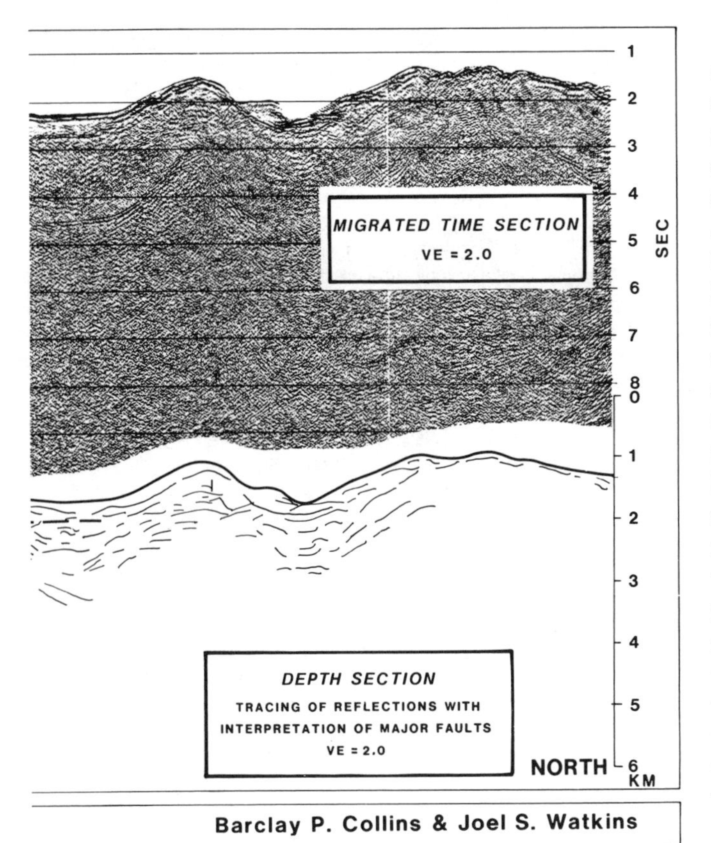

missing Mesozoic and lower Cenozoic active margin rocks that once must have fronted western North America in southern California and the Pacific Northwest where the margin is now formed by a collage of accreted terranes?

The fate of subducted sediment and eroded materials can only be broadly constrained. Volcanic rocks are contaminated by continental rock, but the contaminants constitute a small volume (Kay, 1980). One source of the continental indicators could be erosional products from land that were transported across the shelf and slope to the trench axis and subsequently subducted. Another source could be the subcrustally eroded rocks from the base of the continent. Subcrustal trunction of Cretaceous strata has been imaged in seismic records across the Japan Trench margin (von Huene and others, 1986c), and the 3.5 km of subsidence in 6 to 8 m.y. recorded across the Peru margin indicates subcrustal erosion from beneath the midslope area (von Huene and others, 1987). Most seismic records across the convergent margins of North America that have been processed to research levels show much subducting material. The composition of volcanic rocks indicates that not much material of continental affinity travels down the subduction zone to the source region for volcanic magma. Thus, much material must be underplated beneath the forearc area. Perhaps some areas of uplift are an isostatic adjustment to underplated sediment. Considerable underplating was inferred in the A-2 corridor to account for continued uplift well landward of the compressional deformation that might cause uplift from thickening of the crust, yet well seaward of the arc where intrusion might cause uplift. A similar situation was noted beneath Vancouver Island along the B-2 transect.

Active margin processes seen in the transects include a spectrum from accretion to erosion. Erosion is represented by gaps in the temporal record of a margin. Such gaps are documented convincingly only by dating rock bodies. The addition of materials to the edges of a continent occurs with major breaks; interpretive models of constant accretion, so often applied to active margins, may only be valid for time spans on the order of a few millions of years.

REFERENCES CITED

Barnard, W. D., 1978, The Washington continental slope; Quaternary tectonics and sedimentation: Marine Geology, v. 27, p. 79–114.

Bayer, K. C., Mattick, R. E., Plafker, G., and Bruns, T. R., 1977, Refraction studies between Icy Bay and Kayak Island, Eastern Gulf of Alaska: U.S. Geological Survey Open-File Report 77–550, 29 p.

Bruns, T. R., 1983a, Model for the origin of the Yakutat block; An accretionary terrane in the northern Gulf of Alaska: Geology, v. 11, p. 718–721.

—— , 1983b, Structure and petroleum potential of the Yakutat segment of the northern Gulf of Alaska continental margin: U.S. Geological Survey Miscellaneous Field Studies Map MF–1480, 22 p., 3 sheets, scale 1:500,000.

—— , 1985, Tectonics of the Yakutat block; An allochthonous terrane in the northern Gulf of Alaska: U.S. Geological Survey Open-File Report 85–13, 112 p.

Bruns, T. R., and Carlson, P., 1987, Geology and petroleum potential of the southeastern Alaska continental margin, *in* Scholl, D. W., Grantz, A., and Vedder, J. G., eds., Geology and resource potential of the continental margin of western North America and adjacent ocean basins—Beaufort Sea to Baja California: Houston, Texas, Circum-Pacific Council for Energy and Mineral Resources.

Bruns, T. R., von Huene, R., Culotta, R. C., Lewis, S. D., and Ladd, J. W., 1987, Geology and petroleum potential of the Shumagin Margin, Alaska, *in* Scholl, D. W., Grantz, A., and Vedder, J. G., eds., Geology and resource potential of the continental margin of western North America and adjacent ocean basins—Beaufort Sea to Baja California: Houston, Texas, Circum-Pacific Council for Energy and Mineral Resources Earth Science Series, v. 6, p. 157–189.

Clowes, R. M., and 6 others, 1987, LITHOPROBE–Southern Vancouver Island; Cenozoic subduction complex imaged by deep seismic reflections: Canadian Journal of Earth Sciences, v. 24, p. 31–51.

Collins, B. P., and Watkins, J. S., 1986, The Middle America Trench, *in* von Huene, R., ed., Seismic images of modern convergent margin tectonic structure: American Association of Petroleum Geologists Studies in Geology 26, p. 30–33.

Cowan, D. S., and Potter, C. J., compilers, 1986, B-3, Juan de Fuca Spreading Ridge to Montana Thrust Belt: Boulder, Colorado, Geological Society of America, Centennial Continent-Ocean Transect no. 9, scale 1:500,000.

Davis, D. M., and von Huene, R., 1987, Inferences on sediment strength and fault friction from structures at the Aleutian Trench: Geology, v. 15, p. 517–522.

Engebretson, D. C., Cox, A., and Gordon, R. G., 1985, Relative motions between oceanic and continental plates in the Pacific basin: Geological Society of America Special Paper 206, 59 p.

Fisher, M. A., and von Huene, R., 1980, Structure of upper Cenozoic strata beneath Kodiak Shelf, Alaska: American Association of Petroleum Geologists Bulletin, v. 64, no. 7, p. 1014–1033.

Gawthrop, W. H., Page, R. A., Reichle, M., and Jones, A., 1973, The southeast Alaska earthquake of July 1973: EOS Transactions of the American Geophysical Union, v. 54, no. 11, p. 1136.

Grantz, A., Moore, T. E., and Roeske, S., compilers, 1991, A-3, Gulf of Alaska to Arctic Ocean: Boulder, Colorado, Geological Society of America, Centennial Continent-Ocean Transect no. 15, scale 1:500,000.

Hyndman, R. D., and Ellis, R. M., 1981, Queen Charlotte Fault zone; Microearthquakes from a temporary array of land stations and ocean bottom seismographs: Canadian Journal of Earth Sciences, v. 18, p. 776–788.

Kay, R. W., 1980, Volcanic arc magma genesis; Implications for element recycling in the crust-upper mantle system: Journal of Geology, v. 88, p. 497–522.

Keller, G., von Huene, R., McDougall, K., and Bruns, T. R., 1984, Paleoclimatic evidence for Cenozoic migration of Alaskan Terranes: Tectonics, v. 3, no. 4, p. 473–495.

Kulm, L. D., von Huene, R., eds., 1973, Initial reports of the Deep Sea Drilling Project: Washington, D.C., U.S. Government Printing Office, v. 18, 1077 p.

Lewis, B.T.R., 1984, Deep-tow seismic reflection experiment, Washington Margin, *in* Kulm, L. D., and others, eds., Ocean Margin Drilling Program, Atlas 1; Western North American Continental Margin and Adjacent Ocean Floor off Oregon and Washington: Woods Hole, Massachusetts, Marine Science International.

McCarthy, J., and Scholl, D. W., 1985, Mechanisms of subduction accretion along the central Aleutian trench: Geological Society of America Bulletin, v. 96, p. 691–701.

Minster, J. B., and Jordan, T. H., 1978, Present-day plate motions: Journal of Geophysical Research, v. 63, no. B-11, p. 5331–5354.

Monger, J.W.H., and 5 others, compilers, 1985, B-2, Juan de Fuca Plate to Alberta Plains: Boulder, Colorado, Geological Society of America, Centennial Continent-Ocean Transect no. 7, scale 1:500,000.

Moore, C. J., Watkins, J. S., and Shipley, T. H., 1982, Summary of accretionary processes, Deep Sea Drilling Project Leg 66; Offscraping, underplating, and deformation of the slope apron, *in* Moore, J. C., and Watkins, J. S., eds., Initial reports of the Deep Sea Drilling Project, v. 66: Washington, D.C., U.S. Government Printing Office, p. 825–836.

Naughler, F. P., and Wageman, J. W., 1973, Gulf of Alaska; Magnetic anomalies, fracture zones, and plate interactions: Geological Society of America Bulletin, v. 84, no. 5, p. 1575.

Perez, O. J., and Jacob, K. H., 1979, Tectonic model and seismic potential of the eastern Gulf of Alaska and Yakataga seismic gap: Journal of Geophysical Research, v. 85, no. B-12, p. 7132–7150.

Plafker, G., 1983, The Yakutat block; An actively accreting tectonostratigraphic terrane in southern Alaska: Geological Society of America Abstracts with Programs, v. 15, no. 5, p. 406.

—— , 1987, Regional geology and petroleum potential of the northern Gulf of Alaska continental margin, *in* Scholl, D. W., Grantz, A., and Vedder, J. G., eds., Geology and resource potential of the continental margin of western North America and adjacent ocean basins—Beaufort Sea to Baja California: Houston, Texas, Circum-Pacific Council for Energy and Mineral Resources Earth Science Series, v. 6, p. 229–268.

Riddihough, R. P., and Hyndman, R. D., 1989, The Queen Charlotte Island Margin, *in* Winterer, E. L., Hussong, D. L., and Decker, R. W., eds., The eastern Pacific Ocean and Hawaii: Boulder, Colorado, Geological Society of America, The Geology of North America, v. N, p. 403–412.

Scholl, D. W., Vallier, T. L., and Stevenson, A. J., 1987, Geologic evolution and petroleum geology of the Aleutian Ridge, *in* Scholl, D. W., Grantz, A., and Vedder, J. G., eds., Geology and resource potential of the continental margin of western North America and adjacent ocean basins—Beaufort Sea to Baja California: Houston, Texas, Circum-Pacific Council for Energy and Mineral Resources Earth Science Series, v. 6, p. 123–155.

Schwab, W. C., Bruns, T. R., and von Huene, R., 1980, Maps showing structural interpretation of magnetic lineaments in the northern Gulf of Alaska: U.S. Geological Survey Miscellaneous Field Studies Map MF-1245, 12 p., 3 maps, scale 1:500,000.

Seely, D. R., 1977, The significance of landward vergence and oblique structural trends on trench inner slopes, *in* Talwani, M., and Pitman, W. C., eds., Island arcs, deep-sea trenches, and back-arc basins: American Geophysical Union, Maurice Ewing Series no. 1, p. 187–198.

Shipley, T. H., 1982, Seismic facies and structural framework of the southern Mexico continental margin, *in* Watkins, J. S., Moore, J. C., eds., Initial reports of the Deep Sea Drilling Project, v. 66: Washington, D.C., U.S. Government Printing Office, p. 775–790.

Snaveley, P. D., Jr., 1987, Geologic framework, neotectonics, and petroleum geology of the Oregon-Washington continental margin, *in* Scholl, D. W., Grantz, A., and Vedder, J. G., eds., Geology and resource potential of the continental margin of western North America and adjacent ocean basins—Beaufort Sea to Baja California: Houston, Texas, Circum-Pacific Council for Energy and Mineral Resources, v. 6, p. 305–335.

Snaveley, P. D., Jr., von Huene, R., Mann, D., and Miller, J., 1986, The central Oregon continental margin, lines W076-4 and W076-5, *in* von Huene, R., ed., Seismic images of modern convergent margin tectonic structure: American Association of Petroleum Geologists Studies in Geology no. 26, p. 24–29.

van Alstine, D. R., Bazard, D. R., and Whitney, J. W., 1985, Paleomagnetism of cores from the Yakutat well, Gulf of Alaska [abs.]: EOS Transactions of the American Geophysical Institute, v. 66, no. 46, p. 865.

von Huene, R., and Kulm, L. D., 1973, Tectonic summary of Leg 18, *in* Kulm, L. D., and von Huene, R., eds., Initial Reports of the Deep Sea Drilling Project, v. 18: Washington, D.C., U.S. Government Printing Office, p. 961–976.

von Huene, R., Shor, G., Jr., and Wageman, J., 1979, Continental margins of the eastern Gulf of Alaska and boundaries of tectonic plates, *in* Watkins, J., and Montedart, L., eds., Geological and geophysical investigations of continental margins: American Association of Petroleum Geologists Memoir 29, p. 273–290.

von Huene, R., Box, S., Detterman, B., Fisher, M., Moore, C., and Pulpan, H., 1985, A-2, Kodiak to Kuskokwim, Alaska: Boulder, Colorado, The Geological Society of America, Centennial Continent/Ocean Transect no. 6, scale, 1:500,000.

von Huene, R., Bruns, T. R., and Childs, J., 1986a, Aleutian Trench, Shumagin segment, seismic section 104, part 1, *in* von Huene, R., ed., Seismic images of modern convergent margin tectonic structure: American Association of Petroleum Geologists Studies in Geology 26, p. 15–17.

von Huene, R., Miller, J., and Fisher, M., 1986b, The Eastern Aleutian continental margin, *in* von Huene, R., ed., Seismic images of modern convergent margin tectonic structure: American Association of Petroleum Geologists Studies in Geology 26, p. 20–23.

von Huene, R., Nasu, N., Culotta, R., and Aoki, Y., 1986c, The Japan Trench; Line ORI 78-4, *in* von Huene, R., ed., Seismic images of modern convergent margin tectonic structure: American Association of Petroleum Geologists Studies in Geology 26, p. 57–60.

von Huene, R., Suess, E., and Emeis, K., 1987, Convergent tectonics and coastal upwelling; A history of the Peru continental margin: Episodes, v. 10, no. 2, p. 87–93.

von Huene, R., Fisher, M. A., and Bruns, T. R., 1987, Geology and evolution of the Kodiak margin, Gulf of Alaska, *in* Scholl, D. W., Grantz, A., and Vedder, J. G., eds., Geology and resource potential of the continental margin of western North America and adjacent ocean basins—Beaufort Sea to Baja California: Houston, Texas, Circum-Pacific Council for Energy and Mineral Resources Earth Science Series, v. 6, p. 191–212.

Watkins, J. S., and Moore, J. C., eds., 1982, Initial reports of the Deep Sea Drilling Project, leg 66: Washington, D.C., U.S. Government Printing Office, v. 66, 864 p.

Yorath, C. J., and Hyndman, R. D., 1983, Subsidence and thermal history of Queen Charlotte Basin: Canadian Journal of Earth Sciences, v. 20, p. 135–159.

Yorath, C. J., and 7 others, compilers, 1985, B-1, Intermontane Belt (Skeena Mountains) to Insular Belt (Queen Charlotte Islands): Boulder, Colorado, Geological Society of America, Continent-Ocean Transect no. 8, scale 1:500,000.

Manuscript Accepted by the Society December 14, 1987

NOTES ADDED IN PROOF

Since we wrote this chapter, some of the inferences made during construction of the transects have been substantiated through new seismic reflection data acquisition. The EDGE line, transecting the northern Kodiak shelf and Aleutian Trench, was acquired in 1988 with seismic reflection techniques used for deep crustal profiling. Continuous reflections from the subducting oceanic crust beneath the continental slope and shelf were observed after initial processing (Moore and others, 1991). Beneath the lower continental slope, the upper detached kilometer of trench sediment is faulted and rotated until the frontally accreted mass is about 2.5 km thick. Underlying the accreted mass is a 2-km thick subducting section. Mass balance calculations indicate that about 25% is frontally accreted and 75% underthrust, a partitioning also recorded in other Aleutian transects (Scholl and others, 1990). Of the underthrust sediment, 40% underplates the prism beneath the lower slope. Farther landward beneath the shelf, and at depths of 9 to 35 km, are semihorizontal parallel reflectors presumed to be from sediment underplated in Eocene and Oligocene time. The area of reflections that represent underplating greatly exceeds the area of landward-dipping reflections from frontally accreted deposits. Apparently the rapid seaward growth by frontal accretion during Eocene and Oligocene time required extensive underplating farther landward to maintain the tapered crustal structure of the continental margin (Moore and others, 1991).

A series of lines across the shelf and slope of the Oregon margin shot with the same seismic system (MacKay and others, 1989) reveals in greater detail structures observed in earlier reprocessed seismic records (Snavely and others, 1986). They indicate contemporary underthrusting of thick sediment sections beneath planes of detachment 1 to 2 km above igneous oceanic crust. Areas with elevated planes of detachment are found adjacent to areas where detachment is near the base of the sediment section. Most lines off southern Oregon illustrate landward-dipping structure of the accretionary prism above horizontal underthrust strata 1 to 2 km thick. In contrast are the lower slopes of northern segments of the margin where landward-vergent overthrusts extend nearly to the top of the igneous oceanic crust (Snavely and others, 1986; MacKay and others, 1989). Lower-slope landward vergence reverses again and the upper slope is composed of seaward-verging thrust slices beneath which little structure is resolved. The cause of these sharp structural reversals is still problematic as are the effects on tectonism farther landward of the alternate thick and thin subducted sediment sequences.

Off Vancouver Island a series of lines acquired with a similar seismic system have been variously interpreted. The earlier reported results indicating considerable underthrusting and underplating in the accretionary prism have been followed by later interpretation of little underplating (Davis and Hyndman, 1989; Hyndman and others, 1990). Despite the indication of earlier underplating of a vast area beneath Vancouver Island, these authors suppose that all late Cenozoic sediment is accreted at the front of the margin with little underthrust sediment to nourish thickening of the continental crust landward of the zone where frontal accretion is currently active. The results of further processing of these seismic data will be followed with interest.

Data published since our chapter was written strengthen earlier observation that a great spectrum of tectonic processes and varied geological histories characterize the active and convergent margins of the Pacific rim along North America. As exemplified above, the majority of the terrigenous sediment filling the trench is subducted and underplated along the Alaska margin whereas most of the incoming sediment is thought to be detached and frontally accreted along the Cascadia margin off Vancouver Island and locally off Oregon. One implication may be that in sediment-flooded trenches relatively rapid plate convergence is associated with sediment subduction and frequent large earthquakes whereas slower convergence tends to promote frontal accretion and fewer large quakes. Another implication may be that the age and buoyancy of the subducting plate may alter the critical taper sufficiently to affect the efficiency of accretionary processes. If earlier Pacific margins were similar to the contemporary ones, their remnants, which are thought to build much of the continental crust, must also be diversely structured.

ADDITIONAL REFERENCES

Davis, E. E., and Hyndman, R. D., 1989, Accretion and recent deformation of sediments along the northern Cascadia subduction zone: Geological Society of America Bulletin, v. 110, p. 1465–1480.

Hyndman, R. D., and Yorath, C. J., 1990, The northern Cascadia subduction zone at Vancouver Island: seismic structure and tectonic history: Canadian Journal of Earth Sciences, v. 27, p. 313–329.

MacKay, M., Cochrane, G., Moore, G. F., Moore, J. C., and Kulm, L. D., 1989, High-resolution seismic study of the Oregon accretionary prism: EOS, Transactions of American Geophysical Union, v. 70, p. 1345.

Moore, J. C., and 11 others, 1991, EDGE deep seismic reflection transect of the eastern Aleutian arc-trench layered lower crust reveals underplating and continental growth: Geology, v. 19, p. 420–424.

Scholl, D. W., von Huene, R., and Dieffenbach, H. L., 1990, Rates of sediment subduction and subduction erosion; implications for growth of terrestrial crust: EOS, Transactions of the American Geophysical Union, v. 71, p. 1576.

Snaveley, P. D., Jr., von Huene, R., Mann, D. M., and Miller, J., 1986, The central Oregon margin, line WO 76-4; seismic images of modern convergent margin tectonic structure: American Association of Petroleum Geologists Studies in Geology, v. 26, p. 26–27.

Printed in U.S.A.

Index

[Italic page numbers indicate major references]

C

N

Q

R

Typeset by WESType Publishing Services, Inc., Boulder, Colorado
Printed in U.S.A. by Malloy Lithographing, Inc., Ann Arbor, Michigan

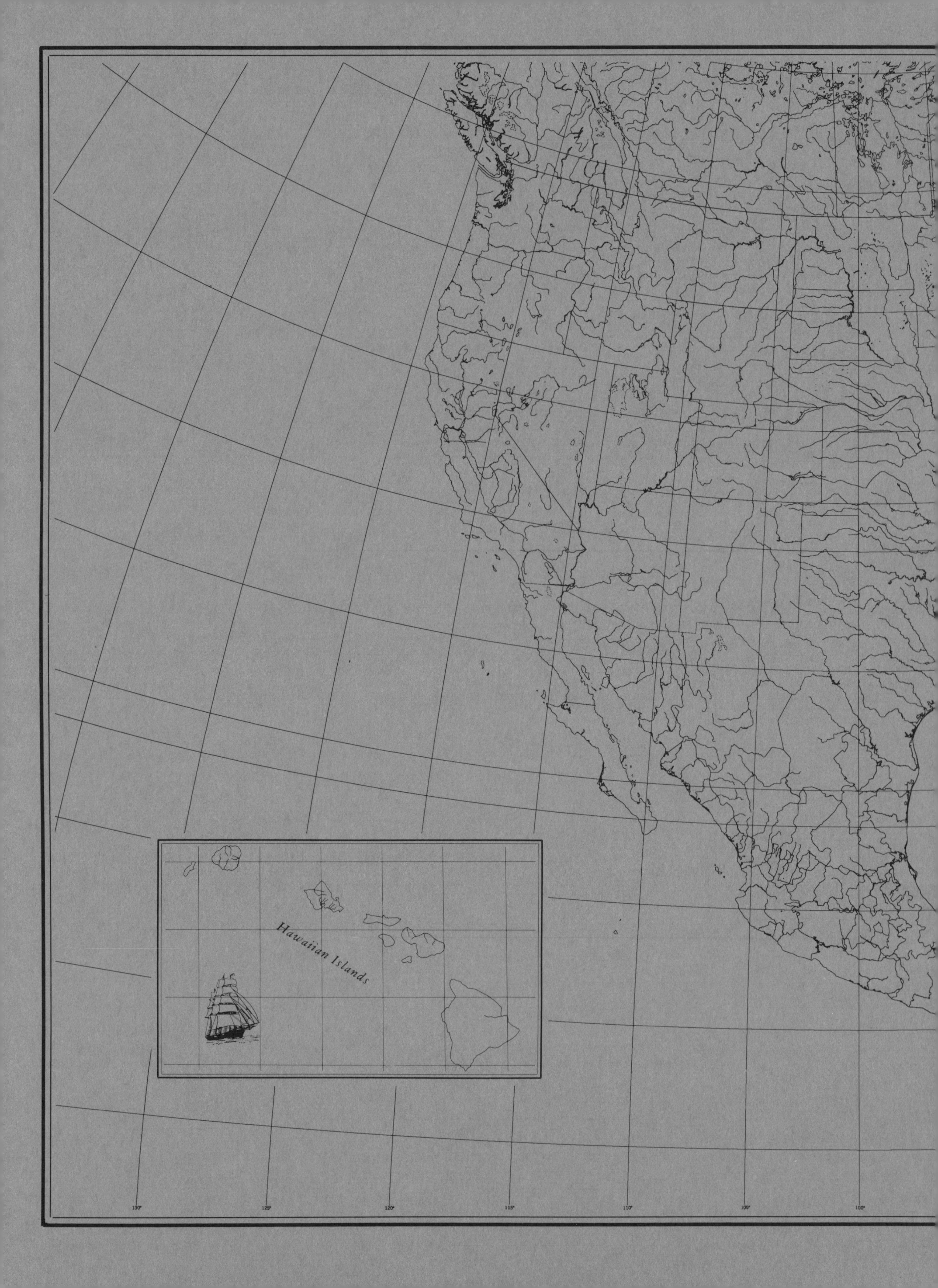
Hawaiian Islands
130°
125°
120°
115°
110°
105°
100°